Urban Engineering

Urban engineers provide a physical definition of the urban habitat by planning, designing, building and constructing, operating, and maintaining infrastructure, applying the tools of engineering, science, and good management to address the complex problems associated with infrastructure, services, buildings, environment, and land use generally encountered in cities.

Urban Engineering serves as a textbook to support a range of undergraduate courses in civil and environmental engineering, urban planning, and related areas. It is broad and inclusive, and takes a modular approach, where each theme is discussed comprehensively from the macro to the micro level. Highlights include urban design, housing, wastewater systems, transportation systems, smart cities, and urban agriculture.

The textbook has a particular emphasis on engineering solutions in sustainability.

Urban Engineering

Ariva Sugandi Permana

Routledge
Taylor & Francis Group
NEW YORK AND LONDON

First edition published 2026
by Routledge
605 Third Avenue, New York, NY 10518

and by Routledge
4 Park Square, Milton Park, Abingdon, Oxon, OX14 4RN

Routledge is an imprint of Taylor & Francis Group, LLC

Library of Congress Cataloging-in-Publication Data
Names: Permana, Ariva Sugandi author
Title: Urban engineering / Ariva Sugandi Permana.
Description: First edition. | Boca Raton, FL : CRC Press, 2026. |
Includes bibliographical references and index.
Identifiers: LCCN 2025009326 (print) | LCCN 2025009327 (ebook) |
ISBN 9781032664606 hbk | ISBN 9781032658018 pbk | ISBN 9781032664613 ebk
Subjects: LCSH: Municipal engineering–Textbooks |
Sustainable urban development–Textbooks | Sustainable engineering–Textbooks
Classification: LCC TD160 .P47 2026 (print) | LCC TD160 (ebook) |
DDC 628–dc23/eng/20250709
LC record available at https://lccn.loc.gov/2025009326
LC ebook record available at https://lccn.loc.gov/2025009327

ISBN: 978-1-032-66460-6 (hbk)
ISBN: 978-1-032-65801-8 (pbk)
ISBN: 978-1-032-66461-3 (ebk)

DOI: 10.1201/9781032664613

Typeset in Times
by Newgen Publishing UK

Contents

List of Abbreviations and Acronyms

3Rs	Reduce, reuse and recycle
ASCT	Adaptive signal control technology
ARM	Adjustable-rate mortgages
AMI	Advance metering infrastructure
AfriNIC	African Network Information Center
AI	Artificial Intelligence
ANSI	American National Standards Institute
ARIN	American Registry for Internet Numbers, North America
ASHRAE	American Society of Heating, Refrigerating, and Air-Conditioning Engineers
ASME	American Society of Mechanical Engineers
APNIC	Asia-Pacific Network Information Center
AV	Autonomous vehicles
BSC	Base Station Controller
BTS	Base Transceiver Station
BOD	Biological Oxygen Demand
BREEAM	Building Research Establishment Environmental Assessment Method
BIPV	Building-Integrated Photovoltaic
BI	Business Intelligence
CCPA	California Consumer Privacy Act
PCI DSS	Payment Card Industry Data Security Standard
CIDA	Canadian International Development Agency
CAHF	Center for Affordable Housing Finance Africa
CBD	Central Business District
CPU	Central Processing Unit
COD	Chemical Oxygen Demand
	Chemical parameters, 292, 294, 295
CCTV	Closed-circuit Television
CASB	Cloud Access Security Brokers
CSPM	Cloud Security Posture Management
CWPP	Cloud Workload Protection Platforms
CCR	Coal Combustion Residual
CFR	Code of Federal Regulation
CSO	Combined Sewer Overflow
CRA	Community Reinvestment Act
CS	Community-supported Agriculture
CFL	Compact Fluorescent Lamp
CRAC	Computer Room Air Conditioners

CRAH	Computer Room Air Handlers
CNN	Convolutional Neural Network
CRM	Customer Relationship Management
DSM	Demand-side Management
DNA	Deoxyribonucleic Acid
VA	US Department of Veterans Affairs
DSL	Digital Subscriber Line
NOC	Network Operation Center
DBP	Disinfection By-product
DO	Dissolved Oxygen
DDoS	Distributed denial-of-service
DNS	Domain Name System
eePRO	Environmental e-Professional
EV	Electric Vehicle
EDI	Electrodeionization
EMF	Electromagnetic Field
HER	Electronic Health Records
eSIM	Embedded Subscriber Identity Module
EDW	Enterprise Data Warehouse
EIA	Environmental Impacts Assessment
EMA	Environmental Management Act
EPANET	Environmental Protection Agency Network Evaluation Tool
ESG	Environmental, Social, and Governance
EMC	Event Mean Concentration
EPR	Extended Producer Responsibility
EOS	Extensible Operating System
	Externalities, 54, 124, 125, 741
EHVAC	Extra High Voltage Alternating Current
ETL	Extract, Transform, and Load
FAPR	Floor Area to Plot Ratio) 54, 55
FCC	Federal Communications Commission
FERC	Federal Energy Regulatory Commission
FHLB	Federal Home Loan Bank
FHA	Federal Housing Administration
FTC	Federal Trade Commission
FDA	Food & Drug Administration
GDPR	General Data Protection Regulation (GDPR)
GIS	Geographic Information System
GTO	Geostationary Transfer Orbit
GEO	Geosynchronous Equatorial Orbit
	Geothermal, 59, 83, 181, 187, 528, 536, 542, 549, 553, 556, 596, 597, 897, 923

GAC	Granular Activated Carbon
GHG	Greenhouse Gas
GDP	Gross Domestic Product
GPP	Gross Primary Productivity
HDD	Hard Disk Drives
HV-AC	High Voltage Alternating Current
HVAC	Heating, Ventilation, and Air Conditioning
HV-DC	High Voltage Direct Current
HDPE	High-density Polyethylene
	High-efficiency appliance, 549, 551
HAWT	Horizontal Axis Wind Turbines
HFC	Housing Finance Companies
HEV	Hybrid Electric Vehicles
HEC-RAS	Hydrologic Engineering Center's River Analysis System
ATMS	Advanced Traffic Management Systems
IEQ	Indoor Environmental Quality
ICT	Information and Communication Technology
IT	Information Technology
IrDA	Infrared Data Association
IM	Instant messaging
IEEE	Institute of Electrical and Electronics Engineers
IBMS	Integrated Building Management Systems
IWMF	Integrated Waste Management Facility
ITS	Intelligent Transportation System
IDF	Intensity, Duration, and Frequency
IPCC	Intergovernmental Panel on Climate Change
IEA	International Energy Agency
ISO	International Standardization Organization
ITU	International Telecommunication Union
IANA	Internet Assigned Numbers Authority
ICANN	Internet Corporation for Assigned Names and Numbers
IP	Internet Protocol
ISP	Internet Service Providers
IOS	Internetwork Operating System
KPI	Key Performance Indicators
LACNIC	Latin American and Caribbean Internet Addresses Registry, Latin America
LEED	Leadership in Energy and Environmental Design
LMS	Learning Management Systems
LOS	Level of Services
LCA	Life Cycle Assessment
LED	Light Emitting Diodes

LRT	Light Rail Transit
LNG	Liquefied Natural Gas
LDD	London Development Database
LIHTC	Low Income Housing Tax Credit
ML	Machine Learning
MRF	Materials Recovery Facility
MEO	Medium Earth Orbit
MNO	Mobile Network Operators
MSC	Mobile Switching Center
MaaS	Mobility as a Service
MBS	Mortgage-backed Securities
MU-MIMO	Multi-User Multiple-Input Multiple Output
MSWLF	Municipal Solid Waste Landfill
MSWM	Municipal Solid Waste Management
NEP	Net Ecosystem Productivity
NPP	Net Primary Productivity
NAS	Network-attached Storage
NLP	Neural Language Processing
NBFC	Non-banking Financial Companies
NMT	Non-motorized Transport
NIMBY	Not In My Backyard
Ofcom	UK Office of Communications
OS	Operating System
OFDMA	Orthogonal Frequency-Division Multiple Access
PPE	Personal Protective Equipment
PV	Photovoltaic
POTS	Plain Old Telephone Services
PHEV	Plug-in Hybrid Electric Vehicles
POE	Point-of-entry Systems
	Polychlorinated Biphenyls
PAH	Polycyclic Aromatic Hydrocarbons
PVC	Polyvinyl Chloride
PE	Population Equivalent
PSU	Power Supply Units
PMAY	Pradhan Mantri Awas Yojana
PST	Primary Sedimentation
PPP	Public-Private Partnerships
QoS	Quality of Service
RF	Radio frequencies
RAM	Random Access Memory
RIR	Regional Internet Registries
R&D	Research & Development

RIPE NCC	Réseaux IP Européens Network Coordination Center, Europe
RCRA	Resource Conservation and Recovery Act
RO	Reverse Osmosis
RNA	Ribonucleic Acid
ROW	Right of Way
RPA	Robotic Process Automation
SDG	Sustainable Development Goals
SCC	Smart Community Capabilities
SRM	Smart Resource Management
SDN	Software-defined Networking
SSD	Solid State Drives
SWM	Solid waste management
SAPC	Static Aerated Pile Composting
SPC	Static Pile Composting
SAN	Storage Area Networks
SWMM	Storm Water Management Model
SWMM-CAT	Storm Water Management Model Climate Adjustment
SRO	Surface runoff
SEI	Swedish Environmental Institute
TWT	Target Wake Time
TFC	Thin-film composite (TFC) membrane
ToU	Time-of-Use
TSS	Total Suspended Solids
TDM	Traffic Demand Management
TRX	Transceiver units
TOD	Transit-oriented Development
TRB	Transportation Research Board
TELUM	Transportation, Economic, and Land-Use Model
TSP	Traveling Salesman Problem
THMs	Trihalomethanes
UHV-DC	Ultra-High Voltage Direct Current
UV	Ultraviolet (UV) light
UVGI	Ultraviolet Germicidal Irradiation
UPS	Uninterruptible Power Supply
UCLA	University of California at Los Angeles
UHI	Urban Heat Island
USDA	US Department of Agriculture
US-HUD	US Department of Housing and Urban Development (HUD)
US-DOT	US Department of Transportation (DOT)
VMT	Vehicle miles traveled (VMT)
VRP	Vehicle Routing Problem
V2X	Vehicle-to-Everything

VAWT	Vertical Axis Wind Turbines
VM	Virtual Machines
VPN	Virtual Private Networks
VR/AR	Virtual Reality/Augmented Reality
VoIP	Voice over Internet Protocol
VOC	Volatile Organic Compounds
WtE	Waste-to-Energy
WPA3	Wi-Fi Protected Access 3
Wi-Fi	Wireless Fidelity
WLAN	Wireless Local Area Networks
WHO	World Health Organization
ZLD	Zero Liquid Discharge

Preface

This *Urban Engineering* textbook has been meticulously designed to cater to both bachelor's and master's degree students in urban planning and related fields. Spanning 15 chapters, it aligns perfectly with a typical 15-week semester structure, ensuring comprehensive coverage of the essential topics within an academic term. The primary aim is to address the most pertinent engineering challenges faced in urban environments, equipping students with the knowledge and tools necessary to navigate and resolve these issues effectively. Urban engineering is a dynamic and evolving field that addresses the complexities of urbanization in the modern world. It encompasses a wide range of disciplines, including transportation systems, water resources management, waste management, environmental sustainability, and infrastructure planning. This textbook aims to cover these diverse topics comprehensively, ensuring that students are well-prepared to tackle the multifaceted challenges of urban development.

Given that this is our inaugural edition, I acknowledge the possibility of shortcomings and areas for improvement, particularly in terms of illustration content. I admit that while our resources for illustrations are limited, acquiring the required illustrations from third parties has been quite challenging. However, I wholeheartedly welcome constructive feedback to refine and enhance future editions.

I extend my heartfelt thanks to Mr. Kaung Htet, an undergraduate student of civil engineering at King Mongkut's Institute of Technology Ladkrabang, for his invaluable contributions in providing illustrations for this book. Mr. Htet's dedication and artistic skills have greatly enriched the visual aspects of our text, making complex concepts more accessible and engaging for our readers. I would like to express our deep appreciation to the creators who have generously made their CC-BY illustrations available through Wikipedia and Wikimedia. These contributions have significantly enhanced the visual quality of this text, and I am grateful for their openness and generosity.

My hope is that this textbook serves as a valuable resource throughout your academic journey and beyond. I aim to provide clarity, insight, and practical knowledge essential for aspiring urban engineering professionals. The integration of theoretical foundations with practical applications is intended to foster a deeper understanding of urban engineering's role in shaping sustainable and resilient urban environments.

I am excited to present this first edition of our *Urban Engineering* textbook. I am confident that it will provide a solid foundation for students pursuing careers in urban planning and related disciplines. While I recognize the areas for improvement, I believe that the core content will significantly benefit our readers. I encourage you, the users and the readers, to share your feedback and suggestions, which will be invaluable in shaping future editions of this book.

I hope you find this textbook to be an informative and engaging resource that supports your academic and professional growth in urban engineering.

AS Permana
Bangkok, May 2025

About the Author

Ariva Sugandi Permana is a distinguished scholar in the field of urban planning and environmental management, holding both an MSc and a PhD from the Asian Institute of Technology (AIT), Thailand. His academic journey has been marked by a rigorous commitment to understanding and solving urban and environmental challenges. With a comprehensive background in civil engineering, Ariva has seamlessly integrated his technical expertise with advanced urban planning and environmental management principles, making significant contributions to these interdisciplinary fields.

Before delving into urban planning, Ariva spent over 15 years honing his skills as a civil engineer, specializing in water resources engineering and management. His extensive practical experience has provided him with a profound understanding of the complexities of urban infrastructure and the importance of sustainable water management. This hands-on expertise has enriched his academic pursuits, enabling him to bridge the gap between theory and practice effectively. His dedication to improving urban environments is evident through his various projects and research endeavors that aim to create resilient and sustainable cities.

Ariva's passion for global environmental issues and sustainable development has driven him to explore the intricate relationship between urban planning and environmental management. This curiosity led to his recognition as an outstanding graduate at AIT, where he received the prestigious John H Hrones prize. Additionally, his exceptional academic achievements were further acknowledged through the award of a scholarship from Canadian International Development Agency (CIDA). These accolades underscore his commitment to excellence and his significant contributions to the field. Ariva continues to inspire future urban planners and environmental managers with his innovative approaches and unwavering dedication to creating sustainable urban futures.

1 General Overview

1.1 INTRODUCTION

As the name implies, urban engineering is a branch of engineering that applies science, art, and engineering tools in an urban environment and development. It is concerned with city infrastructure, which involves planning, specifying, designing, constructing, and maintaining various aspects of the built environment. Some of the areas that urban engineering covers include streets, sidewalks, water supply networks, sewers, street lighting, municipal solid waste management and disposal, storage depots for various bulk materials used for maintenance and public works, landscape design, public parks, and cycling infrastructure. In the case of underground utility networks, it may also include the civil portion of the local distribution networks of electrical and telecommunications services. Urban engineering can also involve optimizing garbage collection and bus service networks. Some of these disciplines overlap with other civil engineering specialties; however, urban engineering focuses on the coordination of these infrastructure networks and services.

Urban engineering may also be a field that deals with the design, construction, and maintenance of the infrastructure of urban areas. This includes roads, bridges, water supply systems, sewage systems, and public transportation systems. Urban engineers, therefore, work to ensure that the infrastructure of cities is safe, efficient, and sustainable. One of the main challenges that urban engineers face is the need to balance the competing demands of different stakeholders. For example, urban engineers must consider the needs of residents, businesses, and government agencies when designing transportation systems. They must also consider environmental concerns and the need to preserve historic buildings and landmarks. Urban engineers need to use a variety of tools and techniques to address these challenges. For instance, they may use computer simulations to model traffic flow and predict the impact of different transportation options. They may also employ geographic information systems (GIS) to map out the location of various infrastructure elements and identify potential conflicts. In addition to technical skills, urban engineers must also have effective communication and collaboration skills. They often work in teams with other engineers, architects, planners, and government officials. They must be able to explain complex technical concepts to nonexperts and build consensus among diverse stakeholders. By this state, urban engineering is a dynamic and challenging field that plays a critical role in shaping the future of our cities. Designing safe, efficient, and sustainable infrastructure enables urban engineers to help improve the quality of life for millions of people worldwide. The profession of urban engineers is irreplaceable by the current trend of predicted diminishing professions in the future due to technological advancements such as artificial intelligence. These gradually diminishing jobs that are most likely to be replaced by technology include accountants/auditors, telemarketers, technical writers, word processors, real estate agents, retail salespersons, retail workers, medical diagnosticians, proofreaders, couriers,

DOI: 10.1201/9781032664613-1

compensation and benefits managers, taxi drivers, construction workers, inventory managers and stockists, and some other professionals who are fading or gone already such as structural engineers, telephone operators, travel agents, cashiers, bank tellers, mail carriers (package carriers will still be there unless teleportation technology does exist), printing press operators, and many more.

Urban engineering is one of the main supporters of city development as it is inseparable from infrastructure. As infrastructure and urban development are closely intertwined, infrastructure is a key component of urban development, providing the physical, social, institutional, and conceptual elements necessary for the growth and functioning of cities. The development of infrastructure systems can attract private investments for housing and economic development, stimulating the growth of human settlements in urban areas.

The World Economic Forum recognizes the importance of infrastructure and urban development, with its Infrastructure and Urban Development Industry community providing leadership and driving change in the industry. The community has identified collaborative project delivery mechanisms and public–private cooperation as priority challenges to develop collaborative mechanisms that deliver major infrastructure and urban development projects on time and budget.

The World Bank also recognizes the challenges brought by the speed and scale of urbanization, including meeting accelerated demand for affordable housing and viable infrastructure such as transport systems, basic services, and jobs. The expansion of urban land consumption outpaces population growth by as much as 50%, putting pressure on land and natural resources. Cities represent two-thirds of global energy consumption and account for more than 70% of greenhouse gas emissions.

With the above arguments, it safely says that infrastructure is a crucial component of urban development, providing the necessary elements for the growth and functioning of cities. The development of infrastructure systems requires the support of urban engineering disciplines and urban engineers. From an economic viewpoint, urban development can attract various kinds of private investments and stimulate growth in urban areas. Organizations such as the World Economic Forum and the World Bank recognize the importance of infrastructure and urban development, working to address challenges such as meeting demand for affordable housing, viable infrastructure, basic services, and jobs.

Now, let us observe where urban engineering can be posed, given that urban infrastructure and urban development are two interrelated concepts that shape the physical and social aspects of cities. Urban engineering connects urban infrastructure and urban development as illustrated in Figure 1.1.

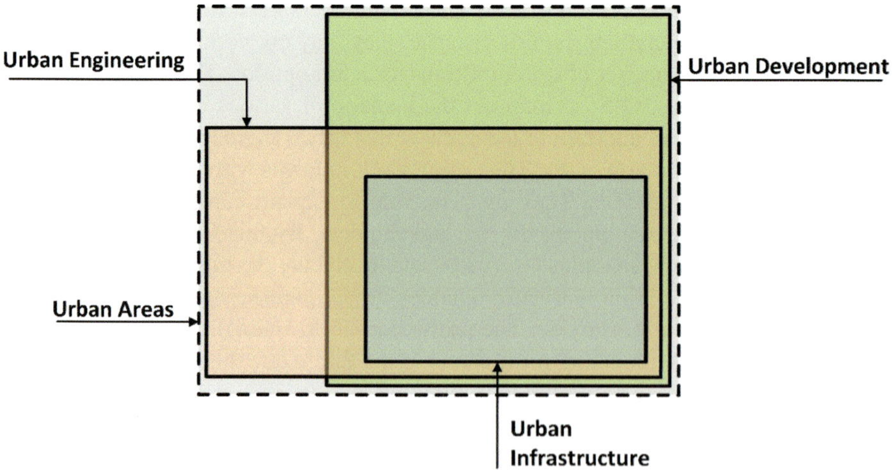

FIGURE 1.1 The connection between urban engineering, urban infrastructure, and urban development.

We recognize the notion that urban infrastructure refers to the systems and facilities that provide essential services and support to urban populations, and urban development refers to the processes and outcomes of planning, designing, building, constructing, and managing urban spaces, including land use, housing, economic activity, social services, and environmental quality. The comprehension of the intertwin or cross-section between urban infrastructure and urban development can be understood from different perspectives. One perspective is that urban infrastructure is a precondition and a driver for urban development, as it enables the movement of people, goods, and information; facilitates the provision of basic needs and amenities; and creates opportunities for economic growth and innovation. Another perspective is that urban development is a determinant and a challenge for urban infrastructure, as it shapes the demand for and the impact of infrastructure services, requires coordination and integration of different infrastructure sectors, and poses trade-offs and conflicts between competing interests and goals. Both perspectives highlight the importance of aligning urban infrastructure and urban development to achieve sustainable and resilient urban outcomes. This alignment involves considering the interdependencies and synergies between different infrastructure systems and urban functions, as well as the trade-offs and conflicts between different stakeholders and objectives. It also involves adopting a long-term and holistic view of urban planning and management that balances the needs of current and future generations, as well as the local and global impacts of urbanization.

1.2 WORKING DEFINITIONS

As urban engineering is strongly correlated with three entities of city, infrastructure, and engineering, we define urban engineering as a discipline in engineering that focuses on the planning, designing, constructing, operating, and maintaining infrastructure those are required for the development and advancement of the cities. The goal of urban engineering is the welfare of urban citizens. With the emergence of sustainable urban development, urban engineering scope might be expanded to social, economic, and environmental sustainability, and not only be limited to urban infrastructure. Therefore, urban engineering should be able to ensure social sustainability by designing and implementing infrastructure that caters to the needs of all members of the community, including vulnerable groups such as the elderly and disabled.

While economic sustainability could be achieved through urban engineering by creating efficient and cost-effective solutions for transportation, energy, and waste management systems that support local businesses and industries, to accomplish environmental sustainability, urban engineering could offer solutions such as the promotion of green building design, stormwater management, and renewable energy systems being implemented to reduce carbon emissions and promote sustainable living. Urban engineering could also play a role in promoting sustainable urban development by collaborating with other professionals such as urban planners, architects, and policymakers to create integrated solutions that address the complex challenges facing cities today.

1.3 CITIES AND URBAN AREAS

Cities and urban areas are related but distinct concepts. A city is a legally defined entity with a specific geographic boundary, usually with a large population and a high degree of economic, social, and cultural activities. A city can also be distinguished by its function in governmental and other services, while urban areas do not include this quality. Urban areas, on the other hand, refer to densely populated regions that may or may not coincide with city boundaries. Urban areas can include cities as well as surrounding suburbs and other developed areas. In other words, a city is a specific type of urban area that has been formally recognized by a government or other entity, while an urban area refers more broadly to any densely populated region.

From the geographical viewpoint, cities are the epicenters of human activity and are often the engines of economic growth. Understanding the geography of cities is essential for policymakers, urban planners, and businesses seeking to thrive in these dynamic environments. In this chapter, we will explore the key components of city geography and how they shape the urban landscape. One of the most important aspects of city geography is its physical environment. Topography, climate, and natural resources all play a role in shaping a city's development. For example, cities located near waterways or natural harbors often become important centers of trade and commerce. Similarly, cities located in areas with fertile soil or abundant natural resources may develop strong agricultural or mining industries.

Another key component of city geography is its infrastructure. Transportation networks, such as roads, highways, and public transit systems, are essential for connecting people and goods within and between cities. Additionally, access to utilities such as electricity, water, and sewage systems is critical for ensuring a high quality of life for residents. The built environment is also a crucial component of city geography. The design and layout of buildings, streets, and public spaces can shape the character of a city and influence how people use and interact with urban spaces. For example, cities with dense, walkable neighborhoods and vibrant public spaces may foster a keen sense of community and social interaction.

Population demographics are another crucial factor in city geography. The size, age, income level, and cultural background of a city's population can all impact its development trajectory. For example, cities with large populations of young professionals may attract new businesses and investments, while cities with aging populations may face challenges related to healthcare and social services.

Political and economic factors also play a role in shaping city geography. Government policies related to zoning, taxation, and regulation can influence the location and type of development that occurs within a city. Economic trends such as globalization and automation can also impact the types of industries that thrive in urban areas. The geography of cities is a complex and multifaceted topic that encompasses a range of physical, social, and economic factors. By understanding these key components of city geography, policymakers, urban planners, and businesses can make informed decisions that support the continued growth and development of our urban centers.

It is easily recognizable and observable whether a city has undergone an appropriate planning process – in which the city has been, is being, and will be continuously planned – or if a city is a naturally grown city, without undergoing a proper planning process, or simply a city with an afterthought planning, where the city planning was heavily dictated by many interests, such as economy and politics. In the latter case, urban engineering would not be able to boost what a city could offer to the citizens. A city could offer optimal services to its citizens if the city is thoughtfully planned and the city planning is well implemented. To this notion, city planning has been a crucial aspect of human civilization since the dawn of urbanization. As cities continue to grow and evolve, the need for effective city planning has become more pressing than ever. One of the most important considerations in city planning is the concept of organically grown cities as a disadvantageous kind of city development. An organically grown city has developed over time in a natural, unplanned way. These cities often have a unique character and charm that is difficult to replicate in planned cities. However, they can also present challenges for city planners, as they may lack the infrastructure and amenities necessary to support a growing population. Great city planning requires a wide knowledge of either excellent or disgraceful planning examples of city planners.

In many developing countries, cities have undergone an organically grown pattern, because a city with natural growth is easy to accomplish. The city manager or city authority needs nothing to do. However, one of the key benefits of organically grown cities is their ability to adapt and evolve. As the needs of the population change, these cities can respond flexibly and dynamically, because no fixed city plan is in place. This can lead to the development of new neighborhoods, and it might lead to the development of urban sprawl. Organically grown cities may lack the infrastructure necessary to support modern transportation systems, such as highways and mass transit. They may also lack the zoning regulations necessary to prevent incompatible land uses from coexisting nearby.

To address these challenges, city planners must strike a balance between preserving the unique character of organically grown cities and providing the infrastructure and amenities necessary to support a growing population. This can be a delicate balancing act, as changes to these cities must be made carefully to preserve their unique character. This notion can easily be said than done, as the facts show that many cities in developing countries are unable to escape from this inexcusable problem.

To achieve this balance in case a city cannot escape from unplanned growth because the growth is beyond retractable to a proper path of the planned city, the city could adopt a "smart growth" approach to city planning. This approach emphasizes the importance of compact, walkable neighborhoods that are well-connected to public transit and other amenities. By promoting sustainable development patterns, smart growth can help preserve the unique character of organically grown cities while also providing the infrastructure necessary to support a growing population.

Another important consideration in city planning is the need to balance economic development with environmental sustainability. Many cities have embraced sustainable development practices in recent years, recognizing that economic growth must be balanced with environmental protection. This can include initiatives such as green building standards, renewable energy development, and sustainable transportation systems.

As cities continue to expand and grow, the lack of proper planning can negatively impact the environment, economy, and quality of life for residents. However, certain infrastructure developments can help minimize these negative impacts and create a more sustainable and livable city. One key infrastructure development is the implementation of public transportation systems. By providing efficient and accessible public transportation options such as buses, trains, and subways, cities can reduce the number of cars on the road and promote a more sustainable mode of transportation. This not only benefits the environment but also improves accessibility for residents who may not have access to a personal vehicle.

Another important infrastructure development is the establishment of green spaces and parks. With urbanization comes the loss of natural habitats and green spaces, leading to a decrease in air quality and biodiversity. By creating parks and green spaces within the city, residents can enjoy the benefits of nature while also improving air quality and reducing the urban heat island effect. Additionally, green spaces can serve as a gathering place for community events and activities, promoting social cohesion and a sense of community.

In addition to public transportation and green spaces, another important infrastructure development is the implementation of sustainable building practices. With a lack of planning, cities may be filled with buildings that are not energy-efficient or environmentally friendly. By implementing sustainable building practices such as green roofs, solar panels, and energy-efficient appliances, cities can reduce their carbon footprint and promote a more sustainable way of living. Furthermore, the establishment of waste management systems is crucial in minimizing the negative impacts of unplanned cities. With a growing population comes an increase in waste production, leading to issues with sanitation and pollution. By implementing effective waste management systems such as recycling programs and waste-to-energy facilities, cities can reduce their environmental impact while also promoting a more circular economy.

The development of smart city technologies can also help minimize the negative impacts of unplanned cities. With the use of sensors and data analysis, cities can optimize their infrastructure and services to improve efficiency and reduce waste. For example, smart lighting systems can reduce energy consumption by adjusting lighting levels based on occupancy, while smart traffic management systems can reduce congestion by optimizing traffic flow.

However, the best way to cope with naturally grown cities is to bring them back on the right track toward planned cities before urban engineering can play its role in developing the city and offering welfare to the citizens. It is because a planned city is a solid foundation for the city to further develop. Urban engineering works in different conditions including naturally grown cities, but the best outcomes will be obtained when urban engineering is applied in well-planned cities.

1.4 URBAN ENGINEERING IN BRIEF

As we have discussed earlier, urban engineering is the discipline that applies scientific and technological principles to design, construct, operate, and maintain the physical infrastructure, facilities, and utilities that support the functioning of a city. Thus, urban engineering is equivalent to promoting urban infrastructure that encompasses various aspects such as transportation, water supply, wastewater treatment, solid waste management, energy systems, communication networks, urban planning, and environmental protection.

A city with sufficient urban infrastructure can provide a high quality of life for its residents and visitors, as well as economic opportunities, social cohesion, and environmental sustainability. On the other hand, a city without sufficient urban infrastructure or significant involvement in urban engineering can face many challenges such as traffic congestion, pollution, flooding, disease outbreaks, power outages, crime, poverty, and social unrest. Therefore, urban engineering involvement in city development is essential for creating livable, resilient, and smart cities that can meet the needs and aspirations of current and future generations.

Some differences between a city with sufficient infrastructure and a city with insufficient infrastructure are:

- A city with sufficient infrastructure ensures that its residents have access to basic public utilities, such as clean water, stable electricity, efficient transportation, quality healthcare, adequate education, and proper waste management. A city with insufficient infrastructure fails or struggles to provide these services, resulting in low quality of life, health hazards, and social disparities.
- A city with sufficient infrastructure can support economic growth and development, by attracting investments, businesses, and skilled workers. A city with insufficient infrastructure faces challenges in creating jobs, generating income, and reducing poverty.
- A city with sufficient infrastructure can cope better with environmental and social challenges, such as climate change, natural disasters, urbanization, and migration. A city with insufficient infrastructure is more vulnerable to these challenges and may experience conflicts, violence, and instability.

Another significant role of urban engineering (i.e., sufficient infrastructure) is to enhance disaster resilience in cities. Disaster resilience is the ability of a city to withstand, adapt, and recover from natural and human-made hazards, such as earthquakes, floods, fires, pandemics, and terrorism. Urban engineering can contribute to disaster resilience in several ways, as shown in Table 1.1.

Some examples of urban engineering in disaster resilience are:

- The city of Ahmedabad, India, is implementing a few projects to make the city more resilient to climate change and natural disasters. These projects include building a new stormwater drainage system, upgrading the city's water supply system, and planting trees to help absorb stormwater.
- The city of Medellín, Colombia, is known for its innovative approach to urban planning and development. The city has implemented several projects to make it more resilient to climate change, including building a new metro system, upgrading its drainage system, and planting trees to help absorb stormwater.

TABLE 1.1
Roles of Urban Engineering for Disaster Resilience

Roles	Explanation
Mitigating risks	Urban engineering can reduce the exposure and vulnerability of urban systems and populations to hazards by applying risk assessment, hazard mapping, structural design, land use planning, and building codes.
Enhancing preparedness	Urban engineering can improve the readiness and capacity of urban systems and populations to respond to hazards by developing early warning systems, emergency plans, evacuation routes, shelters, and communication networks.
Facilitating recovery	Urban engineering can support the restoration and improvement of urban systems and populations after a hazard by providing damage assessment, debris removal, temporary housing, infrastructure repair, and reconstruction.

- The Sustainable Water Management Project in Hanoi, Vietnam, is aimed at improving the city's water supply and drainage systems, as well as its flood protection measures. The project includes the construction of new reservoirs, water treatment plants, and drainage canals. It also includes the upgrading of existing infrastructure, such as levees and floodwalls.
- The Sustainable Urbanization and Disaster Risk Reduction Project in Manila, Philippines, is aimed at making the city more resilient to the impacts of climate change and natural disasters. The project includes the construction of flood-resilient infrastructure, as well as the development of early warning systems.

By enhancing a city's disaster resilience, urban engineering helps minimize losses, damages, and disruptions caused by hazards. In addition to that, it enables cities to seize opportunities for learning, innovation, and transformation. Therefore, urban engineering plays an essential role in creating and maintaining healthy, prosperous, and sustainable cities.

The MOVE-UP Project in the Philippines, showcased in Box 1.1, exemplifies an urban development approach that aims to enhance the resilience of impoverished communities by addressing both infrastructure and climate adaptation needs in a holistic manner.

BOX 1.1 MOVING URBAN POOR COMMUNITIES TOWARD RESILIENCE PROJECT

The Moving Urban Poor Communities toward Resilience (MOVE-UP) Project was implemented from 15 February 2016 to 14 September 2017. It was a collaborative effort managed by Action Against Hunger, CARE Nederland, and CARE Philippines, along with local partners Assistance and Cooperation for Community Resilience and Development (ACCORD) and Plan International Philippines. The project aimed to enhance resilience, disaster risk reduction, and management in urban areas.

Key Objectives:

- *Urban Development Approaches for Resettlement and Upgrading:* The project developed approaches for resettlement and upgrading based on pilot projects in selected areas of Metro Manila. These approaches were designed to improve the living conditions of urban poor communities.

- *Climate-Adaptive Evaluation and Resettlement Approaches:* The project co-produced climate-adaptive evaluation and resettlement strategies with local residents. These approaches aimed to enhance community livelihoods while addressing climate-related risks.

The MOVE-UP Project drew from the consortium's experience in resilience, disaster risk reduction, and lessons learned from broader engagements with cities in the Philippines. Through such initiatives, sustainable urbanization and disaster risk reduction are critical components for building resilient cities in the face of climate challenges.

Source: Moving Urban Poor Communities toward Resilience (MOVE-UP) Project, www.urban-response.org/system/files/content/resource/files/main/Final_MOVEUP_Philippimes_13122017.pdf

1.5 SUSTAINABLE INFRASTRUCTURE PRINCIPLES

Sustainable infrastructure principles are the guiding values and practices, which ensure the development and maintenance of infrastructure systems that are environmentally friendly, socially inclusive, and economically viable. Sustainable infrastructure principles aim to balance the needs of present and future generations, as well as the natural and human-made environments. Some of the key sustainable infrastructure principles are:

1. Minimizing the environmental impact of infrastructure projects, such as reducing greenhouse gas emissions, enhancing biodiversity, and promoting resource efficiency.
2. Maximizing the social benefits of infrastructure projects, such as improving access to basic services, enhancing the quality of life, and fostering community participation and empowerment.
3. Ensuring the economic viability of infrastructure projects, such as optimizing costs and benefits, increasing resilience and adaptability, and creating opportunities for innovation and value creation.

By applying these principles, sustainable infrastructure can contribute to achieving the Sustainable Development Goals (SDGs) and advancing the global agenda for climate action, human rights, and peace.

There are many ways to minimize the environmental impact of infrastructure projects. One way is to use environmentally responsible and sustainable construction practices and be resource-efficient throughout a building's lifecycle, from siting to design, construction, operation, maintenance, renovation, and deconstruction. Sustainable construction methods span a lot of Ps, such as products, practices, processes, and policies. They begin in a project's design phase and carry through to operation. Some of these methods include:

1. Selective uses of materials, such as locally available, energy-efficient, and renewable materials with minimal ecological footprint, and utilizing reusable building materials.
2. Limiting fuel usage by reducing the burning of fossil fuels like gas and diesel.
3. Reducing noise pollution by minimizing construction noise.
4. Effectively managing and disposing of waste.
5. Utilizing reusable materials and technology.
6. Expediting the project to increase overall project efficiency.

Some infrastructure within the scope of urban engineering, for example, roads, bridges, buildings, dams, power plants, and the like, requires building materials to construct. Within the span of their life cycle, building materials can affect the quality, durability, cost, environmental impact, and social acceptability of infrastructure projects. Therefore, choosing the right building materials is a crucial decision for infrastructure development. Building materials is one of the entry points to accomplish sustainable infrastructure, for instance, using building materials that are environmentally friendly and resource-efficient from the whole life cycle of the materials. These materials can be made from renewable resources or recycled materials.

The life cycle of building materials involves the extraction, production, transportation, use, and disposal of various materials used in construction. Each stage of this process has environmental, social, and economic impacts that need to be considered and minimized to accomplish a sustainable building material. Some examples of the main building materials and their life cycle stages are:

1. *Concrete:* Concrete is a mixture of cement, water, sand, and gravel. Cement is made from limestone, clay, and other minerals that are heated in a kiln and then ground into a fine powder. Water, sand, and gravel are added to the cement to form concrete, which is then transported to the construction site and poured into molds or forms. Concrete is used for foundations, walls, floors, and other structural elements. Concrete can last for decades or even centuries, but it can also crack, deteriorate, or become obsolete. When concrete is no longer needed, it can be demolished and recycled into new concrete or used as aggregate for roads or other purposes.
2. *Steel:* Steel is an alloy of iron and carbon that is produced by smelting iron ore in a blast furnace and then adding carbon and other elements to adjust its properties. Steel is then rolled, shaped, or cast into various forms such as beams, bars, plates, or wires. Steel is used for frames, columns, beams, trusses, and other structural components. Steel is strong, durable, and versatile, but it also requires a lot of energy and resources to produce and transport. When steel is no longer needed, it can be melted and recycled into new steel products or used as scrap metal for other industries.
3. *Wood:* Wood is a natural material that comes from trees that are harvested from forests or plantations. Wood is then processed into lumber, plywood, particleboard, or other products that are used for framing, sheathing, flooring, roofing, or furniture. Wood is renewable, biodegradable, and aesthetically pleasing, but it also requires land, water, and nutrients to grow and can be affected by pests, diseases, or fire. When wood is no longer needed, it can be reused, recycled, or composted.
4. *Glass:* Glass is a transparent material that is made from sand, soda ash, and limestone that are melted in a furnace and then cooled into sheets or shapes. Glass is used for windows, doors, skylights, or facades. Glass is durable and recyclable and allows natural light and views, but it also requires a lot of energy to produce and transport and can break or shatter. When glass is no longer needed, it can be crushed and recycled into new glass products or used as aggregate for concrete or asphalt.
5. *Plastic:* Plastic is a synthetic material that is made from petroleum or natural gas that is processed into polymers that are then molded or extruded into various shapes or forms. Plastic is used for pipes, insulation, siding, roofing, or fixtures. Plastic is lightweight, cheap, and versatile, but it also requires a lot of energy and resources to produce and transport and can degrade or leach chemicals over time. When plastic is no longer needed, it can be reused, recycled, or incinerated.

Sustainable infrastructure is essential for achieving social, economic, and environmental goals, such as reducing poverty, improving health and well-being, enhancing resilience, and mitigating climate change. Sustainable infrastructure can also foster innovation, create jobs, and stimulate growth. Some examples of sustainable infrastructure are renewable energy sources, green buildings, public transportation, waste management, and water conservation. Applying the principles of sustainable infrastructure, some general approaches to achieve this include:

1. Using renewable energy sources such as solar, wind, hydro, and biomass to power the infrastructure and reduce greenhouse gas emissions.
2. Incorporating green design principles such as energy efficiency, water conservation, waste reduction, and biodiversity protection into the planning, construction, and operation of the infrastructure.
3. Applying circular economy concepts such as reuse, repair, refurbish, and recycle to extend the lifespan of the infrastructure and minimize the use of natural resources and the generation of waste.
4. Engaging stakeholders such as communities, governments, businesses, and civil society in the decision-making and implementation of infrastructure projects to ensure social inclusion, equity, and accountability.
5. Evaluating the environmental, social, and economic impacts of the infrastructure throughout its lifecycle and adopting adaptive management strategies to address any challenges or risks that may arise.

Adopting sustainable infrastructure principles, following the path of generic ways toward sustainable infrastructure, and, to some extent, utilizing materials with a minimal environmental footprint in their life cycle are appropriate pathways to accomplish sustainable infrastructure in urban engineering.

1.6 BASIC SERVICES TO ADVANCED INFRASTRUCTURE OF A CITY

Basic infrastructure in a city refers to the essential facilities, services, and organizational structures that are necessary for the functioning of a community and city. The goal of the provision of basic infrastructure is to create a sustainable, accessible, and functional infrastructure that meets the needs of citizens and businesses in their social and economic activities. This requires thoughtful planning and management to ensure that the infrastructure can support the growing population and changing needs of the city. The after-thought planning can therefore be avoided. The development of basic infrastructure in a city also requires significant investment and careful planning. City leaders must work closely with urban planners, engineers, architects, and other experts to design and implement sustainable infrastructure solutions. This involves setting clear objectives, collecting, and analyzing data, forecasting future needs, and engaging with the public to gather feedback and input.

A city is more than just a collection of buildings and streets. It is a complex system that provides assorted services and infrastructure to its residents, businesses, and visitors. A city can have several types of infrastructure from basic to advanced, depending on its size, population, and level of development. Basic infrastructure includes roads, bridges, water supply, sewerage, and electricity. Advanced infrastructure includes public transportation, communication networks, healthcare facilities, education institutions, and cultural venues. Infrastructure is essential for the functioning of a city, as it provides services and amenities to its residents and visitors. Infrastructure also influences the economic, social, and environmental aspects of a city, as it affects its productivity, livability, and sustainability.

The quality of life of the citizens is closely related to the state of the infrastructure in their country. Think about it: How much easier is it to get around, access essential services, and enjoy the beauty of nature and culture, when the roads, bridges, railways, airports, hospitals, schools, parks, museums, and so on are well-maintained and functional? How much happier and healthier are we when we have clean water, reliable electricity, fast internet, and safe buildings? How much more prosperous and innovative are we when we have efficient and sustainable infrastructure that supports our economy and society? These are not just rhetorical questions but proven facts backed by research and experience. Infrastructure is the backbone of our civilization, and we should be proud and grateful for what we have achieved so far and eager and ambitious for what we can do in the future.

Is it true that without sufficient infrastructure we can still have a happy life? Well, that depends on how you define happiness, right? Some people might say that happiness is having access to basic needs like food, water, shelter, healthcare, education, etc. Others might say that happiness is more about personal fulfillment, relationships, hobbies, passions, etc. And then some might say that happiness is a state of mind, not dependent on external factors at all.

However, people, in general, will think that without sufficient infrastructure, it would be extremely hard to have a happy life in the first sense. Infrastructure is what enables us to meet our basic needs and have a decent quality of life. Without it, we would face many challenges and risks, such as hunger, disease, poverty, violence, natural disasters, etc. It would be incredibly stressful and exhausting to live in such conditions.

But what about the second and third senses of happiness? Could we still have a happy life without sufficient infrastructure if we focus on our personal fulfillment, relationships, hobbies, passions, or state of mind? Some of us may think that is possible, but it is not easy. Without sufficient infrastructure, we would have fewer opportunities and resources to pursue our goals and interests. We would also have less time and energy to spend with our loved ones and to enjoy our hobbies. We might face more obstacles and frustrations along the way, and we might have more difficulty maintaining a positive outlook and attitude in the face of adversity. Infrastructure may not be the only factor that affects our happiness, but it is an especially important one. That is why we should invest more in developing and improving our infrastructure, not only for our benefit but also for the benefit of future generations.

How can we improve our infrastructure? That is a good question and one that has many answers. Infrastructure is a broad term that covers many aspects of our society, from transportation to education. Improving our infrastructure means making it more efficient, reliable, accessible, affordable, sustainable, and resilient. There are many ways to improve our infrastructure, depending on the specific needs and challenges of each sector and region. Some general strategies that could help are:

1. Investing more in research and innovation to develop innovative technologies and solutions that can enhance our infrastructure performance and reduce its environmental impact.
2. Increasing public and private funding and partnerships to support infrastructure projects and maintenance, especially in areas that are underserved or vulnerable.
3. Implementing policies and regulations that encourage and incentivize best practices and standards for infrastructure planning, design, construction, operation, and management.
4. Engaging and empowering local communities and stakeholders to participate in infrastructure decision-making and governance, ensuring that their needs and preferences are considered and respected.
5. Promoting education and awareness among the public and policymakers about the importance and benefits of infrastructure, as well as the risks and costs of neglecting it.

By improving our infrastructure, we can improve our quality of life, our economic growth, our social equity, and our environmental sustainability. Infrastructure is not only a means to an end but also an end. It reflects our values and aspirations as a society.

Many cities in developing countries are facing serious challenges due to the lack of adequate infrastructure, as infrastructure is essential for providing basic services. Due to rapid urbanization, population growth, poverty, corruption, and weak governance, many cities are unable to meet the demand for infrastructure and maintain its quality. This leads to various problems such as environmental degradation, public health risks, social inequality, economic inefficiency, and reduced quality of life. Therefore, the governments and stakeholders of these cities must invest in infrastructure development and management, as well as adopt innovative and sustainable solutions that can address the current and future needs of urban dwellers.

Sustainable urban infrastructure planning involves balancing economic growth with environmental protection and social equity. This means that infrastructure projects must be designed to minimize their impact on the environment while also providing benefits to all citizens and communities of the city.

1.7 ENVIRONMENTAL, SOCIAL, AND ECONOMIC CONSIDERATIONS IN URBAN ENGINEERING

Sustainable development is a concept that aims to balance the needs of the present generation with the preservation of future generations. It involves considering the social, economic, and environmental aspects of human activities and their impacts on natural resources and ecosystems. Social aspects of sustainable development include ensuring human rights, equity, justice, participation, education, health, and well-being for all people. Economic aspects of sustainable development include promoting green growth, innovation, efficiency, productivity, competitiveness, and resilience for all sectors and regions. Environmental aspects of sustainable development include protecting biodiversity, climate, water, air, soil, and waste for the benefit of all life forms. By integrating and balancing these three aspects, sustainable development can create a harmonious and prosperous society that respects the planet and its diversity. Sustainable development requires the collaboration and cooperation of all stakeholders, such as governments, businesses, civil society, academia, and individuals. It also requires a long-term vision and a holistic approach that considers the interconnections and trade-offs among different goals and actions.

Sustainable development and urban engineering are closely related fields, as infrastructure development in cities through the employment of urban engineering principles may support the accomplishment of SDGs in cities. While sustainable development is a global vision that seeks to balance social, economic, and environmental goals, urban engineering is a discipline that applies engineering principles and methods to design, construct, operate, and maintain urban systems and infrastructure. Urban engineering can contribute to sustainable development by providing solutions for efficient resource management, resilient infrastructure, low-carbon transportation, green buildings, smart cities, and inclusive communities. Some examples of urban engineering projects that support sustainable development are:

1. The Bus Rapid Transit (BRT) system in Curitiba, Brazil, reduces traffic congestion, air pollution, and greenhouse gas emissions by offering a fast, reliable, and affordable public transport service (see Box 1.2).
2. The Sponge City initiative in China aims to enhance the natural water cycle and reduce urban flooding by creating permeable surfaces, green spaces, and water storage facilities.
3. The Masdar City project in Abu Dhabi, United Arab Emirates, is designed to be a carbon-neutral and zero-waste city powered by renewable energy sources and featured with innovative technologies for energy efficiency and waste management.
4. The Kibera Slum Upgrading Program in Nairobi, Kenya, improves the living conditions and social inclusion of slum dwellers by providing access to basic services such as water, sanitation, electricity, and healthcare.

These projects demonstrate how urban engineering can play a vital role in delivering SDGs using scientific knowledge and experience to turn innovative ideas into sustainability projects for the benefit of all.

BOX 1.2 CURITIBA SUSTAINABLE URBAN TRANSPORT

Curitiba is a city that has transformed its public transportation system into a model of efficiency, sustainability, and social inclusion. The BRT system, which was implemented in 1974, is the backbone of this success story. The BRT system consists of a network of bus routes that operate on dedicated lanes, with tube-shaped stations that allow for fast and easy boarding and alighting. The system also integrates different types of buses, such as express, direct, feeder, and inter-district, to provide a comprehensive service that covers the entire city and metropolitan area. The BRT system is used by about 75% of the population, who pay a flat fare regardless of the distance traveled. The system has reduced traffic congestion, air pollution, greenhouse gas emissions, and travel time for millions of people. It has also stimulated urban development along the corridors, creating mixed-use areas that offer housing, commerce, services, and leisure opportunities. The BRT system has been recognized internationally as a best practice in urban mobility and has inspired many other cities worldwide to adopt similar solutions. The BRT system is not only a transport innovation but also a social one, as it promotes accessibility, equity, and quality of life for all citizens. Curitiba's BRT system is a testament to the vision and leadership of the city's planners and politicians, who have made public transportation a priority and a source of pride for the city.

Source: Terminal Cic. www.urbs.curitiba.pr.gov.br/

Environmental consideration in infrastructure development is essential for achieving the goals of sustainable development and climate action. Infrastructure projects can have significant impacts on the natural environment, such as biodiversity loss, habitat fragmentation, pollution, and greenhouse gas emissions. Therefore, planners and investors must adopt a holistic, systematic, and integrated approach to infrastructure development that considers the long-term impacts, risks, and trade-offs of different options. They must also consider the effects of climate change on the infrastructure's resilience and performance as well as the potential of infrastructure to contribute to climate mitigation and adaptation. Moreover, they must engage local communities and stakeholders at the earliest possible stages of planning to ensure that their needs and preferences are reflected in the design and implementation of the infrastructure. Adhering to these principles, infrastructure development can foster climate-compatible progress and yield optimal social, environmental, and economic outcomes for people and nature.

Similarly, economic consideration in infrastructure development is a crucial factor that influences the planning, design, construction, operation, and maintenance of public works. Infrastructure projects, such as roads, bridges, railways, airports, dams, power plants, water supply, and sanitation systems, have significant impacts on the economic growth, social welfare, and environmental quality of a country. Therefore, it is important to assess the costs and benefits of different alternatives and select the most efficient and effective option that maximizes net social welfare. Economic analysis of infrastructure projects involves estimating the present value of the capital and operating costs as well as the benefits that accrue to the users and society. The benefits may include direct effects, such as reduced travel time, increased safety, improved reliability, and enhanced service quality, as well as indirect effects, such as increased productivity, employment, income, trade, and competitiveness. The costs and benefits are then compared using various criteria, such as benefit-cost ratio, net present value, internal rate of return, and economic rate of return. These criteria help to rank the alternatives and identify the optimal solution that satisfies the technical, financial, and social objectives of the project.

The other aspect that we need to consider in infrastructure development is the social aspect. Why do we need to consider social aspects in infrastructure development?

Infrastructure development is a key factor for economic growth and human well-being. However, infrastructure projects can also have significant social impacts, both positive and negative, on the affected communities and stakeholders. Therefore, it is essential to consider the social aspects of infrastructure development to ensure that the projects are sustainable, inclusive, and beneficial for all.

What are the social aspects of infrastructure development? The social aspects of infrastructure development refer to the several ways that infrastructure projects affect the social environment, such as:

1. The quality of life, health, safety, and well-being of the people who use or live near the infrastructure facilities.
2. The access to and affordability of essential services, such as water, sanitation, electricity, transport, education, and healthcare.
3. The participation and empowerment of the local communities and stakeholders in the planning, design, implementation, and monitoring of the infrastructure projects.
4. The respect for and protection of the human rights, dignity, and culture of the affected people, especially the vulnerable and marginalized groups, such as women, children, Indigenous people, ethnic minorities, and persons with disabilities.
5. The prevention and mitigation of potential social risks and conflicts, such as displacement, resettlement, land acquisition, environmental degradation, social exclusion, and inequality.

Considering the social aspects of infrastructure development is important for several reasons, such as:

1. To enhance the effectiveness and efficiency of the infrastructure projects by ensuring that they meet the needs and expectations of the intended beneficiaries and users.
2. To maximize the positive social impacts and benefits of the infrastructure projects by creating opportunities for social development, poverty reduction, and improved livelihoods.
3. To minimize the negative social impacts and costs of the infrastructure projects by avoiding or reducing adverse effects on the affected people and environment.
4. To promote the sustainability and resilience of the infrastructure projects by ensuring that they are compatible with the local context and culture and that they can cope with future changes and challenges.
5. To foster the social acceptance and legitimacy of the infrastructure projects by engaging and involving the relevant stakeholders and ensuring transparency and accountability.

Some various tools and methods that can help us consider the social aspects of infrastructure development are shown in Table 1.2.

Considering the social aspects of infrastructure development is not only a moral duty but also a strategic advantage. When considering the social aspects of infrastructure projects, we can ensure they are not only technically viable but also socially responsible. We can also create more value for our investments by enhancing the quality of life, well-being, and satisfaction of our customers and beneficiaries. Therefore, we should always strive to integrate the social aspects into our infrastructure development processes and practices. Please take note that the goal of urban infrastructure development is citizens' welfare, and citizens are one of the components of social attributes. It becomes void if infrastructure development ignores the social aspect.

1.8 URBAN DEVELOPMENT AND SUSTAINABILITY

We subscribe to the mainstream concept of sustainability, which is a concept that refers to the ability of a system to meet the needs of the present without compromising the ability of future generations

TABLE 1.2
Tools and Methods of Assessing Social Aspects of Infrastructure Development

Name of Tools or Methods	Explanation
Social impact assessment (SIA)	SIA is a process of identifying, analyzing, evaluating, and managing the potential social impacts of an infrastructure project throughout its life cycle.
Social safeguards policies and standards (SSPSs)	SSPSs are a set of principles and guidelines that aim to prevent or mitigate the negative social impacts of an infrastructure project and to enhance its positive outcomes.
Social development plans (SDPs)	SDPs are a set of actions and measures that aim to address specific social issues and challenges related to an infrastructure project and to promote its social benefits.
Stakeholder engagement (SE)	SE is a process of communicating and consulting with the affected people and other relevant parties about an infrastructure project and its potential impacts.
Grievance redress mechanisms (GRMs)	GRMs are a means of receiving, responding to, and resolving complaints or concerns raised by the affected people or other stakeholders about an infrastructure project or its impacts.

to meet their own needs. Referring to the original concept of sustainable development, sustainability encompasses three dimensions: environmental, social, and economic. In the context of urban development, sustainability implies that cities should be designed and managed in a way that minimizes their negative impacts on the natural environment, promotes social equity and inclusion, and fosters economic growth and innovation. On the other hand, urban development is the process of creating and improving the physical, social, and economic conditions of urban areas. Urban development can involve various aspects such as land use planning, infrastructure provision, housing construction, transportation systems, public services, cultural activities, and governance. Urban development can have significant effects on the environment, society, and economy, both locally and globally.

The relevance of the sustainability aspect in urban development is evident from the fact that cities are facing multiple challenges and opportunities in the 21st century. Challenges and opportunities of cities in the 21st century are shown in Table 1.3.

Opportunities in cities from the perspective of socioeconomics are the chances and possibilities for people to improve their well-being, income, education, health, and quality of life by living in urban areas. Cities offer more opportunities than rural areas because they have higher population density, more economic activity, more infrastructure, more services, and more diversity. However, not all cities are equally attractive or accessible for people seeking opportunities. Some factors that affect the opportunities in cities are:

1. *Accessibility:* This is the ease and convenience of reaching and using the facilities, services, and socioeconomic opportunities in a city. Accessibility depends on the availability, affordability, and quality of transport infrastructure and services, as well as the spatial distribution of opportunities within and between cities. Accessibility can also be influenced by social factors such as safety, security, and inclusion. Higher accessibility can enhance the productivity, efficiency, and competitiveness of a city, as well as the well-being and satisfaction of its residents. However, accessibility can also create challenges such as congestion, pollution, inequality, and sprawl if not managed properly.
2. *Revenue:* This is the income that a city generates from various sources such as service fees, fines, charges, taxes, and assets. Revenue is essential for a city to provide public goods and services, invest in infrastructure and development projects, and support social programs and policies. Revenue can also reflect the economic performance and attractiveness of a city, as well as its fiscal autonomy and capacity. However, revenue can also pose challenges such as inefficiency, corruption, inequality, and dependency if not collected and allocated properly.

TABLE 1.3
Challenges and Opportunities of Cities in the 21st Century

Challenges/Opportunities	Explanations
Rapid urbanization	According to the United Nations, more than half of the world's population lives in urban areas, and this proportion is expected to increase to 68% by 2050. This means that cities will have to accommodate more people and provide more services and resources while dealing with limited space and increasing pressures on the environment.
Climate change	Cities are both contributors and victims of climate change. Cities account for about 70% of global greenhouse gas emissions, from energy use, transport, and industry. At the same time, cities are vulnerable to the impacts of climate change, such as increasing sea levels, extreme weather events, heat waves, droughts, and floods. These impacts can threaten the lives and livelihoods of millions of urban dwellers, especially the poor and marginalized.
Social inequality	Cities are also characterized by prominent levels of social inequality and exclusion. Many urban residents lack access to adequate housing, water, sanitation, healthcare, education, and employment opportunities. Moreover, cities are often divided by spatial segregation, discrimination, and violence along lines of class, ethnicity, gender, and other factors. These conditions can undermine social cohesion and human dignity in urban areas.
Economic uncertainty	Cities are also facing economic uncertainty and instability in the context of globalization, digitalization, and innovation. Cities are hubs of economic activity and innovation, generating about 80% of global GDP. However, cities also face competition from other regions and countries, as well as challenges such as unemployment, poverty, informality, and corruption. These factors can affect the economic performance and resilience of cities in the face of shocks and crises.
Urban potentials	Despite their parasitic characteristics, cities can also be sources of solutions for sustainable development. Cities can leverage their creativity, diversity, and dynamism to develop and implement innovative approaches to address their challenges and improve their conditions. For example, cities can adopt low-carbon technologies; use renewable energy sources, green buildings, smart grids; develop circular economy models; employ efficient and sufficient public transport systems; generate urban agriculture; adopt sustainable waste management systems and other measures to reduce their environmental footprint; and enhance their resource efficiency.
Social inclusion	Cities can also promote social inclusion and equity in their development processes. Cities can foster participatory and inclusive governance structures that enable all urban stakeholders to have a voice and a role in decision-making and implementation. Cities can also ensure that all urban residents have access to basic services and opportunities that improve their quality of life and well-being. Moreover, cities can celebrate their cultural diversity and heritage as assets for social cohesion and identity.
Economic prosperity	Cities can also sustainably pursue economic prosperity and innovation. Cities can enhance their competitiveness and attractiveness by investing in human capital and infrastructure.

3. *Innovation:* This is the creation and adoption of innovative ideas, products, processes, or methods that improve the existing or generate new value for a city. Innovation can be driven by numerous factors such as technology, research, education, entrepreneurship, culture, and collaboration. Innovation can enhance the competitiveness, resilience, and sustainability of a city, as well as the creativity and diversity of its residents. However, innovation can also create challenges such as disruption, uncertainty, risk, and exclusion if not supported and regulated properly.

Urban development and sustainability are essential for the future of our cities. Implementing responsible growth strategies and improving the quality of life for all citizens are essential. We can create vibrant, livable, and sustainable urban environments for generations to come. It is up to city leaders, planners, citizens, and other city stakeholders to work together to achieve this vision. Let us build a better future for our cities.

1.9 NATURAL RESOURCES-BASED DEVELOPMENT

Natural resources-based development is a strategy that aims to use the endowment of natural resources in a country or region to foster economic growth, social development, and environmental sustainability. Natural resources-based development can be a source of comparative advantage, employment, income, public revenue, and foreign exchange for resource-rich countries. However, natural resources-based development also poses significant challenges and risks, such as resource dependence, volatility, depletion, corruption, conflict, environmental degradation, and inequality.

One example of a natural resources-based development is exhibited by Brazil. Brazil, the largest country in South America, is known for its vast agricultural resources, particularly sugarcane. The country has leveraged these resources to become a global leader in the production and use of biofuels, specifically ethanol, as shown in Box 1.3.

BOX 1.3 BIOFUELS FOR SUSTAINABLE DEVELOPMENT IN BRAZIL

The biofuel industry in Brazil began in the 1970s with the launch of the ProAlcool program, a government initiative aimed at reducing dependence on foreign oil. The program encouraged the production of ethanol from sugarcane and its use as a fuel for vehicles. Today, Brazil is the world's second-largest producer of ethanol, and most cars in the country are flex-fuel vehicles that can run on either gasoline or ethanol.

The development of the biofuel industry in Brazil has had significant economic, social, and environmental benefits. It has created jobs, stimulated rural development, reduced air pollution, and contributed to the mitigation of climate change by reducing greenhouse gas emissions.

Here are some quantitative data on Brazil's biofuel industry:

- *Production:* Total Brazilian ethanol production for 2022 is estimated at 31.66 billion liters, an increase of 6% relative to 2021 due to the expected increase in sugarcane production and the steady increase in corn ethanol production.
- *Market:* The Brazil Biofuel Market size is estimated at 46.54 million cubic meters in 2024 and is expected to reach 55.25 million cubic meters by 2029, growing at a compound annual growth rate (CAGR) of 3.49% during the forecast period (2024–2029).
- *Employment:* The biofuel industry in Brazil employs over 1 million people, many of whom are in rural areas.
- *Environmental Impact:* The use of ethanol instead of gasoline in Brazil has reduced the country's carbon dioxide emissions by over 350 million tons since the 1970s.

Brazil's experience with biofuels illustrates how a country can leverage its natural resources for sustainable development. It underscores the importance of government policy, technological innovation, and market development in promoting the sustainable use of natural resources.

Sources: Knowledge.energyinst.org, fas.usda.gov, mordorintelligence.com

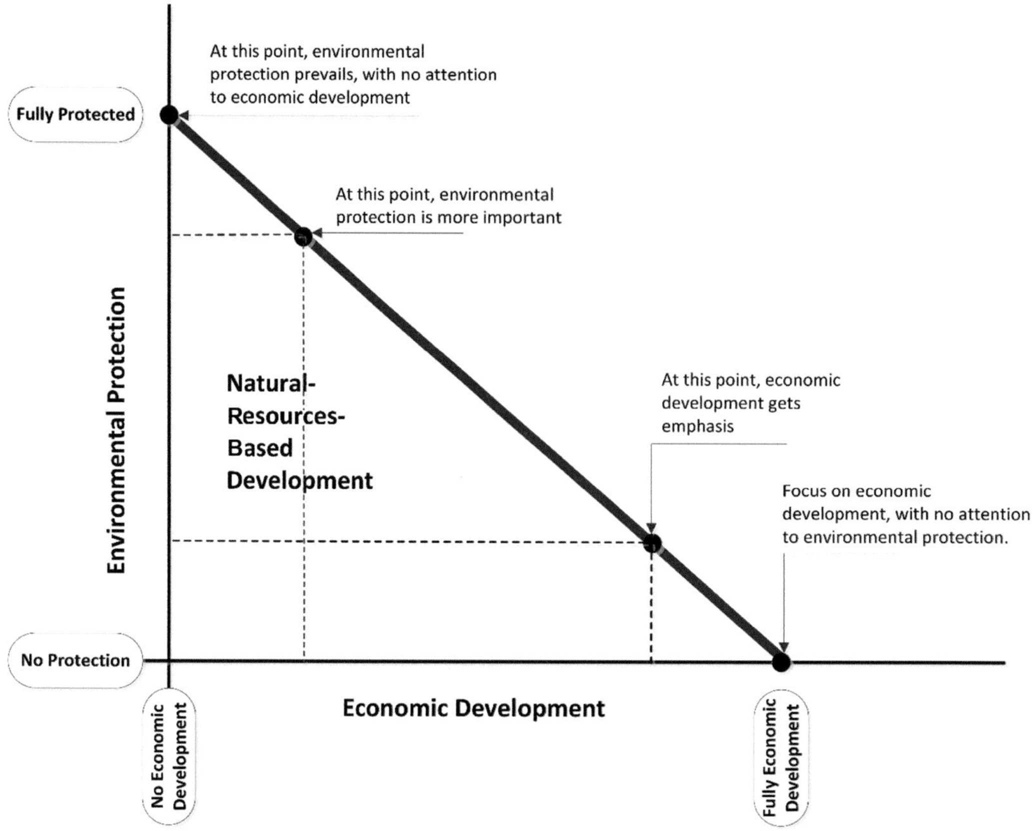

FIGURE 1.2 The dilemma between urban development and environmental protection.

The dilemma of city development and environmental protection is a complex issue that requires careful thinking to deal with. On one hand, city development often requires the use of natural resources such as sand, gravel, boulders, and other various mineral sources, water, and energy. These resources are essential for building infrastructure, providing basic services, and supporting economic growth. On the other hand, environmental protection aims to preserve natural resources and prevent their depletion. This involves conserving ecosystems, reducing pollution, and mitigating the impacts of climate change. Balancing these two objectives can be challenging. City development can have negative impacts on the environment, such as deforestation, habitat destruction, and air and water pollution. At the same time, strict environmental regulations can hinder economic growth and limit the availability of resources for city development. The situation is illustrated graphically in Figure 1.2.

The dilemma between economic development and environmental protection is an intricate issue, despite involving only two opposing sides, which has been debated for many years. Both are important for the well-being of society, but they often are at odds with each other. Economic development is necessary for improving the standard of living and providing opportunities for citizens, while environmental protection is essential for preserving the natural world and ensuring a sustainable future. It is important to realize that both economic development and environmental protection are vital for the well-being of society. The challenge is to find a balance between these two objectives and implement policies that promote sustainable growth while also protecting the environment. This requires thorough planning, collaboration, and a long-term perspective, as we cannot easily find the balance between the two. The balance does not necessarily mean that the proportion is 50% by 50%.

Some argue that economic development should be given priority because it provides immediate benefits to society. Economic growth can create jobs, increase incomes, and improve the standard of living for people. It can also provide the resources necessary to invest in infrastructure, education, and healthcare. These benefits are tangible and can have a significant impact on people's lives. On the other hand, others argue that environmental protection should be given priority because it is essential for the long-term sustainability of our planet. The natural world provides us with the resources we need to survive, such as clean air, water, and food. If we do not protect the environment, these resources could become depleted or polluted, making it difficult for future generations to meet their needs. Environmental protection also helps to preserve biodiversity and maintain the balance of ecosystems.

The decision about whether to prioritize economic development or environmental protection depends on a variety of factors, including the specific context and needs of society. In some cases, economic development may be necessary to address immediate needs such as poverty or unemployment. In other cases, environmental protection may be more important to ensure a sustainable future. It therefore depends on the policies of the city or country, to which side priority should be given. As an illustration, in a developed city, in which economic development is at an advanced stage, priority is supposed to be given to the preservation of natural resources. On the other end, for a developing city or country with a low capacity of human resources, but with abundant natural resources, the use of natural resources for the development of the city or country along with human resources development could not be sidestepped.

Kuznets curve, however, illustrates the correlation between the level of emission production and the income level. Based on global data, the correlation shows an inverted U curve. This means that as income increases, the level of emission generation also increases until a certain point where it reaches its peak (based on the global data for 2022, the peak of the per-capita income level is around 40–50 thousand US dollars annually). After this peak point, as income continues to increase, the level of emission generation starts to decrease. This fact shows that a country, at the initial stage of development, runs on natural resources-based development to progress its economy. While economic development is progressing, the dependence on natural resources lessens and industry shifts to clean and green technology. As a result, the high-income countries, with their technological advancement, were able to limit the use of their natural resources for their development to minimize pollution. The economic development of low-income countries depends on natural resources or natural resources-based economic development. Therefore, the use of natural resources for development cannot be evaded. Consequently, the excessive use of natural resources that degrades the environment takes place. They have little or no industry. If industrial processes take place, the types of machinery used are mostly nongreen and non-clean or polluted industries. This situation is illustrated in Figure 1.3.

City leaders must adopt a sustainable approach to urban development. This involves implementing strategies that promote economic growth while also protecting the environment. For example, cities can invest in renewable energy sources, promote sustainable transportation options, and implement green building practices. By taking a holistic approach to urban development and environmental protection, cities can create a more sustainable future for their residents.

The use of natural resources is essential and cannot be prevented for the development of cities. Natural resources (primary sources) or processed from natural resources (secondary or tertiary sources) such as sand, gravel, boulder, clay, water, steel, cement, and woods are essential for city development. They are used for various purposes such as the construction of various buildings, roads, bridges, dams, tunnels, houses, and other infrastructure. However, the use of these resources also has environmental and social impacts that need to be considered. For example, excessive extraction of sand and gravel can cause erosion, habitat loss, water pollution, and land degradation (Box 1.4). Moreover, the demand for these resources can create conflicts among different stakeholders such as local communities, governments, and private companies. Therefore, it is important to use

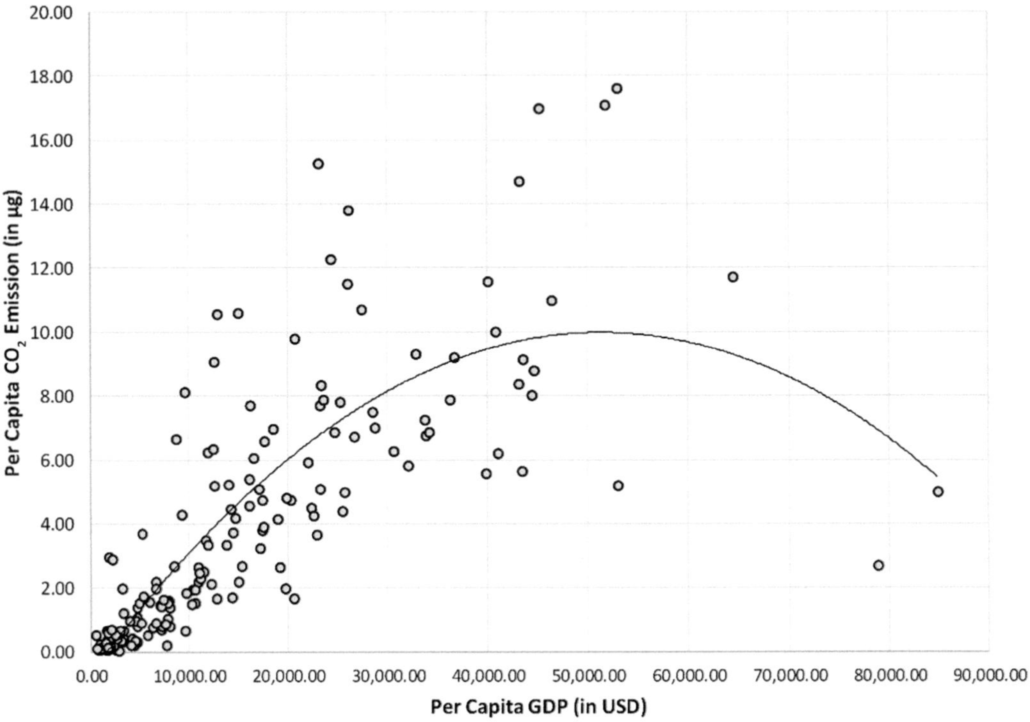

FIGURE 1.3 Inverted U Kuznets curve.

natural resources sustainably and responsibly to balance the needs of city development and the protection of the environment and human rights.

BOX 1.4 SAND EXTRACTION IN INDONESIA FOR SINGAPORE

Sand Extraction Creating Environmental Problems in Indonesia: The recent policy to resume sea sand exports to Singapore has sparked controversy among environmentalists, who warn that it will lead to an increase in sand extraction, which they say will be detrimental to coastal dwellers dependent on ocean ecosystems. They argue that sand mining causes erosion, loss of biodiversity, destruction of coral reefs, and salinization of groundwater. Moreover, they claim that the sand exported to Singapore is used for land reclamation projects that encroach on Indonesia's territorial waters and sovereignty. On the other hand, miners welcomed the policy, saying that it will boost their income and create more jobs in the sector (Asian News Network).

Source: Asian News Network

City development needs natural resources. These resources are not available within the city limits and must be extracted and transported from other regions. This creates an environmental impact, that is, increasing the environmental footprint in the region where the natural resources were extracted and processed. This fact is against the argument that the city–state of Singapore accomplishes sustainable development. It is not. Singapore does not have all its natural resources within the boundaries of Singapore. The natural resources used for the development of Singapore are imported from outside Singapore. Sands, gravel, and boulders were imported from Myanmar and Indonesia; soil comes from the Riau province of Indonesia; and water is supplied by the State of Johor, Malaysia. Singapore itself either does not disturb its natural resources, or it does not have natural resources at all. It brings us to the fact that the development of Singapore is at the price of the environmental degradation in those countries. This fact shows that sustainable development cannot be measured from local boundaries. It must be measured globally. It is therefore the jargon of "think globally, act locally" still very much valid.

Excessive groundwater extraction in Jakarta, Indonesia, for example, has led to severe land subsidence. The city, largely sitting on loose and unconsolidated soils and sediments, is sinking at a rate of up to 26 cm a year, making it one of the fastest-sinking megacities in the world. This over-extraction is primarily driven by the city's growing population and their demand for water (Box 1.5).

BOX 1.5 EXCESSIVE GROUNDWATER EXTRACTION IN JAKARTA

Jakarta is a megacity built on a delta of 13 rivers that flow into the Java Sea. The city has a population of about 10 million people and covers an area of 661 km^2. Jakarta faces many water challenges, such as flooding, pollution, saltwater intrusion, and water scarcity.

Groundwater is one of the main sources of water supply for Jakarta, especially for domestic and industrial uses. However, due to rapid urbanization and population growth, the demand for groundwater has exceeded the natural recharge rate, resulting in excessive groundwater extraction. Excessive groundwater extraction, particularly from deep aquifers, which is prevalent in Jakarta, results in massive pumping of both the water and air. This deep groundwater cannot be replenished immediately as the process is slow and can take up to 100 years. The pumping of water and air creates a void space in the aquifer, which causes the overlying soil layers to compact and settle. This phenomenon is known as land subsidence, which is the sinking of the land surface due to various natural or human-induced factors.

Land subsidence in Jakarta has been observed since the 1970s, but it has accelerated in recent decades due to the increasing rate and extent of groundwater extraction. According to various studies, Jakarta is one of the fastest-sinking megacities in the world, with some areas experiencing land subsidence rates of up to 26 cm/year. Land subsidence has serious consequences for Jakarta, such as increasing flood risk, damaging infrastructure, altering drainage patterns, and reducing land value. Moreover, land subsidence also enhances saltwater intrusion, which is the movement of saline water into freshwater aquifers or surface waters.

Sources: nationalgeographic.com, channelnewsasia.com, thehindu.com, theconversation.com, theconversation.com, cnbctv18.com, earthobservatory.nasa.gov, link.springer.com

Insufficient in both quality and quantity, the water supply service in a metropolitan has generated the excessive extraction of groundwater by the citizens and eventually leads to other environmental problems, that is, seawater intrusion and land subsidence that increase the vulnerability to flooding problems.

Natural resources-based development requires a sound and coherent policy framework that can address these challenges and risks and ensure that natural resources are managed and used in a way that optimizes the balance between natural resources used and economic development as well as maximizes the benefits for current and future generations. Such a policy framework should include the elements as shown in Table 1.4.

TABLE 1.4
Elements in Policy Framework

Element	Explanation
Vision and objective	A sharp vision and objectives for natural resource-based development that is aligned with the national development plan and the 2030 Agenda for Sustainable Development.
Governance	A comprehensive and participatory governance system that ensures transparency, accountability, inclusiveness, and legitimacy in the decision-making and implementation of natural resource-based development policies and projects.
Fiscal policy	A prudent and sustainable fiscal policy that balances the trade-offs between saving, spending, and investing natural resource revenues, and that supports macroeconomic stability, intergenerational equity, and public service delivery.
Economic structure	A diversified and competitive economic structure that reduces dependence on natural resources, fosters innovation and productivity, creates value-added activities and linkages, and enhances human capital and social protection.
Environmental management	Resilient and adaptive environmental management that minimizes the negative impacts of natural resource extraction and use on the ecosystem and human health, promotes the conservation and restoration of natural capital, and addresses the challenges of climate change and disaster risk reduction.

Natural resources-based development is not a curse or a blessing per se but rather an opportunity and a responsibility for resource-rich countries to harness their natural wealth for inclusive and sustainable development. The success of natural resources-based development depends on the quality of policies and institutions that govern the management and use of natural resources. Therefore, policymakers, stakeholders, and partners should work together to design and implement effective policies and practices that can transform natural resources into lasting development outcomes.

1.10 ACTORS AND STAKEHOLDERS OF THE URBAN DEVELOPMENT

Urban development involves a range of actors and stakeholders, who are the people or groups that have an interest or influence in the urban outcomes. The stakeholders include the government, the private sector, the citizens, and the intermediaries. They have different perspectives, needs, and preferences that should be considered and balanced in the planning process. Stakeholder involvement can enhance the quality and legitimacy of urban plans and projects, as it can foster a shared vision, consensus, and ownership among the stakeholders. It can also improve the transparency, accountability, and inclusiveness of the decision-making and implementation processes. Stakeholder involvement can increase the effectiveness and sustainability of urban plans and projects, as it can improve the understanding of urban problems and opportunities, identify the best solutions and alternatives, and mobilize the resources and capacities of the stakeholders. It can also reduce the risks of conflicts, delays, or failures that may arise from stakeholder dissatisfaction or opposition. Therefore, stakeholder involvement is a key factor for successful urban planning and infrastructure planning. It requires a systematic and participatory approach that can identify, engage, and empower the stakeholders throughout the project lifecycle. It also requires a clear communication strategy that can inform, consult, and collaborate with the stakeholders in a timely and appropriate manner.

As mentioned earlier, some of the main actors and stakeholders in urban development are reflected in Table 1.5.

The involvement of all stakeholders in urban planning, which can also be applied to infrastructure planning, is to increase the effectiveness of outcomes of the planning. The outputs of the planning

TABLE 1.5
Stakeholders in Urban Development

Stakeholder	Explanation
Government	The government is the main actor and regulator of urban development, as it sets the vision, policies, plans, and standards for the development of cities. The government also provides public services, infrastructure, and amenities that are essential for the well-being and prosperity of urban residents. The government can be divided into various levels, such as national, regional, and local, each with its roles and responsibilities. The government also interacts with other stakeholders, such as the private sector, civil society, and international organizations, to coordinate and collaborate on urban development issues.
Private sector	The private sector is the main driver and innovator of urban development, as it contributes to the economic growth, job creation, and competitiveness of cities. The private sector includes various actors, such as developers, investors, entrepreneurs, businesses, and industries, which operate in different sectors and markets. The private sector also provides goods and services that meet the needs and demands of urban consumers. The private sector works with the government to comply with the regulations and standards of urban development, as well as to benefit from the incentives and opportunities offered by the government.
Citizens	The citizens are the main beneficiaries and participants of urban development, as they live, work, and play in cities. The citizens include various groups of people, such as residents, workers, commuters, visitors, migrants, and minorities, who have divergent backgrounds, interests, and preferences. The citizens also have rights and duties as urban dwellers, such as the right to access basic services and amenities, and the duty to pay taxes and fees. The citizens engage with the government and the private sector to voice their opinions and concerns, as well as to influence and monitor the decisions and actions of urban development.
Community-based organizations (CBOs) and nongovernmental organizations (NGOs)	While the NGO, in most cases, works for a cause, the CBO works within a particular area for a particular case of its interests. They share a similar motive – performing social service and improving society. However, NGOs are usually larger and have a broader reach than CBOs. NGOs are often funded by international organizations or governments, while CBOs are often funded by local communities or individuals. Their roles in urban planning and development are: 1. Advocating for the rights and interests of marginalized groups, such as the poor, the homeless, people with disabilities, ethnic minorities, and women. NGOs and CBOs can raise awareness, mobilize resources, and influence policies to address the needs and challenges of these groups. 2. Participating in the planning and implementation of urban projects, such as housing, infrastructure, transportation, environment, and social services. NGOs and CBOs can provide valuable input, feedback, and expertise to improve the quality and sustainability of urban development. 3. Promoting community empowerment and participation by organizing workshops, forums, surveys, and campaigns to engage urban residents in the decision-making and problem-solving processes. NGOs and CBOs can also foster social cohesion, trust, and solidarity among urban communities. 4. Monitoring and evaluating the impacts and outcomes of urban policies and programs, by collecting data, conducting research, and producing reports to assess the effectiveness and efficiency of urban interventions. NGOs and CBOs can also hold the government and other stakeholders accountable for their actions and responsibilities.
Intermediaries	Intermediaries are the actors that implement or facilitate urban development policies and projects. Intermediaries include various groups or organizations that have specific functions or expertise in urban issues. For example: 1. Transport operators provide public or private transport services for urban mobility. 2. Infrastructure operators manage or maintain urban infrastructure such as roads or utilities. 3. Media outlets inform or report on urban affairs or events. 4. Interest groups represent or advocate for certain causes or sectors related to urban development. 5. Academic institutions conduct research or education on urban topics.

process will be more effective as the sense of responsibility of all stakeholders exists as they were involved from the beginning of the process, and the outputs are therefore part of their responsibility. The involvement of the stakeholders of urban planning, including citizens, is made possible through bottom-up planning in contrary to top-down planning.

Top-down planning is a method that starts with the big picture and then works its way down to the details. It involves defining the overall objectives and strategies first and then breaking them down into smaller tasks and actions, in which the planning decision is only made by top selected stakeholders, for example, government. Top-down planning is usually done by leaders or managers who have a sharp vision and authority over the project or organization, and the other stakeholders of the project are not involved in the planning process. Bottom-up planning, on the other hand, is a method that starts with the details and builds up to the big picture. It involves identifying the specific needs and preferences of the stakeholders first and then integrating them into a coherent framework or plan. Bottom-up planning is usually done by citizens or participants who have direct experience and knowledge of the project or organization.

Both top-down planning and bottom-up planning have their pros and cons, depending on the situation and context. Here are some of them:

1. Top-down planning can be more efficient, consistent, and organized, as it follows a clear direction and structure. It can also reduce confusion, risk, and conflict, as it eliminates ambiguity and uncertainty. However, top-down planning can also be more rigid, hierarchical, and authoritarian, as it limits the input and involvement of the lower-level stakeholders. It can also be less responsive, adaptable, and innovative, as it may overlook or ignore the changing realities and opportunities.
2. Bottom-up planning can be more flexible, democratic, and participatory, as it allows more input and involvement of the lower-level stakeholders. It can also be more responsive, adaptable, and innovative, as it can capture the changing realities and opportunities. However, bottom-up planning can also be more time-consuming, chaotic, and fragmented, as it lacks a clear direction and structure. It can also increase confusion, risk, and conflict, as it creates ambiguity and uncertainty.

Identifying the stakeholders is a crucial step in the planning process. There are many ways to identify stakeholders for a project, but some common methods are shown in Table 1.6.

After identifying the stakeholders, we should also analyze them based on their power, interest, influence, and expectations. This can help us prioritize and manage them effectively throughout the project lifecycle. We can use tools such as a stakeholder matrix or a stakeholder map to visualize and categorize the stakeholders.

Decision-making by stakeholders in the planning process is an important aspect of any project or organization. Stakeholders can influence the success or failure of a project or organization by providing feedback, support, resources, or opposition. Engaging stakeholders in the planning process can have several benefits, such as:

1. *Sharing Ideas and Solutions:* Stakeholders can help provide diverse perspectives and creative inputs that can improve the quality and feasibility of the plan.
2. *Problem-Solving Resources:* Stakeholders can help identify and resolve potential issues or risks that may arise during the execution of the plan.
3. *Employees as Stakeholders:* Stakeholders who are also employees can increase their motivation, commitment, and performance by being involved in the plan.
4. *Preventing Unforeseen Problems:* Stakeholders can help anticipate and avoid conflicts or challenges that may affect the plan.

TABLE 1.6
Stakeholders Identification Methods for Planning Process

Method	Explanation
Brainstorming	The process can commence by listing down all the possible stakeholders based on the project's scope, objectives, deliverables, and impacts. It can also be discussed with the project team, sponsor, or client for their input. Brainstorming can help to generate a comprehensive list of stakeholders without missing anyone important.
Organizational analysis	This process can start by looking at the organizational structure of the project and identifying the stakeholders based on their roles, responsibilities, and relationships. For example, from the organizational chart, we can see who reports to whom, who has decision-making authority, who has technical expertise, etc. Organizational analysis can help us to understand the power dynamics and communication channels among the stakeholders.
Environmental scanning	We can scan the external environment of the project and identify the stakeholders based on their interests, influences, and impacts. For example, we can use a political, economic, sociological, technological, legal and environmental (PESTLE) analysis to examine the PESTLE factors that may affect the project. Environmental scanning can help us to anticipate the opportunities and threats that may arise from external stakeholders.

However, engaging stakeholders in the planning process also requires some skills and strategies, such as:

1. *Stakeholder Analysis:* This is a process of identifying and prioritizing the stakeholders based on their power, interest, and influence on the project or organization.
2. *Stakeholder Communication:* This is a process of establishing and maintaining effective communication channels with the stakeholders to inform them of the plan's objectives, progress, and outcomes.
3. *Stakeholder Participation:* This is a process of involving the stakeholders in various stages of the planning process, such as defining the scope, setting the goals, developing the alternatives, evaluating the options, and implementing the actions.

Despite their critical roles in urban planning and development, the stakeholders may face several challenges, such as:

1. Lack of coordination and collaboration among stakeholders themselves can lead to conflicts, duplication of efforts, and suboptimal outcomes. The stakeholders may have different interests, priorities, or visions that can hinder their ability to work together effectively.
2. Inadequate legal frameworks and policies can create uncertainties, ambiguities, or contradictions in the roles and responsibilities of the stakeholders. The stakeholders may face legal restrictions, bureaucratic hurdles, or political pressures that prevent them from fulfilling their mandates or objectives.
3. Limited access to funding and resources can constrain the capacity of the stakeholders to carry out their activities and programs. The stakeholders may have to rely on external sources of funding or support that may not be sustainable or sufficient to meet the demands of urban development.
4. Insufficient participation and representation of certain groups in the decision-making processes and benefits of urban development. The stakeholders may not adequately engage or empower the marginalized or vulnerable groups in the urban areas.

5. Inadequate information and data management systems can impede the stakeholders' ability to make informed decisions and proposals. The stakeholders may not have the technical skills, equipment, or networks to collect, analyze, and disseminate data on urban issues.

These challenges are complex and interrelated, requiring a holistic and collaborative approach to address them. The stakeholders need to foster trust, communication, innovation, and accountability among themselves to achieve sustainable and inclusive urban development.

PROBLEMS AND ISSUES FOR DISCUSSION

1. *Urban Development:* How does urbanization impact infrastructure planning and community well-being?
2. *Urban Management:* What strategies can enhance efficient waste management in growing cities?
3. *Economic Development:* How can sustainable economic growth be achieved without compromising environmental quality?
4. *Environmental Protection:* What policies are effective in safeguarding natural resources within urban areas?
5. *Kuznets Curve:* Explain the Kuznets curve hypothesis and its relevance to urban development.
6. *Stakeholders in Urban Development:* Who are the key stakeholders involved in shaping urban landscapes?
7. *Environmental Management:* How can cities balance economic progress with ecological conservation?
8. *Dilemma of Economic Development versus Environmental Protection:* Discuss the trade-offs between economic prosperity and environmental sustainability.
9. *Smart Cities and Sustainability:* How can technology contribute to greener, more resilient urban environments?
10. *Equity in Urban Development:* What measures promote equitable access to resources and opportunities in cities?
11. *Gentrification and Social Equity:* Discuss the impact of gentrification on marginalized communities and potential mitigation strategies.
12. *Gentrification and Social Equity:* Discuss the impact of gentrification on marginalized communities and potential mitigation strategies.
13. Why does urban development in developing countries tend to be influenced more by urban economics than by urban planning, resulting in organic city growth?

MULTIPLE-CHOICE QUESTIONS

1. Urban development primarily involves:
   a. Building skyscrapers
   b. Expanding city limits
   c. Infrastructure development and land use planning
   d. Increasing population density.
2. Which of the following is a key aspect of urban planning?
   a. Predicting population growth
   b. Designing city layouts
   c. Managing public transportation
   d. All of the above.

3. Urban management focuses on:
   a. Building parks and recreational areas
   b. Implementing policies and plans for urban areas
   c. Designing city skylines
   d. Increasing the number of residential buildings.
4. Environmental management in urban areas is important for:
   a. Reducing pollution
   b. Conserving natural resources
   c. Ensuring sustainable development
   d. All of the above.
5. Sustainable urban development aims to:
   a. Increase the city's population
   b. Build more commercial establishments
   c. Balance development with environmental protection
   d. Increase the city's land area.
6. Economic development in a city can be measured by:
   a. The number of skyscrapers
   b. The number of public parks
   c. The number of schools
   d. Job growth and income levels.
7. Environmental protection in urban areas can involve:
   a. Implementing recycling programs
   b. Reducing carbon emissions
   c. Promoting renewable energy
   d. All of the above.
8. Stakeholders in urban development include:
   a. Residents
   b. City officials
   c. Business owners
   d. All of the above.
9. Which of the following is a challenge in urban planning?
   a. Ensuring adequate public transportation
   b. Managing waste disposal
   c. Balancing development with environmental conservation
   d. All of the above.
10. Urban sprawl refers to:
    a. The spread of urban areas into neighboring rural areas
    b. The growth of skyscrapers in a city
    c. The increase in population density in a city
    d. The expansion of public transportation networks.
11. A green city prioritizes:
    a. The construction of parks and green spaces
    b. The use of renewable energy sources
    c. Sustainable waste management practices
    d. All of the above.
12. Gentrification often results in:
    a. Increased property values
    b. Displacement of lower-income residents
    c. Changes in the character of a neighborhood
    d. All of the above.

13. Smart cities use technology to:
    a. Improve infrastructure
    b. Enhance public services
    c. Increase sustainability
    d. All of the above.
14. Urban renewal involves:
    a. The redevelopment of areas in decline
    b. The construction of new buildings
    c. The renovation of existing structures
    d. All of the above.
15. Public participation in urban planning is important because:
    a. It ensures the needs of all residents are considered.
    b. It helps build community support for projects.
    c. It can lead to better decision-making.
    d. All of the above.
16. Which of the following is a principle of sustainable urban development?
    a. Economic growth at all costs
    b. Development without consideration for the environment
    c. Balancing economic, social, and environmental needs
    d. Prioritizing urban development over rural development
17. In the context of urban development, mixed-use development refers to:
    a. The use of a variety of building materials
    b. The combination of residential, commercial, and other uses in a single development
    c. The use of both public and private funding for development projects
    d. The development of both urban and rural areas.
18. Which of the following is a strategy for environmental management in cities?
    a. Increasing the number of cars on the road
    b. Reducing green spaces
    c. Promoting energy efficiency and conservation
    d. Ignoring air and water pollution.
19. Economic development in cities can be promoted by:
    a. Attracting new businesses
    b. Investing in infrastructure
    c. Enhancing education and workforce development
    d. All of the above.
20. Stakeholders in cities can influence urban development by:
    a. Participating in public meetings and hearings
    b. Voting in local elections
    c. Advocating for policies and projects that meet their needs
    d. All of the above.
21. Which of the following is a key principle of urban management?
    a. Prioritizing economic growth over environmental protection
    b. Ignoring the needs of marginalized communities
    c. Ensuring equitable access to urban services and resources
    d. Focusing solely on infrastructure development.
22. What is the main goal of environmental management in urban areas?
    a. To increase the city's population density
    b. To ensure sustainable use and conservation of environmental resources
    c. To build as many skyscrapers as possible
    d. To ignore the impact of urbanization on the environment.

23. Sustainable urban development is characterized by:
    a. Rapid, uncontrolled growth
    b. High levels of pollution and waste
    c. Balanced growth that meets the needs of the present without compromising the ability of future generations to meet their own needs
    d. A focus on economic development at the expense of social equity and environmental protection.
24. Economic development in a city is often associated with:
    a. Decreased employment opportunities
    b. Lower standards of living
    c. Increased poverty and inequality
    d. Improved living standards and prosperity.
25. Environmental protection in urban areas can be achieved through:
    a. Unregulated industrial growth
    b. Overuse of natural resources
    c. Implementation of sustainable practices and policies
    d. Ignoring the impact of urban activities on the environment.
26. Stakeholders in urban development typically include:
    a. Only government officials
    b. Only business owners
    c. Only residents
    d. Government officials, business owners, residents, and other community members.
27. Which of the following is a common challenge in urban planning?
    a. Ensuring adequate housing for all residents
    b. Managing traffic and transportation systems
    c. Balancing development with environmental sustainability
    d. All of the above.
28. Urban sprawl can lead to:
    a. Reduced commuting times
    b. Increased green spaces
    c. Loss of agricultural land and natural habitats
    d. Decreased air and water pollution.
29. A green city is characterized by:
    a. High levels of air and water pollution
    b. Limited access to parks and green spaces
    c. Sustainable practices, renewable energy use, and a focus on environmental conservation
    d. Rapid, uncontrolled urban growth.
30. Gentrification often results in:
    a. Decreased property values
    b. Increased diversity and inclusivity
    c. Displacement of lower-income residents due to increasing property values
    d. Improved living conditions for all residents.
31. Smart cities typically use technology to:
    a. Increase pollution levels
    b. Improve public services and infrastructure
    c. Decrease connectivity and communication
    d. Ignore environmental sustainability.

32. Urban renewal often involves:
    a. The redevelopment of areas in decline
    b. The construction of new skyscrapers
    c. The expansion of city limits
    d. The reduction of green spaces.
33. Public participation in urban planning is important because:
    a. It can lead to better decision-making and community support.
    b. It ensures the needs of all residents are considered.
    c. It allows for the implementation of top-down policies.
    d. Both a and b.
34. Sustainable urban development prioritizes:
    a. Economic growth at all costs
    b. Development without consideration for the environment
    c. Balancing economic, social, and environmental needs
    d. Prioritizing urban development over rural development.
35. Mixed-use development in the context of urban development refers to:
    a. The combination of residential, commercial, and other uses in a single development
    b. The use of a variety of building materials
    c. The use of both public and private funding for development projects
    d. The development of both urban and rural areas.
36. A strategy for environmental management in cities could be:
    a. Increasing the number of cars on the road
    b. Reducing green spaces
    c. Promoting energy efficiency and conservation
    d. Ignoring air and water pollution.
37. Economic development in cities can be promoted by:
    a. Attracting new businesses
    b. Investing in infrastructure
    c. Enhancing education and workforce development
    d. All of the above.
38. Stakeholders in cities can influence urban development by:
    a. Participating in public meetings and hearings
    b. Voting in local elections
    c. Advocating for policies and projects that meet their needs
    d. All of the above.
39. Which of the following is a key aspect of urban management?
    a. Ignoring the needs of marginalized communities
    b. Prioritizing economic growth over environmental protection
    c. Ensuring equitable access to urban services and resources
    d. Focusing solely on infrastructure development.
40. What is the main goal of environmental management in urban areas?
    a. To increase the city's population density
    b. To ensure sustainable use and conservation of environmental resources
    c. To build as many skyscrapers as possible
    d. To ignore the impact of urbanization on the environment.

ANSWERS TO MULTIPLE-CHOICE QUESTIONS

1	c	11	d	21	c	31	b
2	d	12	d	22	b	32	a
3	b	13	d	23	c	33	d
4	d	14	a	24	d	34	c
5	c	15	d	25	c	35	a
6	d	16	c	26	d	36	c
7	d	17	b	27	d	37	d
8	d	18	c	28	c	38	d
9	d	19	d	29	c	39	c
10	a	20	d	30	c	40	b

FURTHER READING

Angel, S., Parent, J., Civco, D. L., & Blei, A. M. (2011). *Atlas of urban expansion*. Lincoln Institute of Land Policy.

Batty, M. (2013). *The new science of cities*. MIT Press.

Breheny, M. (Ed.). (2013). *The compact city: A sustainable urban form?* Routledge.

Carmona, M., Heath, T., Oc, T., & Tiesdell, S. (2010). *Public places, urban spaces: The dimensions of urban design*. Routledge.

Cervantes Puma, G., Salles, A., & Bragança, L. (2024). Nexus between urban circular economies and sustainable development goals: A systematic literature review. *Sustainability*, 16(6), 2500.

Florida, R. (2002). *The rise of the creative class: And how it's transforming work, leisure, community and everyday life*. Basic Books.

Glaeser, E. L. (2011). *Triumph of the city: How our greatest invention makes us richer, smarter, greener, healthier, and happier*. Penguin Books.

Grossman, G. M., & Krueger, A. B. (1991). Environmental impacts of a North American free trade agreement. NBER Working Paper Series, No. 3914.

Hall, P. (2014). *Good cities, better lives: How Europe discovered the lost art of urbanism*. Routledge.

Hardin, G. (1968). The tragedy of the commons. *Science*, 162(3859), 1243–1248.

Hardoy, J. E., Mitlin, D., & Satterthwaite, D. (Eds.). (2001). *Environmental problems in an urbanizing world: Finding solutions for cities in Africa, Asia and Latin America*. Earthscan.

Jacobs, J. (1961). *The death and life of great American cities**. Random House.

Luo, P., Pei, J., Hou, Q., Wang, W., Liu, J., & Hou, J. (Eds.). (2022). *Urban management based on the concept of sustainable development*. MDPI.

Montgomery, C. (2013). *Happy city: Transforming our lives through urban design*. Farrar, Straus and Giroux.

Newman, P., & Thornley, A. (2019). *Urban planning in a changing world: The twentieth century experience*. Routledge.

Pacione, M. (2009). *Urban geography: A global perspective*. Routledge.

Sachs, J. D. (2015). *The age of sustainable development*. Columbia University Press.

Sassen, S. (2018). *The global city: New York, London, Tokyo*. Princeton University Press.

United Nations. (2015). Transforming our world: The 2030 agenda for sustainable development.

United Nations. (2018). The State of the World's Cities 2018: Harmonious cities.

United Nations Environment Programme (UNEP). (2019). *Global Environment Outlook 6: Healthy planet, healthy people.

United Nations Human Settlements Programme (UN-Habitat). (2016). Urbanization and development: Emerging futures.

World Bank. (2019). World Development Report 2019: The changing nature of work.

World Bank. (2020). World Development Report 2020: Trading for Development in the Age of Global Value Chains.

World Commission on Environment and Development. (1987). Our common future (Brundtland Report). Oxford University Press.

World Resources Institute. (2018). World resources report: Towards a more equal city.

World Resources Institute. (2024). Urban heritage, planning and design and development. *Sustainability*, 15(16), 12359.

2 Fundamentals of Urban Planning

2.1 CITIES AND URBAN AREAS

Before we proceed with the discussion on urban planning, let us first organize our understanding of the difference between cities and urban areas. Cities and urban areas are related but distinct concepts can be defined in diverse ways depending on the context and criteria. However, a general way to understand the difference is:

- Cities are human settlements that have a specific administrative, legal, or historical status. Cities usually have a defined boundary, a name, a local government, and a high population density. Cities can be part of larger metropolitan areas or urban regions that include other cities and towns. Examples of cities are New York City, London, Tokyo, Bangkok, and Cairo.
- Urban areas, however, are geographical areas that have a high concentration of people, buildings, infrastructure, and economic activities. Urban areas can be defined by various indicators such as population size, density, land use, or degree of urbanization. Urban areas can span across multiple cities or towns or be contained within a single city. Examples of urban areas are the Greater London Urban Area, the New York Metropolitan Area, the Tokyo–Yokohama Metropolitan Area, the Bangkok Metropolitan Region, and the Cairo Urban Area.

Cities and urban areas are related to each other in that cities are usually part of urban areas, but not all urban areas are cities. Cities are usually the largest and most prominent human settlements within urban areas, with a high population density, a specific administrative or legal status, and a local government. However, some urban areas may not have a city as their core or may have multiple cities within them. The relationship between cities and urban areas can be complex and dynamic, depending on several factors such as history, geography, demography, economy, and culture.

Normal cities or urban areas are usually planned rather than organically grown ones without an appropriate planning process. Therefore, any decent city undergoes a planning process. Urban planning is the process of developing and designing urban areas to meet the needs of a community and country. The practice draws from many disciplines – architecture, engineering, economics, sociology, public health, finance, and more – and strives to prepare cities and towns for the future.

One of the fundamentals of urban planning is to understand the physical, social, and economic environment of the urban area. This involves collecting and analyzing data on the location, climate, population, land use, infrastructure, resources, services, and challenges of the city or town. Urban planners use various tools and methods to gather and visualize this information, such as geographic information systems (GIS), surveys, maps, models, and statistics. Another fundamental of urban planning is to set goals and objectives for the development or revitalization of the urban area. These

DOI: 10.1201/9781032664613-2

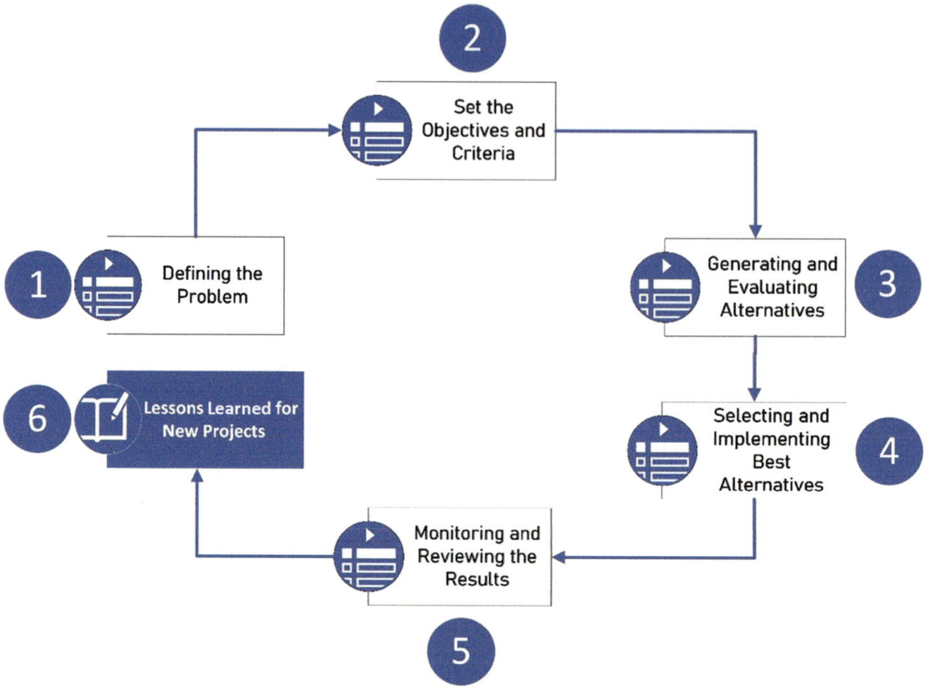

FIGURE 2.1 Rational planning process.

goals and objectives should reflect the vision, values, and needs of the community, as well as the principles of sustainability, equity, efficiency, and livability. Urban planners engage with various stakeholders, such as citizens, the private sector, government agencies, and civil society groups, to identify and prioritize the issues and opportunities that affect the urban area. A third fundamental of urban planning is to design and propose solutions for the improvement of the urban area. These solutions may include master plans, neighborhood plans, project plans, or policy alternatives that address various aspects of the urban environment, such as land use, transportation, housing, public space, utilities, environment, heritage, culture, and social services. Urban planners use various techniques and tools to create and evaluate these solutions. The techniques and tools may include scenario planning, cost–benefit analysis, impact assessment, for example, environmental, and social impacts assessment, design guidelines, and public consultation. A fourth fundamental of urban planning is to implement and monitor the solutions that have been adopted or approved by the relevant authorities. This involves coordinating and collaborating with various actors and institutions that are responsible for the execution and delivery of the plans or policies. Urban planners also monitor and evaluate the outcomes and impacts of the implemented solutions using various indicators and feedback mechanisms. Urban planners use this information to adjust and improve plans or policies as needed. The urban planning process follows the steps of the rational planning process.

A rational planning process (Figure 2.1) is a systematic and logical approach to designing and implementing plans for achieving a desired goal or outcome. A rational planning process typically involves the following steps:

1. Defining the problem or opportunity involves identifying and analyzing the current situation, the needs and preferences of the stakeholders, and the gaps or challenges to be addressed.
2. Setting the objectives and criteria, which involves specifying the desired outcomes, the indicators to measure progress and success, and the standards to evaluate the alternatives.

3. Generating and evaluating alternatives involves brainstorming and researching viable solutions, assessing their feasibility, effectiveness, efficiency, and impacts, and comparing them against the criteria.
4. Selecting and implementing the best alternative involves choosing the most suitable solution based on the evaluation results, developing an action plan, allocating resources, and executing the plan.
5. Monitoring and reviewing the results, which involves collecting and analyzing data on the outputs and outcomes of the plan, identifying any problems or deviations, and adjusting or correcting as needed.

A rational planning process can help to ensure that plans are based on sound evidence, analysis, and reasoning, that they address the real needs and interests of the stakeholders, that they consider the costs and benefits of different options, and that they are adaptable to changing circumstances.

Stakeholders can participate in a rational planning process in several ways, such as:

- Identifying the problems and opportunities that need to be addressed by the plan, based on their knowledge, experience, and interests.
- Providing feedback and suggestions on the objectives, criteria, and alternatives of the plan, based on their needs, preferences, and values.
- Participating in the evaluation and selection of the best alternative for the plan, based on their assessment of the feasibility, effectiveness, efficiency, and impacts of the alternatives.
- Contributing to the implementation and monitoring of the plan, based on their resources, skills, and networks.

Stakeholders can participate in a rational planning process through various channels such as public hearings, surveys, workshops, focus groups, social media, or online platforms. The participation of stakeholders can help to ensure that the plan is relevant, feasible, acceptable, and sustainable. It can also help to build trust, legitimacy, and ownership among the stakeholders. However, stakeholder participation can also pose challenges such as conflicts of interest, power imbalances, or information asymmetry. Therefore, it is important to design and manage stakeholder participation in a way that is inclusive, transparent, accountable, and respectful.

NGOs and CBOs are two distinct types of stakeholders. NGOs (Non-Governmental Organizations) are independent, nonprofit entities that operate locally, nationally, or internationally to address social, humanitarian, or environmental issues. They are not affiliated with governments and rely on funding from donations, grants, and memberships. NGOs play crucial roles in advocacy, disaster relief, education, and development, often working alongside governments and businesses to implement policies and programs that benefit communities.

CBOs (Community-Based Organizations) are grassroots organizations primarily focused on addressing local issues and empowering communities through direct engagement. Unlike NGOs, they are deeply rooted in specific neighborhoods or demographic groups, often managed by community members themselves. CBOs facilitate social services, economic development, health programs, and educational initiatives, fostering self-reliance and strengthening communal bonds through participatory decision-making.

They can play important roles in stakeholder participation in a rational planning process, for instance:

- Representing the interests and needs of the marginalized, vulnerable, or underrepresented groups in the planning process. NGOs and CBOs can provide a voice and platform for these groups to express their concerns, preferences, and values.

- Mobilizing resources and support for the stakeholders to participate effectively in the planning process. NGOs and CBOs can provide information, training, tools, or incentives to help the stakeholders overcome the barriers or challenges to their participation.
- Facilitating communication and collaboration among the stakeholders to enhance their understanding, trust, and cooperation. NGOs and CBOs can organize meetings, workshops, forums, or social media platforms to promote dialogue, feedback, and exchange among the stakeholders.
- Monitoring and evaluating the planning process to ensure that it is inclusive, transparent, accountable, and respectful of the stakeholders' rights and interests. NGOs and CBOs can collect data, conduct research, or produce reports on the quality and outcomes of the planning process.

NGOs and CBOs can also help to ensure that the planning process is based on sound evidence, analysis, and reasoning; that it addresses the real needs and interests of the stakeholders; that it considers the costs and benefits of different options; and that it is adaptable to changing circumstances. NGOs and CBOs can leverage their strengths in advocacy, participation, innovation, and accountability to promote sustainable and inclusive urban development.

2.2 URBAN PLANNING FROM SCRATCH

One of the challenges of urban planning is to deal with the existing constraints and limitations of a city, such as its history, culture, infrastructure, environment, and social dynamics. These factors can often hinder the implementation of innovative and sustainable solutions for urban problems. Therefore, it will be easier if planning a city starts from the blank page, where the planners can design the city from scratch, without any preconceived notions or restrictions. This way, they can optimize the city for the current and future needs of its inhabitants, as well as for the ecological and economic well-being of the region. A blank page approach would allow the planners to create a city that is more efficient, resilient, livable, and adaptable to the changing circumstances. In contrast to planning a city from scratch is retrofitting a city. Retrofitting a city is a huge task that requires careful planning, coordination, and implementation, and huge costs as well. It involves upgrading the existing infrastructure, buildings, and services to make them more efficient, resilient, and sustainable. Retrofitting a city can have multiple benefits, such as reducing greenhouse gas emissions, improving air quality, enhancing livability, and creating new economic opportunities, as the city is aligned to a decent planned city. However, retrofitting a city also poses many challenges, such as securing funding, engaging stakeholders, overcoming technical barriers, and managing social impacts. Therefore, retrofitting a city is not a straightforward process but rather a complex and dynamic one that requires a holistic and adaptive approach. An example of urban planning from scratch is shown in Box 2.1.

BOX 2.1 JAKARTA VIS-À-VIS NEW CAPITAL OF INDONESIA

One of the clearest cases of creating a new urban center from the ground up is the situation of Jakarta and its replacement by the planned Indonesian National Capital (IKN). Jakarta has been facing severe challenges such as overpopulation, traffic congestion, environmental degradation, and flooding. These problems have prompted the government to relocate the administrative functions of the country to a new site in East Kalimantan, on the island of Borneo. The IKN is envisioned as a smart, green, and sustainable city that will showcase the nation's identity and aspirations. The project is expected to cost around $33 billion and take about a decade to complete. However, it also faces many uncertainties and risks, such as land acquisition, financing, governance, social impact, and environmental protection. The IKN is a bold and ambitious attempt to address the urban woes of Jakarta and to create a new symbol of Indonesia's development and progress.

In line with the plan of IKN, Jakarta Governor Anies Baswedan has proposed a plan to retrofit Jakarta as a modern and sustainable city, even as the national government prepares to relocate the capital to Ibu Kota Nusantara (IKN) in East Kalimantan. According to Anies, Jakarta will remain as the center of Indonesia's activity and economy, and therefore needs to improve its infrastructure, environment, and public services. The proposal includes projects such as expanding the mass rapid transit (MRT) system, developing green spaces and parks, reducing air pollution, and flooding, and revitalizing historical and cultural sites. The estimated cost of the proposal is around $40 billion, which will be funded by a combination of public and private sources. Anies said that he will seek cooperation from the central government, local governments, businesses, and civil society to implement his vision for Jakarta. The proposal to retrofit Jakarta has received mixed reactions from critics. The proposal by Jakarta Governor Anies Baswedan to retrofit Jakarta has faced negative criticism from various parties. Some experts and activists have questioned the feasibility, effectiveness, and sustainability of the plan, which aims to transform the capital city into a more livable and resilient place. They have also raised concerns about the potential social and environmental impacts of the retrofitting, such as displacement of low-income residents, loss of green spaces and historical heritage, and increased carbon emissions. The critics have urged the governor to reconsider his proposal and to involve more stakeholders in the decision-making process.

Sources: bbc.com, ikn.go.id, en.tempo.co

The option of whether to build a new city or to retrofit an existing city has its pros and cons. Some of the pros and cons are shown in Table 2.1.

Building a new city is a complex and challenging task that requires careful consideration of a range of factors. Some of the most important aspects to consider are the social, economic, and environmental implications of city design and development.

Social considerations include the needs and preferences of the potential residents, such as their culture, lifestyle, health, education, safety, and social cohesion. A new city should aim to provide a high quality of life for its inhabitants, with diverse and inclusive opportunities for social interaction, recreation, learning, and civic participation. For example, a new city could design public spaces that encourage people to meet and mingle, such as parks, plazas, libraries, and community centers. A new city could also offer a range of educational and cultural facilities that cater to different ages, interests, and backgrounds, such as schools, museums, theaters, and galleries. A new city should also respect and protect the human rights and dignity of all people, regardless of their background, identity, or status. For instance, a new city could implement policies and programs that promote equality, diversity, and inclusion, such as anti-discrimination laws, affirmative action measures, and social welfare schemes.

Economic considerations include the costs and benefits of the city's construction and operation, for instance, the infrastructure, services, employment, trade, and innovation. A new city should aim to be financially sustainable and resilient, with a balanced and diversified economy that supports growth and development. For example, a new city could invest in efficient and reliable infrastructure that facilitates the movement of people, goods, and information, such as roads, railways, airports, and broadband networks. A new city could also provide essential services that enhance the well-being and productivity of its residents, such as healthcare, education, security, and sanitation. A new city should also foster a competitive and creative environment that attracts and retains talent, investment,

TABLE 2.1
Pros and Cons of Building a New City or Retrofitting an Existing City

Build New City	Retrofit Existing City
The Pros	
Fresh Start: Building a new city allows for a fresh start with no pre-existing infrastructure or social systems to work around, including the applications of a sustainable city.	*Preservation of Heritage:* Retrofitting an existing city can preserve cultural heritage and historical landmarks.
Modern Design: A new city can be designed with modern technology and infrastructure in mind. The ideal state of the city is possible to be planned, designed, and implemented.	*Sustainability:* Retrofitting an existing city can promote sustainability by reducing waste and energy consumption.
Economic Development: Building a new city can create jobs and stimulate economic growth, particularly in the new site.	*Community Engagement:* Retrofitting an existing city can involve the community in the planning process.
All Aboard: A new city may represent its stakeholders from the very beginning of the planning process.	
The Cons	
Costly: Building a new city is expensive and requires significant investment.	*Complexity:* Retrofitting an existing city is complex and requires working around pre-existing infrastructure and social systems.
Environmental Impact: Building a new city can have negative environmental impacts such as deforestation, habitat destruction, and pollution.	*Cost:* While being constrained by existing conditions, retrofitting an existing city can be expensive and require significant investment.
Social Disruption: Building a new city can disrupt existing social systems and communities.	*Limited Options:* Retrofitting an existing city may limit options for design and infrastructure due to pre-existing conditions.
	Coverage: Retrofitting will not be able to cover all aspects required by all city stakeholders.

and entrepreneurship. For instance, a new city could establish special economic zones that offer incentives and support for businesses and industries, such as tax breaks, subsidies, and incubators.

Environmental considerations include the impacts and risks of the city on the natural resources, ecosystems, and climate, such as land use, energy consumption, waste management, and greenhouse gas emissions. A new city should aim to be environmentally friendly and responsible, with a low-carbon and circular economy that minimizes waste and maximizes efficiency. For example, a new city could adopt green building standards that reduce the energy demand and environmental footprint of its structures, such as insulation, solar panels, and rainwater harvesting systems. A new city could also implement waste reduction and recycling strategies that prevent pollution and conserve resources, such as composting, biogas production, and material recovery facilities. A new city should also adapt to the changing climate and mitigate the potential hazards and disasters that may affect its people and assets. For instance, a new city could incorporate climate-resilient features that enhance its ability to cope with extreme weather events and natural disasters, such as flood barriers, stormwater drainage systems, and emergency shelters.

By applying a rational planning process, planning, and designing a new city is a complex, challenging, and exhausting task that requires thoughtful and careful consideration of numerous factors that minimize the possibility of after-thought decisions. Some of the most important steps to follow are:

1. Define the vision and goals of the city. What is the purpose and identity of the city? What are the values and principles that guide its development? What are the desired outcomes and impacts of the city?

2. Conduct a comprehensive analysis of the site and context. What are the physical, social, economic, and environmental characteristics of the location? What are the opportunities and constraints that influence city design? What are the needs and expectations of the stakeholders and potential residents?
3. Develop a master plan that outlines the spatial layout and structure of the city. How will the city be organized and divided into different zones, districts, and neighborhoods? How will the city connect and integrate with its surroundings? How will the city accommodate divergent functions and activities, such as housing, commerce, industry, recreation, and culture?
4. Design the urban elements that shape the form and character of the city. How will the city look and feel? What are the architectural styles and features that define the city's identity? What are the public spaces and landmarks that create a sense of place and community? What are the infrastructure and services that support the city's functionality and livability?
5. Implement and evaluate the city design and development. How will the city be built and managed? What are the resources and strategies that ensure the feasibility and sustainability of the city? What are the indicators and methods that measure the performance and impact of the city?

The series of questions can serve as guidance on the design of the city while continuous consultation with all city stakeholders is maintained. Designing a new city is an iterative and collaborative process that involves multiple disciplines, stakeholders, and perspectives. A new city should aim to balance social, economic, and environmental considerations while reflecting its vision, goals, and context.

In the process of city planning, the requirements and qualifications for a planner to be able to do the city planning process vary depending on the context and the jurisdiction. However, most planners have at least a bachelor's degree in urban planning, architecture, public policy, or a related field. Some positions may require a master's degree or additional experience. Planners may also need to be licensed or certified by a professional organization or regulatory body.

A conventional notion asserts that city planning is a process that aims to improve the quality of life of people living in urban areas by creating a sustainable environment that meets their needs. The fundamental objectives of city planning are shown in Table 2.2.

However, from the modern and sustainable viewpoint, the above objectives are only basic objectives centered on anthropogenic and do not cover the contemporary but long-term sustainability objectives.

TABLE 2.2
Fundamental Objectives of City Planning

Objectives	Explanation
Removal of slums	City planning aims to remove slums and provide people with proper housing facilities or accommodation along with basic infrastructure like electricity and water supply.
Provision of basic infrastructure	City planning aims to provide basic infrastructure like roads, water supply, sewage system, and electricity.
Organization and improvement in the means of transport and communication	City planning aims to organize and improve the means of transport and communication.
Creation of open spaces	City planning aims to create open spaces like parks, playgrounds, and recreational areas.
Preservation of natural resources	City planning aims to preserve natural resources like forests, rivers, and lakes.

Smart growth is an urban planning and transportation theory that concentrates growth in compact walkable urban centers to avoid sprawl. It also advocates compact, transit-oriented, walkable, bicycle-friendly land use, including neighborhood schools, complete streets, and mixed-use development with a range of housing choices. Smart growth aims to promote sustainable development by reducing the negative impacts of urbanization such as traffic congestion, air pollution, and habitat destruction. It also aims to promote social equity and economic vitality by creating vibrant and diverse communities that offer a high quality of life for all residents. A smart growth city adopts a set of development and conservation strategies that aim to protect the environment, enhance the quality of life, and foster economic prosperity for its residents. A smart growth city is called smart growth because it uses smart planning and design principles to create compact, walkable, mixed-use, and transit-oriented neighborhoods that offer a range of housing options, preserve open space and natural resources, and promote community engagement and collaboration. A smart growth city is not only smart in terms of environmental sustainability, but also in terms of social equity, cultural diversity, and fiscal responsibility.

The terms smart growth cities and smart cities may be confusing for some. Smart growth is an urban planning and transportation theory that concentrates growth in compact walkable urban centers to avoid sprawl. However, a smart city is a city in which the operations concentrate on real-time actions and use technology to enhance the benefits and diminish the shortcomings of urbanization for its citizens.

Some examples of smart growth cities are shown below:

1. Singapore has a Smart Nation Vision that leverages technology to improve public services, mobility, healthcare, and security.
2. Oslo, Norway, is building a carbon-neutral district with renewable energy, green roofs, and electric vehicles.
3. Portland, OR, United States, has invested in public transportation, bicycle infrastructure, and urban revitalization.
4. San Francisco, CA, United States, has implemented policies to increase affordable housing, reduce greenhouse gas emissions, and support local businesses.

An example of a smart-growth city, which is demonstrated by San Francisco, CA, USA, is shown in Box 2.2. The city has implemented policies to increase affordable housing, reduce greenhouse gas emissions, and support local businesses, creating a more sustainable and equitable urban environment.

BOX 2.2 SAN FRANCISCO SMART GROWTH CITY

San Francisco, a bustling city in California, United States, has been a pioneer in implementing smart growth policies to address urban challenges. These policies focus on increasing affordable housing, reducing greenhouse gas emissions, and supporting local businesses.

Affordable Housing: San Francisco has implemented various policies to increase affordable housing. The city has set a goal of building or rehabilitating 30,000 homes by 2020, with at least one-third of them being permanently affordable to low- and moderate-income families. To achieve this, the city has adopted inclusionary housing policies, which require developers to include a certain percentage of affordable units in new residential developments.

Greenhouse Gas Emissions: San Francisco is committed to reducing its greenhouse gas emissions. The city has set an ambitious goal of reducing emissions to 40% below 1990 levels by 2025 and 80% by 2050. To achieve this, the city has implemented various measures, including promoting public transit, cycling, and walking; improving energy efficiency in buildings; and increasing the use of renewable energy.

Supporting Local Businesses: San Francisco has a strong commitment to supporting local businesses. The city has implemented various programs to provide financial assistance, technical support, and training to local businesses. These programs aim to promote economic development, create jobs, and enhance the vitality of local neighborhoods.

Some quantitative data on San Francisco's smart growth policies:

- *Housing*: As of 2012, San Francisco had a total of 379,597 housing units, with a homeownership rate of 36.5%.
- *Emissions*: San Francisco has reduced its greenhouse gas emissions by 28.5% from 1990 levels, surpassing its 2017 goal of a 25% reduction.
- *Local Businesses*: San Francisco is home to over 120,000 businesses, with small businesses (those with fewer than 100 employees) making up 98% of all businesses in the city.

San Francisco's smart growth policies demonstrate how cities can address urban challenges through strategic planning and policy implementation. These policies not only improve the quality of life for residents but also contribute to sustainable urban development.

Sources: sfpublicpress.org, greenbelt.org

Some examples of smart cities are shown below:

1. Seoul has implemented a smart transportation system that includes a real-time traffic information system and an integrated payment system for public transportation.
2. Hong Kong has implemented a smart lamppost system that provides free Wi-Fi and collects data on air quality and traffic.
3. London has implemented a smart traffic flow management system that can adjust the timing of traffic lights based on the amount of traffic and how well that traffic is flowing.
4. Hangzhou has been described as "the smartest city in China" due to its close ties between large tech companies and public sector organizations which allows for the quick development of innovative solutions and fast implementation.
5. Bangkok is an emerging smart city for the ubiquitous use of cashless transactions, even for street vendor transactions.

A real case study of a smart city's traffic management system is showcased by London's Smart Traffic Flow Management System, which optimizes traffic flow and reduces congestion using advanced sensors, data analysis, and adaptive control algorithms, as shown in Box 2.3.

BOX 2.3 LONDON'S SMART TRAFFIC FLOW MANAGEMENT SYSTEM (AN EXAMPLE OF A SMART CITY)

London, one of the world's busiest cities, has been at the forefront of implementing intelligent traffic management systems to manage its complex road network. One such system is the Real Time Optimizer (RTO), a cutting-edge traffic light control system developed by Transport for London (TfL) and Siemens Mobility Limited. The RTO system revolutionizes the traditional traffic light control system that has been in operation for over 30 years. It optimizes traffic light timings to enable more people and goods to move on the capital's road network with fewer delays. When the road network is disrupted by an unplanned incident, planned work, or events, TfL's 24/7 control center uses the RTO system to return the road network back to normal as quickly as possible. Before the implementation of the RTO system, London used an extensive network of SCOOT-controlled traffic lights, which use sensors to detect traffic and adjust the signal timings to manage queues and give buses priority if they are running late. This system alone kept traffic delay 13% lower than it would otherwise be.

The new RTO system takes this a step further by using a range of new data sources and connecting to new types of sensors. It can even share data with connected vehicles, which could include autonomous vehicles in the future. The data generated by the system can then be used to better inform road users of current or expected road conditions to support improved journey planning. The RTO system is a key part of London's broader Surface Intelligent Transport Systems program, which aims to improve air quality, reduce congestion, and make sustainable travel more appealing and accessible to everyone. The system is set to deliver £1bn of benefits through reducing delays for all road users by 2036.

Sources: tfl.gov.uk, londonroadsafetycouncil.org.uk, data.london.gov.uk

The notion of smart cities has existed for years, but it is not without its complexities and constraints. Smart cities aim to use technology and data to improve the efficiency, livability, and resilience of urban areas; however, they often neglect the environmental and social aspects of sustainability. To address these issues, some experts and practitioners are proposing a shift from smart cities to sustainable cities, which integrate the three dimensions of sustainability: economic, environmental, and social. Sustainable cities not only use technology and data to optimize urban systems but also to enhance the well-being, participation, and empowerment of their citizens, while reducing their ecological footprint and greenhouse gas emissions. Sustainable cities are more than just smart; they are also green, inclusive, and adaptive.

One of the main challenges of the 21st century is to create sustainable cities that can meet the needs of the present without compromising the future. Sustainable cities are those that balance the environmental, social, and economic aspects of urban development while ensuring the well-being and participation of their citizens. Some of the principles that guide the design and management of sustainable cities are:

1. Reducing greenhouse gas emissions and enhancing resilience to climate change impacts.
2. Promoting renewable energy sources and energy efficiency.
3. Improving public transportation and mobility options,
4. Enhancing urban green spaces and biodiversity.
5. Fostering social inclusion and diversity.
6. Encouraging civic engagement and democratic governance.
7. Supporting local innovation and creativity.
8. Strengthening urban–rural linkages and regional cooperation

Infrastructure plays a vital role in creating and maintaining services for sustainable cities. These services must have direct and indirect impacts on the environmental, social, and economic aspects of urban sustainability. Some of the roles of infrastructure in sustainable cities are:

1. Providing efficient and accessible mobility options for people and goods, reducing greenhouse gas emissions, air pollution, and congestion.

2. Enhancing the resilience and reliability of water supply and sanitation, improving public health and hygiene, and reducing water stress and pollution.
3. Promoting the use of renewable and low-carbon energy sources, minimizing energy consumption and waste, and increasing energy security and affordability.
4. Implementing integrated and circular waste management systems, reducing waste generation and disposal, and maximizing resource recovery and reuse.
5. Facilitating the development and adoption of digital technologies, improving connectivity and information access, and fostering innovation and social inclusion.
6. Offering diverse and inclusive public spaces and facilities, enhancing the quality of life and well-being of urban residents, and fostering social cohesion and participation.

To achieve these roles, infrastructure planning and design should be guided by the following foundations. These foundations include:

1. Adopting a holistic and systemic approach, considering the interdependencies and trade-offs among different infrastructure sectors and urban systems.
2. Engaging in participatory and collaborative processes, involving multiple stakeholders and perspectives in decision-making and implementation.
3. Applying adaptive and flexible solutions, responding to the changing needs and conditions of urban contexts.
4. Incorporating environmental and social considerations, minimizing negative impacts, and maximizing positive outcomes for people and the planet.

Infrastructure is a key enabler of sustainable urban development. By sustainably fulfilling its roles, infrastructure can contribute to building cities that are resilient, livable, productive, and equitable. These principles are interrelated and mutually reinforced and require a holistic and integrated approach to urban planning and policymaking. By applying these principles, sustainable cities can become more livable, equitable, and prosperous for all.

2.3 SUSTAINABLE URBAN PLANNING PRINCIPLES

Conventional urban planning, which is urban planning with conventional goals where contemporary concepts such as smart city planning, smart growth, and sustainable city planning are not there yet, is the process of designing and managing the physical and social aspects of a city or a town. It involves various disciplines, such as architecture, engineering, geography, sociology, economics, and politics, to create livable, sustainable, and efficient urban environments.

Some of the main goals of conventional urban planning are quite straightforward, which are to provide adequate housing, transportation, public services, and amenities for the urban population; to preserve the natural and cultural heritage of the city; to promote economic development and social cohesion; and to mitigate the environmental impacts of urbanization.

Conventional urban planning can be divided into different scales and types, depending on the scope and purpose of the plan. For example, regional planning covers a large area that may include multiple cities or towns and aims to coordinate their development and resources. Master planning provides a comprehensive vision and framework for the long-term development of a single city or town. Zoning regulates the use and density of land and buildings within a city or town. Urban design focuses on the esthetic and functional aspects of urban space, such as streetscapes, landmarks, public spaces, and building forms.

Conventional urban planning faces many challenges and criticisms in the 21st century. Some of the challenges are related to rapid urbanization, population growth, climate change, social inequality, and technological innovation that affect urban dynamics and demands. Some of the criticisms are related to the lack of public participation, diversity, flexibility, and innovation in the planning process and outcomes. Therefore, conventional urban planning needs to adapt and evolve to meet the changing needs and aspirations of urban dwellers and stakeholders.

One of the challenges is the emergence of sustainability in the planning process, which is sustainable urban planning. Sustainable urban planning is the process of designing and managing cities that balance the social, economic, and environmental needs of current and future generations. It aims to create livable, resilient, and inclusive urban spaces that minimize the negative impacts of urbanization on natural resources, climate, and human health. Due to the nature of involving sustainability principles, the key principles of sustainable urban planning are directed to accomplish the following:

1. Integrating land use and transport planning to reduce greenhouse gas emissions, air pollution, and congestion, and to promote active mobility, public transit, and mixed-use development.
2. Enhancing and promoting urban greenery and biodiversity to provide ecosystem services, such as cooling, stormwater management, carbon sequestration, and recreation.
3. Promoting social equity and inclusion by ensuring access to affordable housing, basic services, public spaces, and opportunities for all urban dwellers, especially the marginalized and vulnerable groups.
4. Fostering participatory and collaborative governance by engaging diverse stakeholders, such as local communities, civil society, the private sector, and academia, in the planning and decision-making processes.
5. Adopting a long-term and holistic perspective by considering the interconnections and trade-offs among different sectors, scales, and time horizons.
6. Supporting circular economy and low-carbon transition by optimizing resource efficiency, waste reduction, renewable energy, and green innovation despite high consumption in the city.

2.3.1 Integrating Land Use and Transport

One of the most pressing challenges of the 21st century is to mitigate the impacts of climate change and urbanization on the environment and human health. A key strategy to achieve this goal is to integrate land use and transport planning, which can reduce greenhouse gas emissions, air pollution, and congestion, and promote active mobility, public transit, and mixed-use development.

Integrating land use and transport planning means coordinating the location, density, and design of urban development with the provision and management of transport infrastructure and services. This can create more compact, walkable, and transit-oriented cities that offer a variety of housing, employment, and amenities within a short distance. Such cities can also reduce the need for private car use and encourage more sustainable modes of transport, such as walking, cycling, and public transit. These modes can lower greenhouse gas emissions and air pollution, improve public health and well-being, and enhance social equity and inclusion.

However, integrating land use and transport planning is not an easy task. It requires a holistic and long-term vision that aligns with the local context and needs. It also requires effective collaboration and coordination among various levels of government, sectors, and stakeholders, as well as adequate funding and institutional capacity. Moreover, it requires addressing potential trade-offs and barriers, such as land availability, affordability, accessibility, safety, and public acceptance. Therefore, it is essential to develop and implement evidence-based policies and practices that can support the integration of land use and transport planning sustainably and equitably. A case study on this is shown in Box 2.4.

BOX 2.4 INTEGRATING LAND USE AND TRANSPORTATION IN FRESNO, UNITED STATES

Fresno, located in the heart of California's San Joaquin Valley, is a prime example of a city that has integrated land use and transportation policies for sustainable development. The city has long been grappling with the issue of urban sprawl, a common problem in many American cities. To address this, Fresno has turned to a classic gravity model, Transportation, Economic, and Land-Use Model (TELUM), to examine how land-use or transportation policy must be regulated to make urban compaction occur.

Five scenarios were considered in the study: the baseline (BL) is a natural growth scenario, L1 (high-intensity zoning), L2 (growth boundary), T1 (location impedance), and T2 (carbon tax). The results suggest that high-density zoning and growth boundary policies (L1 and L2) could make compaction occur. On the other hand, transportation interventions (T1 and T2) would create barriers among regions/areas and therefore should be carefully used for compaction.

The study not only adds to the literature on urban modeling but also contributes to the practice of smart growth or new urbanism policies for sustainability. It provides valuable insights into how a typical auto-dependent city like Fresno can regulate land use or transportation policy to achieve urban compaction.

Sources: transweb.sjsu.edu, mdpi.com

2.3.2 Enhancing and Promoting Urban Greenery and Biodiversity

Urban greenery and biodiversity are essential components of sustainable urban development. They provide various ecosystem services that enhance the quality of life and well-being of urban residents, as well as mitigate the negative impacts of urbanization on the environment. Some of the key ecosystem services provided by urban greenery and biodiversity are:

1. *Cooling*: Urban green spaces can reduce the urban heat island effect, which is the phenomenon of higher temperatures in urban areas compared to rural areas due to human activities and built structures. Urban green spaces can lower air and surface temperatures by providing shade, evapotranspiration, and albedo effects. This can improve thermal comfort, reduce energy consumption for cooling, and prevent heat-related illnesses and mortality.
2. *Stormwater Management*: Urban green spaces can reduce the runoff and flooding caused by heavy rainfall events, which are expected to increase in frequency and intensity due to climate change. Urban green spaces can increase the infiltration and storage of rainwater, reduce the peak flow and volume of runoff, and improve water quality by filtering pollutants and sediments. This can reduce the pressure on urban drainage systems, prevent erosion and landslides, and enhance water security and availability.
3. *Carbon Sequestration*: Urban green spaces can sequester carbon dioxide from the atmosphere through photosynthesis and store it in their biomass and soil. This can help mitigate climate change by reducing greenhouse gas emissions and enhancing carbon sinks. Urban green spaces can also reduce the carbon footprint of urban areas by reducing the need for fossil fuel-based energy sources for cooling and transportation.
4. *Recreation*: Urban green spaces can provide opportunities for recreation, leisure, physical activity, social interaction, and mental health for urban residents. Urban green spaces can offer diverse amenities and facilities, such as parks, gardens, playgrounds, trails, sports fields,

and community gardens. They can also support various forms of nature-based recreation, such as birdwatching, wildlife observation, gardening, and urban farming. These activities can enhance the physical and mental health of urban residents, as well as foster social cohesion and inclusion.

Therefore, enhancing and promoting urban greenery and biodiversity is a vital strategy for achieving sustainable urban development. It requires a holistic and integrated approach that considers the multiple benefits and trade-offs of urban green spaces, as well as the diverse needs and preferences of urban residents. It also requires collaboration and coordination among various stakeholders, such as governments, planners, developers, researchers, civil society organizations, and communities.

2.3.3 Promoting Social Equity and Inclusion

One of the key challenges of urban development is to ensure that all people, regardless of their socioeconomic status, gender, ethnicity, disability, or other factors, can enjoy the benefits of living in a city. Promoting social equity and inclusion means that urban policies and interventions should aim to reduce inequalities and discrimination, and to empower the marginalized and vulnerable groups to participate in decision-making and access opportunities.

One of the aspects of social equity and inclusion is the provision of affordable housing for all urban dwellers. Housing is not only a basic human right, but also a determinant of health, well-being, and social integration. However, many cities face a shortage of adequate and affordable housing, especially for low-income households, migrants, refugees, women-headed families, and people with disabilities. This leads to overcrowding, slums, homelessness, and forced evictions. To address this challenge, urban planners and policymakers should adopt strategies that increase the supply and diversity of housing options, such as social housing, rental subsidies, cooperative housing, and community-led initiatives. They should also ensure that housing is located near basic services, such as water, sanitation, electricity, healthcare, education, and transport.

Another aspect of social equity and inclusion is the creation and maintenance of public spaces that are accessible, safe, inclusive, and vibrant. Public spaces are essential for social interaction, cultural expression, recreation, and civic engagement. They can also foster social cohesion, identity, and belonging among diverse groups of people. However, many public spaces in cities are underutilized, degraded, privatized, or segregated. To address this challenge, urban planners and policymakers should adopt strategies that enhance the quality and quantity of public spaces, such as parks, squares, streets, markets, libraries, and cultural venues. They should also ensure that public spaces are designed and managed with the participation and needs of all users in mind, especially women, children, elderly people, people with disabilities, ethnic minorities, and other groups that may face barriers or discrimination.

A third aspect of social equity and inclusion is the promotion of opportunities for all urban dwellers to access education, employment, healthcare, culture, and other services that can improve their quality of life and well-being. Opportunities are not equally distributed in cities; often they are concentrated in certain areas or sectors that exclude or disadvantage some groups of people. This leads to spatial and social polarization, poverty traps, and social exclusion. To address this challenge, urban planners and policymakers should adopt strategies that foster spatial justice and social mobility, such as integrated urban development plans, inclusive economic policies, social protection systems, and affirmative action programs. They should also ensure that opportunities are responsive to the diverse needs and aspirations of all urban dwellers, especially those who face multiple forms of discrimination or marginalization.

2.3.4 Fostering Participatory and Collaborative Governance

To ensure that diverse stakeholders are involved in the planning and decision-making processes, fostering participatory and collaborative governance should be encouraged. This is a way to enhance the legitimacy, transparency, accountability, and effectiveness of public policies and actions. By engaging local communities, civil society, the private sector, and academia, participatory and collaborative governance can foster social learning, innovation, empowerment, and ownership of the development agenda. Moreover, it can help to address complex and interrelated issues that require collective action and coordination across different sectors and levels of governance.

To foster participatory and collaborative governance, some of the possible strategies are:

1. Establishing inclusive and representative platforms or mechanisms for dialogue, consultation, and deliberation among diverse stakeholders on relevant issues and policies.
2. Providing adequate information, capacity building, and resources to enable meaningful participation and collaboration of different stakeholders, especially those who are marginalized or vulnerable.
3. Ensuring that the views and inputs of diverse stakeholders are considered and reflected in the planning and decision-making processes and that the outcomes and impacts of these processes are communicated and monitored.
4. Creating an enabling environment that supports the rights, roles, and responsibilities of different stakeholders in the governance process, and that promotes trust, mutual respect, and constructive engagement among them.

2.3.5 Adopting a Long-Term and Holistic Perspective

Adopting a long-term and holistic perspective by considering the interconnections and trade-offs among different sectors, scales, and time horizons is necessary for accomplishing sustainable urban planning. This means that decision-makers need to consider not only the immediate impacts of their actions, but also the potential consequences for future generations, other regions, and the global environment. A long-term and holistic perspective also requires a recognition of the complexity and uncertainty of the systems that shape human and natural well-being and the need for adaptive and participatory approaches to governance. By adopting a long-term and holistic perspective, decision-makers can identify synergies and trade-offs among multiple objectives, such as economic growth, social equity, environmental protection, and human health, and seek solutions that balance these objectives fairly and efficiently.

2.3.6 Supporting Circular Economy and Low-Carbon Transition

A circular economy is an economic system that aims to eliminate waste and pollution, keep products and materials in use, and regenerate natural systems. It is based on three principles, driven by design: eliminate waste and pollution, circulate products and materials (at their highest value), and regenerate nature. A circular economy reduces the consumption of finite resources and the environmental impact of materials while creating new opportunities for innovation, growth, and social well-being. A circular economy is not only a way to manage resources, but also a way to transform our system for a more sustainable and resilient future.

Balancing high consumption levels with environmental sustainability is a strong challenge. A solution is to adopt a circular economy approach, which aims to minimize waste and maximize the value of resources throughout their life cycle. By reusing, repairing, refurbishing, and recycling materials and products, cities can reduce their environmental impact and greenhouse gas emissions, while also creating new economic opportunities and social benefits.

A circular economy also supports the low-carbon transition by promoting the use of renewable energy sources and green innovation. For example, cities can generate energy from organic waste, use biogas for transport and heating, or install solar panels and wind turbines on buildings and public spaces. Green innovation can also help cities develop new products and services that are more efficient, durable, and adaptable to changing needs and preferences.

To achieve a circular economy and a low-carbon transition, cities need to involve various stakeholders, such as businesses, consumers, policymakers, and civil society. They need to create incentives and regulations that encourage circular practices, such as extended producer responsibility, green public procurement, waste prevention and management, or carbon pricing. They also need to raise awareness and educate citizens about the benefits of circular consumption and production patterns, such as saving money, improving health, enhancing social cohesion, and contributing to climate action.

Sustainable urban planning is not a one-size-fits-all approach. It requires context-specific solutions that reflect the local culture, history, geography, and aspirations of each city. It also requires continuous monitoring, evaluation, and adaptation to respond to the dynamic and complex challenges of urbanization. Sustainable urban planning requires a holistic and integrated approach that considers the interconnections between different sectors, scales, and stakeholders. It also requires a long-term vision that anticipates the challenges and opportunities of urbanization in a changing world. By applying the principles of sustainable urban planning, cities can become more livable, resilient, inclusive, and competitive, while contributing to global sustainability goals.

One of the essential components of urban planning is land use planning, which determines how different types of land are allocated and utilized for various purposes, such as residential, commercial, industrial, recreational, environmental, and so on. Land use planning aims to balance the needs and interests of different stakeholders, such as developers, residents, businesses, government agencies, and civil society groups, while ensuring the sustainability and livability of the urban environment. Land use planning also involves setting goals and objectives, developing policies and regulations, conducting analysis and evaluation, implementing and monitoring plans and projects, and engaging with the public and other relevant actors.

2.4 ZONING AND DEVELOPMENT

Land use zoning and development are two interrelated concepts that affect the spatial distribution of human activities and the environmental quality of a region. Zoning is the process of dividing land into various categories, such as residential, commercial, industrial, agricultural, etc., and imposing regulations on the types and intensities of uses that are allowed within each category. Some examples of zoning regulations are:

1. the minimum lot size and setback requirements for buildings;
2. the maximum height and floor area ratio of structures;
3. the permitted and conditional uses within each zone;
4. the parking and landscaping standards; and
5. the design and appearance guidelines.

On the other hand, development is the process of transforming the land from its natural or existing state to a more productive or urbanized state, such as by building roads, houses, factories, parks, etc. However, zoning and development have both positive and negative impacts on various aspects of society and the environment, such as economic growth, social equity, public health, biodiversity, climate change, etc. Therefore, it is important to plan and manage zoning and development in a way that balances the needs and interests of different stakeholders and minimizes the adverse effects on natural resources and ecosystems.

Zoning is a tool of land use planning that regulates the types of activities that can take place on a given parcel of land. Zoning aims to promote orderly development, protect public health and safety, preserve environmental quality, and balance the competing interests of landowners and the community. Zoning typically divides a municipality into districts or zones, such as residential, commercial, industrial, agricultural, and so on, and specifies the permitted and prohibited uses within each zone. For example, zoning may allow single-family homes, duplexes, or apartments in a residential zone, but prohibit factories, warehouses, or junkyards. Zoning may also impose standards on the size, height, density, setback, and appearance of buildings and structures. For instance, zoning may limit the height of buildings to avoid blocking sunlight or views or require a minimum lot size or frontage to maintain open space and privacy. Zoning is usually implemented through a zoning ordinance or by-law, which is a legal document that defines the zoning districts and their regulations. Zoning is often supplemented by other planning tools, such as subdivision regulations, design guidelines, overlay districts, incentive zoning, and variances. These tools provide more flexibility and specificity in regulating land use and development. For example, subdivision regulations control how land is divided into lots and streets, design guidelines provide aesthetic standards for buildings and landscapes, overlay districts apply additional rules to areas with special features or needs, incentive zoning offers bonuses or waivers in exchange for public benefits, and variances grant exceptions to zoning requirements for cases of hardship or unique circumstances.

One of the obvious disadvantages of the zoning system is that zoning can discourage mixed-use development by separating several types of land use into different areas of the city. This can create barriers between different land uses that make it difficult for people to walk or bike between them. Therefore, it is in opposition to the concept of a compact city where the principles of "within walking distance" prevail.

One of the examples of the simplest zoning system is the Burgess model. The Burgess land use model is a theory of urban structure that divides a city into five concentric zones based on the age and density of development (Figure 2.2). The model was proposed by sociologist Ernest Burgess in 1925 and is based on his observation of Chicago. The five zones are:

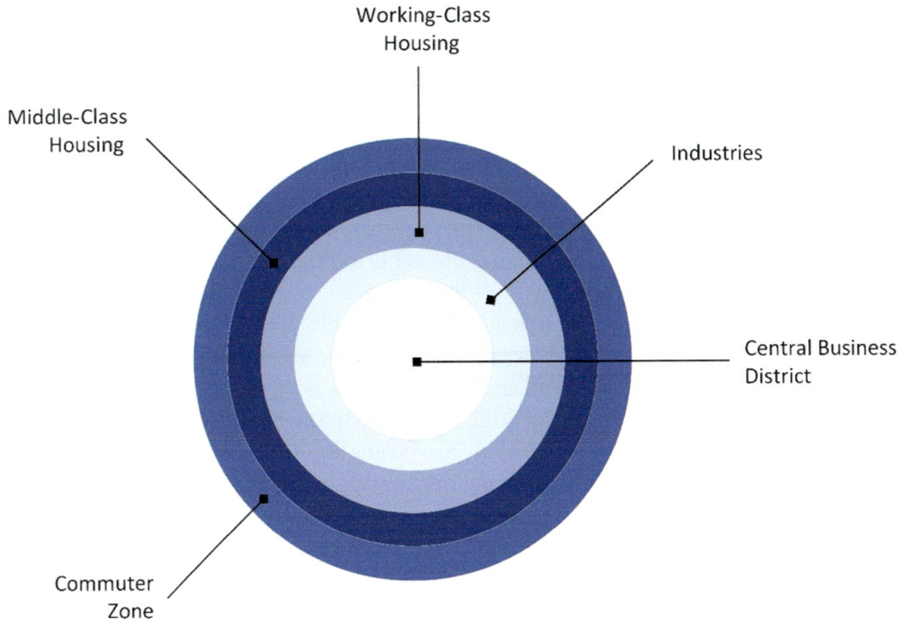

FIGURE 2.2 Burgess land use model.

Zone 1: The central business district (CBD), where most of the economic and commercial activities take place. This zone has the highest land values and the tallest buildings.

Zone 2: The zone of transition, where older residential areas are mixed with industrial and commercial uses. This zone has the lowest quality of housing and the highest rates of social problems, such as crime and poverty.

Zone 3: The zone of independent workers' homes, where newer and more spacious housing is built for the working class. This zone has a more stable population and a higher quality of life than Zone 2.

Zone 4: The zone of better residences, where middle-class and upper-class families live in larger and more expensive houses. This zone has a higher level of social status and a lower population density than Zone 3.

Zone 5: The commuter zone, where people live in suburban areas and commute to the city for work. This zone has the lowest population density and the highest quality of the environment.

The Burgess land use model is one of the earliest and simplest models of urban structure, but it also has some limitations. It assumes that the city is isotropic, meaning that it has a uniform physical and social environment. However, cities are influenced by factors such as topography, transportation, history, and culture, which create variations in land use patterns. It assumes that the city grows outward from the center in a concentric manner. However, cities may grow in different directions due to the location of natural resources, infrastructure, or political boundaries. It assumes that people's residential choices are determined by their income and social status. However, people may have other preferences or constraints, such as ethnicity, lifestyle, or family size, which affect their housing decisions.

Urban structure refers to the spatial patterns of urban areas, such as how land use, population density, transportation networks, and social and economic activities are distributed. An urban structure can be influenced by a range of factors, such as historical development, physical geography, planning policies, market forces, economic forces, and transport infrastructure.

One way to classify urban structures is based on the number and location of centers of activity within a city. For example, a concentric urban structure has one dominant center, usually the CBD, surrounded by concentric rings of residential and commercial areas. A polycentric urban structure has multiple centers of activity, often located in suburban areas or along major transport corridors. A linear urban structure has a single center that extends along a linear axis, such as a river, a coast, or a highway.

Another way to classify urban structures is based on the density and continuity of development within a city. For example, a compact urban structure has high-density and continuous development, with mixed land uses and short distances between different activities. A sprawl urban structure has low-density and discontinuous development, with segregated land uses and long distances between different activities. A hybrid urban structure has elements of both compact and sprawl development, such as high-density nodes surrounded by low-density areas.

Some common types of urban structures are shown in Table 2.3.

Urban structures may have a certain degree of impact on the environment, social equity, and quality of life of urban residents. Some of these impacts are positive, for instance, reducing transport energy consumption and traffic jams due to the creation of walking communities because of the compact structure of the city along with the promotion of excellent public transport; more distributed traffic flows, and eventually reducing traffic congestion in the city because of the quality of poly-centricity of the city. However, some of the impacts are negative, such as contributing to air pollution, greenhouse gas emissions, urban heat island effect, noise, waste generation, and loss of natural habitats. Moreover, urban structures often reflect and reinforce existing social inequalities, such as spatial segregation, gentrification, displacement, and lack of access to affordable housing, public services, and green spaces. Therefore, it is important to design and manage urban structures

TABLE 2.3
Types of Urban Structures

Type	Explanation
Compact city	A high-density urban form that concentrates most of the population and activities within a small area, often with a well-defined core and a clear boundary. Compact cities are usually associated with mixed land use, public transit, walkability, and environmental sustainability.
Sprawl city	A low-density urban form that spreads over a large area, often with no clear center or edge. Sprawl cities are usually characterized by single-use zoning, car dependency, fragmentation, and environmental degradation.
Polycentric city	A multi-nodal urban form that consists of several centers or sub-centers that are connected by transportation corridors. Polycentric cities are usually the result of urban growth, decentralization, or regional integration. Polycentric cities can offer more diversity, accessibility, and resilience than monocentric cities.
Concentric city	A model of urban development that assumes that the city grows outward from a central core. The core is typically the oldest and most densely populated part of the city, where most economic and cultural activities take place. The core is surrounded by concentric rings of land use that reflect distinct stages of urban expansion and function. For example, the first ring may consist of industrial zones and low-income housing, while the outermost ring may contain suburban areas and green spaces.
Linear city	A one-dimensional urban form that follows a linear axis, such as a river, a coast, a road, or a rail line. Linear cities are usually designed to optimize the use of land and infrastructure along the axis while minimizing the distance between the axis and the surrounding areas. Linear cities can facilitate mobility, connectivity, and efficiency.

in a way that minimizes their environmental footprint, maximizes their social benefits, and enhances the well-being of urban residents.

Transport infrastructure plays a key role in shaping and supporting urban structures. It provides accessibility and mobility for people and goods within and between cities. It also affects the location decisions of households and businesses, as well as the environmental and social impacts of urbanization. Different modes of transport have different advantages and disadvantages in terms of speed, capacity, cost, safety, convenience, and environmental performance. Therefore, the choice and design of transport infrastructure can influence the type and quality of urban structures that emerge or evolve.

In the absence of sufficient public transportation systems that would attract most citizens to significantly shift their transport modes from private transport to public transport, there is a strong connection between urban structure and traffic flows. Traffic flows are the movements of vehicles and people on the road network. The urban structure affects the traffic flows by influencing the travel demand, mode choice, route choice, and travel time of the travelers. A monocentric city with a high concentration of jobs and services in the business area may generate more traffic flows toward the center, while a polycentric city with multiple subcenters may reduce the traffic flows between the subcenters. Understanding the connection between urban structure and traffic flows can help planners and engineers design more efficient and sustainable urban transportation systems.

2.5 MONOCENTRIC VIS-À-VIS POLYCENTRIC CITIES

A monocentric urban structure is a model of urban development that assumes a single CBD surrounded by concentric rings of residential and commercial zones. The CBD is the most accessible and attractive location for businesses and services, and it generates the highest land rents and property values. The model has been criticized for its simplicity and inability to account for the diversity and complexity of modern urban forms, such as poly-centricity, edge cities, suburbanization, and

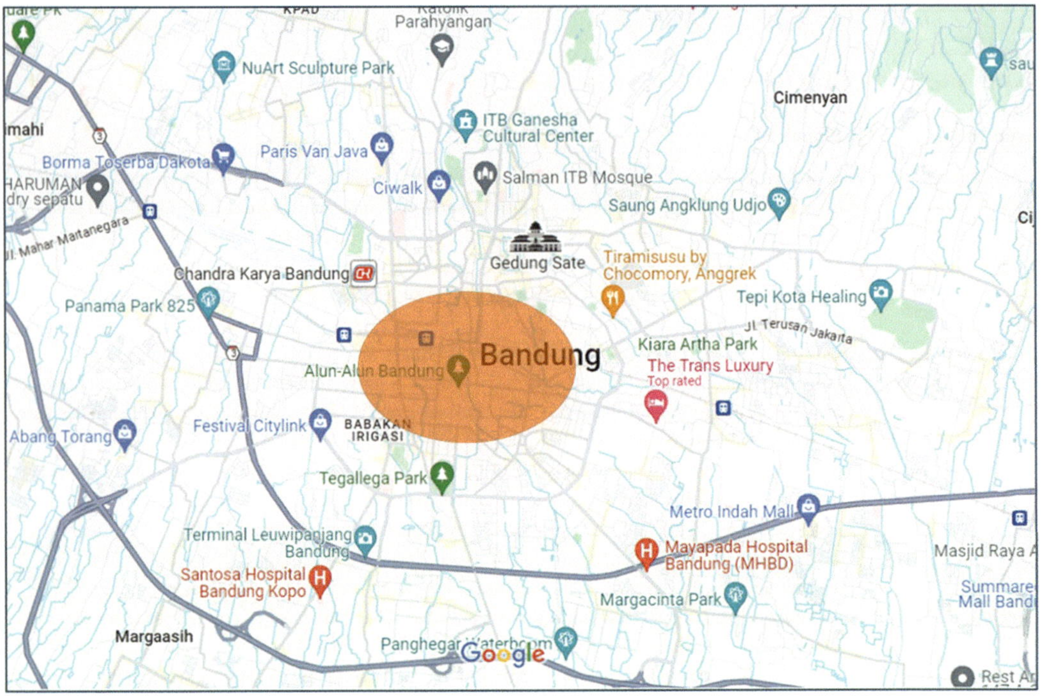

FIGURE 2.3 Monocentric city model of Bandung city.

sprawl. A monocentric city has one main core or CBD where most of the economic activity takes place. The CBD is usually located at the center of the city, and the land rent, population density, and building height tend to decline as the distance from the CBD increases. A monocentric city explains how land use, transportation, and housing are determined by the spatial equilibrium between workers and firms in a city. Some examples of monocentric cities are Chicago, Paris, Tokyo, Bangkok, and Bandung Indonesia (Figure 2.3) which have historically developed around a single dominant center.

A monocentric city has some advantages, such as a clear spatial organization, an intense sense of identity, a high accessibility to public services and amenities, a high agglomeration of human and physical capital, and a high potential for social interaction and cohesion. However, it also has some disadvantages, such as high congestion, high pollution, high inequality, and a low resilience to external shocks. A monocentric city may not be able to cope with the challenges of urbanization, globalization, and technological change, which require more flexibility and diversity in the urban system.

A contrasting quality of a monocentric city is a polycentric city. A polycentric city is a type of urban structure that has multiple centers of employment and activity, which are independent and interconnected. These centers may have divergent functions, sizes, and locations, creating a complex and dynamic spatial pattern. A polycentric city has some advantages, such as lower congestion, lower pollution, higher efficiency, and higher innovation. However, it also has some disadvantages, such as a loss of coherence, a loss of identity, and higher fragmentation. A polycentric city may be more adaptable to the changing economic and social conditions, but it may also face coordination and governance problems among the different centers.

A monocentric city, where most of the population and functions are concentrated in a single core, may not be able to adapt to the dynamic process of urbanization and globalization, which demand more adaptability, flexibility, and diversity in the urban system. Because of its nature, a monocentric city may face problems of congestion, pollution, inequality, social exclusion, and lack of innovation,

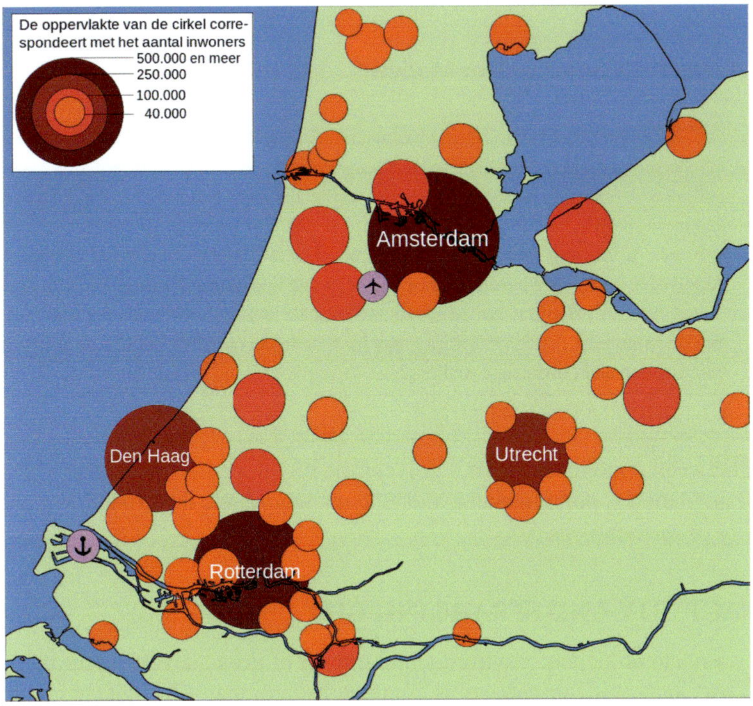

FIGURE 2.4 Polycentric city of Randstad, Netherlands.

Source: Wikipedia.

while a polycentric urban region, where multiple centers of economic and social activity coexist within a larger metropolitan area, may offer more opportunities for mobility, accessibility, and sustainability while maintaining social cohesion, and creativity.

Polycentric city consists of several distinct and independent urban centers, rather than a single dominant core. Although a polycentric city may have various advantages, it may also face challenges and disadvantages in coordinating governance, managing sprawl, ensuring equity, and maintaining identity. A polycentric city is not simply a collection of separate towns, but a networked system that operates at multiple scales and functions.

Some examples of polycentric cities are:

1. The Randstad in the Netherlands (Figure 2.4), which comprises four major cities (Amsterdam, Rotterdam, The Hague, and Utrecht) and several smaller towns, forms a ring-shaped urban region with a population of about 8 million people.
2. The Rhine–Ruhr in Germany consists of several large- and medium-sized cities (such as Cologne, Düsseldorf, Essen, Dortmund, and Duisburg) along the Rhine and Ruhr rivers, with a total population of about 10 million people.
3. The Pearl River Delta in China encompasses nine cities (including Guangzhou, Shenzhen, Hong Kong, and Macau) and several other towns in the Guangdong province, with a population of about seventy million people.

Given the quality of both monocentric cities and polycentric cities, there is no definitive answer to which one is better, as both types of cities have advantages and disadvantages. Some of the factors that may influence the preference for one or the other are shown in Table 2.4.

TABLE 2.4
Factors Influencing the Choice of Two Models

Factor	Explanation
Accessibility	A monocentric city may have better public transportation and walkability, but also more congestion and pollution. A polycentric city may have less traffic and more green spaces, but also longer commutes and more sprawl.
Diversity	A monocentric city may have more cultural and social diversity, but also more inequality and segregation. A polycentric city may have more homogeneity and integration, but also less vibrancy and innovation.
Efficiency	A monocentric city may have more economies of scale and agglomeration, but also more externalities and diseconomies. A polycentric city may have more decentralization and competition, but also more coordination and governance challenges.

The choice between a monocentric or a polycentric city depends on the preferences, values, and goals of the urban planners, policymakers, and citizens who shape the urban form and function, and enjoy the quality of their choices.

2.6 COMPACT CITY AND SPRAWLING CITY

A compact city aims to reduce the environmental impact of urban development by promoting high-density, mixed-use, and transit-oriented development. In a compact city, landed homes may not exist or rarely exist, rather, the residential buildings are in compact form like apartment buildings. They are sometimes shared with offices, shops, and public transport stations, and they are within walking distance of each other, creating a vibrant and diverse urban environment. The cities are characterized by efficient use of land, energy, and resources, as well as reduced greenhouse gas emissions, air pollution, and traffic congestion. A compact city may have lower per capita energy consumption and carbon footprint than a sprawling city, due to less reliance on private cars and more use of renewable energy sources. Compact cities also offer social and economic benefits, such as improved accessibility, livability, diversity, and innovation. A compact city may have better access to education, health care, culture, and employment opportunities than a dispersed city, as well as more opportunities for social interaction and creativity. However, compact cities also carry some disadvantages, such as overcrowding, noise, loss of green space, and social inequality. Therefore, compact city planning requires careful consideration of the trade-offs and synergies between different dimensions of sustainability.

Compact cities may carry a quality that leads to accomplishing sustainable urban development, where the design of a compact city incorporates substantial amounts of open space within its boundaries. This type of urban form can offer several benefits for various planning aspects, such as increasing the greenery coverage in the city, enhancing the efficiency of land use by allowing higher density and intensity of development, and improving the quality of life for the residents.

Compact city promotes a higher floor area to plot ratio (FAPR). The FAPR is a measure of how efficiently a building uses the land on which it is constructed. It is calculated by dividing the total floor area of the building by the area of the plot. A higher FAPR means that the building has more floor space per unit of land, which can indicate higher density, compactness, and profitability. However, a higher FAPR can also have negative impacts on the environment, such as increased energy consumption, reduced green space, and lower air quality. Therefore, FAPR is often regulated by zoning codes and urban planning policies to balance the benefits and costs of development. The ideal FAPR depends on several factors, such as the location, function, design, and demand of the building. There is no universal formula to determine the optimal FAPR, but rather a range of acceptable values that vary according to the context and objectives of each project.

TABLE 2.5
Floor Area to Plot Ratio

Consideration	Explanation
The type of building	Distinct types of buildings have different FAPR requirements. For example, residential buildings may have lower FAPR than commercial or industrial buildings, to ensure adequate living space and privacy for the occupants.
The site characteristics	The physical features of the site, such as the shape, size, slope, orientation, and soil conditions, may affect the FAPR. For example, a narrow or irregular site may require a higher FAPR to achieve a feasible design, while a steep or unstable site may limit the FAPR to reduce the risk of landslides or erosion.
The environmental impact	The FAPR should consider the potential impact of the building on the surrounding environment, such as solar access, ventilation, noise, traffic, stormwater runoff, and biodiversity.
The esthetic quality	The FAPR should also consider the visual appearance and character of the building and its compatibility with the existing or desired urban fabric. For example, a lower FAPR may be desirable to maintain a consistent scale and harmony with the historical or cultural heritage of the area, while a higher FAPR may be acceptable to create a distinctive landmark or focal point.

The FAPR is a measure of how efficiently a building uses the land on which it is constructed. It is calculated by dividing the total floor area of the building by the area of the plot. A higher FAPR means that the building has more floor space per unit of land, which can indicate higher density, compactness, and profitability. However, a higher FAPR can also have negative impacts on the environment, such as increased energy consumption, reduced green space, and lower air quality. Therefore, FAPR is often regulated by zoning codes and urban planning policies to balance the benefits and costs of development.

The ideal FAPR depends on a range of factors, such as the location, function, design, and demand of the building. There is no universal formula to determine the optimal FAPR, but rather a range of acceptable values that vary according to the context and objectives of each project. The range of acceptable FAPR can be influenced by the considerations as shown in Table 2.5.

There is no universal formula to determine the optimal FAPR, but rather a range of acceptable values that vary according to the context and objectives of each project. The FAPR for residential buildings is typically lower than for commercial or industrial buildings. This is because residential buildings require adequate living space and privacy for the occupants, as well as access to natural light and ventilation. The FAPR for residential buildings may vary depending on the type of housing, whether it is single-family or multi-family dwellings, and the density and character of the neighborhood. According to Wikipedia, in New York City, the FAPR for residential buildings ranges from 0.5 to 10, depending on the zoning district and the availability of public amenities. In Singapore, the FAPR for residential buildings ranges from 1.4 to 4.9, depending on the location and type of development. In India, the FAPR for residential buildings varies from 1 to 4, depending on the city and the zone.

From the perspective of different land uses in the city, there is no simple answer to the question of what the ideal mix of land is used in a city. Different cities have different histories, geographies, cultures, and characters that influence their land use patterns. However, some general principles can be applied to achieve balanced and sustainable urban development. A city should have enough residential land to accommodate its population, but also enough parks and open space to provide recreation and environmental benefits. A city should have enough commercial and industrial land to support its economy, but also enough institutional land to provide public services and facilities. A city should also have efficient transportation networks to connect different land uses and reduce congestion and pollution.

TABLE 2.6
Typical Percentage of Land Uses

Land Use Category	Percentage
Residential zone	24–49
Parks and open space	11–28
Institutional zone	8–20
Commercial zone	4–7
Vacant land	4–20
Transportation	14–24
Water bodies	2–5

Sources: Modes and Averages values of 87 cities in the world.

The typical percentage of land use in cities varies widely, based on hundreds of land use data of the city, some common ranges are shown in Table 2.6.

These percentages are not fixed or optimal, but rather indicative of the diversity and complexity of urban land use planning. Each city should consider its context, needs, and goals when deciding how to allocate its land resources. By this notion, there is no limitation on how compact a compact city is.

A key component of compact city design is the specific infrastructure required to accommodate high-density urban functions (Figure 2.5). A compact city necessitates effective and dependable mobility systems, such as mass transit, cycling networks, and walkability, to decrease the reliance on private vehicles and related emissions and congestion. For example, a compact city like Singapore has invested heavily in its public transportation system, which covers 90% of the island and carries 4.5 million passengers daily. A compact city also necessitates green and open spaces, such as parks, gardens, and plazas, to offer recreational and social opportunities for inhabitants and visitors, as well as to improve the environmental quality and resilience of the urban area. Another example is that a compact city like Copenhagen has created a network of green corridors that connect various parts of the city and provide access to nature and leisure activities. Moreover, a compact city necessitates intelligent and sustainable utilities, such as water, energy, and waste management systems, to optimize the use of resources and minimize the environmental impact of urban activities. An example from Asia is that a compact city like Seoul has implemented a smart grid system that integrates renewable energy sources and reduces greenhouse gas emissions by 23%. These specific infrastructure elements are vital to creating a habitable, appealing, and competitive compact city that can satisfy the needs and aspirations of current and future generations.

In contrast with a compact city, a sprawling city is a type of urban development that is characterized by low-density, dispersed, and car-dependent land use (Figure 2.6). Sprawl cities consume more land, energy, and resources than compact cities, and often create negative impacts on the environment, public health, and social equity. Sprawl city is usually associated with suburbanization, decentralization, and fragmentation of urban areas. Some examples of sprawl cities are Atlanta, GA; Los Angeles, CA; Houston, TX; and Phoenix, AZ in the United States, and Melbourne, Sydney, and Brisbane in Australia.

The quality of a sprawling city is characterized by low-density, car-dependent, and dispersed land use patterns. Sprawl city is often associated with negative environmental, social, and economic impacts, such as air pollution, traffic congestion, loss of natural habitats, social isolation, and increased public costs. However, some proponents of sprawl city argue that it can also offer some benefits, such as affordability, personal freedom, and consumer choice. Therefore, the question of how far a sprawl city can accomplish a sustainable city is not easy to answer.

One viable way to approach this question is to consider the three dimensions of sustainability: environmental, social, and economic. Environmental sustainability refers to the ability of a city to maintain or enhance the quality and quantity of natural resources and ecosystems. Social sustainability refers to the ability of a city to foster social cohesion, equity, diversity, and well-being among its residents. Economic sustainability refers to the ability of a city to generate and distribute wealth and opportunities for its residents and businesses.

From an environmental perspective, sprawl cities face many challenges to achieve sustainability. Sprawl city consumes more land and energy than compact city, which leads to higher greenhouse gas emissions and ecological footprint. Sprawl city also reduces the availability and accessibility of green spaces and public transportation, which affects the health and quality of life of its residents. Therefore, sprawl city needs to adopt various strategies to reduce its environmental impacts, such as increasing the density and diversity of land use, promoting mixed-use and transit-oriented development, enhancing the efficiency and renewable share of energy use, preserving, and restoring natural habitats and corridors, and implementing green infrastructure and urban forestry.

From a social perspective, sprawl cities also have some drawbacks to achieving sustainability. Sprawl city tends to create spatial segregation and inequality among diverse groups of people based on income, race, ethnicity, age, and lifestyle. Sprawl city also tends to weaken the sense of community and civic engagement among its residents, as they have fewer opportunities to interact with each other in public spaces and institutions. Therefore, sprawl city needs to adopt various strategies to improve its social cohesion and inclusion, such as ensuring the availability and affordability of housing for all income levels and preferences, providing adequate and accessible public services and facilities for all groups of people, encouraging social interaction and participation in local governance and decision-making processes, and fostering a culture of diversity and tolerance.

From an economic perspective, sprawl cities may have some advantages in achieving sustainability. A sprawling city may offer lower housing costs and taxes for its residents than a compact city, which may attract more people and businesses to locate there. Sprawl city may also provide more consumer choice and market competition for its residents and businesses than compact city, which may stimulate innovation and entrepreneurship. Therefore, sprawl city needs to adopt various strategies to maintain its economic vitality and competitiveness, such as investing in human capital and education for its workforce, supporting the development of local industries and clusters that are compatible with the regional economy, enhancing the connectivity and accessibility of transportation and communication networks within and outside the city, and balancing the fiscal revenues and expenditures of the local government.

A sprawling city can accomplish a sustainable city only if it addresses the environmental, social, and economic challenges that it faces. Sprawl city needs to transform itself into a more compact and mixed-use city to limit the consumption of natural resources by the city, as sustainable cities aim to reduce their environmental impact, enhance their social equity, and improve their economic efficiency. Sustainable cities seek to balance the needs of the present and the future, as well as the needs of humans and nature. To achieve this goal, sprawl cities need to adopt strategies toward sustainable cities, such as:

1. Promoting compact and mixed-use urban forms that reduce land consumption, transportation demand, and infrastructure costs.
2. Enhancing public transportation, walking, and cycling options improves mobility, accessibility, and health outcomes.
3. Implementing renewable energy sources, energy efficiency measures, and waste management systems that lower greenhouse gas emissions and resource consumption.

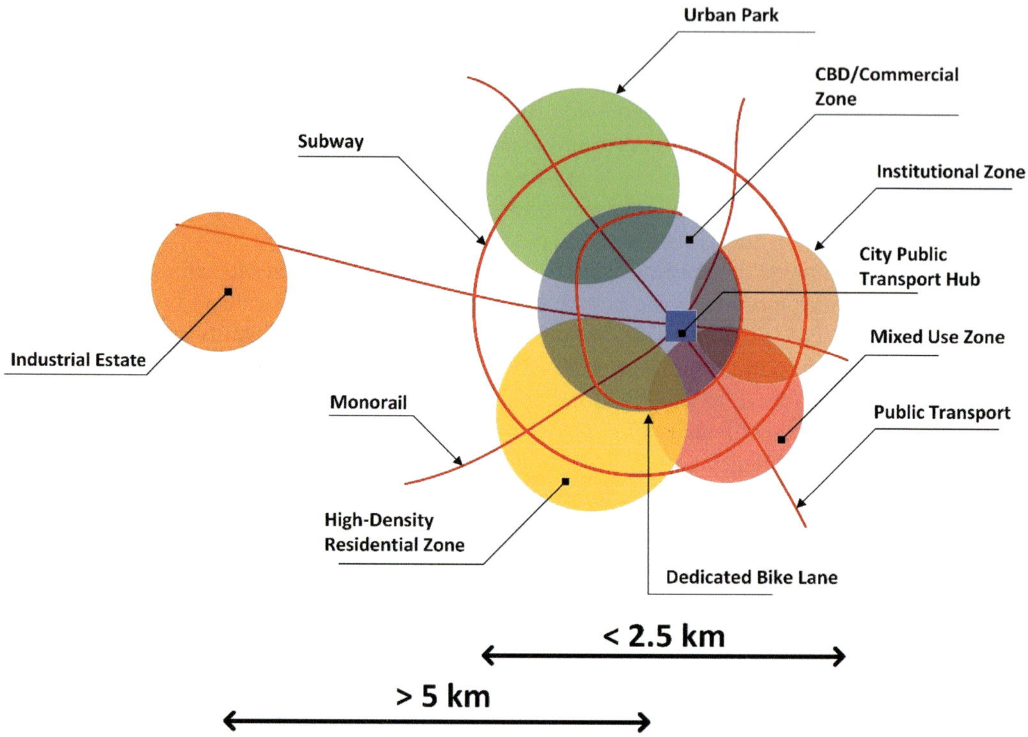

FIGURE 2.5 An idealized compact city.

FIGURE 2.6 A sprawl city.

Source: Wikipedia.

4. Preserving and restoring natural habitats, green spaces, and biodiversity provide ecosystem services and recreational opportunities.
5. Fostering social inclusion, participation, and diversity that strengthen community ties and civic engagement.
6. Supporting local economy, innovation, and culture that create jobs, wealth, and identity.

Retrofitting sprawl cities toward compact cities is not an option to accomplish sustainable cities.

Despite its disadvantages, sprawl cities offer many opportunities and challenges for their citizens. To maximize the services and quality of life for the residents, the city needs to invest in a robust and resilient infrastructure system that can support the growing and changing needs of the population. Some of the key infrastructure components that are required in sprawl city are (these are not specific to sprawl city) reflected in Table 2.7.

Building and maintaining such a system requires a lot of financial resources that may not be readily available. How can we fund these infrastructure components?

One conceivable way is to increase the taxes and fees that the citizens pay for using the infrastructure services, such as water, electricity, transportation, and communication. This would generate more revenue for the city and encourage more efficient and responsible use of the resources. However, this may also impose a burden on the low-income and vulnerable groups in society, who may not be able to afford the higher costs.

Another viable way is to seek external funding from other sources, such as the federal government, private sector, or international organizations. This would reduce the financial pressure on the

TABLE 2.7
Infrastructure Required in a Sprawl City

Type	Explanation
Transportation	The city should provide efficient and accessible public transportation options, such as buses, light rail, subways, and bike lanes, which can connect various parts of the city and reduce traffic congestion and pollution. The city should also maintain and upgrade the roads, bridges, tunnels, and sidewalks that enable the safe and smooth movement of vehicles and pedestrians.
Water supply	The city should ensure an adequate and reliable supply of clean water for domestic, industrial, and agricultural use. The city should also manage the wastewater and stormwater systems that prevent flooding and contamination of the environment. The city should adopt water conservation and reuse practices that reduce water waste and increase water security.
Energy	The city should diversify its energy sources and use more renewable and low-carbon options, such as solar, wind, hydro, and geothermal power. The city should also improve the efficiency and reliability of the power grid that delivers electricity to homes, businesses, and public facilities. The city should encourage energy conservation and efficiency measures that lower energy demand and costs.
Communication	The city should expand and modernize its communication network which enables fast and reliable access to information and services. The city should provide broadband internet, wireless, and fiber optic connections that support digital inclusion, innovation, and economic development. The city should also enhance the emergency communication system that ensures public safety and disaster response.
Social	The city should foster a vibrant and inclusive social infrastructure that promotes community engagement, cultural diversity, and civic participation. The city should provide public spaces, such as parks, libraries, museums, and community centers, which offer recreational, educational, and cultural opportunities for the residents. The city should also support social services, such as healthcare, education, housing, and welfare, which address the needs and challenges of diverse groups in society.

city and allow for more collaboration and innovation in infrastructure projects. Some of the benefits of external funding are:

1. It can provide access to more capital and expertise that may not be available locally.
2. It can help spread the risks and costs of the infrastructure projects among different parties.
3. It can create opportunities for learning and knowledge transfer among different actors.
4. It can enhance the credibility and reputation of the city as a reliable and attractive partner.

However, external funding may also entail some trade-offs and compromises in terms of the autonomy, accountability, and sustainability of the infrastructure system.

A third feasible way is to adopt a mixed approach that combines both internal and external funding sources, as well as several types of financing mechanisms, such as bonds, loans, grants, subsidies, or public–private partnerships. This would enable the city to leverage its resources and capabilities, as well as tap into the expertise and experience of other actors. However, this may also require more coordination and communication among the various stakeholders and partners involved in the infrastructure system.

The next inquiry is how we can ensure sustainability in infrastructure projects. Sustainability is a key aspect of infrastructure projects in the short and long term. Some of the ways to ensure sustainability in infrastructure projects are:

1. Conducting comprehensive feasibility studies that assess the needs, costs, benefits, risks, and alternatives of the projects. For example, a feasibility study may compare different options for expanding the communication network in terms of their technical feasibility, economic viability, social acceptability, and environmental compatibility.
2. Applying life cycle analysis that considers the whole life span of the projects from planning to operation to decommissioning. For example, a life cycle analysis may estimate the total energy consumption, greenhouse gas emissions, waste generation, and resource depletion of a solar power project over its lifetime.
3. Adopting green design principles that minimize the use of natural resources, energy, and emissions in the projects. Green design principles are guidelines or criteria that aim to reduce the environmental impact of buildings or products by incorporating elements such as:
   - *Renewable Energy Sources*: These are sources of energy that are derived from natural processes that are replenished constantly, such as solar, wind, hydro, or geothermal power.
   - *Energy-Efficient Appliances*: These are devices or machines that use less energy than conventional ones to perform the same function or service, such as LED lights, smart thermostats, or low-flow faucets.
   - *Water-Saving Devices*: These are devices or systems that reduce water consumption or waste by using less water or reusing water for different purposes, such as rain barrels, dual-flush toilets, or greywater systems.
   - *Waste-Reduction Measures*: These are actions or practices that prevent or minimize waste generation or disposal by reducing, reusing, recycling, or composting materials or products, such as biodegradable packaging, reusable bags, or worm bins.
   - *Natural Ventilation*: This is a method of cooling or heating buildings by using natural air flow instead of mechanical systems such as fans or air conditioners.

A green design may incorporate renewable energy sources, energy-efficient appliances, water-saving devices, waste-reduction measures, and natural ventilation in a public building project. In addition, implementing participatory approaches by involving the stakeholders and beneficiaries in the decision-making and implementation processes of the projects is necessary. For example, a participatory approach may solicit feedback from the residents on their preferences and expectations for a new park project. Also, establishing monitoring and evaluation systems that track the performance, outcomes, and impacts of the projects should not be skipped. For example, a monitoring and evaluation system may collect data on indicators such as service quality, user satisfaction, cost recovery, social inclusion, environmental protection, etc.

By ensuring sustainability in infrastructure projects, sprawl cities can maximize the services to the citizens and create a more livable, sustainable, and prosperous urban environment.

2.7 OPEN SPACES AND CITIZENS

Open space constitutes an essential element of the urban environment and citizens' well-being. It has multiple advantages for the physical, mental, social, and environmental health of the people and the planet. Open space may encompass parks, gardens, plazas, playgrounds, greenways, trails, natural areas, and other public spaces that enable people to access nature, recreation, leisure, culture, and community. Open space may augment the livability, attractiveness, and resilience of cities by providing opportunities for relaxation, exercise, social interaction, education, creativity, and civic engagement. Open space may also assist in mitigating the impacts of climate change, such as heat islands, air pollution, flooding, and biodiversity loss, by offering ecosystem services such as cooling, shading, filtering, storing, and absorbing water and carbon. Open space may engender a sense of place, identity, and belonging for the diverse urban population and support social inclusion, equity, and justice. Open space is a valuable resource that requires planning, design, management, and protection for the present and future generations.

One of the challenges of urban planning is how to transform urban wasteland into productive and livable spaces. Urban wasteland refers to areas that are abandoned, contaminated, or underutilized, such as brownfields, vacant lots, or industrial sites. These areas often pose environmental and social problems, such as pollution, crime, and blight. However, they also offer opportunities for creative and sustainable urban regeneration. Here are some ways to generate more useful urban spaces from urban wasteland:

1. *Urban Farming:* Urban farming is the practice of growing food in urban areas, using various methods such as hydroponics, aquaponics, vertical farming, or rooftop gardening. Urban farming can provide fresh and healthy food for local communities, reduce food miles and greenhouse gas emissions, and create green jobs and education opportunities. Urban farming can also improve the soil quality and biodiversity of urban wastelands and create attractive and edible landscapes.
2. *Urban Art:* Urban art is the expression of art in public spaces, using various forms such as graffiti, murals, sculptures, installations, or performances. Urban art can beautify and revitalize urban wasteland by adding color, texture, and meaning to the otherwise dull and dreary environment. Urban art can also engage and empower local communities, by reflecting their culture, identity, and aspirations, and by providing a platform for dialogue and social change.
3. *Urban Ecosystems:* Restoring and enhancing the ecological functions and services of urban wastelands, such as water purification, air quality improvement, climate regulation, and habitat provision are necessary. Reviving urban ecosystems can also promote human well-being and quality of life, by creating healthy, livable, and enjoyable spaces that foster physical activity, mental health, and social interaction.
4. *Urban Greenery/Urban Forest:* Urban wastelands can be transformed into urban greenery by active greening of urban wastelands. Urban greenery with its various stages of vegetation can provide many ecosystem services to tackle challenges such as stopping the loss of biodiversity, adapting to climate change, and creating recreational and healthy urban environments. The conservation of spontaneous biotopes offers the potential to develop urban green systems

that provide a range of essential ecological, social, and aesthetic services. Various planning and development approaches can be used to transform urban wastelands into urban greenery, these approaches include:

- the creation of community gardens;
- the creation of green corridors;
- the creation of green roofs;
- the creation of green walls; and
- the creation of urban forests.

5. *Urban Park:* Planning and creativity can transform urban wasteland into urban parks that provide multiple benefits for the community. Some examples of urban parks that were created from wastelands are:
   - The High Line in New York City is a linear park built on an abandoned elevated railway.
   - Gas Works Park in Seattle is a park that incorporates the remains of a former gasification plant.
   - Parc de la Villette in Paris is a park that features cultural and recreational facilities on a former slaughterhouse site.

Here are some generic steps to follow for a successful transformation of urban wasteland:

1. Identify the potential and challenges of the site. Conduct a site analysis to understand the physical, social, and ecological aspects of the wasteland. What are the existing features, such as vegetation, soil, water, wildlife, infrastructure, and pollution? What are the needs and preferences of the local community? What are the opportunities and constraints for development?
2. Develop a vision and a design concept. Based on the site analysis, create a vision statement that summarizes the main goals and values of the project. Then, develop a design concept that illustrates how the wasteland will be transformed into a park. Consider aspects such as layout plans, circulation, amenities, aesthetics, and sustainability.
3. Implement and manage the project. Once the design concept is approved, prepare a detailed plan that specifies the budget, timeline, materials, and methods for implementation. Seek funding sources and partnerships with stakeholders, such as government agencies, NGOs, businesses, and community groups. During implementation, monitor and evaluate the progress and quality of the work. After completion, establish a management plan that ensures the maintenance and operation of the park.

Urban parks are public spaces that provide countless benefits for the well-being of city dwellers. They offer opportunities for recreation, social interaction, physical activity, exposure to nature, and stress relief. There is a correlation between the number of urban parks and the happiness of citizens. This was based on the data from the World Happiness Report and the Urban Green Space Index to measure the levels of happiness and urban park availability. A multiple regression analysis to examine the relationship between these variables has been done. The variables in the analysis include population, income, governance, and environmental quality.

Multiple regression analysis is a statistical technique that allows us to estimate how much each independent variable (such as urban park availability) contributes to the dependent variable (such as happiness) while holding other variables constant. The linear model to express the relationship between the variables has the form:

$$y = b_0 + b_1 x_1 + b_2 x_2 + \ldots + b_n x_n + \varepsilon \quad (2.1)$$

where:

y	the independent variable, in this case, happiness
x_1, x_2, x_n	independent variables, for example, park number, area, population, income, governance
b_0	intercept
b_1, b_2, b_2	coefficients
ε	error

The coefficients indicate the direction and magnitude of the effect of each independent variable on the dependent variable. Many statistical software can be used, for example, SPSS. The results show that there is a positive and significant correlation between the number of urban parks and the happiness of citizens, even after accounting for other determinants of happiness. It is also found that the effect of urban parks on happiness is stronger in developing countries than in developed ones.

This study suggests some ways to improve urban parks and increase their positive impact on happiness. Some of these ways are increasing the size, quality, and accessibility of urban parks; enhancing the diversity and attractiveness of park features and activities; promoting community involvement and participation in park management and maintenance; integrating urban parks with other green spaces and transportation networks; and raising public awareness and appreciation of the value of urban parks. Allocating more resources and funding for urban park development and improvement; establishing standards and guidelines for urban park design and quality; creating incentives and regulations for private sector involvement in urban park provision and maintenance; fostering collaboration and coordination among different stakeholders and agencies involved in urban park management; and developing indicators and monitoring systems to evaluate the impact of urban parks on happiness are required.

2.8 NATURE AND HUMANKIND

The urban environment is the result of the interaction between human activities and natural processes. Anthropocentric activities are those that are centered on human needs and interests, often at the expense of the environment. Some examples of anthropocentric activities in urban areas are:

1. *Urbanization*: This is the process of expanding and densifying urban areas, often accompanied by population growth, economic development, and social change. Urbanization can have negative impacts on the environment, such as land degradation, habitat loss, biodiversity decline, air pollution, water scarcity, and waste generation.
2. *Industrialization:* This is the process of transforming the economy from agriculture-based to manufacturing-based, often involving the use of fossil fuels, chemicals, and machinery. Industrialization can have negative impacts on the environment, such as greenhouse gas emissions, climate change, acid rain, soil contamination, and environmental health risks.
3. *Consumption*: This is the process of using natural resources and goods to satisfy human wants and needs. Consumption can have negative impacts on the environment, such as resource depletion, overexploitation, deforestation, desertification, and marine pollution.

These are some of the anthropocentric activities that affect the urban environment. Urban environmental planning can help to mitigate these impacts by promoting more sustainable and equitable ways of living in urban areas. Urban environmental planning can also help to enhance the positive aspects of urbanization, industrialization, and consumption, such as innovation, diversity, culture, and education. The poor interplay between anthropogenic activities and nature, along with poor urban environmental planning, results in a poor city environment.

One example of a city with poor environmental planning is Beijing, China. Beijing has been facing severe environmental problems due to rapid urbanization and industrialization. The city

TABLE 2.8
Aspects in Urban Environmental Planning

Aspect	Explanation
Land use planning	This refers to the allocation and regulation of land for different purposes, such as residential, commercial, industrial, recreational, and ecological. Land use planning can help to optimize the use of urban space, reduce urban sprawl, preserve natural resources, and enhance the quality of life for urban residents.
Urban greenery	This refers to the vegetation and natural elements that are integrated into urban areas, such as parks, gardens, street trees, green roofs, and green walls. Urban greenery can provide multiple benefits, such as improving air quality, mitigating the urban heat island effect, enhancing biodiversity, reducing stormwater runoff, and promoting human health and well-being.
Municipal waste management	This refers to the collection, treatment, disposal, and recycling of solid and liquid waste generated by urban activities. Urban waste management can have significant impacts on the environment, public health, and resource efficiency of urban areas. Urban environmental planning can help to reduce the amount of waste produced, increase the recovery of valuable materials, and minimize the negative effects of waste disposal.

has prominent levels of air pollution, water pollution, and waste generation, which have negative impacts on public health and the environment. The main causes of these problems are the overuse of fossil fuels, the lack of green spaces, and the inadequate waste management system.

To address these problems, Beijing has implemented various measures, such as promoting renewable energy, improving public transportation, and increasing green spaces. However, these measures are still insufficient to solve the environmental problems in the city. Beijing needs more comprehensive and integrated environmental planning that considers the interrelationships among numerous factors and stakeholders.

Urban environmental planning is the process of designing and managing urban spaces to harmonize nature and humanity. It aims to create livable, resilient, and sustainable cities that balance the needs of human development and environmental protection. Urban environmental planning involves various aspects as shown in Table 2.8.

These are some of the main aspects of urban environmental planning that can help to harmonize nature and humanity in urban areas. Urban environmental planning requires a holistic and participatory approach that considers the interrelationships among several factors and stakeholders. Urban environmental planning can also benefit from the use of innovative technologies and methods that can enhance the data collection, analysis, visualization, and communication of urban environmental issues.

Anthropogenic activities that degrade the urban environment are those human actions that negatively affect the quality of life, health, and well-being of urban dwellers. In a broad sense, the domain of pollution, the pollution caused by human activities include:

1. *Air Pollution:* The emission of harmful gases and particles from vehicles, industries, power plants, and other sources can cause respiratory diseases, cardiovascular problems, and climate change.
2. *Water Pollution:* The discharge of untreated sewage, industrial waste, agricultural runoff, and other contaminants into rivers, lakes, and oceans can harm aquatic ecosystems, reduce water availability, and spread diseases.
3. *Land Pollution:* The accumulation of solid waste, hazardous materials, and construction debris on land can degrade soil quality, increase the risk of fire and flooding, and create aesthetic problems.

4. *Noise Pollution:* Exposure to elevated levels of noise from traffic, machinery, entertainment venues, and other sources can impair hearing, induce stress, and disturb sleep.
5. *Light Pollution:* The excessive use of artificial lighting at night can interfere with the natural cycles of plants, animals, and humans, affecting their biological rhythms, behavior, and health.

The anthropogenic activities that degrade the urban environment have dire consequences for the sustainability and livability of cities. Therefore, it is important to adopt measures to prevent, reduce, and mitigate their impacts, such as:

1. Promoting the use of renewable energy sources and energy efficiency technologies to lower greenhouse gas emissions and air pollution.
2. Implementing proper waste management systems and recycling programs to minimize land pollution and resource consumption.
3. Treating wastewater before discharging it into water bodies and enforcing strict regulations on industrial and agricultural activities to prevent water pollution.
4. Planning urban development in a way that preserves green spaces, reduces urban sprawl, and enhances public transportation and mobility options to lower noise and light pollution.
5. Raising awareness among urban residents and stakeholders about the environmental problems and solutions in their cities and encouraging their participation in environmental initiatives and actions.

The harmonious interplay between the environment and people is the state of mutual benefit and balance between nature and humans. It means that people respect, protect, and restore the environment, while the environment provides resources, services, and beauty for people. The harmonious interplay between the environment and people can enhance the well-being, resilience, and sustainability of both.

To accomplish the harmonious interplay between environment and people, some actions will be necessary. These actions may include the following:

1. Adopting a holistic and participatory approach to environmental planning and management, which considers the interrelationships among several factors and stakeholders.
2. Promoting green and low-carbon development, which reduces greenhouse gas emissions, mitigates climate change, and improves energy efficiency.
3. Conserving and restoring biodiversity and ecosystems, which maintain the functioning and diversity of nature, and provide essential services for people.
4. Enhancing environmental education and awareness fosters a sense of responsibility, appreciation, and stewardship for the environment among people.
5. Encouraging green lifestyles and consumption patterns, which reduce the ecological footprint, waste generation, and resource depletion caused by human activities.

Environmental planning must ensure that the interplay between the environment and people is in harmony as this harmonious interaction may bring multiple benefits to both. This requires a holistic and participatory approach that considers the interrelationships among a range of factors and stakeholders. Environmental planning must also be based on scientific evidence, best practices, and innovative technologies that can enhance the sustainability, resilience, and well-being of urban areas.

Urban environmental planning can help to achieve a harmonious interplay between the environment and people by promoting more sustainable and equitable ways of living in urban areas. Urban environmental planning can also help to enhance the positive aspects of human–nature interactions by fostering a sense of responsibility, appreciation, and stewardship for the environment among people.

The harmonious interplay between the environment and people can provide multiple benefits, such as:

1. *Improved Health and Well-Being*: Exposure to nature has been shown to reduce stress, anxiety, depression, and other mental health problems. Nature can also promote physical health by providing opportunities for exercise, recreation, and social interaction.
2. *Enhanced Resilience and Adaptation*: A healthy environment can provide natural defenses against natural disasters, climate change, and other environmental risks. Nature can also provide resources for food, water, energy, and materials that are essential for human survival and development.
3. *Increased Productivity and Creativity*: Exposure to nature has been shown to enhance cognitive performance, creativity, innovation, and job satisfaction. Nature can also inspire art, culture, and spirituality.
4. *Improved Social Cohesion and Equity*: A healthy environment can provide opportunities for social interaction, community building, and cultural diversity. Nature can also promote social justice by providing equal access to resources and services for all people.
5. *Enhanced Ecological Integrity and Sustainability*: A healthy environment can maintain the functioning and diversity of nature, which is essential for the survival of all species. Nature can also provide essential services for human well-being, such as air purification, water regulation, soil fertility, pollination, and pest control.

Urban environmental planning can help to achieve these benefits by promoting more sustainable and equitable ways of living in urban areas. Urban environmental planning can also help to enhance the positive aspects of human–nature interactions by fostering a sense of responsibility, appreciation, and stewardship for the environment among people. To accomplish a better urban environment, better awareness, and responsibility of the citizens on the environment, and create a harmonious balance between environment and people, some viable policies are, among others, shown in Table 2.9.

These policies require a holistic and participatory approach that considers the interrelationships among a range of factors and stakeholders. These policies also require political will, public support, private investment, and international cooperation to achieve their goals. Some challenges of implementing these policies are:

1. *Political Will:* Environmental policies require political support and leadership at all levels of government. However, political will can be influenced by numerous factors, such as economic interests, public opinion, and international relations. Environmental policies may face opposition or resistance from stakeholders who perceive them as a threat to their interests or values.
2. *Public Participation:* Environmental policies require the participation and engagement of citizens and communities who are affected by them. However, public participation can be hindered by numerous factors, such as lack of awareness, lack of trust, and lack of resources. Environmental policies may also face conflicts or trade-offs among diverse groups of stakeholders who have unique needs or priorities.
3. *Financial Resources:* Environmental policies require financial resources to implement and maintain. However, financial resources can be limited or uncertain due to numerous factors, such as budget constraints, economic downturns, and competing demands. Environmental policies may also face challenges in attracting private investment or international aid that can support their implementation.

TABLE 2.9
Viable Policies for Better Urban Environment

Policy	Explanation
Green infrastructure planning	This refers to the strategic planning and management of natural and semi-natural areas in urban areas, such as parks, gardens, green roofs, green walls, and urban forests. Green infrastructure planning can help to enhance the ecological, social, and economic benefits of urban greenery, such as biodiversity conservation, climate change mitigation, air quality improvement, and human health promotion.
Sustainable transportation planning	This refers to the planning and management of urban transportation systems that are safe, efficient, affordable, and environmentally friendly. Sustainable transportation planning can help to reduce greenhouse gas emissions, air pollution, traffic congestion, and noise pollution caused by urban mobility. Sustainable transportation planning can also promote active modes of transport, such as walking and cycling, which can improve public health and social interaction.
Circular economy policy	This refers to the policy framework that promotes the reduction, reuse, recycling, and recovery of waste materials in urban areas. The circular economy policy can help to reduce the amount of waste generated by human activities, conserve natural resources, and create new economic opportunities for sustainable development. A circular economy policy can also promote the shift from linear consumption patterns to circular consumption patterns that are more resource-efficient and less polluting.
Environmental education program	This refers to the program that provides formal or informal learning opportunities for citizens on environmental issues and solutions. Environmental education programs can help to raise awareness, knowledge, skills, and values among citizens on environmental issues and solutions. Environmental education programs can also promote citizen participation in environmental decision-making processes that are transparent, inclusive, and accountable.
Green lifestyle campaign	This refers to the campaign that promotes green behaviors and attitudes among citizens in their daily lives. Green lifestyle campaigns can help encourage citizens to adopt more sustainable consumption patterns that are less resource-intensive and less polluting. Green lifestyle campaigns can also promote citizen engagement in environmental initiatives that are community-based, collaborative, and innovative.

4. *Technological Innovation:* Environmental policies require technological innovation that can enhance the efficiency, effectiveness, and sustainability of human activities. However, technological innovation can be slow or uncertain due to several factors, such as lack of research funding, lack of skilled workforce, and lack of market demand. Environmental policies may also face challenges in adopting or adapting innovative technologies that can meet local needs or conditions.
5. *International Cooperation:* Environmental policies require international cooperation that can address global environmental challenges that transcend national boundaries. However, international cooperation can be complex or contentious due to several factors, such as geopolitical tensions, cultural differences, and legal frameworks. Environmental policies may also face challenges in aligning with international standards or agreements that can ensure their effectiveness or legitimacy.

One of the major challenges of urban development is the lack of responsibility of the citizens for the impact of their actions on the environment. Urban areas are often characterized by high population density, consumption of resources, generation of waste, and emission of pollutants. These factors contribute to environmental degradation, climate change, and public health problems. However, many citizens are unaware or indifferent to the consequences of their behavior on the urban environment. They may lack the knowledge, awareness, motivation, or incentives to adopt more sustainable practices. Therefore, it is essential to foster a sense of responsibility and civic engagement among urban dwellers and to promote environmental education, awareness, and participation. By doing so, the citizens can become more active agents of change and improvement in their urban communities. For example, they can participate in recycling programs, use public transportation or bicycles, reduce their energy and water consumption, plant trees and gardens, and support local initiatives and organizations that aim to protect and enhance the urban environment.

To push forward the sense of responsibility of the citizens in the urban environment, some strategies to achieve this goal are:

2.8.1 Environmental Education

The main goal of environmental education is to raise awareness and understand the complex and interrelated issues that affect our planet and its inhabitants. Educating the public about the benefits of green spaces, waste management, energy efficiency, and other environmental issues can be done through various means, such as campaigns, workshops, exhibitions, and social media. These methods can help inform, inspire, and empower people to act and make positive changes in their own lives and communities. Another important challenge of our time is to protect the environment and ensure the sustainability of our planet. To achieve this goal, we need to educate the public about the benefits of green spaces, waste management, energy efficiency, and other environmental issues. These benefits include improving the quality of life, health, and well-being of people and communities, reducing greenhouse gas emissions and pollution, preserving biodiversity and natural resources, and creating economic opportunities and social justice.

There are many ways to educate the public about these topics, such as campaigns, workshops, exhibitions, and social media. Campaigns can raise awareness and mobilize people to act for the environment, for instance, recycling, saving water and electricity, and choosing eco-friendly products and services. Workshops can provide practical skills and knowledge on how to implement environmental solutions in everyday life, such as composting, gardening, renewable energy, and green building. Exhibitions can showcase the best practices and innovations in environmental protection and conservation, such as urban farming, electric vehicles, solar panels, and biodegradable materials. Social media can spread the message and engage the audience dynamically and interactively through videos, podcasts, blogs, quizzes, and games.

By educating the public about the benefits of green spaces, waste management, energy efficiency, and other environmental issues, we can inspire them to adopt more sustainable lifestyles and behaviors and to support environmental policies and initiatives. This way, we can contribute to the global efforts to combat climate change and safeguard our future.

These are some of the ways that environmental education can be delivered and received by the public. By engaging people in meaningful and relevant learning experiences, environmental education can foster a sense of responsibility, stewardship, and action for the environment. Improving the sense of responsibility and awareness of the citizens on the city environment can also be done through (1) educating the public about the benefits of green spaces, waste management, energy efficiency, and other environmental issues. (2) Involving the citizens in the decision-making process of urban projects, such as participatory budgeting, public consultations, and feedback mechanisms. This can increase the transparency, accountability, and legitimacy of urban governance. (3) Encouraging the citizens to act for the urban environment, such as volunteering, recycling, composting, planting trees, and using public transportation. This can foster a sense of ownership, empowerment, and civic engagement among urban dwellers. (4) Creating incentives and rewards for the citizens who adopt environmentally friendly behaviors, such as tax breaks, subsidies, discounts, and recognition. This can motivate the citizens to act in the best interest of the urban environment.

Another strategy for creating a sustainable and resilient urban environment is to develop a network of stakeholders and partners who share the same vision and goals. These stakeholders and partners can include non-governmental organizations, businesses, academia, and media, among others. By establishing a network of diverse and committed actors, urban environmental issues can be addressed more effectively and holistically. A network can also facilitate collaboration, innovation, and advocacy for urban environmental solutions, as well as foster mutual learning and capacity building. Developing a network of stakeholders and partners requires building trust, communication, and alignment of interests and values among the different actors. It also requires identifying usual challenges and opportunities, as well as leveraging the strengths and resources of each partner. A network of stakeholders and partners can enhance the impact and sustainability of urban environmental initiatives, as well as create a sense of shared ownership and responsibility for the urban environment.

2.8.2 Involving Citizens in the Decision-Making Process

In progressive governance, inviting citizens to the decision-making process is common. This is to foster a sense of responsibility among citizens for the common good. A viable way to achieve this goal is to involve citizens in the decision-making process, allowing them to participate actively in shaping the policies and actions that affect their lives. By doing so, citizens can gain a deeper understanding of the issues, the trade-offs, the alternatives, and the consequences of their choices. They can also develop a sense of ownership and commitment to the outcomes, as well as a sense of accountability and transparency. Involving citizens in the decision-making process can also enhance the legitimacy and effectiveness of governance, as it can increase the trust and confidence between the public and the authorities and improve the responsiveness and adaptability of policies to the changing needs and preferences of society.

2.8.3 Encouraging the Citizens to Act for the Urban Environment

The urban environment is a vital part of our lives; however, it also faces many challenges such as pollution, congestion, waste, and climate change. These issues affect not only our health and well-being but also our economy and society. That is why we need to act for the urban environment, and we can do so in many ways.

One way is to adopt more sustainable habits, such as using public transportation, biking, or walking instead of driving; recycling and composting our waste; saving energy and water at home and work and buying local and organic products. These actions can reduce our environmental footprint, save us money, and improve our quality of life.

Another way is to get involved in our communities, such as joining or supporting environmental groups, volunteering for green projects, participating in public consultations, or contacting our local representatives. These actions can raise awareness, influence policies, and create positive change in the urban environment.

The urban environment is our shared responsibility, and we all have a role to play in protecting and enhancing it. Acting for the urban environment allows us to make a difference for ourselves, for our future generations, and for the planet.

PROBLEMS AND ISSUES FOR DISCUSSION

1. Discuss in detail the role of urban planning in promoting sustainable development. Consider the various strategies that urban planners can use to ensure that cities grow in a way that is environmentally friendly, socially equitable, and economically viable.
2. Analyze the concept of environmental planning. How does it differ from traditional urban planning? What unique challenges does it address, and how does it integrate environmental considerations into the planning process?

3. Evaluate the importance of environmental management in urban areas. How can effective environmental management contribute to the overall quality of life in cities? Consider factors such as air and water quality, waste management, and green spaces.
4. Compare and contrast the concepts of a compact city and a sprawl city. What are the advantages and disadvantages of each? Consider factors such as population density, transportation, and resource use.
5. Discuss the impact of urban sprawl on the environment. How does sprawl contribute to issues like air pollution, habitat destruction, and increased energy consumption? Consider the long-term implications of these impacts.
6. Analyze the benefits of compact city design. How can it promote sustainability and improve residents' quality of life? Consider factors such as walkability, efficient use of resources, and community cohesion.
7. Discuss the role of public participation in urban planning. How can engaging community members in the planning process lead to better outcomes? Consider the benefits of incorporating local knowledge and addressing community needs.
8. Evaluate the challenges of implementing environmental management strategies in rapidly growing cities. How can these challenges be overcome? Consider factors such as infrastructure development, resource allocation, and policy enforcement.
9. Discuss the concept of "green infrastructure" in urban planning. How can it contribute to environmental sustainability in cities? Consider the benefits of features such as green roofs, rain gardens, and permeable pavements.
10. Analyze the impact of urban design on social equity. How can urban planning promote inclusivity and equal access to resources? Consider the role of affordable housing, accessible transportation, and public spaces.
11. Discuss the role of transportation planning in creating sustainable cities. How can public transit systems reduce environmental impact? Consider the benefits of reducing car dependency and promoting active transportation.
12. Evaluate the importance of open spaces in urban planning. How do parks and green spaces contribute to the overall health of a city? Consider the benefits for mental health, biodiversity, and climate regulation.
13. Discuss the concept of "smart growth" in urban planning. How does it address the challenges of urban sprawl? Consider strategies such as mixed-use development, transit-oriented development, and infill development.
14. Analyze the role of zoning regulations in shaping urban development. How can zoning promote or hinder sustainability? Consider the impact of regulations on land use, building design, and community character.
15. Discuss the impact of climate change on urban planning. How must cities adapt to rising sea levels and increasing temperatures? Consider the need for resilient infrastructure, disaster preparedness, and climate mitigation strategies.
16. Evaluate the role of urban agriculture in promoting food security and sustainability in cities. How can urban farms contribute to local food systems, community engagement, and environmental education?
17. Discuss the concept of "resilient cities." How can urban planning help cities withstand natural disasters and other crises? Consider the role of adaptive design, emergency planning, and community resilience.
18. Analyze the impact of urbanization on biodiversity. How can urban planning help to protect and enhance biodiversity in cities? Consider strategies such as habitat restoration, green corridors, and biodiversity-friendly design.

19. Discuss the role of energy efficiency in urban planning. How can cities reduce their energy consumption and transition to renewable energy sources? Consider the role of building design, transportation planning, and urban form.
20. Evaluate the challenges and opportunities of waste management in cities. How can urban planning contribute to waste reduction and recycling? Consider strategies such as circular economy, waste-to-energy, and community composting.

MULTIPLE-CHOICE QUESTIONS

1. What is the primary goal of urban planning?
   a. To increase the city's population
   b. To ensure the orderly development of urban areas
   c. To increase the city's revenue
   d. To reduce the city's population
   e. None of the above
2. What is a key characteristic of a compact city?
   a. Low population density
   b. High reliance on private vehicles
   c. High population density
   d. Urban sprawl
   e. Large geographical area
3. What is the problem associated with air pollution in cities?
   a. Lack of entertainment options
   b. Water pollution
   c. Traffic congestion
   d. High cost of living
   e. Lack of job opportunities
4. What is a common strategy used in environmental planning to promote sustainability?
   a. Increasing the use of private vehicles
   b. Promoting urban sprawl
   c. Reducing green spaces
   d. Encouraging the use of renewable energy sources
   e. Increasing industrial activities
5. What is the major disadvantage of urban sprawl?
   a. Increased community cohesion
   b. Reduced commuting times
   c. Increased air pollution
   d. Increased access to amenities
   e. Reduced traffic congestion
6. What is the key benefit of a compact city design?
   a. Increased reliance on private vehicles
   b. Reduced access to amenities
   c. Increased air pollution
   d. Reduced commuting times
   e. Increased traffic congestion
7. What is a common tool used in urban planning to control land use?
   a. Zoning regulations
   b. Traffic lights
   c. Building codes
   d. Tax incentives
   e. All of the above

8. What is the key goal of environmental management?
   a. To increase industrial activities
   b. To reduce the use of renewable energy sources
   c. To protect and conserve natural resources
   d. To promote urban sprawl
   e. None of the above
9. What is a common strategy used in environmental planning to reduce air pollution?
   a. Promoting the use of private vehicles
   b. Reducing green spaces
   c. Encouraging the use of public transportation
   d. Increasing industrial activities
   e. Promoting urban sprawl
10. What is a major challenge associated with urban planning in rapidly growing cities?
    a. Reduced population density
    b. Lack of job opportunities
    c. Managing infrastructure development
    d. Reduced traffic congestion
    e. Reduced air pollution
11. What is the key characteristic of a sprawl city?
    a. High population density
    b. High reliance on public transportation
    c. Low population density
    d. Compact development
    e. Reduced commuting times
12. What is a common strategy used in urban planning to promote social equity?
    a. Reducing access to amenities
    b. Promoting urban sprawl
    c. Increasing the cost of living
    d. Ensuring affordable housing
    e. Reducing green spaces
13. What is the key benefit of environmental planning?
    a. Increased air pollution
    b. Reduced reliance on renewable energy sources
    c. Protection and conservation of natural resources
    d. Increased traffic congestion
    e. Increased cost of living
14. What is a common strategy used in environmental management to reduce water pollution?
    a. Increasing industrial activities
    b. Reducing green spaces
    c. Promoting urban sprawl
    d. Implementing wastewater treatment systems
    e. Increasing the use of private vehicles
15. What is the major disadvantage of a compact city design?
    a. Increased air pollution
    b. Reduced access to amenities
    c. Increased traffic congestion
    d. Reduced community cohesion
    e. Increased commuting times

16. What is the key goal of urban planning?
    a. To reduce the city's population
    b. To increase the city's revenue
    c. To ensure the orderly development of urban areas
    d. To increase the city's population
    e. None of the above
17. What is a common tool used in environmental planning to promote sustainability?
    a. Reducing green spaces
    b. Encouraging the use of renewable energy sources
    c. Increasing the use of private vehicles
    d. Promoting urban sprawl
    e. Increasing industrial activities
18. What is a major challenge associated with environmental management in urban areas?
    a. High cost of living
    b. Traffic congestion
    c. Air and water pollution
    d. Lack of entertainment options
    e. Lack of job opportunities
19. What is the key benefit of a compact city design?
    a. Increased traffic congestion
    b. Reduced commuting times
    c. Increased reliance on private vehicles
    d. Reduced access to amenities
    e. Increased air pollution
20. What is the major disadvantage of urban sprawl?
    a. Reduced commuting times
    b. Increased community cohesion
    c. Increased air pollution
    d. Reduced traffic congestion
    e. Increased access to amenities
21. What is a common strategy used in environmental planning to reduce air pollution?
    a. Reducing green spaces
    b. Encouraging the use of public transportation
    c. Promoting the use of private vehicles
    d. Increasing industrial activities
    e. Promoting urban sprawl
22. What is a major challenge associated with urban planning in rapidly growing cities?
    a. Reduced air pollution
    b. Managing infrastructure development
    c. Reduced population density
    d. Lack of job opportunities
    e. Reduced traffic congestion
23. What is the key characteristic of a sprawl city?
    a. Compact development
    b. High reliance on public transportation
    c. Low population density
    d. High population density
    e. Reduced commuting times

24. What is a common strategy used in urban planning to promote social equity?
    a. Reducing green spaces
    b. Promoting urban sprawl
    c. Increasing the cost of living
    d. Reducing access to amenities
    e. Ensuring affordable housing
25. What is the key benefit of environmental planning?
    a. Increased traffic congestion
    b. Protection and conservation of natural resources
    c. Increased air pollution
    d. Reduced reliance on renewable energy sources
    e. Increased cost of living
26. What is a common strategy used in environmental management to reduce water pollution?
    a. Reducing green spaces
    b. Promoting urban sprawl
    c. Increasing the use of private vehicles
    d. Implementing wastewater treatment systems
    e. Increasing industrial activities
27. What is the major disadvantage of a compact city design?
    a. Reduced community cohesion
    b. Increased commuting times
    c. Increased traffic congestion
    d. Increased air pollution
    e. Reduced access to amenities
28. What is a key goal of urban planning?
    a. To increase the city's population
    b. To ensure the orderly development of urban areas
    c. To reduce the city's population
    d. To increase the city's revenue
    e. None of the above
29. What is a common tool used in environmental planning to promote sustainability?
    a. Increasing industrial activities
    b. Encouraging the use of renewable energy sources
    c. Reducing green spaces
    d. Increasing the use of private vehicles
    e. Promoting urban sprawl
30. What is a major challenge associated with environmental management in urban areas?
    a. Lack of job opportunities
    b. Traffic congestion
    c. Air and water pollution
    d. High cost of living
    e. Lack of entertainment options

ANSWERS TO MULTIPLE-CHOICE QUESTIONS

1	b	11	c	21	b
2	c	12	d	22	b
3	c	13	c	23	c
4	d	14	d	24	e
5	c	15	c	25	b
6	d	16	c	26	d
7	a	17	b	27	c
8	c	18	c	28	b
9	c	19	b	29	b
10	c	20	c	30	c

FURTHER READING

Books on Urban Planning:

Garvin, A. (1995). *The American city: What works and what doesn't*. McGraw-Hill.

Hack, G. (Ed.). (2009). *Local planning: Contemporary principles and practice*. ICMA Press.

Jacobs, J. (1961). *The death and life of great American cities*. Random House.

Lynch, K. (1960). *The image of the city*. MIT Press.

Mumford, L. (1972). *The city in history: Its origins, its transformations, and its prospects*. New York: Harcourt, Brace & World.

Olmsted, F. L. (1997). *Civilizing American cities: Writings on city landscapes*. Da Capo Press.

Books on Environmental Planning:

Daniels, T. (2014). *The environmental planning handbook: For sustainable communities and regions*. Routledge.

Books on Environmental Management:

Ghiuzan, L. (2023). *Manufacturing excellence and quality management in sustainable fashion apparel: Strategies for lean production and environmental responsibility*. Springer.

Lehmann, J., & Joseph, S. (2024). *Biochar for environmental management: Science, technology and implementation*. Earthscan.

Scott, T. (2023). *Cambridge International AS level environmental management coursebook with digital access*. Cambridge University Press.

Journal Articles on Urban Planning:

Bontje, M. (2024). Multiscalar governance of shrinkage in the Netherlands: Past, Present… Future? *Urban Planning*, 9,1–18. https://doi.org/10.17645/up.7972

Loewen, B. (2024). From social barriers to transformative planning in energy transitions: Learning from Norwegian planners' perspectives. *Urban Planning*, 9,1–18. https://doi.org/10.17645/up.8158

Olt, G., Csizmady, A., Bagyura, M., & Kőszeghy, L. (2024). Captured by political power: More-Than-Neoliberal urban development and planning in post-socialist hungary. *Urban Planning*, 9, 1–19. https://doi.org/10.17645/up.7708

Journal Articles on Environmental Planning:

Jafari, H. R., Karimi, S., & Alavipoor, F. S. (2019). Environmental Planning and Management. *Environmental Science and Pollution Research*, 26, 24313–24314.

Journal Articles on Environmental Management:

Li, C., Yang, X., & Wang, L. (2024). The impact of renewable energy development on regional carbon emission reduction: Based on the Spatio-Temporal analysis of 30 Provinces in China. **Environmental Management*, 74, 439–460.

Underwood, A., Marcantonio, R., & Crippa, P. (2024). Interstate air pollution governance in the United States: Exploring clean air act section 126. *Environmental Management*, 74, 401–413.

3 Urban Design

3.1 URBAN DESIGN AND URBAN INFRASTRUCTURE

Urban design is the process of shaping the physical features and spaces of cities, towns, and regions. It involves the design of buildings, infrastructure, streets, parks, and public spaces, as well as the consideration of social, economic, and environmental aspects of urban development. Urban design aims to create urban environments that are equitable, beautiful, performative, and sustainable.

Urban infrastructure is the system of facilities and services that support the functioning and well-being of urban populations. It includes transportation, communication, water supply, sanitation, energy, waste management, healthcare, education, security, and recreation. Urban infrastructure is essential for providing access to opportunities, enhancing the quality of life, and reducing poverty and inequality.

Urban design and urban infrastructure are closely correlated, as they both influence and depend on each other. Urban design affects the demand and supply of urban infrastructure, as well as its efficiency and effectiveness. For example, the density, layout, and land use of urban areas determine the need for transportation infrastructure and the mode of travel. Compact and mixed-use urban design can reduce the reliance on private vehicles and encourage walking, cycling, and public transit. This can lower the cost of transportation infrastructure, reduce congestion and pollution, and improve health and safety. Similarly, urban design can affect the demand and supply of water, energy, and waste infrastructure by promoting water conservation, renewable energy sources, and waste reduction.

Urban infrastructure also affects urban design by providing the physical framework and enabling conditions for urban development. For example, transportation infrastructure connects various parts of the city and facilitates the movement of people and goods. It also shapes the spatial structure and growth patterns of urban areas by influencing the location choices of residents and businesses. Transportation infrastructure can also enhance or degrade the aesthetic and social qualities of urban spaces by creating barriers or opportunities for interaction. Similarly, water, energy, and waste infrastructure can affect urban design by providing essential services for human activities and environmental protection.

Therefore, urban design and urban infrastructure are interrelated and interdependent aspects of urban development. They both have significant impacts on the economic, social, and environmental outcomes of cities. A holistic and integrated approach to urban design and urban infrastructure planning is needed to create livable, resilient, and inclusive cities for the present and future generations.

DOI: 10.1201/9781032664613-3

3.2 PRINCIPLES OF URBAN DESIGN

Urban design is the process of shaping the physical environment of cities and towns to create livable, sustainable, and attractive places for people. Urban design involves the integration of various disciplines, such as architecture, landscape architecture, urban planning, transportation, engineering, ecology, and sociology. Urban design principles are the guidelines that inform the decision-making and evaluation of urban design projects. Some of the common urban design principles are shown in Table 3.1.

Among the above principles, what is the most important one? This is a tricky question to answer, as different principles may have diverse levels of importance depending on the context and objectives of each urban design project. However, one conceivable way to approach this question is to consider which principle is most essential for achieving the overall goal of urban design: creating livable, sustainable, and attractive places for people. Based on this criterion, one could argue that human scale is the most important principle of urban design. The human scale ensures that the urban environment is designed with people in mind, not only in terms of their physical needs and comfort but also in terms of their psychological and emotional well-being. Human scale creates a sense of belonging and attachment to a place, as well as a sense of dignity and respect for oneself and others. The human scale also enables people to interact with each other and with their surroundings

TABLE 3.1
Urban Design Principles

Principle	Explanation
Context	Urban design should respect and respond to the existing natural, cultural, and historical context of a place, as well as its future aspirations and challenges. For example, urban design should preserve and enhance the heritage values of a historic district or adapt to the changing needs and preferences of a growing population.
Diversity	Urban design should promote diversity and inclusivity in terms of land uses, activities, people, cultures, and built forms, to create vibrant and dynamic urban environments. For example, urban design should encourage mixed-use development that offers a variety of functions and services or supports social interaction and cultural expression among diverse groups of people.
Connectivity	Urban design should enhance the connectivity and accessibility of different modes of transportation, such as walking, cycling, public transit, and vehicles, as well as the connectivity and legibility of different spaces, such as streets, squares, parks, and buildings. For example, urban design should create a network of safe and attractive pedestrian and bicycle paths or provide clear and consistent signage and wayfinding systems.
Human scale	Urban design should create spaces and structures that are comfortable and appropriate for human perception, movement, and interaction, as well as respect the natural proportions and rhythms of the human body. For example, urban design should ensure adequate lighting and ventilation in indoor spaces or use materials and colors that are harmonious and pleasing to the eye.
Sustainability	Urban design should minimize the negative environmental impacts of urban development, such as greenhouse gas emissions, energy consumption, water consumption, and waste generation, and maximize the positive environmental benefits, such as biodiversity, green infrastructure, and renewable energy sources. For example, urban design should incorporate green roofs and walls that reduce the heat island effect and stormwater runoff or utilize solar panels and wind turbines that generate clean energy.
Quality	Urban design should ensure lofty standards of quality in terms of aesthetics, functionality, durability, and resilience of the urban environment, as well as foster a sense of pride and identity among the users and stakeholders of a place. For example, urban design should employ innovative and elegant design solutions that solve problems and create opportunities or create distinctive landmarks and symbols that represent the character and identity of a place.

in meaningful ways. Without human scale, urban design may fail to achieve its ultimate purpose: enhancing the quality of life for people who live in cities.

Some examples of urban design are:

1. Central Park in New York City, United States, is one of the most famous urban parks in the world. It covers 843 acres of land in the heart of Manhattan and provides a green oasis for millions of visitors and residents. It was designed by Frederick Law Olmsted and Calvert Vaux in the 19th century and features a variety of landscapes, such as meadows, lakes, forests, and gardens.
2. The High Line in New York City, United States, is a linear park that was built on an abandoned elevated railway line. It stretches for 1.45 miles along the west side of Manhattan and offers a unique perspective of the city and the Hudson River. It was designed by James Corner Field Operations, Diller Scofidio + Renfro, and Piet Oudolf and opened to the public in 2009. It features a variety of plants, artworks, and activities, such as walking, biking, and stargazing.
3. The Eiffel Tower in Paris, France, is one of the most iconic landmarks in the world. It stands at 324 m tall and was built for the 1889 World's Fair. It was designed by Gustave Eiffel and his engineers and was initially criticized for its appearance and function. However, it soon became a symbol of Paris and France and attracts millions of visitors every year. It features a restaurant, a museum, and an observation deck, as well as a light show at night.
4. The Forbidden City in Beijing, China, is one of the largest and most well-preserved palace complexes in the world. It covers 72 hectares of land and was the imperial residence of 24 emperors from the Ming and Qing dynasties. It was built between 1406 and 1420 and was designed by architects and craftsmen from various regions of China. It features a symmetrical layout, a hierarchical arrangement, and a rich collection of buildings, gardens, artworks, and artifacts.
5. The Burj Khalifa in Dubai, United Arab Emirates, is the tallest building in the world. It stands at 828 m tall and was completed in 2010. It was designed by Skidmore, Owings and Merrill (SOM) and was inspired by Islamic architecture and geometry. It features a hotel, offices, residences, restaurants, observation decks, and a fountain show.

Urban design is the art of shaping the physical and social characteristics of cities and towns. This involves developing and implementing plans, policies, and initiatives that improve the quality of life, sustainability, and identity of urban areas. The typical steps in the urban design process include the following:

1. *Analysis*: This step involves collecting and analyzing data on the existing conditions, needs, and challenges of the urban area. It also includes identifying the stakeholders, goals, and vision for the project.
2. *Synthesis*: This step involves generating and evaluating alternative concepts and solutions that address the issues and opportunities identified in the analysis. It also includes testing and refining the ideas through feedback, simulation, and prototyping.
3. *Implementation*: This step involves developing and executing the final plan, policy, or project that reflects the best solution from the synthesis. It also includes monitoring and evaluating the outcomes and impacts of the intervention.
4. *Communication*: This step involves communicating and engaging with the public and other relevant parties throughout the process. It also includes disseminating and sharing the results and lessons learned from the project.

The analysis step in urban design is a critical part of the planning and design process. It involves collecting and analyzing data on the existing conditions, needs, and challenges of the urban area. This step is also known as site analysis.

Site analysis is a pre-design step of architectural and urban design that entails the examination of a site's climatic, geographical, historical, legal, and infrastructural contexts. No development takes place in isolation – it will be influenced by the conditions of the site and will have an impact on its context. Good site and context analysis form the foundation of superior design. During this step, it is important to identify the stakeholders, goals, and vision for the project. Stakeholders can include residents, business owners, government officials, and other interested parties. The goals and vision for the project should be developed in collaboration with these stakeholders to ensure that the project meets their needs and expectations.

Some key topics to be considered in the analysis process include site location details (road names, address, major landmarks, etc.), neighborhood context (existing and proposed building uses in the neighborhood), site-specific information (site boundary and dimensions), natural features (topography of the site), man-made features (previous use of the site), and circulation (how visitors/pedestrians/traffic to or near the site flow around or within it).

The synthesis step in urban design is the process of generating and evaluating alternative concepts and solutions that address the issues and opportunities identified in the analysis. This step involves testing and refining the ideas through feedback, simulation, and prototyping. During this step, designers use a variety of techniques to generate and evaluate alternative concepts and solutions. These can include brainstorming, sketching, modeling, and simulation. Designers may also use feedback from stakeholders to refine their ideas and ensure that they meet the needs and expectations of the community.

One method for generating fast conceptual urban design prototypes is to synthesize spatial configurations for street networks, parcels, and building volumes. This involves implementing custom data structures for these configurations and controlling and parameterizing the generation process. By using these components, designers can show use-case applications of the synthesis algorithms.

The implementation step in urban design involves developing and executing the final plan, policy, or project that reflects the best solution from the synthesis. This step also includes monitoring and evaluating the outcomes and impacts of the intervention. During this step, the final plan, policy, or project is developed and executed. This can involve a range of activities, including obtaining necessary approvals and permits, securing funding, and coordinating with contractors and other stakeholders. The implementation process can vary depending on the specific project and its context.

Once the intervention is implemented, it is important to monitor and evaluate its outcomes and impacts. This can involve collecting data on key indicators, such as changes in land use patterns, transportation patterns, or environmental conditions. The data can then be analyzed to assess the effectiveness of the intervention and to identify any areas for improvement.

Stakeholders play an integral part in the implementation of urban design projects. They are people, groups, or individuals who have the power to either affect or be affected by the design project. Stakeholder collaboration is increasingly recognized as an important strategy to address complex global problems. Effective collaboration requires at least a few members of the group to commit to investing their time and resources to strive for collaborative tasks. Stakeholders can be involved in the implementation process in a variety of ways. For example, they can provide feedback on the final plan, policy, or project; help secure necessary approvals and permits; and assist with coordinating with contractors and other stakeholders. Stakeholders can also play a role in monitoring and evaluating the outcomes and impacts of the intervention.

Urban design plays a crucial role in shaping the physical and social aspects of cities and towns. It involves creating and implementing plans, policies, and projects that enhance the quality of life, sustainability, and identity of urban spaces. Urban design determines the physical scale, space, and ambiance of a place and establishes the built and natural forms within which individual buildings and infrastructure are sited. As such, it affects the balance between natural ecosystems and built environments and their sustainability outcomes.

Cities can use a range of principles and approaches to integrate the environment into urban planning and management. It starts with recognizing that action taken in towns and cities is essential for addressing global environmental problems. The report describes the social and economic benefits that can be generated by addressing environmental challenges and provides convincing evidence to support engagement with these issues.

Many urban design projects have started to introduce nature-based solutions (NBS) to provide sustainable management of ecosystems to tackle different environmental challenges. These solutions can reconnect the population with nature, mitigate air pollution, improve thermal comfort in the cities, reduce the effect of urban heat islands, and manage stormwater runoff, among many other benefits to the environment and to the physical and mental health of urban dwellers. Moreover, NBS can also contribute towards creative designs that can cost much less than traditional solutions.

NBS can help cities address urgent and fundamental environmental challenges by bringing ecosystem services back into cities and rebalancing cities' relationships with their surrounding areas. Some examples of NBS in cities include the following.

1. *Green Roofs:* green roofs are roofs that are covered with vegetation, which can help reduce the urban heat island effect, improve air quality, and manage stormwater runoff.
2. *Urban Gardens and Green Spaces:* Urban gardens and green spaces can improve the food security of an area, reduce environmental hazards, and improve the quality of life and well-being of residents.
3. *Trees:* Trees can help clean the air, reduce noise pollution, and cool the air.
4. *Wetlands and Rain Gardens:* Wetlands and rain gardens can help reduce flood risk.
5. *Mangroves and Marshes:* Mangroves and marshes can help buffer storms.
6. *Greenspaces:* Greenspaces can facilitate exercise, spiritual nourishment, and community interaction.
7. *Greenways:* Greenways can support active travel.

While generic steps provide a platform for the urban design process to happen, the following more specific steps attempt to guide the process of urban design itself. These specific steps are (Figure 3.1):

1. Analyzing the context and needs of the site and its users, such as the history, culture, geography, climate, demographics, economy, and transportation. The objective of this analysis is to understand the history, culture, geography, climate, demographics, economy, and transportation of the site and its users. To do the analysis, urban designers can use a variety of methods. These can include conducting surveys or interviews with users, reviewing existing data or research on the site and its users, and observing how people use the site. Designers can also analyze the physical characteristics of the site, such as its topography, climate, and infrastructure.
2. Developing a vision and goals for the urban design project, such as improving livability, accessibility, safety, aesthetics, and functionality. The objective of this step is to establish a clear direction for the project and to identify the desired outcomes, such as improving livability, accessibility, safety, aesthetics, and functionality. It is important to involve a wide range of stakeholders, including residents, business owners, government officials, and other interested parties. These stakeholders can provide valuable input on the needs and desires of the community, which can help shape the vision and goals of the project. The vision and goals should be developed through a collaborative process that involves brainstorming, discussion, and consensus-building. It is important to ensure that the vision and goals are realistic, achievable, and aligned with the needs and desires of the community.

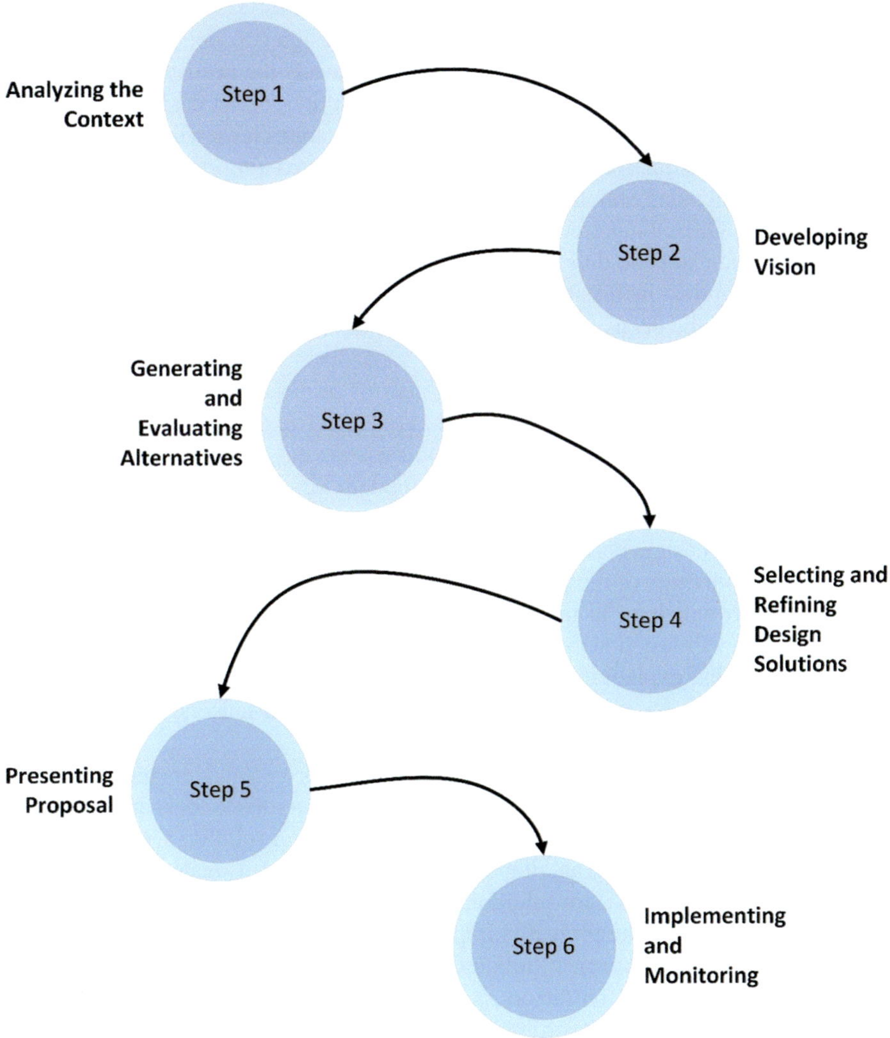

FIGURE 3.1 Specific steps in urban design.

3. Generating and evaluating alternative design concepts and scenarios, such as different land uses, densities, layouts, forms, and styles. The objective of this step is to explore different options for the design of the urban area, such as different land uses, densities, layouts, forms, and styles. To generate and evaluate alternative design concepts and scenarios, urban designers can use a variety of methods. These can include sketching, modeling, and simulation. Designers can also use feedback from stakeholders to refine their ideas and ensure that they meet the needs and expectations of the community. During this step, urban designers can explore different options for the design of the urban area. For example, they can do experiments and modeling with different land uses, such as residential, commercial, or industrial. They can also explore different densities, layouts, forms, and styles. The goal is to identify the best solution that meets the needs and desires of the community.
4. Selecting and refining the preferred design solution, based on the criteria of feasibility, suitability, and desirability. The objective of selecting and refining the preferred design solution

is to choose the best design option based on criteria such as feasibility, suitability, and desirability. This step involves evaluating the alternative design concepts and scenarios generated in the previous step and selecting the one that best meets the needs and desires of the community. To select and refine the preferred design solution, urban designers can use a variety of methods. These can include conducting surveys or interviews with stakeholders to gather feedback on the distinctive design options, reviewing existing data or research on the site and its users, and using simulation tools to test the feasibility and suitability of the different design options. Once the preferred design solution has been selected, it can be refined based on feedback from stakeholders and additional analysis. This can involve making changes to the design to improve its feasibility, suitability, or desirability. The goal is to develop a final design that meets the needs and desires of the community while also being feasible, suitable, and desirable.

5. Preparing and presenting the urban design plan or proposal, which may include drawings, models, simulations, reports, and guidelines. The objective of this step is to communicate the final design solution to stakeholders, decision-makers, and the public. This step involves creating a range of materials, such as drawings, models, simulations, reports, and guidelines, to effectively convey the design vision and its benefits. To prepare and present the urban design plan or proposal, designers can use a variety of methods. These can include creating detailed drawings or renderings of the proposed design, building physical or digital models to show how the design will look in three dimensions, and using simulation tools to demonstrate how the design will function in different scenarios. Urban designers can also prepare reports and guidelines that explain the design vision, its benefits, and how it will be implemented. These materials can be presented to stakeholders, decision-makers, and the public through meetings, presentations, or exhibitions.
6. Implementing and monitoring the urban design plan or project may involve coordination, collaboration, consultation, and communication with various stakeholders and agencies. The primary objective of this step is to ensure that the final design solution is executed according to the plan and that its outcomes and impacts are monitored and evaluated. This step may involve coordination, collaboration, consultation, and communication with various stakeholders and agencies to ensure that the project is implemented successfully. To implement and monitor the urban design plan or project, urban designers can use a variety of methods. These can include coordinating with contractors, government agencies, and other stakeholders to ensure that the project is executed according to the plan. Urban designers can also collaborate with these stakeholders to address any issues or challenges that arise during the implementation process. Consultation and communication with stakeholders are also important during this step. Designers can use a variety of methods, such as meetings, presentations, or surveys, to gather feedback from stakeholders on the progress of the project and its outcomes. This feedback can be used to make any necessary adjustments to the project to ensure that it meets the needs and desires of the community.

Urban designers can ensure that their designs meet the needs of users by involving users in the design process. This can be achieved through a variety of methods, such as conducting surveys or interviews with users to gather feedback on their needs and desires, observing how users interact with the site or product, and involving users in the testing and evaluation of design prototypes. By involving users in the design process, designers can gain valuable insights into the needs and desires of their target audience. This information can help designers make informed decisions about how to design the site or product to meet the needs of its users. Additionally, involving users in the design process can help build a sense of ownership and investment in the final design, which can increase user satisfaction and engagement.

3.3 WATER-SENSITIVE URBAN DESIGN

Climate change is a global challenge that affects the lives and well-being of millions of people, especially in urban areas. Urban design can have a significant impact on how cities cope with and mitigate the effects of climate change, such as rising temperatures, extreme weather events, flooding, droughts, air pollution, and health risks. Therefore, climate change consideration is important in the urban design process for several reasons. A case study is shown in Box 3.1.

BOX 3.1 URBAN DESIGN TO ENHANCE RESILIENCE TO CLIMATE CHANGE IN NEGRIL, JAMAICA

Negril, a coastal area in Jamaica, is highly vulnerable to climate change-related hazards, primarily sea-level rise. To combat this, a sixfold urban design framework was proposed to assess and enhance the resilience of urban form to climate change. The framework considers the patterns of streets, buildings, and land uses. The study revealed that Negril's current adaptation strategies heavily rely on bounce-back resilience measures that predominantly consider the impacts of extreme climatic events rather than slow-onset ones. These strategies support current tourism-driven development patterns while overlooking Negril's inherent abilities for generative transformation and incremental changes to meet climatic uncertainty. The study suggests that generative urban form transformation would better equip Negril to cope with future uncertainty – climatic or other. This transformation involves making changes to the urban form, such as the arrangement of streets and buildings, to enhance its resilience to climate change.

While the study does not provide specific quantitative data, it offers valuable insights into how urban design can be used to combat climate change. It underscores the importance of considering both extreme and slow-onset climatic events in urban design and planning.

Negril's experience demonstrates how cities can use urban design to enhance their resilience to climate change. It highlights the need for a comprehensive approach that considers both the physical aspects of urban form and the socio-ecological dynamics of urban development.

Sources: journal.plos.org, archdaily.com

First, climate change consideration can help urban designers create more resilient and adaptable cities that can withstand and recover from the shocks and stresses caused by climate change. Urban designers can use green infrastructure, such as trees, vegetation, wetlands, and green roofs, to reduce the urban heat island effect, improve air quality, enhance stormwater management, and provide habitats for wildlife. Urban designers can also use low-carbon and renewable energy sources, such as solar panels, wind turbines, and geothermal systems, to reduce greenhouse gas emissions and increase energy efficiency. Urban designers can also use smart technologies, such as sensors, data analytics, and artificial intelligence (AI), to monitor and optimize urban systems and services.

Second, climate change consideration can help urban designers create more equitable and inclusive cities that can address the needs and preferences of diverse and vulnerable groups of people. In this case, urban designers can use participatory and co-design methods to involve local communities and stakeholders in the urban design process and ensure that their voices are heard and respected. Urban designers can also use universal design principles to ensure that urban spaces and facilities are accessible and usable for people of different ages, abilities, genders, cultures, and backgrounds. Urban designers can also use social justice frameworks to address the root causes of inequality and discrimination in urban areas and promote human rights and dignity for all.

Third, climate change considerations can help urban designers create more livable and attractive cities that can enhance the quality of life and happiness of urban dwellers. For this purpose, urban designers can use biophilic design principles to incorporate nature and natural elements into urban environments and foster a sense of connection and belonging among people and nature. Urban designers can also use placemaking strategies to create vibrant and meaningful public spaces that support social interaction, cultural expression, civic engagement, and community identity. Urban designers can also use human-scale design approaches to create walkable and bikeable neighborhoods that encourage active transportation, physical activity, and health.

Water-sensitive urban design (WSUD) is an approach to urban planning and design that integrates the management of the water cycle into the built environment. The goal of WSUD is to minimize the impact of urbanization on the natural water cycle while also providing multiple benefits to the community, such as improved water quality, reduced flood risk, and enhanced urban amenities.

WSUD involves a range of strategies and techniques, including the use of permeable pavements, rain gardens, green roofs, and constructed wetlands to capture, treat, and reuse stormwater runoff. These features help to reduce the volume of stormwater runoff and improve its quality before it is discharged into local waterways.

3.3.1 The Use of Permeable Pavements

Permeable pavement is a type of pavement that allows water to pass through its surface and into the ground below. This can help to reduce the volume of stormwater runoff and improve its quality by filtering out pollutants. Permeable pavement is often used in urban areas as part of a WSUD strategy to manage the water cycle in the built environment, as shown in Figure 3.2.

Permeable pavement can be made from a variety of materials, including porous asphalt, pervious concrete, and permeable interlocking concrete pavers. These materials allow water to pass through their surface while also providing a durable and stable surface for vehicles and pedestrians. The use of permeable pavement can provide multiple benefits to the community, including improved water quality, reduced flood risk, and enhanced urban amenity. By allowing water to infiltrate into the ground, permeable pavement can help to recharge groundwater supplies and support the growth of vegetation. It can also help to reduce the urban heat island effect by cooling the air through evaporation. In addition, permeable pavement can provide aesthetic benefits by creating attractive and functional urban spaces. Factors such as soil type, climate, and topography can affect the performance of permeable pavement systems. It is important to conduct a thorough site analysis to determine the suitability of permeable pavement for a particular location.

In addition to site-specific factors, the design of permeable pavement systems must also consider the needs and desires of the community. This can involve engaging with stakeholders, such as residents, business owners, and government agencies, to gather input on the design and implementation of permeable pavement projects. Once a permeable pavement system has been designed and implemented, it is important to monitor its performance and maintain it properly. This can involve regular inspections to ensure that the system is functioning as intended, as well as routine maintenance activities such as cleaning and repairing damaged sections.

The use of permeable pavement can provide multiple benefits to the community, including improved water quality, reduced flood risk, and enhanced urban amenity. By allowing water to infiltrate into the ground, permeable pavement can help to recharge groundwater supplies and support the growth of vegetation. It can also help to reduce the urban heat island effect by cooling the air through evaporation.

3.3.2 Rain Garden

A rain garden is a type of green infrastructure that is often used in WSUD to manage stormwater runoff (Figure 3.3). Rain gardens are shallow, landscaped depressions that are designed to capture,

FIGURE 3.2 Permeable pavement.

Source: Wikipedia.

filter, and infiltrate stormwater runoff from impervious surfaces such as roofs, driveways, and parking lots. Rain gardens are typically planted with a mix of native vegetation that can tolerate both wet and dry conditions. The plants in a rain garden help to filter pollutants from the stormwater runoff, improving water quality. The soil and gravel layers in a rain garden also help to filter and infiltrate the stormwater, reducing the volume of runoff and recharging groundwater supplies. Rain gardens can provide multiple benefits to the community, including improved water quality, reduced flood risk, and enhanced urban amenities. They can also provide habitat for wildlife and create attractive and functional urban spaces. To implement rain gardens effectively, cities can develop a WSUD strategy that includes the use of rain gardens as a key component. Cities can also engage with stakeholders, such as residents, business owners, and government agencies, to gather input on the design and implementation of rain garden projects. Once rain gardens have been implemented, it is important to monitor their performance and maintain them properly to ensure their ongoing effectiveness.

The success of a rain garden community project can be determined by several factors, including its design, location, and maintenance. Some key factors that can contribute to the success of a rain garden are shown in Table 3.2.

Citizens can provide valuable input on the design and location of rain gardens, as well as feedback on their performance and maintenance. This input can help cities make informed decisions about how to implement rain gardens effectively as part of their WSUD strategies. In addition to providing input, citizens can also play an active role in the maintenance of rain gardens. For example, community groups or individual volunteers can help with activities such as weeding, pruning, and

FIGURE 3.3 Rain gardens.

Source: Wikipedia.

TABLE 3.2
Key Factors in Successful Rain Gardens

Key Factor	Explanation
Design	The design of a rain garden should consider the local climate, soil type, and topography to ensure that it can effectively capture, filter, and infiltrate stormwater runoff. The selection of appropriate plants that can tolerate both wet and dry conditions is also important.
Location	The location of a rain garden should be carefully chosen to maximize its effectiveness. Rain gardens should be placed in areas where they can capture stormwater runoff from impervious surfaces such as roofs, driveways, and parking lots.
Maintenance	Regular maintenance is essential to ensure the ongoing effectiveness of a rain garden. This can include activities such as weeding, pruning, and removing sediment and debris.
Involvement of communities at large	The involvement of citizens is a crucial factor in the success of a rain garden. By engaging with citizens and involving them in the planning, design, and implementation of rain garden projects, cities can ensure that these projects meet the needs and desires of the community.

removing sediment and debris. This involvement can help build a sense of ownership and investment in the rain garden, which can increase its success and sustainability.

3.3.3 Green Roofs

Green roofs are a type of vegetated WSUD system that can cover the already available dense impermeable roof area and provide environmental, economic, and social benefits. Numerous cities have adopted WSUD to mitigate the negative consequences of urbanization and climate change. Green roofs have a few benefits at economic, ecological, and societal levels. Some of the benefits of green roofs are as follows:

1. *Provides a Rainwater Buffer:* A green roof absorbs rainwater by buffering it in the plants, substrate, and drainage layer. This delays the discharge of rainwater to the sewage system, purifies the rainwater, and evaporates water through the plants. This helps to stabilize the groundwater level, reduces the peak load on the sewage system, and reduces the risk of flooding.
2. *Purifies the Air:* The plants in a green roof filter particulate matter from the air and convert CO_2 into oxygen. Green roofs contribute to air purification.
3. *Reduces the Ambient Temperature:* Plants absorb fifty percent of sunlight and reflect thirty percent of it. This helps to create a cooler and more pleasant climate. For the indoor climate, this means that the air conditioning does not have to work so hard, which in turn means energy savings. And this too has an extra positive effect on the climate near the building and the temperature in the city. Overall, it means a three-centigrade temperature reduction in the city.
4. *Increases Solar Panel Efficiency:* A green roof reduces the temperature on the roof. Therefore, it reduces energy costs.
5. *Reduces Ambient Noise Outside and Inside:* A green roof acts as a sound barrier to the buildings. It absorbs sound and thus provides a quieter environment, both inside and outside of the building.
6. *Extends Life Span of Roof:* A green roof protects the roofing material from external influences such as sun, rain, wind, and temperature fluctuations and doubles or triples the life span of the roof to up to 60 years or even longer.
7. *Adds Value to Building:* The natural and sustainable appearance, combined with a reduction in energy costs and an extension of the life span of the roof, means an increase in the value of the property.
8. *Increases Biodiversity:* The green roof that contains sedums, herbs, grasses, or host plants promotes the habitat of birds, butterflies, and insects, especially in city environments which are concrete and asphalt.
9. *Creates Fire-Resistant Layer:* Plants naturally contain a lot of moisture. The green roof creates a natural fire-resistant layer on the house or office building.
10. *Increases Feeling of Well-Being:* Living and working in a green environment has a positive effect on people's well-being. Greenery offers relaxation and reduces stress.

Green roofs (Figure 3.4) have many benefits, as shown above, but there are also some disadvantages to consider before installing one. Some of the disadvantages of green roofs are:

1. *Structural Limitations*: The roof and load-bearing walls of some buildings may not be strong enough to support the added weight of a green roof. Extensive green roofs, which are thin and lightweight, often weigh less than standard gravel and tar roofs. However, intensive green roofs, which are thicker and heavier, may require additional structural support.
2. *Cost*: Installing a green roof can cost up to twice as much as installing a conventional roof. An extensive green roof costs between $100/m^2 and $250/m^2, while intensive green

FIGURE 3.4 Green Roof L'Historial de la Vendée.

Source: Wikipedia.

roofs typically cost twice as much to install due to their thicker, more complex assembly. Maintenance for intensive green roofs can also be labor-intensive and expensive.

3. *Damage from Leaks*: Although most green roofs include a root barrier layer, the roots of plants can sometimes penetrate the waterproof membrane, causing roof leaks that could result in structural damage.
4. *Limited Choice of Plants*: Green roofs require careful plant selection, as not all plants are suitable for growing on a rooftop. The choice of plants may be limited by factors such as the depth of the growing medium, the amount of sunlight and water available, and the climate.

There are several ways to transform an existing roof (a non-green roof) to be more environmentally friendly. Here are some works to be done to achieve this transformation:

1. *Reduce Energy Consumption:* Reducing energy consumption is one of the most effective ways to make the roof more environmentally friendly.
2. *Use Reflective or "Cool" Roofing Materials:* Using reflective roofing materials is a terrific way to reduce the heat of the roof.
3. *Use Sustainable Materials:* To make the roof more environmentally friendly, avoid materials that have a high environmental impact. Instead, opt for sustainable materials like metal, clay, or recycled-content roofing products.
4. *Implement Green Roof Techniques:* Planting a green roof can improve insulation and air quality.
5. *Maintain the Roof:* Inspect the roof regularly for damage and make repairs as needed to ensure it continues to function efficiently.

TABLE 3.3
Types of Green Roofs

Type	Explanation
Intensive green roofs	These are the most expensive and thickest type of green roof, providing the best insulation, drainage management, and temperature regulation. They can replicate a natural landscape and can even be used for agricultural purposes.
Extensive green roofs	These are the cheapest and most lightweight green roofs on the market. They are ideal for flat roofs and are primarily built to promote biodiversity.
Semi-intensive green roofs	These combine elements of extensive and intensive living roofs. They are typically used to harness the environmental benefits associated with an extensive roof, but with a more diverse, structurally sound blend of plants and substrate, you would get with an intensive roof.
Biodiverse green roofs	These are extensive green roofs designed specifically to create habitats for plants and animals. They can aim to recreate the habitat that was lost when the building was erected or even enhance it.
Blue-green roofs	These are extensive green roofs designed specifically to create habitats for plants and animals. They can aim to recreate the habitat that was lost when the building was erected or even enhance it.
Biosolar green roofs	These combine solar panel and green roof technologies to provide renewable energy and deliver biodiversity.

There are several types of green roofs, each with their unique characteristics and benefits. Some of the most common types of green roofs are shown in Table 3.3.

3.3.4 Constructed Wetlands to Capture, Treat, and Reuse Stormwater Runoff

Constructed wetlands are engineered systems that mimic natural wetlands to capture, treat, and reuse stormwater runoff. They consist of shallow basins or channels that are planted with aquatic vegetation and filled with water. The plants, soil, and microorganisms in the wetlands help to remove pollutants such as nutrients, metals, and pathogens from the runoff. Constructed wetlands can also provide habitat for wildlife, aesthetic benefits, and flood control. One of the methods to design constructed wetlands is to use the hydrologic rational formula, which estimates the peak runoff rate from a catchment area based on the rainfall intensity, duration, and runoff coefficient. The formula can be expressed as:

$$Q = \frac{CIA}{360} \tag{3.1}$$

where Q is the peak runoff rate into the wetland (m^3/s), C is the runoff coefficient (dimensionless), I is the average rainfall intensity in the catchment area of the wetland (mm/h), and A is the catchment area (ha). The runoff coefficient reflects the proportion of rainfall that becomes runoff, and it depends on factors such as land use, soil type, slope, and vegetation cover. The hydrologic rational formula can help to determine the size and capacity of the constructed wetlands needed to treat the runoff from a given area. The benefits of constructed wetlands are: (1) Reducing pollutants in the runoff, such as nutrients, metals, and pathogens, which can improve the water quality of downstream water bodies and protect human health. (2) Providing habitat for wildlife, such as birds, fish, amphibians, and insects, which can enhance biodiversity and ecosystem services. (3) Creating aesthetic and recreational opportunities, such as parks, trails, and educational sites, can increase the livability and attractiveness of urban areas. (4) Mitigating flood risks, by storing excess water and releasing it slowly, which can reduce the peak flows and erosion of downstream channels. An example of case study of flood mitigation is shown in Box 3.2.

BOX 3.2 FLOOD MITIGATION IN TULSA, OKLAHOMA

Tulsa, Oklahoma, once identified as the most frequently flooded city in the United States, has transformed into a model city for flood risk reduction. The city, built on the banks of the Arkansas River and located within "Tornado Alley," regularly experiences severe storms. In the 1960s and 1970s, the city flooded every few years, and in 1984, a major flood killed 14 residents, injured nearly 300, and left thousands of damaged or destroyed buildings.

Approximately 40 years ago, the city mayor and activist citizens began a new approach to flood risk reduction. This long-standing, strong support from the government and citizens for flood mitigation has allowed for a continuing, voluntary home buyout program. To date, the city has bought out over a thousand homes and paid the homeowners to move to safer locations.

The city has and maintains a citywide master drainage plan and a hazard mitigation plan, both of which are used to track and measure mitigation projects. Tulsa's hazard mitigation plan also serves as the city's flood management plan and includes comprehensive watershed management, land use controls, voluntary buyouts, and green space that doubles as stormwater detention.

A team dedicated to promoting integrated planning, higher-standard regulations, outreach, and education is the key to Tulsa's floodplain management program and hazard mitigation planning success. Citizen involvement helps encourage sound political leadership and wise long-term actions. For instance, since the 1980s the city has mailed an annual notice to people living in the floodplain to help inform them about flood insurance.

In 2022, the city of Tulsa, Oklahoma received a Class 1 Community Rating System (CRS) rating from FEMA, which is the highest possible level. This recognition is a testament to the city's successful flood mitigation efforts.

Sources: fema.gov, isomitigation.com

Constructed wetlands (Figure 3.5) are artificial systems that consist of three main components: an impermeable layer, a substrate layer, and a vegetation layer. The wastewater flows through the substrate and the root zone, where it undergoes physical, chemical, and biological transformations by the action of plants, microorganisms, and the atmosphere. Constructed wetlands can remove suspended solids, organic matter, nutrients, pathogens, metals, and other pollutants from various sources of wastewater, such as municipal, industrial, agricultural, or stormwater runoff. Constructed wetlands can also provide habitat for wildlife and aesthetic benefits.

There are two main types of constructed wetlands: free water surface (FWS) wetlands and subsurface flow (SSF) wetlands. FWS wetlands have a visible water surface and resemble natural wetlands. They are suitable for treating low-strength wastewater with high solids content, such as stormwater or agricultural runoff. FWS wetlands can also create diverse habitats for aquatic and terrestrial organisms. However, they may require more land area, produce more odors and mosquitoes, and be more susceptible to climatic variations than SSF wetlands. SSF wetlands have water flowing below the surface of the substrate, which can be gravel, sand, or soil. They are suitable for treating high-strength wastewater with low solids content, such as municipal or industrial effluent. SSF wetlands can also reduce odors and mosquitoes, require less land area, and be more resilient to climatic variations than FWS wetlands. However, they may have lower oxygen transfer rates, lower biodiversity, and higher clogging potential than FWS wetlands.

The design, construction, operation, and maintenance of constructed wetlands depend on several factors, such as the type and quality of wastewater, the treatment objectives and standards, the site characteristics and availability, the climate and hydrology, the plant selection and harvesting, and

FIGURE 3.5 Constructed wetland at Flintenbreite near Lübeck, Germany.

Source: **Wikipedia.**

the monitoring and evaluation methods. Constructed wetlands should follow the guiding principles developed by the Interagency Workgroup on Constructed Wetlands to ensure water quality improvement and wildlife habitat provision.

Constructed wetlands are artificial ecosystems that mimic the functions of natural wetlands. They are designed to treat wastewater, stormwater, agricultural runoff, and other sources of water pollution. Constructed wetlands have many advantages and disadvantages for the people and environment.

Some of the advantages of constructed wetlands are as follows. (1) They provide a low-cost and low-maintenance alternative to conventional wastewater treatment systems, especially in rural areas or developing countries. (2) They enhance the biodiversity and habitat of the area, attracting wildlife and creating recreational opportunities for people. (3) They reduce greenhouse gas emissions and nutrient loading to the water bodies, improving the water quality and mitigating climate change impacts. (4) They create aesthetic and educational values, increasing the awareness and appreciation of wetland ecosystems.

Some of the disadvantages of constructed wetlands are as follows. (1) They require a large land area and may compete with other land uses, such as agriculture or urban development. (2) They may pose health risks to humans and animals due to exposure to pathogens, parasites, or toxic substances in the wastewater. (3) They may have negative impacts on the hydrology and ecology of the surrounding areas, altering the water flow and affecting the native wetland species. (4) They may have limited performance and reliability under extreme weather conditions, such as droughts or floods.

3.3.5 Rainwater Harvesting

Rainwater harvesting is a sustainable practice that involves collecting, storing, and reusing rainwater for various purposes, rather than letting it run off and be wasted. Rainwater harvesting can provide a source of clean and fresh water, especially in areas where water is scarce, polluted, or only seasonally available. Rainwater harvesting can also reduce the demand for groundwater extraction, prevent soil erosion and flooding, and lower water bills. The systems typically consist of four main components: a catchment area, such as a roof or a ground surface, where rainwater is collected; a conveyance system, such as pipes or gutters, that transports the rainwater from the catchment to a storage tank; a filter, such as a mesh screen or a sand filter, that removes debris and contaminants from the rainwater; and a storage tank, such as a barrel or a cistern, that holds the rainwater until it is needed.

Rainwater harvesting can be used for various domestic and agricultural purposes, such as drinking, cooking, washing, cleaning, bathing, gardening, watering livestock, or irrigation. Depending on the intended use, the rainwater may need to be treated further to ensure its safety and quality. For example, rainwater used for drinking may need to be disinfected with chlorine or ultraviolet light or filtered through activated carbon or reverse osmosis.

Rainwater harvesting is an ancient practice that has been revived in recent years due to the increasing challenges of water scarcity and climate change. Rainwater harvesting can provide many benefits for individuals, communities, and the environment. However, it also requires careful planning, design, installation, maintenance, and regulation to ensure its effectiveness and sustainability.

Rainwater harvesting is the practice of collecting and storing rain for later use (see Box 3.3). It is a sustainable and cost-effective way of managing water resources, especially in areas where water scarcity or quality is a concern. Rainwater harvesting has many benefits for households, communities, and the environment. Here are some of the main benefits of rainwater harvesting:

1. The rainwater collection is free and simple. Rainwater is a natural resource that falls freely from the sky. Anyone can collect rainwater using simple methods such as buckets, barrels, tanks, or cisterns. Depending on the size and design of the system, rainwater harvesting can provide enough water for various domestic uses such as gardening, washing, flushing toilets, or even drinking.
2. Rainwater harvesting can reduce water bills and save money. By using rainwater instead of mains water for some or all household needs, one can reduce the dependence on the municipal water supply and lower the water bills. According to a study by Texas A&M University, a typical household can save up to 65% of its water consumption by using rainwater for outdoor irrigation. Another study by the University of Arizona estimated that a household can save up to $200 per year by using rainwater for indoor non-potable uses.
3. Rainwater harvesting can save energy and reduce greenhouse gas emissions. The production and distribution of mains water require a lot of energy, which contributes to greenhouse gas emissions and climate change. According to the US Environmental Protection Agency (EPA), the public water supply sector accounts for about 4% of the total electricity use in the United States. By using rainwater instead of mains water, one can reduce the energy demand and the associated environmental impacts.
4. Rainwater harvesting can improve water quality and reduce pollution. Rainwater is clean and free of chemicals such as chlorine, fluoride, or chloramines that are added to mains water for disinfection purposes. These chemicals can have adverse effects on human health and the environment. For example, chlorine can react with organic matter in water to form disinfection by-products (DBPs) such as trihalomethanes (THMs) and haloacetic acids (HAAs), which are carcinogenic and mutagenic. By using rainwater, one can avoid exposure to these harmful substances.

5. Rainwater harvesting can also reduce the pollution of surface water and groundwater by reducing stormwater runoff. Stormwater runoff is the excess water that flows over land after a rainfall event. It can carry pollutants such as pesticides, fertilizers, metals, sediments, bacteria, and viruses from urban and agricultural areas into rivers, lakes, streams, and aquifers. This can degrade the water quality and harm aquatic ecosystems. By collecting rainwater on-site, one can reduce the amount and velocity of stormwater runoff and prevent erosion and contamination.
6. Rainwater harvesting can enhance water security and resilience. Rainwater harvesting can provide a reliable and alternative source of water in times of drought or water scarcity. It can also increase the self-sufficiency and independence of households and communities from external water sources that may be unreliable, expensive, or inaccessible. Rainwater harvesting can also improve preparedness and response to emergencies such as natural disasters or power outages that may disrupt the municipal water supply.
7. Rainwater harvesting can benefit plants and landscapes. Rainwater is ideal for irrigation purposes because it is soft and has a neutral pH. It does not contain salts or minerals that can accumulate in the soil and affect plant growth. It also does not cause scaling or corrosion of pipes and sprinklers. Rainwater can help maintain soil moisture and fertility, promote plant health and diversity, and enhance the aesthetic value of gardens and landscapes.
8. Rainwater harvesting can create social and educational opportunities. Rainwater harvesting can foster community engagement and cooperation among neighbors who share a common interest in water conservation and management. It can also provide educational opportunities for children and adults to learn about the water cycle, hydrology, ecology, engineering, economics, and other related topics.
9. Rainwater harvesting can reduce flooding and erosion. Harvesting rainwater can help mitigate the negative effects of heavy rainfall events that can cause flooding and damage to property and infrastructure. By diverting some of the rainwater from roofs and other impervious surfaces into storage tanks or cisterns, one can reduce the peak flow volume and pressure on storm drains and sewers that may overflow or burst during storms. Rainwater harvesting can also protect soil from erosion by reducing surface runoff that can wash away topsoil and nutrients.
10. Rainwater harvesting can be used as a source of drinking water. Rainwater can be treated and purified to meet the standards for potable water use. This can be done by using filters, disinfectants, ultraviolet light, or boiling. Rainwater can be a safe and healthy source of drinking water, especially in areas where groundwater is contaminated, saline, or depleted. Rainwater can also have a better taste and smell than mains water, as it does not contain chlorine or other additives.

BOX 3.3 RAINWATER HARVESTING IN KOTLA, INDIA

Kotla, a village located at the foothills of the Aravalis in Rajasthan, faced severe water scarcity due to over-exploitation of groundwater and minimal percolation of rainwater. To address this, the Sehgal Foundation implemented a watershed intervention that significantly improved the village's water situation. The intervention involved the construction of check dams to store and percolate rainwater into the ground. The unique ridge-to-valley approach was adopted, which reduced the flow velocity of rainwater, minimized erosion in agricultural fields, and increased the recharging potential. This approach transformed the entire length of the stream into a storage and percolation basin.

The harvested rainwater was more than ten times the storage capacity of a check dam. This significantly increased the percolation of rainwater, leading to a rise in the groundwater level. As a result, the village saw a remarkable improvement in the availability of fresh groundwater.

This successful implementation of rainwater harvesting in Kotla village demonstrates how such interventions can effectively address water scarcity issues in arid regions. It underscores the importance of community participation and the adoption of appropriate technologies in ensuring the sustainability of such initiatives.

Sources: indiawaterportal.org, nationalgeographic.com, cseindia.org

However, there are some situations where rainwater harvesting cannot be done or is not advisable. Some of these situations are as follows. (1) When the rainfall is too low or erratic to provide a reliable source of water. (2) When the quality of the rainwater is contaminated by pollutants, such as acid rain, dust, chemicals, or pathogens. (3) When the cost of installing and maintaining the rainwater harvesting system is too high compared to other sources of water. (4) When the legal or social constraints prevent the use or ownership of rainwater harvesting systems. (5) When the environmental impact of rainwater harvesting is negative, such as reducing the natural recharge of aquifers, altering the hydrological cycle, or affecting biodiversity. In these situations, rainwater harvesting may not be feasible or desirable socially and economically, and alternative sources of water should be considered.

An economic evaluation through cost–benefit analysis of rainwater harvesting (like all environmental projects) is a method of evaluating the economic and environmental impacts of collecting and using rainwater for various purposes. Rainwater harvesting can provide benefits such as reducing the demand for potable water, mitigating floods and droughts, enhancing groundwater recharge, improving water quality, and saving energy. However, rainwater harvesting also involves costs such as installing and maintaining the collection system, treating the harvested water, and complying with regulations and standards.

To conduct a cost–benefit analysis of rainwater harvesting, one needs to identify and quantify the costs and benefits associated with the project and compare them over a certain time horizon. The costs and benefits can be categorized into direct and indirect, tangible and intangible, market and non-market, and private and social. For example, the direct costs of rainwater harvesting include the capital and operational expenses of the collection system, while the indirect costs include the opportunity cost of land use and the environmental impacts of the system. The tangible benefits of rainwater harvesting include savings in water bills and energy consumption, while the intangible benefits include the aesthetic and recreational value of the water source. The market benefits of rainwater harvesting include the increased property value and the potential revenue from selling excess water, while the non-market benefits include the improved health and well-being of the users and the community. The private benefits of rainwater harvesting accrue to the individual or organization that implements the project, while the social benefits accrue to society.

To compare the costs and benefits of rainwater harvesting, one needs to discount them to their present value using an appropriate discount rate. The discount rate reflects the time preference and the opportunity cost of capital for the project. A higher discount rate means that future costs and benefits are valued less than present ones. The net present value (NPV) of rainwater harvesting is the difference between the present value of benefits and the present value of costs. A positive NPV means that the project is economically viable and beneficial. The benefit–cost ratio (BCR) of rainwater harvesting is the ratio of the present value of benefits to the present value of costs. A BCR greater than one means that the project is economically viable and beneficial. The internal rate of return (IRR) of rainwater harvesting is the discount rate that makes the NPV equal to zero. It represents the average annual return on investment for the project. A higher IRR means that the project is more profitable and attractive.

3.3.6 Flood Control with Structural and Non-Structural Measures

Climate change is a global phenomenon that impacts urban life. Rising global temperatures cause sea levels to rise, increase the number of extreme weather events such as floods, droughts, and storms, and increase the spread of tropical diseases. All these have costly impacts on cities' basic services, infrastructure, housing, human livelihoods, and health. For example, heatwaves can threaten tens of thousands in densely populated city centers from Karachi to London, where dark pavements capture and prolong the sun's heat. Changes in the climate are also threatening freshwater supplies and electric grids.

Climate change is expected to increase the frequency and intensity of extreme weather events, including heavy rainfall and flooding. As global temperatures continue to rise, the likelihood of more frequent and severe floods increases. This is because warmer air can hold more moisture, leading to heavier rainfall and increased runoff. In addition, rising sea levels caused by melting polar ice caps can exacerbate coastal flooding.

Urban design can play a crucial role in anticipating and mitigating the consequences of flooding. City leaders are realizing the need to adapt planning and construction to make cities more resistant to extreme floods and heat. Projects in Berlin and London show urban water features can be an effective way of creating heat sinks and reining in flooding. Around the world, cities are developing action plans that combine global thinking in resiliency with local geographic and urban conditions, all with a common goal of defending urban populations from floods. These plans often include measures such as constructing flood barriers, improving drainage systems, and implementing green infrastructure such as green roofs and rain gardens to absorb and retain stormwater. By incorporating these measures into urban design, cities can become more resilient to the impacts of climate change and reduce the risk of flood damage.

One of the consequences of climate change is the increased risk of flooding, especially in urban areas that are densely populated and have inadequate drainage systems. Flooding can have various effects on the urban environment and society, such as:

1. Damage to infrastructure, property, and human lives can result in economic losses and social disruption.
2. Contamination of water sources, soil, and air can pose health risks and environmental degradation.
3. Displacement of people and animals can cause social conflicts and humanitarian crises.
4. Erosion of land and coastlines can alter the natural landscape and biodiversity.

Urban design needs to anticipate and adapt to the potential impacts of flooding and implement strategies to reduce vulnerability and enhance resilience. Some of the possible strategies include the following:

1. Creating green spaces and permeable surfaces that can absorb rainwater and reduce runoff.
2. Building flood barriers and levees that can protect the city from storm surges and river floods.
3. Elevating buildings and infrastructure that are in flood-prone areas or using floating structures that can adapt to water levels.
4. Developing early warning systems and emergency plans that can inform and evacuate residents in case of a flood.
5. Promoting community participation and education can raise awareness and preparedness for flooding.

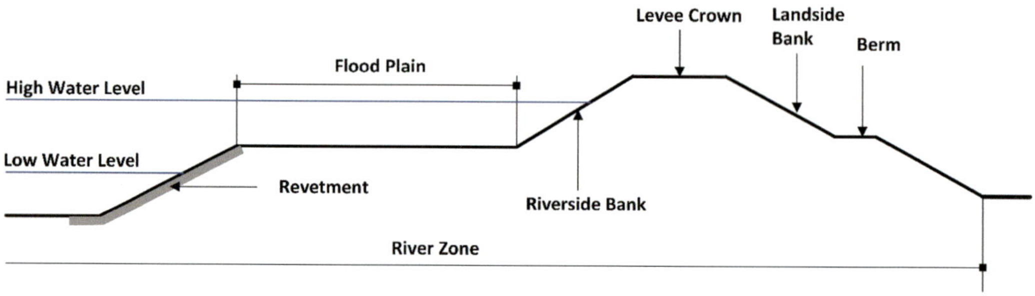

FIGURE 3.6 Levee and its parts.

Flooding problems are a serious threat to many communities around the world, especially in low-lying areas or near rivers and coasts. Flooding can cause damage to infrastructure, property, crops, and human lives. Therefore, it is important to find ways to mitigate flooding problems and reduce their impact.

Flood mitigation measures can be categorized into two types: structural and non-structural measures. Structural measures are physical interventions that aim to modify the flow of water, such as building dams, levees, and increasing channel capacity. On the other hand, non-structural measures aim to reduce the consequences of a flood by reducing exposure, and vulnerability, and increasing resiliency to flooding without modifying the flow of water. These measures include flood forecasting, developing flood early warning systems, establishing reservoir operation rules, land use policy, public education, and emergency services. An integrated approach to flood management requires the co-implementation of both structural and non-structural flood mitigation strategies. For example, watershed management can be considered a structural measure while reoperation of the system of reservoirs can be considered a non-structural measure.

Some examples of mitigation strategies are as follows:

1. Constructing flood barriers, such as levees (Figure 3.6), or seawalls, to prevent water from overflowing into flood-vulnerable areas.
2. Constructing dams to regulate the extent of the flood, and thus the flow rate released downstream can be controlled.
3. Constructing retention basins, and therefore the flood water can be temporarily stored and released gradually to the downstream later.
4. Improving the capacity of flood channels.
5. Enhancing drainage systems, such as storm sewers, culverts, or ditches, to facilitate the flow of excess water away from populated areas.
6. Implementing flood warning systems, such as sensors, sirens, or alerts, to inform people of impending floods and allow them to evacuate or prepare accordingly.
7. Implementing watershed management, such as land management practices, and restoring natural habitats, such as wetlands, forests, or grasslands, to increase the water retention capacity of the land and reduce runoff and erosion.
8. Adopting flood-resilient practices, such as elevating buildings, using waterproof materials, or relocating to safer areas, to minimize the damage caused by floods.

These are some examples of how to mitigate flooding problems, but there may be other options depending on the specific context and needs of each situation. It is advisable to consult with experts and stakeholders to assess the risks and benefits of each strategy and implement them effectively. By incorporating these strategies into urban design, cities can mitigate the negative effects of flooding and improve their sustainability and livability in the face of climate change.

Cities can implement WSUD effectively by adopting a range of strategies and techniques that integrate the management of the water cycle into the built environment. Some steps that cities can take to implement WSUD effectively include the following:

1. *Developing a WSUD Strategy*: Cities can develop a WSUD strategy that outlines their vision, goals, and objectives for managing the water cycle in the urban environment. This strategy can provide a framework for implementing WSUD projects and initiatives.
2. *Engaging with Stakeholders*: Cities can engage with a wide range of stakeholders, including residents, business owners, government agencies, and environmental groups, to gather input on the design and implementation of WSUD projects. This engagement can help ensure that WSUD projects meet the needs and desires of the community.
3. *Investing in WSUD Infrastructure*: Cities can invest in WSUD infrastructure, such as permeable pavements, rain gardens, green roofs, and constructed wetlands, to capture, treat, and reuse stormwater runoff. These features can help reduce the volume of stormwater runoff and improve its quality before it is discharged into local waterways.
4. *Monitoring and Evaluating WSUD Projects*: Cities can monitor and evaluate the performance of WSUD projects to assess their effectiveness in managing the water cycle. This information can be used to make any necessary adjustments to improve the performance of WSUD projects.

One example of a successful WSUD project is the Cheonggyecheon Restoration Project in Seoul, South Korea (Figure 3.7). This project involved the restoration of the Cheonggyecheon stream, which had been covered by an elevated highway for several decades. The highway was removed,

FIGURE 3.7 Cheonggyecheon restoration project.

***Source:* Wikipedia.**

and the stream was daylighted, creating a vibrant public space that has become a popular destination for residents and tourists alike.

The Cheonggyecheon Restoration Project incorporated a range of WSUD features, including rain gardens, permeable pavements, and constructed wetlands. These features help to capture, treat, and reuse stormwater runoff, improving water quality and reducing flood risk. The project has also had a positive impact on the local economy, with increased tourism and business activity in the area.

3.4 MICROCLIMATIC CONSIDERATIONS IN URBAN DESIGN

Urban design and microclimate are closely related concepts that affect the quality of life in cities. Urban design refers to the arrangement and organization of buildings, streets, public spaces, and infrastructure in urban areas. Microclimate refers to the local variations of temperature, humidity, wind, and precipitation within a city, influenced by factors such as topography, vegetation, water bodies, and human activities.

Urban microclimates are the local variations of climate conditions within urban areas, influenced by factors such as building density, layout, height, materials, vegetation, water bodies, and human activities. Urban microclimates can affect the energy performance of buildings, the comfort and health of urban dwellers, and the environmental quality of cities. These factors interact to create complex microclimates that can vary over short distances as exhibited in Table 3.4.

Some of the effects of urban microclimates on cities are:

1. Increased air temperature due to the urban heat island effect, which is the difference between the urban and rural temperatures caused by the absorption and emission of heat by urban surfaces and anthropogenic heat sources. This can lead to higher cooling demand, higher greenhouse gas emissions, lower air quality, and increased heat-related mortality and morbidity.
2. Reduced wind speed due to the aerodynamic drag of buildings and obstacles can alter the ventilation and natural cooling of buildings and urban spaces, as well as the dispersion and dilution of air pollutants.

TABLE 3.4
Factors That Affect Local Microclimate

Factor	Explanation
Building density	The density of buildings in an urban area can affect the local microclimate by altering wind patterns and reducing the amount of sunlight that reaches the ground. Higher building density can reduce wind speed and amplify air temperature.
Layout of building	The layout of buildings and streets can also affect the local microclimate. For example, narrow streets with tall buildings can create urban canyons that trap heat and pollutants.
Height of building	The height of buildings can also affect the local microclimate. Taller buildings can block sunlight and wind, creating cooler and more sheltered areas at ground level.
Materials used	The materials used in the construction of buildings can also affect the local microclimate. For example, dark materials such as asphalt absorb more heat than lighter materials such as concrete, which can contribute to the urban heat island effect.
Vegetation	Vegetation can help mitigate the effects of urbanization on the local microclimate by providing shade, reducing air temperatures, and improving air quality.
Water bodies	Water bodies such as lakes, rivers, and fountains provide an evaporative surface, and can help to moderate the local microclimate by providing a cooling effect through evaporation.
Human activities	Human activities such as driving cars and operating air conditioning units can also affect the local microclimate by releasing heat and pollutants into the air.

3. Modified precipitation patterns due to the changes in atmospheric moisture, pressure, and temperature induced by urbanization. This can result in more frequent and intense rainfall events, flooding, and stormwater runoff in some areas or reduced rainfall and drought in others.
4. Altered solar radiation is due to the shading effects of buildings and vegetation, which can affect daylight availability, thermal comfort, and passive solar heating of buildings and urban spaces.

Water bodies in cities, such as lakes and rivers, may influence the urban microclimate, which is the local atmospheric condition within a city. Water bodies can affect the temperature, humidity, wind, and precipitation patterns of the urban environment. Some of the possible effects are as follows:

1. Water bodies can moderate the temperature extremes in cities by absorbing heat during the day and releasing it at night. This can create a cooling effect in summer and a warming effect in winter.
2. Water bodies can increase the humidity in cities by evaporating water into the air. This can create a more comfortable climate in dry regions but also increases the risk of fog, mist, and smog formation.
3. Water bodies can alter the wind patterns in cities by creating pressure differences between land and water. This can enhance the ventilation and dispersion of pollutants in some cases but also generate strong and gusty winds in others.
4. Water bodies can influence the precipitation patterns in cities by providing moisture and nuclei for cloud formation. This can increase the rainfall and snowfall in some areas but also reduce it in others.

Water bodies in cities are therefore key factors to consider when planning and designing urban spaces. They can provide environmental benefits as well as challenges for urban dwellers. Research on a micro-scale in two university campuses in Thailand with two distinct characteristics pertaining to microclimate, that is, percentages of hardscape, softscape, water body area, and greenery of the total area of the respective campus, shows that average temperatures in the campus with greenery and more water bodies were 2 degrees lower.

Urban greening is one of the strategies that can mitigate some of the negative impacts of urban microclimates on cities by providing shading, evaporative cooling, air purification, noise reduction, stormwater management, biodiversity enhancement, and aesthetic improvement. However, urban greening also poses some challenges and trade-offs, such as water consumption, maintenance costs, allergens, pests, fire hazards, and unintended effects on microclimate and air quality depending on the type, location, and design of green infrastructure. Therefore, urban greening should be carefully planned and implemented in conjunction with other measures to optimize its benefits and minimize its drawbacks for urban microclimates and cities.

Urban design can have a significant impact on microclimates, both positively and negatively. For example, urban design can create or reduce urban heat islands, which are areas of higher temperature in cities compared to their surroundings, caused by the absorption and emission of heat by buildings and pavements. Urban design can also enhance or diminish natural ventilation, which is the movement of air through urban spaces, influenced by the shape and orientation of buildings and streets. Urban design can also promote or hinder urban greening, which is the provision of vegetation and water features in urban areas, which can moderate temperature, increase humidity, reduce air pollution, and provide aesthetic and psychological benefits. Refer to the comparison in Figures 3.8 and 3.9. The former, with less hardscape, shows lower air temperatures, while the latter, with more hardscape, shows higher air temperatures, despite both being situated in nearly the same location.

FIGURE 3.8 Greener and more evaporative surface campus of Asian Institute of Technology (AIT), Bangkok.

Source: **Wikipedia.**

Urban design should consider the effects of microclimate on the comfort, health, and well-being of urban residents and visitors. Urban design should aim to create a balanced and harmonious microclimate that suits the local context and culture, as well as the needs and preferences of different users. Urban design should also adopt strategies to mitigate the negative impacts of climate change on urban microclimates, such as increasing resilience to extreme heat, flooding, and drought. Urban design should also integrate with other disciplines and sectors, such as architecture, engineering, planning, landscape, ecology, energy, transportation, and health, to achieve a holistic and sustainable approach to urban development.

The interplay between urban design and urban microclimate is a complex and dynamic phenomenon that affects the quality of life, health, and sustainability of urban areas. Urban design can influence the urban microclimate by modifying the geometry, materials, vegetation, and activities of the built environment, which in turn affect the solar radiation, air temperature, humidity, wind speed and direction, and air pollution levels. Urban microclimate can also feedback on the urban design by creating thermal comfort or discomfort, encouraging, or discouraging outdoor activities, and providing or reducing ecosystem services. Urban design and urban microclimate mutually influence each other, and urban designers can determine where the city will go to serve the citizens. A holistic and integrated approach that balances the needs and preferences of different stakeholders, respects the local context and culture, and promotes resilience and adaptation to climate change shall not be forgotten by urban designers.

FIGURE 3.9 More hardscape campus of King Mongkut's Institute of Technology Ladkrabang (KMITL), Bangkok.

Source: **Wikipedia.**

3.5 URBAN HEAT ISLAND PHENOMENON

The average temperature in urban areas is commonly warmer than its counterparts in peripheral rural areas because of many factors. It is like a warmer island encircled by cooler surroundings. This phenomenon is known as urban heat island (UHI). With the presence of the UHI phenomenon, cities experience higher air temperatures than the surrounding countryside. This effect is caused by several factors, including the following:

1. *Urban Development:* As cities grow, new developments such as roads and buildings displace natural surfaces like trees, ponds, and soil. This change in the local environment results in a corresponding change in the local climate. That is because these natural surfaces help moderate air temperatures.
2. *Human Activities:* Human activities such as power generation and the use of cars and air conditioners can also contribute to the UHI effect by releasing heat into the air.
3. *The Geometry of Cities:* The geometry of cities also contributes to heat islands: the narrow spaces between tall structures, known as urban canyons, can block wind and trap heat.

These factors interact in complex ways to create unique microclimates in urban areas (Figure 3.10).

UHIs can have a significant impact on human health. The increased temperatures, reduced nighttime cooling, and higher air pollution levels associated with UHIs can contribute to heat-related

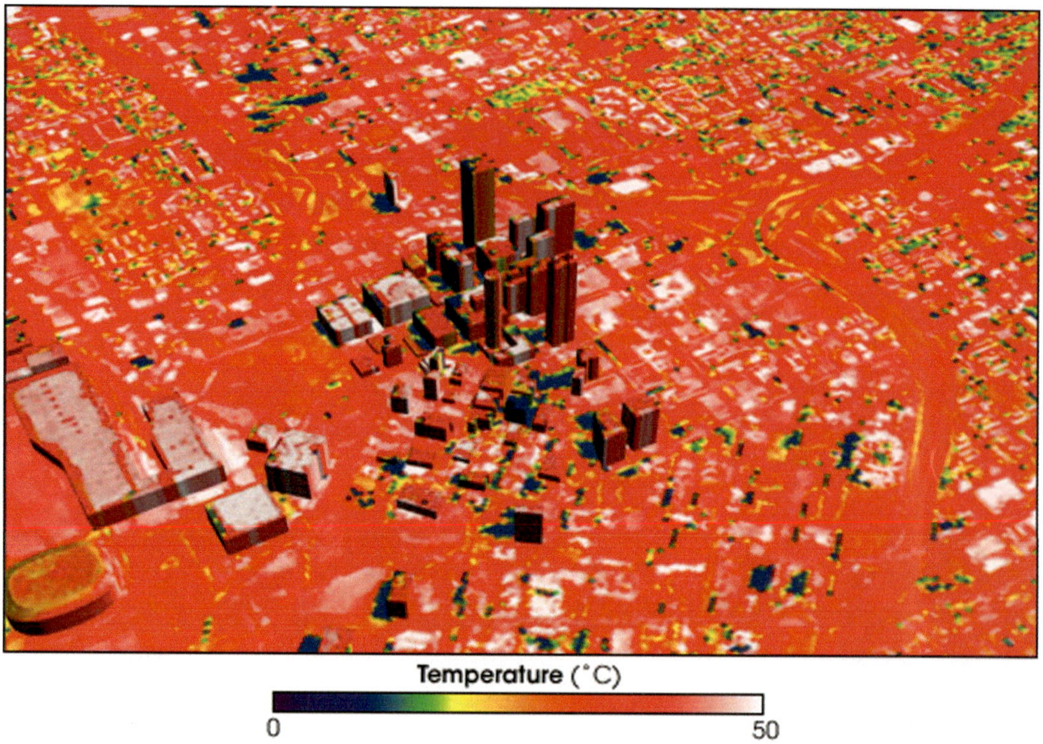

FIGURE 3.10 Urban heat island.

Source: **Wikipedia.**

illnesses such as general discomfort, respiratory difficulties, heat cramps, heat exhaustion, and non-fatal heat stroke. Heat islands can also exacerbate the impact of naturally occurring heat waves, which are periods of abnormally hot and often humid weather. Sensitive populations, such as older adults, are particularly at risk during these events.

In addition to the direct effects of heat on human health, UHIs can also have indirect effects by increasing air pollution levels. As demand for electricity increases due to the need for air conditioning, power plants that rely on fossil fuels release more air pollutants and greenhouse gases into the air. These pollutants can harm human health and contribute to complex air quality problems such as the formation of ground-level ozone (smog), fine particulate matter, and acid rain. UHI can have significant impacts on the energy consumption of cities, as it affects the demand for heating and cooling, the efficiency of power generation and transmission, and the emission of greenhouse gases.

Some of the factors that contribute to UHI are the modification of the land surface, the reduction of vegetation, the emission of waste heat, and the absorption and reflection of solar radiation by buildings and pavements. Because the UHI carries negative impacts on the urban environment and citizens, the effects of UHI should be minimized if cannot be mitigated. There are several ways to mitigate the effects of UHIs, including the following:

1. *Increasing Vegetation:* Planting trees and other vegetation can help to reduce the urban heat island effect by providing shade and cooling the air through evapotranspiration.
2. *Cool Roofs:* Cool roofs are designed to reflect more sunlight and absorb less heat than traditional roofs, which can help to reduce the urban heat island effect.

3. *Using Cool Pavements:* Cool pavements are designed to reflect more sunlight and absorb less heat than traditional pavements, which can help to reduce the urban heat island effect.
4. *Promoting Energy Efficiency:* Promoting energy-efficient buildings and transportation can help to reduce the amount of heat generated by human activities, which can help to reduce the urban heat island effect.

One of the challenges of urban development is to reduce the energy consumption of cities, which can have environmental, economic, and social benefits. Energy consumption in cities is influenced by several factors, such as the UHI phenomenon, the building characteristics, the transportation system, and the behavior of urban dwellers.

UHI is a phenomenon that describes the higher temperature of urban areas compared to their rural surroundings, which can increase the demand for cooling and decrease the efficiency of power generation and transmission. UHI can be mitigated by increasing urban greenery, enhancing natural ventilation, using cool or reflective materials, and improving urban planning and design.

Building characteristics, such as the size, shape, orientation, insulation, and materials of buildings, can affect the energy performance of buildings and the thermal comfort of occupants. Building energy consumption can be reduced by adopting energy-efficient design, construction, operation, and maintenance practices, as well as by using renewable energy sources and smart technologies.

Transportation systems, such as mode, distance, frequency, and time of travel, can affect the energy use and emission of vehicles. Transportation energy consumption can be reduced by promoting public transportation, cycling, walking, and carpooling, as well as by improving traffic management and vehicle efficiency.

The behavior of urban dwellers, such as the choice of appliances, lighting, heating, cooling, and hot water use, can affect the energy demand and consumption of households. Behavioral energy consumption can be reduced by raising awareness, providing incentives, implementing policies, and encouraging behavioral change.

Reducing energy consumption in cities is not only a necessity but also an opportunity to achieve multiple benefits for the environment, the economy, and society. Energy consumption in cities is influenced by numerous factors, such as the UHI phenomenon, building characteristics, the transportation system, and the behavior of urban dwellers. By reducing energy consumption in cities, we can (1) protect the environment by reducing greenhouse gas emissions, air pollution, noise pollution, and waste generation; (2) save money by lowering energy bills, maintenance costs, and operational expenses; (3) improve health and well-being by enhancing thermal comfort, indoor air quality, and access to green spaces; (4) increase productivity and competitiveness by improving energy efficiency, reliability, and security; and (5) promote social equity and inclusion by providing affordable and accessible energy services, creating jobs, and reducing poverty.

To achieve these benefits, we need to adopt a holistic and integrated approach that considers the interrelationships among the varied factors that affect energy consumption in cities. We also need to involve various stakeholders, such as governments, businesses, communities, and individuals, in the planning and implementation of energy-saving strategies. By working together, we can make our cities more sustainable and livable.

3.6 FACILITATING FUTURE DEVELOPMENT IN URBAN DESIGN

3.6.1 Viable Future Development

Facilitating future development in urban design is a topic that has gained more attention and relevance in recent years, especially after the COVID-19 pandemic. Some of the trends and challenges that are influencing the future of urban design are as follows:

1. *Green Planning of Public Spaces*: Cities are being planned and designed for people, with "green" streets, new corridors, and public spaces as centers of social life. Green planning aims to create more natural and healthy environments, reduce pollution, and carbon emissions and enhance social interaction and well-being.
2. *Smart Health Communities*: Cities develop healthcare ecosystems that are focused not only on diagnosing and treating sickness but also on supporting well-being through early intervention and prevention while leveraging digital technologies. Smart health communities use data and analytics, telemedicine, wearable devices, and other innovations to improve access, quality, and efficiency of healthcare services.
3. *15-Minute City*: Cities are being designed in a way that amenities and most services are within a 15-minute walking or cycling distance, creating a new neighborhood approach. The 15-minute city concept promotes local living, reduces car dependency and traffic congestion, and fosters social cohesion and economic vitality.
4. *Mobility*: intelligent, sustainable, and as-a-service: Cities work towards offering digital, clean, intelligent, autonomous, and intermodal mobility, with more walking and cycling spaces, where transport is commonly provided as a service. Mobility as a service (MaaS) integrates different modes of transport into a single platform, offering users convenience, flexibility, and affordability.
5. *Inclusive Services and Planning:* Cities evolve to have inclusive services and approaches, fighting inequalities by providing access to housing and infrastructure, equal rights, and participation, as well as jobs and opportunities. Inclusive cities aim to ensure that no one is left behind or discriminated against based on their gender, race, ethnicity, disability, age, or income.
6. *Digital Innovation Ecosystem:* Cities attract talent, enable creativity, and encourage disruptive thinking, developing themselves through an innovation model approach and a combination of physical and digital elements. Digital innovation ecosystems foster collaboration among different actors, such as universities, businesses, start-ups, governments, and citizens.
7. *Circular Economy and Producing Locally:* Cities adopt circular models based on a healthy circulation of resources, and on principles of sharing, reusing, and restoration, with an emphasis on limiting municipal waste volumes and on producing locally – for instance, by urban farming. Circular economy and local production can reduce environmental impacts, create new jobs, and enhance food security.

The above examples are some of the trends that are shaping the future of urban design. However, each city has its context, needs, and aspirations. Therefore, urban designers need to adopt a participatory and adaptive approach that involves multiple stakeholders and responds to the changing conditions and challenges of urban living. A more detailed discussion of the above concepts is explained below.

3.6.2 Green Planning of Public Spaces

The essence of green planning is to create more natural and healthy environments for vacant spaces, particularly public spaces, under the concept that greenery absorbs CO_2, thus reducing pollution and carbon emissions, enhances social interaction, improves well-being, and generates serendipitous nuances of the city. Cities are being planned and designed for people, with 'green' streets, new corridors, and public spaces as centers of social life. The World Economic Forum's Global Agenda Council on the Future of Cities has included increasing green canopy cover in its list of top ten urban planning initiatives. Green public spaces entail:

1. Promoting a larger number of trees in urban parks, vacant lands, abandoned lands, and wastelands.

2. Creating more and larger public parks and NBS in the urban environment, such as riverbanks, forest areas, and trees in streets, and fostering a closer connection to nature, even in cities with high population density.
3. Increasing walking and cycling facilities instead of car-centric designs and parking areas, with space for children and adults to enjoy outdoor activities.
4. Fostering a sense of security and safety while promoting a shift to mass transit and developing walking and cycling corridors.

Cities around the world are recognizing the benefits of a green approach to urban planning, as it has the potential to lower urban temperatures, mitigate air pollution, and build environmental resilience. The government created green spaces, recreational areas, linear parks along streets, and roof parks. It also encouraged planting on rooftops, and tax rebates were offered to private-sector construction projects that were in alignment with the city's development plan.

3.6.3 Fifteen-Minute City

The concept of a 15-minute city is an urban planning concept that aims to create compact and sustainable urban spaces where most daily necessities and services, such as work, shopping, education, healthcare, and leisure, can be easily reached by a 15-minute walk or bike ride from any point in the city. This approach aims to reduce car dependency, promote healthy and sustainable living, and improve well-being and quality of life for city dwellers. The concept gained significant traction after being advocated by Parisian mayor Anne Hidalgo during her 2020 re-election campaign. Since then, many cities worldwide have adopted the same goal and used the 15-minute model as a spatial analysis tool to evaluate accessibility levels within the urban fabric.

Implementing the 15-minute city concept requires a multidisciplinary approach involving transportation planning, urban design, and policymaking to create well-designed public spaces, pedestrian-friendly streets, and mixed-use development. The lifestyle change may include remote working, which reduces daily commuting and is supported by the recent widespread availability of information and communications technology (ICT).

The concept has its roots in pre-modern urban planning traditions where walkability and community living were the primary focus before the advent of street networks and automobiles. In recent times, it builds upon similar pedestrian-centered principles found in New Urbanism, transit-oriented development, and other proposals that promote walkability, mixed-use developments, and compact, livable communities.

3.6.4 Intelligent, Sustainable, and as-a-Service Mobility

Intelligent, sustainable, and as-a-service mobility is a concept that aims to provide digital, clean, intelligent, autonomous, and intermodal mobility, with more walking and cycling spaces, where transport is commonly provided as a service, for example, convenient, and affordable. This approach aims to reduce traffic congestion, air pollution, and carbon emissions while improving accessibility, convenience, and affordability of transportation services. This system is based on the use of digital technologies such as the Internet of Things (IoT), AI, and big data analytics to optimize transport systems' performance and efficiency. Intelligent mobility can help reduce traffic congestion by providing real-time traffic information, optimizing traffic signals, and predicting travel demand. It should use environmentally friendly modes of transport such as walking, cycling, electric vehicles (EVs), and public transport that emit fewer greenhouse gases and pollutants than conventional vehicles. Sustainable mobility can, therefore, help reduce air pollution, noise pollution, and carbon emissions while promoting healthy living and social interaction.

MaaS is a new business model that integrates different modes of transport such as buses, trains, taxis, car-sharing services, and bike-sharing services into a single platform that offers users convenience, flexibility, and affordability. MaaS can help reduce car dependency by providing alternative modes of transport that are more sustainable and cost-effective than private cars. MaaS can also help improve accessibility to transportation services for people who do not own cars or have limited mobility options.

As-a-service mobility models are becoming increasingly popular in urban areas worldwide. For example, in Helsinki, *Whim-How to get there* is a MaaS platform that allows users to access public transport services such as buses and trains as well as private car-sharing services such as *DriveNow* and *Car2Go* through a single app. *Grab, Bolt, Moovit*, and *In-drive* are ride-hailing platforms that offer various transportation services such as taxis, private cars, bikes, e-scooters, and buses through their apps.

3.6.5 Inclusive Services and Planning

Inclusive services and planning are based on the concept of inclusiveness that aims to fight inequalities by providing access to housing and infrastructure, equal rights, and participation, as well as jobs and opportunities. By this platform, a city ensures that no one is left behind or discriminated against based on their gender, race, ethnicity, disability, age, or income. Inclusive services and planning can be achieved through the following:

1. *Affordable Housing*: Cities can provide affordable housing options for low-income households through rent control policies, social housing programs, and public-private partnerships.
2. *Universal Design*: Cities can adopt universal design principles in public spaces, buildings, and transportation systems to ensure that they are accessible and usable by people of all ages and abilities.
3. *Community Engagement*: Cities can involve local communities in the planning and decision-making process to ensure that their needs and aspirations are considered.
4. *Equitable Access to Services*: Cities can provide equitable access to essential services such as healthcare, education, public transport, and public spaces for all residents.
5. *Diversity and Inclusion Policies*: Cities can adopt diversity and inclusion policies that promote equal opportunities for all residents regardless of their gender, race, ethnicity, disability, age, or income.

Inclusive services and planning can help create more resilient and sustainable cities that are better equipped to face the challenges of the future. By ensuring that everyone has access to basic services and opportunities, cities can foster social cohesion, reduce inequalities, and promote economic growth.

3.6.6 Creating Ecosystems in the City for Digital Innovation

The digital innovation ecosystem promotes the combination of physical and digital elements through creativity, collaboration, and disruptive thinking. Digital innovation ecosystems aim to attract talent, enable creativity, and encourage innovation by providing a supportive environment for start-ups, businesses, universities, governments, and citizens to work together. The common characteristics of digital innovation ecosystems are exhibited in Table 3.5.

Digital innovation ecosystems can take different forms depending on the local context and needs. For example:

1. *Innovation Districts*: Innovation districts are physical locations that bring together start-ups, businesses, universities, research institutions, and other actors in a compact urban area.

TABLE 3.5
Characteristics of Digital Innovation

Characteristics	Explanation
Opennes	Digital innovation ecosystems are open to innovative ideas, technologies, and business models. They encourage experimentation and risk-taking while providing a supportive environment for learning and growth.
Collaboration	Digital innovation ecosystems foster collaboration among different actors such as start-ups, businesses, universities, governments, and citizens. They provide opportunities for networking, knowledge sharing, and co-creation.
Access to resources	Digital innovation ecosystems provide access to resources such as funding, mentorship, incubation facilities, and research infrastructure. They help start-ups and businesses overcome the initial barriers to entry and scale up their operations.
Talent attraction	Digital innovation ecosystems attract talent from diverse backgrounds such as entrepreneurs, researchers, designers, developers, and investors. Some of them provide opportunities for personal and professional growth while fostering a sense of community.
Impact orientation	Digital innovation ecosystems focus on creating solutions that have a positive impact on society and the environment. They promote sustainable development goals such as reducing poverty, improving health outcomes, and mitigating climate change.

Innovation districts aim to create a vibrant ecosystem that fosters creativity, collaboration, and innovation.

2. *Living Labs*: Living labs are real-life environments where users can test new products or services in a controlled setting. Living labs provide opportunities for co-creation between users and producers while generating valuable feedback for product development.
3. *Hackathons*: Hackathons are events where developers, designers, entrepreneurs, and other participants come together to solve specific challenges using technology. Hackathons provide opportunities for learning new skills, networking with peers, and showcasing innovative solutions.

3.6.7 Circular Economy and Producing Locally

Circular models are gaining popularity among cities that want to reduce their environmental impact and enhance their resilience. These models are based on a healthy circulation of resources, where waste is minimized, materials are reused or recycled, and energy is generated from renewable sources. By adopting circular models, cities can improve their livability, competitiveness, and social inclusion, while also contributing to global sustainability goals. Here are some examples of circular models in urban development:

1. *Urban Farming:* Urban farming is a practice that involves growing crops and raising animals within the city limits. Urban farming can help reduce food waste, promote healthy eating habits, and provide opportunities for social interaction and community building. A case study of urban farming is shown in Box 3.4.
2. *Biogas Production:* Biogas production is a process that involves converting organic waste into biogas through anaerobic digestion. Biogas can be used as a renewable energy source for heating, cooking, and electricity generation. Biogas production can help reduce waste volumes, mitigate greenhouse gas emissions, and improve energy security.
3. *Circular Construction:* Circular construction is a practice that involves designing buildings for disassembly and reuse. Circular construction can help reduce waste volumes, conserve natural resources, and promote sustainable building practices.

BOX 3.4 URBAN FARMING IN DETROIT, MI, UNITED STATES

Detroit, once known as the automotive capital of the world, has faced significant economic and social challenges over the past few decades. However, the city has turned to urban farming as a way to revitalize vacant land, provide fresh produce to residents, and create job opportunities. The Detroit Agriculture Network, a coalition of community-based organizations, has been instrumental in promoting urban farming in the city. The network provides resources and support to local residents interested in starting their own urban farms.

One of the most successful urban farming initiatives in Detroit is the Detroit Black Community Food Security Network's D-Town Farm. This seven-acre organic farm produces a variety of fruits, vegetables, and herbs, which are sold at local farmers markets and used in school lunch programs.

Urban farming in Detroit has had significant social, economic, and environmental benefits. It has provided fresh, locally grown produce to residents in a city where access to healthy food can be a challenge. It has also created job opportunities and provided a source of income for local residents.

In terms of environmental benefits, urban farming has helped to improve air and soil quality, reduce the UHI effect, and promote biodiversity. It has also contributed to the city's efforts to reduce greenhouse gas emissions by reducing the need for long-distance transportation of food.

Detroit has over 1,400 urban farms and gardens, covering more than 180 acres of land. D-Town Farm produces over thirty different fruits, vegetables, and herbs. In terms of employment, urban farming initiatives in Detroit have created dozens of jobs and provided volunteer opportunities for hundreds of local residents.

Sources: worldbank.org, mdpi.com, tapipedia.org

Implementing circular models in urban development can be challenging due to assorted reasons. Here are some of the challenges:

1. *Lack of Awareness and Understanding:* Circular models are still a new concept, and many stakeholders may not be aware of their benefits or how to implement them. Therefore, there is a need for awareness-raising campaigns and capacity-building programs to educate stakeholders about circular models and their potential.
2. *Policy and Regulatory Barriers:* Existing policies and regulations may not be conducive to circular models, and there may be a need for new policies and regulations that support circular models. For example, waste management policies may need to be revised to promote waste reduction, reuse, and recycling.
3. *Financial Constraints:* Implementing circular models may require significant investments in infrastructure, technology, and human resources. Therefore, there is a need for innovative financing mechanisms that can mobilize public- and private-sector investments in circular models.
4. *Lack of Collaboration and Coordination:* Implementing circular models requires collaboration and coordination among different stakeholders such as government agencies, businesses, civil society organizations, and citizens. Therefore, there is a need for platforms that can facilitate stakeholder engagement, knowledge sharing, and co-creation.
5. *Technical Challenges:* Implementing circular models may require innovative technologies or modifications to existing technologies. For example, biogas production requires anaerobic

digestion technology that may not be available in all contexts. Therefore, there is a need for research and development programs that can develop innovative technologies or adapt existing ones to local contexts.

From the above discussions, the city is facing a dynamic and challenging situation as it must cope with the existing and emerging development opportunities. To ensure that the city can benefit from these opportunities, it needs to plan and provide the necessary resources such as land, infrastructure, workforce, policies, etc. The city should also adopt a proactive and collaborative approach to engage with various stakeholders and address the potential issues and risks that may arise from the development process.

PROBLEMS AND ISSUES FOR DISCUSSION

1. How can the principles of water-sensitive design be effectively integrated into urban planning to address the challenges of stormwater management, promote biodiversity, and support the broader goals of sustainable development?
2. How does placemaking contribute to the vibrancy and liveliness of urban spaces?
3. Placemaking, involving community engagement, design interventions, and activation of public areas, plays a pivotal role in shaping cities where people feel connected, inspired, and engaged. Discuss its impact on social interactions, cultural identity, and overall urban well-being.
4. What are the advantages and obstacles associated with seamlessly integrating residential, commercial, and recreational areas in an urban context?
5. Consider economic vitality, walkability, traffic management, and community well-being. Explore how balanced zoning fosters vibrant neighborhoods while addressing potential conflicts and trade-offs.
6. How does site analysis contribute to identifying both constraints and opportunities for urban development?
7. Explore the role of site-specific factors such as topography, soil quality, existing infrastructure, and environmental considerations.
8. Discuss how thorough analysis informs design decisions and ensures sustainable, contextually appropriate development.
9. Water-sensitive design aims to mitigate the adverse effects of stormwater runoff. By incorporating permeable surfaces, green infrastructure (such as rain gardens and bioswales), and retention basins, cities can reduce flooding, improve water quality, and recharge aquifers. Discuss how these strategies enhance urban resilience during extreme weather events and contribute to overall urban hydrological balance.
10. Water-sensitive design fosters biodiversity by creating habitats for native flora and fauna. Wetlands, ponds, and vegetated swales not only manage water but also support wildlife. Evaluate the role of urban green spaces in promoting biodiversity and enhancing urban ecosystems.
11. Discuss the impact of involving residents and stakeholders in the design process.
12. Evaluate how transportation choices shape urban form and accessibility.
13. Ponder the role of site analysis in identifying constraints and opportunities for urban development.
14. Think about how green roofs contribute to biodiversity, energy efficiency, and air quality.
15. Discuss the role of constructed wetlands in treating urban runoff and improving water quality.
16. What factors impact the economic feasibility of widespread green roof adoption in urban areas? (Consider installation costs, maintenance expenses, and long-term benefits.)

17. Consider factors like population density, quality of life, ecological sustainability, and equitable access to parks and recreational areas. How do urban planners balance the need for high-density development with preserving green spaces in cities?
18. How does the preservation of heritage buildings and cultural landmarks contribute to the identity and character of a city? In your discussion, you may consider historical significance, architectural value, and community pride.
19. Water-sensitive design aims to mitigate the adverse effects of stormwater runoff. By incorporating permeable surfaces, green infrastructure (such as rain gardens and bioswales), and retention basins, cities can reduce flooding, improve water quality, and recharge aquifers. Discuss how these strategies enhance urban resilience during extreme weather events and contribute to overall urban hydrological balance.
20. Water-sensitive design fosters biodiversity by creating habitats for native flora and fauna. Wetlands, ponds, and vegetated swales not only manage water but also support wildlife. Evaluate the role of urban green spaces in promoting biodiversity and enhancing urban ecosystems.
21. Think about the role of circular economy principles in minimizing waste. How does the circular economy minimize waste?
22. Discuss the benefits of designing neighborhoods where essential services are within a 15-minute walk or bike ride.
23. Analyze the economic shifts toward a circular urban economy (key phrases: resource efficiency, sustainable economic growth, reduced emissions).
24. How did Detroit address the issue of contaminated soil in urban farming?

MULTIPLE-CHOICE QUESTIONS

1. What does context-sensitive urban design emphasize?
   a. Aesthetic appeal
   b. Local context, culture, and history
   c. High-density development
   d. Economic feasibility
   e. None of the above
2. Which principle focuses on creating vibrant and engaging urban spaces?
   a. Placemaking
   b. Zoning regulations
   c. Traffic management
   d. Industrial development
   e. All of the above
3. What is the primary goal of mixed-use zoning?
   a. Separating residential and commercial areas
   b. Enhancing community identity
   c. Promoting green spaces
   d. Reducing traffic congestion
   e. Allowing flexibility in land use
4. Which step involves analyzing site-specific factors such as topography and soil quality?
   a. Community engagement
   b. Conceptual design
   c. Site analysis
   d. Infrastructure planning
   e. None of the above

5. What role does community engagement play in the urban design process?
   a. Identifying constraints
   b. Enhancing aesthetics
   c. Involving stakeholders
   d. Implementing green infrastructure
   e. All of the above
6. Transportation planning influences which aspect of urban form?
   a. Density
   b. Aesthetics
   c. Cultural heritage
   d. Economic growth
   e. None of the above
7. What benefits does WSUD offer in managing stormwater?
   a. Reduced energy consumption
   b. Enhanced biodiversity
   c. Improved water quality
   d. Increased traffic flow
   e. All of the above
8. Which green infrastructure feature helps manage stormwater runoff?
   a. Green roofs
   b. Constructed wetlands
   c. Permeable pavements
   d. Elevated highways
   e. None of the above
9. What challenges are associated with implementing WSUD in densely populated cities?
   a. High installation costs
   b. Limited space availability
   c. Lack of community support
   d. Excessive vegetation
   e. All of the above
10. Which ecological advantage is associated with green roofs?
    a. Increased air pollution
    b. Enhanced biodiversity
    c. Higher energy consumption
    d. Reduced noise levels
    e. None of the above
11. What role do constructed wetlands play in urban environments?
    a. Treating wastewater
    b. Providing recreational spaces
    c. Enhancing visual aesthetics
    d. Supporting wildlife
    e. None of the above
12. Which factor affects the economic viability of widespread green roof adoption?
    a. Maintenance costs
    b. Cultural heritage
    c. Traffic flow
    d. Industrial development
    e. All of the above

13. How does rainwater harvesting contribute to water scarcity mitigation?
    a. By increasing water demand
    b. By reducing stormwater runoff
    c. By promoting water wastage
    d. By depleting groundwater
    e. None of the above
14. What design considerations are crucial for effective rainwater harvesting systems?
    a. Storage capacity
    b. Aesthetics
    c. Cultural significance
    d. Traffic patterns
    e. All of the above
15. Which social equity aspect should be considered in rainwater harvesting implementation?
    a. Access for all income groups
    b. Aesthetic preferences
    c. Traffic management
    d. Industrial needs
    e. None of the above
16. How do urban designers balance aesthetics and functionality in public spaces?
    a. Prioritizing aesthetics
    b. Ignoring functionality
    c. Integrating both aesthetics and functionality
    d. Focusing solely on economic factors
    e. None of the above
17. Which of the following is a consideration when preserving historic buildings in urban areas?
    a. Maximizing floor area ratio
    b. Adapting for modern use
    c. Demolition for new development
    d. Ignoring cultural significance
    e. All of the above
18. What trade-offs might arise when balancing density and open space in urban planning?
    a. Reduced property values
    b. Increased traffic congestion
    c. Limited green spaces
    d. Enhanced social interactions
    e. None of the above
19. What causes the urban heat island effect?
    a. Excessive tree cover
    b. High population density
    c. Absorption of solar radiation by buildings
    d. Cool ocean breezes
    e. None of the above
20. Which strategy helps mitigate the urban heat island effect?
    a. Installing reflective pavements
    b. Increasing impervious surfaces
    c. Removing green spaces
    d. Encouraging industrial emissions
    e. All of the above

21. How do urban green spaces contribute to reducing UHI intensity?
    a. By trapping heat
    b. By providing shade and evaporative cooling
    c. By increasing building density
    d. By promoting concrete surfaces
    e. None of the above
22. What is the central idea behind the 15-minute city concept?
    a. Reducing commute times
    b. Expanding suburban areas
    c. Focusing on long-distance transportation
    d. Enhancing regional connectivity
    e. Prioritizing local amenities
23. Which of the following aligns with the 15-minute city model?
    a. Large-scale industrial zones
    b. Mixed-use neighborhoods
    c. Car-dependent suburbs
    d. Highways connecting distant cities
    e. None of the above
24. How does walkability impact residents' quality of life in a 15-minute city?
    a. Increases stress levels
    b. Enhances physical health
    c. Decreases social interactions
    d. Encourages sedentary behavior
    e. All of the above
25. What is the primary goal of a circular urban economy?
    a. Maximizing waste generation
    b. Linear resource consumption
    c. Promoting reuse and recycling
    d. Ignoring environmental impact
    e. None of the above
26. Which sector benefits from circular practices in construction and infrastructure?
    a. Real estate speculation
    b. Transportation
    c. Energy production
    d. Building materials
    e. All of the above
27. How can cities encourage circular economy principles among businesses and residents?
    a. Implementing strict regulations
    b. Providing incentives and education
    c. Ignoring sustainability goals
    d. Promoting linear production models
    e. None of the above

ANSWERS FOR MULTIPLE-CHOICE QUESTIONS

1	b	11	d	21	b
2	a	12	a	22	e
3	e	13	b	23	b
4	c	14	a	24	b
5	c	15	a	25	b
6	a	16	c	26	d
7	e	17	b	27	b
8	c	18	e		
9	b	19	c		
10	b	20	a		

FURTHER READING

Batty, M. (2013). *The new science of cities*. MIT Press.

Biddulph, M. (2007). *Introduction to residential layout*. Routledge.

Biddulph, M. (2012). Radical Streets? The Impact of Restrictive Street Standards on Urban Form. *International Journal of Urban and Regional Research*, 36(2), 326–341. https://doi.org/10.1111/j.1468-2427.2011.01100.x

Calthorpe, P. (1993). *The next American metropolis: Ecology, community, and the American dream*. Princeton Architectural Press.

Carmona, M. (2014). The place-shaping continuum: A theory of urban design process. *Journal of Urban Design*, 19(1), 2–36. https://doi.org/10.1080/13574809.2013.854695

Cuthbert, A. R. (2007). Urban Design: Requiem for an Era – Review and Critique of the last 50 Years. *Urban Design International*, 12(4), 177–223. https://doi.org/10.1057/palgrave.udi.9000200

Dovey, K. (2016). *Urban design thinking: A conceptual toolkit*. Bloomsbury Academic.

Dovey, K., & Pafka, E. (2014). The Urban Densification Equation: Density, Urban Form and Sustainable Transport. *Urban Policy and Research*, 32(1), 7–21. https://doi.org/10.1080/08111146.2013.833705

Dunnett, N., & Kingsbury, N. (2004). *Planting green roofs and living walls*. Timber Press.

Ellen MacArthur Foundation. (2013). Towards the Circular Economy: Economic and Business Rationale for an Accelerated Transition.

Ellin, N. (1999). *Postmodern urbanism*. Princeton Architectural Press.

Forsyth, A. (2015). What is a Walkable Place? The Walkability Debate in Urban Design. *Urban Design International*, 20(4), 274–292. https://doi.org/10.1057/udi.2015.22

Frey, H. (1999). *Designing the city: Towards a more sustainable urban form*. E & FN Spon.

Gehl, J. (2010). *Cities for people*. Island Press.

Gould, J., & Nissen-Petersen, E. (1999). *Rainwater catchment systems for domestic supply: Design, construction and implementation*. ITDG Publishing.

Grant, J. (2006). Planning the good community: New urbanism in theory and practice. *Journal of the American Planning Association*, 72(4), 503–504. https://doi.org/10.1080/01944360608976766

Habe, R. (2019). Aesthetic Dimensions of Urban Design: Artistic Elements and their Role in Urban Design Practice. *Journal of Urban Design*, 24(3), 393–411. https://doi.org/10.1080/13574809.2018.1549052

Hall, P. (1988). *Cities of tomorrow: An intellectual history of urban planning and Design in the twentieth century*. Blackwell.

Hough, M. (1984). *City form and natural process: Towards a new urban vernacular*. Routledge.

Jacobs, A. B. (1993). *Great streets*. MIT Press.

Kostof, S. (1991). *The city shaped: Urban patterns and meanings through history*. Thames & Hudson.

Kropf, K. (2017). *The handbook of urban morphology*. Wiley Blackwell.

Lynch, K. (1981). *A theory of good city form*. MIT Press.

Madanipour, A. (1996). *Design of urban space: An inquiry into a ocio-spatial process*. Wiley.

Németh, J., & Langhorst, J. (2014). Rethinking Urban Transformation: Temporary uses for Vacant Land. *Cities*, 40, 143–150. https://doi.org/10.1016/j.cities.2013.04.007

Oke, T. R. (1982). The Energetic Basis of the Urban Heat Island. *Quarterly Journal of the Royal Meteorological Society*, 108(455), 1–24.

Sharifi, A. (2016). From garden city to eco-urbanism: The quest for sustainable neighborhood development. *Sustainable Cities and Society*, 20, 1–16. https://doi.org/10.1016/j.scs.2015.09.002

Tan, X., Zhou, J., & Tan, Y. (2018). Evaluating Urban Design from the Perspective of Walkability: A Case Study in Nanjing, China. *Journal of Urban Design*, 23(5), 677–698. https://doi.org/10.1080/13574809.2018.1457003

Tribby, C. P., Zimring, C. M., & Winters, M. (2020). Advancing Health and the Built Environment through Collaborative Cross-Disciplinary Research. *Journal of Transport & Health*, 19, 100950. https://doi.org/10.1016/j.jth.2020.100950

Vymazal, J. (2013). *Constructed wetlands for wastewater treatment*. Springer.

Walljasper, J. (2007). *The great neighborhood book: A do-it-yourself guide to placemaking*. New Society Publishers.

4 Urban Economy

4.1 ISSUES IN URBAN ECONOMY

If we cover core and collateral issues of the urban economy, it may have an infinite number of them. However, some common issues, that is, may not be an exhaustive list, may include (1) congestion that brings economic, social, and environmental impacts; (2) higher crime rates as a social problem that can bring a negative impact on city economics; (3) pollution as an environmental problem and economic problem as well; (4) Increased levels of inequality and social exclusion; (5) excessive size of city influences transport costs for citizens; (6) overcrowding is not socially desirable and leads to economic problem; (7) urban expansion and loss of agriculture land; (8) shortage of urban services; (9) slum and squatter settlements; (10) traffic congestion; (11) unemployment and underemployment; (12) social and racial issues; (13) fiscal problems; and (14) housing and homelessness. However, we do not discuss these issues individually, rather discussing sectors in the urban economy and associate these sectors with the issues listed above.

Urban economy refers to the economic activity that takes place in cities and metropolitan areas. We subscribed to this definition. Urban economies are characterized by their prominent levels of productivity, innovation, and specialization, which are facilitated by the concentration of people, firms, and institutions in urban areas. Urban economies are also more diverse and dynamic than rural economies, with a wider range of economic activities and opportunities. The urban economy can be divided into different sectors as shown in Table 4.1.

The urban economy is an important arena for policymakers, planners, and researchers who seek to understand the dynamics of economic growth and development in cities. By promoting innovation, entrepreneurship, and investment in urban areas, policymakers can help create more vibrant and sustainable urban economies that benefit all residents.

The formal sector in the urban economy, or economy in general, has overwhelmingly been discussed in many economic books, as its roles and contributions to urban development or vice versa are clear. However, the informal sector is sometimes neglected whereas, in cities in developing countries, its contributions to overall economic activities were even larger than the formal counterpart, due to its characteristics of easy entry.

The informal sector in the urban economy is the part of the economy that is not regulated by the government or formal institutions. It includes activities such as street vending, domestic work, waste picking, and transport services. The most important characteristic of the informal sector is easy entry because unregulated as an opposition to the formal sector that is many barriers to entry. Anyone can engage in informal sector business almost immediately. Because of this characteristic, the informal sector is more prevalent in low- and middle-income countries than in high-income countries. The informal sector is often characterized by low wages, poor working conditions, and a lack of social protection. However, the informal sector also provides livelihoods for millions of

DOI: 10.1201/9781032664613-4

TABLE 4.1
Sectors in Urban Economy

Sector	Explanation
Manufacturing	Manufacturing is an important sector in many urban economies, especially in developing countries. Manufacturing activities can range from small-scale production of artisanal goods to large-scale production of industrial goods.
Services	Services are the largest sector in most urban economies, accounting for a significant share of employment and output. Service activities can range from low-skilled jobs such as retail and hospitality to high-skilled jobs such as finance and professional services.
Creative industries	Creative industries are a growing sector in many urban economies, driven by the demand for cultural and creative products and services. Creative industries can include sectors such as advertising, architecture, design, fashion, film, music, and software development.
Knowledge-based industries	Knowledge-based industries are a key driver of innovation and growth in many urban economies. Knowledge-based industries can include sectors such as biotechnology, information technology, and research and development.

urban workers, especially in developing countries. Some of the main points about the informal sector in the urban economy are:

1. The informal sector accounts for 80% of jobs and is the backbone of economic activity in urban Africa.
2. The informal sector is the product and driver of advanced capitalism and the site of the most entrepreneurial aspects of the urban economy, led by creative professionals such as artists, architects, designers, and software developers.
3. The informal sector drives urbanization through secondary sector employment, which is easy to engage, as a pull factor of a city amid the persistent presence of the rural push factor of no job or no income.
4. The informal sector faces many challenges such as a lack of access to finance, infrastructure, markets, and legal recognition.
5. The informal sector can benefit from coordinated policies with social protection, such as cash transfers, health insurance, and skills training.
6. The informal sector can also contribute to urban resilience and recovery in times of crises, such as civil war or natural disasters.
7. The informal sector is a global and pervasive phenomenon that affects both developed and developing economies.

Urban planners should accommodate the informal sector and informal economy in their planning rather than prevent them from flourishing for their significant contribution to the urban economy while keeping the characteristic of easy entry. Catering to the informal sector in urban planning has two prong objectives, providing easy jobs for citizens and keeping the urban environment under control. Figures 4.1 and 4.2 show the informal sector in Bangkok.

Policymakers can promote innovation and entrepreneurship in urban areas by adopting policies and programs that create a supportive environment for start-ups, businesses, and investors. Some strategies that policymakers can use to promote innovation and entrepreneurship are shown below.

1. Investing in research and development: Policymakers can invest in research and development programs that support innovation and entrepreneurship. These programs can provide funding, mentorship, and networking opportunities for start-ups and businesses.

FIGURE 4.1 Semi-informal sector at Jodd Fair in Bangkok.

Source: **Photograph by Author.**

2. Creating incubators and accelerators: Policymakers can create incubators and accelerators that provide start-ups with affordable office space, shared resources, and access to mentors and investors. These programs can help start-ups overcome the initial barriers to entry and scale up their operations.
3. Streamlining regulations: Policymakers can streamline regulations that may hinder innovation and entrepreneurship. For example, they can reduce bureaucratic red tape, simplify licensing procedures, and provide tax incentives for start-ups and businesses.
4. Promoting public-private partnerships: Policymakers can promote public-private partnerships that leverage the strengths of both sectors to support innovation and entrepreneurship. For example, they can create innovative districts that bring together start-ups, businesses, universities, research institutions, and government agencies in a compact urban area.
5. Encouraging diversity and inclusion: Policymakers can encourage diversity and inclusion in the innovation ecosystem by promoting equal opportunities for all entrepreneurs regardless of their genders, races, ethnicities, or socioeconomic backgrounds. They can also provide training programs that help entrepreneurs develop the skills they need to succeed.
6. Fostering a culture of innovation: Policymakers can foster a culture of innovation by promoting creativity, risk-taking, and experimentation. They can also celebrate successful entrepreneurs as role models for future generations.

These are just a few strategies that policymakers can use to promote innovation and entrepreneurship in urban areas. By creating a supportive environment for start-ups and businesses, policymakers can help create more vibrant and sustainable urban economies that benefit all residents.

FIGURE 4.2 Informal sector business at Indra Bazaar, Pratunam Bangkok.

Source: **Photograph by Author.**

Incubators and accelerators are programs that provide support and resources to start-ups and entrepreneurs to help them grow and succeed. Although the terms are often used interchangeably, there are some differences between them.

Incubators are programs that provide start-ups with affordable office space, shared resources, and access to mentors and investors. Incubators typically focus on early-stage start-ups that are still developing their products or services. Incubators can help start-ups overcome the initial barriers to entry and provide them with the resources they need to develop their business models and prototypes. Accelerators are programs that provide start-ups with intensive mentoring, training, and networking opportunities to help them scale up their operations. Accelerators typically focus on more mature start-ups that have already developed their products or services and are looking to expand their customer base or enter new markets. Accelerators can help start-ups refine their business models, develop their marketing strategies, and connect with potential investors.

Both incubators and accelerators can play a vital role in promoting entrepreneurship by providing start-ups with the resources they need to succeed. By providing affordable office space, shared resources, mentoring, training, and networking opportunities, incubators, and accelerators can help start-ups overcome the initial barriers to entry and scale up their operations.

There are many successful incubators and accelerators around the world that have helped start-ups grow and succeed. Here are some examples:

1. Y-Combinator: Y-Combinator is a start-up accelerator based in Silicon Valley that has helped launch many successful start-ups such as Airbnb, Dropbox, and Reddit. Y-Combinator

provides start-ups with seed funding, mentorship, and networking opportunities to help them grow and succeed.
2. Techstars: Techstars is a global start-up accelerator that provides start-ups with mentorship, training, and networking opportunities to help them scale up their operations. Techstars has helped launch many successful start-ups such as SendGrid, DigitalOcean, and ClassPass.
3. Five Hundred Startups: Five Hundred Startups is a global venture capital firm and start-up accelerator that provides seed funding, mentorship, and training to early-stage start-ups. Five Hundred Startups has helped launch many successful start-ups such as Udemy, Canva, and Talkdesk.
4. Seedcamp: Seedcamp is a European seed-stage venture capital firm and start-up accelerator that provides funding, mentorship, and training to early-stage start-ups. Seedcamp has helped launch many successful start-ups such as TransferWise, Revolut, and UiPath.
5. Startupbootcamp: Startupbootcamp is a global network of industry-focused start-up accelerators that provide mentorship, training, and networking opportunities to early-stage start-ups. Startupbootcamp has helped launch many successful start-ups such as Lendix, Bitwala, and Voi.

Many other programs provide support and resources to start-ups and entrepreneurs around the world. A success story of Airbnb that starts from zero, even under zero, is shown Box 4.1.

BOX 4.1 THE STORY OF AIRBNB

In 2008, founders Brian Chesky and Joe Gebbia faced rent struggles in San Francisco. Their solution? Rent out their living space to travelers attending a design conference. Armed with air mattresses, they launched "Air Bed and Breakfast" The response was immediate bookings poured in. Airbnb was born. Beyond quick cash, they envisioned a platform connecting travelers with unique, local accommodations. Legal battles, regulatory hurdles, and occasional horror stories tested their resilience. Yet, they persisted. They refined their platform, introduced professional photography, and implemented a review system. By 2011, Airbnb spanned 13,000 cities. They expanded beyond air mattresses to entire homes and introduced "Experiences" COVID-19 hit hard in 2020, but adaptability saved the day. Airbnb went public, defying skeptics, and now stands as a testament to disruptive thinking, perseverance, and creating extraordinary experiences.

Sources: blog.12min.com, and fastcompany.com

Policymakers can streamline regulations that may hinder innovation and entrepreneurship. By reducing bureaucratic red tape, simplifying licensing procedures, and providing tax incentives for businesses that invest in the city, policymakers can create a business-friendly environment that supports economic growth and development. Streamlining regulations can also help attract talent, foster innovation, and create jobs. By creating a supportive environment for businesses, entrepreneurs, and investors, policymakers can help create more vibrant and sustainable cities that benefit all residents.

Policymakers face the challenge of balancing the need for regulation with the need to promote innovation and entrepreneurship. On the one hand, regulations are necessary to protect public health, safety, and welfare, and to ensure a level playing field for businesses. On the other hand, regulations can also be a barrier to entry for start-ups and entrepreneurs, especially those with limited resources. Some strategies that policymakers can use to balance the need for regulation with the need to promote innovation and entrepreneurship are, among others:

1. *Risk-Based Approach:* Policymakers can adopt a risk-based approach to regulation that focuses on high-risk activities and sectors while reducing the regulatory burden on low-risk activities and sectors. This approach can help ensure that regulations are proportionate to the risks they aim to address.
2. *Regulatory Sandboxes:* Policymakers can create regulatory sandboxes that provide a safe space for start-ups and entrepreneurs to test new products or services without being subject to full regulatory requirements. Regulatory sandboxes can help start-ups overcome regulatory barriers while ensuring public safety and consumer protection. (*Note:* Sandboxes are isolated environments where software or code can be executed without affecting the rest of the system. They are often used for testing, debugging, or security purposes. Sandboxes can prevent malicious or faulty code from harming the system or accessing sensitive data. Sandboxes can also allow users to experiment with atypical features or settings without risking permanent changes.)
3. *Collaboration with Stakeholders:* Policymakers can collaborate with stakeholders such as businesses, entrepreneurs, and consumers to develop regulations that are effective, efficient, and responsive to their needs. This approach can help ensure that regulations are evidence-based and reflect the realities of the market.
4. *Periodic Review of Regulations:* Policymakers can periodically review existing regulations to ensure that they remain relevant, effective, and efficient. This approach can help identify regulatory barriers that may hinder innovation and entrepreneurship and remove them where appropriate.
5. *International Cooperation:* Policymakers can cooperate with other countries to develop international standards and best practices for regulation. This approach can help reduce regulatory fragmentation and promote cross-border innovation and entrepreneurship.

These are just a few strategies that policymakers can use to balance the need for regulation with the need to promote innovation and entrepreneurship. By adopting a flexible, evidence-based, and collaborative approach to regulation, policymakers can create a supportive environment for businesses, entrepreneurs, and investors while ensuring public safety and consumer protection.

Regulatory sandboxes are a new concept that provides a safe space for start-ups and entrepreneurs to test new products or services without being subject to full regulatory requirements. Here are some examples of regulatory sandboxes in urban areas:

1. *Singapore:* The Monetary Authority of Singapore (MAS) has launched a regulatory sandbox that allows fintech start-ups to test their products or services in a controlled environment. The sandbox provides start-ups with a safe space to experiment with new business models and technologies while ensuring that they comply with regulatory requirements.
2. *United Kingdom:* The Financial Conduct Authority (FCA) has launched a regulatory sandbox that allows fintech start-ups to test their products or services in a controlled environment. The sandbox provides start-ups with access to regulatory expertise, feedback, and support while ensuring that they comply with regulatory requirements.
3. *United States:* The Consumer Financial Protection Bureau (CFPB) has launched a regulatory sandbox that allows fintech start-ups to test their products or services in a controlled environment. The sandbox provides start-ups with access to regulatory expertise, feedback, and support while ensuring that they comply with consumer protection laws.

Participating in a regulatory sandbox program can provide several benefits for start-ups and entrepreneurs. The potential benefits of the program are shown in Table 4.2.

Creating a supportive environment for startups and entrepreneurs, along with regulatory sandboxes, can foster vibrant and sustainable urban economies that benefit all residents. Policymakers can

TABLE 4.2
Potential Benefits of the Sandbox Program

Potential Benefit	Explanation
Reduced regulatory burden	Regulatory sandboxes provide a safe space for start-ups to test new products or services without being subject to full regulatory requirements. This can help reduce the regulatory burden on start-ups and provide them with the flexibility they need to innovate.
Access to regulatory expertise	Regulatory sandboxes provide start-ups with access to regulatory expertise, feedback, and support. This can help start-ups understand the regulatory requirements they need to comply with and ensure that their products or services are safe and compliant.
Networking opportunities	Regulatory sandboxes provide start-ups with networking opportunities with other start-ups, investors, and regulators. This can help start-ups build relationships, gain exposure, and attract investment.
Brand recognition	Participating in a regulatory sandbox program can help start-ups build their brand recognition and reputation. This can help start-ups attract customers, partners, and investors.
Innovation	Regulatory sandboxes can foster innovation by providing start-ups with the resources they need to develop new products or services. This can help start-ups create new markets, disrupt existing markets, and drive economic growth.

TABLE 4.3
Possible Strategies for Effective Sand Boxes

Strategy	Explanation
Undertaking periodic review	Policymakers can periodically review the regulatory sandbox program to ensure that it remains relevant, effective, and efficient. This can help identify areas for improvement and ensure that the program continues to meet the needs of start-ups and entrepreneurs.
Showing transparency and accountability	Policymakers can ensure that the regulatory sandbox program is transparent and accountable. This can help build trust among stakeholders and ensure that the program is perceived as fair and impartial.
Carrying out evaluation	Policymakers can evaluate the impact of the regulatory sandbox program on start-ups, entrepreneurs, and the wider economy. This can help identify the strengths and weaknesses of the program and inform future policy decisions.
Engaging collaboration	Policymakers can collaborate with stakeholders such as businesses, entrepreneurs, investors, and regulators to develop regulations that are effective, efficient, and responsive to their needs. This approach can help ensure that regulations are evidence-based and reflect the realities of the market.
Promoting international cooperation	Policymakers can cooperate with other countries to develop international standards and best practices for regulatory sandboxes. This approach can help reduce regulatory fragmentation and promote cross-border innovation and entrepreneurship.

ensure that regulatory sandboxes remain effective and efficient over time by adopting several strategies as shown in Table 4.3.

Policymakers can create a supportive environment for businesses, entrepreneurs, and investors while ensuring public safety and consumer protection by adopting a flexible, evidence-based, and collaborative approach to regulation,

Urban development can promote the urban economy in several ways. By creating a supportive environment for businesses, entrepreneurs, and investors, urban development can help attract talent, foster innovation, and create jobs. Some strategies that urban development can do to promote urban economy are shown in Table 4.4.

TABLE 4.4
Strategies to Promote Urban Economy

Strategy	Explanation
Investing in infrastructure	Urban development can invest in infrastructure such as transportation, energy, and communication systems that support economic growth and development. For example, investing in public transportation can improve accessibility to jobs and markets, while investing in broadband networks can enhance digital connectivity and innovation.
Creating business-friendly policies	Urban development can create policies and regulations that support businesses and entrepreneurs. For example, they can reduce bureaucratic red tape, simplify licensing procedures, and provide tax incentives for businesses that invest in the city.
Promoting innovation and entrepreneurship	Urban development can promote innovation and entrepreneurship by creating incubators, accelerators, and other programs that provide support and resources to start-ups and entrepreneurs. These programs can help start-ups overcome the initial barriers to entry and scale up their operations.
Fostering a diverse and inclusive community	Urban development can foster a diverse and inclusive community by promoting equal opportunities for all residents regardless of their gender, race, ethnicity, or socioeconomic background. They can also provide training programs that help residents develop the skills they need to succeed.
Encouraging sustainable development	Urban development can encourage sustainable development by promoting green technologies, circular models of production and consumption, and other practices that reduce environmental impact while promoting economic growth.

Through the creation of a supportive environment for businesses, entrepreneurs, and investors, urban development can help create more vibrant and sustainable cities that benefit all residents.

4.2 MARKET FORCES IN CITY DEVELOPMENT

An economy is a complex system of interrelated production, consumption, and exchange activities that determine how resources are allocated among all the participants. The production, consumption, and distribution of goods and services combine to fulfill the needs of those living and operating within the economy. An economy may represent a nation, a region, a single industry, or even a family. There are four main types of economies: traditional economies, command economies, market economies, and mixed economies, with the following explanations.

1. Traditional economies are based on customs, beliefs, and traditions that have been passed down from generation to generation. They are often found in rural areas with low levels of technology and are characterized by subsistence farming and bartering.
2. Command economies are regulated by a central government that determines the goods that are produced, their quantities, and the price paid for them. The government also decides how resources are allocated and how much is produced.
3. Market economies, also called free market economies, are self-regulated, allowing goods to be produced and distributed in response to demand from consumers. Producers determine what is sold and produced, and what prices to charge. If they expect to succeed, they will produce what consumers want and charge what consumers are willing to pay. Through these decisions, the laws of supply and demand determine prices and total production. If consumer's demand for a specific product increases, production tends to increase to satisfy the demand. The increased demand causes prices to rise until consumers balk and cut back on their purchases. Demand for the product will then decline and prices will decline with it. This

constant tug of supply and demand allows a market economy a tendency to naturally balance itself.

4. Mixed economies combine elements of both market-based and command-based systems. In mixed economies, the government intervenes in the market by regulating certain industries or providing public goods such as education or healthcare. However, most economic decisions are made by individuals or businesses operating within the market.

Within the spectrum of economic systems or schools of thought mentioned above, we find the communist economic system, exemplified by countries like North Korea. Additionally, there exists another school of thought: pure competition or a free-market economy, to which the United States belongs. Figure 4.3 illustrates this range.

A market economy is a system in which economic decisions and prices are determined by the interactions of buyers and sellers in a free market. For example, in a market economy, the price of a product is determined by how much consumers are willing to pay for it and how much producers are willing to supply it. In a market economy, individuals and firms have the freedom to own and exchange private property, subject to minimal government intervention. For instance, in a market economy, a farmer can decide what crops to grow and sell them to whoever offers the best price.

A market economy is based on the principles of supply and demand, which reflect the preferences and choices of consumers and producers. For example, if consumers demand more of a good or service, the price will rise, and producers will increase their output. Conversely, if consumers demand less of a good or service, the price will fall, and producers will reduce their output. A market economy tends to allocate resources efficiently, as goods and services are produced at the lowest possible cost and sold at the highest possible price. For example, in a market economy, a firm that can produce a good or service at a lower cost than its competitors will have an advantage and earn higher profits. A market economy also encourages innovation and entrepreneurship, as firms compete to gain market share and profits. For instance, in a market economy, a firm that can create a new or improved product or service that meets the needs of consumers will have an edge over its rivals and attract more customers. However, a market economy also has some drawbacks, such as inequality, externalities, market failures, and public goods. These issues may require government intervention to correct market inefficiencies and provide social welfare. For example, in a market economy, some people may earn much more income than others due to differences in skills, education, or luck. This may create social problems such as poverty, crime, or resentment.

Similarly, in a market economy, some activities may generate negative effects on others that are not reflected in the market price. This may lead to the overproduction or underproduction of certain goods or services. For instance, in a market economy, a factory that pollutes the air or water may not bear the full cost of its environmental damage. This may result in too much

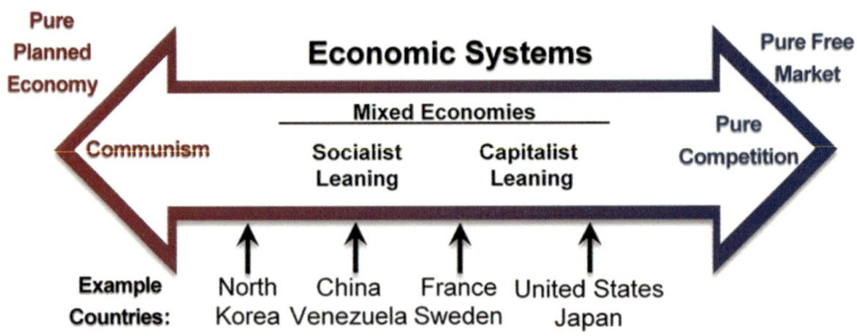

FIGURE 4.3 Communism versus pure competition economy.

Source: **Wikimedia.**

pollution and harm the health of people and wildlife. Likewise, in a market economy, some goods or services may not be provided by the private sector because they are not profitable or excludable. This may lead to under-provision or no provision of certain goods or services. For example, in a market economy, a public park or a national defense system may not be provided by the private sector because they are non-rivalrous and non-excludable. This means that one person's use of them does not reduce their availability to others and that no one can be prevented from using them. Therefore, these goods or services may require government provision or regulation to ensure their availability and quality.

Market forces in the economy may be translated broadly as the interaction between supply and demand. Similarly in urban development, the market force is the interaction between the supply and demand of the urban elements such as land, housing, infrastructure, and other urban amenities. These forces shape the spatial patterns, density, diversity, and affordability of urban areas. Market forces are influenced by a range of factors, such as population growth, income levels, preferences, regulations, policies, and externalities. Understanding how market forces operate and affect urban development is crucial for urban planners, policymakers, developers, and residents who seek to create livable, sustainable, and equitable cities. One of the important economic infrastructures is the market, a place where demand and supply meet (Figure 4.4).

Economic forces play a significant role in shaping the spatial patterns, density, diversity, and affordability of urban areas. For instance, the availability of jobs, wages, and the cost of living can influence the spatial patterns, density, diversity, and affordability of urban areas. High-paying jobs in a particular area can lead to an increase in demand for housing in that area, which can drive up prices and make it less affordable for low-income residents. In addition to economic factors, social factors such as demographics, culture, and lifestyle can also influence the spatial patterns, density, diversity, and affordability of urban areas. For example, areas with a high concentration of young

FIGURE 4.4 A market where supply and demand meet.

Source: **Wikipedia.**

professionals may have a different spatial pattern and density than areas with a high concentration of retirees.

Social factors such as demographics, culture, and lifestyle can also influence the spatial patterns, density, diversity, and affordability of urban areas. For example, areas with a high concentration of young professionals may have a different spatial pattern and density than areas with a high concentration of retirees.

Political factors such as zoning laws, building codes, and land-use regulations can also influence the spatial patterns, density, diversity, and affordability of urban areas. For example, zoning laws that restrict the height of buildings can limit the density of an area.

Environmental factors such as climate change and natural disasters can also influence the spatial patterns, density, diversity, and affordability of urban areas. For example, areas that are prone to flooding may have lower property values due to the increased risk of damage from natural disasters.

Urbanization can have a significant impact on economic forces in cities. As more people move to urban areas, the demand for goods and services increases, which can lead to economic growth. Urbanization can also lead to increased specialization and division of labor, which can increase productivity and efficiency. However, urbanization can also lead to negative economic effects. For example, rapid urbanization can lead to overcrowding, which can put a strain on infrastructure and public services. This can lead to increased costs for businesses and residents, which can negatively impact the economy. Urbanization can also lead to income inequality, as high-paying jobs may be concentrated in certain areas while low-paying jobs are concentrated in others. This can make it difficult for low-income residents to afford housing and other necessities, which can negatively impact their quality of life and the overall economy. Policymakers need to consider these impacts when making decisions about urban planning and development.

Urban planning can have a significant impact on economic forces in cities. Urban planning can influence the spatial patterns, density, diversity, and affordability of urban areas. For example, zoning laws that encourage mixed-use development can lead to more affordable housing options and increased economic activity in an area. Urban planning can also help to address issues such as income inequality and environmental degradation. For example, urban planners can work to create more equitable access to public transportation and green spaces, which can improve the quality of life for residents and reduce the negative impacts of urbanization on the environment. In addition, urban planning can help to promote economic growth by attracting new businesses and industries to an area. By creating a favorable business environment through zoning laws and other regulations, urban planners can help to create jobs and stimulate economic activity.

Urban planning is essential in addressing environmental issues in a city. For example, urban planners can work to create more green spaces and parks in urban areas, which can help mitigate the effects of climate change and improve air quality. In addition, urban planners can work to promote sustainable transportation options such as bike lanes, public transportation, and pedestrian walkways. This can help to reduce traffic congestion and air pollution, which can have a positive impact on the environment. Urban planners can also work to promote sustainable building practices such as green roofs and solar panels. These practices can help to reduce energy consumption and greenhouse gas emissions, which can have a positive impact on the environment.

Overall, urban planning is also important in shaping the economic forces that drive cities. Urban planners must prioritize the needs of residents, businesses, and the environment. By doing so, they contribute to the creation of vibrant, sustainable communities that benefit everyone.

Economic efficiency refers to the ability of an economy to produce goods and services with the least number of resources possible. In the context of city development, economic efficiency can be achieved by promoting sustainable practices and creating a favorable business environment. Sustainable practices such as green building design, energy-efficient transportation, and renewable energy sources can help to reduce costs and increase productivity. For example, green building design can help to reduce energy consumption and waste, which can lead to lower operating costs

TABLE 4.5
Potential Benefits of Sustainable Building

Potential Benefit	Explanation
Reduced energy consumption	Sustainable buildings are designed to use less energy than traditional buildings. This can lead to lower energy bills for residents and businesses, as well as a reduced carbon footprint.
Improved indoor air quality	Sustainable buildings are designed to promote good indoor air quality, which can improve the health and well-being of residents and workers.
Reduced water consumption	Sustainable building design can help to reduce waste by using materials that are recyclable or biodegradable. This can help to reduce the amount of waste that ends up in landfills.
Reduced waste	Sustainable building design can help to reduce waste by using materials that are recyclable or biodegradable. This can help to reduce the amount of waste that ends up in landfills.
Improved quality of life	Sustainable building design can help to create more livable and healthy communities. For example, green spaces and parks can improve mental health and well-being, while bike lanes and pedestrian walkways can promote physical activity.

TABLE 4.6
Benefits of Implementing Water Conservation

Benefit	Explanation
Reduced water bills	Water conservation can help to reduce water bills for residents and businesses, which can lead to cost savings.
Conservation of natural resources	Water conservation can help to conserve natural resources such as rivers, lakes, and groundwater.
Reduced energy consumption	Water conservation can help to reduce energy consumption by reducing the amount of energy needed to pump, treat, and distribute water.
Improved water quality	Water conservation can help to improve water quality by reducing the number of pollutants that enter waterways.

for businesses and residents. Creating a favorable business environment through zoning laws and other regulations can help to attract new businesses and industries to an area. This can lead to job creation and increased economic activity, which can benefit both businesses and residents.

Sustainable building design can benefit cities in several ways. The potential benefits of sustainable building are shown in Table 4.5.

Sustainable building design and green building design are often used interchangeably, but there are some differences between the two. Sustainable building design is a broader concept that encompasses all aspects of building design, including energy efficiency, water conservation, and waste reduction. Sustainable building design aims to create buildings that are environmentally responsible, economically viable, and socially equitable. Green building design, on the other hand, is a subset of sustainable building design that focuses specifically on reducing the environmental impact of buildings. Green building design aims to create buildings that are energy-efficient, water-efficient, and use environmentally friendly materials.

Water conservation and waste reduction can benefit cities in several ways. The benefits of implementing water conservation are shown in Table 4.6.

Similarly, waste reduction can also benefit cities in several ways. The benefits of accomplishing waste reduction are shown in Table 4.7.

Promoting economic efficiency is a crucial aspect of city development. By adopting sustainable practices and creating a favorable business environment, cities can become more efficient,

TABLE 4.7
Benefits of Accomplishing Waste Reduction

Benefit	Explanation
Reduced landfill costs	Waste reduction can help to reduce landfill costs for cities by reducing the amount of waste that needs to be disposed of.
Conservation of natural resources	Waste reduction can help conserve natural resources by reducing the need for raw materials.
Reduced greenhouse gas emissions	Waste reduction can help to reduce greenhouse gas emissions by reducing the amount of waste that is burned or decomposed in landfills.
Creation of jobs	Waste reduction can create jobs in industries such as recycling and composting.

productive, and sustainable promoting economic efficiency through sustainable practices. A favorable business environment is essential for cities to become more efficient, productive, and sustainable. By doing so, cities can improve their quality of life for residents while also attracting new businesses and investments.

4.3 ALLOCATION OF RESOURCES

Resource allocation in urban development is the process of distributing and managing the available resources for planning, designing, and implementing urban projects. These resources may include land, capital, labor, materials, energy, information, and technology. Resource allocation involves balancing the needs and preferences of different stakeholders, such as residents, businesses, governments, and environmental groups while considering the constraints and opportunities of the urban context. For example, resource allocation may involve deciding how much land to allocate for housing, commerce, industry, recreation, and green spaces; how to finance urban infrastructure and services; how to ensure fair access and participation for diverse groups; and how to minimize environmental impacts and risks. Resource allocation can be influenced by several factors, such as political agendas, economic conditions, social equity, environmental sustainability, and technical feasibility. These factors may create conflicts or trade-offs among different objectives and interests. Resource allocation in urban development requires a strategic vision, a collaborative approach, and a flexible framework that can adapt to changing circumstances and uncertainties. A strategic vision defines the long-term goals and directions for urban development; a collaborative approach involves engaging and coordinating with multiple actors and sectors; and a flexible framework allows for learning and feedback throughout the process.

Optimizing available resources in the city is a crucial aspect of urban development. Here are some ways to accomplish this:

1. *Land Resources:* Cities can optimize land use by adopting smart growth principles, such as compact development, mixed-use zoning, and transit-oriented development. This can reduce urban sprawl, preserve open spaces, and improve access to public transportation.
2. *Capital:* Cities can optimize capital by investing in infrastructure that supports sustainable development, such as renewable energy, green buildings, and public transportation. This can reduce energy consumption, greenhouse gas emissions, and air pollution.
3. *Labor:* Cities can optimize labor by promoting entrepreneurship, innovation, and skills development. This can create new job opportunities, increase productivity, and enhance competitiveness.
4. *Materials:* Cities can optimize materials by adopting circular economy principles, such as reducing waste generation, reusing materials, and recycling waste. This can conserve natural resources, reduce landfill waste, and create new business opportunities.

5. *Energy:* Cities can optimize energy by adopting energy-efficient technologies and practices, such as light emitting diodes (LED) lighting, smart grids, and green roofs. This can reduce energy consumption, lower energy bills for households and businesses, and improve air quality.
6. *Information:* Cities can optimize information by adopting smart city technologies that enable data collection, analysis, and sharing. This can improve decision-making processes, enhance public service delivery, and increase transparency and accountability.
7. *Technology:* Cities can optimize technology by investing in research and development of innovative technologies that support sustainable development goals. This can create new business opportunities, increase competitiveness, and improve the quality of life for residents.

Citizens can play a crucial role in optimizing resources for urban development. By engaging in meaningful forms of citizen engagement and social accountability, citizens can contribute innovative solutions to complex development challenges, advocate for more transparent and effective public institutions, and strengthen citizen-state relations. Citizens can participate in community-based initiatives that promote sustainable practices, such as waste reduction, energy conservation, and green transportation. They can also provide feedback on public service delivery, report problems in real time, and collaborate with city officials to identify areas for improvement. Moreover, citizens can support entrepreneurship and innovation by investing in local businesses, sharing knowledge and skills, and creating networks that foster creativity and collaboration. They can also advocate for policies that promote social inclusion, gender equality, and human rights, which are essential for building resilient and equitable cities.

The private sector can play a significant role in optimizing resources for urban development. By investing in sustainable infrastructure, adopting green technologies, and promoting social responsibility, private companies can contribute to building more efficient, productive, and sustainable cities. For instance, private companies can invest in renewable energy projects, such as solar and wind power, which can reduce greenhouse gas emissions and improve air quality. They can also adopt green building practices that promote energy efficiency, water conservation, and waste reduction. Moreover, private companies can promote social responsibility by supporting local communities, creating job opportunities, and respecting human rights. Private companies can also collaborate with public institutions and civil society organizations to identify areas for improvement and develop innovative solutions to complex development challenges. For example, they can participate in public–private partnerships that leverage the strengths of both sectors to achieve common goals.

Can we model the optimization of resource allocation in a simple representation? Theoretically, yes, we can, but the solutions may not be as simple as on paper, as myriad real factors may not be able to be expressed in mathematical variables easily. For instance, how can we express policy in our mathematical equation for optimization? However, to give a small picture of how the optimization for resource allocation works, we will provide a simple example of this process.

A company produces two types of products, A and B, using three types of resources, P, Q, and R. The company has 100 units of P, 90 units of Q, and 120 units of R available. The company wants to maximize its profit subject to resource constraints. The constraints given are: (1) The company has 100 units of resource P available. Each unit of product A requires 2 units of resource P, and each unit of product B requires 1 unit of resource P. (2) The company has 90 units of resource Q available. Each unit of product A requires 1 unit of resource Q, and each unit of product B requires 2 units of resource Q. (3) The company has 120 units of resource R available. Each unit of product A requires 3 units of resource R, and each unit of product B requires 2 units of product R.

The objective of the production is to maximize profit. The profit of a unit of product A is 10 $/unit, and product B is 12 $/unit.

There are several methods to solve linear programming problems, which are mathematical problems that involve optimizing a linear objective function subject to linear constraints. Some of the most common methods are:

- *Graphical Method:* This method involves graphing the feasible region defined by the constraints and finding the optimal solution by inspecting the corner points of the region. This method is suitable for problems with two variables and can provide a visual understanding of the problem.
- *Simplex Method:* This method is an iterative algorithm that starts from a basic feasible solution and moves along the edges of the feasible region until it reaches the optimal solution or proves that the problem is unbounded or infeasible. This method can handle problems with more than two variables and is very efficient in practice.
- *Interior Point Methods:* These methods are a class of algorithms that start from an interior point of the feasible region and move toward the optimal solution by following a certain direction. These methods can handle large-scale problems and have strong theoretical guarantees of convergence.
- *Software Tools:* Various software tools can solve linear programming problems using different methods and techniques. Some examples are R, OpenSolver, Excel Solver, MATLAB, etc. These tools can handle complex and realistic problems with ease and provide numerical solutions.

As we have only two variables to solve, which are the numbers of product A, and the numbers of product B, we can conveniently use the graphical method.

Let X be the number of units of product A, and the profit per unit of product A is \$10. Let Y be the number of units of product B, and the profit per unit of product B is \$12.

We can set up the problem in standard maximization form as follows: The profit to be maximized can be represented as:

$$10X + 12Y$$

10 and 12 are the profit per unit of products A and B, respectively.

The constraints: The company has 100 units of resource P available. Each unit of product A requires 2 units of resource P, and each unit of product B requires 1 unit of resource P. We can write:

$$2X + Y \leq 100$$

The company has 90 units of resource Q available. Each unit of product A requires 1 unit of resource Q, and each unit of product B requires 2 units of resource Q. We can write:

$$X + 2Y \leq 90$$

The company has 120 units of resource R available. Each unit of product A requires 3 units of resource R, and each unit of product B requires 2 units of product R. We can write:

$$3X + 2Y \leq 120$$

We can now plot these constraints into a graph with x–y coordinates (Figure 4.5), by changing the inequality to equality, and find the feasible region. The feasible region is the area that satisfies all

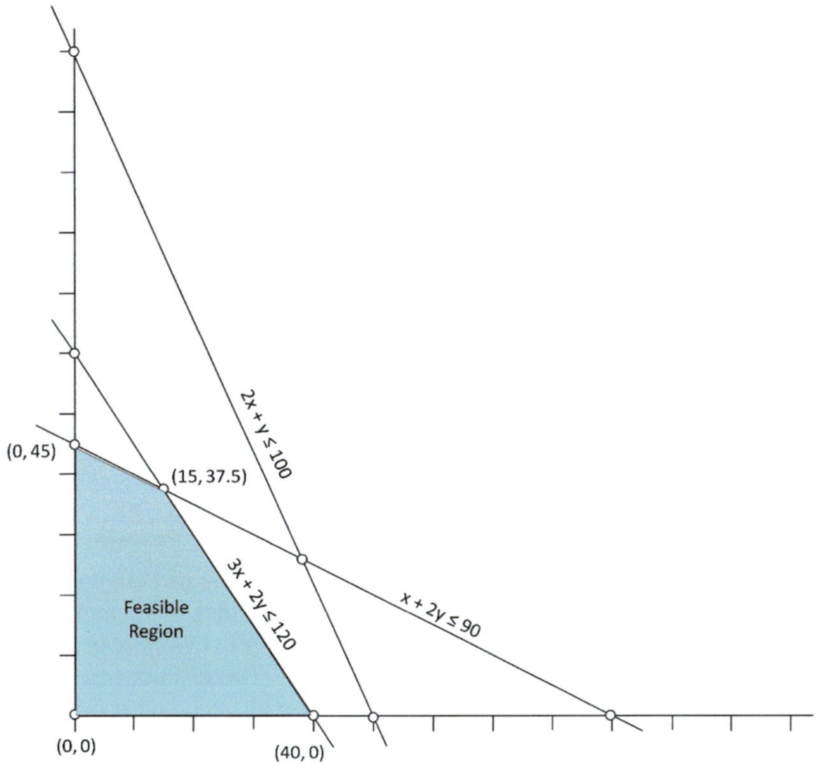

FIGURE 4.5 Linear programming.

constraints. The corner points of the feasible region are (0, 0), (0, 45), (15, 37.5), and (40, 0), and the profits for each corner are at (0, 0) = \$0, at (0.45) = \$540, at (15, 37.5) = \$600, and (40, 0) = \$400.

Through sustainable practices and innovative technologies, cities can achieve their urban development goals while simultaneously enhancing residents' quality of life.

4.4 RESIDENTIAL LOCATION FROM CITY CENTER

The dilemma of residential location whether in the city center or the urban periphery is a classic trade-off between the cost of housing and the cost of transportation. Living in the city center has the advantage of being close to work, which reduces transportation costs and saves time. However, housing prices and rental costs in the city center are higher than in the periphery of the city, which can be a significant financial burden. On the other hand, living in the periphery of the city has the advantage of lower housing prices and rental costs, and a better environment, that is, less pollution. However, this comes at the expense of increased transportation costs and time spent commuting to work. The cost of transportation can be significant, especially if we need to use a car or public transportation.

One way to approach this dilemma is to find a balance between housing costs and transportation costs that fit your budget and lifestyle. For example, you could consider living in a neighborhood that is close to public transportation or within walking or biking distance to work. This could reduce your transportation costs while still allowing you to live in a more affordable area. Figure 4.6 shows this dilemma. We assumed that both housing and commuting costs are linear with respect to distance from the city center toward urban periphery. However, there are no optimal solutions for this

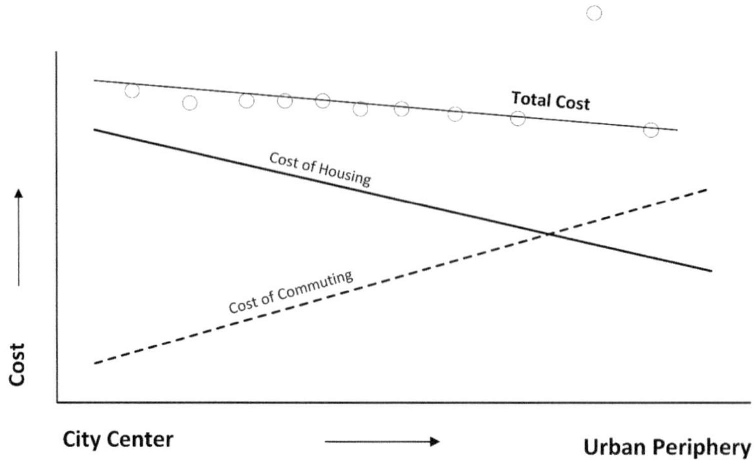

FIGURE 4.6 Trade-off residential location.

tradeoff, very much depending on maximum and minimum values of both commuting cost and housing cost. This figure is simplified, in reality, it is hard to illustrate this complex correlation.

One of the trade-offs that many urban dwellers face is whether to live close to their place of work in the city center or farther away in the periphery of the city. The main factors that influence this decision are the costs of housing and transportation. However, there are other factors to consider as well, such as:

1. *Personal Preferences:* Some people may prefer a quieter and more spacious living environment in the periphery, while others may enjoy the convenience and vibrancy of the city center.
2. *Lifestyle:* Some people may have dissimilar needs and preferences for their daily activities, such as shopping, entertainment, education, health care, and socializing, which may be either more or less accessible in various locations.
3. *Amenities:* Some people may value certain amenities that are available in distinct locations, such as parks, libraries, museums, restaurants, and cultural venues.
4. *Environmental Impact:* Some people may be concerned about the environmental impact of their living and transportation choices, such as greenhouse gas emissions, air pollution, noise pollution, and land use.

The optimal choice depends on how one values these costs and benefits, as well as the concept of opportunity cost, which is the value of the next best alternative that is forgone because of making a choice. The opportunity cost of living in the city center is the lower housing cost that one could have enjoyed in the periphery, while the opportunity cost of living in the periphery is the lower transportation cost that one could have enjoyed in the city center. We should calculate the total cost of living in each location, considering all relevant expenses such as rent/mortgage payments, utilities, food, entertainment, and transportation costs. We should also consider other factors such as quality of life, safety, access to amenities, and proximity to friends and family.

There are several conventional ways to reduce transportation costs:

1. *Use Public Transportation:* Public transportation such as buses, trains, and subways can be a cost-effective way to get around. Many cities offer discounted fares for students, seniors, and low-income individuals.

2. *Carpool:* Carpooling with friends, family, or coworkers can help reduce transportation costs by sharing the cost of gas and parking. Some employers offer incentives for carpooling such as preferred parking spots or cash rewards.
3. *Bike or Walk:* Biking or walking to work or school can be a healthy and cost-effective alternative to driving. Many cities have bike lanes and pedestrian paths that make it safe and easy to get around.
4. *Drive a Fuel-Efficient Car:* If you need to drive, consider purchasing a fuel-efficient car that gets good gas mileage. This can help reduce your fuel costs over time.
5. *Plan Your Trips:* Planning your trips in advance can help you avoid unnecessary driving and reduce transportation costs. For example, you could combine multiple errands into one trip or plan your route to avoid traffic congestion.
6. *Maintain Your Car:* Keeping your car well-maintained can help improve its fuel efficiency and reduce the need for costly repairs. Regular oil changes, tire rotations, and tune-ups can help keep your car running smoothly.

The environmental impact of urban living and transportation is significant and varies depending on the location and mode of travel. According to the US Environmental Protection Agency (EPA), transportation is estimated to be responsible for a quarter of global energy-related CO_2 emissions. In addition, transportation interacts with and depends on other sectors of the economy, such as water resources and energy. Therefore, climate impacts on these sectors can also affect the transportation system. For example, rising sea levels can damage coastal roads and bridges, while droughts can reduce hydropower generation and increase fuel prices.

Moreover, transportation is a major source of air pollution, especially in urban areas where traffic congestion is high. Air pollution can have adverse effects on human health and the environment, such as respiratory diseases, cardiovascular problems, acid rain, and smog. The United Nations Environment Programme (UNEP) estimates that air pollution causes more than 7 million premature deaths each year. Therefore, reducing air pollution from transportation is crucial for improving public health and environmental quality.

The urban periphery is often seen as a desirable location for residential and commercial development, but it also poses significant challenges for sustainability and livability. One of the main problems associated with living in the urban periphery is the transformation of land use patterns, which can result in increased traffic generation, congestion, and air pollution. These effects can have negative consequences for the health and well-being of the residents, as well as for the quality of the urban environment. Another problem related to living in the urban periphery is the encouragement of urban sprawl, which can lead to the loss of natural resources and environmental degradation. Urban sprawl can reduce the availability of green spaces, biodiversity, and ecosystem services, and increase the vulnerability to natural hazards and climate change. Living in the urban periphery, therefore, may not be as beneficial as it seems, and may require careful planning and management to avoid or mitigate its adverse impacts. These factors must be considered when the preference is to live in the urban periphery.

One example of a mathematical model that can be used to analyze the trade-off between residential location and transportation costs is the residential location choice model. This model assumes that individuals choose their residential location based on a trade-off between housing costs and transportation costs.

The model can be formulated as follows:

$$U = V - C \tag{4.1}$$

where U is the utility of living in a certain location, V is the value of the location (e.g., proximity to work, access to amenities), and C is the cost of living in that location (e.g., housing costs, transportation costs). The cost component C can be further decomposed into two parts: $C = C_h + C_t$, where C_h is the cost of housing and C_t is the cost of transportation. The cost of housing can be modeled as a function of housing prices or rents in various locations, while the cost of transportation can be modeled as a function of commuting distance, travel time, and transportation mode. The value component V can also be modeled in numerous ways depending on the specific context. For example, it could include factors such as proximity to work, access to amenities (e.g., grocery stores, parks), quality of schools, and safety.

The utility function U can then be used to estimate the probability that an individual will choose a certain location given their preferences and constraints. This can help policymakers and urban planners make informed decisions about land use, transportation infrastructure, and housing policies.

A numerical example of a residential location choice model is shown as the following:

Suppose we are considering two locations: the city center and the urban periphery. The cost of housing in the city center is \$2,000 per month, while the cost of housing in the urban periphery is \$1,000 per month. The cost of transportation to work in the city center is \$100 per month, while the cost of transportation to work in the urban periphery is \$300 per month. You value living in the city center at \$1,500 per month, while you value living in the urban periphery at \$1,000 per month. The value of living is intangible; thus, it needs to be converted into monetary terms through, for example, valuation.

Using the residential location choice model, we can calculate the utility of living in each location as follows:

$$U(\text{city center}) = V(\text{city center}) - C(\text{city center}) = \$1{,}500 - (\$2{,}000 + \$100) = -\$600$$

$$U(\text{urban periphery}) = V(\text{urban periphery}) - C(\text{urban periphery}) = \\ \$1{,}000 - (\$1{,}000 + \$300) = -\$300$$

The negative values indicate that both locations have a negative utility, which means that neither location is optimal. However, we can still compare the two locations based on their relative utilities. In this case, living in the urban periphery has a higher utility than living in the city center. This simple example illustrates how the residential location choice model can be used to analyze the trade-off between housing costs and transportation costs when choosing a place to live.

Some assumptions of the residential location choice model are: (1) The model assumes that individuals choose their residential location based on a trade-off between housing costs and transportation costs. (2) The model assumes that individuals have a utility function that depends on the value and cost of living in a certain location. (3) The model assumes that individuals have preferences and constraints that affect their choice set and decision process. (4) The model assumes that individuals are rational and consistent in their choices.

Policymakers can use the residential location choice model to inform their decisions in several ways. For example:

1. *Land Use Planning:* The model can be used to identify areas that are most suitable for residential development based on factors such as proximity to employment centers, access to public transportation, and availability of affordable housing.
2. *Transportation Planning:* The model can be used to evaluate the impact of transportation policies and infrastructure investments on residential location choices. For example, policymakers can use the model to estimate the effect of building a new transit line on housing prices and transportation costs in different areas.
3. *Housing Policy:* The model can be used to evaluate the effectiveness of housing policies such as rent control, affordable housing subsidies, and inclusionary zoning. Policymakers can use

the model to estimate the impact of these policies on housing prices, housing quality, and residential location choices.

4. *Environmental Policy:* The model can be used to evaluate the environmental impact of residential location choices. For example, policymakers can use the model to estimate the effect of promoting denser development patterns on transportation emissions and air quality.

By using the residential location choice model, policymakers can make more informed decisions about land use, transportation, housing, and environmental policies that affect where people live and work.

The residential location choice model can be used to optimize the trade-off between housing costs and transportation costs. By estimating the utility of living in various locations, policymakers, urban planners, and citizens can identify the optimal location that maximizes the value of living in a certain area while minimizing the cost. From the urban planning viewpoint, one way to optimize the trade-off is to promote location-efficient development. This approach involves building compact, mixed-use communities that are designed to be walkable, bikeable, and transit friendly. By reducing the need for driving, location-efficient development can help reduce transportation costs and improve access to amenities. Another way to optimize the trade-off is to promote infill development. This approach involves redeveloping underutilized or vacant land in existing urban areas. By increasing the supply of housing in desirable locations, infill development can help reduce housing costs and improve access to jobs and amenities.

Policymakers can also use various policy tools such as incentives, regulations, and pricing mechanisms to encourage location-efficient and infill development to take place. For example, policymakers can offer tax incentives for developers who build location-efficient developments or require developers to include affordable housing units in their projects. Pricing mechanisms can be used to encourage location efficiency and infill development by creating financial incentives for developers and property owners. Examples of pricing mechanisms that can be used to promote location efficiency are shown in Table 4.8.

TABLE 4.8
Pricing Mechanism to Promote Location Efficiency

Mechanism	Explanation
Impact fees	Impact fees are charges that are imposed on new developments to help pay for the cost of infrastructure improvements such as roads, water and sewer lines, and schools. By charging higher impact fees for developments in suburban or rural areas, policymakers can create a financial incentive for developers to build in more location-efficient areas.
Tax increment financing	Tax increment financing (TIF) is a financing mechanism that allows local governments to use the future tax revenues generated by a development to pay for the cost of infrastructure improvements. By using TIF to fund infrastructure improvements in location-efficient areas, policymakers can create a financial incentive for developers to build in those areas.
Density bonuses	Density bonuses are incentives that are offered to developers who build more densely than what is allowed by zoning regulations. By offering density bonuses for developments in location-efficient areas, policymakers can create a financial incentive for developers to build in those areas.
Parking pricing	Parking pricing is a mechanism that charges users for the use of parking spaces. By charging higher prices for parking in suburban or rural areas, policymakers can create a financial incentive for people to use alternative modes of transportation such as walking, biking, or public transit.
Value capture	Value capture is a mechanism that captures the increase in property values that results from public investments such as transit lines or parks. By using value capture to fund infrastructure improvements in location-efficient areas, policymakers can create a financial incentive for developers to build in those areas.

These pricing mechanisms can help create financial incentives for developers and property owners to build more location-efficient and infill developments. Creating location-efficient areas and infill development enables the residential location dilemma to be detached or minimized. By optimizing the trade-off between housing costs and transportation costs, policymakers can create more sustainable, livable, and equitable communities that meet the needs of all residents.

4.5 URBAN POVERTY AND INFORMAL ECONOMY

Urban poverty is a complex and multidimensional phenomenon that affects millions of people around the world. It is characterized by low income, inadequate housing, poor health, limited access to education and social services, vulnerability to crime and environmental hazards, and marginalization from political and economic opportunities. Urban poverty has various causes, such as rural–urban migration, population growth, unequal distribution of resources, lack of employment, inadequate governance, and social exclusion. Urban poverty also has various effects, such as increased crime and violence, reduced human capital, lower productivity and economic growth, social unrest and instability, and environmental degradation. A reflection of urban poverty is shown in Figure 4.7.

Statistical data can illustrate the magnitude and diversity of urban poverty. According to the World Bank, around 24% of the urban population in developing countries lived below the national poverty line in 2012, and 13% lived below the international poverty line of $1.90 a day. The International Institute for Environment and Development estimates that around a billion urban dwellers live in informal settlements, most of which lack basic services and security of tenure. The US Census Bureau reports that the poverty rate for people in urban areas was 16% in 2015, compared to 17.2% in rural areas. *Lifewater* states that a third of the entire urban population is living in a slum, which are

FIGURE 4.7 Urban poverty.

Source: **Wikipedia.**

unsafe or unhealthy homes in a crowded city. The World Bank also notes that urban poverty varies across regions and countries. For instance, poverty in urban areas fell from 26% to 15% in Ethiopia between 2011 and 2016, while it decreased from 30% to 26% in rural areas.

Poverty brings various negative impacts. One of the main drivers of rural-to-urban migration is poverty in rural areas. Poverty in rural areas is often caused by factors such as low agricultural productivity, lack of access to markets and services, environmental degradation, and limited opportunities for non-farm employment. These factors create a push effect that forces rural dwellers to seek better living conditions elsewhere. On the other hand, urban areas offer a pull effect that attracts rural migrants. Urban areas are perceived as places of economic opportunity, social mobility, and cultural diversity. Urban areas also provide better access to education, healthcare, infrastructure, and public services. However, urban areas also have their challenges, such as overcrowding, pollution, crime, and inequality. Therefore, rural-to-urban migration is not a simple solution to poverty, but a complex process that involves trade-offs and risks for both migrants and their places of origin and destination.

Urban poverty and homelessness are major challenges faced by many cities around the world. Poverty is often concentrated in urban areas due to factors such as a lack of affordable housing, limited job opportunities, and inadequate social services. Homelessness is a visible manifestation of urban poverty and can have serious health and safety implications for individuals and communities. The impacts of urban poverty and homelessness on the urban environment can be significant. Homeless encampments can create unsanitary conditions and contribute to the spread of disease. Poverty can also lead to environmental degradation through activities such as illegal dumping, deforestation, and overfishing.

Despite being complex and multifaceted, the solution to the urban homelessness problem is simple, which is housing. However, housing alone may not be enough for some people who face multiple barriers and challenges, such as disabilities, mental health issues, substance use disorders, criminal justice involvement, or lack of income. Therefore, some people may need longer-term rental assistance and service support to achieve stability.

One of the most effective strategies to cope with urban homelessness is to implement a Housing First-like approach, which prioritizes providing permanent housing to people experiencing homelessness, thus ending their homelessness, and serving as a platform from which they can pursue personal goals and improve their quality of life. Housing First programs do not require people to demonstrate readiness for housing or to comply with treatment or service plans before moving into their own homes. Instead, they offer voluntary and flexible services that are tailored to the needs and preferences of everyone.

Another strategy is to strengthen the rehousing system, which includes rapid re-housing and permanent supportive housing programs. Rapid re-housing helps people who are experiencing homelessness to quickly exit shelters or the streets and move into permanent housing, usually with short-term rental assistance and case management services. Permanent supportive housing provides long-term rental assistance and intensive services for people who have chronic patterns of homelessness and complex needs that require ongoing support. Both programs aim to reduce the length of time people spend homeless and prevent their return to homelessness.

A third strategy is to expand the affordable housing supply, especially for the lowest-income renters who are most at risk of homelessness. The American Rescue Plan Act of 2021 provides an unprecedented $10 billion investment in ending homelessness through the Department of Housing and Urban Development's (HUD) Emergency Housing Voucher (EHV) and HOME Investment Partnerships (HOME-ARP) programs. These resources can be used to create new affordable housing units, preserve existing ones, or provide rental subsidies for people experiencing or at risk of homelessness. Additionally, communities can support innovation in development, such as modular housing, tiny homes, or shared housing models, which can lower costs and increase the speed of producing affordable housing.

To implement these strategies effectively, communities need to set community-specific goals, cultivate political will and partnerships, ensure racial equity in all decision-making, and reduce administrative and regulatory barriers. Moreover, communities need to coordinate federal, state, and local housing resources and leverage support services from various sectors, such as healthcare, behavioral health, education, employment, and criminal justice. By doing so, communities can create a comprehensive and coordinated system that can prevent and end homelessness for everyone.

Sufficient urban infrastructure, particularly affordable housing, plays a critical role in addressing homelessness. Affordable or free-of-charge public transportation and social services such as healthcare and education can help reduce poverty and homelessness. Infrastructure investments such as parks, green spaces, and bike lanes can also help improve the quality of life in urban areas.

Policymakers can use various policy tools such as zoning regulations, tax incentives, and public–private partnerships to promote sustainable and equitable urban development. By investing in infrastructure that supports affordable housing, public transportation, and social services, policymakers can help reduce poverty and homelessness while improving the health and well-being of urban residents.

Homelessness requires a sustainable solution, which may not be possible to be solved merely by providing permanent housing. One of the possible ways to address this issue is to provide the homeless with a source of income that can help them meet their basic needs and improve their quality of life. The informal sector, which includes activities such as street vending, recycling, or domestic work, can offer some opportunities for the homeless to earn money and gain skills, amid many challenges and risks, such as low wages, lack of social protection, exploitation, and harassment. The solution, which involves the informal sector economy, should not generate a vicious circle for the second generation of homeless people. Therefore, any solution that involves the informal sector should also consider the rights and welfare of the homeless workers and provide them with adequate support and resources. Whenever possible, the city authority should gradually convert the engagement of homeless people into more permanent and formal or even professional jobs.

In line with the strategies to solve homelessness, we must also adopt strategies to solve urban poverty. Some of the possible strategies to address this challenge are shown in Table 4.9.

TABLE 4.9
Strategies to Address Homelessness

Strategies	Explanation
Inclusive and sustainable urban development	Promoting inclusive and sustainable urban development, which creates opportunities for the poor to access decent housing, basic services, public spaces, and social protection.
Productivity and competitiveness	Enhancing the productivity and competitiveness of the urban informal sector, which employs a large share of the urban poor, by providing them with skills training, free education, financial inclusion, legal recognition, and social security.
Participation and empowerment	Strengthening the participation and empowerment of the urban poor in local governance and decision-making processes, as well as in the design and implementation of urban policies and programs that affect their lives.
Cohesion and solidarity	Fostering social cohesion and solidarity among the urban poor and other stakeholders, such as civil society organizations, the private sector, academia, and media, to build trust, collaboration, and mutual support.
Job opportunities in rural areas	It creates more job opportunities in rural areas, where rural-to-urban migration takes place. City authorities could offer assistance and collaboration with their surrounding rural counterparts, to create opportunities in the rural areas, and therefore suppress the persistent issue of migration.

Urban poverty and homelessness are a challenge for both developed and developing countries and require coordinated and comprehensive policies and interventions to address their root causes and alleviate their consequences.

4.6 TRAFFIC CONGESTION: THE UNNECESSARY ECONOMIC LOSSES

Traffic congestion is a serious problem that affects the quality of life, productivity, and environment of many urban areas. It occurs when the demand for road space exceeds the available capacity, resulting in delays, frustration, and increased fuel consumption and emissions. Traffic congestion has significant negative impacts on the economy, as it reduces the efficiency of transportation systems, increases the cost of travel and logistics, and lowers the competitiveness of businesses and regions. According to a study by INRIX, a global leader in transportation analytics, traffic congestion cost the US economy $88 billion in 2019, equivalent to $1,377 per driver. The study also estimated that drivers in the most congested cities spent an average of 99 hours/year stuck in traffic. These figures indicate the urgent need for effective solutions to reduce traffic congestion and its associated economic losses.

However, the United States is not the only country that suffers from traffic congestion. According to the TomTom Traffic Index, which ranks 390 cities across 56 countries on six continents by their travel time, fuel costs, and CO_2 emissions, Moscow was the most congested city in the world in 2020, with a congestion level of 54%. This means that drivers in Moscow spent 54% more time in traffic than they would in free-flowing conditions. The second most congested city was Bengaluru in India, with a congestion level of 51%, followed by Manila in the Philippines, with a congestion level of 49%. The top 10 most congested cities also included Bogota, Mumbai, Pune, Istanbul, Kyiv, New Delhi, and Novosibirsk.

Traffic congestion also has a significant environmental impact, as it increases the fuel consumption and emissions of vehicles. According to Numbeo, which provides data on traffic index by country based on travel time, fuel costs, and CO_2 emissions, Nigeria had the highest traffic index in 2020, with a score of 310.6. This means that Nigeria had the worst traffic conditions among the 77 countries surveyed. Nigeria also had the highest CO_2 emission index, with a score of 7,657.8, which indicates the amount of CO_2 emissions caused by one passenger traveling an average distance of 10 km by car. The second-highest traffic index was Sri Lanka, with a score of 269.1, followed by Bangladesh, with a score of 262.3. The top 10 countries with the highest traffic index also included Kenya, Egypt, Peru, Iran, India, Colombia, and Indonesia.

Traffic congestion is not only a nuisance for drivers but also a threat to the environment. When vehicles are stuck in traffic, they consume more fuel and emit more pollutants than when they are moving smoothly. This contributes to the global problem of climate change, as well as to the local problem of air pollution, which can harm the health of people and ecosystems.

According to a study by the International Energy Agency (IEA), road transport accounted for about 24% of global CO_2 emissions from fuel combustion in 2019, making it the second largest source after electricity and heat generation. The study also estimated that road transport emissions increased by 0.5% in 2019, despite improvements in vehicle efficiency and fuel quality. The main drivers of this increase were the growing demand for mobility, especially in developing countries, and the shift to larger and heavier vehicles, such as sports utility vehicles (SUVs).

Traffic congestion also worsens the air quality in urban areas, where most of the population lives and works. According to the World Health Organization (WHO), ambient air pollution caused an estimated 4.2 million premature deaths worldwide in 2016, mainly due to cardiovascular and respiratory diseases. The main sources of ambient air pollution are transport, industry, power generation, and household energy use. Among these sources, transport is the most directly exposed to people, as it occurs near ground level and in densely populated areas.

Congestion on roads can lead to significant economic losses. Two primary factors contribute to these losses: fuel costs and travel time costs. Fuel costs refer to the amount of money spent on fuel due to congestion. When traffic is heavy, vehicles move slowly and consume more fuel than they would on a clear road. This increased fuel consumption leads to higher costs for drivers and ultimately results in a loss for the economy. Travel time costs refer to the amount of time lost due to congestion. When traffic is congested, it takes longer for people to reach their destinations. This wasted time can have a significant impact on the economy, as people are unable to use that time for productive activities such as work or leisure.

To estimate the costs of fuel and travel time in congested traffic, we can use the following steps:

1. Determine your vehicle's fuel efficiency in miles per gallon (MPG) or liters per 100 kilometers (L/100 km).
2. Check the current price of fuel per gallon or liter in the area.
3. Estimate the distance we will be traveling and calculate the total fuel cost by multiplying the distance by your vehicle's fuel efficiency and the current fuel price.

The above steps are valid for running vehicles (not idling). In case the cars stop because of congestion, we need to estimate fuel consumption during idling. Fuel consumption during idling due to congestion can be challenging as it depends on various factors such as the vehicle's make and model, engine size, and idling duration. However, you can use a general rule of thumb that most vehicles consume about 0.38–1.39 liters of fuel per hour of idling. To estimate the fuel consumption during idling, we can follow these steps:

1. Determine the average idling time you expect to encounter during your trip.
2. Multiply the average idling time in hours by the estimated fuel consumption rate per hour (e.g., 0.38–1.39 L/h).
3. Multiply the result by the current fuel price to estimate the cost of fuel consumed during idling.

Keep in mind that this is a rough estimate, and actual fuel consumption may vary based on individual vehicle characteristics and driving conditions.

It is not easy to estimate the cost of travel time in congestion. One of the methods is to use an average hourly wage or an estimated value of our time. Multiply the estimated time spent in congestion by your hourly wage or estimated value to get the total cost of travel time.

Example

We travel from A to B within a city. At the non-congestion state, the travel time is 40 minutes, which consists of 25 minutes running and 15 minutes idling (due to traffic lights and minor congestion). The distance between A and B at the shortest route is 12 km. The car's fuel consumption is 14 L/100 km in running mode and 3.5 L/h in idling mode. In a congested state for the same travel route, the travel time is 70 minutes, and the average hourly wage is USD 50/h. The fuel price is USD 1.1 per liter. What is (1) the cost of congestion? (2) the losses due to congestion?

Solutions:

1. The cost of congestion consists of the cost of fuel and the cost of travel time.

In a non-congested state:

The cost of fuel is 12 km × 14 L/100 km × 1.1 USD/L = USD 1.85

The cost of congestion is 15 minutes × 3.5 L/h × 1 hour/60 min × USD 1.1/L = USD 0.97

Total costs: USD 1.85 + USD 0.97 = USD 2.82

In a congested state:

The cost of fuel is the same: USD 1.85

The cost of congestion is (70 – 25) minutes × 3.5 L/h × 1 hour/60 min × USD 1.1/L = USD 0.97 = USD 2.25

Total costs: $\text{USD } 1.85 + \text{USD } 2.25 = \text{USD } 4.10$

The total cost is the per-car cost of traveling in a congested state. If we would like to know the passengers in the car, this figure should be divided by the number of passengers.

2. The losses due to congestion are $\text{USD } 4.10 - \text{USD } 2.82 = \text{USD } 1.28$ per car.

For the other routes, we can do a similar estimation and estimate the number of cars for some routes by undertaking a traffic survey. Then, we can estimate the total losses due to congestion in a day in the city. This is only an estimate given no information about the number of different cars, the distance of origin-destination, exact travel time, and others.

Note that these estimates may vary based on individual circumstances and factors such as traffic conditions, route choice, and vehicle type.

Reducing traffic congestion is beneficial not only for the economy and the quality of life but also for the environment and public health. Some possible solutions to reduce traffic congestion include improving public transport, promoting active mobility, such as walking and cycling, implementing road pricing and parking policies, encouraging teleworking and flexible schedules, and supporting low-carbon and zero-emission vehicles.

Various solutions can be implemented to reduce traffic congestion and its associated costs. Some of these solutions include:

1. Improving public transportation systems, such as buses, trains, subways, and light rail, to provide more efficient, reliable, and affordable alternatives to driving.
2. Encouraging active transportation modes, such as walking, cycling, and scooting, to reduce the number of cars on the road and promote health and environmental benefits.
3. Implementing smart traffic management systems, such as adaptive traffic signals, real-time traffic information, and dynamic tolling, optimizes the flow of traffic and reduces bottlenecks and collisions.
4. Promoting telecommuting and flexible work schedules, to allow workers to work from home or other locations and avoid peak travel times.
5. Developing compact and mixed-use urban planning, to reduce the need for long-distance commuting and foster more livable and sustainable communities.

Traffic congestion is not an inevitable consequence of urbanization. It is a solvable challenge that requires coordinated actions from governments, businesses, and individuals. By adopting these solutions, we can improve the quality of life for millions of people and enhance the economic competitiveness of our cities.

Individual professionals and citizens can actively contribute to reducing traffic congestion. One of them is telecommuting, also known as remote work or popularly known as work from home during COVID-19. It allows professionals to carry out their jobs from a location outside of the traditional office setting. Telecommuting can offer benefits for both employers and employees, such as flexibility, productivity, morale, and cost savings. Telecommuting can also help employees avoid commuting time and challenges and improve their work–home balance. Telecommuting can also be a great benefit for recruitment and retention in some industries. Some of the top benefits experienced by companies that switch to telecommuting include more dedicated workers, increased productivity, and higher quality of work.

However, telecommuting has several disadvantages as well. Managing a telecommuting staff can be difficult, and professional isolation can have negative effects on well-being and career development. Telecommuting may also lead to a more sedentary lifestyle and physical problems such as back pain and increase the risk of not disconnecting and working more hours than usual. Telecommuting only works when employees are self-disciplined, and it does not always promote extra productivity. Working from home offers many distractions that can take people away from their jobs, and parents may be forced to stop working to address issues with their children. In a telecommuting environment, there is less personal contact with the leadership team, which can lead to issues with communication. Employers must find people who can make decisions without the need for constant feedback from others.

4.7 ENVIRONMENTAL POLLUTIONS

Environmental pollution is a major economic problem because it can cause billions of dollars in damage to the environment and public health. Pollution can stunt economic growth, exacerbate poverty and inequality in both urban and rural areas, and significantly contribute to climate change. Poor people, who cannot afford to protect themselves from the negative impacts of pollution, end up suffering the most. Pollution is the largest environmental cause of disease and premature death. Air pollution, for example, is one of the biggest pollution problems, killing 7 million people each year, according to the World Health Organization. In 2013, those deaths cost the global economy $225 billion in lost labor. Around the world, 9 out of 10 people breathe highly polluted air. In the United States, air pollution caused $131 billion in damages in 2011.

Cities are often the source of significant pollution due to their high population density and industrial activity. Some common types of pollution in urban areas include air pollution, water pollution, noise pollution, and light pollution. Air pollution in urban areas is often caused by emissions from vehicles, factories, and power plants, as well as by the burning of fossil fuels for heating and cooking. This can lead to respiratory problems, heart disease, and other health issues for residents. The other common pollution in cities is water pollution. Water pollution in urban areas can result from industrial discharges, sewage overflows, and runoff from roads and other impervious surfaces. This can contaminate drinking water supplies and harm aquatic ecosystems. The third type of most common pollution in urban areas is noise pollution. This can result from traffic, construction, and other sources. This can cause stress, sleep disturbances, and other health problems for residents. The fourth one, which has never been questioned by citizens is pollution by light.

4.7.1 Water Pollution

Water pollution is the contamination of water bodies, such as lakes, rivers, oceans, and groundwater, by harmful substances. It can have a wide range of causes, including industrial discharges, agricultural runoff, sewage and wastewater treatment plant discharges, and oil spills. Water pollution can have serious effects on the environment and human health. It can harm aquatic ecosystems, reduce biodiversity, and contaminate drinking water supplies. In addition, water pollution can contribute to the spread of waterborne diseases and reduce the availability of clean water for irrigation and other uses.

Water pollution can have significant impacts on the economy. It can reduce the availability of freshwater by reducing its quality for drinking water and domestic use, water for irrigation, industrial processes, and other uses, which can affect productivity and economic growth. In addition, water pollution can harm fisheries and other aquatic industries, leading to job losses and reduced income. The costs of treating water to make it safe for drinking and other uses can also be significant. Furthermore, water pollution can have negative effects on tourism and recreation, as people may be less likely to visit areas with polluted water. Overall, the economic costs of water pollution can be substantial, and efforts to prevent and reduce water pollution can provide significant economic benefits.

There are several ways to address water pollution. One approach is to reduce the sources of pollution through regulations and incentives that encourage industries and individuals to reduce their emissions. Another approach is to improve wastewater treatment and sewage systems to prevent pollutants from entering water bodies. In addition, efforts can be made to restore damaged ecosystems and protect sensitive areas from further pollution. Education and awareness campaigns can also help to reduce water pollution by encouraging individuals to take actions such as properly disposing of hazardous waste and reducing their use of harmful chemicals.

One approach to estimating economic losses is to calculate the direct costs of water pollution. These costs may include expenses related to healthcare, property damage, and cleanup efforts. For example, if a factory discharges toxic chemicals into a river, the costs of treating illnesses caused by exposure to those chemicals, repairing damaged property, and cleaning up the river may be calculated.

Another approach is to estimate the indirect costs of water pollution. These costs may include lost income and productivity due to reduced agricultural yields, decreased tourism revenue, and business closures. For example, if a beach is closed due to high levels of pollution, the local businesses that rely on tourism may suffer significant losses.

To estimate economic losses accurately, it is essential to consider both direct and indirect costs and quantify them as accurately as possible. This may involve conducting surveys or analyzing existing data to determine the economic impact of water pollution on various industries and communities.

Water pollution can have some negative effects on human health. Contaminated water can contain harmful bacteria, viruses, and parasites that can cause a range of illnesses, including diarrhea, cholera, and typhoid fever. Chemical pollutants in water can also be harmful to human health. For example, exposure to high levels of nitrates in drinking water can cause methemoglobinemia, or "blue baby syndrome," which reduces the ability of blood to carry oxygen. In addition, exposure to heavy metals such as lead and mercury in drinking water can cause neurological damage and other health problems. Governments and communities need to take steps to prevent water pollution and ensure that everyone has access to clean and safe drinking water.

There are several ways to reduce water pollution. The methods are shown in Table 4.10.

TABLE 4.10
Methods of Reducing Water Pollution

Method	Explanation
Regulating industrial discharges	Governments can implement regulations to limit the amount of pollutants that industries are allowed to discharge into water bodies. This can help to reduce the amount of harmful substances that enter the water.
Improving wastewater treatment	Upgrading wastewater treatment plants and sewage systems can help to prevent pollutants from entering water bodies. This can include implementing advanced treatment technologies and increasing the capacity of treatment plants.
Reducing agricultural runoff	Agricultural runoff, which can contain fertilizers, pesticides, and animal waste, is a major source of water pollution. Measures such as implementing best management practices, reducing fertilizer use, and properly managing animal waste can help to reduce agricultural runoff.
Protecting sensitive areas	Sensitive areas such as wetlands and riparian zones can help to filter pollutants and protect water quality. Efforts to protect these areas from development and pollution can help to reduce water pollution.
Promoting individual actions	Individuals can also take actions to reduce water pollution, such as properly disposing of hazardous waste, reducing their use of harmful chemicals, and conserving water.

4.7.2 Air Pollution

Air pollution can have significant impacts on economic growth. It can reduce productivity by causing health problems and premature deaths, which can lead to increased healthcare costs and reduced labor force participation. In addition, air pollution can harm industries such as agriculture, fisheries, and tourism, leading to job losses and reduced income. The costs of mitigating air pollution, such as implementing emissions controls and transitioning to cleaner energy sources, can also be significant. However, these costs are often outweighed by the economic benefits of improved public health and environmental quality. Overall, efforts to reduce air pollution can provide significant economic benefits by improving public health, protecting natural resources, and promoting sustainable economic growth.

Reducing air pollution can provide significant economic benefits. Some potential benefits of the reduction of air pollution are shown in Table 4.11.

There are several ways to reduce air pollution in urban areas. Some of conventional methods to reduce air pollution in urban areas are reflected in Table 4.12.

In cities, public transport is a prominent strategy to cope with urban air pollution issues. Public transportation provides many benefits to urban areas. It can help reduce congestion by decreasing

TABLE 4.11
Benefits of Air Pollution Reduction

Benefit	Explanation
Improved public health	Reducing air pollution can lead to improved public health by reducing the incidence of respiratory and cardiovascular diseases, which can result in lower healthcare costs and increased productivity.
Increased labor force participation	By improving public health, reducing air pollution can also increase labor force participation, leading to higher economic output and growth.
Enhanced agricultural productivity	Air pollution can harm crops and reduce agricultural yields. Reducing air pollution can therefore improve agricultural productivity and increase food security.
Boosted tourism	Air pollution can deter tourists from visiting certain areas, leading to reduced income from tourism. Reducing air pollution can therefore boost tourism and support local economies.
Reduced environmental damage	Air pollution can harm natural resources such as forests, water bodies, and wildlife. Reducing air pollution can therefore reduce environmental damage and support sustainable economic growth.

TABLE 4.12
Methods to Reduce Air Pollution in Urban Areas

Method	Explanation
Promoting public transportation	Encouraging the use of public transportation, such as buses, trains, and subways, can help reduce the number of cars on the road and decrease emissions from vehicles.
Encouraging active transportation	Promoting walking, cycling, and other forms of active transportation can also help to reduce air pollution by decreasing the number of cars on the road.
Implementing emissions controls	Governments can implement regulations to limit the amount of pollutants that industries and vehicles are allowed to emit. This can help to reduce air pollution from these sources.
Promoting energy efficiency	Encouraging the use of energy-efficient technologies, such as LED lighting and efficient appliances, can help to reduce emissions from power generation.
Transitioning to cleaner energy sources	Switching to cleaner energy sources, such as wind and solar power, can help to reduce emissions from power generation.

the number of cars on the road, leading to faster travel times. Public transportation can also improve air quality by reducing air pollution. It increases accessibility by providing access to jobs, education, healthcare, and other services for people who do not have access to a car. Public transportation is generally more energy-efficient than driving and can help reduce greenhouse gas emissions and combat climate change. Additionally, public transportation can support local economies by providing access to jobs and services and reducing the costs of congestion and air pollution.

Air pollution can also have significant costs on the economy. It can reduce productivity by causing health problems and premature deaths, which can lead to increased healthcare costs and reduced labor force participation. In addition, air pollution can harm industries such as agriculture, fisheries, and tourism, leading to job losses and reduced income. The costs of mitigating air pollution, such as implementing emissions controls and transitioning to cleaner energy sources, can also be significant. However, these costs are often outweighed by the economic benefits of improved public health and environmental quality.

4.7.3 Noise Pollution

Noise pollution is the excessive or harmful level of noise in the environment. It can be caused by a variety of sources, including transportation, construction, industrial activity, and household appliances. Noise pollution can have negative effects on human health, including hearing loss, stress, and sleep disturbances. It can also harm wildlife by interfering with communication and behavior. Efforts to reduce noise pollution include implementing noise barriers, enforcing noise regulations, and promoting the use of quieter technologies.

Table 4.13 shows the noise levels in decibels (dB) and their corresponding impacts.

The noise is measured in dB which is a relative unit of measurement equal to one-tenth of a bel (B). It expresses the ratio of two values of a power or root-power quantity on a logarithmic scale. Two signals whose levels differ by one decibel have a power ratio of $10^{(1/10)}$ (approximately 1.26) or root-power ratio of $10^{(1/20)}$ (approximately 1.12). Decibels are commonly used in various fields, including acoustics, electronics, and control theory. In acoustics, it is used to measure sound pressure levels and express the relative loudness of sounds. In electronics, it is used to describe the gains of amplifiers, attenuation of signals, and signal-to-noise ratios. The decibel is also used to quantify the intensity of sounds and their potential impact on human hearing.

Noise is any unwanted sound that can cause adverse effects on human and animal health, such as hearing loss, stress, hypertension, etc. According to the NIOSH Recommended Exposure Limit (REL), the maximum permissible exposure time for noise is 8 hours/day at 85 dBA. For every 3 dBA increase in noise level, the exposure time must be halved to prevent hearing damage. This is called the exchange rate.

TABLE 4.13
Noise Level and Every Day's Sounds

Noise Level (dB)	Everyday Sounds	Situation
0–30	Whispering, falling leaves	Quiet
30–60	Normal conversation	Normal
60–70	Busy traffic	Loud
70–90	Vacuum cleaner	Very loud
>90	Rock concert	Extremely loud
Appr. 110	Gunshot	The painful, immediate risk of hearing damage
Appr. 140	Jet engine at take-off	Painful and damaging

TABLE 4.14
Methods to Reduce Noise Pollution in Urban Areas

Method	Explanation
Installing noise barriers	Noise barriers, such as walls and berms, can be used to block or absorb sound and reduce noise levels in urban areas.
Enforcing noise regulations	Governments can implement and enforce regulations to limit the amount of noise that is allowed in urban areas. This can include setting maximum noise levels for different times of day and different types of activities.
Promoting the use of quieter	Encouraging the use of quieter technologies, such as electric vehicles and low-noise construction equipment, can help to reduce noise pollution in urban areas.
Designing urban spaces to reduce noise	Urban planners can design cities to reduce noise pollution by incorporating features such as green spaces, water features, and sound-absorbing materials into the urban landscape.

TABLE 4.15
Methods to Promote the Use of Quieter Technologies

Method	Explanation
Providing incentives	Governments and other organizations can provide financial incentives, such as tax breaks or grants, to encourage construction companies to invest in quieter technologies.
Implementing regulations	Governments can implement regulations that require construction companies to use quieter technologies or limit the amount of noise that construction activities are allowed to produce.
Raising awareness	Raising awareness about the benefits of quieter technologies, such as reduced noise pollution and improved public health, can help to encourage their adoption by construction companies.
Supporting research and development	Supporting research and development into quieter technologies can help to improve their performance and reduce their cost, making them more attractive to construction companies.

For example, if a worker is exposed to 88 dBA of noise, the maximum exposure time is 4 hours/day. If the noise level is 91 dBA, the exposure time is reduced to 2 hours/day, and so on. Therefore, it is important to protect your ears from excessive noise by using appropriate hearing protection devices, such as earplugs or earmuffs, and avoiding prolonged exposure to loud noises.

There are several methods to reduce noise pollution in urban areas as shown in Table 4.14.

The use of quieter technology in transportation has been in place for many years, for example, electric trains and cars. There are several methods to promote the use of quieter technologies in the construction industry, as shown in Table 4.15.

Noise barrier technology is one of the methods to reduce noise pollution from various sources, such as road, rail, and air traffic, construction, and industrial activities. Noise barriers are structures that block or deflect the sound waves from reaching the receivers, such as people, animals, or buildings. Noise barriers can be made of different materials, such as concrete, metal, wood, or plastic, and can have different shapes and sizes, depending on the source and type of noise. Noise barriers can also be combined with sound absorption materials, such as mineral wool, foam, or rubber, to increase their effectiveness. Noise barriers can reduce the noise level by up to 10 dB, depending on the frequency and direction of the sound. However, noise barriers also have some limitations, such as cost, maintenance, aesthetic impact, and environmental effects. Therefore, noise barrier technology should be used in conjunction with other noise reduction measures, such as noise-reducing road surfaces, lower speed limits, quieter vehicles and machinery, and noise regulations and standards.

Building a noise barrier is one way to reduce noise pollution from sources outside your home, such as traffic, construction, or neighbors. A noise barrier is a structure that blocks or deflects the sound waves from reaching your ears, creating a quieter environment. To build a noise barrier, you will need some materials, tools, and steps. Here are some general guidelines:

1. *Materials:* We can use different materials such as wood, metal, concrete, or plastic. The material should be dense and heavy enough to block the sound. You can also add some sound absorption materials, such as mass-loaded vinyl, foam, or vegetation, to enhance the noise reduction. Mass-loaded vinyl is a special material that can dampen the sound vibrations and make the barrier more effective.
2. *Tools:* We will need some tools to measure, cut, drill, nail, or screw the materials together. Some common tools are tape measure, a saw, a drill, a hammer, nails, screws, and brackets. You may also need some safety equipment, such as gloves, goggles, and ear protection.
3. Add some vegetation or other decorative elements to improve the appearance of the barrier.

A noise barrier can help us create a more peaceful and comfortable home by blocking out unwanted sounds from outside. However, we should also check with local authorities before building a noise barrier to make sure we comply with any regulations or restrictions.

4.7.4 Pollution by Lights

Light pollution (Table 4.16) is the excessive or misdirected artificial light produced by human activity that can have negative effects on the environment, wildlife, and human health. It is a growing problem in urban areas, where artificial lighting is used extensively for streetlights, advertising, and buildings. Light pollution can disrupt ecosystems by interfering with the natural patterns of light and dark, affecting the behavior of nocturnal animals and plants. It can also have negative effects on human health by disrupting circadian rhythms and sleep patterns. In addition, light pollution can obscure the night sky, making it difficult to see stars and other celestial objects. Efforts to reduce light pollution include using more efficient lighting technologies, shielding outdoor lights to direct light downward, and reducing unnecessary lighting.

Light pollution can have several negative effects on human health. One of the main concerns is the disruption of our circadian rhythms, which are the natural cycles of sleep and wakefulness that our bodies follow. Exposure to artificial light at night can interfere with the production of melatonin, a hormone that regulates sleep, and can lead to sleep disorders, such as insomnia. In addition, studies

TABLE 4.16
Light Pollution Level and Conditions

Light Pollution Level	Conditions
Class 1: Excellent night sky visibility	Minimal light pollution, excellent visibility of celestial objects
Class 2: Typical dark sky site	Minor light pollution, good visibility of celestial objects
Class 3: Rural sky	Moderate light pollution, fair visibility of celestial objects
Class 4: Suburban transition	Significant light pollution, poor visibility of celestial objects
Class 5: Suburban sky	Severe light pollution, very poor visibility of celestial objects
Class 6: Bright suburban sky	Extreme light pollution, no visible stars
Class 7: Suburban/urban transition	Severe light pollution, sky glow visible over the entire sky
Class 8: City sky	Extreme light pollution, few stars visible
Class 9: Inner-city sky	Severe light pollution, no stars visible

Note: The impacts of light pollution can vary depending on the location and context. However, in general, light pollution can have negative effects on human health, wildlife, and the environment. It can disrupt natural circadian rhythms, alter animal behavior and migration patterns, and waste energy and resources.

have shown that exposure to light at night can increase the risk of certain health conditions, such as obesity, diabetes, and even some types of cancer. Light pollution can also cause glare and discomfort, leading to eye strain and headaches. Urban planners and policymakers need to consider the effects of light pollution on human health when designing lighting systems for cities.

There are various ways to measure light pollution, but one common method is by using a device called a Sky Quality Meter (SQM). An SQM measures the brightness of the night sky in units of magnitude per square arcsecond. This measurement can then be used to determine the level of light pollution in a particular area. Another method is by using satellite imagery to analyze the amount of artificial light emitted by a region. One commonly used measure is the Bortle Scale, which categorizes the level of light pollution in a particular area based on the visibility of celestial objects. The scale ranges from Class 1, which represents areas with very little light pollution and excellent night sky visibility, to Class 9, which represents areas with severe light pollution and no visible stars.

Addressing water pollution, air pollution, noise pollution, and light pollution requires collective efforts from individuals, communities, governments, and industries. By raising awareness, implementing sustainable practices, and supporting environmental initiatives, we can work toward a cleaner, healthier, and more sustainable future for our planet.

To tackle the challenges of water, air, noise, and light pollution, we need to join forces across different levels of society. We can educate ourselves and others about the causes and effects of these environmental issues, adopt eco-friendly habits and technologies, and participate in green campaigns and movements. This way, we can contribute to the protection and restoration of our natural resources and ecosystems and ensure a better quality of life for ourselves and future generations.

4.8 SUSTAINABLE ECONOMIC GROWTH

Sustainable economic growth is economic development that attempts to satisfy the needs of humans but in a manner that sustains natural resources and the environment for future generations. It is the practice of supporting long-term economic growth without depleting resources. Sustainable economic growth implies that the growth rate can be maintained over the long term without creating significant economic problems, especially for future generations. It involves both environmentally sustainable growth (not exploiting scarce resources) and sustainable growth in terms of low inflation and a balanced economy. The ecosystem provides the factors of production that fuel economic growth: land, natural resources, labor, and capital (which are created by labor and natural resources). Sustainable economic growth is managing these resources in a manner that they will not be depleted and will remain available for future generations.

While many economists and people disagree about the importance of the environment regarding economic activity, the following facts are seldom disputed:

1. The extraction and depletion of natural resources, as well as pollution and permanent changes made to the landscape, are caused by economic activities and can harm the environment.
2. Many of the costs of the harm created by economic activities are not borne by those who cause it but by other people who neither obtain the benefits from the economic activity nor agree to pay the costs related to it.
3. Humans live in an ecosystem and cannot survive without it. If we destroy the environment, we will eventually destroy ourselves.

Economic growth of a city is often seen as a desirable goal, but it can also pose significant challenges for sustainability. Sustainability refers to the ability of a system to meet the needs of the present without compromising the ability of future generations to meet their own needs. A city that pursues economic growth without considering its environmental and social impacts may face problems such as pollution, congestion, inequality, poverty, and depletion of natural resources. Therefore, it is important for a city to balance its economic objectives with its sustainability goals, and to adopt policies and practices that promote both.

Urban economic growth must be sustainable for several reasons:

1. *Population Growth:* By 2030, 6 in 10 people will live in urban areas. This rapid urbanization poses significant challenges and opportunities for sustainable development and economic activities must be able to promote this sustainability.
2. *Economic Impact:* Cities are drivers of economic growth and contribute more than 80% of global GDP. Sustainable urban development can generate inclusive prosperity.
3. *Environmental Considerations:* Cities account for 60–80% of energy consumption and 75% of carbon emissions. Sustainable urban development can help mitigate these environmental impacts.
4. *Inequality and Poverty:* If urban growth is not sustainable and inclusive, we risk exacerbating global challenges like inequality and poverty. Sustainable economic growth can lead to new and better employment opportunities and provide greater economic security for all.
5. *Infrastructure Needs:* Rapid urbanization increases the demand for affordable housing, viable infrastructure, basic services, and jobs. Sustainable urban development ensures these needs are met without compromising future generations.
6. *Health Concerns:* Urban areas often face significant health challenges, such as air pollution and disease outbreaks. Sustainable urban development prioritizes public health and creates resilient cities.

A city can achieve sustainable economic growth by aligning its economic objectives with its environmental and social goals. This means that the city should pursue economic activities that generate value and income, while also minimizing negative impacts on the natural environment and the urban society. To do this, the city should implement policies and practices that support the following principles:

1. Green and circular economy, which aims to reduce waste, reuse materials, and minimize environmental damage.
2. Public transportation, renewable energy, and energy efficiency can lower greenhouse gas emissions and improve air quality.
3. Social inclusion, participation, and empowerment can enhance the well-being and resilience of the urban population.
4. Innovation, creativity, and diversity can stimulate new ideas and solutions for sustainable development.

A green and circular economy is a model of production and consumption that aims to reduce waste, reuse materials, and minimize environmental damage. It is based on the principles of designing out waste and pollution, keeping products and materials in use, and regenerating natural systems. A green and circular economy can help address some of the major challenges facing humanity, such as climate change, biodiversity loss, resource depletion, and social inequality. By decoupling economic growth from environmental impact, a green and circular economy can foster innovation, competitiveness, and well-being for people and the planet.

One of the ways to achieve a more sustainable and resource-efficient economy is to adopt the principles of the circular economy, which aims to reduce waste, reuse materials, and regenerate natural systems. Here are some examples of green and circular economy initiatives from different sectors and regions:

1. *Industrial Symbiosis:* a model for growth in the circular economy that involves sharing resources, such as energy, water, materials, and by-products, among different industries to create mutual benefits and reduce environmental impacts. For example, in Kalundborg, Denmark, a network of companies exchanges various streams of waste, such as steam, sludge, fly ash, and gypsum, to create value-added products and services.
2. *The Danish Deposit and Return System:* a system for recycling drink cans and bottles that incentivizes consumers to return their used containers to reverse vending machines or retailers in exchange for a small refund. The system has achieved a high return rate of over 90% and has reduced littering, greenhouse gas emissions, and resource consumption.
3. *Denmark's First Circular Social Housing Project:* a project that involves renovating an old building in Copenhagen with recycled materials and energy-efficient solutions. The project aims to demonstrate how circular principles can be applied to the construction sector, which is responsible for a large share of waste generation and resource use.
4. *Recycling of Artificial Turf:* a solution that enables the recovery and reuse of valuable materials from old synthetic grass fields that would otherwise end up in landfills. The process involves separating the different components of the artificial turf, such as rubber granules, sand, and plastic fibers, and turning them into new products or raw materials.
5. *Closed Loop in Reuse Packaging-as-Service:* a service that provides reusable packaging solutions for e-commerce and food delivery businesses. The service replaces single-use packaging with durable containers that can be returned, cleaned, and reused multiple times. This reduces waste generation, carbon footprint, and packaging costs.
6. *Potentials in Potato Starch Production:* a project that explores how potato starch, a by-product of potato processing, can be used as a biodegradable material for various applications, such as packaging, textiles, cosmetics, and pharmaceuticals. The project also aims to optimize the production process of potato starch to reduce water and energy consumption.
7. *Re-Using Old Bricks to Build a Greener Future:* a company that collects and cleans used bricks from demolition sites and sells them to new construction projects. The company uses patented technology that removes mortar and cement from bricks without damaging them. This preserves the quality and aesthetic value of the bricks and saves natural resources.
8. *Circular Economy Examples* – how IKEA, Burger King, Adidas, and more are investing in a circular economy: Some of the well-known brands that are implementing circular strategies in their products and services. For example, IKEA has launched a buy-back scheme that allows customers to sell their old furniture back to the store for store credit or donation. Burger King has introduced reusable packaging for its food items that can be returned by customers for cleaning and reuse. Adidas has partnered with Parley for the Oceans to create shoes and clothing from recycled plastic waste collected from the oceans.
9. *Korean Nuclear Fusion Reactor Achieves 100 Million Degrees for 30 Seconds:* An example of a breakthrough in nuclear fusion technology that could potentially provide a clean and abundant source of energy for the future. Nuclear fusion is the process of fusing light atoms to release energy, similar to what happens in the Sun. However, achieving nuclear fusion on Earth requires extremely high temperatures and pressures that are difficult to sustain. The Korean experiment has demonstrated that it is possible to maintain a stable fusion reaction at temperatures higher than the core of the Sun for a significant duration.
10. *Thousand Fell:* an example of a circular fashion brand that produces shoes from sustainable materials and offers incentives for customers to recycle their old shoes. The shoes are made from materials such as coconut husk, sugar cane, recycled plastic bottles, and natural rubber. The customers can send their worn-out shoes back to the company for free and receive a $20 voucher for their next purchase. The returned shoes are then recycled into new materials or donated to non-profit organizations.

Public transportation, renewable energy, and energy efficiency are three key strategies that can help reduce greenhouse gas emissions and improve air quality. Public transportation reduces the number of vehicles on the road, which lowers the amount of tailpipe emissions and congestion. Renewable energy sources, such as solar, wind, and hydro, do not emit greenhouse gases or other pollutants when they generate electricity. Energy efficiency measures, such as LED lighting, smart thermostats, and insulation, can reduce the amount of energy needed to power homes and businesses, which reduces the demand for fossil fuels and their associated emissions. By implementing these strategies, we can protect the environment and our health, while also saving money and creating jobs.

Social inclusion, participation, and empowerment are crucial factors that contribute to the overall well-being and resilience of the urban population. In today's rapidly evolving society, it is essential to ensure that every individual feels included and valued, regardless of their background or circumstances. Social inclusion refers to the process of creating a society where everyone has equal opportunities to participate and contribute. It involves breaking down barriers and promoting diversity, ensuring that no one is left behind. By fostering social inclusion, cities can create a sense of belonging and unity among their residents.

Participation is another vital aspect of urban life. It involves engaging individuals in decision-making processes that affect their lives and the communities they belong to. When people have a voice and are actively involved in shaping their environment, they feel a sense of ownership and responsibility, leading to increased satisfaction and well-being. Empowerment goes hand in hand with social inclusion and participation. It involves providing individuals with the tools, resources, and opportunities they need to take control of their lives and make positive changes. Empowered individuals are more likely to overcome challenges, adapt to new situations, and contribute to the development of their communities.

The benefits of social inclusion, participation, and empowerment extend beyond individual well-being. They also contribute to the overall resilience of urban populations. Resilience refers to the ability to withstand and recover from shocks and stresses, such as natural disasters or economic downturns. When communities are socially inclusive and empowered, they are better equipped to respond to challenges collectively, supporting each other and finding innovative solutions. Inclusive cities are more resilient because they tap into the diverse skills, knowledge, and experiences of their residents. By embracing diversity and promoting social cohesion, cities can harness the collective intelligence and creativity of their populations. This leads to more effective problem-solving, increased innovation, and improved overall resilience.

Furthermore, social inclusion, participation, and empowerment have a positive impact on various aspects of urban life. They contribute to improved health outcomes by reducing inequalities in access to healthcare and promoting healthy lifestyles. They enhance educational opportunities by ensuring equal access to quality education for all. They also foster economic development by enabling individuals to participate in the labor market and contribute to the local economy.

To achieve social inclusion, participation, and empowerment in urban areas, governments, policymakers, community organizations, and individuals need to work together. This requires creating inclusive policies and practices, providing access to resources and opportunities, and promoting dialogue and collaboration among different stakeholders.

Sustainable development requires the utilization of the three pillars of innovation, creativity, and diversity. Innovation facilitates the generation of new ideas and approaches that can respond to the evolving challenges and opportunities in the global context. Creativity enables the application of unconventional perspectives and problem-solving techniques that can create value and impact. Diversity enhances the diversity of perspectives and experiences by incorporating people from different backgrounds and contexts, resulting in a more cooperative and inclusive development process. These three pillars are interdependent and synergistic, creating a virtuous cycle that improves the ability to find sustainable solutions. By capitalizing on the potential of innovation, creativity, and diversity, a more sustainable future can be achieved for the current and future generations.

Economic growth in the city cannot ignore the urban informal economy. In some developing countries, the urban informal economy is far more important than the formal economy for its free entry characteristics and thus becomes a lifesaver for most low-income citizens. The urban informal economy is a term that refers to the economic activities and workers that operate outside the formal regulations and institutions of the state. It includes street vendors, waste pickers, domestic workers, home-based workers, and many others who provide essential goods and services to urban residents, especially the poor. The urban informal economy is often seen as a survival strategy for the marginalized and excluded segments of society, who face barriers to accessing formal employment, education, healthcare, and social protection. However, the urban informal economy is also a source of dynamism, innovation, and resilience for urban development, as it contributes to the livelihoods of millions of people, generates income and taxes for the city, and reduces waste and environmental impacts.

The sustainability of the urban informal economy is a complex and contested issue, as it depends on various factors such as the legal and policy frameworks, the social and political context, the environmental conditions, and the preferences and aspirations of the informal workers themselves. Some argue that the urban informal economy is inherently unsustainable, as it perpetuates poverty, inequality, exploitation, and informality, and undermines the rule of law, the quality of public services, and urban planning and governance. Others contend that the urban informal economy is a vital and viable part of urban development, as it provides opportunities for economic inclusion, social empowerment, and environmental stewardship, and responds to the needs and demands of the urban population, especially in times of crisis and uncertainty.

There is no definitive answer to whether the urban informal economy is sustainable or not, as it depends on how one defines and measures sustainability, and what are the goals and values that guide urban development. However, there is a growing consensus that the urban informal economy cannot be ignored or eradicated, as it is an integral part of urban reality and potential. Therefore, there is a need for more inclusive and participatory approaches to recognize, protect, and support the urban informal economy and its workers, to enhance their rights, capacities, and contributions to sustainable urban development.

Supporting the urban informal economy is a key challenge and opportunity for sustainable urban development, as it can enhance the economic, social, and environmental well-being of the city and its inhabitants. There are various methods to support the urban informal economy as shown in Table 4.17.

We can create more inclusive and participatory cities that harness the potential of the urban informal economy and its workers for sustainable urban development. By integrating economic

TABLE 4.17
Methods to Support the Urban Informal Economy

Method	Explanation
Recognition	Recognizing the urban informal economy as a legitimate and valuable part of urban development and respecting the rights and dignity of the informal workers.
Protection	Protecting the urban informal economy from harassment, eviction, confiscation, and violence, and ensuring their access to justice and legal remedies.
Support of infrastructure	Supporting the urban informal economy with adequate infrastructure, services, and facilities, such as water, sanitation, electricity, waste management, transport, healthcare, and education.
Access to formal markets	Enabling the urban informal economy to access formal markets, finance, technology, and information, and facilitating their integration into value chains and supply networks.
Empowerment	Empowering the urban informal economy to organize themselves into associations, cooperatives, unions, or networks, and involving them in decision-making and policymaking processes that affect their lives and livelihoods.

growth and sustainability, a city can create a more livable, prosperous, and equitable environment for its current and future inhabitants.

PROBLEMS AND ISSUES FOR DISCUSSION

1. Define urban economy with your own words.
2. What are the policy interventions and strategies that can be implemented to integrate the informal sector with the formal urban economy?
3. The informal sector plays a crucial role in the urban economy, providing employment opportunities and contributing to economic growth. Discuss the significance of this sector, its challenges, and its impact on the overall urban economic landscape.
4. Analyze the challenges faced by informal entrepreneurs in accessing formal financial services and potential solutions.
5. Explore the intricate relationship between urban planning policies and the growth of the informal economy. Analyze how factors like zoning, infrastructure, and land use regulations impact the development and integration of informal economic activities in urban areas.
6. Evaluate the effectiveness of government policies and regulations in addressing issues in the urban informal sector.
7. Compare the contribution of different sectors (e.g., manufacturing, services, trade) to the overall urban economy.
8. Explore the potential of entrepreneurship in driving innovation and economic development in urban areas.
9. Globalization and technological innovations are transforming the urban economic landscape. Discuss how these factors are reshaping the structure, composition, and dynamics of the urban economy, including the emergence of new industries, shifting employment patterns, and changing consumer behaviors.
10. Examine how public–private partnerships can be leveraged to address the pressing challenges of urban infrastructure development and service delivery, fostering more efficient and effective solutions.
11. Assess the effectiveness of resource allocation strategies in addressing the needs of different socioeconomic groups in urban areas.
12. Examine the influence of land use policies and zoning regulations on the distribution of economic activities in cities.
13. The intricate relationship between urban housing policies and the growth of the informal economy has been recognized for a long time. Analyze how factors like affordable housing, gentrification, and residential zoning impact the ability of informal entrepreneurs and workers to access suitable spaces and opportunities within the urban economic landscape.
14. Assess the critical role of the informal sector in providing affordable and accessible goods and services to urban residents, particularly those with limited financial resources. Discuss the impact of this sector on the overall well-being and quality of life in cities.
15. Urban agriculture is a potential source to enhance food security and sustainability in cities. Analyze how the integration of agricultural activities within urban areas can contribute to local food production, reduce reliance on external food sources, and promote environmental sustainability.
16. Urban transportation systems, including public transit, roads, and infrastructure, impact the accessibility and mobility of are economic sectors within the city that play very crucial roles in urban economy. Evaluate how efficient and equitable transportation networks can facilitate the integration and growth of different industries and economic activities in urban areas.
17. Explore the challenges faced by small- and medium-sized enterprises (SMEs) in navigating the urban economic landscape.

18. Evaluate the effectiveness of skills development programs in supporting the growth and formalization of the urban informal sector.
19. Analyze how training, capacity building, and access to resources can empower informal workers and entrepreneurs to enhance their productivity, competitiveness, and integration into the formal economy.
20. Discuss the role of digital technologies in enabling the formalization and integration of the informal economy.
21. Evaluate the impact of urban renewal and gentrification on the displacement of informal economic activities.
22. Analyze the potential of urban tourism in generating employment and economic opportunities in cities.
23. Investigate the relationship between urban environmental policies and the development of sustainable economic activities.
24. Discuss the role of social enterprises in addressing the needs of marginalized communities in urban areas.
25. Assess the impact of urban infrastructure investments on the productivity and competitiveness of different economic sectors.
26. Explore the potential of circular economy models in promoting sustainable resource utilization in urban areas.
27. Investigate how urban governance structures, such as local policies, regulatory frameworks, and decision-making processes, influence the distribution of economic resources, investments, and opportunities within a city.
28. Discuss the potential impact of inclusive and transparent governance on promoting equitable economic development and addressing disparities in urban areas.
29. Discuss the challenges faced by urban residents in accessing affordable and quality healthcare services.
30. Evaluate the role of urban economic development strategies in promoting inclusive and equitable growth.
31. Investigate the potential of urban agriculture in generating income and employment for urban residents.
32. Explore how urban mobility solutions, such as public transportation, ridesharing, and pedestrian-friendly infrastructure, can impact the accessibility and integration of different economic activities within a city.
33. Assess how efficient and inclusive mobility systems can facilitate the connection and collaboration among various sectors, promoting economic growth and development.
34. Discuss the role of urban cultural and creative industries in driving economic diversification and innovation.
35. Assess the potential of urban–rural linkages in enhancing the resilience and sustainability of the urban economy.

MULTIPLE-CHOICE QUESTIONS

1. What is the primary role of the informal sector in the urban economy?
   a. Providing employment opportunities
   b. Generating tax revenue
   c. Promoting environmental sustainability
   d. Improving urban infrastructure

2. Which of the following factors contributes to the growth of the informal economy in urban areas?
   a. Strict regulatory environment
   b. Lack of access to formal financial services
   c. High levels of urbanization
   d. All the above
3. Which sector plays a significant role in the urban economy?
   a. Agriculture
   b. Manufacturing
   c. Services
   d. Tourism
4. Which of the following is a challenge faced by informal entrepreneurs in urban areas?
   a. Lack of access to education
   b. Difficulty in obtaining business licenses
   c. Shortage of skilled labor
   d. All the above
5. Which of the following is a strategy to address issues in the urban informal sector?
   a. Deregulation
   b. Strengthening social protection
   c. Increasing taxation
   d. Promoting informal activities
6. How can entrepreneurship contribute to the development of the urban economy?
   a. Promoting innovation
   b. Creating job opportunities
   c. Diversifying economic activities
   d. All of the above
7. Which factor has the most significant impact on the structure and dynamics of the urban economy?
   a. Technological advancements
   b. Globalization
   c. Political stability
   d. Availability of natural resources
8. What is the primary purpose of public–private partnerships in urban areas?
   a. Reducing government expenditure
   b. Addressing infrastructure and service delivery challenges
   c. Promoting privatization of public assets
   d. Enhancing environmental sustainability
9. Which factor is most influential in the distribution of economic resources in urban areas?
   a. Land use policies
   b. Income inequality
   c. Availability of skilled labor
   d. Access to transportation
10. How can urban agriculture contribute to the urban economy?
   a. Enhancing food security
   b. Promoting sustainable resource utilization
   c. Generating employment opportunities
   d. All the above

11. Which of the following is a challenge faced by SMEs in the urban economy?
    a. Access to funding
    b. Regulatory compliance
    c. Competition from larger firms
    d. All the above
12. What is the primary benefit of digital technologies for the informal economy in urban areas?
    a. Increased tax revenue
    b. Improved worker protection
    c. Formalization and integration
    d. Enhanced environmental sustainability
13. How can urban renewal and gentrification impact the informal economy?
    a. Promote the growth of informal activities
    b. Displace informal economic activities
    c. Increase the accessibility of informal services
    d. None of the above
14. Which sector has the greatest potential for generating employment in urban areas?
    a. Manufacturing
    b. Healthcare
    c. Retail
    d. Construction
15. What is the primary challenge faced by urban residents in accessing affordable healthcare services?
    a. Lack of insurance coverage
    b. Shortage of healthcare facilities
    c. High cost of medical treatments
    d. All the above
16. Which of the following is a key strategy for promoting inclusive and equitable urban economic development?
    a. Privatization of public assets
    b. Targeted investment in specific sectors
    c. Decentralization of economic activities
    d. Concentration of resources in urban centers
17. How can urban–rural linkages contribute to the resilience and sustainability of the urban economy?
    a. Enhance food security
    b. Promote resource sharing
    c. Diversify economic activities
    d. All the above
18. Which factor is most influential in shaping the urban informal economy?
    a. Government regulations
    b. Availability of infrastructure
    c. Access to education and skills
    d. Cultural and social norms
19. What is the primary role of social enterprises in urban areas?
    a. Generating profits for private investors
    b. Addressing the needs of marginalized communities
    c. Promoting environmental sustainability
    d. Influencing urban governance

20. How can urban cultural and creative industries contribute to economic diversification and innovation?
    a. Attract tourism
    b. Generate employment opportunities
    c. Promote entrepreneurship
    d. All the above
21. Which of the following is a key challenge in the urban informal sector?
    a. Lack of access to healthcare
    b. Inadequate access to finance
    c. Poor working conditions
    d. All the above
22. What is the primary driver of the growth of the urban informal economy?
    a. Lack of employment opportunities in the formal sector
    b. High levels of urbanization
    c. Insufficient government regulation
    d. All the above
23. Which sector has the most significant impact on the urban economy in terms of economic output?
    a. Manufacturing
    b. Services
    c. Agriculture
    d. Construction
24. How can public–private partnerships contribute to urban infrastructure development?
    a. Improve resource allocation
    b. Enhance service delivery
    c. Promote cost-efficiency
    d. All the above
25. Which factor is most influential in the distribution of economic opportunities in urban areas?
    a. Access to education and skills
    b. Availability of transportation
    c. Land ownership patterns
    d. Government policies
26. What is the primary challenge faced by urban entrepreneurs in the informal sector?
    a. Accessing formal financial services
    b. Obtaining necessary business licenses
    c. Competing with larger formal businesses
    d. All the above
27. How can urban agriculture contribute to sustainable resource utilization?
    a. Reduce food waste
    b. Promote circular economy practices
    c. Enhance urban biodiversity
    d. All the above
28. Which factor has the most significant impact on the integration of the informal economy into the formal urban economy?
    a. Digitalization and technology
    b. Government policies and regulations
    c. Access to education and skills
    d. Availability of infrastructure

29. What is the primary role of urban governance structures in the distribution of economic resources?
    a. Ensuring equitable access
    b. Promoting economic growth
    c. Regulating market activities
    d. All the above
30. How can urban mobility solutions contribute to the accessibility and integration of different economic activities?
    a. Improve transportation infrastructure
    b. Enhance connectivity between urban centers
    c. Facilitate the movement of goods and services
    d. All the above

ANSWERS TO MULTIPLE-CHOICE QUESTIONS

1	a	11	d	21	d
2	d	12	c	22	d
3	c	13	b	23	b
4	d	14	d	24	d
5	b	15	d	25	c
6	d	16	c	26	d
7	a	17	d	27	d
8	b	18	d	28	a
9	a	19	b	29	d
10	d	20	d	30	d

FURTHER READING

Acs, Z. J., Audretsch, D. B., & Feldman, M. P. (1994). R&D spillovers and recipient firm size. *The Review of Economics and Statistics*, 76(2), 336–340.

Audretsch, D. B., & Feldman, M. P. (1996). R&D spillovers and the geography of innovation and production. *The American Economic Review*, 86(3), 630–640.

Bettencourt, L. M. (2013). The origins of scaling in cities. *Science*, 340(6139), 1438–1441.

Bettencourt, L. M., Lobo, J., Helbing, D., Kühnert, C., & West, G. B. (2007). Growth, innovation, scaling, and the pace of life in cities. *Proceedings of the National Academy of Sciences*, 104(17), 7301–7306.

Camagni, R., Capello, R., & Caragliu, A. (2015). The rise of second-rank cities: What role for agglomeration economies? *European Planning Studies*, 23(6), 1069–1089.

Carlino, G. A., Chatterjee, S., & Hunt, R. M. (2007). Urban density and the rate of invention. *Journal of Urban Economics*, 61(3), 389–419.

Castells, M. (1996). *The rise of the network society* (Vol. 1 of The information age: Economy, society and culture). Blackwell.

Duranton, G., & Puga, D. (2001). Diversity and specialisation in cities: Why, where and when does it matter? *Urban Studies*, 38(8), 1303–1320.

Duranton, G., & Puga, D. (2001). Nursery cities: Urban diversity, process innovation, and the life cycle of products. *American Economic Review*, 91(5), 1454–1477.

Duranton, G., & Puga, D. (2004). Micro-foundations of urban agglomeration economies. In J. V. Henderson & J. F. Thisse (Eds.), *Handbook of regional and urban economics* (Vol. 4, pp. 2063–2117). Elsevier.

Duranton, G., & Turner, M. A. (2011). The fundamental law of road congestion: Evidence from US cities. *American Economic Review*, 101(6), 2616–2652.

Ewing, R., & Cervero, R. (2010). Travel and the built environment: A meta-analysis. *Journal of the American Planning Association*, 76(3), 265–294.

Florida, R. (2002). *The rise of the creative class: And how it's transforming work, leisure, community and everyday life*. Basic Books.
Florida, R. (2005). *Cities and the creative class*. Routledge.
Fujita, M., Krugman, P., & Venables, A. J. (1999). *The spatial economy: Cities, regions, and international trade*. MIT Press.
Fujita, M., & Thisse, J. F. (2002). *Economics of agglomeration: Cities, industrial location, and regional growth*. Cambridge University Press.
Glaeser, E. L. (2008). *Cities, agglomeration, and spatial equilibrium*. Oxford University Press.
Glaeser, E. L. (2011). *Triumph of the city: How our greatest invention makes us richer, smarter, greener, healthier, and happier*. Penguin.
Glaeser, E. L., & Gottlieb, J. D. (2009). The wealth of cities: Agglomeration economies and spatial equilibrium in the United States. *Journal of Economic Literature*, 47(4), 983–1028.
Glaeser, E. L., & Kahn, M. E. (2004). Sprawl and urban growth. In J. V. Henderson & J. F. Thisse (Eds.), *Handbook of regional and urban economics* (Vol. 4, pp. 2481–2527). Elsevier.
Glaeser, E. L., Kallal, H. D., Scheinkman, J. A., & Shleifer, A. (1992). Growth in cities. *Journal of Political Economy*, 100(6), 1126–1152.
Jacobs, J. (1961). *The death and life of great American cities*. Random House.
Jacobs, J. (1969). *The economy of cities*. Vintage.
Kaplan, D. H., Wheeler, J. O., & Holloway, S. R. (2009). *Urban geography*. Wiley.
Katz, B., & Bradley, J. (2013). *The metropolitan revolution: How cities and metros are fixing our broken politics and fragile economy*. Brookings Institution Press.
Katz, B., & Wagner, J. (2014). *The rise of innovation districts: A new geography of innovation in America*. Brookings Institution.
Krugman, P. (1991). Increasing returns and economic geography. *Journal of Political Economy*, 99(3), 483–499.
Lucas, R. E., Jr. (1988). On the mechanics of economic development. *Journal of Monetary Economics*, 22(1), 3–42.
Melo, P. C., Graham, D. J., & Noland, R. B. (2009). A meta-analysis of estimates of urban agglomeration economies. *Regional Science and Urban Economics*, 39(3), 332–342.
Moretti, E. (2012). *The new geography of jobs*. Houghton Mifflin Harcourt.
O'Sullivan, A. (2019). *Urban economics* (9th ed.). McGraw-Hill Education.
Romer, P. M. (1990). Endogenous technological change. *Journal of Political Economy*, 98(5, Part 2), S71–S102.
Rosenthal, S. S., & Strange, W. C. (2003). Geography, industrial organization, and agglomeration. *Review of Economics and Statistics*, 85(2), 377–393.
Rosenthal, S. S., & Strange, W. C. (2004). Evidence on the nature and sources of agglomeration economies. In J. V. Henderson & J. F. Thisse (Eds.), *Handbook of regional and urban economics* (Vol. 4, pp. 2119–2171). Elsevier.
Sassen, S. (2001). *The global city: New York, London, Tokyo*. Princeton University Press.
Stolarick, K., & Florida, R. (2006). Creativity, connections and innovation: A study of linkages in the Montréal region. *Environment and Planning A*, 38(10), 1799–1817.
Storper, M. (2013). *Keys to the city: How economics, institutions, social interaction, and politics shape development*. Princeton University Press.
Storper, M., & Venables, A. J. (2004). Buzz: Face-to-face contact and the urban economy. *Journal of Economic Geography*, 4(4), 351–370.

5 Urban Housing

5.1 SOME ISSUES IN URBAN HOUSING

Urban housing is an essential component of urban development and planning, as it provides shelter, security, and comfort for the residents of cities. However, urban housing also faces many issues and challenges in the context of rapid urbanization, population growth, environmental degradation, social inequality, and economic instability. Some of the major issues and challenges include the following.

5.1.1 Affordability

The demand for urban housing often exceeds the supply, leading to high prices and rents that exclude many low-income and vulnerable groups from accessing adequate and safe housing. Moreover, the cost of housing maintenance, utilities, and services can also pose a heavy burden on the household budget, especially for those living in informal settlements or slums. The affordability issue of urban housing might lead to evictions, squatters, and homelessness, and eventually generate social and environmental problems. This is not only taking place in developing countries but also in developed countries.

Squatters and homelessness are two interrelated phenomena that pose significant challenges to urban planning and social policy. Squatters are people who occupy vacant or abandoned buildings or land without the permission of the owner or the legal authority. Homelessness is the condition of lacking a fixed, regular, and adequate nighttime residence. Both squatters and homeless people face various forms of exclusion, discrimination, and marginalization from mainstream society.

Squatters and homelessness have multiple causes and consequences, such as poverty, unemployment, lack of affordable housing, mental health issues, substance abuse, domestic violence, and natural disasters. Some squatters and homeless people may choose to live in alternative spaces as a form of resistance or protest against the dominant system, while others may have no other option. Squatters and homelessness also have different impacts on the environment, the economy, public health, and the human rights of the affected individuals and communities. This requires comprehensive and coordinated responses from different levels of government, civil society, and the private sector. Some possible solutions include providing adequate and accessible social services, creating more affordable and inclusive housing options, promoting community participation and empowerment, respecting the dignity and autonomy of squatters and homeless people, and addressing the root causes of social inequality and exclusion.

The affordability of urban housing varies by income, race and ethnicity, and age. Lower-income households, racial and ethnic minorities, and younger adults are more likely to face difficulties in finding affordable housing in cities than their counterparts. The affordability of urban housing is

DOI: 10.1201/9781032664613-5

not only a social problem, but also an economic one. It affects the productivity and competitiveness of cities, as well as the well-being and quality of life of their residents. To address this challenge, governments, private sector actors, and civil society organizations need to work together to increase the supply of affordable housing, improve the efficiency and transparency of the housing market, and provide adequate subsidies and support for low-income households.

The affordability of urban housing is a major challenge that affects millions of people around the world. One of the main factors that contributes to the housing affordability crisis is the mismatch between demand and supply. As more people move to cities in search of better opportunities, the housing stock fails to keep up with the growing demand, leading to higher prices and lower vacancy rates. The issue is especially acute in urban areas, where 63% of residents say affordable housing is a major problem, compared with 46% of suburban residents and 40% of those living in rural areas. The affordability of urban housing also varies by income, race and ethnicity, and age. Lower-income households, racial and ethnic minorities, and younger adults are more likely to face difficulties in finding affordable housing in cities than their counterparts. The affordability of urban housing is not only a social problem, but also an economic one. It affects the productivity and competitiveness of cities, as well as the well-being and quality of life of their residents.

To address this challenge, various solutions have been proposed and implemented by different stakeholders. Some of these solutions include:

1. Increasing the supply of affordable housing by building more public and social housing units, incentivizing private developers to include affordable units in their projects, and relaxing zoning and land use regulations that limit the density and diversity of housing types.
2. Improving the efficiency and transparency of the housing market by providing better information and guidance to buyers and renters, enforcing fair housing laws and anti-discrimination policies, and regulating rent increases and evictions.
3. Providing adequate subsidies and support for low-income households by expanding rental assistance programs, offering tax credits and vouchers, and facilitating access to credit and mortgages.

Public housing could be one of the solutions to affordability. It is a form of housing tenure in which the property is usually owned by a government authority, either central or local. Public housing is intended to provide affordable and decent housing for low-income or vulnerable people who cannot access the private housing market. Public housing may also have social, economic, and environmental benefits, such as reducing homelessness, improving social cohesion, enhancing urban regeneration, promoting sustainability, increasing access to education and health services, and supporting community development. However, public housing also faces many challenges, such as insufficient funding, poor maintenance, stigma, segregation, and crime. Therefore, public housing policies and programs need to be carefully designed and implemented to ensure that they meet the diverse needs and preferences of the residents and the communities.

A unique initiative program to give a breakthrough on the vicious circle of housing affordability in a province in Indonesia, the Governor of Central Java has launched a program called "Buy Land Get Home." The story is given in Box 5.1.

BOX 5.1 THE BUY LAND GET HOME PROGRAM (A JAVANESE UNIQUE APPROACH TO HOUSING AFFORDABILITY)

The "Tuku Lemah Oleh Omah" (literally means: Buy Land Get Home) program is an initiative by the Governor of Central Java and the PDIP Presidential Candidates 2024–2029, Ganjar Pranowo, to help people realize their dream of owning a home without a down payment and without installments. This program has been rolled out since 2020 and has already provided hundreds of homes through this scheme.

The program is intended for individuals and communities who are already married, do not yet own a home, and are registered in the Integrated Social Welfare Data (DTKS). The beneficiaries only need to own land, and the government will build a house worth IDR 35 million with an additional IDR 1.8 million for labor.

In the initial year of 2020, the program built 200 units. The following year, 2021, saw the construction of 186 units, and in 2022, 253 units were built. For the year 2023, it has been budgeted to build 615 units.

The program has been well received by the community as it has helped many people achieve their dream of owning a home. The houses can be designed according to the community's self-help capabilities. The program is not only limited to Pemalang but also extends aid to less fortunate residents in other areas such as Brebes District, Magelang City, and Wonosobo District.

Sources: humas.jatengprov.go.id, regional.kompas.com

5.1.2 Quality

The quality of urban housing can vary significantly depending on the location, design, construction, materials, and management of the buildings. Poor quality housing can have negative impacts on the health, well-being, and productivity of the occupants, as well as on the environment and the urban landscape. Some of the common problems include overcrowding, lack of ventilation, insulation, sanitation, drainage, waste management, and fire safety.

Unlike the rural housing system, urban housing in cities, particularly megacities and metropolitan areas in which the land is scarce, and the land price is soaring, offers a variety of housing options for people with different needs, preferences, and budgets. Some of the most common types of housing in cities are shown in Table 5.1.

5.1.3 Accessibility

There are two issues on accessibility centered on urban housing. They are accessibility to urban housing, and the other one is accessibility from urban housing. The first one refers to how easily individuals can access and obtain housing in urban areas. It encompasses factors such as affordability, availability, proximity to amenities and transportation, and inclusivity for people with disabilities. Improving the accessibility of urban housing can create equitable and sustainable cities. The second one refers to the physical and social proximity of the housing to the essential urban services and facilities, such as transportation, education, health care, employment, recreation, and culture. In many cities, urban housing is located far from the urban centers or in peripheral areas that lack adequate infrastructure and connectivity. This can result in long commuting times, increased traffic congestion, air pollution, and greenhouse gas emissions, as well as reduced social interaction and integration.

Accessibility of urban housing affects the quality of life and well-being of the citizens. Urban housing is often expensive, overcrowded, poorly maintained, or located in areas with high levels of pollution, noise, crime, or social exclusion. These factors can have negative impacts on the physical, mental, and social health of urban residents, especially those who belong to vulnerable groups such as low-income families, elderly people, people with disabilities, or ethnic minorities. To improve the accessibility of urban housing, some possible solutions include providing subsidies or incentives for affordable and social housing projects, enforcing regulations and standards for housing quality and

TABLE 5.1
Common Types of Urban Housing

Type	Explanation
Apartments	These are units in a building that share walls, floors, and ceilings with other units. Apartments can range from studios to penthouses and can be rented or owned. Apartments are usually cheaper than houses but may have less space, privacy, and amenities. The building of the apartment is commonly owned by a person. Apartments may also have access to shared facilities such as pools, gyms, and laundry rooms. However, apartments may have less space, privacy, and amenities than houses. Apartments may have restrictions on pets, noise, and visitors. Apartments may also have less security and stability than houses, as renters may face eviction or rent increases.
Condominiums	These are like apartments, but the units are individually owned and managed by a homeowner's association. Condominiums can offer more security, convenience, and services than apartments, but may also have higher fees and regulations. Condominiums can offer more control and equity than apartments, as owners can customize their units and benefit from property appreciation. Condominiums may also have better security, management, and maintenance than apartments. On the other hand, condominiums may have higher upfront and ongoing costs than apartments, such as mortgage payments, property taxes, insurance, and association fees. Condominiums may also have more rules and regulations than apartments, such as on parking, landscaping, and renovations.
Townhouses	These are attached houses that share walls with other houses but have their entrances and yards. Townhouses can offer more space and comfort than apartments but may have less maintenance and responsibility than detached houses. Townhouses can offer more living space and outdoor space than apartments, as well as more privacy and independence. Townhouses may also have lower costs and obligations than detached houses, as some exterior maintenance and repairs may be covered by the association or the landlord. However, townhouses may have less privacy and freedom than detached houses, as they may still share walls, roofs, and common areas with other houses. Townhouses may also have higher fees and regulations than detached houses, depending on the association or the lease agreement.
Detached houses	These are standalone houses that do not share walls with other houses and have their own lot and driveway. Detached houses can offer more privacy, freedom, and customization than other types of housing, but may also have higher costs and obligations. Detached houses can offer more space, comfort, and personalization than other types of housing, as well as more options for landscaping, gardening, and entertainment. Detached houses may also have more value and appreciation than other types of housing, as they are usually in high demand. However, detached houses may have higher upfront and ongoing costs than other types of housing, such as mortgage payments, property taxes, insurance, and maintenance. Detached houses may also have more responsibility and liability than other types of housing, as owners have to take care of all aspects of their property.

safety, promoting mixed-use and inclusive urban planning and design, enhancing public transportation and mobility options, and fostering community participation and empowerment.

5.1.4 Diversity

The diversity of urban housing reflects the diversity of the urban population in terms of their needs, preferences, lifestyles, and cultures. Urban housing should be able to accommodate and respond to the different and changing demands of urban dwellers, while also respecting their rights, dignity, and identity. However, in many cases, urban housing is standardized, uniform, and homogeneous, failing to recognize and celebrate the diversity and creativity of the urban communities.

Standardized urban housing is a concept that aims to provide affordable, efficient, and sustainable living spaces for people in densely populated cities. Standardized urban housing refers to the design and construction of modular units that can be easily assembled, disassembled, and relocated

FIGURE 5.1 A condominium – the unit in the building owned by individual. Credit: Niall Sohan, Wikipedia.

according to the changing needs and preferences of the residents. Standardized urban housing can offer several benefits, such as (1) reducing the cost and time of building and maintaining housing units, as well as the environmental impact of construction materials and waste; (2) increasing the availability and accessibility of housing options for low-income and marginalized groups, as well as the diversity and inclusivity of urban communities; (3) enhancing the flexibility and adaptability of housing units to different climates, terrains, and lifestyles, as well as the resilience and security of urban areas in the face of natural disasters and social unrest; and (4) promoting the innovation and creativity of architectural design and urban planning, as well as the collaboration and participation of various stakeholders in the development and management of urban housing.

Standardized urban housing is not a one-size-fits-all solution, but rather a framework that can be customized and optimized for different contexts and purposes. Standardized urban housing requires careful consideration of the social, cultural, economic, and environmental factors that affect the quality of life and well-being of urban dwellers. Standardized urban housing also poses some challenges, such as (1) balancing the standardization and customization of housing units, as well as the trade-offs between efficiency and comfort, functionality and aesthetics, individuality, and community; (2) ensuring the quality and safety of housing units, as well as compliance with building codes and regulations, human rights and ethical standards, environmental protection, and sustainability goals; and (3) addressing the potential issues of social isolation, alienation, or stigma that may arise from living in standardized housing units, as well as the possible conflicts or disputes that may occur among residents or between residents and authorities.

FIGURE 5.2 An apartment – the whole building owned by individual owner.

Source: **Wikipedia.**

Standardized urban housing is a promising idea that can transform the way people live in cities. Standardized urban housing can provide more than just shelter, but also opportunities for growth, development, and empowerment. Standardized urban housing can be a catalyst for positive change in urban society.

The question of whether urban housing should be standardized or diverse involves many factors. Some of the arguments in favor of standardization are that it can reduce construction costs, increase efficiency, create a sense of order and harmony, and facilitate urban planning and management. For example, standardized housing can allow for the mass production of building materials and prefabricated units, which can lower the expenses and time required for building. Standardized housing can also optimize the use of space and energy, which can benefit urban residents and the environment. Standardized housing can also promote a uniform and coherent image of the city, which can enhance its identity and attractiveness. Standardized housing can also simplify the tasks of urban planners and managers, who can apply consistent rules and regulations to the housing sector.

Some of the arguments in favor of diversity are that it can foster creativity, reflect cultural and individual differences, enhance the aesthetic appeal and livability of the urban environment, and support social diversity and inclusion. For example, customized and diverse housing can encourage architectural innovation and experimentation, which can enrich the urban landscape and culture, and the sense of place. Diverse housing can also accommodate the varied needs and preferences of the residents, which can improve their satisfaction and well-being. Diverse housing can also create

FIGURE 5.3 Shophouses.

Source: Wikipedia.

a more dynamic and vibrant atmosphere in the city, which can attract more visitors and investors. Diverse housing can also support the social diversity and inclusion of different groups of people, who can find suitable and affordable housing options that match their lifestyles and identities.

These issues and challenges require concerted efforts from various stakeholders, such as governments, the private sector, civil society, academia, and international organizations, to develop and implement effective policies, strategies, and interventions that can improve the quantity, quality, accessibility, and diversity of urban housing in cities.

5.2 URBAN VIS-À-VIS SUBURBAN HOUSING

Urban and suburban housing are two types of locations of residential areas that differ in many aspects, such as population density, land use, infrastructure, amenities, and environmental impact. Urban housing refers to dwellings that are in densely populated areas, typically near the center of a city or a town. Suburban housing refers to dwellings that are in less populated areas, typically on the outskirts of a city or a town.

Urban housing tends to be more diverse and compact than suburban housing. Urban dwellers can enjoy the benefits of living close to various services, facilities, and cultural activities. However, urban housing also faces challenges such as traffic congestion, noise pollution, crime, and lack of green space. Suburban housing tends to be more spacious and homogeneous in comparison to urban housing. Suburban dwellers can enjoy the benefits of living in a quieter, safer, and greener

TABLE 5.2
Factors Influencing the Selection of Housing Location

Factor	Explanation
Cost	Urban areas tend to have higher housing prices and taxes than suburban areas, but they may also offer more public services and amenities. Suburban areas may have lower costs, but they may also require more expenses for transportation and maintenance.
Lifestyle	Urban areas may offer more opportunities for social, cultural, and professional activities, as well as more diversity and convenience. Suburban areas may provide more space, privacy, and tranquility, as well as more family-friendly and community-oriented environments.
Commute	Urban areas may have better access to public transportation, which can reduce the time and stress of commuting. Suburban areas may have less traffic congestion, but they may also require longer distances and more dependence on personal vehicles.
Safety	Urban areas may have higher crime rates and pollution levels than suburban areas, but they may also have more security and emergency services. Suburban areas may have lower risks, but they may also have less awareness and preparedness for potential threats.
Environment	Urban areas are predominantly built environments with possible higher pollution levels, that is, air pollution, noise pollution, light pollution, etc. Suburban areas are natural environments, with more serendipity and less pollution levels.
Facilities/amenities	Urban areas have superiority in facilities and amenities in comparison to suburban or rural areas.

environment. However, suburban housing also faces challenges such as sprawl, car dependency, social isolation, and environmental degradation.

One of the criteria that affects the selection of housing is the trade-off between proximity to services in the city and a superior natural environment in the suburban area. Residing in the city offers numerous benefits, such as access to public transportation, healthcare, education, entertainment, and cultural diversity. Nevertheless, residing in an urban area also entails higher expenses, noise, pollution, and congestion. Conversely, residing in a suburban area provides more space, privacy, safety, and greenery. Nevertheless, residing in a suburban area also requires longer commutes, car dependency, lower social interaction, and less diversity. Therefore, the determination of where to reside depends on the preferences, needs, and values of each individual or household.

There are many factors to consider when choosing a housing location (assuming the allocated budget to buy the property was not a problem), whether in urban or suburban areas. Some of the factors that may influence this decision are shown in Table 5.2.

Despite obvious factors, there is no definitive answer to this issue, as different people may have different priorities, preferences, and expectations. However, it is important to weigh the pros and cons of each option and make an informed decision based on your situation and goals. An illustrative example of selecting the housing location based on the Pair-wise Comparison method (a real case of developing countries of Indonesia and Thailand), as we have only two entities to compare. However, other methods can also be used.

The objective of comparing conditions in two different locations is to acquire an economically profitable, environmentally friendly, and socially acceptable solution for the benefit of the citizens. By this objective, the selection is based on factors calculated by the following formula:

$$F_a = a/(a+b) \quad \text{and} \quad F_b = b/(a+b)\,\text{Unit} \tag{5.1}$$

where a and b are quantitative factors of living in. In the above formula, factors a and b represent the relative attractiveness or utility of two residential locations. These factors can be determined based on various criteria that influence residential preferences. Here are some common considerations, as similarly reflected in Table 5.2:

- *Proximity to Workplace:* Distance or commute time to work can be a critical factor. The shorter the commute, the higher the attractiveness.
- *Access to Amenities:* Availability of schools, healthcare facilities, parks, shopping centers, and recreational areas can enhance the attractiveness.
- *Cost of Living:* Housing prices, property taxes, and overall cost of living can influence the decision. Lower costs may increase attractiveness.
- *Safety and Security:* Crime rates and general safety perceptions of the area can significantly impact the attractiveness.
- *Environmental Quality:* Air quality, noise levels, and green spaces contribute to the desirability of a location.
- *Community and Social Factors:* Presence of family, friends, and social networks can also play a role.
- *Public Transport Access:* Availability and convenience of public transportation options can increase the attractiveness.
- *Future Development:* Planned or ongoing development projects that might increase the area's future desirability.

Property price for approximately equivalent quality:

- In the city 75,000 USD
- In suburban 25,000 USD,
- The weighting factor of the property price in the city is $1/(75{,}000/100{,}000) = 1.33$ and in the suburban is $1/(25{,}000/100{,}000) = 4.0$

Living cost per month:

- In the city: 800 USD
- In suburban 500 USD
- The weighting factor of the city: $1/(800/1300) = 1.63$ and suburban: $1/(500/1300) = 2.6$

5.2.1 Lifestyle

Lifestyle has intangible value. As lifestyle is an intangible value, the comparison of city living and the suburban lifestyle is opinionated; therefore the comparison is based on the quantification of their advantages and disadvantages. Living in the city and the suburbs both have their advantages and disadvantages. Here is an example of the comparison:

Advantages of living in the city: (1) access to a variety of amenities such as restaurants, entertainment, and cultural activities; (2) better public transportation; (3) more job opportunities; (4) more opportunities to meet new people, try different activities and explore the latest dining options; and (5) better and more public schools. Total: 5 points. The disadvantages are: (1) higher cost of living; (2) more traffic that creates traffic congestion; (3) more noise, more pollution in general; (4) higher crime rates; and (5) faster pace of life and higher stress levels. Total: 5 points.

Advantages of suburban living: (1) lower housing prices; (2) lower crime rates; (3) less crowding and more free space; and (4) quieter, more relaxed lifestyle, less stressful. Total: 4 points. The disadvantages of suburban living: (1) a slower-paced lifestyle may not suit everyone, (2) may require a car for commuting, and (3) fewer amenities compared to the city. Total: 3 points.

This factor can be estimated by the total number of things associated with leisure, relaxation, affluence, and prosperity activities that benefit the citizens. For each element, another sub-weighting factor could also apply.

- In the city: 5 points
- Suburban: 3 points
- The weighting factor of the city: 5/8 = 0.63 and suburban: 3/8 = 0.37

5.2.2 Commuting

Commuting to work, school, shopping, etc. could be another factor for consideration. For instance, the cost of commuting, time spent, and safety during traveling etc. In the case of living in a city versus living in suburban areas, the factors could be:

- For the cost of commuting 0.7 for living in the city, and 0.3 for living in suburban areas.
- The time spent factor is 0.7 for living in the city, and 0.3 for living in suburban areas.
- The factor of safety during traveling is about the same for both, which is 0.5.
- From these three sub-factors, the factor of commuting is 1.9 for living in cities and 1.1 for living in suburban areas.

5.2.3 Safety of Living

It is generally safer to live in suburban areas than in a city, regarding crime incidents and disasters. Therefore, the factor can be:

- In the city: 0.4
- In suburban areas: 0.6
- As the total of the factor for safety living is a unit, it can be applied directly in the calculation of the total factor.

5.2.4 Environment

The consideration of the environment is based on the level of pollution, the possibility of pollution incidents, and the environmental footprint.

- In the city: 0.2
- In suburban areas: 0.8
- The factor is idiosyncratic; therefore, it depends on the preference of the assessor.

Infrastructure and services:

- In the city: 7 points
- In suburban areas: 1 point
- The weighting factor for living in the city: 7/8 = 0.88, and for suburban areas: 1/8 = 0.12.

From each element of the weighting factor, the total pair-wise comparison between living in the city versus living in suburban areas is shown in Table 5.3.

Even though the above example is based on personal preference and experience in selecting two locations for living, it is important to note that these are general observations, and the experience

TABLE 5.3
Pair-Wise Comparison Method with Weighting Factors to Select the Location

Location	Factors							
	Property Price	Living cost	Lifestyle	Commute	Safety	Environment	Infrastructure/ Services	Total
Urban	1.33	1.63	0.63	1.90	0.40	0.20	0.88	6.97
Suburban	4.00	2.60	0.37	1.10	0.60	0.80	0.12	9.59

Note: The weight of individual factor is determined by individual assessor.

can vary greatly depending on the specific city or suburb. Ultimately, the choice between city and suburban living depends on personal preference, lifestyle, and needs.

5.3 SHELTER AND LAND DEMAND

Urbanization has become a pressing issue for cities around the world. As populations continue to grow, cities are facing the challenge of accommodating this rapid urbanization within limited urban land. The consequences of this pressure are manifold. One of the primary concerns associated with urbanization is the strain it puts on existing infrastructure and housing. As more people move into cities, there is an increased demand for housing, transportation, and public services. This puts a burden on the city's resources and can lead to overcrowding, traffic congestion, and inadequate access to essential services.

In urban areas, the demand for land is constantly increasing due to a variety of factors. One of the main causes is population growth. As more people move to cities, the need for housing, commercial buildings, and infrastructure increases, putting pressure on available land. Another cause of land demand in cities is economic development. As cities grow and become more prosperous, businesses are attracted to the area, leading to an increase in demand for commercial and industrial space. This can also lead to gentrification, where low-income neighborhoods are transformed into higher-end areas, further increasing land demand. Transportation is also a major factor in land demand in cities. As more people rely on cars and public transportation, the need for roads, parking lots, and public transit infrastructure increases, taking up valuable land. Zoning laws and regulations can also contribute to land demand in cities. Certain areas may be designated for specific uses, such as residential or commercial, which can limit the available land for other purposes. City regulations, such as minimum lot sizes or building height restrictions, can limit the amount of usable space in a given area.

Urban sprawl could also be a cause of land demand. As cities expand horizontally, they consume more land and resources. Urban sprawl is often driven by the preference for low-density, single-family housing, car-oriented transportation, and decentralized development patterns. Urban sprawl can lead to environmental degradation, social segregation, loss of agricultural land and cultural heritage, and increased infrastructure costs.

Land speculation is a silent but imminent land pressure. Speculators may create land scarcity and soaring land prices in cities, some actors may engage in speculative practices to profit from the rising land prices. Land speculation involves buying, holding, or selling land with the expectation of future capital gains, rather than using it for productive purposes. Land speculation can inflate land prices, reduce land availability, distort land markets, and discourage investment in urban development.

Several strategies can be implemented to cope with land scarcity in cities. One approach is to promote compact and efficient land use. This can be achieved by encouraging the development of multistory buildings and high-density housing. By building verticals rather than horizontals, cities can make more efficient use of limited land resources. Another strategy is to promote mixed-use development. This involves integrating residential, commercial, and recreational spaces within the same area. By doing so, cities can reduce the need for long commutes and create vibrant, walkable neighborhoods. The mixed-use development also helps to maximize the use of available land, as different functions can be accommodated within the same space.

Cities can also implement policies that prioritize the use of vacant or underutilized land. This could involve providing incentives for property owners to develop or lease their land or implementing regulations that discourage land speculation. Making better use of existing land enables cities to reduce the pressure to develop greenfield sites and protect valuable natural areas. Cities can adopt innovative design solutions to make the most of limited space. This could involve the use of vertical gardens, rooftop parks, and underground infrastructure. By incorporating green spaces into urban areas, cities can improve livability and create a more sustainable environment. Promoting sustainable transportation can also be an option. Investing in public transportation systems allows cities to reduce the need for private vehicles and free up valuable land that would otherwise be used for parking. Additionally, promoting cycling and walking infrastructure can further reduce the demand for cars and create more space for other uses.

Cities can also explore the potential of redeveloping brownfield sites. These are areas that have been previously used for industrial or commercial purposes and are now vacant or underused. By revitalizing these sites, cities can reclaim valuable land and transform them into vibrant, mixed-use developments. Cities can implement policies that encourage the preservation and enhancement of green spaces. Parks, gardens, and other natural areas not only provide recreational opportunities but also help to improve air quality and mitigate the effects of climate change. Protecting and expanding green spaces allows cities to create a more sustainable and livable environment.

Limited urban land further exacerbates the challenges posed by urbanization. With limited space available for expansion, cities must find innovative ways to accommodate their growing populations. This often means intensifying land use through high-rise buildings, mixed-use developments, and compact urban design. However, these solutions come with their own set of challenges, including increased density, reduced green spaces, and potential impacts on the quality of life for residents.

The environmental impact of urbanization is another significant concern. As cities expand, natural habitats are often destroyed or fragmented, leading to a loss of biodiversity. Additionally, increased energy consumption and pollution from transportation and industrial activities contribute to climate change and air pollution. Cities must prioritize sustainable development practices to mitigate these environmental impacts and ensure a livable and healthy environment for their residents.

The social implications of urbanization cannot be overlooked either. As cities become more densely populated, social inequalities can be exacerbated. Limited affordable housing options can lead to gentrification and the displacement of low-income communities. Access to quality education, healthcare, and other essential services may also be compromised for marginalized populations. Cities must implement inclusive policies that address these social disparities and promote equitable access to resources and opportunities for all residents.

To effectively manage the pressures of urbanization, cities must adopt comprehensive urban planning strategies. This includes long-term land use planning, transportation planning, and infrastructure development. Prioritizing sustainable and resilient urban design principles enables cities to create livable environments that can accommodate growing populations while minimizing negative impacts.

Investment in research and innovation is another key aspect of managing urbanization pressures. Leveraging technology and data-driven approaches allows cities to optimize resource allocation, improve infrastructure efficiency, and enhance the overall quality of life for their residents. Smart city initiatives, for example, can help cities become more sustainable, resilient, and responsive to the needs of their inhabitants.

Urbanization and increasing population pose significant challenges for cities around the world. Limited urban land exacerbates these challenges, requiring cities to adopt innovative solutions to accommodate their growing populations. By prioritizing sustainable development practices, inclusive policies, comprehensive urban planning, collaboration among stakeholders, and investment in research and innovation, cities can effectively manage the pressures of urbanization and create livable environments for their residents. Cities must address these challenges proactively to ensure a sustainable and prosperous future for all.

In urban areas where land scarcity is a pressing issue, it becomes crucial to optimize the solutions for urban housing. With limited available land, it is important to consider various factors to effectively address the housing needs of the growing population. In such situations, certain types of urban housing can prove to be more suitable and efficient.

One approach to optimizing urban housing in the face of land scarcity is through vertical expansion. This involves constructing high-rise buildings that can accommodate a larger number of residents within a smaller land area. By utilizing vertical space, cities can maximize their housing capacity without the need for extensive land resources. High-rise buildings can be designed to include multiple-floor condominiums and apartments, providing housing options for a diverse range of residents.

Another strategy is to focus on compact housing solutions. Compact housing refers to the design and construction of smaller living spaces that are still functional and comfortable. This approach allows for increased housing density within a limited land area. Compact housing can be achieved through the development of micro-apartments, studio units, or even modular housing systems (Figure 5.4). These smaller living spaces can be efficiently designed to maximize functionality and optimize the use of space.

FIGURE 5.4 Modular housing systems.

Source: **Wikipedia.**

Mixed-use developments can be a valuable solution for urban housing in land-scarce cities. By combining residential spaces with commercial, retail, or recreational areas, mixed-use developments create vibrant and self-sustaining communities. This not only optimizes land use but also enhances the quality of life for residents by providing easy access to amenities and services. Mixed-use developments can be designed to incorporate green spaces, pedestrian-friendly areas, and efficient transportation systems, further enhancing the overall livability of urban areas.

Adaptive reuse of existing structures can be an effective way to optimize urban housing in land-scarce cities. This involves repurposing underutilized or abandoned buildings into residential spaces. Renovating and retrofitting these structures enables cities to make use of existing infrastructure instead of allocating additional land for new construction. Adaptive reuse not only preserves the architectural heritage of a city but also promotes sustainable development by reducing waste and energy consumption.

To optimize urban housing in land-scarce cities, it is essential to prioritize affordability and accessibility. The housing solutions implemented should cater to the needs of different income groups and ensure that housing is within reach for all residents. This can be achieved through the provision of subsidized housing, rent control policies, or public-private partnerships that promote affordable housing initiatives.

Maximizing the floor-to-land plot ratio is a crucial aspect of urban planning and development. It refers to the efficiency with which a building utilizes the available land area. By maximizing this ratio, developers can make the most out of limited space, allowing for more efficient land use and increased building density. One of the primary reasons for maximizing the floor-to-land plot ratio is to address the growing demand for housing and commercial spaces in densely populated areas. As cities become more crowded, it becomes essential to optimize land usage to accommodate the increasing population. Constructing taller buildings or utilizing multi-story structures allows developers to create more living or working spaces within a limited land area. Maximizing the floor-to-land plot ratio also has economic benefits. Utilizing space efficiently allows developers to maximize their return on investment. Higher building density allows for more units or commercial spaces, which translates to increased revenue potential. This aspect is especially crucial in areas where land prices are high, as maximizing the floor-to-land plot ratio can help offset the cost of land acquisition. Maximizing the floor-to-land plot ratio also promotes sustainable development. Utilizing space efficiently enables developers to contribute to minimizing urban sprawl and preserving open green spaces. This approach is particularly important in urban areas where land is scarce and valuable. Building vertically instead of horizontally allows developers to reduce the environmental impact of construction and preserve natural habitats.

It is crucial to ensure that buildings are designed and constructed in a way that does not compromise safety and structural integrity. Adequate infrastructure and facilities should be in place to support increased building density, including transportation systems, utilities, and public amenities. It is important to strike a balance between maximizing the floor-to-land plot ratio and maintaining quality living and working environments. High-density developments should incorporate elements such as green spaces, recreational areas, and community facilities to enhance the overall livability of the area. This approach helps create vibrant and sustainable communities that cater to the needs of residents and workers.

Of course, it is also important to consider the needs and desires of the people who will be using the building. This includes everything from the number of bedrooms and bathrooms to the overall style and aesthetic of the building. Considering all these factors, a building can be designed to be both efficient and functional, while also being aesthetically pleasing and comfortable to live in. Maximizing the plot-to-floor ratio is all about finding creative ways to make the most of the available space. Whether through innovative design techniques or careful consideration of every aspect of the building, many different strategies can be employed to create a building that is both efficient and beautiful. Taking the time to carefully consider all these factors allows you to create a space that maximizes its potential in terms of both functionality and overall appeal.

5.4 URBAN MIGRATION AND HOMELESSNESS

Urban migration, driven by the search for better opportunities, has led to significant population shifts toward cities. However, this influx often exceeds the capacity of urban infrastructure and housing, resulting in increased homelessness. Affordable housing shortages, coupled with inadequate support systems, exacerbate the issue, leaving many without stable shelter. Homelessness in urban areas highlights the disparity between economic growth and social equity, as cities struggle to accommodate the growing needs of their residents. Addressing this challenge requires comprehensive policies that include affordable housing development, social services, and economic support to ensure sustainable and inclusive urban growth.

5.4.1 Rural Push Factors and Urban Pull Factors

Urban migration is a phenomenon that has been observed in many countries across the world. It refers to the movement of people from rural areas to urban centers in search of better economic opportunities and a better standard of living. There are several causes of urban migration, including push factors such as job scarcity in rural areas and pull factors such as the economic opportunities in cities.

One of the primary causes of urban migration is the lack of job opportunities in rural areas. Many people living in rural areas are engaged in agriculture, which is often seasonal and dependent on weather conditions. This can lead to a lack of consistent income, making it difficult for people to support their families. These become rural push factors. In contrast, urban centers offer a range of job opportunities in various sectors such as manufacturing, services, and technology. These jobs provide a more stable income and better prospects for career growth.

Another factor that contributes to urban migration is the pull of attractions in cities. Urban centers offer a range of amenities and services that are not available in rural areas. These include better healthcare facilities, educational institutions, entertainment options, and cultural experiences. Cities also offer a more diverse population, which can be attractive for people looking for new experiences and opportunities. However, urban migration also has its downsides. The rapid growth of urban centers can lead to overcrowding, increased pollution, and a strain on infrastructure and resources. This can result in a decline in the quality of life for both urban and rural residents.

Creating job opportunities in rural areas and improving the quality of life for rural residents is a direct and straightforward strategy. This can be achieved through investments in infrastructure, education, and healthcare. At the same time, efforts should be made to manage the growth of urban centers by implementing sustainable development practices and improving the efficiency of public services.

The phenomenon of urban migration, where individuals from rural areas move to urban centers in search of better economic opportunities, can have unintended consequences if not properly managed. If these migrants lack the necessary skills and qualifications required by urban jobs, it can lead to an increase in urban unemployment and homelessness.

The root cause of this problem lies in the mismatch between the skills and qualifications of rural migrants and the demands of urban employers. Many rural migrants may have experience in traditional agricultural or manual labor jobs but lack the technical skills and education required for more specialized urban jobs. As a result, they may struggle to find employment in the city, leading to increased competition for limited job opportunities and a rise in urban unemployment. This unemployment can, in turn, lead to homelessness as individuals struggle to afford housing and necessities without a steady income. The strain on social services and infrastructure can also increase, as more individuals require assistance from government programs and non-profit organizations.

From the rural perspective, to reduce urban migration, several key steps should be taken. These measures aim to address the underlying factors that drive individuals to leave rural areas in search of better opportunities in urban centers. Focusing on improving rural infrastructure, enhancing economic prospects, and providing access to quality education and healthcare enables us to create an environment that encourages people to stay in rural areas and contribute to their development.

Investing in rural infrastructure is crucial. This includes improving road networks, transportation systems, and basic amenities such as electricity, water supply, and sanitation facilities. By ensuring that rural areas have adequate infrastructure, we can enhance the quality of life and create an enabling environment for economic growth. This, in turn, will reduce the need for individuals to migrate to urban areas in search of better living conditions. Promoting economic opportunities in rural areas is essential. This can be achieved by supporting agricultural development, encouraging entrepreneurship, and attracting investment in sectors such as manufacturing and services. Providing financial assistance, technical training, and market linkages to rural entrepreneurs can help create sustainable livelihoods and income-generating activities. Diversifying the rural economy and creating job opportunities allows us to provide an alternative to urban migration.

Access to quality education and healthcare in rural areas is vital in reducing urban migration. Improving the availability and quality of schools and healthcare facilities in rural areas will not only benefit the local population but also attract individuals from urban areas seeking a better environment in rural areas. Investing in education and healthcare infrastructure and ensuring the availability of trained professionals enables us to create an environment that provides essential services and opportunities for personal growth in rural areas. This includes promoting community engagement, preserving local traditions and cultural heritage, and fostering a sense of pride and belonging among rural residents. When a strong sense of community and identity is created, individuals may be more inclined to stay in rural areas rather than seek opportunities elsewhere.

5.4.2 Social and Environmental Problems Brought by Migrants

Rural-to-urban migration has become a significant phenomenon in many countries, resulting in a range of social and environmental problems. As individuals and families leave their rural communities in search of better opportunities, they often face challenges that can have far-reaching consequences for both the migrants themselves and the urban areas they settle in.

One of the most tenacious social problems associated with rural-to-urban migration is the strain it places on urban housing, infrastructure, and services. As migrants flood into cities, there is often an insufficient supply of housing, healthcare facilities, and educational resources to meet the increased demand. This can lead to overcrowding, inadequate living conditions, and a lack of access to essential services for both migrants and existing urban residents. Another social problem that arises from rural-to-urban migration is the breakdown of social networks and support systems. In rural areas, communities are often tightly knit, with strong social ties and mutual support among residents. However, when individuals migrate to urban areas, they often leave behind their social networks, leading to feelings of isolation and disconnection. This can have negative effects on mental health and well-being, as individuals struggle to adapt to their new environment without the support of familiar faces and relationships.

Environmental problems are also a significant concern associated with rural-to-urban migration. As more people move to cities, there is increased pressure on natural resources such as water, energy, and land. Urban areas often struggle to provide adequate infrastructure to support the growing population, leading to issues such as water scarcity, pollution, and deforestation. The increased demand for housing and transportation can destroy natural habitats and ecosystems. Rural-to-urban migration can also exacerbate income inequality and social disparities. Many migrants are forced to take on low-paying jobs in the informal sector, which often lacks job security and fair wages. This can perpetuate a cycle of poverty and marginalization, as migrants struggle to access the same opportunities and benefits as urban residents. The resulting income inequality can lead to social unrest and tensions within communities.

Rural-to-urban migration has profound effects on urban society, shaping its demographic, economic, social, and cultural landscape. As individuals and families move from rural areas to cities in search of better opportunities and improved living conditions, the urban structure undergoes significant transformations.

Demographically, rural-to-urban migration alters the population composition of cities. The influx of migrants from rural areas can lead to rapid population growth, straining urban infrastructure and services. This can result in overcrowding, increased demand for housing, and pressure on healthcare, education, and transportation systems. Consequently, urban authorities must adapt to these changes by expanding and upgrading infrastructure and public services to accommodate the growing population.

Economically, rural-to-urban migration has both positive and negative implications. On one hand, it can stimulate urban economies by increasing the labor supply and consumer demand. Migrants often bring new skills and expertise, contributing to the diversification and specialization of urban industries. Moreover, their consumption patterns can drive local businesses and spur economic growth. On the other hand, the rapid influx of migrants can also lead to unemployment and income inequality. The competition for jobs in urban areas can be fierce, particularly for low-skilled workers, resulting in wage stagnation and precarious employment conditions.

Socially, rural-to-urban migration brings about changes in social dynamics and community structures. As migrants settle in urban areas, they form new social networks and communities, often based on shared rural origins or cultural backgrounds. These communities provide support systems and a sense of belonging for migrants in unfamiliar urban environments. However, they can also lead to the formation of parallel societies or ethnic enclaves, potentially exacerbating social divisions and cultural tensions. Integration efforts are crucial to foster social cohesion and ensure that all residents have equal access to opportunities and resources.

Culturally, rural-to-urban migration influences the urban cultural landscape through the introduction of new traditions, languages, and beliefs. Migrants bring their unique cultural practices and customs, enriching the diversity of urban society. This cultural exchange can foster creativity, innovation, and intercultural understanding. However, it can also pose challenges in terms of preserving local traditions and identities. Urban areas must find a balance between embracing cultural diversity and maintaining their cultural heritage.

The question of whether rural-to-urban migration should be allowed to flourish is a complex and multifaceted one. There are arguments both in favor of and against such migration, each with its own merits and considerations. To fully understand the implications and consequences of allowing rural-to-urban migration to flourish, it is important to examine the various factors at play.

One of the key arguments in favor of allowing rural-to-urban migration to flourish is the potential economic benefits it can bring. Urban areas often offer greater job opportunities and higher wages compared to rural areas. By allowing individuals to migrate from rural to urban areas, it can help stimulate economic growth and development in both regions. This can lead to increased productivity, improved living standards, and a reduction in poverty levels.

Rural-to-urban migration can also contribute to the overall development and modernization of rural areas. As individuals migrate to urban areas and acquire new skills and knowledge, they can bring these skills back to their rural communities. This can help diversify the local economy, improve infrastructure, and enhance the quality of education and healthcare services available in rural areas.

However, there are also valid concerns and arguments against allowing rural-to-urban migration to flourish. One of the main concerns is the strain it can put on urban infrastructure and resources. As more individuals migrate to urban areas, there is an increased demand for housing, transportation, healthcare, and other public services. If not properly managed, this can lead to overcrowding, inadequate infrastructure, and a decline in the quality of life for both urban and rural residents.

Rural-to-urban migration can also contribute to social and cultural challenges. As individuals migrate to urban areas, they often leave behind their communities, families, and traditional ways of life. This can lead to a loss of cultural identity and social cohesion in rural areas. It can also result in social inequality and marginalization in urban areas, as migrants often face challenges in accessing education, healthcare, and other social services.

5.4.3 Viable Social and Engineering Solutions

Several viable social and engineering solutions can be implemented to address the overall issues of rural-to-urban migration. These solutions should be able to mitigate the challenges faced by individuals and communities during the process of migration while also promoting sustainable development and improved quality of life in both cities and rural areas.

One social solution is the establishment of comprehensive support systems for migrants. This includes the provision of information and guidance on employment opportunities, housing options, and social services available in urban areas. Additionally, programs can be developed to offer language and vocational training to help migrants acquire the necessary skills for urban employment. By providing these support systems, migrants can have a smoother transition into urban life and be better equipped to contribute to the local economy. Another social solution is the promotion of community engagement and integration. This can be achieved through the creation of community centers or spaces where migrants and residents can interact and participate in cultural exchange activities. By fostering social cohesion and understanding between different groups, tensions and conflicts arising from migration can be reduced. Additionally, initiatives can be implemented to encourage migrants to participate in local decision-making processes, ensuring their voices are heard and their needs are met.

On the engineering front, one solution is the development of sustainable urban infrastructure. This involves the planning and construction of affordable housing, transportation systems, and public amenities that can accommodate the growing population in urban areas. By investing in sustainable infrastructure, cities can provide adequate living conditions for migrants and alleviate strain on existing resources. Furthermore, the integration of smart technologies can enhance the efficiency and effectiveness of urban services, improving the overall quality of life for both migrants and existing residents. Another solution is the decentralization of economic opportunities. This can be achieved by promoting the establishment of industries and job opportunities in rural areas, thereby reducing the need for migration to urban centers. This can be facilitated through government incentives, such as tax breaks or subsidies, to attract businesses to rural areas. By creating economic opportunities in rural regions, individuals are more likely to choose to stay or return to their hometowns, reducing the pressure on urban areas.

Promoting small and micro enterprises in rural areas has emerged as a crucial strategy to suppress the push factors that drive rural-to-urban migration. This approach recognizes the need to create economic opportunities and improve livelihoods in rural communities, thereby reducing the allure of urban areas and stemming the tide of migration.

By focusing on promoting small and micro enterprises in rural areas, policymakers and development practitioners are tackling the root causes of rural-to-urban migration. Small and micro enterprises play a vital role in generating employment opportunities, stimulating economic growth, and improving living standards in rural communities. They provide an avenue for local entrepreneurship, harnessing the potential of rural populations and creating sustainable livelihoods.

One of the key advantages of promoting small and micro enterprises in rural areas is their ability to capitalize on local resources and knowledge. Rural communities often possess unique skills, traditions, and natural assets that can be leveraged to create niche markets and competitive advantages. By supporting these enterprises, policymakers can tap into these local resources, fostering economic diversification and reducing dependence on agriculture as the sole source of income. Promoting small and micro enterprises in rural areas can also help address the issue of youth unemployment. In many rural communities, young people face limited job prospects, leading them to migrate to urban centers in search of employment opportunities. Creating an enabling environment for entrepreneurship and providing targeted support to young entrepreneurs, policymakers can empower youth to stay in their communities and contribute to local economic development.

However, promoting small and micro enterprises in rural areas is not without its challenges. Limited access to finance, inadequate infrastructure, and a lack of business development services are common barriers that entrepreneurs face in rural settings. Therefore, policymakers must address these constraints by implementing targeted interventions such as providing microcredit facilities, improving rural infrastructure, and offering training programs to enhance entrepreneurial skills.

Promoting small and micro enterprises in rural areas can be a challenging task, especially amidst rural poverty and the insufficient skills of rural laborers. However, it is crucial to address these challenges to foster economic growth and development in rural areas. One way to promote small and micro enterprises is by providing access to finance. Rural entrepreneurs often lack collateral, credit history, and financial literacy, making it difficult for them to access finance. Therefore, providing financial support in the form of grants, loans, or subsidies can help stimulate entrepreneurship and innovation in rural areas.

Another way to promote small and micro enterprises is by providing training and capacity-building programs. Rural entrepreneurs often lack the necessary skills and knowledge to run successful businesses. Therefore, training programs that focus on business management, marketing, and financial management can help equip rural entrepreneurs with the skills they need to succeed. In addition to financial support and training programs, creating a supportive business environment can also help promote small and micro enterprises in rural areas. This can be achieved through policies that encourage entrepreneurship, such as tax incentives, streamlined business registration processes, and reduced regulatory burdens.

Promoting collaboration and networking among rural entrepreneurs can also help foster innovation and growth. Creating opportunities for rural entrepreneurs to connect and share ideas enables them to learn from each other's successes and failures and collaborate on projects that benefit the entire community. Promoting small and micro enterprises in rural areas requires a multi-pronged approach that addresses the unique challenges faced by rural entrepreneurs. Providing access to finance, training programs, a supportive business environment, and opportunities for collaboration can stimulate entrepreneurship and innovation in rural areas, contributing to their economic growth and development.

Promoting small and micro enterprises in rural areas is an effective strategy to suppress the push factors of rural-to-urban migration. Creating economic opportunities, improving livelihoods, and empowering rural communities can help policymakers reduce the allure of urban areas and foster sustainable development in rural areas. However, it is essential to address the challenges faced by entrepreneurs in rural settings through targeted interventions that enhance access to finance, infrastructure, and business development services. With concerted efforts and effective policies, promoting small and micro enterprises can be a catalyst for transforming rural economies and creating a brighter future for rural communities.

Rural-to-urban migration brings with it a range of social and environmental problems. From the strain on urban infrastructure and services to the breakdown of social networks, these challenges can have long-lasting effects on both migrants and the urban areas they settle in. Policymakers and stakeholders must address these issues through comprehensive strategies that focus on providing adequate resources, support systems, and sustainable development practices. This approach helps us work towards creating cities that are more inclusive and resilient, benefiting both migrants and existing urban residents.

5.5 INNOVATIVE AND AFFORDABLE HOUSING SYSTEM

Innovative and affordable housing systems are a crucial need in today's world. With the ever-increasing population and limited resources, it is essential to develop housing systems that are not only cost-effective, that is, affordable, but also sustainable. One of the most innovative approaches to affordable housing is the use of modular construction. Modular construction involves building housing units in a factory and then transporting them to the site for assembly. This method reduces construction time and labor costs, making it an affordable option for developers. Additionally, modular construction is environmentally friendly, as it reduces waste and improves energy efficiency. Another innovative approach to affordable housing is the use of tiny homes. Tiny homes are small, portable houses that are usually less than 500 ft^2 (46 m^2). They are a cost-effective solution for individuals or families who want to reduce their living expenses or live a minimalist lifestyle. Tiny homes can be built using sustainable materials, such as recycled wood or metal, making them an eco-friendly option.

Co-housing is another affordable housing system that is gaining popularity. Co-housing involves a group of people living together in a shared space, such as a house or apartment complex. Each resident has their own private space, but they also share common areas, such as kitchens and living rooms. This approach reduces living expenses, as residents share the cost of utilities and other expenses.

In contrast, affordable housing is only a dream for many citizens. The current state of high housing prices in cities can be attributed to the high demand for housing. As more people flock to urban areas in search of better job opportunities and a higher standard of living, the demand for housing has increased significantly. This has led to a shortage of available housing, which in turn has driven up prices. The situation is further exacerbated by the limited amount of land available for development in cities. With space at a premium, developers must pay a premium to acquire land and build new housing units. This cost is then passed on to consumers in the form of higher prices. While high housing prices may be good news for property owners and investors, they pose a significant challenge for those looking to buy or rent a home. The high cost of housing can make it difficult for individuals and families to find affordable and suitable accommodation, especially in areas with high demand.

There are other factors contributing to the high demand for housing in cities. For example, many young professionals are delaying marriage and children, which means they are staying in cities longer than previous generations. Furthermore, immigration to cities from other countries is also contributing to the demand for housing.

Amidst the escalating housing prices in cities, the promotion of affordable housing has become a persistent issue. To address this challenge, it is crucial to consider the design aspect of housing. By focusing on innovative housing design solutions, we can create more affordable and sustainable housing options for urban dwellers. One approach to promoting affordable housing through design is by incorporating efficient space utilization techniques. Many cities face a scarcity of land, making it essential to maximize the use of available space. This can be achieved through the implementation of compact and multifunctional design concepts. For instance, the integration of flexible living spaces that can serve multiple purposes, such as transforming from a bedroom to a workspace, can help optimize the use of limited square footage. The use of modular construction techniques can significantly contribute to affordable housing. Modular housing involves the construction of prefabricated units that can be easily assembled on-site. This method reduces construction time and costs, making it an attractive option for affordable housing projects. Moreover, modular designs allow for customization and scalability, enabling residents to adapt their homes to their changing needs over time.

Sustainable design practices should be prioritized in the development of affordable housing. By integrating energy-efficient technologies and materials, we can reduce long-term operational costs for residents. This includes incorporating renewable energy sources, such as solar panels, and implementing water-saving measures like rainwater harvesting systems. Additionally, the use of eco-friendly building materials can enhance indoor air quality and reduce environmental impact.

To foster a sense of community and promote social interaction, the design of affordable housing should also prioritize communal spaces. Shared amenities such as gardens, playgrounds, and community centers can create opportunities for residents to connect and engage with one another. These spaces not only enhance the quality of life for residents but also contribute to a more vibrant and cohesive neighborhood. It is also essential to consider the accessibility and inclusivity of affordable housing designs. Incorporating universal design principles ensures that housing units are accessible to people of all ages and abilities. Features such as wider doorways, step-free entrances, and accessible bathrooms can improve the usability and comfort of these homes. Additionally, attention should be given to the provision of adequate parking spaces and public transportation accessibility to cater to the diverse needs of residents.

There are different ways to approach this problem from various perspectives, such as urban planning, economics, and social justice. Some possible solutions are (1) increasing the supply of housing units by relaxing zoning regulations, incentivizing mixed-use development, and supporting public housing projects; (2) reducing the demand for housing by encouraging telecommuting, improving public transportation, and promoting regional cooperation; and (3) enhancing the affordability of housing by providing subsidies, vouchers, tax credits, and rent control measures for eligible households.

While there is no one-size-fits-all solution to this affordable housing problem, there are several alternative approaches that cities can explore to make housing more affordable for their residents. One approach is the implementation of inclusionary zoning policies. Inclusionary zoning requires developers to include a certain percentage of affordable units in new housing developments. This helps to ensure that there is a mix of housing options available at various price points, making it more feasible for low and moderate-income individuals to find suitable housing. In addition to creating affordable units, inclusionary zoning can also help to promote economic diversity within neighborhoods and prevent the concentration of poverty.

Another strategy that cities can pursue is the development of community land trusts. Community land trusts are non-profit organizations that acquire and hold land to create affordable housing. By taking land out of the speculative market and placing it under community control, land trusts can help to stabilize housing prices and ensure long-term affordability. This model has been successful in cities such as Burlington, Vermont, where the Champlain Housing Trust has created hundreds of permanently affordable homes.

Cities can also explore the option of implementing rent control policies. Rent control places limits on how much landlords can increase rent each year, providing stability and predictability for tenants. While rent control has its critics, proponents argue that it helps to prevent displacement and allows long-term residents to remain in their communities, even as housing costs rise. However, it is important to strike a balance with rent control to avoid unintended consequences such as reduced investment in rental housing or a decline in the quality of existing units. In addition to these policy approaches, cities can also leverage public–private partnerships to create more affordable housing. Collaborating with developers and investors allows cities to access additional resources and expertise, thereby increasing the supply of affordable housing units. This can take the form of tax incentives or subsidies for developers who include affordable housing in their projects, or the creation of dedicated funding streams for affordable housing initiatives. Cities can prioritize the preservation of existing affordable housing stock. This can be achieved through the rehabilitation and renovation of older buildings, as well as the acquisition of properties at risk of being converted to market-rate housing. Preserving existing affordable units helps cities prevent the loss of affordable housing and ensures that current residents are not displaced. Cities can explore innovative construction methods and designs that can help lower the cost of building new housing. This includes modular construction, which involves the fabrication of building components off-site and their assembly on-site.

Modular construction can help to reduce construction time and costs, making it a more affordable option for developers. Cities can also encourage the use of sustainable and energy-efficient building materials and practices, which can help to lower operating costs for residents.

With increasing numbers of people struggling to find homes that fit their budgets, the affordable housing issue becomes imperative. In response, governments have implemented various measures to provide subsidies for affordable housing. These subsidies can take many forms, including tax incentives, grants, and low-interest loans. The goal of these subsidies is to make it easier for developers to build affordable housing projects, which can then be rented or sold at lower prices. This not only helps people find homes they can afford, but it also helps to address issues of homelessness and overcrowding.

One of the most common ways that governments provide subsidies for affordable housing is through tax incentives. These incentives can take many forms, such as credits for developers who build affordable housing units or tax breaks for individuals who invest in affordable housing projects. These incentives can help to offset some of the costs associated with building affordable housing, making it more financially feasible for developers.

Another way that governments provide subsidies for affordable housing is through grants. These grants can be used to fund various aspects of affordable housing projects, such as land acquisition, construction costs, or infrastructure improvements. Providing these grants allows governments to reduce the overall cost of building affordable housing, making it more accessible to those in need. Governments can also provide low-interest loans to developers who are building affordable housing projects. These loans can help to lower the cost of financing, making it easier for developers to build and maintain affordable housing units. In some cases, these loans may also include provisions for forgiveness or deferment if certain conditions are met.

The possible solution to affordable housing in urban areas is to apply engineering and design principles to create more efficient, sustainable, and livable housing units that can accommodate more people with fewer resources. Some of the aspects that engineering and design can address are as follows: (1) Space optimization: using modular, prefabricated, or adaptable components that can fit different needs and preferences, such as movable walls, foldable furniture, or multifunctional spaces. (2) Energy efficiency: using renewable energy sources, such as solar panels, wind turbines, or geothermal systems, to reduce the reliance on fossil fuels and lower utility bills. Also, incorporating passive design strategies, such as natural ventilation, insulation, or shading, to reduce the heating and cooling loads. (3) Environmental impact: using recycled, biodegradable, or low-impact materials, such as bamboo, hemp, or straw, to reduce the waste and emissions generated by the construction and operation of the housing units. Also, implementing green features, such as rainwater harvesting, greywater recycling, or composting, reduces water consumption and improves soil quality.

Conventional construction methods of affordable housing can be time-consuming and expensive, making it difficult to keep costs down. However, with innovative engineering design, we can explore alternative construction techniques that are both efficient and cost-effective. Other than modular construction, that is, prefabricating housing units in a factory setting, the use of sustainable materials. Incorporating reused and eco-friendly materials into the construction process can reduce the environmental impact of housing and lower costs. For example, using recycled materials such as reclaimed wood or recycled plastic can significantly reduce construction expenses without compromising quality. Energy-efficient design principles can also play a crucial role in achieving affordable housing. Incorporating features such as proper insulation, efficient heating and cooling systems, and energy-saving appliances can significantly reduce utility costs for residents. This not only makes housing more affordable but also contributes to a greener and more sustainable future.

In conclusion, innovative engineering design holds great potential in addressing the challenge of affordable housing for citizens. By embracing modular construction, sustainable materials, energy-efficient design principles, and community engagement, we can create housing solutions that are both affordable and environmentally friendly. It is through these innovative approaches that we can make a meaningful difference in the lives of citizens by providing them with safe, comfortable, and affordable housing options.

5.6 GREEN BUILDING PRINCIPLES IN URBAN HOUSING

Adoption of green building principles in urban housing is a crucial strategy to mitigate the environmental impacts of urbanization and improve the quality of life for urban residents. Green building principles include energy efficiency, water conservation, waste reduction, the use of renewable materials, and integration of natural elements. These principles can be applied to various aspects of urban housing, such as design, construction, operation, maintenance, and renovation. Adopting green building principles, urban housing can reduce greenhouse gas emissions, lower utility costs, enhance indoor comfort and health, and create more livable and resilient communities.

Green building principles are a set of guidelines that aim to reduce the environmental impact of the construction and operation of buildings. Some of the aspects that are considered in green building principles are:

1. *Energy Efficiency:* This refers to the use of renewable and low-carbon sources of energy, as well as the optimization of energy consumption and performance of buildings. Examples of energy efficiency measures include solar panels, insulation, natural ventilation, and smart lighting systems.

   Energy efficiency in buildings is a crucial aspect of reducing greenhouse gas emissions and mitigating climate change. There are several ways to achieve energy efficiency in buildings, such as:

   a. Designing buildings with passive solar features, such as orientation, insulation, shading, ventilation, and thermal mass, reduces the need for artificial heating and cooling.
   b. Installing energy-efficient appliances and lighting systems, such as light emitting diodes (LEDs) bulbs, smart thermostats, and solar water heaters, that consume less electricity and generate less heat.
   c. Implementing renewable energy sources, such as solar panels, wind turbines, and geothermal heat pumps, that provide clean and low-carbon energy for buildings.
   d. Adopting smart building technologies, such as sensors, meters, and controllers, which monitor and optimize the energy performance of buildings and enable demand response and load shifting.
   e. Promoting behavioral changes among building occupants, such as turning off lights and appliances when not in use, adjusting thermostats to comfortable levels, and using natural ventilation and daylighting when possible.
2. *Water Conservation:* This refers to the reduction of water use and waste, as well as the protection and enhancement of water quality and resources. Examples of water conservation measures include rainwater harvesting, low-flow fixtures, greywater recycling, and green roofs.

   Water conservation is the practice of using water efficiently and reducing unnecessary waste. In buildings, water conservation can be achieved by implementing various strategies, such as:

   a. Installing low-flow fixtures and appliances, such as faucets, showerheads, toilets, dishwashers, and washing machines, that use less water per use.
   b. Repairing leaks and drips in pipes, faucets, and toilets can waste a significant amount of water over time.
   c. Harvesting rainwater and using it for non-potable purposes, such as irrigation, flushing toilets, and washing cars.

d. Reusing greywater, which is the wastewater from sinks, showers, and laundry, for similar purposes as rainwater.
e. Educating occupants and tenants about the benefits of water conservation and encouraging them to adopt water-saving behaviors, such as taking shorter showers, turning off the faucet when not in use, and washing only full loads of dishes and clothes.

Water conservation in buildings can have multiple benefits, such as: (a) saving money on water bills and reducing the need for costly infrastructure upgrades; (b) preserving freshwater resources and protecting the environment from water pollution and depletion; (c) enhancing the performance and durability of plumbing systems and appliances; and (d) improving the health and comfort of occupants and tenants by providing adequate water quality and quantity.

Water conservation in buildings is a smart and sustainable practice that can benefit both the building owners and managers and society at large.

3. *Material Selection:* This refers to the use of materials that are durable, recyclable, biodegradable, or have low embodied energy and emissions. Examples of material selection measures include bamboo, cork, recycled steel, and natural paints.

Material selection is an important aspect of sustainable design, as it involves choosing materials that have minimal environmental impact throughout their life cycle. Some of the criteria for material selection are durability, recyclability, biodegradability, and low embodied energy and emissions. These criteria help reduce waste, pollution, and greenhouse gas emissions, as well as conserve natural resources and energy. Some examples of materials that meet these criteria are bamboo, cork, recycled steel, and natural paints. Here are some more examples:

a. *Bamboo:* Bamboo is a rapidly renewable resource with various benefits as a building material. It has high tensile strength, high fire resistance, good elastic properties suitable for earthquake-prone areas, and is easy to cut, handle, repair, reposition, and maintain. It is also versatile and can bend into any shape. Bamboo is non-polluting and does not have crusts or parts that can be considered waste. It's sustainable, eco-friendly, and proliferates.
b. *Cork:* Cork is light, resilient, elastic, wear-resistant, boasts insulating properties, is naturally fire retardant, hypoallergenic, highly flexible, and compressible. It is resistant to moisture, friction, and abrasion. It provides excellent thermal insulation and acoustic insulation. It is highly impermeable and has low embodied energy.
c. *Recycled Steel:* Recycled steel can be used in construction projects to reduce waste and the need for new raw materials. Recycled steel is eco-friendly, retains high quality after recycling, can be recycled easily, and recycling metal reduces waste.
d. *Natural Paints:* Natural paints derived from plant and mineral materials have subtle, soothing colors, pleasant scents, and help create a healthy environment. They avoid indoor air pollution and can have an unusual, attractive appearance. The oils and resins used in natural paints penetrate deep within the timber fibers, giving long-lasting, deep protection. Also, natural paints and finishes are biodegradable, and the manufacturing process involves minimal energy, waste, and pollution.
e. *Hemp:* Hemp is a versatile plant that can be used for textiles, paper, insulation, and biofuel. Hemp fibers are strong, durable, and biodegradable. Hemp also absorbs carbon dioxide and does not require pesticides or herbicides to grow.
f. *Straw Bale:* Straw bale is a building material that consists of compressed straw covered with plaster or lime. Straw bale is a good insulator, fire-resistant, and inexpensive. Straw bale also uses agricultural waste that would otherwise be burned or discarded.
g. *Rammed Earth:* Rammed earth is a technique that involves compacting soil mixed with water and a stabilizer such as cement or lime. Rammed earth creates solid walls that are durable, thermally efficient, and aesthetically pleasing. Rammed earth also uses local materials and reduces the need for transportation and processing.

Through conscious choices of materials like these, we can construct buildings that are not only functional and aesthetically pleasing but also environmentally friendly.

4. *Indoor Environmental Quality (IEQ):* This refers to the provision of a healthy and comfortable indoor environment for the occupants of buildings. Examples of IEQ measures include natural lighting, ventilation, air filtration, noise control, and ergonomic design.

   IEQ is a critical aspect of building design and operation that focuses on creating healthy and comfortable indoor environments for occupants. It encompasses a wide range of elements, including:

   a. *Natural Lighting:* Adequate natural lighting is essential for the well-being of building occupants. It not only reduces the need for artificial lighting but also helps regulate circadian rhythms, improve mood, and increase productivity.
   b. *Ventilation:* Proper ventilation ensures a continuous supply of fresh air, dilutes indoor pollutants, and helps control temperature and humidity levels. It can be achieved through mechanical systems or natural means, such as windows and vents.
   c. *Air Filtration:* High-quality air filtration systems can remove particulates, allergens, and other pollutants from indoor air, contributing to better respiratory health.
   d. *Noise Control:* Acoustic comfort is another important aspect of IEQ. Excessive noise can cause stress and hinder concentration. Noise control measures include sound-absorbing materials and strategic space planning.
   e. *Ergonomic Design:* This involves designing the workspace to fit the user's needs, reducing the risk of strain or injury. It can include adjustable furniture, proper placement of computer screens, and consideration of factors like reach and viewing angles.

      Measuring IEQ involves assessing several key components: (a) *Building Type and Construction Materials:* The type of building and the materials used during its construction can significantly impact the indoor environment. (b) *Ventilation System:* The ventilation system plays a crucial role in maintaining good air quality by ensuring a continuous supply of fresh air. (c) *Occupancy:* The number of people residing or working in the area can affect the indoor environment, particularly in terms of air quality. (d) *Health-Related Problems and Disease History:* Any health-related problems or disease history of people residing or working in the area can provide insights into potential IEQ issues. (e) *Activities and Equipment:* Activities or types of equipment in the vicinity can contribute to indoor pollution. (f) *Cleanliness:* The cleanliness of the area can influence IEQ, particularly in terms of dust and mold. (g) *Air Quality Monitors:* Air quality can be measured using an in-home monitor that gives real-time readings. (h) *Mold Tests:* Mold tests come in various formats, including swabs, tape strips, air pumps, and petri dish tests. (i) *Radon Tests:* Radon is a colorless, odorless, radioactive gas that can be measured using inexpensive devices. (j) *Carbon Monoxide Detectors:* These devices can be installed on every level of your home to detect unhealthy levels of carbon monoxide.

      Emphasizing these aspects in the design and operation of buildings allows us to foster indoor environments that enhance health, comfort, and productivity. Nonetheless, it is crucial to understand that optimal IEQ can only be achieved through a comprehensive approach that integrates all these factors.

5. *Site Selection and Planning:* This refers to the consideration of the location, orientation, and layout of buildings with the natural and built environment. Examples of site selection and planning measures include brownfield redevelopment, urban infill, passive solar design, and landscaping.

   Site selection and planning play a crucial role in sustainable building design. It involves careful consideration of the location, orientation, and layout of buildings in the natural and built environment. Various measures are involved in site selection and planning.

One such measure is brownfield redevelopment, which involves the redevelopment of previously used or contaminated land. This approach helps reduce urban sprawl, preserve green space, and remediate environmental hazards. Another measure is urban infill, which involves developing vacant or underused parcels within existing urban areas. This practice helps make efficient use of existing infrastructure, reduces commuting distances, and revitalizes neighborhoods.

Passive solar design is another approach that involves using the sun's energy for heating and cooling living spaces. Proper orientation and design of a building can take advantage of natural light and heat, reducing the need for artificial lighting and heating/cooling systems. Landscaping is also an important factor to consider, as it can enhance a building's energy performance. Trees can provide shade in summer, reducing cooling needs, while allowing sunlight in winter to help heat the building. Landscaping can also contribute to stormwater management and create habitats for local wildlife.

Building orientation is another critical consideration that impacts energy consumption. For example, in cooler climates, a building with its long axis running east–west can maximize southern exposure, allowing for more solar heat gain. Layout planning is also essential, considering factors like prevailing winds (for natural ventilation), views, privacy, and noise. Placing living areas on the south side of a building can take advantage of daylight.

5.7 DAYLIGHTING, VENTILATION, AND THERMAL COMFORT

Daylighting and ventilation are two aspects of green building design that aim to reduce energy consumption. In tropical countries, these aspects are more critical and require special attention, as tropical countries have great potential to maximize their use of them. Therefore, the aspects of daylighting and ventilation are discussed separately.

Daylighting and ventilation are essential elements in the design and construction of buildings, playing a crucial role in reducing home energy consumption and creating comfortable and healthy indoor environments. Daylighting refers to the practice of utilizing natural light to illuminate the interior spaces of a building, while ventilation involves the movement of fresh air to maintain air quality and regulate temperature. Both daylighting and ventilation have numerous benefits, ranging from energy efficiency and cost savings to improved occupant well-being.

One of the primary advantages of incorporating daylighting into a building is its potential for reducing energy consumption. By maximizing the use of natural light, artificial lighting needs can be minimized, resulting in lower electricity usage. This not only reduces the building's environmental footprint but also translates into significant cost savings for the occupants or owners. Moreover, daylighting can contribute to a more visually appealing and comfortable environment, as natural light tends to be softer and more pleasing to the eye compared to artificial lighting. In addition to energy efficiency, daylighting has been linked to various health benefits for building occupants. Exposure to natural light has been shown to improve mood, increase productivity, and enhance overall well-being. Studies have indicated that access to daylight in educational settings can enhance students' academic performance and concentration levels. Similarly, in office environments, workers exposed to natural light tend to experience higher job satisfaction and lower levels of stress. Furthermore, daylighting has been associated with faster recovery times in healthcare facilities, as patients exposed to natural light often experience reduced pain and anxiety.

To effectively incorporate daylighting into a building, several design strategies can be employed. The orientation and placement of windows and skylights are crucial factors in maximizing the amount of natural light entering a space. South-facing windows receive the most sunlight throughout the day, while north-facing windows provide more consistent but less intense illumination. The size and positioning of windows should be carefully considered to balance the desired amount of daylight with potential glare issues. Additionally, the use of light shelves, light tubes, or reflective surfaces can help distribute daylight deeper into a building's interior.

While daylighting focuses on harnessing natural light, ventilation is concerned with providing fresh air circulation within a building. Proper ventilation is essential for maintaining good indoor air quality, removing pollutants, regulating temperature, and preventing the buildup of excess moisture. Insufficient ventilation can lead to a range of health issues, including respiratory problems, allergies, and the growth of mold and mildew.

There are various methods of achieving effective ventilation in buildings. Natural ventilation relies on the use of windows, doors, and vents to allow air movement. This approach takes advantage of natural pressure differences created by wind or temperature variations to drive airflow. Mechanical ventilation systems, on the other hand, utilize fans or ductwork to circulate air. These systems can be particularly useful in buildings with limited access to natural ventilation or in areas with high pollution levels. Also, hybrid ventilation systems combine both natural and mechanical methods to optimize airflow based on external conditions.

In recent years, there has been a growing emphasis on designing buildings that integrate both daylighting and ventilation strategies. This approach recognizes the interconnectedness of these elements and seeks to create synergies that enhance overall building performance. For example, windows designed for daylighting can also be designed to allow for effective natural ventilation. This integration not only improves energy efficiency but also provides occupants with greater control over their indoor environment.

Thermal comfort is a crucial aspect of building design and construction, referring to the state of mind that expresses satisfaction with the thermal environment. Achieving thermal comfort involves creating a space that maintains a temperature range that is neither too hot nor too cold, and where the humidity and air movement are at appropriate levels. Thermal comfort is essential for occupant well-being, productivity, and overall satisfaction with the indoor environment.

The human body has a natural ability to regulate its internal temperature, and thermal comfort is achieved when the body's heat production and dissipation are in balance. The factors that influence thermal comfort include air temperature, humidity, air velocity, and radiant temperature. These variables can be measured and evaluated using various tools, such as thermal comfort models and sensors.

The American Society of Heating, Refrigerating, and Air-Conditioning Engineers (ASHRAE) has developed a set of standards for thermal comfort, which consider the interactions between environmental factors and human physiology. The ASHRAE standards define a range of acceptable indoor temperatures, humidity levels, and air velocities that are considered optimal for thermal comfort. These standards also consider factors such as clothing insulation, metabolic rate, and air movement preferences.

To achieve thermal comfort in a building, several design strategies can be employed. These include the use of insulation to reduce heat loss or gain, the installation of energy-efficient heating and cooling systems, and the use of shading devices to control solar heat gain. In addition, the use of natural ventilation systems can help regulate indoor air quality and temperature, while also reducing energy consumption.

One of the primary benefits of achieving thermal comfort is improved occupant well-being. Maintaining a comfortable indoor environment can reduce stress levels, improve sleep quality, and enhance overall health. In contrast, exposure to extreme temperatures can lead to a range of health issues, including dehydration, heat exhaustion, and hypothermia.

Thermal comfort also has implications for productivity and performance. Studies have shown that workers in comfortable indoor environments tend to be more productive and exhibit higher levels of job satisfaction. Conversely, workers in uncomfortable environments may experience reduced concentration and motivation levels. In educational settings, thermal comfort has been linked to improved academic performance and reduced absenteeism rates.

Achieving thermal comfort can have significant energy-saving benefits as well. By optimizing heating and cooling systems to maintain an appropriate temperature range, energy consumption can be minimized. This not only reduces the environmental footprint of a building but also translates into significant cost savings for occupants or owners.

5.8 THE MINIMUM ENVIRONMENTAL FOOTPRINT OF BUILDINGS AND BUILDING MATERIALS

The environmental footprint refers to the impact that human activities have on the environment, particularly in terms of resource consumption, pollution generation, and contribution to climate change. It is a measure of the overall environmental impact of a product, service, or activity throughout its life cycle, including the extraction of raw materials, production, use, and disposal. The environmental footprint encompasses various aspects, including energy consumption, water usage, waste generation, and greenhouse gas emissions. These factors are interconnected and contribute to environmental degradation and the depletion of natural resources. Minimizing the environmental footprint is crucial for achieving sustainability and preserving the planet for future generations.

The environmental footprint of a building exhibits the impacts that building construction, operation, and demolition have on the environment. Buildings are responsible for a significant portion of global energy consumption, resource depletion, and greenhouse gas emissions. The environmental footprint of a building encompasses various factors, including energy consumption, water usage, waste generation, and material usage.

To minimize the environmental footprint of a building, it is essential to adopt sustainable practices throughout its life cycle. This includes designing buildings for energy efficiency, water conservation, and waste reduction. Implementing renewable energy sources such as solar panels or geothermal systems can significantly reduce energy consumption and associated greenhouse gas emissions.

The choice of building materials also plays a crucial role in minimizing the environmental footprint of a building. Utilizing sustainable materials such as recycled or renewable materials can help reduce the embodied carbon of a building. Embodied carbon refers to the carbon emissions generated during the production and transportation of building materials. By reducing embodied carbon, it is possible to minimize the environmental impact of a building's construction. Designing buildings for durability and adaptability can help extend their lifespan and reduce the need for new construction. This can significantly reduce waste generation and promote a circular economy approach. Implementing waste management strategies such as recycling and reusing construction materials can also help minimize waste generation and promote sustainability by minimizing the environmental footprint of buildings, including the use of building materials.

The concept of minimizing the environmental footprint of buildings and building materials has gained significant attention in recent years as the construction industry seeks to mitigate its impact on the environment. Buildings are responsible for a substantial portion of global energy consumption, resource depletion, and greenhouse gas emissions. By adopting sustainable practices and utilizing environmentally friendly materials, it is possible to reduce the environmental footprint of buildings and promote a more sustainable built environment.

One of the key aspects of minimizing the environmental footprint of buildings is improving energy efficiency. Buildings consume vast amounts of energy for heating, cooling, lighting, and operating various systems. Implementing energy-efficient design strategies and utilizing advanced technologies can significantly reduce energy consumption and associated greenhouse gas emissions. This can be achieved through measures such as optimizing insulation, using high-performance windows, implementing efficient heating, ventilation, and air conditioning (HVAC) systems, and incorporating renewable energy sources like solar panels or geothermal systems.

Water conservation plays a critical role in minimizing the environmental footprint of buildings. Traditional buildings consume large quantities of water for various purposes, including sanitation, irrigation, and cooling systems. Implementing water-efficient fixtures, like low-flow toilets and faucets, and utilizing rainwater harvesting systems for non-potable water needs can reduce buildings' water consumption and alleviate the strain on local water resources.

The choice of building materials has a significant impact on the environmental footprint of a building. Traditional construction materials, such as concrete and steel, have high carbon footprints due to their manufacturing processes and the extraction of raw materials. To minimize the environmental impact, it is essential to consider alternative materials that are renewable, recycled, or have lower embodied carbon. For example, using sustainably sourced timber as a structural material can help sequester carbon dioxide from the atmosphere and reduce reliance on energy-intensive materials.

Another important aspect of minimizing the environmental footprint of buildings is waste reduction and management. Construction and demolition activities generate substantial amounts of waste that often end up in landfills. Implementing waste management strategies, like recycling and reusing construction materials, can minimize waste generation and promote a circular economy approach. Also, designing buildings for adaptability and disassembly facilitates easier renovation or deconstruction in the future, reducing the need for new construction and further minimizing waste.

Beyond the operational phase, the lifespan of a building also plays a role in its environmental footprint. Designing buildings for durability and longevity can help reduce the need for frequent renovations or demolitions. Additionally, considering the potential for adaptive reuse or repurposing of existing structures can further extend their lifespan and minimize resource consumption.

To assess and quantify the environmental footprint of buildings, various tools and certifications are available. Life Cycle Assessment (LCA) is commonly used to evaluate the environmental impact of a building throughout its entire life cycle, including materials extraction, construction, operation, maintenance, and end of life. LCA takes into account factors such as energy consumption, greenhouse gas emissions, water usage, and waste generation to provide a comprehensive analysis of a building's environmental performance.

Several green building certifications exist to recognize buildings that meet specific sustainability criteria. Leadership in Energy and Environmental Design (LEED) is one of the most widely recognized certifications globally. LEED evaluates buildings based on criteria such as energy efficiency, water conservation, indoor environmental quality, and sustainable materials. Other certifications include Building Research Establishment Environmental Assessment Method (BREEAM), Green Star, and WELL Building Standard, which focuses on occupant health and well-being.

Selecting building materials that minimize the environmental footprint of a building involves considering various factors, including the embodied carbon of the materials, their sustainability, and their potential for recycling or reuse. Here are factors to consider when selecting building materials that minimize the environmental footprint of buildings, as shown in Table 5.4.

A life cycle assessment (LCA) of a building is a comprehensive assessment that evaluates the environmental impact of a building throughout its entire life cycle, from raw material extraction to construction, operation, and eventual demolition or reuse. LCA takes into account various factors and stages involved in the building's life cycle to provide a holistic understanding of its environmental footprint.

The first stage of the LCA process involves defining the goal and scope of the study. This includes determining the functional unit, which is the quantifiable measure of the building's performance or service provided. For example, the functional unit could be the square meters of usable space or the number of occupants supported by the building. Once the goal and scope have been established, the inventory analysis stage begins. This involves identifying and quantifying all inputs and outputs associated with each life cycle stage of the building. Inputs include materials, energy, and water, while outputs include emissions and waste. The inventory analysis covers all stages of the building's life cycle, including raw material extraction, manufacturing, transportation, construction, operation, maintenance, and end-of-life.

TABLE 5.4
Factors for Selecting Building Materials

Factor	Explanation
Embodied carbon	Embodied carbon refers to the carbon emissions generated during the production and transportation of building materials. Materials with high embodied carbon, such as concrete and steel, have a significant environmental footprint. Consider using materials with lower embodied carbon, such as sustainably sourced timber or recycled materials.
Sustainability	Consider the sustainability of the materials used in construction. Look for materials that are renewable, such as bamboo or cork, or materials that can be recycled or reused.
Durability	Durability is an essential factor in selecting building materials that minimize the environmental footprint of a building. Durable materials can help extend the lifespan of a building and reduce the need for new construction.
Energy efficiency	Energy-efficient materials can help reduce energy consumption and associated greenhouse gas emissions. Look for materials that have high insulation values or reflective properties to reduce heating and cooling needs.
Water usage	Water-efficient materials can help reduce water consumption and alleviate the strain on local water resources. Look for materials that require less water in their production or can be recycled.
Waste management	Consider the waste generated during the production and disposal of building materials. Choose materials that can be recycled or reused and avoid materials that generate significant waste during production.
Certifications	Look for materials that have been certified by third-party organizations, such as the Forest Stewardship Council (FSC) or Cradle to Cradle (C2C), which verify their sustainability and environmental impact.

The third stage of LCA is impact assessment. This involves evaluating the potential environmental impacts resulting from the inputs and outputs identified in the inventory analysis. The impact assessment considers various categories such as greenhouse gas emissions, energy consumption, water usage, air pollution, land use, and waste generation. The results of the impact assessment help identify significant environmental hotspots and potential areas for improvement. The interpretation stage involves analyzing and interpreting the results of the impact assessment to understand the significant environmental hotspots and potential areas for improvement. This includes comparing different scenarios or alternative designs to identify which options have a lower environmental impact. The interpretation stage helps stakeholders develop strategies to minimize the identified environmental impacts.

The fifth stage of LCA is improvement strategies. Based on the findings from the LCA, stakeholders can develop strategies to minimize the identified environmental impacts. This may involve optimizing energy efficiency, selecting sustainable materials, implementing water conservation measures, improving waste management practices, or incorporating renewable energy sources.

Presenting and communicating the results of the LCA study clearly and transparently to stakeholders, including architects, engineers, builders, and policymakers, are essential. The findings can be used to inform decision-making processes and promote sustainable design and construction practices.

The inventory analysis step in LCA is a crucial phase where data collection and calculation procedures are carried out to quantify relevant inputs and outputs of a product system. For a building construction project, this inventory analysis could involve the factors as shown in Table 5.5.

The data collected in this step is usually extensive and detailed to ensure a comprehensive analysis. It is important to note that data quality can significantly affect the results of an LCA. Therefore, reliable data sources and rigorous data collection methods are crucial in this step.

TABLE 5.5
Factors of Inventory Analysis in LCA

Factor	Explanation
Material inputs	This includes all the raw materials used in the construction process. It could range from concrete, steel, and wood to smaller components like nails and screws. The quantity of each material used is recorded.
Energy inputs	This involves the amount of energy consumed during the construction process. It could include electricity used by machinery, fuel for transportation, and heating, among others.
Water inputs	The amount of water used in the construction process is also considered. This could be water used in mixing concrete, cleaning, and other construction processes.
Outputs	These are the emissions and waste produced during the construction process. It could include CO_2 emissions from machinery, waste materials from construction, and wastewater.

5.9 HOUSING FINANCE

Housing finance covers various aspects; however, the discussion is limited only to some important topics, including viable housing finance schemes, the involvement of banks in housing finance, government funding schemes, the presence of low-interest financing systems, insurance schemes, and attempts to discuss the institutional aspects as well.

5.9.1 Viable Housing Finance Schemes and Institutions Involved

A viable housing finance scheme should be present in making housing affordable and accessible to all, especially for low- and middle-income households. These schemes can take various forms, including loans, grants, subsidies, and guarantees, provided by a range of institutions such as banks, non-banking financial companies (NBFCs), housing finance companies (HFCs), and government bodies.

One of the most common types of housing finance schemes is mortgage loans offered by banks and HFCs. These loans are typically long-term, allowing borrowers to repay the loan over an extended period, often up to 30 years. The interest rates for these loans can be either fixed or variable, depending on the terms of the loan agreement.

Another viable scheme is the government-backed housing programs. These programs aim to support low-income households in purchasing their own homes. They often provide subsidies or low-interest loans to make housing more affordable. For instance, the Pradhan Mantri Awas Yojana (PMAY) in India offers credit-linked subsidies on home loans for eligible beneficiaries. Several government-backed housing programs aim to make homeownership more accessible and affordable. In the United States, some government-backed housing programs are also running, for example:

1. *Federal Housing Administration (FHA) Loans:* The FHA manages the FHA loan program. It helps buyers by insuring their loans so lenders can give them lower down payments and closing costs. The home must be within FHA loan limits, depending on its location, and you must qualify for a loan with an FHA-approved lender.
2. *Veteran and Service Member Programs:* These programs offer special home-buying assistance to veterans and service members.
3. *Rural Home Buyer Assistance Programs:* These programs assist home buyers in rural areas.
4. *Indian Home Loan Guarantee Program:* This program is specifically for the Native American community.

5. *Homeownership Vouchers for First-Time Home Buyers:* If you have a low income and want to buy your first home, the Housing Choice Voucher homeownership program could help.
6. *US Department of Agriculture (USDA) Loans:* These loans are designed to help low- to moderate-income households purchase homes in rural areas.
7. *Department of Veterans Affairs (VA) Loans:* These loans provide long-term financing to eligible American veterans or their surviving spouses.

There are also many housing finance programs offered by non-governmental organizations and international organizations around the world. Here are some examples:

1 *HUD-Approved Non-Profit Organizations and Government Entities:* These are organizations that partner with the US Department of Housing and Urban Development (HUD) to provide affordable housing opportunities to low- and moderate-income families in the US. They can participate in various programs such as HUD Homes, FHA Mortgage, and Secondary Financing.
2. *International Grants:* These are grants provided by various foundations, corporations, and agencies to support non-profit organizations working on social development, human rights, environmental protection, and other issues in different regions of the world. Some examples are the Aga Khan Foundation, Asia Foundation, Canadian Government Grants and Loans, and Central and Eastern Europe-Focused Grantmakers.
3. *Asian Development Bank:* This is a multilateral development bank that supports economic and social development in Asia and the Pacific. It provides loans, grants, technical assistance, and policy advice to its member countries and partner organizations.
4. *International Corporate Foundations and Grants Programs:* These are foundations and programs established by multinational corporations to support projects and programs that align with their corporate social responsibility goals. They often focus on areas such as education, health, environment, and innovation. Some examples are Google.org, Microsoft Philanthropies, Coca-Cola Foundation, and Ford Foundation.

Private international organizations have contributed globally to promoting housing programs. They are shown in Table 5.6.

TABLE 5.6
International Organizations Involve in Housing Programs

Organization	Explanation
World Housing	This organization receives donations and ensures that 100% of the donation goes to provide safe housing for those in need. They work with World Housing Ambassadors to provide homes to the world's poorest families.
World Bank Group	The World Bank Group works on a variety of housing finance development challenges by providing client countries with the tools to build and expand housing finance markets, fund housing finance (often through the mobilization of long-term mortgage securities where possible), develop sustainable and affordable housing finance markets, and respond effectively to housing finance crises.
HUD Awards	HUD was awarded $36 million for local economic and housing programs. Three national groups were selected to distribute the funds to local initiatives that support affordable housing and community development – Local Initiatives Support Corp. (LISC), Enterprise Community Partners, and Habitat for Humanity International.
Center for Affordable Housing Finance Africa (CAHF)	This organization is working in Nigeria. They also mention real estate investment platforms that offer investors the opportunity for fractional ownership of rental income-generating properties.

One international organization that works in housing development, Habitat for Humanity, has been working in the northeastern provinces of Thailand since 2007. The case study is presented in Box 5.2.

BOX 5.2 HABITAT FOR HUMANITY IN THAILAND

Habitat for Humanity began working in Thailand in the northeastern province of Udon Thani in 1998. Since July 2007, Habitat Thailand has been operating through its national office in Bangkok and zone offices located in Chiang Mai (North), Rayong (Northeast), Patumthani (Central), and Phangnga (South).

The organization's aim is to improve the quality of life of the Thai people through building homes and transforming communities. It works with like-minded partners to achieve a lasting, positive impact through holistic development and sustainable change.

Habitat Thailand has partnered with 11,500 families or over 46,000 people to build, repair, and rehabilitate their homes. Based on the needs of families and communities in Thailand, Habitat and its partners also implement community development programs that provide training in livelihood options and assistance in setting up home-based businesses.

In addition to housing, health centers, multipurpose community centers, school libraries, vegetable greenhouses, and water and sanitation facilities are also built. Corporate partners include Korean Reinsurance, Destination Asia, Thrivent, Bain & Company, Dow, HMC Polymer, Cargill, Ananda Development & Partners, Siam Commercial Bank, Seagate, Caterpillar, Nissan, Cigna, Bank of America, P&G, Credit Suisse, HMC Polymer, Hilton Pattaya, Benefit Cosmetics, and HSBC. Local and International schools and universities also lend a hand.

Sources: habitat.org, habitatthailand.org, Wikipedia.org

5.9.2 The Involvement of Local Banks in House Financing

Banks have long been integral to the housing finance sector, playing a multifaceted role that encompasses various aspects of mortgage lending and investment. These financial institutions serve as critical intermediaries in the housing market, connecting prospective home buyers, real estate developers, and builders with the capital they need to purchase or construct residential properties. The array of financial instruments and services offered by banks in this domain is both diverse and complex, with the overarching goal of facilitating homeownership and ensuring the smooth functioning of the housing market.

One of the primary functions of banks in the realm of housing finance is the provision of loans to individuals looking to purchase homes. These loans, commonly referred to as mortgages, come in a variety of forms to cater to the diverse financial situations and preferences of borrowers. Some of the most common types of mortgages offered by banks include fixed-rate mortgages, adjustable-rate mortgages (ARMs), and interest-only loans. Each of these mortgage types has distinct characteristics that make them suitable for different borrowers and circumstances.

Fixed-rate mortgages, for instance, provide borrowers with a stable and predictable monthly payment, as the interest rate remains constant over the life of the loan. This type of mortgage is especially attractive to those who seek financial security and want to avoid the uncertainty associated with interest rate fluctuations. In contrast, ARMs have interest rates that can change periodically, typically after an initial fixed-rate period. ARMs often offer lower initial interest rates, making them appealing to borrowers who anticipate interest rates to remain stable or decline in the future. Interest-only loans, on the other hand, allow borrowers to pay only the interest for a specified period before they begin repaying the principal. This can be advantageous for individuals with irregular income streams or those who want to minimize their initial monthly payments.

Beyond direct lending, banks also play a crucial role in the secondary mortgage market. In this capacity, banks have the option to sell the mortgages they originate from to investors or to securitize them into mortgage-backed securities (MBS). The secondary mortgage market is a key mechanism for banks to manage their risk and liquidity. When banks sell loans to investors, they receive a lump sum of cash, which they can then use to create new loans. This process is essential in ensuring the continuous flow of funds into the housing market. Alternatively, when banks securitize mortgages into MBS, they bundle a group of mortgages together and issue securities that represent a claim on the cash flows generated by the underlying loans. These MBS can be sold to investors, such as pension funds and other financial institutions, providing banks with another source of capital to fund additional loans.

In addition to selling their loans, banks can also purchase MBS from other issuers or invest in mortgage-related derivatives. This practice allows banks to diversify their portfolios and manage their exposure to various risks associated with the housing market. The ability to buy and sell MBS and derivatives gives banks flexibility in their investment strategies and helps them optimize their overall financial performance.

Banks' involvement in housing finance is not limited to their interactions with individual homebuyers and investors. They are integral members of the Federal Home Loan Bank (FHLB) system, a network of government-sponsored banks established to provide low-cost funding to financial institutions for housing finance. Banks can access the FHLB system by pledging their mortgages or other assets as collateral, which enables them to borrow funds at favorable terms. This system plays a pivotal role in ensuring the availability of capital for housing finance, particularly in times of economic turbulence or when credit markets are constrained. The FHLB system effectively complements banks' resources, supporting their ability to extend loans and sustain a robust housing finance market.

Banks actively contribute to affordable housing programs and initiatives, which are essential for addressing the housing needs of low- and moderate-income communities. One of the most prominent mechanisms through which banks promote affordable housing is the Community Reinvestment Act (CRA). Enacted in 1977, the CRA mandates that banks meet the credit needs of the communities in which they operate, particularly low- and moderate-income neighborhoods. Banks are evaluated on their CRA performance, which can influence regulatory approvals for mergers, acquisitions, and branch expansions. To comply with the CRA, banks often engage in activities such as providing affordable mortgage loans, supporting community development projects, and offering financial services to underserved populations.

Apart from the CRA, banks actively participate in various federal, state, and local programs that are designed to promote and finance affordable housing. These programs may offer grants, tax credits, or subsidies to incentivize the construction or rehabilitation of housing units that are affordable to low- and moderate-income households. Such initiatives are essential for addressing the affordability challenges that many communities face, ensuring that housing remains accessible to a wide range of income groups.

One significant federal program aimed at increasing affordable housing options is the Low-Income Housing Tax Credit (LIHTC) program. Administered by the Internal Revenue Service (IRS), LIHTC provides tax incentives to investors who contribute equity to finance low-income housing developments. Banks frequently invest in LIHTC projects to benefit from the tax credits and to fulfill their obligations under the CRA.

State and local governments also offer a variety of housing finance programs tailored to the specific needs of their communities. These programs may include down payment assistance, homebuyer education, and subsidies for affordable rental housing. Banks collaborate with these entities to support and implement these initiatives, further underscoring their commitment to making housing more accessible and affordable.

In addition to their active participation in affordable housing programs, many banks have established their initiatives and partnerships aimed at addressing local housing challenges. These initiatives can encompass a wide range of activities, including financial literacy programs, support for non-profit organizations dedicated to housing solutions, and grants for community development projects. Engaging in such activities enables banks to directly foster housing stability and economic growth within the communities they serve.

It is important to recognize that the multifaceted role banks play in housing finance extends beyond profit-seeking activities. While they certainly aim to generate revenue through lending and investment, they also have a significant social and economic responsibility. Housing is not only a fundamental human need but also a cornerstone of economic stability and growth. Providing financing options, participating in the secondary mortgage market, and supporting affordable housing programs allows banks to contribute to the overall health and vibrancy of the housing market, and by extension, the broader economy.

One way by which banks contribute to the housing market's stability and efficiency is by managing risk. Mortgages are long-term loans, typically spanning 15 to 30 years, and involve substantial amounts of capital. Banks need to ensure that they can meet their financial obligations over these extended periods. Interest rate risk is one of the key challenges they face. If banks hold a portfolio of fixed-rate mortgages and interest rates rise, the value of those mortgages may decrease. Conversely, if interest rates fall, banks may face prepayment risk, as borrowers refinance their mortgages to secure lower rates. To mitigate these risks, banks can use financial derivatives like interest rate swaps and options, which allow them to hedge against unfavorable interest rate movements. These tools enable banks to maintain financial stability and continue offering competitive mortgage products.

The secondary mortgage market serves as an essential component of the housing finance ecosystem, with banks at its core. The ability to securitize mortgages into MBS provides banks with a mechanism to free up capital and allocate it to new loans. By doing so, they facilitate the recycling of funds in the housing market, which is vital for its continuous operation. This practice is particularly important in situations where housing markets experience increased demand, such as during periods of economic growth or low-interest rate environments. When housing markets enter periods of increased demand, there is a surge in the number of individuals seeking mortgages to purchase homes. Banks, as primary lenders, can find themselves inundated with loan applications, requiring significant capital to fund these loans. This is where the secondary mortgage market plays a pivotal role.

During such periods of heightened demand, banks can securitize a portion of their existing mortgages into MBS. These securities are then sold to investors, providing banks with an infusion of cash. This influx of funds allows banks to continue extending new mortgages to borrowers without depleting their available capital reserves. In essence, it is a process that promotes the efficient circulation of capital within the housing market, ensuring that the momentum of the market is not hampered by capital constraints.

The practice of securitization not only benefits banks but also offers advantages to investors seeking stable returns. MBSs are known for their relatively predictable cash flows, making them attractive investments for various entities, including pension funds, insurance companies, and other financial institutions. Investing in MBS allows these entities to diversify their portfolios and generate a steady stream of income, often with a degree of risk management due to the inclusion of numerous mortgages in a single MBS pool.

The MBS investments indirectly influence interest rates in the broader financial market. When banks securitize mortgages and sell MBS to investors, they are essentially offloading some of the interest rate risk associated with the loans. This can influence the supply and demand dynamics for mortgages, which, in turn, can impact interest rates for borrowers. In times of increased demand for MBS, the prices of these securities may rise, causing their yields to fall. Lower yields on MBS can lead to reduced interest rates for borrowers, making homeownership more affordable and stimulating further demand in the housing market.

Conversely, during periods of economic expansion and increased competition among lenders, banks may offer a wider array of mortgage products to cater to diverse borrower needs. This could include specialized mortgages designed for specific demographics, such as first-time homebuyers, veterans, or low-income individuals. In this environment, banks may leverage their resources and expertise to provide customized loan options and financial solutions that align with the evolving demands of the housing market.

As banks continue to evolve and adapt to changing economic and regulatory landscapes, technology plays a pivotal role in reshaping the housing finance industry. In recent years, the rise of fintech (financial technology) companies and online lenders has introduced a new dimension to the way mortgages are originated, processed, and serviced. These digital disruptors leverage advanced algorithms, data analytics, and user-friendly interfaces to streamline the mortgage application and approval process.

Many traditional banks have also embraced technology to enhance their housing finance operations. Online and mobile banking platforms have made it easier for consumers to apply for mortgages, track the progress of their applications, and access information about the housing market. Automated underwriting systems have sped up the approval process, making it more efficient and reducing the time it takes for borrowers to secure a mortgage.

Moreover, data analytics and machine learning have become valuable tools for banks to assess creditworthiness and mitigate risks. By analyzing vast amounts of data, banks can make more informed lending decisions, which can benefit both borrowers and lenders. Accurate risk assessment allows banks to offer competitive interest rates to borrowers with strong credit profiles while managing their risks effectively.

In addition to using technology to streamline operations, banks are increasingly focused on digital marketing and customer engagement strategies. They employ data-driven marketing campaigns and personalized customer experiences to attract potential homebuyers and retain existing mortgage customers. These efforts contribute to the overall growth and competitiveness of banks in the housing finance sector.

The evolving role of banks in housing finance also extends to their involvement in sustainable and responsible lending practices. In recent years, there has been a growing emphasis on environmental, social, and governance (ESG) considerations in the financial industry, and housing finance is no exception. Banks are under increasing pressure to integrate ESG principles into their lending and investment activities, with a focus on promoting affordable, energy-efficient, and environmentally sustainable housing options.

To align with ESG goals, banks are exploring ways to finance green and energy-efficient homes, as well as developments that promote sustainability. Green mortgages, for example, are designed to incentivize homeowners to make energy-efficient improvements to their properties by offering favorable terms and rates for such upgrades. Banks are actively developing and promoting these financial products to encourage homeowners to reduce their carbon footprint while simultaneously improving the energy efficiency and value of their homes.

Furthermore, banks are increasingly considering the social aspect of ESG in their housing finance efforts. This includes initiatives aimed at increasing access to housing for underserved and marginalized communities. Banks may offer special loan programs and community development projects that prioritize affordable housing, particularly in neighborhoods that have historically faced housing discrimination or disinvestment.

The governance component of ESG is also relevant to banks' involvement in housing finance. Responsible governance practices encompass fair lending, transparency, and ethical decision-making. Banks are continually refining their lending practices to ensure that they are fair and unbiased, free from discriminatory practices, and in compliance with regulations such as the

Community Reinvestment Act (CRA). Transparency in lending terms and conditions is crucial to ensure borrowers fully understand their mortgage agreements, and ethical decision-making is fundamental to maintaining the trust of both borrowers and the broader community.

Banks' adaptation to technological advancements has streamlined the mortgage application and approval process, making it more efficient and user-friendly. Data analytics and machine learning have improved risk assessment and credit underwriting, benefiting both borrowers and lenders. Moreover, the integration of ESG principles into housing finance reflects a broader commitment to sustainability, inclusivity, and ethical governance.

As banks continue to evolve and respond to the ever-changing landscape of housing finance, they remain key stakeholders in the pursuit of affordable and accessible housing, underpinning the foundation of homeownership and economic growth. Their role in housing finance is not only about profit but also about facilitating social responsibility, economic stability, and sustainable housing solutions.

5.9.3 Low-Interest Home Financing

The dream of homeownership is a fundamental aspiration for people around the world. Owning a home not only provides a sense of stability and security but also serves as a significant investment for the future. However, one of the primary barriers to achieving this dream for many individuals and families is the availability of affordable financing options. Low-interest home financing has emerged as a crucial issue in the modern world, reflecting the pressing need for accessible and affordable housing. This article delves into the complexities surrounding this issue, exploring the current state of low-interest home financing and the various factors contributing to its importance.

Low-interest home financing refers to the availability of mortgage loans with interest rates that are substantially lower than the prevailing market rates. The primary goal of such financing is to make homeownership more accessible and affordable for a broader segment of the population. In recent years, low-interest home financing has become a topic of increased concern due to the rising cost of housing and the associated challenges faced by potential homebuyers.

In many countries, central banks and financial institutions play a significant role in determining interest rates. Low-interest home financing often depends on the monetary policies and lending practices in place. In economies with high interest rates, securing a mortgage can be prohibitively expensive, making it difficult for many people to enter the housing market.

The importance of low-interest home financing cannot be overstated in the realm of housing and real estate. It serves as a cornerstone of financial accessibility, transforming the dream of homeownership into a tangible reality for a broader cross-section of society. By mitigating the prohibitive costs associated with high-interest mortgages, low-interest home financing opens doors for countless individuals and families, enabling them to secure a stable and prosperous future. Some benefits of low-interest home financing include the following: (1) *Reduced Monthly Payments:* Lower interest rates can significantly reduce your monthly mortgage payments, making homeownership more affordable and easing your budget. (2) *Long-Term Savings:* With lower interest, we will pay less over the life of your mortgage, potentially saving you tens of thousands of dollars or more compared to a higher interest rate. (3) *Higher Purchasing Power:* Lower interest rates may allow you to qualify for a larger loan, which means you can consider homes in a higher price range and potentially secure a more spacious or desirable property. (4) *Equity Build-Up:* Lower interest rates help you build home equity faster because a larger portion of your monthly payment goes toward the principal rather than interest. (5) *Financial Stability:* Reduced interest rates make your housing costs more predictable and manageable, which can lead to greater financial stability and peace of mind. (6) *Refinancing Opportunities:* If interest rates drop further after you've secured your mortgage, you may have the opportunity to refinance at an even more favorable rate, saving you additional money. (7) *Economic Stimulus:* Low-interest home financing can stimulate the housing market and overall economy by encouraging home purchases and investments. (8) *Diversification of Investment Portfolio:* Homeownership is an investment that can diversify your financial portfolio and potentially provide long-term value. (9) *Tax Benefits:* Some homeowners may be eligible for tax deductions and credits related to mortgage interest payments, which can further reduce the cost of homeownership. (10) *Improved Credit Opportunities:* Successfully managing a mortgage with a low interest rate can improve your credit score and open doors to other favorable credit opportunities.

These benefits make low-interest home financing an attractive option for individuals and families looking to purchase a home and secure their financial future. However, it is important to consider your financial situation and consult with a financial advisor or lender to determine the best financing option for your needs.

The demand for low-interest home financing stems from various contributing factors. First and foremost, escalating housing prices, particularly in urban areas, are driven by factors like limited housing supply, population growth, and urbanization. In these markets, accessible low-interest home financing becomes crucial for aspiring homeowners. Second, income inequality prevails, affecting people's home-buying capacity. Those with higher incomes secure traditional mortgages with ease, while lower-income individuals grapple with down payment and interest rate requirements. Third, many, particularly young adults, face hurdles in qualifying for mortgage loans due to limited credit history, high student loan debt, or financial constraints, where low-interest financing offers relief. Fourth, economic conditions, including unemployment rates and income growth, profoundly affect housing market entry. In economic downturns, low-interest home financing provides stability. Lastly, demographic shifts, such as an aging population and increased single-parent households, have reshaped housing needs, making low-interest home financing vital in adapting to these evolving requirements.

Low-interest home financing has gained prominence as a vital solution to the growing challenges of housing affordability, wealth building, and social stability. It offers numerous benefits, including increased access to affordable housing, wealth accumulation, economic growth, social stability, and a reduction in income inequality. However, addressing this issue requires a multi-pronged approach that includes government-backed programs, community land trusts, affordable housing initiatives, interest rate subsidies, and public–private partnerships.

PROBLEMS AND ISSUES FOR DISCUSSION

1. What are the main factors driving the increased demand for urban housing in developed countries, such as population growth, urbanization trends, economic opportunities, and the rise of small households?
2. How can local governments effectively address the issue of affordable housing in densely populated cities?
3. What are the potential benefits and drawbacks of high-rise residential developments in densely populated urban areas, including their impact on housing supply, community integration, resource efficiency, and the overall livability of the urban environment?
4. What innovative design strategies, such as modular construction, flexible floor plans, and the incorporation of multifunctional spaces, can be implemented to maximize the efficient use of limited living space in compact urban housing developments?
5. How do widening income disparities and the process of gentrification in urban areas negatively impact the accessibility and affordability of housing for low-income residents, often leading to displacement and the loss of diverse and inclusive communities?
6. What role can public–private partnerships play in the development of sustainable and inclusive urban housing?
7. How can the integration of smart technologies improve the efficiency and livability of urban housing?

8. What are the potential environmental consequences of urban housing development, such as increased energy consumption, greenhouse gas emissions, and habitat fragmentation, and how can sustainable design strategies, renewable energy integration, and nature-based solutions help mitigate these impacts?
9. How do zoning regulations and land-use policies determine the quantity, type, and pricing of urban housing units available to residents? In what ways do these policies shape the accessibility and affordability of housing in metropolitan areas?
10. What demographic shifts and economic factors contribute to the challenges of ensuring sufficient and appropriate housing options for the aging urban population? How can zoning, design, and social policies address the unique housing needs of the elderly in cities?
11. How can urban planning and design promote the creation of diverse and mixed-income housing communities?
12. What economic, demographic, and lifestyle trends are driving the growth of micro-apartments and co-living arrangements in densely populated urban areas?
13. How do these innovative housing models address the challenges of limited space and rising rents in metropolitan centers?
14. How can urban housing developments be designed to foster a stronger sense of community and social cohesion?
15. What are the potential benefits and drawbacks of incentivizing developers to include affordable housing units in new projects?
16. How can the integration of renewable energy and energy-efficient technologies impact the sustainability of urban housing?
17. What are the unique housing needs and challenges faced by marginalized communities in urban areas?
18. How can the preservation of historic buildings and neighborhoods be balanced with the need for urban housing development?
19. What are the implications of remote work and flexible living arrangements for the future of urban housing demand?
20. How can urban housing policies and programs address the issue of housing insecurity and homelessness in cities?
21. What are the potential benefits and challenges of incorporating urban agriculture and green spaces into high-density housing developments?
22. How do transportation accessibility and connectivity influence the desirability and affordability of urban housing?
23. What are the potential social and economic impacts of short-term rental platforms on the urban housing market?
24. How can the integration of universal design principles improve the accessibility and inclusivity of urban housing?
25. What are the key considerations for developing sustainable and resilient urban housing in the face of climate change?
26. How can the utilization of prefabricated and modular construction techniques impact the delivery and cost of urban housing?
27. What are the potential barriers to the widespread adoption of co-living and communal housing models in urban areas?
28. How can the incorporation of mixed-use development in urban housing projects enhance the overall livability of a neighborhood?
29. What are the unique housing needs and challenges faced by young professionals and families in urban areas?

30. How can urban housing policies and programs support the creation of intergenerational and age-friendly communities?
31. What are the potential benefits and drawbacks of implementing rent control or stabilization measures in urban housing markets?
32. How can the integration of smart home technologies and automation enhance the efficiency and convenience of urban housing?
33. What are the potential implications of the rise of remote work on the future demand and design of urban housing?
34. How can the development of mixed-tenure housing projects contribute to the creation of more inclusive and equitable urban communities?
35. What are the key factors that influence the decision-making process of individuals and families when choosing urban housing options?
36. How can the incorporation of biophilic design elements in urban housing projects improve the overall well-being of residents?
37. What are the potential social and economic consequences of the growing trend of corporate investment in urban rental housing?
38. How can the integration of public transportation and active mobility infrastructure impact the accessibility and desirability of urban housing?
39. What are the potential environmental benefits and challenges of integrating urban farming and food production into housing developments?
40. How can the development and implementation of co-living and shared housing models in urban areas address the pressing issue of housing affordability, considering factors such as cost of living, population density, socioeconomic diversity, and the need for community building and social interaction?
41. What are the potential implications of the shift towards remote and hybrid work on the future design and layout of urban housing?
42. What are the specific challenges that urban areas face in providing adequate and accessible housing for the rapidly growing elderly population, considering factors such as accessibility, safety, affordability, proximity to healthcare services, and the need for community support and social interaction?

MULTIPLE-CHOICE QUESTIONS

1. Which of the following is the primary goal of urban zoning regulations?
    a. Maximizing developer profits
    b. Increasing housing density
    c. Preserving neighborhood character
    d. Reducing infrastructure costs
2. What is a key factor contributing to the shortage of affordable housing in many cities?
    a. Declining population
    b. Overregulation of construction
    c. Low mortgage interest rates
    d. Increasing wages
3. Which type of housing policy is designed to set aside a certain percentage of new residential units for low-income households?
    a. Rent control
    b. Inclusionary zoning
    c. Tax abatement
    d. Urban renewal

4. What is a key factor that has contributed to the rise of micro-apartments and co-living spaces in urban areas?
   a. Declining population
   b. Increasing land values
   c. Improved infrastructure
   d. Falling construction costs
5. Which of the following is a potential challenge associated with providing adequate housing for the growing elderly population in urban areas?
   a. Declining property values
   b. Lack of accessibility features
   c. Oversupply of housing units
   d. Decreasing demand for senior living
6. Which of the following is a common strategy used by cities to incentivize the development of affordable housing?
   a. Rent control
   b. Property tax increases
   c. Zoning for larger lot sizes
   d. Density bonuses
7. Which of the following is a common challenge faced by urban residents when seeking affordable housing?
   a. Abundant housing supply
   b. Declining construction costs
   c. Lack of access to transportation
   d. Increasing household incomes
8. What is a key strategy used by cities to address the housing needs of the homeless population?
   a. Rent control
   b. Inclusionary zoning
   c. Transitional housing
   d. Tax incentives for developers
9. Which of the following is a potential consequence of restrictive zoning regulations in urban areas?
   a. Increased housing affordability
   b. Reduced urban sprawl
   c. Improved public transportation
   d. Limited housing supply and higher prices
10. What is a common approach used by cities to revitalize blighted or underutilized urban areas?
    a. Demolition and redevelopment
    b. Increasing minimum lot sizes
    c. Restricting new construction
    d. Providing tax incentives for homeowners
11. Which of the following is a factor contributing to the rise of co-living arrangements in urban centers?
    a. Declining housing costs
    b. Aging population
    c. Increasing household sizes
    d. Desire for shared amenities and community
12. What is a potential benefit of implementing inclusionary zoning policies in urban areas?
    a. Increased development of luxury housing
    b. Reduced demand for affordable housing
    c. Gentrification of low-income neighborhoods
    d. Greater availability of affordable housing units

13. Which of the following is a common strategy used by cities to address the housing needs of the elderly population?
    a. Restricting new construction
    b. Promoting aging-in-place initiatives
    c. Increasing minimum unit sizes
    d. Reducing public transportation access
14. What is a potential downside of urban renewal efforts that involve the demolition of existing housing?
    a. Increased housing supply
    b. Improved neighborhood aesthetics
    c. Displacement of low-income residents
    d. Reduced infrastructure costs
15. Which of the following is a strategy used by cities to encourage the development of affordable housing units?
    a. Requiring large minimum lot sizes
    b. Establishing rent control policies
    c. Providing zoning incentives for developers
    d. Increasing building height restrictions
16. What is a key factor contributing to the high cost of urban housing in many cities?
    a. Declining population
    b. Overregulation of the housing market
    c. Decreasing construction costs
    d. Improved public transportation
17. Which of the following is a potential benefit of mixed-use development in urban areas?
    a. Reduced housing density
    b. Increased segregation of land uses
    c. Decreased access to amenities and services
    d. Improved walkability and reduced car dependence
18. What is a common challenge faced by urban planners when addressing the housing needs of low-income residents?
    a. Excessive supply of affordable housing units
    b. Lack of community opposition to new development
    c. Sufficient funding for housing subsidies
    d. Resistance to changes in zoning and land-use policies
19. Which of the following is a strategy used by cities to encourage the development of affordable housing units?
    a. Increasing minimum unit sizes
    b. Providing tax incentives for developers
    c. Establishing strict height limits
    d. Requiring large minimum lot sizes
20. What is a potential consequence of the high cost of urban housing in many cities?
    a. Increased economic diversity
    b. Reduced demand for rental housing
    c. Improved job opportunities
    d. Displacement of low-income residents
21. Which of the following is a strategy used by cities to address the housing needs of the elderly population?
    a. Reducing accessibility requirements
    b. Promoting age-restricted communities
    c. Increasing density in urban centers
    d. Providing subsidies for home modifications

22. What is a key factor contributing to the rise of micro-apartments and co-living spaces in urban areas?
    a. Declining land values
    b. Increasing household sizes
    c. Improved public transportation
    d. Limited availability of land for development
23. Which of the following is a potential benefit of implementing inclusionary zoning policies?
    a. Increased segregation of income groups
    b. Reduced housing supply and higher prices
    c. Gentrification of low-income neighborhoods
    d. Greater socioeconomic diversity in neighborhoods
24. What is the common challenge faced by cities in providing affordable housing for the growing elderly population?
    a. Declining demand for senior living
    b. Lack of accessibility features in housing
    c. Oversupply of age-restricted communities
    d. Decreasing construction costs for senior-friendly homes
25. Which of the following is a strategy used by cities to encourage the development of affordable housing units?
    a. Increasing minimum parking requirements
    b. Establishing rent control policies
    c. Providing zoning bonuses for affordable units
    d. Requiring large minimum unit sizes
26. What is the potential consequence of restrictive zoning regulations in urban areas?
    a. Increased housing affordability
    b. Reduced urban sprawl
    c. Improved public transportation
    d. Limited housing supply and higher prices
27. Which of the following is a common approach used by cities to revitalize blighted or underutilized urban areas?
    a. Demolition and redevelopment
    b. Increasing minimum lot sizes
    c. Restricting new construction
    d. Providing tax incentives for homeowners
28. What is a factor contributing to the rise of co-living arrangements in urban centers?
    a. Declining housing costs
    b. Aging population
    c. Increasing household sizes
    d. Desire for shared amenities and community
29. What is the potential benefit of implementing inclusionary zoning policies in urban areas?
    a. Increased development of luxury housing
    b. Reduced demand for affordable housing
    c. Gentrification of low-income neighborhoods
    d. Greater availability of affordable housing units
30. Which of the following is a common strategy used by cities to address the housing needs of the elderly population?
    a. Restricting new construction
    b. Promoting aging-in-place initiatives
    c. Increasing minimum unit sizes
    d. Reducing public transportation access

31. What is a potential downside of urban renewal efforts that involve the demolition of existing housing?
    a. Increased housing supply
    b. Improved neighborhood aesthetics
    c. Displacement of low-income residents
    d. Reduced infrastructure costs
32. Which of the following is a strategy used by cities to encourage the development of affordable housing units?
    a. Requiring large minimum lot sizes
    b. Establishing rent control policies
    c. Providing zoning incentives for developers
    d. Increasing building height restrictions
33. What is a key factor contributing to the high cost of urban housing in many cities?
    a. Declining population
    b. Overregulation of the housing market
    c. Decreasing construction costs
    d. Improved public transportation
34. Which of the following is a potential benefit of mixed-use development in urban areas?
    a. Reduced housing density
    b. Increased segregation of land uses
    c. Decreased access to amenities and services
    d. Improved walkability and reduced car dependence
35. What is a common challenge faced by urban planners when addressing the housing needs of low-income residents?
    a. Excessive supply of affordable housing units
    b. Lack of community opposition to new development
    c. Sufficient funding for housing subsidies
    d. Resistance to changes in zoning and land-use policies
36. Which of the following is a strategy used by cities to encourage the development of affordable housing units?
    a. Increasing minimum unit sizes
    b. Providing tax incentives for developers
    c. Establishing strict height limits
    d. Requiring large minimum lot sizes
37. What is a potential consequence of the high cost of urban housing in many cities?
    a. Increased economic diversity
    b. Reduced demand for rental housing
    c. Improved job opportunities
    d. Displacement of low-income residents
38. Which of the following is a strategy used by cities to address the housing needs of the elderly population?
    a. Reducing accessibility requirements
    b. Promoting age-restricted communities
    c. Increasing density in urban centers
    d. Providing subsidies for home modifications
39. What is a key factor contributing to the rise of micro-apartments and co-living spaces in urban areas?
    a. Declining land values
    b. Increasing household sizes
    c. Improved public transportation
    d. Limited availability of land for development

40. Which of the following is a potential benefit of implementing inclusionary zoning policies?
    a. Increased segregation of income groups
    b. Reduced housing supply and higher prices
    c. Gentrification of low-income neighborhoods
    d. Greater socioeconomic diversity in neighborhoods
41. What is a common challenge faced by cities in providing affordable housing for the growing elderly population?
    a. Declining demand for senior living
    b. Lack of accessibility features in housing
    c. Oversupply of age-restricted communities
    d. Decreasing construction costs for senior-friendly homes
42. Which of the following is a strategy used by cities to encourage the development of affordable housing units?
    a. Increasing minimum parking requirements
    b. Establishing rent control policies
    c. Providing zoning bonuses for affordable units
    d. Requiring large minimum unit sizes
43. What is a potential consequence of the declining population in some urban areas?
    a. Increased housing affordability
    b. Reduced demand for affordable housing
    c. Improved public infrastructure
    d. Gentrification of neighborhoods
44. Which of the following is a common urban planning strategy to address the lack of affordable housing?
    a. Promoting single-family home development
    b. Increasing minimum lot sizes
    c. Restricting high-density residential zoning
    d. Incentivizing the construction of multi-family housing
45. What is a key factor contributing to the rise of co-living arrangements in urban areas?
    a. Decreasing household sizes
    b. Aging population
    c. Declining construction costs
    d. Desire for shared living spaces and amenities
46. Which of the following is a potential downside of urban renewal efforts that involve the demolition of existing housing?
    a. Increased housing supply
    b. Improved neighborhood aesthetics
    c. Displacement of low-income residents
    d. Reduced infrastructure costs
47. What is a common strategy used by cities to address the housing needs of the homeless population?
    a. Rent control
    b. Inclusionary zoning
    c. Transitional housing
    d. Tax incentives for developers
48. Which of the following is a potential benefit of mixed-use development in urban areas?
    a. Increased segregation of land uses
    b. Reduced housing density
    c. Decreased access to amenities and services
    d. Improved walkability and reduced car dependence

ANSWER TO MULTIPLE-CHOICE QUESTIONS

1	c	11	d	21	d	31	c	41	b
2	b	12	d	22	d	32	c	42	c
3	b	13	b	23	d	33	b	43	a
4	b	14	c	24	b	34	d	44	d
5	b	15	c	25	c	35	d	45	d
6	d	16	b	26	d	36	b	46	c
7	c	17	d	27	a	37	d	47	c
8	c	18	d	28	d	38	d	48	d
9	d	19	b	29	d	39	d		
10	a	20	d	30	b	40	d		

FURTHER READING

Angotti, T. (2013). *New York for sale: Community planning confronts global real estate*. MIT Press.

Breuste, J. H., Artmann, M., Ioja, C., & Qureshi, S. (2015). Hybrid governance for sustainable urban green infrastructure. *Sustainability*, 7(3), 3955–3973.

Bulkeley, H., & Betsill, M. M. (2013). Revisiting the urban politics of climate change. *Environmental Politics*, 22(1), 136–154.

Carpenter, J., & Lees, L. (1995). Gentrification in New York, London and Paris: an international comparison. *International Journal of Urban and Regional Research*, 19(2), 286–303.

Catte, E., Price, J., Serratore, A., & Thompson, C. (2018). The value of black neighborhoods. *Boston Review*, 43(1), 65–77.

Chaskin, R. J., & Joseph, M. L. (2015). Contested space: Design principles and regulatory regimes in mixed-income communities replacing public housing. *Annals of the American Academy of Political and Social Science*, 660(1), 136–153.

Checa-Artasu, M. M. (2011). Gentrifying the Mexican metropolis: The case of Polanco, Mexico City. *Geographical Review*, 101(4), 517–536.

Coaffee, J., & Healey, P. (2003). 'My voice: my place': Tracking transformations in urban governance. *Urban Studies*, 40(10), 1979–1999.

Crang, M. (1998). *Cultural geography*. Routledge.

Desmond, M. (2016). *Evicted: Poverty and profit in the American city*. Crown.

Doucet, B. (2014). A process of change and a changing process: Introduction to the special issue on contemporary gentrification. *Tijdschrift voor economische en sociale geografie*, 105(2), 125–139.

Evans, G. L. (2011). A novice researcher's first walk through the maze of grounded theory: Rationalization for classical grounded theory. *The Grounded Theory Review*, 10(1), 23–41.

Ewing, R., & Cervero, R. (2010). Travel and the built environment: A meta-analysis. *Journal of the American Planning Association*, 76(3), 265–294.

Florida, R. (2002). *The rise of the creative class*. Basic Books.

Hackworth, J. (2002). Postrecession gentrification in New York City. *Urban Affairs Review*, 37(6), 815–843.

Harvey, D. (1989). From managerialism to entrepreneurialism: The transformation in urban governance in late capitalism. *Geografiska Annaler. Series B. Human Geography*, 71, 3–17.

Hossain, M. S., & Nuruddin, A. (2019). A systematic review of urban housing policy in Bangladesh. *ICTD,* 2019, 265.

Iveson, K. (2013). Cities within the city: Do-it-yourself urbanism and the right to the city. *International Journal of Urban and Regional Research*, 37(3), 941–956.

Jacobs, J. (1961). *The death and life of great American cities*. Random House.

Kemeny, T. (2020). Immigrant diversity and economic development in cities: a critical review. *Geography Compass*, 14(4), e12480.

Lees, L. (2008). Gentrification and social mixing: Towards an inclusive urban renaissance. *Urban Studies*, 45(12), 2449–2470.

Lees, L., Slater, T., & Wyly, E. (2013). *Gentrification*. Routledge.

Lefebvre, H. (1991). *The production of space* (Vol. 142). Blackwell.

Logan, J. R., & Molotch, H. L. (2007). *Urban fortunes: The political economy of place*. University of California Press.
Marcuse, P. (1985). Gentrification, abandonment, and displacement: Connections, causes, and policy responses in New York City. *Washington University Journal of Urban & Contemporary Law*, 28, 195.
Mayer, M. (2009). The 'Right to the City'in the context of shifting mottos of urban social movements. *City*, 13(2–3), 362–374.
Newman, K., & Ashton, P. (2004). Neoliberal urban policy and new paths of neighborhood change in the American inner city. *Environment and Planning A*, 36(7), 1151–1172.
Paccoud, A. (2017). Planning law, power, and practice: Haussmann in Paris (1853–1870). *Planning Perspectives*, 32(1), 41–63.
Sassen, S. (2001). *The global city: New York, London, Tokyo*. Princeton University Press.
Storper, M., & Manville, M. (2006). Behaviour, preferences and cities: Urban theory and urban resurgence. *Urban Studies*, 43(8), 1247–1274.
Turner, M. A., & Rawlings, L. A. (2009). *Promoting neighborhood diversity: Benefits, barriers, and strategies*. Urban Institute.

6 Urban Hydrology

6.1 INTRODUCTION

Hydrology is the scientific study of water in the Earth's atmosphere, on its surface, and underground. It encompasses the distribution, movement, and properties of water in various forms, including liquid, solid (ice and snow), and vapor. Hydrology plays a crucial role in understanding and managing water resources, including issues related to precipitation, evaporation, runoff, groundwater, and water quality.

Hydrologists, professionals who specialize in hydrology, analyze the natural processes involved in the water cycle or hydrological cycle, such as precipitation, infiltration, percolation, surface runoff (SRO), and the replenishment of aquifers. They also study the impact of human activities on water resources and the environment, including water pollution, flood management, irrigation, and the design of water-related infrastructure like dams and reservoirs.

The field of hydrology is interdisciplinary, drawing upon principles from meteorology, geology, geography, environmental science, and engineering to gain comprehensive understanding of how water behaves and moves within the Earth's various reservoirs and systems. This knowledge is essential for sustainable water resource management and environmental protection.

With the above definitions, urban hydrology refers to the study and analysis of hydrological processes within urban areas, encompassing a comprehensive examination of how water interacts within the urban environment. This includes an exploration of the precipitation, runoff, infiltration, and drainage patterns specific to urban landscapes, as well as the associated consequences and implications for urban activities, residents, infrastructure, and the overall urban ecosystem. The key focus of urban hydrology is to understand and manage the various ways in which water interacts with and influences the dynamics of cities, addressing issues like flood risk, water quality, sustainable water management, and the overall well-being of urban populations and their built environment.

While urban hydrology aims to understand the quantity and quality of water resources in urban environments, as well as the effects of urbanization on the natural hydrological processes, urban drainage is the design and management of systems that collect, transport, and dispose of stormwater and wastewater in urban areas. Urban drainage systems include pipes, sewers, culverts, channels, detention basins, wetlands, and other structures that convey water from urban surfaces to receiving water or treatment facilities. The systems aim to prevent flooding, protect public health, and enhance environmental quality.

Urban hydrology and urban drainage are, therefore, closely connected, as they both deal with the effects of urbanization on the water cycle. Urban hydrology provides the input data and models for urban drainage design, such as rainfall intensity, runoff coefficients, and infiltration rates. Urban drainage affects urban hydrology by altering the flow paths, volumes, and quality of water in urban

DOI: 10.1201/9781032664613-6

areas. Urban drainage can also provide opportunities for water reuse, groundwater recharge, and ecological restoration.

In this chapter, we aim to provide a discussion of the urban drainage system, which is essential for preventing floods and maintaining water quality in the city. We will explain how the excess water from rainfall is collected, transported, treated, and discharged to the sea through various components of the drainage system, such as pipes, channels, pumps, detention basins, and outfalls. The process of urban drainage starting from rainfall can be summarized as follows:

Rainfall: Rainfall is the main source of water that enters the urban drainage system. Rainfall intensity, duration, and frequency (IDF) determine the amount and rate of runoff that is generated. Rainfall data can be obtained from historical records, gauges, or models.

Runoff: Runoff is the portion of rainfall that flows over the land surface and reaches the drainage system. Runoff depends on the characteristics of the urban area, such as land use, soil type, slope, vegetation, and imperviousness. Runoff can be estimated using methods such as the rational method or the Colorado Urban Hydrograph Procedure.

Collection: Collection is capturing and conveying the runoff from the urban area to the drainage system. Collection can be done using natural or artificial channels, such as gutters, gullies, pipes, sewers, culverts, or ditches. Collection can also include devices that control the flow rate, such as weirs, orifices, or valves.

Discharge: Discharge is releasing the runoff from the drainage system to the receiving water body, such as a stream, river, lake, or ocean. Discharge can be done using outlets, such as outfalls, diffusers, or spillways. Discharge can also include devices that reduce the impact of the runoff, such as detention basins, wetlands, or filters.

Treatment: Treatment improves the quality of the runoff before or after it is discharged to the receiving water body. Treatment can be done using physical, chemical, or biological methods, such as sedimentation, filtration, coagulation, disinfection, or biodegradation. Treatment can also include facilities that remove or recover pollutants, such as wastewater treatment plants, stormwater treatment plants, or resource recovery plants.

One of the concerns of urban hydrology in urban engineering is dealing with urban drainage, which is to prevent urban areas from flood problems because of stormwater. Therefore, urban areas must be adequately facilitated by urban drainage with sufficient capacity. Discussing urban drainage would not be comprehensive if only discussing drainage for stormwater management without the drainage system for wastewater treatment as well. The wastewater treatment, however, is discussed in a separate chapter of wastewater management. Therefore, in this chapter, the focus will be on drainage stormwater management. These two systems, stormwater and wastewater drainages, are separated for the difference of water quality managed.

6.2 URBAN FLOOD MECHANISMS

Rainwater that falls on the earth's surface can cause urban floods or other types of floods. The amount of rainwater that becomes SRO depends on how well the soil layers can absorb it. SRO is also known as overland flow, and it occurs when the rainwater flows over the land instead of infiltrating into the soil.

When the rainwater infiltrates into the upper soil layers, it increases the soil saturation. Soil saturation is the degree to which the soil pores are filled with water. Soil pores are tiny spaces between soil particles that can hold air or water. When the soil pores are mostly filled with water, the soil is said to be saturated. When the upper layer of the soil reaches its maximum saturation level, it cannot absorb any more rainwater. The overland flow begins.

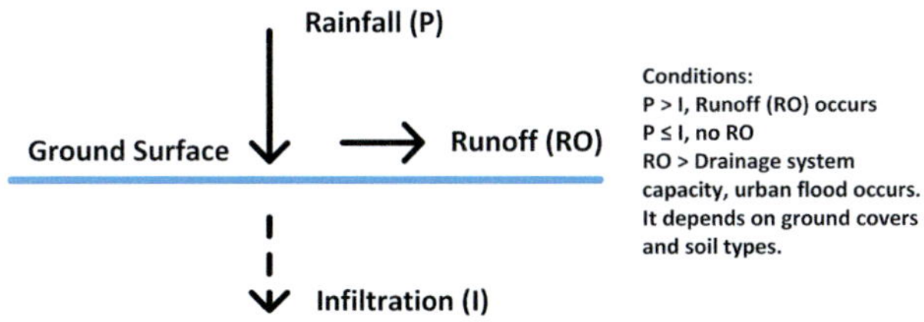

FIGURE 6.1 Urban flood (oversimplified).

The overland flow can vary in size depending on how long and how much it rains. The longer and heavier the rainfall, the larger and faster the overland flow will be. Other factors can also affect the overland flow such as slope, vegetation, land use, and soil type. Urban floods happen when the overland flow exceeds the capacity of the urban drainage facilities. Urban drainage facilities are structures such as pipes, channels, culverts, and storm drain that are designed to collect and transport excess water from urban areas. If these facilities are inadequate, damaged, blocked, or overwhelmed by too much water, they cannot drain the overland flow quickly and efficiently enough. As a result, the water accumulates in low-lying areas and causes urban floods.

The process of interplay between rainfall, infiltration, and runoff that generate urban floods can be simplified in the illustration as shown in Figure 6.1. When overland flow exceeds the capacity of drainage systems in the city along with other factors, the city floods.

We have no control over the amount of rainwater that falls, as rainfall is a purely natural phenomenon. However, we can influence the volume of runoff by modifying the surface of the ground. It is known that the ground surface plays a crucial role in determining runoff (RO), as demonstrated in the rational formula for maximum flood calculation:

$$Q_p = 0.00278\ C \times i \times A \tag{6.1}$$

where:

Q_p denotes peak discharge (m^3/s)

C is runoff coefficient, which depends on the ground cover.

i is rainfall intensity (mm/h).

A represents catchment area (km^2).

The runoff coefficient stands for the ratio of the volume of water that runs off a surface to the volume of rainfall that falls on the ground surface. The runoff coefficient is dimensionless and usually varies between 0 and 1. A value of 0 indicates that all the surface absorbs the rainfall and none of it runs off, while a value of 1 indicates that all the rainfall runs off and none of it is absorbed.

Several factors influence the runoff coefficient. The most significant of these are the characteristics of the surface, including its permeability, roughness, slope, and the cover of the ground surface. For instance, a smooth, impermeable surface like concrete will have a high runoff coefficient because most of the rainfall will run off. On the other hand, a rough, permeable surface like a grassy field will have a low runoff coefficient because much of the rainfall will be absorbed.

The runoff coefficient also depends on the antecedent moisture condition of the surface, which refers to the amount of moisture in the surface before the rainfall event. If the surface has already saturated with water, it cannot absorb much more, and the runoff coefficient will be high. Conversely, if the surface is dry, it will absorb more water, and the runoff coefficient will be low.

TABLE 6.1
Runoff Coefficients (C) for Various Land Uses

Types of Uses	Descriptions	Approximate C
Commercial	Downtown, with more hardscape	0.70–0.95
	Neighborhood, predominantly hardscape with some softscape	0.50–0.70
Residential	Single unit, for example, predominantly bungalow	0.30–0.50
	Multi-unit detached	0.40–0.60
	Multi-unit attached, for example, row house	0.60–0.70
Industrial	Light industries, for example, garments	0.50–0.80
	Heavy industries, for example, machinery	0.60–0.90
Suburban	Single-unit residential plus woods	0.25–0.40
Apartment dwelling	Residential plus hardscape parking lots	0.50–0.70
Park and cemeteries	Lawn, grass, and woods	0.10–0.25
Unimproved areas	Predominantly natural areas	0.10–0.30

The runoff coefficient is used in several hydrological models to estimate runoff from rainfall. One of the most common of these is the Rational Method, which estimates the peak runoff rate from a rainfall event. The formula for the Rational Method. Despite its widespread use, the runoff coefficient has several limitations. It assumes that the runoff process is linear, and that the runoff coefficient is constant for a given surface. However, in reality, the runoff process is highly nonlinear, and the runoff coefficient can vary with the intensity and duration of the rainfall.

In recent years, there has been a move towards more physically based models that do not rely on the concept of a runoff coefficient. These models simulate the physical processes of infiltration and runoff more realistically. However, the runoff coefficient remains a useful concept, particularly for preliminary design and planning purposes.

Table 6.1 displays the magnitude of runoff coefficients, or 'C' values, for various land uses and surface conditions in typical urban areas. These 'C' values apply to areas with relatively flat terrain and typical soil structures. However, in areas where the terrain is predominantly steep, these 'C' values tend to be higher.

When calculating the peak discharge of a catchment area using the rational formula, it is assumed that the rainfall intensity across the area accumulated instantly. However, rainfall requires time to converge at a specific outlet for the rational formula to be applied correctly.

Therefore, the time of concentration should be considered when determining the peak discharge. If the duration of the rainfall is shorter than the time of concentration, not all the watersheds will contribute to the peak discharge, resulting in a lower value. Conversely, if the rainfall duration is at least equal to the time of concentration, the entire watershed will contribute to the rainfall intensity.

There are numerous formulas to approximate the time of concentration, but Kirpich proposed a simple and reliable one in 1940. The Kirpich formula is a method used to estimate the time of concentration of a catchment area, which is then used to determine the critical storm duration during runoff and drainage design1. The time of concentration (T_c) is the time required for runoff to travel from the hydraulically most distant point in the watershed to the outlet. The Kirpich formula is used for natural basins with well-defined routes for overland flow along bare earth or mowed grass roadside channels. It is important to note that the Kirpich formula might give shorter times compared to other methods.

Factors that can affect the time of concentration include surface roughness, channel shape, and slope. In watersheds with low (flat) topographic slopes, the calculation of T_c using commonly accepted equations with slope in the denominator often results in unreasonably large values.

Therefore, an adjustment to the slope is sometimes recommended to allow more realistic results for low topographic slope watersheds.

The formula is as follows:

$$t_c = 0.0195L^{0.77}S^{-0.385} \quad (6.2)$$

where:

t_c is time of concentration, in minute
L denotes length of the channel, in meter
S represents slope of the channel

6.3 PRECIPITATIONS

Precipitation refers to the process where condensed water descends from the atmosphere to the Earth's surface in various forms, such as rainfall, snowfall, hail, or sleet, depending on atmospheric conditions. In this urban drainage chapter, we will focus on rainfall only as the most fundamental element of hydrology. The hydrological cycle, often referred to as the water cycle, is a fundamental process that regulates the distribution and movement of water on our planet (Figure 6.2). It is an intricate and continuous cycle that sustains life, supports ecosystems, and shapes the Earth's landscapes. The hydrological cycle encompasses various components and is essential for a multitude of reasons, from providing freshwater resources to supporting agriculture, industry, and the very existence of life itself.

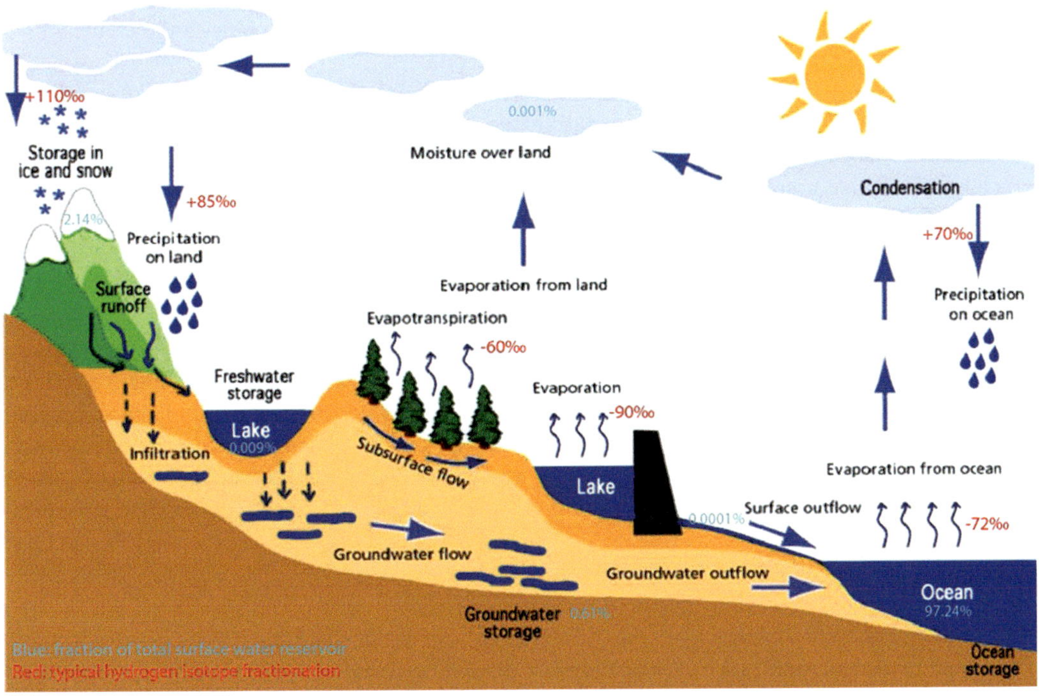

FIGURE 6.2 Hydrological cycle.

Source: **Wikimedia Commons.**

The hydrological cycle consists of several key components. Evaporation is the transformation of liquid water into water vapor because of solar radiation, primarily occurring in oceans, lakes, rivers, and moist soil surfaces. Transpiration, the release of water vapor from plants, particularly through tiny openings in their leaves, transfers water from the land back into the atmosphere. Condensation sees water vapor rise into the atmosphere, cool, and condense into clouds. Precipitation occurs when water droplets combine in clouds and fall as rain, snow, sleet, or hail. SRO carries precipitation over land, eventually flowing into rivers, streams, and oceans. Infiltration involves some precipitation seeping into the ground, replenishing groundwater reservoirs.

The hydrological cycle is essential for numerous reasons, including its role in providing freshwater for drinking, agriculture, and industrial use, supporting ecosystems, regulating global climate patterns, and enabling food production through agriculture. It also plays a critical role in providing water for industrial processes and energy production. The cycle ensures human survival by providing clean, potable water and is the foundation for human existence and societal progress. However, the dependency of people on the hydrological cycle is not uniform worldwide, leading to disparities in water availability, which can result in conflicts over water resources. As the world faces increasing population growth, urbanization, and climate change, wise water management, conservation efforts, and adaptation to evolving climate challenges are essential to ensure a sustainable future. The hydrological cycle serves as a constant reminder of water's vital role in our lives, emphasizing the need for responsible stewardship of this precious resource for the well-being of present and future generations.

Among the components of the hydrological cycle, rainfall is one of the most essential components. Rainfall analysis holds a pivotal role in the realm of urban hydrology, where the focus is on understanding and managing water dynamics within urban environments. Urbanization has reshaped the natural landscape, replacing permeable surfaces with impermeable ones and altering the way rainfall is absorbed and managed. In this context, rainfall analysis becomes paramount, as it helps urban planners, engineers, and water resource managers grapple with the challenges of increased runoff, flooding, and water quality issues in cities.

Urban areas are particularly vulnerable to the effects of rainfall, primarily because of the proliferation of impervious surfaces like roads, buildings, and parking lots, which hinder natural infiltration and lead to greater SRO. Understanding rainfall patterns and their temporal and spatial variations is crucial for effective urban hydrology. The analysis of rainfall data provides essential insights into the IDF of precipitation events, which are vital for designing stormwater management systems that can handle the unique demands of urban areas.

Rainfall analysis within the context of urban hydrology also plays a significant role in flood prediction and prevention. The rapid flow of rainwater over impermeable surfaces can lead to flash floods in urban settings, making it imperative to monitor rainfall events closely and to design drainage systems capable of mitigating the risk. Additionally, as urban areas expand and face the challenges of climate change, the role of rainfall analysis in promoting resilient and sustainable urban planning becomes increasingly evident. By comprehending rainfall behavior and integrating this knowledge into infrastructure design and urban development, we can work toward creating cities that are better equipped to manage their water resources efficiently while safeguarding against the perils of urban flooding.

6.3.1 Types of Precipitation

Precipitation is a term that describes the different forms of water that fall from the sky to the earth. Precipitation can be liquid, such as rain or drizzle, or solid, such as snow, sleet, or hail. Precipitation occurs when water vapor in the atmosphere condenses into droplets or crystals and becomes too heavy to stay suspended in the air. The type and amount of precipitation depend on various factors, such as temperature, humidity, wind, and air pressure. Precipitation is an important part of the water cycle, as it replenishes the freshwater supply on the planet.

Based on the shape, size, and phase of the water particles. The main types of precipitation are:

- *Drizzle:* Drizzle is a type of precipitation that consists of very small droplets of liquid water, usually less than 0.5 mm in diameter. Drizzle falls from low clouds, such as stratus or stratocumulus, and is often associated with foggy or misty weather. Drizzle is usually light and gentle and does not make much noise when it hits the ground.
- *Rain:* Rain is a type of precipitation that consists of larger droplets of liquid water, usually more than 0.5 mm in diameter. Rain falls from various types of clouds, such as cumulus, nimbostratus, or cumulonimbus, and is often associated with stormy or windy weather. Rain can vary in intensity and duration, from light showers to heavy downpours.
- *Sleet:* Sleet is a type of precipitation that consists of small pellets of ice, usually less than 5 mm in diameter. Sleet forms when raindrops freeze as they fall through a layer of cold air near the ground. Sleet falls from high clouds, such as cirrus or altostratus, and is often associated with cold or freezing weather. Sleets can be slippery and dangerous and can cause damage to plants and crops.
- *Snow:* Snow is a type of precipitation that consists of ice crystals that form in the atmosphere and stick together in clusters of snowflakes. Snow forms when water vapor freezes directly into solid ice, without passing through the liquid phase. Snow falls from cold clouds, such as cirrostratus or cirrocumulus, and is often associated with winter or polar weather. Snow can vary in shape and size, from fine powder to large flakes.
- *Graupel:* Graupel is a type of precipitation that consists of soft balls of ice, usually less than 5 mm in diameter. Graupel forms when snowflakes are coated with supercooled water droplets that freeze on contact. Graupel falls from convective clouds, such as cumulus or cumulonimbus, and is often associated with thunderstorms or hailstorms. Graupel can be white or gray and can bounce or break when it hits the ground.
- *Hail:* Hail is a type of precipitation that consists of hard balls of ice, usually more than 5 mm in diameter. Hail forms when graupel or ice pellets are lifted and lowered by strong updrafts and downdrafts in a thunderstorm, accumulating layers of ice. Hail falls from severe clouds, such as cumulonimbus or supercell, and is often associated with violent or destructive weather. Hail can be clear or opaque, and can cause injury or damage to people, animals, and property.

However, precipitation can also be classified into different forms based on the way it develops in the atmosphere. Some of the common forms are:

1. *Cyclonic Precipitation:* This form of precipitation occurs when two air masses of different temperatures and humidity collide, forming a front. The warmer, lighter air mass rises over the colder, denser air mass, creating a slope. As the warm air rises, it cools and condenses, forming clouds and precipitation. Cyclonic precipitation can be either warm or cold, depending on the type of front. Warm fronts produce steady, light to moderate precipitation, while cold fronts produce sudden, heavy precipitation.
2. *Convective Precipitation:* This form of precipitation occurs when warm, moist air rises and cools in the atmosphere. As the air cools, it becomes denser and sinks, creating a cycle of rising and sinking air currents. This cycle is called convection. Convective precipitation is often associated with thunderstorms, which can produce heavy rainfall, lightning, and hail.
3. *Orographic Precipitation:* This form of precipitation occurs when moist air is forced to rise over a mountain range. As the air rises, it cools and condenses, forming clouds and precipitation. The amount of precipitation depends on the height, shape, and orientation of the mountain range, as well as the speed and direction of the wind.

4. *Monsoon:* A monsoon is a seasonal reversal of wind direction that brings heavy rainfall to some regions, especially in Asia and Africa. Monsoons are caused by the difference in temperature and pressure between the land and the ocean. During the summer, the land heats up faster than the ocean, creating a low-pressure zone that attracts moist air from the ocean. This results in heavy rainfall over the land. During the winter, the land cools down faster than the ocean, creating a high-pressure zone that pushes dry air from the land to the ocean. This results in dry conditions over the land.
5. *Frontal:* Frontal precipitation is the rain or snow that falls when two air masses of different temperatures and humidity meet. A front is the boundary between the two air masses. The warmer, lighter air mass rises over the colder, denser air mass, creating a slope. As the warm air rises, it cools and condenses, forming clouds and precipitation. Frontal precipitation can be either warm or cold, depending on the type of front. Warm fronts produce steady, light to moderate precipitation, while cold fronts produce sudden, heavy precipitation. Frontal precipitation is common in areas such as the mid-latitudes, where warm and cold air masses collide.

6.3.2 Precipitation Patterns

Precipitation patterns in urban areas are a critical aspect of understanding the local climate and environmental conditions. Urban environments have unique characteristics that can significantly influence the distribution, intensity, and frequency of precipitation within city limits. This discussion will delve into the particularities of precipitation in urban areas, highlighting the factors that shape these patterns and their implications for urban planning and sustainability.

The assessment of precipitation distribution patterns in urban areas involves the examination of two primary aspects: spatial distribution and temporal distribution. Spatial distribution pertains to the geographical variation of precipitation across the urban area, while temporal distribution involves the temporal variation of precipitation over time. By analyzing both aspects, researchers and urban planners can gain a comprehensive understanding of how precipitation is distributed within the city and how it changes over different time intervals. This information is crucial for addressing various urban hydrological challenges, including flood risk management, urban drainage design, and the sustainable use of water resources within urban environments.

Given that cities typically have relatively small spatial extents, with linear dimensions typically less than 30 km, it is common to observe a relatively uniform or homogeneous distribution of rainfall depth within the city. This means that, over the city's area, the amount of rainfall tends to be consistent, and there are no significant variations or gradients in precipitation levels across different parts of the city. This uniform rainfall pattern is often a result of the relatively small size of the urban area and the local-scale meteorological factors that influence precipitation. Understanding this pattern is important for urban hydrology and infrastructure planning, as it can help inform decisions related to stormwater management and flood mitigation within the city.

Urban areas may also exhibit unique precipitation distribution patterns. The presence of buildings and impervious surfaces can disrupt the natural flow of air masses and the formation of rain clouds. In some cases, urban canyons (the tall buildings on either side of a street) can channel moist air and create "urban corridors" where precipitation is concentrated. On the other hand, some urban areas may experience reduced rainfall because of the urban heat island effect, as warm air rises and suppresses cloud formation.

Local factors can also influence the frequency of precipitation events in urban areas. Urbanization can lead to changes in local climate patterns and weather systems, affecting when and how often it rains. Some cities experience changes in the frequency of precipitation events, with an increase in intense, sporadic rainfall rather than consistent, steady rain. This can be attributed to the altered microclimates within cities.

Urban development can disrupt natural landscapes and vegetation, reducing the ability of the land to absorb and store rainwater. This can result in more frequent, localized flooding during heavy rainfall events. In response, many cities implement stormwater management systems to control the flow of water and mitigate flood risks.

Besides the physical characteristics of urban areas, human activities play a significant role in shaping precipitation patterns. Urban air pollution can influence cloud formation and rainfall. Pollutants such as particulate matter can act as cloud condensation nuclei, potentially leading to changes in the size and number of cloud droplets and affecting precipitation. The presence of pollutants in the atmosphere can either enhance or suppress rainfall, depending on the specific atmospheric conditions.

Urbanization can also have indirect effects on regional precipitation patterns. Large metropolitan areas often lead to deforestation and land-use changes in surrounding regions. This can disrupt natural ecosystems and potentially alter the sources of moisture that contribute to precipitation in urban areas. Changes in land use, such as the conversion of forests to agriculture or urban development, can influence regional climate and precipitation patterns.

Urban areas can experience localized weather phenomena, such as "urban thunderstorms" or "urban flash floods." These intense, short-duration storms are more common in cities because of the complex interplay of factors, including the urban heat island effect, the presence of impervious surfaces, and the altered microclimates within cities. Urban flash floods can lead to rapid inundation of streets, basements, and transportation infrastructure, posing risks to public safety and property damage.

City planners and urban developers are increasingly recognizing the importance of understanding and managing urban precipitation patterns. The sustainable urban design incorporates strategies to mitigate the negative effects of UHIs and precipitation-related challenges. For example, green infrastructure solutions, such as green roofs, permeable pavements, and urban parks, are implemented to enhance rainwater absorption, reduce SRO, and cool the urban environment.

Effective stormwater management is another crucial aspect of urban planning. Cities install drainage systems, retention ponds, and flood control measures to manage rainfall and reduce the risk of flooding. Proper stormwater management not only protects the city from flooding but also helps maintain water quality by preventing the transport of pollutants into water bodies.

Climate-resilient urban planning is becoming increasingly important in the face of global climate change. As the planet warms, extreme weather events, including heavy rainfall and intense storms, are expected to become more frequent and severe. Urban areas must adapt to these changes by improving infrastructure, implementing sustainable urban planning strategies, and enhancing emergency preparedness to mitigate the effects of extreme precipitation events.

6.3.3 Instruments of Precipitation Data Collection

Precipitation is a generic term that can be used for both rainfall and snowfall. As snowfall will not be discussed in this chapter, therefore the precipitation hereinafter will be referred to as rainfall. Rainfall measurement is a crucial aspect of meteorology and hydrology, as it provides valuable information for weather forecasting, flood prediction, and water resource management. Various methods are used to measure rainfall, ranging from traditional manual techniques to modern automated systems. Here are some of the most common methods for measuring rainfall:

1. *Rain Gauge:* From the way the rain gauge works to measure the rainfall depth, the gauge can be divided into two types, they are, ordinary rainfall instrument (non-recording rain gauge), in which the measurement of the rainfall depth collected by the instrument is carried out manually by an operator, and the automatic rainfall recorder (recording rain gauge), in this case, the measurement is done automatically by the instrument. The operator collects the results.

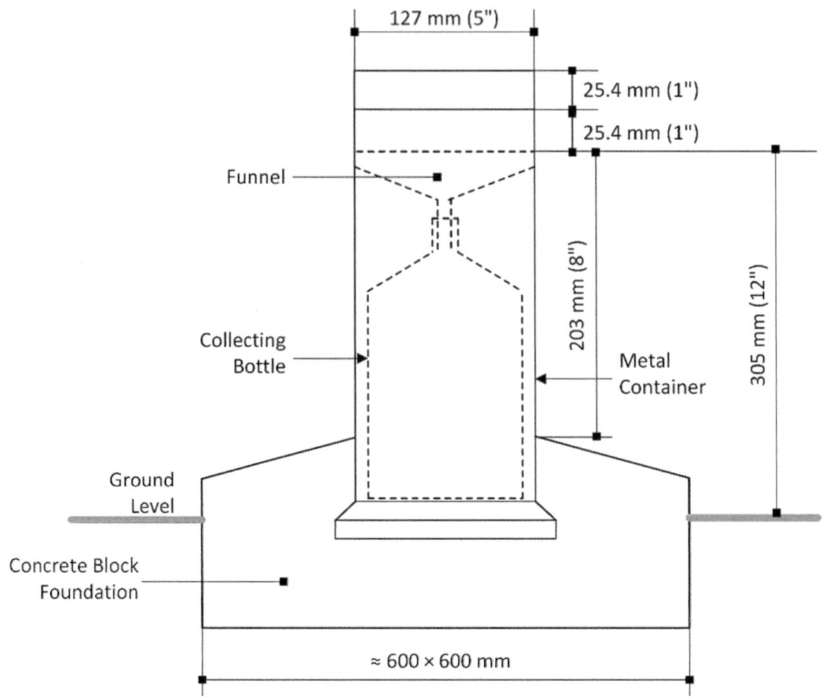

FIGURE 6.3 Symons manual rainfall recorder.

This instrument (Figure 6.3) is designed to collect and measure the amount of rainfall that falls within a specific area. Here is a breakdown of the components and their functions: (a) *Collector Rim:* The rim is set at a fixed height of 12 inches (305 mm) above the ground level. It acts as a reference point for the collection of rainfall and ensures that the collected water flows into the funnel. (b) *Funnel:* The funnel is designed to direct rainwater into a receiving vessel. It typically has a wide opening at the top to capture raindrops and channels the water into the vessel, ensuring accurate measurement. (c) *Receiving Vessel:* This is the container that collects the rainwater funneled from the rim. It is securely housed within a metallic container to protect it from external factors that might interfere with the measurement. (d) *Metallic Container:* The metallic container protects the receiving vessel and the measurement apparatus from external elements, such as wind, debris, or temperature fluctuations. (e) *Measuring Glass:* The measuring glass is used to measure the amount of rainfall collected. It has an accuracy of up to 0.1 mm, which allows for precise measurements of the accumulated rainfall.

2. *Tipping Bucket Rain Gauge:* Tipping bucket rain gauges are often used in automated weather monitoring systems. A Tipping Bucket Rain Gauge (Figure 6.4) is a specific rain measurement instrument to quantify precipitation over a predetermined area during a specified timeframe. The device operates via a funnel that gathers rainwater and directs it into a small bucket. Once this bucket reaches its capacity, it tips over, releasing the water, and concurrently activates a switch that registers the rainfall quantity. Subsequently, the bucket reverts to its initial position, and the cycle recommences. When rainwater falls into the funnel, it flows into the bucket. The bucket has a pivot point at its center, and when it fills with water, it tips over, emptying the water into a container below. As the bucket tips, it triggers a switch that records the amount of rainfall. The bucket then returns to its original position, ready to collect more rainwater. The amount of rainfall is calculated by counting the number of times

FIGURE 6.4 Automatic rainfall recorder – tipping bucket.

Source: **Wikipedia.**

the bucket tips over and empties. This rain gauge type is renowned for its capacity to provide precise and dependable measurements. The number of bucket tips provides an accurate measurement of rainfall, and the data can be recorded electronically.

3. *Weighing Rain Gauge:* A weighing rain gauge is a type of rain gauge that operates on the principle of measuring rainfall by weighing the collected precipitation. Its parts and the way it works can be briefly described as follows: (a) *Collection Surface:* The rain gauge consists of a collection surfacc, often a funnel or a wider opening, designed to efficiently collect rainwater. This surface is exposed to the open air to receive rainfall. (b) *Funneling:* The collected rainwater is funneled into a container, typically a container with a known weight. This container is suspended or supported by a mechanism that allows it to move freely. (c) *Weight Measurement:* As rainwater accumulates in the container, its weight increases. This weight increase is continuously monitored using a precision weighing scale or load cell that is sensitive enough to detect even small weight changes. (d) *Data Recording:* The changes in weight because of the added rainwater are recorded by a data recording system, which could be manual or automated, depending on the specific design of the rain gauge. (e) *Calibration:* The system needs to be carefully calibrated to convert the changes in weight into an accurate measurement of rainfall. This calibration accounts for factors like the container's weight, the rain gauge's design, and any buoyancy effects. (f) *Accumulated Rainfall:* By continuously monitoring the weight change in the container, the rain gauge can accurately measure the accumulated rainfall over time. The amount of precipitation is typically recorded in units of millimeters or inches, representing the depth of rainfall that has been collected.

Weighing rain gauges are known for their accuracy and precision in measuring rainfall, as they directly measure the amount of water accumulated. They are commonly used in research, meteorology, and hydrology, providing valuable data for weather forecasting, flood monitoring, and water resource management.

There are also instruments to measure the quantity of rainfall indirectly. Unlike conventional rain gauges, in which the measurement is carried out directly in contact between precipitation and the instrument, these rainfall measurement instruments employ a non-contact method by using ultrasonic or microwave pulses to estimate the rainfall depth. The apparent and significant advantages of these instruments are real time and cover large areas at once, and therefore they can be used as an early warning system in a flood prevention project. These instruments are:

4. *Acoustic Rain Gauge:* An acoustic rain gauge is a weather instrument designed to measure and record the amount of rainfall by using sound waves. Instead of traditional rain gauges that collect rainwater directly, acoustic rain gauges work based on the principle of measuring the impact of raindrops on a surface and converting the sounds of that impact into a measurement of precipitation. The acoustic rain gauge works as follows: (a) *Microphones or Sensors:* Acoustic rain gauges are equipped with microphones or sensors that are sensitive to vibrations or sound waves. (b) *Raindrop Impact:* When raindrops fall onto a surface, such as a rooftop, the ground, water body, or any surface, they generate sounds. The microphone, or a sensor specifically designed for this purpose captures the sound, they create vibrations or acoustic signals upon impact. (c) *Sound Wave Analysis:* The microphones or sensors in the acoustic rain gauge capture these acoustic signals and analyze them. The characteristics of the sound waves, such as their frequency, intensity, and duration, can provide information about the size and number of raindrops that hit the surface. (d) *Precipitation Measurement*: By processing the data collected from the sound waves, the acoustic rain gauge can calculate the amount of precipitation that has fallen over a given period. This measurement is usually expressed in millimeters or inches of rainfall.

   Acoustic rain gauges offer certain advantages, such as the ability to measure precipitation with no physical collection containers and the potential for continuous monitoring. They are often used in research, hydrology, and meteorology to study rainfall patterns and collect data for weather and climate studies. However, the accuracy and reliability of acoustic rain gauges can be influenced by factors like wind, surface characteristics, and the type of raindrops (e.g., heavy or light rain). Calibration and careful installation are important to ensure accurate measurements.

5. *Precipitation Radar (Telemetering):*
   Precipitation radar, also known as weather radar or meteorological radar, is a specialized radar system used to detect and monitor various forms of precipitation in the atmosphere, such as rain, snow, sleet, and hail. It plays a crucial role in weather forecasting, severe weather monitoring, and climate research. Here are the key features and functions of precipitation radar: (a) *Detecting Precipitation:* Precipitation radar operates by emitting radio waves, typically in the microwave or radio frequency bands, into the atmosphere. When these waves encounter precipitation particles, they scatter in various directions. (b) *Measuring Return Signals:* The radar system has a receiver that collects the signals reflected or scattered by the precipitation particles. The radar measures the strength of these return signals, which is related to the intensity of the precipitation. (c) *Displaying Reflectivity:* The data collected by the radar is processed to create a reflectivity map, often displayed in color on weather radar screens. Reflectivity is a measure of the radar signal's intensity after interacting with precipitation. Different colors indicate different precipitation rates, with lighter colors representing light precipitation and darker colors indicating heavy precipitation. (d) *Identifying Precipitation Types:* Precipitation radar can help meteorologists distinguish between different types

of precipitation, such as rain, snow, sleet, or hail, based on the characteristics of the radar returns and the temperature profiles in the atmosphere. (e) *Monitoring Storms:* Precipitation radar is an essential tool for monitoring severe weather events, including thunderstorms and hurricanes. It can detect features like heavy rain, severe wind, and the potential for tornadoes. (f) *Tracking Movement:* By capturing successive radar scans over time, meteorologists can track the movement and evolution of weather systems and precipitation patterns. This information is critical for weather forecasting and issuing weather warnings. (g) *Measuring Doppler Velocity:* Many modern precipitation radars also incorporate Doppler radar technology to measure the velocity of precipitation particles. This provides valuable data on wind patterns and can help identify the presence of rotation in severe thunderstorms, which may indicate the potential for tornadoes. (h) *Network of Radars:* National and regional weather agencies often deploy networks of radar stations to provide comprehensive coverage of an area. The data from multiple radar sites can be combined to create a detailed and real-time view of precipitation across a region.

6. *Weather Satellites:* Weather satellites equipped with radiometers can provide estimates of rainfall on a global scale. The noticeable disadvantage of this system is that weather satellites cannot be used to measure the exact quantity of precipitation. It can only be used to estimate the precipitation over large areas and for weather forecasts.

Each method of measuring rainfall has its advantages and limitations, and the choice of method depends on factors such as the required accuracy, the scale of measurement (local or regional), and the available resources. Combining multiple measurement methods can provide a more comprehensive understanding of rainfall patterns and their impact on the environment and society. In a city or urban areas, as the size is normally not large, the use of large-scale measurements with low accuracy is not necessary.

Rainfall measurements, which are typically recorded over a specific area and time period using various instruments, are commonly expressed in units of millimeters (mm) or inches (in), representing the equivalent depth of water. These measurements can be reported at different intervals – hourly, daily, weekly, monthly, or annually – depending on the specific requirements.

Hourly Rainfall Measurements are particularly useful for monitoring short-term events such as storms or flash floods. They can also aid in weather forecasting and issuing public warnings about potential hazards.

Daily Rainfall Measurements are typically used to assess the water balance and soil moisture content of a region. This data can inform the planning of irrigation and drainage systems, including agricultural and urban drainage systems.

Weekly Rainfall Data provides insights into the seasonal patterns and trends of precipitation. This information can enhance our understanding of the hydrological cycle and the potential effects of climate change.

Monthly Rainfall Measurements can help identify periods of drought and flooding and their respective effects on ecosystems and society. This data is crucial for environmental management and disaster mitigation.

Annual Rainfall Measurements are valuable for evaluating long-term averages and variability of precipitation. This data can inform climate studies and long-term water resource management.

For urban drainage design, however, the intensity of rainfall is a critical factor. This is typically derived from the analysis of IDF of the rainfall. The IDF analysis provides a statistical representation of the likelihood of the occurrence of different rainfall intensities based on historical data, which is crucial for designing effective drainage systems.

6.3.4 Transferring Individual Station Point Rainfall to Basin Rainfall

To estimate the average rainfall over a river basin, different methods can be applied based on the spatial distribution of the rainfall stations. Three common methods are the isohyet method, the Thiessen polygon method, and the hypsometric method. The isohyet method involves drawing lines of equal rainfall on a map and calculating the area and the mean rainfall for each zone. The Thiessen polygon method assigns a weight to each station based on the area of influence of the station, which is determined by drawing perpendicular bisectors of the lines joining adjacent stations. The hypsometric method uses elevation zones to divide the basin and assigns a rainfall value to each zone based on the average elevation of the stations in that zone.

- *Isohyet Method:* This method involves drawing lines, called isohyets, on a map. Each isohyet represents an area with equal rainfall. The area and the mean rainfall for each zone between the isohyets are calculated. The average rainfall over the basin is then determined by taking a weighted average of the mean rainfall in each zone, with the weights being the areas of the zones.
- *Thiessen Polygon Method:* In this method, each rainfall station is assigned a weight based on its area of influence. The area of influence is determined by drawing perpendicular bisectors of the lines joining adjacent stations, which form polygons around each station. The rainfall measurement at each station is then multiplied by the area of its corresponding polygon, and these products are summed and divided by the total area of all polygons to get the average rainfall.
- *Hypsometric Method:* This method uses elevation zones to divide the basin. A rainfall value is assigned to each zone based on the average elevation of the rainfall stations in that zone. The average rainfall over the basin is then calculated by taking a weighted average of the rainfall values in each zone, with the weights being the areas of the zones.

These methods consider the spatial distribution of rainfall stations and the variability of rainfall regarding location and elevation, providing a more accurate estimate of average rainfall over a river basin. However, the average rainfall can also be derived by using arithmetic mean, in which the spatial distribution of rainfall stations is ignored.

Examples of the calculation with hypothetical illustrations are given in the following paragraphs.

6.3.4.1 Isohyet Method

An isohyetal line is a line on a map that connects points that have the same amount of rainfall in a given period of time or for a specific storm. Isohyetal lines are used to show the spatial distribution and variation of precipitation over an area. The closer the isohyetal lines are, the higher the rainfall gradient is, meaning that the rainfall changes rapidly over a short distance. The farther the isohyetal lines are, the lower the rainfall gradient is, meaning that the rainfall changes slowly over a long distance. To make an isohyetal map, we need to have data on the rainfall depths at different stations, measured by rain gauges or weather stations. Then, we can plot the rainfall values on a base map and draw lines that connect points with equal rainfall. We need interpolation or estimation used to determine the points where a particular isohyetal line is located. For practical purposes and easiness of development, the isohyetal lines are normally designated as an even number with a last digit of 0 or 5, such as 100 or 105, not 103 or 107.

An example of developing isohyetal lines is illustrated below. Suppose we are going to analyze the average weekly rainfall depth of a river basin, given the maximum weekly rainfall depth at each station available within and near the river basin. We have five rainfall stations A, B, C, D, and E with respective rainfall depths as shown in Figure 6.5.

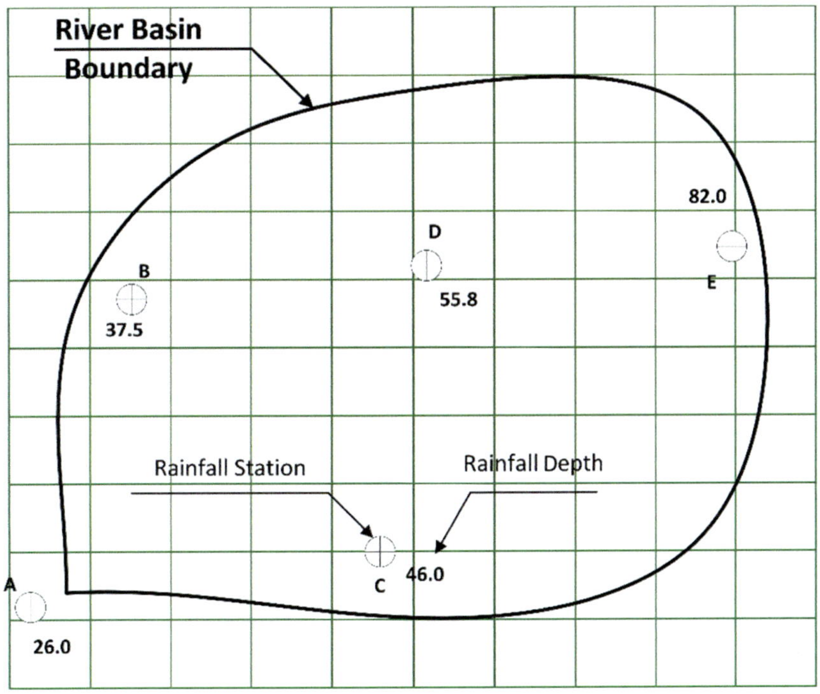

FIGURE 6.5 Rainfall stations in a river basin.

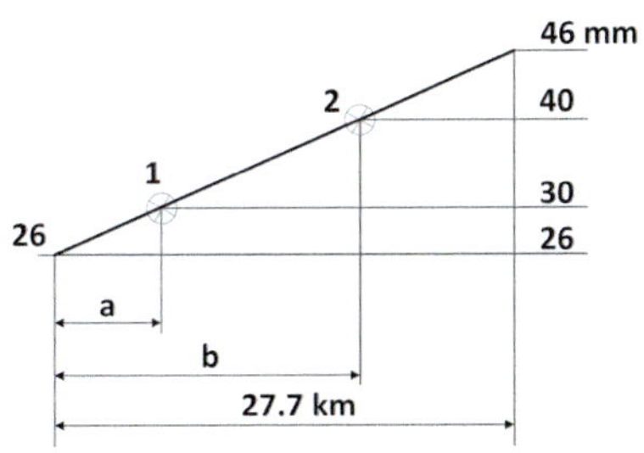

FIGURE 6.6 Interpolation to determine the locations of 1 and 2.

We draw an imaginary line between each pair of neighboring stations to represent the rainfall distribution, such as A–B, A–C, B–C, B–D, C–D, D–E, and C–E. For example, in line of A–C, the rainfall depth at station A is 26 mm, and at station C is 46 mm. There are also two intermediate points on this line with rainfall depths of 30 mm and 40 mm, respectively. The question is, how can we determine the locations of these two points?

We can determine these two points by using interpolation that follows the rule of the right triangle that is congruent to another right triangle. This is illustrated in Figure 6.6.

To determine the location of point 1, that is, the horizontal distance of point 1 from station A, can be found by using the rule of straight triangle, as follows:

$$\frac{a}{27.7} = \frac{(30-26)}{(46-26)} \rightarrow a = 5.54 \text{ km}$$

We can find the point on the map where the isohyetal line of 30 mm intersects line A–C by measuring the horizontal distance from station A. This distance is 5.54 km, which we can easily mark on the map. Similarly, we can approximate the location of the point where the isohyetal line of 40 mm intersects line A–C by measuring the horizontal distance from station A. This distance is b, which we can also mark on the map.

$$\frac{b}{27.7} = \frac{(40-26)}{(46-26)} \rightarrow b = 19.39 \text{ km}$$

For other locations, the approximation is the same. Through this process, we can draw the isohyetal lines for the whole river basin. The isohyetal map of this river basin is shown in Figure 6.7.

The following equation calculates the average basin rainfall:

$$P_{avg} = \frac{\sum_{i=1}^{n} A_i \times P_i}{\sum_{i=1}^{n} A_i} \tag{6.3}$$

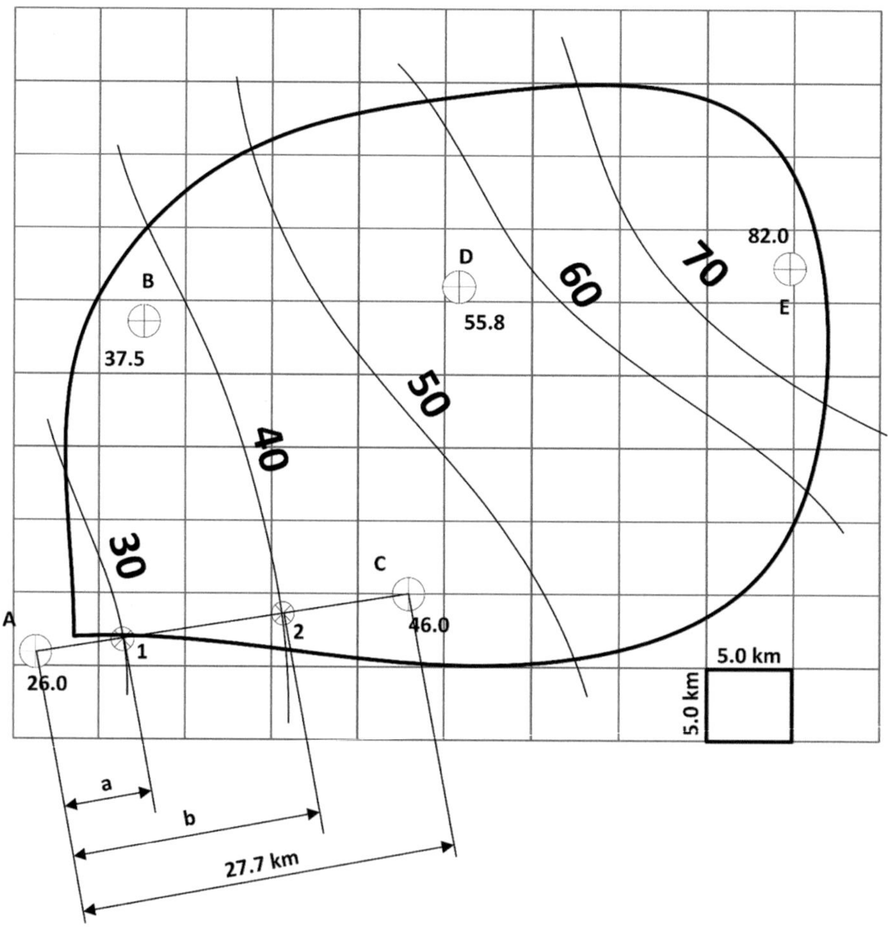

FIGURE 6.7 The isohyetal map.

TABLE 6.2
Basin Rainfall Calculation Using Isohyet Method

Isohyet	Area, A_i, km²	Rainfall, P_i, mm	$A_i \times P_i$, mm . km²	Notes
<30	16	26	416	$P_{avg} = \frac{63861}{1184} = 53.9$ mm
30–40	153	35	5,355	
40–50	280	45	12,600	
50–60	458	55	25,190	
60–70	142	65	9,230	
>70	135	82	11,070	
Total, £	1,184		63,861	

where:

A_i is area of the ith isohyetal zone.

P_i denotes average precipitation, that is, depth of rainfall at the ith zone.

The next step in using the isohyetal method to convert point rainfall to basin rainfall is to estimate the area of each isohyetal zone in the river basin. The total area of the river basin is the sum of the areas of all the isohyetal zones. We can use any online or offline app that can calculate the area of irregular shapes to find the area between two isohyetal lines. Until this point, the above process is summarized in Table 6.2.

The average rainfall depth of 53.9 mm in a watershed refers to the average amount of rainfall that the watershed receives. It is a measure of the volume (or depth) of rain that falls, spread evenly across the entire area of the watershed. This measurement is usually taken over a specific period of time, such as a month or a year. It is a useful metric for understanding the water availability in the area, which can be critical for things like agriculture, water supply, and flood management. To put it in perspective, if you imagine the entire watershed as a flat surface, the average rainfall depth of 53.9 mm means that if all the rain that fell over this area stayed where it landed, without running off or evaporating, it will form a layer of water 53.9 mm deep across the entire watershed. In reality, water does not stay evenly distributed – it runs off slopes and into rivers, and some of it evaporates back into the atmosphere. However, the average rainfall depth gives us a useful way to quantify and compare rainfall amounts.

6.3.4.2 Thiessen Polygon Method

The Thiessen polygon method is a common method of estimating the average rainfall over a certain area based on point rainfall values measured by rain gauges. The method involves the following steps:

1. Draw a map of the basin and locate all the rain gauges in and around the basin.
2. Construct a Delaunay triangulation by connecting each rain gauge with its nearest neighbors. A Delaunay triangulation is a net of non-overlapping triangles such that no rain gauge is inside the circle that passes through the vertices of any triangle.
3. Bisect each segment of the triangulation by a perpendicular line that passes through the midpoint of the segment. Extend the bisectors beyond the segments to form polygons.
4. Assign each polygon to the rain gauge that is contained within it. The area of each polygon represents the influence area of the corresponding rain gauge.
5. Calculate the average rainfall over the basin by multiplying the rainfall value at each rain gauge by the area of its polygon and dividing the sum by the total area of the basin, as shown in the following formula:

$$P_{avg} = \frac{\sum_{i=1}^{n} A_i \times P_i}{\sum_{i=1}^{n} A_i} \tag{6.4}$$

where:

A_i represents area of the ith polygon zone (influence area).

P_i denotes average precipitation, that is, depth of rainfall at the ith influence area.

Equations (6.3) and (6.4) are identical, but A_i and P_i are different. Therefore, we prefer to write twice.

For the same example of a river basin, the use of the Thiessen polygon is illustrated in Figure 6.8.

The Thiessen polygon method assumes that the rainfall at any point within a polygon is equal to that of the rain gauge inside it. This method is also called the weighted mean method because it weighs the rain gauge observations according to their influence areas. The calculation is shown in Table 6.3.

By using Thiessen Polygon, the average basin rainfall is 53.0 mm. The result shows an insignificant difference with the isohyet method.

6.3.4.3 Hypsometric Method

The hypsometric method is based on the hypsometric curve of a river basin, which is a graphical representation of the distribution of elevations in a given area. It shows the cumulative area or percentage of land at or below a specific elevation. Although hypsometric parameters do not have a

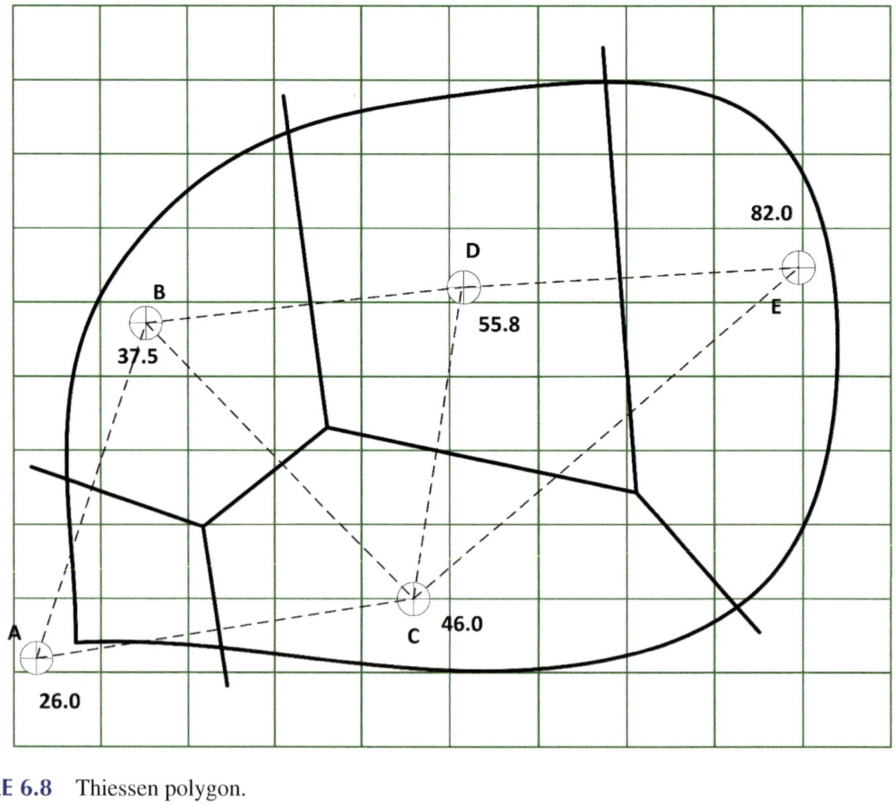

FIGURE 6.8 Thiessen polygon.

TABLE 6.3
Basin Rainfall Calculation Using Thiessen Polygon Method

Station	Area, A_i, km²	Rainfall, P_i, mm	$A_i \times P_i$, mm . km²	Notes
A	81	26.0	2,106	$P_{avg} = \frac{62703}{1184} = 53.0$ mm
B	192	37.5	7,200	
C	340	46.0	15,640	
D	346	55.8	19,307	
E	225	82.0	18,450	
Total, £	1,184		62,703	

direct correlation with rainfall depth, they are related to topography, hypsometry, and climate, which can affect rainfall patterns.

Topography, as represented by the hypsometric curve, can affect rainfall through orographic effects. Mountains can cause moist air masses to rise, leading to more rainfall on windward slopes and less rainfall on leeward slopes. Elevation, as indicated by the hypsometric curve, can affect temperature, which affects rainfall patterns. Temperature decreases with increasing elevation, and this can affect the distribution and types of precipitation.

Therefore, it is reasonable to say that there is a correlation between hypsometric parameters and precipitation. Some suggested that the hypsometric method can also be used as a technique for estimating the average rainfall over a basin by using the relationship between elevation and precipitation. The method assumes that the precipitation varies linearly or nonlinearly with the elevation and that the basin can be divided into elevation zones with known areas. The method involves the following steps:

1. Obtain the elevation data for the basin and construct a hypsometric curve, which is a plot of the cumulative percentage of basin area versus the relative elevation.
2. Obtain the precipitation data for the basin and develop a mathematical equation expressing the relationship between precipitation and elevation. This can be done by using regression analysis or other methods.
3. Apply the equation to each elevation zone and calculate the average precipitation for each zone.
4. Multiply the average precipitation for each zone by the percentage of the basin area in that zone and sum up the products to obtain the basin's average rainfall.

The hypsometric method accounts for the topographic effects on precipitation and can be used for basins with complex terrain and sparse rain gauges. However, the method also has some limitations, such as:

- The method requires accurate elevation and precipitation data, which may not be available or reliable for some regions.
- The method assumes a uniform distribution of precipitation within each elevation zone, which may not be realistic in some cases.
- The method may not capture the spatial variability of precipitation caused by factors other than elevation, such as wind direction, moisture availability, orographic effects.

While there may be some correlations between hypsometric curve characteristics and local rainfall patterns because of topographic influences, the direct relationship is complex and depends on various factors such as geography, atmospheric circulation patterns, and climate types specific to

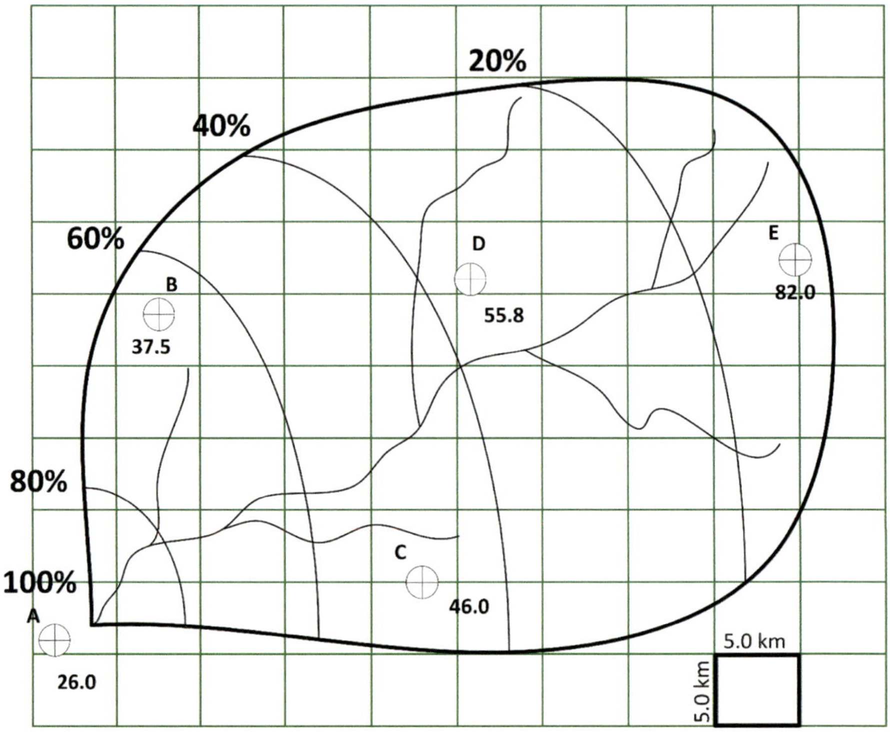

FIGURE 6.9 Hypsometric method.

the region in question. It's important to consider the broader climatic and geographic context when exploring these relationships.

We did not apply any formulas to find the depth of rain from the hypsometric parameters, which show the relationship between the area and height of a watershed. We used the hypsometric curves instead, which we obtained from Geographic Information Systems (GIS) data on the area and ground surface elevation of the watershed, as in Figure 6.9. The hypsometric curves let us estimate the depth of rain without measuring it directly.

We selected four hypsometric lines, corresponding to the 20%, 40%, 60%, and 80% percentiles of the area. The selection was based on the spatial distribution of the rainfall stations in the basin so that the stations were evenly distributed along the hypsometric lines. This ensured that the rainfall data reflected the variability of the elevation and area of the basin. Figure 6.10 illustrates the selected hypsometric lines and the rainfall stations for the river basin.

The hypsometric configuration of the river basin is shown in Figure 6.11.

Like the Isohyet method and Thiessen polygon method, the average basin rainfall is calculated by using either Equation (6.3) or (6.4). The equation is:

$$P_{avg} = \frac{\sum_{i=1}^{n} A_i \times P_i}{\sum_{i=1}^{n} A_i}$$

where:

A_i is area of the ith hypsometric zone.

P_i denotes average precipitation, that is, depth of rainfall in the ith zone.

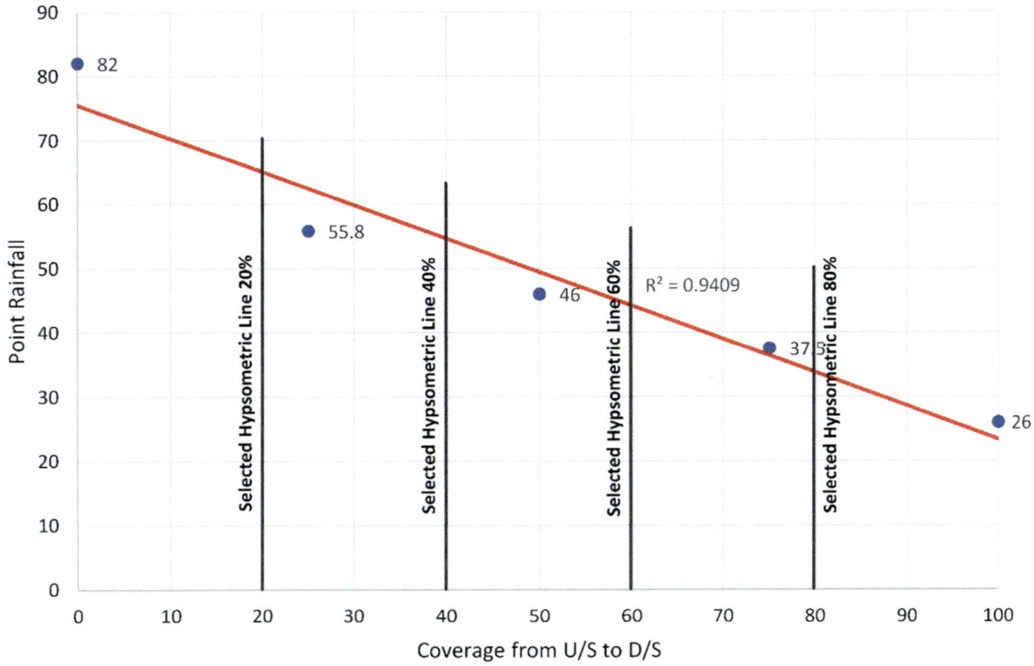

FIGURE 6.10 Selected hypsometric lines.

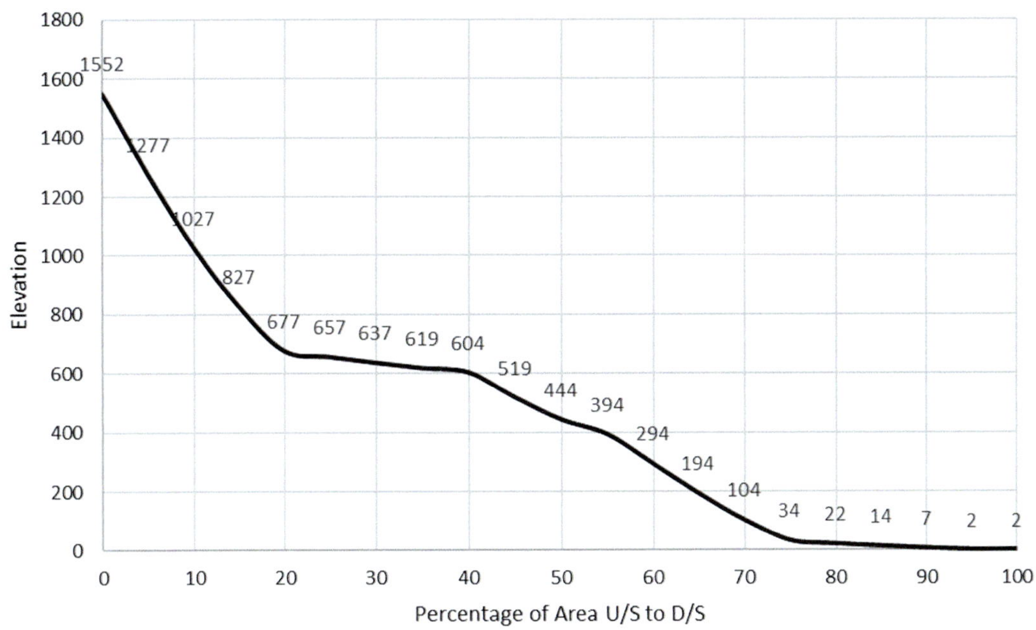

FIGURE 6.11 The hypsometric configuration of the river basin.

TABLE 6.4
Basin Rainfall Calculation Using Hypsometric Method

Hypsometric Lines	Area, A_i, km²	Rainfall, P_i, mm	$A_i \times P_i$, mm . km²	Notes
0-20	202	82.0	16,564.0	$P_{avg} = \frac{64,537.2}{1184} = 54.5$ mm
20-40	529	55.8	29,518.2	
40-60	220	46.0	10,120.0	
60-80	198	37.5	7,425.0	
80-100	35	26.0	910.0	
Total, £	1,184		64,537.2	

The calculation of the average precipitation in the river basin is shown in Table 6.4.

This chapter covers urban hydrology and urban drainage, with a focus on cities or urban areas as the study area. These areas differ from natural river basins, which are larger and more complex in terms of spatial extent, land use, and hydrological characteristics. Therefore, the methods for estimating the average precipitation over an urban area are simpler than those for natural catchments. The average precipitation is crucial for the design and management of the urban drainage system. A city or urban area is usually a small part of the river basin, and it may have only one rainfall station, which can represent the whole urban area or city.

6.4 RAINFALL IDF

The IDF of the rainfall are intercorrelated and an essential element of the information in the hydrologic design of urban drainage systems.

The are numerous formulas of rainfall intensity. However, a simple general formula to determine rainfall intensity was presented by Talbot (1881):

$$i = \frac{a}{t + b} \tag{6.5}$$

where:

- i is rainfall intensity (mm/h)
- t denotes duration of rainfall (hour)
- a, b are constant that may be derived from the regression analysis of the observed past data of i and t. The values of a and b are specific to the watershed.

The following formulas can be used to obtain the values of a and b, by taking note:

$$a = \frac{[\Sigma i \cdot t][\Sigma i^2] - [\Sigma i^2 \cdot t][\Sigma i]}{n[\Sigma i^2] - [\Sigma i]^2} \tag{6.6}$$

and

$$b = \frac{[\Sigma i][\Sigma i \cdot t] - n[\Sigma i^2 \cdot t]}{n[\Sigma i^2] - [\Sigma i]^2} \tag{6.7}$$

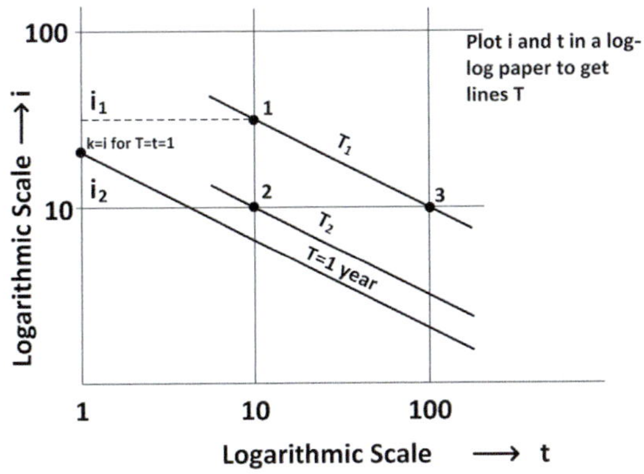

FIGURE 6.12 Derivation of rainfall intensity.

Another method to compute rainfall intensity can be derived from the recorded rainfall intensity. By using the generic formula of Sherman which is

$$i = \frac{kT^a}{t^b} \tag{6.8}$$

where:

i is rainfall intensity, for example, in mm/h
T is average rate of rainfall occurrence, in year
t denotes duration of rainfall, in hour
k, a, b are constants

The procedure of the derivation of the constants and therefore the development of the formula for rainfall intensity follows this step.

Observe Figure 6.12, and Sherman's generic equation to derive rainfall intensity.

From Equation (6.6):

$$i = \frac{kT^a}{t^b}$$

Taking a log of both sides, we have:

$$\log i = \log k + a \log T - b \log t$$

Take any two values of T_1 and T_2 for the same t, for instance for t = 10, we get point 1 for T_1 and point 2 for T_2, and therefore:

[1] $\log i_1 = \log k + a \log T_1 - b \log t$
[2] $\log i_2 = \log k + a \log T_2 - b \log t$

Deduct [1] by [2]:

$$\log i_1 - \log i_2 = a \log T_1 - a \log T_2$$

or

$$\log i_1 - \log i_2 = a\left(\log T_1 - \log T_2\right)$$

$$a = \frac{\Delta \log i}{\Delta \log T} \tag{6.9}$$

Now, to obtain the value of constant b, we do the similar way:

Take any two values of t for the same T, for example, take the T_1 line for t = 10 and t = 100, which are point 1 and point 3 in Figure 6.2.

[3] $\log i_1 = \log k + a \log T - b \log t_1$
[4] $\log i_2 = \log k + a \log T - b \log t_2$

Deduct [3] by [4]:

$$\log i_1 - \log i_2 = -b \log t_1 + b \log t_2$$

or

$$\log i_1 - \log i_2 = -b(\log t_1 - \log t_2)$$

$$b = -\frac{\Delta \log i}{\Delta \log t} \tag{6.10}$$

To find the constant k, construct a line of T = 1 year (T = 1), from existing lines. Line T = 1 year is supposed to be parallel to line T = 10 years, and the same distance with lines T = 100 years and T = 10 years. We can write:

$$\log i = \log k + a \log T - b \log t$$

For T = 1, and t = 1, we have:

$$\log i = \log k + a \log 1 - b \log 1$$

$$k = i \tag{6.11}$$

Equation (6.9) tells us that for T = t = 1 year, the constant k equals i.

By knowing the constants, a, b, and k, the rainfall intensity based on Sherman's generic equation can be established.

Example of the Application

This is hypothetical data of rainfall intensity based on the duration of 5, 15, 30, 60 minutes. The data records were obtained from four independent events of the rainfall, as shown in Table 6.5.

The actual data as shown in Table 6.2 is random. However, the data shown in the table was arranged in descending order where the smallest data (smallest figure) has the lowest rank (greater

TABLE 6.5
Hypothetical Data of 20 Years of Rainfall Intensity for a Given Duration

	T = m/n,	Independent Rainfall Event, with Duration (mm/h)							
Rank, m	Year	5 minutes	1 hour	15 minutes	1 hour	30 minutes	1 hour	60 minutes	1 hour
1	2	3	4	5	6	7	8	9	10
20	1	4	48	10	40	15	30	20	20
19	2	5	60	11	44	17	34	22	22
18	3	6	72	12	48	18	18	24	24
17	4	7	84	13	52	19	38	25	25
16	5	8	96	14	56	20	40	26	26
15	6	8	96	15	60	21	42	27	27
14	7	9	108	16	64	22	44	28	28
13	8	9	108	17	68	23	46	29	29
12	9	10	120	18	72	24	48	30	30
11	10	11	132	18	72	25	50	31	31
10	11	12	144	19	78	26	52	32	32
9	12	12	144	20	80	27	54	34	34
8	13	13	156	20	80	28	56	36	36
7	14	14	168	21	84	29	58	38	38
6	15	15	180	23	92	30	60	40	40
5	16	16	192	24	96	32	64	42	42
4	17	17	204	25	100	34	68	44	44
3	18	18	216	26	104	36	72	46	46
2	19	19	228	27	108	38	76	48	48
1	20	20	240	28	112	40	80	50	50

TABLE 6.6
Selected Data of 1 Year, 10 Years, and 20 Years

	Independent Rainfall Event, with Duration (mm/h)			
T = q m/n, Year	t = 5 minutes	t = 15 minutes	t = 30 minutes	t = 60 minutes
1	48	40	30	20
10	132	72	50	31
20	240	112	80	50

figure), and the average return period T is calculated by using the simplest California method with $T = m/n$, where m is rank order, and n is the number of data. In this example, n = 20. To convert the duration rainfall data expressed in minutes to be in 1-hour duration, the data must be multiplied by a factor of 1 hour/duration in minute . For example, to convert 5-minute duration data to 1-hour data, the multiplication factor is $60/5 = 12$.

To construct the Sherman's intensity of rainfall we can, in principle, select any T. However, for our convenience, we selected only three data sets of T = 1 year, 10 years, and 20 years. This selection is shown in Table 6.6.

The data are then plotted in a semi-logarithmic scale of plotting paper, with the horizontal axis of t (logarithmic scale), and vertical axis of i (linear scale), as shown in Figure 6.13.

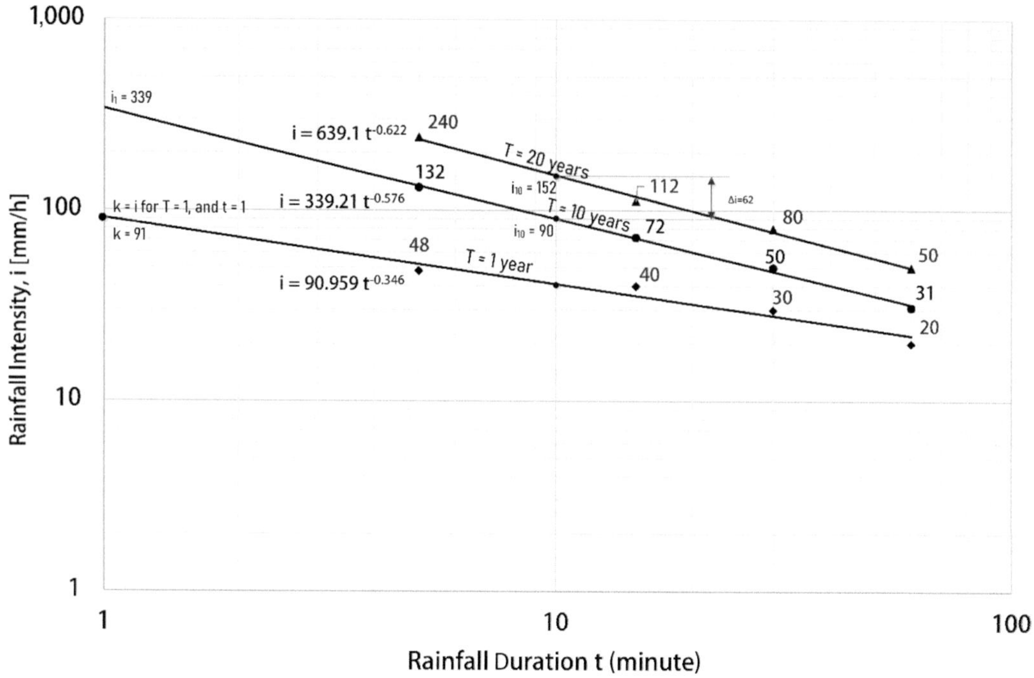

FIGURE 6.13 Rainfall duration and intensity.

From Figure 6.13, we can get the following information:

$$\text{Equations}\,(6.8)\,\mathrm{i} = \frac{\mathrm{kT^a}}{\mathrm{t^b}}\,\text{and}\,(6.11)\;\mathrm{k} = \mathrm{i}$$

- The value of k is obtained from the linear equation T = 1 year (computed by using Excel), which is $\mathrm{i} = 90.959\ \mathrm{t}^{-346}$ for t = 1, then $\mathrm{k} = 90.9 \sim 91.0$.
- The value of a can be obtained from Equation (6.7) by taking the same t = 10 minutes, for T = 20 years, i10 = 152 mm, and for T = 10 years, i10 = 90 mm, therefore $\Delta\mathrm{i} = \log 152 - \log 90 = 0.23$, and $\Delta\mathrm{t} = \log 20 - \log 10 = 0.30$, and $\mathrm{a} = \frac{\Delta \log \mathrm{i}}{\Delta \log \mathrm{T}} = \frac{0.23}{0.30} = 0.77$.
- Similarly, for the value of b, $\mathrm{b} = -\frac{\Delta \log \mathrm{i}}{\Delta \log \mathrm{t}} = -\frac{\log 339 - \log 90}{\log 10 - \log 1} = 0.57$.

The rainfall intensity is therefore $\mathrm{i} = \frac{91\mathrm{T}^{0.77}}{\mathrm{t}^{0.57}}$ In this case, the units are the same as the unit of the seed data, which are, i in mm/h, T in year, and t in mm/h.

Rainfall intensity, duration, and frequency (IDF) equations are usually specific to a certain watershed, so they cannot be applied to different watersheds without adjustments. Alternatively, we can use a regionalization method to transfer the IDF equation from one watershed to another. This method involves finding a regional equation that relates the IDF parameters to some geographical or climatic variables, such as latitude, longitude, elevation, and mean annual precipitation. If two watersheds have similar climatic and geographic variables, the constants or parameters used in the IDF equations are likely to be similar as well.

6.5 RUNOFF AND DRAINAGE SYSTEMS

SRO is the flow of water that occurs when excess rainwater, meltwater, or other sources of water cannot infiltrate the soil. When this excess water accumulates, a drainage system may be required to properly drain the excess water before it generates flooding problems that may hamper urban life. A drainage system is a network of pipes, channels, ditches, or streams that collect and transport SRO to a common outlet, such as a river, lake, or ocean.

SRO and drainage systems have important implications for water quality, flood control, erosion prevention, and urban planning itself, as SRO may carry pollutants, sediments, nutrients, as well as pathogens from various land uses and activities, such as agriculture, industry, mining, and construction. On the other hand, drainage systems mitigate the negative effects of SRO by confining the runoff into a proper drainage facility, avoiding uncontrollable runoff, and therefore reducing its volume of overland flow, filtering out contaminants, and enhancing infiltration and groundwater recharge.

SRO can overwhelm the capacity of a drainage system and cause flooding, especially in urban areas where impervious surfaces such as roads, roofs, and parking lots prevent water from infiltrating the soil. Drainage systems can alter the natural hydrological cycle and affect the ecological functions of wetlands, streams, and rivers. Drainage systems can also create new pathways for invasive species, pests, and diseases to spread.

6.5.1 Surface Runoff

6.5.1.1 The ϕ-Index

SRO is the water that flows over the land surface when the rainfall rate exceeds the infiltration capacity of the soil. The excess rainfall, which is the difference between the rainfall intensity and the infiltration rate, is the driving force for SRO generation. However, not all of the excess rainfall contributes to SRO, as some of it may be stored in surface depressions or intercepted by vegetation. Only effective rainfall contributes to the SRO. Therefore, effective rainfall is the portion of excess rainfall that reaches the stream channel and causes runoff. One way to estimate the effective rainfall is to use the ϕ-index, which is a measure of the average infiltration rate of the soil over a given period. The ϕ-index can be calculated from the rainfall hyetograph and the runoff hydrograph, using the following formula:

$$\phi = \frac{P - SRO}{t} \tag{6.12}$$

where P is the total rainfall depth, SRO is the total runoff volume, and t is the duration of the rainfall event. The ϕ-index represents the constant infiltration rate that would produce the same runoff volume as the actual variable infiltration rate. Therefore, the effective rainfall can be obtained by subtracting the ϕ-index from the excess rainfall at each time step. This method assumes that the infiltration rate is uniform over the catchment area and does not change with time or soil moisture conditions.

Example

Determine the ϕ-index, given catchment area of 50 km^2, rainfall intensity $i_1 = 3$ cm/h, $i_2 = 5$ cm/h, and $i_3 = 4$ cm/h, and hydrograph (shown in Figure 6.14).

The volume of direct runoff in the hydrograph is 3,000,000 m^3 = 3×10^{12} cm^3.

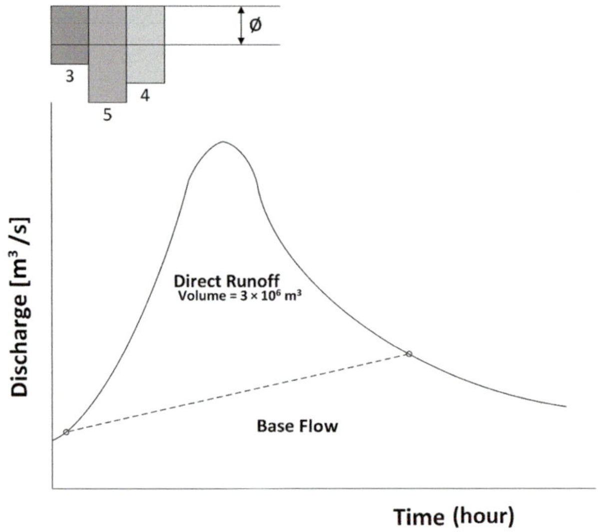

FIGURE 6.14 ϕ-Index.

The area of the catchment = 50 km^2 = 50 $\times 10^{10}$ cm^2.

Assume the ϕ-index less than the smallest rainfall, which is 3 cm. Also, the volume of excess rainfall over the catchment area must be equal to the volume of direct runoff:

$$\left\{(3-\phi)\frac{\text{cm}}{\text{h}}\times 1\text{ h}+(5-\phi)\frac{\text{cm}}{\text{h}}\times 1\text{ h}+(4-\phi)\frac{\text{cm}}{\text{h}}\times 1\text{ h}\right\}$$
$$\times 50\times 10^{10}\text{ cm}^2 = 3\times 10^{12}\text{ cm}^3$$

$$12-3\phi = 6$$

$$\phi - \text{index} = 2.0\text{ cm/h}$$

If the ϕ-index is larger than the smallest rainfall, then the above calculation should be repeated with $3-\phi$ term of the above calculation substituted by $\phi-3$.

Excess rainfalls are i_{e1} = 3.0–2.0 = 1.0 cm/h, i_{e2} = 3.0 cm/h, and i_{e3} = 2.0 cm/h respectively. These excess rainfalls generated direct runoff.

The effective rainfall can be summed up throughout the event to obtain the total effective rainfall, which should be equal to the total volume of runoff. The effective rainfall hyetograph can be plotted along with the excess rainfall hyetograph and the runoff hydrograph to visualize the relationship between them.

6.5.1.2 Factor Affecting SRO

SRO is the unconfined flow of water over the ground surface, in contrast to channel runoff or stream flow. It occurs when excess rainwater, stormwater, meltwater, or other sources can no longer sufficiently rapidly infiltrate the soil. SRO is a major component of the water cycle and a primary agent of soil erosion by water. It is also a potential source of water pollution, as it can carry human-made contaminants or natural forms of pollution. SRO is affected by both meteorological factors and the physical geology and topography of the land. Some factors affecting SRO are:

Type and Intensity of Precipitation: Rain, snow, sleet, or hail can have different effects on runoff generation. Rainfall intensity is the rate at which rain falls, and it influences how much water can infiltrate into the soil. High-intensity rainfall can produce more runoff than low-intensity rainfall. Snowmelt can also contribute to runoff, especially in spring when the temperature rises and the snowpack melts.

Soil Characteristics: The physical properties of the soil, such as texture, structure, porosity, permeability, and moisture content, affect how much water can infiltrate into the soil and how much water is stored in the soil. Sandy soils tend to have high infiltration rates and low water-holding capacity, while clayey soils tend to have low infiltration rates and high water-holding capacity. Soil moisture is the amount of water present in the soil, and it influences the soil can absorb how much additional water. Saturated soils have no more space to store water, so any excess water becomes runoff.

Land Use and Vegetation Cover: Land use reflects the human activities on a particular land area that modify the natural environment, such as because of agricultural activities, or the development of a built environment, that is, residential areas, industrial areas, commercial areas, or other anthropogenic activities that modify the landscape such as logging, mining, etc. The type and extent of land use and vegetation cover affect how much water can infiltrate into the soil and how much water is intercepted by the plants, and therefore determines the magnitude of the SRO. Vegetation cover refers to the plants that grow on the land surface, such as grasses, shrubs, trees, etc. Land use and vegetation cover can alter the soil characteristics, such as organic matter content, compaction, erosion, etc., and affect the runoff generation. Natural vegetation tends to reduce runoff by increasing infiltration and interception, while human activities such as in urban development tend to increase runoff by decreasing infiltration and interception, as schematically shown in Figure 6.14.

The impact of ground cover variations during rainfall events is vividly illustrated in the runoff hydrograph. This graphical representation provides a clear picture of how different types of ground cover influence the amount and speed of water runoff during and after a rainfall event.

In urban areas, where the built environment is predominant, the ground surface is often covered with impervious materials such as concrete and asphalt. These are common in industrial and commercial areas and are referred to as 'hardscape'. Hardscape surfaces prevent the infiltration of water into the ground, leading to an increased volume of SRO.

This phenomenon is quantified using a parameter known as the runoff coefficient. The runoff coefficient is a dimensionless indicator that represents the ratio of the amount of rainfall that becomes runoff to the total amount of rainfall. In areas where hardscape is prevalent, the runoff coefficient is extraordinarily high. This means that a large proportion of the rainfall is immediately converted into runoff.

The high runoff coefficient in urban areas has significant implications for the hydrological phenomenon. One of the most notable consequences is the increase in peak discharge, which refers to the maximum rate of flow at a particular point in a river or stream following a rainfall event. As shown in Figures 6.16 and 6.17, the peak discharge in areas with a high runoff coefficient is significantly higher than in areas with a lower runoff coefficient.

On the other hand, in areas where the natural elements of ground cover predominate, the situation is markedly different. These areas, which include forests, grasslands, and agricultural fields, are characterized by permeable surfaces that allow water to infiltrate into the ground. This natural ground cover significantly reduces the volume of water runoff, resulting in a lower runoff coefficient compared to areas with predominantly impervious surfaces.

The lower runoff coefficient in areas with natural ground cover has a mitigating effect on the hydrological effects of rainfall events. One of the key benefits is the reduction in peak discharge. As shown in Figure 6.15, the peak discharge in areas with a high proportion of natural ground cover could be substantially lower than in areas with a high proportion of impervious surfaces.

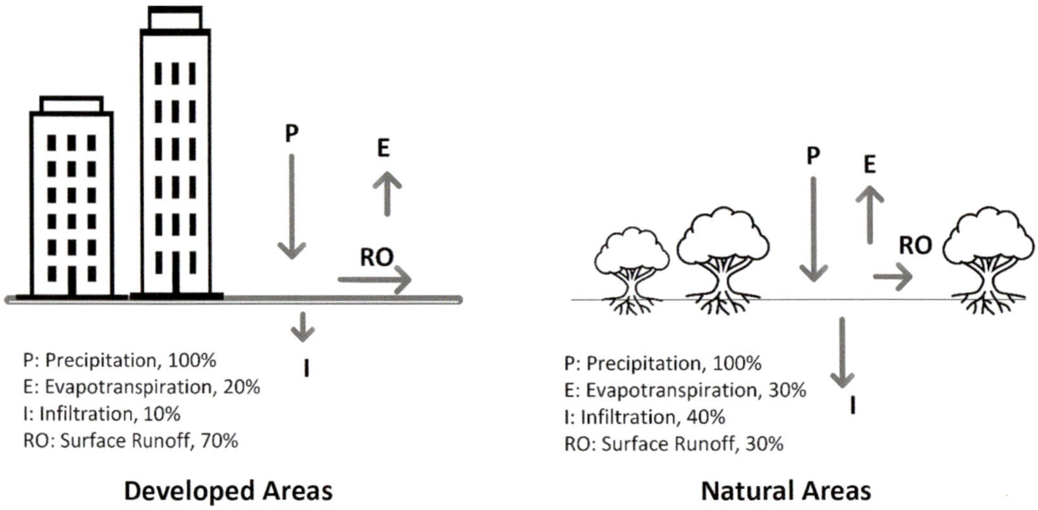

FIGURE 6.15 Runoff at developed and natural areas.

The explanation confirms that the type of ground cover plays a crucial role in determining the hydrological response to rainfall events. Urban areas, with their high proportion of impervious surfaces, tend to have a high runoff coefficient and consequently a high peak discharge. In contrast, areas with a high proportion of natural ground cover have a lower runoff coefficient, leading to a lower peak discharge. These differences underscore the importance of sustainable urban planning and land management practices in mitigating the hydrological effects of urbanization.

Figure 6.16 depicts an SRO hydrograph, highlighting the rapid increase and peak flow of runoff in urban areas due to impervious surfaces, followed by a quick decrease as water drains.

Figure 6.17 illustrates SRO in natural areas. Due to lower runoff coefficients compared to urban areas, less water flows rapidly over the surface, leading to lower peak discharge levels, highlighting the natural landscape's capacity to absorb rainwater effectively.

Topography: Topography refers to the elevation, relief, aspect, and drainage pattern of the land surface. The shape and slope of the land surface affect how fast water flows over the land and how much water can infiltrate into the soil. According to Manning, the overland flow velocity depends on the slope.

$$v = \frac{1}{n} R^{2/3} S^{1/2} \tag{6.13}$$

where:

n denotes Manning's roughness coefficient

R is radius hydraulic of the channel, for overland flow, equals the width of the flow, as the depth of flow is very much smaller than the flow width

S stands for longitudinal slope of the channel

The larger the slope of the land, the larger the velocity of the SRO, and the larger the velocity, the smaller the possibility of the runoff infiltrating the soil. This linear correlation confirms that topography determines the runoff.

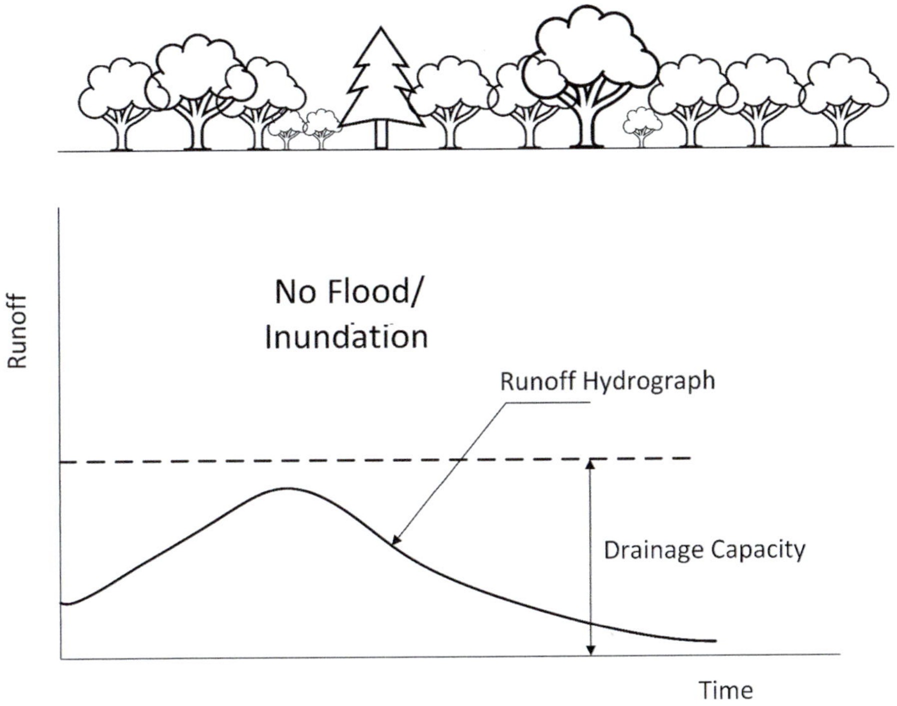

FIGURE 6.16 SRO hydrograph on urban areas.

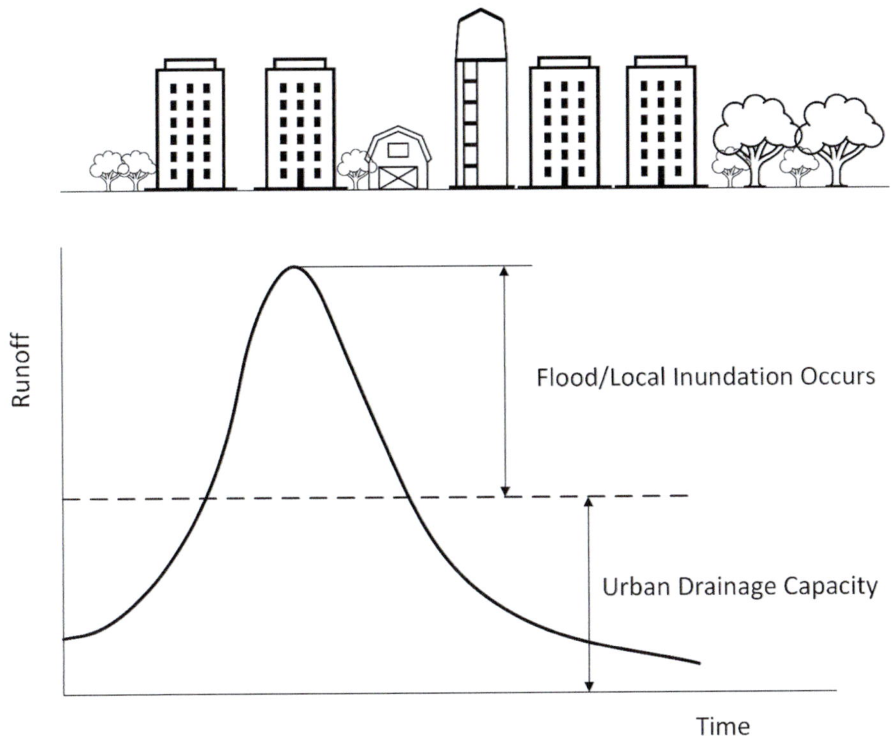

FIGURE 6.17 SRO on natural areas.

6.5.2 Stormwater Drainage Networks

Stormwater drainage systems are designed to collect, convey, and treat stormwater runoff from urban areas. They consist of various components, such as sewers, culverts, and retention ponds, which have different functions and design criteria. The stormwater drainage system is commonly separated from the wastewater system for substantial differences in the quality of the water they handle. The stormwater is better in quality, and therefore it can be directly discharged into the public water body such as rivers, lakes, or oceans. On the other hand, the wastewater effluents that come from kitchens, toilets, and bathrooms need to be treated by a wastewater treatment plant before being discharged to the public water body. The stormwater drainage network is aimed at managing excess rainfall and preventing inundation. This network is designed to follow the areas that need to be protected from flooding, ensuring that water is efficiently drained away from these regions.

The design and construction of a stormwater drainage network involve significant costs. To optimize the network and minimize these costs, the network is typically subdivided into three levels: the tertiary, secondary, and primary systems.

The tertiary system is the first level of the network. It consists of tertiary canals, which are small channels designed to collect runoff directly from the fields. These canals are strategically placed to capture and direct the flow of water toward the secondary system. The placement and design of these canals are crucial in ensuring efficient water collection and minimizing the risk of flooding in the fields.

The secondary system is the next level of the network. It consists of secondary canals, which are larger than the tertiary canals. These canals receive the flow of water from the tertiary network. The secondary canals are designed to handle larger volumes of water and to transport this water to the primary system. The design of these canals is crucial in ensuring that the water collected by the tertiary canals is efficiently transported away from the fields.

The primary system is the final level of the network. It consists of the primary canal, which is the largest in capacity. The primary canal can be either an artificial canal or a natural river. This canal receives the flows from the secondary system and discharges the water load to a public water body, such as a river, lake, or ocean, as illustrated in Figure 6.18.

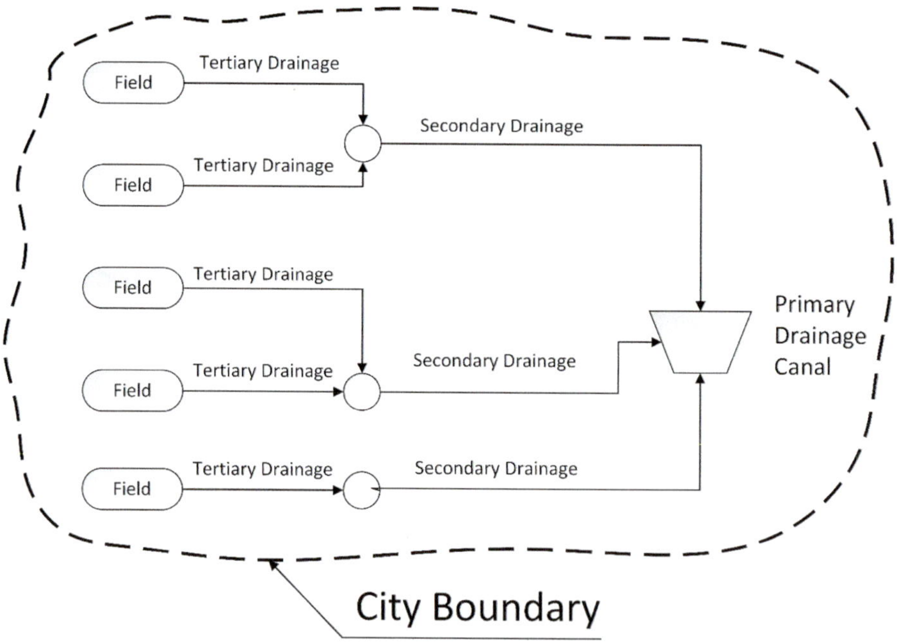

FIGURE 6.18 Hierarchy of a drainage system.

By subdividing the network in this manner, each level of the network plays a specific role in managing stormwater. The tertiary network, being the highest order, directly receives the flows from the fields. The secondary system then receives the flows of water from the tertiary network, and subsequently, the primary system receives the flows from the secondary system.

This hierarchical structure of the stormwater drainage network ensures that water is efficiently collected, transported, and discharged, thereby minimizing the risk of flooding. It also allows for the optimization of the network design, ensuring that each level of the network is appropriately sized and positioned to handle the water flows effectively.

There are two main types of drainage channels: open channel flow and pressure flow. Open channel flow means that the water flows freely without any external pressure. The water surface in an open channel flow is exposed to the atmosphere, and the pressure at any point is equal to the atmospheric pressure. Pressure flow means that the water flows under a higher or lower pressure than the atmospheric pressure. The water surface in a pressure flow is not exposed to the atmosphere, and the pressure at any point differs from the atmospheric pressure. Pressure flows can occur in pipes, hoses, or closed conduits. Open channel flows can have different shapes, depending on the design and the terrain. Some of the common shapes are circular, V-shaped, or trapezoidal. Circular channels have a circular cross-section, and they are usually used for pipes or tunnels. V-shaped channels have a V-shaped cross-section, and they are often used for ditches or canals. Trapezoidal channels have a trapezoidal cross-section, with a flat bottom and sloping sides. They are widely used for irrigation or drainage channels, as they can accommodate different flow rates and depths. These types of channels are very common to use as stormwater drainage channels. The hydrologic and hydraulic design examples will use these channel shapes in either pipe in the form of pressurized flows or open channel flows.

The design and analysis of stormwater drainage systems involve the following steps:

1. *Hydrologic Analysis:* This step estimates the amount and distribution of rainfall and runoff in the drainage area, using, for example, the rational method with derived rainfall intensity.
2. *Hydraulic Analysis:* This step calculates the flow characteristics and maximum capacity, that is, flow rates of the drainage system by applying methods such as the Manning equation and the continuity equation if the drainage exhibits open channel characteristics. If the channels show pressurized pipe flow, the energy approach can be used.
3. *Water Quality Analysis:* This step evaluates the pollutant loadings and concentrations in the runoff and the drainage system, using methods such as the event mean concentration (EMC), the buildup-wash-off model, or the SWMM water quality module. The water quality analysis also assesses the performance of the treatment facilities, such as retention ponds, that are designed to remove pollutants from the runoff. This section is particularly used for wastewater management, which will be deliberated in the Chapter on Wastewater System.
4. *Design Optimization:* This step selects the optimal size, shape, and location of the drainage system components, using methods such as cost-benefit analysis, multi-objective optimization, or the SWMM-CAT tool. The design optimization aims to minimize the construction and maintenance costs, maximize the hydraulic and water quality performance, and satisfy the regulatory and environmental requirements.

Here is a numerical example of the design and analysis of a stormwater drainage system for a small urban catchment.

Given: The catchment area is 10 ha, with 70% imperviousness and 0.20 runoff coefficient. The design rainfall intensity is 100 mm/h, with a duration of 1 hour and a return period of 10 years. The catchment is drained by a circular sewer with a diameter of 0.6 m and a slope of 0.01. The sewer discharges into a rectangular retention pond with a length of 50 m, a width of 20 m, and a depth of 2 m. The pond has an outlet weir with a crest length of 5 m and a coefficient of discharge of 0.6. The

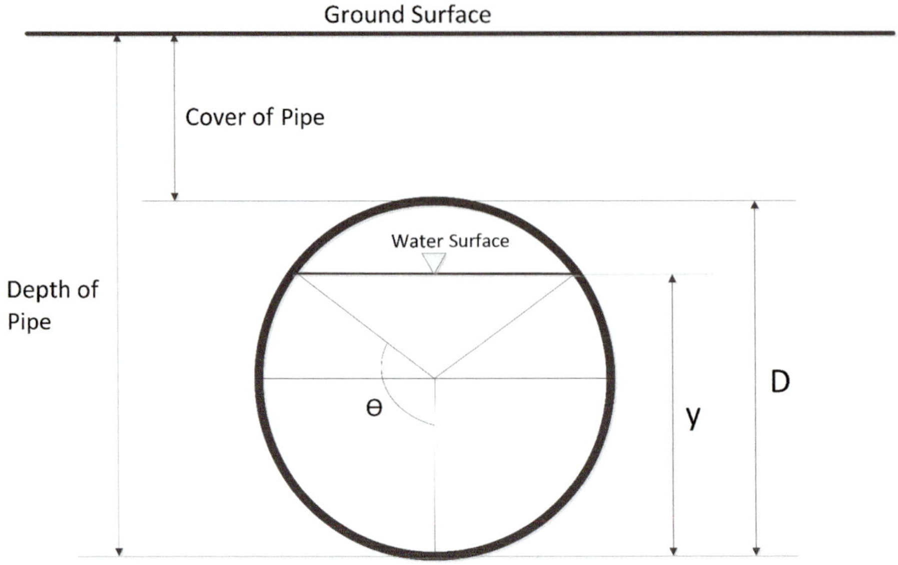

Flow Area $A = \frac{1}{8}(2\theta - \sin 2\theta) D^2$

Hydraulic Radius $R = \frac{1}{4}\left(1 - \frac{\sin 2\theta}{2\theta}\right) D$

FIGURE 6.19 Circular shape drainage channel.

EMC, which is the average concentration of pollutants of total suspended solids (TSS) in the runoff is 100 mg/L, and the removal efficiency of the pond is 50%.

Find:

1. The peak runoff rate and volume from the catchment.
2. The flow depth and velocity in the sewer.
3. The water level and detention time in the pond.
4. The TSS load and concentration at the pond outlet.

Solution (see Figure 6.19):

1. The peak runoff rate and volume from the catchment.
   The peak runoff rate from the catchment is calculated by the rational formula.

$$Q_p = C \times i \times A = 0.20 \times 100\left[\frac{\text{mm}}{\text{hour}} \times \frac{1\ \text{hour}}{3{,}600\ \text{s}} \times \frac{1\ \text{m}}{1{,}000\ \text{mm}}\right] \times 10\left[\text{ha} \times \left(\frac{10{,}000\ \text{m}^2}{1\ \text{ha}}\right)\right]$$

$$Q_p = 0.6\ \text{m}^3/\text{s}$$

The volume of runoff during the time of concentration when rainfall in all parts of the catchment contributes to the peak runoff. Assume that t_c is equal to the duration of the rainfall, which is 1 hour or 3,600 s.

$$V = Q_p \times t_c = 0.6\left(m^3/s\right) \times 3{,}600\ (s) = 2{,}160\ m^3$$

2. The flow depth and velocity in the sewer are calculated by the Manning equation, as the flow is an open channel flow.

$$Q = \frac{A}{n} R^{2/3} S^{1/2}$$

where A is the cross-sectional area, R is the hydraulic radius, S is the longitudinal slope of the channel, and n is the Manning roughness coefficient.
Assuming n = 0.013, and the flow depth y = 0.45 m.

Flow area, A:

$$A = \frac{1}{8}(2\theta - \sin 2\theta) D^2 \ \text{ with D = 0.6 m, and y = 0.45 m}$$

Then $\theta = 90^\circ + \text{asin}\left(\frac{2y}{D} - 1\right)$

$$\theta = 90^\circ + \text{asin}\left(\frac{2y}{D} - 1\right) = 90^\circ + \text{asin}\left(\frac{2(0.45)}{0.60} - 1\right)$$
$$= 125.7^\circ = 2.19 \text{ rads}$$

Thus:

$$A = \frac{1}{8}\left((2 \times 2.19) - \sin(2 \times 125.7)\right)(0.6)^2 = 0.19\ m^2$$

and

$$R = \frac{1}{4}\left(1 - \frac{\sin 2\theta}{2\theta}\right) D = \frac{1}{4}\left(1 - \frac{\sin 250.14}{4.38}\right) = 0.30$$

$$Q = \frac{0.19}{0.013} \times (0.30)^{2/3} (0.01)^{1/2} = 0.65\ m^3/s$$

$$v = Q/A = 0.65/0.19 = 3.42\ m/s$$

3. The water level and detention time in the pond.
Use continuity equation:

$$Q_{in} = Q_{out} + \frac{dV}{dt}$$

where Q_{in} is the inflow rate, Q_{out} is the outflow rate, and dV/dt is the change of volume in the pond. Assuming a steady state, dV/dt = 0, and Q_{out} is given by the weir equation as:

$$Q_{out} = C_d L \sqrt{2g} H^{3/2}$$

or

$$H = \left(\frac{Q}{C_d L \sqrt{2g}} \right)^{2/3} = \left(\frac{0.65}{(0.6)(5)\sqrt{2(9.81)}} \right)^{2/3} = \left(\frac{0.65}{13.29} \right)^{2/3} = 0.13 \text{ m}$$

where C_d is the coefficient of discharge, L is the crest length, g is the gravitational acceleration, and H is the head over the weir. Solving for H, we get:

$$H = 0.13\text{m}$$

The water level in the pond is equal to the depth plus the head over the weir:

$$Z = d + H = 2 + 0.13 = 2.13\text{m}$$

Detention time, T:

$$T = \frac{V}{Q} = \frac{L \times W \times d}{Q} = \frac{(50)(20)(2)}{0.65} = 0.85 \text{ hours}.$$

4. The TSS load and concentration at the pond outlet.
   The TSS load and concentration at the pond outlet are calculated by multiplying the inflow load and concentration by the removal efficiency:

$$L_{out} = L_{in}(1 - E) = Q_{in} C_{in}(1 - E)$$

$$L_{out} = \left(0.65 \left[\frac{m^3}{s} \right] \left[\frac{1{,}000 \text{ L}}{m^3} \right] \right) \left(100 \left[\frac{mg}{L} \right] \left[\frac{kg}{1{,}000 \text{ mg}} \right] \right) (1 - 0.5)$$
$$= 32.5 \text{ kg/s}$$

Concentration at outlet:

$$C_{out} = \frac{L_{out}}{Q_{out}} = \frac{32.5 \text{ kg/s}}{0.65 \text{ m}^3/\text{s}} = 50 \text{ kg/m}^3 = 50 \text{ mg/L}$$

As it has been briefly discussed, from a hydraulic perspective, the flow of water in drainage channels can be classified into three types, depending on the pressure and cross-sectional shape of the channel. These types are: (1) *Pipe flow*, which occurs when the channel is completely filled with water and the pressure is greater than the atmospheric pressure. Pipe flow is also known as pressurized flow or closed conduit flow. (2) *Open channel flow*, which occurs when the channel is

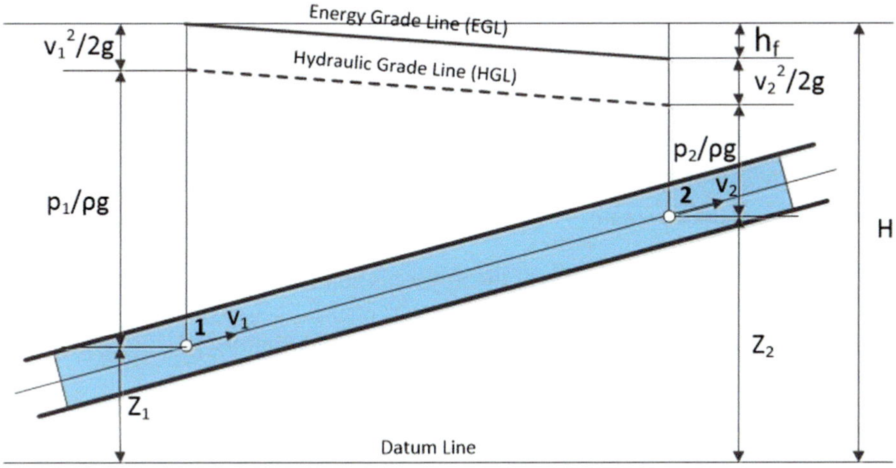

FIGURE 6.20 Bernoulli equation in pipe flow.

partially filled with water and the pressure is equal to the atmospheric pressure. Open channel flow is also known as free surface flow or open conduit flow. (3) *Transitional flow* occurs when the channel alternates between pipe flow and open channel flow, depending on the flow conditions and channel geometry. Transitional flow is also known as mixed flow or partially full flow. The approach to the hydraulic calculations of these types of flow is different.

In pipe flow, the water is under pressure that is greater than the atmospheric pressure. This means that the flow rate is determined by the pipe diameter, the pipe roughness, and the pressure difference between the ends of the pipe. The longitudinal slope of the channel does not affect the flow rate, unlike in open channel flow, where the water flows under the influence of gravity. Therefore, to calculate the flow rate in pipe flow, we cannot use the formulas for open channel flow, such as the Manning equation or the Chezy equation. Instead, we need to use the energy flow principle, which states that the total energy of the water (including the pressure energy, the kinetic energy, and the potential energy) is conserved along the pipe, except for the losses because of friction and other factors. One of the most common formulas that apply the energy flow principle is the Hazen-Williams equation, which estimates the velocity and the discharge of the water in pipe flow, as shown in Figure 6.20.

Consider particle flow of 1 and 2 in a pipe. Each particle has three components of energy, which are potential energy (Z), pressure head $(p/\rho g)$, and kinetic energy $v^2/2g$. The total head at points 1 and 2 is given by:

$$H = z_1 + \frac{p_1}{\rho g} + \frac{v_1^2}{2g} = z_2 + \frac{p_2}{\rho g} + \frac{v_2^2}{2g} + h_f \tag{6.14}$$

For horizontal same-diameter pipe, $z_1 = z_2$ and $v_1 = v_2$:

$$h_f = \frac{p_1}{\rho g} - \frac{p_2}{\rho g} = \frac{p_1 - p_2}{\rho g} \tag{6.15}$$

The friction loss in the pipe is the reduction of pressure or head because of the resistance of fluid flow in the pipe. It depends on factors such as pipe roughness, diameter, length, flow velocity, and fluid properties. The Darcy–Weisbach equation is a widely used formula to calculate the friction loss in a pipe.

Darcy–Weisbach introduced a formula for friction losses in pipe:

$$h_f = f\frac{L}{D}\frac{v^2}{2g} \tag{6.16}$$

where h_f is the friction loss (in m), f is the Darcy friction factor (dimensionless), L is the pipe length (in m), D is the pipe diameter (in m), v is the flow velocity (in m/s), and g is the gravitational acceleration (in m/s^2).

The Darcy friction factor is a function of the Reynolds number (Re) and the relative roughness of the pipe. The relative roughness is the ratio of the pipe roughness height (ϵ) to the pipe diameter (D). It reflects the effect of the pipe surface on the flow. The Reynolds number is a dimensionless number that characterizes the flow regime (laminar or turbulent). It is given by:

$$R_e = \frac{\rho v D}{\mu} \tag{6.17}$$

where ρ is the fluid density in kilograms per cubic meter (kg/m^3), v is the fluid velocity in meters per second (m/s), D is the characteristic length in meters (m), and μ is the dynamic viscosity of the fluid in pascal-seconds (Pa·s).

Another method to estimate the friction in a pipe is by using the Colebrook–White equation. This is an implicit equation that relates the Darcy friction factor to the Reynolds number and the relative roughness. It is given by:

$$\frac{1}{\sqrt{f}} = -2\log_{10}\left(\frac{\varepsilon/D}{3.7} + \frac{2.51}{R_e\sqrt{f}}\right) \tag{6.18}$$

This equation cannot be solved directly for f so an iterative method or an explicit approximation is needed. Some explicit solutions for the Colebrook–White equation are available. One of the simplest approaches is shown below by Serghide (1984):

If

$$A = -2\log_{10}\left(\frac{\varepsilon/D}{3.7} + \frac{12}{R_e}\right) \tag{6.19}$$

$$B = -2\log_{10}\left(\frac{\varepsilon/D}{3.7} + \frac{2.51\times A}{R_e}\right) \tag{6.20}$$

$$C = -2\log_{10}\left(\frac{\varepsilon/D}{3.7} + \frac{2.51\times B}{R_e}\right) \tag{6.21}$$

then

$$f = A - \left[\frac{(B-A)^2}{(C-2B+A)}\right]^{-2} \tag{6.22}$$

Serghide's solution can be used for various ranges of Moody's diagram, with acceptable accuracy.

6.5.3 Design and Analysis of Stormwater Drainage Systems

The design and analysis of stormwater drainage systems are the most essential aspects of urban drainage, in which the stormwater drainage systems are designed to manage the flow of rainwater and prevent flooding in urban and suburban areas. These systems involve a necessary comprehensive understanding of hydrology, hydraulics, and environmental engineering principles.

Stormwater drainage systems are designed to collect, convey, and manage rainwater runoff from impervious surfaces such as roads, parking lots, and rooftops. The goal is to safely transport the stormwater to natural water bodies or designed retention areas without causing damage to infrastructure or the environment. The design process considers factors such as rainfall intensity, land use, soil type, and topography to determine the appropriate size and layout of the drainage system.

The analysis of stormwater drainage systems involves assessing the performance of the system under various rainfall conditions. This includes evaluating the capacity of pipes, channels, and storage facilities to handle different flow rates and volumes. Hydraulic modeling and computer simulations are often used to analyze the behavior of stormwater within the drainage network and predict potential flooding or overflow scenarios.

One of the key considerations in the design and analysis of stormwater drainage systems is the concept of sustainable urban drainage. This approach emphasizes use of green infrastructure, such as permeable pavements, rain gardens, and vegetated swales, to mimic natural hydrological processes and reduce the impact of urban development on the water cycle. Sustainable drainage systems aim to enhance water quality, promote groundwater recharge, and provide habitat for wildlife while effectively managing stormwater.

Besides technical considerations, the design and analysis of stormwater drainage systems also involve compliance with regulatory requirements and environmental standards. Engineers and planners must adhere to local, state, and federal regulations related to stormwater management, erosion control, and water quality protection. This often requires the integration of best management practices and innovative design solutions to achieve sustainable and resilient stormwater infrastructure.

The integration of climate change considerations is becoming increasingly important in the design and analysis of stormwater drainage systems. As extreme weather events become more frequent and intense, engineers must account for changing precipitation patterns and rising sea levels in their designs. This may involve incorporating adaptive measures, such as increased storage capacity and floodplain mapping, to mitigate the impact of climate-related challenges on stormwater management.

This section discusses the design methods of essential stormwater drainage infrastructure. Urban drainage infrastructure includes various types of systems and components designed to manage stormwater and wastewater in urban areas. Some common types of urban drainage infrastructure include:

- *Stormwater Drains:* These are underground pipes designed to collect and convey rainwater from streets, sidewalks, and other paved surfaces to prevent flooding and waterlogging.
- *Sewer Systems:* These systems collect and convey wastewater from residential, commercial, and industrial areas to treatment facilities. There are separate sewer systems for stormwater and wastewater, as well as combined sewer systems that handle both.
- *Culverts:* These are structures that allow water to flow under roads, railways, and embankments. They help to maintain the natural flow of water and prevent flooding.
- *Detention Basins:* These are constructed ponds or reservoirs designed to temporarily store stormwater runoff, allowing sediment to settle and pollutants to be removed before the water is released into the environment.

- *Green Infrastructure:* This includes natural or nature-based systems such as green roofs, permeable pavements, rain gardens, and vegetated swales that help manage stormwater and reduce the impact of urban development on the natural water cycle.
- *Pumping Stations:* These are used to lift wastewater or stormwater from a lower to a higher elevation when gravity flow is not possible.
- *Flood Barriers:* These structures are designed to protect urban areas from flooding by diverting or containing floodwater.

Despite the above types of infrastructure, the discussion on design and analysis is focused only on stormwater drainage infrastructure, as wastewater infrastructure is included in the chapter on wastewater management. Here are some examples of hydrologic and hydraulic design, with hypothetical data.

6.5.3.1 Stormwater Drains

We are designing a drainage system for the part of a city, given the catchment area, which is the land we are going to drain, is 50 ha. The approximate time of concentration for this area is 30 minutes. Based on the available IDF curve in the city, it is found that for the time concentrations of 30 minutes, the rainfall intensity for a 25-year recurrence interval is 80 mm/h. The ground surface of the area is mostly paved and urbanized with C = 0.8. Determine the diameter of the pipe drain.

6.5.3.2 Solution

We assume that the flow is an open channel flow, where the flow pressure is 0, even though the channel is made from pipes. One possible way to calculate the size of pipe drains for a stormwater drainage system is to use the Rational Method. This method assumes that the peak runoff rate from a catchment area is proportional to the rainfall intensity and the catchment characteristics. The rational formula (6.1) for the peak runoff rate is:

$$Q_p = 0.00278\ C \times i \times A$$

where: Q_p is peak discharge (m^3/s), C is runoff coefficient, which depends on the ground cover, i is rainfall intensity (mm/h), A is catchment area (km^2).

For the given information, the runoff coefficient C is 0.8, i = 80 mm/h, and A = 50 ha = 0.5 km^2, and the peak discharge is:

$$Q_p = 0.00278 \times (0.8) \times (80) \times (0.5) = 0.09\ m^3/s$$

Next, we need to determine the size of the pipe drains that can convey this flow rate. We can use the **Manning Equation (6.11)** to relate the flow rate, pipe diameter, pipe slope, and pipe roughness. The equation is:

$$Q = \frac{A}{n} R^{2/3} S^{1/2}$$

where Q is the flow rate in m^3/s, n is the Manning roughness coefficient, which is a dimensionless factor that depends on the pipe material and condition, A is the cross-sectional area of the pipe in m^2, R is the hydraulic radius of the pipe in m, which is equal to the cross-sectional area divided by the wetted perimeter, S is the slope of the pipe in m/m

Assume that the pipe drains (made from concrete, n = 0.013) are circular with diameter D and have a uniform slope of 0.01 m/m. Plugging the values we have, we can simplify the equation as:

$$Q = \frac{A}{n} R^{2/3} S^{1/2} = \frac{\pi D^2}{4n} \left(\frac{D}{4}\right)^{2/3} (0.01)^{1/2} = 1.93\ D^{8/3}$$

$$D = \left(\frac{0.09}{1.93}\right)^{3/8} = 0.32\ \text{m}$$

Therefore, the pipe drains should have a diameter of at least 0.32 m, say 0.40 m to convey the peak runoff rate from the catchment area.

6.5.3.3 Culverts

A culvert is a structure that allows water to flow under a road, railway, trail, or similar obstruction from one side to the other. It is designed to collect and lead water across the road so as not to cause damage to the road bank or the stream bed by scouring. It also allows sufficient waterways to prevent the heading up of water above the road surface. Culverts are preferred to minor bridges as culverts are always beneath the carriageway, whereas bridge surfaces form a part of the carriageway.

From a hydraulic viewpoint, culverts may have different shapes and hydraulic characteristics, that is, pressure-flow (pipe flow) or non-pressure flow (open channel flow). Culverts mainly consist of three parts: culvert barrel, inlet structure, and outlet structure. The alignment of the culvert should conform closely to that of the natural waterway regarding plan and profile. Therefore, the culvert may be skewed and right-angled concerning highway alignment. However, the culvert slope should confirm the natural bed slope of the waterways. There are different types of culverts such as pipe culverts, box culverts, slab culverts, arch culverts, etc.

Figure 6.21 shows various hydraulic situations where the hydraulic calculation of the culverts is different.

There are four hydraulic conditions of culverts, in which the ways we design the culverts are also different. Given the depth of flow in the upstream part of the culvert is less than 1.2 diameters of the barrel, or larger than 1.2 diameters of the barrel of the culvert, the hydraulic conditions of culverts are determined by the downstream water level relative to the height of the barrel of the culvert.

In the situation as shown by Figure 6.21(a) when the hydrostatic head at the entrance is less than 1.2 diameters of the culvert, the open channel will dictate the condition, and the culvert is considered as open channel flow. This is because the current rate of flow is much lower than the hydraulic capacity of the culvert, as a result, atmospheric pressure affects the entire flow system in the culvert and therefore the flow is an open channel flow. Here, the hydraulic design is based on the open channel flow design, that is, using Manning's formula of flow.

Figure 6.21(b) shows a different hydraulic condition, where the upstream flow depth is larger than 1.2 diameters of the culvert, the flow is considered as an orifice, in which the discharge through the culvert is calculated by:

$$Q = C_d A \sqrt{2gh} \tag{6.23}$$

where A is the cross-sectional area of the culvert, h is the hydrostatic head above the center of the pipe opening (orifice), and C_d is the discharge coefficient, which is 0.6 (for square entrance), and 0.95 (for rounded).

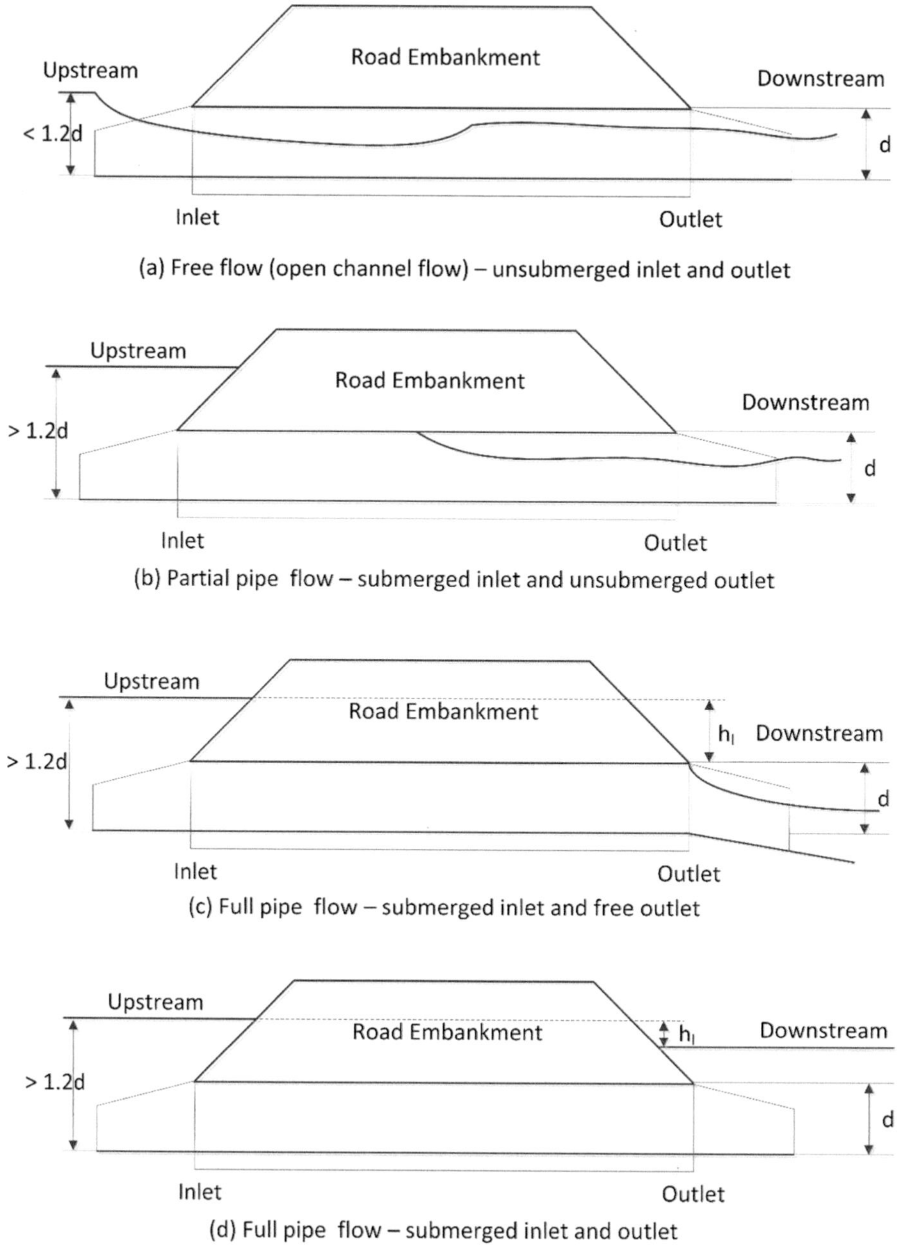

FIGURE 6.21 Various hydraulic conditions of culverts.

With the situations, as shown in Figure 6.21(c) and 6.21(d), the hydraulic calculations are based on pipe flow. The culvert is treated as a pressure pipe, and head loss is the sum of entrance loss (h_e), friction loss (h_f), and exit loss (h_d):

$$h_L = h_e + h_f + h_d \tag{6.24}$$

or

$$h_L = k_e\left(\frac{V^2}{2g}\right) + \frac{n^2V^2L}{R_h^{4/3}} + \frac{V^2}{2g} \tag{6.25}$$

$$h_L = \left[k_e + \left(\frac{n^2L}{R_h^{\frac{4}{3}}}\right)(2g) + 1\right]\frac{8Q^2}{\pi^2 g D^4} \tag{6.26}$$

Example: Hydraulic Analysis of a Culvert

The design flow for a proposed corrugated steel culvert is 5.25 m³/s. The maximum available headwater (H) is 3.2 m above the culvert invert (inside bottom) (Figure 6.22). The culvert is 40 m long and has a square-edged entrance and slope of 0.003. The outlet is not submerged (free discharge). Determine the required diameter.

Solution:

The outlet of the culvert is not submerged; therefore, it is not fully pressurized pipe flow, and a fully open channel is also not possible as the entrance is likely to be submerged. Therefore, partially full pipe flow and full pipe flow with free outlet are possible.

Assumption 1: assuming full pipe flow, the energy balance for this culvert is:

$$H + S_b L = d + h_L$$

$$h_L = H + S_b L - d$$

$$h_L = 3.2 + 0.003(40) - d$$

$$h_L = 3.32 - d$$

The tailwater is assumed to be equal to d:

$$h_L = \left(K_e + \frac{n^2L}{R_h^{4/3}}(2g) + 1\right)\frac{8Q^2}{\pi^2 g d^4}$$

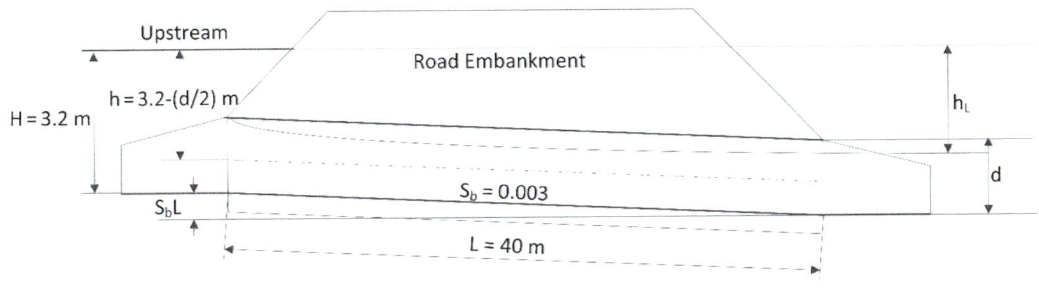

FIGURE 6.22 Partial pipe flow culvert.

$$h_L = \left(0.5 + \frac{(0.024)^2(40)}{(d/4)^{4/3}} 2(9.81) + 1\right) \frac{8(5.25)^2}{\pi^2(9.81)d^4}$$

$$d + \left(1.5 + \frac{2.87}{d^{4/3}}\right)\left(\frac{2.28}{d^4}\right) = 3.32$$

It yields d = 1.41 m.

Assumption 2: Assuming partially full pipe flow, the discharge is controlled by the entrance condition only in this case, the head is measured above the centerline of the pipe. Thus:

$$h + \frac{d}{2} = 3.2$$

and

$$Q = C_d A\sqrt{2gh} = C_d(\pi d^2/4)\sqrt{2gh}$$

$$5.25 = 0.60(\pi d^2/4)\sqrt{2(9.81)\left(3.2 - \frac{d}{2}\right)}$$

It yields d = 1.25 m.

We got two different diameters of culvert. The diameter that safely conveys the discharge of 5.25 m^3/s is d = 1.41 m. If we select d = 1.25 m, at a condition of fully pressurized pipe flow, the culvert will be incapable of passing the discharge of 5.25 m^3/s.

6.5.3.4 Detention Basins

A detention basin is a type of stormwater management facility that temporarily stores runoff and releases it at a controlled rate. The volume of a detention basin depends on the design storm, the watershed characteristics, and the outlet structure. Here are some examples of how to calculate the volume of a detention basin using different methods:

6.5.3.5 Hydrograph Volume

We assume that the objective of constructing a detention basin is only to divert temporarily the SRO that generates urban floods. It is therefore the volume of the detention basin should at least be equal to the volume of SRO. The volume of SRO can be approximated from the existing hydrograph. This is the simplest method to approximate the volume of the detention basin, as the function of the detention basin is to keep the city safe from floods generated by SRO (Figure 6.23).

The illustration consists of a hydrograph on the left, showing discharge over time, and a schematic of a detention basin on the right. The hydrograph depicts the peak discharge from SRO and indicates that the detention basin's capacity is designed to handle this peak runoff. The detention basin schematic shows the process of water inflow (Q_{det}) into the basin and the controlled outflow ($Q_{flood} - Q_{det}$) into the river, effectively managing and temporarily storing excess runoff to prevent flooding downstream. This setup highlights the importance of detention basins in urban water management by reducing peak discharge into rivers.

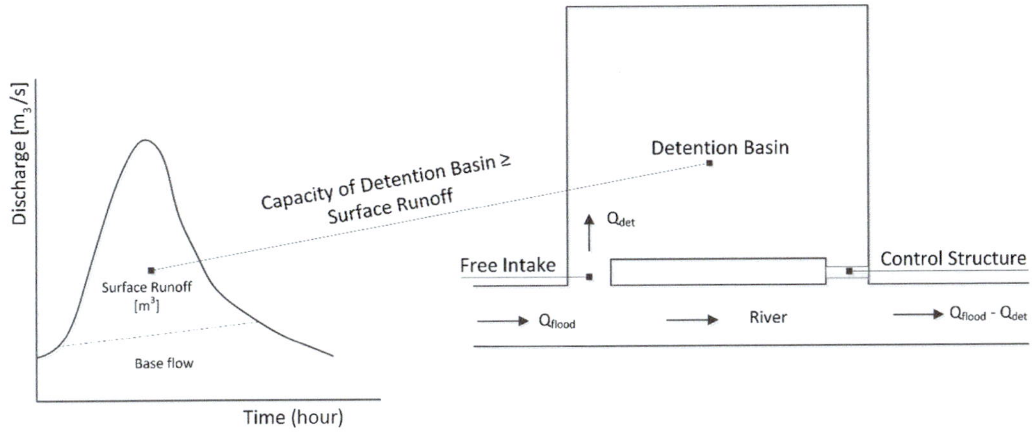

FIGURE 6.23 Detention basin volume equals to SRO.

The volume of the detention basin is:

$$V_s \geq \text{Volume of SRO} \tag{6.27}$$

The limited urban land may hamper the city's construction of the detention basin. It can be solved by acquiring land in the other administrative regions and constructing the basin in the upstream part of and not so far from the city to increase the effectiveness of the detention function.

6.5.3.6 Generalized Model

This method assumes a linear relationship between the runoff depth and the peak runoff rate for both pre-development and post-development conditions. The formula gives the volume of the detention basin in m^3:

$$V_s = 10A\left(h_a - h_b\right) \tag{6.28}$$

where V is the volume of the detention basin in m^3, A is the watershed area in hectare, h_a and h_b are the runoff depths in mm for post-development and pre-development conditions, respectively, the volume of the detention basin is to keep the excess of urbanization.

6.5.3.7 Rational Hydrograph Model

This method uses the rational method to estimate the peak runoff rate and the triangular hydrograph shape to estimate the runoff volume. The formula gives the volume of the detention basin:

$$V_s = Ah_a\left[\frac{\gamma + \alpha + \alpha\gamma\left(g + \alpha - 4\right)}{\gamma - \alpha}\right] \tag{6.29}$$

where:

V_s is the detention basin volume (mm),
A is the area of the watershed (hectare),
h_a is the runoff depth after development (mm),

γ is the time of concentration ratio for pre-development and after-development $(\gamma = t_{cb}/t_{ca})$, α is the runoff rate ratio before and after development $(\alpha = q_{pb}/q_{pa})$ and g is the runoff depth ratio for pre and after-development $(g = h_b/h_a)$.

Example

Given the drainage area of the watershed is 50 ha, the runoff coefficient before development is 0.3 and after development is 0.8. The time of concentration before development is 2 hours and after development is 1 hour. The design storm has a return period of 10 years and a duration of 2 hours. The rainfall intensity for the design storm is 50 mm/h. The peak runoff rate before development is 75 m^3/h/ha.

Find:

- The peak runoff rate after development
- The required detention basin volume

Solution:

The rational formula can calculate the peak runoff rate after development:

$$q_p = CIA$$

where q_p is the peak runoff rate (m^3/h), C is the runoff coefficient, I is the rainfall intensity (mm/h), and A is the drainage area (ha).

Plugging in the given values, we get:

$$q_p = 0.8 \times 50 \times 50 = 2{,}000 \text{ m}^3/\text{h}$$

The generalized model equation can calculate the required detention basin volume:

$$V_s = Ah_a\left[\frac{\gamma + \alpha + \alpha\gamma(g + \alpha - 4)}{\gamma - \alpha}\right]$$

where:

V_s is the storage volume (mm),
A is the area of the drainage system,
h_a is the runoff depth after development (mm),

γ is the time of concentration ratio before and after development, which can be calculated by:

$$\gamma = \frac{t_{cb}}{t_{ca}}$$

$(\gamma = t_{cb}/t_{ca})$, α is the runoff rate ratio $(\alpha = q_{pb}/q_{pa})$, and g is the runoff depth ratio $(g = h_b/h_a)$.

To find h_a, we use the equation below with t_{ca} as the time of concentration after development (h), and substitute them with known values:

$$h_a = q_{pa}t_{ca} = 2{,}000 \times 1 = 2000 \text{ mm}$$

To find q_b, we use the equation below, given the known values:

$$h_b = q_{pb} t_{cb} = 75 \times 2 = 150 \text{ mm}$$

where t_{cb} is the time of concentration before development.

To find γ, we use the equation $\gamma = t_{cb}/t_{ca}$ and given a known value, we get $\gamma = 2$.

Next, find $\alpha = q_{pb}/q_{pa}$, with the known values we have

$$\alpha = 75/2000 = 0.0375$$

and

$$g = q_{pb}/q_{pa} = 150/2000 = 0.075$$

Now we can plug in all the values into the generalized model equation:

$$V_s = Ah_a \left[\frac{\gamma + \alpha + \alpha\gamma(g + \alpha - 4)}{\gamma - \alpha} \right]$$

$$V_s = 5 \times 10^5 [m^2] \times 2[m] \times \left[\frac{2 + 0.0375 + (0.0375 \times 2)(0.075 + 0.0375 - 4)}{2 - 0.0375} \right]$$

$$V_s = 8.9 \times 10^5 \, m^3$$

Therefore, the required detention basin volume is $8.9 \times 10^5 \, m^3$.

The other essential drainage infrastructure such as green infrastructure, pumping stations, flood barriers, and the infrastructure of low-affected development are planned in line with the overall plan of urban drainage.

6.6 WATER QUALITY ASPECTS OF URBAN DRAINAGE SYSTEMS

Stormwater drainage systems are designed to collect and convey runoff from the area where precipitation occurred. This precipitation may generate inundation in that area and the surroundings when the stormwater drainage systems do not exist or are inadequate to swiftly drain the runoff generated by the precipitation. On the other hand, stormwater runoff can carry various pollutants, such as sediments, nutrients, metals, bacteria, and chemicals. These pollutants may originate from urban areas, agricultural lands, industrial sites, and construction activities. They can degrade the quality of the receiving water bodies and pose risks to human health and aquatic ecosystems. For this reason, it is essential to discuss the relevant water quality aspects in this chapter. However, some other aspects of water quality, such as water treatment, are discussed in another chapter.

6.6.1 Sources of Pollution

The water quality of stormwater is associated with the physical, chemical, and biological characteristics of the runoff and the receiving water bodies that affect their suitability for different uses and their ecological functions. Water pollutants carried by runoff come from various sources, such as agricultural fields, urban areas, industrial sites, and mining operations. These pollutants can include nutrients, pesticides, metals, pathogens, and sediment. They can contaminate surface water and groundwater, causing eutrophication, algal blooms, fish kills, disease outbreaks, and reduced water quality.

To minimize the pollution carried by runoff, several measures can be taken at different levels. At the source level, organic or less harmful alternatives can reduce or replace the use of fertilizers and pesticides. The management of waste and hazardous materials can be improved to prevent leaks and spills. The erosion and sedimentation of soil can be controlled by planting vegetation, using mulch, and building terraces or contour plowing. At the transport phase, the volume and velocity of runoff can be reduced by increasing the infiltration capacity of the soil, using permeable pavements, rain barrels, green roofs, and rain gardens. The treatment of runoff can be enhanced by using natural or constructed wetlands, buffer strips, filter strips, bioretention basins, and detention ponds. These practices can help remove pollutants from runoff before they reach the water bodies.

Some of the essential water quality parameters that are measured or monitored in stormwater drainage systems are:

Turbidity: A measure of the cloudiness or clarity of water caused by suspended solids, such as soil particles, organic matter, or algae. Turbidity can reduce light penetration and photosynthesis in aquatic plants, impair the visibility and feeding of fish and other organisms, and increase the transport and deposition of pollutants.

Temperature: A measure of the heat content of water that affects the metabolic rates and oxygen demand of aquatic organisms. Temperature can also influence the solubility and toxicity of some pollutants, such as metals and ammonia. Temperature can vary because of seasonal changes, solar radiation, shading, evaporation, or thermal discharges from industrial or power plants.

pH: A measure of the acidity or alkalinity of water that affects the availability and toxicity of some nutrients and metals. pH can also affect the growth and survival of aquatic organisms. pH can vary because of natural processes, such as photosynthesis and respiration, or human activities, such as acid rain or wastewater discharges.

Dissolved Oxygen (DO): A measure of the amount of oxygen dissolved in water that is essential for the respiration of aquatic organisms. DO can also affect the oxidation and reduction reactions of some pollutants, such as nitrate and iron. DO can vary because of biological processes, such as photosynthesis and decomposition, or physical processes, such as aeration and mixing.

Nutrients: A group of elements or compounds that are required for the growth and development of living organisms. The most common nutrients in stormwater runoff are nitrogen and phosphorus, which can originate from fertilizers, animal wastes, sewage, or atmospheric deposition. Excessive nutrients can cause eutrophication, which is the enrichment of water bodies with organic matter that leads to algal blooms, oxygen depletion, fish kills, and loss of biodiversity.

Metals Contents: A group of elements that have metallic properties and can be toxic to aquatic organisms at high concentrations. The most common metals in stormwater runoff are copper, zinc, lead, cadmium, chromium, nickel, and mercury. These metals can originate from various sources, such as vehicles (e.g., brake pads), roofs (e.g., galvanized steel), pipes (e.g., copper), batteries (e.g., lead), paints (e.g., cadmium), industrial processes (e.g., chromium), or mining activities (e.g., mercury). Metals can accumulate in sediments or bioaccumulate in aquatic food webs, posing risks to wildlife and human health.

Bacteria and Viruses Contents: A group of microscopic organisms that can cause diseases or infections in humans and animals. The most common bacteria in stormwater runoff are fecal coliforms and *Escherichia coli* (*E. coli*), which indicates the presence of fecal contamination from human or animal sources. Other bacteria that can be found in stormwater runoff are *Salmonella*, *Shigella*, *Campylobacter*, *Vibrio cholerae* (cholera), *Legionella pneumophila* (Legionnaires' disease), *Leptospira interorgan* (leptospirosis), or *Mycobacterium avium* complex (MAC).

Bacteria can enter stormwater drainage systems through direct discharges from sanitary sewers (e.g., overflows or leaks), septic systems (e.g., failures or malfunctions), animal wastes (e.g., pets or wildlife), or illicit connections (e.g., illegal dumping).

Chemicals Contents: a group of substances that have various effects on water quality and aquatic life. Some of the common chemicals in stormwater runoff are pesticides (e.g., herbicides, insecticides), polycyclic aromatic hydrocarbons (PAHs) (e.g., oil spills), polychlorinated biphenyls (PCBs) (e.g., transformers), pharmaceuticals (e.g., drugs), personal care products (e.g., cosmetics), endocrine disrupting compounds (EDCs) (e.g., hormones), or emerging contaminants (e.g., nanomaterials). These chemicals can originate from various sources, such as agricultural lands, urban areas, industrial sites, or household wastes. These chemicals can have acute or chronic effects on the health and reproduction of aquatic organisms, as well as human health.

Water quality aspects in stormwater drainage systems are also influenced by several factors, such as the type, intensity, and duration of precipitation events, the characteristics and land uses of the drainage area, the design and operation of the stormwater drainage system, and the quality and quantity of the receiving waters.

6.6.2 Impacts of Urbanization on Water Quality

Urbanization significantly impacts water quality, primarily through the transformation of landscapes and hydrological systems. As cities expand, natural vegetation is replaced by impervious surfaces such as roads, parking lots, and buildings, leading to increased runoff during rainfall events. This surge in runoff not only elevates the volume of water entering drainage systems but also enhances the transport of pollutants, including sediments, heavy metals, and chemicals, into nearby water bodies. Additionally, urban development alters the natural flow of water, disrupting established hydrological patterns and potentially leading to increased flooding and erosion. The alteration of these natural flows can result in sedimentation and habitat destruction, further degrading water quality. Moreover, urban areas often experience higher temperatures due to the heat island effect, which can raise the temperature of runoff and receiving waters, negatively impacting aquatic ecosystems and promoting harmful algal blooms. Lastly, the use of fertilizers and landscaping practices in urban settings contributes to nutrient loading in drainage systems, leading to eutrophication and deteriorating water quality. Collectively, these issues underscore the urgent need for effective urban planning and sustainable drainage solutions to mitigate the adverse effects of urbanization on water quality.

6.6.2.1 Increased Runoff

Urbanization leads to the proliferation of impervious surfaces, such as asphalt and concrete, which prevent water from infiltrating into the ground. This results in higher volumes of SRO during rainfall events. As the amount of runoff increases, so does the potential for pollutants to be transported into drainage systems. Pollutants such as oils, heavy metals, and sediments are often washed off these surfaces and carried into stormwater drains, which can discharge directly into rivers, lakes, and oceans without treatment. This runoff can lead to significant degradation of water quality, harming aquatic habitats and posing risks to public health.

6.6.2.2 Alteration of Natural Water Flows

Changes in land use due to urban development disrupt the natural hydrological cycle. The replacement of vegetation with buildings and roads alters how water flows through the landscape, leading to increased runoff velocities and reduced time for water to percolate into the soil. This alteration affects sediment transport dynamics, often resulting in increased sedimentation in water bodies. Higher sediment loads can smother aquatic habitats, reduce light penetration, and carry additional pollutants, which can concentrate in receiving waters. The disruption of natural flows can also lead to flooding, as drainage systems become overwhelmed by the sudden influx of water.

6.6.2.3 Temperature Changes

Urban areas typically experience higher temperatures than surrounding rural areas, a phenomenon known as the urban heat island effect. This increase in temperature can significantly affect the thermal regime of nearby water bodies. Warmer water temperatures can accelerate chemical reactions, leading to increased rates of evaporation and altered dissolved oxygen levels, which are critical for aquatic life. Elevated temperatures can also promote the growth of harmful algal blooms, which can produce toxins detrimental to both aquatic ecosystems and human health. Additionally, higher temperatures can stress aquatic organisms, particularly those sensitive to temperature changes, disrupting local biodiversity.

6.6.2.4 Nutrient Loading

Urbanization often involves extensive landscaping practices that utilize fertilizers and pesticides, contributing to nutrient loading in stormwater runoff. When it rains, these nutrients – primarily nitrogen and phosphorus – are washed off into drainage systems and ultimately into receiving waters. This nutrient enrichment can lead to eutrophication, a process characterized by excessive growth of algae and aquatic plants. Eutrophication can result in oxygen depletion in the water as decomposing algae consume oxygen, leading to hypoxic conditions that can cause fish kills and loss of biodiversity. Furthermore, the proliferation of harmful algal blooms can produce toxins that pose serious health risks to humans and wildlife, making nutrient management a critical issue in urban water quality.

6.7 URBAN FLOOD AND RISK ASSESSMENT AND MANAGEMENT

Risk assessment and management of urban floods or any floods attempt to understand and address the potential effects of flooding in urban areas. Various factors can cause flooding such as heavy rainfall, storm surges, river overflow, or drainage failure, along with insufficient infrastructure and facilities to control and prevent floods. As floods can pose serious threats to the lives, livelihoods, and assets of urban residents, especially the poor and vulnerable groups, and therefore flood events must be prevented, and the threats must be minimized and controlled. Flooding can also damage urban infrastructure, disrupt essential services, and affect environmental and social well-being. We cannot avoid the presence of floods, as floods are natural random processes, what we can do is just to minimize the losses because of floods by providing a flood control and prevention system that can cope with the floods to a certain degree that we can afford.

Urban flood and risk assessment and management aims to reduce the exposure and vulnerability of urban communities to flood hazards and to enhance their resilience and adaptive capacity. It involves identifying and analyzing the sources, characteristics, and consequences of urban flooding, and evaluating and implementing different measures to prevent, mitigate, prepare for, respond to, and recover from flood events. Urban flood and risk assessment and management requires an approach that considers the physical, social, economic, and environmental dimensions of urban flooding and involves multiple stakeholders, such as local governments, urban planners, engineers, disaster managers, community members, and other relevant actors.

Some of the key steps and components of urban flood and risk assessment and management are:

Hazard Assessment: This assessment involves estimating the frequency, intensity, duration, and extent of flood events, based on historical data, hydrological and hydraulic models, climate projections, and other sources of information. Hazard assessment helps to identify the areas and assets that are prone to flooding, and to define the scenarios and indicators for risk analysis.

Exposure and Vulnerability Assessment: This activity involves assessing the characteristics and conditions of the people, infrastructure, and environment that are exposed to flood hazards, and their susceptibility to harm, loss, or damage. Exposure and vulnerability assessment helps to quantify the potential effects and losses of flooding and to identify the factors and drivers that increase or reduce the flood risk.

Risk Assessment: To estimate the likelihood and severity of flood damage and losses, and to assess the extent and variation of flood risk, the risk assessment involves combining the assessments of hazard, exposure, and vulnerability. Risk assessment assists in identifying the areas and sectors that require action, and in establishing the goals and standards for risk mitigation and management.

Risk Reduction and Management: Designing and implementing different measures to reduce flood risk and enhance the resilience of urban communities is what this entails. Risk reduction and management can include structural measures, such as building or improving flood control and drainage infrastructure, and non-structural measures, such as developing or enforcing land use and building regulations, raising public awareness and preparedness, establishing early warning and emergency response systems, and providing financial protection and social support. Risk reduction and management requires a participatory and adaptive process that involves stakeholder consultation, cost–benefit analysis, feasibility study, monitoring and evaluation, and feedback and learning.

6.7.1 Flood Risks Mapping

One of the most important activities of risk assessment is creating flood hazard maps. This map shows the areas that are likely to be affected by floods of different probabilities, based on historical data, hydrological and hydraulic models, climate projections, and other sources of information. Flood hazard maps provide valuable information about the predicted flood depth and flow velocity, which can help to identify the exposure and vulnerability of people, infrastructure, and environment to flood hazards. With this information, citizens may avoid undertaking permanent activities in the area marked by having risks on the map.

Flood hazard maps can also support the design and implementation of risk reduction and management measures, such as flood control and drainage infrastructure, land use and building regulations, public awareness and preparedness, early warning and emergency response systems, and financial protection and social support. Flood hazard maps are one of the key instruments for flood risk management, and they require a participatory and adaptive process that involves stakeholder consultation, data collection and analysis, map validation and dissemination, and feedback and learning.

A flood risk map can be made by conducting a ground survey, and then marking floodwater elevation and coordinates of the points and plotting them in the existing map with appropriate scale. The map's scale depends on the need. The map can also be prepared by combining a digital elevation model data map with a computer model of a 2-D or 3-D rainfall-runoff model like HEC-RAS, which is freely available for download at the Hydraulic Engineering Center of the US Army Corps of Engineers.

A flood risk map typically consists of two components: a flood hazard map and a flood consequence map. A flood hazard map identifies areas affected by floods of different probabilities, such as a 1% annual chance (also known as a 100-year flood) or 0.2% annual chance (also known as a 500-year flood). It also provides information about the predicted flood depth and flow velocity, which can affect the severity of flood damage. A flood consequence map, meanwhile, presents the potential impact of floods on human health and life, economic activity and infrastructure, cultural heritage,

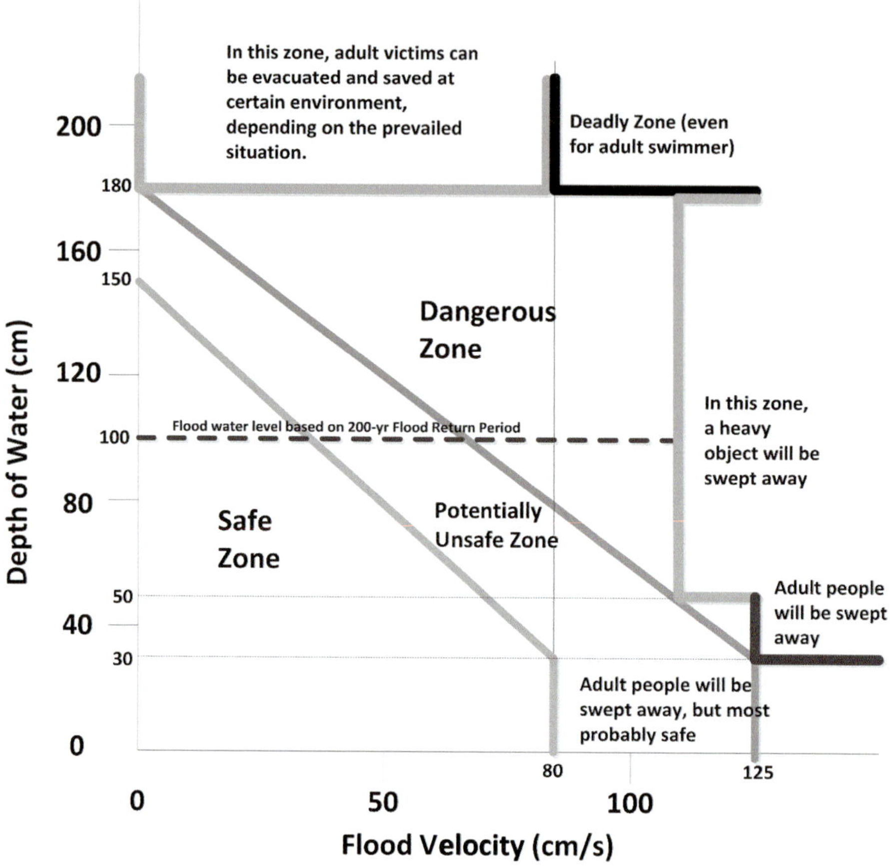

FIGURE 6.24 Flood velocity and depth zones.

and the environment. It can show the number of people, buildings, roads, bridges, crops, and other assets that are exposed or vulnerable to floods, as well as the estimated cost of damage and recovery.

With a flood hazard map, the unsafe or safe zone for citizens based on the correlation between the velocity of floodwater and flood depth can also be determined, as shown in Figure 6.24. The figure applies particularly to Asian adults with average height of 165–170 cm and with basic skill of swimming.

By combining the flood hazard and consequence maps, a flood risk map can help you assess the level of risk for different scenarios and locations. You can use a flood risk map to compare your risk with other communities, plan for emergency response and evacuation, identify areas for mitigation and adaptation, and make informed decisions about insurance and investment.

6.7.2 Benefits of Preventing Floods

Preventing floods can have a multitude of benefits. First, it can help reduce the risk of loss of life and property. Floods can cause significant damage to homes, businesses, and infrastructure, leading to financial losses and displacement of people. Second, it can improve public health by reducing the risk of waterborne diseases and other health hazards that can arise from contaminated floodwaters. Third, it can protect the environment by preserving natural habitats and ecosystems that are often destroyed by floods. Fourth, it can boost the economy by reducing the cost of flood damage and increasing the availability of resources for other purposes. Finally, it can promote social stability by reducing the risk of social unrest and displacement that can arise from floods. There are many ways to prevent floods, including building flood barriers, improving drainage systems, and restoring wetlands and other natural habitats. It is important to take action to prevent floods and protect communities from the devastating effects of flooding.

In planning a flood control and prevention system, we usually do an economic analysis to ensure that the system has optimum economic viability. In doing so, all tangible and intangible benefits of flood prevention are considered and counted. While counting the tangible benefits is relatively easy to do, identifying the intangible benefits is otherwise. What we need to do is develop some transfer functions that can translate the intangible benefits into countable ones.

Transfer functions are mathematical functions that represent how inputs are transformed into outputs by a system. The functions take in incomplete or unprocessed inputs of intangible quantities and produce tangible outputs. For instance, in flood prevention works, injury due to flood is a loss that can be avoided by preventing flood occurrence, and thus this loss can be converted into a benefit. However, some processes are not so simple and smooth that they can be done by using direct transfer functions, some processes need external tools to obtain input variables, such as perceptions and willingness to pay for events that are not measurable but imaginable. Here are some possible and usually applicable transfer functions in evaluating the intangible benefits and the methods of flood prevention.

6.7.2.1 Benefits of Avoiding Loss of Life

This type of loss is the most difficult to quantify as somebody's life is invaluable for whatever reasons. However, along with the other quantified losses, there must be a way to approximate the losses in monetary terms. By employing the economic logic in which every single man is a profit taker and always tries to take advantage to meet his/her hedonic life the loss of life could be approximated by using the comparison of his/her desired salary against the life and death of working environment. A straightforward example is the willingness to work in Afghanistan (rank 163/163 in Global Peace Index, the most hostile country in the world) with salary X against working in Iceland (rank 1/163 in GPI, the most peaceful country in the world) with salary Y, in which X is multiple times of Y. It is then reasonable to measure the loss of life from this simple perception as the respondent fully understood and was aware of the situation, including the risk of life.

The benefits, B_l, from avoiding loss of life of one person during a flood event can be expressed in the following transfer function:

$$B_l = \left(A_l - A_p\right) n_1 \times S_{ann} \tag{6.30}$$

for the age of victims younger than productive age.

$$B_l = \left(A_l - A_v\right) n_2 \times S_{ann} \tag{6.31}$$

for the age of the victim older than or same as productive age.

Where:

A_l is national life expectancy (year)
A_p is national productive age (year)
A_v is age of the victim at the time of loss of life happened because of flood event (year)
n_1 is number of victims of the younger-than-productive-age group (person)
n_2 is number of victims of the older-than-or-same-as-productive-age group (person)
S_{ann} is average national annual salary (currency unit/person/year)

Alternately, whichever is higher, it can also be approximated by using the life insurance claims for the death multiplied by a factor (k) determined by the assessor, multiplied by number of victims (n):

$$B_l = k \times n \times \text{Insurance claims} \tag{6.32}$$

The assessor may determine and adjust the k-factor based on the current and local situation at the time of flood events.

6.7.2.2 Absence from Work because of Traffic Disruptions

Flood events, to some extent, may disrupt traffic and further prevent citizen's activities of studying, working, and other business activities. This disruption certainly causes economic losses, in terms of the length of interruption of the activities. The following transfer function can compute the economic losses or the benefits of preventing and evading such a loss. Elmasry et al. (2018) also asserted about similar direction in estimating the losses because of traffic disruptions.

$$B_w = V_T \times 5 \times W_{day} \times N \tag{6.33}$$

where:

- B_w denotes economic benefits of preventing traffic disruption by flood control works (currency unit)
- V_T represents daily traffic volume at a particular disrupted road by flood in equivalent passenger car unit (unit)
- W_{day} is daily wage of the worker (currency unit/day)
- N denotes number of days of disrupted traffic, integer (day)

If the disruption rather occurred at workplaces, in which the workers could come to the workplaces, but the workplace itself could not operate because of being submerged by flood waters or inundated, the benefits from avoiding this distractor by flood work prevention could be approximated by the following function.

$$B_p = \left(\frac{N}{365} \times \text{Annual Benefit of the Company}\right) + \left(n \times N \times W_{day}\right) \tag{6.34}$$

The transfer functions (equation 6.34) can also be applied to industrial and commercial activities interrupted by flood events.

6.7.2.3 School Disruption by Flood

School disruption by flood events may void the teaching and learning activities. The disruption may hamper the advancement of the students toward their future to whatever the level or scale. The losses of the school disruptions or the benefits of preventing school disruptions by flood control works can be computed in two perspectives (1) from the teaching viewpoint, and (2) from the learning viewpoint. From the teaching viewpoint, the benefit of evading school disturbance from flood events can be seen in the losses suffered by the teacher, which is the losses of the teachers when he/she is absent from the teaching activities, or the benefits gained if such losses can be avoided. From this perspective, the following can express the transfer function of losing equivalent earnings.

$$B_t = n_e \times N \times W_t \tag{6.35}$$

where:
W_t refers to equivalent daily wage of a teacher
n_e denotes number of teachers and supporting staff of the schools

From the perspective of the learning process, the transfer function is formulated according to the duration of education up to the undergraduate level, which is 6 years of primary education plus 6 years of secondary education plus 4 years of university education is equal to 16 years of education or equivalent to 5,844 days of educational duration. The transfer function from the learning viewpoint, B_e, is therefore expressed by the following.

$$B_e = \frac{N}{5844} \times \left(A_l - A_p\right) n \times S_{ann} \tag{6.36}$$

where:
N is number of disrupted learning activities (da)]
n denotes number of students impeded by flood events (person)
A_l represents national life expectancy (year)
A_p is national productive age (year)
S_{ann} refers to average national annual salary (currency unit/person/year)

The physical losses of the school's buildings, facilities, equipment, furniture, books, etc. must be calculated separately in tangible economic losses, which is out of the scope of this study.

6.7.2.4 Long-Term Sickness

If the flood generates a long-term sickness that causes a person to be disabled to sustain his/her own life, then the transfer function is equal to the loss of life plus the medical cost of the person until she/he recovers. The following equation expresses the function:

$$B_s = \left(A_l - A_p\right) n_1 \times S_{ann} + C_{med} \tag{6.37}$$

for the age of victims younger than productive age:

$$B_s = \left(A_l - A_v\right) n_2 \times S_{ann} + C_{med} \tag{6.38}$$

for the age of the victim older than or the same as productive age.

Where:
B_s is benefit of avoiding long-term sickness because of flood of one victim (currency unit)
A_l refers to national life expectancy (year)
A_p is national productive age (year)
A_v denotes age of the victim at the time of loss of life because of flood event (year)
n_1 is number of victims of the younger-than-productive-age group (person)
n_2 is number of victims of the older-than-or-same-as-productive-age group (person)
S_{ann} is average national annual salary (currency unit/person/year)
C_{med} refers to total medical cost until recovery (currency unit)

If the long-term sickness does not generate the lifelong disability of the victims, the benefits of preventing long-term sickness are expressed by the following transfer function.

$$B_s = C_{med} \tag{6.39}$$

6.7.2.5 Damage to Natural Environment

Floods may create devastation to the natural environment. This intangible loss could be transferred to intangible benefits of preventing this loss because of flood control works. There are several components of benefits of preventing natural environmental damage: (1) aesthetical component, (2) loss of habitat, and (3) cost of rehabilitation. The aesthetical component and loss of habitat could not be estimated straightforwardly, but rather through the valuation process by exercising the willingness to pay of the citizens. The cost of rehabilitation of the natural environment damage could either be combined with components (1) and (2) to become one complete package of valuation survey, or as a separate element as this component is a tangible component. To get maximum benefit, in the valuation survey, the willingness to pay for aesthetical components and loss of habitat can be separated in a separate survey with different respondents. The transfer function of the combination of all components because of the willingness to pay from the citizens in the valuation survey, the benefit of avoiding this damage, B_n, is expressed by:

$$B_n = N_a \times W_{tp} \tag{6.40}$$

where:

N_a is number of populations in the city and surroundings those aware of the issues (person)
W_{tp} denotes willingness to pay from the citizens of one person (currency unit/person)

In case the valuation survey separates the tangible component, the transfer function of the benefits can be computed according to the following equation:

$$B_n = \left(N_a \times W_{tp}\right) + C_{rehab} \tag{6.41}$$

where:

C_{rehab} is rehabilitation cost of the environmental damage (currency unit)

6.7.2.6 Post-Disaster Indoor Pollution

After the flood occurs, particularly when the inundation takes a long time, the indoor high humidity may generate indoor air pollution, for example, because of fungus that generates bad odor in the room. Here, the following equation can approximate the benefit of preventing this pollution from taking place.

$$B_i = C_{med} + C_{rehab} + \left(N \times C_d\right) \tag{6.42}$$

where:

B_i is benefit from preventing indoor air pollution (currency unit)
C_{med} denotes medical costs because of indoor pollution, that is, hospital visits and cost of medicines (currency unit)
C_{rehab} is rehabilitation cost of the building for eliminating the pollution (currency unit)

N stands for number of days of the incidence of indoor pollution (day)
C_d is daily cost of grievance because of indoor air pollution (currency unit/day)

6.7.2.7 Water Sources Pollution

In some cases, the flood waters pollute the other water sources, for example, dug wells of the individuals or communities. Groundwater pollution because of flood water rarely takes place because the soil itself may act as the groundwater filter unless the flood water carries hazardous pollutants that may not be screened by the soil. This issue will be discussed separately. If floodwaters contaminate commonly used domestic water sources, the benefits of pollution prevention through flood control measures can be represented by the following transfer function.

$$B_{ws} = n\left(N \times C_{al}\right) + C_{rehab} \tag{6.43}$$

where:

B_{ws} is benefit of preventing water source pollution by flood prevention works (currency unit)
n is number of people affected (person)
N is number of days one person spends money for alternate domestic water (day)
C_{cal} is cost of buying alternate domestic water (currency unit/person/day)
C_{rehab} is rehabilitation cost of the water source to return to normal function (currency unit).

6.7.2.8 Groundwater Contamination

Flood waters can also contaminate the groundwater, particularly when the flood waters carry hazardous contaminants. When the situation is reversible then the benefits of preventing such groundwater contamination are equal to the cost of rehabilitation plus the alternate cost during the contamination (if the groundwater is also the source of domestic water). When the situation is irreversible, then the benefit is equal to the alternate replacement plus bequest and existence costs. The bequest and existence costs are approximated from the valuation survey. The transfer function of the reversible condition is given by:

$$B_{gw} = C_{rehab} + \left(N \times C_a\right) \tag{6.44}$$

where:

B_{gw} is benefit of preventing groundwater contamination (currency unit)
C_{rehab} denotes rehabilitation cost to return the groundwater to its original function (currency unit)
N is number of days of the duration of contamination (day)
C_a is alternate cost during the contamination (currency unit/day)

When the contamination is irreversible, then the transfer function is approximated by the following transfer functions.

$$B_{gw} = C_r + C_{bequest} + C_{existence} \tag{6.45}$$

$$C_{bequest} = N_a \times W_{tpb} \tag{6.46}$$

$$C_{existence} = N_a \times W_{tpe} \tag{6.47}$$

where:

C_r is replacement/alternate cost (currency unit)
$C_{bequest}$ is cost of the lost heritage resulting from the valuation survey (currency unit)
$C_{existence}$ is cost of the environment as a result of the inexistence of the environment, which previously existed.
N_a denotes number of populations in the city and surroundings those aware of the issues (person)
W_{tpb} stands for willingness to pay for the bequest value of the environment
W_{tpe} refers to willingness to pay for the existence value of the environment

6.7.2.9 Aesthetic Damage

Damage to the aesthetics of the environment can be measured by using perceptions if the aesthetics are irreplaceable because of something abstract. If the aesthetic damage is repairable or replaceable, then the benefit of preventing aesthetic damage is equal to the substitute cost of the damage. In case the aesthetic is irreplaceable, the method to value this intangible benefit can be done through environmental valuation, and the benefits can be approximated by the following transfer function.

$$B_{ae} = N_a \times W_{tpa} \tag{6.48}$$

where:

B_{ae} is benefit from preventing aesthetic damage (currency unit)
N_a refers to number of populations in the city and surroundings those aware of the issues (person)
W_{tpa} refers to Willingness to pay the citizens of one person from a valuation survey of aesthetic damage (currency unit/person)

If the aesthetic is replaceable or repairable by the assessment of the expert, then the benefits from avoiding aesthetic damage from flood prevention works can be directly approximated by the following transfer function.

$$B_{ae} = C_r \tag{6.49}$$

where C_r is replacement cost to return the aesthetics to the original function.

6.7.2.10 Total Intangible Benefits

The total intangible benefits due to flood prevention works are equal to the sum of all intangible benefits as discussed above and mathematically expressed in the following equation.

$$B_{int} = \sum_{i=1}^{n} B_i \tag{6.50}$$

where:

B_{int} is total intangible benefits of preventing losses by flood control works (currency unit)
B_i denotes individual benefits as per the above discussion (currency unit)

The intangible benefits are, in monetary terms, more complex to approximate in comparison to the direct tangible damages. The intangible benefits approximation sometimes needs perceptions to estimate through the willingness to pay of the citizens, which is largely based on perceptions. However, this study has attempted to present the rational transfer functions, which translate the complicated states into reasonable and simple inputs of the transfer functions to obtain the approximation monetary values of the benefits. By understanding the overall benefits and costs of the flood control works, a more reasonable result of economic analysis of the flood control works can be obtained.

PROBLEMS AND ISSUES FOR DISCUSSION

Note: You can observe that most numerical calculation questions lack more information, you may determine other factors required for solving the problems by your own knowledge and creativity.

1. A specific catchment area has a runoff coefficient of 0.6 and a rainfall intensity of 50 mm/h, how would you calculate the peak discharge using the Rational method?
2. If a detention basin has a storage volume of 5,000 m^3 and is designed to control the runoff from a 100-year storm event, how long would it take to empty the basin if the outflow rate is 50 m^3/s?
3. A drainage pipe made from a certain material, has a diameter of 1 m and a slope of 0.01, how would you calculate the flow rate using Manning's equation, assuming a roughness coefficient of 0.013?
4. If a permeable pavement system is designed to infiltrate a rainfall event with a return period of 10 years, what would be the required infiltration rate given a rainfall intensity of 100 mm/h?
5. Given a catchment area of 2 km^2 and a rainfall event with a total rainfall depth of 50 mm, how would you calculate the total volume of runoff, assuming a runoff coefficient of 0.7?
6. If a bio-retention system is designed to treat the runoff from a 1-year storm event, what would be the required surface area of the system given a rainfall intensity of 20 mm/h and a catchment area of 0.5 ha?
7. Given a drainage channel with a width of 2 m, a depth of 1 m, and a slope of 0.005, how would you calculate the velocity of flow using the Chezy's equation, assuming a roughness coefficient of 60?
8. If a rainwater harvesting tank is designed to supply water for a household with a daily water demand of 200 L, how many days can the tank provide water given a tank volume of 5,000 L?
9. A catchment area has a time of concentration of 30 minutes and a rainfall intensity-duration-frequency (IDF) curve, how would you calculate the peak runoff for a 5-year storm event?
10. If a swale is designed to convey a peak discharge of 10 m^3/s, what would be the required cross-sectional area of the swale given a flow velocity of 1 m/s?
11. A small catchment area of 2 ha has a runoff coefficient of 0.8 and a rainfall intensity of 75 mm/h, how would you calculate the peak discharge using the Rational method?
12. If a detention basin has a storage volume of 10,000 m^3 and is designed to control the runoff from a 50-year storm event, how long would it take to empty the basin if the outflow rate is 100 m^3/s?
13. Given a drainage pipe with a diameter of 1.5 m and a slope of 0.02, how would you calculate the flow rate using the Manning's equation, assuming a roughness coefficient of 0.015?
14. If a permeable pavement system is designed to infiltrate a rainfall event with a return period of 20 years, what would be the required infiltration rate given a rainfall intensity of 150 mm/h?
15. Given a catchment area of 3 km^2 and a rainfall event with a total rainfall depth of 100 mm, how would you calculate the total volume of runoff, assuming a runoff coefficient of 0.9?

16. If a bio-retention system is designed to treat the runoff from a 2-year storm event, what would be the required surface area of the system given a rainfall intensity of 30 mm/h and a catchment area of 1 ha?
17. Given a drainage channel with a width of 3 m, a depth of 1.5 m, and a slope of 0.01, how would you calculate the velocity of flow using Chezy's equation, assuming a roughness coefficient of 80?
18. If a rainwater harvesting tank is designed to supply water for a household with a daily water demand of 300 L, how many days can the tank provide water given a tank volume of 10,000 L?
19. Given a catchment area with a time of concentration of 45 minutes and a rainfall IDF curve, how would you calculate the peak runoff for a 10-year storm event?
20. If a swale is designed to convey a peak discharge of 20 m^3/s, what would be the required cross-sectional area of the swale given a flow velocity of 2 m/s?
21. What are the key factors that influence the design and implementation of an effective urban drainage system, and why are they important?
22. How does the process of urbanization impact the natural hydrological cycle, and what are the potential consequences for local ecosystems?
23. Why is it crucial to manage stormwater in urban areas, and what strategies can be employed to mitigate the negative impacts of stormwater runoff?
24. Can you analyze the role of green infrastructure in urban hydrology and explain how it contributes to sustainable water management?
25. Evaluate the effectiveness of permeable pavements as a stormwater management strategy. What are their benefits and limitations?
26. How do detention basins function in flood control, and what considerations need to be considered when implementing them in an urban setting?
27. What are the factors that need to be considered when designing urban drainage systems, and why are they significant?
28. Why is hydraulic modeling an essential component in the design of urban drainage systems, and what insights can it provide?
29. Given a specific catchment area and runoff coefficient, how would you calculate the runoff volume for a 1-hour, 100-year storm event?
30. How do urban drainage systems interact with the broader urban water cycle, and why is this interaction important for urban water management?
31. What are the potential impacts of climate change on urban hydrology and how can urban drainage systems be adapted to cope with these changes?
32. How can sustainable urban drainage systems (SUDS) contribute to the improvement of water quality in urban areas?
33. Evaluate the role of wetlands in urban hydrology. How can they be integrated into urban planning for better water management?
34. How does soil type affect the design of urban drainage systems and why is it important to consider in hydrological calculations?
35. Given a rainfall IDF curve, how would you calculate the peak runoff for a specific storm event in an urban area?
36. What are the challenges in managing urban flood risks and how can such challenges be addressed through effective urban drainage design?
37. Analyze the impact of land use changes on urban hydrology. How can urban planners mitigate the negative effects of these changes?
38. How can the principles of water-sensitive urban design (WSUD) be applied to improve urban hydrology and drainage?

39. Given the dimensions and slope of a drainage channel, how would you calculate its hydraulic capacity?
40. What are the benefits and challenges of using rainwater harvesting systems in urban areas, and how can they contribute to sustainable water management?
41. How does the design of stormwater detention facilities affect their performance in controlling peak runoff?
42. Evaluate the role of bio-retention systems in managing urban stormwater. What are their benefits and limitations?
43. Given the properties of a catchment area, how would you calculate the time of concentration for stormwater runoff?
44. How can floodplain mapping be used in urban planning to reduce flood risks?
45. What are the potential impacts of sea-level rise on urban drainage systems in coastal cities and how can they be mitigated?
46. How does urbanization affect the groundwater recharge process and what are the potential consequences for urban water management?
47. Evaluate the effectiveness of green roofs in reducing stormwater runoff. What are their benefits and limitations?
48. Given the rainfall data and the properties of a catchment area, how would you calculate the annual runoff volume?
49. How can the principles of low impact development (LID) be applied to improve urban drainage and hydrology?
50. What are the potential impacts of urban heat islands on urban hydrology and how can they be mitigated?
51. Given the dimensions and roughness of a pipe, how would you calculate the flow rate using the Manning's equation?
52. How can the principles of integrated water resources management (IWRM) be applied in urban areas to improve water management?
53. Evaluate the role of constructed wetlands in treating urban stormwater. What are their benefits and limitations?
54. Given the rainfall data and the properties of a soil, how would you calculate the infiltration rate using the Horton's equation?
55. How can urban forestry contribute to the improvement of urban hydrology and drainage?
56. What are the potential impacts of urban pollution on urban water bodies and how can they be mitigated through effective urban drainage design?
57. Given the dimensions and slope of a drainage channel, how would you calculate the velocity of flow using the Chezy's equation?
58. How can the principles of ecosystem-based management (EBM) be applied in urban areas to improve water management?
59. Evaluate the role of swales in managing urban stormwater. What are their benefits and limitations?
60. Given the rainfall data and the properties of a catchment area, how would you calculate the peak runoff using the Rational method?
61. How can urban agriculture contribute to the improvement of urban hydrology and drainage?
62. What are the potential impacts of urban sprawl on urban hydrology and how can they be mitigated through effective urban drainage design?
63. Given the dimensions and roughness of a pipe, how would you calculate the head loss using the Darcy–Weisbach equation?
64. How can the principles of adaptive management be applied in urban areas to improve water management?

65. Evaluate the role of rain gardens in managing urban stormwater. What are their benefits and limitations?
66. Given the rainfall data and the properties of a soil, how would you calculate the overland flow using the kinematic wave equation?
67. How can urban parks and green spaces contribute to the improvement of urban hydrology and drainage?
68. What are the potential impacts of urban development on river ecosystems and how can they be mitigated through effective urban drainage design?
69. Given the dimensions and slope of a drainage channel, how would you calculate the sediment transport capacity using the Einstein-Brown equation?
70. How can the principles of circular economy be applied in urban areas to improve water management?
71. Consider a catchment area of 5 km^2 with varying land use types: 50% impervious surfaces with a runoff coefficient of 0.95, and 50% pervious surfaces with a runoff coefficient of 0.3. If a storm event with a rainfall intensity of 75 mm/h occurs. Calculate the peak discharge using the Rational method? Consider the weighted average of the runoff coefficients for different land use types.
72. A detention basin with a storage volume of 10,000 m^3 is designed to control the runoff from a 100-year storm event. The outflow rate is controlled by an orifice at the bottom of the basin with a diameter of 0.5 m. Assuming the orifice flow follows Torricelli's Law $\sqrt{2gh}$, where v is the velocity, g is the acceleration due to gravity, and h is the height of water above the orifice), how long would it take to empty the basin?
73. A circular drainage pipe with a diameter of 1.5 m is operating at three-quarters full. The pipe is laid on a slope of 0.02. Calculate the flow rate using the Manning's equation, assuming a roughness coefficient of 0.013? Consider the cross-sectional area and hydraulic radius for a circular pipe operating at three-quarters full.
74. A permeable pavement system is designed to infiltrate a rainfall event with a return period of 20 years. The rainfall intensity for this event is 150 mm/h. The pavement is made up of a 0.3 m thick layer of permeable concrete with a porosity of 0.2, underlain by a 1.5 m thick gravel layer with a porosity of 0.4. What would be the required surface area of the system to ensure that no runoff occurs?
75. A catchment area of 3 km^2 experiences a storm event with a total rainfall depth of 100 mm. The catchment has a time of concentration of 1 hour. The rainfall intensity-duration-frequency (IDF) curve for the area is given by the equation $i = 1{,}000/(T + 10)$, where i is the rainfall intensity in mm/h and T is the duration in hours. Calculate the peak runoff using the Rational method, assuming a runoff coefficient of 0.9?
76. Consider a mixed land use urban catchment of 15 km^2 with 40% impervious surfaces (runoff coefficient of 0.9), 30% semi-pervious surfaces (runoff coefficient of 0.6), and 30% pervious surfaces (runoff coefficient of 0.3). If a storm event with a rainfall intensity of 120 mm/h occurs, calculate the peak discharge using the Rational method?
77. A stormwater detention pond with a storage volume of 30,000 m^3 is designed to control the runoff from a 100-year storm event. The outflow rate is controlled by a circular orifice at the bottom of the pond with a diameter of 1.5 m. Assuming the orifice flow follows Torricelli's Law, how long would it take to empty the pond?
78. A trapezoidal open channel with a bottom width of 2 m, side slopes of 2:1 (horizontal:vertical), a depth of 1.5 m, and a slope of 0.005 is designed to convey stormwater runoff. How would you calculate the flow rate using the Manning's equation, assuming a roughness coefficient of 0.035?

79. A permeable pavement system is designed to infiltrate a rainfall event with a return period of 50 years. The rainfall intensity for this event is 250 mm/h. The pavement is made up of a 0.5-m thick layer of permeable asphalt with a porosity of 0.3, underlain by a 2.5-m thick gravel layer with a porosity of 0.4. What would be the required surface area of the system to ensure that no runoff occurs?
80. A catchment area of 10 km^2 experiences a storm event with a total rainfall depth of 200 mm. The catchment has a time of concentration of 3 hours. The rainfall intensity-duration-frequency (IDF) curve for the area is given by the equation i = 3,000/(T + 30), where i is the rainfall intensity in mm/h and T is the duration in hours. How would you calculate the peak runoff using the Rational method, assuming a runoff coefficient of 0.95?

MULTIPLE-CHOICE QUESTIONS

1. What is the main purpose of urban drainage systems?
   a. To collect and dispose of solid waste.
   b. To control and manage stormwater runoff.
   c. To provide water supply for urban areas.
   d. To treat wastewater.
2. Which of the following is a key component of an urban drainage system?
   a. Water treatment plant.
   b. Sanitary sewer system.
   c. Storm sewer system.
   d. All of the above.
3. The capacity of an urban drainage system is typically determined by:
   a. The average daily water consumption.
   b. The peak stormwater flow rate.
   c. The population density of the area.
   d. The annual rainfall.
4. Which of the following is a technique used to reduce stormwater runoff in urban areas?
   a. Permeable pavement.
   b. Groundwater recharge.
   c. Detention ponds.
   d. All of the above.
5. The concept of "green infrastructure" in urban drainage refers to:
   a. Concrete-lined channels and pipes.
   b. Underground storage tanks.
   c. Vegetated swales and bioretention areas.
   d. Impervious surfaces.
6. Which of the following factors can influence the hydrological response of an urban catchment?
   a. Soil type.
   b. Land use.
   c. Slope.
   d. All of the above.
7. What is the primary purpose of a detention pond in an urban drainage system?
   a. To store and treat wastewater.
   b. To recharge groundwater.
   c. To attenuate peak stormwater flows.
   d. To provide water supply.

8. The term "first flush" in urban hydrology refers to:
   a. The initial high-intensity rainfall event.
   b. The concentrated pollutant load in the initial stormwater runoff.
   c. The groundwater recharge during the wet season.
   d. The overtopping of drainage infrastructure.
9. Which of the following is a common method for reducing combined sewer overflows (CSOs) in urban areas?
   a. Increasing the capacity of the storm sewer system.
   b. Implementing green infrastructure practices.
   c. Constructing additional wastewater treatment plants.
   d. Both b and c.
10. The concept of "low impact development" in urban drainage aims to:
    a. Minimize the use of impervious surfaces.
    b. Maximize the use of natural drainage systems.
    c. Reduce the volume and rate of stormwater runoff.
    d. All of the above.
11. What is the primary function of a sump pump in an urban drainage system?
    a. To remove groundwater from building foundations.
    b. To pump stormwater from detention ponds.
    c. To divert wastewater to the sanitary sewer system.
    d. To treat stormwater before discharge.
12. Which of the following is a common issue associated with the design of urban drainage systems?
    a. Insufficient capacity to manage heavy rainfall events.
    b. Lack of maintenance and sediment buildup.
    c. Conflicts with underground utility infrastructure.
    d. All of the above.
13. The term "hydraulic gradient" in urban hydrology refers to:
    a. The slope of the ground surface.
    b. The slope of the water surface in a pipe or channel.
    c. The difference in water pressure between two points.
    d. The rate of groundwater flow.
14. What is the primary purpose of a storm sewer outfall in an urban drainage system?
    a. To discharge treated wastewater.
    b. To release stormwater runoff into a receiving water body.
    c. To collect and convey groundwater.
    d. To provide access for maintenance and inspection.
15. Which of the following is a common method for controlling runoff from impervious surfaces in urban areas?
    a. Constructing detention basins.
    b. Installing infiltration trenches.
    c. Implementing green roofs.
    d. All of the above.
16. The concept of "water quality treatment" in urban drainage systems refers to:
    a. Removing pollutants from stormwater runoff.
    b. Disinfecting wastewater before discharge.
    c. Reducing the turbidity of surface water bodies.
    d. All of the above.

17. What is the primary function of a catch basin in an urban drainage system?
    a. To collect and convey stormwater runoff.
    b. To remove sediment and debris from the stormwater.
    c. To prevent the backflow of wastewater into the storm sewer.
    d. Both a and b.
18. Which of the following is a common technique used to model urban drainage systems?
    a. Hydrologic modeling.
    b. Hydraulic modeling.
    c. Geographic information systems (GIS).
    d. All of the above.
19. The term "overland flow" in urban hydrology refers to:
    a. The flow of water through the soil.
    b. The flow of water across the ground surface.
    c. The flow of water in open channels and pipes.
    d. The flow of water in the groundwater system.
20. What is the primary purpose of a retention pond in an urban drainage system?
    a. To store and treat stormwater runoff.
    b. To recharge groundwater.
    c. To provide water supply for urban areas.
    d. To attenuate peak flows during heavy rainfall events.
21. Which of the following is a common factor that can impact the performance of an urban drainage system?
    a. Climate change and increasing rainfall intensity.
    b. Urbanization and changes in land use.
    c. Inadequate maintenance and sediment buildup.
    d. All of the above.
22. The concept of "detention time" in urban drainage systems refers to:
    a. The time it takes for stormwater to travel through the system.
    b. The time that stormwater is stored in a detention or retention facility.
    c. The time required for wastewater treatment.
    d. The time between consecutive rainfall events.
23. What is the primary purpose of a culvert in an urban drainage system?
    a. To convey stormwater under a roadway or other obstruction.
    b. To discharge treated wastewater into a receiving water body.
    c. To provide access for maintenance and inspection.
    d. To recharge groundwater.
24. Which of the following is a common technique used to reduce the volume of stormwater runoff in urban areas?
    a. Increasing the capacity of the storm sewer system.
    b. Implementing rain gardens and bioretention areas.
    c. Constructing additional wastewater treatment plants.
    d. Both a and b.
25. The term "hydraulic residence time" in urban hydrology refers to:
    a. The time it takes for water to travel through a drainage system.
    b. The time that water is stored in a detention or retention facility.
    c. The time required for groundwater recharge.
    d. The time between consecutive rainfall events.

26. What is the primary purpose of a stormwater pump station in an urban drainage system?
    a. To remove groundwater from building foundations.
    b. To convey stormwater from low-lying areas to higher elevations.
    c. To divert wastewater to the sanitary sewer system.
    d. To treat stormwater before discharge.
27. Which of the following is a common approach to reducing the risk of urban flooding?
    a. Increasing the capacity of the storm sewer system.
    b. Implementing green infrastructure practices.
    c. Constructing additional wastewater treatment plants.
    d. Both a and b.
28. The term "hydrograph" in urban hydrology refers to:
    a. The relationship between water depth and flow rate.
    b. The plot of flow rate or water level over time.
    c. The distribution of water across a watershed.
    d. The relationship between rainfall and groundwater recharge.
29. What is the primary purpose of a detention basin in an urban drainage system?
    a. To store and treat stormwater runoff.
    b. To recharge groundwater.
    c. To attenuate peak stormwater flows.
    d. To provide water supply for urban areas.
30. Which of the following is a common technique used to reduce the impact of urban heat islands on stormwater runoff?
    a. Increasing the use of impervious surfaces.
    b. Implementing green infrastructure practices.
    c. Constructing additional wastewater treatment plants.
    d. Both a and c.
31. The concept of "runoff coefficient" in urban hydrology refers to:
    a. The ratio of stormwater runoff to rainfall.
    b. The ratio of groundwater recharge to rainfall.
    c. The ratio of wastewater discharge to water consumption.
    d. The ratio of detention time to residence time.
32. What is the primary function of a storm drain inlet in an urban drainage system?
    a. To collect and convey stormwater runoff.
    b. To remove sediment and debris from the stormwater.
    c. To prevent the backflow of wastewater into the storm sewer.
    d. All of the above.
33. Which of the following is a common technique used to manage the quality of stormwater runoff in urban areas?
    a. Oil-water separators.
    b. Constructed wetlands.
    c. Bioretention cells.
    d. All of the above.
34. The term "first flush" in urban hydrology refers to:
    a. The initial high-intensity rainfall event.
    b. The concentrated pollutant load in the initial stormwater runoff.
    c. The groundwater recharge during the wet season.
    d. The overtopping of drainage infrastructure.

35. What is the primary function of a sump pump in an urban drainage system?
    a. To remove groundwater from building foundations.
    b. To pump stormwater from detention ponds.
    c. To divert wastewater to the sanitary sewer system.
    d. To treat stormwater before discharge.
36. Which of the following is a common issue associated with the design of urban drainage systems?
    a. Insufficient capacity to manage heavy rainfall events.
    b. Lack of maintenance and sediment buildup.
    c. Conflicts with underground utility infrastructure.
    d. All of the above
37. The term "hydraulic gradient" in urban hydrology refers to:
    a. The slope of the ground surface.
    b. The slope of the water surface in a pipe or channel.
    c. The difference in water pressure between two points.
    d. The rate of groundwater flow.
38. What is the primary purpose of a storm sewer outfall in an urban drainage system?
    a. To discharge treated wastewater.
    b. To release stormwater runoff into a receiving water body.
    c. To collect and convey groundwater.
    d. To provide access for maintenance and inspection.
39. Which of the following is a common method for controlling runoff from impervious surfaces in urban areas?
    a. Constructing detention basins.
    b. Installing infiltration trenches.
    c. Implementing green roofs.
    d. All of the above.
40. The concept of "water quality treatment" in urban drainage systems refers to:
    a. Removing pollutants from stormwater runoff.
    b. Disinfecting wastewater before discharge.
    c. Reducing the turbidity of surface water bodies.
    d. All of the above.
41. What is the primary function of a catch basin in an urban drainage system?
    a. To collect and convey stormwater runoff.
    b. To remove sediment and debris from the stormwater.
    c. To prevent the backflow of wastewater into the storm sewer.
    d. Both a and b.
42. Which of the following is a common technique used to model urban drainage systems?
    a. Hydrologic modeling.
    b. Hydraulic modeling.
    c. Geographic Information Systems (GIS).
    d. All of the above.
43. The term "overland flow" in urban hydrology refers to:
    a. The flow of water through the soil.
    b. The flow of water across the ground surface.
    c. The flow of water in open channels and pipes.
    d. The flow of water in the groundwater system.

44. What is the primary purpose of a stormwater retention pond in an urban drainage system?
    a. To store and treat stormwater runoff.
    b. To recharge groundwater.
    c. To provide water supply for urban areas.
    d. To attenuate peak flows during heavy rainfall events.

ANSWERS TO MULTIPLE-CHOICE QUESTIONS

1	b	11	a	21	d	31	a	41	d
2	c	12	d	22	b	32	d	42	d
3	b	13	b	23	a	33	d	43	b
4	a	14	b	24	b	34	b	44	d
5	c	15	d	25	b	35	a		
6	d	16	a	26	b	36	d		
7	c	17	d	27	b	37	b		
8	b	18	d	28	b	38	b		
9	b	19	b	29	c	39	d		
10	d	20	a	30	b	40	a		

FURTHER READING

Akan, A. O. (2006). *Urban stormwater hydrology: A guide to engineering calculations*. CRC Press.

Andrieu, H., & Chocat, B. (2004). Introduction to the special issue on urban hydrology Part I. *Journal of Hydrology*, 299(3-4), 163–165.

Baraniecki, V., Heal, K. V., & Wallis, S. (2021). Performance assessment of sustainable urban drainage systems using a combination of laboratory experiments and field monitoring. *Water*, 13(15), 2119.

Bolognesi, A., Casadio, A., Piccinini, L., & Savage, N. (2012). Experimental results on sediment transport in urban drainage pipes. *Water Science and Technology*, 66(12), 2738–2746.

Butler, D., & Davies, J. W. (2011). *Urban drainage*. CRC press.

Campisano, A., Cutore, P., & Modica, C. (2011). Reclaimed water distribution for toilet flushing: A case study to evaluate the budget and the environmental impact. *Water Science and Technology*, 63(12), 2919–2926.

Charlesworth, S. M. (2010). A review of the adaptation and mitigation of global climate change using sustainable drainage in cities. *Journal of Water and Climate Change*, 1(3), 165–180.

Charlesworth, S. M., & Booth, C. A. (Eds.). (2016). *Sustainable surface water management: A handbook for SuDS*. Wiley.

Charlesworth, S. M., Harker, E., & Rickard, S. (2003). A review of sustainable drainage systems (SuDS): A soft option for hard drainage questions? *Geography*, 88(2), 99–107.

Chocat, B., Ashley, R., Marsalek, J., Matos, M. R., Rauch, W., Schilling, W., & Urbonas, B. (2007). Toward the sustainable management of urban storm-water. *Indoor and Built Environment*, 16(3), 273–285.

Chocat, B., Krebs, P., Marsalek, J., Rauch, W., & Schilling, W. (2001). Urban drainage redefined: from stormwater removal to integrated management. *Water Science and Technology*, 43(5), 61–68.

Dotto, C. B., Mannina, G., Kleidorfer, M., Vezzaro, L., Henrichs, M., McCarthy, D. T., ... & Rauch, W. (2012). Comparison of different uncertainty techniques in urban stormwater quantity and quality modelling. *Water Research*, 46(8), 2545–2558.

Elliott, A. H., & Trowsdale, S. A. (2007). A review of models for low impact urban stormwater drainage. *Environmental Modelling & Software*, 22(3), 394–405.

Fletcher, T. D., Shuster, W., Hunt, W. F., Ashley, R., Butler, D., Arthur, S., ... & Viklander, M. (2015). SUDS, LID, BMPs, WSUD and more – The evolution and application of terminology surrounding urban drainage. *Urban Water Journal*, 12(7), 525–542.

Freni, G., Mannina, G., & Viviani, G. (2010). Urban water quality modelling: A parsimonious holistic approach for a complex real case study. *Water Science and Technology*, 61(2), 521–536.

Garofalo, G., Palermo, S., Principato, F., Theodosiou, T., & Santamouris, M. (2016). The impact of pavement material on the urban heat island in southern Italy. *Sustainability*, 8(10), 989.

Guan, M., Sillanpää, N., & Koivusalo, H. (2015). Storm runoff response to rainfall pattern, magnitude and urbanization in a developing urban catchment. *Hydrological Processes*, 29(10), 2396–2409.

Guan, M., Sillanpää, N., & Koivusalo, H. (2016). Modelling and assessment of hydrological changes in a developing urban catchment. *Hydrological Processes*, 30(4), 603–619.

Hatt, B. E., Fletcher, T. D., & Deletic, A. (2009). Hydrologic and pollutant removal performance of stormwater biofiltration systems at the field scale. *Journal of Hydrology*, 365(3–4), 310–321.

Hatt, B. E., Fletcher, T. D., Walsh, C. J., & Taylor, S. L. (2004). The influence of urban density and drainage infrastructure on the concentrations and loads of pollutants in small streams. *Environmental Management*, 34(1), 112–124.

Hoang, L., & Fenner, R. A. (2016). System interactions of stormwater management using sustainable urban drainage systems and green infrastructure. *Urban Water Journal*, 13(7), 739–758.

Holman-Dodds, J. K., Bradley, A. A., & Potter, K. W. (2003). Evaluation of hydrologic benefits of infiltration-based urban stormwater management. *Journal of the American Water Resources Association*, 39(1), 205–215.

House, M. A., Ellis, J. B., Herricks, E. E., Hvitved-Jacobsen, T., Seager, J., Lijklema, L., ... & Pontier, H. (1993). Urban drainage–impacts on receiving water quality. *Water Science and Technology*, 27(12), 117–158.

Huber, W. C., & Dickinson, R. E. (1988). *Stormwater management model, version 4: User's manual (No. EPA/600/3-88/001a)*. Environmental Protection Agency, Athens, GA (USA). Environmental Research Lab.

Integrated stormwater management (iSWMTM) design manual for construction and development. (2010). North Central Texas Council of Governments.

Jia, H., Lu, Y., Yu, S. L., & Chen, Y. (2012). Planning of LID–BMPs for urban runoff control: The case of Beijing Olympic Village. *Separation and Purification Technology*, 84, 112–119.

Jones, P. S., & Hunt, W. F. (2010). Hydrologic and water quality performance of four bioretention cells in central North Carolina. *Journal of Irrigation and Drainage Engineering*, 136(3), 200–208.

Jones, P. S., & Hunt, W. F. (2010). Performance of rainwater harvesting systems in the southeastern United States. *Resources, Conservation and Recycling*, 54(10), 623–629.

Kazemi, F., Beecham, S., & Gibbs, J. (2011). Streetscape biodiversity and the role of bioretention swales in an Australian urban environment. *Landscape and Urban Planning*, 101(2), 139–148.

Kerkez, B., Gruden, C., Lewis, M., Montestruque, L., Quigley, M., Wong, B., ... & Jacques, C. (2016). Smarter stormwater systems. *Environmental Science & Technology*, 50(14), 7267–7273.

Kim, L. H., Zoh, K. D., Jeong, S. M., Kayhanian, M., & Stenstrom, M. K. (2007). Estimating pollutant mass accumulation on highways during dry periods. *Journal of Environmental Engineering*, 133(9), 882–891.

Kleidorfer, M., Moderl, M., Urich, C., Rauch, W., & Sitzenfrei, R. (2014). A case independent approach on the impact of climate change effects on combined sewer system performance. *Water Science and Technology*, 69(3), 646–654.

Konrad, C. P. (2003). Effects of urban development on floods (No. 3-C). US Geological Survey.

Kuichling, E. (1889). The relation between the rainfall and the discharge of sewers in populous districts. *Transactions of the American Society of Civil Engineers*, 20(1), 1–56.

Lee, J. G., & Heaney, J. P. (2003). Estimation of urban imperviousness and its impacts on storm water systems. *Journal of Water Resources Planning and Management*, 129(5), 419–426.

Lee, J. M., Hyun, K. H., Choi, J. S., & Yoon, Y. N. (2012). Assessing the impact of climate change on urban flooding in South Korea. *Journal of the Korean Society of Civil Engineers*, 32(2B), 111–119.

Liu, Y., Ahiablame, L. M., Bralts, V. F., & Engel, B. A. (2015). Enhancing a rainfall-runoff model to assess the impacts of BMPs and LID practices on storm runoff. *Journal of Environmental Management*, 147, 12–23.

Locatelli, L., Mark, O., Mikkelsen, P. S., Arnbjerg-Nielsen, K., Bergen Jensen, M., & Binning, P. J. (2014). Modelling of green roof hydrological performance for urban drainage applications. *Journal of Hydrology*, 519, 3237–3248.

Macdonald, E. (2010). *Green infrastructure for landscape planning: Integrating human and natural systems*. Routledge.

Marsalek, J., Jiménez-Cisneros, B., Karamouz, M., Malmquist, P. A., Goldenfum, J., & Chocat, B. (2006). *Urban water cycle processes and interactions*. IWA Publishing.

Mays, L. W. (Ed.). (2001). *Stormwater collection systems design handbook*. McGraw-Hill.

Moshiri, G. A. (Ed.). (1993). *Constructed wetlands for water quality improvement*. CRC Press.

Nehrke, S. M., & Roesner, L. A. (2004). Effects of design practice for flood control and best management practices on the flow-frequency curve. *Journal of Water Resources Planning and Management*, 130(4), 311–319.

Nicholson, A. R., Wilkinson, M. E., O'Donnell, G. M., & Quinn, P. F. (2012). Runoff attenuation features: A sustainable flood mitigation strategy in the Belford catchment, UK. *Area*, 44(4), 463–469.

Niemczynowicz, J. (1999). Urban hydrology and water management–present and future challenges. *Urban Water*, 1(1), 1–14.

Palla, A., & Gnecco, I. (2015). Hydrologic modeling of Low Impact Development systems at the urban catchment scale. *Journal of Hydrology*, 528, 361–368.

Paterson, E., Uzomah, V. C., Scholz, M. (2014). Soil infiltration, water quality, and leak detection using induced gradient arrays. *Water*, 6(7), 2042–2062.

Pitt, R., Chen, S., Clark, S., Swenson, J., & Ong, C. (2008). Compaction's impacts on urban storm-water infiltration. *Journal of Irrigation and Drainage Engineering*, 134(5), 652–658.

Rossman, L. A. (2015). Storm water management model user's manual version 5.1. National Risk Management Research Laboratory, Office of Research and Development, US Environmental Protection Agency.

Rowe, A. A., Rector, P., & Bright, T. M. (2021). Evaluation of the hydrologic performance of bioretention systems under varying soil media depths. *Journal of Sustainable Water in the Built Environment*, 7(2), 04021003.

Scholz, M., & Uzomah, V. C. (2013). Rapid decision support tool based on novel ecosystem service variables for retrofitting of permeable pavement systems in the presence of trees. *Science of the Total Environment*, 458, 486–498.

Schueler, T. R. (1987). Controlling urban runoff: a practical manual for planning and designing urban BMPs (No. 87703). Metropolitan Washington Council of Governments.

Semadeni-Davies, A., Hernebring, C., Svensson, G., & Gustafsson, L. G. (2008). The impacts of climate change and urbanisation on drainage in Helsingborg, Sweden: Combined sewer system. *Journal of Hydrology*, 350(1–2), 100–113.

Stovin, V. R., Jorgensen, A., & Clayden, A. (2008). Street trees and stormwater management. *Arboricultural Journal*, 30(4), 297–310.

Suarez, J. L., Puertas, J., Anta, J., Jácome, A., & Taboada, J. (2015). Stormwater management in Santiago de Compostela, Spain. *Journal of Water Management Modeling*, 23, C392.

Suds, U. K. (2015). *The SuDS manual*. CIRIA.

Tsihrintzis, V. A., & Hamid, R. (1998). Runoff quality prediction from small urban catchments using SWMM. *Hydrological Processes*, 12(2), 311–329.

Urbonas, B. R., & Stahre, P. (1993). *Stormwater best management practices and detention for water quality, drainage, and CSO management*. Prentice Hall.

Walsh, C. J., Roy, A. H., Feminella, J. W., Cottingham, P. D., Groffman, P. M., & Morgan, R. P. (2005). The urban stream syndrome: Current knowledge and the search for a cure. *Journal of the North American Benthological Society*, 24(3), 706–723.

Wang, M., Sweetapple, C., Fu, G., Farmani, R., & Butler, D. (2017). A framework to support decision making in the selection of sustainable drainage system design alternatives. *Journal of Environmental Management*, 201, 145–152.

Wanielista, M. P., & Yousef, Y. A. (1993). *Stormwater management*. Wiley.

Williams, J. R., & Hann, R. W. (1978). *Optimal operation of large irrigation systems*. Texas Water Resources Institute.

Woods-Ballard, B., Kellagher, R., Martin, P., Jefferies, C., Bray, R., & Shaffer, P. (2007). *The SUDS manual*. CIRIA.

7 Urban Water Supply Systems

7.1 WATER DEMAND AND CONSUMPTION

Water is essential for life, health, and well-being. It is also a vital resource for various human activities, such as agriculture, industry, energy, and transportation. However, water is not evenly distributed or accessible across the globe. Freshwater, which is suitable for drinking and other domestic uses, constitutes only about 2.5% of the total water on earth, and most of it is locked in glaciers and ice caps. Moreover, the availability and quality of freshwater are affected by natural factors, such as climate, hydrology, and geography, as well as human factors, such as population, consumption, pollution, and management. Water scarcity and water security are increasingly becoming major challenges for many regions and countries in the world.

To address these challenges, water supply systems are developed to provide water for various purposes and needs. A water supply system consists of three main components: the source of supply, the treatment or purification of water, and the distribution of water to consumers. The source of supply can be either surface water (such as rivers, lakes, and reservoirs) or groundwater (such as wells and springs). The treatment or purification of water involves removing or reducing contaminants, such as sediments, microorganisms, chemicals, and salts, to make water safe and acceptable for a specific use. The distribution of water to consumers involves a network of pipes, pumps, valves, meters, and other devices that deliver water to homes, businesses, industries, and public facilities.

Water supply systems are crucial for the functioning of modern societies. They provide water for drinking, cooking, washing, sanitation, hygiene, and other domestic needs. They support water for economic activities, such as agriculture, industry, energy, and transportation. They also enable water for environmental and social benefits, such as recreation, wildlife, culture, and education. Water supply systems may face various challenges and risks, due to aging infrastructure, increasing demand, climate change, water quality deterioration, water losses, energy consumption, and governance issues.

Water is a precious and finite resource that we cannot live without. It is also a complex and dynamic resource that we need to manage wisely. Water supply systems are how we harness and utilize water for our diverse and evolving needs. They are the lifelines that connect us to water and each other. They are the foundations that support our development and well-being.

Water is a vital resource for human life and well-being, especially in urban areas where millions of people live and work. This resource is essential for various purposes, such as drinking, cooking, washing, sanitation, hygiene, and health. Unfortunately, not all cities have access to adequate and reliable water supply systems that can provide water that is safe and acceptable for these uses. Many cities, especially in developing countries, face challenges and risks related to water scarcity and water security. These include factors such as population growth, urbanization, climate change,

DOI: 10.1201/9781032664613-7

pollution, overexploitation, and poor management of water resources and infrastructure. As a result, many urban dwellers suffer from the lack of water that is clean, let alone drinkable. This can have negative impacts on their health, livelihoods, environment, and quality of life. Therefore, water is indispensable in the cities, but many cities are struggling to meet the potable water demand. Let us focus on the domain of water supply, demand, and consumption in the cities, not in other areas.

To meet the demand for potable water, the total water consumption in a city must be estimated. A common measure of water consumption is the amount of water supplied per population. For large cities, the total population cannot be based merely on the registered population. During the daytime, the city accommodates additional populations, including laborers and workers who commute from suburban areas, as well as visitors. These groups also contribute to the city's water consumption. The number can be neglected and must be estimated appropriately.

7.1.1 Demands and Consumptions

The aim of developing an urban water supply system is to meet the demand for water in the city, which may include various categories of water consumers. In a city, the categories of water consumers/consumptions are shown in Table 7.1.

The supply must, therefore, meet all legitimate demand (domestic, public, commercial, industrial, institutional, and agricultural) plus losses. The total water demand depends on various factors such as population, income, climate, land use, and technology. The water supply system must be able to satisfy the current and future demand while ensuring the quality, quantity, reliability, and affordability of the water. It consists of several components, such as sources, treatment plants, distribution networks, storage facilities, and meters. The water sources for raw water supply purposes can be surface water (rivers, lakes, and reservoirs), groundwater (wells, springs, and aquifers), or nonconventional water (rainwater, desalination, and wastewater reuse).

Water treatment plants are responsible for removing contaminants and pathogens from raw water to make it safe for consumption and other uses. The water distribution networks then transport the treated water from the plants to the consumers through pipes, pumps, valves, and hydrants. To ensure that the water supply is consistent and reliable, water storage facilities provide buffer capacity and pressure regulation for the distribution networks. Finally, water meters measure the water consumption and billing for the customers.

TABLE 7.1
Water Consumption Categories

Categories	Explanation
Domestic consumptions	They include (1) in-house consumption such as drinking, cooking, sanitation, laundry, house-cleaning, and car washing; (2) out-house consumption such as watering gardens, lawn sprinkling, and swimming pools; (3) standpipes and public fountains.
Public consumptions	Public uses include firefighting, public parks, green areas, street watering, sewer flushing, and street watering.
Commercial, industrial, and institutional consumptions	These types of consumptions may include (1) commercial uses such as shops, offices, hotels, supermarkets, traditional markets, restaurants, airports, harbors, railway stations, and bus stations; (2) industrial uses such as factories, various industries, power stations, shipyards, and docks; and (3) government offices, hospitals, health centers, community centers, public libraries, universities, schools, and military establishments.
Agricultural consumptions	Agricultural consumptions include farming, livestock, dairies, and greenhouses.
Losses	It covers various types of losses, among others: (1) distribution losses such as leakage from the distribution network; (2) wastage: losses from consumer's premise; and (3) other losses: meter errors and unrecorded consumptions.

One of the most important parts of a water supply system is to estimate the water demand of a given area or population. Water demand is the total amount of water used by an area or population per unit of time, usually expressed in liters per capita per day (LCD) or gallons per capita per day (GCD). Knowing the water demand can help policymakers and water authorities plan and manage water resources effectively and sustainably.

There are different methods to approximate the water demand of an area or population. Still, one of the simplest and most common methods to approximate the average daily water demand (ADD) is to use the following formula:

$$\text{ADD} = \text{LCD} \times (1 + \text{f})\text{P} \tag{7.1}$$

where:

ADD is the average daily water demand in liters or gallons.
LCD or GCD is the average daily water usage per person in liters or gallons.
P is the size of the population in the area.
f is a factor to account for any potential water waste.

Example

If we want to estimate the water demand of a city with a population of 500,000 people and an average daily water usage per person of 225 L (60 gallons), we can use the above formula (7.1) as follows: $\text{ADD} = 225(\text{LCD}) \times (1 + 0.1) \times 500{,}000$ (persons).

$$\text{ADD} = 123{,}750{,}000 \text{ L/day} = 123{,}750 \text{ m}^3\text{/day}$$

This means that the city needs about 123.75 million liters (32.74 million gallons) of water per day to meet its domestic, commercial, and industrial needs.

However, this formula only gives an approximation of the ADD. It does not account for variations in water usage due to seasonal changes, weather conditions, economic activities, conservation measures, or other factors. Therefore, it is important to use other methods to refine the estimate and adjust it according to the specific characteristics and conditions of the area or population.

Some of these methods include:

- Using historical data on water consumption and population trends to project future water demand,
- Using surveys or metering to measure actual water usage by different sectors and users,
- Using models or simulations to analyze different scenarios and factors that affect water demand, and
- Using benchmarks or standards to compare water demand with similar areas or populations.

By using these methods, we can obtain a more accurate and reliable estimate of the water demand of an area or population. This can help us design and implement effective strategies to manage water resources efficiently and sustainably.

According to the US Environmental Protection Agency, the average American family used more than 300 gallons of water per day at home, of which 70% occurred indoors. The main indoor uses of water were toilets (24%), showers (20%), faucets (19%), clothes washers (17%), leaks (12%), and other (8%). The main outdoor uses of water were lawn watering (32%), pools (25%), car washing (7%), and others (36%).

To calculate the water consumption per capita, we need to divide the total water consumption by the population size. Water consumption per capita is typically measured in LCD. The average daily water consumption per person can vary depending on several factors such as the region, climate, and lifestyle of the people.

It is important to note that this is only an approximation of the real water consumption per capita. The actual water consumption can vary depending on several factors such as the season, climate, and lifestyle of the people. Therefore, it is essential to incorporate these factors when calculating the water consumption per capita.

7.1.1.1 Domestic Water Demand

Domestic water demand is to meet the demand for drinking, cooking, sanitation, laundry, housecleaning, car washing, watering gardens, lawn sprinkling, swimming pools, standpipes, and public fountains. It commonly shows that household consumption of water depends on personal habits, social statuses such as lifestyle and income, local climatic conditions such as the hotter climate that leads to higher consumption, and customs of the local people such as showering once or twice a day or as required. The typical (average) domestic water consumption is shown in Table 7.2.

Water demand should be able to predict future demand. As the demand for domestic water consumption is highly correlated with the number of populations, we should estimate the future projection based primarily on the number of populations. This is the simplest but acceptable method for planning purposes.

There are some methods to forecast the future population, such as (1) the geometrical increase method, (2) the arithmetical increase method, and (3) the exponential growth method.

7.1.1.1.1 Geometrical Increase

The geometrical increase method, also known as the geometric growth method, assumes that the rate of growth of the population is proportional to the population at present. This method is often used for young and developing towns or regions.

The formula for the geometrical increase method is given by:

$$P_n = P_0 (1+g)^n \tag{7.2}$$

where:

P_n is the predicted population after n years,

P_o is the initial population (number of populations at $t = 0$),

TABLE 7.2
Typical Domestic Water Consumption in LCD

No.	Type of Consumption	Quantity (LCD)
1	Bathing/showering	60
2	Toilet flushing	30
3	Cloth-washing	20
4	House-cleaning	10
5	Utensil and cutlery-washing	10
6	Cooking and drinking	10
	Total (LCD)	140
	Other consumption	
1	Car washing (small-medium)	200–250 L per car
2	Garden-watering, landscaping	1500 L per 100 m²/day
3	Swimming pool	Varies
4	Standpipe	1500–4000 L/min

g is the rate of population growth per year (expressed as a decimal), and
n is the number of years into the future you want to predict.

This formula assumes that the population grows at a constant rate, which may not always be the case in real-world scenarios. Factors such as migration, changes in birth and death rates, and policy changes can affect the rate of population growth. Therefore, it is important to use this method as a starting point and adjust as necessary based on additional data and factors.

7.1.1.1.2 Arithmetical or Linear Increase

The other method is the arithmetical increase method. This formula assumes that net population growth, migration, birth rate, and death rate are reflected in the figure. The formula of the arithmetical or linear increase method is given by:

$$P_n = P_0 + g \cdot n \tag{7.3}$$

The accuracy of the arithmetical increase method can vary depending on the context in which it is used. This method is most suitable for large cities that have reached a saturation point in their population growth. It is mainly adopted for old and developed towns, where the rate of population growth is nearly constant. This method also assumes a constant rate of population growth, which may not always be the case. Therefore, the population predicted by this method is usually the lowest of all.

In comparison with other methods like the geometrical increase method and the incremental increase method, the arithmetical increase method might not always provide the most accurate results. For instance, the geometrical increase method is found to be more accurate in some cases.

7.1.1.1.3 Exponential Growth

The exponential growth method, also known as the geometrical increase method, assumes that the population is growing at a constant rate. This method is often used for young and developing towns, where the rate of population growth is proportional to the current population. The population of a species that grows exponentially over time can be modeled by the formula:

$$P_n = P_0 e^{gn} \tag{7.4}$$

The accuracy of the exponential growth method can vary. It is generally used to make short-term forecasts, but longer-term forecasts using this technique can be quite unreliable. More recent observations are given larger weights by exponential smoothing methods, and the weights decrease exponentially as the observations become more distant.

This method is suitable for expanding communities unaffected by any constraints. It is best used for short-term forecasts and in the absence of seasonal or cyclical variations. It is also suitable for young and developing towns, where the rate of growth of population is proportional to the population at present.

The exponential growth method has several weaknesses. It produces forecasts that lag behind the actual trend. This lag is a side effect of the smoothing process, which neglects the difficulties associated with random variation. Moreover, this method cannot manage trends well. Forecasts are not accurate when data with cyclical or seasonal variations are present. This kind of averaging will not work well if there is a trend in the series. Over the long run, both the exponential growth method and other methods were found to suffer problems, such as negative populations and trend reversals.

7.1.1.2 Water Demand for Public Use

Public water demands encompass a wide range of uses that contribute to the functioning and well-being of a community. Here are some of the key areas where public water is essential.

7.1.1.2.1 Firefighting

Water is crucial for firefighting efforts. Fire departments require a reliable water supply to combat fires and protect communities.

Estimating water demand for firefighting involves several steps and considerations:

1. *Identify the Protected Area:* Determine what exactly is being protected and how it is being protected. Depending on the occupancy, commodity, or fire hazard classifications, different amounts of water will be required to control a fire effectively.
2. *Understand the Sprinkler Systems:* Understand the sprinkler systems/sprinklers to be used to protect against the hazard. Adjustments may be required to the design area, depending on the type selected.
3. *Identify the Most Demanding Area:* Identify what is the most hydraulically remote/demanding area. If it is unclear, then multiple areas may need to be iteratively analyzed to determine which one is the most demanding.
4. *Estimate Required Water Flow Demand:* To estimate the flow demand, identify the number of fire sprinklers required to be calculated within the design area. This is the number of sprinklers anticipated to operate during a fire.
5. *Use a Formula:* Several empirical formulas are available to estimate the water demand:

   Kuichling's formula:

$$Q = 3182\sqrt{P} \tag{7.5}$$

   where:
   Q is the water required in liters/minute, and
   P is the population in thousands.

   Freeman's formula:

$$Q = 1136\left[\left(P/10\right) + 10\right] \tag{7.6}$$

   Engineering toolbox formula:

$$Q = 3.7854\left[1020P^{0.5}\left(1 - 0.01P^{0.5}\right)\right] \tag{7.7}$$

   All the above formulas are empirical; therefore, some adjustments may be necessary.

7.1.1.2.2 Public Parks

Water is used for the maintenance of public parks, including watering lawns, trees, and flower beds. It is also used in fountains and other water features.

There are several factors to be considered while estimating water demand for public parks:

1. *Size of the Park:* The larger the park, the more water it will generally require. This includes the area of lawns, flower beds, trees, and other plantings.
2. *Type of Vegetation:* Different plants have different water needs. For example, lawns typically require more water than xeriscaping.
3. *Climate and Weather:* Parks in hot, dry climates will require more water than those in cooler, wetter climates. Seasonal changes also affect water demand.
4. *Irrigation Efficiency:* The type of irrigation system used can greatly affect water usage. Drip irrigation systems are typically more efficient than sprinkler systems.

5. *Maintenance Practices:* Practices such as mulching and proper plant selection can reduce water demand.

The procedure to estimate the water demand can involve the following steps, such as for irrigation water demand:

1. *Calculate the Evapotranspiration Rate (ET):* This is the amount of water that is evaporated from the land and transpired by plants. It can be calculated using local weather data.
2. *Determine the Plant Water Requirement (PWR):* This is the amount of water needed by the specific types of plants in the park. It can be calculated as a percentage of the ET.
3. *Calculate the Irrigation Requirement (IR):* This is the amount of water needed to supplement natural rainfall. It can be calculated by subtracting the effective rainfall from the PWR.
4. *Consider Peak Demand:* Peak demand is the maximum amount of water required during the hottest part of the year. It is often calculated as a multiple of the average demand.
5. *Adjust for Irrigation Efficiency:* Not all water applied through irrigation is used by plants. Some are lost to runoff, evaporation, or deep percolation. Adjust the irrigation requirement (IR) based on the efficiency of the irrigation system.

7.1.1.2.3 Green Areas

Green spaces such as community gardens, green roofs, and green walls all require water for irrigation.

Because of the similar nature of public parks and green areas, the estimation of water demand for green spaces like community gardens, green roofs, and green walls is similar and involves several factors and steps:

Community gardens/green roofs/green walls:

1. *Size of the Garden/Roofs/Walls:* The larger the size, the more water it will require.
2. *Type of Vegetation:* Different plants have different water needs.
3. *Climate and Weather:* Gardens in hot, dry climates will require more water than those in cooler, wetter climates.
4. *Irrigation Efficiency:* The type of irrigation system used can affect water usage.
5. *Maintenance Practices:* Practices such as mulching and proper plant selection can reduce water demand.
6. *Water Demand Calculator:* You can use a water demand calculator to quickly find the basic parameters for estimating the average water usage in a given area or population.

7.1.1.2.4 Street Watering and Cleaning

Water is used to clean and cool down streets, especially during hot weather. It helps to control dust and maintain a pleasant environment.

Estimating water demand for street watering involves several factors, including the size of the area, the frequency of watering, and the local climate. These are some steps to be taken:

1. *Determine the Area:* Measure the total area of the streets that need to be watered. This could be in square meters or square feet.
2. *Identify Watering Frequency:* Determine how often the streets need to be watered. This could depend on local weather conditions, dust levels, and regulations.
3. *Estimate Water Requirement:* This needs an experiment to determine various water demands for cleaning or cooling down the streets.
4. *Consider Local Climate:* Local temperature, humidity, and air quality are, among others, important variables to be considered in estimating water requirements.

5. *Use Catch Cans:* You can also use catch cans to measure the amount of water being distributed over a specific area. Lay out low-catch cans in a rectangular or triangular pattern. Run sprinklers for a certain amount of time and then measure the water in each can.

7.1.1.2.5 Sewer Flushing

Water is used to flush sewers and maintain the sanitation system. Regular flushing helps to prevent blockages and ensure the system operates efficiently. We need to take some steps to determine water demand for this purpose.

1. *Determine the Size of the Sewer System:* Measure the total length and diameter of the sewer pipes that need to be flushed. This will give you the total volume of the sewer system.
2. *Identify Flushing Frequency:* Determine how often the sewers need to be flushed. This could depend on local sanitation requirements and the amount of waste that the system handles.
3. *Estimate Water Requirement:* The amount of water required for flushing can be estimated using the formula: Volume of water = Volume of sewer pipe × Flushing frequency.
4. *Consider Local Sanitation Requirements:* The local sanitation requirements can significantly impact the water demand. Some areas might require more frequent flushing due to higher amounts of waste or stricter sanitation standards.
5. *Use a Water Demand Calculator:* A water demand calculator can help estimate the average water usage in a given area or population.

7.1.2 Industry and Commercials

Industries and commercial sectors stand as two pivotal pillars in the landscape of water consumption, with their roles underscored by the sheer volume of their water demand. These sectors, each with their unique needs and processes, draw heavily upon our planet's water resources, shaping the ebb and flow of water usage on a global scale.

In the realm of industries, water is the lifeblood that fuels a myriad of processes. From the cooling towers of power plants to the assembly lines of manufacturing units, water is an indispensable resource. It is used in copious amounts for various purposes such as cooling, cleaning, diluting, and even as a raw material in certain products. The scale of industrial water consumption underscores the sector's critical role in the overall water demand.

On the other hand, the commercial sector, with its vast array of establishments – offices, restaurants, hotels, and retail spaces, to name a few – also contributes significantly to water demand. Here, water serves a multitude of purposes, from maintaining hygiene in restrooms to landscaping aesthetics, from culinary requirements in restaurants to laundry needs in hotels. The diversity and extent of water usage in this sector are a testament to its importance in the water consumption matrix.

However, this significant demand for water by industries and commercial sectors also brings to light the pressing need for sustainable water management practices. As we continue to draw upon our water resources, we must also strive to replenish and conserve them. The journey toward water sustainability is a shared responsibility, and these sectors, given their substantial water demand, are key players in this endeavor. Their actions can significantly influence our collective water footprint, shaping a future where water, our most precious resource, is used judiciously and sustainably.

While residential water use certainly holds significance, the industrial and commercial sectors play an indispensable contribution in shaping water demand. Let us delve deeper into how each of these sectors influences our water resources.

7.1.2.1 Industries

Industrial processes, particularly in sectors like manufacturing, mining, and energy production, necessitate substantial water usage for functions such as manufacturing, cooling, and cleaning. These water-intensive industries often grapple with issues of water scarcity, quality, and sustainability, making efficient water management practices vital to minimize waste and ensure long-term availability. To mitigate their impact on freshwater resources, some industries have embraced innovative solutions, including water recycling, rainwater harvesting, and process optimization.

Estimating water demand for industries involves understanding the specific water requirements of the industry's processes. Here are some general steps and methods we can follow:

1. *Understand the Industrial Processes:* Identify the processes within the industry that require water. This could include manufacturing, cooling, cleaning, and other operations.
2. *Determine Water Consumption per Unit of Production:* This is the amount of water used per unit of product produced. It can vary greatly depending on the industry and the efficiency of the processes.
3. *Calculate Total Water Demand:* Multiply the production rate by the water consumption per unit of production to get the total water demand.
4. *Consider Other Uses:* Do not forget to account for other uses of water in the industry, such as boiler feed, cooling and condensing, and sanitary uses.
5. *Project Future Demand:* Consider factors like projected growth, changes in processes, or efficiency improvements that could affect future water demand.

Estimating water consumption in industries involves a combination of direct measurement and calculation based on known parameters. Here are some methods that can be used:

1. *Direct Methods/Field Methods:* These involve measuring the actual water use in the industry. This could be done using water meters or other measurement devices.
2. *Empirical Methods:* These are based on data from similar industries or processes. They use known relationships between water use and other variables (like production volume) to estimate water consumption.
3. *Batch Processes/Manufacturing:* Estimate the water used per unit of product or per batch of process. This requires knowledge of the manufacturing process and its water requirements.
4. *Evaporative Cooling Systems:* These systems use water to cool industrial processes. The water consumption can be estimated based on the cooling requirements of the process.
5. *Plumbing Fixtures, Steam Heating Systems, Vehicle Wash Systems:* These are common in many industries, and their water use can be estimated based on standard values or measurements.

Let us consider the livestock industry as an example. The water used per unit of product in this industry can be estimated by considering the entire life cycle of the livestock. For instance, the water used to produce 1 kg of beef has been estimated to range from 3 to 540 liters of water or water equivalents for the life cycle assessment approach and from 10,000 to 200,000 liters of water for the water footprint.

Water consumption per unit of production can vary widely among different industries due to the diverse nature of their operations. Based on many online sources, some typical rates of water use for various industries in some industrial countries are shown in Table 7.3.

However, some selected industries have typical water consumption per unit of product, as shown in Table 7.4.

7.1.2.2 Commercials

Commercial establishments include offices, restaurants, hotels, and retail spaces. Their water consumption varies based on the type of business and daily activities. Restaurants and hotels use water for cooking, cleaning, laundry, and guest services. Efficient fixtures (like low-flow faucets) can help reduce consumption. Office buildings consume water for restrooms, landscaping, and heating,

TABLE 7.3
Water Consumption Per Unit of Production

Type of Industry	Water Consumption
Energy production	Approximately 70% of industrial water use in high-income countries is for energy production. In the United States, about 4.8 billion gallons of water are used by power plants every day.
Paper industry	Among major industries, the paper industry is the largest consumer of industrial water, followed by food and coal industries globally.
Textile industry	About 19% of China's industrial water is used by the textile industry. The global fashion industry consumes about 79 billion cubic meters of water per year.
Mining industry	Around 3.8% of Europe's total water is used by its mining industry. Over 17.4% of South Africa's water is used for mining purposes.
Chemical and petrochemical sector	This sector is responsible for about 30% of the industrial consumption of water in the European Union.
Beverage industry	The beverage industry in India uses around 2% of the country's total industrial water use.
Knit mills	Water use can vary widely between similar operations as well. For example, knit mills average 10 gallons of water per pound of production, yet water use ranges from a low of 2.5 gallons to a high of 45.2 gallons.

TABLE 7.4
Typical Water Consumption for Different Industries

No.	Type of Industries	In m^3 per ton of Product
1	Canned food	1,000–6,000
2	Chemical industry: Cardon dioxide	14,000–20,000
3	Chemical industry: Sulfur	1,900–2.500
4	Chemical industry: Ammonia	24,000–70,000
5	Beer	2,400–3,800
6	Bread	450–1,000
7	Meat packaging	3,500–4,800
8	Milk products	2,400–4,800
9	Pulp and paper	60,000–190,000
10	Textile	40,000–70,000
11	Steel/metal works	22–28
12	Underground mining	2-5 m^3 per 10^{12} J

ventilation, and air conditioning (HVAC) systems. Implementing water-saving technologies can lead to significant reductions. Retail stores may require water for cleaning floors and restrooms and maintaining a pleasant shopping environment.

However, generic steps to determine to estimate water demand for commercial purposes are as follows:

1. *Understand the Commercial Activities:* Identify the activities within the commercial sector that require water. This could include cleaning, cooking, laundry, irrigation, and other operations.
2. *Determine Water Consumption per Unit of Activity:* This is the amount of water used per unit of activity. It can vary greatly depending on the type of commercial establishment and the efficiency of its operations.

3. *Calculate Total Water Demand:* Multiply the activity rate by the water consumption per unit of activity to get the total water demand.
4. *Consider Other Uses:* Do not forget to account for other uses of water in the commercial sector, such as landscaping, HVAC systems, and sanitary uses.
5. *Project Future Demand:* Consider factors like projected growth, changes in operations, or efficiency improvements that could affect future water demand.

Based on many sources, the water demand for some selected commercial purposes is shown in Table 7.5.

Urban water management requires balancing the needs of various sectors. Besides catering to the water demand for households, public facilities, industries, and commercial establishments, city waterwork authorities also need to ensure a minimum urban low flow. Urban low flow, also known as base flow or ecological flow, is the minimum amount of water flow that must be always maintained in the city's water distribution system. This is crucial for several reasons:

1. *System Integrity:* Maintaining a minimum flow helps ensure the integrity of the water distribution system. It prevents pressure drops that could lead to pipe collapses or the intrusion of contaminants.
2. *Firefighting Needs:* A certain level of water flow is necessary to meet emergency needs, such as firefighting. Fire departments rely on always having sufficient water pressure and flow in the system.

TABLE 7.5
Average Water Demand for Selected Commercial Purposes

No	Type of Establishment	Quantity	Unit
1	Hotels	10	Liter/day/m^2
2	Motels	9	Liter/day/m^2
3	Barber shops	200	Liter/day/chair
4	Beauty salons	1000	Liter/day/station
5	Restaurants	90	Liter/day/seat
6	Hospitals	1300	Liter/day/bed
7	Nursing homes	500	Liter/day/bed
8	Medical offices	25	Liter/day/m^2
9	Laundry	10	Liter/day/m^2
10	Laundromats	88	Liter/day/washing machine
11	Retail	4	Liter/day/m^2
12	Schools (not boarding)	22	Liter/day/student
13	Bus/train depots	136	Liter/day/m^2
14	Car washes	195	Liter/day/m^2
15	Churches	1	Liter/day/member
16	Bowling alleys	503	Liter/day/alley
17	Golf/swim clubs	117	Liter/day/member
18	Residential college	400	Liter/day/student
19	Offices	4	Liter/day/m^2
20	Theaters	12	Liter/day/seat
21	Apartments	820	Liter/day/occupied unit
22	Fast food restaurants	6800	Liter/day/establishment

3. *Ecological Health:* If the city has water bodies like rivers or streams, maintaining a minimum flow is essential for the health of aquatic ecosystems. Adequate water flow supports the survival of various species and maintains the ecological balance.
4. *Water Quality:* Continuous flow helps prevent stagnation, reducing the risk of water quality degradation due to factors like bacterial growth.

7.2 WATER SUPPLY SOURCES

Urban water supply can come from a variety of sources, each with its own advantages and challenges. Some common water sources for urban water supply are shown in Table 7.6.

It is important to understand that each water source carries its own considerations regarding cost, sustainability, and infrastructure needs. As a result, urban water management typically employs a mix of these sources to guarantee a dependable and sustainable water supply.

7.2.1 Surface Water: Rivers, Lakes, and Reservoirs

Rivers, with their inherent abundance, provide an economical solution for water supply, requiring only a modest capital investment for the establishment of intake structures. They serve as the bedrock of diverse ecosystems, fostering a wide range of plant and animal life. However, the delicate balance of these habitats can be disrupted by excessive water extraction, leading to a series of environmental consequences such as reduced flow, habitat degradation, and biodiversity loss.

The reliability of river flow as a water source is a delicate balance with nature, influenced by seasonal variations, climate change impacts, and upstream water usage. In regions where rainfall is sporadic, rivers may not consistently offer a reliable water source.

TABLE 7.6
Water Sources for Urban Water Supply

Water Source	Explanation
Surface water	This includes rivers, lakes, and reservoirs. Surface water is usually treated in a water treatment plant before being distributed to consumers. Surface water is a very common source of water supply in large cities.
Groundwater	This includes wells and boreholes. Groundwater is a common source for single homes and small towns. Groundwater is considered an improved source of water if it is constructed and used in such a way that it adequately protects the water from contamination.
Surface water and groundwater conjunctive use	Conjunctive use of surface water and groundwater for water supply involves strategically managing both resources to maximize the reliability and sustainability of the water supply. Surface water, from rivers and lakes, is typically used when available, while groundwater from wells and aquifers serves as a reliable backup during dry periods. This approach helps maintain water levels, prevent land subsidence, and protect against drought. It requires careful planning and infrastructure but can significantly enhance water security in urban and agricultural areas.
Rainwater harvesting	This involves collecting and storing rainwater for later use. It is a sustainable and cost-effective way to supplement urban water supply, especially in areas with sufficient rainfall.
Reclaimed water	This is wastewater that has been treated and purified so that it can be reused. It is a nonconventional water resource that can be used for non-potable purposes, such as irrigation, flushing toilets, and cooling water for power plants.
Desalinated water	This is seawater or brackish water that has been treated to remove salts and other impurities. It is an important source of water in coastal areas or regions with limited freshwater resources.
Imported water	In some cases, water may be transferred from other regions or countries through pipelines or other means. This is often necessary in areas where local water resources are insufficient to meet demand.

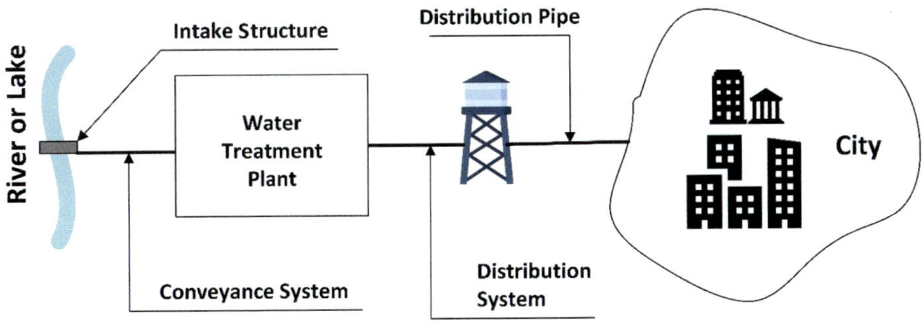

FIGURE 7.1 Intake system (from river to city).

In most cases, rivers require development before they can be fully utilized as water sources for urban areas. The financial expenditure per cubic meter of harnessing river water includes the costs of constructing intake structures, treatment facilities, and distribution networks. This cost is variable, fluctuating based on local conditions and the level of treatment required. One of the methods the rivers are utilized as a source of water supply is through the construction of an intake structure, which is schematically shown in Figure 7.1.

Water conservation is of paramount importance. By encouraging water conservation practices across various sectors – from industries and agriculture to households – we can significantly reduce the demand for river water, thereby helping to maintain ecological balance. The enforcement of stringent regulations on river water extraction is vital to prevent overexploitation. Regular monitoring can ensure adherence to these regulations, preserving our rivers for future generations.

Measures to control pollution are essential to protect our rivers from contamination. This includes treating industrial waste before it is discharged into rivers and advocating for the use of eco-friendly pesticides and fertilizers in agriculture. Additionally, undertaking river restoration projects can aid in repairing damaged ecosystems. Such initiatives can include riverbank reforestation to prevent erosion, removal of invasive species, and reintroduction of native species.

Involving local communities in the management and protection of river resources is another effective approach. This not only instills a sense of ownership and responsibility but also taps into the valuable traditional knowledge that local communities often have about sustainable resource management.

Lakes, another form of surface water resources, have some distinctions from rivers. Both rivers and lakes are significant bodies of water, but they possess unique characteristics.

Rivers are flowing bodies of water that traverse a defined path, linking different areas. Conversely, lakes are bodies of water encircled by land and characterized by still or slow-moving water. Rivers are formed when water flows from elevated points, such as mountains or hills, to lower points, such as lakes or oceans, due to gravitational force. On the other hand, lakes are formed when precipitation accumulates in a low-lying area.

Rivers are connected to seas or oceans. Lakes, however, are not connected to any sea or ocean. Also, rivers often have a serpentine appearance, while lakes resemble large ponds. A river is not surrounded by land on all sides, while a lake is completely enclosed by land. Lakes are vital components of the global water supply system and could potentially serve as sources of urban water supply. They function as natural reservoirs that store water, making it available for various uses such as drinking, irrigation, industrial processes, and more. Lakes can contribute to the urban water supply system through the following mechanisms:

1. *Source of Freshwater:* Lakes are a significant source of freshwater. They store rainwater and runoff from surrounding land, making it available for use.

2. *Water Treatment:* Lakes naturally filter water, improving its quality before it is used. As water flows into a lake, sediments and contaminants settle to the bottom. This process, known as sedimentation, helps to purify the water. Additionally, biological processes within the lake can further improve water quality by breaking down certain types of pollutants.
3. *Storage and Regulation:* Lakes help to regulate the water supply by storing excess water during periods of heavy rainfall and releasing it during dry periods. This function is critical for preventing floods and ensuring a steady supply of water.
4. *Recreation and Tourism:* While not directly related to the water supply system, lakes also support recreation and tourism, which can have economic benefits for local communities. Activities such as fishing, boating, and swimming all rely on clean, plentiful lake water.
5. *Biodiversity:* Lakes support a wide range of plant and animal life, contributing to local and global biodiversity. This biodiversity can have indirect benefits for the water supply system. For example, certain species help to maintain water quality by filtering pollutants or breaking down organic matter.

Reservoirs (the artificial ones) can also be a source of water supply. They are man-made lakes, typically formed by constructing a dam across a river, that store water for various uses. There are also other functions of the reservoirs along with their contribution to providing water sources for the water supply system:

1. *Water Storage:* Reservoirs store large volumes of water, ensuring a steady supply even during periods of low rainfall. This is particularly important in regions with seasonal rainfall patterns.
2. *Flood Control:* By controlling the flow of water, reservoirs can help prevent flooding. During periods of heavy rainfall, reservoirs can store excess water, releasing it gradually to prevent sudden, large volumes of water from causing floods downstream.
3. *Water Treatment:* Like natural lakes, reservoirs can also help improve water quality through natural processes. As water sits in the reservoir, particles and sediments settle to the bottom, a process known as sedimentation. This can help reduce the amount of treatment required before the water is safe for use.
4. *Recreation and Tourism:* While not directly related to the water supply, reservoirs often support recreational activities such as fishing, boating, and swimming. These activities can have economic benefits for local communities.
5. *Hydroelectric Power:* Many reservoirs are part of hydroelectric power facilities. The stored water can be released to turn turbines and generate electricity, providing a renewable source of energy.

The intake structure for the water supply system can be placed directly in the reservoir as an appurtenant structure of the dam; therefore, the water supply system receives water from the reservoir directly through this intake structure. Alternatively, the reservoir maintains the downstream discharge by delivering a certain level of flow to the downstream of the reservoir where the intake structure of the water supply system is placed. By this arrangement, the demand of water for the water supply system is secured, unless an extreme condition of the climate, like a prolonged drought, happens.

7.2.2 Groundwater

Groundwater is a critical source of water for urban supply systems. It is the water found underground in the cracks and spaces in soil, sand, and rock. It is stored in and moves slowly through geologic formations of soil, sand, and rocks called aquifers.

One of the main advantages of groundwater for urban water supply is its availability and reliability. Unlike surface water sources such as rivers and lakes, groundwater is less vulnerable to drought conditions and can provide a steady supply of water throughout the year. This makes it a reliable source of water for cities, especially those located in regions with variable rainfall patterns.

Groundwater is also often of high quality, requiring less treatment than surface water before it is safe to drink. This can result in cost savings for cities that use groundwater as their primary water source. However, groundwater can be contaminated by human activities, such as industrial waste disposal or agricultural runoff, so regular monitoring and treatment may still be necessary.

Despite these advantages, the use of groundwater for urban water supply also presents challenges. One of the main issues is over-extraction, which can lead to a decrease in the water table, land subsidence, and saltwater intrusion in coastal areas. Groundwater is typically extracted using wells. There are some methods to explore the groundwater:

1. *Hand-Dug Wells:* If the water table is relatively close to the surface, a well can be dug by hand or with an excavator.
2. *Drilled Wells:* In most cases, a drill is used to go down deep enough. The well must be drilled at least as deep as the water table but in fact must go much deeper; first, because the water table may change from season to season and from year to year, and second, because when water is being pumped, the water level will drop, at least temporarily.
3. *Well Casing and Screening:* When a well is drilled in unconsolidated sediments or relatively weak rock, it has to be lined with casing (steel pipe in most cases) to ensure that it does not cave in. A specially designed well screen is installed at the bottom of the casing. The size of the holes in the screen is carefully chosen to make sure that it allows the water to move into the well freely but prevents aquifer particles from entering the well.
4. *Submersible Pumps:* A submersible pump is typically used to lift water from within the well up to the surface where it is needed.
5. *Cone of Depression:* Pumping water from the well removes water from inside the well at first. This lowers the water level inside the well. It means that water will flow from the surrounding aquifer toward the pumping well where the water level is now lower. This creates a shape known as a cone of depression surrounding the well.

In case the use of surface water is insufficient to meet the demand for water, the conjunctive use of surface water and groundwater is carried out. The conjunctive use of groundwater and surface water is primarily done to optimize the use of water resources, enhance water reliability, and improve resilience against drought. This strategic approach to water management integrates the use of both resources to maximize the total water supply in a region. This approach is particularly useful in situations where surface water alone cannot meet the entire water demand.

In many urban areas, surface water sources such as rivers and lakes are the primary sources of water. However, during periods of drought or high demand, these sources may not be sufficient to meet the entire water demand. In such cases, groundwater can serve as a valuable supplemental source of water.

The conjunctive use of groundwater and surface water involves the coordinated and planned management and use of these two resources. During periods of high rainfall or low demand, surface water is used as the primary source of water, and excess surface water can be used to recharge groundwater aquifers. During periods of drought or high demand, when surface water supplies are low, groundwater can be extracted to supplement the water supply.

This approach provides several benefits. It allows for a more reliable water supply, as it is less dependent on variable weather conditions. It can also help to reduce the impacts of drought and climate change on water availability. Furthermore, it can help to prevent the over-extraction of groundwater, which can lead to problems such as land subsidence and saltwater intrusion.

The conjunctive use of groundwater and surface water also requires careful planning and management to ensure the sustainable use of both resources. Excessive extraction of groundwater may lead to land subsidence or sinkholes. Sinkholes are depressions or holes in the ground caused by some form of collapse of the surface layer. They occur when the ground below the land surface cannot support the land surface. Sinkholes are common where the rock below the land surface is limestone, carbonate rock, salt beds, or rocks that can naturally be dissolved by groundwater circulating through them. As the rock dissolves, spaces and caverns develop underground. If there is not enough support for the land above the spaces, then a sudden collapse of the land surface can occur.

7.3 WATER QUALITY

The availability of a water supply, both in terms of quantity and quality, is indeed essential to human existence. Water is a fundamental resource that underpins all aspects of human life, from personal hygiene and health to agriculture and industry.

Water is essential for preserving human health. It is required for hydration, a key factor in almost all bodily functions such as blood flow, digestion, and the elimination of waste. Furthermore, uncontaminated water is indispensable for food preparation and sanitation, aiding in disease prevention and hygiene maintenance.

Playing a pivotal role in food production, water is heavily utilized in agriculture for crop irrigation. A significant fraction of the world's water consumption is attributed to this sector. In the absence of a sufficient water supply, the impact on food production could be severe, potentially leading to food scarcity and a surge in food prices.

Many industries rely heavily on water. It is integral to manufacturing processes, the production of energy, and waste disposal, among other applications. Consequently, the availability of a dependable water supply is a critical factor for economic productivity and expansion.

It is not just the quantity of water that matters, but also its quality. Contaminated water can lead to a range of health problems, from diarrhea to more serious diseases such as cholera and dysentery. Ensuring access to clean water is therefore a key public health priority.

7.3.1 Drinking Water Quality Standards

Drinking water standards are crucial for ensuring the safety and quality of the water that we consume. These standards are typically categorized into physical, chemical, biological, and bacteriological parameters:

1. *Physical Parameters:* Physical standards pertain to the sensory and observable characteristics of water. These include aspects such as color, odor, taste, temperature, and turbidity. For instance, water should ideally be clear (low turbidity), colorless, and free from any objectionable taste or odor.
2. *Chemical Parameters:* Chemical standards relate to the concentration of various chemical substances in water. These include parameters such as pH level, hardness, alkalinity, and the presence of various ions like calcium, magnesium, and potassium. Additionally, chemical standards also cover harmful substances like lead, arsenic, and other heavy metals, ensuring their concentrations remain below harmful levels.
3. *Biological Parameters:* Biological standards are concerned with the presence of microorganisms in water. These standards typically specify acceptable levels for different types of microorganisms to prevent waterborne diseases. For example, coliform bacteria are commonly used as an indicator organism to detect fecal contamination.
4. *Bacteriological Parameters:* Bacteriological standards specifically deal with the presence of bacteria in drinking water. These standards set limits for pathogenic bacteria such as *Escherichia coli*, *Salmonella*, and *Shigella*, which can cause serious illnesses if ingested.

7.3.1.1 Physical Parameters

Concerning the physical parameters of drinking water quality, here are some of the key physical parameters of water quality standards:

1. *Electrical Conductivity:* It measures the water's ability to conduct electricity, which provides a measure of what is dissolved in water. High conductivity in water indicates the presence of dissolved salts and other inorganic chemicals. This can affect the taste of drinking water and the health of aquatic life. The acceptable range for drinking water is 50–1500 μS/cm.
2. *Salinity:* It is the saltiness or dissolved salt content of a body of water. High salinity can affect the water's suitability for drinking, irrigation, and industrial uses. It can also impact the biodiversity of aquatic ecosystems. For drinking water, the total dissolved solids (TDS), which is a measure of salinity, should be less than 1000 mg/L.
3. *TDS:* It measures the combined content of all inorganic and organic substances contained in a liquid. High TDS levels can affect the taste of water and may indicate the presence of harmful minerals. It can also affect the health of aquatic organisms. The World Health Organization (WHO) guideline for TDS in drinking water is less than 1000 mg/L.
4. *Turbidity:* It measures the degree to which the water loses its transparency due to the presence of suspended particulates. High turbidity can affect the aesthetic quality of water and interfere with disinfection processes in water treatment. It can also reduce light penetration in water, affecting aquatic plant growth. For drinking water, the turbidity should be less than 5 NTU and ideally below 1 NTU.
5. *Temperature:* It refers to the heat or coldness of water. Temperature can affect the solubility and reaction rates of chemicals in water. Temperature affects the solubility of gases in water, the rates of chemical reactions, and the life cycles of aquatic organisms. While there is no set standard for temperature, it can affect other water quality parameters. The typical range for freshwater systems is 6–30°C.
6. *Color:* The color of water can reveal the presence of certain minerals or chemicals. Unusual color in water can indicate the presence of pollutants or high concentrations of dissolved organic matter. For drinking water, the color should be less than 15 color units.
7. *Taste and Odor:* These are aesthetic parameters of water quality. They can be indicators of water quality issues. Unpleasant tastes and odors can make water unsuitable for drinking and cooking. They can also indicate the presence of pollutants. These are aesthetic parameters. While not health-threatening, noticeable taste or odor can make water unpalatable.
8. *pH:* It measures how acidic/basic water is. The range goes from 0 to 14, with 7 being neutral. The pH of water can affect the solubility and toxicity of chemicals in water. It can also affect the health of aquatic life. The acceptable range for drinking water is 6.5–8.5.
9. *Dissolved Oxygen (DO):* It is the amount of oxygen dissolved in the water. DO is a very important indicator of water quality and its capacity to support a balanced aquatic ecosystem. DO is essential for the survival of aquatic organisms. Low DO levels can lead to the death of fish and other aquatic life. For aquatic life, the minimum DO concentration should be 5 mg/L.
10. *Oxidation–Reduction Potential (ORP):* It measures the ability of a lake or river to cleanse itself or break down waste products, such as contaminants and dead plants and animals. ORP affects the ability of water bodies to self-purify and break down waste products. Changes in ORP can indicate changes in water quality. There is no set standard for ORP as it can vary greatly depending on the water source and treatment process.

These parameters are often measured in quantitative terms, such as mg/L for TDS, nephelometric turbidity units (NTU) for turbidity, °C for temperature, and so on. The specific standards for these parameters can vary depending on the intended use of the water (e.g., drinking, recreation, agriculture, and industry).

7.3.1.2 Chemical Parameters

As chemical substances are mostly harmful to humans and the environment, their presence in drinking water must be limited. Some of the key chemical parameters of drinking water quality standards include:

1. *Arsenic:* The WHO recommends a limit of 10 μg/L. Arsenic in water can lead to various health issues, including skin damage, problems with the circulatory system, and an increased risk of cancer.
2. *Barium:* The recommended limit is 10 μg/L. High levels of barium in drinking water can increase blood pressure.
3. *Boron:* The recommended limit is 2400 μg/L. High levels of boron can lead to reproductive and developmental health issues.
4. *Chromium:* The recommended limit is 50 μg/L. Chromium can cause liver and kidney damage, lung cancer, and dermatitis.
5. *Fluoride:* The recommended limit is 1500 μg/L. While fluoride can help prevent dental cavities, too much fluoride can lead to dental fluorosis or skeletal fluorosis.
6. *Selenium:* The recommended limit is 40 μg/L. High levels of selenium can cause hair loss, fatigue, neurological damage, and circulatory problems.
7. *Uranium:* The recommended limit is 30 μg/L. Uranium is a radioactive element that can increase the risk of cancer and kidney toxicity.
8. *Lead:* Lead can affect almost every organ and system in your body. Children younger than 6 years are most susceptible to lead poisoning, which can severely affect their mental and physical development. The US Environmental Protection Agency (EPA) and the WHO have set the maximum allowable concentration of lead in public drinking water at 15 μg/L.
9. *Copper:* Copper is an essential nutrient, but in excess, it can cause vomiting, diarrhea, stomach cramps, and nausea. It has also been associated with liver damage and kidney disease. The US EPA has set the maximum contaminant level (MCL) goal for copper in drinking water at 1.3 mg/L or 1.3 ppm. The WHO recommends no more than 2 mg of copper per liter of water.
10. *Nitrate:* High levels of nitrate in water can cause methemoglobinemia or "blue baby" disease. The US EPA has set the MCL for nitrates in drinking water at 10 mg/L or 10 ppm.

The specific standards for these parameters can vary depending on the intended use of the water (e.g., drinking, recreation, agriculture, and industry).

7.3.1.3 Biological Parameters

Some of the key biological parameters of drinking water quality standards include:

1. *Bacteria:* Bacteria such as *E. coli* are commonly used as indicators of fecal contamination. Drinking water should be free from *E. coli* to prevent diseases like diarrhea. The WHO recommends a limit of 0 *E. coli* per 100 mL of water.
2. *Viruses:* Viruses in drinking water can cause a variety of diseases, including gastroenteritis and hepatitis. Drinking water should be free from viruses.
3. *Parasites:* Parasites such as *Giardia* and *Cryptosporidium* can cause gastrointestinal diseases. Drinking water should be free from disease-causing parasites.
4. *Algae:* Some types of algae can produce toxins that are harmful to human health. While there is no set standard for algae, drinking water should be monitored for harmful algal blooms.
5. *Protozoa:* For drinking water, protozoa such as *Giardia* and *Cryptosporidium* should be absent.

6. *Fungi:* There is no specific limit for fungi in drinking water, but their presence could indicate a problem with the water supply.
7. *Phytoplankton:* While there is no set standard for phytoplankton, harmful algal blooms (which can be caused by certain types of phytoplankton) are usually indicated by a cell count of 5000 cells/mL or a chlorophyll concentration of 20 μg/L.
8. *Zooplankton:* There is no specific limit for zooplankton in drinking water, but their presence could indicate a problem with the water supply.
9. *Biochemical Oxygen Demand (BOD):* For freshwater ecosystems, a BOD less than 5 mg/L is considered good, while a BOD greater than 5 mg/L could indicate pollution.
10. *Chemical Oxygen Demand (COD):* For freshwater ecosystems, a COD of less than 20 mg/L is considered good, while a COD greater than 20 mg/L could indicate pollution.
11. *Nutrients:* For freshwater ecosystems, a total phosphorus concentration of less than 0.1 mg/L and a total nitrogen concentration of less than 1.0 mg/L are generally considered good.

The specific standards for these parameters can vary depending on local regulations and the intended use of the water.

7.3.1.4 Bacteriological Parameters

Most of the bacteria are harmful to human beings, and they, therefore, must be absent or limited in their presence in drinking water. Some of the key bacteriological parameters of drinking water quality standards (as also shown in Table 7.7) are:

TABLE 7.7
Typical Drinking Water Quality Standards

Physical Parameters		Chemical Parameters		Biological Parameters		Bacteriological Parameters	
Electrical conductivity	50–1,500 μS/cm	Arsenic	<10 μg/L	Bacteria	Absent	Total coliforms	Absent
Salinity	<1,000 mg/L	Barium	<10 μg/L	Viruses	Absent	Fecal coliforms/*E. coli*	Absent
Total dissolved solids (TDS)	<1,000 mg/L	Boron	<2400 μg/L	Parasites	Absent	*Enterococci*	Absent
Turbidity	<5 NTU	Chromium	<50 μg/L	Algae	Monitor	*Pseudomonas aeruginosa*	Absent
Temperature	6–30°C	Fluoride	<1500 μg/L	Protozoa	Absent	Heterotrophic plate count (HPC)	
Color	<15 color units	Selenium	<40 μg/L	Fungi	Monitor		
Taste and odor	Tasteless and odorless	Uranium	<30 μg/L	Phytoplankton	<5,000 cells/mL		
pH	6.5–8.5	Lead	<15 μg/L	Zooplankton			
Dissolved oxygen (DO)	>5 mg/L	Copper	<1.3 mg/L	BOD	<5 mg/L		
Oxidation-reduction Potential (ORP)	Varies	Nitrate	<10 mg/L	COD	<20 mg/L		
				Nutrients (total P)	<0.1 mg/L		

1. *Total Coliforms:* Coliforms are bacteria that are always present in the digestive tracts of animals, including humans, and are found in their wastes. They are also found in plants and soil materials. The presence of coliform bacteria in drinking water may indicate a possible presence of harmful, disease-causing organisms. The WHO recommends a limit of 0 total coliforms per 100 mL of water.
2. *Fecal Coliforms/E. coli:* Fecal coliform bacteria are a kind of total coliform. The feces (or stool) and digestive systems of humans and warm-blooded animals contain millions of fecal coliforms. *E. coli* is part of the fecal coliform group and may be tested for by itself. Fecal coliforms and *E. coli* are usually harmless. However, a positive test may mean that feces and harmful germs have found their way into your water system. These harmful germs can cause diarrhea, dysentery, and hepatitis. The WHO recommends a limit of 0 *E. coli* or thermotolerant coliform bacteria per 100 mL of water.
3. *Enterococci:* Enterococci are a type of bacteria that live in the gastrointestinal tract of humans and other warm-blooded animals. Enterococci can survive in salt water, and this makes them important bacteria to study when testing beach water. High counts show there may be sewage contamination. The WHO recommends a limit of 0 enterococci per 100 mL of water.
4. *Pseudomonas aeruginosa: Pseudomonas aeruginosa* is a common bacterium that can cause disease in animals, including humans. It is found in soil, water, skin flora, and most man-made environments throughout the world. It thrives not only in normal atmospheres but also in low-oxygen atmospheres and has, thus, colonized many natural and artificial environments. The WHO recommends a limit of 0 *P. aeruginosa* per 100 mL of water.
5. *Heterotrophic Plate Count (HPC):* HPC measures a range of bacteria that are naturally present in the environment. They are used as an indicator of general water quality. The US EPA recommends a limit of no more than 500 bacterial colonies per milliliter.
6. *Nutrients:* For freshwater ecosystems, a total phosphorus concentration of less than 0.1 mg/L and a total nitrogen concentration of less than 1.0 mg/L are generally considered good.

Again, it should be noted that the specific standards for these parameters can vary depending on local regulations and the intended use of the water.

7.3.2 Non-drinking Water Quality Standards

Water quality standards for non-drinking water, such as recreational water (beaches, lakes), wastewater, and industrial water, vary widely depending on the specific use of the water. Some general criteria for non-drinking water and their quality standards are explained in the following:

1. *Recreational Water:* Recreational water quality standards are designed to protect the health of swimmers and other recreational users. These standards often focus on the levels of bacteria (such as *E. coli* and *Enterococci*), algae, and other potential pathogens.
2. *Wastewater:* Wastewater quality standards are designed to protect the environment and public health when wastewater is discharged into the environment. These standards often focus on parameters such as BOD, COD, pH, temperature, and the levels of nutrients (such as nitrogen and phosphorus), pathogens, and toxic substances.
3. *Industrial Water:* Industrial water quality standards depend on the specific industrial use. For example, water used in food and beverage processing must meet high standards for purity and safety, while water used in cooling towers might have different standards.
4. *Agricultural Water:* Agricultural water quality standards focus on parameters that can affect plant growth and livestock health, such as salinity, pH, hardness, and levels of specific ions (such as sodium, chloride, and boron).

5. *Aquatic Life:* Water quality standards for protecting aquatic life often focus on parameters such as temperature, pH, dissolved oxygen, and levels of specific pollutants that can harm aquatic organisms.

The specific standards can vary greatly depending on local regulations, which will not be discussed in this section, the specific type of water body, and the intended use of the water.

7.4 WATER SUPPLY NETWORK

An urban water supply network is a complex system designed to provide water to the population in a city and surrounding areas. The key components of an urban water supply network include:

1. *Water Source:* This could be a river, lake, reservoir, or groundwater source. The choice of water source depends on local availability and quality of water.
2. *Treatment Facilities:* Before the water is distributed, it needs to be treated to ensure it is safe for consumption. This usually involves processes like filtration, disinfection, and sometimes desalination.
3. *Storage Facilities:* Treated water is stored in facilities like water tanks or reservoirs. These facilities help maintain a steady supply of water, even during times of high demand.
4. *Distribution System:* This is the network of pipes that delivers water from the storage facilities to the consumers. The distribution system needs to be designed and optimized appropriately to deliver water at the required pressure and discharge. It often includes pumps and pressure control valves.
5. *Intelligent Water Networks:* With advancements in technology, urban water supply networks are becoming more intelligent. They use smart sensors, flow meters, and integrated supervisory control and data acquisition (SCADA) systems for efficient management and operation.
6. *Waste Water Management:* After the water has been used, it needs to be collected, treated, and either discharged safely or reused. This is an important part of the urban water supply network.

The exact facilities and their configuration can vary depending on the specific needs and constraints of the urban area.

7.4.1 Types of Pipe Networks

In an urban water supply system, the pipe network is the most essential component in distributing water from the source to the consumers. Here are the main types of pipe networks:

1. *Serial or Sequential Pipes:* In this type of network, pipes are connected in a series, meaning the output of one pipe is the input to the next. This is often used in situations where the water needs to pass through several processes or stages before it reaches the consumer. This is shown in Figure 7.2.
2. *Parallel Pipes:* In this type of network, several pipes run parallel to each other, carrying water from the same source to different locations. This is often used to distribute water to different parts of the city or to provide redundancy in case one pipe fails. The parallel pipe is shown in Figure 7.3.
3. *Branching Pipes:* In this type of network, the pipes branch out from a main pipe, like the branches of a tree. This is often used to distribute water within a specific area or neighborhood. An illustration of this network is shown in Figure 7.4.

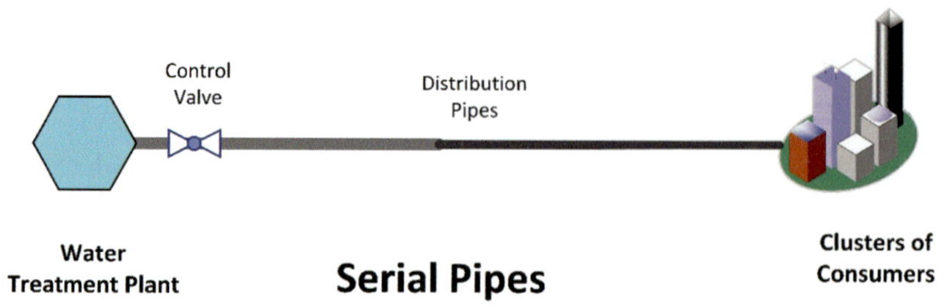

FIGURE 7.2 Serial pipes.

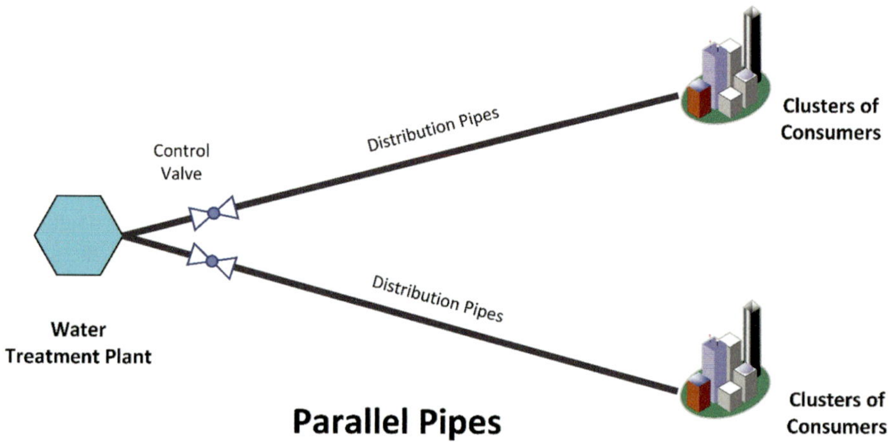

FIGURE 7.3 Parallel pipes.

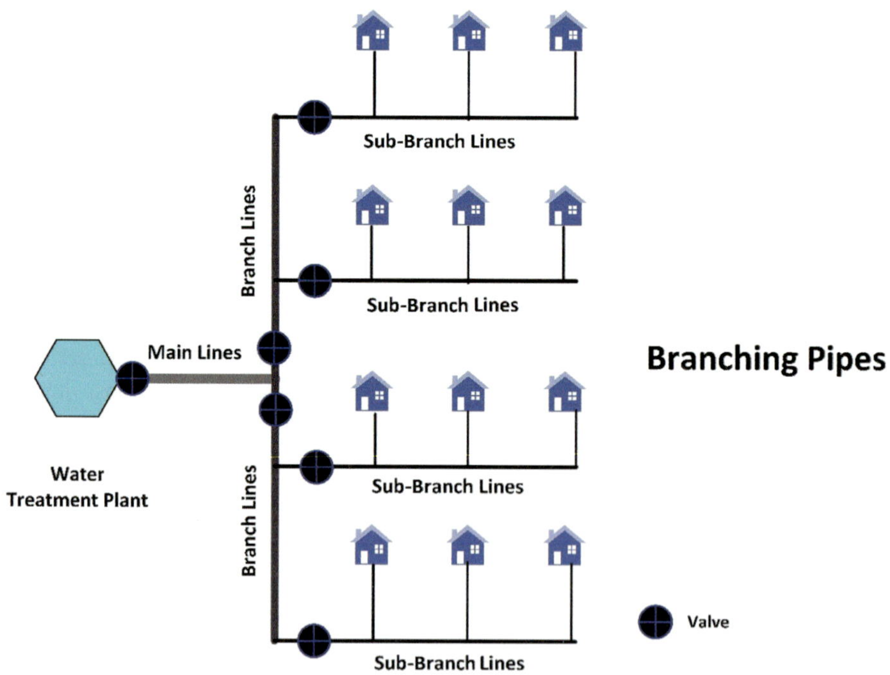

FIGURE 7.4 Branching Pipes.

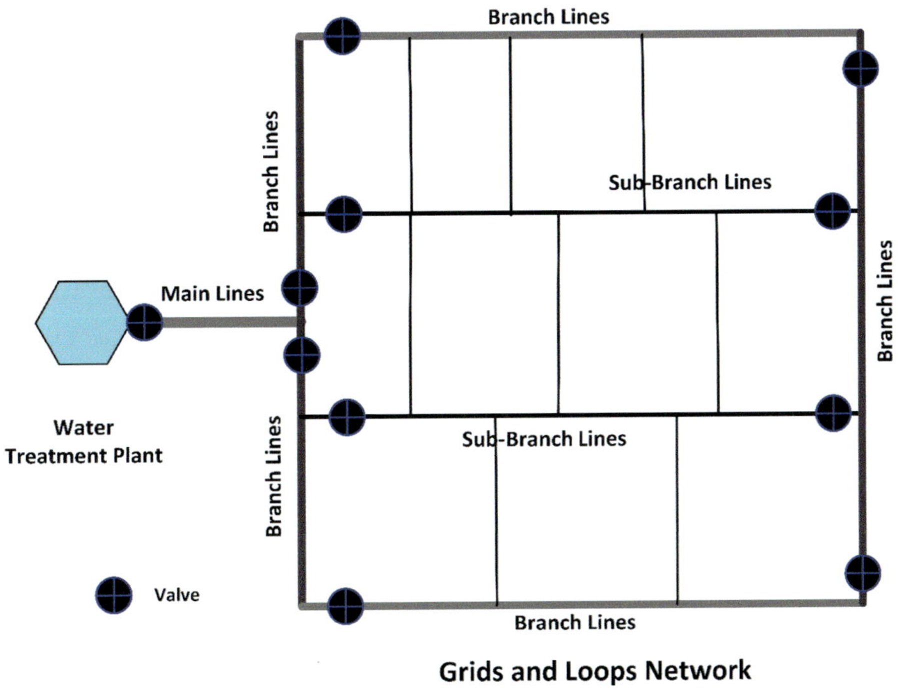

FIGURE 7.5 Grids and loops network.

4. *Grid-and-Loop Networks:* In this type of network, the pipes form a grid or loop, allowing water to flow in multiple paths. This is often used in urban areas to ensure a reliable supply of water even if one part of the network fails. It is shown in Figure 7.5.
5. *Open-Loop (Pseudoloop) Networks:* In this type of network, each open-loop makes the connection between a node with a known piezometric head (reservoir) or with a determined relation discharge-piezometric head (pump station) and another node with a known piezometric head or a determined relation discharge-piezometric head. It is shown in Figure 7.6.
6. *Hybrid Networks:* These are a combination of two or more of the systems.

Each type of network has its own advantages and is used in different situations depending on factors like the layout of the city, the source of the water, and the needs of the consumers.

In addition to these, with advancements in technology, urban water supply networks are becoming more intelligent. They use smart sensors, flow meters, and integrated SCADA systems for efficient management and operation.

Serial pipes, also known as branching patterns with dead ends, are a type of water distribution system. Here are some of their advantages and disadvantages:

Advantages:

1. It is a very simple method of water distribution.
2. Calculations are easy and simple to do.
3. The required dimensions of the pipes are economical.
4. This method requires comparatively a smaller number of cut-off valves.

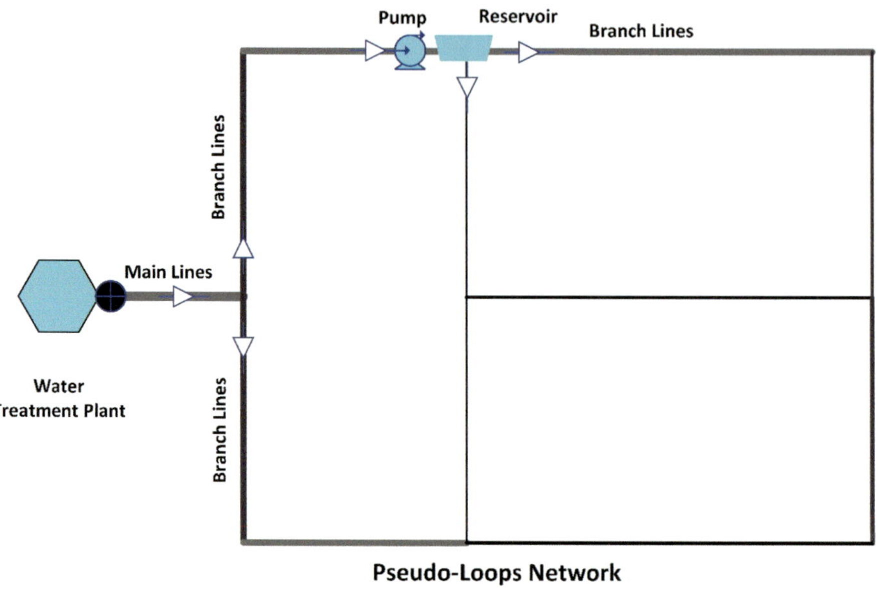

FIGURE 7.6 Pseudoloops network.

Disadvantages:

1. The area receiving water from a pipe under repair is without water until the work is completed.
2. In this system, there are large numbers of dead ends where water does not circulate but remains static.
3. Sediments accumulate due to stagnation of the dead end, and bacterial growth may occur at these points.
4. It is difficult to maintain chlorine residual at the dead ends of the pipe.
5. Water available for firefighting will be limited as it is being supplied by only one water main.

For the serial pipes arrangement, the advantages of using this distribution method in a water distribution system include:

1. *Flexibility:* Serial pipes allow for more flexibility in designing the layout of the distribution system to accommodate various terrains and obstacles.
2. *Cost-Effectiveness:* Using serial pipes can sometimes be more cost-effective than other distribution system configurations, especially in specific scenarios where it reduces the need for extensive excavation or construction.
3. *Ease of Maintenance:* As each pipe segment is separate, it can be easier to isolate and repair a particular section without disrupting the entire system.

The disadvantages of using serial pipes in a water distribution system are:

1. *Increased Risk of Leakage:* With multiple connections between pipes, there is a higher risk of leaks occurring at these joints, potentially leading to water loss and infrastructure damage.
2. *Complexity:* The design and operation of a system with serial pipes can be more complex compared to other configurations, requiring careful planning and maintenance.

3. *Pressure Drops:* Due to the increased number of connections, there can be a higher pressure drop along the system, affecting the efficiency of water distribution and potentially requiring additional pumping.

Advantages of using branching pipes in a water distribution system include:

1. *Efficient Water Distribution:* Branching pipes can help distribute water more evenly throughout the system, ensuring adequate water supply to different areas.
2. *Redundancy:* Having branching pipes can provide redundancy in the system, allowing for alternative pathways for water flow in case of pipe failures or maintenance needs.
3. *Ease of Expansion:* Branching pipes make it easier to expand the water distribution system by adding new branches to accommodate growing demand or new developments.

The disadvantages of using branching pipes in a water distribution system are:

1. *Complexity:* The design and management of a system with branching pipes can be more complex, requiring careful planning to ensure proper flow and pressure throughout the network.
2. *Increased Risk of Contamination:* With multiple branches, there is a higher risk of contamination entering the system through various entry points, necessitating stricter monitoring and maintenance protocols.
3. *Higher Construction and Maintenance Costs:* Building and maintaining a system with branching pipes can be more expensive due to the additional materials, fittings, and maintenance requirements associated with a more intricate network.

Advantages of using grid-and-loop pipes in a water distribution system include:

1. *Redundancy and Reliability:* Grid-and-loop systems provide redundancy and multiple pathways for water flow, which can enhance system reliability by reducing the risk of service disruptions in case of pipe failures or maintenance needs.
2. *Consistent Water Pressure:* The interconnected grid-and-loop layout helps maintain more consistent water pressure throughout the system, ensuring adequate supply to all areas and reducing the likelihood of pressure fluctuations.
3. *Improved Water Quality:* The continuous circulation of water in a grid-and-loop system can help prevent stagnation and improve water quality by reducing the risk of bacterial growth and contamination.

The disadvantages of using grid-and-loop pipes in a water distribution system are:

1. *Complexity and Planning:* Designing and managing a grid-and-loop system can be more complex and require careful planning to ensure optimal flow patterns, pressure regulation, and water quality maintenance.
2. *Higher Initial Cost:* Implementing a grid-and-loop system may involve higher initial construction costs due to the need for more pipes, fittings, and valves compared to simpler distribution layouts.
3. *Increased Maintenance Requirements:* The interconnected nature of grid-and-loop systems may result in higher maintenance requirements, as issues in one section of the network can potentially affect multiple areas, necessitating more extensive monitoring and repair efforts.

Advantages of open-loop (pseudoloop) networks in a water distribution system include:

1. *Redundancy:* Open-loop networks provide a level of redundancy by allowing water to flow in multiple directions, which can help maintain water supply in case of pipe failures or maintenance activities.
2. *Improved Water Quality:* Continuous water circulation in an open-loop network can help prevent stagnation and improve water quality by reducing the risk of bacterial growth and contamination.
3. *Flexibility:* Open-loop networks offer flexibility in water distribution, allowing for adjustments and modifications to the system layout to accommodate changes in demand or infrastructure requirements.

Disadvantages of open-loop (pseudoloop) networks in a water distribution system are:

1. *Complexity:* Designing and managing an open-loop network can be more complex compared to simpler distribution layouts, requiring careful planning to ensure optimal flow patterns and pressure regulation.
2. *Higher Initial Cost:* Implementing an open-loop network may involve higher initial construction costs due to the need for additional pipes, fittings, and valves compared to more straightforward distribution configurations.
3. *Maintenance Challenges:* The interconnected nature of open-loop networks may lead to increased maintenance requirements, as issues in one part of the network can potentially impact multiple areas, necessitating more extensive monitoring and repair efforts.

Several key aspects that should be considered in the selection of a pipe distribution network in a water supply system are shown in Table 7.8.

TABLE 7.8
Key Aspects for Consideration in the Selection of Pipe Distribution Network

Aspect	Explanation
Water demand	Understanding the current and future water demand of the area served by the distribution network is crucial in determining the size and capacity requirements of the pipes.
Topography and terrain	The geographical features of the area, such as elevation changes, soil conditions, and obstacles like rivers or buildings, play a significant role in selecting the most suitable pipe distribution network layout.
Water quality requirements	Considerations for maintaining water quality, such as preventing stagnation, minimizing contamination risks, and ensuring adequate disinfection, should influence the choice of distribution network design.
System reliability	Assessing the need for system redundancy, backup options, and emergency response capabilities is essential to ensure reliable water supply under various operating conditions.
Cost	Balancing initial construction costs, ongoing maintenance expenses, energy requirements, and potential future expansion needs is crucial in selecting a cost-effective pipe distribution network.
Operational efficiency	Factors such as energy efficiency, pressure management, flow control, and water loss reduction should be considered to optimize the operational efficiency of the distribution network.
Regulatory compliance	Compliance with local regulations, standards, and codes related to water distribution systems should be a key consideration in selecting the appropriate pipe network design.
Environmental impact	Evaluating the environmental impact of the chosen distribution network in terms of energy consumption, carbon footprint, material usage, and potential ecosystem effects is important for sustainable water management.

7.4.2 Engineering Principles of Pipe Networks

The engineering principles of pipe networks in a water supply system involve various considerations to ensure the efficient and reliable distribution of water. Some key engineering principles include:

1. *Hydraulic Design:* This involves determining the optimal pipe sizes, layout, and configurations to ensure adequate water flow, pressure, and velocity throughout the system. Hydraulic calculations consider factors such as demand patterns, elevation changes, friction losses, and pipe materials to design a network that meets the required performance criteria.
2. *Pressure Management:* Maintaining appropriate pressure levels within the network is crucial to ensure consistent water supply to consumers. Engineers must design the system to manage pressure variations effectively, considering factors like elevation changes, pipe diameters, pump locations, and pressure control devices.
3. *Water Quality Considerations:* Engineers must design pipe networks to minimize the risk of water quality degradation, stagnation, and contamination. Proper design features such as adequate disinfection, flushing programs, and strategic pipe materials selection help maintain water quality standards throughout the distribution system.
4. *System Resilience and Reliability:* Designing for system resilience involves incorporating redundancy, backup options, and emergency response plans to ensure uninterrupted water supply even during pipe failures, maintenance activities, or emergencies. Engineers consider factors like looped configurations, isolation valves, and alternative supply sources to enhance system reliability.
5. *Energy Efficiency:* Engineers aim to design pipe networks that minimize energy consumption by optimizing pump operations, reducing pressure losses, and selecting energy-efficient components. Efficient designs help reduce operational costs and environmental impact while maintaining system performance.
6. *Material Selection:* Choosing appropriate pipe materials based on factors like durability, corrosion resistance, cost-effectiveness, and environmental impact is essential in designing a reliable and long-lasting water distribution network. Engineers consider factors such as pipe material properties, lifespan, maintenance requirements, and compatibility with water quality standards.
7. *Monitoring and Control:* Incorporating monitoring systems, control devices, and automation technologies in the design of pipe networks enables real-time monitoring of water flow, pressure, quality parameters, and system performance. These tools help optimize operations, detect issues promptly, and facilitate timely interventions to maintain system efficiency and reliability.

By applying these engineering principles in the design and operation of pipe networks in water supply systems, engineers can create a robust, efficient, and sustainable infrastructure that meets the water demands of communities while ensuring reliable and high-quality water supply.

7.4.3 Hydraulic Computation of the Pipe Network

Pipe networks are an essential part of water supply systems to distribute water from sources to consumers. The hydraulic computation of pipe networks involves determining the flow rates, pressures, and head losses in the pipes to ensure efficient and reliable water supply. For a limited number of nodes and loops, the hydraulic analysis can be done manually, as the number of iterations will also be limited, and it is therefore manageable. However, in the real world, to serve a city or community in which the number of nodes and loops is tens of thousands, the manual process is almost impossible. Therefore, it is done using hydraulic analysis software based on the principles of fluid mechanics and network theory.

Some software is available for hydraulic analysis of pipe networks, either commercial or freeware, for example:

1. *Pipe Flow Expert Software:* This is a premier software program for piping design and pipe system modeling. It calculates fluid flow in open or closed-loop pipe networks with multiple supply and discharge tanks, multiple pumps in series or parallel, and multiple pipe sizes and fittings.
2. *KYPipe:* It provides the capability to handle transient, unsteady flow calculations in large distribution systems.
3. *Pipeflowcalculations.com:* This is a free online software for the calculation of pipe networks, flows in pipes, and pressures in nodes.
4. *PIPENET Standard:* This software is used for steady-state flow analysis of pipe and duct networks with compressible and incompressible fluids. It is ideal for hydraulic calculations of complex networks, pipe and pump sizing, pressure drop, flow rates, and cavitation analysis.
5. *Environmental Protection Agency Network Evaluation Tool (EPANET):* This is a public domain software developed by the US Environmental Protection Agency. It is used throughout the world to model water distribution systems. Environmental Protection Agency Network Evaluation Tool (EPANET) can perform extended-period simulation of the hydraulic and water quality behavior within pressurized pipe networks. The networks can consist of pipes, nodes (pipe junctions), pumps, valves, storage tanks, and reservoirs. It can track the flow of water in each pipe, the pressure at each node, and the height of the water in each tank throughout the network. It can also be used to model contamination threats and evaluate resilience to security threats or natural disasters. EPANET is widely used by engineers and consultants to design and size new water infrastructure, retrofit existing aging infrastructure, optimize operations of tanks and pumps, reduce energy usage, investigate water quality problems, and prepare for emergencies. It is a powerful tool for the hydraulic analysis of pipe networks.

As EPANET is a public domain and everyone can download, install, and use the software, it would be useful to briefly explain the procedure for performing hydraulic calculations of a pipe network using EPANET:

1. *Draw a Network Representation:* Create a diagram of your distribution system in EPANET. This includes all pipes, nodes (junctions), pumps, valves, and storage tanks or reservoirs.
2. *Edit the Properties:* Input the properties of the objects that make up the system. This includes the physical properties of the fluid (like density and viscosity), the dimensions of the pipes (diameter and length), and the pressure conditions at the start and end of each pipe.
3. *Describe the System Operation:* Define how the system is operated. This could include the operation of pumps and valves, the demand at each node, and the initial water levels in tanks.
4. *Select Analysis Options:* Choose the options for the hydraulic and water quality analysis. This includes the time step for the simulation, the accuracy of the hydraulic analysis, and the type of water quality analysis.
5. *Run the Analysis:* Execute the hydraulic/water quality analysis. EPANET will calculate the flow of water in each pipe, the pressure at each node, the height of water in each tank, and the concentration of chemical species throughout the network.
6. *Review the Results:* After the simulation, review the results. EPANET provides various ways to view the results, including color-coded maps, time-series graphs, and tables.

7.4.3.1 Hydraulic Computation of Pipeline

The hydraulic computation of pipelines, particularly when the pipes are arranged in series, relies primarily on the application of Bernoulli's equation, which elucidates the relationship between fluid velocity, pressure, and elevation along the pipeline. In this computational framework, the determination of head loss, a crucial factor in assessing the efficiency and performance of the pipeline system, is facilitated through the utilization of the Darcy–Weisbach equation. This equation accounts for various parameters such as pipe roughness, flow rate, and pipe diameter, offering a comprehensive understanding of the energy dissipation within the pipeline network. Through the integration of Bernoulli's equation and the Darcy–Weisbach equation, we can model and optimize the hydraulic behavior of interconnected pipelines, ensuring the effective conveyance of fluids while minimizing energy losses and operational inefficiencies.

Bernoulli's equation is a fundamental principle in fluid mechanics that describes the conservation of energy in a flowing fluid. It states that the total mechanical energy per unit mass remains constant along a streamline. The general form of Bernoulli's equation is:

$$P + \frac{1}{2}\rho v^2 + \rho g h = \text{constant} \tag{7.8}$$

It can also be expressed in a different form:

$$\frac{P}{\rho g} + \frac{v^2}{2g} + h = \text{constant} \tag{7.9}$$

where:

P is the pressure of the fluid,
ρ is the density of the fluid,
v is the velocity of the fluid,
g is the acceleration due to gravity, and
h is the height above a reference point.

Bernoulli's equation helps in understanding the relationship between pressure, velocity, and elevation along a streamline in a flow field. Therefore, head loss along the pipelines can be identified.

Assume sections 1 and 2 along the pipeline as shown in Figure 7.7. The average velocity of flow at sections 1 and 2 are v_1 and v_2, respectively. According to Bernoulli:

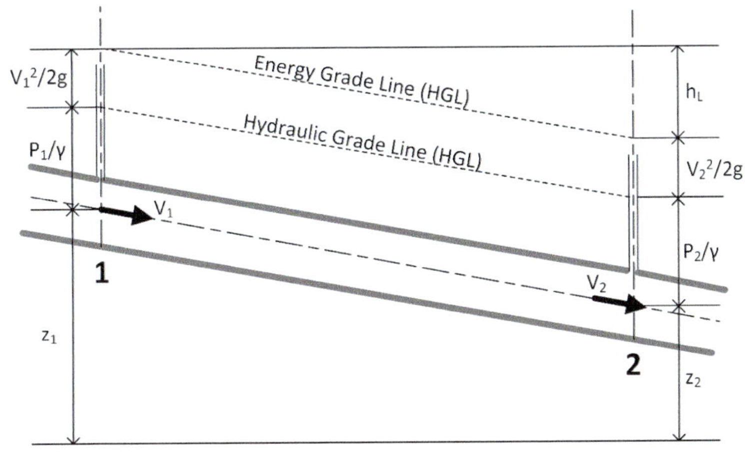

FIGURE 7.7 Bernoulli equation.

$$\frac{V_1^2}{2g} + \frac{P_1}{\gamma} + z_1 = \frac{V_2^2}{2g} + \frac{P_2}{\gamma} + z_2 + h_L \qquad (7.10)$$

Based on the Darcy–Weisbach formula, the head loss due to friction of the flow and internal pipe:

$$h_f = f\left(\frac{L}{D}\right)\frac{V^2}{2g} \qquad (7.11)$$

In this case,

f is the friction factor,
L is the pipe length, and
D is the pipe diameter.

The friction factor f for turbulent flow can be approximated by the Colebrook–White semi-empirical formula:

$$\frac{1}{\sqrt{f}} = -2.0\log\left(\frac{e/D}{3.7} + \frac{2.51}{R_e\sqrt{f}}\right) \qquad (7.12)$$

Formula (7.12) is not an explicit equation. The value of f cannot be easily obtained. To solve this problem, a diagram, which was developed semi-empirically by Moody in 1944 can help solve the friction factor value during hydraulic calculation of the pipe flow.

When we do the hydraulic calculation of pipe flow using the Moody diagram, we need information about Reynold's number of the flow and the relative roughness of the pipe. Reynold's number can be found by using this formula:

$$R_e = \frac{VD}{\nu} \qquad (7.13)$$

where:

V is the flow velocity,
D is the diameter of the pipe, and
ν is the kinematic viscosity of the water.

The relative roughness, on the other hand, can be found by calculating e/D, where e is the roughness height of the internal pipe wall. The value of e can be found in Table 7.9.

During the calculation of flow in the pipe, there are various minor losses. The minor losses are head losses caused by inlets, outlets, bends, and other appurtenances such as fittings, valves, expansions, and contractions. These losses referred to as minor losses, form losses, or secondary losses are caused by flow separation and the generation of turbulence. The minor losses can be expressed by:

$$h_{Lm} = K\frac{V^2}{2g} \qquad (7.14)$$

where K is the loss coefficient (see Table 7.10).

TABLE 7.9
Values of e for Different Types of Pipes

Pipe Materials	e (mm)	Pipe Materials	e (mm)
Brass	0.0015	PVC	0.0015
Concrete, smooth with steel forms	0.18	Polyethylene, HDPE	0.0015
Concrete, average	0.36	Steel, enamel-coated	0.0048
Concrete, rough	0.60	Steel, riveted	0.9 ~ 9.0
Copper	0.0015	Steel, seamless	0.004
Corrugated metal	45	Steel, commercial	0.045
Iron with asphalt-lined	0.12		
Iron, cast	0.26		
Iron, with cement mortar lined	0.12		
Iron, galvanized	0.15		
Iron, wrought	0.045		

TABLE 7.10
Minor Loss Coefficient, K $\left[\frac{V^2}{2g}\right]$

Type of Minor Loss	K	Type of Minor Loss	K
Contraction	0.04–0.50	Pipe fittings, 90 degrees elbow-long radius	0.14–0.23
Expansion	0.03–1.00	Pipe fittings, 45 degrees elbow regular	020
Entrance, bell-mouthed	0.04	Sluice gate, submerged	0.8
Entrance, slightly taunted	0.23	Sluice gate, contraction in conduit	0.5
Entrance, square-edged	0.50	Sluice gate, width equal to conduit width	0.2
Entrance, projecting	1.00	Valve = 0	0.3–1.3
Exit, bell-mouthed	$0.10\left(\frac{V_1^2}{2g}-\frac{V_2^2}{2g}\right)$	Valve, $\theta = 30$	3.6–3.9
Exit, submerged	1.00	Valve, $\theta = 60$	90–120
Pipe fittings, 90 degrees elbow regular	0.21–0.30		

EXAMPLE 7.1

Water flows in a 1,000-m-long pipeline with a diameter of 200 mm at a velocity of 5 m/s. The pipeline is a new cast iron pipe. Determine the head loss in the pipeline. The kinematic viscosity of water is $\nu = 1.007 \times 10^{-6}$ m²/s.

Solution:
The pipe is made from cast iron, from Table 7.5. The e for cast iron is 0.26 mm = 0.00026 m.

The head loss is computed using the Darcy – Weisbach formula: $h_L = f\frac{L}{D}\frac{V^2}{2g}$

To determine the friction factor f, the Reynold's number must be computed:

$$R_e = \frac{VD}{\nu} = \frac{5\text{ m/s} \times \left(\frac{200}{1000}\text{m}\right)}{1.007 \times 10^{-6}\text{ m}^2/\text{s}} = 9.93 \times 10^5$$

The relative roughness $\frac{e}{D} = \frac{0.00026}{0.200} = 0.0013$. From Moody diagram, $f = 0.021$

$$h_L = 0.021 \frac{1{,}000}{0.200} \frac{(5)^2}{2(9.81)} = 133.8\text{ m}$$

The head loss in the pipe is 133.8 m.

EXAMPLE 7.2

A 10-cm diameter, 2000-m-long pipeline connects two reservoirs open to the atmosphere. What is the discharge in the pipeline if the water surface elevation difference of the reservoirs is 50 m? Assume a smooth pipe, and $\nu = 1.02 \times 10^{-6}\text{ m}^2/\text{s}$.

Solution:
The problem can be schematically depicted as shown in Figure 7.8.

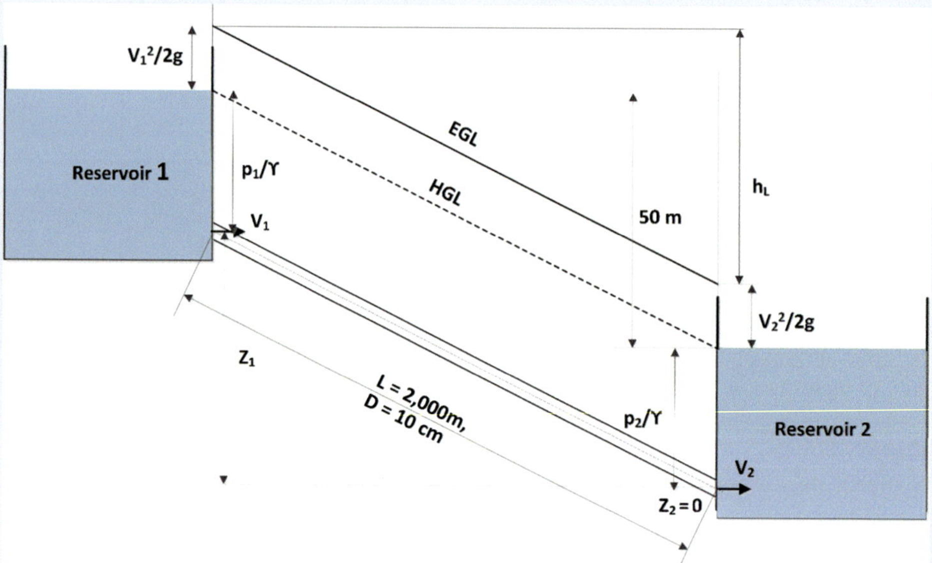

FIGURE 7.8 Two reservoirs.

The energy equation between the reservoir surfaces is used to determine the head loss due to friction, which is then used to determine the velocity of flow in the pipeline.

From the drawing, the total energy at 1 and 2: $\frac{p_1}{\gamma}+\frac{v_1^2}{2g}+z_1=\frac{p_2}{\gamma}+\frac{v_2^2}{2g}+z_2+h_L$

Minor losses at Entrance 1 and Exit 2 are ignored.

Apply continuity equation: Velocity of flow along the pipe is the same:

$\frac{p_1}{\gamma}=\frac{p_2}{\gamma}$ and $\frac{v_1^2}{2g}=\frac{v_2^2}{2g}$ and, therefore, they are canceled.

Head loss:

$$h_L = z_1 - z_2 = 50 \text{ m} \rightarrow h_L = f\frac{L}{D}\frac{V^2}{2g} = f\frac{2000}{0.1}\frac{V^2}{2\times 9.81} = 50$$

$$V = 0.2215/\sqrt{f}$$

The value of f cannot be solved directly; therefore, it is solved by trial and error: (1) assume f, (2) compute V, and (3) compute Re and repeat until the value of f or V at steps n and n + 1 is the same. Then calculation can be stopped.

Assume $f = 0.02$, $V = \frac{0.2215}{\sqrt{0.02}} = 1.566$ m/s

Compute $R_e = \frac{VD}{\nu} = \frac{1.566(0.1)}{1.02\times 10^{-6}} = 1.54\times 10^5$ from Moody diagram $f = 0.016$,

$V = \frac{0.2215}{\sqrt{0.016}} = 1.751$ m/s.

Repeat $R_e = \frac{VD}{\nu} = \frac{1.751(0.1)}{1.02\times 10^{-6}} = 1.72\times 10^5$ from Moody diagram f = 0.016 close enough.

Thus, V = 1.751 m/s. Stopped.

Discharge $Q = AV = \frac{\pi(0.1)^2}{4}(1.751) = 0.0137 \text{ m}^3/\text{s}$.

7.4.3.2 Hydraulic Computation of Pipe Network

Hydraulic computation of pipe network in this section employs only a very popular and easy method, which is the Hardy cross method. The theory of this pipe network calculation is described in the following.

Water supply distribution systems in municipal districts are usually constructed of a large number of pipes interconnected to form loops and branches. Calculation of flow in a pipe network involves a large number of pipes and may become tedious work. The solution scenario is based on the same principles that govern flow in pipelines and branching pipes.

At this state, a series of simultaneous equations must be satisfied:

At any junction:

$$Q_{in} = Q_{out} \quad \text{Kirchhoff's rule} \tag{7.15}$$

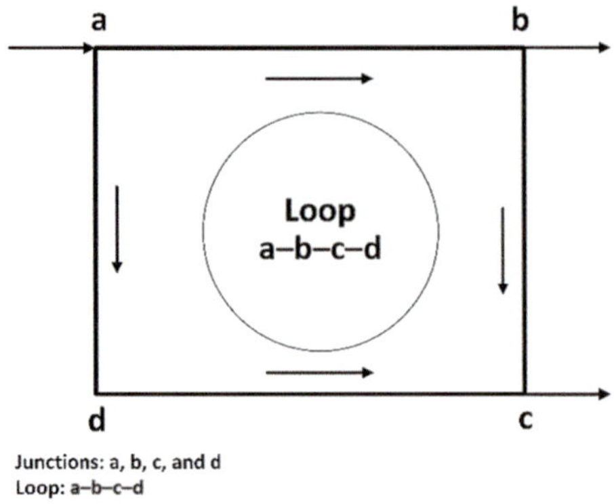

FIGURE 7.9 Junction and loop.

or

$$\Sigma Q = 0 \tag{7.16}$$

Between any two junctions, the head loss is independent of the path taken. For the above example (Figure 7.9), the head loss path a–b–c is equal to the head loss path a–d–c.

The Hardy cross method utilizes successive flow approximations based on two conditions stated previously for each junction and loop in the network. In loop **a–b–c–d** (Figure 7.9), the arrowheads indicate the presumed flow direction, and the loop must satisfy the conditions of:

- At each junction a, b, c, and d, the total inflow must be equal to the total outflow (Equation (7.15)).
- The head loss from flow in the counterclockwise direction a–b–c along pipe ab and bc must equal the head loss in the clockwise direction a–d–c along pipe ad and dc.

To start the process, we estimate the flow distribution in each pipe so that the total inflow matches the total outflow at every junction in the network. In a network with "n" junctions, we can establish "(n – 1)" junction equations to determine the system's flow rates. After establishing flows for the first "(n – 1)" junctions, the flows to and from the final junction become fixed.

We need the estimated flows, pipe diameters, lengths, friction coefficients, and other network data such as connectivity and junction elevations. After that, we compute the head loss for the network based on the estimated flow rates in all pipes. We then adjust the estimated pipe flows until the head losses in each loop are balanced, that is, the head losses in the clockwise direction are equal to those in the counterclockwise direction.

Knowing the diameter, length, and roughness of a pipe, the head loss is a function of Q. Applying Darcy–Weisbach equation:

$$h_f = f\left(\frac{L}{D}\right)\frac{V^2}{2g} = \left[f\left(\frac{L}{D}\right)\frac{1}{2gA^2}\right]Q^2 = KQ^2 \tag{7.17}$$

In any network loop, the total head loss in the *clockwise direction* (subscript *c*) is the sum of the head loss in all pipes that carry flow in the clockwise direction around the loop:

$$\Sigma h_{fc} = \Sigma K_c Q_c^2 \tag{7.18}$$

Similarly, the head loss in the *counterclockwise direction* (subscript *cc*):

$$\Sigma h_{fcc} = \Sigma K_{cc} Q_{cc}^2 \tag{7.19}$$

By using the assumed Q, we do not expect that these two values (Equations (7.18) and (7.19)) will be equal during the first trial. The difference is therefore:

$$\Sigma K_c Q_c^{\ 2} - \Sigma K_{cc} Q_{cc}^{\ 2} = \text{closure error} = \Delta Q \tag{7.20}$$

The error is then distributed into both flows in clockwise and counterclockwise directions. While the surplus (the larger) flow is deducted by the closure error, the deficit (the smaller) flow is added by closure error, as demonstrated by Equation (7.21). In this case, the clockwise flow is the larger one; therefore, the flow is deducted by closure error.

$$\Sigma K_c \left(Q_c - \Delta Q\right)^2 = \Sigma K_{cc} \left(Q_{cc} + \Delta Q\right)^2 \tag{7.21}$$

Expanding the terms on both sides:

$$\Sigma K_c \left(Q_c^{\ 2} - 2Q_c \Delta Q + \Delta Q^2\right) = \Sigma K_{cc} \left(Q_{cc}^{\ 2} + 2Q_{cc} \Delta Q + \Delta Q^2\right) \tag{7.22}$$

Assume that the correction term ΔQ^2 is small compared with Q_c and Q_{cc}; thus, we can ignore this term, and Equation (7.22) reduces to:

$$\Sigma K_c \left(Q_c^{\ 2} - 2Q_c \Delta Q\right) = \Sigma K_{cc} \left(Q_{cc}^{\ 2} + 2Q_{cc} \Delta Q\right) \tag{7.23}$$

Simplify, we get:

$$\Delta Q = \frac{\Sigma K_c Q_c^{\ 2} - \Sigma K_{cc} Q_{cc}^{\ 2}}{2\left(\Sigma K_c Q_c + \Sigma K_{cc} Q_{cc}\right)} \tag{7.24}$$

By substituting Equations (7.17) through (7.19) to Equation (7.24), the error can be expressed in terms of head loss:

$$\Delta Q = \frac{\left(\Sigma h_{fc} - \Sigma h_{fcc}\right)}{2\left(\Sigma \frac{h_{fc}}{Q_c} + \Sigma \frac{h_{fcc}}{Q_{cc}}\right)} \tag{7.25}$$

EXAMPLE 7.3

The water supply distribution system has pipe diameters and lengths specified inFigure 7.10, with a flow rate Q measured in liters per second (l/s). The water surface elevation in the storage tank is 50 m above A. Elevation at all junctions is the same as A. All pipes are made of ductile iron with $e = 0.26$ mm. Determine the flow rate in each pipe, and check whether the pressure at junction F satisfies the requirement of 185 kPa.

Solution:

The layout of the distribution system is shown in Figure 7.10. The assumed flow rate and direction in each pipe are also shown. This assumption will change in the final results. For this assumed flow rate, $Q_{in} = Q_{out}$ must be ensured and checked. *Note:* (Q = 200) is the assumed flow rate and flow direction.

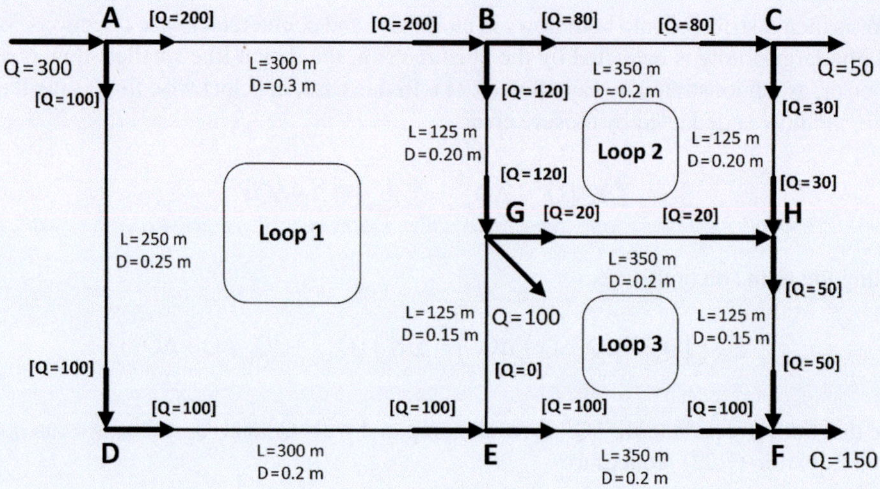

FIGURE 7.10 Pipe distribution system layout.

The assumed flow rate and direction, pipe length and diameter, e and e/D of each pipe are given in Table 7.11.

TABLE 7.11
Information on Pipe Distribution System

Pipe	Assumed Flow rate (m^3/s)	Pipe Length (m)	Pipe Diameter (m)	Rel. Roughness e/D	Friction factor, f	K (s^2/m^5)
AB	0.20	300	0.30	0.00087	0.019	194
AD	0.10	250	0.25	0.00104	0.020	423
BC	0.08	350	0.20	0.00130	0.021	1,900
BG	0.12	125	0.20	0.00130	0.021	678
GH	0.02	350	0.20	0.00130	0.021	1,900
CH	0.03	125	0.20	0.00130	0.021	678
DE	0.10	300	0.20	0.00130	0.021	1,630
GE	0.00	125	0.15	0.00173	0.022	2,990
EF	0.10	350	0.20	0.00130	0.021	1,900
HF	0.05	125	0.15	0.00173	0.022	2,990

Note: $K = f\left(\frac{L}{D}\right)\frac{V^2}{2g} = \left[f\left(\frac{L}{D}\right)\frac{1}{2gA^2}\right]$

Start with loop 1:

- Clockwise direction flows are at pipes: AB–BG–GE.
- Counterclockwise direction flows are at pipes: AD–DE. The numbers are put in parentheses.

Enter the results of the calculation in tabular format (Table 7.12).

TABLE 7.12
Iteration 1, Loop 1

Loop No.	Pipe	Q (m^3/s)	K (s^2/m^5)	h_f (m)	h_f/Q (s/m^2)	New Q
1	2	3	4	5	6	7
	AB	0.200	194	7.76	38.8	0.205
1	BG	0.120	678	9.76	81.3	0.125
	GE	0.000	2,990	0.00	0.0	0.005
	AD	(0.100)	423	(4.23)	(42.3)	(0.095)
	DE	(0.100)	1,630	(16.30)	(163.0)	(0.095)

Note to Table 7.12.

- The flows listed in Column 3 are original estimates.
- Counterclockwise flows in loop 1 are placed in parentheses.
- (Column 4): K is the original estimate based on Q (Column 3).
- (Column 5): $h_f = KQ^2$.
- (Column 6): h_f/Q .
- Compute $\Delta Q = \dfrac{\left(\sum h_{fc} - \sum h_{fcc}\right)}{2\left(\sum \dfrac{h_{fc}}{Q_c} + \sum \dfrac{h_{fcc}}{Q_{cc}}\right)} = \dfrac{(7.76 + 9.76) - (4.23 + 16.3)}{2\left[(38.8 + 81.3) + (42.3 + 163)\right]} = -0.005\ m^3/s$.
- The negative sign on the flow $\Delta Q = -0.005\ m^3/s$ indicates that counterclockwise head losses dominate ($\sum h_{fcc} > \sum h_{fc}$). Therefore, the flow correction of –0.005 m^3/s is applied in the clockwise direction flows, and –0.005 m^3/s is applied in the counterclockwise direction (Column 7, New Q). Therefore, all discharges in a counterclockwise direction should be deducted by 0.005 m^3/s. It will help equalize the losses in the next iteration.

Iteration 1, loop 2 (Table 7.13):

- Clockwise direction flows are at pipes: BC–CH.
- Counterclockwise direction flows are at pipes: BG–GH. The numbers are put in parentheses.

As pipe BG is shared by loops 1 and 2, the revised Q from the calculation in loop 1 is used here. In loop 1, the flow in BG is clockwise; in loop 2, the flow is counterclockwise. The flow correction ΔQ is:

$$\Delta Q = \frac{\left(\sum h_{fc} - \sum h_{fcc}\right)}{2\left(\sum \frac{h_{fc}}{Q_c} + \sum \frac{h_{fcc}}{Q_{cc}}\right)} = \frac{(12.2 + 0.61) - (10.61 + 0.76)}{2\left[(152.2 + 20.3) + (84.8 + 38.0)\right]} = +0.002\ m^3/s$$

TABLE 7.13
Iteration 1, Loop 2

Loop No.	Pipe	Q (m^3/s)	K (s^2/m^5)	h_f (m)	h_f/Q (s/m^2)	New Q
1	2	3	4	5	6	7
	BC	0.080	1,900	12.2	152.5	0.078
2	CH	0.030	678	0.61	20.3	0.028
	BG	(0.125)	678	(10.6)	(84.8)	(0.127)
	GH	(0.020)	1,900	(0.76)	(38.0)	(0.022)

Note to Table 7.13.

TABLE 7.14
Iteration 1, Loop 3

Loop No.	Pipe	Q (m^3/s)	K (s^2/m^5)	h_f (m)	h_f/Q (s/m^2)	New Q
1	2	3	4	5	6	7
	GH	0.022	1,900	0.92	41.8	0.035
3	HF	0.050	2,990	7.48	149.6	0.063
	GE	(0.005)	2,990	(0.07)	(14.0)	0.008
	EF	(0.100)	1,900	(19.0)	(190.0)	(0.087)

Note to Table 7.14.

Clockwise flow dominates the head loss, so the correction of 0.002 m^3/s is added in the counterclockwise direction, and the correction of –0.002 m^3/s is applied to the counterclockwise direction (Column 7: New Q).

Iteration 1, Loop 3 (Table 7.14):

- Clockwise direction flows are at pipes: GH–HF.
- Counterclockwise direction flows are at pipes: GE–EF. The numbers are put in parentheses.

The flow correction at this step:

$$\Delta Q = \frac{\left(\Sigma h_{fc} - \Sigma h_{fcc}\right)}{2\left(\Sigma \dfrac{h_{fc}}{Q_c} + \Sigma \dfrac{h_{fcc}}{Q_{cc}}\right)} = \frac{(0.92 + 7.48) - (0.07 + 19.0)}{2\left[(41.8 + 149.6) + (14.0 + 190.0)\right]} = -0.013 \text{ m}^3/\text{s}$$

Counterclockwise flow dominates the head loss. Therefore, the correction of 0.013 m^3/s is added in the clockwise direction, and the correction of –0.013 m^3/s is applied to the counterclockwise direction (Column 7: New Q). For pipe GE, the new Q is changing the direction as 0.005–0.013 m^3/s = –0.008 m^3/s (in the opposite direction with the previous step).

As the last error $\Delta Q = -0.013 \text{ m}^3/\text{s}$ is considered as high, the iterative process continues with the second iteration.

TABLE 7.15
Iteration 2, Loop 1

Loop No.	Pipe	Q (m³/s)	K (s²/m⁵)	h_f (m)	h_f/Q (s/m²)	New Q
1	2	3	4	5	6	7
	AB	0.205	194	8.15	39.8	0.205
1	BG	0.127	678	10.9	85.8	0.127
	AD	(0.095)	423	(3.82)	(40.2)	(0.095)
	DE	(0.095)	1,630	(14.7)	(154.7)	(0.095)
	EG	(0.008)	2,990	(0.19)	(23.8)	(0.008)

Note to Table 7.15.

TABLE 7.16
Iteration 2, Loop 2

Loop No.	Pipe	Q (m³/s)	K (s²/m⁵)	h_f (m)	h_f/Q (s/m²)	New Q
1	2	3	4	5	6	7
	BC	0.078	1,900	11.6	148.7	0.080
2	CH	0.028	678	0.53	18.9	0.030
	BG	(0.127)	678	(10.9)	(85.8)	(0.125)
	GH	(0.035)	1,900	(2.33)	(66.6))	(0.033)

Iteration 2, Loop 1 (Table 7.15):
Based on the previous result, the flow direction changes:

- Clockwise direction flows are at pipes: AB–BG.
- Counterclockwise direction flows are at pipes: AD–DE–EG. The numbers are put in parentheses, as usual.

The flow correction:

$$\Delta Q = \frac{\left(\Sigma h_{fc} - \Sigma h_{fcc}\right)}{2\left(\Sigma \dfrac{h_{fc}}{Q_c} + \Sigma \dfrac{h_{fcc}}{Q_{cc}}\right)} = \frac{(8.15 + 10.9) - (3.82 + 14.7 + 0.19)}{2\left[(39.8 + 85.8) + (40.2 + 154.7 + 23.8)\right]}$$
$$= +0.000 \text{ m}^3/\text{s}$$

The correction is very small (<0.0005 m³/s). However, as this is not the last loop, the iterative process continues to loop 2.

Iteration 2, Loop 2 (Table 7.16):
The flow direction:

- Clockwise direction flows are at pipes: BC–CH.

TABLE 7.17
Iteration 2, Loop 3

Loop No.	Pipe	Q (m^3/s)	K (s^2/m^5)	h_f (m)	h_f/Q (s/m^2)	New Q
1	2	3	4	5	6	7
	EG	0.008	2,990	0.19	23.8	0.008
3	GH	0.033	1,900	2.07	62.7	0.033
	HF	0.063	2,990	11.9	188.9	0.063
	EF	(0.087)	1,900	(14.4)	(165.5)	(0.087)

Note to Table 7.17.

- Counterclockwise direction flows are at pipes: BG–GH. The numbers are put in parentheses, as usual.

The flow correction:

$$\Delta Q = \frac{\left(\Sigma h_{fc} - \Sigma h_{fcc}\right)}{2\left(\Sigma \frac{h_{fc}}{Q_c} + \Sigma \frac{h_{fcc}}{Q_{cc}}\right)} = \frac{(11.6+0.53)-(10.9+2.33)}{2\left[(148.7+18.9)+(85.8+66.6)\right]} = -0.0002\ m^3/s$$

The process continues to loop 3.

Iteration 2, Loop 3 (Table 7.17):
The flow direction:

- Clockwise direction flows are at pipes: EG–GH–HF.
- Counterclockwise direction flows are at pipes: EF. The numbers are put in parentheses, as usual.

The flow correction:

$$\Delta Q = \frac{\left(\Sigma h_{fc} - \Sigma h_{fcc}\right)}{2\left(\Sigma \frac{h_{fc}}{Q_c} + \Sigma \frac{h_{fcc}}{Q_{cc}}\right)} = \frac{(2.07+11.9+0.19)-(14.4)}{2\left[(62.7+188.9+23.8)+(165.5)\right]} = -0.000\ m^3/s$$

As the correction is very small on all three loops, the flow rates are accepted, and the process ends. The final flow rate is shown in Table 7.18.
The final flow of the network is shown in Figure 7.11.
Note to Table 7.17:

- The head loss is calculated by:

$$h_f = f\left(\frac{L}{D}\right)\frac{V^2}{2g} = \left[f\left(\frac{L}{D}\right)\frac{Q^2}{2gA^2}\right]$$

TABLE 7.18
Final Flow Rate

Pipe	Flow rate, Q (L/s)	Length (m)	Diameter (cm)	Head loss h_f (m)	Pressure difference, ΔP (kPa)
AB	205	300	30	8.2	80.3
AD	95	250	25	3.8	37.2
BC	80	350	20	12.2	119.4
BG	125	125	20	10.6	103.8
GH	33	350	20	2.1	20.6
CH	30	125	20	0.6	5.9
DE	95	300	20	14.7	143.9
EG	8	125	15	0.2	2.0
EF	87	350	20	14.4	141.0
HF	63	125	15	11.9	116.5

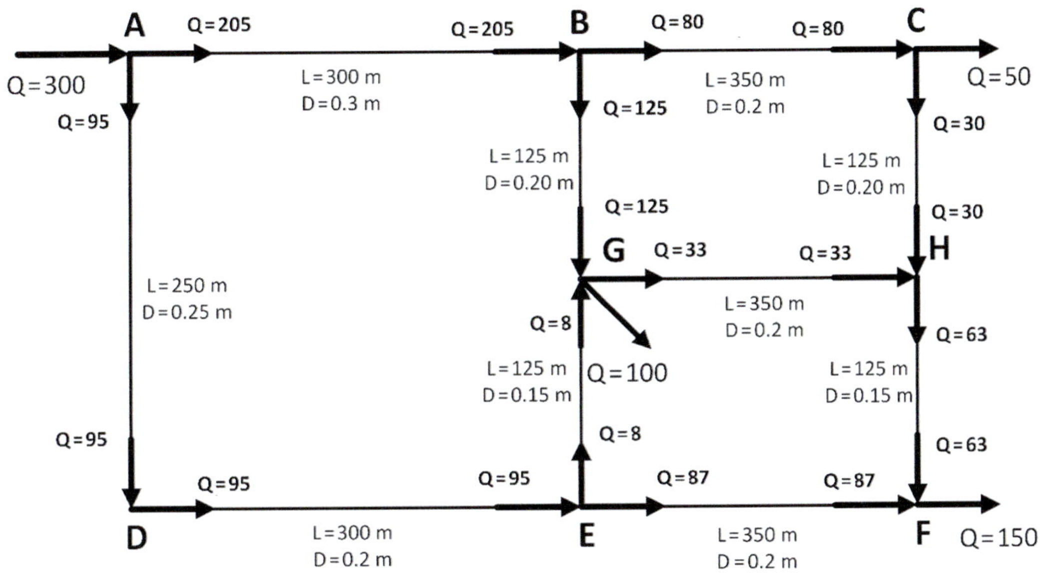

FIGURE 7.11 The final flow of the network.

- The head loss h_f is converted to a pressure drop in the last column ($\Delta P = \gamma h_f$) where $\gamma = 9790\ \text{N/m}^3$.
- Now, we determine whether the pressure at F is large enough to satisfy the customer at that location.
- Using energy balance, the water surface elevation in the tank is 50 m above junction A. It equals $P_A = \gamma h = \left(9790 \dfrac{\text{N}}{\text{m}^3}\right)(50\ \text{m}) = 489.5\ \text{kPa}$.
- To arrive at the pressure at F, given pressure at A, we can traverse through different paths: A–B–C–C–F, A–B–G–H–F, or A–D–E–F. The shortest route is A–D–E–F. Using this route, the pressure at F: $P_F = P_A - \Delta P_{AD} - \Delta P_{DE} - \Delta P_{EF} = 489.5 - 37.2 - 143.9 - 141 = 167.4\ \text{kPa} < 185\ \text{kPa}$. This

TABLE 7.19
Results by EPANET

Pipe	Length (m)	Diameter (mm)	Flowrate (l/s)	Velocity (m/s)	Head Loss (m/km)	Friction Factor
AB	300	300	205.03	2.90	37.14	0.026
BC	350	200	79.75	2.54	48.00	0.029
BG	125	200	125.28	3.99	118.02	0.029
GE	125	150	7.63	0.43	2.10	0.033
EF	350	200	87.35	2.78	57.53	0.029
DE	300	200	94.97	3.02	67.97	0.029
AD	250	250	94.97	1.93	20.96	0.027
GH	350	200	32.90	1.05	8.28	0.030
CH	125	200	29.75	0.95	6.78	0.030
HF	125	150	62.65	3.55	135.81	0.032

pressure is below the pressure required by the customers. Therefore, the pressure must be increased, at least, by 185 – 167.4 kPa =17.6 kPa. This is equivalent to the increase of the surface water at the water tank of at least: $50\text{ m} + \dfrac{17.6\text{ kPa}}{9.79\text{ kN/m}^3} = 52.0\,\text{m}$.

Hydraulic computation using EPANET 2.2:
The above manual calculation is compared with the hydraulic computation using EPANET. EPANET does not really fit into the information available in the problem.

Input to the EPANET system:

- Set the unit into SI unit by changing the flow unit into liter per second (LPS) in the Project > Analysis Option > LPS.
- When the flow unit is in LPS, the other unit will change, by default, to metric units. For example, the diameter pipe is in millimeters, the length of the pipe is in meters, the elevation of the junction is in meters, the hydraulic head is in meters, and the pressure is in meters.
- Also important to note is to set the head loss formula into Darcy–Weisbach. There are three options in the EPANET: Hazen–William (H-W), Darcy–Weisbach (D-W), and Chezy–Manning (C-M). To set the head loss into D-W: Project > Analysis Option > Head Loss Formula.
- Place all the junctions into the editor, and put the coordinates of each junction. Place the pipe link into all junctions and enter length (in meters) and diameter (in millimeters).
- There must be a tank or a reservoir at junction A with a 50-m head (in this problem).

Run the analysis. Table 7.19 shows the results of EPANET.

The analysis using EPANET (Figure 7.12) and manual calculation using the Hardy cross method (Figure 7.11) gave the same results.

While manual calculation took 4 h to complete (from the preparation until getting the final result), using EPANET took an hour to complete. However, the process of calculating itself by EPANET took a few seconds.

7.5 WATER TREATMENT SYSTEM

Water treatment systems represent the cornerstone of public health infrastructure, serving as the frontline defense against waterborne diseases and contaminants. These systems are intricate and

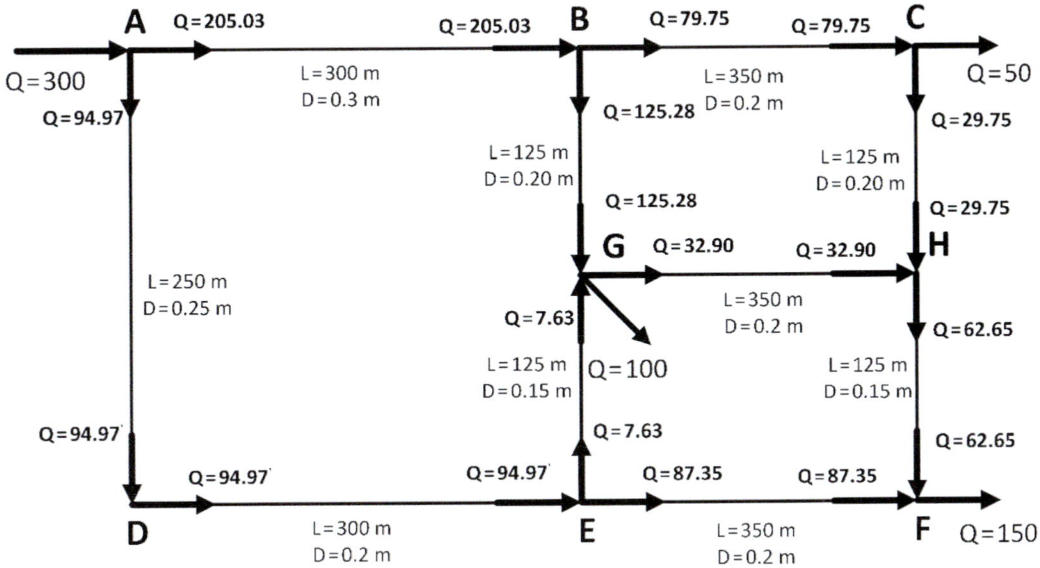

FIGURE 7.12 Results of EPANET computation.

multifaceted, encompassing a series of interdependent processes aimed at purifying raw water from natural sources such as rivers, lakes, or groundwater. By employing a combination of physical, chemical, and biological mechanisms, water treatment facilities ensure that the water supplied to communities meets stringent quality standards, thus safeguarding public health and well-being. The treatment process of the drinking water is commonly following these phases:

1. *Raw Water Intake:* Raw water is collected from natural sources such as rivers, lakes, or groundwater. Intake structures and pipelines transport the raw water to the treatment facility.
2. *Screening:* Large debris and objects such as leaves, twigs, and aquatic organisms are removed from the raw water through screening processes. Screens or grates prevent these materials from entering the treatment system, thus protecting downstream equipment.
3. *Pretreatment:* Pretreatment processes may include aeration and pre-chlorination to address specific water quality issues. Aeration improves taste and odor by removing dissolved gases, while pre-chlorination helps control algae and bacterial growth.
4. *Coagulation and Flocculation:* Chemical coagulants such as aluminum sulfate (alum) or ferric chloride are added to the water to destabilize suspended particles and colloids. Flocculation promotes the aggregation of these particles into larger, denser flocs through gentle mixing.
5. *Sedimentation:* The water enters sedimentation basins, where the flow velocity is reduced, allowing suspended particles and flocs to settle out of the water under gravity. Clarified water rises to the surface, while settled particles form a sludge layer at the bottom of the basin.
6. *Filtration:* The clarified water undergoes filtration to remove remaining suspended particles and microorganisms. Various filtration media such as sand, gravel, and activated carbon trap impurities as the water passes through.
7. *Disinfection:* Disinfection processes such as chlorination, ultraviolet (UV) irradiation, or ozonation are employed to kill or deactivate pathogenic microorganisms. This step ensures that the treated water is safe for consumption and meets regulatory standards for microbial quality.
8. *pH Adjustment and Chemical Addition:* pH adjustment may be necessary to achieve the desired pH level for corrosion control and taste improvement. Additional chemicals such as fluoride may be added for dental health benefits, while corrosion inhibitors protect distribution system pipes.

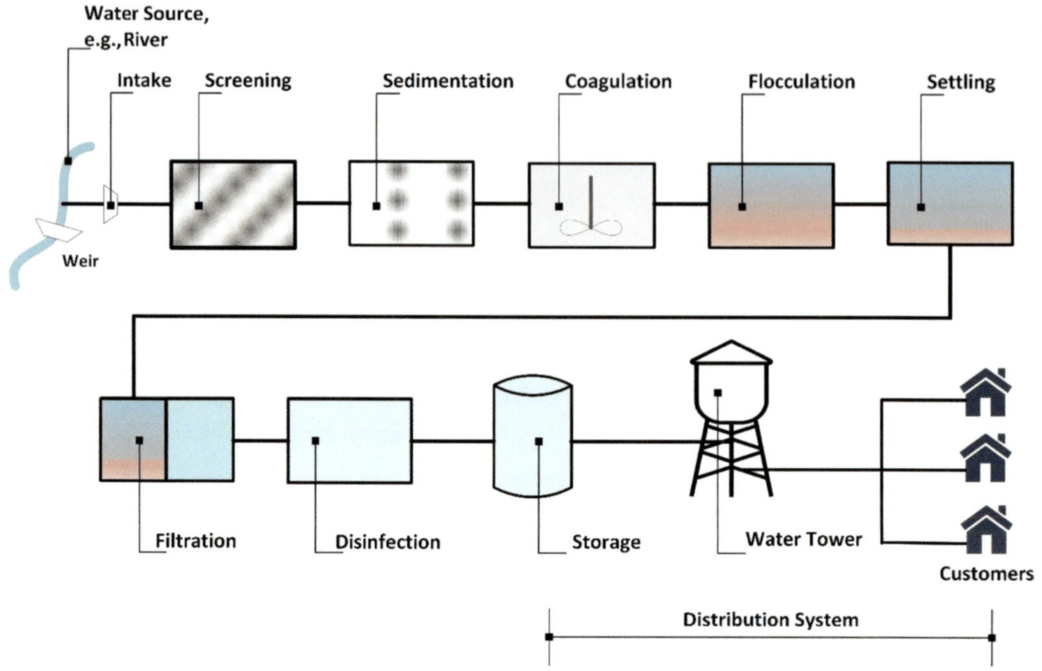

FIGURE 7.13 Schematic diagram of the water treatment plant.

9. *Storage:* Treated water is stored in reservoirs or tanks to provide a buffer against fluctuations in demand and to ensure a reliable water supply. Storage also allows for the mixing and blending of treated water from multiple sources to maintain consistent water quality.
10. *Distribution:* Treated water is distributed through a network of pipelines to consumers, including homes, businesses, and industrial facilities. Distribution systems may include pumps, valves, and storage tanks to regulate flow and pressure and to maintain water quality during transport.

These 10 stages collectively ensure that raw water from natural sources is transformed into clean, safe drinking water that meets regulatory standards and is readily available to communities for consumption and various other purposes.

The stages are schematically shown in Figure 7.13, and the following sections discuss the process.

7.5.1 Raw Water Sources

Raw water sources for drinking water treatment can be drawn from various natural sources, including rivers, lakes, or groundwater aquifers. Each of these sources presents unique challenges and considerations in terms of water quality and treatment requirements.

Rivers: Rivers are dynamic bodies of water that can vary significantly in flow rate, sediment load, and pollutant levels depending on factors such as weather patterns, land use activities, and upstream inputs. The quality of river water can be influenced by industrial discharges, agricultural runoff, urban pollution, and natural processes. As a result, river water may contain suspended solids, organic matter, nutrients, pathogens, and various contaminants that need to be addressed during the treatment process.

Lakes: Lakes are large bodies of standing water that can serve as important sources of drinking water for communities. Unlike rivers, lakes are generally less influenced by fluctuations in flow and can provide a more stable water supply. However, lakes may still be subject to contamination from sources such as stormwater runoff, shoreline development, and recreational activities. The quality of lake water can vary depending on factors such as nutrient levels, algal blooms, sediment accumulation, and aquatic ecosystems.

Groundwater: Groundwater is water that is stored beneath the earth's surface in aquifers, which are layers of permeable rock or sediment. Groundwater is often considered a relatively clean and reliable source of drinking water, as it is naturally filtered as it percolates through the soil and rock layers. However, groundwater can still be susceptible to contamination from sources such as leaking underground storage tanks, septic systems, agricultural chemicals, and industrial pollutants. Additionally, groundwater quality can vary depending on factors such as geology, hydrology, and land use practices.

The quality of water at these raw water sources directly impacts the complexity and cost of the treatment process. Higher levels of contamination or pollutants in the raw water may require more extensive treatment processes, additional treatment steps, or specialized treatment technologies to meet regulatory standards and produce safe drinking water. Consequently, the costs associated with water treatment can vary depending on the source water quality and the specific treatment requirements.

In many cases, the intake of raw water from rivers, lakes, or groundwater sources may require the construction of intake structures to capture and convey water to the treatment facility. Intake structures are designed to regulate the flow of water, minimize the intake of sediment and debris, and protect aquatic habitats. Depending on the location and characteristics of the raw water source, intake structures may include intake pipes, screens, pumps, diversion channels, or dam structures.

The quality of raw water sources is a critical factor in the provision of safe and reliable drinking water. Raw water sources, such as rivers, lakes, and groundwater, can vary significantly in their quality due to natural processes and human activities. The quality of raw water as the source of water supply determines the effectiveness and complexity of the treatment processes required and ensures the water is economically viable for further treatment processes and safe for consumption. The presence of contaminants such as bacteria, viruses, heavy metals, and organic compounds in the raw water can necessitate more advanced and costly treatment methods. Furthermore, the quality of raw water can impact the taste, odor, and color of the treated water. Therefore, protecting the quality of raw water sources is crucial for maintaining the efficiency, cost-effectiveness, and reliability of drinking water supply systems. It also underscores the importance of sustainable water management practices and policies.

Some factors may influence the quality of raw water sources. These factors include the following:

1. *Natural Processes:* Natural processes can influence the quality of raw water sources. For example, weather patterns can affect the amount of rainfall, which can influence the flow rates and turbidity of surface water sources. Geological formations can also impact water quality. For instance, limestone can contribute to water hardness, while volcanic rocks can increase the concentration of certain minerals.
2. *Human Activities:* Human activities can also significantly impact the quality of raw water sources. Industrial discharges, agricultural runoff, and domestic wastewater can introduce a range of pollutants into water sources. These can include nutrients (such as nitrogen and phosphorus), pathogens, heavy metals, and synthetic organic compounds.

Conservation of fresh water sources in the environment, monitoring, and assessment of raw water quality are essential components of drinking water supply systems. Regular testing can help identify potential issues and inform treatment processes. Key parameters that are often monitored include pH, temperature, turbidity, conductivity, and concentrations of various chemical constituents.

As there are no natural water sources, in which the quality meets the safe standard of drinking water, that is, without the necessity of further treatment, treatment processes will always be required. Treatment processes are designed to improve the quality of raw water to meet drinking water standards. These can include coagulation, flocculation, sedimentation, filtration, and disinfection. The specific processes used can depend on the quality of the raw water and the contaminants present.

Depending on the situation, the intake of water sources for a water supply system commonly involves several facilities and structures designed to ensure the efficient and safe collection of water. These facilities and structures are crucial in maintaining the continuous supply and quality of the water and preventing any potential contamination.

1. *Intake Structure and Weir (Figure 7.14):* This structure houses and supports the intake conduit, valves, and pumps. It permits the withdrawal of water from the source. The design of the intake structure and weir should consider factors such as external forces caused by waves and currents, the impact of floating and submerged bodies, and ice pressures.
2. *Intake Conduit:* This is a pipe or open channel that carries water from the source to the treatment facility. The inlet end of the conduit is usually fitted with a screen to prevent the entry of large debris and aquatic life. A valve is also present to control the flow of water.

FIGURE 7.14 Intake structure.

Source: **Wikipedia.**

3. *Pumping Stations:* These are facilities that house pumps and equipment needed to transport the water from the source to the treatment plant or distribution system.
4. *Screen or Grating:* This is placed at the inlets of the intake structure to prevent the entrance of large and medium objects and fishes.
5. *Valves:* These are used to control the flow of water through the intake conduit.
6. *Wells:* In the case of groundwater sources, wells are constructed to extract water from the aquifer.

The design and construction of these facilities and structures depend on several factors, including the type of water source (e.g., lake, river, reservoir, or canal), the quality of the water, and the specific requirements of the water supply system. Regular maintenance and monitoring are also essential to ensure the continued efficiency and safety of these facilities.

Choosing the right intake structure for the water supply system depends on several factors:

1. *Type of Water Source:* The type of water source (e.g., lake, river, reservoir, or groundwater) influences the design of the intake structure.
2. *Water Availability:* The availability of water at the proposed location is a critical factor. Certainly, the intake structures will not be constructed at a point where no sources of water are available for the whole period of a year.
3. *Topography:* The topography of the site plays a vital role in the selection process. In most cases, the gravitational flow from intake structures to the water treatment plant is desirable.
4. *Geology:* The geological conditions at the site must be carefully evaluated to ensure the stability and longevity of the intake structure.
5. *Water Quality:* The quality of the water source can affect the design and location of the intake structure. This is to minimize the costs of water treatment.
6. *Quantity of Water Required:* The amount of water needed for the supply system can influence the size and design of the intake structure.
7. *Environmental Considerations:* The surrounding environment, including potential impacts on local ecosystems and navigation channels, should be considered.
8. *Future Demand:* The intake should be in a place where it can draw more water in the future if demand increases.
9. *Proximity to Treatment Plant:* The water intake should be as close as possible to the treatment plant to reduce the cost and complexity of conveying the water.

7.5.2 Screening

A screen or trashrack is an essential component of the intake structure in a water supply system. It is typically made of metals and is often supported by masonry. The primary function of a trashrack is to prevent waterborne debris, such as logs, boats, animals, and masses of cut waterweed, from entering the intake of a water mill, pumping station, or water conveyance. This helps to protect the mechanical equipment inside the intake from damage and ensures that the water supply remains uncontaminated and free from large debris.

In terms of design, an intake should be robust enough to withstand external forces like heavy waves, currents, and the impact of floating or submerged bodies. It should also have sufficient self-weight to prevent it from being washed away by the current. To avoid the entrance of large and medium objects and fishes, screens are provided on the inlets. The inlets should be of sufficient size to allow the required quantity of water to enter and be positioned in such a way that they can admit water in all seasons near the surface where the quality of water is good.

The difference between a screen and a trashrack in a water supply system lies primarily in their design and the size of debris they are intended to filter out.

TrashRacks:

1. Trashracks, also known as bar screens, are heavy-duty water screen systems designed to prevent large and potentially harmful debris such as trees, branches, and trash from entering a water intake system.
2. They have larger openings, typically ranging from 36 to 144 mm (1.5 to 6 inches), to stop big objects while maintaining the required flow rate.

Screens:

1. Screens are used to remove smaller objects that could damage influent pumps or block flow in sewage channels and piping systems.
2. They can be classified into coarse and fine screens, with coarse screens having openings from 6 to 36 mm (0.25 to 1.5 inches) to remove items like rags, sticks, and other debris.

Common materials used for screens and trashracks (Figure 7.15) in water supply systems include aluminum, which is known for its strength and cost-effectiveness. Aluminum is a popular choice for stormwater debris screens. It is noncorrosive, which is essential for the longevity of the screens in a water environment. Another material is stainless steel. This material is also noncorrosive and is used for both screens and trashracks. Stainless steel is durable and can withstand the harsh conditions

FIGURE 7.15 Trashracks.

Source: **Sandro Senn – Wikipedia.**

within waterway systems. The screens or trashracks can also be made from welded bar stock. For trashracks, which require robust construction to handle larger debris, welded bar stock is a common material. It is strong enough to resist the force of debris in the applications where the trashrack is placed. These materials are selected for their durability, resistance to corrosion, and ability to withstand the physical demands of filtering debris from water systems.

Maintaining trashrack or screen is crucial for ensuring they function effectively. These activities can be done to maintain the function of trashracks or screens:

1. *Regular Inspection:* Check the trashrack or screen periodically for any signs of damage, corrosion, or clogging,
2. *Cleaning:* Remove debris accumulation regularly. The frequency of cleaning depends on the rate at which trash accumulates. Ensure that no more than 33% of the trashrack space is clogged at any given time.
3. *Deep Cleaning:* Conduct an annual deep clean, paying close attention to bar spacing and removing build-ups of silt and algae that can reduce flow through the trashrack.
4. *Repair and Replacement:* Repair or replace rusted, broken, or worn sections as needed to maintain structural integrity.
5. *Storm Checks:* After storm events, check the trashrack or screen more frequently to ensure they are functioning properly and to clear any accumulated debris.

A specific maintenance requirement can vary based on the design and material of your trashrack or screen, as well as the environmental conditions they are exposed to. It is always a good idea to refer to the manufacturer's guidelines or consult with a professional for a maintenance plan tailored to a particular system.

7.5.3 Pretreatment Process

The pretreatment process in a water supply system is a series of procedures designed to prepare raw water for further treatment and use. It involves removing contaminants and conditioning the water so that it is suitable for the specific requirements of the subsequent treatment processes, such as reverse osmosis (RO) systems. Pretreatment methods can include screening, filtration, coagulation, flocculation, aeration, and chemical treatment. These steps help improve the efficiency and lifespan of the RO system by preventing fouling or damage to the membranes and ensuring that the water meets the necessary quality standards.

The pretreatment process involves several activities designed to prepare the water for further treatment and use. Here is a detailed explanation of the steps involved:

1. *Screening:* This is the first step where screens are used to remove large debris such as sticks, leaves, and other large particles from the water.
2. *Coagulation and Flocculation:* Chemicals with a positive charge are added to the water. The positive charge of these chemicals neutralizes the negative charge of dirt and other dissolved particles in the water. Once the charge is neutralized, the particles bind with the chemicals and form larger particles, known as floc.
3. *Sedimentation:* During sedimentation, floc settles to the bottom of the water supply, due to its weight. This settling process is called sedimentation.
4. *Filtration:* Once the floc has settled to the bottom, the clear water on top will pass through filters of varying compositions (sand, gravel, and charcoal) and pore sizes to remove dissolved particles, such as dust, parasites, bacteria, viruses, and chemicals.
5. *Aeration:* This step involves exposing the water to air, which can help to remove dissolved gases (like hydrogen sulfide) and volatile organic compounds. Aeration also helps in the oxidation of iron and manganese, making them easier to filter out.

6. *Chemical Addition:* Chemicals may be added to the water to adjust pH or to remove specific contaminants like iron, manganese, and hardness. This step is crucial for protecting the downstream processes, such as RO systems, from fouling or scaling.
7. *Pre-chlorination:* In some cases, chlorine may be added to the water at this stage to control algae and other biological growths in the tanks and pipelines.

Steps 1 to 4 are common in a pretreatment process. These steps are essential for protecting downstream processes and ensuring the water meets quality standards before further treatment. However, in some instances, the pretreatment process may also include steps 5 to 7. Each of these steps is collectively to ensure that the water is adequately prepared for the subsequent treatment processes, which may include RO, UV disinfection, or chlorination. The specific activities and their order can vary depending on the quality of the source water and the treatment goals.

The purpose of pretreatment in water supply systems is to prepare the water for further treatment processes, such as RO, by removing contaminants that could damage equipment or affect the efficiency of subsequent treatment stages. The pretreatment process is essential due to the following reasons:

1. *Protects RO Membranes:* Pretreatment extends the life of RO plant membranes by removing particles that could cause fouling.
2. *Ensures Water Quality:* It ensures that the water delivered to the RO or electrodeionization (EDI) systems meets the required quality standards.
3. *Continuous Supply:* Supports the delivery of a continuous and dependable water supply.
4. *Cost Efficiency:* Maximizes the overall water processing plant lifecycle, thus reducing costs.

In the water supply system, a pretreatment process is a common process for most freshwater sources available in nature, unless an extraordinary quality of fresh water is available, which is usually very rare. To determine if the water needs pretreatment, we should consider the following:

1. *Water Quality Testing:* Conduct a comprehensive water analysis to identify contaminants present in the water supply.
2. *System Requirements:* Check the specifications of your water treatment system, such as an RO system, which may require certain levels of water quality to operate efficiently.
3. *Regulatory Standards:* Ensure your water meets local and national drinking water standards or other regulatory requirements.
4. *Visual Inspection:* Look for signs of sediment, turbidity, or color changes in your water, which can indicate the presence of contaminants.
5. *Equipment Performance:* Monitor the performance of your water-using equipment; scaling, corrosion, or biofouling can signal the need for pretreatment.
6. *Professional Assessment:* Consult with a water treatment specialist who can provide a detailed assessment based on your water source and usage.

After the pretreatment process in a water supply system, the water undergoes further treatment to ensure it is safe for consumption and use. The further treatment typically includes the following steps:

1. *Filtration:* After sedimentation, the water passes through filters to remove any remaining particles and microorganisms.
2. *Disinfection:* Chemicals like chlorine or chloramine are added to kill any remaining bacteria, viruses, and other pathogens.
3. *pH Adjustment:* The pH of the water is adjusted to prevent corrosion in the distribution system and to improve the taste of the water.

4. *Fluoridation:* Some water treatment facilities add fluoride to the water to help reduce tooth decay.
5. *Additional Treatment:* Depending on the source water quality and the intended use of the water, additional treatment processes such as activated carbon adsorption, ion exchange, or advanced oxidation may be used.

These steps ensure that the water meets the required safety standards and is suitable for distribution to homes, businesses, and industries. A specific treatment process may be required. However, it may vary depending on local water quality and regulatory requirements.

7.5.4 Coagulation and Flocculation

Coagulation and flocculation are vital processes in water treatment due to their effectiveness in improving water quality. These two processes remove suspended solids, including organic and inorganic materials, from water, which significantly reduces turbidity. This is particularly important for wastewater treatment, where these processes can decrease organic loads and suspended solids by up to 90%. By aggregating fine particles into larger flocs, coagulation and flocculation simplify subsequent treatment stages, especially disinfection, making them more effective. Moreover, these processes are cost-effective and essential for producing treated water that meets health and safety standards. They serve as an indispensable first step in the water treatment sequence, setting the stage for filtration, disinfection, and other purification methods that follow.

Coagulation is the process where chemicals, known as coagulants, are added to water. These coagulants are usually organic or inorganic substances that carry a positive charge. When they are added to the water, they neutralize the negative charges of particles suspended in the water, such as dirt, bacteria, and other pathogens. This neutralization causes the particles to bind together into larger particles, called microflocs.

Flocculation follows coagulation and involves gentle mixing of the water to encourage the microflocs to come together to form larger, settleable particles known as flocs. During this stage, flocculant aids, which are long-chain polymers, may be added to enhance the process. The flocs grow in size and weight until they are large enough to settle out of the water by gravity in a sedimentation basin or be removed by filtration.

The main purpose of these processes is to treat turbid water, thus improving water clarity and quality. This is particularly important for wastewater treatment, where it can reduce organic loads and suspended solids by up to 90%. The removal of these particles simplifies later stages of water treatment, especially disinfection, making it more effective.

Some common coagulants used in water treatment include:

1. *Alum (Aluminum Sulfate, $Al_2(SO_4)_3$):* Alum is one of the most common coagulants used in water treatment. It is effective for a wide range of pH levels and is often used in conjunction with cationic polymers.
2. *Polyaluminum Chloride (PAC)* $[Al_2(OH)_n Cl_{6-n}]_m$: PAC is efficient in some waters, requiring less pH adjustment and producing less sludge compared to alum.
3. *Ferric Chloride ($FeCl_3$):* Ferric chloride can be more effective than alum in some applications.
4. *Ferric Sulfate ($Fe_2(SO_4)_3$):* This coagulant is effective in some waters and can be more economical in certain locations.
5. *Ferrous Sulfate ($FeSO_4$):* Although it may cost more compared to ferric chloride, the results can be better.
6. *Sodium Aluminate ($NaAlO_2$):* Sodium aluminate is another option used in water treatment processes.

7. *Cationic Polymers:* These can be used alone as the primary coagulant or in conjunction with aluminum or iron coagulants. They are effective over a wider pH range and produce less sludge, but they are more expensive compared to inorganic coagulants.

The choice of coagulant depends on the nature of the suspended solids to be removed, the raw water conditions, the facility design, and the cost of the chemical necessary to produce the desired result. It is important to conduct a jar test or pilot study with different coagulants to determine which one is most effective for your specific water treatment needs.

On the other hand, flocculants are substances that promote the clumping of particles, which helps in the separation of solids from liquids in water treatment processes. They are typically added to water after the coagulation process. The most common flocculants used in water treatment are polyacrylamides, which are synthetic polymers that have a high molecular weight.

There are three key classes of flocculants within the water treatment industry:

Powder Flocculants: Powder flocculants are long-chain molecules that produce highly charged ions when dissolved in water. They are used in water treatment to help remove suspended solids. By introducing these substances into a solution, particles are grouped together to form easily removable "flocs." This allows for easy extraction of harmful impurities like suspended solids, dissolved solids, organic and inorganic elements, and metals.

Emulsions: Emulsions in water treatment refer to the process where water forms a stable emulsion with another substance, such as oil. When this emulsion cannot be removed in conventional storage tanks, emulsion-treating methods must be used. These methods involve the use of coagulants and flocculants to promote the clumping of particles, making them easier to remove. Emulsions are particularly challenging to treat due to their stability and the small size of the dispersed particles.

Water-Based Dispersion Flocculants: Water-based dispersion flocculants are substances introduced into a solution to promote the formation of "flocs." These flocs are larger groupings of particles that can be easily removed, allowing for the extraction of harmful impurities like suspended solids, dissolved solids, organic and inorganic elements, and metals. This type of flocculant is particularly useful in enhancing the removal of suspended particles and impurities from water.

Each of these flocculants has its own advantages and disadvantages, and the choice of flocculant depends on the specific requirements of the water treatment process.

7.5.5 Sedimentation

Sedimentation is a physical water treatment process used to settle out colloidal and suspended solids in water under the influence of gravity. It is one of the widely used conventional water treatment methods. The process involves removing suspended particles from water by encouraging them to settle at the bottom of a tank, where they can be removed periodically.

The sedimentation process is essential in water treatment plants for several reasons:

1. *Particle Settling:* Sedimentation allows for the gravitational settling of suspended particles, as their density exceeds that of the surrounding liquid.
2. *Enhanced Particle Aggregation:* During sedimentation, coagulants are often added to facilitate particle aggregation or flocculation.
3. *Removal of Organic Matter:* Sedimentation also aids in reducing organic matter content present in raw water sources.

4. *Improving Water Quality and Clarity:* Both drinking water treatment and wastewater treatment plants use sedimentation to separate suspended particles from water, significantly improving water quality and clarity.
5. *Minimizing Chemical Use:* Sedimentation reduces the chemicals – and their associated costs – required at later treatment stages.

The predecessor steps to sedimentation usually include coagulation and flocculation. As it has been discussed earlier, coagulation involves adding a coagulant to the water to neutralize the charge of the particles, which allows them to stick together. Flocculation is the process where the particles form larger clumps, or flocs, making them easier to remove.

After sedimentation, the successor steps typically include filtration and disinfection. Filtration involves passing the water through a filter to remove any remaining particles. Disinfection is the final step, where chemicals (like chlorine) or other methods (like UV light) are used to kill any remaining bacteria or other microorganisms.

The sedimentation process is done in a particular tank. There are several types of sedimentation tanks used in water treatment plants, each with their own design and purpose. The types of sedimentation tanks are shown in Table 7.20.

As additional notes to primary and secondary sedimentation tanks. Primary and secondary sedimentation tanks serve different purposes in the water treatment process and are particularly used in the wastewater treatment process.

TABLE 7.20
Types of Sedimentation

Type	Explanation
Rectangular sedimentation tanks	These tanks are rectangular in shape and are widely used in water and wastewater treatment plants. They are built with either concrete or steel and have a series of baffles or plates installed in them to facilitate the settling of suspended solids.
Circular sedimentation tanks	This type of sedimentation tank is circular in shape and is typically used in larger treatment plants. They are more efficient in removing suspended solids compared to rectangular tanks due to their design, which promotes the formation of a circular flow pattern.
Inclined plate settlers	These tanks consist of a series of inclined plates installed at an angle of 45-60 degrees inside a rectangular tank. The plates are spaced a few centimeters apart to provide a large surface area for particles to settle.
Upflow sedimentation tanks	In this type of sedimentation tank, the influence flows in from the bottom and moves upward to the surface.
Hopper bottom tanks	These tanks are generally in circular shape and flow takes place in a vertical direction. A hopper bottom is provided at the bottom of the tank to dispose the collected sludge.
Primary sedimentation tanks	Primary sedimentation tanks, also known as primary clarifiers, are the first stage of treatment in a water or wastewater treatment plant.
Secondary sedimentation tanks	These tanks are used for further treatment after primary sedimentation.
Tertiary sedimentation tanks	These are used for additional treatment after secondary sedimentation.
Swimming pool sedimentation tanks	These are used specifically for treating water in swimming pools.
Hydrodynamic sedimentation tanks	This tank is designed to enhance the settling process by creating a flow pattern that helps to separate particles more efficiently.

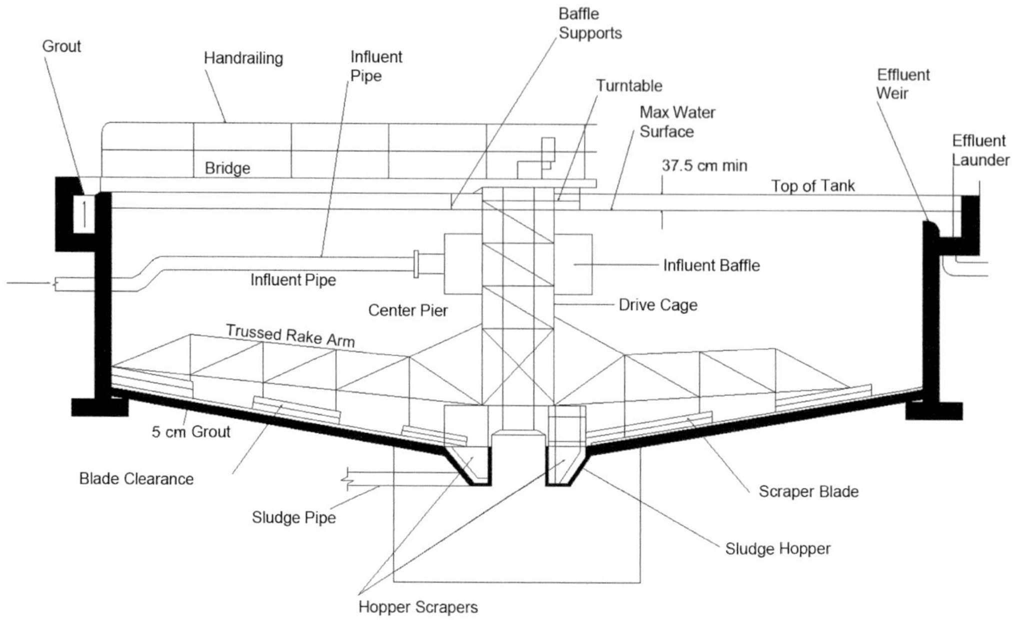

FIGURE 7.16 Sedimentation tank.

Primary Sedimentation Tanks: These tanks are the first step in the water treatment process, where they remove suspended solids, oil, and grease. The solids settle at the bottom of the tank and are collected by a rake and removed by a sludge removal system. Meanwhile, oil and grease float to the surface and are skimmed off. A typical primary clarifier removes 60% of suspended solids and 30%–40% of biological oxygen demand (BOD).

Secondary Sedimentation Tanks: These tanks follow the biological treatment process with the main goal of returning activated sludge. During this process, the biomass from microorganisms settles to the bottom in the form of activated sludge. After settling over a period of time, the biomass of microorganisms is returned to the aeration tank with the cycle repeating until the effluent is clean before being sent for filtration and/or disinfection. Waste sludge is removed and thickened before the digestion process.

Figure 7.16 illustrates a circular sedimentation tank used in urban water treatment processes. It clarifies water by allowing suspended particles to settle at the bottom, improving overall water quality.

Each type of sedimentation tank has its own advantages and is used based on the specific requirements of the water treatment process.

7.5.6 Filtration

The filtration process, along with other processes, in a water treatment plant is to ensure the delivery of safe and clean drinking water. It involves the removal of suspended solids, microorganisms, and other contaminants from water through various chemicals and processes.

The first step in the conventional water treatment process is coagulation, which involves adding iron or aluminum salts, such as aluminum sulfate, ferric sulfate, ferric chloride, or polymers, to the water. These chemicals, known as coagulants, have a positive charge that neutralizes the negative charge of dissolved and suspended particles in the water. When this reaction occurs, the particles

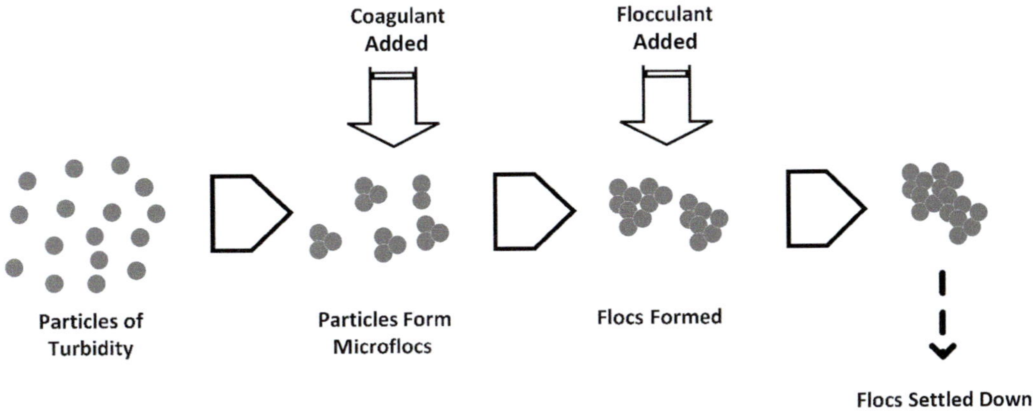

FIGURE 7.17 Coagulation, flocculation, and sedimentation processes.

bind together, or coagulate, forming larger particles, or floc. This process is sometimes also referred to as flocculation.

The floc is heavy and quickly settles to the bottom of the water supply in a process called sedimentation. The coagulated water can either be filtered directly through a medium filter (such as sand and gravel), a microfiltration or ultrafiltration membrane, or it can be moved to a settling tank. In a settling tank, or clarifier, the heavy particles settle to the bottom and are removed, and the water moves on to the filtration step of the treatment process.

Figure 7.17 illustrates the water treatment processes of coagulation, flocculation, and sedimentation, where particles are combined, formed into larger clusters, and then settled out to clarify the water.

The second step in a conventional water treatment system is filtration, which removes particulate matter from water by forcing the water to pass through porous media. The filtration system consists of filters with varying sizes of pores and is often made up of sand, gravel, and charcoal. These filters remove dissolved particles and germs, such as dust, chemicals, parasites, bacteria, and viruses. Activated carbon filters also remove any bad odors.

There are different types of filtration methods, such as rapid sand filtration, activated carbon filtration, and membrane filtration. Each technique has its own set of advantages and disadvantages depending on the specific needs of the water source being treated. For example, rapid sand filtration is a process that uses gravity to move water through a bed of sand, trapping and removing particles. Activated carbon filtration uses a bed of activated carbon to remove contaminants and impurities, using chemical adsorption. Membrane filtration, on the other hand, uses a semipermeable membrane to separate particles, unwanted molecules, and ions from the water.

The filtration process removes harmful substances from the water supply while retaining essential minerals and nutrients necessary for human consumption. Filtration offers promise in providing access to clean drinking water for communities worldwide. However, achieving optimal results requires careful consideration of appropriate filters and accompanying chemical treatments tailored to specific contamination profiles.

There are two common methods used in water treatment, which are conventional and membrane filtration, but they operate on different principles and are suited to different applications.

7.5.6.1 Conventional Filtration

Conventional filtration typically involves the use of media such as sand, gravel, or activated carbon. The process works by having water pass over and through the media multiple times. Pollutants become trapped in the voids of the media particles and on the surfaces of the media particles. In some cases, pollutants are removed through chemical and molecular action or cation exchange. This method is commonly used to capture large particles or remove pollutants such as pathogens, chlorine, iron, and other metals and impurities through adsorption.

Conventional filtration in a drinking water treatment plant involves a combination of coagulation, sedimentation, filtration, and disinfection. This method has been widely used since the early twentieth century.

In the coagulation process, chemicals like iron or aluminum salts, which are positively charged, are added to the water. These coagulants neutralize the negative charge of dissolved and suspended particles in the water, causing the particles to bind together or coagulate. The larger particles, or floc, are heavy and quickly settle to the bottom of the water supply in a process called sedimentation.

The coagulated water is then either filtered directly through a medium filter (such as sand and gravel), a microfiltration or ultrafiltration membrane, or it can be moved to a settling tank. In a settling tank, the heavy particles settle to the bottom and are removed, and the water moves on to the filtration step of the treatment process.

During filtration, the clear water passes through filters that have different pore sizes and are made of different materials (such as sand, gravel, and charcoal). These filters remove dissolved particles and germs, such as dust, chemicals, parasites, bacteria, and viruses.

On the other hand, membrane filtration requires a permeable layer or sheet of material that the water passes through to remove the pollutants or targeted impurities. Membrane filtration offers a variety of uses with its four categories: microfiltration, ultrafiltration, nanofiltration, and RO filtration.

The advantages of membrane filtration include its accuracy and sensitivity, as it tests the entire sample, not just part of it. It is also simple and rapid, and due to low installation and energy costs, it is usually more cost-effective than many other alternatives.

However, membrane filtration also has its disadvantages. The processes are prone to membrane fouling effects, which lead to a decrease in permeate flux. Expensive cleaning and regeneration schemes may be necessary. The high flow rates used in crossflow feed can damage shear-sensitive materials, and equipment costs can be high.

In comparison, conventional filtration is a physical process that does not require a chemical process to filter water. It can successfully remove a large amount of organic compounds, including some dissolved organic material. However, it may not be as effective as membrane filtration in removing certain types of contaminants, especially those that are very small or dissolved in water.

7.5.6.2 Membrane Filtration

Membrane filtration, on the other hand, uses a permeable layer or sheet of material that the water passes through to remove the pollutants or targeted impurities. Like media filtration, membrane filtration has two removal mechanisms: physical blockage of particles by the membrane material and the removal of substances that are attached to the particles captured during the primary removal process. Membrane filters are manufactured using a wide variety of materials, with pore size and durability being key factors for selecting a material. Membrane filtration offers a variety of uses with its four categories, such as microfiltration, ultrafiltration, nanofiltration, and RO filtration.

Membrane filtration methods have significant efficiency in removing fine particles and organic matter from water. Thin-film composite (TFC) membranes have better removal capacity than conventional membranes, although they require higher operating pressures, which can limit their use.

TFC membranes are semipermeable membranes that are manufactured to provide selectivity with high permeability. They are often used in water purification or water desalination systems but also have applications in chemical processes such as gas separations, dehumidification, batteries, and fuel cells.

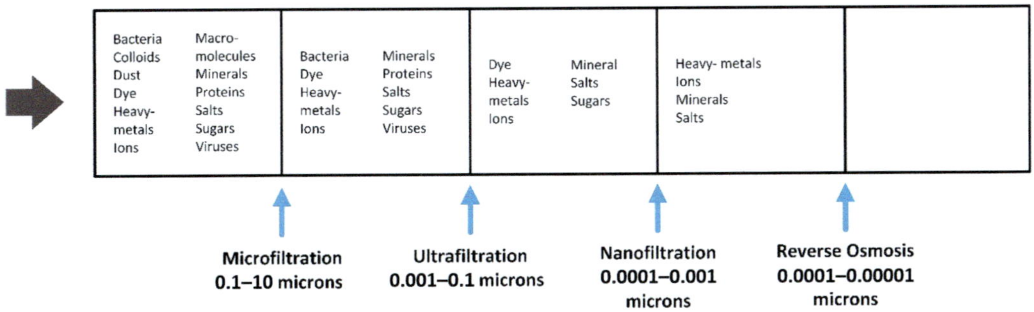

FIGURE 7.18 Membrane filtration.

A TFC membrane can be considered a molecular sieve constructed in the form of a film from two or more layered materials. The additional layers provide structural strength and a low-defect surface to support a selective layer that is thin enough to be selective but not so thick that it causes low permeability.

TFC membranes for water treatment are commonly classified as nanofiltration (NF) and RO membranes. Both types are typically made of a thin polyamide layer (<200 nm) deposited on top of a polyethersulfone or polysulfone porous layer (about 50 microns) on top of a non-woven fabric support sheet. The three-layer configuration gives the desired properties of high rejection of undesired materials (like salts), high filtration rate, and good mechanical strength. The polyamide top layer is responsible for the high rejection and is chosen primarily for its permeability to water and relative impermeability to various dissolved impurities including salt ions and other small, unfilterable molecules.

Although not fully commercialized yet, TFCs are also used in other water treatment technologies, including forward osmosis, membrane distillation, and electrodialysis.

The first viable RO membrane was made from cellulose acetate as an integrally skinned asymmetric semipermeable membrane. This membrane was made by Loeb and Sourirajan at UCLA in 1959 and patented in 1960. In 1972, John Cadotte of North Star Technologies (later FilmTec Corporation) developed the first interfacial polyamide (IP) TFC membrane. The current generation of RO membrane materials is based on a composite material patented by FilmTec Corporation in 1970 (now part of DuPont). Today, most such membranes for RO and nanofiltration use a polyamide active layer.

Figure 7.18 illustrates membrane filtration processes, including microfiltration, ultrafiltration, nanofiltration, and RO, each filtering out progressively smaller particles and sediments to purify water in various treatment stages.

As is suggested by the name, TFC membranes are composed of multiple layers. Membranes designed for desalination use an active thin-film layer of polyamide layered with polysulfone as a porous support layer. The active layers tend to be extremely thin and relatively nonporous. The chemistry of these layers often imparts selectivity. Meanwhile, the support layers tend to need to be both extremely porous and robust to higher pressures. Other materials, usually zeolites, are also used in the manufacture of TFC membranes.

7.5.7 Disinfection

The disinfection process in a water treatment plant is a process that ensures the water we consume is safe and free from harmful microorganisms. This process involves the use of various methods to kill or deactivate pathogens present in the water.

Water disinfection begins after the water has undergone preliminary treatment stages such as coagulation, flocculation, sedimentation, and filtration. At this point, the water is largely free of physical impurities, but it may still contain harmful microorganisms. To ensure these pathogens are eliminated, disinfectants are added to the water.

The most common disinfectants used in water treatment include chlorine and its compounds, ozone, and UV light. Each of these methods has its own advantages and disadvantages.

Chlorination is a widely accepted method of disinfection. It involves the addition of chlorine or its compounds to the water. The chlorine reacts with the water to form hypochlorous acid and hypochlorite ions, which are capable of oxidizing and thus killing many types of bacteria and viruses. However, chlorination has some drawbacks, such as the potential formation of harmful by-products if organic matter is present in the water. Chlorination, the most used water treatment method, is very effective at killing harmful bacteria and viruses. However, it frequently causes both an off-taste and odor. This is because chlorine can react with naturally occurring organic compounds in the water, producing potentially harmful disinfection by-products. These by-products can affect the taste and odor of the water.

The addition of ammonia to produce monochloramine (NH_2Cl), which lasts longer in the pipes, can make the taste and odor even worse. Chemical reactions can also occur between chlorine and chemicals in kettles, rubber, or plastic components within your domestic plumbing system, leading to antiseptic and medicinal tastes. The taste and smell may be stronger if the water has been boiled or left to stand in a glass.

When a system first starts chlorinating, it is normal for people to say they can taste and/or smell the chlorine. Over time, the system stabilizes, and any tastes or smells will decrease or go away. People also usually get used to chlorine in water over time.

However, there are several health risks associated with chlorinated water. While chlorine is effective at killing many types of pathogens in water, it can react with organic compounds in the water to produce harmful by-products, called trihalomethanes (THMs). These by-products can produce free radicals in the body, which can trigger cell damage.

Some scientific studies have linked chlorine and chlorination by-products to various health issues, including:

1. Increased risk of cancer, specifically bladder, liver, rectal, and colon cancer.
2. Reproductive and developmental effects.
3. Problems with the immune system.
4. Heart disease.
5. Atherosclerosis (hardening of the arteries).
6. Anemia.
7. High blood pressure.
8. Allergic reactions.

However, it is also important to remember that the risk must be balanced against the benefits of using chlorine to disinfect water and protect against waterborne diseases. If you have specific health concerns related to chlorinated water, it is best to consult with a healthcare provider.

Ozonation is another effective method of disinfection. Ozone is a powerful oxidant that can destroy a wide range of pathogens. It also improves the taste and odor of water. However, ozone is unstable and must be generated on-site, which can be costly.

UV disinfection is a physical process that involves exposing the water to UV light. This light damages the DNA of the microorganisms, preventing them from reproducing. UV disinfection is effective against many types of pathogens and does not introduce any chemicals into the water. However, the water must be relatively clear for the UV light to penetrate effectively.

After the primary disinfection stage, secondary disinfection may be carried out to provide further protection against pathogens as the water travels through the distribution system to consumers. This ensures that the water remains safe to drink even if it becomes contaminated during distribution.

Secondary disinfection in water treatment refers to the process of adding disinfectants to the water after it has undergone primary disinfection. This process is designed to provide longer-lasting water treatment as the water moves through pipes to consumers.

The purpose of secondary disinfection is to maintain water quality by killing potentially harmful organisms that may get into the water as it moves through the distribution system. This ensures that the water remains safe to drink even if it becomes contaminated during distribution.

Chloramines, formed when ammonia is added to chlorine, are commonly used as a secondary disinfectant. They provide longer-lasting disinfection as the water moves through pipes to consumers. This type of disinfection is known as secondary disinfection.

In some cases, effluent from the sedimentation tank is disinfected with chlorine before being discharged into receiving waters to complete secondary treatment. Done properly, chlorination will kill more than 99% of the harmful bacteria in an effluent.

Several potential issues can arise during the disinfection process in a water treatment plant:

1. *Formation of Disinfection By-products (DBPs):* When disinfectants like chlorine react with organic and inorganic matter in the water, they can form DBPs. Some DBPs have been linked to adverse health effects.
2. *Interference from Other Substances:* The presence of substances such as ferrous and manganese ions, nitrites, sulfides, and organic substances can interfere with the disinfection process. These substances can reduce the concentration of oxidizing disinfectants, leading to a decrease in the inactivation of microorganisms.
3. *Lack of Residual Disinfectant:* Some disinfection methods, like UV disinfection, do not leave a residual disinfectant in the water. This can be a problem when water must be piped or stored over significant distances and time, as re-growth of the microbial population is considered a risk.
4. *Inadequate Contact Time:* The effectiveness of the disinfection process depends on the contact time between the disinfectant and the water. If the contact time is too short, the disinfectant may not fully inactivate all pathogens.
5. *Changes in Water Quality:* Changes in water quality, such as pH and temperature, can affect the effectiveness of the disinfection process.
6. *Cost and Infrastructure:* Advanced disinfection technologies like ozonation and UV disinfection can be costly and require significant infrastructure.
7. *Taste and Odor:* Some disinfectants, like chlorine, can affect the taste and odor of the water.

These issues highlight the complexity of water disinfection and the need for careful monitoring to ensure the safety and quality of treated water.

7.5.8 pH Adjustment

The pH level of water is a critical factor that affects its quality and suitability for various uses, including drinking. The term "pH" stands for "potential of hydrogen," and it measures the concentration of hydrogen ions in a solution. The pH scale ranges from 0 to 14, with 7 being neutral. A pH less than 7 indicates acidity, while a pH greater than 7 indicates alkalinity.

In the context of drinking water, the US EPA recommends that the pH of water should be between 6.5 and 8.5. This range is considered safe for human consumption and helps to minimize the corrosion of pipes and plumbing fixtures.

In the water treatment process, the pH value of water is adjusted to optimize the effectiveness of the treatment process and to ensure the treated water is safe for consumption. The optimal pH value for drinking water is typically between 6.5 and 8.5. Adjusting the pH value involves adding chemicals known as pH adjusters or neutralizers. If the water is too acidic (pH less than 7), a base is added to increase the pH. Common bases used for this purpose include soda ash (sodium carbonate) and sodium hydroxide. On the other hand, if the water is too alkaline (pH greater than 7), an acid is added to decrease the pH. Acids commonly used for this purpose include sulfuric acid and hydrochloric acid. The amount of acid or base added depends on the initial pH of the water and the desired pH level. This requires careful monitoring and control to ensure the pH is adjusted accurately and maintained within the desired range.

After the pH adjustment, the water undergoes further treatment processes such as coagulation, flocculation, sedimentation, filtration, and disinfection. The adjusted pH helps to optimize these processes. For example, maintaining the pH within a specific range (generally between 6 and 8) enhances the performance of the coagulant and improves the formation of larger, more settleable flocs.

The process of pH adjustment must be carried out carefully to avoid overshooting the desired pH level. This requires accurate measurement of the initial pH, careful calculation of the amount of acid or base needed, and thorough mixing to ensure even distribution of the added substance.

The pH of water can also be affected by various factors such as temperature, dissolved solids, and gases. Therefore, continuous monitoring and adjustment may be necessary to maintain the desired pH level.

While pH adjustment is a common practice in water treatment, it is not without challenges. For instance, the substances used for pH adjustment can introduce additional contaminants into the water. Also, the process can be complex and require specialized equipment and trained personnel.

While it might seem logical to adjust the pH after sedimentation to remove any particles that could interfere with the pH measurement, doing so could compromise the effectiveness of the coagulation, flocculation, and sedimentation processes. Therefore, pH adjustment is typically done before these processes in a water treatment plant.

The pH adjustment is typically done before coagulation, flocculation, and sedimentation for several reasons:

Optimizing Coagulation: The effectiveness of the coagulation process is highly dependent on the pH of the water. Certain coagulants work best at specific pH levels. Therefore, adjusting the pH before coagulation helps optimize the performance of the coagulant and enhances the formation of larger, more settleable flocs.

Preventing Precipitation of Coagulants: Some coagulants, like aluminum sulfate, can precipitate out of solution at certain pH levels. By adjusting the pH before the coagulation process begins, the water is treated to have a pH that exceeds the solubility of aluminum, reducing the residuals that the process of coagulation leaves behind.

Improving Taste and Reducing Corrosion: Adjusting the pH improves taste, reduces corrosion (breakdown) of pipes, and ensures chemical disinfectants continue killing germs as the water travels through pipes.

The processes of storage and distribution, represented by steps 9 and 10, do not fall within the technical scope of water treatment, but they are essential for ensuring a reliable supply to the customers.

7.5.9 Storage

After the water has been treated completely in a water treatment plant and is safe for consumption, a storage system is necessary in the water supply system for several reasons:

1. *Balancing Supply and Demand:* Storage helps balance the supply of treated water with the demand. The quantity of water people use, or would like to use, is called the demand. What is actually available is the supply. On a daily scale, some sources, such as springs and rivers, flow constantly, day and night. Treatment works, pumping stations, and water mains are designed to produce an average daily supply, operating 24 hours a day. It would be inefficient to build them larger to meet the demand during the day but then not use them much at night. Using these constant 24-hour flows, a service reservoir fills up at night and releases water during the day when it is needed.
2. *Ensuring Constant Supply:* Storage provides a constant source of supply to the treatment plant. These "reservoirs" (lakes) provide an area for large amounts of water to collect so the treatment plant has a constant source of supply.
3. *Emergency Preparedness:* Storage provides some safety if a pump breaks down, a power supply fails, or a pipe bursts. More storage means you have longer to repair the system. There can be unexpected increases in demand, such as water needed for firefighting or hot weather. The risk of running out of water during an emergency must be balanced with the saving in capital expenditure.

The volume of storage and requirements of the storage are determined by several factors:

1. *Flow Rate:* The flow rate is the amount of wastewater that is expected to be treated in the plant.
2. *Peak Flow Rate:* The peak flow rate is the maximum flow rate that can occur during a specific period.
3. *Retention Time:* The retention time is the time required for wastewater to be treated in the tank. It is determined by dividing the tank's volume by the wastewater flow rate.
4. *Operational Storage (OS):* OS is the volume of the reservoir devoted to supplying the water system while, under normal operating conditions, the source(s) of supply is in "off" status.
5. *Equalizing Storage (ES):* ES is designed to provide enough water to system customers while the source is off.
6. *Standby Storage (SB):* SB is the volume of the reservoir devoted to emergency or breakdown situations.
7. *Fire Suppression Storage (FSS):* FSS is the volume of the reservoir devoted to meeting the fire flow and duration requirements for the service area.
8. *Dead Storage (DS):* DS is the volume of the reservoir devoted to storing water that cannot be drained by gravity.
9. *Location of the Storage:* The amount of effective storage may also be dependent upon the location of the storage relative to the place of its use.
10. *Regulatory Requirements:* There are regulations within the Safe Drinking Water Act called the Surface Water Treatment Rule (SWTR), which specifically addresses surface water treatment requirements.

These factors help ensure that the storage facilities are designed to meet the specific needs of the water treatment plant and the communities they serve.

There are several types of storage facilities used in water treatment:

1. *Elevated Tanks:* These are water storage tanks that are elevated above the ground, usually on a stand or tower. The height provides the necessary pressure to distribute the water through the system.
2. *Standpipes:* Standpipes are tall, narrow tanks used for water storage. They also provide pressure in the water supply system due to their height.

3. *Reservoirs:* Reservoirs are large natural or artificial lakes used for water storage. They can store large volumes of water and are often used in conjunction with damming projects.
4. *Hydro-Pneumatic Tanks:* These are pressure vessels that store and regulate water in a pressurized environment. They are often used in small water systems and pump stations.
5. *Point-of-Use (POU) Systems:* These include water pitchers, faucet filters, and RO systems, etc.
6. *Whole-House/Point-of-Entry (POE) Systems:* These include municipal systems, pressurized storage tanks, UV microbiological systems, water softeners, etc.

Balancing the supply and demand for treated water in a water supply system is mandatory. In many cases, a service reservoir is necessary. During periods of low demand, such as at night, the water treatment plant continues to treat water at a constant rate. This treated water is stored in the service reservoir. During peak demand periods, such as in the morning and evening when water usage is high, water is supplied from both the treatment plant and the service reservoir. This ensures that the demand is met without overloading the treatment plant or depleting the water source.

The service reservoir also provides a buffer against fluctuations in demand. For example, in the event of an emergency such as a fire, there may be a sudden increase in demand for water. The water stored in the service reservoir can be used to meet this additional demand.

In addition to balancing supply and demand, the service reservoir also helps to maintain pressure in the distribution system. This is particularly important in hilly areas, where the reservoir can be placed at a high point to create a gravity-fed system.

An example of how to calculate the storage capacity for a water supply system in metric units is given below:

Suppose a small community has an ADD of 150 m^3 and a peak hourly demand of 25 m^3. The water treatment plant operates for 20 hours each day.

The volume of water produced per hour when the plant is operating can be calculated as follows:

Volume per hour = Daily demand/Operating hours.

Volume per hour = 150 m^3/20 h = 7.5 m^3/h.

During peak demand, the treatment plant cannot produce enough water to meet the demand (7.5 m^3/h < 25 m^3/h). Therefore, we need to have enough storage to supply the deficit.

The volume of storage required can be calculated as follows:

Storage volume = (Peak hourly demand – Volume per hour) × Peak demand duration.

Assuming the peak demand lasts for 3 hours, the storage volume required would be:

Storage volume = (25 m^3/h – 7.5 m^3/h) × 3 hours = 52.5 m^3.

Therefore, the community would need a storage tank with a capacity of at least 52.5 m^3 to meet the peak demand.

7.5.10 Distribution

The distribution process of the drinking water supply system has a goal to ensure the delivery of safe and clean water from the treatment plants to the consumers' taps. This process generally involves several stages, each of which plays a specific role in maintaining the quality and safety of the water.

1. *Water Treatment:* The first step in the distribution process is the treatment of raw water at the water treatment plant. This involves several stages, including coagulation, flocculation, sedimentation, filtration, and disinfection. The goal of these processes is to remove impurities and pathogens from the water, making it safe for consumption.

2. *Storage:* After treatment, the water is stored in reservoirs or tanks. This serves several purposes, including balancing supply and demand, maintaining pressure in the distribution system, and providing a buffer against fluctuations in demand or interruptions in the supply.
3. *Distribution Mains:* From the storage facilities, the water is distributed to consumers through a network of pipes known as distribution mains. These mains are typically made of durable materials such as cast iron, ductile iron, or PVC and are designed to withstand the pressure of the water and the stresses of the surrounding environment.
4. *Service Connections:* Branching off from the distribution mains are smaller pipes known as service connections. These pipes deliver water directly to individual homes, businesses, and other buildings.
5. *Water Meters:* At each service connection, a water meter is installed to measure the amount of water used by the consumer. This information is used for billing purposes and to monitor water usage patterns.
6. *Maintenance and Monitoring:* Throughout the distribution process, regular maintenance and monitoring are carried out to ensure the integrity of the system and the quality of the water. This includes inspecting and repairing pipes and other infrastructure, testing the water for contaminants, and monitoring water pressure and flow rates.

The maintenance and monitoring process in a drinking water distribution system is a vital aspect of ensuring the delivery of safe and clean water. This process involves several key steps. Regular inspection of the distribution system is crucial to identify any potential issues early, such as inspecting pipes for leaks or damage, checking storage tanks for structural integrity, and assessing pumps and other equipment for proper operation.

Preventive maintenance involves regularly scheduled tasks to keep equipment and infrastructure in good working condition. This can include cleaning and disinfecting storage tanks, flushing pipes to remove sediment, and servicing pumps and valves. Over time, parts of the distribution system may need to be repaired or replaced. This can include fixing leaks in pipes, replacing worn-out pumps, or upgrading old infrastructure.

Regular testing of the water is conducted to ensure it meets health and safety standards. This can involve testing for various parameters such as pH, turbidity, and levels of disinfectant. In addition, tests for harmful bacteria or other pathogens are conducted. Maintaining the right pressure in the distribution system is important to prevent leaks and ensure water is delivered effectively. Pressure can be monitored using pressure gauges installed at various points in the system.

Monitoring the flow rate of water in the distribution system helps to manage supply and demand. It can also help detect leaks or blockages in the system. Data from inspections, maintenance activities, and monitoring are collected and analyzed. This information can help identify trends, predict future issues, and inform decision-making. Having a plan in place for responding to emergencies such as main breaks, contamination incidents, or natural disasters is an important part of system maintenance and monitoring. By carrying out these steps, water utilities can ensure the reliability of the distribution system, maintain the quality of the water, and provide a consistent supply to consumers.

PROBLEMS AND ISSUES FOR DISCUSSION

1. Can you explain the basic process of urban water supply and how it ensures the delivery of clean water to households?
2. What steps are involved in the treatment of water before it is supplied to urban households, and why are these steps important?
3. What are the key standards that must be met for urban water supply, and how do they ensure the safety and health of the population?

4. What types of equipment are essential in an urban water supply system, and what are their specific roles?
5. How are water supply facilities maintained in urban areas to ensure continuous and reliable water supply?
6. What role do water towers play in urban water supply, and why are they a common sight in many cities?
7. How is water pressure managed in high-rise buildings to ensure that water reaches all floors efficiently?
8. What measures are taken to ensure the quality of water in urban supply systems, and how are these measures enforced?
9. How is water supply managed during periods of drought, and what strategies are used to conserve water?
10. What are the common challenges faced in urban water supply, and how can they be addressed?
11. What role does technology play in urban water supply management, and how has it improved efficiency and reliability?
12. How are leaks detected and repaired in urban water supply systems, and why is this important for water conservation?
13. What is the impact of climate change on urban water supply, and how are cities adapting to these changes?
14. How is the demand for water forecasted in urban areas, and what factors are considered in these forecasts?
15. What are the environmental considerations in urban water supply, and how do they influence the design and operation of water supply systems?
16. How is water rationing implemented in urban areas during shortages, and what impact does it have on residents?
17. What are the health implications of not meeting water supply standards, and how can these risks be mitigated?
18. How is the cost of water supply calculated in urban areas, and what factors influence these costs?
19. How do urban water supply systems interact with other urban systems, such as energy and waste management systems?
20. What are the primary sources of water that are commonly used to supply water to communities and how are these sources selected and developed?
21. How do water treatment plants utilize various physical, chemical, and biological processes to purify and disinfect raw water to ensure it meets drinking water quality standards before distribution to consumers?
22. What is the purpose of water storage tanks in a water supply network, and how do they help regulate water pressure, store emergency reserves, and accommodate fluctuations in water demand throughout the day and over time?
23. How are water distribution networks designed to efficiently and reliably deliver water from treatment plants to households, businesses, and other end-users while maintaining appropriate water pressure and flow rates?
24. What are the common types of pipes used in water supply systems, and how do their material properties, durability, and installation methods affect the overall performance and longevity of the water distribution infrastructure?
25. How do water supply systems utilize pressure-regulating valves, booster pumps, and other technologies to maintain appropriate water pressure and flow rates throughout the distribution network, ensuring that all consumers receive adequate water service and pressure?

26. What is the role of water metering in water supply systems, and how do water meters installed at individual properties help manage water consumption, detect leaks, and support fair and accurate billing for water services provided to customers?
27. How do water supply systems implement various water quality monitoring, testing, and treatment processes to ensure the safety and compliance of drinking water with regulatory standards for parameters such as microbial content, chemical composition, and aesthetic properties?
28. What are the key challenges faced by water supply system operators, such as aging infrastructure, water scarcity, climate change impacts, and increasing demand, and how do they develop strategies to maintain a reliable and sustainable water supply?
29. How do water supply systems adapt to changes in water demand, population growth, and urbanization by expanding or upgrading their infrastructure, implementing water conservation measures, and exploring alternative water sources to meet the evolving needs of the community?
30. What are the potential environmental impacts of water supply systems associated with water treatment and distribution, and how do water managers work to mitigate these impacts and promote sustainable practices?
31. How do water supply systems incorporate energy-efficient technologies, renewable energy sources, and other sustainable practices to reduce their carbon footprint and environmental impact while also ensuring the cost-effective and reliable operation of the water infrastructure?
32. What are the common water quality issues that can arise in water supply systems, and how do water treatment processes and distribution system management strategies address these challenges to maintain the quality of drinking water?
33. How do water supply systems utilize advanced monitoring and control technologies to continuously monitor water flows, pressures, and quality and quickly identify and respond to any issues or inefficiencies within the distribution network?
34. What is the importance of water supply system cybersecurity and resilience planning, and how do water utilities implement measures to protect their critical infrastructure from cyber threats, natural disasters, and other disruptions to ensure the continuity of water services?
35. How do water supply systems adapt to the impacts of climate change and the occurrence of extreme weather events?
36. What are the different types of water treatment technologies used in water supply systems, and how do water managers select and integrate these technologies to effectively remove contaminants and produce high-quality drinking water?
37. How do water supply systems work to protect their water sources from contamination and pollution to ensure the long-term sustainability of the water resources?
38. What is the role of water supply system automation and smart technologies in improving the efficiency, responsiveness, and resilience of water distribution networks, and how do utilities leverage these technologies to enhance their water service delivery?
39. How do water supply systems integrate the management of wastewater and stormwater to promote a more holistic and sustainable approach to urban water management?
40. What are the key considerations that water managers take into account when planning and designing the expansion or upgrading of water supply infrastructure to meet the evolving needs of a community?
41. How do water supply systems ensure equitable access to safe and affordable water services for all consumers, including low-income households, marginalized communities, and underserved areas?

42. What are the key factors that water supply system designers and planners consider when selecting the most appropriate technologies, materials, and configurations for the various components of the water infrastructure?
43. How do water supply systems integrate and coordinate with other urban infrastructure systems, such as transportation, energy, and telecommunications networks, to optimize the efficient and sustainable delivery of water services while minimizing disruptions and maximizing the benefits of cross-system synergies?
44. What is the role of water pricing and tariff structures in water supply systems, and how do utilities balance the need for revenue generation, cost recovery, water conservation incentives, and social equity considerations when designing and implementing their water pricing policies?
45. How do water supply systems incorporate water conservation and efficiency measures to reduce overall water demand and promote sustainable water use?
46. What are the common operational and maintenance challenges faced by water supply system operators, and how do they develop strategies to address these challenges?
47. How do water supply systems respond to natural disasters to ensure the continuity of water services and the protection of public health and safety?
48. What is the importance of water infrastructure rehabilitation and asset management in water supply systems, and how do utilities prioritize and execute projects to maintain the integrity, performance, and longevity of their water distribution networks, treatment facilities, and other critical components?
49. How do water supply systems incorporate public engagement, stakeholder participation, and community outreach strategies to foster transparency, build trust, and collaborate with citizens, businesses, and other stakeholders in the planning, decision-making, and management of water services?

MULTIPLE-CHOICE QUESTIONS

1. What is the primary source of urban water supply?
   a. Rainwater
   b. Groundwater
   c. Rivers and lakes
   d. Desalinated seawater
   e. All of the above
2. What is the first step in water treatment?
   a. Disinfection
   b. Coagulation
   c. Sedimentation
   d. Filtration
   e. Storage
3. What is the purpose of chlorination in water treatment?
   a. Remove solid particles
   b. Improve taste
   c. Kill bacteria and other pathogens
   d. Remove color
   e. Soften the water
4. What is a water distribution system?
   a. A network of pipes, pumps, and storage facilities
   b. A water treatment plant
   c. A system for collecting rainwater
   d. A system for desalinating seawater
   e. A system for heating water

5. What is the role of a water tower in a water distribution system?
    a. Purify water
    b. Store water
    c. Heat water
    d. Cool water
    e. Desalinate water
6. What is hard water?
    a. Water that has high mineral content
    b. Water that is difficult to treat
    c. Water that is frozen
    d. Water that is contaminated with bacteria
    e. Water that is salty
7. What is a common method for softening hard water?
    a. Chlorination
    b. Filtration
    c. Ion exchange
    d. Sedimentation
    e. Coagulation
8. What is the main disadvantage of hard water?
    a. It is harmful to health.
    b. It causes pipes to corrode.
    c. It is not suitable for drinking.
    d. It causes scale buildup in pipes and appliances.
    e. It has a bad taste.
9. What is the purpose of a water meter?
    a. Measure the amount of water used
    b. Detect leaks in the water supply system
    c. Control the flow of water
    d. Purify water
    e. Store water
10. What is a common method for detecting leaks in a water supply system?
    a. Visual inspection
    b. Pressure testing
    c. Acoustic leak detection
    d. Chemical testing
    e. Thermal imaging
11. What is the main cause of water supply interruptions?
    a. Power outages
    b. Equipment failure
    c. Water shortage
    d. Pipe leaks or bursts
    e. All of the above
12. What is the main advantage of using underground pipes for water distribution?
    a. They are cheaper to install.
    b. They are easier to maintain.
    c. They are less likely to freeze.
    d. They can carry more water.
    e. They are more durable.

13. What is the purpose of a water treatment plant?
    a. Store water
    b. Distribute water
    c. Collect water
    d. Treat water to make it safe for use
    e. Heat water
14. What is the main disadvantage of using chlorine for water disinfection?
    a. It is expensive.
    b. It is difficult to handle.
    c. It can react with organic matter to form harmful by-products.
    d. It is not effective against all types of pathogens.
    e. It has a bad taste.
15. What is the main advantage of using UV light for water disinfection?
    a. It is cheap.
    b. It is easy to handle.
    c. It does not leave any residual disinfectant in the water.
    d. It is effective against all types of pathogens.
    e. It improves the taste of water.
16. What is the main disadvantage of using UV light for water disinfection?
    a. It is expensive.
    b. It is difficult to handle.
    c. It can react with organic matter to form harmful by-products.
    d. It is not effective against all types of pathogens.
    e. It has a bad taste.
17. What is the main advantage of using ozone for water disinfection?
    a. It is cheap.
    b. It is easy to handle.
    c. It does not leave any residual disinfectant in the water.
    d. It is effective against all types of pathogens.
    e. It improves the taste of water.
18. What is the main disadvantage of using ozone for water disinfection?
    a. It is expensive.
    b. It is difficult to handle.
    c. It can react with organic matter to form harmful by-products.
    d. It is not effective against all types of pathogens.
    e. It has a bad taste.
19. What is the main advantage of using chloramines for water disinfection?
    a. They are cheap.
    b. They are easy to handle.
    c. They do not leave any residual disinfectant in the water.
    d. They are effective against all types of pathogens.
    e. They improve the taste of water.
20. What is the main disadvantage of using chloramines for water disinfection?
    a. They are expensive.
    b. They are difficult to handle.
    c. They can react with organic matter to form harmful by-products.
    d. They are not effective against all types of pathogens.
    e. They have a bad taste.

21. What is the main advantage of using activated carbon for water treatment?
    a. It is cheap.
    b. It is easy to handle.
    c. It does not leave any residual disinfectant in the water.
    d. It is effective against all types of pathogens.
    e. It improves the taste of water.
22. What is the main disadvantage of using activated carbon for water treatment?
    a. It is expensive.
    b. It is difficult to handle.
    c. It can react with organic matter to form harmful by-products.
    d. It is not effective against all types of pathogens.
    e. It has a bad taste.
23. What is the main advantage of using reverse osmosis for water treatment?
    a. It is cheap.
    b. It is easy to handle.
    c. It does not leave any residual disinfectant in the water.
    d. It is effective against all types of pathogens.
    e. It improves the taste of water.
24. What is the main disadvantage of using reverse osmosis for water treatment?
    a. It is expensive.
    b. It is difficult to handle.
    c. It can react with organic matter to form harmful by-products.
    d. It is not effective against all types of pathogens.
    e. It has a bad taste.
25. What is the main advantage of using distillation for water treatment?
    a. It is cheap.
    b. It is easy to handle.
    c. It does not leave any residual disinfectant in the water.
    d. It is effective against all types of pathogens.
    e. It improves the taste of water.
26. What is the main disadvantage of using distillation for water treatment?
    a. It is expensive.
    b. It is difficult to handle.
    c. It can react with organic matter to form harmful by-products.
    d. It is not effective against all types of pathogens.
    e. It has a bad taste.
27. What is the main advantage of using aeration for water treatment?
    a. It is cheap.
    b. It is easy to handle.
    c. It does not leave any residual disinfectant in the water.
    d. It is effective against all types of pathogens.
    e. It improves the taste of water.
28. What is the main disadvantage of using aeration for water treatment?
    a. It is expensive.
    b. It is difficult to handle.
    c. It can react with organic matter to form harmful by-products.
    d. It is not effective against all types of pathogens.
    e. It has a bad taste.

29. What is the main advantage of using flocculation for water treatment?
    a. It is cheap.
    b. It is easy to handle.
    c. It does not leave any residual disinfectant in the water.
    d. It is effective against all types of pathogens.
    e. It improves the taste of water.
30. What is the main disadvantage of using flocculation for water treatment?
    a. It is expensive.
    b. It is difficult to handle.
    c. It can react with organic matter to form harmful by-products.
    d. It is not effective against all types of pathogens.
    e. It has a bad taste.
31. What is the main advantage of using sedimentation for water treatment?
    a. It is cheap.
    b. It is easy to handle.
    c. It does not leave any residual disinfectant in the water.
    d. It is effective against all types of pathogens.
    e. It improves the taste of water.
32. What is the main disadvantage of using sedimentation for water treatment?
    a. It is expensive.
    b. It is difficult to handle.
    c. It can react with organic matter to form harmful by-products.
    d. It is not effective against all types of pathogens.
    e. It has a bad taste.
33. What is the main advantage of using filtration for water treatment?
    a. It is cheap.
    b. It is easy to handle.
    c. It does not leave any residual disinfectant in the water.
    d. It is effective against all types of pathogens.
    e. It improves the taste of water.
34. What is the main disadvantage of using filtration for water treatment?
    a. It is expensive.
    b. It is difficult to handle.
    c. It can react with organic matter to form harmful by-products.
    d. It is not effective against all types of pathogens.
    e. It has a bad taste.
35. What is the main advantage of using disinfection for water treatment?
    a. It is cheap.
    b. It is easy to handle.
    c. It does not leave any residual disinfectant in the water.
    d. It is effective against all types of pathogens.
    e. It improves the taste of water.
36. What is the main disadvantage of using disinfection for water treatment?
    a. It is expensive.
    b. It is difficult to handle.
    c. It can react with organic matter to form harmful by-products.
    d. It is not effective against all types of pathogens.
    e. It has a bad taste.

37. What is the main advantage of using storage for water treatment?
    a. It is cheap.
    b. It is easy to handle.
    c. It does not leave any residual disinfectant in the water.
    d. It is effective against all types of pathogens.
    e. It improves the taste of water.
38. What is the main disadvantage of using storage for water treatment?
    a. It is expensive.
    b. It is difficult to handle.
    c. It can react with organic matter to form harmful by-products.
    d. It is not effective against all types of pathogens.
    e. It has a bad taste.
39. What is the main advantage of using distribution for water treatment?
    a. It is cheap.
    b. It is easy to handle.
    c. It does not leave any residual disinfectant in the water.
    d. It is effective against all types of pathogens.
    e. It improves the taste of water.
40. What is the main disadvantage of using distribution for water treatment?
    a. It is expensive.
    b. It is difficult to handle.
    c. It can react with organic matter to form harmful by-products.
    d. It is not effective against all types of pathogens.
    e. It has a bad taste.

ANSWERS TO MULTIPLE-CHOICE QUESTIONS

1	e	11	e	21	e	31	a
2	b	12	c	22	a	32	b
3	c	13	d	23	d	33	d
4	a	14	c	24	a	34	a
5	b	15	c	25	d	35	d
6	a	16	a	26	a	36	c
7	c	17	d	27	e	37	a
8	d	18	a	28	a	38	a
9	a	19	e	29	a	39	a
10	c	20	d	30	b	40	a

FURTHER READING

Ahuja, S., & Ergas, S. J. (Eds.). (2016). *Water reuse: An international survey of current practice, issues and needs*. IWA Publishing.

Alcamo, J., & Henrichs, T. (Eds.). (2002). *World water in 2025: Global modeling and scenario analysis for the World Commission on Water for the 21st Century*. Springer Science & Business Media.

Allan, J. A. (2011). *Virtual water: Tackling the threat to our planet's most precious resource*. I.B.Tauris.

Bartram, J., & Ballance, R. (Eds.). (1996). *Water quality monitoring: A practical guide to the design and implementation of freshwater quality studies and monitoring programs*. CRC Press.

Biswas, A. K. (2010). *Water management in 2020 and beyond*. Springer.

Biswas, A. K., Tortajada, C., & Braga, B. (Eds.). (2005). *Integrated water resources management in South and South-East Asia*. Oxford University Press.

Charalambous, B., Graham, N. J., & Makropoulos, C. (Eds.). (2014). *Decision support tools for sustainable urban development*. Springer.
Chaudhry, M. Q. (2011). *Water supply systems security*. Springer Science & Business Media.
Dandy, G., Maier, H., & Ravalico, J. K. (2016). *Decision making in water resources policy and management: An Australian perspective*. Elsevier.
DeOreo, W. B., Mayer, P. W., Dziegielewski, B., Kiefer, J. C., & Nelson, J. O. (2016). *Residential end uses of water, version 2*. Water Research Foundation.
Fewkes, A., Butler, D., & Memon, F. A. (2018). *Water demand management in urban areas*. IWA Publishing.
Gleick, P. H. (2010). *Bottled and sold: The story behind our obsession with bottled water*. Island Press.
Gupta, A., & Gupta, A. (2014). *Water supply engineering: Environmental engineering series. I.K.* International Publishing House Pvt Ltd.
Gupta, R. P., & Gupta, S. K. (2012). *Water resources systems planning and management*. PHI Learning Pvt. Ltd.
Hammer, M. J. (2015). *Water and wastewater technology*. Pearson.
Huisman, L., Olsthoorn, T., & Van de Ven, F. H. M. (Eds.). (2007). *Water governance in the Netherlands: Fit for the future?* CRC Press.
Liu, G., & Zhang, Y. (2018). *Sustainable water management and technologies*. CRC Press.
Mays, L. W. (2012). *Water distribution systems handbook*. McGraw-Hill Education.
Pescod, M. B., & Bosher, L. S. (2017). *Water supply and pollution control* (8th ed.). Pearson.
Postel, S. L. (2015). *Replenish: The virtuous cycle of water and prosperity*. Island Press.
Salgot, M., Folch, M., & Huertas, E. (Eds.). (2016). *Water distribution systems*. CRC Press.
Sharma, S. K., & Goyal, R. K. (2018). *Water supply engineering*. S. Chand Publishing.
Sharma, S. K., & Sanghi, R. (2017). *Water supply and sanitation*. S. Chand Publishing.
Swain, R. B., & Saha, N. C. (2016). *Water management in 21st century: Challenges and opportunities*. Springer.
Swarzenski, P. W., & Izbicki, J. A. (2018). Groundwater availability of the Central Valley Aquifer, California: US Geological Survey Professional Paper 1766.
Tchobanoglous, G., & Schroeder, E. D. (2013). *Water supply* (6th ed.). Pearson.
US Environmental Protection Agency. (2018). Drinking water standards and health advisories. EPA.
Viessman, W., & Hammer, M. J. (2012). *Water supply and pollution control*. Pearson.
Ward, S., & Butcher, D. (Eds.). (2006). *Water recycling in Australia: A guide to understanding and practice*. CSIRO Publishing.
Ward, S., & Pulido-Velazquez, M. (Eds.). (2008). *Water conservation, reuse, and recycling: Proceedings of an Iranian-American workshop*. National Academies Press.

8 Wastewater System

8.1 GENERAL OVERVIEW

Wastewater systems, also known as sewerage systems, are a network of pipes, pumps, and force mains designed for the collection, treatment, and disposal of wastewater. Wastewater is the water-borne waste material of a community that includes all normal wastes from human activities and residences. These systems can be categorized into domestic and industrial sewers and storm sewers.

A wastewater system is essential in maintaining the health and well-being of communities around the world. These systems are designed to collect, treat, and dispose of water that has been used in homes, businesses, and industries. This water, known as wastewater, contains a variety of pollutants that can be harmful to both humans and the environment if not properly managed.

The first stage in a wastewater system is the collection of used water. This is typically done through a network of pipes and pumps that transport the wastewater from its source to a treatment facility. The design and maintenance of this collection system is a complex task that requires careful planning and engineering.

Once the wastewater has been collected, it undergoes a series of treatment processes to remove the pollutants. These processes can be broadly categorized into primary, secondary, and tertiary treatment. Primary treatment involves the removal of large solids from the wastewater, while secondary treatment uses biological processes to break down organic matter. Tertiary treatment, also known as advanced treatment, involves further processes to remove nutrients and other contaminants.

The treated wastewater is then discharged back into the environment, usually into a body of water such as a river or ocean. The discharge of treated wastewater is regulated by environmental laws to ensure that it does not harm the ecosystem.

8.2 PLANNING AND DESIGN OF WASTEWATER TREATMENT SYSTEMS

When planning and designing wastewater treatment systems, several important factors need to be considered:

1. *Understanding the Local Environment:* The climate, geography, and geology of the area must be considered to create an effective and efficient plant.
2. *Analyzing the Wastewater:* This includes testing the water for various contaminants, such as bacteria, viruses, fungi, and other microorganisms.
3. *Designing the Treatment Process:* This includes choosing the right type of equipment and layout for the plant. It is important to consider the plant's size and the type of wastewater that will be treated.

DOI: 10.1201/9781032664613-8

4. *Building the Wastewater Treatment Plant:* After the design process, the next step is to build the wastewater treatment plant.
5. *Flow Rate, Temperature, pH, Organic and Inorganic Content:* These factors are crucial to analyze the characteristics of the wastewater to be treated thoroughly.
6. *Presence of Toxic Substances:* The presence of toxic substances in the wastewater must also be considered.

A detailed explanation of the above factors is given in the following sections.

8.2.1 Local Environment

Understanding the local environment in planning and design of a wastewater system, particularly a wastewater treatment system, is necessary for the high effectiveness and efficiency of the system. Let us break it down:

Climate: The climate of the area where the plant will be located can significantly impact the design and operation of the wastewater treatment plant. For example, in colder climates, the plant may need to be designed to handle the freezing of wastewater, which can affect the treatment process. On the other hand, in warmer climates, evaporation rates may be higher, which can affect the volume of wastewater that needs to be treated. Additionally, rainfall patterns can influence the amount of stormwater entering the system, which can lead to overflow events if not properly managed.

Geography: The geographical location of the plant can also have significant implications. For instance, the topography of the land can affect the gravity flow of wastewater through the system, potentially reducing the need for energy-intensive pumping. The proximity to residential areas, businesses, and sensitive ecological sites may also influence the design of the plant to minimize odor, noise, and environmental impact.

Geology: The underlying geology of the site is another important factor. The type of soil and bedrock can affect the construction of the plant and the design of the wastewater disposal system. For example, in areas with permeable soils, treated wastewater might be discharged through land application systems, where it can percolate through the soil for additional treatment. However, in areas with impermeable soils or high groundwater tables, alternative disposal methods may need to be considered.

By considering the climate, geography, and geology of the area, engineers can design a plant that is both effective in treating wastewater and efficient in its use of resources.

8.2.2 Analyzing the Characteristics of Wastewater

Analyzing the wastewater includes testing the water for various contaminants, such as bacteria, viruses, fungi, and other microorganisms. It refers to the process of examining the wastewater to identify and quantify the presence of various pollutants. This helps determine the appropriate treatment methods and assess the effectiveness of the treatment process. Let us delve into this issue in more detail:

Testing for Contaminants: Wastewater typically contains a wide range of contaminants, including organic and inorganic substances, suspended solids, nutrients like nitrogen and phosphorus, heavy metals, and various microorganisms. These contaminants can originate from various sources, including domestic sewage, industrial effluents, and stormwater runoff. Testing the wastewater involves collecting samples and analyzing them in a laboratory using various

TABLE 8.1
Methods for Testing Wastewater

Method	Explanation
Colorimetric methods	These methods involve the use of color reactions to determine the concentration of a particular substance in the wastewater.
Titrimetric methods	These methods involve the use of titration, a process where a solution of known concentration is used to determine the concentration of an unknown solution.
Electrometric methods	These methods involve the use of meters and probes to measure various properties of the wastewater, such as pH, conductivity, and dissolved oxygen.
Turbimetric and nephelometric methods	These methods involve the measurement of the turbidity or cloudiness of the wastewater, which can indicate the amount of suspended solids.
Biochemical oxygen demand (BOD)	This test measures the amount of oxygen required by bacteria to decompose the organic matter in the wastewater over a certain period.
Chemical oxygen demand (COD)	This test measures the total quantity of oxygen required to oxidize all organic material into carbon dioxide and water.
Total organic carbon (TOC)	This test measures the total amount of carbon in the wastewater, including both inorganic and organic carbon.
Oil and grease (O&G):	This test measures the concentration of oil and grease in the wastewater.

analytical techniques. The results of these tests provide valuable information about the composition of the wastewater and the levels of different contaminants.

Microorganisms in Wastewater: One of the key groups of contaminants in wastewater is microorganisms, which include bacteria, viruses, and fungi. These microorganisms can cause a variety of diseases if they are not effectively removed during the treatment process. Therefore, wastewater is routinely tested for the presence of pathogenic (disease-causing) microorganisms. This is usually done by culturing the microorganisms on a nutrient medium and then identifying them under a microscope. More advanced techniques, such as molecular methods, can also be used to detect specific types of microorganisms.

Importance of Wastewater Analysis: Analyzing wastewater is not only important for protecting public health and the environment, but it is also a regulatory requirement in many jurisdictions. The results of the wastewater analysis are used to ensure that the treatment process is working effectively and that the treated wastewater meets the required standards before it is discharged back into the environment. Furthermore, by identifying the types and concentrations of contaminants in the wastewater, it is possible to identify potential sources of pollution and implement measures to reduce these at the source.

There are several common methods for testing wastewater, each designed to detect and measure different types of contaminants as shown in Table 8.1.

These methods provide valuable information about the quality of the wastewater and the effectiveness of the treatment process. They help ensure that the treated wastewater meets the required standards before it is discharged back into the environment.

8.2.3 Designing the Treatment Process

Designing the treatment process for a wastewater treatment plant involves several important considerations to ensure the plant operates efficiently and effectively. This process entails the following steps:

Choosing the Right Type of Equipment: Choosing the right type of equipment is a critical step in ensuring the efficient and effective treatment of wastewater. The selection process involves identifying and acquiring the necessary machinery and technology essential for treating wastewater, such as pumps, filters, and reactors. This equipment is essential in various stages of the treatment process, from removing impurities to facilitating chemical reactions that help purify the water.

One of the key factors influencing the choice of equipment is the specific contaminants present in the wastewater being treated. Different contaminants require specific treatment methods and equipment to effectively remove or neutralize them. By understanding the composition of the wastewater and the types of contaminants it contains, treatment facilities can select the most suitable equipment to achieve optimal results.

Treatment facilities must also consider the desired quality of the treated water and the regulatory standards that govern wastewater treatment. The quality of the treated water often dictates the level of treatment required, which in turn influences the selection of equipment. Furthermore, adherence to regulatory standards is essential to ensure that the treatment plant operates in compliance with environmental regulations and safeguards public health. By carefully evaluating these factors, treatment facilities can make informed decisions when choosing the right type of equipment for their wastewater treatment processes.

Plant Layout: Plant layout is a component that may impact the efficiency and functionality of a wastewater treatment facility. It encompasses the spatial arrangement of equipment and facilities within the plant, determining how wastewater flows through different treatment stages. A well-thought-out layout is essential for ensuring the smooth movement of wastewater from one process to another, minimizing bottlenecks, and optimizing the overall treatment process. Additionally, a well-designed layout provides easy access for maintenance personnel to perform routine inspections, repairs, and equipment upgrades efficiently.

An effective plant layout allows for flexibility and scalability in plant operations. By considering future expansion or upgrading needs during the initial design phase, the layout can accommodate changes in treatment processes or capacity requirements without significant disruptions to plant operations. This adaptability is necessary for ensuring the long-term sustainability and viability of the treatment facility in response to evolving regulatory standards or increased wastewater loads.

The plant layout must also address logistical considerations such as waste sludge removal and the integration of control systems. Proper planning for waste sludge management ensures efficient handling and disposal of by-products generated during the treatment process, minimizing environmental impact, and ensuring regulatory compliance. Integration of control systems within the layout allows for centralized monitoring and control of plant operations, enhancing operational efficiency and facilitating timely adjustments to optimize treatment performance.

Size of the Plant: The size of a treatment plant determines its ability to effectively treat wastewater. It should be determined based on the volume of wastewater that needs to be processed, which is influenced by factors such as the population size it serves and the industrial or commercial activities in the surrounding area. A plant that is too small relative to the wastewater volume may struggle to handle the influx, leading to overloading and inadequate treatment capacity. On the other hand, an excessively large plant can result in unnecessary costs both during construction and operation, as resources are underutilized.

Ensuring that the size of the treatment plant is proportionate to the wastewater volume is essential for maintaining operational efficiency and effectiveness. By accurately assessing the expected volume of wastewater based on population and industrial activities, treatment facilities can design a plant that meets the demand without being overwhelmed or underutilized. Careful consideration helps to optimize treatment processes, ensuring that wastewater is adequately treated to meet regulatory standards and environmental requirements.

An appropriate size of a treatment plant is important for ensuring long-term sustainability and cost-effectiveness. A plant that matches the wastewater volume allows for efficient operation and maintenance, minimizing the risk of system failures or inefficiencies. This approach not only enhances the treatment plant's performance but also helps to control operational costs by avoiding unnecessary expenses associated with oversized infrastructure or emergency repairs due to overload situations.

Therefore, striking the right balance in determining the size of a treatment plant is essential for achieving optimal performance and cost-efficiency. By aligning the plant size with the expected wastewater volume based on population and industrial activities, treatment facilities can ensure reliable and effective wastewater treatment while managing operational costs effectively. Proper sizing considerations contribute to the overall success and sustainability of the treatment plant, enabling it to meet current and future wastewater treatment needs efficiently.

Type of Wastewater: The type of wastewater to be treated determines the design and implementation of an effective treatment process. Wastewater can originate from various sources, including domestic, industrial, or commercial activities, each contributing unique characteristics and levels of pollutants to the water. Understanding the origin and composition of the wastewater is essential for customizing the treatment approach to address the specific contaminants present in the water.

Domestic wastewater, for example, typically contains organic matter, nutrients, and pathogens from household activities, while industrial wastewater may include heavy metals, chemicals, and other industrial by-products depending on the industry. Commercial wastewater, on the other hand, can vary widely based on the nature of the business and activities conducted on the premises. By identifying the source and composition of the wastewater, treatment facilities can tailor their processes to effectively target and remove the specific pollutants present in the water.

The treatment process must, therefore, be designed to meet discharge or reuse standards set by regulatory authorities. By customizing the treatment approach to address the unique characteristics of the wastewater, treatment facilities can ensure that the treated water meets environmental quality guidelines and can be safely discharged into water bodies or reused for various purposes. Tailoring the treatment process to the type of wastewater being treated is essential for achieving efficient and effective removal of contaminants, ultimately contributing to environmental protection and public health.

8.2.4 Building the Wastewater Treatment Plant

Constructing a wastewater treatment plant is a step following the design process when the design was approved including the approval of socio-environmental impact assessments (EIAs). This marks the transition from planning to implementation in the journey toward efficient wastewater management. This phase involves the physical construction of the treatment facility based on the detailed design specifications developed earlier. The construction process typically encompasses various stages, including site preparation, foundation laying, structural erection, equipment installation, and system integration, all aimed at bringing the treatment plant from concept to reality.

Site preparation is the initial stage of building a wastewater treatment plant and involves clearing and leveling the designated area where the facility will be constructed. This step is essential for creating a suitable foundation for the plant structures and ensuring proper drainage to prevent flooding or waterlogging issues. The site preparation phase also includes soil testing to assess the ground conditions and determine the necessary measures for ensuring structural stability and integrity during construction.

Once the site is prepared, the foundation laying phase begins, where the structural base of the treatment plant is established. This involves excavating trenches, pouring concrete footings, and setting the foundation for various components of the facility, such as tanks, reactors, and buildings. The foundation serves as the structural support for the treatment plant, ensuring stability and durability over the plant's operational lifespan. Proper foundation construction is critical for withstanding environmental factors, such as seismic activity or soil settlement, and ensuring the long-term functionality of the treatment facility.

Following foundation laying, the structural erection phase involves assembling the main structural components of the treatment plant, including tanks, pipelines, and buildings. This phase requires skilled labor and adherence to engineering specifications to ensure that the structures are built according to design requirements. Structural erection plays a significant role in shaping the physical layout of the treatment plant and creating the framework for housing essential equipment and systems necessary for wastewater treatment. Attention to detail and precision during this phase are essential for ensuring that the treatment plant is constructed in a manner that optimizes operational efficiency and functionality.

Equipment installation is another critical aspect of building a wastewater treatment plant, involving the setup and integration of various machinery, technology, and systems required for treating wastewater. This phase includes installing pumps, filters, reactors, control systems, instrumentation, and other equipment essential for carrying out treatment processes effectively. Proper installation and calibration of equipment are essential for ensuring optimal performance and adherence to design specifications. Additionally, system integration involves connecting and testing the interdependent components of the treatment plant to ensure seamless operation and functionality across all treatment stages. By meticulously installing and integrating equipment and systems, treatment facilities can ensure that their wastewater treatment plant operates efficiently and meets regulatory standards for environmental protection and public health.

The EIA process is unavoidable in the design of a water treatment plant to evaluate and mitigate potential environmental effects associated with the project. The stages of EIA typically include scoping, baseline data collection, impact prediction, mitigation measures, and monitoring.

8.2.4.1 Scoping

The scoping stage involves defining the scope and boundaries of the EIA. This includes identifying the key environmental issues and concerns that need to be addressed during the design and construction of the water treatment plant. Stakeholder engagement must be undertaken in this stage to gather input from local communities, regulatory agencies, and other relevant parties to ensure that all potential impacts are considered.

8.2.4.2 Baseline Data Collection

Baseline data collection involves gathering information on the existing environmental conditions in the project area before the construction of the water treatment plant. This includes assessing factors such as air quality, water quality, soil conditions, biodiversity, and socioeconomic aspects. The collected baseline data serves as a reference point for evaluating changes that may occur as a result of the project.

8.2.4.3 Impact Prediction

In this stage, the potential environmental impacts of the water treatment plant project are identified and predicted based on the collected baseline data and project activities. This involves assessing both direct and indirect impacts on the environment, such as changes in water quality, habitat disruption, air emissions, noise pollution, and socioeconomic effects. Impact prediction helps in understanding the magnitude and significance of potential impacts to inform decision-making and planning for mitigation measures.

8.2.4.4 Mitigation Measures

After identifying potential environmental impacts, mitigation measures are developed to minimize or eliminate adverse effects on the environment. These measures aim to reduce impacts to acceptable levels through design modifications, technology improvements, operational practices, or compensatory actions. Examples of mitigation measures in the design of a water treatment plant may include incorporating green infrastructure for stormwater management, implementing noise reduction measures, or restoring habitat areas affected by construction activities.

8.2.4.5 Monitoring

Monitoring is a critical stage that involves ongoing assessment and evaluation of environmental impacts throughout the life cycle of the water treatment plant project. Monitoring helps ensure that mitigation measures are effective in reducing impacts as predicted and allows for adaptive management to address any unforeseen environmental consequences. Regular monitoring also facilitates compliance with regulatory requirements and demonstrates a commitment to environmental stewardship and sustainability.

By doing the EIA in the design of a water treatment plant, project developers can proactively identify, assess, and address potential environmental impacts to ensure sustainable development and minimize harm to the surrounding environment and communities.

8.2.5 Wastewater Sample Contents

To ensure the successful design and construction of a wastewater treatment plant, it is imperative to thoroughly understand the characteristics of the wastewater it will process. This understanding is foundational to the plant's efficiency and effectiveness, as it informs every aspect of its operation, from the selection of treatment methods to the sizing of equipment and the overall layout of the facility. The goal of this sampling is to obtain a sample that accurately reflects the wastewater's characteristics for reliable analysis and treatment design.

Sufficient sampling of wastewater is the cornerstone of this understanding. By collecting and analyzing a representative range of samples, engineers and scientists can gain insights into the variability and composition of wastewater. This includes determining the concentration of organic and inorganic matter, the presence of hazardous substances, and the physical properties such as temperature and pH. These factors are critical in selecting the most appropriate and cost-effective treatment technologies.

Moreover, regular sampling allows for the monitoring of influent and effluent quality, ensuring that the plant meets regulatory standards and environmental protection goals. It also aids in identifying any changes in the wastewater's characteristics over time, which could necessitate adjustments in the treatment process.

The practice of taking sufficient samples is not merely a procedural step; it is a strategic approach that underpins the entire lifecycle of a wastewater treatment plant. It ensures that the plant is designed with a deep understanding of the challenges it will face, equipped to handle the demands of the wastewater it treats, and capable of delivering treated water that is safe for discharge or reuse.

Sampling is a worthy investment in the plant's future performance and reliability. It is through this diligent process that we can guarantee the creation of wastewater treatment facilities that are not only efficient and effective but also sustainable and adaptable to the evolving needs of the communities they serve. Here is a guidance for taking sufficient samples of wastewater:

1. *Determine the Sampling Method:*
   a. *Grab Sampling:* This method involves collecting a single sample at a specific moment in time. It is useful for analyzing pollutants that vary significantly over short periods.

b. *Composite Sampling:* This method combines multiple grab samples over a set period, providing an average representation of the wastewater characteristics.

2. *Select the Sampling Point:* Choose locations that are representative of the wastewater stream, such as the influent or effluent of the treatment plant.
3. *Use Proper Equipment:*
   a. Utilize clean, sterilized containers to avoid contamination.
   b. For composite samples, automatic samplers can be used to collect samples at predetermined intervals.
4. *Follow Safety Procedures:*
   a. Wear appropriate personal protective equipment (PPE) such as gloves and safety glasses.
   b. Ensure that the sampling procedures do not pose any risk to the sampler or the environment.
5. *Preserve the Sample:*
   a. Keep the samples at a controlled temperature, usually around 4°C, to prevent biological activity from altering the sample composition.
   b. Transport the samples to the laboratory as soon as possible for analysis.

8.2.6 Presence of Toxic Substances

From the samples we took and tested in the laboratory, the sample may reflect the presence of toxic substances in wastewater. This is a concern for wastewater treatment processes. These substances can include a variety of pollutants such as heavy metals, pesticides, pharmaceuticals, and industrial chemicals. Here are the explanations:

1. *Impact on Treatment Processes:* Toxic substances can disrupt the biological treatment processes that are central to many wastewater treatment systems. For instance, heavy metals can inhibit the growth and function of microorganisms that break down organic matter. Some toxic substances may not be effectively removed through standard treatment processes and require specialized treatment methods, such as advanced oxidation or adsorption.
2. *Environmental and Health Risks:* If not properly removed, toxic substances can pose significant risks to the environment and public health. They can contaminate water bodies, accumulate in the food chain, and cause acute or chronic health effects in humans and wildlife. The treatment of wastewater aims to reduce these pollutants to levels that nature can handle without causing harm.
3. *Regulatory Compliance:* Wastewater treatment plants must comply with stringent regulatory standards for the discharge of treated effluent. These regulations often set limits on the concentration of toxic substances to protect water quality. Failure to meet these standards can result in legal penalties and increased scrutiny from regulatory agencies.
4. *Challenges in Detection and Analysis:* Detecting and quantifying toxic substances can be challenging due to their diverse nature and the complexity of wastewater matrices. Accurate analysis is essential for understanding the extent of contamination and for designing treatment processes that can effectively remove these substances.
5. *Advancements in Treatment Technologies:* Research and development are leading to more effective treatment technologies that can address the challenges posed by toxic substances in wastewater. Innovations such as membrane filtration, advanced biological treatment, and chemical precipitation are improving the ability of treatment plants to remove a wider range of contaminants.

Considering the presence of toxic substances is essential for the design and operation of wastewater treatment plants. It ensures the protection of environmental and public health, compliance with regulations, and the advancement of treatment technologies to meet the demands of modern wastewater management.

8.3 COLLECTION AND CONVEYANCE

Effective wastewater management is a critical component of modern urban infrastructure. As populations continue to grow and concentrate in cities, the collection and conveyance of wastewater become increasingly complex and vital. Proper wastewater handling is essential for protecting public health, preserving the environment, and supporting economic development.

The wastewater collection and conveyance system in a typical city consists of three main elements: the sewer network, pumping stations, and wastewater treatment plants.

The foundation of the wastewater system is the sewer network – the underground pipes that collect and transport wastewater from homes, businesses, and other facilities to the treatment plant. Sewer networks can be designed as either combined systems, which carry both stormwater and sanitary sewage, or separate systems, with distinct pipes for stormwater and wastewater.

Sewer pipes are constructed from materials like concrete, polyvinyl chloride (PVC), or clay, and are sized and sloped to allow gravity-driven flow. The network is organized hierarchically, with smaller lateral pipes collecting wastewater from individual properties and feeding into larger trunk lines and interceptor sewers. Manholes provide access points for inspection, maintenance, and cleaning of the sewer system.

While gravity is the primary driving force, pumping stations are often necessary to lift wastewater and overcome changes in elevation or other hydraulic constraints within the sewer network. These stations use electric pumps to boost the flow of wastewater through the pipes, ensuring continuous conveyance to the treatment plant.

Pumping stations are strategically placed at low points in the system, where wastewater naturally accumulates. They are equipped with backup power sources, overflow storage, and other redundancies to maintain operations during power outages or periods of high flow.

The final destination for the collected wastewater is the treatment plant, where physical, biological, and chemical processes remove contaminants and pollutants. Treatment plants are designed based on the anticipated volume and composition of the incoming wastewater, as well as the desired quality of the treated effluent.

The treatment process typically begins with screening and grit removal to extract large solids and debris. Primary clarification then settles out suspended organic matter. Secondary biological treatment, often using activated sludge, degrades the remaining dissolved and particulate organic material. Advanced tertiary treatment, including disinfection, may also be employed to meet stringent water quality standards before the treated effluent is discharged or reused. The solids removed during treatment, known as sludge, undergo further processing, such as thickening, digestion, and dewatering, before being disposed of or beneficially reused, such as in land application as a soil amendment.

Many cities struggle with aging, deteriorating wastewater infrastructure that was built decades or even centuries ago. Crumbling pipes, outdated pumping equipment, and capacity limitations can lead to sewer overflows, infiltration and inflow, and other operational issues. Addressing this infrastructure gap requires significant investment in rehabilitation, replacement, and system upgrades. Municipalities must carefully plan and prioritize capital improvements to maintain reliable service and compliance with environmental regulations.

Heavy rainfall and snowmelt can overwhelm combined sewer systems, causing untreated wastewater and stormwater to overflow into local waterways. This is a significant environmental and public health concern, as these combined sewer overflows (CSOs) can introduce pathogens, nutrients, and other pollutants into rivers, lakes, and coastal areas.

Strategies to mitigate CSOs include separating combined sewers, constructing storage tanks or tunnels to hold excess flows, and implementing green infrastructure solutions like permeable surfaces and stormwater detention basins.

The wastewater stream increasingly contains trace concentrations of pharmaceuticals, personal care products, microplastics, and other emerging contaminants that traditional treatment plants are not designed to remove. The potential environmental and human health impacts of these substances are not yet fully understood but are an area of growing concern and research. Addressing emerging contaminants may require advanced treatment technologies, such as membrane filtration, ozonation, or activated carbon adsorption, which can be costly to implement at the municipal scale.

As wastewater management practices evolve, there is growing interest in transitioning toward more sustainable, resource–recovery–focused approaches. This includes exploring opportunities to extract energy, nutrients, and other valuable materials from the wastewater stream, rather than simply treating and disposing of it. Innovations like anaerobic digestion, nutrient recovery, and water reuse can transform wastewater facilities into resource recovery centers, contributing to the circular economy and reducing the environmental footprint of urban water systems.

The collection and conveyance of wastewater is a fundamental component of modern urban infrastructure, protecting public health and the environment. However, aging systems, wet weather challenges, emerging contaminants, and the need for greater sustainability require cities to continuously invest in, optimize, and innovate their wastewater management practices.

8.3.1 Types of Collection Systems

This subsection can discuss the different types of collection systems, such as combined sewer systems, sanitary sewer systems, and storm sewer systems. Each type can be explained in detail, including its design, operation, and typical applications.

Municipalities typically employ one of three main types of wastewater collection systems:

1. combined sewer systems,
2. sanitary sewer systems, and
3. storm sewer systems.

Each system has distinct design characteristics, operational considerations, and applications, as discussed in the subsequent sections.

8.3.1.1 Combined Sewer Systems

Combined sewer systems are designed to collect and convey both sanitary wastewater and stormwater runoff through a single set of underground pipes. This integrated approach was common in many older cities, particularly those developed prior to the mid-20th century (Figure 8.1).

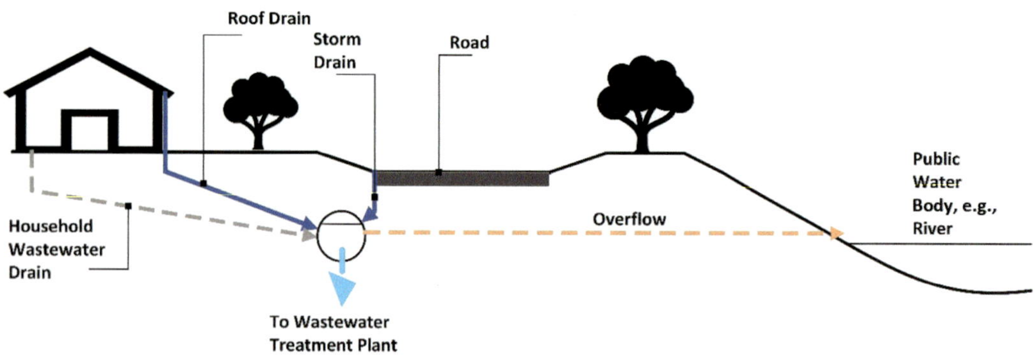

FIGURE 8.1 Combined sewer systems.

8.3.1.1.1 Design and Operation

In a combined system, stormwater from streets, roofs, and other impervious surfaces flows into the same sewer network that collects wastewater from homes, businesses, and other facilities. During dry weather, the combined flow is directed to a wastewater treatment plant for full treatment. However, during heavy rain or snowmelt events, the volume of stormwater can exceed the capacity of the sewer system and treatment plant. In these situations, CSOs are designed to discharge the excess flow directly into nearby waterways, bypassing treatment. This prevents backups and flooding within the collection system.

8.3.1.1.2 Typical Applications

Combined sewer systems are most common in older, densely populated urban areas where the original infrastructure was built before the separation of sanitary and storm sewers became standard practice. Cities like New York, Chicago, and Boston have extensive combined sewer networks.

8.3.1.1.3 Advantages and Disadvantages

The primary advantage of combined systems is their simplicity and cost-effectiveness, as a single set of pipes and a single treatment plant can handle both wastewater streams. However, the major downside is the environmental and public health impact of CSOs, which can introduce untreated sewage, debris, and other pollutants into local water bodies.

8.3.1.2 Sanitary Sewer Systems

Sanitary sewer systems are designed to collect and convey only domestic, commercial, and industrial wastewater, keeping it separate from stormwater runoff (Figure 8.2).

8.3.1.2.1 Design and Operation

In a sanitary sewer system, wastewater from toilets, sinks, showers, and other plumbing fixtures flows into a network of underground pipes that transport it to a treatment plant. Stormwater is handled through a separate storm sewer system or other drainage infrastructure. Sanitary sewers are sized and sloped to maintain sufficient flow velocity for self-cleaning, and are equipped with manholes, cleanouts, and other access points for maintenance. Lift stations or pumping stations may be required to overcome changes in elevation and convey the wastewater to the treatment facility.

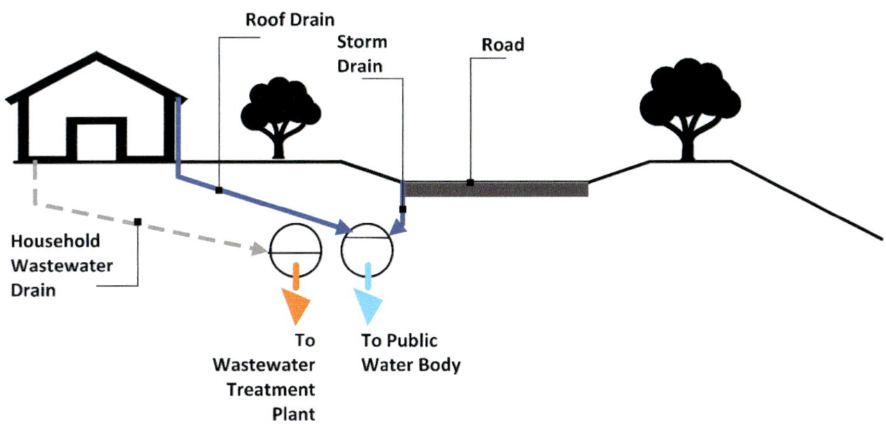

FIGURE 8.2 Separated sewer systems (sanitary and storm).

8.3.1.2.2 *Typical Applications*

Sanitary sewer systems are the predominant wastewater collection approach in most modern cities and suburbs. They are the preferred option in areas with separate storm drainage infrastructure or where stormwater is managed through other means, such as surface drainage or infiltration.

8.3.1.2.3 *Advantages and Disadvantages*

The key advantage of a sanitary sewer system is the elimination of CSOs and the associated environmental and public health risks. However, the need for a dedicated network of sanitary sewer pipes, as well as separate stormwater infrastructure, can result in higher construction and maintenance costs compared to combined systems.

8.3.1.3 Storm Sewer Systems

Storm sewer systems are designed solely to collect and convey stormwater runoff from streets, parking lots, roofs, and other impervious surfaces. They do not handle any sanitary wastewater (Figure 8.2).

8.3.1.3.1 *Design and Operation*

Storm sewer networks typically consist of catch basins, inlets, and underground pipes that transport stormwater to a nearby water body, such as a river, lake, or the ocean. The pipes are sized and sloped to accommodate the expected peak flows during heavy rainfall or snowmelt events. Unlike sanitary sewers, storm sewers are not required to maintain self-cleaning velocities, as the intermittent nature of stormwater flow means that debris and sediment can accumulate between storm events. Regular maintenance, including street sweeping and catch basin cleaning, is necessary to prevent blockages and ensure proper drainage.

8.3.1.3.2 *Typical Applications*

Storm sewer systems are commonly found in modern suburban and urban developments, where they work in tandem with sanitary sewer networks to manage stormwater runoff separately from wastewater. They are also used in some older cities that have transitioned from combined to separate sewer systems.

8.3.1.3.3 *Advantages and Disadvantages*

The primary advantage of a dedicated storm sewer system is the elimination of CSOs and the associated environmental impacts. It also allows for more targeted treatment and management of stormwater, which may contain different pollutants than sanitary wastewater. However, the need for a separate network of storm sewer pipes and associated drainage infrastructure can result in higher overall costs for the municipal water system.

8.3.2 Components of a Collection System

The wastewater collection system transports used water and other waste products from homes, businesses, and industries to a treatment facility. This comprehensive system comprises several interconnected components, each playing a vital role in the efficient management of wastewater. We will explore the various components that make up a wastewater collection system, including pipes, manholes, pumping stations, and treatment facilities, as well as the materials used and their typical lifespan.

8.3.2.1 Pipes

The backbone of a wastewater collection system is the network of pipes that convey the wastewater from its source to the treatment facility. These pipes come in a variety of materials, each with its own advantages and disadvantages. Common pipe materials used in wastewater collection systems include:

1. *PVC:* PVC pipes are widely used due to their durability, corrosion resistance, and relatively low cost. They have an estimated lifespan of 50–100 years.
2. *Ductile Iron:* Ductile iron pipes are known for their strength and resistance to physical damage. They can last for 75–100 years but may be susceptible to corrosion if not properly coated or protected.
3. *Concrete:* Concrete pipes are often used for large-diameter, gravity-fed wastewater lines. They can have a lifespan of 50–100 years but may be vulnerable to chemical and biological corrosion.
4. *High-Density Polyethylene (HDPE):* HDPE pipes are becoming increasingly popular due to their flexibility, resistance to corrosion, and ease of installation. They can have a lifespan of 50–100 years.

8.3.2.2 Manholes

Manholes are essential components of the wastewater collection system, providing access points for maintenance, inspection, and cleaning. These structures are typically made of concrete, precast concrete, or brick, and can have a lifespan of 50–100 years, depending on the material and environmental conditions (Figure 8.3).

8.3.2.3 Pumping Stations

In areas where the terrain is not conducive to gravity-fed wastewater flow, pumping stations are used to lift the wastewater to a higher elevation, allowing it to continue its journey to the treatment facility. Pumping stations typically consist of the following components:

1. *Wet Well:* The wet well is a submerged chamber that collects the incoming wastewater.
2. *Pumps:* Submersible or dry-pit pumps are used to lift the wastewater from the wet well and discharge it into the next section of the collection system.
3. *Controls and Electrical Systems:* These components regulate the operation of the pumps and monitor the system's performance.

Pumping stations are often constructed using concrete, with the wet well and other components designed to withstand the corrosive nature of wastewater. The lifespan of a pumping station can range from 20 to 50 years, depending on the quality of the equipment, maintenance practices, and environmental factors.

8.3.2.4 Treatment Facilities

The final component of the wastewater collection system is the treatment facility, where the collected wastewater undergoes a series of physical, chemical, and biological processes to remove contaminants and make the water safe for discharge or reuse. The main components of a typical wastewater treatment facility include:

1. *Preliminary Treatment:* This stage removes large objects, grit, and other coarse materials from the wastewater.
2. *Primary Treatment:* In this stage, the wastewater undergoes sedimentation, where solids settle out and are removed.
3. *Secondary Treatment:* Biological processes, such as activated sludge or trickling filters, are used to break down organic matter in the wastewater.

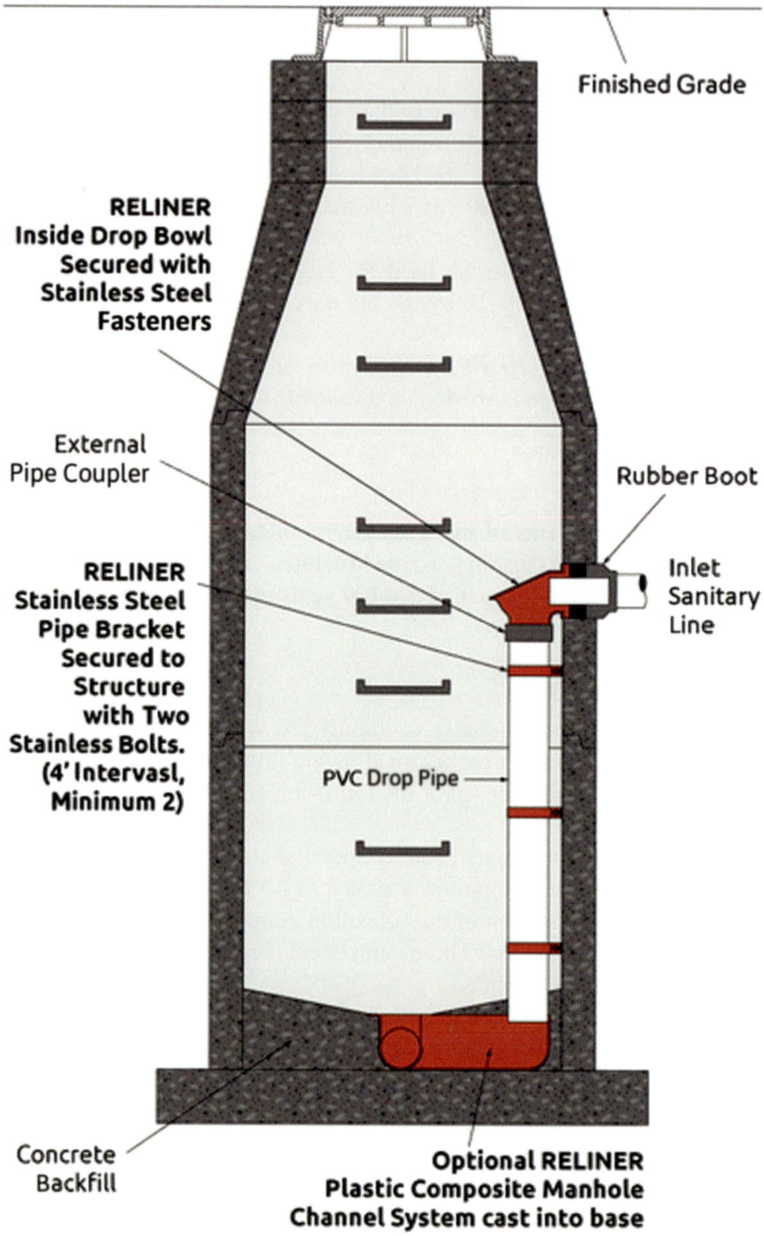

FIGURE 8.3 Manhole and its components.

Source: Opentextbc.ca

4. *Tertiary Treatment:* Additional treatment steps, such as disinfection and nutrient removal, may be employed to further purify the wastewater.
5. *Sludge Treatment and Disposal:* The solid waste (sludge) generated during the treatment process is treated and disposed of or recycled.

Wastewater treatment facilities are typically constructed using a combination of concrete, steel, and specialized equipment, such as pumps, aeration systems, and disinfection units. The lifespan of a treatment facility can vary widely, ranging from 20 to 50 years, depending on the quality of construction, maintenance practices, and technological upgrades.

The wastewater collection system is an interconnected network of pipes, manholes, pumping stations, and treatment facilities, all working together to effectively manage the flow and treatment of used water and waste products. The choice of materials used in these components, such as PVC, ductile iron, concrete, and HDPE, is crucial in ensuring the system's durability and longevity, with typical lifespans ranging from 20 to 100 years, depending on the specific component and environmental conditions. By understanding the various components of a wastewater collection system and their respective materials and lifespans, municipalities and water authorities can plan for the maintenance, repair, and eventual replacement of this critical infrastructure, ensuring the continued provision of essential wastewater management services to their communities.

8.3.3 Design Considerations

In the design of a wastewater collection system usually considers various factors to ensure the efficient and effective management of wastewater. These factors include population size, topography, soil type, climate, and future growth projections. Each of these elements plays a crucial role in shaping the design of the wastewater collection system, as they directly influence the system's capacity, layout, and overall performance.

8.3.3.1 Population Size

The size of the population served by the wastewater collection system is a fundamental factor in its design. The system must be capable of handling the current and projected future wastewater flows generated by the population. Accurate population estimates, including both residential and commercial/industrial users, are essential to determine the required pipe sizes, pumping capacities, and treatment facility dimensions.

8.3.3.2 Topography

The terrain and elevation of the service area significantly impact the design of the wastewater collection system. In areas with a gradual slope, gravity-fed systems can be employed, allowing wastewater to flow naturally toward the treatment facility. However, in areas with uneven or steep terrain, pumping stations may be necessary to lift the wastewater to higher elevations, enabling it to continue its journey to the treatment plant.

8.3.3.3 Soil Type

The composition and characteristics of the soil in the service area can affect the design of the wastewater collection system. Factors such as soil permeability, corrosivity, and stability influence the selection of pipe materials, the depth of pipe installation, and the need for specialized bedding or backfill requirements. For example, in areas with highly corrosive soils, the use of corrosion-resistant pipe materials, such as PVC or HDPE, may be necessary to ensure the long-term integrity of the system.

8.3.3.4 Climate

The local climate, including factors such as temperature, precipitation, and seasonal variations, can impact the design of the wastewater collection system. In colder climates, the system must be designed to prevent freezing of the pipes, which may require deeper burial depths or the use of insulation. In areas with high rainfall or frequent flooding, the system must be designed to withstand the increased wastewater flows and prevent infiltration and inflow of stormwater into the system.

8.3.3.5 Future Growth Projections

Designing a wastewater collection system requires consideration of future population and development growth within the service area. Accurate projections of population growth, changes in land use, and potential industrial or commercial expansions are essential to ensure that the system has sufficient capacity to accommodate future demands. This may involve the installation of larger-diameter pipes, the construction of additional pumping stations, or the expansion of the treatment facility.

8.3.3.6 Influence on System Design

The consideration of these factors directly influences the design of the wastewater collection system. For example, a system serving a large population will require larger-diameter pipes, higher-capacity pumping stations, and a treatment facility with a greater processing capacity compared to a system serving a smaller population.

Similarly, the topography of the service area will dictate the layout of the pipe network, the placement of pumping stations, and the overall gravity-fed or pressure-driven nature of the system. Soil type may necessitate the use of specific pipe materials, specialized bedding, or additional corrosion protection measures.

Climate conditions can influence the depth of pipe installation, the selection of materials that can withstand temperature extremes, and the incorporation of measures to prevent freezing or address increased wastewater flows during wet seasons.

Future growth projections guide the design of the system to ensure that it can accommodate the anticipated changes in wastewater generation and flow patterns over the system's lifespan.

When designing a wastewater collection system, several climatic factors are considered, and they can significantly impact the system's operation. The key climatic factors to be considered are shown in Table 8.2.

When designing a wastewater collection system in areas with high precipitation or heavy snowfall, there are several considerations that design engineers must consider as well. These technical considerations are shown in Table 8.3.

It is important to incorporate various factors into the design process of a wastewater collection system. By doing so, design engineers can create a system that is not only better prepared to withstand the possible and potential constraints, but also ensures reliable and efficient operation throughout its lifespan.

8.3.4 Operation and Maintenance

Proper operation and maintenance of a wastewater collection system are crucial for ensuring its long-term performance, reliability, and compliance with regulatory requirements. The key procedures involved in the operation and maintenance of a wastewater collection system include routine inspections, cleaning, repair, and replacement of components.

8.3.4.1 Routine Inspections

Regular inspections of the wastewater collection system are essential to identify potential issues and address them before they escalate. This includes:

1. Visual inspections of manholes, lift stations, and other access points to check for signs of damage, corrosion, or blockages.
2. Closed-circuit television inspections of the sewer pipes to detect cracks, breaks, root intrusion, or other structural defects.
3. Flow monitoring to identify areas with excessive inflow and infiltration or changes in wastewater flow patterns.

TABLE 8.2
Climatic Factors in Wastewater Collection System

Climatic Factor	Explanation
Temperature	• Extreme temperatures, both high and low, can affect the wastewater collection system. • In cold climates, the system must be designed to prevent freezing of the pipes, which can lead to blockages and damage. This may require deeper burial depths, the use of insulation, or the installation of heat tracing systems. • In hot climates, high temperatures can accelerate the deterioration of pipe materials and increase the risk of corrosion, particularly in the presence of hydrogen sulfide gas.
Precipitation	• The amount, intensity, and distribution of rainfall can impact the wastewater collection system. • Heavy or prolonged rainfall can lead to increased inflow and infiltration (I&I) of stormwater into the system, which can overload the system and cause backups or overflows. • In areas with high rainfall, the system must be designed to accommodate the increased wastewater flows, which may require larger-diameter pipes, additional pumping capacity, or the incorporation of stormwater management features.
Humidity and moisture	• High humidity and moisture levels can contribute to the formation of corrosive environments within the wastewater collection system. • Condensation within the system can lead to the accumulation of water, which can promote the growth of bacteria and the production of hydrogen sulfide gas, further exacerbating corrosion. • The design of the system may need to consider the use of corrosion-resistant materials, proper ventilation, and drainage systems to mitigate the effects of high humidity and moisture.
Seasonal variations	• Seasonal changes in temperature, precipitation, and groundwater levels can impact the performance of the wastewater collection system. • During wet seasons, the system may need to handle increased wastewater flows due to I&I, while during dry seasons, the system may experience lower flows, which can lead to the deposition of solids and the potential for odor issues. • The design of the system should account for these seasonal variations, ensuring that it can adapt to the changing conditions and maintain efficient operation throughout the year.
Natural disasters	• Extreme weather events, such as hurricanes, floods, or earthquakes, can significantly impact the wastewater collection system. • The system must be designed to withstand the forces and stresses associated with these natural disasters, including the potential for damage to pipes, manholes, and other infrastructure. • Resilience planning, such as the use of flexible pipe materials, redundant systems, and emergency backup power, can help mitigate the impact of natural disasters on the wastewater collection system.

4. Smoke testing or dye testing to locate sources of inflow and infiltration (I&I) and identify illegal connections.

8.3.4.2 Cleaning

Routine cleaning of the wastewater collection system is necessary to maintain the system's efficiency and prevent blockages. Cleaning methods may include:

1. *Flushing:* High-pressure water jets are used to remove accumulated debris and sediment from the pipes.
2. *Rodding or Jetting:* Mechanical devices are used to dislodge and remove obstructions in the pipes.
3. *Vacuum Truck Cleaning:* Specialized vacuum trucks are used to remove accumulated sludge and debris from manholes and lift stations.

TABLE 8.3
Technical Consideration for Wastewater Collection System in Areas with High Precipitation

Factor	Explanation
Inflow and infiltration (I&I) management	• High precipitation, whether in the form of rain or snowmelt, can lead to increased inflow and infiltration of stormwater into the wastewater collection system. • This excess water can overload the system, causing backups, overflows, and potential environmental damage. • To address I&I, the system design may include: - Properly sealing and maintaining pipes, manholes, and other infrastructure to prevent infiltration. - Incorporating stormwater diversion systems, such as catch basins and storm sewers, to separate stormwater from the wastewater collection system. - Designing the system with additional capacity to handle the increased flows during wet weather events.
Pipe depth and insulation	• In areas with heavy snowfall, the wastewater collection system must be designed to prevent freezing of the pipes. • This typically requires burying the pipes at a sufficient depth below the frost line, which can vary depending on the local climate. • Additionally, the use of insulation around the pipes, either using specialized pipe materials or external insulation, can help prevent freezing. • The design may also consider the use of heat tracing systems or other heating methods to maintain the temperature of the wastewater within the pipes.
Pumping station design	• In areas with high precipitation or heavy snowfall, the wastewater collection system may rely more heavily on pumping stations to lift the wastewater to higher elevations. • These pumping stations must be designed to handle the increased flows during wet weather events, with sufficient capacity and backup power systems to ensure continuous operation. • The location and elevation of the pumping stations should also be carefully considered to minimize the risk of flooding or damage from snowmelt or heavy rainfall.
Sewer system layout and slope	• The layout and slope of the wastewater collection system can be influenced by the local topography and the need to accommodate high precipitation or snowmelt. • In areas with steep slopes, the system may require the use of inverted siphons or other specialized design features to maintain the necessary flow velocities and prevent the deposition of solids. • In flat or low-lying areas, the system may need to be designed with increased slopes to ensure adequate drainage and prevent the accumulation of water during wet weather events.
Materials selection	• The selection of pipe materials and other infrastructure components must consider the potential for corrosion and deterioration due to the presence of moisture, chemicals, and other environmental factors associated with high precipitation or heavy snowfall. • Corrosion-resistant materials, such as PVC, HDPE, or stainless steel, may be preferred in these environments to ensure the long-term integrity of the system.

8.3.4.3 Repair and Replacement

Over time, components of the wastewater collection system may require repair or replacement due to age, wear, or damage. Common repair and replacement activities include:

1. *Pipe Rehabilitation:* Trenchless technologies, such as pipe relining or pipe bursting, are used to repair or replace deteriorated pipes without the need for extensive excavation.
2. *Manhole Rehabilitation:* Damaged or deteriorated manholes may be repaired or replaced to maintain structural integrity and prevent infiltration.
3. *Pump and Equipment Replacement:* Pumps, valves, and other mechanical components in lift stations are replaced as they reach the end of their useful life.

8.3.4.4 Mitigation Strategies

Operating and maintaining a wastewater collection system can present several challenges, including:

1. *Aging Infrastructure:* As the system ages, the risk of pipe failures, corrosion, and other structural issues increases, requiring proactive maintenance and replacement strategies. Mitigation: Implement a comprehensive asset management plan, prioritize rehabilitation and replacement projects, and allocate sufficient funding for system upgrades.
2. *Inflow and Infiltration:* Excessive I&I can overload the system, leading to backups, overflows, and potential environmental impacts. Mitigation: Conduct regular inspections, implement a program to identify and address sources of I&I, and incorporate I&I reduction measures in the system design and maintenance.
3. *Limited Access and Confined Spaces:* Maintenance activities within the wastewater collection system, such as inspections and cleaning, can be challenging due to the confined spaces and limited accessibility. Mitigation: Provide proper training and safety equipment for personnel, follow confined space entry protocols, and utilize specialized equipment and technologies (e.g., CCTV, robotic inspections) to access and maintain the system.
4. *Odor and Hydrogen Sulfide Issues:* The anaerobic conditions in the wastewater collection system can lead to the production of hydrogen sulfide, which can cause odor problems and contribute to the corrosion of infrastructure. Mitigation: Implement odor control measures, such as chemical treatment, ventilation systems, or the use of corrosion-resistant materials, to mitigate the impact of hydrogen sulfide.
5. *Regulatory Compliance:* Wastewater collection systems must comply with various environmental regulations, which can require extensive monitoring, reporting, and record-keeping. Mitigation: Establish a comprehensive compliance management system, provide training for personnel, and maintain detailed records of system operations and maintenance activities.

A well-planned and proactive approach to the operation and maintenance of the wastewater collection system, utilities can address the challenges, extend the system's lifespan, and ensure the reliable and efficient conveyance of wastewater to the treatment facility.

8.4 PHYSICAL UNIT OF WASTEWATER TREATMENT SYSTEM

A typical wastewater treatment system consists of several unit processes that work together to treat wastewater effectively. Here is a description of the physical unit process components from the beginning (influent) to the end (discharge of effluent):

1. *Screening:* The influent wastewater first passes through screens to remove large debris such as sticks, rags, and plastics. This step helps protect downstream equipment from damage and clogging.
2. *Grit Removal:* After screening, the wastewater undergoes grit removal to separate and remove heavy materials like sand, gravel, and other grit particles. This process prevents abrasion and damage to equipment in subsequent treatment stages.
3. *Primary Sedimentation:* In this step, the wastewater flows into a primary sedimentation tank where heavier solids settle at the bottom (sludge) and oils and greases float to the surface (scum). The clarified water in the middle layer undergoes further treatment.
4. *Aeration:* The clarified water from the primary sedimentation tank enters an aeration tank or basin where air or oxygen is introduced into the wastewater. This aeration process creates an aerobic environment that promotes the growth and activity of aerobic bacteria and other microorganisms responsible for breaking down organic matter in the wastewater.

5. *Secondary Sedimentation (Secondary Clarification):* The aerated wastewater then flows into a secondary clarifier where any remaining solids settle out as sludge, leaving clearer water at the top.
6. *Filtration:* Some treatment plants include a filtration step after secondary clarification to remove any remaining suspended solids and impurities from the water. This step helps improve water quality before disinfection.
7. *Disinfection:* The final effluent undergoes disinfection to kill harmful pathogens and bacteria. Common disinfection methods include chlorination, ultraviolet (UV) disinfection, or ozonation.
8. *Effluent Discharge:* The treated and disinfected effluent is then discharged into receiving water bodies such as rivers or oceans in compliance with environmental regulations. The discharged water should meet specific quality standards to protect human health and the environment.

The schematic typical process in the wastewater treatment plant is shown in Figure 8.4.

The treatment processes in both drinking water supply treatment and wastewater treatment systems share fundamental similarities, with the main distinction lying in the characteristics of the influents introduced into each system. Despite their differences, the core treatment methods employed in both systems aim to purify water and ensure its safety for consumption or environmental discharge. These treatment processes involve a series of physical, chemical, and biological steps designed to remove impurities, contaminants, and pollutants from the water.

In a drinking water treatment system, the influent water typically starts with a higher quality compared to the influent entering a wastewater treatment system. The higher quality influents in drinking water treatment often require less intensive treatment processes to meet regulatory standards for potable water supply. Common treatment steps in drinking water treatment include screening, coagulation, sedimentation, filtration, disinfection, and sometimes additional steps like pH adjustment or fluoridation to enhance water quality.

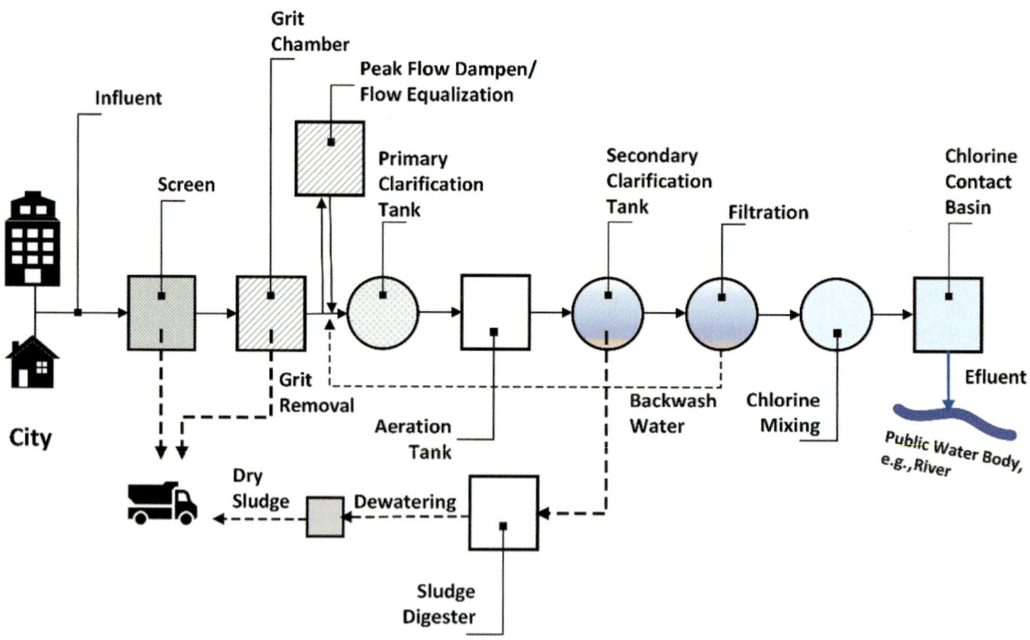

FIGURE 8.4 Typical process in the wastewater treatment system.

On the other hand, wastewater treatment systems deal with influents characterized by a wide range of contaminants, including organic matter, pathogens, nutrients, and pollutants from various sources. The influent in wastewater treatment requires more extensive treatment processes to effectively remove or reduce these contaminants before discharging the treated effluent back into the environment. Wastewater treatment typically involves screening, grit removal, primary sedimentation, biological treatment (aeration), secondary clarification, filtration, disinfection, and other advanced treatment processes depending on the specific requirements and regulations.

Despite the differences in influent quality between drinking water supply and wastewater treatment systems, both play crucial roles in protecting public health and the environment. Employing treatment processes tailored to the characteristics of the influent ensures sustainable management of water resources, providing safe and clean water for various purposes while minimizing environmental impacts.

8.4.1 The Wastewater Influents

Wastewater influent or affluent refers to the incoming water in a wastewater treatment plant, which can be of various types and from different sources. An overview of some characteristics of the wastewater influent can be discussed as follows:

Types and Sources: This can be (1) domestic wastewater, which originates from household activities such as bathing, cooking, and washing, (2) industrial wastewater, which is produced by industrial processes, manufacturing, and mining, and (3) agricultural wastewater, which comes from agricultural activities, including irrigation and livestock farming.

Conduits: The wastewater channels and stormwater channels are commonly separated as the latter will not go to the wastewater treatment system for processing as the quality of stormwater is usually better than wastewater. The channels for wastewater that collect and discharge the effluent from the sources are (1) *Sewers:* Pipes that transport wastewater from its source to treatment facilities, (2) *Drains:* smaller channels that collect wastewater and lead it to larger sewer systems.

Treatment Methods: There are four methods of treatment in the wastewater treatment plant (1) *Physical Processes:* such as screening, sedimentation, and filtration to remove solids; (2) *Biological Processes:* this process utilizes microorganisms to break down organic matter in wastewater; (3) *Chemical Processes:* this process involves coagulation, flocculation, and disinfection to remove contaminants and pathogens; and (4) *Advanced Treatment:* these are the techniques like activated carbon adsorption and membrane bioreactors (MBRs) for further purification.

The treatment process is designed to reduce pollutants in wastewater to a level where it can be safely discharged into the environment or reused. Each method addresses different contaminants and is chosen based on the composition of the influents and the desired quality of the effluent.

8.4.2 Screening Process

The screening process serves as an indispensable initial phase in the wastewater treatment sequence, acting as a critical line of defense that ensures the removal of large and potentially damaging solids, such as rags and debris, from the wastewater stream. This preliminary step is fundamental to safeguarding the intricate array of subsequent treatment stages, which rely on the absence of these materials to operate smoothly and efficiently. By meticulously filtering out these larger items, the screening process not only protects the mechanical integrity of downstream equipment but also significantly enhances the overall effectiveness of the wastewater treatment plant. It is, therefore, essential to grasp the various facets of this process, including the types of screens used, the methodologies employed for cleaning them, and the operational protocols that govern their maintenance and disposal of the extracted waste, to appreciate its role in the comprehensive treatment chain. Here are some key points to understand about this process:

1. *Purpose of Screening:* The primary goal of the screening is to remove large solids, rags, and debris that could damage or clog downstream equipment and piping systems. The screening process also protects the efficiency of the whole system and prevents contamination of waterways by removing large solids and debris at the beginning of the treatment process, screens prevent damage to downstream equipment such as pumps, valves, and biological reactors. This protection reduces maintenance needs and prolongs the lifespan of the equipment, ensuring continuous and efficient operation.
2. *Types of Screens:* Coarse screens, which have openings from 6 to 150 mm, are used to remove large objects like rags and sticks. The fine screens have smaller openings of less than 6 mm and are used to remove fine solids, and the micro screens can further reduce suspended solids to levels near those achieved by primary clarification.
3. *Cleaning Methods:* While manually cleaned screens are typically used in smaller or older facilities or as standby screens during high-flow periods or maintenance, mechanically cleaned screens employ automatic mechanisms to clean the screens, reducing operating and maintenance problems. These screens require less human intervention due to their automated cleaning process, which ensures that the screen remains free from debris. However, regular inspections and maintenance are still necessary to ensure optimal performance and longevity of the equipment.
4. *Screening Equipment:* Equipment like comminutors and grinders are installed to shred larger solids into smaller pieces that can be handled in downstream processes. Comminutors are mechanical devices designed to reduce the size of solid waste materials and debris found in wastewater, making them easier to handle and treat in subsequent stages of the treatment process. Comminutors consist of a rotating slotted cylinder through which wastewater flows. Solids too large to pass through the slots are cut by blades as the cylinder rotates, effectively reducing their size until they can pass through the slot openings.
5. *Regulatory Compliance:* Wastewater treatment plants are subject to stringent regulatory standards that govern the disposal of screenings, which are the materials removed during the initial screening process of wastewater treatment. The regulations ensure that screened materials are handled in a manner that minimizes their impact on the environment, protecting ecosystems and public health.

The disposal methods for screened materials should be outlined by environmental agencies and can include landfilling, incineration, or even composting, depending on the nature of the material. Landfilling must be done in designated areas to prevent any leachate from contaminating soil and groundwater. Incineration, if chosen, must be performed at specific temperatures to ensure complete combustion and minimize the release of harmful emissions. Composting is another option for organic materials, turning them into beneficial soil amendments under controlled conditions.

Treatment plants must maintain meticulous records of their disposal practices and often undergo regular inspections to ensure compliance with these standards. Failure to adhere to these regulations can result in significant fines and legal action, as well as damage to the plant's reputation. Proper disposal practices are part of the broader commitment to environmental stewardship and sustainability, reflecting the treatment plant's role in the community as a protector of natural resources.

As mentioned earlier, the screening process in a wastewater treatment plant is the first unit operation and has several important functions as shown in Table 8.4.

TABLE 8.4
The Function of Screening Process

Function	Explanation
Removal of large objects	As wastewater enters the treatment plant, it flows through a screen that removes large floating objects such as rags and sticks that might clog pipes or damage equipment. This helps to prevent operational issues and maintenance problems.
Protection of downstream processes	By removing these large objects early in the process, the screening process protects the downstream processes from damage and inefficiency.
Types of screens	There are various types of screens used in wastewater treatment. Coarse screens have clear openings ranging from 6 to 150 mm and are used to remove large solids. Fine screens with clear openings less than 6 mm are used to remove smaller solids.
Screening design	The design of the screening process is important. For example, the screen channels should be designed to prevent excessive accumulation of grit and other heavy materials.
Mechanical screens	In larger treatment plants, mechanical screens are often used to automate the screening process and reduce the need for manual cleaning.

The screening process in a wastewater treatment plant helps remove large debris and solid materials from the influent wastewater. Screens are used to capture and separate objects such as sticks, plastics, rags, and other large particles that could potentially damage downstream equipment or hinder the efficiency of the treatment process. By removing these larger materials at the beginning of the treatment train, the screening process helps protect pumps, pipes, and other equipment from clogging or mechanical damage.

Various types of screens are employed in wastewater treatment plants, including bar screens, coarse screens, fine screens, and micro screens. Bar screens consist of vertical or inclined bars spaced apart to allow water to pass through while trapping larger objects. Coarse screens have larger openings to capture bigger debris, while fine screens have smaller openings to target finer particles. Microscreens use fine mesh filters to capture even smaller particles, providing additional levels of filtration.

The effectiveness of the screening process in a wastewater treatment plant is essential for maintaining the overall performance and efficiency of the treatment system. Properly functioning screens help prevent blockages in downstream equipment, such as pumps and pipelines, which can lead to operational disruptions and increased maintenance costs. Regular maintenance and cleaning of screens are necessary to ensure continuous operation and optimal removal of debris from the influent wastewater.

In addition to protecting equipment, the screening process plays a vital role in safeguarding downstream treatment processes by removing large objects that could interfere with subsequent treatment stages. Removing these materials early in the process helps prevent damage to pumps, valves, and other components in the treatment plant, ensuring smooth operation and consistent treatment performance. Effective screening also reduces the risk of blockages in pipes and channels, which can impede the flow of wastewater through the system.

Properly designed and maintained screens are essential for achieving the desired level of treatment efficiency in a wastewater treatment plant. The selection of appropriate screen types and sizes depends on factors such as the characteristics of the influent wastewater, flow rates, and the specific requirements of the treatment process. Regular inspection and cleaning of screens are necessary to prevent buildup and ensure optimal performance. In some cases, automated screen cleaning systems or mechanical rakes are used to remove accumulated debris efficiently.

The screening process in a wastewater treatment plant is not only essential for protecting equipment and ensuring efficient operation but also contributes to environmental protection and

TABLE 8.5
Screen Classification

Factor	Explanation
Based on the size of openings	Screens can be classified as coarse, fine, or micro based on the size of their openings. 1. *Coarse Screens:* These have clear openings ranging from 6 to 150 mm and are used to remove large solids. 2. *Fine Screens:* These have clear openings less than 6mm and are used to remove smaller solids. 3. *Micro Screens:* These are the smallest type of screening in wastewater treatment.
Shape of screens	Screens can be discs, bands, drums, cages, wings, and perforated plates.
Condition of movement	Screens can be fixed or movable.
Method of cleaning	Screens can be hand-cleaned or mechanically cleaned.
Mechanical traits	Screens can also be classified based on other mechanical traits. For example, continuous belt screens are ultra-high-tech, functional, efficient, and self-cleaning.

public health. Removing large debris and solid materials from influent wastewater helps prevent pollutants from entering downstream water bodies and ecosystems. This initial step in the treatment process reduces visible contaminants in the effluent discharged from the plant, contributing to cleaner water quality and minimizing potential impacts on aquatic life and habitats.

Screens used in wastewater treatment are classified based on several factors as shown in Table 8.5.

Choosing the right type of screen for a wastewater treatment plant involves several considerations. (1) *It Depends on Influent Characteristics:* Conduct on-site testing of your plant's influent characteristics. This includes the size and volume of the material that needs to be removed. (2) *Screen Type and Grid Opening:* Identify the best type of screen, grid type, and grid opening based on your influent characteristics. (3) *Screen Sizing:* Optimize your screen sizing to maximize protection and extend the life of your downstream processes, while reducing the amount of equipment maintenance and the chance of equipment failure. (4) *Specific Situation:* Depending on the specific situation, static wedge wire, rotary drum, or step fine screens may be used. (5) *Minimum and Maximum Water Levels:* Consider the minimum and maximum water levels in the area.

8.4.3 Grit Removals

Grit removal is a component of the wastewater treatment process, designed to eliminate sand, gravel, cinders, and other heavy materials from wastewater early in the treatment process. These materials, collectively referred to as "grit" have specific gravities significantly higher than organic matter found in wastewater and can cause substantial issues if not removed.

The primary purpose of grit removal is to protect mechanical equipment from unnecessary abrasion and wear, which can lead to increased maintenance and operational costs. Grit can also settle in pipelines and channels, causing blockages and reducing the capacity of the treatment plant.

There are several methods employed for grit removal, as discussed in the following sections.

8.4.3.1 Gravity Settling

This method reduces the flow velocity of wastewater, allowing grit to settle at the bottom of a grit chamber due to its higher density.

Gravity settling is effective for grit removal because it is a relatively simple and low-cost process that can be easily incorporated into the design of a wastewater treatment plant. It is also a passive process, requiring no energy input other than that needed for the removal and washing of the settled grit. However, the settling tanks must be properly designed to handle the expected flow rates and grit loads to ensure optimal removal efficiency.

Gravity settling is a process in the removal of grit from wastewater that relies on the principle that particles with a higher density than water will settle to the bottom of a tank due to gravitational forces. The way gravity settling works in the context of grit removal is as follows.

8.4.3.1.1 In Settling Tanks

1. Wastewater enters a settling tank or grit chamber where the velocity of the flow is carefully controlled to allow grit to settle by gravity.
2. The design of these tanks is such that the flow velocity is slow enough to permit settling, but not so slow that organic matter settles with the grit.

8.4.3.1.2 Grit Hoppers

1. Grit hoppers are typically located at the bottom of settling tanks or grit chambers and are constructed with steeply sloped sides to facilitate the accumulation and removal of grit.
2. The shape of the hopper is often trapezoidal or V-shaped to ensure that the settled grit moves toward the bottom for easy extraction.

As the wastewater flows through the settling tank, the heavier grit particles settle by gravity into the hopper. The design of the hopper prevents the resuspension of these particles back into the water column.

8.4.3.1.3 Removal of Grit

After separation, the grit is concentrated and washed to remove organic material that may have been captured with it. The collected grit is then removed from the hopper using either a grit pump or an airlift pump. This is done periodically to prevent the accumulation of grit, which can affect the efficiency of the process. The washing process is important for reducing odors and making the grit suitable for disposal or further processing. Disposal methods may include landfilling or reuse as a construction material, depending on the local regulations and the quality of the grit.

The design of the grit removal system must account for the expected grit load, which can vary based on factors like the type of sewer system and the presence of industrial waste. Proper sizing and design ensure that the system can handle peak loads, especially during wet weather events when grit accumulation can be significant.

8.4.3.1.4 Washing and Dewatering

Once removed, the grit is often washed to reduce organic material that may have settled with it. After washing, the grit is dewatered before final disposal.

The washing step is designed to remove organic material that may have settled with the grit. This is important because organic matter can lead to odor issues and may complicate disposal. The degree of washing depends on several factors, including the need for odor control and the requirements for disposal. A thorough washing process can reduce the organic content to less than 5%.

Dewatering reduces the volume and weight of the grit, making it easier and more cost-effective to transport and dispose of. This process involves concentrating the grit slurry to remove as much water as possible. This can be achieved using equipment like centrifuges, belt presses, or drying beds.

Proper washing and dewatering are essential for minimizing the environmental impact of disposing of grit. They ensure that the grit is free of organic material that could decompose and cause odors or attract pests. The grit can be disposed of in landfills or potentially reused in construction or other applications, depending on its quality and local regulations.

S
The treated effluent, now free of grit, exits at the top of the tank and proceeds to the next stage of the wastewater treatment process. As the grit settles at the bottom of the tank, the upper layers of water, now largely free of heavier particles, begin to clear. This treated water, or effluent, is drawn from the top of the tank, avoiding the disturbance to the settled grit. The design of the tank ensures that the cleanest water exists, while the grit remains at the bottom to be removed and processed separately.

From here, the water flows to the next treatment stage, which could involve sedimentation, where finer particles are allowed to settle, or it might go through biological treatment processes where microorganisms break down organic matter. In some cases, it may undergo chemical treatment to remove phosphorus and nitrogen or be filtered to remove additional solids.

8.4.3.2 Centrifugal Separation

This process utilizes rotational forces to separate grit particles from the liquid stream. Centrifugal separation is a method used in the grit removal process of wastewater treatment that leverages the force of gravity and rotational motion to separate particles based on their density.

In centrifugal separation, wastewater is introduced into a cyclone system or a hydro-cyclone, where it is set into a rapid spinning motion. This rotational force, much like a centrifuge, creates a vortex within the system, causing the denser grit particles to move outward toward the walls of the cyclone due to their higher specific gravity.

As the wastewater spirals downwards, the heavier grit particles are forced to the sides and bottom of the cyclone, where they can be collected and removed. The lighter organic material and water, having lower specific gravities, remain closer to the center of the vortex and exit through the top of the cyclone.

Once separated, the grit slurry is extracted, typically with a grit pump or an airlift system, and then further processed through grit concentration, washing, and dewatering steps.

The design of centrifugal grit removal systems can vary, but they often feature a circular chamber for creating the vortex and a bottom grit well for collection.

8.4.3.3 Vortex Chambers

Create a vortex to enhance the settling of grit, often used in modern systems due to their efficiency and smaller footprint. Vortex chambers are a sophisticated component of the grit removal process in wastewater treatment plants. They utilize the principles of rotational flow to separate heavier grit particles from the wastewater.

Vortex chambers typically consist of a cylindrical tank where wastewater enters tangentially, creating a swirling flow pattern. This rotational flow, or vortex, forces the heavier grit particles to move outward and downward due to centrifugal force and gravity, settling into the bottom of the tank. The settled grit accumulates in a grit hopper located at the bottom of the chamber. From here, it can be removed by a grit pump or an airlift pump for further processing.

The design of vortex chambers allows for a high separation efficiency, effectively removing grit while allowing organic material and water to remain suspended and exit the chamber. They are particularly effective at maintaining performance across variable flow rates, ensuring consistent grit removal even during fluctuating wastewater volumes.

The grit that was separated from the wastewater must be removed and disposed of properly. This is typically done using automated systems like inclined screw or tubular conveyors, chain and bucket elevators, clamshell buckets, or pumping.

The removed grit is then washed to reduce organic content and dewatered before disposal. Proper handling and disposal of grit are essential to comply with environmental regulations and prevent potential harm to ecosystems.

8.4.4 Primary Sedimentation

Primary sedimentation is a phase in the wastewater treatment process, which serves as the first major step in separating solids from liquid waste. The process begins when wastewater enters a primary sedimentation tank, also known as a primary clarifier. The breakdown of the process is as follows:

- *Flow into the Tank:* Wastewater is directed into the primary sedimentation tank after it has passed through preliminary treatment steps like screening and grit removal. These steps remove larger solids and inorganic materials like sand and gravel.
- *Settling of Solids:* Inside the tank, the flow of the wastewater is slowed down significantly, allowing gravity to act on the heavier solids. These solids, often referred to as sludge, gradually settle at the bottom of the tank. The sludge consists of a variety of materials, including organic and inorganic matter, which can be further processed or disposed of after removal from the tank.
- *Rising of Oils and Greases:* Concurrently, lighter substances like oils and grease rise to the surface of the tank, forming a layer known as scum. This layer is skimmed off and removed, often requiring additional treatment due to its high concentration of organic material.
- *Clarification Zone:* It lies between the sludge at the bottom and the scum at the top lies the clarified water. This middle layer is relatively free of solids and is carefully extracted from the tank to proceed to the next stage of treatment.
- *Further Treatment:* The clarified water, now significantly cleaner than when it entered, is directed to secondary treatment processes. These may include biological treatment methods that further reduce organic content and other contaminants.

Primary sedimentation tanks are designed to maximize the efficiency of solid removal while minimizing the carryover of solids into the subsequent treatment stages. They play a vital role in the overall wastewater treatment process, ensuring that the effluent moving forward is in a suitable state for more refined treatment methods. The success of primary sedimentation directly impacts the effectiveness and efficiency of the entire wastewater treatment plant.

Some hydraulic parameters used in designing a primary sedimentation tank are given in Table 8.6.

TABLE 8.6
Hydraulic Parameters Used in Design of Primary Sedimentation Tank

Parameter	Value
Hydraulic retention time (HRT) for plain sedimentation	3–4 hours
HRT for coagulated sedimentation	2–2.5 hours
Velocity of flow	<30 cm/min
Tank dimensions (length to breadth ratio)	3 to 5:1
Tank length	30 m (max 100 m)
Tank breadth	6–10 m
Circular tank diameter	<60 m Generally, 20–40 m
Average overflow rate	30–50 m^3/m^2/day Typically, 40 m^3/m^2/day
Surface overflow rate	600–900 L/m^2/day (for average flow) 81,000–122,000 L/m^2/day (for peak flow)

The values in Table 8.6 are general guidelines and the actual design parameters can vary based on the specific requirements of the wastewater treatment process.

Here is a simple example of hydraulic design of a primary sedimentation tank.

EXAMPLE 8.1

Detention period or hydraulic retention time (HRT) is calculated by using the following formula:

$$HRT = \frac{V}{Q} \tag{8.1}$$

V is the volume of the treatment unit, for example, tank
Q is the flow rate of the wastewater

Detention period: Let us say we are designing a tank for plain sedimentation. The detention period would be typically 3–4 hours. If we choose a detention period of 3 hours, that is 180 minutes.

Flow velocity: The velocity of flow should not be greater than 30 cm/min. So, if we have a tank that is 30 m long, it will take 100 minutes for water to flow from one end to the other.

Tank dimensions: Let us assume we are designing a rectangular tank with a length to breadth ratio (L:B) of 3:1. If we choose a length of 30 m, then the breadth would be 10 m.

Surface area: The surface area of the tank would be the length multiplied by the breadth. So, for our example, it would be 30 m × 10 m = 300 m^2.

Hydraulic loading rate (HLR) is given by the following formula:

$$HLR = \frac{Q}{A} \tag{8.2}$$

where:
HLR is the hydraulic loading rate, typically measured in $m^3/m^2/day$.
Q is the flow rate, typically measured in m^3/day.
A is the surface area of the tank, typically measured in m^2.

This formula helps to determine the rate at which water is applied to the surface of the sedimentation tank. It is a crucial parameter in the design and operation of a primary sedimentation tank.

Let us assume a hydraulic loading rate of 200 $m^3/m^2/day$. To calculate the required flow rate (Q), we can rearrange the formula (8.2) for HLR to get $Q = HLR \times A$. Substituting the values, we get Q = 200 $m^3/m^2/day$ × 300 m^2 = 60,000 m^3/day.

Volume of the tank: The volume of the tank would be the surface area multiplied by the depth. If we choose a depth of 3 m, then the volume would be 300 m^2 × 3 m = 900 m^3.

In designing secondary primary sedimentation tank, we need also to understand the organic loading rate (OLR). OLR is used to assess the impact of pollutants on treatment plants. For instance, OLR can be used to size the infiltrative surface for high-strength septic tank effluent (STE). A numerical example of calculation of OLR is given below.

EXAMPLE 8.2

Let us consider a hypothetical example for calculating the OLR in the design of a primary sedimentation tank.

Flow Rate (Q): Let us assume the flow rate of the wastewater entering the tank is 5,000 m³/day.

Biochemical Oxygen Demand (BOD) Concentration: The BOD is a measure of the quantity of oxygen used by microorganisms while decomposing organic matter under aerobic conditions. Let's assume the BOD concentration of the wastewater is 200 mg/L.

Volume of the Tank (V): Let's assume the volume of the tank is 2,500 m³.

The OLR is calculated using the formula:

$$\text{OLR} = \frac{Q \times \text{BOD}}{V} \tag{8.3}$$

where:

OLR is the organic loading rate, typically measured in g/m³/day.
Q is the flow rate, typically measured in m³/day.
BOD is the biochemical oxygen demand, typically measured in mg/L or g/m³.
V is the volume of the tank, typically measured in m³.

Substituting the values into the formula, we get:

$$\text{OLR} = \frac{5{,}000\ \text{m}^3/\text{day} \times 200\ \text{mg/L}}{2{,}500\ \text{m}^3}$$

Please note that 1 m³ = 1,000 L, so we can convert the units to be consistent:

$$\text{OLR} = \frac{5{,}000\ \text{m}^3/\text{day} \times 200\ \text{g/m}^3}{2{,}500\ \text{m}^3} = 4{,}000\ \text{g/m}^3/\text{day}$$

So, the OLR for this example would be 4,000 g/m³/day.

These are general guidelines, and the actual design parameters can vary based on the specific requirements of the wastewater treatment process.

There are some common design mistakes to avoid when designing primary sedimentation tanks:

1. *Oversizing the Tanks:* One common mistake is oversizing the tanks. This is often done as a response to poor design, which can lead to system disorders and turbulences.
2. *Ignoring the Accumulation of Solids:* The accumulation of solids in the tank can reduce its effectiveness in removing solids from wastewater and can lead to issues in subsequent treatment processes.

3. *Ignoring the Formation of Scum:* The formation of scum is another issue that can reduce the effectiveness of the tank.
4. *Ignoring Algae Growth:* The growth of algae in the tank can also reduce its effectiveness.
5. *Ignoring Turbulence Flow Fields:* PSTs contain turbulence flow fields that need to be considered in the design.

The above mistakes can lead to reduced efficiency and capability of a primary sedimentation tank. Therefore, it is important to consider these factors during the design process to ensure the effectiveness of the tank.

The primary sedimentation process may also encounter several common issues that affect its efficiency. The issues include the following:

1. *Inadequate Removal of Solids:* Sometimes, the primary sedimentation tanks do not remove enough solids. This can be due to high flow rates, which do not allow sufficient time for solids to settle.
2. *Scum Accumulation:* Oils and grease can accumulate on the surface, forming scum. If not regularly removed, this layer can become too thick and hinder the sedimentation process.
3. *Sludge Buildup:* The sludge that settles at the bottom of the tank must be periodically removed. If this is not done efficiently, the accumulated sludge can resuspend and reduce the quality of the effluent.
4. *Tank Design Issues:* Inadequate design of the sedimentation tanks can lead to poor distribution of flow or dead zones where solids do not settle properly.
5. *Fluctuating Influent Characteristics:* Variations in the characteristics of the incoming wastewater, such as changes in flow rate or solids content, can disrupt the sedimentation process.
6. *Operational Issues:* Operational issues such as improper adjustment of weirs and baffles, or mechanical failures, can also affect the performance of primary sedimentation tanks. Addressing these issues involves optimizing flow rates, regular maintenance, and potentially redesigning tank components to improve the efficiency of the process. Continuous monitoring and adjustment of the process are crucial for ensuring that primary sedimentation functions effectively within the wastewater treatment plant.

8.4.5 Aeration

The clarified water from the primary sedimentation tank enters an aeration tank or basin where air or oxygen is introduced into the wastewater. This aeration process creates an aerobic environment that promotes the growth and activity of aerobic bacteria and other microorganisms responsible for breaking down organic matter in the wastewater.

In the aeration tank, the aerobic bacteria and microorganisms, also known as activated sludge, consume the organic pollutants, converting them into carbon dioxide, water, and energy for their own growth and reproduction. This biological process is known as aerobic respiration. The vigorous mixing and aeration also ensure that the activated sludge is kept in suspension, contact between the microorganisms and the organic material in the wastewater is maximized, and anoxic or anaerobic conditions are prevented.

After aeration, the wastewater is directed into a secondary sedimentation tank, also known as a clarifier. Here, the activated sludge settles down due to gravity, forming a concentrated layer of sludge at the bottom of the tank. Some of this sludge is recycled back into the aeration tank to maintain a high population of beneficial microorganisms, while the excess is removed for further treatment and disposal.

The water that leaves the secondary clarifier is typically free of harmful substances and can be discharged into a body of water or undergo further treatment if necessary. This entire process, from the primary sedimentation tank to the secondary clarifier, is a crucial part of the wastewater treatment process, ensuring that the water we return to the environment is clean and safe.

8.4.6 Secondary Sedimentation or Secondary Clarification

The aerated wastewater then flows into a secondary clarifier where any remaining solids settle out as sludge, leaving clearer water at the top. This clearer water, also known as effluent, is then subjected to disinfection processes, such as chlorination or UV treatment, to kill any remaining pathogens and ensure the water is safe for discharge into the environment or for reuse. The settled sludge is often returned to the aeration tank to maintain a high concentration of microorganisms for the degradation process, or it may be further treated and disposed of in a responsible manner. This entire process is a testament to the effectiveness of modern wastewater treatment methods in protecting our water resources and the environment.

UV disinfection, specifically Ultraviolet Germicidal Irradiation (UVGI), is a disinfection method that uses UV light to kill or inactivate microorganisms. This method is commonly used in various applications, including air, surface, food, and water disinfection. The process works by exposing microorganisms to UV light, particularly UV-C light, which has a wavelength range of 180–280 nm. This UV-C light can penetrate the cells of microorganisms and damage their DNA or RNA. This damage inhibits the microorganisms' ability to reproduce and carry out vital functions, effectively killing or inactivating them.

In the context of water treatment, UV disinfection is often used as a final step after other treatment processes. The treated water is exposed to UV light, which inactivates any remaining microorganisms, making the water safe for discharge or reuse.

One specific application of UVGI is the Upper-Room UVGI system. In this system, a disinfection zone of UV energy is created above the people in the room they occupy. Air passes through this disinfection zone, and airborne pathogens are killed once they receive an appropriate amount of UV energy. This method is effective at reducing exposures to airborne pathogens in certain group settings.

UV disinfection has several advantages over traditional chemical disinfection methods:

1. *Non-Chemical Treatment:* UV disinfection is a non-chemical process, meaning it does not require the addition of harmful chemicals. This reduces the risk of chemical exposure, which can damage the lungs, skin, and eyes.
2. *Effectiveness:* UV disinfection is effective against a wide range of pathogens, including those resistant to chlorination, such as certain protozoa like Cryptosporidium and Giardia. It can eliminate almost all known viruses, bacteria, and fungi.
3. *No Byproducts:* Unlike chemical disinfection methods, UV disinfection does not produce harmful byproducts. Chemical disinfection methods can change the taste and odor of water and produce byproducts.
4. *Economical and Energy Efficient:* UV disinfection systems have low power requirements, making them economical and energy efficient.
5. *Easy Installation and Maintenance:* UV systems are easy to install and maintain.
6. *Safe for Users:* There's no chemical handling involved, making it safer for users.
7. *Compact and Fast Treatment:* UV systems are compact and provide fast treatment without the need for a holding tank or reaction time.
8. *Approved Technology:* UV disinfection is recognized by regulatory agencies, including the United States Environmental Protection Agency (USEPA), and is approved by the FDA (Food & Drug Administration in the United States) and ÖNORM in Europe.

9. *Ecological Technology:* UV disinfection has a smaller carbon footprint, lower energy consumption, and almost no safety concerns.
10. *Reliable:* UV systems workday and night, providing continuous disinfection.

However, while UV disinfection has many advantages, it does have some limitations. For example, UV light can only eliminate the microorganisms present in the water and does not remove other contaminants such as heavy metals, salts, chlorine, or artificial substances such as petroleum or pharmaceutical products. Therefore, other filtration methods should be used in conjunction with UV to ensure that all contaminants are removed from the water.

8.4.7 Filtration

Some treatment plants include a filtration step after secondary clarification to remove any remaining suspended solids and impurities from the water. This step helps improve water quality before disinfection. The filtration process typically involves passing the water through layers of sand, gravel, or other filter media that trap and remove particles. Some systems may also use activated carbon filters to remove organic compounds and improve taste and odor. After filtration, the water is usually disinfected using methods such as chlorination, ozonation, or UV light to kill any remaining pathogens. The result is clean, safe water that can be returned to the environment or reused for various purposes.

Activated carbon filters are a popular method for purifying both air and water. They work on the principle of adsorption, which is a chemical reaction where certain particles are attracted to activated carbon and bond with it. How they work and their applications:

1. *What Is Activated Carbon:* Activated carbon is produced by taking a carbon-containing material, like coconut shells or coal, and putting it through a thermochemical process. This process creates a char that is then chemically treated and heated again, resulting in activated carbon. The activation process creates millions of tiny pores on the carbon's surface, increasing its surface area significantly.
2. *How Do Activated Carbon Filters Work:* When water or air passes over the activated carbon, the contaminants in them are attracted to the carbon surface and get trapped in its pores, a process known as adsorption. This makes activated carbon filters very effective at removing a variety of pollutants.
3. *Applications of Activated Carbon:* Activated carbon filters are used widely in water and air purification systems. They can remove volatile organic compounds (VOCs), odors, chlorine, and other chemicals from water. In air purifiers, they can adsorb gases, chemical vapors, and odor molecules.
4. *Types of Activated Carbon Filters:* There are several types of activated carbon filters, including granular activated carbon (GAC) filters and block carbon filters. GAC filters use granules of activated carbon and are common in water filtration systems.

While activated carbon filters are highly effective at removing many types of pollutants, they do not remove all types of contaminants. For example, they are not effective at removing minerals, salts, and certain inorganic compounds. Therefore, they are often used in combination with other types of filters to ensure comprehensive water or air purification.

Regenerable activated carbon filters are a type of activated carbon filter that can be regenerated or renewed, extending their lifespan, and making them more cost-effective. Here how they work and their applications:

1. *Regeneration Process:* The regeneration process typically involves soaking the activated carbon in water while it is still in the filter. This process helps to remove the adsorbed

pollutants from the carbon, renewing its adsorption capacity. Some systems automatically detect when the filter is saturated and launch the regeneration process.

2. *Water Regenerable Carbon Filters:* These filters are designed to allow for in-situ water regeneration of the carbon. This eliminates the downtime and cost associated with replacing impregnated carbon on a regular basis.
3. *Photocatalytic Regeneration:* Some regenerable activated carbon filters incorporate photocatalytic particles. These particles can regenerate the filter under UV light, making them renewable indoor VOC filters. For example, TiO_2-bearing activated carbon filters can be regenerated by up to 99% under UV-C irradiation for 22 hours.

Regenerable activated carbon filters are used in various applications, including odor control and VOC removal. They are particularly useful in settings where the filters need to be used continuously and replacing them frequently would be impractical or costly.

8.4.8 Disinfection

The final effluent undergoes disinfection to kill harmful pathogens and bacteria. Common disinfection methods include chlorination, UV disinfection, or ozonation.

1. Chlorination involves adding chlorine or chlorine compounds to the water. This method is effective and relatively cheap, but it can leave residual chlorine in the water and form disinfection byproducts.
2. UV disinfection uses UV light to inactivate microorganisms by damaging their DNA, preventing them from reproducing. This method is effective and does not introduce any chemicals into the water, but it requires electricity and does not provide residual disinfection.
3. Ozonation involves injecting ozone gas into the water. Ozone is a powerful oxidant that can effectively kill a wide range of microorganisms. However, ozonation systems can be more expensive to install and operate, and like UV, do not provide residual disinfection.

Advantages, disadvantages, and safety considerations of above disinfection methods are as follows:

Advantages of Chlorination:

1. It is effective against almost all types of harmful bacteria and viruses.
2. Chlorination inactivates microorganisms.
3. It keeps the water safe.
4. It is easy to use.

Disadvantages of Chlorination:

1. Some microbes are resistant to its effects.
2. It can produce harmful byproducts.
3. It can change the taste and odor of water.

Safety Considerations:

1. Chlorine solutions should be prepared fresh daily, as they are light sensitive.
2. Most chlorine products require the removal of organic soil prior to their use (removing bulk spill matter, cleaning prior to disinfecting).
3. Disinfection requires a wet contact time (amount of time the disinfectant is required to be left on the surface to be effective).
4. Do not attempt to combine different types of chlorine.
5. Do not inhale any of the fumes.

6. Do not contact chlorine with bare hands.
7. Do not combine chlorine with additional substances.

Advantages of UV Disinfection:

1. It is a non-chemical treatment.
2. It is easy to install and maintain.
3. It is economical and energy efficient.
4. It is effective on a wide range of pathogens.
5. It is recognized by regulatory agencies.
6. It is ecological technology.
7. It contains no disinfection by-products.
8. It is reliable.
9. It is cost-effective.
10. It is compact and fast treatment.

Disadvantages of UV Disinfection:

1. The initial cost of some UV applications is somewhat higher than chlorination.
2. UV light can only eliminate the microorganisms present in the water.
3. If the water is cloudy, a pre-filter should be used.
4. UV water systems require electricity to operate.
5. UV disinfection does not offer the persistence of some other chemicals.

Safety Considerations:

1. Avoid working in or around the safety cabinet when the UV light is on or avoid using the room when UV light is on.
2. Always close the sash completely when the UV light is on.
3. Even a small opening of the sash can cause skin damage and other biological effects.
4. Ensure the UV light is off prior to working in the cabinet.
5. Remove any objects or surfaces that could block UV light from reaching all areas to be disinfected.
6. Ensure that reflective surfaces, such as mirrors or stainless steel, are covered to prevent light reflection.
7. UV disinfection should only be performed in unoccupied areas.

Advantages of Ozonation:

1. Ozone is more effective than chlorine in destroying viruses and bacteria.
2. The ozonation process utilizes a short contact time.
3. There are no harmful residuals that need to be removed after ozonation because ozone decomposes rapidly.

Disadvantages of Ozonation:

1. Low dosages may not effectively inactivate some viruses, spores, and cysts.
2. Ozonation is more complex than other disinfection technologies.
3. Ozone does not provide residual protection against recontamination during distribution.

Safety Considerations:

1. The ozonation of bromide-containing waters can form bromate, a known carcinogen, though usually at concentrations well below the health concern threshold.
2. Techniques to control bromate formation involve ozonation at slightly acidic pH values, multi-stage ozonation and the use of ammonia or chlorine.

3. While maintaining the basic safety rules and provided the rooms are properly prepared, disinfecting the apartment with ozone is completely safe.
4. To take the plants and animals out of the room before starting ozonation.

Each method has its own strengths and weaknesses, and the choice of method can depend on a variety of factors, including the specific needs of the water treatment system, the quality of the water being treated, and regulatory requirements.

8.4.9 Effluent Discharge

The treated and disinfected effluent is then discharged into receiving water bodies such as rivers or oceans in compliance with environmental regulations. The discharged water should meet specific quality standards to protect human health and the environment.

These standards typically include limits on the concentration of harmful substances such as heavy metals, organic compounds, and pathogens. Regular monitoring and testing are conducted to ensure that these standards are met. If the effluent does not meet the required standards, it must be further treated until it does.

In some cases, the treated water can also be reused for non-potable purposes, such as irrigation, industrial cooling, or toilet flushing, which can help conserve freshwater resources. This process, known as water reclamation, is becoming increasingly important in regions where water scarcity is a concern.

Water quality standards for discharged effluent can vary depending on the specific regulations of a country or region. However, some common parameters that are often regulated are shown in Table 8.7.

These standards are typically enforced through regular monitoring and testing. If the effluent does not meet the required standards, it must be further treated until it does.

TABLE 8.7
Common Parameters of Effluent Standards

Parameter	Explanation
pH	The pH value of the effluent should typically be between 6 and 9[3]. This ensures that the water is neither too acidic nor too alkaline, which could harm aquatic life or corrode pipes.
Temperature	The temperature of the effluent should not exceed certain limits (e.g., 45°C) to prevent thermal pollution, which can harm aquatic ecosystems.
Biological oxygen demand (BOD)	BOD measures the amount of oxygen needed by bacteria to break down organic material present in the water. High BOD indicates high levels of organic pollution.
Chemical oxygen demand (COD)	COD measures the total quantity of oxygen required to oxidize all organic material into carbon dioxide and water.
Total suspended solids (TSS)	TSS measures the weight of solids suspended in water, and high levels can cause problems for aquatic life and for water treatment processes.
Nutrients	Levels of nutrients like nitrogen and phosphorus are often regulated because excessive amounts can cause eutrophication, a process that depletes oxygen in water bodies and harms aquatic life.
Pathogens	The presence of disease-causing organisms such as bacteria, viruses, and parasites is often regulated to protect public health.
Toxic substances	Levels of toxic substances such as heavy metals and certain organic compounds are often regulated to protect both public health and the environment.

The safe discharge of wastewater effluent is a critical environmental consideration. Here are some common practices:

1. *Surface Water Bodies:* Treated and disinfected effluent is often discharged into receiving water bodies such as rivers, lakes, or oceans. This is done in compliance with environmental regulations to protect water quality and public health.
2. *Sanitary Sewer Systems:* In some cases, wastewater is discharged into sanitary sewer systems, where it is transported to a Publicly Owned Treatment Works (POTW) for further treatment.
3. *Land Application:* Treated wastewater can also be applied to land for irrigation or to recharge groundwater. This must be done in a manner that prevents runoff and protects groundwater resources.
4. *Zero Effluent Discharge Approaches:* Some facilities aim to achieve zero effluent discharge, meaning they recycle all their wastewater for reuse within the facility.

Regardless of the discharge method, the effluent must meet specific quality standards set by regulatory authorities, as discussed earlier. These standards typically regulate parameters such as pH, temperature, BOD, Chemical Oxygen Demand (COD), Total Suspended Solids (TSS), nutrients, pathogens, and toxic substances.

There are some current thoughts instead of discharging treated water, recycle and recovery the treated water through what so called Zero Liquid Discharge (ZLD). ZLD systems offer several environmental benefits, among others: (1) ZLD systems help conserve water resources by recycling and reusing wastewater. (2) ZLD systems prevent the discharge of pollutants into the environment by treating wastewater to remove contaminants. This helps protect water bodies and ecosystems from pollution. (3) ZLD systems can recover valuable resources from wastewater, such as minerals and other substances. These resources can be reused or sold, reducing the need for new resource extraction. (4) Some ZLD systems can recover energy from the treatment process, improving overall energy efficiency. (5) By minimizing the volume of wastewater entering the main flow of water, ZLD reduces the energy required for treatment and minimizes greenhouse gas emissions. (6) ZLD systems can improve a company's environmental performance and regulatory risk profile for future permitting.

ZLD is a water treatment process designed to remove all the liquid waste from a system. The goal of ZLD is to reduce wastewater and produce clean water that is suitable for reuse. Some important points about ZLD are shown in Table 8.8.

ZLD is not a new concept, but its adoption and implementation have been gaining momentum in recent years due to increasing water scarcity and stricter environmental regulations. The demand for ZLD is already strong and is expected to grow in the future. In 2018, European industries and countries purchased ZLD systems worth more than $1.5 billion annually, and globally, the current revenue for ZLD is nearing $6 billion.

Industries and government regulatory agencies spanning from Europe, North America to China and India are quickly turning to ZLD systems to recover all the liquid from wastewater effluent and return it to the system as clean, reusable water. The market for ZLD systems is projected to grow to at least $9 billion by the year 2025.

However, ongoing research and technological advancements of ZLD are continually improving the efficiency and affordability of ZLD systems. This includes the development of more energy-efficient treatment processes, better waste management strategies, and methods to recover valuable resources from the waste stream.

While beneficial, do come with a set of challenges that need to be addressed for effective implementation. Some key challenges associated with ZLD approaches are shown in Table 8.9.

Addressing the challenges associated with ZLD systems requires innovative solutions, and research and experiments should be done.

TABLE 8.8
Important Point of ZLD

The Point	Explanation
Process	ZLD systems employ various wastewater treatment technologies and desalination to purify and recycle virtually all wastewater. The technologies used in ZLD facilities can include sedimentation, dissolved air flotation (DAF), biological treatment, media filtration, activated carbon filtration, ultrafiltration, and reverse osmosis.
Benefits	ZLD systems help industrial facilities meet discharge and water reuse requirements, enabling them to comply with government discharge regulations, achieve higher water recovery, and treat and recover valuable materials from the wastewater streams.
Technologies	ZLD in wastewater treatment applications can be achieved by combining various technologies, such as thermal processes like heat pump evaporators, reverse osmosis (membrane technology), electrodialysis, falling film brine concentrators, forced circulation crystallizer, horizontal spray film evaporator, and solids waste handling.
Challenges	Despite the advantages such as reduced water pollution and resource recovery from waste, there are several challenges to overcome prior to wider applications of ZLD. These include high energy consumption, high capital and operating costs, and the management of the solid waste produced.
Future research	To fulfill the sustainable vision, future research that can bridge the gap between theoretical study and industrial practice is highlighted.

TABLE 8.9
Challenges in ZLD

Challenge	Explanation
High energy consumption	ZLD systems often require significant amounts of energy, particularly for processes like evaporation and crystallization. This can lead to high operating costs and a larger carbon footprint.
High capital and operating costs	The installation and maintenance of ZLD systems can be expensive. This includes the costs of the advanced treatment technologies used in ZLD, as well as ongoing costs for energy, labor, and waste disposal.
Management of solid waste	ZLD systems produce a concentrated solid waste stream that must be managed properly. This waste often requires further treatment before it can be disposed of, which can add to the complexity and cost of the system.
Technical complexity	ZLD systems can be technically complex to design, operate, and maintain. They require skilled operators and regular maintenance to ensure they function effectively.
Water quality	While ZLD systems are designed to produce high-quality water suitable for reuse, the quality of the treated water can vary depending on the specific treatment technologies used and the characteristics of the incoming wastewater.

8.5 HYDRAULICS OF WASTEWATER SYSTEMS

Hydraulics of wastewater systems involves the study of the behavior of water and other liquids in wastewater infrastructure, including pipes, channels, and treatment facilities. Unlike water supply systems, which focus on delivering clean water to consumers, wastewater systems deal with the collection, transportation, and treatment of used water and sewage. One key difference between the two is the nature of the fluid being conveyed: while water supply systems handle clean water, wastewater systems must contend with a more complex and variable mixture of water, organic matter, and solids.

In wastewater systems, hydraulics are influenced by the presence of solid particles, organic materials, and various chemicals, which can affect the flow characteristics and the performance of the system. Additionally, the presence of air in wastewater systems can also impact the hydraulics, as it can lead to the formation of air pockets or air binding in the pipes, affecting the flow and efficiency of the system. In contrast, water supply systems typically deal with a more consistent and predictable flow of clean water, without the presence of solid waste or the need to manage air in the same way.

Despite these differences, there are also several similarities between the hydraulics of wastewater systems and both pipe flow and open channel hydraulics. For instance, the fundamental principles of fluid mechanics apply to all three areas, including concepts such as continuity, energy, and momentum equations. In all cases, the behavior of the fluid, whether it is wastewater in pipes, clean water in supply systems, or open channel flow, is governed by these underlying principles.

The design and analysis of wastewater systems often draw upon the same principles and methodologies used in pipe flow and open channel hydraulics. Engineers and researchers rely on similar mathematical models and computational tools to predict and optimize the performance of wastewater systems, just as they do for pipe flow and open channel hydraulics. This includes the use of equations such as the Manning equation for open channel flow, the Darcy–Weisbach equation for pipe flow, and similar empirical and theoretical approaches to characterize and analyze the behavior of fluids in these different systems.

Another important similarity is the consideration of factors such as friction, turbulence, and flow regime, which are relevant to both wastewater systems and traditional hydraulics. These factors influence the resistance to flow, the distribution of velocities within the system, and the overall efficiency and performance of the infrastructure. Therefore, engineers and practitioners in the field of wastewater hydraulics often leverage the knowledge and techniques developed for pipe flow and open channel hydraulics to address similar challenges and optimize the design and operation of wastewater systems.

While there are distinct differences between the hydraulics of wastewater systems and water supply systems, there are also significant similarities with pipe flow and open channel hydraulics. Understanding the unique characteristics of wastewater, including the presence of solids, organic matter, and air, is crucial for effectively managing the hydraulics of wastewater systems. At the same time, the fundamental principles and analytical tools developed for pipe flow and open channel hydraulics provide valuable insights and methods for addressing the challenges and optimizing the performance of wastewater infrastructure. By leveraging this shared knowledge and expertise, engineers and researchers can continue to advance the field of wastewater hydraulics and contribute to the sustainable management of water resources and environmental protection.

The hydraulic principles of a wastewater network revolve around the behavior of water and other liquids as they move through the collection, conveyance, and treatment infrastructure. These principles are crucial for ensuring the efficient and effective operation of the network. The key hydraulic principles of a wastewater network are shown in Table 8.10.

Engineers and practitioners can effectively design, operate, and maintain wastewater networks by incorporating hydraulic principles. This approach ensures the reliable conveyance and treatment of wastewater while minimizing operational issues and environmental impacts.

8.6 URBAN WASTEWATER SYSTEM NETWORKS

A wastewater systems network (Figure 8.5) is an infrastructure designed to collect, convey, and treat used water and sewage from homes, businesses, and industrial facilities. The network consists of a series of interconnected components, including pipes, pumping stations, treatment plants, and storage facilities, all working together to manage the flow of wastewater. Here is an overview of the key aspects of a wastewater systems network:

TABLE 8.10
Key Hydraulic Principles of Wastewater Network

Hydraulic Principle	Explanation
Flow characteristics	Understanding the flow characteristics of wastewater within the network is essential. This includes analyzing factors such as flow rates, velocities, and the distribution of flow throughout the system. Hydraulic principles help in predicting how the wastewater will behave as it moves through pipes, channels, and treatment facilities.
Pressure and head loss	The principles of pressure and head loss are important in determining the energy requirements for moving wastewater through the network. Frictional losses, changes in elevation, and other factors affect the pressure and head within the system, and hydraulic principles are used to calculate and manage these losses.
Pressure and head loss	The principles of pressure and head loss are important in determining the energy requirements for moving wastewater through the network. Frictional losses, changes in elevation, and other factors affect the pressure and head within the system, and hydraulic principles are used to calculate and manage these losses.
Pipe sizing and design	Proper sizing and design of pipes within the network are critical for maintaining hydraulic efficiency. Hydraulic principles guide engineers in determining the appropriate pipe diameters, gradients, and layouts to ensure optimal flow and performance.
Pump selection and operation	In cases where pumping is necessary to move wastewater through the network, hydraulic principles are used to select suitable pumps and to optimize their operation. This involves considering factors such as flow rates, head requirements, and the characteristics of the wastewater being pumped.
Hydraulic modeling	The use of hydraulic modeling software and techniques is integral to the design and analysis of wastewater networks. Hydraulic principles are applied to develop accurate models that simulate the behavior of wastewater under various conditions, aiding in network optimization and troubleshooting.
Flow control and regulation	Hydraulic principles are utilized to implement flow control and regulation devices within the network, such as valves and regulators. These devices help manage the flow of wastewater, maintain system pressures, and facilitate the proper functioning of treatment processes.
Hydraulic performance assessment	Regular assessment of the hydraulic performance of the network is essential. Hydraulic principles are employed to analyze factors such as flow capacity, hydraulic gradients, and the impact of blockages or obstructions on the system's operation.
Sustainable hydraulics	With a growing emphasis on sustainability, hydraulic principles are increasingly applied to optimize the energy efficiency of wastewater networks, minimize environmental impacts, and enhance the overall sustainability of the infrastructure.

1. *Collection System:* The network begins with the collection system, which comprises a network of underground pipes that capture wastewater from individual buildings and transport it to larger sewer lines. This system also includes manholes and inspection chambers for maintenance and monitoring purposes.
2. *Conveyance Infrastructure:* Once collected, the wastewater is conveyed through a network of gravity-driven and pressurized pipes to transport it to treatment facilities. This infrastructure may also include pumping stations to lift the wastewater to higher elevations when necessary.
3. *Treatment Facilities:* Wastewater treatment plants are integral components of the network, where the collected wastewater undergoes physical, chemical, and biological processes to remove contaminants and pollutants before being discharged back into the environment or reused for various purposes.
4. *Storage and Equalization:* Some wastewater systems include storage and equalization facilities to manage fluctuations in wastewater flow, ensuring that treatment plants can handle varying loads and preventing overflows during peak periods.

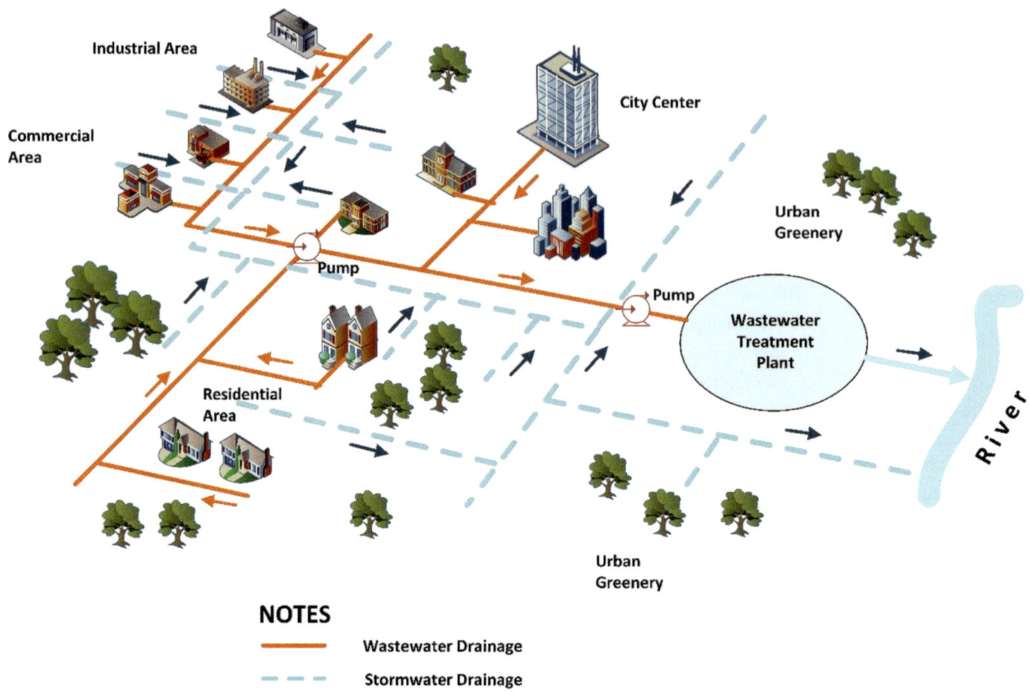

FIGURE 8.5 Separated wastewater drainage network.

5. *Monitoring and Control:* Wastewater systems are equipped with monitoring and control systems to track flow rates, water quality parameters, and operational status. These systems help operators manage the network efficiently and respond to any issues that may arise.
6. *Regulatory Compliance:* Compliance with environmental regulations and standards is a critical aspect of wastewater systems. The network must meet stringent requirements to protect public health and the environment, and it must be designed and operated in accordance with these regulations.
7. *Hydraulic Considerations:* Hydraulic principles play a crucial role in the design and operation of wastewater systems. Engineers must consider factors such as flow rates, pressures, and the behavior of wastewater within the network to ensure efficient conveyance and treatment.
8. *Asset Management:* Effective asset management is essential for the long-term performance of the network. This includes regular maintenance, rehabilitation of aging infrastructure, and strategic planning to optimize the use of resources and investments.
9. *Resilience and Emergency Response:* Wastewater systems must be designed to withstand and recover from natural disasters, infrastructure failures, and other emergencies. Resilience measures, such as backup systems and emergency response plans, are vital for maintaining service continuity.
10. *Sustainability and Innovation:* As the focus on sustainability grows, wastewater systems are incorporating innovative technologies and practices to improve energy efficiency, reduce environmental impact, and promote resource recovery from wastewater.

A wastewater systems network is a vital part of modern urban infrastructure, playing a crucial role in public health, environmental protection, and sustainable development. The effective design, operation, and management of these networks are essential for ensuring the safe and reliable conveyance and treatment of wastewater while meeting the evolving needs and challenges of urban communities.

8.6.1 Collection Systems

The collection system in a wastewater network is a component responsible for gathering used water and sewage from residential, commercial, and industrial sources and transporting it to treatment facilities. This system plays a crucial role in maintaining public health and environmental protection by effectively managing the flow of wastewater. There are some important sub-components of the collection system. These sub-components may include:

1. *Underground Pipes:* The collection system consists of a vast network of underground pipes that form the primary conduit for transporting wastewater. These pipes are typically made of durable materials such as PVC, concrete, or vitrified clay, and they are designed to withstand the corrosive nature of wastewater and the external environment.
2. *Service Laterals:* Service laterals are smaller pipes that connect individual buildings to the main sewer lines. These laterals collect wastewater from homes, businesses, and other properties and convey it to the larger sewer network for further transport.
3. *Manholes and Inspection Chambers:* Throughout the collection system, manholes and inspection chambers are strategically placed to provide access for maintenance, inspection, and cleaning activities. These structures allow workers to enter the sewer network for repairs and to monitor the condition of the pipes.
4. *Gravity-Driven Flow:* In many areas, the collection system relies on gravity to move wastewater through the network. The system is designed with carefully calculated slopes and elevations to ensure that wastewater flows from higher to lower points, guiding it toward treatment facilities without the need for additional pumping.
5. *Inflow and Infiltration Management:* Collection systems must address issues related to inflow and infiltration, which involve excess water entering the sewer network from sources such as stormwater or groundwater. Proper management of inflow and infiltration is essential for preventing overloading of the system and ensuring efficient treatment.
6. *Combined Sewer Systems:* In some older urban areas, combined sewer systems are still in use, where stormwater and sanitary sewage are conveyed through the same network. These systems present unique challenges in managing the flow of wastewater during periods of heavy rainfall and require specialized infrastructure and management practices.
7. *Sewer Odor Control:* To minimize unpleasant odors and potential health hazards, collection systems often incorporate measures for odor control, such as ventilation systems, chemical treatments, and sealing techniques to prevent the escape of noxious gases.

The collection system is a component of a wastewater network, serving as the initial stage in the journey of wastewater from its source to treatment facilities. Effective design, operation, and maintenance of the collection system are essential for ensuring the safe and efficient conveyance of wastewater while protecting public health and the environment.

8.6.2 Conveyance Infrastructure

The conveyance infrastructure of a wastewater system encompasses the network of pipes, pumping stations, and associated structures designed to transport collected wastewater from the point of origin to treatment facilities. This critical component of the wastewater system ensures the efficient conveyance of sewage and used water, contributing to public health and environmental protection. The conveyance infrastructure includes gravity-driven sewer lines that rely on carefully engineered slopes and elevations to guide wastewater flow, as well as pressurized sewer lines that utilize pumping stations to lift wastewater to higher elevations when necessary. These systems are designed to accommodate varying flow rates and hydraulic conditions, with consideration for factors such as pipe materials, diameters, and gradients. Pumping stations play a key role in overcoming elevation differences and conveying wastewater over long distances, and they are equipped with pumps, control systems, and monitoring equipment to ensure reliable operation. Additionally, the conveyance infrastructure may incorporate storage and equalization facilities to manage fluctuations in wastewater flow, helping to prevent overloading of treatment plants during peak periods. Effective design, operation, and maintenance of the conveyance infrastructure are essential for ensuring the safe and reliable transport of wastewater to treatment facilities while minimizing environmental impact and public health risks.

The separation of conveyance infrastructure for wastewater and stormwater involves the distinct routing of these two types of water within a community's infrastructure. Wastewater, which includes sewage and used water from homes, businesses, and industries, is conveyed through a separate network of pipes and pumping stations designed specifically for its transport to treatment facilities. On the other hand, stormwater, which consists of rainwater and surface runoff, is conveyed through a separate system of pipes, ditches, and culverts to manage its flow and mitigate flooding, without being mixed with wastewater. This separation is important because the water quality of both is different. There are also other reasons for this separation.

First, separating wastewater and stormwater helps prevent the overloading of treatment facilities during heavy rainfall events. By keeping stormwater out of the wastewater system, the treatment plants can focus on processing sewage and used water without being overwhelmed by excessive inflows of stormwater. This separation also reduces the risk of untreated wastewater being discharged into the environment during periods of heavy rainfall when combined systems might overflow.

Second, the separation of conveyance infrastructure for wastewater and stormwater allows for more efficient and cost-effective treatment of each type of water. Wastewater treatment plants can be designed and operated to effectively treat sewage and used water without the added burden of managing large volumes of stormwater. Similarly, stormwater management facilities can be optimized to handle runoff and reduce the risk of urban flooding without the need to address the complexities of treating wastewater.

Third, separating wastewater and stormwater conveyance infrastructure can contribute to improved water quality in receiving bodies of water. By preventing the mixing of untreated sewage and stormwater, the risk of pollution in rivers, lakes, and coastal areas can be significantly reduced, benefiting both the environment and public health.

The separation of conveyance infrastructure for wastewater and stormwater is an important in urban water management, providing numerous benefits such as enhanced treatment efficiency, reduced environmental pollution, and improved resilience to extreme weather events. This approach reflects a commitment to sustainable and effective water infrastructure planning and management.

The conveyance infrastructure is designed to address challenges related to inflow and infiltration, which involve the entry of excess water into the sewer network from sources such as stormwater or groundwater. Effective management of inflow and infiltration is crucial for preventing overloading of the conveyance system and treatment facilities, as well as minimizing the risk of system backups and overflows during periods of heavy rainfall or snowmelt. In addition to the conveyance of domestic wastewater, the infrastructure also handles industrial wastewater and commercial effluent, requiring careful consideration of the diverse characteristics and composition of the conveyed fluids. This may involve the implementation of specialized piping and treatment requirements to accommodate the unique properties of industrial and commercial wastewater, ensuring that it is safely and effectively conveyed to treatment facilities.

The conveyance infrastructure maintains the overall hydraulic performance of the wastewater system and ensures that the conveyance of wastewater is efficient and reliable. Hydraulic principles are applied to optimize the design of the conveyance infrastructure, considering factors such as flow velocities, pressure requirements, and the behavior of wastewater within the network. Regular monitoring and maintenance of the conveyance infrastructure are essential to address issues such as pipe blockages, structural deterioration, and hydraulic inefficiencies, helping to uphold the integrity and functionality of the system. By effectively managing the conveyance infrastructure, wastewater utilities can ensure the safe and sustainable transport of wastewater to treatment facilities, contributing to the protection of public health and the environment.

8.6.3 Treatment Facilities

Wastewater treatment facilities are primary components of the overall wastewater system, responsible for the purification and removal of contaminants from collected sewage and used water before it is discharged back into the environment or reused for various purposes. These facilities employ a combination of physical, chemical, and biological processes to treat wastewater, ensuring that it meets environmental standards and poses no harm to public health. Please refer to Section 8.2. Physical Unit of Wastewater Treatment System.

8.7 POPULATION EQUIVALENTS

Population equivalent (PE) in the context of wastewater monitoring and treatment refers to the amount of oxygen-demanding substances (measured in BOD or BOD5) present in wastewater. Specifically, it represents the average oxygen demand of the wastewater produced by one person for 1 day. For practical calculations, it is commonly assumed that one PE equals 54 grams of BOD per 24 hours.

When planning wastewater treatment systems, understanding the PE helps determine the capacity needed to effectively treat the wastewater generated by a specific population. By considering this metric, authorities can design and implement appropriate treatment processes to manage the organic load and ensure environmental protection.

PE serves several important purposes in the context of wastewater management and treatment:

1. *Designing Wastewater Treatment Systems:*
   - PE helps determine the capacity needed for wastewater treatment plants.
   - By understanding the average organic load generated by a specific population, authorities can design treatment processes effectively.
   - It ensures that treatment facilities can handle the organic pollutants produced by industries or communities.
2. *Designing Wastewater Treatment Systems:*
   - When assessing the impact of wastewater discharges, PE provides a standardized measure.
   - It allows us to compare an industry's polluting potential (in terms of biodegradable organic matter) with the equivalent pollution load produced by a population.
   - This assessment helps safeguard the environment and public health.
3. *Designing Wastewater Treatment Systems:*
   - Municipal sewage systems often charge fees based on the PE of industrial wastewater discharges.
   - Regulatory compliance and permit requirements are often tied to PE values.
   - Industries must adhere to PE-based guidelines to manage their wastewater appropriately.
4. *Determining Loadings and Treatment Efficiency:*
   - PE helps estimate the total waste loadings that need treatment.
   - It allows us to express the industrial wastewater load in terms of an equivalent number of people.
   - Treatment efficiency can be evaluated by comparing actual pollutant loads to PE-based estimates.

PE is calculated by taking the wastewater flow (Q) and multiplying it by the BOD5 Concentration in the wastewater. PE represents the polluting potential of an industry (in terms of biodegradable organic matter) compared to a population (or a certain number of people) that would produce an equivalent polluting load.

A commonly used base value for PE is 54 grams of BOD per person per day. However, other values are also in use. In Europe, 1 PE equates to 60 grams of BOD per person per day and equals 200 liters of sewage per day. In the United States, a figure of 80 grams BOD per day is typically used.

PE is calculated by multiplying the wastewater flow (Q) by the BOD5 concentration in the wastewater, and then dividing the result by 60:

$$PE = \frac{\left(Q\left[m^3/day\right] \times C\left[mg/L\right]\right)}{60} \tag{8.4}$$

where:

PE: It is a relative measure used in environmental engineering. In this context, it refers to the amount of BOD5 (biochemical oxygen demand over 5 days) that a population would produce, given average per capita BOD5 production.

Q: This is typically measured in cubic meters per day (m^3/day) or liters per day (L/day). It represents the volume of wastewater being produced.

C (BOD5 Concentration in Wastewater): This is typically measured in milligrams per liter (mg/L). It represents the amount of oxygen required to break down organic material present in the wastewater over a 5-day period.

By understanding PE, we can design and size wastewater treatment plants effectively. It helps ensure that treatment processes can handle the organic load generated by a specific population or industry.

EXAMPLE 8.3

Suppose we have a wastewater flow (Q) of 5,000 m^3/day and a BOD5 concentration in the wastewater (C) of 300 mg/L. We can substitute these values into the equation to calculate the PE.

Solution:

Convert the units for Q and C to be compatible. In this case, they already are, so no conversion is necessary.

Substitute the values into Equation (8.1):

$$PE = \frac{\left(5{,}000\left(m^3/day\right) \times 300\left(mg/L\right)\right)}{60}$$

$$PE = 250{,}000 \text{ g/day}$$

Therefore, the PE in this example would be 250,000. This means that the pollution load of this wastewater is equivalent to the waste produced by a population of 250,000 people.

When planning wastewater treatment systems, understanding the PE is important. PE estimates the usage made of sewage facilities and helps determine the capacity needed for effective treatment. The PE values for various premises are summarized in Table 8.11.

TABLE 8.11
PE Values for Various Premises

Premise	PE Value
Residential areas	• PE: 5 per house • Direct measurement of the population in an area.
Commercial areas (e.g., offices, shopping complexes, restaurants)	• PE: Calculated from the floor area (proportional to the number of people using the premises during the day). • Does not reflect the population living in the area. • Typically, 3 per 100 m² of gross area.
Schools/Educational institutions	• Day schools / Institutions: 0.2 per student. • Fully residential: 1 per student. • Partial residential: 0.2 per non-residential student and 1 per residential student.
Hospital	4 per bed
Hotels with dining and laundry facilities	4 per room
Factories (excluding process water)	0.3 per staff
Market	• Wet type: 3 per stall • Dry type: 1 per stall
Petrol kiosks/service stations	15 per toilet
Transportation terminals	• Bus terminal: 4 per bus bay. • Taxi terminal: 4 per taxi bay. • Airport: 0.2 per passenger/day and 0.3 per employee.
Public facilities	• Mosque/Church/Temple: 0.2 per person. • Stadium: 0.2 per person. • Swimming pool or sports complex: 0.5 per person. • Public toilet: 15 per toilet.
Other premises	• Laundry: 10 per machine. • Prison: 1 per person. • Golf course: 20 per hole.

PE values guide the design and efficiency of wastewater treatment systems, ensuring environmental protection.

EXAMPLE 8.4

Suppose a small town has a PE of 10,000. This means that the pollution load of the wastewater produced by this town is equivalent to that produced by a population of 10,000 people.

The design of a wastewater treatment plant typically considers a per capita wastewater generation rate. Let us assume this rate is 200 liters per person per day (L/person/day). This rate can vary depending on the specific conditions and water usage habits of the population.

We can calculate the total daily wastewater flow (Q) for the town as follows:

Q = PE multiplied by per capita wastewater generation rate
Q = 10,000 people × 200 L/person/day
Q = 2,000,000 L/day

Therefore, the wastewater treatment plant for this town should be designed to treat 2,000,000 liters of wastewater per day.

This is a simplified example. The actual design of a wastewater treatment plant would need to consider many other factors, such as peak flow rates, the specific composition of the wastewater, local regulations, future population growth, and more.

PROBLEMS AND ISSUES FOR DISCUSSION

1. What are the primary objectives of a wastewater treatment system, and how do they differ from a water treatment system?
2. Explain the purpose and importance of the preliminary treatment stage in a wastewater treatment process.
3. Describe the role of primary sedimentation in a wastewater treatment plant, and how it contributes to the overall treatment process.
4. What are the key differences between aerobic and anaerobic biological treatment processes used in wastewater treatment?
5. How does the activated sludge process work, and what are the main components of this biological treatment method?
6. Explain the purpose and operation of a secondary clarifier in a wastewater treatment plant.
7. Discuss the advantages and disadvantages of using trickling filters as a biological treatment method for wastewater.
8. What is the purpose of disinfection in a wastewater treatment system, and what are the common disinfection methods used?
9. Describe the role of nutrient removal (nitrogen and phosphorus) in wastewater treatment, and the various techniques employed.
10. Explain the concept of sludge management in a wastewater treatment plant, and the various sludge handling and disposal methods.
11. How does the hydraulic retention time (HRT) affect the performance of a wastewater treatment process, and how is it calculated?
12. Discuss the importance of process control and monitoring in a wastewater treatment system, and the key parameters that are typically monitored.
13. What are the common sources of inflow and infiltration (I&I) in a wastewater collection system, and how can they be mitigated?
14. Explain the concept of MBRs and their advantages over conventional activated sludge systems.
15. Describe the role of anaerobic digestion in the treatment and stabilization of wastewater sludge.
16. What are the potential environmental impacts of untreated or inadequately treated wastewater, and why is proper treatment essential?
17. Discuss the importance of energy efficiency and sustainability in the design and operation of wastewater treatment systems.
18. How does the type of wastewater (domestic, industrial, or a combination) affect the design and operation of a treatment plant?
19. Explain the concept of biological nutrient removal (BNR) and its advantages over conventional nutrient removal methods.
20. Describe the role of constructed wetlands in the treatment of wastewater, and the key design considerations.
21. What are the common methods used for the disposal or beneficial reuse of treated wastewater effluent?

22. Discuss the challenges associated with the treatment of emerging contaminants, such as pharmaceuticals and personal care products, in wastewater.
23. Explain the concept of water reclamation and the various applications of reclaimed water in urban and agricultural settings.
24. How does the geographical location and climate affect the design and operation of a wastewater treatment system?
25. What are the key considerations in the design and construction of wastewater pumping stations within a collection system?
26. Discuss the importance of preventive maintenance and asset management in ensuring the long-term reliability of a wastewater treatment system.
27. Explain the concept of resource recovery from wastewater, such as the production of biogas or the extraction of nutrients.
28. How can the integration of renewable energy sources, such as solar or wind power, contribute to the sustainability of wastewater treatment operations?
29. Describe the role of public education and stakeholder engagement in the successful implementation and acceptance of wastewater treatment projects.
30. What are the common methods used for the treatment and disposal of hazardous or toxic waste generated in industrial wastewater?
31. Discuss the importance of odor control and the various strategies employed to mitigate odor issues in wastewater treatment facilities.
32. Explain the concept of decentralized or distributed wastewater treatment systems and their potential benefits in certain applications.
33. How do regulatory requirements, such as discharge standards and environmental regulations, influence the design and operation of wastewater treatment systems?
34. Describe the role of advanced oxidation processes (AOPs) in the removal of recalcitrant or difficult-to-treat contaminants in wastewater.
35. What are the key considerations in the design and operation of wastewater treatment systems for small communities or rural areas?
36. Discuss the potential impacts of climate change on the performance and resilience of wastewater treatment infrastructure, and strategies for adaptation.
37. Explain the concept of water–energy–nutrient nexus and its relevance in the design and optimization of wastewater treatment systems.
38. How can the integration of digital technologies, such as sensors and data analytics, improve the efficiency and optimization of wastewater treatment processes?
39. Discuss the role of public–private partnerships and innovative financing models in the development of sustainable wastewater infrastructure projects.
40. Describe the key design considerations and operational challenges associated with the treatment of high-strength industrial wastewater, such as from food processing, chemical manufacturing, or pharmaceutical industries.
41. Explain the concept of membrane filtration technologies, such as microfiltration, ultrafiltration, and reverse osmosis, and their applications in advanced wastewater treatment processes for the removal of suspended solids, pathogens, and dissolved contaminants.
42. Discuss the role of life cycle assessment (LCA) in the design and evaluation of wastewater treatment systems, and how it can be used to optimize environmental, economic, and social sustainability throughout the system's life cycle.
43. Describe the potential benefits and limitations of using natural treatment systems, such as constructed wetlands or lagoons, for wastewater treatment, and the key design parameters and operational considerations for their effective implementation.
44. Explain the concept of co-digestion, where organic waste streams from various sources (e.g., municipal, agricultural, industrial) are combined and anaerobically digested to enhance biogas production and improve the overall efficiency of sludge management.

45. Discuss the challenges and best practices associated with the operation and maintenance of wastewater collection systems, including the inspection, cleaning, and rehabilitation of sewer pipes, manholes, and lift stations to prevent infiltration, exfiltration, and system failures.
46. Describe the role of mathematical modeling and simulation tools in the design, optimization, and performance evaluation of wastewater treatment processes, and how they can be used to predict the impacts of process changes, influent variations, and environmental conditions.
47. Explain the importance of employee training, safety protocols, and emergency response planning in the effective and responsible operation of wastewater treatment facilities, ensuring the protection of both workers and the surrounding community.
48. Discuss the potential opportunities and barriers for the integration of resource recovery technologies, such as the production of biofuels, the extraction of nutrients, or the reuse of treated water, within the overall wastewater treatment system design and operation.
49. Describe the key considerations and best practices for the decommissioning and demolition of outdated or obsolete wastewater treatment infrastructure, ensuring the safe and environmentally responsible disposal or repurposing of materials and minimizing the impact on the surrounding environment.

MULTIPLE-CHOICE QUESTIONS

1. What is the typical detention period for plain sedimentation in a primary sedimentation tank?
   a. 1–2 hours
   b. 2–2.5 hours
   c. 3–4 hours
   d. 4–5 hours
   e. 5–6 hours
2. What is the maximum flow velocity in a primary sedimentation tank?
   a. 10 cm/min
   b. 20 cm/min
   c. 30 cm/min
   d. 40 cm/min
   e. 50 cm/min
3. What is the typical length to breadth ratio (L:B) for a primary sedimentation tank?
   a. 1:1
   b. 2:1
   c. 3:1 to 5:1
   d. 6:1
   e. 7:1
4. What is the formula for Hydraulic Loading Rate (HLR)?
   a. HLR = Q/V
   b. HLR = Q × A
   c. HLR = Q + A
   d. HLR = Q/A
   e. HLR = Q – A
5. What is the formula for Organic Loading Rate (OLR)?
   a. OLR = Q/BOD
   b. OLR = Q × BOD
   c. OLR = Q + BOD
   d. OLR = Q – BOD
   e. OLR = Q × BOD/V

6. What is the typical detention period for coagulated sedimentation in a primary sedimentation tank?
   a. 1–2 hours
   b. 2–2.5 hours
   c. 3–4 hours
   d. 4–5 hours
   e. 5–6 hours
7. What is the typical surface overflow rate for a primary sedimentation tank?
   a. 100–200 $m^3/m^2/day$
   b. 200–300 $m^3/m^2/day$
   c. 300–400 $m^3/m^2/day$
   d. 600–1200 $m^3/m^2/day$
   e. 1200–1500 $m^3/m^2/day$
8. What is the typical average overflow rate for a primary sedimentation tank?
   a. 10–20 $m^3/m^2/day$
   b. 20–30 $m^3/m^2/day$
   c. 30–40 $m^3/m^2/day$
   d. 30–50 $m^3/m^2/day$
   e. 50–60 $m^3/m^2/day$
9. What is the typical depth of a primary sedimentation tank?
   a. 1–2 m
   b. 2–3 m
   c. 3–4 m
   d. 4–5 m
   e. 5–6 m
10. What is the maximum diameter for a circular primary sedimentation tank?
    a. 20 m
    b. 30 m
    c. 40 m
    d. 50 m
    e. 60 m
11. The primary purpose of wastewater treatment is to:
    a. Recycle water for reuse
    b. Reduce the amount of pollutants in the water
    c. Produce energy from the waste
    d. All of the above
12. Which of the following is not a typical stage in a wastewater treatment process?
    a. Preliminary treatment
    b. Primary treatment
    c. Secondary treatment
    d. Tertiary treatment
    e. Quaternary treatment
13. The process of removing large objects, such as rags, sticks, and plastics, from wastewater is known as:
    a. Screening
    b. Sedimentation
    c. Aeration
    d. Disinfection

14. The process of allowing suspended solids to settle out of the wastewater is called:
    a. Screening
    b. Sedimentation
    c. Aeration
    d. Disinfection
15. The process of adding air or oxygen to the wastewater to promote the growth of aerobic bacteria is called:
    a. Screening
    b. Sedimentation
    c. Aeration
    d. Disinfection
16. Which of the following is a common method of disinfecting wastewater?
    a. Chlorination
    b. Ozonation
    c. UV radiation
    d. All of the above
17. The process of removing nutrients, such as nitrogen and phosphorus, from wastewater is known as:
    a. Nutrient removal
    b. Denitrification
    c. Nitrification
    d. All of the above
18. The process of converting ammonia to nitrite and then to nitrate is called:
    a. Nutrient removal
    b. Denitrification
    c. Nitrification
    d. All of the above
19. The process of converting nitrate to nitrogen gas is called:
    a. Nutrient removal
    b. Denitrification
    c. Nitrification
    d. All of the above
20. Which of the following is a common method of sludge disposal?
    a. Landfilling
    b. Incineration
    c. Land application
    d. All of the above
21. The process of removing dissolved organic matter from wastewater using microorganisms is known as:
    a. Preliminary treatment
    b. Primary treatment
    c. Secondary treatment
    d. Tertiary treatment
22. Which of the following is a common type of secondary treatment process?
    a. Activated sludge
    b. Trickling filter
    c. Rotating biological contactor
    d. All of the above

23. The process of removing specific pollutants, such as heavy metals or pharmaceuticals, from wastewater is known as:
    a. Preliminary treatment
    b. Primary treatment
    c. Secondary treatment
    d. Tertiary treatment
24. Which of the following is a common method of tertiary treatment?
    a. Reverse osmosis
    b. Membrane filtration
    c. Ion exchange
    d. All of the above
25. The process of converting organic matter in wastewater into methane gas is called:
    a. Anaerobic digestion
    b. Aerobic digestion
    c. Sludge dewatering
    d. Sludge incineration
26. Which of the following is a benefit of using constructed wetlands for wastewater treatment?
    a. High energy consumption
    b. Low maintenance requirements
    c. Inability to remove nutrients
    d. Lack of natural habitat creation
27. The process of removing suspended solids from wastewater using a physical barrier, such as a membrane, is called:
    a. Membrane filtration
    b. Reverse osmosis
    c. Ion exchange
    d. Adsorption
28. Which of the following is a common method of sludge dewatering?
    a. Centrifugation
    b. Belt filtration
    c. Drying beds
    d. All of the above
29. The process of using electricity to remove dissolved ions from wastewater is called:
    a. Membrane filtration
    b. Reverse osmosis
    c. Ion exchange
    d. Adsorption
30. Which of the following is a common method of wastewater reuse?
    a. Irrigation
    b. Groundwater recharge
    c. Industrial process water
    d. All of the above
31. The process of removing dissolved organic compounds from wastewater using a porous material is called:
    a. Membrane filtration
    b. Reverse osmosis
    c. Ion exchange
    d. Adsorption

32. Which of the following is a common type of wastewater treatment plant?
    a. Centralized treatment plant
    b. Decentralized treatment system
    c. On-site treatment system
    d. All of the above
33. The process of removing dissolved salts from wastewater using a semi-permeable membrane is called:
    a. Membrane filtration
    b. Reverse osmosis
    c. Ion exchange
    d. Adsorption
34. Which of the following is a common method of wastewater disinfection?
    a. Chlorination
    b. Ozonation
    c. UV radiation
    d. All of the above
35. The process of converting organic matter in wastewater into carbon dioxide and water in the presence of oxygen is called:
    a. Anaerobic digestion
    b. Aerobic digestion
    c. Sludge dewatering
    d. Sludge incineration
36. Which of the following is a common type of wastewater treatment lagoon?
    a. Aerated lagoon
    b. Stabilization pond
    c. Facultative lagoon
    d. All of the above
37. The process of removing dissolved metals from wastewater using a chemical process is called:
    a. Chemical precipitation
    b. Membrane filtration
    c. Ion exchange
    d. Adsorption
38. Which of the following is a common method of sludge disposal?
    a. Landfilling
    b. Incineration
    c. Land application
    d. All of the above
39. The process of converting organic matter in wastewater into methane gas in the absence of oxygen is called:
    a. Anaerobic digestion
    b. Aerobic digestion
    c. Sludge dewatering
    d. Sludge incineration
40. Which of the following is a common type of secondary treatment process?
    a. Activated sludge
    b. Trickling filter
    c. Rotating biological contactor
    d. All of the above

41. The process of removing dissolved organic compounds from wastewater using a chemical or physical process is called:
    a. Chemical precipitation
    b. Membrane filtration
    c. Adsorption
    d. Ion exchange
42. Which of the following is a common method of wastewater disinfection using ultraviolet (UV) light?
    a. Low-pressure UV
    b. Medium-pressure UV
    c. High-intensity UV
    d. All of the above
43. The process of removing nitrogen from wastewater through the conversion of nitrate to nitrogen gas is called:
    a. Nitrification
    b. Denitrification
    c. Nutrient removal
    d. Biological nutrient removal
44. Which of the following is a common method of sludge stabilization?
    a. Anaerobic digestion
    b. Aerobic digestion
    c. Lime stabilization
    d. All of the above
45. The process of removing phosphorus from wastewater through chemical or biological means is called:
    a. Nitrification
    b. Denitrification
    c. Phosphorus removal
    d. Biological nutrient removal
46. Which of the following is a common type of wastewater treatment plant that treats wastewater from multiple sources?
    a. Centralized treatment plant
    b. Decentralized treatment system
    c. On-site treatment system
    d. Satellite treatment plant
47. Which of the following is a common method of wastewater reuse for agricultural irrigation?
    a. Drip irrigation
    b. Spray irrigation
    c. Subsurface irrigation
    d. All of the above
48. The process of removing dissolved salts from wastewater using an ion exchange resin is called:
    a. Membrane filtration
    b. Reverse osmosis
    c. Ion exchange
    d. Adsorption
49. Which of the following is a common method of wastewater reuse for industrial processes?
    a. Cooling water
    b. Boiler feed water
    c. Process water
    d. All of the above

ANSWERS TO THE MULTIPLE-CHOICE QUESTIONS

1	c	11	d	21	c	31	d	41	c
2	c	12	e	22	d	32	d	42	d
3	c	13	a	23	d	33	b	43	b
4	d	14	b	24	d	34	d	44	d
5	e	15	c	25	a	35	b	45	c
6	b	16	d	26	b	36	d	46	a
7	d	17	d	27	a	37	a	47	d
8	d	18	c	28	d	38	d	48	c
9	c	19	b	29	c	39	a	49	d
10	e	20	d	30	d	40	d		

FURTHER READING

Abdel-Shafy, H. I., & Mansour, M. S. (2018). A review on polycyclic aromatic hydrocarbons: Source, environmental impact, effect on human health and remediation. *Egyptian Journal of Petroleum*, 27(4), 839–848.

Ahn, Y. H. (2006). Sustainable nitrogen elimination biotechnologies: A review. *Process Biochemistry*, 41(8), 1709–1721.

Akpor, O. B., & Muchie, M. (2011). Environmental and public health implications of wastewater quality. *African Journal of Biotechnology*, 10(13), 2379–2387.

Arden, S., & Ma, X. (2018). Constructed wetlands for greywater recycle and reuse: A review. *Science of the Total Environment*, 630, 587–599.

Barker, D. J., & Stuckey, D. C. (1999). A review of soluble microbial products (SMP) in wastewater treatment systems. *Water Research*, 33(14), 3063–3082.

Bhatnagar, A., & Sillanpää, M. (2010). Utilization of agro-industrial and municipal waste materials as potential adsorbents for water treatment – A review. *Chemical Engineering Journal*, 157(2-3), 277–296.

Cai, T., Park, S. Y., & Li, Y. (2013). Nutrient recovery from wastewater streams by microalgae: Status and prospects. *Renewable and Sustainable Energy Reviews*, 19, 360–369.

Canizares, P., Martínez, F., Jiménez, C., Sáez, C., & Rodrigo, M. A. (2008). Coagulation and electrochemical processes in municipal wastewater treatment. *Journal of Chemical Technology & Biotechnology*, 83(8), 1204–1212.

Chong, M. N., Jin, B., Chow, C. W., & Saint, C. (2010). Recent developments in photocatalytic water treatment technology: A review. *Water Research*, 44(10), 2997–3027.

Demirbas, A. (2010). Treatment of biodiesel waste for recovery and reuse. *Energy Conversion and Management*, 51(12), 2391–2398.

Eckenfelder, W. W. (2000). *Industrial water pollution control*. McGraw-Hill Education.

Fang, H. H., & Jia, X. S. (1996). Autotrophic denitrification and its effect on metal speciation. *Water Research*, 30(6), 1573–1578.

Foley, J., de Haas, D., Hartley, K., & Lant, P. (2010). Comprehensive life cycle inventories of alternative wastewater treatment systems. *Water Research*, 44(5), 1654–1666.

Gao, D. W., Wen, Z. D., Li, B., & Liang, H. (2017). Microbial community structure characteristics associated with membrane fouling in a submerged membrane bioreactor process for municipal wastewater treatment. *Bioresource Technology*, 240, 196–203.

Ghangrekar, M. M. (2022). *Wastewater to water: Principles, technologies and engineering design*. Springer.

Gray, N. F. (2004). *Biology of wastewater treatment*. Imperial College Press.

Gude, V. G. (2015). Energy and water autarky of wastewater treatment and power generation systems. *Renewable and Sustainable Energy Reviews*, 45, 52–68.

Henze, M., Gujer, W., Mino, T., & van Loosdrecht, M. (2000). *Activated sludge models ASM1, ASM2, ASM2d, and ASM3*. IWA Publishing.

Henze, M., Harremoës, P., Jansen, J. L. C., & Arvin, E. (2002). *Wastewater treatment: Biological and chemical processes*. Springer Science & Business Media.

Hu, Z., Lotti, T., van Loosdrecht, M., & Kartal, B. (2013). Nitrogen removal with the anaerobic ammonium oxidation process. *Biotechnology Letters*, 35(8), 1145–1154.

Ivanovic, I., & Lebow, P. K. (2009). A review of studies relevant to the environmental fate of triclocarban in the United States. *Environmental Toxicology and Chemistry*, 28(1), 1–9.

Joss, A., Siegrist, H., & Ternes, T. A. (2008). Are we about to upgrade wastewater treatment for removing organic micropollutants? *Water Science and Technology*, 57(2), 251–255.

Katsoyiannis, I. A., & Samara, C. (2007). The fate of dissolved organic carbon (DOC) in the wastewater treatment process and its importance in the removal of wastewater contaminants. *Environmental Science and Pollution Research*, 14(5), 284–292.

Khiewwijit, R., Temmink, H., Rijnaarts, H., & Keesman, K. J. (2015). Energy and nutrient recovery for municipal wastewater treatment: How to design a feasible plant layout? *Renewable Energy*, 75, 374–386.

Kujawa-Roeleveld, K., & Zeeman, G. (2006). Anaerobic treatment in decentralised and source-separation-based sanitation concepts. *Reviews in Environmental Science and Biotechnology*, 5(1), 115–139.

Lateef, A., Chaudhry, M. N., & Ilyas, S. (2013). Biological treatment of dairy wastewater using activated sludge. *The Science World Journal*, 2013.

Lettinga, G. (1995). Anaerobic digestion and wastewater treatment systems. *Antonie van Leeuwenhoek*, 67(1), 3–28.

Li, W. W., Yu, H. Q., & Rittmann, B. E. (2015). Chemistry: Reuse water pollutants. *Nature*, 528(7580), 29–31.

Liang, S., Liu, C., & Song, L. (2010). Soluble microbial products in membrane bioreactor operation: Behaviors, characteristics, and fouling potential. *Water Research*, 44(20), 5750–5760.

Liu, Y., & Tay, J. H. (2004). State of the art of biogranulation technology for wastewater treatment. *Biotechnology Advances*, 22(7), 533–563.

Luo, Y., Guo, W., Ngo, H. H., Nghiem, L. D., Hai, F. I., Zhang, J., ... & Wang, X. C. (2014). A review on the occurrence of micropollutants in the aquatic environment and their fate and removal during wastewater treatment. *Science of the Total Environment*, 473, 619–641.

Metcalf & Eddy, Inc. (2014). *Wastewater engineering: Treatment and resource recovery* (5th ed.). McGraw-Hill Education.

Muga, H. E., & Mihelcic, J. R. (2008). Sustainability of wastewater treatment technologies. *Journal of Environmental Management*, 88(3), 437–447.

Oller, I., Malato, S., & Sánchez-Pérez, J. A. (2011). Combination of advanced oxidation processes and biological treatments for wastewater decontamination – A review. *Science of the Total Environment*, 409(20), 4141–4166.

Peavy, H. S., Rowe, D. R., & Tchobanoglous, G. (1985). *Environmental engineering*. McGraw-Hill Education.

Qasim, S. R. (1999). *Wastewater treatment plants: Planning, design, and operation*. CRC Press.

Radjenović, J., Petrović, M., Ventura, F., & Barceló, D. (2008). Rejection of pharmaceuticals in nanofiltration and reverse osmosis membrane drinking water treatment. *Water Research*, 42(14), 3601–3610.

Reed, S. C., Crites, R. W., & Middlebrooks, E. J. (1995). *Natural systems for waste management and treatment*. McGraw-Hill Education.

Rittmann, B. E., & McCarty, P. L. (2001). *Environmental biotechnology: Principles and applications*. McGraw-Hill.

Saeed, T., & Sun, G. (2012). A review on nitrogen and organics removal mechanisms in subsurface flow constructed wetlands: Dependency on environmental parameters, operating conditions and supporting media. *Journal of Environmental Management*, 112, 429–448.

Spellman, F. R. (2013). *Handbook of water and wastewater treatment plant operations*. CRC Press.

Tchobanoglous, G., Burton, F. L., & Stensel, H. D. (2003). *Wastewater engineering: Treatment and reuse* (4th ed.). McGraw-Hill.

Verstraete, W., & Vlaeminck, S. E. (2011). ZeroWasteWater: Short-cycling of wastewater resources for sustainable cities of the future. *International Journal of Sustainable Development & World Ecology*, 18(3), 253–264.

Vymazal, J. (2010). Constructed wetlands for wastewater treatment. *Water*, 2(3), 530–549.

Wintgens, T., Salehi, F., Hochstrat, R., & Melin, T. (2008). Emerging contaminants and treatment options in water recycling for indirect potable use. *Water Science and Technology*, 57(1), 99–107.

9 Municipal Solid Waste Management

9.1 SUSTAINABLE WASTE MANAGEMENT

Sustainable waste management is a crucial aspect of environmental conservation and public health. It involves the efficient handling, disposal, and recycling of waste materials to minimize negative impacts on the environment and human health. Implementing sustainable waste management practices is essential for reducing pollution, conserving resources, and promoting a cleaner and healthier living environment for current and future generations.

One key aspect of sustainable waste management is waste reduction at the source. This involves minimizing the amount of waste generated through practices such as reusing items, reducing packaging, and employing sustainable manufacturing processes. By reducing the amount of waste produced, the overall environmental impact can be significantly decreased.

Another important component of sustainable waste management is recycling. Recycling allows materials such as paper, glass, plastics, and metals to be reprocessed and reused, reducing the need for raw materials and cutting down on energy consumption. By encouraging recycling initiatives and providing accessible recycling facilities, communities can significantly reduce the amount of waste sent to landfills.

Composting is also an integral part of sustainable waste management. Organic waste, such as food scraps and yard trimmings, can be composted to create nutrient-rich soil amendments for gardening and agriculture. By diverting organic waste from landfills, composting helps reduce greenhouse gas emissions and enriches soil health.

Furthermore, sustainable waste management involves proper disposal of hazardous waste. Hazardous materials, such as batteries, electronic waste, and chemicals, must be handled and disposed of in a manner that prevents environmental contamination and protects public health.

In addition to waste reduction, recycling, composting, and hazardous waste disposal, sustainable waste management also encompasses the concept of waste-to-energy (WtE). This involves converting waste materials into energy through processes such as incineration, anaerobic digestion, and gasification. These methods can help generate electricity and heat while reducing the volume of waste sent to landfills.

Public awareness and education play a vital role in promoting sustainable waste management practices. By informing and engaging the community about the importance of waste reduction, recycling, and proper disposal, individuals can actively contribute to sustainable waste management efforts.

Governments and local authorities also play a crucial role in implementing sustainable waste management policies and regulations. By establishing comprehensive waste management plans, promoting sustainable practices, and investing in infrastructure for waste treatment and recycling, authorities can contribute to a cleaner and healthier environment.

DOI: 10.1201/9781032664613-9

Furthermore, businesses and industries have a responsibility to adopt sustainable waste management practices. By implementing efficient production processes, reducing packaging waste, and investing in recycling initiatives, businesses can minimize their environmental footprint and contribute to a more sustainable future.

Technological advancements also play a significant role in sustainable waste management. Innovations in waste sorting, recycling technologies, and WtE processes continue to improve the efficiency and effectiveness of waste management practices.

Community involvement and participation in waste management initiatives are essential for achieving sustainable outcomes. Engaging local communities in cleanup campaigns, waste separation programs, and environmental awareness activities can foster a sense of responsibility and stewardship toward the environment.

Sustainable waste management is closely linked to the concept of a circular economy, where resources are used efficiently, and waste is minimized through reuse, recycling, and regeneration. By embracing the principles of a circular economy, societies can move toward a more sustainable and resource-efficient model of waste management.

The significant point of sustainable waste management, when compared to conventional waste management, lies in its focus on long-term environmental preservation and resource conservation. Unlike conventional waste management, which primarily emphasizes the disposal of waste materials, sustainable waste management places a strong emphasis on minimizing waste generation, promoting recycling and composting, and harnessing WtE technologies. By prioritizing the reduction, reuse, and recycling of materials, sustainable waste management aims to minimize the environmental impact of waste and contribute to a more sustainable and resource-efficient approach to handling waste materials.

Sustainable waste management takes a holistic approach by considering the entire life cycle of products and materials, from production to disposal. This approach emphasizes the importance of sustainable manufacturing processes, eco-friendly packaging, and promoting the concept of extended producer responsibility (EPR), where producers take responsibility for the environmental impact of their products throughout their life cycle. In contrast, conventional waste management often focuses solely on the end-of-pipe solutions, such as landfilling and incineration, without addressing the root causes of waste generation.

Sustainable waste management recognizes the interconnectedness of environmental, social, and economic factors. It acknowledges that effective waste management is not only about reducing environmental pollution but also about creating employment opportunities, fostering innovation in recycling technologies, and promoting sustainable consumption patterns. In contrast, conventional waste management may overlook these broader societal and economic implications, focusing primarily on waste disposal without considering the potential benefits of a more sustainable approach.

Moreover, sustainable waste management encourages collaboration and partnerships among various stakeholders, including governments, businesses, communities, and non-governmental organizations. By fostering a multi-stakeholder approach, sustainable waste management seeks to leverage diverse expertise and resources to address waste management challenges comprehensively. This collaborative approach sets it apart from conventional waste management, which often relies heavily on centralized waste disposal facilities and may not actively engage with the broader community in waste reduction and recycling efforts.

9.2 WASTE CATEGORIES

Waste refers to any material or substance that is discarded, unwanted, or no longer serves its original purpose. It encompasses a wide range of materials, including, but not limited to, household garbage, industrial byproducts, construction debris, agricultural waste, and hazardous substances. Waste can take various forms, such as solid, liquid, or gaseous, and may pose environmental, health, and aesthetic challenges if not managed properly.

Solid waste is one of the most common forms of waste and includes everyday items such as packaging materials, food scraps, paper, plastics, glass, and textiles. This type of waste is generated from households, businesses, industries, and construction activities. Improper disposal of solid waste can lead to pollution of land, water, and air, as well as contribute to the depletion of natural resources.

Liquid waste, on the other hand, consists of wastewater from domestic, industrial, and agricultural sources. It includes sewage, industrial effluents, and agricultural runoff. Improper disposal of liquid waste can contaminate water bodies, affecting aquatic ecosystems and posing risks to human health through the spread of waterborne diseases and chemical pollutants.

Gaseous waste encompasses emissions released into the atmosphere as a result of industrial processes, transportation, and energy production. These emissions can include pollutants such as carbon dioxide, sulfur dioxide, nitrogen oxides, and volatile organic compounds. Gaseous waste contributes to air pollution, climate change, and respiratory health issues, highlighting the importance of effective waste management strategies to mitigate these impacts.

Hazardous waste represents a category of waste that poses substantial risks to human health and the environment due to its toxic, flammable, reactive, or corrosive nature. Examples of hazardous waste include batteries, electronic waste, chemical solvents, pesticides, and medical waste. Proper handling and disposal of hazardous waste are critical to prevent contamination of soil, water, and air, as well as to protect the health and safety of communities and ecosystems.

Waste is a by-product of human activities and consumption patterns, and its proper management is essential for safeguarding the environment, public health, and natural resources. Effective waste management practices, including waste reduction, recycling, proper disposal, and WtE initiatives, are essential for minimizing the negative impacts of waste and fostering a more sustainable approach to resource utilization.

Solid waste refers to a broad range of discarded materials that are not in a liquid or gaseous state. It encompasses various types of waste generated from residential, commercial, industrial, and construction activities. Solid waste can take the form of everyday items such as packaging materials, food scraps, paper, plastics, glass, and textiles, as well as bulky items like furniture and appliances. The management of solid waste is a critical aspect of environmental conservation and public health, as improper handling and disposal can lead to pollution, resource depletion, and health hazards.

The composition of solid waste can vary significantly depending on the source and location. Household solid waste typically includes organic materials, paper, cardboard, plastics, glass, metals, and miscellaneous items. Commercial and industrial solid waste may consist of manufacturing byproducts, packaging materials, office waste, and construction debris. Understanding the composition of solid waste is essential for developing effective waste management strategies tailored to specific waste streams.

Proper management of solid waste involves a combination of waste reduction, recycling, composting, and responsible disposal. Waste reduction practices aim to minimize the amount of waste generated by promoting reuse, repair, and the adoption of sustainable consumption habits. Recycling involves the collection and processing of materials such as paper, glass, plastics, and metals to produce new products, reducing the demand for raw materials and conserving energy.

Composting is another important aspect of solid waste management (SWM), particularly for organic waste such as food scraps, yard trimmings, and other biodegradable materials. Composting involves the natural decomposition of organic matter to produce nutrient-rich soil amendments, which can be used in gardening, landscaping, and agriculture. By diverting organic waste from landfills, composting helps reduce greenhouse gas emissions and enriches soil health.

Effective SWM also involves responsible disposal of non-recyclable and hazardous materials. This includes the proper operation of landfills, WtE facilities, and hazardous waste treatment plants to ensure that waste is managed in a manner that minimizes environmental impacts and protects public health. Overall, sustainable SWM practices are essential for reducing environmental pollution, conserving resources, and promoting a cleaner and healthier living environment for current and future generations.

Waste comes in many different forms and can be categorized in a variety of ways. Some common categories of waste may include the following:

1. *Agricultural Waste:* Waste generated from farming activities.
2. *Animal Byproducts:* Waste from slaughterhouses and related industries.
3. *Biodegradable Waste:* Waste that can be broken down by microorganisms.
4. *Biomedical Waste:* Waste from healthcare facilities, such as hospitals.
5. *Bulky Waste:* Large items, such as furniture and appliances, that are discarded.
6. *Business or Commercial Waste:* Waste produced by commercial activities.
7. *Chemical Waste*: Waste that contains or is contaminated by chemicals.
8. *Construction and Demolition Waste:* Waste generated from building or tearing down structures.
9. *Electronic Waste (E-Waste):* Discarded electronic devices or parts.
10. *Food Waste:* Unconsumed food and food preparation waste from residences and commercial establishments.
11. *Green Waste:* Biodegradable waste that can be composed of garden or park waste, such as grass or flower cuttings and hedge trimmings.
12. *Hazardous Waste:* Waste that poses substantial or potential threats to public health or the environment.
13. *Household Waste:* Waste generated by households.
14. *Industrial Waste:* Waste produced by industrial activity, such as factories.
15. *Municipal Solid Waste (MSW):* Waste collected by municipalities from households, institutions, and small businesses.
16. *Nuclear Waste:* Waste that contains radioactive material.
17. *Organic Waste:* Material that is biodegradable and comes from either a plant or an animal.
18. *Packaging Waste:* Waste generated by packaging materials.
19. *Recyclable Waste:* Waste that can be converted into reusable material.

Each type of waste requires a different method of disposal and treatment. It is important to classify waste correctly to ensure it is handled and disposed of properly. For example, hazardous waste must be dealt with differently than organic waste to prevent harm to people and the environment.

The Mekong Region, which includes countries like Cambodia, Lao PDR, Thailand, and Vietnam, has a simple yet effective way of categorizing waste. This categorization is based on the nature of the waste and how it can be managed or disposed of. Here are the five categories:

1. *Biodegradable Waste:* This type of waste can be broken down naturally over time by biological processes. It typically includes organic materials such as food scraps, paper, and wood. In the Mekong Region, this waste can be managed through composting or other biological treatment methods.
2. *Non-Biodegradable Waste:* This category includes waste that cannot be broken down naturally. Examples include plastic, glass, and metal items. These materials often end up in landfills or are incinerated.
3. *Hazardous/Toxic Waste:* This category includes waste that is dangerous or potentially harmful to human health or the environment. It can include certain types of industrial waste, batteries, paints, and medical waste. Special handling and disposal methods are required for this type of waste.
4. *Reusables:* These are items that might be considered a waste but can be reused for the same or a different purpose. Examples include glass containers, clothing, furniture, and electronic

equipment. Reusing items can significantly reduce the amount of waste that ends up in landfills.

5. *Recyclable Waste:* This category includes waste that can be converted into new products. Common recyclable materials include paper, cardboard, glass bottles, and certain types of plastic. Recycling helps to conserve resources and reduce the amount of waste that goes to landfills.

These categories help in effective management and disposal of waste in the Mekong Region. Again, each category requires a different method of waste management and treatment, which can range from composting for biodegradable waste to special disposal facilities for hazardous waste. The success of waste management in any region depends on the cooperation of every individual in properly sorting and disposing of their waste. Figure 9.1 shows the from-cradle-to-grave of the five categories of waste and the typical percentage of waste in the Mekong Region countries.

Figure 9.1 shows the present situation in the Mekong Region countries, that is, Thailand, Lao, Cambodia, and Vietnam, less the accomplishment of zero waste to the landfill site. The willingness of the local governments in this region to attain zero waste at the landfill site was high with the assistantships provided by CIDA Canada, SEI Sweden, and the Government of Finland. The pilot projects of zero waste were more than enough, but when it comes to the scale-up, replication, and real implementation (not a pilot project), the curse of developing nations of "good in planning but worse in implementing" hampered the program. As shown in Figure 9.1, a high percentage (70–80%) of biodegradable waste in the region is present in the Southeast Asia region.

Waste can also be categorized based on its composition into various types. Based on their composition, the categories of waste may include the following:

1. *Garbage:* This refers to animal and vegetable wastes resulting from the handling, sale, storage, preparation, cooking, and serving of food.

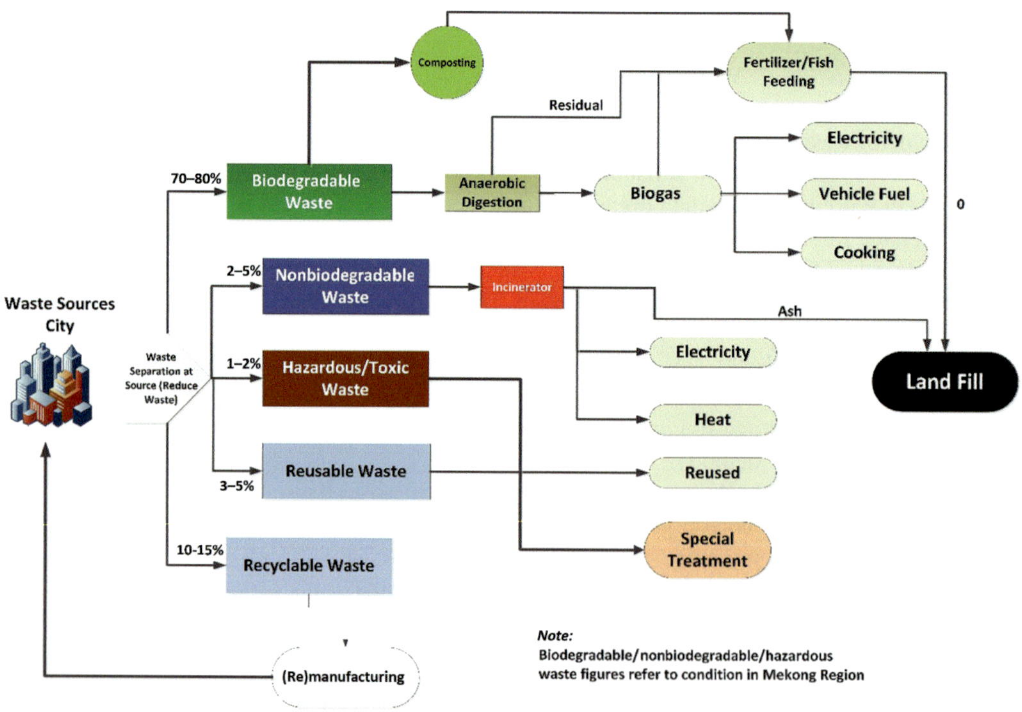

FIGURE 9.1 From source to landfill.

2. *Ashes and Residues:* These are the remains after the combustion of wood, coal, charcoal, or other materials.
3. *Combustible Waste:* These are wastes that can catch fire and burn, such as paper, wood, and certain types of plastic.
4. *Non-combustible Wastes:* These are wastes that do not burn, such as glass, metals, ceramics, etc.
5. *Bulky Waste:* These are large items that are not usually collected with regular household waste, such as furniture, appliances, etc.
6. *Street Wastes:* These include leaves, dirt, and other organic materials that are collected from street cleaning.
7. *Dead Animals:* These are animals that have died and are collected by animal control services.
8. *Construction and Demolition Wastes:* These are wastes generated from building or tearing down structures.
9. *Hazardous Wastes:* These are wastes that pose a substantial or potential threat to public health or the environment.

The number of categories of waste based on its composition can increase or decrease depending on the level of granularity one wants to achieve in categorizing the waste. For example, one could further divide electronic waste into categories like discarded computers, discarded televisions, discarded mobile phones, etc.

9.3 SOLID WASTE MANAGEMENT SYSTEM

While it may seem that waste is an insignificant part of our lives, the reality is quite the contrary. Waste management is a crucial aspect of our daily existence, and neglecting it can lead to catastrophic consequences. This paradoxical situation is particularly evident in many cities across Indonesia, where the implementation of sustainable SWM has been less than successful due to a reliance on traditional, outdated methods.

In many cities in developing countries, the approach to SWM has remained unchanged, adhering to a routine process of collection, transportation, and disposal, with no intermediate processing of the waste before it reaches the landfill. This method, while straightforward, has led to an unsustainable situation where the volume of solid waste being transported to landfill sites surpasses their capacity.

The problem is further exacerbated by the fact that the amount of waste generated is continually on the rise. This traditional approach to waste management needs a radical overhaul, shifting toward a more comprehensive and sustainable model of municipal solid waste management (MSWM) that incorporates unconventional options. Without such a shift, landfills will continue to be filled to capacity on a regular and ongoing basis, leading to the creation of "Not In My Back Yard" (NIMBY) situations. This relentless cycle of SWM needs to be broken to prevent the escalating and snowballing problems associated with waste management.

9.3.1 Sustainable versus Pragmatic Options

Pragmatic or conventional waste handling is a traditional approach to waste management that focuses on four main stages: waste production, waste collection, waste transport, and waste disposal. This flow is maintained for a long time and becomes a routine, business-as-usual, and traditional way of waste management in many developing cities.

- *Waste Production:* This is the initial stage where waste is generated. It can come from various sources, including households, industries, businesses, and more.

- *Waste Collection:* Once the waste is produced, it needs to be collected. This is usually done by local municipalities or private waste management companies. The waste is gathered from different sources and then taken to a central location for further processing.
- *Waste Transport:* After collection, the waste is transported to a disposal site. This could be a landfill, an incineration plant, or a recycling facility. The mode of transport can vary depending on the type and amount of waste and the distance to the disposal site.
- *Waste Disposal:* The final stage is disposal. This is where the waste is finally discarded. Depending on the type of waste and the disposal facility, this could involve burying the waste in a landfill, burning it in an incinerator, or processing it at a recycling facility.

This pragmatic approach is straightforward and effective in dealing with waste in the short term. However, it does not consider the long-term sustainability of waste management. It often leads to environmental issues such as pollution and resource depletion, as well as health risks for communities near disposal sites.

In contrast, sustainable waste management approaches aim to minimize waste production, maximize recycling, reuse, and reduce the environmental impact of waste disposal. They consider the entire lifecycle of products and materials, aiming to create a circular economy where waste is seen not as a problem, but as a resource.

Sustainable SWM is supposed to be able to accomplish the situation in which the waste hierarchy of waste reduction, reuse, recycle, WtE conversion, and waste disposal are in order of preference. This means the hierarchy must be largest to smallest with respect to percentage of contribution. The waste reduction must therefore be the largest, and the waste disposal must be the smallest and zero (Figure 9.2(a)). By this goal, the perpetual NIMBY could be avoided, particularly when a city accomplishes zero waste, that is, all waste is properly managed and utilized for further useful products. No waste is produced.

The approach to MSWM in many developing cities still adheres to a traditional, non-progressive business-as-usual process. This approach, which is primarily centered on the collection, transportation, and disposal of waste without much consideration for sustainability or waste reduction, is depicted in Figure 9.2(a). However, there are a few local authorities that have begun to shift their perspective and are making strides toward a more sustainable and progressive waste management model. This transition is represented in Figure 9.2(b). Notable examples of such cities in Indonesia include Surabaya, Bandung, Surakarta, and Semarang. These cities have started to implement more advanced SWM systems and technologies. They are exploring waste reduction strategies, enforcing

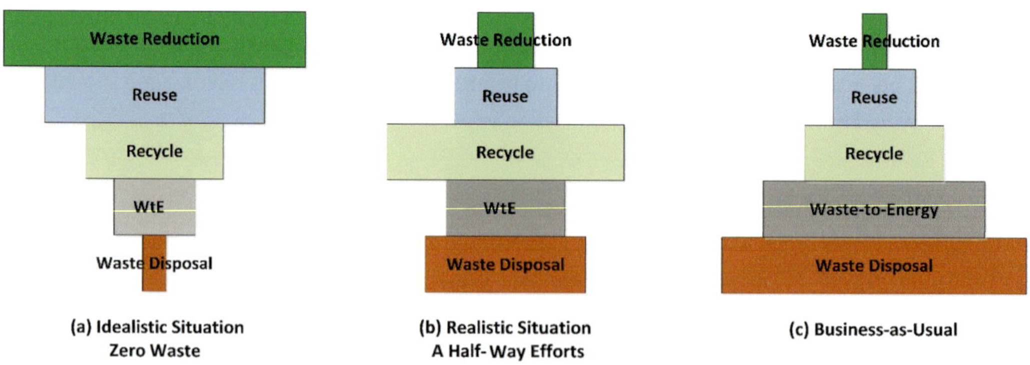

FIGURE 9.2 Waste hierarchy.

at-source segregation, adjusting local budgets to cover waste management costs, upgrading waste collection systems, and applying advanced SWM systems as appropriate.

These efforts are highly commendable and represent important steps toward achieving sustainable waste management. However, for these initiatives to be truly effective, they need to be accompanied by large-scale and community-based non-conventional approaches. Non-conventional approaches could include promoting waste reduction and recycling at the community level, encouraging the use of composting for organic waste, and implementing WtE technologies. Public education and awareness campaigns about the importance of waste reduction and proper waste disposal are also crucial.

9.3.2 Conventional Process of MSWM

Conventional MSWM encompasses a series of interconnected processes. It begins with waste generation at residential, commercial, and institutional sources, followed by waste collection by dedicated vehicles and personnel. The collected waste is then transferred to intermediate facilities for consolidation before being transported to processing facilities, such as recycling centers, composting sites, or WtE plants. At these facilities, the waste undergoes various treatment methods, including sorting, recycling, composting, or incineration, depending on its composition and characteristics. Finally, any residual waste that cannot be further processed is disposed of in landfill sites, where it is compacted and covered to minimize environmental impact. This conventional approach to waste management aims to reduce the volume of waste sent to landfills, maximize resource recovery, and ensure the safe and responsible disposal of residual waste. The order of the process of municipal (solid) waste management is shown in Table 9.1.

The flow of discussion will generally follow the order of the process of conventional SWM, starting from waste generation and collection, then progressing to waste transfer, transportation, processing or disposal, and concluding with landfill disposal. Each stage in the waste management process plays a critical role in minimizing environmental impact, conserving resources, and promoting sustainable practices. By understanding the interconnected nature of these processes, stakeholders can work toward implementing comprehensive and efficient waste management strategies that prioritize environmental stewardship and public health.

TABLE 9.1
The Order of Conventional SWM

1	*Waste Generation:* The process begins at the household level where waste is generated from various activities such as cooking, cleaning, and other daily tasks. The waste generator can be households, industries, commercials, institutional, etc.
2	*Waste Collection:* Local waste collection services collect the waste from households. This is typically done using specialized vehicles that travel along designated routes.
3	*Waste Transfer:* The collected waste is then transported to a transfer station. A transfer station is a facility where municipal solid waste is unloaded from collection vehicles and briefly held while it is reloaded onto larger long-distance transport vehicles for shipment to landfills or other treatment or disposal facilities.
4	*Waste Transportation:* The waste is then transported to a waste processing or disposal facility. This could be a recycling center, a composting facility, a waste-to-energy plant, or a landfill.
5	*Waste Processing or Disposal:* Depending on the type of waste and the facilities available, the waste may be recycled, composted, converted into energy, or disposed of in a landfill.
6	*Landfill Disposal:* If the waste is not recycled, composted, or converted into energy, it is disposed of in a landfill. A landfill is a carefully designed and monitored structure that isolates waste from the surrounding environment. This is the final step in the waste management process.

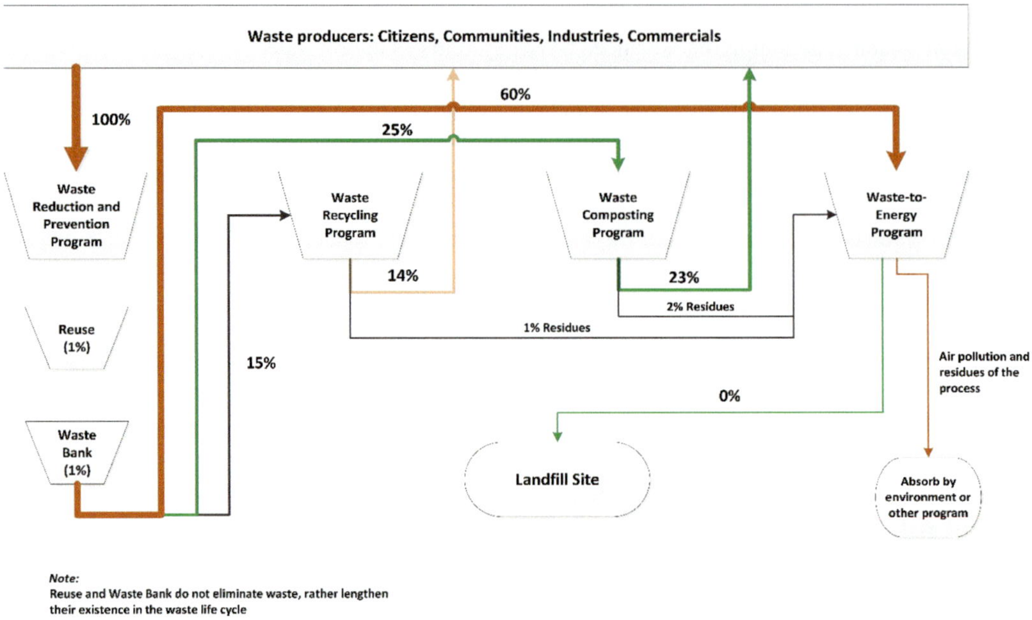

FIGURE 9.3 Waste: from production to landfill site.

9.3.3 Waste Generations/Productions

Everyone produces waste, whether it is from daily activities like cooking and cleaning, or larger-scale operations such as industrial manufacturing and construction. This waste, if not managed properly, can have significant environmental impacts. Therefore, each of us needs to understand our role in waste production and take steps toward reducing, reusing, and recycling wherever possible. By doing so, we can contribute to a more sustainable future and help preserve our planet for generations to come. It is not just about managing the waste we produce but also about minimizing it and finding ways to turn it into a resource. Figure 9.3 illustrates the flow of waste production to the landfill. Waste producers are entities that generate waste because of their activities. They can be broadly categorized into the following groups:

1. *Households:* Every home generates waste, from food scraps to packaging materials and old appliances.
2. *Industries:* Various industries produce waste as a byproduct of their operations. This includes manufacturing waste, chemical waste, and more.
3. *Businesses:* Retail stores, offices, restaurants, and other businesses generate waste, including paper, food waste, and electronic waste.
4. *Construction Sites:* Building and demolition activities produce a significant amount of waste, including wood, metal, concrete, and more.
5. *Healthcare Facilities:* Hospitals and other healthcare facilities generate medical waste, which can be hazardous.
6. *Agricultural Operations:* Farms produce waste in the form of crop residues, animal manure, and more.
7. *Municipal Services:* City services like street cleaning and public facility maintenance also generate waste.

In developing countries, biodegradable waste is the largest predominant quantity of waste. It could not be wholly absorbed by the waste composting program. On the other hand, the WtE program is practically able to absorb any kind of waste. It is, therefore, more efficient if the WtE program is placed as the last resort of the entire municipal waste management program. The WtE program could, to some extent, eliminate the hazardous waste, given the hazardous waste incinerator-friendly waste, that is, without creating severe by-products. Another consideration is that the waste elimination and transformation process would be more efficient if the process follows practicality and the pre-processing cost minimization. For example, it would be more efficient if unsorted waste produced by individual households were separated in advance.

By these considerations, the sequence of waste elimination and transformation should follow the order as exhibited in Figure 9.3. The authority fosters waste prevention and reduction program for the waste producers, that is, citizens, communities, industries, and commercials. Within the program, reuse, waste bank, and another community-based program can be included. The waste producers should also have done waste separation, which segregates recyclables, compostable wastes, and others. Separated waste could help transfer the waste to recycling centers, composting facilities, and WtE plants. The transfer process would be smoother with the interconnectivity of those centers, facilities, and plants coordinated under local authority.

Waste production varies greatly around the world depending on their culture, lifestyle, and gross domestic product (GDP). For example, each US citizen currently produces 811 kg of waste a year, with about half of that ending up in landfills. On the other hand, Turkey, Latvia, and New Zealand have been named the top 3 biggest waste producers in the 2019 Global Waste Index. Effective waste management strategies are needed to handle the waste produced by these various entities, with an emphasis on reducing waste production, reusing materials where possible, and recycling.

Countries like Canada, the United States, Japan, Germany, and Sweden, despite being high-income nations, have managed to keep their per capita waste production low. This is a significant achievement, as it demonstrates that economic prosperity does not necessarily have to result in high levels of waste production. These countries are represented in the shaded region (lower-right part) of Figure 9.4, which symbolizes the ideal condition for sustainable development. In this scenario, economic growth is balanced with environmental considerations, as evidenced by the low per capita waste production.

On the other hand, countries such as Malawi, Vietnam, Indonesia, Thailand, Guyana, and Oman, which have lower waste production, face a linear, one-dimensional challenge on their path to sustainable development. Their primary task is to ensure that waste production remains low as their economies grow and develop. This is a delicate balancing act, as economic advancement often leads to increased consumption and, consequently, more waste.

Lao PDR faces a more complex, two-dimensional challenge. Not only does it need to advance its economy, but it also needs to reduce its waste production to achieve a state of sustainable development. This requires a comprehensive approach that integrates economic planning with waste management strategies.

However, the journey toward sustainable development is fraught with challenges. Many city governments have struggled to implement effective waste management strategies while fostering economic growth. This failure can be attributed to a variety of factors, including lack of resources, inadequate infrastructure, and lack of public awareness about waste management.

Lifestyle has a significant impact on waste production. The choices we make in our daily lives, from the products we buy to the food we eat, all contribute to the amount of waste we generate. Here is how different aspects of our lifestyle can affect waste production:

- *Consumption Habits:* The more we consume, the more waste we produce. This includes everything from food to clothing to electronics. For example, a lifestyle that involves frequent shopping and the use of disposable products will generate more waste than a minimalist lifestyle.

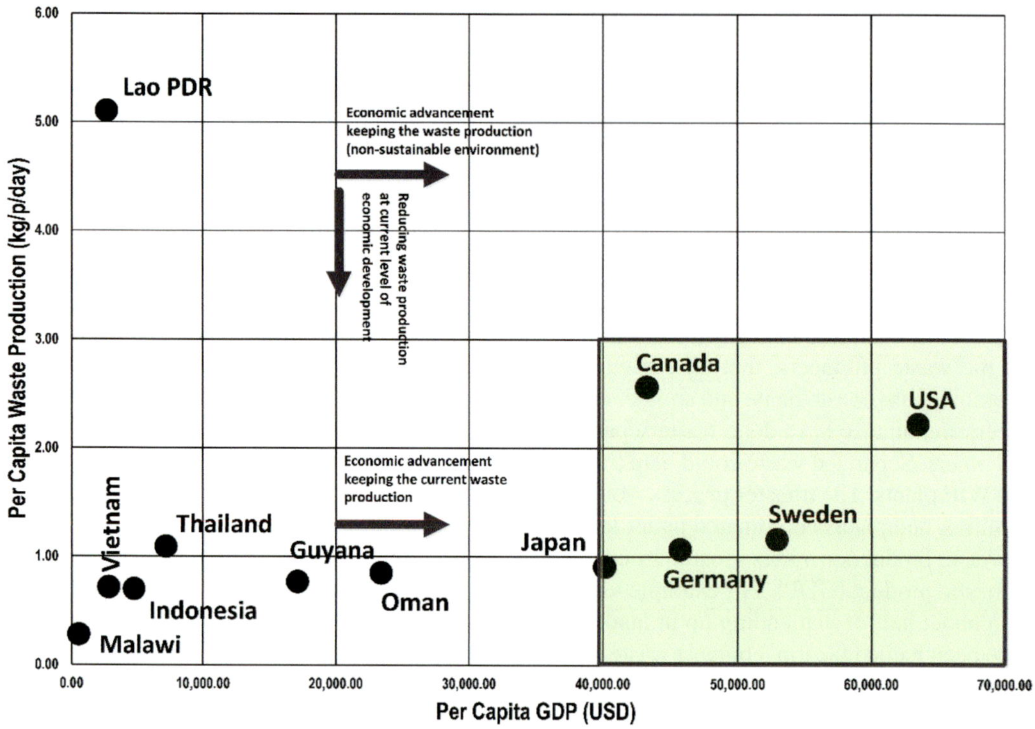

FIGURE 9.4 Waste production versus per capita GDP.

- *Food Waste:* Our eating habits also play a role. If we buy more food than we can consume, the leftovers end up as waste. Similarly, dining out often can lead to more waste due to the disposable containers and cutlery commonly used in the food service industry.
- *Packaging:* The preference for convenience has led to an increase in packaged goods. These packages often end up as waste. Choosing products with less packaging or reusable packaging can significantly reduce waste.
- *Recycling Habits:* Whether we choose to recycle can greatly impact the amount of waste we produce. Effective recycling can turn waste into valuable resources, reducing the need for raw materials.
- *Product Lifespan:* Frequently replacing items contributes to waste production. A lifestyle that values durability and repairability over disposability can help reduce waste.
- *Digital Lifestyle:* The digital age has also influenced waste production. While it has reduced the need for physical goods (like CDs and books), it has increased the production of electronic waste.

By making conscious decisions to reduce, reuse, and recycle, we can significantly decrease our waste production and contribute to a more sustainable world.

Balancing economic growth and environmental protection in developing nations is a complex task that requires a multifaceted approach. Some strategies that can help achieve this balance may include

- *Sustainable Production and Consumption:* Shifting how and where food and energy are produced can help meet global growth projections while achieving national habitat protection commitments, reducing greenhouse gas emissions, and improving air and water quality.

This involves promoting sustainable agricultural practices, increasing energy efficiency, and encouraging the use of renewable energy sources.

- *Waste Management:* Implementing effective waste management strategies is crucial. This includes promoting recycling and composting, improving waste collection and disposal systems, and encouraging waste reduction at the source.
- *Policy and Regulation:* Governments can play a key role by implementing policies and regulations that promote sustainable practices. This could include enforcing environmental standards, providing incentives for green businesses, and investing in sustainable infrastructure.
- *Education and Awareness:* Raising public awareness about the importance of environmental protection is essential. This can be achieved through education programs, public campaigns, and community engagement initiatives.
- *International Cooperation:* Developing nations can benefit from international cooperation. This can involve sharing of technology, knowledge, and resources, as well as collaboration on global environmental issues.
- *Economic Incentives:* Economic tools such as subsidies for sustainable practices, tax incentives for green businesses, and penalties for environmental violations can also be effective.
- *Investment in Research and Development:* Investing in research and development can lead to new technologies and practices that boost economic growth while minimizing environmental impact.

The zero-waste lifestyle represents a fundamental shift in the way we approach consumption and waste generation. At its core, it embodies a philosophy that emphasizes the importance of taking responsibility for the environmental impact of our daily choices. By actively seeking to minimize the waste we produce, individuals who embrace this lifestyle are making a commitment to reduce their ecological footprint and contribute to a more sustainable future for the planet.

Embracing a zero-waste lifestyle entails a holistic approach to living, encompassing various aspects of daily life such as shopping, meal planning, and waste management. It involves a conscious effort to prioritize products and packaging that are reusable, recyclable, or compostable, thereby minimizing the amount of waste destined for landfills. This often requires a reevaluation of consumer habits and a shift toward more sustainable alternatives, such as opting for package-free items, utilizing reusable containers, and choosing products with minimal or biodegradable packaging.

Furthermore, the zero-waste philosophy encourages individuals to adopt mindful consumption practices, fostering a deeper awareness of the life cycle of products and their environmental implications. By making informed decisions about the items they purchase and the resources they use, proponents of the zero-waste lifestyle actively contribute to the preservation of natural resources and the reduction of pollution. This conscientious approach extends beyond individual actions, serving as a catalyst for advocating systemic changes and promoting sustainable practices within communities and industries.

The zero-waste lifestyle represents a commitment to redefining our relationship with consumption and waste, striving to create a more harmonious coexistence with the environment. Through deliberate choices and mindful actions, individuals can play a pivotal role in shaping a future where sustainability and responsible resource management are integral components of everyday life.

Achieving a zero-waste lifestyle involves a variety of strategies, among others. (1) Consume Less: The less we consume, the less waste we produce. (2) Reuse What You Have: Instead of buying new things all the time, try to reuse what you already have. (3) Avoid Single-Use Items: Items that are used once and then thrown away contribute significantly to waste. (4) Always Go for Reusable Items: Choose items that can be used multiple times over disposable ones. (5) Buy in Bulk: Buying in bulk can reduce the amount of packaging waste.

In a zero-waste lifestyle, the focus is on minimizing waste as much as possible, but it is recognized that some waste may be unavoidable. When it comes to handling unavoidable waste, there are several strategies that can be employed to mitigate its impact:

- *Minimize Unavoidable Waste:* The first step is to keep unavoidable waste to an absolute minimum. For example, if certain medications are necessary and only available in plastic bottles, the aim would be to purchase larger quantities less frequently to reduce the amount of plastic waste.
- *Repurpose or Reuse:* Whenever possible, unavoidable waste should be repurposed or reused. For instance, glass jars can be used for storage, and cardboard boxes can be used for organizing.
- *Recycling and Composting:* Items that cannot be avoided, such as certain packaging materials or food scraps, can often be recycled or composted. By separating these materials from general waste, they can be diverted from landfills and repurposed into new products or used to enrich soil.
- *Upcycling and Reusing:* Finding creative ways to repurpose items or packaging can help reduce the amount of waste sent to landfills. Upcycling involves transforming waste materials into new products or finding alternative uses for them, while reusing involves using items multiple times before discarding them.
- *Supporting EPR:* Encouraging manufacturers and producers to take responsibility for the end-of-life disposal of their products can help reduce the burden of waste on individuals. This can involve advocating for more sustainable packaging and product design, as well as supporting initiatives that promote the circular economy.
- *Community Initiatives:* Participating in or establishing community initiatives such as swap meets, repair cafes, or bulk buying cooperatives can help reduce waste by promoting the reuse of items and minimizing the need for new purchases.
- *Waste Reduction Advocacy:* Engaging in advocacy efforts to promote policies and practices that minimize waste at the local, regional, and national levels can help address systemic issues related to waste management and encourage a more sustainable approach to resource use.

While the goal of a zero-waste lifestyle is to minimize waste to the greatest extent possible, addressing unavoidable waste requires a combination of individual actions, community involvement, and advocacy for systemic change.

9.3.4 Waste Collection

Waste collection systems manage the disposal of waste generated by households, businesses, and public institutions. These systems encompass a range of processes and infrastructure designed to gather, transport, and process waste materials in a safe and sustainable manner. Waste collection systems typically involve the regular collection of household waste from residential areas. This often involves the use of specialized garbage trucks that make scheduled rounds to pick up waste from designated collection points, such as curbside bins or communal dumpsters. The collected waste is then transported to appropriate facilities for further processing.

In a waste collection system, pickup locations refer to the designated points where waste is collected for disposal or recycling. These locations can vary depending on the type of waste being collected. For residential waste, pickup locations are often the curbside in front of homes or designated communal bins within a neighborhood. In commercial areas, pickup locations may be specific dumpsters or waste collection points within the premises.

The waste collection system is designed to efficiently gather waste from these pickup locations and transport it to appropriate disposal or recycling facilities. It is important for these pickup locations to be easily accessible for waste collection vehicles and convenient for waste generators to deposit their waste.

In densely populated urban and suburban communities, the primary method of waste collection is through curbside or alley pickup services. These services are typically provided by municipal waste management departments or private waste collection companies. In some municipalities, these services are extended to larger residential complexes such as apartment buildings, as well as commercial establishments like offices, restaurants, and retail stores. The waste generated by these buildings is usually stored in large, specialized containers such as dumpsters or roll-off containers, which are designed for easy pickup and transport.

On the other hand, waste collection in rural areas often operates differently due to the dispersed nature of residences and the larger distances involved. Residents in these areas are often required to place their waste containers near their mailboxes or at designated pickup points along major routes. This method allows waste collection trucks to cover large areas efficiently.

In addition to curbside pickup, some municipalities offer a drop-off arrangement. Under this system, residents are responsible for transporting their waste to a designated facility known as a transfer station. Transfer stations serve as central collection points where waste is temporarily stored before it is transported to a landfill, recycling center, or WtE facility.

While a drop-off system can be more cost-effective for the municipality, as it reduces the need for collection trucks and personnel, it can be less convenient for residents, particularly those who live far from the transfer station or have large amounts of waste to dispose of.

The effectiveness of different waste collection methods can vary widely depending on a variety of factors, including the local population density, geography, waste generation rates, and local regulations. Therefore, municipalities often employ a mix of different methods to provide the most effective and efficient waste management services to their residents. Table 9.2 provides a

TABLE 9.2
Advantages and Disadvantages of Waste Collection Points

Curbside Collection System

Advantages

- *Convenience:* Curbside collection is convenient for residents as waste is collected directly from their homes.
- *Efficiency:* It can be more efficient in densely populated areas where homes are close together.
- *Reduced Environmental Impact:* One truck collecting everything means fewer miles driven and less gas used, which reduces the environmental impact.

Disadvantages

- *Cost:* The cost of purchase and operation, including fuel and maintenance, can be high.
- *Unpredictable Pauses:* Garbage collection can sometimes stop a program for a bit to clean up unused data, which can make the program's speed seem uneven.
- *Additional System Resources:* It needs extra power from the computer to work, which can take away from what is available for the actual program.

Drop-Off Collection System

Advantages

- *Cost-Effective:* Drop-off systems can be more cost-effective for the municipality, as it reduces the need for collection trucks and personnel.
- *Increased Waste Diversion:* The benefits of moving to centralized collection points include increased waste diversion.
- *Environmentally Friendly:* Every single day trees are cut down to make paper and more paper.

Disadvantages

- *Less Convenient:* While a drop-off system is more cost-effective than a collection service, it is less convenient for residents.
- *Health Hazard:* Insects, rodents, etc. can cause damage due to air pollution.
- *Clutter in Homes:* Getting rid of waste helps keep our homes clean and tidy, making room for the things we need or want.

comprehensive overview of various waste collection methods, detailing their respective advantages and disadvantages, and offering insights into their suitability for different contexts. This information can be invaluable for decision-makers in designing and implementing effective waste management strategies.

Many waste collection systems also include provisions for the separate collection and recycling of materials such as paper, plastic, glass, and metal. This often involves the establishment of recycling centers or the implementation of separate collection schemes to encourage residents to segregate recyclable materials from general waste. Furthermore, waste collection systems may also incorporate special provisions for the collection and disposal of hazardous waste materials, such as chemicals, batteries, and electronic waste. Due to the potential environmental and health risks associated with these materials, they require specialized handling and disposal methods to ensure their safe management.

Local regulations and guidelines frequently govern waste collection systems, with the aim of encouraging practices that support sustainable waste management. These rules may specify the kinds of waste eligible for collection, how often collections occur, and the appropriate disposal methods. The primary objective of these regulations is to reduce environmental harm and foster the conservation of resources.

Public education and outreach initiatives are often incorporated into waste collection systems to increase awareness about correct waste disposal methods, the significance of recycling, and the need to minimize waste production. Such endeavors can cultivate a sense of environmental stewardship and stimulate community involvement in waste management tasks.

Waste collection systems are constantly adapting to include cutting-edge technologies and innovative approaches with the goal of boosting efficiency, lessening environmental harm, and optimizing resource recovery. This could involve the application of intelligent waste management technologies, like bins equipped with sensors and software for optimizing routes, to make collection processes more efficient and reduce operational expenses.

A waste collection system requires a variety of equipment to efficiently gather, transport, and manage waste materials. Some essential equipment includes

1. *Garbage Trucks:* These specialized vehicles are used for collecting and transporting general household waste from designated collection points. They come in various types, including rear-loading, side-loading, and front-loading trucks, each suited for different collection methods.
2. *Recycling Containers:* Different containers are needed for collecting recyclable materials such as paper, plastic, glass, and metal. These containers can range from curbside bins to larger communal recycling dumpsters, each designed to facilitate the separation and collection of recyclables.
3. *Hazardous Waste Containers:* For the safe collection and disposal of hazardous waste materials, specialized containers are required to ensure proper containment and handling of substances such as chemicals, batteries, and electronic waste.
4. *Compactors and Balers:* These machines are used to compress and package waste materials, reducing their volume for efficient transportation and storage. Compactors are particularly useful for reducing the size of general waste, while balers are essential for processing recyclable materials.
5. *Waste Transfer Stations:* These facilities serve as intermediate collection points where waste is consolidated from smaller collection vehicles before being transported to larger disposal or processing facilities. Transfer stations may include compactors, sorting equipment, and storage areas for different types of waste.
6. *Collection Bins and Containers:* Various types of bins and containers are needed for different waste streams, including general waste, recyclables, organic waste, and hazardous materials.

These can range from small curbside bins for household waste to larger roll-off containers for bulk waste.

7. *Personal Protective Equipment (PPE):* Workers involved in waste collection require appropriate PPE, such as gloves, safety vests, and, in some cases, respiratory protection, to ensure their safety while handling waste materials.
8. *Sorting and Processing Equipment:* Facilities involved in waste processing and recycling require equipment for sorting, shredding, and processing different types of waste materials. This may include conveyor belts, sorting screens, shredders, and other specialized machinery.

These are just a few examples of the equipment needed for a waste collection system. The specific equipment requirements can vary depending on the scale of the operation, the types of waste being managed, and the overall waste management strategy in place.

The procedure for collecting waste from various sources, such as households, industries, and commercial establishments, involves several key steps to ensure the efficient and safe management of waste materials.

1. *Waste Generation:* Waste is initially generated at its source, which can include households, businesses, industrial facilities, and public institutions. This can encompass a wide range of materials, including general household waste, recyclables, organic waste, and potentially hazardous materials.
2. *Segregation:* At the point of generation, waste materials are ideally segregated into different categories to facilitate proper handling and disposal. This may involve separating recyclable materials from general waste, as well as identifying and setting aside hazardous waste for specialized handling.
3. *Collection Scheduling:* Waste collection services typically operate on a scheduled basis, with designated collection days for different types of waste. For households, this may involve regular curbside pickup of general waste and recyclables, while businesses and industries may have specific arrangements for waste collection based on their operational needs.
4. *Collection Process:* On the scheduled collection days, waste collection vehicles, such as garbage trucks, are dispatched to the designated collection points at each source. Trained personnel are responsible for loading the waste into the collection vehicles, ensuring that the materials are handled safely and efficiently.
5. *Transportation:* Once the waste is collected, it is transported to appropriate facilities for further processing. This may involve transporting general waste to landfills, recyclables to recycling centers, and hazardous waste to specialized treatment or disposal facilities.
6. *Processing and Disposal:* At the receiving facilities, the collected waste undergoes processing and disposal based on its type. General waste may be compacted and disposed of in landfills, while recyclables are sorted, processed, and prepared for recycling. Hazardous waste is subject to specialized treatment to mitigate its environmental and health risks.
7. *Monitoring and Compliance:* Throughout the entire collection and disposal process, regulatory compliance and environmental monitoring are key considerations. Waste management authorities may conduct inspections and audits to ensure that waste collection and disposal activities adhere to relevant regulations and environmental standards.
8. *Public Education and Participation:* In many waste collection systems, public education and participation are integral components. Outreach programs and educational initiatives are often implemented to raise awareness about proper waste disposal practices, recycling, and the importance of reducing waste generation.

These steps guide the waste collection services to gather, transport, and manage waste materials from their sources, contributing to the overall goal of SWM and environmental stewardship.

Technology can significantly improve solid waste collection systems by enhancing efficiency, reducing environmental impact, and maximizing resource recovery. There are some technologies being used:

- *Smart Waste Bins:* A smart waste bin is equipped with various sensors, connectivity features, and data analytics capabilities. They are designed to make the waste collection and recycling process more efficient, convenient, and sustainable for both businesses and consumers. These bins use artificial intelligence (AI)-based object recognition to automatically sort recyclables into separate compartments. After sorting, the machine compresses the waste and monitors how full each bin is. This can lower waste management costs by as much as 80% and drastically improve employee efficiency. Smart waste bins can help in cleaning up urban landscapes and streamlining waste management systems. They can transform operational costs and waste management benchmarks, offering deeper insights into the lifecycle of waste.
- *Waste Level Sensors:* These sensors monitor trash levels in real-time and schedule collections to maximize efficiency.
- *AI Recycling Robots:* These robots can sort and separate waste with a high level of precision, resulting in increased material purity for recycling.
- *Garbage Truck Weighing Mechanisms:* These mechanisms weigh the waste as it is collected, providing real-time data on waste volumes.
- *Pneumatic Waste Pipes:* These pipes transport waste quickly and efficiently from buildings to waste collection points.
- *Solar-Powered Trash Compactors:* These compactors use solar power to compact waste, reducing its volume and making collection more efficient.
- *E-Waste Kiosks:* These kiosks provide a convenient drop-off point for electronic waste, ensuring it is properly recycled.
- *Recycling Apps:* These apps provide information on what can be recycled and where, making it easier for individuals and businesses to recycle.
- *Route Optimization Software:* This software uses real-time data to plan the most efficient routes for waste collection vehicles, reducing fuel consumption and emissions.

These technologies not only make waste collection more efficient but also help in achieving sustainability goals by reducing environmental impact and promoting resource conservation.

Route optimization in waste collection involves the strategic planning and organization of collection routes to maximize efficiency, minimize costs, and reduce environmental impact. This process utilizes advanced technologies and algorithms to optimize the scheduling and routing of waste collection vehicles, leading to improved operational performance and resource utilization. The components of route optimization in waste collection include

- *Vehicle Scheduling:* Route optimization software considers factors such as vehicle availability, capacity, and maintenance schedules to efficiently allocate collection vehicles to specific routes and tasks.
- *Dynamic Routing:* Advanced algorithms are used to dynamically adjust collection routes based on real-time factors such as traffic conditions, weather, and unexpected events, ensuring that routes remain optimized throughout the day.
- *Geospatial Analysis:* Geographic information systems (GIS) are leveraged to analyze spatial data and optimize route planning based on factors such as the location of collection points, road network characteristics, and the distribution of waste sources.
- *Bin Monitoring and Data Analysis:* Some route optimization systems integrate data from smart waste bins equipped with sensors to provide real-time information on fill levels, allowing for more accurate route planning and collection frequency adjustments.

- *Cost and Time Optimization:* Route optimization algorithms aim to minimize travel time, fuel consumption, and vehicle wear and tear, ultimately reducing operational costs and environmental impact.
- *Service Level Optimization:* The software considers service level agreements, regulatory requirements, and customer preferences to ensure that waste collection services are delivered in accordance with established standards.
- *Integration with Fleet Management:* Route optimization systems may be integrated with fleet management solutions to provide real-time tracking of vehicles, monitor driver performance, and facilitate communication between the operations center and collection crews.

The adoption of route optimization in waste collection can lead to substantial advantages for municipalities and waste management companies. These benefits encompass a decrease in fuel usage, a reduction in emissions, a cutback in operational expenses, and an enhancement in service dependability. Furthermore, the optimization of routes aids in alleviating traffic congestion and bolsters the overall sustainability of urban areas.

Route optimization can have a significant impact on waste collection systems, offering numerous benefits:

- *Efficiency:* Route optimization can make waste collection more efficient by minimizing travel distance and time. This allows each waste collection route to maximize the use of resources and minimize the costs involved.
- *Cost Reduction:* By reducing the distance traveled and time spent on each route, route optimization can help lower operational costs. This includes savings on fuel, vehicle maintenance, and labor costs.
- *Environmental Impact:* Route optimization can also reduce the environmental impact of waste collection. By minimizing the distance traveled, it can help reduce fuel consumption and, consequently, lower greenhouse gas emissions.
- *Service Reliability:* Optimized routes can improve service reliability by ensuring that all areas are serviced on schedule. This can lead to increased customer satisfaction.
- *Traffic Congestion:* By optimizing routes, waste collection vehicles can avoid contributing to traffic congestion, enhancing overall urban sustainability.
- *Adaptability:* Route optimization software can adjust routes in real-time based on various factors such as traffic conditions, road closures, and changes in waste pickup locations.

Route optimization in waste collection systems often involves solving complex mathematical problems known as Vehicle Routing Problems (VRP). These problems aim to find the most efficient routes for a fleet of vehicles to travel between various locations.

One common approach to solving VRP is using Dijkstra's algorithm, which finds the shortest path between two points in a graph. The formula for Dijkstra's algorithm is shown in Table 9.3.

Another common approach is the Traveling Salesman Problem (TSP), which aims to find the shortest possible route that visits each location once and returns to the origin location. The TSP can be solved using various methods, including the Haversine formula, which calculates the shortest distance between two points on the surface of a sphere.

The Haversine formula is given by:

$$a = \sin^2\left(\frac{\Delta\phi}{2}\right) + \cos\phi_1 \cdot \cos\phi_2 \cdot \sin^2\left(\frac{\Delta\lambda}{2}\right) \tag{9.1}$$

$$c = 2 \cdot \text{atan2}\left(\sqrt{a}, \sqrt{1-a}\right) \tag{9.2}$$

TABLE 9.3
Code for Dijkstra Algorithm

```
function Dijkstra(Graph, source):
  create vertex set Q
  for each vertex v in Graph:
    distance[v] ← INFINITY
    previous[v] ← UNDEFINED
    add v to Q
  distance[source] ← 0
  while Q is not empty:
    u ← vertex in Q with min distance[u]
    remove u from Q
    for each neighbor v of u:
      alt ← distance[u] + length(u, v)
      if alt < distance[v]:
        distance[v] ← alt
        previous[v] ← u
  return distance[], previous[]
```

$$d = R \cdot c, \tag{9.3}$$

where

ϕ_1, ϕ_2 are the latitude of point 1 and latitude of point 2 (in radians),
$\Delta\phi$ is the difference of latitudes,
$\Delta\lambda$ is the difference of longitudes,
R is the Earth's radius (mean radius = 6,371 km),
d represents the actual distance between two points on the surface of the sphere (i.e., along the sphere's surface). This is calculated by multiplying R, the radius of the sphere, by c, the angular distance in radians.

These formulas and algorithms are used to calculate the most efficient routes for waste collection vehicles, reducing operational costs and environmental impact.

EXAMPLE 9.1

A simplified numerical example of the VRP to illustrate the concept:

Let us consider a small fleet of vehicles that needs to serve a set of delivery points. The vehicles start and end their routes at a central depot. The goal is to minimize the total distance traveled by the vehicles while ensuring that each delivery point is visited exactly once and that the total demand served by each vehicle does not exceed its capacity.

Parameters:
Number of vehicles: 2. Depot location: (0, 0)

Delivery points:
A: (2, 3), Demand = 4
B: (5, 7), Demand = 7

C: (8, 1), Demand = 5
D: (3, 6), Demand = 3

Vehicle capacity: 10 units. This problem is shown in Table 9.4.
Distance between two points A and B is calculated by:

$$d = \sqrt{(x_B - x_A)^2 + (y_B - y_A)^2} \quad (9.4)$$

Solution:
In the numerical example provided, we have a simplified instance of the VRP involving a small fleet of vehicles and a set of delivery points. The goal is to determine the optimal routes for the vehicles to minimize the total distance traveled while satisfying capacity constraints and visiting each delivery point exactly once.

A step-by-step explanation of the numerical example:

Problem Setup:

- We have two vehicles available to serve the delivery points.
- The depot, where the vehicles start and end their routes, is located at coordinates (0, 0).
- There are four delivery points labeled A, B, C, and D, each with specific demand (the amount of goods to be delivered) and coordinates.
- The order of priority: vehicle must deliver the demanded goods, then calculate distance.

Distance Matrix:

- A distance matrix is provided, which represents the pairwise distances between the depot and the delivery points, as well as between the delivery points themselves. In this example, we use Euclidean distances for simplicity.

Vehicle Capacity:

- Each vehicle has a capacity constraint, meaning it can only carry a certain total demand of goods. In this example, the vehicle capacity is set to 10 units. So, we assign the routes A–C with demand equal to 9 units, and B–D with demand is 10 units.

TABLE 9.4
Distance Matrix (Euclidean Distance)

	Depot	A	B	C	D
Depot	0	3.6	8.6	8.0	6.7
A	3.6	0	5.0	6.4	3.0
B	8.6	5.0	0	7.2	2.2
C	8.0	6.4	7.2	0	7.6
D	6.7	3.0	2.2	7.6	0

Optimization Process:

- The goal is to find the optimal routes for the vehicles to serve all delivery points while minimizing the total distance traveled. This involves assigning delivery points to the vehicles, determining the sequence of visits for each vehicle, and ensuring that the capacity constraints are satisfied.

Final Step:

- Based on the distance matrix and the demand at each delivery point, an optimization algorithm (such as a VRP solver) is used to determine the optimal routes for the vehicles.
- In the provided example, the optimal solution assigns delivery points to the vehicles in a way that minimizes the total distance traveled while ensuring that the capacity constraints are not exceeded.

Optimal Routes:

The optimal solution for the example is as follows:

The routes satisfy the demand constraints and are a feasible solution to the problem. The total distance for these routes:

- *Vehicle 1:* Depot to A then to C and returned to Depot (Depot–A–C–Depot):
  Total Demand: 4 (A) + 5 (c) = 9 < 10 (Vehicle Capacity)
  Total Distance: 3.6 (Depot–A) + 6.4 (A–C) + 8.0 (C–Depot) = 18.0
- *Vehicle 2:* Depot–B–D–Depot
  Total Demand: 7 (B) + 3 (D) = 10 = 10 (Vehicle Capacity)
  Total Distance: 8.6 (Depot–B) + 2.2 (B–D) + 6.7 (D–Depot) = 17.5
  The total distance traveled by both vehicles would be 18.0 + 17.5 = 35.5 units. This is a solution for the VRP given the constraints.

The numerical process above demonstrates the application of the VRP to a simple example involving a small fleet of vehicles and a set of delivery points. The goal is to efficiently assign delivery points to the vehicles and determine their routes to minimize the total distance traveled while satisfying capacity constraints, ultimately optimizing the logistics and transportation operations. In practice, VRP instances can involve larger numbers of vehicles, delivery points, and more complex constraints, requiring sophisticated optimization algorithms to find the best solutions.

In reality, the VRP might not be practical because 101 situations could be there, and the situation is dynamic. There should be some practical and logical rules for optimizing waste collection vehicle routes. The objective of the rules is to reduce the costs, but at the same time, to improve their environmental performance. For this purpose, the rules may include the items as shown in Table 9.5.

The potential benefits of the rules of waste collection vehicles routes are shown in Table 9.6.

However, there are possible constraints and challenges to accomplish the objectives of the route's optimization. They are shown in Table 9.7.

Considering, to some extent, the rules, benefits, and challenges of the route's optimization, there are some practical routes that minimize the distance of the waste collection vehicle traveled. The routes are shown in Figure 9.5. There are two options, depending on the traffic conditions, that is, one-way traffic or restrictions; the waste collection vehicles may select the most suitable route for them, for example, option 1 or option 2. This example is only for a particular situation. We may encounter various conditions and city configurations.

TABLE 9.5
The Rules of Waste Collection Vehicles Routes

Objectives: Reduce operational costs and increase the service performance.

Minimize Distance: The primary goal should be to minimize the total distance traveled by all vehicles. This can be achieved by using route optimization algorithms that consider factors such as road network, traffic conditions, and vehicle load capacity.

Maximize Load: Each vehicle should collect as much waste as possible on each trip to maximize efficiency. This means that routes should be planned in such a way that vehicles reach their maximum load capacity before returning to the depot.

Prioritize High Waste Areas: Areas that generate more waste should be prioritized. This means that these areas should be serviced more frequently or at the beginning of the route.

Consider Traffic and Road Conditions: Traffic conditions and road quality should be considered when planning routes. Routes should be planned to avoid heavy traffic and poor road conditions, which can slow down collection and increase fuel consumption.

Use Real-Time Data: Use real-time data to adjust routes dynamically based on current conditions. For example, if a particular area generates more waste than expected, the route can be adjusted to service that area sooner.

Regularly Review and Update Routes: Routes should be regularly reviewed and updated based on changes in waste generation patterns, road conditions, and other relevant factors.

Use Route Optimization Software: Use route optimization software that can automatically generate the most efficient routes based on the above factors. This software can also provide real-time updates and adjustments based on current conditions.

Train Drivers: Drivers should be trained to follow the optimized routes and to understand the importance of route optimization. They should also be encouraged to provide feedback on the routes, which can be used to further improve route planning.

TABLE 9.6
Potential Benefits of Routes Optimization

Operational Efficiency: Route optimization makes waste collection as efficient as possible for the waste vehicle, which can lead to increased operational efficiency.

Cost Reduction: By reducing the distance traveled and the time spent on each route, route optimization can significantly reduce fuel consumption and maintenance costs. It can also reduce labor costs by enabling drivers to complete their routes more quickly.

Environmental Impact: Route optimization can help reduce the environmental impact of waste collection by reducing fuel consumption and associated greenhouse gas emissions.

Improved Service: Route optimization can lead to improved service for customers by ensuring that waste is collected in a timely and efficient manner.

Data-Driven Decision-Making: Route optimization provides valuable data that can be used to make informed decisions about waste collection operations. This can include decisions about vehicle allocation, route planning, and scheduling.

Adaptability: Route optimization allows for real-time adjustments to routes based on current conditions, such as traffic or weather. This can help ensure that waste collection continues to operate efficiently even under changing conditions.

The waste collection vehicles are equipped with GPS and real-time traffic data, allowing the drivers to select the most suitable route based on the current conditions. By having these two options, the waste collection service can ensure that it operates efficiently under a variety of conditions, thereby reducing fuel consumption, minimizing wear and tear on the vehicles, and ensuring timely waste collection.

TABLE 9.7
Constraints and Challenges of Route's Optimization

Drivers' Knowledge: Drivers often believe they know the best route based on their experience, which can make them resistant to following optimized routes.
Loss of Knowledge: When drivers leave the company, their knowledge of the routes and specific customer preferences can be lost.
Incomplete Notes: Driver notes, including bin weight and dispatch notes, may not be comprehensive or easily accessed by field teams.
Too Many Planning Meetings: Team meetings may be required to plan routes and pass on information, but too many meetings can be time-consuming.
Proof of Collection: Proof of collection may not be taken or may be difficult to obtain.
Customer Expectations: Customers may expect a collection ETA or time window, which can be challenging to provide with route optimization.
Last-Minute Changes: Customers often come up with last-minute changes, which can disrupt a day's schedule.
Bad Routes: On a map, it may be clear if there is a stop that is costing money to service or if drivers are deviating from their route.

These routes are not fixed and can be adjusted over time based on changes in waste generation patterns, road conditions, and other relevant factors. Regular review and update of these routes are crucial to maintaining the efficiency of the waste collection service. The ultimate goal is to provide reliable waste collection service while minimizing the environmental impact and operational costs.

Road closures can occur due to a variety of reasons, ranging from planned events to unexpected emergencies. Planned road closures often stem from events such as parades, marathons, street festivals, and construction projects. These closures are typically scheduled in advance and communicated to the public through official channels, allowing motorists to plan alternate routes. Additionally, road maintenance and repair work, including resurfacing, bridge maintenance, and utility repairs, can necessitate temporary closures to ensure the safety of both workers and the traveling public.

On the other hand, unexpected emergencies, such as traffic accidents, hazardous material spills, severe weather conditions, and natural disasters like floods or landslides, can lead to immediate and unplanned road closures. In these cases, the primary concern is public safety, and the closure may remain in effect until the situation is resolved and the road is deemed safe for travel. Emergency services and transportation authorities work swiftly to manage these closures and provide timely updates to minimize disruptions and ensure the swift reopening of affected roadways.

Facing the possibility of road closures, the waste collection vehicles can adapt to road closures by leveraging a combination of technology, planning, and communication, as shown in Table 9.8.

Adjusting to road closures can ensure that waste collection services will not be disrupted and the service continues to operate efficiently and effectively, even when faced with unexpected road closures.

9.3.5 Waste Separation

Waste separation, also known as waste sorting or waste segregation, is the process of dividing waste into different categories based on their characteristics and properties. This typically involves separating waste into recyclable materials, organic materials, and non-recyclable waste. The purpose of waste separation is to make it easier to recycle and process different types of waste, reducing the amount of waste sent to landfills and incinerators. Sorting waste at the source, whether in households or businesses, enables the recovery of valuable resources, minimizes environmental impact, and promotes a more sustainable approach to waste management.

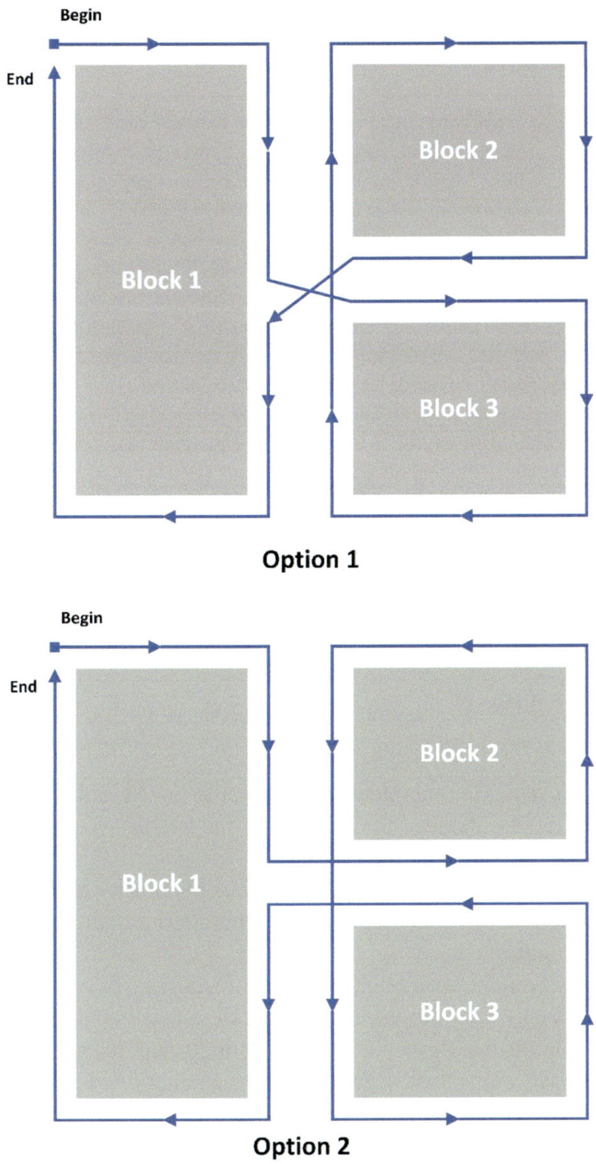

FIGURE 9.5 The route for three-block configuration.

Waste separation, also known as waste sorting, is the process by which waste is divided into different categories. This can be done manually at the household level and collected through curbside collection schemes, or it can be automatically separated in materials recovery facilities (MRFs) or mechanical biological treatment systems.

Some common categories for waste separation are as follows:

- Paper,
- Cardboard (including packaging for return to suppliers),
- Glass (clear, tinted no light bulbs or windowpanes, which belong with residual waste),
- Plastics,

TABLE 9.8
Adaptation to Road Closures

Dynamic Route Planning: This involves adjusting routes in real-time based on current traffic conditions, including road closures. Advanced route optimization software can automatically update routes to avoid closed roads.
Use of GPS and Real-Time Traffic Data: Modern waste collection vehicles are often equipped with GPS and real-time traffic data systems. These systems can provide up-to-date information about road closures and suggest alternative routes.
Driver Training: Drivers should be trained to adapt to unexpected situations, including road closures. This includes understanding how to use onboard navigation systems and how to safely navigate alternative routes.
Communication: There should be clear lines of communication between drivers, dispatchers, and route planners. If a driver encounters a road closure, they should immediately inform the dispatcher, who can then update the routes for all other vehicles.
Pre-Planning: Route planners should have contingency plans in place for known road closures (e.g., due to construction or events). These plans can be activated as soon as the road closure is confirmed.
Automated Vehicles: In the future, autonomous waste collection vehicles could be used to further improve adaptability to road closures. These vehicles could use advanced sensors and AI algorithms to detect road closures and automatically reroute.

- Textiles,
- Wood, leather, rubber,
- Scrap metal,
- Compost,
- Special/hazardous waste, and
- Residual waste.

Waste separation is a critical component of sustainable SWM, and it determines the successful recycling management as it enhances the quality of recyclables, reduces municipal solid waste (MSW), and optimizes incineration.

These days, automatic waste segregators are gaining popularity and are already being used in many parts of the world. The key is to make waste separation a habit and part of our daily routine for a more sustainable future.

Waste separation can be done manually and using machinery or equipment. Manual separation involves individuals sorting waste into different categories such as paper, plastic, glass, and organic waste. This method is labor-intensive but allows for high accuracy in waste separation. On the other hand, automatic waste separation uses advanced machinery and technology to sort waste. This includes conveyor belt systems, magnetic separators for metals, optical sorters for plastics, and even AI-powered robots for more complex sorting tasks. While this method can handle large volumes of waste efficiently, it requires significant initial investment and ongoing maintenance.

Regardless of the method used, the goal of waste separation is to reduce the amount of waste that ends up in landfills, promote recycling, and contribute to a more sustainable environment. It is important to note that the success of both methods relies heavily on public awareness and participation. Therefore, education and awareness campaigns are crucial components of effective waste management strategies. The best place to do waste separation is waste separation at source, which is at the waste producers.

Waste separation has several benefits, both environmental and economic:

Environmental Benefits:

- *Reducing Landfill Waste:* Waste separation can significantly reduce the amount of waste sent to landfills. Overfilled and poorly managed landfills are a significant source of environmental pollution.

- *Lowering Greenhouse Gas Emissions:* Decomposition of garbage in landfills produces methane, a potent greenhouse gas. When organic waste is separated and composted, it decomposes aerobically, significantly reducing methane emissions.
- *Preserving Natural Resources:* Recycling materials like paper, glass, and plastic reduces the demand for raw materials and energy-intensive manufacturing processes. This helps protect forests, reduce energy consumption, and lower the carbon footprint associated with the production of new materials.
- *Protecting Wildlife:* Non-recyclable waste can be hazardous to wildlife when improperly disposed of in natural habitats. By ensuring that non-recyclable materials are appropriately managed and disposed of, we can help safeguard our planet's diverse ecosystems.

Economic Benefits:

- Communities that implement effective waste separation programs often save money on waste management and disposal costs.
- Recycling and composting programs can create jobs and stimulate local economies.

Both manual and automatic waste separation methods have their own strengths and weaknesses.

Manual Waste Separation:

- Manual waste separation is inherently flexible and gives control to collectors.
- It demands more from residents and is a more costly service.
- Manual collection allows more waste per cycle.
- However, it requires frequent scooping and waste disposal.

Automatic Waste Separation:

- Automatic waste separation can provide a high service level to residents at a low cost.
- It requires a major process change and capital commitment.
- It is relatively inflexible for service adjustment and takes away some collection control.
- Automated collection is slightly more frequent.
- Automatic systems may require more frequent cleaning of the waste receptacle and mechanical components, as well as regular maintenance and possible repairs.

The choice between manual and automatic waste separation depends on various factors such as the scale of waste, time, effort, cost, and the specific waste management goals of the community. Some cities even use a combination of both methods. For instance, a smart robot is being developed to automate the process of urban solid waste separation.

Automatic waste separation requires a variety of facilities and technologies. Here are some key components:

- *Automated Sorting System:* This system uses conveyor belts, screens, magnets, and lasers to separate materials. The system can be designed to separate waste into different categories such as plastics, metals, mixed paper, and residual waste.
- *Robotic Arms:* These are used to pick and sort waste. For example, the robotic sorting station supplied by ZenRobotics can perform up to 6,000 picks per hour.
- *AI:* AI-powered robots can sort several fractions of industrial waste and are also trained to recognize new ones. This opens new business opportunities for the waste industry.
- *Near-Infrared (NIR) Optical Sorters:* These are used to separate waste into different types of plastic, mixed paper, metals, and residual waste.

- *Anaerobic Digestion Facility:* This is used to turn green bags of food waste into biogas, which can fuel waste collection trucks.
- *Waste Compaction and Storage Facilities:* The separated waste is compacted and stored in fully concealed containers, which are transported to landfill or other waste handling facilities for further treatment.
- *Waste Collection Trucks:* These are used to collect household and food waste from municipalities.
- *Waste Incineration Facilities:* The residual waste is incinerated and used for nearby heating and electricity applications.

These facilities and technologies can be combined in various ways to create a fully automated waste separation system. The specific configuration will depend on the scale of waste, the specific waste management goals of the community, and other factors.

MRFs serve as centralized facilities where various types of recyclable materials are sorted, processed, and prepared for recycling and reuse. When integrated with waste separation facilities, MRFs can handle mixed waste streams, contributing to resource conservation, waste reduction, and sustainable material management within the circular economy framework. Here is how they work:

- *Resource Recovery:* MRFs are designed to efficiently separate different types of recyclable materials, such as plastics, metals, glass, and paper, from mixed waste.
- *Waste Diversion:* MRFs divert a significant portion of waste from going to landfills or incineration facilities.
- *Quality Improvement:* MRFs employ advanced sorting technologies, including optical sorting, magnetic separation, and eddy current separation, to enhance the quality of recovered materials.
- *Closed-Loop Systems:* The materials recovered from MRFs are often used as feedstock for new products.
- *Energy Savings:* Recovering materials through MRFs generally requires less energy than producing materials from raw resources.
- *Economic Benefits:* MRFs create economic opportunities by generating jobs in the waste management and recycling industries.
- *Education and Awareness:* MRFs often provide opportunities for public engagement and education about recycling practices and the importance of waste separation.
- *Innovation and Research:* MRFs drive innovation in waste sorting and recycling technologies.

An Integrated Waste Management Facility (IWMF) is a large-scale waste processing plant that combines several waste management technologies such as MRFs, composting, WtE, and landfilling. It can handle up to 500–1,000 tons of waste per day.

Operating MRFs can present several challenges:

- *Contamination Issues:* One of the biggest challenges that MRFs face today is contamination in the recycling load. This can occur when non-recyclable items, such as plastic bags and straws, are mixed in with recyclable materials. Contamination can reduce the quality of the recovered materials and make recycling more difficult.
- *Heterogeneous Nature of Waste:* The heterogeneous nature of non-biodegradable materials is a significant challenge in managing them after their intended use. The random appearance of specific grades of material among heaps of non-biodegradable waste makes resource recovery unattractive to scrap dealers.

- *Market Challenges:* In general, recycling commodity markets have had significant challenges in the past few years due to low oil prices, high contamination rates, and reduced export market options.
- *Technological Challenges:* The increasing use of state-of-the-art technology like image processing systems and electromagnetic sensing has contributed to the rising recovery rates of MRFs. However, these technologies require significant investment and ongoing maintenance.
- *Source Segregation:* The real driver for recovery rate success lies in segregation at the source. In many regions, especially in the Middle East, source segregation needs to be better implemented in order for facilities to increase efficacy.
- *Consumer Education:* There is a need for improved consumer education about recycling practices and the importance of waste separation. Without proper education and awareness, the effectiveness of MRFs can be limited.

Waste separation at the source is widely recognized as one of the most effective strategies for efficient waste management and environmental sustainability. By segregating waste at the point of origin, such as households, businesses, and institutions, it becomes possible to divert recyclable materials, organic waste, and hazardous substances away from landfills, thereby reducing the overall environmental impact. However, the successful implementation of waste separation at the source relies heavily on raising awareness and fostering active participation among all waste producers, including householders.

Householders play a pivotal role in waste separation as they are directly responsible for generating a significant portion of municipal waste. Through education and awareness campaigns, individuals can be empowered to properly sort their waste into distinct categories, such as recyclables, organic waste, and non-recyclable materials. This not only facilitates the recycling process but also minimizes the volume of waste destined for landfills, ultimately conserving natural resources and mitigating environmental pollution. By promoting an understanding of the benefits of waste separation and providing clear guidelines for proper waste disposal, communities can encourage widespread adoption of this practice, leading to lasting positive impacts on the environment and public health.

In addition to raising awareness, successful waste separation at the source also requires the provision of accessible and well-maintained infrastructure to support the process. This includes the availability of clearly labeled recycling bins, composting facilities, and hazardous waste collection points in residential areas and public spaces. Furthermore, local authorities and waste management organizations can play a crucial role in facilitating waste separation by implementing effective collection systems, providing educational materials, and offering incentives for active participation, such as reward programs for proper waste sorting.

The integration of modern technologies, such as smart waste bins equipped with sensors and communication capabilities, can streamline the waste separation process and provide valuable data for optimizing collection routes and resource allocation. By embracing innovative solutions and fostering a culture of environmental responsibility, communities can harness the collective efforts of waste producers to achieve significant reductions in landfill waste and promote a more sustainable approach to waste management. Ultimately, waste separation at the source, supported by comprehensive awareness initiatives and robust infrastructure, represents a fundamental step toward building cleaner, healthier, and more resilient communities for the future.

Waste separation at the source presents several challenges that can hinder its effective implementation. One significant challenge is the lack of awareness and understanding among waste producers, including households, businesses, and institutions, regarding the importance and methods of waste separation. Without proper education and outreach, individuals may struggle to grasp the distinctions between recyclable materials, organic waste, and non-recyclable items, leading to contamination of separated waste streams and reducing the overall effectiveness of the process.

Another obstacle is the availability and accessibility of infrastructure to support waste separation. In some areas, the absence of adequate recycling facilities, composting sites, and hazardous waste collection points can discourage individuals from engaging in proper waste sorting. Limited access to convenient and clearly labeled waste disposal options may lead to apathy or noncompliance with separation guidelines, resulting in mixed waste streams that are more challenging and costly to process.

Cultural and behavioral factors can pose challenges to waste separation at the source. Long-standing habits and attitudes toward waste disposal may need to be shifted through targeted awareness campaigns and community engagement initiatives. Overcoming resistance to change and fostering a sense of responsibility for environmental stewardship among diverse populations requires sustained effort and collaboration among stakeholders.

Economic considerations, such as the cost of implementing and maintaining separation infrastructure, as well as market fluctuations in the demand for recyclable materials, can impact the viability of waste separation programs. Without sustainable funding and viable markets for recycled materials, the long-term success of waste separation initiatives may be jeopardized.

Addressing these challenges requires a multifaceted approach that encompasses public education, infrastructure development, community involvement, and supportive policies. By tackling these obstacles, communities can work toward establishing effective waste separation practices that yield tangible environmental and social benefits.

9.3.6 Waste Transfer and Transportation

Waste transfer and transportation play a crucial role in the management of municipal, industrial, and hazardous waste. Waste transfer typically involves the consolidation of waste from smaller collection vehicles into larger transport vehicles, which then convey the waste to processing facilities, recycling centers, landfill sites, or other disposal locations. This process helps optimize the efficiency of waste collection and reduces the number of individual trips required for waste disposal, thereby minimizing fuel consumption and associated emissions.

9.3.6.1 Waste Transfer

This process involves moving waste from the collection vehicle to a transport vehicle, perhaps via a processing facility. Transfer stations usually act as spaces for the waste collector to discharge their load before reloading it into a larger vehicle for long-haul waste transportation. A transfer station also plays a critical role that directly links the waste collection stream to the waste treatment facilities or final disposal site.

9.3.6.2 Waste Transportation

This involves moving waste significant distances from a transfer location to processing, treatment, or disposal facilities. The option of any waste transportation should be a compromise in terms of effective operation in the waste collection and transfer to the transfer station, treatment facilities, and landfill. The distance from a collection area and discharge location influences the waste transport size and speed.

These processes are designed to ensure that waste is collected, transferred, and transported in a safe, efficient, and environmentally friendly manner. They require careful planning and coordination, as well as the use of specialized vehicles and facilities.

Transporting waste from transfer stations to its final destination involves a range of vehicles, including specialized trucks, trailers, and containers designed to accommodate various types of waste. These vehicles must adhere to safety and environmental regulations to prevent spillage, minimize odors, and mitigate potential hazards associated with transporting hazardous materials.

The logistics of waste transportation require careful planning to optimize routes, minimize travel distances, and avoid congestion. Utilizing advanced route planning software, GPS tracking systems, and real-time monitoring tools can enhance the efficiency of waste transportation operations, leading to cost savings and reduced environmental impact.

Waste transfer and transportation activities should prioritize safety, environmental protection, and compliance with regulations to ensure the secure and responsible movement of waste materials. Additionally, efforts to promote the use of alternative fuel vehicles and the adoption of sustainable transportation practices can contribute to reducing the carbon footprint associated with waste transport operations.

A transfer station, can also be a resource recovery center, is a building or processing site for the temporary deposition, consolidation, and aggregation of waste. They are often used as places where local waste collection vehicles will deposit their waste cargo prior to loading into larger vehicles.

Transfer stations vary significantly in size and function. Some transfer stations allow residents and businesses to drop off small loads of waste and recycling and may perform some preliminary sorting of material. Other transfer stations are places where local waste collection vehicles will deposit their waste cargo prior to aggregation and loading into larger vehicles. These larger vehicles will transport the waste to the end point of disposal in an incinerator, landfill, or hazardous waste facility, or for recycling. Transfer stations can be publicly or privately owned. They vary in size, from small regional sites managing less than 1,000 tons/year to large sites managing over 200,000 tons/year.

Transfer stations can be categorized based on their capacity and the volume of waste they are designed to handle. Table 9.9 shows the common types of transfer stations based on capacity.

TABLE 9.9
Types and Characteristics of Transfer Stations

Types	Characteristics
Small-scale transfer stations	These transfer stations are designed to handle relatively lower volumes of waste. They are typically found in smaller communities or rural areas where the waste generation rate is moderate. Small-scale transfer stations may serve as collection points for waste consolidation before transfer to larger regional facilities or processing sites. Small-scale transfer stations typically have a capacity ranging from 50 to 200 tons of waste per day. These facilities are designed to handle moderate volumes of waste and serve as collection and consolidation points for smaller communities or rural areas.
Medium-scale transfer stations	Medium-scale transfer stations can handle a higher volume of waste compared to small-scale facilities. They are often located in suburban or semi-urban areas where waste generation is more substantial. These transfer stations play a crucial role in consolidating waste from multiple collection routes before onward transportation to larger processing or disposal facilities. Medium-scale transfer stations are capable of handling higher volumes of waste, with capacities ranging from 200 to 500 tons per day. These facilities are commonly found in suburban or semi-urban areas and play a pivotal role in consolidating waste from multiple collection routes before onward transportation.
Large-scale transfer stations	Large-scale transfer stations are designed to handle significant volumes of waste from densely populated urban areas, industrial zones, or regions with high waste generation rates. These facilities are equipped with advanced waste-handling equipment, such as conveyor systems, sorting machinery, and compaction technology, to efficiently manage large quantities of waste before transfer to distant processing or disposal sites. Large-scale transfer stations are designed to manage substantial volumes of waste, with capacities typically ranging from 500 to 2,000 tons per day or more. These facilities are in densely populated urban areas or industrial zones and are equipped with advanced waste handling technology to efficiently process and transfer large quantities of waste.
Regional transfer stations	Regional transfer stations serve as central hubs for consolidating waste from multiple smaller transfer stations and collection points within a broader geographic area. These facilities are designed to accommodate substantial waste volumes and play a critical role in optimizing the transportation and transfer of waste to regional or national processing and disposal facilities. Regional transfer stations serve as central hubs for consolidating waste from multiple smaller transfer stations and collection points within a broader geographic area. Their capacities can range from 2,000 tons per day to several thousand tons per day, depending on the scale of the region's waste management needs.

The classification of transfer stations based on capacity reflects the diverse needs of different communities and regions, as well as the scale of waste management operations. The design and operation of transfer stations are tailored to the specific waste management requirements and logistical considerations of each area, ensuring efficient and sustainable waste transfer and transportation.

The main purpose of a transfer station is to act as an intermediate station where waste from smaller collection vehicles is unloaded and reloaded back into large vehicles for transport to a disposal or processing site. This process helps to increase the efficiency of waste collection and transportation, reduce costs, and minimize the environmental impact.

After waste is transferred at a transfer station, it is typically transported to its next destination, which can vary based on the types of waste received in the transfer station, specific waste management practices, and infrastructure in place. Some possible routes for waste after leaving a transfer station:

1. *Recycling Facilities:* If the waste includes recyclable materials, it may be transported to recycling facilities where sorting, processing, and recovery of recyclable materials such as paper, cardboard, glass, metals, and plastics take place. These materials are then prepared for reuse in manufacturing processes.
2. *Composting Sites:* Organic waste, such as food scraps and yard waste, may be transported to composting sites where it undergoes decomposition to produce nutrient-rich compost, which can be used in agriculture, landscaping, and soil improvement.
3. *WtE Plants:* Some waste may be transported to WtE plants where it is incinerated to generate electricity or heat. Advanced technologies are used to capture energy from the combustion process while minimizing emissions and environmental impact.
4. *Landfill Sites:* Residual waste that cannot be recycled, composted, or processed further may be transported to landfill sites for final disposal. At the landfill, the waste is compacted and covered with soil to reduce odor, prevent litter, and minimize environmental impact.
5. *Hazardous Waste Treatment Facilities:* Hazardous waste, such as chemicals, batteries, and electronic waste, may be transported to specialized treatment facilities equipped to handle and process hazardous materials in a manner that ensures environmental protection and public safety.

The specific destination for waste after leaving a transfer station depends on factors such as local regulations, waste management policies, available infrastructure, and the composition of the waste stream. Each destination represents a different stage in the waste management process, aimed at maximizing resource recovery and minimizing the environmental impact of waste disposal.

In some cities of Southeast Asian countries, transfer stations are found in very simple form, which is a dumpster or a bit bigger than a dumpster, and these dumpsters are transported directly to the landfill site without undergoing waste separation or the like. As a result, the lack of proper waste separation and processing at these rudimentary transfer stations leads to significant challenges, including increased environmental pollution, reduced resource recovery, and heightened public health risks. The direct transport of unsorted waste to landfill sites contributes to the accumulation of non-biodegradable materials, soil and water contamination, and the release of greenhouse gases, exacerbating the environmental impact of landfill disposal. The absence of waste diversion practices hinders the recovery of recyclable materials and organic waste, limiting opportunities for sustainable resource management and circular economy initiatives within these communities. Addressing these challenges requires comprehensive strategies to improve waste management infrastructure, promote waste reduction and recycling, and enhance environmental stewardship practices within the region.

If a transfer station does not exist, and to avoid further environmental consequences, alternative strategies can be employed to manage solid waste effectively. Some possible solutions for consideration to solve this problem may include the following:

- *Direct Transport to Processing Facilities:* In the absence of a transfer station, waste collection vehicles can transport waste directly to processing facilities, such as recycling centers, composting sites, or WtE plants. This approach requires careful coordination and planning to ensure that the appropriate vehicles and equipment are available at the processing facilities to receive and handle the incoming waste.
- *On-Site Sorting and Processing:* Waste management practices can be adapted to incorporate on-site sorting and processing at the point of waste collection. This may involve providing separate collection bins for recyclables, organic waste, and non-recyclable materials at the source, allowing for immediate segregation and processing without the need for an intermediate transfer station.
- *Mobile Processing Units:* Mobile processing units, such as portable recycling and composting facilities, can be deployed to different locations to process waste directly at the source. This approach can be particularly useful in areas where centralized processing facilities are not readily accessible.
- *Community-Based Solutions:* Community involvement and decentralized waste management initiatives, such as neighborhood composting programs and local recycling centers, can help mitigate the need for a centralized transfer station. By empowering communities to manage their own waste, the burden on centralized waste management infrastructure can be reduced.
- *Investment in Infrastructure:* In the long term, consideration can be given to the establishment of new transfer stations or the expansion of existing waste management infrastructure to better accommodate the needs of the community. This may involve collaboration with local authorities, private sector investment, and comprehensive planning to address the gaps in the waste management system.

In the absence of a transfer station, a combination of these approaches can be tailored to the specific needs and resources of the community to ensure efficient and sustainable SWM. However, it is essential to prioritize waste reduction, recycling, and resource recovery while exploring innovative solutions to address the challenges associated with the absence of a transfer station.

9.3.7 Waste Disposal

Waste disposal is the final stage of the whole process of the SWM, and the effective operation of recycling facilities is instrumental in diverting recyclable materials from landfill disposal, promoting resource recovery, and contributing to the development of a circular economy. Recovering and reprocessing valuable materials allows recycling facilities to significantly reduce the environmental impact of waste disposal, conserve natural resources, and mitigate the need for energy-intensive extraction and production of raw materials. The sale of recycled materials provides economic incentives for recycling, fostering sustainable practices and supporting the development of a closed-loop system where materials are continuously reused and recycled, minimizing the reliance on finite resources, and reducing the overall environmental footprint of waste management. Let us start from the transfer station, which is one step before the disposal.

9.3.7.1 Transfer Stations

As we are aware, a transfer station is a facility where waste is temporarily stored before it is transported to a recycling facility. Transfer stations serve as the link between local waste collection facilities and waste disposal facilities like landfills, WtE facilities, and other important elements of the process like the MRFs and recycling centers.

Transfer stations are designed to consolidate waste from multiple collection routes, which can make transportation more efficient. They are places where local waste collection vehicles will deposit their waste cargo prior to aggregation and loading into larger vehicles. These larger vehicles will transport waste to the end point of disposal in an incinerator, landfill, or hazardous waste facility, or for recycling.

9.3.7.2 To Recycling Facilities

9.3.7.2.1 Transportation

From the transfer station, the waste is transported to a recycling facility. This is usually done using larger vehicles, which can carry larger loads of waste. The transportation of waste is a critical step in the waste management process.

If the waste transfer station is connected to a MRF, waste will be sorted and screened to collect any recyclable items and remove them from the waste stream. Workers will also do a screen for any hazardous waste or unauthorized items to dispose of separately.

At the recycling facility, the waste is further sorted into different categories such as paper, plastic, glass, and metal. This is often done using both manual labor and automated machinery. Once the waste is sorted, it is processed into a form that can be used as raw material to produce new goods. For example, paper is pulped, glass is crushed, and plastic is shredded or pelletized.

The processed materials are then sold to manufacturers who use them to produce new products. This helps to reduce the demand for virgin materials, conserve resources, and reduce environmental impact.

9.3.7.2.2 Cleaning

The cleaning process involves removing any contaminants or impurities from the waste materials. This is a crucial step, as it ensures that the waste is safe to handle and process further. It also improves the quality of the recyclable materials and increases their value.

9.3.7.2.3 Processing and Reprocessing

Once the waste materials are cleaned, they are processed into a form that can be used as raw material to produce new goods. For example, paper is pulped, glass is crushed, and plastic is shredded or pelletized. The processed materials are then sold to manufacturers who use them to produce new products.

Reprocessing involves the cleaning, disinfection, and sterilization of reusable medical equipment/devices in healthcare and personal service settings. Following reprocessing guidelines and proper procedures reduces the spread of bacterial infection through reusable equipment and devices.

9.3.7.2.4 Quality Control

Quality control is an essential part of the waste management process. It involves checking the quality of the waste materials at various stages of the process to ensure that they meet the required standards. This includes checking the cleanliness of the waste materials, the efficiency of the processing methods, and the quality of the final recycled products.

9.3.7.2.5 Impacts

The impacts of waste management processes can be categorized into social, environmental, and economic impacts.

Social Impacts: Proper waste management practices can lead to improved public health outcomes by reducing the spread of diseases. They also create jobs and contribute to the local economy.

Environmental Impacts: Effective waste management practices help to reduce the amount of waste that ends up in landfills, thereby reducing greenhouse gas emissions. Recycling also conserves natural resources by reducing the demand for virgin materials.

Economic Impacts: Waste management and recycling can generate revenue through the sale of recycled materials. They also create jobs and contribute to the local economy. Moreover, efficient waste management practices can lead to cost savings in the long run by reducing the amount of waste that needs to be disposed of in landfills.

9.3.7.3 To Composting Site

Organic waste, including food scraps and yard waste, can be taken to composting sites, where it goes through a decomposition process to create nutrient-rich compost. This compost can then be utilized for agricultural purposes, landscaping, and soil enhancement.

Composting is a natural process that transforms organic material like kitchen scraps and yard waste into a nutrient-rich soil conditioner. Composting in composting sites, also known as community composting, involves several steps, as shown in Table 9.10.

Composting can be done through community composting or larger-scale industrial composting facilities, providing sustainable solutions for managing organic waste and reducing environmental impact. Community composting involves individuals or small groups managing organic waste in their local area. This can be done through backyard composting, where households compost their food scraps and yard waste to create nutrient-rich soil for gardening. Community composting also includes shared composting sites where people in a neighborhood collectively contribute their organic waste to be composted together. These community-based approaches promote local sustainability and reduce the amount of organic waste sent to landfills.

On the other hand, larger-scale industrial composting facilities are equipped to handle significant volumes of organic waste. These facilities use specialized machinery and controlled environments to efficiently break down organic materials into compost. Industrial composting is often used by municipalities, businesses, and agricultural operations to manage large quantities of organic waste in a

TABLE 9.10
Composting Process in Composting Sites

Collection of Organic Waste: The first step in the composting process is the collection of organic waste. This can include fruit and vegetable scraps, coffee grounds, eggshells, grass clippings, leaves, and other yard waste.
Building the Compost Pile: The collected organic waste is then arranged in a compost pile at the composting site. A good compost pile should have a balance of "green" materials (which provide nitrogen) like fruit and vegetable scraps, and "brown" materials (which provide carbon) like leaves and twigs.
Collection of Organic Waste: The first step in the composting process is the collection of organic waste. This can include fruit and vegetable scraps, coffee grounds, eggshells, grass clippings, leaves, and other yard waste.
Turning the Compost Pile: The compost pile needs to be turned regularly to provide oxygen, which is essential for the composting process. Turning the pile also helps to speed up the decomposition process and prevents the pile from developing an unpleasant odor.
Maturation: Over time, the organic waste in the compost pile breaks down into a rich, dark soil-like substance known as compost. This process can take anywhere from a few weeks to several months, depending on the materials used and the conditions at the composting site.
Use of Compost: Once the compost has fully matured, it can be used to enrich the soil in gardens and landscapes. Compost improves soil structure, provides essential nutrients to plants, and helps retain moisture in the soil.

sustainable manner. By diverting organic waste from landfills, industrial composting helps reduce greenhouse gas emissions and minimizes the environmental impact of waste disposal.

Community composting and industrial composting facilities both contribute to a circular economy by converting organic waste into a valuable resource. The resulting compost can be used to enrich soil for agricultural purposes, landscaping, and horticulture. This closed-loop approach reduces the need for chemical fertilizers, conserves natural resources, and promotes soil health. Additionally, the use of compost helps sequester carbon in the soil, contributing to climate change mitigation efforts.

One of the key benefits of composting is its ability to reduce methane emissions from landfills. When organic waste decomposes in anaerobic conditions, such as those found in landfills, it produces methane, a potent greenhouse gas. By composting organic waste instead of sending it to landfills, the production of methane is significantly reduced, thus mitigating the impact on climate change. This highlights the importance of composting as a sustainable waste management solution with positive environmental implications.

Furthermore, composting plays a crucial role in promoting sustainable food systems. By returning organic matter to the soil in the form of compost, the nutrient cycle is closed, supporting the growth of healthy and nutritious crops. This is particularly important in the context of sustainable agriculture, where the use of compost can improve soil structure, fertility, and water retention, leading to higher crop yields and reduced reliance on synthetic fertilizers.

In addition to its environmental benefits, composting also has economic advantages. By diverting organic waste from landfills, communities and businesses can reduce their waste management costs. Instead of paying for disposal fees at landfills, they can redirect organic waste toward composting, creating a valuable resource that can be used locally. This not only saves money but also contributes to the creation of green jobs in the composting industry, further enhancing the economic sustainability of composting initiatives.

Composting also fosters a sense of environmental stewardship and community engagement. When individuals and communities participate in composting initiatives, they become more aware of the environmental impact of their actions and gain a deeper appreciation for the natural cycles of waste and resource management. This heightened awareness often leads to broader sustainable practices, such as reducing overall waste generation, recycling more materials, and supporting local environmental conservation efforts.

Moreover, composting can serve as an educational tool, especially in schools and community organizations. By incorporating composting into educational curricula and community programs, participants can learn about the environmental benefits of composting, the science of decomposition, and the importance of waste reduction. This hands-on approach not only instills valuable knowledge but also empowers individuals to take active roles in environmental conservation and sustainable living.

Whether through community-based initiatives or large-scale facilities, composting offers a versatile and effective solution for managing organic waste and reducing environmental impact. By embracing composting practices, individuals, communities, and businesses can contribute to a more sustainable and resilient future while simultaneously addressing environmental, economic, and social challenges associated with waste management.

Community composting sites are local and often dispersed throughout a town, city, or neighborhood. They source organic materials locally and engage the community in the composting process. The compost produced is used in local soils, contributing to local sustainability and resilience. Some methods in community composting are shown in Table 9.11.

Amid all composting can be done through both community composting and large-scale industrial composting, Table 9.12 displays various methods that are suitable for large-scale composting.

Each method has its own advantages and is suitable for different types of waste and conditions. The choice of method depends on the amount and type of waste, available space, and personal preference.

TABLE 9.11
Methods in Community Composting

Methods	Explanations
Backyard composting	This is a common method of composting where organic waste is collected in a bin or heap in your backyard. It can be further divided into hot (with oxygen) and cold (without oxygen) methods.
Three-bin system composting	This method involves three bins where one is used for adding new waste, the second is for turning and aerating the compost, and the third is for the finished compost.
Windrows	This method involves creating long narrow piles of organic waste. The piles are turned to improve oxygenation.
Aerated static piles	This method involves creating piles of organic waste and using a system to push air up through the pile.
Vermicomposting	This method uses worms to consume organic waste and convert it into nutrient-rich compost.
Bokashi composting	This is a method where kitchen waste is fermented in a sealed bucket using a special ingredient called Bokashi.

TABLE 9.12
Methods in Large-scale Composting

Methods	Explanations
Static pile composting (SPC)	This method involves creating large piles of organic material. These may be turned from time to time, often with heavy machinery.
Static aerated pile composting (SAPC)	This method uses a compost aerator to force air through piles of organic material.
In-vessel composting	This method involves composting organic materials within a drum, silo, bin, or similar equipment.
Tunnel composting	This method involves composting organic materials in a tunnel-like structure.
Anaerobic digestion	This method involves breaking down organic material in the absence of oxygen.

TABLE 9.13
Types of Combustible Waste

Municipal Solid Waste (MSW): This is mainly household waste and commercial waste. It is the most common type of waste incinerated in waste-to-energy plants.

Sewage Waste: Sewage waste can be placed into an incinerator and is sometimes mixed with other waste for incineration, such as municipal waste and biowaste.

Clinical Waste: There are other subcategories related to each individual waste stream.

Industrial Effluents: Some industrial effluents may be suitable for waste-to-energy plants, such as activated sludge, food scraps, grease, pulp, paper waste, etc.

Biomass: Biomass waste can also be incinerated in waste-to-energy plants.

9.3.7.4 To WtE Plants

A WtE plant, also known as an energy-from-waste (EfW) plant, is a facility that converts municipal and industrial solid waste into electricity and/or heat. There are some types of waste that can be incinerated and sent to WtE plants to be converted to energy, as shown in Table 9.13.

Not all waste is suitable for incineration. Some waste materials, such as those containing heavy metals or other toxic substances, may need to be treated differently to prevent environmental contamination.

9.3.7.5 To Landfill Sites

A landfill site, also known as a tip, dump, rubbish dump, garbage dump, trash dump, or dumping ground, is a site for the disposal of waste materials. Dumping waste in landfills is often considered the easiest method of waste disposal for several reasons: (1) *Simplicity:* The process of landfilling is relatively straightforward. It involves collecting waste, transporting it to the landfill site, and depositing it there. (2) *Low Cost:* Compared to other waste disposal methods such as incineration or recycling, landfilling is often more cost-effective. (3) *Versatility:* Landfills can accommodate a wide variety of waste types, including MSW, construction and demolition waste, and certain types of industrial waste. (4) *Availability of Land:* In many countries, especially those with large land areas, there is sufficient space available for landfill sites. (5) *Lack of Infrastructure:* In some cases, countries may lack the necessary infrastructure or resources for more complex waste management strategies, making landfilling a more feasible option. However, while landfilling is a common method of waste disposal, it is not without its drawbacks. Landfills can have significant environmental impacts, including air and water pollution, and they contribute to the production of greenhouse gases. Therefore, many countries are working toward reducing their reliance on landfills and implementing more sustainable waste management practices, such as recycling and composting.

The environmental problems associated with landfill sites include the release of greenhouse gases, such as methane, into the atmosphere as organic waste decomposes in anaerobic conditions. Methane is a potent greenhouse gas that contributes to climate change, and its release from landfills can have significant environmental implications. Additionally, the leachate produced because of waste decomposition can contaminate soil and water, posing risks to ecosystems and human health. Landfill sites also occupy valuable land space, and their expansion can lead to habitat destruction and loss of biodiversity.

From a social perspective, the development of landfill sites can negatively impact nearby communities. Issues such as odors, noise, and traffic associated with waste disposal activities can diminish the quality of life for residents living in the vicinity of landfills. Moreover, the stigma attached to living near a landfill can affect property values and community well-being, leading to social disparities and inequities. Therefore, while landfill disposal may seem convenient, it is crucial to consider the long-term social and environmental consequences of this approach.

Considering the huge impacts of the development of landfill sites as part of MSWM, there are several ways to reduce reliance on landfilling by doing the following things:

- *Reducing Waste:* Reducing waste involves reusing materials, reducing consumption, and increasing public awareness about sustainable waste management practices. Policies can also drive waste reduction.
- *Recycling:* Recycling breaks down materials into their components and converts them into reusable products. This not only reduces the amount of waste in landfills but also helps conserve resources and energy.
- *Composting:* Organic waste like food scraps and yard waste can be composted at home or in industrial facilities. This reduces the amount of organic waste that ends up in landfills and produces a valuable soil amendment.
- *WtE:* Some waste can be used to generate energy. For example, organic waste can be converted into biogas through anaerobic digestion. This not only reduces the amount of waste going to landfills but also produces a renewable source of energy.
- *Reuse and Reduction:* Reusing materials and products, using energy-efficient green technologies, and reducing everyday material consumption can all help reduce waste.

- *Education and Awareness:* Educating the public about the importance of waste reduction and proper waste management can also help reduce reliance on landfills.
- *Regulations and Policies:* Policies such as landfill bans and taxes, and incentives for recycling and recycling infrastructure have proven successful in reducing landfilling.

Every little bit helps. Even small changes in our daily habits can add up to making a big difference in reducing waste and our reliance on landfills. Some habits, like buying in bulk, reusing bottles and cups, using cloth bags, avoiding plastic-wrapped produce, supporting local farmers, avoiding single-use plastics, composting, reducing energy use, thinking before buying, and donating or selling unused items, can help reduce daily waste.

9.4 HAZARDOUS WASTE MANAGEMENT

9.4.1 General Overview

Hazardous waste is defined as waste with properties that make it dangerous or capable of having a harmful effect on human health or the environment. The Environmental Protection Agency (EPA) developed a regulatory definition and process that identifies specific substances known to be hazardous and provides objective criteria for including other materials in the regulated hazardous waste universe. There are four types of hazardous wastes based on their key characteristics: ignitability, corrosivity, reactivity, and toxicity, according to the Resource Conservation and Recovery Act (RCRA) and the EPA:

- *Ignitability:* Ignitable wastes can create fires under certain conditions, are spontaneously combustible, or have a flash point less than 60°C (140°F). Examples include waste oils and used solvents. They can cause fire through friction, absorption of moisture, or spontaneous chemical changes. Ignitable wastes are identified by the EPA hazardous waste number D001.
- *Corrosivity:* Corrosive wastes are acids or bases (pH less than or equal to 2, or greater than or equal to 12.5) that are capable of corroding metal containers, such as storage tanks, drums, and barrels. They include strong acidic or alkaline substances. Corrosive wastes are identified by the EPA hazardous waste number D002.
- *Reactivity:* Reactive wastes are unstable under "normal" conditions. They can cause explosions, toxic fumes, gases, or vapors when heated, compressed, or mixed with water. They are chemically unstable and react violently with air or water. Reactive wastes are identified by the EPA hazardous waste number D003.
- *Toxicity:* Toxic waste is harmful when ingested or absorbed. They present a concern as they may be able to leach from waste and pollute groundwater. Waste is considered toxic if it is poisonous, radioactive, explosive, or bioaccumulative (i.e., increasing in concentration at the higher ends of food chains) or if it causes cancer, damage to chromosomes, or birth defects.

These characteristics help in the identification and proper management of hazardous waste to prevent harm to human health and the environment.

Management of hazardous waste entails the procedures for handling materials classified as hazardous to prevent environmental contamination and protect human health. It involves systematic steps, which consist of collection, recycling, treatment, transportation, disposal, and monitoring of waste disposal sites. These practices are essential to manage the risks associated with hazardous waste, ensuring it is dealt with in a manner that complies with environmental regulations and safeguards both the ecosystem and public welfare. The process of the collection, treatment, transportation, disposal, and recycling of waste may cause substantial harm to humans and the environment. The RCRA, passed in 1976, was established to set up a framework for the proper management of hazardous waste. Hazardous waste is generated from many sources, ranging from industrial manufacturing process wastes to batteries, and may come in many forms, including liquids, solids, gases, and sludges. In the current scenario of developing countries, hazardous wastes are often disposed of directly into the environment, posing health and environmental risks.

9.4.2 Regulatory Framework

The regulatory framework of hazardous waste is a set of laws and policies designed to manage and control the generation, transportation, treatment, storage, and disposal of hazardous waste. The key elements of this framework include the following:

1. *Waste Framework Directive:* This is a directive by the European Commission that sets the basic concepts and definitions related to waste management, including definitions of waste, recycling, and recovery. It introduces the "polluter pays principle" and the "EPR." It also establishes a five-step "waste hierarchy" for managing and disposing of waste.
2. *RCRA:* This is a US federal law that gives the EPA the authority to control hazardous waste from the "cradle-to-grave." This includes the generation, transportation, treatment, storage, and disposal of hazardous waste.
3. *Institutional and Policy Framework:* This framework reviews key results from the Environmental Performance Reviews on the legal and policy frameworks for waste management and the circular economy in various countries. It includes legislation, strategies, and planning for waste management, coordination between national and sub-national planning, and management of waste services at the local level.
4. *Hazardous Waste Program:* Under RCRA Subtitle C, this program establishes a system for controlling hazardous waste from the time it is generated until its ultimate disposal.

These frameworks ensure that waste is managed without endangering human health and harming the environment and without causing a nuisance through noise or odors. They also provide additional labeling, record keeping, monitoring, and control obligations from the "cradle to the grave" for hazardous waste. This concept refers to the responsibility of waste generators for their waste from the moment it is created until its final disposal or destruction. This concept is a key component of the RCRA in the United States.

The cradle-to-grave concept covers (1) *Generation:* The "cradle" refers to the generation stage where it all begins. Once a company generates waste, it must identify whether it is hazardous or regular waste. Once the hazardous material is determined, it should be documented and properly labeled. (2) *Transportation:* The waste must be transported to the proper treatment facility for safe disposal. The transportation methods must meet the requirements of the Department of Transportation (DOT). The material must be properly labeled, stored in appropriate containers, and sealed to prevent leakage. (3) *Treatment, Storage, and Disposal:* The "grave" refers to the final disposal or destruction of the waste. Even after the waste leaves the generator's site, the generator is still responsible for the waste. If an accidental spill or release of hazardous material occurs during transport, the generator is responsible for the cleanup and disposal of the material. (4) *Post-Disposal:* Even after the waste has been disposed of, the generator's responsibility continues. If the waste is not managed properly at the disposal facility, the generator may be held responsible. The cradle-to-grave system aims to ensure that hazardous waste is managed safely and responsibly, protecting both the environment and human health.

9.4.3 Hazardous Waste Identification and Classification

Hazardous waste identification is a crucial step in the waste management process. The RCRA regulations require that any person who produces or generates waste must determine if that waste is hazardous. This identification process can be complex and involves several steps, including determining if the waste is a solid waste, if it is excluded from regulation, and if it exhibits any characteristics of hazardous waste.

Hazardous waste can be classified as either listed or characteristic waste. Listed wastes are wastes from common manufacturing and industrial processes, and specific industries and can be generated from discarded commercial products. There are four lists (the F, K, P, and U lists) found in title 40 of the Code of Federal Regulations (CFR) in part 261 that specifically list hazardous wastes. Characteristic wastes, on the other hand, are wastes that exhibit any one or more of the following characteristic properties: ignitability, corrosivity, reactivity, or toxicity.

There is also a category called mixed waste, which has both a hazardous component and a radioactive component. This type of waste is regulated under both the RCRA and the Atomic Energy Act. Proper identification and classification of hazardous waste are essential for its safe and effective management, ensuring protection of both human health and the environment.

Countries enforce hazardous waste regulations through a combination of legal frameworks, policy measures, and enforcement mechanisms. For instance, countries establish legal frameworks that define what constitutes hazardous waste, how it should be managed, and the penalties for noncompliance. These laws often include requirements for permits, licenses, inspections, and enforcement. For example, the RCRA in the United States establishes a system for controlling hazardous waste from the time it is generated until its ultimate disposal.

Then, countries implement policy measures to encourage compliance with these laws. These measures can include "negative incentives" that impose costs on the waste generator for noncompliance. For instance, companies may face fines or other penalties if they fail to properly manage their hazardous waste.

Finally, enforcement mechanisms are put in place to ensure that these laws and policies are followed. This can include regular inspections, audits, and monitoring of waste management practices. In some cases, the enforcement of these regulations is carried out by specific government departments or local authorities.

The effectiveness of these enforcement mechanisms can vary widely between countries, and there can be challenges in implementing and enforcing these regulations, particularly in developing countries. Despite these challenges, the goal of these regulations is to ensure the safe and responsible management of hazardous waste, protecting both human health and the environment.

Hazardous waste materials can come from various sources and take many forms. Here are some common examples:

- *Industrial By-products:* The industrial sector, including the petroleum industry, is a significant contributor to hazardous waste generation. Different industries produce a variety of by-products, depending on their operations.
- *Solvents:* These are often used in industrial processes and can become hazardous waste when they are spent or no longer usable.
- *Electronics Waste (E-Waste):* This includes discarded electronic devices and components, which can contain hazardous materials such as lead, mercury, and cadmium.
- *Batteries:* Both household and industrial batteries can be hazardous due to the heavy metals and corrosive materials they contain.
- *Pesticides and Herbicides:* These garden chemicals can be hazardous, especially in large quantities.
- *Cleaning Agents:* Certain cleaning fluids and chemicals can be hazardous, particularly if they are corrosive or contain toxic substances.
- *Pharmaceutical Waste:* This includes expired or unused medications, which can be hazardous if not disposed of properly.

TABLE 9.14
Some Hazardous Waste Treatment Methods

Physical Methods
Landfills: Specifically designed for hazardous waste, these offer a controlled environment to dispose of solid hazardous waste materials.
Incineration: This method involves burning the waste at high temperatures to convert it into residue and gaseous products.
Deep-Well Injection: This involves injecting liquid waste deep into the ground below the level of groundwater.
Physical Treatment: This includes methods like solidification, flotation, sedimentation, evaporation, or filtration.
Encapsulation: This involves sealing waste in a stable material to prevent it from contaminating the environment.
Thermal Desorption: This method uses heat to vaporize contaminants, which are then collected and treated.
Recycling: Some hazardous wastes can be recycled into new products.
Plasma Arc Recycling: This method uses a plasma torch to break down waste into basic elements.
Chemical Methods
Chemical Stabilization: This involves adding chemicals to the waste to stabilize it and make it less hazardous.
Biological Methods
Biological Treatment: This method uses microorganisms to break down hazardous organic compounds into less harmful substances.

- *Paints and Oils:* These can include waste solvents, paints, oils, and polychlorinated biphenyls (PCBs).
- *Mercury-Containing Devices:* Devices like thermometers, thermostats, and compact fluorescent lamps (CFLs) contain mercury, which is a hazardous material.

Hazardous waste needs to be treated properly to prevent harm to both the environment and human health. Improper disposal can lead to soil, water, and air pollution, causing significant damage to ecosystems and potentially leading to health issues in humans and wildlife. It is important that hazardous waste is identified correctly at the source and separated from regular waste to ensure it does not contaminate nonhazardous materials.

Once identified, hazardous waste should be handled and transported by trained professionals to prevent accidents and spills. Treatment methods vary depending on the type of waste and can include incineration, chemical treatment, and biological treatment. Some hazardous waste can even be recycled into new products, reducing the need for raw materials and conserving resources.

Public awareness and education are also key components in ensuring hazardous waste is treated properly. By understanding the risks and responsibilities associated with hazardous waste, individuals and businesses can make informed decisions and contribute to a safer and cleaner environment.

Hazardous waste treatment methods can be broadly categorized into physical, chemical, and biological methods, as shown in Table 9.14.

9.4.4 Hazardous Waste Handling and Storage

Handling hazardous waste requires careful planning and adherence to regulations to ensure safety and environmental protection. Some steps to handle hazardous waste according to ISO 14001 and other sources are as follows.

9.4.4.1 Evaluate the Waste

Determine whether the waste is hazardous or not and whether handling of that waste is regulated by legislation. Determining whether a waste is hazardous involves a series of steps. The first step is to determine if the material is a solid waste. If it is, the next step is to investigate whether the waste is specifically excluded from regulation as a solid or hazardous waste. Once these steps are completed, the generator must determine if the waste is a listed or characteristic hazardous waste. Listed wastes are those that are specifically listed by regulatory authorities as hazardous. Characteristic wastes are those that exhibit one or more of the four characteristics of hazardous waste: ignitability, corrosivity, reactivity, and toxicity.

The handling of hazardous waste is regulated by various legislations. In the United States, the RCRA regulates the management of hazardous waste. Businesses and individuals have a legal duty of care when it comes to any waste, especially waste that is hazardous. They must ensure they handle all hazardous waste responsibly, correctly, and safely. If they do not, it can result in prosecution, fines, and even imprisonment.

In addition to federal laws, there may be state and local regulations that also apply. For example, the Environmental Management Act (EMA) prohibits the introduction of waste into the environment in a way that will cause pollution, except in accordance with a regulation, permit, approval, or code of practice issued under the Act. It is important for waste generators to be aware of and comply with all applicable regulations to ensure safe and legal waste management.

9.4.4.2 Store the Waste

Use Sturdy, Leakproof Containers: Hazardous waste must always be kept in receptacles constructed of materials that are appropriate for the waste they contain. These containers should be sturdy and leakproof to prevent any accidental spillage or leakage. Containers used for household detergent or those in the foodservice industry are unacceptable for use. It is preferable to use the original chemical receptacle for hazardous waste disposal.

Keep Containers Closed: Containers used to store hazardous material must be free of leaks and exterior residue. Using a screw-type lid or another suitable mechanism, containers should be properly sealed. The use of improvised covers such as aluminum foil or plastic wrap is prohibited. When not adding or removing garbage, a container storing hazardous waste must always be sealed.

Use Compatible Containers: Different kinds of waste may require different types of storage containers. For example, corrosive waste should not be stored in metal containers, as it can corrode the metal and cause leaks. Similarly, flammable waste should be stored in fire-resistant containers.

Store in Dry, Cool, and Ventilated Areas: Hazardous waste must be stored in dry, cool, and ventilated areas. This is to prevent any chemical reactions that could be triggered by heat or moisture. Ventilation is important to prevent the buildup of harmful fumes.

Label Your Containers: As soon as the container is used to collect hazardous waste, hazardous chemical waste tags must be labeled on the container irrespective of whether the receptacle is filled with hazardous waste. The laboratory must supply a variety of details on the tags, including the person's name, phone number, address, and room number. When the container is full, enter the date in the appropriate field.

Segregate Different Types of Waste: When possible, segregate different types of hazardous waste from each other. When hazardous trash is combined with nonhazardous garbage, the total amount of hazardous waste that must be disposed of increases. Combining hazardous wastes might raise disposal costs as different hazardous wastes have different disposal strategies.

9.4.4.3 Label the Waste

Nonhazardous waste does not have to be labeled in any special way. However, hazardous waste labeling is often prescribed by law. The label for marking packed hazardous waste usually contains the following information:

- The words “Hazardous Waste”,
- A clear description of the contents,
- The date when the waste is first placed in the container.

9.4.4.4 Segregate Incompatible Wastes

Segregating incompatible waste is necessary. Incompatible wastes are those that, when mixed, can cause dangerous reactions such as heat generation, fires, explosions, or the release of toxic gases. The process of segregating incompatible wastes involves the following:

Identify Incompatible Wastes: The first step is to identify which waste is incompatible. This can be done by referring to safety data sheets (SDS) or other resources that provide information on chemical incompatibilities.

Use Separate Storage Areas or Containers: Once the incompatible wastes have been identified, they should be stored separately. This can be done by placing them in different physical locations or using separate containers. The containers used should be suitable for the type of waste they are storing.

Use Physical Barriers: In some cases, it may be necessary to use a physical barrier, such as a wall or berm, to separate incompatible wastes. This can provide an extra level of protection in case of accidental mixing.

Follow Local Regulations: It is important to follow local regulations and guidelines when segregating waste. These regulations are designed to ensure safety and prevent environmental harm.

9.4.4.5 Maintain Proper Ventilation and Temperature

Maintaining proper ventilation and temperature inside facilities where hazardous waste is stored is of paramount importance. Proper ventilation is crucial as it helps to regulate air quality levels, prevent the buildup of combustible fumes, and ensure compliance with health and safety regulations. When hazardous waste is stored, certain chemicals can emit vapors that can ignite or explode when they accumulate to specific concentrations. Ventilation systems help mitigate these risks by continually circulating air and dissipating the concentration of these harmful fumes. These systems also remove heat and moisture, reducing the risk of spontaneous combustion and maintaining an environment conducive to safe storage. On the other hand, maintaining an appropriate temperature is key to preventing unwanted chemical reactions or degradation of chemicals over time. Certain hazardous chemicals are sensitive to heat and moisture, which can lead to these unwanted reactions. Therefore, ventilation and temperature control are key to maintaining optimal storage conditions, preventing damage to stored materials, and ensuring the safety of the facility.

9.4.4.6 Ensure All Hazardous Materials Are Properly Marked

Ensuring that all hazardous materials (hazmat) containers are adequately labeled and that all chemicals are stored in the appropriate containers is a critical aspect of hazardous waste management. Proper labeling provides essential information about the contents of the container, including the type of waste, its hazards, and how to handle it safely. This helps to prevent accidents, facilitates proper waste handling, and is often required by law. Similarly, storing chemicals in appropriate containers is crucial to prevent leaks, spills, and chemical reactions. Containers should be compatible with their contents, sturdy, and leakproof. They should also be kept closed when not in use to prevent evaporation and dispersion of harmful substances.

9.4.5 Hazardous Waste Disposal

Proper disposal of hazardous waste is inevitable to prevent harm to human health and the environment. There are some methods for properly disposing of hazardous waste:

9.4.5.1 Incineration

This method involves burning hazardous waste at high temperatures in specially designed incinerators. The heat breaks down organic compounds and reduces the volume of the waste. Incineration is effective for destroying certain types of hazardous waste, such as organic solvents and some types of medical waste. However, it is important to ensure that the incineration process is carried out in compliance with environmental regulations to minimize air emissions.

9.4.5.2 Secure Landfill Disposal

Hazardous waste that cannot be incinerated or recycled may be disposed of in secure landfills designed to prevent the release of hazardous substances into the environment. These landfills have protective liners and monitoring systems to prevent groundwater contamination. Hazardous waste is carefully placed in designated areas within the landfill and covered with multiple layers of soil to prevent exposure.

9.4.5.3 Chemical Treatment

Some hazardous waste can be treated chemically to neutralize its hazardous properties or to separate it into less harmful components. This may involve processes such as oxidation, reduction, precipitation, or chemical stabilization. Chemical treatment can reduce the toxicity of the waste, making it safer for disposal or recycling.

9.4.5.4 Physical Treatment

Physical treatment methods include techniques such as filtration, sedimentation, and evaporation to separate hazardous components from the waste stream. Physical treatment can be used to remove contaminants or to reduce the volume of the waste before disposal.

9.4.5.5 Recycling

Certain types of hazardous waste, such as used oil, batteries, and electronic waste, can be recycled to recover valuable materials and reduce the amount of waste requiring disposal. Recycling hazardous waste reduces the need for raw materials and minimizes environmental impact.

9.4.5.6 Biological Treatment

Biological treatment methods, such as bioremediation, use microorganisms to break down hazardous substances in the waste. This method is particularly effective for treating certain types of organic waste, petroleum products, and contaminated soil.

9.4.5.7 Neutralization

Hazardous waste containing acids or bases may be neutralized to adjust its pH and reduce its corrosive or reactive properties. This process can make the waste less hazardous and safer for disposal.

The selection of the disposal method should be based on the specific characteristics of the hazardous waste and must comply with local, national, and international regulations. Proper disposal methods should be carried out by trained personnel using appropriate equipment and facilities to ensure the safety of both the environment and individuals involved.

9.5 SANITARY LANDFILL

A sanitary landfill is a waste management system designed to safely and efficiently dispose of solid waste. By layering waste between compacted soil, sanitary landfills minimize environmental contamination and health hazards. These sites are engineered with protective liners and leachate collection systems to prevent groundwater pollution. Gas collection systems capture methane emissions for energy use, reducing greenhouse gas release. Properly managed, sanitary landfills offer a more controlled and sustainable waste disposal method compared to open dumps, ensuring that waste is isolated from the environment and public health is safeguarded.

9.5.1 Types of Landfills

A landfill site is a location that has been prepared for the purpose of dumping waste, rubbish, or debris. It is a site for the disposal of waste materials. Landfills are sites designed to store garbage and are designed to minimize the effects of the trash on human health and the environment. The waste is deposited in thin layers and promptly compacted by heavy machinery.

There are several types of landfills based on types of waste received by the landfill or dumped to the landfill, or types of waste they are designed to handle. These landfills are shown in Table 9.15.

Each type of landfill has specific design, operation, and closure requirements to ensure environmental safety and compliance with regulations. It is important to note that landfilling is just one part of an integrated waste management system, and efforts should always be made to reduce, reuse, and recycle waste where possible, and the final disposal does not necessarily have to be a landfill site.

A sanitary landfill (Figure 9.6), on the other hand, is a pit with a protected bottom where trash is buried in layers, compacted, and covered. It is a waste management method that involves the disposal of solid waste into specially designed areas known as sanitary landfills. It is designed to cause the least amount of environmental harm and inconvenience. A sanitary landfill can reduce harm from waste that has been collected and is safer than an open dumping site. It has specialized structures to reduce the contamination of the surrounding soil or water.

The difference between a landfill site and a sanitary landfill lies in their design and purpose. While both are used for waste disposal, a sanitary landfill is specifically designed to minimize environmental impact and protect public health. It uses engineering methods to protect the environment from contamination by solid or liquid wastes. In contrast, a landfill site, while also used for waste disposal, may not have these protective measures in place. Therefore, sanitary landfills are more environmentally friendly and safer for human health compared to regular landfill sites.

9.5.2 Landfill Operations

Operators of well-run landfills confine waste to as small an area as possible, compact waste to reduce volume, and cover waste (usually daily) with layers of soil or other types of material. This process is crucial in maintaining the efficiency and safety of a landfill.

The first step in the operation of a landfill is the confinement of waste. This involves the careful placement of waste in a designated area within the landfill. The goal is to use the available space

TABLE 9.15
Types of Landfills Based on Waste to Handle

1	*Municipal Solid Waste Landfills (MSWLFs):* These are designed to receive household waste, as well as other types of nonhazardous wastes.
2	*Bioreactor Landfills:* A type of MSWLF that operates to rapidly transform and degrade organic waste.
3	*Industrial Waste Landfill:* Designed to collect commercial and institutional waste, which is often a significant portion of solid waste.
4	*Construction and Demolition (C&D) Debris Landfill:* Designed exclusively for construction and demolition materials.
5	*Coal Combustion Residual (CCR) Landfills:* Used to manage and dispose of coal combustion residuals (CCRs or coal ash).

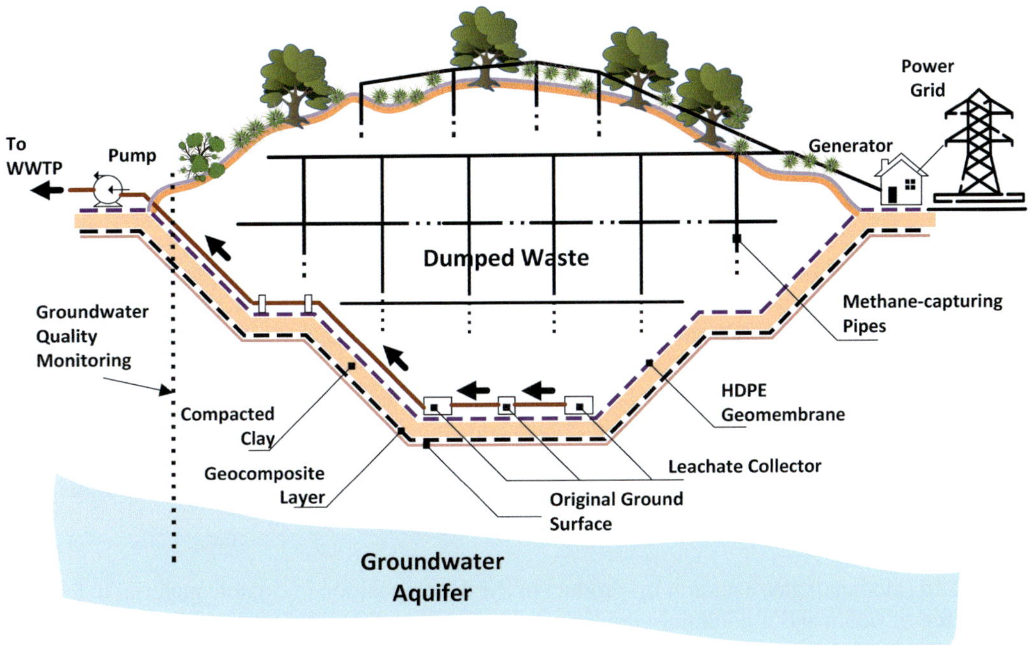

FIGURE 9.6 Sanitary landfill.

as efficiently as possible, which is achieved by compacting the waste. Compaction reduces the volume of the waste and increases the landfill's capacity. Heavy machinery, such as compactors and bulldozers, is typically used for this purpose.

Covering the waste is another critical aspect of landfill operations. This is usually done daily to minimize odors and keep waste from being carried away by wind. The cover material, often soil, also deters pests like rodents and birds and helps to prevent the spread of disease. In some cases, alternative daily cover materials such as foam, tarps, or compost materials may be used instead of soil.

In addition to these daily operations, landfill operators also must manage the landfill gas and leachate produced by the decomposing waste. Landfill gas, a natural by-product of the decomposition of organic material in landfills, is a source of odors and a potential greenhouse gas. It is typically collected and either flared (burned off) or used to generate electricity. Leachate, the liquid that drains or "leaches" from a landfill, can contaminate groundwater and surface water and must be carefully collected and treated.

Overall, the operation of a landfill involves a complex set of procedures and regulations designed to protect human health and the environment. It requires careful planning, skilled management, and rigorous monitoring to ensure compliance with all relevant regulations. Despite the challenges, when operated correctly, landfills can provide a safe and effective method for the disposal of many types of waste. However, it is always important to remember disposing of waste to landfill sites should not be the final goal of waste management; rather, we should always think that reducing, reusing, and recycling waste can significantly decrease the amount of waste that needs to be landfilled. This not only conserves landfill space but also helps to preserve our natural resources and protect the environment.

There are two significant substances that may have significant impacts on the environment if not properly handled. They are leachate and methane gas. Landfills manage leachate and gas through a combination of containment, treatment, and conversion methods.

9.5.2.1 Leachate Management

Leachate, the liquid that drains or "leaches" from a landfill, can contaminate groundwater and surface water and must be carefully collected and treated. Managing the leachate in a landfill site can be done in the following ways:

- *Containment:* Landfills are designed to prevent leachate from escaping into the surrounding environment. This is typically achieved using liners and collection systems that capture the leachate and direct it to a treatment facility.
- *Treatment:* Various treatment methods can be used to remove contaminants from leachate. These include biological treatments, chemical treatments, and physical processes. The choice of treatment method depends on the composition of the leachate and local environmental regulations.
- *Recirculation:* In some cases, leachate is recirculated back into the landfill. This can enhance waste decomposition, increase landfill gas yield, and accelerate airspace recovery.
- *Evaporation:* Powered by landfill gas, an evaporator can heat up the leachate, driving off the water vapor and leaving behind sludge and solids that can be safely disposed of on the landfill active face.

Landfill (methane) gas, a natural by-product of the decomposition of organic material in landfills, is a source of odors and a potential greenhouse gas. The landfill methane gas management can be done simultaneously with the leachate management.

- *Collection:* Landfills are equipped with gas collection systems that capture the gas produced by the decomposing waste.
- *Treatment:* The collected gas is then treated to remove contaminants and moisture, making it suitable for use or safe for release into the atmosphere.
- *Conversion:* In many cases, the treated landfill gas is used as a source of renewable energy. It can be burned in a flare to produce heat, used to generate electricity, or processed into pipeline-quality gas.
- *Commercialization:* The process of commercializing methane capture technologies can be complex and time-consuming, often taking up to 20 years to get a product out of the lab and onto the market. However, with strategic collaboration, clear identification of the product's market position, and access to excellent manufacturing and engineering capabilities, a new technology is more likely to cross the finish line.

These management practices are essential for minimizing the environmental impact of landfills and ensuring compliance with waste management regulations. However, they also highlight the importance of reducing waste and increasing recycling and composting efforts to minimize the amount of waste that ends up in landfills in the first place.

The methane gas produced in landfill sites is a pollutant if not properly managed and utilized. Methane gas can be converted into energy through a process that involves collection, treatment, and conversion.

The first step in the process is to collect the landfill gas. This is done by drilling wells into the landfill and using a network of pipes to transport the gas to a central location. The gas extraction system runs under a vacuum created by blowers at the facility, which causes landfill gas to flow from the wells.

Once collected, the landfill gas is monitored for its methane concentration. If there is insufficient methane present for combustion in the energy process, the landfill gas is simply sent to a flare. This is usually corrected by throttling back the extracted gas flow or rebalancing the valves at various locations on the collection system pipework. If the landfill gas is within the range of methane content needed for power generation, it is filtered or "scrubbed."

The treated landfill gas is then used to generate electricity. This is typically done using internal combustion engines. These engines have a shaft that rotates and is coupled to turn large magnetic windings in the electricity generators to make electricity. The electricity produced is an alternating current that flows through transformers that raise the voltage up to about 11,000 volts, sometimes higher, and through an exported power meter.

This process not only helps to manage the landfill gas produced by decomposing waste but also provides a source of renewable energy. It is a great example of how waste management can contribute to sustainability and energy production.

Converting landfill gas (LFG) into energy has several significant benefits:

- *Reduce Greenhouse Gas Emissions:* MSW landfills are a major source of methane emissions, a potent greenhouse gas. Using LFG for energy generation can capture and destroy methane, reducing greenhouse gas emissions.
- *Reduce Air Pollution:* By offsetting the use of non-renewable resources like coal, oil, or natural gas, LFG energy projects can reduce air pollution. This includes emissions of CO_2, sulfur dioxide, particulate matter, and nitrogen oxides.
- *Create Health and Safety Benefits:* The reduction in air pollution can lead to improved public health outcomes. Additionally, the proper management of LFG reduces the risk of explosions caused by methane accumulation.
- *Benefit the Community and Economy:* LFG energy projects can create local jobs and stimulate the local economy. They also generate a source of clean, cost-effective power for the local community.
- *Reduce Environmental Compliance Costs:* Using LFG for energy generation can help landfill operators comply with environmental regulations and avoid potential penalties.
- *Sustainable Community Planning:* LFG utilization projects create partnerships among citizens, nonprofit organizations, local governments, and industry in sustainable community planning.

Harnessing the energy potential of LFG can turn a waste product into a valuable resource while also protecting the environment and public health.

Besides the benefits of capturing methane gas (LFG), there are also challenges, as shown in Table 9.16.

Despite these challenges, methane capture technologies hold great promise for reducing greenhouse gas emissions and mitigating climate change. Continued research and development, along with supportive policies and investment, will be key to overcoming these challenges and realizing the potential of these technologies.

9.5.3 Environmental Considerations

Environmental considerations are the first and foremost in the construction of landfill sites to ensure that potential environmental impacts are minimized and that the surrounding ecosystems and communities are protected. Here are some key considerations to minimize environmental impacts and NIMBYs (Not-In-My-Backyard issues) when constructing landfill sites:

1. *Site Selection:* Choosing an appropriate location for a landfill is critical. Environmental impact assessments are conducted to identify suitable sites that minimize potential harm to sensitive ecosystems, water sources, and human populations. Factors such as proximity to residential areas, wetlands, floodplains, and environmentally sensitive areas are carefully evaluated.
2. *Land Use and Zoning:* Compliance with local zoning regulations and land use planning is essential to ensure that the landfill site is located in an area designated for waste management activities. This helps prevent conflicts with other land use and protects nearby communities from potential hazards.

TABLE 9.16
Challenges in Capturing LFG

Technical Efficiency: These technologies often require high temperatures or other stringent conditions to operate. They may also contain expensive metals like platinum and do not work well at the concentrations of methane found in ambient air.

Scaling Up: Scaling up these technologies to a level where they can have a significant impact on global methane levels is a major challenge. The volume of air that would need to be processed would be enormous.

Cost: The cost of implementing these technologies can be high, which can be a barrier to widespread adoption.

Resource Usage: These technologies can require significant amounts of energy, land, and water, which could have other environmental impacts.

Social Barriers: There can be social barriers to the implementation of these technologies, such as public acceptance and regulatory hurdles.

Uncertainties in Emissions and Sinks: There are large uncertainties that exist in both anthropogenic and natural emissions and sinks, that is, large differences in estimates of methane emissions and sinks between different inventories, models, and atmospheric inversions.

Commercialization: The process of commercializing methane capture technologies can be complex and time-consuming, often taking up to 20 years to get a product out of the lab and onto the market. Most prototypes fail at the first hurdle – getting industry backing.

3. *Geomorphology and Hydrogeology:* Understanding the local geomorphology and hydrogeology is vital to prevent groundwater contamination and soil erosion. Proper site grading, drainage systems, and impermeable liners are implemented to contain leachate and prevent its migration into groundwater sources.
4. *Water Management:* Effective stormwater management and erosion control measures are put in place to prevent surface water runoff from carrying pollutants into adjacent water bodies. Sedimentation ponds, vegetative buffers, and engineered drainage systems are often utilized to minimize the impact on water quality.
5. *LFG Management:* Landfills produce landfill gas, primarily methane and carbon dioxide, which can contribute to air pollution and climate change if not properly managed. Gas collection systems are installed to capture and treat landfill gas, reducing its release into the atmosphere.
6. *Waste Minimization and Recycling:* Incorporating waste minimization and recycling programs can reduce the volume of waste requiring disposal, thereby extending the lifespan of the landfill and minimizing its environmental footprint.
7. *Vegetation and Habitat Preservation:* Efforts are made to preserve natural vegetation and wildlife habitats in and around the landfill site. Restoration of the site after closure is also considered to mitigate the long-term environmental impact.
8. *Monitoring and Compliance:* Regular monitoring of environmental parameters such as groundwater quality, air emissions, and surface water quality is essential to ensure compliance with environmental regulations and to detect and address any potential environmental impacts.

The above considerations during the construction and operation of landfill sites are expected to mitigate potential negative effects on the environment and surrounding communities while ensuring the safe and responsible management of solid waste.

9.5.4 Regulations

Regulations associated with the construction and operation of a landfill site typically include a range of environmental, safety, and operational requirements. Here are some common regulations that may apply:

9.5.4.1 Environmental Impact Assessment

Before construction, an environmental impact assessment (EIA) may be required to evaluate the potential environmental effects of the landfill site. EIA is a comprehensive process that involves the systematic evaluation of the potential environmental, social, and economic impacts of the proposed landfill site before the commencement of construction activities. This assessment typically includes the identification, prediction, and assessment of potential impacts on various environmental components such as air quality, water resources, soil quality, biodiversity, as well as the potential effects on nearby communities and cultural heritage. The EIA process also involves the identification of potential mitigation measures to minimize adverse impacts and may require public consultation and the preparation of an Environmental Impact Statement (EIS) to provide detailed information to regulatory authorities and stakeholders. The goal of the EIA is to ensure that decision-makers have a thorough understanding of the potential environmental consequences of the landfill project and to facilitate informed decision-making regarding its approval and implementation.

An EIA needs to be taken before the construction of a landfill site. This assessment is a systematic process that identifies, predicts, and evaluates the potential environmental effects of the proposed landfill site. The EIA process involves several steps:

1. *Screening:* This initial step determines whether the proposed project requires an EIA and identifies the key environmental issues that need to be addressed.
2. *Scoping:* This step identifies the key environmental issues, the information needed, and the boundaries of the assessment.
3. *Impact Analysis:* This involves predicting and identifying the potential environmental impacts of the proposed landfill site. It includes assessing the magnitude, extent, and duration of these impacts.
4. *Mitigation Measures:* This step involves identifying measures to avoid, reduce, or offset the significant adverse impacts of the landfill site.
5. *EIA Report:* This is a document that presents the findings of the EIA process. It includes a description of the environment, an explanation of the methodology used for the assessment, a description of the proposed project, an outline of the alternatives considered, a summary of the likely significant effects of the project on the environment, a description of the mitigation measures, and a non-technical summary.
6. *Review of EIA Report:* This involves a technical evaluation of the EIA report to ensure that the assessment has been carried out satisfactorily and that the information provided is complete and reliable.
7. *Decision-Making:* The competent authority will make a decision on whether to approve the project and what conditions to apply.
8. *Monitoring, Compliance, Enforcement, and Environmental Auditing:* If the project is approved, it is necessary to ensure that the mitigation measures are implemented and have the desired effect.

The EIA ensures that potential environmental impacts are considered before the landfill site is constructed. It helps decision-makers understand the potential environmental implications of the proposed landfill site, thereby aiding in the decision-making process. It also helps to minimize adverse impacts and enhance positive impacts, contributing to sustainable development.

9.5.4.2 Permitting and Zoning

Obtaining permits and complying with zoning regulations are essential. This includes obtaining permits for construction and waste disposal and complying with local zoning laws.

Permitting and zoning are essential in the process of establishing a landfill site. They ensure that the proposed site complies with local, regional, and national regulations and standards.

1. Permitting involves obtaining necessary permissions from relevant authorities before starting the construction of the landfill site. There are several types of permits that may be required:
   a. *Construction Permits:* These are required to ensure that the construction of the landfill site complies with local building codes and regulations. They cover aspects such as site preparation, excavation, and the construction of infrastructure like roads and utilities.
   b. *Waste Disposal Permits:* These are specific to the operation of a landfill site. They regulate the types of waste that can be accepted, the methods of waste treatment and disposal, and the monitoring and management of the site. These permits are usually issued by environmental protection agencies and require detailed plans for waste management and site operation.
2. *Zoning:* This refers to the regulation of land use within a specific area. Zoning laws dictate what types of activities can take place in different areas of a city or county. For a landfill site, it is essential to comply with these laws. Zoning regulations may cover aspects such as the following:
   a. *Land Use:* The proposed landfill site must be in an area that is zoned for waste management or industrial use. This is to ensure that the landfill does not negatively impact residential areas or other sensitive land uses.
   b. *Setbacks:* These are minimum distances that must be maintained between the landfill site and other features such as property lines, water bodies, and residential areas. Setbacks are designed to protect these features from potential impacts of the landfill.
   c. *Height and Volume Restrictions:* These regulate the maximum size of the landfill. They are designed to ensure that the landfill does not become a visual nuisance or pose a risk to air traffic.

Permits and complying with zoning regulations in the establishment of a landfill site ensure that the site is constructed and operated in a manner that minimizes its impact on the environment and the surrounding community. Failure to comply with these requirements can result in penalties, delays, and even the denial of permission to operate the landfill site. Therefore, one needs to understand and adhere to these regulations when planning for a landfill site.

9.5.4.3 Liner and Leachate Collection

Sanitary landfills require impermeable liners and leachate collection systems to prevent groundwater contamination. This becomes an integral component of modern landfill design, aimed at protecting the environment and public health from potential contamination.

Liners are barriers that are installed at the bottom and sides of the landfill to prevent the escape of leachate (the liquid that is generated within the landfill) into the surrounding environment. The liner is typically made of a synthetic material such as high-density polyethylene (HDPE) that is highly resistant to puncture, tear, and chemical degradation. It is designed to be impermeable, meaning it does not allow liquids to pass through. The liner is laid on a compacted clay base, providing a double layer of protection. The purpose of the liner is to contain the waste and the leachate within the landfill, preventing them from contaminating the soil and groundwater.

Leachate Collection Systems are systems designed to collect and manage the leachate that is generated within the landfill. Leachate is a liquid that forms when water (from rainfall or from the waste itself) percolates through the waste, picking up dissolved or suspended contaminants. The leachate collection system typically consists of a network of perforated pipes, gravel packs, and a layer of sand or gravel placed above the liner. The pipes collect the leachate that accumulates at the bottom of the landfill and transport it to a leachate treatment facility. The treatment process removes harmful substances from the leachate before it is discharged or reused.

These measures are crucial because by containing the waste and leachate within the landfill, liners and leachate collection systems prevent the contamination of soil and groundwater. This helps to preserve the quality of the environment and protect ecosystems. Contaminated groundwater can pose serious risks to public health if it enters the drinking water supply. By preventing the contamination of groundwater, liners and leachate collection systems help to safeguard public health.

Regulatory Compliance: Many jurisdictions require landfills to have liners and leachate collection systems as part of their waste management regulations. These measures help landfill operators to comply with these regulations and avoid penalties.

9.5.4.4 Methane Gas (LFG) Management

Regulations may require the implementation of methane gas management systems to control and monitor the release of methane gas generated by the landfill.

Methane gas is a potent greenhouse gas that is produced as a by-product of the decomposition of organic waste in landfills. It is approximately 25 times more potent at trapping heat in the atmosphere than carbon dioxide, making its management a critical aspect of landfill operations.

Various local, national, and international regulations may be in place to govern the management of methane gas from landfills. These regulations are designed to minimize the environmental impact of landfills and ensure the safety of nearby communities. They may specify standards for the design, operation, and monitoring of methane gas management systems.

To comply with these regulations, landfill operators may be required to implement comprehensive methane gas management systems. These systems typically involve the installation of gas collection wells throughout the landfill, which capture the methane gas as it is produced. The collected gas can then be flared (burned off), used on-site for energy generation, or processed and sold as a renewable energy source.

The purpose of methane gas management systems is to control and monitor the release of methane gas generated by the landfill. This involves regularly measuring the amount of gas being produced and ensuring that it is being effectively captured and managed. Monitoring can help identify any issues with the system, such as leaks, which can then be addressed to prevent uncontrolled releases of methane gas.

The methane gas that these systems manage is generated by the landfill itself. As organic waste in the landfill decomposes, it produces a variety of gases, including methane. The amount of methane gas produced can vary depending on factors such as the type and amount of waste, the age of the landfill, and environmental conditions.

9.5.4.5 Operational Standards

Operational regulations may cover waste acceptance criteria, waste compaction, cover material requirements, and daily cover procedures.

These are rules and guidelines set by environmental agencies or governing bodies that dictate how a landfill or waste management facility should operate. They are designed to minimize the environmental impact, ensure public health and safety, and promote efficient waste management practices.

Waste Acceptance Criteria refer to the types of waste that a landfill or waste management facility is permitted to accept. The criteria can be based on the nature of the waste (e.g., hazardous, nonhazardous), its source (e.g., industrial, residential), and its physical and chemical properties. The purpose is to prevent the disposal of inappropriate waste that could pose risks to the environment and public health.

Waste Compaction involves the use of heavy machinery to compress the waste and reduce its volume. Compaction increases the landfill's capacity, improves stability, and reduces the amount of space that waste takes up in the landfill. It also helps to minimize the production of leachate and methane gas.

Landfills are required to cover deposited waste with layers of soil or other approved materials. The cover material helps to control odors, prevent wind-blown litter, discourage pests, and reduce the infiltration of rainwater. The type, thickness, and frequency of the cover are usually specified in the operational regulations.

Daily cover procedures refer to the practice of covering the waste deposited in the landfill at the end of each operating day. The daily cover helps to minimize odors, prevent the spread of disease by vectors such as birds and rodents, and reduce the risk of fires. It also enhances public acceptance by improving the aesthetics of the landfill site.

9.5.4.6 Monitoring and Reporting

Regular monitoring of groundwater, surface water, air quality, and leachate may be required, along with reporting to regulatory authorities.

Monitoring and Reporting involves the regular collection and analysis of data to assess the impact of operations on the environment. The results are then reported to regulatory authorities to demonstrate compliance with environmental regulations.

Regular Monitoring involves systematic checking or tracking of environmental parameters over time. It is done to understand the ongoing effects of an operation or process on the environment. Regular monitoring can help identify trends, measure the effectiveness of mitigation measures, and detect any potential issues early.

Groundwater refers to the water found underground in the cracks and spaces in soil, sand, and rock. It is stored in and moves slowly through geologic formations of soil, sand, and rocks called aquifers. Monitoring of groundwater involves checking for any contaminants that may have leached into the groundwater from the operation.

Surface Water includes bodies of water such as rivers, lakes, and reservoirs. Monitoring of surface water usually involves testing various physical, chemical, and biological parameters to assess water quality.

Air Quality monitoring involves monitoring the levels of pollutants in the air. These can include gases (like carbon monoxide, sulfur dioxide, and nitrogen oxides) and particulate matter. Poor air quality can have significant impacts on human health and the environment.

Leachate is the liquid that drains or "leaches" from a landfill. It usually contains both dissolved and suspended materials. Leachate can pose a significant environmental hazard if it contaminates groundwater or surface water, so it is important to monitor its composition and manage its disposal carefully.

Reporting to regulatory authorities involve compiling the data collected during monitoring into reports and submitting them to the relevant environmental authorities. These reports help the authorities to verify that the operation is complying with environmental regulations and to act if necessary.

9.5.4.7 Closure and Post-Closure Care

Regulations often specify the requirements for landfill closure, including capping, landscaping, and ongoing post-closure care and monitoring.

Closure and post-closure care refers to the activities that are required when a landfill reaches its maximum capacity and is no longer in operation. The aim is to minimize the potential environmental impact of the landfill after it has been closed.

Regulations are rules set by environmental agencies or governing bodies that specify the procedures for closing a landfill and the post-closure care that must be taken. These regulations are designed to protect the environment and public health.

When a landfill is closed, it must be properly sealed off to prevent any further waste from being deposited and to minimize the release of leachate and landfill gas. This usually involves several steps:

1. *Capping:* This involves covering the landfill with a layer of soil and then a layer of clay or synthetic material. The cap helps to prevent rainwater from infiltrating the waste and creating leachate. It also helps to control the release of landfill gas.

2. *Landscaping:* After capping, the landfill is usually landscaped to blend in with the surrounding environment. This can involve planting grass or trees on the cap and shaping the landfill so that it looks like a natural hill or mound.

Ongoing post-closure care and monitoring is still necessary to be looked after even after a landfill has been closed. This is known as post-closure care and can last for several decades. It involves the following:

1. *Monitoring:* Regular checks are made to ensure that the cap is intact and that the landfill gas and leachate control systems are working properly. Groundwater and surface water around the landfill may also be tested to check for any signs of contamination.
2. *Maintenance:* Any necessary repairs or maintenance are carried out. This could involve repairing the cap, maintaining the landfill gas and leachate control systems, or dealing with any issues that arise from the monitoring.

The closure and post-closure care of a landfill is a complex process that requires careful planning and ongoing commitment. It is governed by regulations to ensure that it is carried out in a way that protects the environment and public health. It is an essential part of responsible waste management.

9.6 ZERO WASTE

The concept of zero waste has been a visionary target for MSWM in numerous cities, gaining prominence in urban development discourse during the 2000–2010 period. The zero-waste movement began in the 1990s when environmental activists in Europe advocated for the complete elimination of waste. The focus during this time was on environmental sustainability, emphasizing sustainable consumption and production. The goal was not to completely eliminate waste at the source, but rather to minimize waste production by reducing consumption.

The motivation behind zero waste was the recognition that land resources, particularly in urban areas, are limited and cannot continually accommodate the increasing demand for landfill sites. This is coupled with the fact that no one willingly wants to live next to a landfill site, a phenomenon known as Not-In-My-Backyard (NIMBYism). This led to efforts to minimize waste disposal to landfill sites to zero, thereby eliminating the constant need for landfill sites and the potential social and environmental impacts associated with them. While landfill sites may provide income for scavengers and recyclable collectors, it appears that the negative effects generated by these activities far outweigh the benefits gained by the scavengers.

Zero waste is a set of principles focused on waste prevention. The goal is to minimize waste production as much as possible and then compost, reuse, or recycle any other waste that is generated. This concept encourages the redesign of resource life cycles so that all products are repurposed (also known as "up cycled") and/or reused. Zero waste does not necessarily mean we produce no waste. It is impossible as long as human beings need to produce and consume something. Therefore, zero waste would rather produce no waste to landfill sites.

The zero-waste movement aims to avoid sending trash to landfills, incinerators, oceans, or any other part of the environment. It is a whole-system approach that aims for a massive change in the way materials flow through society, resulting in no waste.

The framework for zero waste includes the 5Rs: refuse, reduce, reuse, recycle, and rot, as shown in Table 9.17.

These principles were proposed by Bea Johnson in her book "Zero Waste Home" and have become the rallying cry of zero wasters around the globe. By implementing these principles, we can reduce the negative impact on the planet, conserve resources, and reduce the presence of toxins. The goal is to do the best that you can with whatever resources you have.

TABLE 9.17
The Zero-Waste Framework

Refuse What You Do Not Need: This prevents unwanted items from coming into your home and applies to all those promotional items you are offered, along with things like junk mail and plastic straws.
Reduce What You Do Use: This equals less waste overall.
Reuse Whatever You Can: Can you extend the lifespan of something by mending, handing down, or repairing? Can you buy or sell second hand? Reusing also means swapping disposable products for reusable ones that can easily be laundered instead of thrown away.
Recycle What You Cannot Refuse or Reduce: Saving resources from the waste stream by recycling is one option, but keep in mind that some items will not be recycled indefinitely and will end up in the landfill at the end of their (often short) lifespan. These are the items that zero wasters try to avoid.
Rot What Is Left Over: Composting food scraps, paper pieces, and wooden or bamboo toothbrushes returns nutrients and fiber back to the earth.

TABLE 9.18
Benefits and Challenges of Zero Waste

Benefits	Challenges
1. *Improved Physical Health:* Zero waste often leads to healthier eating habits as packaged food often means a lot of packaging.	1. *Political Barriers:* There can be political barriers in the macro environment that make it difficult to implement zero waste practices.
2. *Improved Well-Being:* Mindfulness in consumption can lead to stress reduction, increased working memory and focus, less emotional reactivity, and more.	2. *Cultural Barriers:* Cultural norms and attitudes toward waste can also pose a challenge.
3. *Fewer Toxic Chemicals:* Zero waste cleaning and self-care products tend to have fewer potentially damaging chemicals.	3. *Stakeholder Resistance:* Stakeholders, industries, and municipalities may resist changes required for zero waste.
4. *Reduced Unnecessary Spending:* Many people participate in emotional spending, which is when you buy something you do not want or need because of negative (or sometimes positive) emotions.	4. *Infrastructure:* The lack of necessary infrastructure for waste reduction, recycling, and composting can be a significant challenge.
5. *Environmental Conservation:* By minimizing waste generation, zero waste living significantly contributes to environmental conservation.	

Zero waste goes beyond just reducing, reusing, and recycling. It should focus on restructuring distribution and production systems to reduce waste. This approach is not only sustainable but also considerate of life and future generations.

The potential benefits and challenges of implementing zero waste are shown in Table 9.18.

The zero-waste lifestyle may look different depending on where we live, but everyone can participate on some level. Whether we reside in a bustling city or a rural community, embracing a zero-waste mindset can involve simple yet impactful changes. In urban areas, individuals can opt for reusable alternatives to single-use items, such as water bottles, shopping bags, and coffee cups. They can also support businesses that offer package-free products and bulk purchasing options. In contrast, those in rural settings may focus on reducing food waste through composting, growing their own produce, and repairing or repurposing items rather than discarding them. Additionally, advocating for sustainable practices within local government and community organizations can amplify the impact of zero waste efforts. Regardless of the specific actions taken, the collective commitment to minimizing waste fosters a healthier environment for all.

9.7 REDUCE, REUSE, AND RECYCLE

Reduce, reuse, and recycle are the core activities of the zero-waste program, which are now becoming 5Rs by adding refuse and rot. The principles of reduce, reuse, and recycle, often referred to as the 3Rs, are the core of sustainable waste management.

1. *Reduce:* This is the most effective of the 3Rs. The idea is to use less, thereby generating less waste. It is about being mindful of our resources and not buying or using unnecessary items. For instance, using a reusable water bottle instead of single-use plastic bottles is an example of reducing.
2. *Reuse:* This involves using items as many times as possible before replacing them. This can be done by repairing, donating, selling, or repurposing items. For example, we can reuse old cardboard boxes for storage or DIY projects.
3. *Recycle:* This is the process of collecting and processing materials that would otherwise be thrown away and turning them into new products. While recycling is beneficial, it should be the last resort after reducing and reusing, as recycling uses more energy and resources.

9.7.1 Waste Reductions

9.7.1.1 Understanding Reduction

Reduction, in the context of waste management, refers to the strategies and actions taken to lessen the amount of waste that is produced. It is the first step in the waste hierarchy, which is often represented as "Reduce, Reuse, Recycle." The idea is to minimize waste at its source by making conscious decisions that lead to the creation of less waste.

Reduction aims to prevent the generation of waste at its source. This could be achieved by choosing products with less packaging, opting for durable goods over disposable ones, or simply buying only what is necessary. Additionally, embracing a "repair, reuse, recycle" mentality can further contribute to waste reduction by extending the lifespan of items and diverting them from landfills. Encouraging businesses to implement sustainable packaging and production practices, as well as advocating for policies that promote waste reduction and resource conservation, are also crucial steps in achieving this goal. By prioritizing reduction, individuals and organizations can significantly minimize their environmental impact and pave the way for a more sustainable future.

By reducing waste, we can significantly decrease the amount of waste that ends up in landfills or incinerators. This can help conserve natural resources, reduce pollution, and mitigate the effects of climate change. Furthermore, waste reduction can lead to energy savings, as the production and disposal of goods often require significant amounts of energy. Additionally, minimizing the amount of waste sent to landfills can alleviate the release of greenhouse gases, such as methane, which are produced during the decomposition of organic waste. By implementing effective waste reduction strategies, such as composting organic materials and promoting the reuse of items, communities and businesses can play a pivotal role in fostering a more sustainable and environmentally conscious society for future generations.

Reduction can also lead to cost savings. By buying less and choosing durable goods, consumers can save money, as they are purchasing fewer items and investing in products that have a longer lifespan. Businesses can also cut costs by improving their production processes to generate less waste, reducing the need for raw materials and disposal expenses. Additionally, implementing efficient packaging and distribution methods can result in significant savings. By prioritizing waste reduction, both individuals and organizations can not only contribute to environmental conservation but also benefit economically through reduced expenses and improved resource management.

Producing new items requires energy, often derived from burning fossil fuels. By reducing consumption, we can decrease the demand for new items, thereby conserving energy and reducing greenhouse gas emissions. This shift toward reduced consumption also lessens the environmental

TABLE 9.19
Reduction Strategies by Businesses and Commercials

1	*Efficient Production Processes:* Businesses can optimize their production processes to use resources more efficiently, thereby reducing waste. This could involve using advanced technology or rethinking the production process to minimize waste.
2	*Sustainable Procurement:* Businesses can choose to purchase goods and services that are environmentally friendly. This could mean choosing suppliers who also practice waste reduction or opting for products that have less packaging.
3	*Product Design:* Designing products to use fewer materials or to be more durable can significantly reduce waste. For example, a business could design a product to be easily disassembled for repair, which would extend the product's life and reduce the need for replacement.
4	*Employee Training:* Businesses can train their employees on the importance of waste reduction and how to practice it in their work. This could involve training on efficient use of resources, proper waste segregation, and recycling.
5	*Waste Audits:* Regular waste audits can help businesses identify where waste is being generated and how it can be reduced. This involves tracking waste generation and disposal over a certain period and then analyzing the data to identify opportunities for reduction.
6	*Reuse and Recycling Programs:* Implementing programs for reusing and recycling materials within the business can help reduce waste. This could involve setting up recycling stations, composting organic waste, or finding ways to reuse materials in the production process.
7	*Green Marketing:* Businesses can also use their waste reduction efforts as a marketing tool. By communicating their commitment to sustainability to customers, businesses can not only reduce waste but also attract environmentally conscious consumers.

impact of extracting raw materials, such as minerals and timber, further preserving natural habitats and biodiversity. Additionally, extending the lifespan of products through repair and reuse diminishes the need for manufacturing new goods, ultimately contributing to a more sustainable and resource-efficient economy. Embracing a culture of mindful consumption can lead to significant environmental benefits, including reduced carbon emissions and minimized ecological degradation.

Practicing reduction promotes a more sustainable lifestyle. It encourages mindful consumption and the efficient use of resources, which are key to achieving sustainability. By embracing a mindset of "less is more," individuals can prioritize the long-term health of the planet over short-term convenience, leading to a reduced ecological footprint. Furthermore, this approach fosters a culture of appreciation for the intrinsic value of goods and resources, inspiring responsible consumption habits that align with the principles of environmental stewardship and conservation. The practice of reduction plays a fundamental role in shaping a more sustainable and harmonious relationship between humanity and the natural world.

Businesses and commercials can incorporate reduction strategies in several ways to minimize waste and promote sustainability, as seen in Table 9.19.

9.7.1.2 Benefits of Reduction

Reducing consumption means reducing waste. It is not just good for the environment but also beneficial from an economic perspective. The environmental and economic benefits of reducing consumption and therefore waste reduction are shown in Table 9.20.

It is well-known that with current consumption, theoretically we need 7 earths; it means our life is in an unsustainable state. There was a radical thought that was born in the 1970s. It broadly means shrinking rather than growing economies, so we use less of the world's energy and resources and put well-being ahead of profit. The idea is that by pursuing degrowth policies, economies can help themselves, their citizens, and the planet by becoming more sustainable. This economic school of thought was the degrowth theory.

TABLE 9.20
Environmental and Economic Benefits of Reducing Consumption

Environmental Benefits	Economic Benefits
• *Resource Conservation*: Recycling reduces the need for raw material extraction, preserving natural resources such as forests, minerals, and water. • *Energy Savings*: Processing recycled materials often require less energy than producing new materials, leading to lower greenhouse gas emissions. • *Pollution Reduction*: Proper waste management minimizes air, water, and soil pollution by reducing landfill waste and harmful emissions. • *Biodiversity Protection*: Less resource extraction means fewer disruptions to ecosystems and wildlife habitats. • *Landfill Reduction*: Diverting waste from landfills decreases methane emissions and extends landfill lifespan.	• *Cost Savings*: Individuals and businesses save money by purchasing fewer goods and reducing waste disposal costs. • *Efficiency Gains*: Companies that optimize resource use and streamline production benefit from lower operational costs. • *Circular Economy Growth*: Encourages sustainable business models that focus on reuse, repair, and recycling. • *Long-Term Economic Stability*: Reducing reliance on finite resources helps create a more resilient economy. • *Innovation Opportunities*: Drives the development of sustainable technologies and alternative materials.

Here are some key points about degrowth:

- The term "degrowth" was coined in 1972 by an Austrian-French social philosopher, André Gorz. As a movement, degrowth started to take off in the early 2000s.
- Practical degrowth actions might include buying less stuff, growing your own food, and using empty houses instead of building new ones.
- Modern degrowth protagonists include French economist Serge Latouche, who argues that society's current model of economic growth is unsustainable.
- Degrowth theory's main argument is that an infinite expansion of the economy is fundamentally contradictory to the finiteness of material resources on Earth. It argues that economic growth measured by GDP should be abandoned as a policy objective.
- Policy should instead focus on economic and social metrics such as life expectancy, health, education, housing, and ecologically sustainable work as indicators of both ecosystems and human well-being.
- Degrowth theory is highly critical of free market capitalism, and it highlights the importance of extensive public services, care work, self-organization, commons, relational goods, community, and work sharing.

9.7.1.3 Impact on Waste Reduction

The concept of reduction will be directly decreasing the amount of waste that ends up in landfills.

When we reduce our consumption, there is less demand for goods. This leads to less production and, consequently, less waste generated from the production process. Reducing the volume of goods manufactured conserves resources such as raw materials, water, and energy, contributing to a more sustainable use of natural resources. Moreover, reduced production results in lower emissions of greenhouse gases and pollutants associated with manufacturing processes, thereby mitigating environmental impact. Embracing a culture of reduced consumption not only minimizes waste but also fosters a more balanced and responsible approach to resource utilization, aligning with the principles of sustainable development and environmental stewardship.

A lot of the waste that ends up in landfills comes from packaging. By choosing products with less packaging or reusable packaging, we can significantly reduce the amount of waste we produce. This shift toward minimal and reusable packaging not only lessens the burden on landfills but also conserves resources and reduces the environmental impact of manufacturing and disposal. Also, supporting businesses that prioritize sustainable packaging practices can encourage industry-wide adoption of eco-friendly alternatives and promote a more circular economy, where materials are reused and recycled to minimize waste generation.

By buying and cooking only what we need, we can significantly reduce food waste, a large part of which ends up in landfills. This approach not only minimizes the environmental impact of food production and disposal but also conserves resources such as water and energy that are used in the cultivation, processing, and transportation of uneaten food.

When we choose high-quality products and take good care of them, they last longer. This means we do not need to throw them away and replace them as often. Investing in durable and well-made items reduces the frequency of disposal, thereby decreasing the amount of waste sent to landfills. In addition, extending the lifespan of products through proper maintenance and repair conserves resources and reduces environmental impact. This approach promotes more sustainable consumption, aligning with the principles of responsible stewardship and resource conservation.

Many products, especially electronic goods, contain hazardous materials that can be harmful if they end up in landfills. By reducing our consumption of these products, we can decrease the amount of hazardous waste in landfills. This reduction in hazardous waste not only protects the environment and human health but also lessens the need for specialized disposal methods and the associated environmental risks. Furthermore, embracing a mindset of responsible consumption and proper disposal of electronic waste through recycling and proper e-waste management can contribute to a healthier and more sustainable approach to technology use and disposal.

Reduction also promotes the repair and maintenance of items, which extends their life and prevents them from becoming waste. Embracing a culture of repair and maintenance encourages individuals to value and care for their possessions, fostering a more sustainable approach to consumption. Repairing and prolonging the lifespan of goods reduces the volume of discarded items, conserves resources, and minimizes the environmental impact associated with manufacturing and disposal. This approach contributes to a more circular and resource-efficient economy.

Reduction encourages mindful consumption, where we think about the lifecycle of a product before we buy it. This can lead to choices that generate less waste. By considering factors such as durability, recyclability, and environmental impact, individuals can make informed decisions that prioritize sustainability. This approach fosters a more conscientious relationship with consumer goods, prompting consideration of the long-term consequences of purchases and encouraging the selection of items with minimal environmental impact throughout their lifecycle, ultimately contributing to a more sustainable and responsible consumption pattern.

9.7.1.4 Role of Consumers

Consumers are the focus of practicing consumption reduction and making sustainable choices. Their roles and contributions are as follows.

- *Mindful Consumption:* Consumers can practice mindful consumption by considering the environmental impact of their purchases. This involves thinking about the lifecycle of a product, from production to disposal, before making a purchase.
- *Choosing Sustainable Products:* Consumers can choose to buy products that are made sustainably or ethically. This could mean choosing products made from recycled materials, products that are designed to be durable and long-lasting, or products made by companies that follow sustainable practices.

- *Reducing Waste:* Consumers can reduce waste by buying only what they need, reusing items as much as possible, and recycling. They can also choose products with less packaging or packaging that is recyclable.
- *Supporting Responsible Companies:* By choosing to buy from companies that prioritize sustainability, consumers can encourage more businesses to adopt sustainable practices.
- *Educating Themselves and Others:* Consumers can educate themselves about the environmental impact of different products and practices and share this knowledge with others. This can help create a culture of sustainability.
- *Participating in Community Initiatives:* Consumers can participate in community initiatives such as local cleanup drives, recycling programs, and other environmental projects.
- *Advocacy:* Consumers can advocate for policies and regulations that promote sustainability and hold companies accountable for their environmental impact.
- *Energy Conservation:* Consumers can reduce their energy consumption by using energy-efficient appliances, reducing their use of heating and cooling, and using public transportation or carpooling.
- *Water Conservation:* Consumers can conserve water by fixing leaks, installing water-efficient appliances, and using water-saving techniques in their gardens.
- *Food Choices:* Consumers can make sustainable food choices by reducing their consumption of meat, choosing locally grown produce, and reducing food waste.

9.7.2 Reuse

9.7.2.1 Concept of Reuse

Reuse refers to the process of using an item more than once, either for the same purpose or for a different one. This could involve using a glass jar to store leftovers, using old T-shirts as rags, or donating unwanted clothes and furniture for others to use. Reuse helps extend the life of an item and delay or avoid its disposal.

On the other hand, recycling involves the process of collecting and processing materials (like paper, glass, metal, and plastic) that would otherwise be thrown away as trash and turning them into new products. This often involves breaking down the original product and re-manufacturing it into a new product, which can require significant energy.

The key difference between the two lies in the fact that reuse involves using an item as is without breaking it down, while recycling involves breaking down an item to its raw materials and then using these materials to create new items. Both are important strategies in waste management and contribute to the conservation of resources, but reuse is often considered more environmentally friendly because it requires less energy than recycling.

9.7.2.2 Benefits and Challenges of Reuse

While there are environmental and economic benefits to reusing items, there are also challenges that need to be addressed. The economic and environmental benefits of reusing items and some challenges are shown in Tables 9.21 and 9.22.

Overcoming these challenges requires efforts from individuals, businesses, and governments to promote and facilitate reuse.

9.7.2.3 Promoting Reuse

Businesses and consumers can play pivotal roles in promoting reuse, fostering a more sustainable and resource-efficient economy. Firstly, businesses can implement strategies to encourage reuse by offering reusable alternatives to single-use products, such as refillable containers, packaging-free options, and durable goods designed for long-term use. Incorporating reusable packaging and products into their offerings allows businesses to reduce the volume of disposable items and

TABLE 9.21
Environmental and Economic Benefits of Reuse

Environmental Benefits	Economic Benefits
• *Conservation of Resources:* Reusing items means less new products are needed, which reduces the demand for raw materials and helps conserve natural resources. • *Reduced Waste:* Reuse extends the life of products and prevents them from becoming waste, reducing the amount of waste that ends up in landfills or incinerators. • *Energy Savings:* Manufacturing new products requires energy. By reusing items, we can save the energy that would have been used to produce new ones. • *Lowered Pollution:* The production of new items often involves processes that release pollutants into the environment. Reuse can help reduce this pollution.	• *Cost Savings:* Reusing items can save money. This applies to individuals who reuse items instead of buying new ones, as well as businesses that can save on material costs. • *Job Creation:* Reuse can create jobs, for example in repair services, secondhand sales, and rental services. • *Market Opportunities:* The demand for reused and refurbished products can open new market opportunities and stimulate economic activity.

TABLE 9.22
Challenges of Reusing Items

Perception and Behavior: One of the main challenges is changing people's perceptions and behaviors. Many people prefer new items and may perceive reused items as inferior or unhygienic.
Health and Safety Concerns: Some items may not be safe to reuse due to health and safety concerns, such as certain types of food containers or medical equipment.
Lack of Standards: There may be a lack of standards or guidelines for reusing certain items, which can make it difficult for businesses and consumers to know what is safe and effective to reuse.
Logistical Challenges: Collecting, sorting, and distributing reusable items can present logistical challenges, particularly for businesses.

minimize waste generation. Also, businesses can establish take-back programs for products, enabling customers to return items for refurbishment, repair, or repurposing, thereby extending their lifespan and reducing the need for new manufacturing.

Businesses can collaborate with consumers to promote reuse through educational initiatives and awareness campaigns. By highlighting the environmental benefits of reusing products and packaging, businesses can empower consumers to make informed choices that support a circular economy. This can involve providing guidance on proper product maintenance, repair services, and encouraging the reuse of packaging materials. By engaging with consumers and fostering a culture of reuse, businesses can contribute to a more sustainable and environmentally conscious society.

On the consumer side, individuals can actively seek out and support businesses that prioritize reusable products and packaging. By consciously choosing reusable alternatives and favoring products with minimal packaging, consumers can directly influence market demand and encourage businesses to expand their offerings of reusable options. Additionally, consumers can engage in practices such as upcycling, repurposing, and sharing items within their communities, contributing to the extension of product lifespans and the reduction of waste. Embracing a mindset of reuse and advocating for sustainable practices enables consumers to drive positive change and contribute to a more circular and resource-efficient economy.

When we reuse items, we extend their life, delaying the need for disposal. For example, reusing a glass jar for storage or an old t-shirt as a cleaning rag means these items are not being thrown away. This practice not only conserves resources by reducing the demand for new products but also minimizes waste generation, contributing to a more sustainable and environmentally responsible approach to consumption and resource utilization.

Reuse decreases the demand for new items. When we reuse, we do not need to buy new items as frequently, which means fewer items are produced and eventually disposed of. This reduction in consumption not only conserves resources and minimizes the environmental impact of manufacturing processes but also contributes to a more sustainable and circular economy. By extending the lifespan of products through reuse, we lessen the burden on natural resources, reduce energy consumption, and mitigate the emissions and waste associated with the production and disposal of goods, fostering a more responsible and environmentally conscious approach to consumption.

Instead of throwing items away when they are no longer needed, we can find new uses for them. This prevents these items from becoming waste and ending up in landfills. Reimagining items for new purposes and finding alternative uses not only shrinks the mountain of waste heading to landfills, preserving valuable space, but also mitigates the environmental impact of disposal. Embracing a reuse culture champions sustainability – reducing the need for new products, conserving raw materials, and significantly lowering the energy and emissions associated with manufacturing. These efforts collectively pave the way for a more circular and resource-efficient economy.

Producing new items requires raw materials and energy. By reusing items, we reduce the need for these resources and thus conserve them. This approach not only lessens the demand for finite resources, such as timber, minerals, and fossil fuels, but also minimizes the environmental impact of extraction and processing. Additionally, by extending the lifespan of products through reuse, we decrease the amount of waste generated from manufacturing processes, further contributing to resource conservation and environmental sustainability. Embracing a culture of reuse is integral to fostering a more sustainable and responsible approach to resource utilization, aligning with the principles of environmental stewardship and conservation.

By reusing items, we reduce the need for new items to be produced, thereby reducing pollution. This practice lessens the emissions of greenhouse gases, air pollutants, and water contaminants associated with manufacturing processes, contributing to improved air and water quality. Moreover, reducing the demand for new production through reuse plays a vital role in lessening the environmental impact of industrial activities. This approach fosters a more sustainable and environmentally responsible method for resource utilization and waste management.

Reuse can be economically efficient. Buying used items, repairing broken items, or repurposing items can often be cheaper than buying new ones. This approach not only saves money for individuals and businesses but also contributes to a more sustainable economy by extending the usefulness of products. Additionally, embracing a culture of reuse stimulates secondary markets for used goods, fostering economic activity and providing affordable options for consumers. Focusing on reuse empowers individuals and businesses to save money, ease financial pressure, and promote a more resource-efficient and financially sustainable way of consuming and producing.

Reusing items is a powerful strategy for waste reduction. It not only reduces the amount of waste that needs to be disposed of but also has numerous environmental and economic benefits. It is a practice that everyone can adopt to contribute to a more sustainable world.

9.7.3 Recycle

9.7.3.1 Understanding Recycling

Recycling is a primary component of modern waste management and zero waste. It involves converting waste materials into reusable materials and objects. The recycling process can be broken down into a few steps:

1. *Collection and Sorting:* The first step in the recycling process is the collection of recyclable materials. These materials are then sorted based on their type, such as paper, plastic, glass, and metal.
2. *Cleaning:* Once sorted, the recyclable materials are cleaned to remove any impurities, such as food residue on a plastic container.
3. *Processing:* The cleaned materials are then processed to break them down into their base materials. For example, paper is broken down into pulp, and plastic is broken down into pellets.
4. *Manufacturing:* These base materials are then used to manufacture new products. For example, the pulp from recycled paper can be used to make new paper products, and the pellets from recycled plastic can be used to make new plastic products.
5. *Distribution and Sale:* The newly manufactured products are then distributed and sold, completing the recycling loop.

Recycling minimizes the quantity of waste that ultimately reaches landfills or incinerators. By engaging in the process of recycling, we actively contribute to the transformation of waste into new products, thereby effectively diminishing the overall volume of waste. This practice not only aids in conserving natural resources but also significantly reduces the strain on landfill capacities and minimizes the environmental impact of incineration. Moreover, recycling serves as a sustainable solution to manage waste, fostering a circular economy where materials are reused, thus decreasing the demand for raw resources. Embracing recycling as a fundamental aspect of our daily lives promotes environmental sustainability and fosters a more responsible approach to consumption and waste management, ultimately contributing to the well-being of our planet and future generations.

Recycling is instrumental in the conservation of natural resources by utilizing recycled materials in lieu of new ones. This practice allows for substantial savings on raw materials such as timber, water, and minerals, thereby lessening the strain on our natural resources. By incorporating recycled materials into the production process, we effectively reduce the need for resource extraction, subsequently mitigating the environmental impact associated with mining, logging, and other extraction activities. Embracing recycling as a fundamental component of resource management promotes sustainable utilization of our planet's finite resources, fostering a more environmentally conscious approach to material consumption and production.

The manufacturing of products from recycled materials frequently consumes less energy compared to manufacturing the same products from new materials. This is primarily due to the reduced energy requirements for processing recycled materials, which typically undergo less extensive refining and processing than raw materials. By utilizing recycled materials, manufacturers can significantly decrease energy consumption, thereby lessening the environmental impact associated with production processes. Additionally, the use of recycled materials in manufacturing contributes to a more sustainable and eco-friendly approach, aligning with global efforts to reduce carbon emissions and promote energy efficiency in industrial operations. Embracing recycled materials in manufacturing supports the transition toward more environmentally responsible production practices.

By diminishing the necessity for extracting and processing raw materials, recycling plays a pivotal role in reducing pollution. Through the utilization of recycled materials, there is a notable reduction in the environmental impact associated with resource extraction and manufacturing processes, consequently mitigating pollution levels. This approach aligns with sustainable practices aimed at curbing air, water, and soil pollution, thereby contributing to a cleaner and healthier environment. Embracing recycling as a fundamental component of resource management fosters a more environmentally conscious approach, ultimately aiding in pollution reduction.

Recycling holds the potential for substantial economic benefits by fostering job creation within the recycling and manufacturing sectors. The establishment and expansion of recycling facilities, coupled with the demand for processing and sorting recyclable materials, contribute to the generation of employment opportunities. Furthermore, the manufacturing industry benefits from the use of recycled materials, leading to the development of additional jobs in the production of goods utilizing these resources. The economic impact extends beyond job creation, as recycling also presents opportunities for entrepreneurship and innovation, driving economic growth through the establishment of new businesses and technologies focused on sustainable practices. Additionally, the recycling industry contributes to the overall economy by reducing waste management costs for municipalities and businesses, thus freeing up resources for other critical needs. Embracing recycling as an economic driver not only promotes job growth but also fosters a more sustainable and resilient economy.

9.7.3.2 Types of Recyclable Materials

Recyclable materials (or recyclables) encompass a wide range of materials that can be reprocessed and reused to create new products. Common recyclables include paper, cardboard, glass, plastic, and metal. By collecting and processing these materials, recyclables contribute to resource conservation, energy savings, and pollution reduction. Embracing the recycling of these materials is crucial for sustainable waste management and environmental preservation. An example of recyclable materials is shown in Table 9.23.

Not all materials are recyclable in all areas, so it is important to check with local recycling facilities to see what they accept. Also, recycling is just one part of waste management – reducing consumption and reusing items are equally important for environmental sustainability.

9.7.3.3 Benefits of Recycling

Recycling offers a multitude of environmental and economic benefits by conserving natural resources, reducing pollution, and minimizing energy consumption. Simultaneously, it fosters job creation, stimulates innovation, and contributes to cost savings, positioning recycling as a key driver of sustainable economic growth. The highlight of this is shown in Table 9.24.

TABLE 9.23
Recyclable Materials

Paper: This includes newspapers, office paper, magazines, and cardboard without corrugation (like cereal boxes). Even gift-wrapping paper can be recycled if it does not have a plastic film coating on the top.
Metal: There are two major categories of metal recycling: aluminum and steel. Aluminum is often found in soda cans, while steel cans often store soups, sauces, beans, or fruit.
Glass: Many glass products, such as bottles and jars, are recyclable materials. However, ceramics, heat-resistant glass (like Pyrex), or mirror glass are not recyclable.
Plastic: Plastics can be confusing due to the different types and recycling restrictions. The most commonly seen and recyclable plastics are #1 (PET: clear plastics, like water and soda bottles), #2 (HDPE: usually more opaque plastics, like laundry soap and milk jugs), and #5 (yogurt, butter, sour cream containers).
Batteries: Batteries contain toxins like silver, zinc, or even mercury. It is important to recycle batteries to prevent them from leaking into the ground in landfills.
Electronics: More than 130,000 computers are thrown away in the United States every day. Yet almost all parts of a computer can be recycled: plastic, metal, and glass.
Textiles: Textiles, including clothes, can often be recycled or donated for reuse.
Wood: Wood can be recycled into a variety of products, including paper, furniture, and more.
Rubber: Rubber, especially from tires, can be recycled into a variety of products, including asphalt for roads.

TABLE 9.24
Environmental and Economic Benefits of Recycling

Environmental Benefits	Economic Benefits
• *Conservation of Resources:* Recycling helps conserve natural resources by reducing the need for raw materials. For example, making paper from recycled materials rather than from virgin wood uses 30–55% less water and 40–70% less energy. • *Energy Savings:* Manufacturing products from recycled materials typically requires less energy than making them from new materials. This can result in significant energy savings, which also means fewer greenhouse gas emissions. • *Reduced Pollution:* The process of extracting and processing raw materials generates substantial air and water pollution. By using recycled materials instead, we can reduce this pollution. • *Waste Reduction:* Recycling turns materials that would have been thrown away into valuable resources. This not only reduces the amount of waste going to landfills and incinerators but also prevents the environmental damage associated with waste disposal.	• Job Creation: The recycling industry plays a significant role in job creation. According to some estimates, recycling and reuse activities provide numerous jobs worldwide. • Economic Efficiency: Recycling can often be more economically efficient than waste disposal. The revenues from selling recycled materials, along with the societal benefits of recycling, can outweigh the costs of recycling. • Market Development: Recycling can contribute to the development of new markets for recyclable materials, which can stimulate economic activity and innovation. • Cost Savings: Using recycled materials in the manufacturing process can often be cheaper than extracting and processing new materials.

9.7.3.4 Challenges in Recycling

Recycling programs may face challenges including the quality of recyclables, insufficient recycling capacity, and lack of public awareness. Solutions involve educating the public on proper recycling practices, implementing advanced sorting technologies, and expanding recycling facilities. Collaboration between governments, businesses, and communities is essential to address these challenges and ensure the success of recycling programs. The challenges and possible solutions are shown in Table 9.25.

PROBLEMS AND ISSUES FOR DISCUSSION

Here are 40 essay questions on municipal solid waste management, each within the specified word count:

1. Discuss the impact of urbanization on municipal solid waste generation. How does increasing population density influence waste management practices in cities?
2. Evaluate the effectiveness of recycling programs in reducing municipal solid waste. What challenges do cities face in implementing successful recycling initiatives?
3. Analyze the role of public awareness campaigns in promoting responsible waste disposal. How can education influence community participation in waste management?
4. Explore the relationship between municipal solid waste management and public health. What are the potential health risks associated with poor waste management practices?
5. Examine the concept of zero waste and its applicability in urban settings. What strategies can municipalities adopt to move toward a zero-waste goal?
6. Investigate the environmental impacts of landfilling municipal solid waste. How does landfill leachate affect local ecosystems and groundwater quality?
7. Discuss the economic implications of municipal solid waste management. How can cities balance cost-efficiency with effective waste management practices?

TABLE 9.25
Challenges and Possible Solutions in Recycling

Challenges
Consumer Behavior: One of the main challenges is consumer behavior. Many people are unaware of what can be recycled and how to do it properly.
Lack of Access to Recycling: In some areas, especially rural ones, access to recycling facilities or services can be limited.
Sorting of Challenging Plastics: Sorting of so-called challenging plastics, which refers to non-bottle packaging materials such as films and flexible items, foams, and small-format items, is a significant challenge.
Insufficient Recycling Capacity: The capacity to recycle certain materials, particularly certain types of plastic, is often insufficient.
Quality of Recyclable Materials: The quality of recyclable materials can be a challenge, as contamination or mixing of different types of materials can make recycling more difficult.
Regulations: Sometimes laws and regulations can be an impediment to recycling success.
Possible Solutions
Education and Awareness: Educating consumers about what can be recycled and how to do it properly can significantly improve recycling rates.
Improved Access to Recycling: Providing more recycling bins and facilities, particularly in areas where access is currently limited, can help increase recycling.
Advanced Recycling Technologies: New technologies, known as "advanced recycling," are being developed to recycle challenging plastics.
Increase Recycling Capacity: Investments in recycling infrastructure can help increase the capacity to recycle more materials.
Improve Quality of Recyclable Materials: Implementing better sorting and collection practices can help improve the quality of recyclable materials.
Simplifying and Standardizing Regulations: Simplifying and standardizing regulations, including definitions of recyclable and compostable materials, can enhance recycling practices.

8. Assess the role of technology in modern waste management systems. What innovations are emerging to improve waste collection, sorting, and recycling?
9. Analyze the challenges of hazardous waste management within municipal solid waste streams. How can municipalities ensure the safe disposal of hazardous materials?
10. Explore the significance of composting in municipal solid waste management. How can cities promote composting as a viable waste reduction strategy?
11. Evaluate the impact of legislation on municipal solid waste management practices. How do regulations shape the waste management landscape in urban areas?
12. Discuss the role of private sector involvement in municipal solid waste management. What are the advantages and disadvantages of privatizing waste services?
13. Investigate the effectiveness of waste-to-energy technologies in managing municipal solid waste. What are the environmental and economic trade-offs?
14. Examine the challenges of managing electronic waste (e-waste) in urban areas. How can municipalities develop effective e-waste recycling programs?
15. Assess the role of community engagement in successful waste management initiatives. How can municipalities foster collaboration with residents to improve waste practices?
16. Discuss the impact of climate change on municipal solid waste management. How can cities adapt their waste management strategies to mitigate climate-related challenges?
17. Analyze the influence of cultural factors on waste disposal behaviors. How do cultural attitudes shape municipal solid waste management practices?
18. Explore the concept of extended producer responsibility (EPR) in waste management. How can EPR policies incentivize manufacturers to reduce waste generation?

19. Evaluate the effectiveness of landfill diversion strategies in reducing municipal solid waste. What methods can municipalities employ to divert waste from landfills?
20. Investigate the role of data collection and analysis in improving municipal solid waste management. How can data-driven approaches enhance waste management efficiency?
21. Discuss the challenges of managing organic waste in urban areas. What strategies can municipalities implement to effectively handle food waste?
22. Assess the impact of illegal dumping on municipal solid waste management. What measures can cities take to combat illegal waste disposal practices?
23. Explore the benefits and challenges of implementing pay-as-you-throw (PAYT) programs. How can these programs encourage waste reduction among residents?
24. Analyze the role of social media in promoting sustainable waste management practices. How can municipalities leverage social platforms to engage the community?
25. Discuss the importance of lifecycle assessments in evaluating waste management options. How can lifecycle thinking inform better waste management decisions?
26. Investigate the effects of globalization on municipal solid waste management. How do international trade practices influence local waste management strategies?
27. Explore the potential of circular economy principles in municipal solid waste management. How can cities transition toward a circular economy model?
28. Assess the role of local governments in addressing food waste within municipal solid waste. What initiatives can be implemented to minimize food waste generation?
29. Discuss the implications of climate justice in municipal solid waste management. How can waste management practices address social equity issues?
30. Analyze the effectiveness of educational programs in changing waste disposal behaviors. What methods can municipalities use to teach residents about waste management?
31. Explore the relationship between tourism and municipal solid waste generation. How can cities manage waste effectively in high-tourism areas?
32. Assess the challenges of managing construction and demolition debris in urban environments. What strategies can municipalities employ to handle this type of waste?
33. Discuss the role of innovation in developing sustainable waste management solutions. What emerging technologies show promise in enhancing waste management practices?
34. Evaluate the impact of transportation on municipal solid waste management efficiency. How can optimizing waste collection routes improve overall waste management?
35. Investigate the effects of socioeconomic factors on waste generation and management. How do income levels and education influence waste disposal behaviors?
36. Explore the significance of stakeholder collaboration in municipal solid waste management. How can partnerships enhance waste management outcomes in communities?
37. Discuss the potential of behavioral economics in influencing waste management practices. What psychological strategies can motivate individuals to reduce waste?
38. Assess the impact of climate change adaptation strategies on municipal solid waste management. How can cities prepare for the effects of climate change on waste systems?
39. Explore the role of grassroots movements in shaping municipal solid waste policies. How can community activism drive change in waste management practices?
40. Analyze the implications of international agreements on municipal solid waste management. How do global treaties influence local waste management policies?

MULTIPLE-CHOICE QUESTIONS

1. What is the primary goal of municipal solid waste management?
   a. To increase waste generation
   b. To minimize waste and protect public health
   c. To promote illegal dumping
   d. To reduce recycling rates
   e. To create more landfills
2. Which of the following is a common method of waste disposal?
   a. Incineration
   b. Composting
   c. Landfilling
   d. Recycling
   e. All of the above
3. What does the term "zero waste" refer to?
   a. Creating no waste at all
   b. Landfilling all waste
   c. Recycling 100% of waste
   d. Reducing waste to a minimum
   e. Disposing of waste in oceans
4. Which of the following is NOT a component of municipal solid waste?
   a. Food waste
   b. Yard waste
   c. Industrial waste
   d. Paper products
   e. Plastics
5. What is the main benefit of recycling?
   a. Increases waste volume
   b. Conserves resources
   c. Creates pollution
   d. Increases landfill use
   e. Reduces public awareness
6. Which of the following is a type of hazardous waste?
   a. Glass bottles
   b. Batteries
   c. Food scraps
   d. Cardboard
   e. Yard trimmings
7. What is composting primarily used for?
   a. Reducing greenhouse gases
   b. Decomposing organic waste
   c. Storing waste
   d. Burning waste
   e. Creating landfill space
8. Which agency typically oversees waste management in municipalities?
   a. Department of Transportation
   b. Environmental Protection Agency (EPA)
   c. Department of Education
   d. Department of Health
   e. Department of Energy

9. What does the term "landfill diversion" mean?
   a. Increasing landfill capacity
   b. Reducing the amount of waste sent to landfills
   c. Burning waste
   d. Transporting waste to other countries
   e. Storing waste underground
10. Which of the following practices can help reduce food waste?
    a. Buying more than needed
    b. Composting leftovers
    c. Throwing away expired food
    d. Ignoring food storage guidelines
    e. None of the above
11. What is the primary source of municipal solid waste?
    a. Industrial activities
    b. Residential households
    c. Agricultural practices
    d. Construction projects
    e. Medical facilities
12. Which of the following is an example of a pay-as-you-throw (PAYT) program?
    a. Flat fee for waste collection
    b. Charging based on weight of waste
    c. Free waste disposal
    d. Unlimited waste collection
    e. None of the above
13. What is the purpose of extended producer responsibility (EPR)?
    a. To increase production costs
    b. To hold manufacturers accountable for waste
    c. To reduce consumer choices
    d. To promote illegal dumping
    e. To limit recycling
14. Which of the following materials is most commonly reused?
    a. Plastic bags
    b. Glass bottles
    c. Food waste
    d. Yard waste
    e. Electronics
15. What is a significant challenge in electronic waste (e-waste) management?
    a. High recycling rates
    b. Toxic materials
    c. Easy disposal methods
    d. Lack of consumer interest
    e. Abundant resources
16. What does "waste-to-energy" refer to?
    a. Burning waste to produce energy
    b. Recycling waste into new products
    c. Composting organic waste
    d. Storing waste for future use
    e. None of the above

17. Which of the following is a benefit of incineration?
    a. Produces large amounts of waste
    b. Reduces waste volume significantly
    c. Increases landfill use
    d. Generates toxic emissions
    e. Requires extensive land
18. What is the primary goal of public awareness campaigns in waste management?
    a. To increase waste generation
    b. To educate the public on waste reduction
    c. To promote illegal dumping
    d. To decrease recycling
    e. To ignore waste issues
19. What is a common method to incentivize recycling?
    a. Increasing landfill fees
    b. Offering financial rewards for recyclables
    c. Reducing waste collection services
    d. Banning recycling altogether
    e. None of the above
20. Which of the following materials is NOT recyclable in most programs?
    a. Aluminum cans
    b. Glass bottles
    c. Plastic containers
    d. Pizza boxes
    e. Paper products
21. What is the primary disadvantage of landfilling?
    a. High cost
    b. Land consumption and pollution
    c. Easy to manage
    d. Generates energy
    e. Reduces waste volume
22. Which of the following is a method for reducing plastic waste?
    a. Using single-use plastics
    b. Promoting reusable bags
    c. Ignoring recycling guidelines
    d. Increasing plastic production
    e. None of the above
23. What is the purpose of a waste audit?
    a. To increase waste generation
    b. To analyze waste composition and management
    c. To promote illegal dumping
    d. To ignore waste issues
    e. To increase landfill space
24. Which of the following practices can help reduce construction waste?
    a. Over-ordering materials
    b. Implementing recycling programs on-site
    c. Ignoring waste management plans
    d. Disposing of all materials in landfills
    e. None of the above

25. What is the primary challenge of managing yard waste?
    a. High recycling rates
    b. Seasonal surges in volume
    c. Easy disposal
    d. Lack of public interest
    e. Abundant resources
26. Which of the following is a method of waste reduction?
    a. Composting
    b. Buying in bulk
    c. Reusing items
    d. Recycling
    e. All of the above
27. What is the main environmental concern associated with landfills?
    a. Air pollution
    b. Water contamination
    c. Habitat destruction
    d. Noise pollution
    e. None of the above
28. Which of the following is a benefit of community composting programs?
    a. Increases waste volume
    b. Reduces greenhouse gas emissions
    c. Requires extensive land
    d. Generates toxic emissions
    e. None of the above
29. What is the primary purpose of a waste management plan?
    a. To ignore waste issues
    b. To outline strategies for waste reduction and disposal
    c. To increase waste generation
    d. To promote illegal dumping
    e. None of the above
30. Which of the following technologies is used in waste sorting?
    a. Manual sorting only
    b. Optical sorting systems
    c. Incineration
    d. Landfilling
    e. None of the above
31. What is the primary goal of waste minimization?
    a. To increase waste generation
    b. To reduce the amount of waste produced
    c. To promote illegal dumping
    d. To ignore waste issues
    e. To increase landfill use
32. Which of the following is a key principle of the circular economy?
    a. Linear production processes
    b. Waste as a resource
    c. Increased consumption
    d. Landfilling all waste
    e. None of the above

33. What is a common challenge in implementing recycling programs?
    a. High public participation
    b. Contamination of recyclables
    c. Abundant recycling facilities
    d. Easy public awareness
    e. None of the above
34. Which of the following is an example of a biodegradable material?
    a. Glass
    b. Plastic
    c. Food scraps
    d. Aluminum
    e. Styrofoam
35. What is the primary environmental benefit of recycling?
    a. Increased waste volume
    b. Conservation of natural resources
    c. More landfills
    d. Higher emissions
    e. None of the above
36. Which of the following is a method to combat illegal dumping?
    a. Ignoring the issue
    b. Increasing penalties and fines
    c. Reducing waste collection services
    d. Decreasing public awareness
    e. None of the above
37. What is an advantage of using incineration for waste management?
    a. Generates large amounts of waste
    b. Reduces waste volume significantly
    c. Increases landfill use
    d. Creates toxic emissions
    e. Requires extensive land
38. Which of the following is a common strategy for reducing packaging waste?
    a. Using single-use plastics
    b. Promoting bulk purchasing
    c. Ignoring waste management
    d. Increasing packaging materials
    e. None of the above
39. What is the primary purpose of waste segregation at the source?
    a. To increase contamination
    b. To facilitate recycling and composting
    c. To ignore waste issues
    d. To promote illegal dumping
    e. None of the above
40. Which of the following is a key component of a successful recycling program?
    a. Public education and awareness
    b. Ignoring contamination
    c. Reducing participation
    d. Increasing landfill use
    e. None of the above

41. What is the role of municipal solid waste management in public health?
    a. To increase waste generation
    b. To minimize health risks from waste
    c. To promote illegal dumping
    d. To ignore waste issues
    e. None of the above
42. Which of the following is a disadvantage of composting?
    a. Reduces landfill waste
    b. Requires space and management
    c. Produces organic material
    d. Is environmentally friendly
    e. None of the above
43. What is a common feature of successful waste management policies?
    a. Limited public participation
    b. Comprehensive planning and community involvement
    c. Ignoring recycling
    d. Increasing waste generation
    e. None of the above
44. Which of the following is a method for reducing textile waste?
    a. Discarding old clothes
    b. Donating or recycling garments
    c. Ignoring textile waste
    d. Increasing clothing production
    e. None of the above
45. What is the main purpose of a landfill gas collection system?
    a. To increase emissions
    b. To capture methane for energy
    c. To ignore landfill issues
    d. To promote illegal dumping
    e. None of the above
46. Which of the following is an effective way to engage the community in waste management?
    a. Ignoring public input
    b. Hosting educational workshops
    c. Reducing recycling options
    d. Increasing landfill fees
    e. None of the above
47. What is the primary objective of waste reduction initiatives?
    a. To increase waste generation
    b. To minimize waste produced
    c. To promote illegal dumping
    d. To ignore waste issues
    e. None of the above
48. Which of the following is the benefit of using biodegradable materials?
    a. Increases landfill waste
    b. Breaks down naturally in the environment
    c. Requires more resources
    d. Promotes pollution
    e. None of the above

49. What is a common challenge in implementing composting programs?
    a. High public interest
    b. Lack of space and resources
    c. Easy management
    d. High recycling rates
    e. None of the above
50. Which of the following is the goal of sustainable waste management?
    a. To increase waste generation
    b. To minimize environmental impact
    c. To promote illegal dumping
    d. To ignore waste issues
    e. None of the above

ANSWERS FOR MULTIPLE-CHOICE QUESTIONS

1	b	11	b	21	b	31	b	41	b
2	e	12	b	22	b	32	b	42	b
3	d	13	b	23	b	33	b	43	b
4	c	14	b	24	b	34	c	44	b
5	b	15	b	25	b	35	b	45	b
6	b	16	a	26	e	36	b	46	b
7	b	17	b	27	b	37	b	47	b
8	b	18	b	28	b	38	b	48	b
9	b	19	b	29	b	39	b	49	b
10	b	20	d	30	b	40	a	50	b

FURTHER READING

Awasthi, A., & Zeng, X. (2020). Municipal solid waste management in developing countries: A review of challenges and opportunities. *Waste Management*, 102, 1–12. https://doi.org/10.1016/j.wasman.2019.10.022.

Beigl, P., & Huber, J. (2018). The role of waste management in a circular economy. *Waste Management*, 74, 1–12. https://doi.org/10.1016/j.wasman.2017.11.015.

Bhatia, S., & Jain, P. (2016). Municipal solid waste management in India: Current practices and future challenges. *Waste Management & Research*, 34(8), 743–755. https://doi.org/10.1177/0734242X16647773.

Chandra, P., & Sharma, S. (2019). Solid waste management in urban areas: A review of recent developments. *Environmental Science and Pollution Research*, 26, 4321–4335. https://doi.org/10.1007/s11356-019-04345-5.

Chen, Y., & Xu, Y. (2020). The impact of urbanization on municipal solid waste generation: Evidence from China. *Sustainability*, 12(4), 1445. https://doi.org/10.3390/su12041445.

Das, S., & Kumar, A. (2019). A comprehensive review on municipal solid waste management: Challenges and solutions. *Journal of Cleaner Production*, 221, 1–12. https://doi.org/10.1016/j.jclepro.2019.02.006.

De Feo, G., & De Gisi, S. (2018). Municipal solid waste management in developing countries: A review of the literature. *Waste Management*, 77, 1–12. https://doi.org/10.1016/j.wasman.2018.04.007.

Dhir, A., & Gupta, A. (2021). Role of technology in solid waste management: A review. *Resources, Conservation and Recycling*, 167, 105419. https://doi.org/10.1016/j.resconrec.2020.105419.

Geng, Y., & Doberstein, B. (2008). Developing the circular economy in China: Challenges and opportunities for the industrial sector. *Journal of Cleaner Production*, 16(1), 1–8. https://doi.org/10.1016/j.jclepro.2006.07.004.

Hoornweg, D., & Bhada-Tata, P. (2012). *What a waste: A global review of solid waste management*. World Bank. https://doi.org/10.1596/978-0-8213-8812-4.

Iyer, R. H., & Gupta, A. (2016). Municipal solid waste management in India: A review. *International Journal of Environmental Science and Technology*, 13(4), 1109–1120. https://doi.org/10.1007/s13762-015-0851-5.

Kaza, S., Yao, L. C., & Bhada-Tata, P. (2018). *What a waste 2.0: A global snapshot of solid waste management to 2050.* World Bank. https://doi.org/10.1596/978-1-4648-1329-0.

Khan, M. I., & Ghosh, S. (2020). Waste management practices in urban India: A review. *Waste Management & Research*, 38(5), 489–496. https://doi.org/10.1177/0734242X20916156.

Kumar, S., & Singh, R. (2020). Sustainable municipal solid waste management: A review of current practices and future challenges. *Environmental Science and Pollution Research*, 27, 1–15. https://doi.org/10.1007/s11356-020-09109-8.

Lemaire, R., & Renaud, J. (2016). Waste management in urban areas: An overview of the challenges. *Waste Management*, 54, 1–13. https://doi.org/10.1016/j.wasman.2016.05.014.

Li, J., & Wang, Y. (2018). Municipal solid waste management in China: A review of the current situation and future challenges. *Waste Management*, 74, 1–11. https://doi.org/10.1016/j.wasman.2017.10.015.

Mazzotta, M., & Ciriello, R. (2019). Circular economy in solid waste management: Challenges and opportunities. *Waste Management,* 95, 1–11. https://doi.org/10.1016/j.wasman.2019.06.017.

Mehta, R., & Hossain, M. (2019). Solid waste management in urban areas: A review of challenges and strategies. *Environmental Monitoring and Assessment*, 191(7), 431. https://doi.org/10.1007/s10661-019-7647-6.

Nascimento, L. F., & Silva, C. C. (2018). A review of solid waste management in Brazil: Current practices and challenges. Waste Management, 77, 1–9. https://doi.org/10.1016/j.wasman.2018.04.004.

Pires, A., & Martinho, G. (2019). Municipal solid waste management: A review of the literature. *Waste Management*, 87, 1–10. https://doi.org/10.1016/j.wasman.2019.01.017.

Qu, X., & Zhang, Y. (2020). Solid waste management in China: Current situation and future prospects. *Resources, Conservation and Recycling*, 159, 104822. https://doi.org/10.1016/j.resconrec.2020.104822.

Rathi, S. (2006). Alternative approaches for better municipal solid waste management in Mumbai, India. *Waste Management*, 26(10), 1192–1200. https://doi.org/10.1016/j.wasman.2005.08.013.

Rojas, C., & Tello, A. (2018). Municipal solid waste management in Latin America: Current challenges and future directions. *Waste Management*, 74, 1–9. https://doi.org/10.1016/j.wasman.2017.12.017.

Singh, R. P., & Kumar, S. (2020). Sustainable solid waste management: A review of current practices and future directions. *Journal of Cleaner Production*, 263, 121490. https://doi.org/10.1016/j.jclepro.2020.121490.

Sudhakar, K., & Reddy, K. (2019). Waste management practices in India: A review. *Environmental Science and Pollution Research*, 26, 12345–12358. https://doi.org/10.1007/s11356-019-05236-5.

Tchobanoglous, G., & Kreith, F. (2002). *Handbook of solid waste management* (2nd ed.). McGraw-Hill.

Tchobanoglous, G., Theisen, H., & Vigil, S. (1993). *Integrated solid waste management: Engineering principles and management issues*. McGraw-Hill.

Wang, X., & Zhang, Y. (2021). The role of community participation in municipal solid waste management: A case study from China. *Resources, Conservation and Recycling,* 167, 105419. https://doi.org/10.1016/j.resconrec.2020.105419.

Zeng, X., & Awasthi, A. (2019). Municipal solid waste management in developing countries: A review of challenges and opportunities. *Waste Management*, 87, 1–12. https://doi.org/10.1016/j.wasman.2019.01.025.

10 Low Impact Development PLUS (LID+) in Urban Areas

10.1 WHAT IS LOW IMPACT DEVELOPMENT?

Low impact development (LID) is a sustainable land development strategy that aims to manage stormwater runoff and promote environmental conservation. LID practices focus on mimicking natural hydrological processes to reduce the impact of urbanization on the environment. One key principle of LID involves managing stormwater at its source through techniques such as permeable pavements, green roofs, and rain gardens. These methods help to infiltrate, filter, and store stormwater, reducing the volume and velocity of runoff. By integrating LID into urban planning, communities can mitigate flooding, protect water quality, and preserve natural hydrological systems.

LID is a term used in Canada and the United States to describe a land planning and engineering design approach to manage stormwater runoff as part of green infrastructure. It emphasizes conservation and use of on-site natural features to protect water quality. This approach implements engineered small-scale hydrologic controls to replicate the pre-development hydrologic regime of watersheds through infiltrating, filtering, storing, evaporating, and detaining runoff close to its source.

The foundational concept of LID initially centered on stormwater management, which may seem incongruent with the term "low impact." However, to align more closely with the principles of sustainable development, LID should encompass a broader scope, integrating green infrastructure and sustainable practices. By expanding its focus beyond stormwater management, LID can embrace a comprehensive approach that incorporates elements such as permeable pavements, green roofs, rain gardens, and preservation of natural landscapes. This evolution allows LID to promote not only effective stormwater control but also ecological preservation, resource conservation, and the creation of resilient, environmentally responsible communities. We called this expansion LID+ and used it as the title of this chapter.

LID is not only about stormwater management, but LID also emphasizes the preservation of natural landscapes and open spaces within developed areas. This approach involves minimizing land disturbance, preserving existing vegetation, and incorporating green infrastructure elements such as vegetated swales and bioretention areas. By preserving natural features, LID helps to maintain biodiversity, enhance wildlife habitat, and promote ecological balance within urban environments.

LID also promotes the use of sustainable materials and construction techniques to minimize the environmental impact of development. This includes utilizing recycled materials, implementing energy-efficient designs, and reducing impervious surfaces. Incorporating sustainable practices, LID strives to minimize resource consumption, cut down on pollution, and lower the overall environmental footprint of development projects.

DOI: 10.1201/9781032664613-10

Another fundamental aspect of LID is the integration of community engagement and education. Encouraging public participation and awareness regarding sustainable land development practices is essential for the successful implementation of LID initiatives. Community involvement can foster a sense of stewardship, promote responsible land use, and ensure the long-term sustainability of LID projects.

LID emphasizes the importance of adaptive and flexible design approaches to accommodate changing environmental conditions. By incorporating resilient design features such as green infrastructure adaptable to varying precipitation patterns and climate change impacts, LID strives to create durable and effective stormwater management systems.

LID represents a holistic approach to sustainable land development, integrating stormwater management, ecological preservation, sustainable design, community engagement, and adaptive strategies to create resilient and environmentally responsible built environments. Embracing LID principles allows communities to mitigate the adverse effects of urbanization, protect natural resources, and promote a more sustainable and harmonious relationship between human development and the natural environment.

There are several types of LID practices, including:

- Sustainable use of materials,
- Bioretention cell,
- Green roof,
- Infiltration trench,
- Permeable pavement,
- Rain barrel or cistern,
- Rooftop disconnection, and
- Vegetative swale.

Each type of LID has its own principles, design, performance, advantages, disadvantages, and costs. Although LID has been quite broadly applied and has demonstrated success in urban stormwater management in many countries, there are still some main challenges during the implementation, such as clogging and water quality.

10.2 LOW-IMPACT HOUSING

Low-impact housing aims to minimize environmental footprints through sustainable design and construction practices. It promotes sustainable living by reducing energy consumption and carbon emissions, and often uses natural materials like straw, timber, and lime. Energy-efficient features such as solar roofs and super insulation are incorporated to reduce the need for artificial heating and cooling. The unique financing model of low-impact housing ensures its affordability for people on low or variable incomes. It often involves the creation of communities where residents share financial responsibility, land, and the management of the project. LID strategies can lead to reduced infrastructural costs and increased property values due to the desirability of sustainable and energy-efficient features. Achieving these objectives allows low-impact housing to provide a sustainable and affordable solution to housing while minimizing its environmental impact.

10.2.1 Sustainable Living

Low-impact housing is a revolutionary concept that aims to promote sustainable living by significantly reducing energy consumption and carbon emissions. These houses are specifically designed to be environmentally friendly, with a primary focus on achieving carbon negativity. By generating more energy than they consume, low-impact housing sets a new standard for eco-friendly living.

This innovative approach not only benefits the environment but also contributes to a more sustainable future for generations to come.

The core principle of low-impact housing revolves around the reduction of energy consumption and carbon emissions. Through thoughtful and innovative design, these houses are engineered to minimize their environmental impact. Utilizing renewable energy sources like solar power and implementing energy-efficient technologies, low-impact housing sets a new benchmark for sustainable residential living. This forward-thinking approach not only addresses current environmental challenges but also serves as a model for future developments in the housing industry.

One of the most remarkable features of low-impact housing is its carbon-negative design. By generating surplus energy, these houses actively contribute to reducing carbon emissions, creating a positive environmental impact. This innovative approach helps combat climate change while demonstrating the potential for sustainable living without compromising comfort or convenience. Low-impact housing represents a significant step toward a greener and more sustainable future, setting an example for environmentally conscious construction and design practices.

The concept of low-impact housing embodies a holistic approach to sustainable living. Beyond energy efficiency, these houses often incorporate eco-friendly materials, water conservation systems, and sustainable landscaping practices. This comprehensive strategy aims to minimize the ecological footprint of residential living, making a significant contribution to environmental preservation. Embracing low-impact housing enables individuals and communities to actively contribute to the global effort to mitigate climate change and promote a more sustainable way of life.

Low-impact housing represents a groundbreaking approach to sustainable living, emphasizing the reduction of energy consumption and carbon emissions. By achieving carbon negativity and generating surplus energy, these houses set a new standard for eco-friendly residential living. Through innovative design, renewable energy utilization, and a holistic approach to sustainability, low-impact housing serves as a beacon of hope for a greener and more sustainable future. This concept not only addresses the urgent need to combat climate change but also inspires a shift toward more environmentally conscious living practices on a global scale.

Sustainable living encompasses a wide range of practices and principles that aim to minimize environmental impact and promote long-term ecological balance. In addition to reducing energy consumption and carbon emissions, there are several other important aspects to consider when embracing sustainable living:

1. *Resource Conservation:* Sustainable living involves the conscientious use of natural resources such as water, land, and raw materials. This includes practices such as water conservation, responsible land use, and sustainable forestry to ensure that resources are utilized in a manner that supports their renewal and long-term availability.
2. *Waste Reduction and Recycling:* A key component of sustainable living is the reduction of waste generation and the promotion of recycling. This involves minimizing single-use products, repurposing materials, and adopting recycling practices to divert waste from landfills and reduce the overall environmental impact.
3. *Sustainable Transportation:* Sustainable living encourages the use of eco-friendly transportation options such as walking, cycling, public transit, and electric vehicles. By reducing reliance on fossil fuel-powered vehicles, individuals can significantly decrease their carbon footprint and contribute to improved air quality.
4. *Supporting Local and Organic Agriculture:* Embracing sustainable living often involves supporting local and organic food production. This includes purchasing locally grown produce, supporting sustainable farming practices, and reducing reliance on industrially produced and transported food items, which can contribute to lower carbon emissions and support local economies.

5. *Community Engagement:* Sustainable living extends beyond individual actions and often involves community engagement and advocacy. This can include participating in local environmental initiatives, supporting sustainable businesses, and advocating for policies and practices that promote sustainability at the community and governmental levels.

By considering these additional aspects of sustainable living, individuals and communities can make meaningful contributions to environmental preservation and long-term ecological sustainability. Embracing a holistic approach to sustainable living can lead to positive impacts on both the environment and the well-being of present and future generations.

10.2.2 Energy Efficiency

Energy consumption has a direct impact on our carbon footprint. Efficient energy use reduces greenhouse gas emissions and reliance on fossil fuels.

Energy-efficient housing is designed to reduce energy consumption and greenhouse gas emissions, potentially offering healthier, cleaner living conditions. Here are some key features of energy-efficient homes:

10.2.2.1 Efficient Home Design

Energy-efficient homes are designed and constructed with a comprehensive whole-house systems approach, considering all variables, details, and interactions that influence energy consumption within the home. By considering every aspect of a home's energy usage – from appliances to insulation, lighting, space heating and cooling, water heating, and windows – this approach aims to optimize energy efficiency and minimize environmental impact. By adopting a holistic perspective, energy-efficient homes aim to create a harmonious balance between various components, ultimately achieving the highest level of energy performance.

Electrical appliances at home contribute significantly to energy consumption, and therefore, to the energy efficiency of a home, and the whole-house systems approach ensures that energy-efficient models are selected to minimize electricity consumption. From refrigerators and washing machines to dishwashers and air conditioning units, these appliances are carefully chosen to meet strict energy efficiency standards, thereby reducing overall energy demand within the home. In addition, the approach considers the integration of smart technologies and energy management systems to further enhance the efficiency of these appliances.

Home insulation is also a fundamental component of energy-efficient homes, as it maintains indoor thermal comfort and reduces the need for excessive heating or cooling. By selecting and installing high-quality insulation materials, the whole-house systems approach ensures that heat loss and gain are minimized, resulting in reduced energy consumption for space heating and cooling. Attention is given to air sealing to prevent unwanted drafts and heat loss, contributing to a more comfortable and energy-efficient living environment.

In the realm of lighting, the whole-house systems approach focuses on integrating energy-efficient lighting solutions such as light emitting diode (LED) bulbs and fixtures, which consume significantly less energy than traditional incandescent or fluorescent lighting. By strategically planning the placement of lighting fixtures and maximizing the use of natural light, energy-efficient homes can significantly reduce electricity usage for lighting, leading to overall energy savings. Moreover, sensors and smart lighting controls optimize lighting based on occupancy and natural light levels, further enhancing energy efficiency. In tropical countries, leveraging daylight can notably reduce the need for artificial lighting, contributing to substantial energy savings.

Space heating and cooling systems should be selected and designed to operate efficiently within energy-efficient homes. This includes the use of high-efficiency heating and cooling equipment, proper sizing of heating, ventilation, and air conditioning (HVAC) systems, and the implementation of advanced thermostat and zoning technologies to optimize energy usage based on specific heating and cooling needs. Considering these factors within the whole-house systems approach, energy-efficient homes can achieve superior thermal comfort while minimizing energy consumption and environmental impact.

10.2.2.2 Size and Shape

The size and shape of a home determine its energy consumption and efficiency, as they directly impact the amount of energy required for heating and cooling. Larger homes, characterized by their expansive floor area and volume, often demand higher energy consumption to maintain comfortable indoor temperatures due to their larger surface area. This increased surface area results in greater heat loss during colder months and heat gain during warmer months, necessitating more energy for space heating and cooling to offset these effects. As a result, the size of a home has a direct correlation with its energy usage, making it a crucial consideration in the pursuit of energy-efficient residential design.

In addition to size, the shape of a home also influences its energy efficiency. The architectural layout and form of a house can impact its exposure to external elements such as sunlight, wind, and outdoor temperatures, thereby affecting its thermal performance. For instance, homes with elongated or irregular shapes may experience uneven heat distribution and air circulation, leading to potential energy inefficiencies. Conversely, well-designed and streamlined home shapes can optimize solar gain, natural ventilation, and thermal comfort, contributing to reduced energy consumption for heating and cooling.

The orientation of a home in relation to the sun and prevailing winds is a crucial factor in its energy efficiency. Properly orienting the home to maximize solar exposure in winter and minimize it in summer can significantly impact its heating and cooling requirements. Strategically positioning windows, shading elements, and outdoor landscaping allows energy-efficient homes to harness the benefits of passive solar heating and natural ventilation. This reduces reliance on mechanical heating and cooling systems, enhancing overall energy performance.

The layout and distribution of interior spaces within a home can influence its energy efficiency. Efficient floor plans that minimize circulation space and optimize room layout can improve the flow of natural light and air, reducing the need for artificial lighting and mechanical ventilation. In addition, thoughtful zoning of living areas and bedrooms can facilitate targeted heating and cooling, allowing for more precise temperature control and energy conservation.

The size and shape of a home become critical determinants of its energy efficiency, directly impacting its heating and cooling requirements. Taking these factors into account in residential design and construction allows homeowners and builders to optimize energy performance, reduce environmental impact, and enhance the overall sustainability of residential dwellings. Thoughtful consideration of size, shape, orientation, and interior layout ensures energy-efficient homes achieve superior thermal comfort while minimizing energy consumption and operational costs.

Energy consumption in a house can be influenced by its size, which includes the floor area or volume of the house. However, this is a simplification, and actual energy consumption can be affected by many other factors such as insulation, number of occupants, efficiency of appliances, etc.

10.2.2.3 Window Direction

The orientation of windows plays a pivotal role in influencing the average temperature and natural light levels within a room. The direction in which windows face determines the amount of sunlight that enters the space throughout the day, consequently affecting the room's thermal conditions. Depending on the latitude of the house location, for instance, houses located in the northern hemisphere at high latitudes, windows facing south tend to receive a significant amount of direct sunlight, especially in the northern hemisphere, resulting in increased solar heat gain during the winter months. This can contribute to higher average temperatures within the room and reduce the need for additional heating, thereby enhancing thermal comfort and energy efficiency.

Conversely, windows facing north typically receive less direct sunlight and may experience minimal solar heat gain. As a result, rooms with north-facing windows may exhibit lower average temperatures, particularly during colder seasons. Although this may necessitate additional heating to maintain comfortable conditions, it also reduces the risk of overheating during warmer months. Understanding the impact of window orientation on natural light and temperature allows homeowners and designers to strategically position windows for optimal daylighting and thermal performance. This approach promotes a more comfortable and energy-efficient indoor environment.

In addition to thermal considerations, the orientation of windows profoundly influences the quality and quantity of natural light within a room. South-facing windows, for example, allow for ample natural light penetration, creating a bright and inviting atmosphere. This not only reduces the reliance on artificial lighting during daylight hours but also contributes to a more visually appealing and uplifting indoor environment. Conversely, north-facing windows may provide softer, more diffused light, which can be advantageous for spaces where glare and excessive sunlight penetration are undesirable, such as work areas or entertainment spaces.

Careful selection and placement of windows based on their orientation can have a substantial impact on overall energy usage and environmental sustainability. Harnessing natural light and optimizing solar heat gain allows homeowners to reduce their reliance on artificial lighting and mechanical heating systems, thereby lowering energy consumption and reducing carbon emissions. This highlights the importance of considering window orientation as a critical element in designing and utilizing interior spaces, with the potential to enhance both comfort and energy efficiency.

10.2.2.4 Insulation

Effective insulation serves as a crucial component in regulating indoor temperatures, contributing to a comfortable living environment throughout the year. By minimizing heat transfer, insulation plays a pivotal role in reducing the amount of heat that escapes from the interior of the house during winter and inhibiting the entry of external heat during summer. This results in more stable and comfortable indoor temperatures, enhancing the overall livability of the home and reducing the reliance on heating and cooling systems to maintain thermal comfort.

During the colder months, well-installed insulation acts as a thermal barrier, preventing the escape of heated air from the interior of the house. This helps to retain warmth within the living spaces, reducing the need for excessive heating and promoting a cozy and inviting atmosphere. In addition, effective insulation minimizes cold spots and drafts, creating a more uniform distribution of heat throughout the home, further enhancing comfort and reducing energy consumption.

In the summer, proper insulation functions as a shield against external heat, limiting the transfer of heat from the outdoors into the interior of the house. Insulation prevents the infiltration of warm air, helping to maintain cooler indoor temperatures and reducing the need for air conditioning systems. This mitigation of outdoor heat impact not only promotes a more pleasant living environment during hot weather but also enhances energy efficiency by lowering the demand for mechanical cooling, ultimately reducing energy costs and environmental impact.

10.2.2.5 Energy Star-Certified Appliances

Appliances bearing the Energy Star certification emblem are meticulously engineered to operate with significantly reduced energy consumption, representing a hallmark of energy efficiency and environmental responsibility. The Energy Star program, established by the US Environmental Protection Agency and the Department of Energy, aims to promote the widespread adoption of energy-efficient products by setting stringent criteria for performance and energy usage. Appliances that meet these rigorous standards not only deliver superior energy efficiency but also contribute to cost savings for consumers and reduced environmental impact.

Energy Star-certified appliances encompass a wide range of household products, including refrigerators, dishwashers, washing machines, air conditioners, televisions, and more. Each appliance category is subject to specific energy performance criteria, ensuring that certified products meet or exceed the established efficiency benchmarks. By incorporating advanced technologies and innovative design features, Energy Star appliances are engineered to optimize energy usage. This results in reduced electricity consumption without compromising performance or functionality.

The implementation of energy-efficient technologies in Energy Star-certified appliances encompasses various aspects of their operation. For instance, refrigerators and freezers utilize advanced insulation materials and compressor designs to minimize energy loss and reduce electricity consumption. Similarly, washing machines and dishwashers integrate features such as improved water circulation, sensor-based load detection, and optimized wash cycles to achieve energy savings without sacrificing cleaning efficacy.

The Energy Star certification program places emphasis on standby power consumption, also known as "phantom load," which refers to the energy consumed by appliances while in standby mode. Certified products are required to meet strict standby power requirements, ensuring that they consume minimal energy when not in active use. This focus on standby power reduction contributes to overall energy savings and aligns with the program's commitment to promoting energy-efficient practices across all aspects of appliance operation.

Energy Star-certified appliances often incorporate environmentally friendly features such as water conservation, non-toxic materials, and recyclability. By prioritizing sustainability throughout the product life cycle, these appliances minimize their environmental impact and uphold the principles of eco-conscious manufacturing and consumption. This holistic approach highlights the broader commitment of the Energy Star program to advancing sustainable practices and reducing the ecological footprint of household appliances.

The widespread adoption of Energy Star-certified appliances has significant implications for energy conservation and environmental stewardship on a global scale. Choosing these energy-efficient products allows consumers to reduce electricity consumption, lower greenhouse gas emissions, and decrease the demand for fossil fuels. In addition, the cost savings from reduced energy usage can lead to long-term financial benefits, making the transition to Energy Star-certified appliances a wise investment both environmentally and economically.

The Energy Star certification serves as a symbol of energy efficiency and environmental stewardship, indicating that appliances are engineered to function with minimal energy usage. With strict performance standards and innovative technologies, Energy Star-certified appliances provide exceptional energy efficiency, decreased electricity consumption, and financial savings for consumers. In addition, the program's emphasis on sustainability and environmental responsibility highlights its overarching dedication to advocating for eco-friendly practices and lessening the environmental footprint of household appliances. Selecting Energy Star-certified products allows consumers to actively promote energy conservation and support the establishment of a more sustainable future for upcoming generations.

10.2.2.6 Energy-Efficient Lighting

Energy-efficient lighting, exemplified by the utilization of LED bulbs, represents a significant advancement in lighting technology, offering a multitude of benefits over traditional incandescent bulbs. LED bulbs are engineered to consume significantly less electricity than their incandescent counterparts, making them a compelling choice for both residential and commercial lighting applications. This heightened energy efficiency stems from the fundamental operational principles of LED technology, which enable these bulbs to convert a greater proportion of electrical input into visible light while minimizing wasted energy in the form of heat.

LED bulbs achieve superior energy efficiency through their utilization of solid-state lighting technology, which enables them to produce light with minimal energy loss. Unlike traditional incandescent bulbs, which rely on heating a filament to produce light, LED bulbs operate by passing an electrical current through a semiconductor, generating light through a process known as electroluminescence. This efficient conversion of electrical energy into visible light allows LED bulbs to deliver a higher lumen output per watt, translating to reduced electricity consumption for the same level of illumination compared to incandescent bulbs.

The energy efficiency of LED lighting extends beyond reduced electricity consumption to encompass an extended operational lifespan. LED bulbs are engineered to provide significantly longer service life compared to traditional incandescent bulbs, thereby reducing the frequency of bulb replacements and the associated energy and material costs. This longevity not only contributes to energy conservation by minimizing the resources required for bulb manufacturing and disposal but also results in long-term cost savings for consumers and businesses through reduced maintenance and replacement expenses.

LED bulbs offer enhanced versatility and control, further contributing to their appeal as a sustainable lighting solution. LED technology enables the production of bulbs with adjustable color temperatures, dimmable capabilities, and directional lighting, providing users with greater flexibility in creating customized lighting environments while optimizing energy usage. This adaptability aligns with the broader objective of energy-efficient lighting to promote sustainable practices and reduce overall electricity demand, contributing to environmental preservation and resource conservation.

By embracing energy-efficient lighting solutions such as LED bulbs, individuals and organizations can actively participate in the global effort to mitigate climate change, reduce electricity consumption, and promote sustainable living practices. The widespread adoption of LED technology not only contributes to lower energy bills and reduced environmental impact but also represents a tangible step toward a more energy-efficient and environmentally conscious future.

10.2.2.7 Renewable Energy Systems

Ultra-efficient homes represent a pinnacle of sustainable residential design, integrating energy-efficient construction practices with advanced renewable energy systems to achieve unparalleled levels of energy performance. By combining innovative building techniques with renewable energy technologies like solar water heating and solar electricity, these homes establish a new standard for environmental responsibility and energy conservation. The synergy between energy-efficient construction and renewable energy systems allows ultra-efficient homes to significantly reduce their reliance on traditional energy sources, leading to substantial cost savings and environmental benefits.

Energy-efficient construction forms the foundation of ultra-efficient homes, encompassing a range of design and building practices aimed at minimizing energy consumption and maximizing thermal performance. This includes the use of high-performance insulation, airtight building envelopes, energy-efficient windows, and advanced HVAC systems, all of which contribute to reduced energy demand and enhanced indoor comfort. Incorporating these elements allows ultra-efficient homes to optimize energy usage and minimize heat loss, establishing a foundation for a sustainable and energy-conscious living environment.

In conjunction with energy-efficient construction, ultra-efficient homes integrate renewable energy systems such as solar water heating and solar electricity to further enhance their energy performance. Solar water heating systems utilize solar thermal collectors to harness the sun's energy and heat water for domestic use, reducing the need for conventional water heating methods and lowering overall energy consumption. This technology not only contributes to energy savings but also aligns with the overarching goal of reducing reliance on non-renewable energy sources, thereby promoting environmental sustainability.

The integration of solar electricity systems, commonly known as photovoltaic (PV) systems, allows ultra-efficient homes to generate clean, renewable electricity from sunlight. By converting solar energy into electricity, PV systems allow homeowners to offset a portion of their electricity consumption with a sustainable and environmentally friendly energy source. This approach not only reduces the environmental impact of electricity generation but also provides homeowners with greater energy independence and resilience against fluctuating energy costs and grid disruptions.

Ultra-efficient homes represent a harmonious integration of energy-efficient construction and renewable energy systems, exemplified by technologies such as solar water heating and solar electricity. Through the combination of advanced building techniques and renewable energy solutions, these homes achieve exceptional levels of energy performance, cost savings, and environmental sustainability.

10.2.2.8 Cool Roofs

Cool roofs represent an innovative approach to sustainable building design, using highly reflective materials to minimize heat absorption from sunlight and maintain cooler indoor temperatures during hot weather. By leveraging the reflective properties of specialized roofing materials, cool roofs effectively reduce the amount of solar heat gain absorbed by buildings. This results in numerous benefits in terms of energy efficiency, thermal comfort, and environmental impact.

The key characteristic of cool roofs lies in their ability to reflect a greater proportion of sunlight, particularly in the visible and near-infrared spectrum, while simultaneously absorbing less heat. This reflective capability allows cool roofs to mitigate the transfer of solar radiation into the building, thereby reducing the thermal load on the interior spaces and lessening the need for mechanical cooling. As a result, homes with cool roofs experience reduced indoor temperatures, leading to enhanced comfort for occupants and decreased reliance on air conditioning systems, especially during periods of intense heat.

The use of highly reflective materials in cool roof construction plays a pivotal role in minimizing solar heat gain by redirecting a significant portion of the incident sunlight back into the atmosphere. This reflective action effectively reduces the amount of solar energy absorbed by the building, thus limiting the transfer of heat to the interior spaces. Adopting cool roofs allows homeowners and building occupants to benefit from cooler indoor environments, improved thermal comfort, and reduced energy consumption for space cooling. This ultimately translates into lower utility costs and decreased environmental impact.

The implementation of cool roofs contributes to the mitigation of urban heat islands, a phenomenon characterized by elevated temperatures in densely populated areas due to the absorption and retention of solar heat by buildings and paved surfaces. By incorporating highly reflective materials in roofing systems, cool roofs help counteract the urban heat island effect by reducing the amount of heat absorbed and re-radiated by buildings. This contributes to lower ambient temperatures and improved outdoor comfort in urban environments.

Cool roofs advance environmental sustainability by reducing the demand for mechanical cooling, decreasing greenhouse gas emissions, and promoting energy conservation. The reflective nature of cool roofs contributes to lower energy consumption for space cooling, leading to reduced electricity usage and associated carbon emissions, which aligns with broader efforts to mitigate climate change and promote sustainable building practices.

The construction of cool roofs can lead to financial benefits for homeowners and building owners by lowering energy costs associated with space cooling. With reduced indoor temperatures and decreased reliance on air conditioning, homes with cool roofs can realize substantial savings on cooling-related energy expenses, contributing to long-term cost efficiency and economic sustainability. This cost-effectiveness, coupled with the environmental advantages of reduced energy consumption, positions cool roofs as a compelling investment for sustainable building design and operation.

Cool roofs, characterized by their use of highly reflective materials to minimize solar heat gain, promote energy efficiency, thermal comfort, and environmental sustainability in building design. Through their ability to reflect more light and absorb less heat from sunlight, cool roofs contribute to cooler indoor temperatures, reduced energy consumption for space cooling, mitigation of urban heat islands, and financial savings for homeowners and building owners. By embracing cool roof technology, individuals and communities can make meaningful contributions to energy conservation, environmental preservation, and the advancement of sustainable building practices.

10.2.3 Materials

The selection of building materials significantly influences the environmental footprint of a structure, encompassing impacts that span from the extraction of resources to the disposal of materials at the end of their life cycle. The entire life cycle of building materials, including extraction, manufacturing, transportation, installation, use, and eventual disposal, contributes to their overall environmental impact. Opting for sustainable, low-impact materials and considering factors such as embodied energy, recyclability, and longevity allows builders and designers to minimize the environmental consequences of construction and promote eco-friendly and resource-efficient building practices. Natural materials are often used in low-impact housing due to their lower environmental impact and biodegradability. Examples of sustainable materials are shown in Table 10.1.

Other materials like sheep's wool, cellulose, cork, and wood fiber are also used. These materials are renewable, biodegradable, and often found on-site or nearby, reducing the need for transportation.

When evaluating materials for sustainable housing, it is crucial to consider a range of key factors that can significantly influence the environmental footprint and long-term viability of the construction. These factors encompass various elements such as embodied energy, renewable and recycled materials, durability and longevity, recyclability and reusability, low toxicity and indoor air quality, local sourcing, and energy efficiency, as shown in Table 10.2. Carefully assessing these aspects allows homeowners, architects, and builders make informed decisions that prioritize sustainability, minimize environmental impact, and promote the resilience and eco-friendliness of the built environment.

Selecting materials and conducting a life cycle analysis of building materials, homeowners, architects, and builders can make informed decisions that prioritize sustainability, environmental responsibility, and long-term resilience in housing construction.

The concept of embodied energy underscores the importance of considering the full life-cycle energy requirements of building materials. Using materials with lower embodied energy, construction projects can reduce their environmental impact, conserve energy, and promote sustainability. This approach not only aligns with environmental objectives but also holds potential economic benefits, highlighting the multifaceted impact of embodied energy considerations in the construction industry.

When choosing materials with lower embodied energy, such as recycled or locally sourced materials, the environmental impact of construction can be significantly reduced. Recycled

TABLE 10.1
Examples of Sustainable Building Materials

Timber: It is a renewable resource that can be locally sourced and has good insulation properties.
Clay: Used in construction for its excellent thermal mass, helping regulate indoor temperatures.
Lime: Often used in mortars and plasters, it is less energy-intensive to produce than cement.
Rammed Earth: A technique where natural raw materials like earth, chalk, lime, or gravel are compressed to form a solid wall.
Cob: A mixture of clay, sand, straw, water, and earth, cob is sculpted into walls while still wet.
Straw: Used in bale construction for its excellent insulation properties.
Hemp: Can be mixed with lime to create a lightweight, insulating material called hempcrete.
Stone: Locally sourced stone can have a lower environmental impact and provides excellent thermal mass

TABLE 10.2
Factors for Selection of Materials

Embodied Energy: This refers to the total energy required for the extraction, processing, manufacturing, transportation, and installation of building materials. Choosing materials with lower embodied energy, such as recycled or locally sourced materials, can help reduce the environmental impact of construction.

Renewable and Recycled Materials: Utilizing renewable materials, such as sustainably harvested wood or bamboo, as well as recycled materials like reclaimed wood or recycled metal, can help minimize the depletion of natural resources and reduce waste.

Durability and Longevity: Selecting materials with high durability and a long lifespan can reduce the need for frequent replacements and repairs, ultimately reducing the consumption of resources over the building's lifetime.

Recyclability and Reusability: Opting for materials that are recyclable at the end of their life cycle or can be easily repurposed contributes to a more circular approach to construction, minimizing waste and promoting a sustainable, closed-loop system.

Low Toxicity and Indoor Air Quality: Prioritizing materials with low emissions of volatile organic compounds (VOCs) and other harmful chemicals can help maintain good indoor air quality, promoting the health and well-being of occupants.

Local Sourcing: Using locally sourced materials can reduce the environmental impact of transportation and support local economies, while also ensuring that the materials are well-suited to the local climate and conditions.

Energy Efficiency: Incorporating materials that contribute to the energy efficiency of the building envelope, such as high-performance insulation and windows, can enhance the overall sustainability of the structure by reducing energy consumption for heating and cooling.

materials, for example, require less energy for processing and manufacturing compared to their non-recycled counterparts. The use of locally sourced materials can minimize the energy expended on transportation, as they do not need to be transported over long distances. Using materials with lower embodied energy, construction projects can contribute to overall energy conservation and environmental sustainability.

Reducing embodied energy in building materials can lead to a decrease in greenhouse gas emissions and resource depletion. By opting for materials with lower embodied energy, construction practices can align with principles of sustainable development and environmental stewardship. This approach not only benefits the immediate construction project but also contributes to broader efforts aimed at mitigating climate change and preserving natural resources for future generations.

The consideration of embodied energy in construction materials can also have economic implications. By selecting materials with lower embodied energy, construction projects can potentially lower overall costs associated with energy consumption, transportation, and manufacturing. This can result in more cost-effective and sustainable construction practices, demonstrating the interconnectedness of environmental and economic factors in decision-making processes.

Utilizing renewable materials, such as sustainably harvested wood or bamboo, as well as recycled materials like reclaimed wood or recycled metal, can help minimize the depletion of natural resources and reduce waste. Sustainable harvesting practices for wood and bamboo involve carefully managing forests to ensure that the rate of harvesting does not exceed the rate of regrowth, thus maintaining the long-term health and productivity of the ecosystems.

The use of recycled materials like reclaimed wood or recycled metal offers a valuable opportunity to reduce waste and minimize the environmental impact of construction projects. Through repurposing materials, which would otherwise end up in landfills, the demand for new raw materials can be reduced, leading to lower resource depletion and decreased energy consumption associated with extraction and processing. This approach aligns with the principles of the circular economy, where materials are reused and recycled to extend their lifespan and minimize the generation of waste.

The utilization of renewable and recycled materials in construction can contribute to the development of a more sustainable and resilient built environment. The construction industry can move toward a more regenerative model that seeks to minimize its ecological footprint while promoting the efficient use of resources. This shift not only addresses immediate environmental concerns but also fosters long-term sustainability by encouraging the adoption of practices that support the health and well-being of both human communities and the natural world.

10.2.4 Housing Affordability

Affordability is not directly linked to the environmental aspects of low-impact housing. However, it is an essential component for social sustainability and inclusivity. Affordable housing access for individuals and families is a key element of social sustainability, promoting diverse, vibrant, and inclusive communities. Low-impact housing, which strives to reduce environmental footprints and resource use, must also cater to the essential need for affordable housing to be truly equitable and effective.

In this scenario, the overlap of environmental and social sustainability becomes clear. Housing affordability significantly affects the capacity of individuals from various socioeconomic backgrounds to inhabit environmentally friendly spaces. If low-impact housing costs surpass most people's financial capabilities, the goals of establishing sustainable and inclusive communities could be jeopardized. Hence, it is an important factor in both environmental and social aspects when developing and implementing low-impact housing solutions, acknowledging their interrelated nature.

Tackling housing affordability within the scope of low-impact housing can yield wider societal advantages, such as enhanced public health, economic stability, and community resilience. Affordable housing allows individuals and families to devote more resources to other vital needs like healthcare, education, and personal growth, thereby enhancing overall well-being and social equity. Making low-impact housing accessible to a diverse population, communities can encourage greater social unity and economic variety, setting the stage for enduring sustainability and prosperity.

Incorporating affordable housing into low-impact housing initiatives can boost the overall efficacy and influence of environmental preservation efforts. Low-impact housing can attract more community support and involvement if it is affordable, leading to a rise in the adoption of eco-friendly practices and technologies. This inclusive strategy not only aids individuals and families seeking affordable housing but also furthers broader environmental goals, such as energy consumption reduction, waste minimization, and ecological guardianship.

10.3 GREEN ROOFS

Green roofs, also known as living roofs or eco-roofs, are innovative roofing systems designed to support the growth of vegetation on the top of buildings. These roofs are constructed with a waterproofing membrane, a root barrier, a drainage system, and a growing medium that can support a variety of plants, ranging from grasses and sedums to small trees and shrubs. Green roofs provide numerous environmental benefits, including reducing urban heat island effect by absorbing and deflecting solar radiation, improving air quality by capturing airborne pollutants, and reducing stormwater runoff by retaining and filtering rainwater. In addition, green roofs can enhance the aesthetic appeal of urban areas, provide habitat for wildlife, and contribute to energy efficiency by insulating buildings and reducing the need for heating and cooling.

From a practical standpoint, green roofs offer a sustainable solution for utilizing underutilized space in urban environments, creating opportunities for urban agriculture, recreational areas, and even rooftop gardens. In addition to their environmental and social benefits, green roofs can also extend the lifespan of the roofing membrane by protecting it from the elements, thereby reducing maintenance costs and contributing to the overall resilience and longevity of buildings. Green roofs represent a versatile and effective strategy for promoting sustainable urban development and

TABLE 10.3
Green Roofs' Contributions to Low Impact Development

1	*Stormwater Management:* Green roofs absorb and retain rainwater, which reduces the volume of stormwater runoff and delays the time at which runoff occurs, resulting in decreased stress on sewer systems at peak flow periods.
2	*Temperature Regulation:* By providing shade and removing heat from the air through evapotranspiration, green roofs can significantly reduce the temperatures of roofs and the surrounding air, combating the urban heat island effect.
3	*Energy Efficiency:* Green roofs act as insulators for buildings, reducing the amount of energy needed to provide heating and cooling, thereby contributing to energy conservation and climate change mitigation.
4	*Biodiversity and Habitat:* Green roofs can serve as habitat for a variety of species, contributing to biodiversity. They can provide a refuge for species that have been displaced from their natural habitats due to urban development.
5	*Air Quality:* The vegetation on green roofs can capture airborne pollutants and atmospheric deposition, and can also filter noxious gases, improving air quality.
6	*Aesthetics and Mental Health:* Green roofs can significantly improve a building's appearance and can provide recreational space, which can contribute to the mental well-being of the inhabitants.
7	*Longevity of Roof Membranes:* By protecting the roof material from extreme temperature fluctuations, ultraviolet radiation, and physical damage, green roofs can extend the lifespan of the roof.
8	*Noise Reduction:* The soil and plants on a green roof can provide sound insulation, reducing noise levels inside buildings

TABLE 10.4
The Benefits of Green Roofs

1	*Absorb Rainfall:* Green roofs can absorb rainfall, reducing the amount of stormwater runoff and easing the burden on city drainage systems.
2	*Reduce Urban Runoff:* They help to decrease the amount of stormwater that runs off roofs by absorbing water like sponges.
3	*Improve Runoff Quality*: The plants on green roofs take up and transpire rainwater and help process pollutants such as nutrients.
4	*Reduce Urban Heat Island Effect:* During the heat of summer, the surface temperature of a vegetated rooftop can be cooler than the air temperature, significantly reducing the urban heat island effect of conventional roofs.
5	*Add Green Spaces and Wildlife Habitat:* Green roofs add natural landscapes and wildlife habitat to dense urban areas.
6	*Provide Building Insulation:* They provide excellent insulation and can reduce indoor heating and cooling costs.
7	*Reduce Noise Pollution:* Green roofs can help reduce noise pollution.
8	*Increase Roof Lifespan:* The layer of soil and plants is more resistant than conventional roofing materials to the damaging effects of ultraviolet light and temperature extremes, helping to prolong the life of the roof.

mitigating the environmental impact of buildings in densely populated areas. The ways green roofs contribute to LID is shown in Table 10.3.

Green roofs are crucial in LID due to their multiple benefits, such as stormwater management, temperature regulation, and biodiversity enhancement, as shown in Table 10.4.

Green roofs help decrease energy costs by providing thermal mass and evaporative cooling. They are visually attractive and provide increased wildlife habitat and urban green space. By making sustainable living options accessible to a broader demographic, low-impact housing can garner greater support and participation from the community, leading to increased adoption of environmentally friendly practices and technologies. This inclusive approach not only benefits individuals and families seeking affordable housing but also contributes to the advancement of broader environmental objectives, such as reducing energy consumption, minimizing waste, and promoting ecological stewardship.

10.4 GREEN GARDENS AND BOX PLANTERS

Green gardens and box planters offer a versatile and space-saving way to bring nature into urban environments. They enhance aesthetic appeal, improve air quality, and provide a haven for relaxation. Whether on a balcony or in a small yard, these solutions make gardening accessible to everyone.

10.4.1 Green Gardens

Green gardens, also known as rain gardens, are a crucial element of LID. They are designed to capture, filter, and infiltrate stormwater runoff from impervious surfaces such as roofs, roads, and parking lots. This helps reduce the volume and speed of stormwater that reaches streams and rivers, thereby minimizing erosion and pollution.

The design of a green garden is simple yet effective. It is typically a shallow, vegetated basin that includes a variety of plants, from grasses to trees. These plants are chosen for their ability to thrive in both wet and dry conditions. The garden is usually located at a slight depression, allowing it to collect runoff from surrounding areas.

The benefits of green gardens extend beyond stormwater management. They provide habitat for local wildlife, increase green space, and enhance the aesthetic appeal of urban areas. Furthermore, by reducing the amount of runoff that needs to be managed by traditional stormwater infrastructure, they can result in significant cost savings. However, the effectiveness of green gardens depends on proper design and maintenance. They need to be sized and located correctly to handle the expected volume of runoff. Regular maintenance, such as removing sediment and replacing dead plants, is also crucial to ensure they continue to function properly.

Green gardens are a versatile and sustainable tool for managing stormwater in urban areas. They exemplify the principles of LID, which seeks to manage rainfall at the source using decentralized and nature-based solutions. As cities continue to grow and climate change increases the frequency of heavy rainfall events, the role of green gardens and other LID practices will only become more important.

Green gardens contribute to enhancing biodiversity and providing habitat for wildlife within urban environments. By incorporating native plants and creating diverse ecological niches, these gardens support the presence of pollinators, birds, and other beneficial organisms. This not only adds to the aesthetic appeal of the surroundings but also fosters ecological resilience and contributes to the overall health of the local ecosystem. An example of a rain garden is shown in Figure 10.1.

Green gardens can also serve as educational and community engagement tools, raising awareness about sustainable water management practices and environmental stewardship. Their visible presence in public spaces can inspire and educate residents and visitors about the importance of LID principles and the role of green infrastructure in creating more sustainable and resilient cities.

A green garden, or rain garden, is composed of several key components that work together to manage stormwater runoff. These components are shown in Table 10.5.

Each of these components plays a different role in the function of a green garden, contributing to its ability to effectively manage stormwater, create habitat, and enhance urban landscapes.

Overflow management is a critical aspect of a green garden's design, as it plays a fundamental role in safeguarding the garden against the adverse impacts of heavy rainfall events and preventing potential flooding. During periods of excessive rainfall, the overflow management system within the green garden is designed to efficiently divert excess water away from the garden, thereby preventing inundation and ensuring that the garden can continue to effectively manage stormwater without becoming overwhelmed (Figure 10.1). This proactive approach to overflow management helps maintain the garden's functionality, protects surrounding areas from waterlogging, and contributes

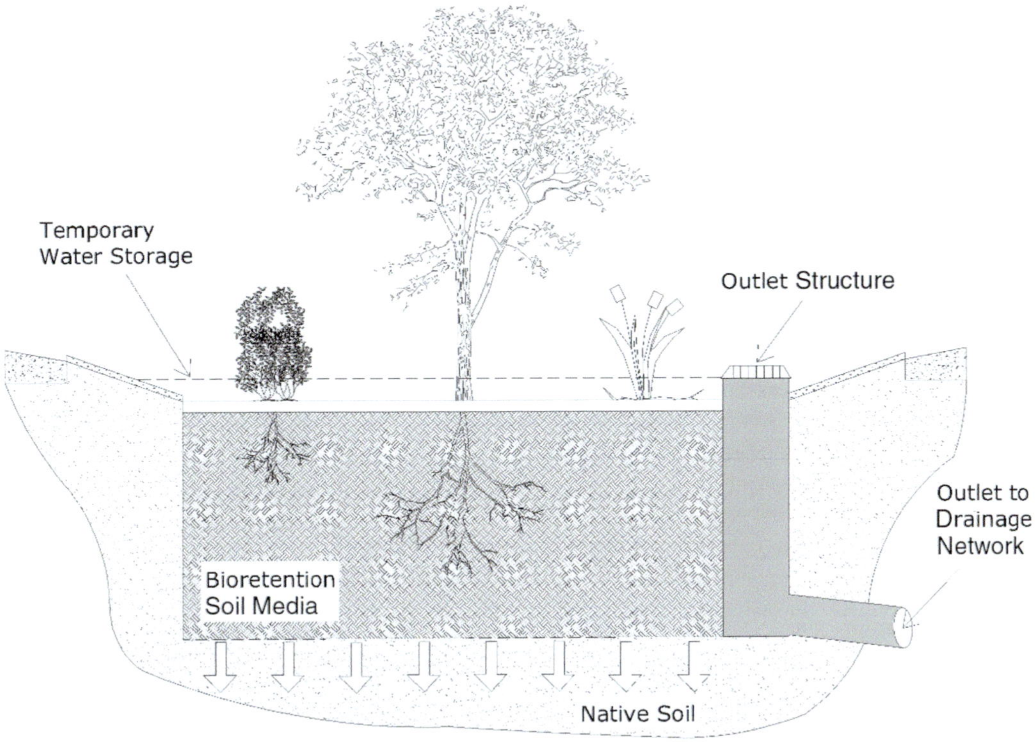

FIGURE 10.1 Rain garden with overflow management.

TABLE 10.5
The Main Components of Green Gardens

1	*Garden Location:* The garden is usually located in a slight depression and in an area where it can collect runoff from impervious surfaces such as roofs, driveways, or streets.
2	*Soil Mix:* The soil in a green garden is typically a mix of sand, compost, and topsoil. This special blend improves water infiltration and provides necessary nutrients for plant growth.
3	*Plants:* The garden is vegetated with a variety of plants that can tolerate both wet and dry conditions. These may include native grasses, flowers, shrubs, and trees. The roots of these plants help to absorb water and filter pollutants.
4	*Mulch Layer:* A layer of mulch is often applied on top of the soil. This helps to retain moisture, suppress weeds, and prevent erosion.
5	*Overflow:* An overflow area is designed for times when the volume of water exceeds the garden's capacity. This could be a grassy area or a dry creek bed that can safely convey excess water to the storm drain system.
6	*Maintenance:* Regular maintenance is a crucial component of a green garden. This includes tasks such as removing sediment, replacing dead plants, and ensuring the overflow area is clear.

to the overall resilience of the stormwater management system. The components of overflow management are shown in Table 10.6.

An effective overflow management of green gardens can handle large volumes of water, protect property from water damage, and contribute to the overall health of the local watershed. Overflow management is a key component in making the gardens a sustainable solution for stormwater management.

TABLE 10.6
The Components of Overflow Management

1	*Design:* The garden is designed with an overflow area, which is a lower section of the garden where excess water can flow out safely. This could be a grassy area, a dry creek bed, or a direct connection to the storm drain system.
2	*Capacity:* The size and location of the overflow area are determined based on the garden's capacity and the expected volume of runoff. It should be large enough to handle water from heavy rain events that exceed the garden's infiltration capacity.
3	*Routing:* The overflow area is typically located at the lowest point of the garden, allowing water to naturally flow toward it. Sometimes, a swale or a shallow ditch may be used to direct water to the overflow area.
4	*Protection:* The overflow area is designed to prevent erosion. This could involve using erosion control measures such as rocks or riprap to slow down the flow of water.
5	*Maintenance:* Regular checks are needed to ensure the overflow area is clear of debris and functioning properly. Any blockages could cause the garden to flood during heavy rain.

FIGURE 10.2 A box planter.

Source: Jensen Precast.

10.4.2 Box Planters

Box planters (Figure 10.2), also known as raised beds or container gardens, are a popular choice for both novice and experienced gardeners. They provide a controlled environment for plants to grow in, making it easier to manage soil conditions and improve plant health.

TABLE 10.7
The Components/Parts of Box Planters

1	*Planter Box:* This is the actual container in which the plants grow. It can be made from a variety of materials, including wood, plastic, or metal. The size and depth of the box can vary depending on what you plan to grow.
2	*Soil:* The soil within the box planter is usually a mix of garden soil, compost, and other organic matter. This provides a nutrient-rich environment for plants to thrive.
3	*Drainage System:* Proper drainage is crucial in a box planter to prevent waterlogging. This can be achieved through holes at the bottom of the box or using a layered system with gravel or clay pebbles at the bottom.
4	*Plants:* The choice of plants can vary widely depending on the size of the box and the gardener's preferences. You can grow anything from herbs and vegetables to flowers and small shrubs.
5	*Mulch:* A layer of mulch on top of the soil can help retain moisture, suppress weeds, and regulate soil temperature.
6	*Support Structures:* For certain plants like tomatoes or climbing vines, you might need to add support structures like stakes or trellises within the box planter.

The design of a box planter is quite simple. It is essentially a large box that can be filled with soil and plants. The box can be made from a variety of materials, including wood, metal, or plastic. It is typically raised off the ground, which helps to prevent soil compaction and improve drainage. A box planter consists of several key parts that contribute to its functionality, as shown in Table 10.7.

Regular maintenance such as watering, weeding, and replenishing the soil nutrients is also a key part of keeping a box planter healthy and productive.

One of the main advantages of box planters is their versatility. They can be used to grow a wide range of plants, from flowers and herbs to vegetables and small fruit trees. They are also ideal for people with limited space, as they can be placed on patios, balconies, or rooftops. However, box planters do require some maintenance. The soil in the box will need to be replenished regularly, as it can become depleted of nutrients over time. It is also important to ensure that the box has adequate drainage to prevent waterlogging.

Box planters are a practical and flexible gardening solution. They offer a range of benefits, including improved soil conditions, easier maintenance, and greater versatility. Whether you are a seasoned gardener or just starting out, a box planter could be the perfect addition to your outdoor space.

10.5 INFILTRATION TRENCH

An infiltration trench (Figure 10.3) is a type of stormwater management system designed to capture and manage stormwater runoff. It consists of a shallow trench filled with gravel, stone, or other porous materials that allow water to seep into the ground. The trench is typically located underground and is often situated near impervious surfaces such as parking lots, roads, or buildings to capture and treat stormwater before it enters natural water bodies.

The main purpose of an infiltration trench is to reduce the volume of stormwater runoff and to improve water quality by allowing it to naturally infiltrate into the soil. As stormwater flows into the trench, it percolates through the porous materials, where it undergoes filtration and biological processes, removing pollutants and contaminants. This helps to mitigate the impacts of urbanization on natural hydrology and reduces the risk of flooding and erosion in downstream areas.

Infiltration trenches are designed to be sustainable and low-impact stormwater management solutions. They can be integrated into green infrastructure designs to mimic natural hydrological processes and promote groundwater recharge. By allowing stormwater to infiltrate into the ground, infiltration trenches help replenish aquifers and maintain base flow in streams, which is essential for supporting aquatic ecosystems and maintaining water supply.

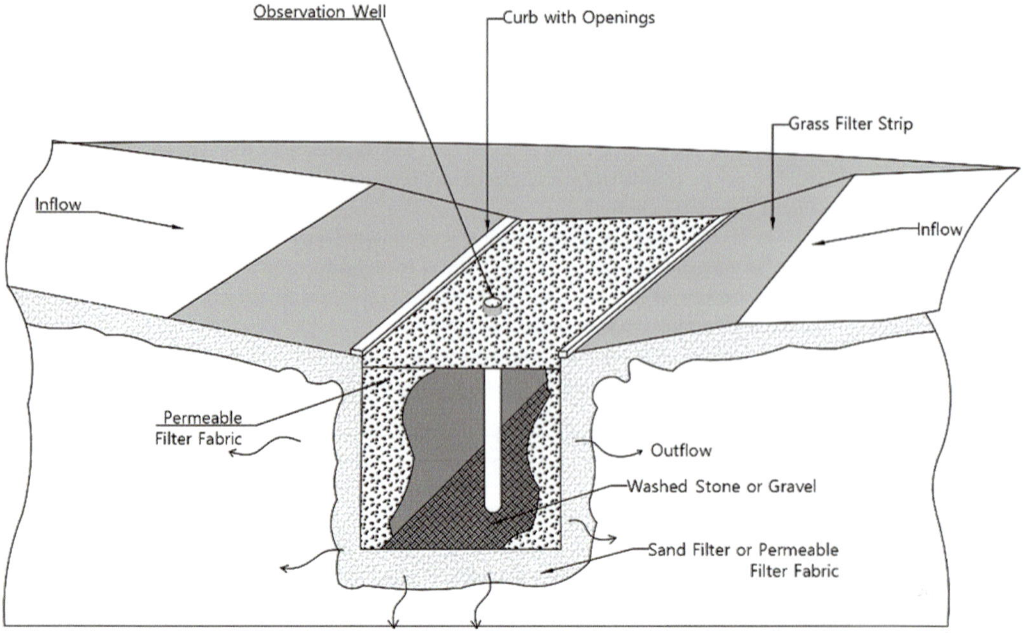

FIGURE 10.3 Infiltration trench.

Proper design and maintenance of infiltration trenches are crucial to ensure their effectiveness and longevity. Factors such as soil type, groundwater levels, and the volume and intensity of rainfall must be considered during the design phase. Regular inspection and maintenance, including sediment removal and vegetation management, are necessary to prevent clogging and to preserve the infiltration capacity of the trench.

Infiltration trenches contribute to sustainable stormwater management by reducing the impact of urban development on natural water systems. Their ability to capture, treat, and infiltrate stormwater helps to protect water quality, reduce flooding, and support the ecological health of watersheds. As cities continue to face challenges related to stormwater management, infiltration trenches offer an effective and environmentally friendly solution for mitigating the adverse effects of urbanization on the water cycle.

Geotextile materials can be utilized in the construction of infiltration trenches. They are typically made from natural or synthetic fibers, such as polypropylene, polyester, and nylon. These fibers are woven or joined to form a permeable fabric. Geotextile is used to wrap the aggregate fill material in the trench. It surrounds all the aggregate fill material except for the top 1 ft, which is placed over the geotextile. This fabric serves several important functions. It separates the aggregate from the surrounding soil, preventing the migration of soil particles into the trench, which could lead to clogging. It also allows water to pass through while keeping the aggregate and soil layers separate.

The selection of geotextile fabric needs to have acceptable properties to avoid plugging. The fabric should be selected based on its physical and hydraulic properties, including its permeability, strength, and durability. Geotextile products may also be used in other stormwater management facilities, such as bioretention facilities. The designer determines the appropriate geotextile to be used in stormwater management applications.

The geotextile materials help maintain the trench's effectiveness in managing stormwater runoff, thereby contributing to the overall success of LID strategies.

EXAMPLE 10.1: NUMERICAL EXAMPLE OF INFILTRATION TRENCH CALCULATION

Let's say the volume of runoff that needs to be managed by the trench during the duration of storm event is 14.16 m^3. We need to calculate the volume of the trench, given that the void space of the trench is 40%.

Solution:
The volume of the runoff must be absorbed by the trench safely. This runoff will be stored temporarily by the trench in the available void space, which is 40%. Therefore, the total volume of the trench will be:

$$V = 14.16 \text{ m}^3/0.4 = 35.4 \text{ m}^3$$

If the trench's width, w, is 1.5 m, and the trench's depth, d, is 2.0 m, the trench's length is, therefore, $L = V/(w \times d) = 35.4 \text{ m}^3/(1.5 \times 2) = 11.8 \text{ m}$, say 12.0 m.

The type of soil determines its infiltration rate, which is the speed at which water can soak into the soil. For example, sandy soils have high infiltration rates, while clay soils have low infiltration rates. The infiltration rate should ideally range between 12 and 75 mm/h. The soil type and infiltration rate influence the design of the trench, including its size, depth, and the type of aggregate used. For instance, a trench in a sandy soil area may need to be deeper to prevent quick drainage and ensure adequate water treatment.

Infiltration rate can be determined in the geotechnical laboratory or can also be done onsite. In case the test is done onsite, the procedure of the test is as follows:

1. Choose a location for the test that is representative of the area where the infiltration trench will be installed. We will need a ring or pipe that can be driven into the ground. The ring should be about 20 cm in height.
2. Drive the ring halfway into the ground using a hammer. This should leave about 10 cm of the ring standing above the ground, while the remaining 10 cm is below the ground surface.
3. Pour water into the ring to a depth of 10 cm. If we are using a 3″ ring, we will need approximately 107 mL of water. For a 6″ ring, we will need about 444 mL.
4. Start a stopwatch immediately after adding the water. Record the time it takes for the water to infiltrate the soil. The average infiltration rate is equal to the water depth in the ring, that is, 10 cm divided by time (in minute or hour).
5. Conduct the test at several locations to get a representative measurement.

Never leave excavations unsecured and unmarked, and always notify all applicable authorities prior to any work. Also, keep in mind that this is a simplified explanation, and actual procedures may vary based on local regulations and site-specific conditions.

10.6 PERMEABLE PAVEMENTS

Permeable pavements, also known as pervious or porous pavements, are a type of surface that allows water to pass through it. They are made of either porous material or nonporous blocks spaced so that water can flow between the gaps. Permeable pavements can include a variety of surfacing techniques for roads, parking lots, and pedestrian walkways.

The main purpose of permeable pavements is to manage stormwater runoff. Unlike traditional impervious paving materials such as concrete and asphalt, permeable paving systems allow

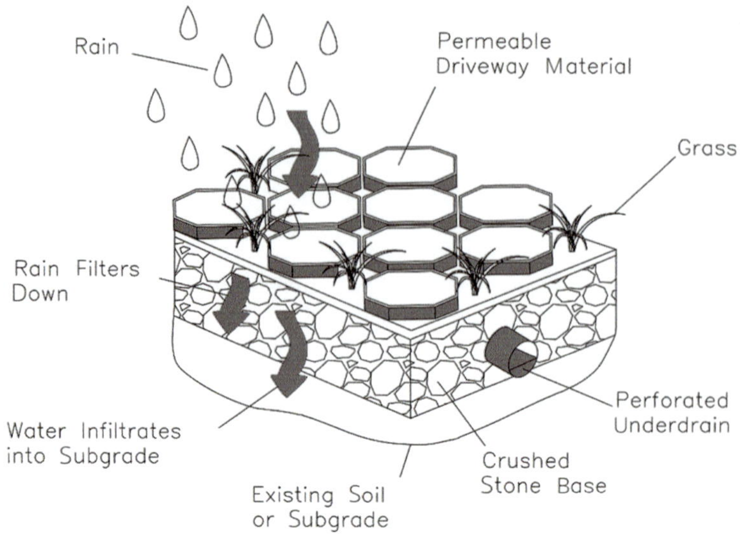

FIGURE 10.4 How does permeable pavement work.

stormwater to percolate and infiltrate through the pavement and into the aggregate layers and/or soil below. This reduces surface runoff, thereby minimizing flooding and erosion.

Permeable pavements (Figure 10.4) can be composed of various materials, including pervious concrete, porous asphalt, paving stones, or interlocking pavers. These materials are designed to be porous, allowing water to flow through them. They can also trap suspended solids, filtering pollutants from stormwater.

Permeable pavements are designed to allow water to pass through, reducing runoff and helping manage stormwater. Here is how the permeable pavements work.

1. *Surface Layer:* The surface layer of permeable pavement is porous, allowing water to flow through it. This can be achieved using pervious concrete, porous asphalt, or interlocking pavers with spaces between them.
2. *Subsurface Layer:* Beneath the surface layer is a subsurface layer, often made up of crushed stone or gravel. This layer acts as a reservoir, temporarily storing the water that has passed through the surface layer.
3. *Infiltration:* The water stored in the subsurface layer then slowly infiltrates into the ground. This process helps recharge groundwater resources and reduces the volume of runoff that ends up in storm drains.
4. *Filtration:* As water passes through the pavement and the subsurface layer, pollutants are filtered out. This improves the quality of the water that eventually reaches the groundwater or local waterways.
5. *Overflow:* During heavy rain events, when the volume of water exceeds the pavement's capacity, the excess water will flow over the surface and into storm drains. This is similar to what happens with traditional, non-permeable pavements.

Permeable pavements are commonly used on roads, paths, and parking lots subject to light vehicular traffic. They are also suitable for different conditions, including hot climates, high-speed traffic areas, and less-traveled areas. The three common types of permeable pavement include traditional concrete/asphalt, plastic pavers, and concrete pavers. Permeable pavements can be installed

TABLE 10.8
Permeable Pavement Installation

Parking Lots: Permeable pavements are commonly used in parking lots. They can effectively manage the large amounts of runoff generated by these impervious surfaces.
Sidewalks and Walkways: Sidewalks and pedestrian walkways are also good candidates for permeable pavements. They can improve pedestrian safety by reducing surface water and ice formation.
Driveways: Residential driveways can benefit from permeable pavements. They can manage runoff and add aesthetic appeal to the property.
Low-Traffic Roads: Roads with low traffic volumes are suitable for permeable pavements. They can handle the weight of vehicles while effectively managing stormwater.
Garden Areas: Garden areas and other landscaped spaces can also benefit from permeable pavements. They can help manage irrigation runoff and contribute to the health of plants.
Swimming Pool Decks: Permeable pavements can be used around swimming pools to manage splash water and provide a slip-resistant surface.

in a variety of places where runoff from rainwater is an issue. Table 10.8 shows some suitable places for the installation of permeable pavements.

The suitability of permeable pavements also depends on the local climate, soil type, and maintenance considerations.

Traditional concrete/asphalt is the standard mix minus the fine particles, which are left out to make it more porous. Plastic pavers have a honeycomb shape that allows vegetation, such as grass, to grow through the holes. Concrete pavers have spaces between the blocks, which aid better drainage and water permeability.

One of the advantages of permeable pavements is that they do not produce a "heat island" – an area on the pavement that is much warmer than the surroundings. They can also be made using recycled materials, which puts less pressure on the environment. As water and precipitation seep through the pavement, there will not be any ice formation on it. However, permeable pavements also have some disadvantages. They are more expensive to install compared to traditional pavements. The maintenance requirements of permeable pavement are quite different. They are prone to clogging if the water in the reservoir is not drained out properly.

The sand and fine particles that can block the space between the pavers must be removed using an industrial vacuum. It can even clog when you sand for ice during the winter. If you do not cater to clogging quickly, it will cause the water and pollutants to run off the surface, defeating the purpose of installing permeable pavement. Permeable pavements are an effective tool for managing stormwater runoff. They help prevent flooding, improve water quality, and contribute to groundwater recharge. However, careful design and regular maintenance are crucial to their effectiveness.

There are several types of porous materials that can be used in permeable pavements, as shown in Table 10.9.

Materials #5 to #8 are considered innovative materials, as they have just emerged recently. Please take note that each of these materials has its own advantages and disadvantages, and the choice of material will depend on the specific requirements of the project. Factors to consider in selecting the correct materials for permeable pavement are shown in Table 10.10.

Calculating the life cycle cost for different permeable pavement material options involves identifying initial installation costs, estimating ongoing maintenance expenses and replacement costs, and considering environmental impact. Factors such as durability, traffic resistance, water quality impact, and long-term societal benefits should be assessed to make informed decisions aligning with sustainability goals and optimizing cost-effectiveness. Estimating the life-cycle costs involves the steps shown in Table 10.11.

TABLE 10.9
Porous Materials

1	*Porous Asphalt:* This is a type of asphalt that has been engineered to have more void spaces than traditional asphalt. It allows water to drain through it into the underlying layers.
2	*Pervious Concrete:* Pervious concrete is a special type of concrete with a high porosity that allows water to pass through it. It is often used in parking lots and on sidewalks.
3	*Permeable Interlocking Concrete Pavers (PICP):* These are concrete blocks with spaces between them that allow water to infiltrate. They are often used in driveways, patios, and walkways.
4	*Gravel, Stone Aggregate, and Shells:* On porous foundations, stone aggregate or gravel with an average grain diameter that is consistent could be used. These materials could require more maintenance when they are used on less solid foundations.
5	*Plastic Grids:* Plastic grids have a honeycomb shape that allows vegetation, such as grass, to grow through the holes. They are often used in parking lots and driveways.
6	*Woodchips and Pine Bark:* These natural materials can also be used as a loose covering layer or to be embedded in clay or sand.
7	*Grass-Concrete Pavers:* These are concrete blocks with large holes that can be filled with soil and seeded with grass. They provide a sturdy surface that can handle vehicle traffic while still allowing water to infiltrate.
8	*Composite Resin:* This is a type of plastic that can be molded into a variety of shapes and sizes. It is often used in combination with other materials to create permeable surfaces.

TABLE 10.10
Factors to Consider in Selecting Materials

1	*Project Requirements:* Consider the specific needs of your project. For instance, for paving a high-traffic area, we might need a material that can withstand heavy use.
2	*Climate:* The local climate can influence your choice of material. Some materials may be more suitable for certain climates than others.
3	*Cost:* The cost of the material can be a significant factor. While some innovative materials may offer superior performance, they may also be more expensive.
4	*Maintenance:* Consider the maintenance requirements of the material. Some materials may require more frequent maintenance than others.
5	*Environmental Impact:* If sustainability is a priority, consider the environmental impact of the material. Some materials may be more eco-friendly than others.
6	*Aesthetics:* The appearance of the material can also be a factor, especially for projects where aesthetics are important.
7	*Local Regulations:* Check local regulations and guidelines. Some materials may not be allowed in certain areas.

Once we have estimated all of these costs and benefits, we can compare the total life-cycle costs of different material options. This can help us make an informed decision about which material is the most cost-effective choice for our specific project.

10.7 VEGETATIVE SWALE

Vegetative swales are an innovative stormwater management practice designed to effectively manage runoff and promote environmental sustainability. These landscaped channels are engineered to slow, collect, and filter stormwater, reducing the risk of flooding and erosion while improving water quality. Typically integrated into urban and suburban settings, vegetative swales are characterized by gently sloping, vegetated channels that effectively capture and treat stormwater runoff. Vegetative

TABLE 10.11
Steps in Calculating Life-Cycle Costs

1	*Initial Construction Costs:* This includes the costs of materials, labor, and equipment needed for the initial construction of the pavement. It is important to consider all aspects of construction, including excavation, fill, subbase, base, permeable surface, line painting, drainage system, etc.
2	*Maintenance and Rehabilitation Costs:* These are the costs associated with maintaining the pavement over its lifetime. This can include routine maintenance tasks such as vacuuming, as well as major rehabilitation activities such as resurfacing or replacing worn and cracked pavers.
3	*Replacement Costs:* Eventually, the pavement will reach the end of its service life and need to be replaced. The cost of replacement should be included in the life-cycle cost analysis.
4	*Environmental and Social Costs/Benefits:* These are the costs and benefits associated with environmental and social impacts. For example, permeable pavements can reduce stormwater runoff, which can lead to cost savings in terms of reduced stormwater management infrastructure. They can also provide social benefits such as improved water quality and reduced flooding.
5	*Discount Rate:* The discount rate is used to convert future costs and benefits into present value terms. This allows for a fair comparison of costs and benefits that occur at different times.

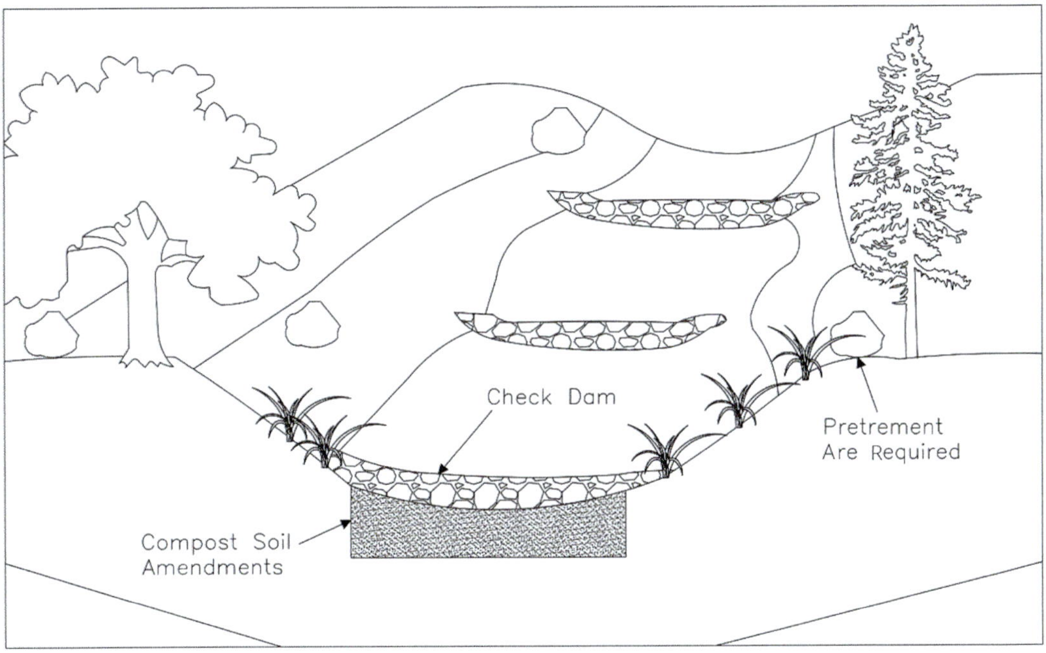

FIGURE 10.5 Vegetative swales.

swales incorporate a variety of native plants and grasses; these swales enhance infiltration and pollutant removal, contributing to the overall health of the surrounding ecosystem. An example of the vegetative swale is shown in Figure 10.5.

One of the primary functions of vegetative swales is to mitigate the impact of stormwater runoff by promoting infiltration and reducing peak flow rates. As stormwater flows into the swale, the vegetation and soil help to slow down the water, allowing sediment and pollutants to settle out and be filtered. This process helps to prevent erosion and reduces the amount of contaminants entering nearby water bodies, ultimately improving water quality. In addition, the vegetation within the swale plays a crucial role in enhancing infiltration, allowing stormwater to percolate into the soil, replenishing groundwater, and reducing the burden on traditional stormwater infrastructure.

Vegetative swales contribute to the overall aesthetic and ecological value of the surrounding environment and provide habitat for wildlife, support biodiversity, and enhance green spaces within urban areas, by incorporating diverse plant species into their elements. The visual appeal of the swales can also contribute to the creation of attractive, sustainable landscapes that benefit the community and improve the overall quality of life. In addition, the presence of vegetative swales can serve as an educational tool, raising awareness about sustainable stormwater management practices and fostering a sense of environmental stewardship within the community.

The implementation of vegetative swales supports sustainable development and provide a cost-effective and environmentally friendly approach to stormwater management. Compared to traditional stormwater infrastructure, such as concrete channels or underground pipes, vegetative swales offer a natural and low-impact solution that can be integrated into various land uses. This approach not only reduces the reliance on conventional stormwater systems but also contributes to the conservation of natural resources and the preservation of ecological functions within urban and suburban areas. The long-term maintenance requirements of vegetative swales are often minimal, making them a sustainable and practical choice for stormwater management in diverse settings.

Vegetative swales enhance community resilience to climate change and extreme weather events. These swales help to reduce the risk of localized flooding and erosion, contributing to the overall resilience of urban and suburban areas. The increased infiltration and groundwater recharge facilitated by vegetative swales can help mitigate the impacts of drought and support water resource management in regions prone to water scarcity. As a result, these swales contribute to building more resilient and sustainable communities that are better equipped to adapt to the challenges posed by a changing climate.

Vegetated swales, also known as bioswales, are carefully engineered landscape features that serve as effective stormwater management tools. These contoured land areas or ditches are strategically planted with a variety of vegetation to facilitate the collection, conveyance, and treatment of stormwater runoff. Vegetated swales promote infiltration and pollutant removal, enhancing water quality and reducing the risk of flooding and erosion. The design of these swales is tailored to redirect stormwater away from impermeable surfaces such as roads, parking lots, and rooftops, allowing it to be naturally absorbed into the soil.

The diverse appearance of vegetated swales reflects their adaptability to different environments and land uses. While some swales may resemble straight channels, others are designed to mimic the characteristics of meandering natural streams. This variability in design allows for the customization of swales to suit specific site conditions and aesthetic preferences. Straight channels are often used in urban settings to efficiently convey stormwater, while meandering swales are utilized in more natural or park-like environments to create visually appealing features that enhance the overall landscape. Regardless of their appearance, both types of swales serve the common purpose of managing stormwater runoff and promoting environmental sustainability.

Along with their functional role in stormwater management, the design of vegetated swales also contributes to their ecological and aesthetic value. Selected vegetation within the swales not only aids in stormwater treatment but also provides habitat for wildlife, supports biodiversity, and enhances the visual appeal of the surrounding area. The integration of native plants and grass helps to create naturalistic landscapes that blend seamlessly with the built environment, enriching urban and suburban spaces with green infrastructure that benefits both the community and the local ecosystem.

The design of vegetated swales aligns with principles of LID and sustainable urban design. By incorporating vegetated swales into the built environment, cities and communities can manage stormwater while minimizing the reliance on traditional, energy-intensive stormwater infrastructure.

TABLE 10.12
Contributing Process of Vegetative Swales to Groundwater

1	*Stormwater Collection:* Vegetated swales are designed to collect stormwater runoff from impervious surfaces like roads, parking lots, and rooftops. This water would otherwise flow directly into storm drains or water bodies, carrying pollutants with it.
2	*Infiltration:* The design of vegetated swales slows down the flow of water, allowing more time for it to infiltrate, or soak into the ground. The soil and plants in the swale help to filter out pollutants, and the clean water then percolates down to recharge the groundwater.
3	*Soil and Vegetation:* The type of soil and vegetation used in the swale can also affect the rate of infiltration. Soils with higher permeability allow water to infiltrate more quickly. Similarly, certain types of vegetation can enhance infiltration rates by creating channels in the soil for water to flow through.
4	*Check Dams:* In some cases, structures known as check dams are installed in the swale to further slow down water flow and promote infiltration. These small dams create pooling areas that increase the residence time of water in the swale, allowing more of it to infiltrate.
5	*Water Table Recharge:* The water that infiltrates the ground in vegetated swales eventually reaches the water table, recharging the groundwater system. This process is vital in areas where groundwater is a significant source of water supply.
6	*Sustainable Water Management:* By promoting groundwater recharge, vegetated swales contribute to sustainable water management. They help to reduce overland flow and therefore prevent floods, maintain water balance in the local ecosystem, reduce the impact of droughts and water shortages, and support the health of local water bodies and wetlands.

This approach not only reduces the burden on municipal drainage systems but also contributes to the conservation of natural resources and the enhancement of urban resilience. The thoughtful design of vegetated swales reflects a commitment to sustainable development practices that prioritize environmental stewardship, community well-being, and long-term ecological health.

The design and function of vegetated swales represent a harmonious integration of practical stormwater management with ecological and aesthetic considerations. Their ability to collect and convey stormwater runoff, combined with their capacity to support biodiversity and enhance the visual appeal of urban and suburban landscapes, underscores the multifaceted benefits of these green infrastructure features.

Vegetative swales can contribute to groundwater recharge, and therefore reduce surface runoff. This process is shown in Table 10.12.

Implementing vegetated swales for groundwater recharge can pose challenges, including space requirements, maintenance aspects, generate standing water, and other challenges. These challenges need to be addressed for effective long-term functionality and success. The challenges are shown in Table 10.13.

Incorporating innovative designs and technologies, such as engineered media for improved pollutant removal, advanced monitoring systems for real-time performance evaluation, and innovative plant species selection for enhanced biodiversity, can significantly elevate the effectiveness and environmental impact of vegetated swales.

Engineered media for improved pollutant removal:

- Customized media blends are designed to effectively capture and retain pollutants from stormwater runoff.
- Incorporation of specific filter materials, such as activated carbon or zeolite, to target and remove contaminants like heavy metals and hydrocarbons.
- Implementation of biofiltration technologies within the media to enhance biological degradation of pollutants, promoting cleaner stormwater discharge.

TABLE 10.13
Challenges in Implementing Vegetative Swales

1	*Space Requirements:* Vegetated swales require more room than traditional stormwater management systems like curb and gutter systems. This can be a challenge in urban areas where space is often limited.
2	*Maintenance:* Vegetated swales require regular maintenance to ensure their effectiveness. This includes removing sediment and trash, managing vegetation, and repairing any damage. Lack of maintenance can lead to overgrowth of vegetation and reduced functionality.
3	*Standing Water:* If not properly designed or maintained, vegetated swales can result in standing water, which can pose safety risks and create potential mosquito problems.
4	*Biological Evolution:* The biological growth occurring in the surface layer of vegetated swales, as well as the biological evolution taking place in them, can introduce further difficulties. This can affect the swale's performance over time.
5	*Climate Change and Urbanization:* Rapid urban development and climate change can impact the effectiveness of vegetated swales. Increased runoff from impervious surfaces and more frequent intense rainfall can challenge the capacity of swales.
6	*Design Modifications:* Significant modifications might be needed in the existing nature-based treatment solutions to control floods and remove pollutants in rapid urban landscapes and extreme climate conditions.

Advanced monitoring systems for real-time performance evaluation:

- Integration of sensor networks to continuously monitor flow rates, water quality parameters, and sediment accumulation within vegetated swales.
- Utilization of remote telemetry systems for real-time data collection and analysis, enabling prompt response to changing environmental conditions.
- Adoption of automated monitoring technologies to track the effectiveness of stormwater treatment and identify maintenance needs.

Innovative plant species selection for enhanced biodiversity:

- Utilization of native plant species adapted to local climate and soil conditions to support diverse wildlife habitats and ecological functions.
- Incorporation of vegetation with varying root structures to improve soil stability, enhance water infiltration, and promote microbial activity.
- Integration of flowering plants and grasses to attract pollinators and contribute to the overall ecological richness of the swale environment.

Soil amendment to improve the infiltration rate:

- Application of permeable amendments, such as compost or biochar, to enhance soil porosity and water retention capacity.
- Incorporation of organic matter to promote microbial activity and improve soil structure, facilitating increased infiltration and groundwater recharge.
- Implementation of soil aeration techniques to alleviate compaction and enhance the permeability of the soil within vegetated swales.

Vegetated swales are a sustainable and effective solution for stormwater management. They not only enhance the aesthetic appeal of urban landscapes, but also contribute to biodiversity, improve water quality, and promote groundwater recharge. Their low maintenance and cost-effectiveness make them an excellent choice for urban planning.

Vegetated swales stand as a testament to the integration of nature-based solutions within urban and suburban landscapes. Their multifaceted benefits, from stormwater management and water quality improvement to biodiversity support and community resilience, underscore their pivotal role in sustainable development, fostering healthier and more resilient environments for generations to come.

10.8 CONSTRUCTED WETLANDS

A constructed wetland is an artificial wetland created as a new or restored habitat for native and migratory wildlife, which mimics the functions of natural wetlands. Constructed wetlands are designed to treat diverse types of wastewater, such as sewage, greywater, stormwater runoff, and industrial wastewater. They can also be designed for land reclamation after mining, or as a mitigation step for natural areas lost to land development.

Constructed wetlands use the natural functions of vegetation, soil, and organisms to provide secondary treatment to wastewater. They are engineered systems that can improve water quality. The design of the constructed wetland must be adjusted according to the type of wastewater to be treated.

There are two main types of constructed wetlands: subsurface flow and surface flow:

- *Subsurface Flow Constructed Wetlands (SF CWs)*: In these wetlands, the water level is kept below the top of the rock or gravel media. The water flows through a bed or channel containing appropriate media, such as coarse rock, gravel, sand, and other soils. The medium is typically planted with the same types of emergent vegetation present in marshes. The main advantages of this subsurface water level are the prevention of mosquitoes and odors, and the elimination of the risk of public contact with the partially treated wastewater.
- *Surface Flow Constructed Wetlands (FWS CWs)*: In these wetlands, water flows above ground. They aim to replicate the naturally occurring processes of a natural wetland, marsh, or swamp. As water slowly flows through the wetland, pollutants are removed through various biological, chemical, and physical processes. The water is exposed to the atmosphere, allowing for oxygen transfer. These wetlands are often used to treat stormwater runoff, wastewater, and other water sources that may contain pollutants. However, they require more land area compared to subsurface flow constructed wetlands.

Constructed wetlands can remove a range of pollutants from the water, such as suspended solids, organic matter, nutrients (nitrogen and phosphorus), and all types of pathogens (i.e., bacteria, viruses, protozoans, and helminths). They are cost-effective, require low maintenance, and provide significant water quality benefits.

Constructed wetlands are an essential component of LID strategies, designed to manage stormwater and improve water quality in urban and suburban areas. These engineered ecosystems mimic the natural functions of wetlands, utilizing vegetation, soil, and microbial processes to treat and retain stormwater runoff. One key feature of constructed wetlands is their ability to capture and filter pollutants from stormwater, such as sediment, nutrients, and heavy metals. This natural filtration process helps to improve water quality and reduce the negative environmental effects of urban runoff on receiving water bodies.

The vegetation in constructed wetlands contributes to the uptake and transformation of pollutants, further enhancing the treatment efficiency of these systems. Constructed wetlands also replenish groundwater by facilitating the infiltration of stormwater into the soil. This helps to recharge aquifers and maintain base flow in streams, which is particularly important in regions experiencing water scarcity.

The presence of wetlands in urban areas can provide valuable habitat for wildlife, contributing to biodiversity conservation and ecological balance within the built environment. In addition to their

TABLE 10.14
Design Considerations in Constructed Wetlands

1	*Conservation of Ecological Character:* The design should aim to conserve and enhance the ecological character of the area.
2	*Innovative Approach:* Look for innovative approaches to integrate the wetland with the surrounding environment.
3	*Local Culture and Lifestyle:* The design should consider the culture and lifestyle of local people.
4	*Economic Benefits:* Try to integrate the economic benefits of wetlands with urban design.
5	*Landscape Integration:* The landscape of wetlands should be integrated into urban design.
6	*Species-Specific Design:* The design must be species-specific, catering to the needs of local flora and fauna.
7	*Hydrodynamic Design:* The hydrodynamics of the wetland, including water flow and water level, are crucial factors in the design.
8	*Ease of Maintenance:* The wetland should be designed in a way that makes maintenance easy.
9	*Water Quality Characterization:* Understanding the quality of the water to be treated is crucial for the design.
10	*Wetland Inflow Estimation:* The volume and rate of inflow into the wetland need to be estimated accurately.
11	*Wetland Sizing and Configuration:* The size and configuration of the wetland should be based on the volume of water to be treated, the type of pollutants present, and the desired quality of the treated water.
12	*Inlet and Outlet Structures:* The design of the inlet and outlet structures is important for controlling the flow of water through the wetland.

environmental benefits, constructed wetlands also offer aesthetic and recreational value to communities. These green spaces can serve as natural amenities, providing opportunities for passive recreation, wildlife observation, and educational activities.

Constructed wetlands represent a versatile and effective solution for managing stormwater in urban and suburban environments. As a component of LID, these engineered ecosystems play a critical role in promoting water quality, mitigating the impacts of urbanization, and enhancing the overall sustainability of communities. By harnessing the natural processes of wetlands, LID strategies can achieve multiple benefits, ranging from pollution reduction and groundwater recharge to ecological restoration and community well-being.

When designing a constructed wetland, it is crucial to consider various factors such as ecological conservation, technological innovation, cultural relevance, and lifestyle integration. These considerations ensure that the wetland effectively harmonizes with its surroundings while fulfilling its environmental and societal functions. The design considerations are shown in Table 10.14.

Following these considerations during the design and implementation of constructed wetlands may ensure that the constructed wetland is effective in treating wastewater, sustainable, and beneficial to the local ecosystem.

Construction of artificial wetlands generates a multitude of social, ecological, and economic benefits. These benefits are shown in Table 10.15.

The presence of diverse and resilient vegetation within well-designed wetlands is essential for enhancing the ecological functions, water treatment capabilities, and overall success of these engineered ecosystems in managing stormwater and promoting environmental sustainability. The roles of vegetation are shown in Table 10.16.

The soil in a well-functioning wetland plays a multifaceted role critical to its success. It serves as a provider of essential nutrients, offers habitat for diverse organisms, regulates water flow and quality, and contributes to the overall ecological balance and functionality of the wetland ecosystem. Its roles are shown in Table 10.17.

TABLE 10.15
The Benefits of Artificial Wetlands

1	*Water Quality Improvement:* Wetlands act as natural filters, improving water quality by absorbing excess nutrients and trapping suspended sediments. They can remove pollutants before water leaves a wetland, making it healthier for drinking, swimming, and supporting plants and animals.
2	*Flood Control:* Wetlands can reduce the frequency and intensity of floods by acting as natural buffers, soaking up and storing a significant amount of floodwater. This can help prevent property damage and loss of life.
3	*Erosion Control:* Wetlands help control erosion by stabilizing shorelines and riverbanks. They can also trap sediment that would otherwise contribute to downstream sedimentation.
4	*Habitat Provision:* Wetlands provide a home to a vast array of wildlife, including many threatened and endangered species. They serve as habitats for a variety of plants and animals, contributing to biodiversity.
5	*Carbon Sequestration:* Wetlands play a crucial role in climate regulation by sequestering carbon, thus helping to mitigate climate change.
6	*Recreational Opportunities:* Wetlands offer opportunities for recreation and aesthetic appreciation, such as bird-watching, hiking, and photography.
7	*Economic Benefits:* Wetlands contribute to the economy by producing resources, enabling recreational activities, and providing other benefits, such as pollution control and flood protection.

TABLE 10.16
The Roles of Vegetations in Constructed Wetlands

1	*Water Purification:* Wetland vegetation, such as reeds, grasses, and trees, plays a crucial role in water purification. The extensive root systems of these plants provide habitat for microorganisms that participate in the degradation of contaminants. In addition, wetland plants absorb nutrients, effectively preventing their entry into water bodies.
2	*Sediment Trapping and Retention:* Plants enhance the overall resilience of coastal wetlands to sea-level rise by improving mineral sediment trapping and retention. This is especially important when and where supplies are low.
3	*Contribution to Soil Volume:* Plants contribute organic matter to soil volume, which allows continued vertical expansion, within limits, even when mineral sediment is deficient.
4	*Resistance to Erosion:* Plants aid in resistance to compaction and erosion of deposited sediment by increasing soil shear strength.
5	*Habitat Provision:* The vegetation in wetlands provides a habitat for a variety of wildlife, contributing to biodiversity.
6	*Hydroclimatic Regime:* Understanding how flow regime drives the spatiotemporal dynamics of wetland habitats is fundamental to effective conservation practices.

10.9 RAIN BARRELS AND RAIN HARVESTING

Rain barrels collect and store rainwater from rooftops, providing a sustainable source of water for activities such as watering gardens, washing cars, and other non-potable uses. Rainwater harvesting systems capture and store rainwater for various purposes, including landscape irrigation, toilet flushing, and even potable water use in some cases, thereby reducing the demand on municipal water supplies and mitigating stormwater runoff. These methods help to conserve water resources, minimize urban flooding, and alleviate pressure on traditional water sources, contributing to more sustainable and resilient water management practices in urban and suburban areas.

Rain barrels are containers designed to collect and store rainwater that accumulates on rooftops. They are typically connected to the gutters and downspouts of a building, allowing them to efficiently capture rainwater runoff. By intercepting and storing rainwater, rain barrels play a crucial role in reducing the volume of runoff, which, if left unchecked, can lead to landscape erosion and the transportation of pollutants into nearby waterways.

TABLE 10.17
The Roles of Soil in Well-Functioning Artificial Wetlands

1	*Nutrient Provision:* Soils provide nutrients for plants, creating a fertile environment for vegetation growth.
2	*Habitat for Organisms:* Soils offer habitats for a variety of organisms, including animals and microbes. After death, all organisms are decomposed in the soil.
3	*Water Regulation:* Soils play a key role in regulating water storage and flows, and consequent flood risk. The capacity for water storage within wetland soils is limited to those periods when soil pores may not be fully charged with water.
4	*Soil Formation:* Soils are formed by long-term interactions between external environmental factors. Like wetlands themselves, soils can take a long time to form, though almost invariably their damage or destruction can be much more rapid, particularly in the case of peat soils.
5	*Hydromorphic Soils:* The main characteristic that wetlands share is their prolonged saturation with water, and soils influenced by these conditions are termed hydromorphic soils. Depending on features of the local geology, topography, and climate, the source of this water will vary.
6	*Quality Protection:* Protection of the soil should be an integral part of wetland conservation and play a key role in wetland management.

The stored rainwater in these barrels can be utilized for a variety of purposes, such as watering gardens and plants, or even for flushing toilets. By providing an alternative source of water for non-potable uses, rain barrels help to alleviate the demand for treated water, thereby promoting water conservation and reducing the strain on municipal water supplies. This sustainable practice not only contributes to the efficient use of water resources but also aids in minimizing the environmental impact of stormwater runoff on urban and suburban areas.

Rainwater harvesting encompasses a wider scope than rain barrels, as it involves the collection of rainwater from diverse surfaces beyond rooftops. This collected water can be stored in larger systems such as cisterns or utilized directly for irrigation purposes. By adopting rainwater harvesting systems, communities can effectively reduce runoff and mitigate the adverse impacts of stormwater on the environment.

Similar to rain barrels, rainwater harvesting systems play a vital role in managing stormwater by intercepting and storing rainwater, thereby decreasing the volume of runoff and its associated negative effects. By implementing these systems, communities can enhance their resilience to urban flooding, reduce erosion, and minimize the transport of pollutants into water bodies, contributing to improved water quality and sustainable water management practices.

Both rain barrels and rainwater harvesting systems can play a significant role in managing stormwater, reducing overland flow, and contributing to water conservation. They are particularly beneficial in urban areas where impervious surfaces are abundant, and overland flow can quickly become a problem.

Rain barrels and rainwater harvesting systems are, to some extent, effective in reducing overland flow, depending on the volume of rainwater captured by the system. The ways rain barrels and rainwater harvesting systems are shown in Table 10.18.

The percentage of overland flow reduction by rain barrels and rainwater harvesting can vary based on the specific design and implementation of the system, that is, the volume of rainwater to be captured, as well as the local climate and rainfall patterns. However, several studies provide some estimates. A study conducted at Purdue University found that the deployment of large residential rain barrels resulted in up to a 12% reduction in predicted combined sewer overflow volume.

Implementing rain barrels and rainwater harvesting can face several challenges and constraints, as shown in Table 10.19.

TABLE 10.18
Rain Barrels and Rainwater Harvesting Systems

Reduction in Runoff and Erosion: These systems capture rainwater that would otherwise contribute to runoff, thereby reducing soil erosion and flooding.
Conservation of Water Resources: By storing rainwater for later use, these systems reduce the demand for municipal water supply, thereby conserving precious natural resources.
Protection of Waterways: Reduced runoff means fewer pollutants are carried into our waterways, which can harm aquatic life and human health.
Energy Conservation: Less demand for municipal water supply reduces the energy used for water treatment and distribution.
Energy Conservation: Less demand for municipal water supply reduces the energy used for water treatment and distribution.
Climate Change Mitigation: By managing stormwater, these systems can help mitigate the impacts of increased rainfall intensity due to climate change.

TABLE 10.19
Challenges in Rainwater Harvesting Implementation

Financial Constraints: Adequate funding is essential for successful rainwater harvesting projects. The costs of installation and maintenance can be high, especially in developing countries.
Technical Complexity: Rainwater harvesting systems can be intricate and require technical expertise for design, construction, and maintenance.
Lack of Acceptance and Motivation: There can be a lack of acceptance, motivation, and involvement among potential users.
Missing Hydrological Data: There can be a lack of hydrological data and information for confident planning, design, and implementation of rainwater harvesting systems.
Inadequate Policy Support: A conducive regulatory framework is vital for promoting rainwater harvesting practices on a larger scale.
Space Availability: The availability of space for installing rainwater harvesting systems can be a challenge, especially in urban areas.
Unpredictability of Rainfall: The unpredictability of rainfall can affect the efficiency and reliability of rainwater harvesting systems.
Quality of Collected Water: There can be issues related to the quality of collected water, which may require additional treatment before use.

PROBLEMS AND ISSUES FOR DISCUSSION

1. How does LID contribute to sustainable urban planning, and what are the key principles that differentiate it from traditional stormwater management practices?
2. Discuss the role of green infrastructure in LID. How do features like green roofs, permeable pavements, and rain gardens enhance urban resilience against flooding and water quality issues?
3. Analyze the economic benefits of implementing LID strategies in urban areas. How can these practices lead to cost savings in infrastructure and maintenance over the long term?
4. Evaluate the effectiveness of LID in mitigating the impacts of climate change. In what ways can LID practices help communities adapt to changing weather patterns and increased precipitation?

6. Compare and contrast LID with conventional development methods. What are the environmental, social, and economic implications of choosing one approach over the other in urban settings?
7. How can community engagement and education enhance the implementation of LID practices? Discuss strategies for involving local residents in LID initiatives to promote sustainable practices.
8. Investigate the regulatory challenges faced by municipalities in adopting LID practices. What policy changes are necessary to facilitate the widespread use of LID in urban planning?
9. Explore the relationship between LID and ecosystem services. How do LID practices contribute to the preservation and enhancement of natural habitats within urban environments?
10. What role do native plants play in LID? Discuss how the use of indigenous vegetation can improve biodiversity and support local ecosystems in urban landscapes.
11. Examine case studies of successful LID projects. What lessons can be learned from these examples, and how can they inform future LID initiatives in different contexts?
12. How does LID address water quality issues in urban runoff? Discuss the mechanisms by which LID practices filter pollutants and improve the health of local waterways.
13. Assess the impact of LID on urban heat islands. In what ways can LID strategies help mitigate temperature increases in densely populated areas?
14. Discuss the importance of site design in LID. How can thoughtful planning and layout contribute to the success of LID practices in managing stormwater effectively?
15. Analyze the role of technology in advancing LID practices. What innovations are emerging to enhance the effectiveness and efficiency of LID solutions in urban environments?
16. How can LID contribute to social equity in urban areas? Discuss the potential benefits of LID for underserved communities and the importance of inclusive planning processes.
17. Explore the challenges of retrofitting existing urban areas with LID features. What strategies can be employed to integrate LID into established neighborhoods without significant disruption?
18. How do LID practices align with the principles of sustainable development? Discuss the interconnectedness of LID with broader sustainability goals in urban planning.
19. Evaluate the role of stormwater management in LID. How can effective stormwater strategies reduce flooding risks and improve overall community resilience?
20. What are the potential drawbacks or limitations of LID? Discuss scenarios where LID may not be the most effective solution for managing stormwater and environmental concerns.
21. How can public–private partnerships enhance the implementation of LID initiatives? Explore the benefits of collaboration between government agencies and private entities in promoting LID.
22. Discuss the significance of maintenance and monitoring in LID. How can ongoing care for LID features ensure their longevity and effectiveness in managing stormwater?
23. Analyze the role of educational institutions in promoting LID. How can schools and universities contribute to research, advocacy, and practical applications of LID practices?
24. How does LID intersect with public health concerns? Discuss the potential health benefits of LID practices in urban areas, particularly in relation to air and water quality.
25. Explore the concept of "living shorelines" as an LID strategy. How do these approaches differ from traditional shoreline stabilization methods, and what benefits do they offer?
26. Discuss the importance of hydrology in the design and implementation of LID. How does understanding local water cycles inform effective LID strategies?
27. Evaluate the role of community gardens within the framework of LID. How can these spaces contribute to urban sustainability and enhance community engagement?
28. How can LID practices be integrated into transportation planning? Discuss the potential for LID features to manage stormwater in roadways and parking areas effectively.

29. What are the implications of climate change for LID strategies? How can LID practices be adapted to address the increasing frequency and intensity of extreme weather events?
30. Discuss the importance of interdisciplinary collaboration in advancing LID. How can professionals from various fields work together to create effective LID solutions?
31. Analyze the potential for LID to enhance property values in urban areas. How can LID features contribute to the attractiveness and marketability of residential and commercial properties?
32. How can social media and digital platforms be utilized to promote awareness and education about LID? Discuss strategies for engaging the public in LID initiatives through technology.
33. Explore the relationship between LID and urban biodiversity. How can LID practices support the conservation of wildlife and plant species in metropolitan areas?
34. Assess the role of funding and financial incentives in promoting LID. What strategies can municipalities employ to encourage private investment in LID initiatives?
35. How can LID practices be tailored to different climatic regions? Discuss the importance of local context in designing effective LID solutions for diverse environments.
36. Evaluate the potential for LID to influence future urban policies. How can successful LID initiatives serve as models for broader changes in urban planning and development practices?

MULTIPLE-CHOICE QUESTIONS

1. What is the primary goal of LID?
   a. Increase urban density
   b. Enhance aesthetic appeal
   c. Manage stormwater sustainably
   d. Promote industrial growth
   e. Reduce construction costs
2. Which of the following is a common feature of LID?
   a. Concrete barriers
   b. Green roofs
   c. Asphalt roads
   d. Retention basins
   e. High-rise buildings
3. LID primarily aims to reduce which environmental issue?
   a. Urban sprawl
   b. Air pollution
   c. Stormwater runoff
   d. Noise pollution
   e. Waste generation
4. Which of the following practices can be classified as LID?
   a. Conventional stormwater drains
   b. Permeable pavements
   c. Deforestation
   d. Landfilling
   e. Industrial waste disposal
5. What is the role of vegetation in LID?
   a. To increase impervious surfaces
   b. To absorb and filter rainwater
   c. To enhance urban heat
   d. To promote invasive species
   e. To block sunlight

6. Which type of infrastructure is considered green infrastructure in LID?
   a. Parking lots
   b. Stormwater retention ponds
   c. Highways
   d. Landfills
   e. Power plants
7. LID can help mitigate which of the following impacts of urbanization?
   a. Increased biodiversity
   b. Urban heat islands
   c. Improved air quality
   d. Increased wildlife habitats
   e. Decreased energy consumption
8. Which of the following is NOT a benefit of LID?
   a. Improved water quality
   b. Increased flooding risk
   c. Enhanced community aesthetics
   d. Reduced urban runoff
   e. Increased groundwater recharge
9. Which LID practice involves the use of soil and plants to manage stormwater?
   a. Rain gardens
   b. Concrete channels
   c. Underground tanks
   d. Retention basins
   e. Paved surfaces
10. In LID, what is the purpose of using permeable materials?
    a. To increase runoff
    b. To allow water infiltrations
    c. To create barriers
    d. To prevent vegetation growth
    e. To enhance soil compaction
11. Which of the following is a challenge in implementing LID?
    a. High initial costs
    b. Increased biodiversity
    c. Reduced maintenance needs
    d. Enhanced community engagement
    e. Improved public health
12. What is the purpose of a bioswale in LID?
    a. To store waste
    b. To filter and convey stormwater
    c. To increase impervious surfaces
    d. To create recreational areas
    e. To promote industrial activities
13. Which of the following is a significant advantage of implementing green roofs?
    a. Increased building weight
    b. Enhanced biodiversity
    c. Higher energy costs
    d. Increased runoff
    e. Decreased air quality

14. What is the main purpose of using native plants in LID?
    a. To require more water
    b. To improve soil quality
    c. To support local wildlife
    d. To increase maintenance costs
    e. To enhance aesthetic appeal
15. Which of the following is a key component of a successful LID strategy?
    a. Increased impervious surfaces
    b. Community involvement
    c. Limited vegetation
    d. High water usage
    e. Industrial development
16. What is a common misconception about LID?
    a. It enhances stormwater management
    b. It is cost-effective in the long run
    c. It only applies to new developments
    d. It improves water quality
    e. It promotes sustainability
17. Which of the following practices can help manage stormwater on-site?
    a. Detention basins
    b. Paved parking lots
    c. Conventional gutters
    d. Stormwater pipes
    e. Landfills
18. How does LID contribute to sustainable urban growth?
    a. By increasing the need for concrete
    b. By promoting green spaces and reducing runoff
    c. By encouraging industrial expansion
    d. By limiting community participation
    e. By increasing pollution
19. Which regulatory aspect is important for the implementation of LID?
    a. Restrictive zoning laws
    b. Lack of funding
    c. Supportive policies and incentives
    d. Increased taxes
    e. Limited public awareness
20. What is the primary function of a rain barrel in LID?
    a. To increase runoff
    b. To store rainwater for later use
    c. To promote flooding
    d. To block vegetation
    e. To enhance soil erosion
21. Which of the following is the benefit of using bioswales?
    a. Increased impervious surfaces
    b. Improved water quality
    c. Higher maintenance costs
    d. Reduced biodiversity
    e. Increased flooding

22. What type of development does LID primarily focus on?
    a. Rural development
    b. Industrial development
    c. Urban development
    d. Agricultural development
    e. Mining development
23. Which of the following is NOT typically a part of LID?
    a. Green roofs
    b. Rain gardens
    c. Conventional drainage systems
    d. Permeable pavements
    e. Bioswales
24. What is the role of soil in LID?
    a. To increase runoff
    b. To filter and absorb rainwater
    c. To promote erosion
    d. To block vegetation
    e. To create impervious surfaces
25. Which of the following is a key principle of LID?
    a. Maximize impervious surfaces
    b. Minimize land disturbance
    c. Increase water consumption
    d. Promote urban sprawl
    e. Limit green spaces
26. How can LID practices help reduce the urban heat island effect?
    a. By increasing concrete surfaces
    b. By incorporating vegetation and green spaces
    c. By promoting industrial activities
    d. By reducing public transportation
    e. By limiting green roofs
27. What is a common method for assessing the effectiveness of LID practices?
    a. Visual inspections only
    b. Monitoring water quality and quantity
    c. Ignoring local conditions
    d. Relying solely on anecdotal evidence
    e. Limiting community feedback
28. Which of the following is a potential drawback of LID?
    a. Increased biodiversity
    b. Higher initial costs
    c. Improved water quality
    d. Enhanced community engagement
    e. Reduced flooding
29. What is the primary environmental benefit of using rain gardens?
    a. Increase in runoff
    b. Reduction of stormwater pollution
    c. Increase in construction costs
    d. Decrease in local wildlife
    e. Increase in impervious surfaces

30. Which of the following is a characteristic of permeable pavements?
    a. They are entirely impermeable
    b. They allow water to infiltrate through the surface
    c. They increase runoff
    d. They are made of solid concrete
    e. They require no maintenance
31. What is the impact of LID on groundwater recharge?
    a. It decreases groundwater levels
    b. It has no effect
    c. It promotes groundwater recharge
    d. It creates barriers to recharge
    e. It increases evaporation
32. Which of the following is essential for the success of LID?
    a. Community support and involvement
    b. High construction costs
    c. Limited vegetation
    d. Increased impervious surfaces
    e. Minimal maintenance
33. What is a common source of funding for LID projects?
    a. Private donations only
    b. Government grants and incentives
    c. Increased property taxes
    d. Corporate sponsorships only
    e. No funding is required
34. Which of the following best describes the concept of "living shorelines"?
    a. Concrete barriers along coastlines
    b. Natural approaches to shoreline stabilization
    c. Industrial development near water bodies
    d. Urban sprawl along coastlines
    e. Deforestation near shorelines
35. How does LID contribute to climate resilience?
    a. By increasing greenhouse gas emissions
    b. By reducing the need for traditional infrastructure
    c. By promoting deforestation
    d. By ignoring local climate conditions
    e. By limiting green spaces
36. Which of the following is a method of community engagement in LID projects?
    a. Public meetings and workshops
    b. Ignoring local opinions
    c. Limiting information access
    d. Exclusive decision-making by officials
    e. Avoiding feedback mechanisms
37. What is the function of a rain garden?
    a. To store rainwater indefinitely
    b. To channel water away from the site
    c. To absorb and filter rainwater runoff
    d. To increase soil erosion
    e. To block sunlight

38. Which of the following is a key outcome of successful LID implementation?
    a. Increased urban flooding
    b. Improved stormwater management
    c. Decreased biodiversity
    d. Higher maintenance costs
    e. Reduced community engagement
39. What is the primary reason for using green roofs in urban areas?
    a. To increase building weight
    b. To provide insulation and reduce energy costs
    c. To block sunlight
    d. To promote concrete usage
    e. To enhance air pollution
40. What is the main focus of LID in relation to water management?
    a. To increase water consumption
    b. To manage stormwater on-site
    c. To divert water to rivers
    d. To ignore water quality
    e. To promote industrial water use

ANSWERS TO MULTIPLE-CHOICE QUESTIONS

1	c	11	a	21	b	31	c
2	b	12	b	22	c	32	a
3	c	13	b	23	c	33	b
4	b	14	c	24	b	34	b
5	b	15	b	25	b	35	b
6	b	16	c	26	b	36	a
7	b	17	a	27	b	37	c
8	b	18	b	28	b	38	b
9	a	19	c	29	b	39	b
10	b	20	b	30	b	40	b

FURTHER READING

Ahiablame, L. M., Engel, B. A., & Chaubey, I. (2012). Effectiveness of low impact development practices: Literature review and suggestions for future research. *Water, Air, & Soil Pollution*, 223(7), 4253–4273.

Bichai, F., & Ashbolt, N. J. (2017). Public health and water quality management in low-impact development. *Environmental Science & Technology*, 51(19), 10718–10719.

Dhakal, K. P., & Chevalier, L. R. (2015). Managing urban stormwater for urban sustainability: Barriers and policy solutions for green infrastructure adoption. *Journal of Environmental Management*, 203, 171–181.

Dillman, K., McMahon, M., & Sanicola, O. (2018). Urban stormwater management in the 21st century: A review of the current state of the science. *Journal of Hydrology*, 558, 327–339.

Efstathiou, M. N., & Varotsos, C. A. (2012). Application of detrended fluctuation analysis to global climate system and stormwater forecast. *Environmental Research Letters*, 7(3), 034017.

Fairlie, S. (1996). *Low impact development: Planning and people in a sustainable countryside*. Permanent Publications.

Kim, H., & Shafique, M. (2017). Low impact development practices: A review of current research and future directions. *Water, Air, & Soil Pollution*, 228(8), 1–15.

Liu, T., Lawluvy, Y., Shi, Y., & Yap, P. S. (2021). Low impact development (LID) practices: A review on recent developments, challenges, and prospects. *Water, Air, & Soil Pollution*, 232(344), 1–20.

MacMullan, E., & Reich, S. (2007). *The economics of low-impact development: A literature review*. ECONorthwest.

McMahon, M., Dillman, K., & Sanicola, O. (2018). The influence of human activities on urban stormwater management. *Journal of Environmental Management*, 224, 1–10.

National Research Council. (2009). *Urban stormwater management in the United States*. National Academies Press.

Shafique, M., & Kim, H. (2017). The role of low impact development in urban stormwater management. *Environmental Science & Technology*, 51(19), 10718–10719.

United States Environmental Protection Agency. (2000). *Low impact development (LID): A literature review*. EPA.

University of Arkansas Community Design Center. (2011). *Low impact development: A design manual for urban areas*. University of Arkansas Press.

Varotsos, C. A., & Efstathiou, M. N. (2013). Non-linear processes in urban stormwater management. *Environmental Research Letters*, 8(4), 044017.

Wang, M., Feng, S., Ikram, R. M. A., Chen, T., Sun, C., Chen, B., Rao, Q., Jin, H., & Li, J. (2023). Assessing the performance and challenges of low-impact development under climate change: A bibliometric review. *Sustainability,* 15(18), 13616.

Wong, T. H. F., & Brown, R. R. (2009). The water sensitive city: Principles for practice. *Water Science and Technology*, 60(3), 673–682.

Yang, B., & Li, S. (2013). Green infrastructure design for stormwater runoff and water quality: Empirical evidence from field monitoring and simulation. *Journal of Environmental Management*, 129, 577–585.

Zhang, K., & Chui, T. F. M. (2018). Evaluating the benefits of low impact development practices in urban stormwater management. *Journal of Hydrology*, 560, 1–10.

Zhou, Q., & Wang, R. (2014). Low impact development and urban stormwater management: A review. *Environmental Science & Technology*, 48(16), 8911–8920.

11 Urban Energy Supply Systems

11.1 AN OVERVIEW OF ENERGY SUPPLY SYSTEMS

In the modern world, energy supply systems are among the most essential requirements for human existence, playing a vital role in meeting the diverse and increasing energy needs of society. These systems encompass a wide range of technologies and infrastructure designed to generate, transmit, and distribute energy to end-users. One key component of energy supply systems is the power generation sector, which includes traditional sources such as coal, natural gas, and nuclear, as well as renewable sources like solar, wind, and hydroelectric power. Each of these sources has its own unique advantages and challenges, contributing to the diversity of energy supply systems.

In addition to power generation, energy supply systems also involve the transmission and distribution of electricity. Transmission lines carry high-voltage electricity over long distances, connecting power plants to local distribution networks. These networks, in turn, deliver electricity to homes, businesses, and industries. The efficiency and reliability of transmission and distribution infrastructure are critical for ensuring a stable energy supply.

Energy supply systems are increasingly integrating advanced technologies to enhance efficiency and sustainability. Smart grid technologies, for example, enable real-time monitoring and control of electricity flow, improving reliability and enabling the integration of renewable energy sources. Energy storage systems, such as batteries and pumped hydro storage, are also becoming more prominent, helping to balance supply and demand and supporting the integration of intermittent renewables.

The transition toward cleaner and more sustainable energy supply systems is a key focus in many regions. Governments and industry stakeholders are investing in renewable energy projects, energy efficiency measures, and carbon capture technologies to reduce environmental impact and combat climate change. This shift toward cleaner energy sources is reshaping the landscape of energy supply systems and driving innovation in the sector.

Nevertheless, energy supply systems face various challenges, including aging infrastructure, cybersecurity threats, and the need for significant investment to modernize and expand capacity. Balancing the growing energy demand with environmental concerns and the need for resilience in the face of natural disasters and other disruptions is an ongoing challenge for energy supply systems.

Energy supply systems are complex and dynamic, reflecting the intricate interplay of various elements. These systems encompass a diverse array of technologies, each with its unique strengths and challenges. From traditional fossil fuels to renewable energy sources like wind and solar, the technological landscape of energy supply systems is vast and varied. Additionally, these systems are underpinned by extensive infrastructure, including power plants, transmission lines, and distribution networks, which enable the delivery of energy to end users.

As the world continues to evolve, these energy supply systems will play a pivotal role in meeting the growing demand for energy. The need for reliable and affordable energy is more pressing than ever, given the rapid pace of industrialization and urbanization. Moreover, as societies become more

DOI: 10.1201/9781032664613-11

digital and interconnected, the importance of energy security and resilience cannot be overstated. Therefore, the evolution and optimization of energy supply systems are crucial for supporting economic growth and social development.

Embracing innovation is key to shaping the future of energy supply systems. This involves not only the adoption of advanced technologies but also the development of innovative business models and regulatory frameworks. Investing in infrastructure is equally important, as it lays the foundation for the efficient and sustainable operation of energy supply systems. Promoting responsible energy consumption is essential for mitigating environmental impacts and achieving sustainability goals. Through these concerted efforts, we can ensure that our energy supply systems are fit for the future and capable of delivering clean, affordable, and reliable energy for all.

11.2 ENERGY SYSTEMS IN CITIES

Energy systems in cities encompass energy sources, a complex network of infrastructure, technologies, and policies designed to meet the diverse energy needs of urban populations. These systems involve the generation, transmission, distribution, and consumption of electricity, as well as the provision of heating, cooling, and transportation fuels. Urban energy systems are constantly evolving to incorporate a mix of traditional and renewable energy sources, advanced grid technologies, energy-efficient buildings, and sustainable transportation solutions. The integration of smart grid technologies, energy storage systems, and demand-side management strategies plays a critical role in enhancing the resilience, efficiency, and sustainability of urban energy supply. As cities continue to grow and face the challenges of climate change and environmental sustainability, the development of innovative and integrated energy systems is essential for ensuring a reliable, affordable, and environmentally responsible energy supply for urban communities.

11.2.1 Energy Sources in Urban Areas

In this exploration of energy sources in urban areas, we delve into the utilization of various energy sources, including traditional fossil fuels and nuclear energy, as well as the increasing prominence of renewable sources like solar and wind power. Urban energy landscapes are diverse and dynamic, making it crucial to understand the interplay of these sources in shaping sustainable, resilient, and environmentally responsible energy supply systems in our cities. The various sources of energy used in urban areas may include the following.

11.2.1.1 Fossil Fuels

Fossil fuels have long been the cornerstone of energy production in urban areas. These fuels, which include coal, natural gas, and petroleum, are derived from the remains of ancient plants and animals buried and subjected to heat and pressure over millions of years. They have powered the Industrial Revolution and continue to support modern economies, providing a reliable and readily available source of energy.

Coal, one of the earliest sources of fossil fuel, has been used for centuries for heating and cooking. In the context of urban energy systems, coal is primarily used for electricity generation. Power plants burn coal to produce steam, which then drives turbines to generate electricity. Despite its widespread use, coal is the most carbon-intensive fossil fuel, releasing a high amount of carbon dioxide when burned.

Natural gas, another key fossil fuel, is primarily composed of methane. It is used in urban areas for a variety of purposes, including heating homes and businesses, cooking, and generating electricity. Natural gas power plants work by burning natural gas to produce steam for electricity generation, similar to coal power plants. However, natural gas is cleaner than coal, emitting less carbon dioxide and other pollutants when burned.

Petroleum, also known as crude oil, is a liquid fossil fuel that is used extensively in transportation. It is refined into various products, including gasoline, diesel, and jet fuel. In urban areas, petroleum fuels cars, buses, and other vehicles play a crucial role in enabling mobility and connectivity. Additionally, petroleum is used in power plants for electricity generation, although to a lesser extent compared to coal and natural gas. The fossil-based fuels are still the main source of urban energy in many urban areas (Figures 11.1 and 11.2).

Despite their widespread use, fossil fuels are nonrenewable resources. This means that they cannot be replenished on a human timescale. The extraction and use of fossil fuels have led to a significant depletion of these resources, raising concerns about energy security and sustainability. As the world's population and energy demand continue to grow, the reliance on finite fossil fuels becomes increasingly problematic.

The use of fossil fuels contributes significantly to air pollution. When burned, fossil fuels release a variety of pollutants, including sulfur dioxide, nitrogen oxides, and particulate matter. These pollutants can have detrimental effects on air quality and human health, leading to respiratory diseases and other health problems. In urban areas, where population density is high, the impact of air pollution from fossil fuel combustion can be particularly severe.

The burning of fossil fuels is also the main driver of climate change. It releases large amounts of carbon dioxide, a greenhouse gas (GHG) that traps heat in the atmosphere and causes global warming. The consequences of climate change, including rising temperatures, changing weather patterns, and sea-level rise, pose significant challenges to urban areas. These challenges range from heat stress and water scarcity to increased risk of extreme weather events and flooding.

As we move forward, transitioning to cleaner and renewable sources of energy will be crucial for mitigating the impacts of air pollution and climate change and for ensuring the sustainability of our urban energy systems. In recent years, there has been a growing emphasis on diversifying urban energy portfolios by incorporating renewable energy sources. Solar power, for instance, has become increasingly prevalent in urban settings, with photovoltaic (PV) panels being installed on rooftops and in solar farms to harness the sun's energy. This clean and abundant energy source offers a promising alternative to traditional fossil fuels, helping to reduce reliance on nonrenewable resources and mitigate environmental impacts.

FIGURE 11.1 Fossil-based fuels station.

Source: **Uwe Aranas, Wikipedia.**

FIGURE 11.2 A refueler, fuel tanker.

Source: **Wikipedia.**

Wind power has emerged as a viable renewable energy option for urban areas, with wind turbines being strategically deployed to capture the kinetic energy of the wind and convert it into electricity. The scalability and relatively low environmental impact of wind energy make it an attractive complement to traditional energy sources, contributing to a more sustainable urban energy mix. As technology and infrastructure continue to advance, the integration of renewable energy sources like solar and wind power holds great potential for reducing the carbon footprint of urban energy systems.

Nuclear energy has also played a role in urban energy supply, offering a low-carbon alternative to fossil fuels. Nuclear power plants harness the energy released from nuclear reactions to generate electricity, providing a reliable and consistent source of power. While concerns about safety, waste management, and the potential for nuclear accidents exist, advancements in reactor design and stringent safety protocols have contributed to the continued consideration of nuclear energy as part of the urban energy landscape.

Considering these developments, urban areas are increasingly exploring energy efficiency measures and smart grid technologies to optimize energy use and enhance the resilience of energy supply systems. The implementation of energy-efficient building designs, advanced lighting systems, and smart meters contributes to reducing energy consumption and lowering environmental impact. Smart grid infrastructure enables more efficient monitoring, management, and distribution of electricity, supporting the integration of diverse energy sources and enhancing the reliability of urban energy supply.

As urban populations continue to grow and the demand for energy escalates, the transition toward cleaner, more sustainable energy sources becomes imperative. The adoption of renewable energy technologies, coupled with energy efficiency initiatives and responsible consumption practices, is pivotal in shaping urban energy systems that are resilient, environmentally responsible, and capable of meeting the evolving needs of urban communities. These efforts are essential for mitigating the environmental impact of energy use in urban areas and fostering a more sustainable and livable urban environment for current and future generations.

11.2.1.2 Nuclear Energy

Nuclear power plants are a key component of the energy infrastructure in many urban areas around the world. These facilities generate a significant amount of electricity, powering homes, businesses, and public services. The energy produced by nuclear power plants is not only substantial but also reliable, providing a steady supply of electricity that is not subjected to fluctuations in weather or fuel prices. The efficiency and reliability of nuclear power plants have made them a key component of urban energy supply, providing a consistent source of low-carbon electricity.

The operation of a nuclear power plant involves a process known as nuclear fission. In this process, the nucleus of a heavy atom, such as uranium or plutonium, is split into two or more smaller nuclei. This splitting releases a large amount of energy in the form of heat. The heat generated by nuclear fission is then used to heat water, turning it into steam. The steam produced in a nuclear power plant is the key element in electricity generation. It is directed onto the blades of a turbine, causing the turbine to spin. The spinning turbine is connected to a generator, which converts the mechanical energy of the spinning turbine into electrical energy. This electricity is then transmitted through power lines to homes and businesses.

One of the main advantages of nuclear energy is that it is a low-carbon source of energy. Unlike fossil fuels, which release carbon dioxide when burned, nuclear power plants emit virtually no GHGs during operation. This makes nuclear energy a potentially important tool in efforts to combat climate change and reduce the carbon footprint of our energy systems. However, despite its benefits, the use of nuclear energy is not without controversy. Safety is a major concern when it comes to nuclear power plants. Accidents, while rare, can have devastating consequences, as seen in the Chernobyl and Fukushima disasters. These incidents have underscored the importance of stringent safety measures and robust regulatory oversight in the operation of nuclear power plants, as well as the need for continuous improvements in reactor design and emergency preparedness.

Waste disposal is another significant challenge associated with nuclear energy. Nuclear power plants produce radioactive waste, which remains hazardous for thousands of years. Managing and disposing of this waste in a safe and secure manner is a complex and costly process. The management and disposal of nuclear waste present significant challenges. Radioactive waste generated from nuclear power plants requires careful handling and long-term storage to prevent environmental contamination and safeguard public health. Developing secure and sustainable solutions for the storage and disposal of nuclear waste is essential for ensuring the responsible use of nuclear energy in urban areas and beyond. There are concerns about the potential for nuclear waste to be used in the production of weapons. The potential for weaponization of nuclear materials is a further concern that has limited the adoption of nuclear energy. The same technology used to produce electricity in a nuclear power plant can also be used to produce nuclear weapons. This dual-use nature of nuclear technology has led to concerns about nuclear proliferation, particularly in regions of the world with political instability.

The term "dual use" refers to technologies that have both peaceful and military applications. In the context of nuclear technology, this dual-use nature is particularly significant.

Nuclear technology, which is used for generating power in many parts of the world, can also be used for the creation of nuclear weapons. This is because many of the technologies and materials associated with the creation of a nuclear power program have a dual-use capability. For instance, several stages of the nuclear fuel cycle allow for the diversion of nuclear materials for nuclear weapons.

This dual-use nature of nuclear technology is often seen as a lamentable fact, as states can pursue nuclear weapons under the guise of peaceful intentions. The same technology that can be used to produce electricity in a nuclear power plant can also be used to produce nuclear weapons. However, there is an interesting aspect to this dual-use dilemma. The deniable nature of dual-use technology makes it more amenable to coercive counterproliferation. In other words, if a state is caught proliferating, it is more likely to come into compliance if it can deny that it was ever out of compliance. This means that the dual-use dilemma is both a challenge and an opportunity for the nuclear non-proliferation regime.

One key issue is how the commercialization of certain nuclear technologies, such as laser enrichment, would affect its spread globally. If a technology like laser enrichment becomes profitable, it could inspire other states to develop it for themselves. There are also concerns about whether classified details of certain processes can be kept secret for decades, potentially against repeated attempts by foreign governments to acquire them.

Despite these challenges, ongoing advancements in nuclear technology and safety practices continue to shape the landscape of nuclear energy. Innovations in reactor design, such as the development of next-generation nuclear reactors, aim to enhance safety, efficiency, and waste management capabilities. These advanced reactor designs incorporate inherent safety features and novel fuel cycles, offering potential improvements in both operational safety and waste reduction.

Research and development efforts are focused on exploring advanced nuclear fuel technologies, such as thorium-based and molten salt reactors, which have the potential to address some of the challenges associated with traditional uranium-based nuclear power. These innovative approaches seek to optimize fuel utilization, reduce waste generation, and enhance the sustainability of nuclear energy as a long-term energy option for urban areas.

In parallel, the integration of small modular reactors (SMRs) is gaining attention as a promising avenue for expanding the deployment of nuclear power in urban settings. SMRs offer scalability, enhanced safety features, and potential cost efficiencies, making them well-suited for diverse urban energy needs and applications. The compact size and modular nature of these reactors provide flexibility in deployment and can support the electrification of urban transportation and heating systems.

As urban areas strive to meet their energy demands while addressing climate change and air quality concerns, the role of nuclear power in the urban energy landscape remains subject to ongoing debate and scrutiny. Balancing the potential benefits of low-carbon electricity generation with the associated safety, waste management, and proliferation risks requires a comprehensive and nuanced approach to the consideration of nuclear energy in urban energy planning and policy-making. Collaborative efforts among stakeholders, including governments, industry, and the public, are essential for fostering informed dialogue and decision-making regarding the role of nuclear power in urban energy supply.

11.2.1.3 Solar Power

Solar energy is a form of renewable energy that is harnessed using PV cells (Figure 11.3). These cells, which are typically made of silicon, work by absorbing sunlight and converting it directly into electricity. This process, known as the PV effect, involves the creation of an electric field that pushes electrons freed by sunlight toward the front of the cell, creating a flow of electricity.

In urban areas, solar panels, which are composed of many PV cells, can be installed in a variety of locations. One common place for installation is rooftops. Rooftop solar panels can be a practical and efficient way to utilize space that is otherwise often unused. They can be installed on homes, businesses, and other buildings, providing a source of clean, renewable energy right where it is needed. Parking structures are another potential location for solar panel installation in urban areas. Solar canopies, which are essentially solar panels mounted on structures that provide shade, can be installed over parking lots. These canopies not only generate electricity but also provide shade for vehicles, reducing the heat island effect common in urban areas. Other available spaces in urban

FIGURE 11.3 PV panels.

Source: **AleSpa, Wikipedia.**

areas, such as vacant lots or unused land, can also be utilized for solar energy generation. Large-scale solar installations, known as solar farms or solar parks, can be set up in these spaces. These installations can generate a significant amount of electricity, contributing to the energy supply of the urban area.

One of the key advantages of solar power is that it is renewable. Unlike fossil fuels, which are finite and will eventually run out, the sun is a virtually inexhaustible source of energy. If the sun shines, we can continue to harness its energy. Solar power produces no emissions during operation, making it a clean source of energy that does not contribute to air pollution or climate change. However, the generation of solar power is dependent on weather conditions. Solar panels need sunlight to generate electricity, so their output can be affected by cloudy or rainy weather. The angle of the sun changes with the seasons, which can also affect the amount of electricity generated by solar panels. Therefore, while solar power can be a significant source of energy, it is not always a consistent one.

Another challenge with solar power is that it requires significant space for installation. Each solar panel only produces a certain amount of electricity, so to generate large amounts of power, many panels are needed. This can be a challenge in urban areas, where space is often at a premium. However, with the ongoing development of more efficient solar cells and the creative use of space, such as installing solar panels on rooftops and parking structures, this challenge can be mitigated. As technology continues to advance and the efficiency of solar cells improves, we can expect to see an increasing reliance on solar power in our cities.

There are several innovative ways to integrate solar panels into urban design.

11.2.1.3.1 Building Rooftops and Structures

Building rooftops represent a significant but underutilized resource in urban areas. Roofs can house PV solar panels, transforming buildings into power generators. Notably, commercial buildings with large flat rooftops are ideal for large-scale solar installation. In some cases, solar panels are being incorporated into the very material of rooftops, through solar shingles and tiles, offering an aesthetic yet functional alternative to traditional roofing.

11.2.1.3.2 Building-Integrated PVs

Building-integrated photovoltaics (BIPV) involves integrating solar panels into the very fabric of building structures. This innovative design strategy can include:

- *Solar Windows:* These can generate electricity while allowing natural light to filter through.
- Solar Cladding Systems: These act as a building's exterior protective layer while generating power.
- Solar Facade Systems: These convert the building's external surface into a solar energy-producing system.

BIPV contributes toward energy-neutral or even energy-positive buildings, playing a significant role in urban sustainability.

11.2.1.3.3 Solar Parking Lots

Urban areas abound with large open spaces in the form of parking lots, which can be transformed into solar farms. Installing solar canopies over parking areas can provide shade for vehicles, reduce heat islands, and generate substantial amounts of electricity. These solar structures also have the potential to be paired with electric vehicle charging stations, creating a synergistic relationship that promotes cleaner transportation.

11.2.1.3.4 Solar Panel-Roofed Structures

In addition to conventional rooftop installations, solar panels can be seamlessly integrated into garden structures, pergolas, and various outdoor architectural elements. This innovative approach not only enhances the aesthetic appeal of outdoor spaces and provides valuable shade but also leverages underutilized areas to generate clean, renewable electricity. Combining functionality with sustainability, this integration demonstrates the versatility of solar technology in enhancing both the visual and environmental aspects of urban and suburban landscapes.

11.2.1.3.5 Camouflaging Solar Panels with Art

Solar panels offer the potential to seamlessly blend with their surroundings through artistic camouflage, incorporating colors, patterns, and designs that harmonize with the aesthetic of the environment. This innovative approach not only enhances the visual appeal of solar installations but also integrates renewable energy generation into the fabric of urban and rural landscapes. By fusing functionality with artistry, camouflaged solar panels demonstrate the creative potential of sustainable technology in enhancing the visual harmony and cultural integration of renewable energy solutions.

11.2.1.3.6 Solar Fabrics

Solar fabrics, featuring integrated PV cells, represent a cutting-edge innovation in renewable energy technology. These versatile materials can be seamlessly incorporated into a wide range of applications, including awnings, tents, and architectural structures, to harness solar energy and generate electricity. Transforming everyday textiles into sources of clean, sustainable power, solar fabrics offer a promising avenue for integrating renewable energy generation into the fabric of urban and outdoor environments, enhancing both functionality and environmental sustainability.

11.2.1.3.7 Solar Panel Arrays in Shapes

Solar panels present a versatile canvas for creative design, allowing them to be arranged in diverse shapes and patterns that not only enhance their functionality but also contribute to the visual aesthetics of their surroundings. This flexibility in design opens up opportunities for integrating solar technology into architectural elements and urban landscapes in a visually appealing and innovative manner.

11.2.1.3.8 Wall-Mounted Solar Panels

Mounting solar panels on walls offers a practical solution for maximizing solar energy generation in urban areas with limited roof space. This approach not only optimizes the utilization of available surfaces for solar installation but also contributes to the integration of renewable energy into the built environment. Leveraging vertical spaces, such as building facades and solar panels, can expand the capacity of urban solar power generation.

11.2.1.3.9 Integrating Solar Panels into Windows

The integration of solar panels into windows represents a dual-purpose innovation that combines energy generation with architectural functionality. Incorporating PV technology into window design, buildings can harness sunlight for both natural illumination and electricity production. This innovative approach not only optimizes the use of available space for solar energy generation but also enhances the sustainability and energy efficiency of urban and architectural environments.

These innovative strategies not only enhance the aesthetic appeal of urban areas but also contribute to energy efficiency and sustainability.

11.2.1.4 Wind Power

Wind turbines are a marvel of modern engineering, designed to harness the power of the wind and convert it into usable energy (Figure 11.4). These structures work by capturing the kinetic energy in wind and converting it into mechanical power. The blades of a wind turbine are designed to capture

FIGURE 11.4 Wind power, Xinjiang China.

Source: **Chris Lim, Wikipedia.**

the wind's kinetic energy. As the wind blows, it applies force to the blades of the turbine, causing them to spin. This spinning motion is the mechanical power that is then converted into electricity.

The conversion of mechanical power into electricity happens in the generator of the wind turbine. The generator is connected to the rotor, which is set in motion by the spinning blades. As the rotor spins, it creates a rotating magnetic field in the generator, which induces a flow of electrical current. This electricity can then be used directly or sent to the grid for distribution.

While wind farms, which consist of many large wind turbines, are typically located in rural areas, wind power is not exclusive to these locations. This is because wind speeds in rural areas are often higher and more consistent, making them ideal for large-scale wind energy production. These wind farms can cover large areas and generate significant amounts of electricity. However, smaller-scale wind turbines can also be used in urban settings. These turbines, often referred to as "urban wind turbines" or "small wind turbines," are designed to operate in lower wind speeds and turbulent conditions typical of urban environments. They can be installed on rooftops or other open spaces in the city, providing a source of renewable energy that is locally produced.

Like solar power, wind energy is a renewable source of energy. This means that it is derived from a resource that is not depleted when used, in this case, the wind. If the wind continues to blow, we can continue to harness its energy to generate electricity. This makes wind power a sustainable choice for energy production, contributing to the diversification of our energy mix and the reduction of our reliance on fossil fuels. Wind energy is emission-free. Unlike burning fossil fuels, generating electricity with wind turbines does not release GHGs or other pollutants. This makes wind power a clean source of energy, contributing to the mitigation of climate change and the improvement of air quality.

However, the generation of wind power is variable and dependent on wind conditions. Wind speeds are not constant – they can vary from moment to moment and from place to place. This means that the amount of electricity a wind turbine can generate can also vary. To manage this variability, wind power is often used in conjunction with other forms of renewable energy, such as solar power, and energy storage solutions.

Wind turbines can also generate negative visual impacts, particularly when the surrounding areas of wind farms are tourist destinations. Addressing the visual impact of wind turbines in urban areas involves a combination of thoughtful planning, innovative design, and community engagement. The strategies to cope with visual impacts are shown in Table 11.1.

TABLE 11.1
Strategies to Cope with Visual Impacts of Wind Turbines

1	*Visual Impact Analyses:* Visual impact analyses can accurately portray how wind turbines affect aesthetics and help local residents better understand what a project might look like when completed. This can involve creating visual simulations or mock-ups of proposed wind turbine installations.
2	*Location Planning:* The placement of wind turbines can significantly influence their visual impact. By carefully selecting locations that minimize visibility, such as lower-lying areas or sites screened by buildings or vegetation, the visual impact can be reduced.
3	*Innovative Turbine Design:* Adopting advanced turbine technologies, such as blade designs that reduce noise and improve aesthetics, can mitigate visual concerns Turbines can be designed with sleeker, more aesthetically pleasing profiles, and can even be colored to blend in with the urban landscape.
4	*Size and Scale:* The size and scale of the turbines should be appropriate for the urban context. Smaller-scale turbines may be more suitable for urban settings, as they can be less visually intrusive than larger models.
5	*Community Engagement:* Engaging the community in the planning and design process can help address visual concerns. This can involve public meetings, workshops, or surveys to gather community input and feedback.
6	*Regulations and Guidelines:* Development regulations and guidelines can help mitigate any local viewshed impacts. These can establish criteria for turbine siting, design, and operation to minimize visual impact.
7	*Landscaping and Screening:* Landscaping and other screening methods can be used to partially conceal turbines and integrate them more harmoniously into the urban environment.

Wind turbines can have both direct and indirect effects on local wildlife. They are:

Direct Effects: One of the most direct ways wind turbines can affect wildlife is through collisions. Birds and bats can be at risk of colliding with the turbine blades. It is estimated that tens to hundreds of thousands of bats die at wind turbines each year in North America alone. It is possible that wind turbines interfere with seasonal migration and mating patterns in some species of bats.

Indirect Effects: Indirect effects of wind turbines on wildlife include habitat disruption and noise pollution. The construction and operation of wind turbines can lead to changes in land use, potentially disrupting the habitats of various wildlife species. Noise from the turbines can also disturb wildlife, potentially affecting their behavior and survival.

During Construction: The greatest impact of wind farms on wildlife is often during construction. For example, noise and dust could disturb wildlife in surrounding habitats, while construction during breeding seasons could affect some species.

Despite these potential impacts, many strategies are being implemented to mitigate these effects. These include thoughtful site selection to avoid wildlife areas, implementing "shutdown-on-demand" systems that halt turbines when birds or bats are detected nearby, and ongoing research to better understand and minimize impacts.

11.2.1.5 Other Renewable Sources

Other renewable energy sources utilized in urban areas play a crucial role in diversifying the energy mix and reducing reliance on fossil fuels. Hydropower, one of the prominent renewable energy sources, harnesses the energy of flowing or falling water to generate electricity. This method involves the construction of dams or diversion structures to capture the energy of water, and it has been widely adopted in various urban settings due to its reliability and sustainability. In addition to its electricity generation capabilities, hydropower also provides benefits such as flood control, irrigation, and water supply.

Geothermal energy (Figure 11.5), another significant renewable energy source, taps into the heat stored within the earth's crust. In urban areas, geothermal energy is utilized for heating and cooling purposes through geothermal heat pumps, which transfer heat between buildings and the ground. This technology offers a reliable and efficient way to regulate indoor temperatures while reducing GHG emissions, making it an attractive option for urban sustainability initiatives. Furthermore, geothermal power plants can also be found in some urban areas, where they harness the earth's heat to generate electricity, contributing to the diversification of the energy supply.

Bioenergy, derived from organic materials such as plant and animal waste, represents a versatile renewable energy source with various applications in urban settings. In urban areas, bioenergy is commonly used to produce heat, electricity, and biofuels. For instance, organic waste materials can be processed through anaerobic digestion to produce biogas, which can then be utilized for heating or to generate electricity. Additionally, biofuels derived from organic sources can be used to power vehicles, thereby reducing the carbon footprint of urban transportation. Moreover, bioenergy projects in urban areas often contribute to waste management and environmental sustainability by diverting organic waste from landfills and reducing methane emissions.

The integration of these renewable energy sources into urban environments not only reduces GHG emissions and mitigates climate change but also fosters energy independence and resilience. By diversifying the energy mix, urban areas can enhance their energy security and reduce vulnerability to supply disruptions and price fluctuations associated with conventional energy sources. Furthermore, the deployment of renewable energy technologies in urban settings creates opportunities for job creation, economic growth, and technological innovation, driving the transition toward

FIGURE 11.5 Geothermal energy.

Source: **Gretar Ivarsson, Wikipedia.**

a more sustainable and low-carbon future. As urban populations continue to grow, the adoption of renewable energy sources becomes increasingly vital in ensuring the long-term sustainability and livability of cities around the world. The shift toward more sustainable and renewable sources of energy is a key trend in urban energy systems. This transition is driven by the need to reduce GHG emissions, improve air quality, and enhance energy security. However, it also requires addressing challenges related to energy storage, grid integration, and the need for significant infrastructure investments.

11.2.1.6 Waste-to-Energy

Cities are significant waste producers, and many cities cannot fully and appropriately handle their municipal solid waste management. One of the strategies to cope with waste generation is the waste-to-energy program. Waste-to-energy projects using bioenergy are an innovative approach to managing waste and producing energy. They involve the conversion of organic waste materials, also known as biomass, into energy in the form of heat, electricity, or biofuels.

One common method of waste-to-energy conversion is through a process called anaerobic digestion. In this process, organic waste is broken down by microorganisms in an oxygen-free environment, producing biogas. This biogas, which is primarily composed of methane, can then be burned to generate heat and electricity. Anaerobic digestion is commonly used for treating municipal solid waste, agricultural waste, and wastewater sludge.

Another method involves the production of liquid biofuels from waste. For example, waste from food processing can be converted into bioethanol or biodiesel, which can be used as a renewable alternative to gasoline or diesel fuel. This not only provides a source of renewable energy but also helps reduce the environmental impact of waste disposal.

There are also projects that focus on producing energy from specific types of waste. For instance, one project in Kenya uses waste from avocado processing to produce biogas. Another project in Brazil generates electricity from food production residues. These projects demonstrate the potential of waste-to-energy technologies to address both waste management and energy production challenges.

Implementing waste-to-energy projects involves several challenges, including technical, economic, and regulatory hurdles. Despite these challenges, waste-to-energy projects using bioenergy offer a promising pathway toward sustainable waste management and renewable energy production.

11.2.2 Energy Infrastructure and Distribution

Energy infrastructure in cities is a vital system that ensures the reliable delivery of energy to homes, businesses, and public facilities. This infrastructure encompasses a wide range of components, each playing a crucial role in the generation, transmission, and distribution of energy. Understanding this infrastructure is key to appreciating how energy flows from its source to end users.

Power plants are a central part of the urban energy infrastructure. These facilities are where energy is generated. They can be fueled by a variety of energy sources, including fossil fuels like coal and natural gas, nuclear energy, and renewable resources like wind, solar, and hydropower. The type and number of power plants in a city depend on various factors, including the availability of energy resources, the city's energy demand, and environmental and economic considerations.

Once energy is generated at power plants, it needs to be transmitted to where it will be used. This is where transmission lines come in. These high-voltage lines carry electricity over long distances, from power plants to substations located near demand centers. Transmission lines are a critical link in the energy chain, ensuring that the electricity generated at power plants reaches consumers. Figure 11.6 presents a simplified grid system for electricity generation and transmission.

Substations are another key component of the energy infrastructure. They serve as hubs in the power grid, receiving high-voltage electricity from transmission lines and converting it to a lower voltage suitable for distribution. Substations are equipped with transformers and other equipment to carry out this voltage conversion. They also have protective devices to detect and respond to faults in the system, helping to maintain the reliability of the power supply. From substations, electricity is distributed to end users through local distribution networks. These networks consist of a web of lower-voltage lines that deliver electricity to homes, businesses, and other facilities. Distribution networks are designed to provide a reliable and safe supply of electricity to all connected customers.

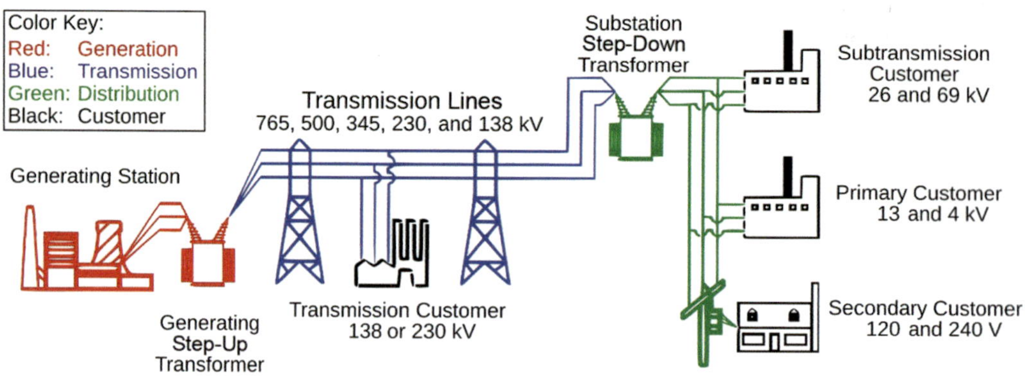

FIGURE 11.6 Simplified grid system of North America.

Source: **US DoE, Wikipedia.**

The energy infrastructure also includes the systems and technologies for managing and operating the power grid. This includes control centers that monitor and control the flow of electricity in real time, advanced metering systems that measure electricity use, and communication networks that enable information exchange between different parts of the grid.

The design and operation of the energy infrastructure need to consider various factors. These include the geographical layout of the city, the pattern of energy demand, the availability and mix of energy sources, and environmental and regulatory constraints. Balancing these factors is a complex task that requires careful planning and coordination. The energy infrastructure also needs to be resilient and adaptable. It needs to withstand and recover from various disturbances, from extreme weather events to equipment failures. It also needs to adapt to changes in energy demand and supply, technological advancements, and shifts in energy policy and regulation.

To enhance the resilience and adaptability of energy infrastructure in cities, several measures can be implemented, as shown in Table 11.2.

Investments in energy infrastructure are crucial for the sustainable development of cities. They can enhance the reliability and efficiency of the energy supply, facilitate the integration of renewable energy, and support the transition to a low-carbon energy system. However, these investments need to be guided by a long-term vision and strategy, considering not only the current needs but also the future challenges and opportunities.

Investments in energy infrastructure are vital for cities' sustainable development, enhancing reliability, efficiency, and renewable energy integration. However, these investments require a long-term vision, considering current needs and future challenges. A smart grid utilizes digital technology, sensors, advanced metering, and communication networks to monitor and manage electricity flow from diverse sources to meet varying demands.

A smart grid is essentially an electricity network that uses digital technology to monitor and manage the transport of electricity from all generation sources to meet the varying electricity demands of end users. It involves the use of sensors, advanced metering systems, and communication networks that enable real-time monitoring and control of the electricity flow.

TABLE 11.2
Strategies to Improve Resilient and Adaptable Energy Infrastructure

1	*Diversification of Energy Sources:* By diversifying the sources of energy, cities can reduce their reliance on a single energy supply, making the infrastructure more resilient to disruptions in any one source.
2	*Integration of Renewable Energy:* Incorporating renewable energy sources such as solar, wind, and hydropower into the energy mix can enhance resilience by reducing dependence on finite fossil fuels and mitigating the impact of supply chain disruptions.
3	*Implementation of Smart Grids:* Smart grid technologies enable real-time monitoring, automation, and optimization of energy distribution, enhancing the system's ability to adapt to changing conditions and improving overall resilience.
4	*Energy Storage Solutions:* Deploying energy storage systems such as batteries or pumped hydro storage can help cities store excess energy during periods of low demand and utilize it during peak times or in the event of power outages.
5	*Enhanced Infrastructure Planning:* Long-term infrastructure planning that considers climate change impacts, extreme weather events, and evolving energy demands can ensure that energy systems are designed to withstand and adapt to future challenges.
6	*DSM:* Encouraging energy efficiency measures and DSM programs can help reduce overall energy consumption and peak demand, thereby improving the resilience of the energy infrastructure.
7	*Robust Emergency Response Plans:* Developing and regularly testing emergency response plans for energy infrastructure can help cities effectively manage and recover from disruptions, ensuring continuity of essential services during crises.

One of the key features of a smart grid is its two-way communication capability. This allows for dynamic interaction between the utility and its customers. For example, during periods of high electricity demand, a utility could send signals to customers encouraging them to reduce their usage, perhaps in return for a discount on their electricity bill.

Smart grids also enable better integration of renewable energy sources, such as solar and wind power. They can manage the variable and intermittent nature of these sources, ensuring a stable and reliable power supply. Moreover, smart grids can support the deployment of electric vehicles by managing the additional electricity demand and utilizing the vehicles' batteries for energy storage.

Another important aspect of smart grids is their ability to quickly detect and respond to power outages. Through real-time monitoring, a smart grid can identify the location of a fault in the network and isolate it, preventing the outage from spreading and enabling faster restoration of service. Smart grids can also provide consumers with more information about their energy usage, helping them to manage their consumption more effectively. This can lead to energy savings and lower electricity bills.

The implementation of smart grid technologies also presents challenges. These include the need for significant infrastructure investments, the management of large volumes of data, and concerns about data security and privacy. Despite these challenges, smart grid technologies hold great promise for transforming urban energy systems. They can enhance the efficiency, reliability, and sustainability of these systems, supporting the transition to a low-carbon and resilient energy future.

Cities can encourage the adoption of smart grid technologies through a variety of strategies, as shown in Table 11.3.

Data analytics plays a crucial role in optimizing smart grids and enhancing their efficiency and sustainability.

Smart grids, a modern advancement in energy distribution, generate a vast amount of data from a variety of sources including smart meters, grid sensors, and other connected devices. This wealth

TABLE 11.3
Strategies for Adoption of Smart Grid Technology

1	*Policy and Regulation:* Governments can create favorable policies and regulations that incentivize the adoption of smart grid technologies. This could include financial incentives, such as tax credits or grants, for utilities that invest in smart grid infrastructure.
2	*PPPs:* Collaborations between public entities and private companies can help drive the adoption of smart grid technologies. These partnerships can leverage the resources and expertise of both sectors to develop and implement smart grid projects.
3	*Education and Awareness:* Raising awareness about the benefits of smart grids can help encourage their adoption. This can involve public education campaigns, workshops, and other initiatives to inform consumers and stakeholders about the advantages of smart grids.
4	*Investment in Research and Development:* Investing in research and development can lead to the creation of more efficient and cost-effective smart grid technologies. This can make smart grids more attractive to utilities and consumers.
5	*Infrastructure Upgrades:* Upgrading existing energy infrastructure to incorporate smart grid technologies can be an effective way to encourage their adoption. This can involve replacing outdated equipment with smart meters and other smart grid devices.
6	*Cybersecurity Measures:* Ensuring the security of smart grids is crucial to gaining public trust and encouraging their adoption. This can involve implementing robust cybersecurity measures and standards to protect against cyber threats.
7	*Demonstration Projects:* Implementing demonstration projects can provide a practical way to showcase the benefits of smart grids. These projects can serve as real-world examples of how smart grids work and the benefits they offer.
8	*Stakeholder Engagement:* Engaging stakeholders, including utilities, consumers, and regulators, in the planning and implementation of smart grid projects can help ensure their success. This can involve regular consultations, feedback sessions, and other forms of engagement.

of data serves as a treasure trove of information, offering valuable insights into patterns of energy consumption across different times and locations. It also aids in the early detection of equipment failures, thereby preventing potential disruptions. Furthermore, this data plays a pivotal role in optimizing energy usage and ensuring efficient and sustainable energy distribution.

Big data analytics techniques are employed to process and analyze the vast and diverse data generated by smart grids. This data, which can be complex and varied, is meticulously sifted through using these advanced techniques. When combined with machine learning (ML) algorithms, these techniques become a powerful tool, playing an indispensable role in electricity transmission and distribution networks. They facilitate efficient data collection, secure storage, and insightful analysis, thereby enhancing the overall performance and reliability of these networks. Through the analysis of this data, utilities can optimize power flows, effectively decreasing peak demands. This also facilitates the integration of renewable energy sources into the grid. Consequently, energy can be delivered more efficiently and cost-effectively, leading to improvements in customer service and contributing positively to societal welfare.

In addition to its current applications, data analytics can play a vital role in predictive forecasting and proactive system maintenance. Through predictive analytics, utilities can accurately forecast energy demand, enabling efficient resource management and optimization. Furthermore, the use of analytics to predict equipment failures allows for proactive maintenance, minimizing downtime and enhancing the overall reliability and performance of energy systems. Data analytics can support the implementation of demand response programs. These programs encourage consumers to adjust their energy usage in response to real-time price signals, helping to balance supply and demand.

Despite holding great promise for transforming urban energy systems and making them more efficient, reliable, and sustainable, the use of data analytics in smart grids also presents challenges. These challenges are shown in Table 11.4.

TABLE 11.4
Challenges of the Use of Data Analytic in Smart Grids

1	*Computational Complexity:* The large volume and high velocity of data generated by smart grids can lead to significant computational challenges. Analyzing this data in real time requires advanced computational methods and high-performance computing infrastructure.
2	*Data Quality and Quantity:* Ensuring the quality and quantity of the data collected from various sensors and devices within the smart grid infrastructure can be challenging, as it requires reliable data acquisition and management processes.
3	*Data Security and Privacy:* Protecting the vast amount of sensitive data generated by smart grid systems from cybersecurity threats and ensuring the privacy of consumers' information are significant concerns that need to be addressed.
4	*Integration of Diverse Data Sources:* Smart grids generate data from a wide range of sources, including sensors, meters, and other devices. Integrating and analyzing this diverse data effectively can be complex and require robust data integration and processing capabilities.
5	*Scalability and Flexibility:* As smart grid systems expand and evolve, ensuring that data analytics solutions are scalable and flexible enough to accommodate increasing data volumes and new technologies is a challenge that needs to be managed.
6	*Interoperability and Standards:* Ensuring interoperability between different components and systems within the smart grid, as well as adhering to industry standards for data formats and communication protocols, is essential for effective data analytics.
7	*Development of Robust ML-Based Algorithms:* The development of more robust ML-based algorithms is needed to effectively analyze the complex and large-scale data generated by smart grids.
8	*Skilled Workforce:* The need for a skilled workforce capable of managing and analyzing the complex data generated by smart grid systems is a challenge, as it requires expertise in data science, analytics, and domain-specific knowledge of energy systems.

TABLE 11.5
Strategies to Address Data Security

1	*Secure Communication Protocols:* These protocols can safeguard AMI from remote attacks. The most successful protocols offer end-to-end security with additional layers in protocol stacks like authentication and encryption.
2	*Unified Security Measures:* Implementing a comprehensive security strategy that includes various measures to mitigate risks and privacy issues pertaining to smart grid customers and uses of their data.
3	*Better Regulatory Framework:* Establishing a robust regulatory framework can help ensure that all stakeholders adhere to the necessary security standards.
4	*Data Loss Prevention Techniques:* These techniques can help prevent unauthorized access to data and protect against data breaches.
5	*Blockchain:* Blockchain technology can provide a decentralized and secure method for data management, enhancing the security of smart grid data.
6	*AI and ML:* AI and ML can be used to detect anomalies and potential security threats in real-time.
7	*Encryption Algorithms, Key Management, and Access Control Models:* These techniques can help preserve privacy and enhance data security in smart grid networks.
8	*Firmware and Software Updates:* Regular updates can help protect against known vulnerabilities.
9	*Privacy-Preserving Data Analytics:* This involves analyzing data in a way that preserves the privacy of individuals.

The utilization of data analytics in smart grids, while promising, presents several challenges, including data security. This concern is paramount as it involves the protection of sensitive information and system integrity. To address this, a multifaceted approach is required, encompassing secure communication protocols, unified security measures, and the application of advanced technologies like artificial intelligence (AI), ML, and blockchain. This comprehensive strategy ensures the secure and efficient operation of smart grids, as shown in Table 11.5.

These methods, when implemented effectively, are expected to enhance the security of smart grids. However, specific security measures may be needed, and they may vary depending on the specific characteristics and requirements of each smart grid system.

11.2.3 Energy Efficiency in Urban Settings

Energy efficiency in urban settings is a fundamental aspect of sustainable development. It involves implementing strategies and technologies that reduce energy consumption and promote the use of renewable energy sources. Here are some strategies and technologies for improving energy efficiency in urban environments.

11.2.3.1 Equitable City-Wide Electrification

Equitable city-wide electrification aims to ensure that all areas of a city, including underserved communities, have reliable access to electricity. This initiative is essential for promoting social equity, economic development, and environmental sustainability. By providing universal access to electricity, cities can enhance energy efficiency and address disparities in energy access, thereby improving the overall quality of life for residents.

One of the key benefits of equitable city-wide electrification is the promotion of energy efficiency across all areas of the city. Access to electricity enables the deployment of energy-efficient technologies and appliances, which can lead to reduced energy consumption and lower utility costs for residents. Additionally, electrification allows for the adoption of modern energy systems and infrastructure, contributing to improved energy productivity and sustainability.

Equitable electrification can support economic development by creating opportunities for businesses, industries, and entrepreneurs to thrive. Access to reliable electricity is crucial for powering commercial activities, supporting small-scale enterprises, and attracting investment in urban areas. By ensuring that all neighborhoods have access to electricity, cities can foster economic growth and job creation, contributing to a more equitable and prosperous urban environment. Equitable city-wide electrification also plays a vital role in enhancing public health and well-being. Access to electricity enables the provision of essential services such as lighting, heating, and cooling, which are critical for maintaining safe and healthy living conditions. Reliable electricity supply in healthcare facilities, schools, and public spaces can improve access to healthcare, education, and community services, ultimately contributing to the well-being of residents.

Universal electrification supports the transition to cleaner and more sustainable energy sources. Through the expansion of access to electricity, cities can facilitate the adoption of renewable energy technologies, energy-efficient lighting, and low-carbon transportation options. This can lead to a reduction in GHG emissions, improved air quality, and a more sustainable urban environment, benefiting both current and future generations.

Equitable electrification is also essential for promoting social equity and addressing energy poverty within urban areas, by ensuring that all communities have access to electricity. By this, cities can reduce disparities in energy access and alleviate the burden of energy costs on low-income households. This can contribute to greater social inclusion, empowerment, and resilience among vulnerable populations, ultimately fostering a more equitable and cohesive urban society. Access to electricity is critical for emergency response, disaster preparedness, and the provision of essential services during crises. By extending reliable electricity access to all neighborhoods, cities can improve their capacity to respond to emergencies and ensure the safety and well-being of residents in challenging circumstances.

11.2.3.2 Adaptive Public Lighting

Adaptive public lighting systems represent a significant advancement in urban lighting technology, offering the capability to automatically adjust the intensity of light based on various factors such as the time of day, ambient light levels, and the presence of people. By dynamically responding to environmental conditions and human activity, these systems can achieve significant energy savings and contribute to enhanced sustainability in urban environments.

The ability of adaptive lighting systems to modulate light intensity in response to natural light levels throughout the day allows for optimal energy utilization. During daylight hours, when natural light is abundant, the system can dim or even turn off certain lighting fixtures, minimizing energy consumption. As the natural light diminishes during dusk and nighttime, the system can gradually increase the light intensity to ensure visibility and safety while still maintaining energy efficiency. Adaptive public lighting systems can incorporate motion sensors and occupancy detection technology to adjust lighting levels based on the presence of people in specific areas. By illuminating spaces only when they are in use, these systems can further reduce energy consumption and light pollution, while also enhancing safety and security for pedestrians and commuters in urban areas.

The implementation of adaptive lighting not only contributes to energy savings but also supports the creation of more comfortable and inviting urban environments. By adjusting the lighting intensity based on the surrounding conditions, these systems can create a more pleasant and visually appealing atmosphere for residents, workers, and visitors, enhancing the overall urban experience. Adaptive public lighting systems can promote safety and security in urban settings by dynamically adjusting light levels based on pedestrian and vehicular activity. These systems can improve visibility and deter potential security threats, contributing to the well-being and confidence of individuals moving through urban spaces.

In addition to energy efficiency and safety benefits, adaptive lighting systems can have positive environmental impacts by reducing light pollution. Minimizing unnecessary light emissions during periods of low activity or when ample natural light is available helps these systems preserve the natural darkness of nighttime environments. This supports the health of ecosystems and wildlife while also reducing energy waste.

Adaptive public lighting systems offer urban areas a substantial prospect to boost energy efficiency, safety, aesthetics, and ecological sustainability. Utilizing cutting-edge technologies to smartly modulate lighting in response to varying conditions allows cities to foster urban environments that are not only more efficient and inviting but also mindful of environmental sustainability, benefiting both residents and visitors.

11.2.3.3 Municipal Green Bonds

Municipal Green Bonds are a form of debt security issued by municipalities, cities, or other local government entities. They are specifically designed to raise capital for projects that have environmental benefits, such as the development of energy-efficient infrastructure, renewable energy installations, waste management systems, and public transportation improvements. These bonds provide a way for cities to invest in sustainable initiatives while offering investors a chance to contribute to environmentally friendly projects.

The funds raised from these bonds are typically used to finance a variety of green projects. For instance, they might be used to upgrade buildings to make them more energy-efficient, install solar panels on public buildings, or develop public transportation systems that reduce carbon emissions. Financing these types of projects allows municipalities to reduce their environmental impact, improve the quality of life for their residents, and potentially save money in the long term through reduced energy costs.

Investing in municipal Green Bonds not only provides financial returns but also contributes to environmental sustainability. For investors, these bonds offer the opportunity to invest in projects that align with their values and commitment to environmental stewardship. Furthermore, as awareness and concern about climate change and environmental issues continue to grow, the demand for Green Bonds is likely to increase, making them an increasingly important tool for financing sustainable urban development.

11.2.3.4 Future-Ready Net-Zero Buildings

Future-ready net-zero buildings represent a significant advancement in sustainable building design, aiming to achieve a balance between energy consumption and energy production. These buildings are designed to minimize their energy demand through efficient design and technologies while offsetting their remaining energy needs by generating renewable energy on-site or through other means. The goal is to ensure that the building's net energy consumption over a defined period, typically a year, is zero, resulting in minimal environmental impact and enhanced sustainability.

Future-ready net-zero buildings are a revolutionary concept in the field of sustainable architecture and urban planning. These buildings are designed with the aim of achieving a net-zero energy balance, meaning they generate as much energy as they consume. This is primarily achieved using renewable energy sources, such as solar panels or wind turbines, integrated into the building design. The goal is to reduce the building's carbon footprint and contribute to a sustainable future.

The design of these buildings involves careful planning and the use of advanced technologies. Energy-efficient materials and construction techniques are used to minimize energy loss. For instance, high-quality insulation, energy-efficient windows, and optimized building orientation are used to reduce heating and cooling needs. Additionally, energy-efficient appliances and lighting are used to minimize electricity consumption.

The energy that is needed for these buildings is generated on-site through renewable energy sources. Solar panels are commonly used, often installed on rooftops or facades. Wind turbines, geothermal energy systems, and even rainwater harvesting systems can also be part of a net-zero building's energy solution. The idea is to create a building that is not only self-sufficient but also contributes to the overall grid by feeding excess energy back into it.

In addition to renewable energy generation, future-ready net-zero buildings prioritize energy efficiency through advanced building design, insulation, high-performance windows, energy-efficient appliances, and lighting systems. Minimizing energy demand through these measures enables net-zero buildings to optimize energy performance and reduce their environmental footprint, while also providing comfortable and healthy indoor environments for occupants.

Future-ready net-zero buildings often incorporate innovative technologies such as smart energy management systems, energy storage solutions, and advanced building automation to optimize energy use and generation. These technologies enable real-time monitoring, control, and optimization of energy flows within the building, ensuring that energy is utilized efficiently and that any surplus energy generated is stored for later use or fed back into the grid. However, creating future-ready net-zero buildings is not without challenges. The challenges are that the initial cost of installing renewable energy systems can be high, and the availability of renewable energy can be inconsistent geographically, requiring the need for energy storage solutions or backup systems. Despite these challenges, the benefits of future-ready net-zero buildings are also significant. They reduce reliance on fossil fuels, lower energy costs, and contribute to environmental sustainability. They represent a forward-thinking approach to architecture and urban planning, one that aligns with global efforts to combat climate change and promote sustainable living. As technology advances and the cost of renewable energy systems continues to decrease, it is expected that more and more buildings in the future will be designed to be net-zero.

11.2.3.5 Solar Passive Architecture

Designing buildings to take advantage of solar energy for heating and lighting is a key aspect of sustainable architecture. This approach, often referred to as passive solar design or solar passive approach, involves the collection and distribution of solar energy using natural means. When we implement the passive solar design, we must consider the factors listed in Table 11.6.

From the viewpoint of energy-efficient buildings, the passive solar design is intended to use less energy for heating, cooling, and running appliances and electronics. This can be achieved through various means, such as improved insulation, energy-efficient lighting, and the use of energy-efficient appliances. Moreover, energy-efficient buildings are also better equipped to switch to renewable energy. The operations of buildings account for a significant portion of global final energy

TABLE 11.6
Factors for Consideration in Passive Solar Design

1	*Orientation and Siting:* Buildings are oriented to ensure a long side faces the sun, minimizing east and west-facing walls and windows to reduce excessive summer heat gain. This allows for maximum exposure to sunlight, especially during the winter months when the sun is lower in the sky.
2	*Material Selection:* Different materials absorb sun radiation in different ways. For instance, dark surfaces absorb heat in larger amounts than light ones. Some surfaces will transmit radiation and store it at the same time, creating what we know as the greenhouse effect.
3	*Direct, Indirect, and Isolated Gain:* Direct gain is when the object is heated and energized by direct sunlight reaching its surface (through a south-facing glass facade, for instance). Indirect gain involves absorbing solar radiation on a surface (like a wall or floor), storing it, and then distributing it to the building. Isolated gain involves systems like sunspaces or solariums that collect solar energy and then distribute it to the building.
4	*Energy-Efficient Landscaping:* Designers can also incorporate energy-efficient landscaping strategies. For example, deciduous trees can be planted on the south side of a building. These trees will provide shade in the summer, reducing cooling needs, and will lose their leaves in the winter, allowing more sunlight to reach the building.
5	*Balance of Energy Efficiency and Material Use:* Creating energy-efficient buildings can sometimes increase the energy-intensive materials required to build them. Therefore, architects and designers around the world have taken steps to minimize the amount of energy that new buildings use, and the emphasis on solar energy and sustainable practices has extended to embodied carbon.

TABLE 11.7
Key Aspects of Energy-Efficient Landscaping

1	*Climate Zones:* Understanding our climate zone can help us determine the best energy-saving landscaping strategies for your home.
2	*Shade and Windbreaks:* A well-placed tree, shrub, or vine can deliver effective shade, act as a windbreak, and reduce your energy bills. Carefully positioned trees can save up to 25% of the energy a typical household uses.
3	*Microclimate:* The climate immediately surrounding our home is called its microclimate. When landscaping for energy efficiency, it is important to consider our microclimate as well as your regional climate.
4	*Water Management:* Energy-efficient landscaping also involves managing water effectively. This includes using plants or construction materials that cost less energy.
5	*Renewable Energy Sources:* Energy-efficient landscaping maximizes outdoor spaces' use and aesthetic appeal while minimizing the detrimental effects of outdoor spaces on the environment via careful planning and design. Water conservation, plant selection, and the use of renewable energy sources are some of its components.

consumption and emissions. Therefore, improving the energy efficiency of buildings can have a significant impact on reducing global energy consumption and emissions.

Solar passive architecture also includes energy-efficient landscaping, as the landscape is an integral part of the building function as a whole. Energy-efficient landscaping is a design strategy that uses natural elements like trees and vegetation to reduce reliance on artificial climate control. Some important aspects of energy-efficient landscaping are shown in Table 11.7.

11.2.4 Urban Energy Policy and Regulation

A government policy refers to a set of guidelines or principles that guide decision-making and actions. Policies are often the starting place for the law. They are typically informed by how people would like to see space defined and are often aspirational in nature. Policies do not typically have legal implications if they are not followed. They are found everywhere we go, whether it be the statehouse, work, or the gym. If we want to create mandates for the policy to be followed and implications if the policy is not followed, we can work with our district representatives and senators to draft bill language that can begin the process of transforming policy into law.

A government regulation, on the other hand, refers to a set of rules or laws that govern conduct. These are rules or mechanisms that limit, steer, or otherwise control social behavior. Regulation has a variety of meanings that are not reducible to a single concept. In the field of public policy, regulation refers to the promulgation of targeted rules, typically accompanied by some authoritative mechanism for monitoring and enforcing compliance. Regulations are vital to the implementation of the law. Regulations can be modified over time, and governments are often required to seek feedback from citizens through a public comment period.

Urban energy systems are networks that serve as the backbone of modern cities, supplying them with the essential power they need to function effectively. These systems encompass a broad range of components, including power generation facilities, transmission lines, distribution networks, and end-use technologies. They are responsible for powering everything from homes and businesses to transportation systems and public services. Urban energy systems embody a wide range of elements, from power generation facilities and transmission lines to end-use technologies and consumers. The functioning of these systems is influenced by a multitude of factors, including technological advancements, market dynamics, and consumer preferences. However, among these various influences, energy pricing stands out as a particularly significant factor due to its direct impact on the economic viability of different energy sources and technologies.

The design and operation of urban energy systems are influenced by a multitude of factors. Technological advancements play a significant role, with innovations continually reshaping the energy landscape. For instance, the advent of renewable energy technologies such as solar panels and wind turbines has opened new possibilities for sustainable power generation. Similarly, advancements in energy storage technologies, such as batteries, are enabling more flexible and resilient energy systems. Economic factors are another key influence on urban energy systems. The costs of different energy sources and technologies can greatly impact the choices made by both energy providers and consumers. For example, if the price of fossil fuels is low, it may discourage the adoption of renewable energy. On the other hand, if renewable energy becomes more cost-competitive, it could lead to a significant shift toward cleaner energy sources.

Social norms and consumer preferences also shape urban energy systems. People's attitudes toward different energy sources and their willingness to adopt energy-saving practices can have a substantial impact on energy demand. For instance, if there is a strong societal push toward sustainability, it could drive greater demand for renewable energy and energy-efficient technologies.

However, among these various influences, policy and regulation stand out as particularly significant. Government policies and regulations can steer the development of urban energy systems in specific directions. They can set standards for energy efficiency, mandate the use of certain energy sources, provide incentives for renewable energy, and impose penalties for excessive emissions. For example, a government might implement policies that favor renewable energy, such as subsidies or feed-in tariffs. These can make renewable energy more financially attractive, leading to increased investment and deployment. Similarly, regulations that set strict emissions standards can push energy producers to adopt cleaner technologies.

11.2.4.1 Energy Pricing

Energy pricing policy is a policy that has a very wide impact on the lives of people. Because of this characteristic, the tool brings significant influence in shaping urban energy systems. It is a mechanism that can be used to influence the behavior of both energy producers and consumers, playing a central role in determining the overall structure and operation of the energy system.

Energy prices can be used to encourage or discourage certain types of energy use. This is because the cost of energy is a major consideration for both energy producers and consumers. For energy producers, the price of energy can affect decisions about which energy sources to use. For example, if the price of fossil fuels is high, it may be more economically viable for energy producers to invest in renewable energy technologies. On the other hand, if the price of renewable energy is low, it may encourage energy producers to continue relying on fossil fuels.

For consumers, the price of energy can influence decisions about energy consumption. High energy prices can encourage consumers to reduce their energy use or to invest in energy-efficient appliances and technologies. Conversely, low energy prices may lead to increased energy consumption, which could have negative environmental impacts.

Energy pricing can also have significant implications for energy efficiency. Energy efficiency refers to the use of less energy to provide the same level of energy service. High prices for energy, particularly for fossil fuels, can provide a strong incentive for energy efficiency. This is because high energy prices can make energy-saving measures more economically attractive. For example, if the price of electricity is high, it may be more cost-effective for a business to invest in energy-efficient lighting or heating systems.

Conversely, low prices for renewable energy can make these sources more competitive with traditional energy sources. This can lead to increased investment in renewable energy technologies and a greater share of renewable energy in the overall energy mix. For example, if the price of solar power is low, it may encourage homeowners to install solar panels on their roofs. However, while energy pricing can be a powerful tool for shaping urban energy systems, it brings also in challenges. One of the main challenges is the volatility of energy prices, which can create uncertainty for both energy

TABLE 11.8
Strategies to Accomplish Balance between Affordable Energy and Sustainability

1	*Implementing Energy-Efficient Technologies:* The use of energy-efficient technologies in buildings, transportation systems, and urban infrastructure can significantly reduce energy consumption, thereby making energy more affordable and reducing environmental impact.
2	*Promoting Renewable Energy:* Renewable energy sources such as solar and wind power are not only environmentally friendly but can also be cost-effective in the long run. Policies that provide incentives for the use of renewable energy can help make these sources more competitive with traditional energy sources.
3	*Encouraging Behavioral Change:* Educating citizens about the importance of energy conservation and providing them with the tools and information they need to reduce their energy consumption can also contribute to both energy affordability and environmental sustainability.
4	*Implementing Progressive Energy Pricing:* Progressive energy pricing, where energy prices increase with consumption, can encourage energy conservation while protecting low-income households from high energy costs.
5	*Investing in Public Transportation:* Investing in public transportation can reduce the reliance on private vehicles, thereby reducing energy consumption and emissions.
6	*Urban Planning:* Urban planning strategies that promote compact, mixed-use development can reduce the need for transportation, thereby reducing energy use.
7	*Government Policies and Regulations:* Government policies and regulations play a crucial role in shaping urban energy systems. They can set standards for energy efficiency, mandate the use of certain energy sources, provide incentives for renewable energy, and impose penalties for excessive emissions.
8	*PPPs:* PPPs can facilitate the implementation of energy-efficient technologies and the development of renewable energy projects.
9	*Research and Development:* Investing in research and development can lead to the discovery of new, more efficient, and more sustainable energy technologies.
10	*Community Engagement:* Engaging with communities in the decision-making process can help ensure that energy policies and initiatives are well-received and effective.

producers and consumers. Additionally, while high energy prices can encourage energy efficiency and the use of renewable energy, they can also place a financial burden on low-income households.

Energy pricing must be carefully managed to ensure fairness. This means that energy prices should reflect the true cost of energy production, including the environmental and social costs. Additionally, energy pricing policies should be designed in a way that protects vulnerable groups from energy poverty. Energy pricing can drive the transition toward more sustainable and efficient energy systems. The use of energy pricing as a policy tool must be carefully managed to balance the need for economic efficiency, environmental sustainability, and social equity.

Given that eco-friendly energy typically costs more than nonrenewable energy and considering the unsustainable nature of non-renewable energy sources, it becomes imperative to find a balance between cost-effective energy and environmental sustainability in urban areas. The strategies to accomplish this equilibrium are outlined in Table 11.8.

11.2.4.2 Emission Standards

Emissions standards serve as a regulatory instrument in the realm of environmental policy. These standards are designed to set specific limits on the amount of pollutants that can be released into the environment during the process of energy production and consumption. By establishing stringent emissions standards, policymakers can stimulate the adoption of cleaner energy sources and technologies.

The concept of emissions standards is rooted in the fundamental principle of environmental protection. The primary objective of these standards is to control and reduce the release of harmful pollutants into the environment, thereby mitigating their detrimental effects on air quality, climate, and overall ecosystem health. These pollutants can include a wide range of substances, such as carbon dioxide, sulfur dioxide, nitrogen oxides, and particulate matter, all of which can have significant impacts on both the environment and human health.

Emissions standards are typically enforced through a variety of mechanisms, including permits, fines, and penalties for non-compliance. These enforcement mechanisms provide a strong incentive for energy producers and consumers to comply with the standards and to continually strive to reduce their emissions. In many cases, the cost of noncompliance, both in terms of financial penalties and reputational damage, can far outweigh the cost of investing in cleaner technologies and practices.

One of the key ways in which emissions standards can shape urban energy systems is by encouraging the adoption of cleaner energy sources. For example, if emissions standards make it more costly to burn coal, energy producers may be more inclined to invest in renewable energy sources such as wind and solar power. Similarly, strict emissions standards for vehicles can stimulate the adoption of electric cars and other forms of clean transportation. In addition to promoting cleaner energy sources, emissions standards can also drive technological innovation. To comply with strict emissions standards, companies may need to invest in the development of new technologies and practices that can reduce their emissions. This can lead to the emergence of innovative solutions that not only reduce emissions but also improve efficiency and reduce costs.

Emissions standards are a powerful tool for promoting environmental sustainability, and the standards must be systematically designed to ensure their effectiveness. This includes setting standards that are achievable with current technology and practices, providing sufficient time for companies to adapt to new standards, and ensuring that standards are enforced consistently and fairly.

Emissions standards vary significantly across different countries, reflecting the diverse environmental, economic, and political contexts in which they are implemented. Table 11.9 shows emission standards in some prominent countries.

Emissions standards have a significant impact on the non-electric-vehicle automotive industry. For example, tightening emissions standards can result in higher rates of automaker noncompliance. Stricter emissions standards not only fail to curtail on-road emissions but increase the likelihood

TABLE 11.9
Emission Standards in Some Prominent Countries

United States	The United States has a long history of implementing vehicle emissions standards, with mandatory standards operating for at least a decade. The US standards are more flexible yet more administratively complex, including trading or pooling arrangements, and have financial penalties for noncompliance.
European Union (EU)	The EU has focused on reducing emissions as part of a broader climate change strategy and introduced mandatory targets in 2009 after previous voluntary targets were not met. The EU and Indian standards focus on CO_2 emissions.
Japan	Japan has been operating mandatory vehicle emissions standards for at least a decade, like the United States and China. Soon, Japan will take a flexible, corporate-average approach to standards, with Japan switching to fleet averaging for its 2020 target.
China	China, like the United States and Japan, has been operating mandatory vehicle emissions standards for at least a decade. China is one of the four major markets (the United States, the EU, China, and Japan) that will take a flexible, corporate-average approach to standards.
Canada	Canada has standards that aim to reduce emissions. The Canadian standard includes restrictions on all GHGs.
India	India, like the EU, has standards that focus on CO_2 emissions.
Sweden, Switzerland, and Australia	These countries had separate emissions standards for many years but have since adopted the European standards.

of noncompliance by automakers. This is because economically, automakers have little financial incentive to abide by the tight emissions standards. Higher noncompliance rates in automakers face greater substitution pressure from competitors, as well as those with less advanced emissions control technology.

Emissions standards can drive technological innovation in the automotive industry. To comply with these standards, automakers may need to invest in the development of new technologies and practices that can reduce their emissions. This can lead to the emergence of innovative solutions that not only reduce emissions but also improve efficiency and reduce costs. Emissions laws also affect a car maker's bottom line. Devices designed to reduce a car's emissions, such as catalytic converters, cost money to develop, test, and mass-produce. The most common ways to meet emissions thresholds are to install expensive catalytic converters or reduce the weight or horsepower of the vehicle. Noncompliant automakers are left with an unappealing set of options: increase the sticker price (or take a hit on profitability), reduce the perceived safety, exit the market, or cheat.

11.2.4.3 Renewable Energy Incentives

Incentives for renewable energy, including mechanisms like tax credits or feed-in tariffs, are essential in shaping urban energy systems. These incentives are designed to make renewable energy more financially appealing, which can lead to increased investment in renewable technologies and a higher rate of utilization.

Renewable energy sources, such as solar, wind, hydro, and geothermal, offer a sustainable alternative to traditional fossil fuels. They provide a means to generate electricity with significantly lower emissions, contributing to the mitigation of climate change. However, despite their environmental benefits, the initial costs associated with the installation and implementation of renewable energy technologies can be a barrier to their widespread adoption. This is where incentives for renewable energy come into play. Reducing the financial burden associated with renewable energy projects makes these incentives more accessible and appealing to investors and consumers alike.

Tax credits are a common form of renewable energy incentive. These provide a reduction in the tax liability for individuals or companies that invest in renewable energy projects. For example, a company that installs solar panels on its buildings may receive a tax credit that reduces its corporate tax bill. This not only makes the project more financially viable but also provides a direct financial return on the investment. Feed-in tariffs, on the other hand, guarantee a fixed price for energy generated from renewable sources. Under a feed-in tariff scheme, energy producers are paid a set rate for the electricity they feed back into the grid. This provides a stable income stream for renewable energy projects, making them more attractive to investors. These incentives can have a significant impact on urban energy systems. Making renewable energy more financially attractive can stimulate the growth of renewable energy infrastructure within cities. This leads to a greater diversity of energy sources in the urban energy mix, reducing reliance on fossil fuels and lowering GHG emissions.

The increased use of renewable energy can have several additional benefits for urban areas. It can lead to the creation of new jobs in the renewable energy sector, from the manufacturing of renewable energy technologies to their installation and maintenance. It can also lead to improvements in air quality, as renewable energy sources do not emit the pollutants associated with the burning of fossil fuels. However, while incentives for renewable energy can play a significant role in shaping urban energy systems, they are not without challenges. The effectiveness of these incentives can depend on a range of factors, including the level of the incentive, the cost of renewable energy technologies, and the price of electricity from the grid. Policymakers must therefore carefully design these incentives to ensure they are effective in promoting the uptake of renewable energy.

11.2.4.4 Building Codes

Building codes and standards are sets of guidelines and regulations that dictate the design, construction, alteration, and maintenance of buildings. They cover a wide range of aspects, including structural integrity, fire safety, health and hygiene, accessibility, and, importantly, energy efficiency. The primary objective of these codes and standards is to ensure the safety, health, and welfare of the occupants and the broader community. Building codes and standards represent the area where policy and regulation can exert a significant influence. These codes and standards can mandate the incorporation of energy-efficient design and construction practices, thereby reducing the energy demand of the built environment.

The impact of building codes and standards on urban energy systems can be profound. Buildings account for a significant proportion of total energy consumption in cities. Requiring energy-efficient design and construction practices, building codes and standards can substantially reduce the energy demand of the built environment. This reduces the strain on urban energy systems and helps mitigate climate change by lowering greenhouse gas emissions associated with energy production.

The implementation of energy-efficient building codes and standards can lead to several benefits. For building owners and occupants, energy efficiency can result in substantial cost savings over the building's lifecycle due to reduced energy consumption. For the community and the environment, energy-efficient buildings contribute to the reduction of greenhouse gas emissions and the mitigation of climate change. They also reduce the demand for energy, which can alleviate pressure on energy infrastructure and reduce the need for costly upgrades or expansions. However, the development and implementation of energy-efficient building codes and standards are not without challenges. One of the key challenges is ensuring compliance. This requires effective enforcement mechanisms, as well as education and training for those involved in the design and construction of buildings. Additionally, there can be upfront costs associated with energy-efficient design and construction. While these costs can often be recouped over time through energy savings, they can still pose a barrier to the adoption of energy-efficient practices.

Another challenge is the need for ongoing adaptation and improvement of building codes and standards. As technology advances and our understanding of energy efficiency improves, building codes and standards need to be updated to reflect these changes. This requires ongoing research and development, as well as mechanisms for regularly reviewing and updating the codes and standards. Despite these challenges, the potential benefits of energy-efficient building codes and standards are significant. Through their ability to reduce energy consumption, these codes and standards are indispensable in determining sustainable urban energy systems. They represent a tool for policymakers aiming to reduce greenhouse gas emissions, mitigate climate change, and improve the sustainability of our cities.

11.2.4.5 Transportation Policies

Urban energy systems are significantly influenced by transportation policies. These policies, when designed with a focus on sustainability, can promote public transit and active transportation modes such as walking and biking. This, in turn, can lead to a reduction in the reliance on personal vehicles, thereby decreasing energy use and emissions.

Due to the nature of one of the largest energy consumptions in a city, transportation policies play a fundamental role in influencing the energy consumption patterns of a city. They determine the types of transportation that are promoted and supported, which directly impacts the energy demand of the urban population. For instance, policies that encourage the use of public transportation can lead to a decrease in the number of private vehicles on the road. This not only reduces traffic congestion but also lowers overall energy consumption as public transit is generally more energy-efficient than private vehicles.

Active transportation, such as walking and biking, is another area that can be promoted through effective transportation policies. These modes of transport require no fuel and produce no emissions,

TABLE 11.10
Challenges in Implementing Transport Policies

Unclear objectives	Clear objectives are crucial for everything including transport policies. It is important to distinguish between objectives, strategy, and the selection of policy measures.
No infrastructure focus	There is often an undue focus on infrastructure provision. It is essential to adopt a multimodal approach to avoid this.
Optimism bias	This is particularly evident in the financing of infrastructure. There is a tendency to underestimate costs and overestimate benefits.
Lack of performance monitoring	Ensuring that the performance of a national transport plan is regularly monitored against its objectives is a challenge.
Unclear responsibility	There can be uncertainty over who is responsible for implementing travel plans.
Lack of ownership	A general lack of ownership can hinder the implementation of travel plans.
Reduced funding	Transit agencies often face reduced or cut funding, which can limit their ability to implement policies.
Changing demographics and travel markets	*Changing Demographics and Travel Markets:* Responding to these changes can be a challenge for transit agencies.
Budget-services mismatch	It is difficult to keep a high standard of services within tight budget constraints.

making them the most sustainable form of transportation. Policies that create safe and accessible walking and biking paths can encourage more people to choose these active modes of transport over driving. This can lead to significant reductions in energy use and emissions. Transportation policies can also influence urban planning decisions. For example, policies that prioritize public transit can lead to the development of transit-oriented developments. These are compact, walkable, mixed-use communities centered around high-quality public transportation. Such developments can further reduce the need for private vehicles, leading to additional energy savings.

However, the success of these policies in determining urban energy systems depends on several factors. These include the availability and reliability of public transit, the safety and accessibility of walking and biking paths, and the public's willingness to change their transportation habits. Therefore, it is important for policymakers to consider these factors when designing transportation policies.

There are several challenges in implementing transportation policies, as shown in Table 11.10.

11.2.4.6 Utility Regulations

Utility regulations can dictate the types of energy sources that utilities must use, thereby influencing the overall energy landscape of the urban environment. For instance, utilities might be mandated to source a certain percentage of their energy from renewable sources. This requirement can significantly alter the energy mix of a city, pushing it toward a more sustainable and environmentally friendly future. The energy mix of a city is the combination of different energy sources used to meet the city's energy needs. This can include traditional fossil fuels like coal, oil, and natural gas, as well as renewable energy sources such as wind, solar, hydro, and bioenergy. The composition of this mix is influenced by a variety of factors, including the availability of resources, economic considerations, and government regulations.

Government regulations are particularly influential in shaping the energy mix. These regulations can set requirements for utilities regarding the types of energy sources they must use. For example, a regulation might require utilities to source a certain percentage of their energy from renewable sources. This type of regulation, often referred to as a renewable portfolio standard (RPS), is designed to increase the use of renewable energy and reduce reliance on fossil fuels. An RPS can have a significant impact on the energy mix of a city. By requiring utilities to source a certain percentage of their energy from renewable sources, an RPS can drive investment in renewable energy technologies and infrastructure. This can lead to an increase in the share of renewable energy in the city's energy mix, reducing its carbon footprint and contributing to environmental sustainability.

Utilities may face technical and economic hurdles in integrating large amounts of renewable energy into their systems. For instance, renewable energy sources like wind and solar are intermittent, meaning they do not produce energy consistently throughout the day. This can pose challenges for grid stability and reliability. Moreover, while the costs of renewable energy technologies have been decreasing, they can still be higher than traditional fossil fuels. This can make it economically challenging for utilities to meet renewable energy requirements without raising energy prices for consumers. To overcome these challenges, regulations may need to be accompanied by supportive policies and incentives. This could include financial incentives to offset the costs of renewable energy technologies, technical assistance to help utilities integrate renewable energy into their systems, and policies to encourage energy efficiency and reduce overall energy demand.

11.2.4.7 Land Use Policies

Land use policies have a profound impact on urban energy systems. These policies, which govern how land is used and developed within a city, can shape the city's energy consumption patterns in significant ways. One of the key ways in which land use policies can influence urban energy systems is through the promotion of compact, mixed-use development. This type of development, which combines residential, commercial, and other uses in close proximity, can reduce the need for transportation, thereby reducing energy use.

Compact, mixed-use development is a form of urban planning that seeks to create dense, walkable neighborhoods where people can live, work, and play without needing to travel long distances. This type of development can include a mix of housing types, retail shops, offices, schools, parks, and other amenities, all located within a short walk or bike ride of each other. By reducing the need for long commutes and car trips, compact, mixed-use development can significantly reduce a city's transportation-related energy use.

Transportation is one of the largest sources of energy used in cities. Cars, trucks, and other vehicles consume large amounts of gasoline and diesel fuel, contributing to air pollution and greenhouse gas emissions. By reducing the need for these vehicles, compact, mixed-use development can help to lower a city's overall energy consumption. In addition to reducing transportation-related energy use, compact, mixed-use development can also lead to more efficient use of infrastructure, further reducing energy consumption. Dense, mixed-use neighborhoods can make more efficient use of public transit, water and sewer systems, and other infrastructure, compared to sprawling, low-density developments. This can result in significant energy savings.

Implementing compact, mixed-use development requires thoughtful planning and no-afterthought planning as well as coordination among various stakeholders, including city planners, developers, residents, and businesses. Zoning laws and building codes may need to be updated to allow for higher densities and a mix of uses. Infrastructure may need to be upgraded to support increased densities. Community engagement is crucial to ensure that new developments meet the needs and preferences of residents. Despite these challenges, many cities around the world are recognizing the benefits of compact, mixed-use development and are implementing policies to promote it. These policies can include changes to zoning laws to allow for higher densities and a mix of uses, incentives for developers to build compact, mixed-use projects, and investments in infrastructure to support these developments. In terms of urban energy consumption, the benefits and drawbacks of compact cities and mixed-use development are shown in Table 11.11.

TABLE 11.11
The Benefits and Drawbacks of Compact City and Mixed-Use Development

Benefits	Drawbacks
Efficient Land Consumption: Compact cities excel in efficient land consumption, more flexible land use patterns, and cost efficiency of development and maintenance.	*Implementation Challenges:* Implementing compact, mixed-use development requires careful planning and coordination among various stakeholders, including city planners, developers, residents, and businesses.
Reduced Reliance on Fossil Fuels and Motor Vehicles: Compact cities reduce reliance on fossil fuels and motor vehicles.	*Zoning Laws and Building Codes:* Zoning laws and building codes may need to be updated to allow for higher densities and a mix of uses.
Energy Efficiency: Mixed-use developments are more energy efficient and reduce emissions.	*Infrastructure Upgrade:* Infrastructure may need to be upgraded to support increased densities.
Promotion of Sustainable Transport: Mixed-use development makes it easy to use public transport, walk, or bike, helping to efficiently connect the city's neighborhoods through sustainable transport.	*Community Engagement:* Community engagement is crucial to ensure that new developments meet the needs and preferences of residents.
Economic and Social Benefits: Compact, mixed-use developments yield economic and social benefits.	*Trade-Offs:* Residents and businesses have to make compromises or trade-offs to sustain mixed uses.
Improved Health, Environment, and Economies: Connected communities improve health, environment, and economies.	*Unsustainability:* There is a point at which such trade-offs become unsustainable.

11.3 URBAN ENERGY PLANNING AND INNOVATION

Urban energy planning and innovation is a rapidly evolving field that is becoming increasingly important as cities around the world grapple with the challenges of climate change, population growth, and urbanization. This field encompasses a wide range of strategies and technologies aimed at improving energy efficiency, reducing greenhouse gas emissions, and promoting sustainable development in urban areas. The sections on urban energy planning and innovation in the following are not stand-alone. They are deeply intertwined and influence each other in various ways, as given below.

11.3.1 Spatial and Urban Form

The spatial and urban form of a city directly impacts its energy demand. Compact cities with efficient public transportation and mixed-use developments can significantly reduce energy consumption. Therefore, understanding the energy demand of urban areas is crucial in shaping the spatial and urban form. It is a critical determinant of its energy demand. This form refers to the physical layout and design of a city, including its building density, transportation infrastructure, and land use patterns. The way a city is structured can significantly influence the amount of energy it consumes.

Compact cities, characterized by high-density development, mixed land use, and efficient public transportation systems, are often more energy-efficient than sprawling cities. High-density development reduces the distance people need to travel for work or leisure, reducing the reliance on private vehicles and thus lowering energy consumption. Mixed land use, where residential, commercial, and recreational areas are closely integrated, further reduces the need for long-distance travel. Efficient public transportation systems provide a low-energy alternative to private vehicles, especially in densely populated areas. Compact cities often have a higher proportion of multi-family residential buildings, which are typically more energy-efficient than single-family homes due to shared walls reducing heat loss. They also facilitate district heating systems, which are more efficient than individual heating systems.

However, creating compact cities is not just about increasing building density. It is also about designing buildings and infrastructure that are energy efficient. This includes everything from the orientation of buildings to take advantage of natural light and ventilation, to the use of energy-efficient materials and technologies, to the integration of renewable energy sources into the urban grid.

Understanding the energy demand of urban areas is therefore crucial in shaping the spatial and urban form. Assessing the energy consumption of different types of buildings and infrastructure, and predicting future energy demand based on trends in population growth, economic development, and technological change, allows urban planners to make informed decisions about the spatial and urban form of a city.

11.3.2 Energy Demand Assessment

Understanding the energy demand of urban areas is a fundamental aspect of urban energy planning. It involves a comprehensive analysis of the energy consumption patterns of different types of buildings and infrastructure, as well as forecasting future energy demand based on various factors such as population growth, economic development, and technological advancements.

11.3.2.1 Assessing Energy Consumption

The first step in understanding energy demand is to assess the current energy consumption of the city. This involves collecting data on the energy use of different types of buildings (residential, commercial, industrial, etc.) and infrastructure (transportation, public lighting, water and waste management systems, etc.). The energy consumption of a building can be influenced by various factors, including its size, design, construction materials, insulation, heating and cooling systems, and the behavior of its occupants. Similarly, the energy consumption of infrastructure depends on factors such as its efficiency, usage patterns, and maintenance practices.

11.3.2.2 Predicting Future Energy Demand

Once the current energy consumption is understood, the next step is to predict future energy demand. This process involves carefully analyzing various trends and projecting how they will impact future energy requirements.

One of the key factors to consider is population growth and urbanization. As the global population continues to rise, more people will require access to energy for their homes, businesses, and transportation. This increased demand can be especially pronounced in rapidly growing urban areas, where the concentration of people and infrastructure tends to be higher. According to the United Nations, the world's urban population is expected to grow from 4.2 billion in 2018 to 6.7 billion by 2050, with much of this growth occurring in developing countries. This substantial increase in the number of people living in cities will undoubtedly drive up energy consumption as more buildings, roads, and other urban facilities are constructed to accommodate the growing population.

In addition to population growth, economic development also plays a crucial role in shaping future energy demand. As countries and regions experience economic growth, their energy needs often change. For example, as living standards improve, there may be a greater demand for air conditioning, appliances, and other energy-intensive technologies in residential and commercial buildings. Furthermore, the types of industries and manufacturing processes employed can significantly impact energy consumption patterns. Heavier industries, such as steel production or chemical manufacturing, tend to be more energy-intensive than lighter industries, such as electronics or software development. Understanding the trajectory of a region's economic development and the resulting changes in its industrial and commercial landscape is essential for accurately forecasting future energy demand.

Technological advancements can also significantly influence energy demand. The introduction of more energy-efficient technologies and systems can help mitigate the increase in energy consumption that may accompany population growth and economic development. For instance, the adoption of LED lighting, high-efficiency HVAC (heating, ventilation, and air conditioning) systems, and advanced building insulation can reduce the energy required for powering and cooling buildings. Similarly, the widespread use of electric vehicles and the integration of renewable energy sources, such as solar and wind power, can help offset the growing energy demands of the transportation and electricity sectors. Furthermore, the development of smart grid technologies and energy storage solutions can enable better management and optimization of energy resources, leading to more efficient energy consumption.

However, the impact of technological advancements on energy demand can be complex. While some technologies may reduce overall energy consumption, others may enable new energy-intensive applications or alter consumer behavior in ways that increase energy demand. For example, the rise of energy-hungry data centers and cryptocurrency mining operations has been driven by the rapid growth of digital technologies and the Internet. Carefully considering both the potential benefits and unintended consequences of technological progress is crucial for accurately forecasting future energy demand.

Predicting future energy demand involves a comprehensive analysis of population growth, economic development, and technological change. By understanding these key drivers and their interplay, policymakers, energy planners, and industry leaders can make more informed decisions about the long-term energy needs of a region or country, enabling the development of sustainable and resilient energy systems for the future.

Understanding energy demand can help cities identify opportunities for energy savings. For instance, if a city finds that a significant portion of its energy is consumed by inefficient buildings, it can prioritize investments in energy-efficient building design and retrofits. Similarly, if a city finds that a large amount of energy is used for transportation, it can explore options for improving public transportation, promoting active transportation (walking and cycling), and supporting electric vehicles.

11.3.2.3 Prioritizing Investments in Energy-Efficient Technologies and Infrastructure

Understanding energy demand can help cities prioritize investments in energy-efficient technologies and infrastructure. By identifying the areas with the highest energy consumption and the greatest potential for energy savings, cities can allocate their resources more effectively and achieve greater energy efficiency.

One of the key areas where cities can focus their efforts is the building sector. Buildings account for a significant portion of urban energy consumption, often ranging from 30% to 50% of a city's total energy use. By analyzing energy usage patterns in different building types, such as residential, commercial, and industrial, cities can identify the most energy-intensive structures and target them for upgrades and retrofits. For example, older commercial buildings with outdated HVAC systems and poor insulation may be consuming far more energy than necessary, presenting an opportunity for significant energy savings through the installation of energy-efficient equipment and building envelope improvements.

Another area of focus for cities is the transportation sector, which can account for a substantial portion of urban energy consumption, especially in car-dependent communities. Understanding the travel patterns and energy usage of different modes of transportation allows cities to prioritize investments in public transit, electric vehicle charging infrastructure, and active transportation options like walking and cycling. This can help reduce the reliance on personal vehicles, which tend to be energy-intensive, and shift the transportation mix toward more efficient and sustainable modes.

In addition to buildings and transportation, cities can also analyze the energy consumption of their public infrastructure, such as street lighting, water treatment and distribution systems, and municipal buildings. By identifying inefficient or outdated systems, cities can prioritize upgrades and retrofits that can significantly reduce energy usage and costs. For example, the replacement of traditional streetlights with energy-efficient LED lighting can result in substantial energy savings, while investments in water pumping and treatment technologies can optimize the energy consumption of the water infrastructure.

Understanding energy demand can help cities identify opportunities for renewable energy integration and distributed energy generation. Analyzing the energy consumption patterns and the availability of local renewable resources, such as solar, wind, or geothermal, allows cities to strategically deploy clean energy technologies. This offsets the demand for grid-supplied electricity, reducing the overall carbon footprint and energy costs.

To effectively prioritize investments in energy-efficient technologies and infrastructure, cities often rely on comprehensive energy assessments and audits. These assessments involve collecting and analyzing detailed data on energy consumption, building characteristics, transportation patterns, and other relevant factors. This data-driven approach allows cities to identify the areas with the greatest potential for energy savings and target their investments accordingly.

Cities can leverage advanced analytics and modeling techniques to simulate the impact of different energy efficiency measures and infrastructure investments. Using these tools, city planners can evaluate the potential energy savings, cost-effectiveness, and environmental benefits of various strategies. This enables them to make more informed decisions about their energy efficiency priorities.

By understanding energy demand and using this knowledge to guide their investments, cities can not only reduce their energy consumption and associated costs but also contribute to broader sustainability goals, such as reducing greenhouse gas emissions and addressing climate change. This strategic approach to energy efficiency can help cities become more resilient, livable, and environmentally responsible, ultimately enhancing the quality of life for their residents.

11.3.3 Integration of Energy-Efficient Design

Incorporating energy-efficient design principles into urban planning and building design is a multifaceted strategy that can yield significant reductions in energy consumption. One of the key components of this approach is the utilization of passive design strategies that leverage natural resources, such as sunlight and wind, to reduce the need for artificial heating, cooling, and lighting.

Passive design techniques can include the strategic orientation of buildings to maximize solar gain in the winter and minimize it in the summer, the use of high-performance insulation and windows to maintain thermal comfort, and the incorporation of natural ventilation systems that harness prevailing winds to circulate air throughout the building. These passive strategies not only reduce energy demand but also contribute to the overall comfort and well-being of building occupants.

The use of high-efficiency appliances and building systems can further enhance the energy performance of urban structures. This includes the installation of energy-efficient HVAC systems, LED lighting, and water-efficient plumbing fixtures, all of which can significantly reduce the operational energy consumption of a building.

The integration of renewable energy technologies, such as solar PV panels and wind turbines, into urban planning and building design is becoming increasingly prevalent. These renewable energy systems can be integrated into the built environment, either as stand-alone installations or as integrated components of the building's infrastructure, allowing for on-site energy generation and reduced reliance on traditional grid-supplied electricity.

Many cities around the world have recognized the importance of energy-efficient design and have implemented building codes and regulations that either require or incentivize the incorporation of these principles. For example, some cities have adopted mandatory energy performance standards for new construction, while others have implemented tax credits or rebate programs to encourage the retrofitting of existing buildings with energy-efficient technologies.

These regulatory measures not only drive the adoption of energy-efficient design but also help to establish a market demand for sustainable building practices. As a result, the construction industry has been increasingly responsive to these demands, developing new technologies, materials, and design approaches that further enhance the energy performance of buildings.

The benefits of incorporating energy-efficient design principles into urban planning and building design are multifaceted. In addition to reducing energy consumption and associated greenhouse gas emissions, these strategies can also lead to significant cost savings for building owners and occupants through reduced utility bills. The improved thermal comfort and indoor air quality provided by energy-efficient buildings can have positive impacts on the health and well-being of the building's occupants.

As cities continue to grow and urbanization trends accelerate, the need for sustainable and energy-efficient design solutions will only become more pressing. Through these principles, urban planners, architects, and policymakers can play a pivotal role in shaping more resilient, livable, and environmentally responsible cities of the future.

11.3.4 Smart Energy Systems

The concept of smart energy systems is becoming a cornerstone in urban energy planning. These systems, which leverage advanced technologies and data analytics, are designed to optimize energy production, distribution, and consumption, thereby enhancing efficiency, reducing costs, and minimizing environmental impact.

Smart energy systems represent a paradigm shift in the way we manage and use energy. Unlike traditional energy systems, which are often characterized by one-way flows of energy and information, smart energy systems are interactive, multidirectional, and highly responsive. They integrate various types of energy resources, including conventional fossil fuels and renewable energy sources, and use information and communication technologies to monitor, control, and optimize their operation.

11.3.4.1 Smart Grids: Balancing Supply and Demand

One of the key components of smart energy systems is the smart grid. A smart grid is an electricity network that uses digital technology to monitor and manage the transport of electricity from all generation sources to meet the varying electricity demands of end users. Smart grids coordinate the needs and capabilities of all generators, grid operators, end-users, and electricity market stakeholders to operate all parts of the system as efficiently as possible, minimizing costs and environmental impacts while maximizing system reliability, resilience, and stability.

Smart grids can balance supply and demand in real time. They can adjust to changes in electricity demand throughout the day and respond to fluctuations in electricity production, especially from renewable energy sources like wind and solar power, which can be highly variable. Smart grids can reduce the need for excess generation capacity, which saves costs and lowers greenhouse gas emissions. They also minimize energy losses in electricity distribution, which are typically around 6% in traditional grids, by implementing better voltage regulation, more accurate metering, and detecting and reducing faults and outages.

11.3.4.2 Integration of Renewable Energy

Another major advantage of smart grids is their ability to integrate a high share of renewable energy. Renewable energy sources, such as wind and solar power, are often decentralized and variable. Smart grids can accommodate these characteristics by providing flexible demand response, energy storage, and advanced forecasting techniques. This enables a higher penetration of renewable energy in the energy mix, which is crucial for reducing greenhouse gas emissions and combating climate change.

11.3.4.3 Smart Buildings and Homes

Smart energy systems also encompass smart buildings and homes. These are buildings and homes equipped with advanced automation systems that control and manage building energy consumption. They can automatically adjust their energy usage based on real-time price signals, weather forecasts, and user preferences, thereby optimizing energy efficiency and comfort. For example, a smart building can adjust its HVAC system based on the outside temperature and the occupancy of the building. It can also adjust its lighting system based on the amount of natural light available. Similarly, a smart home can adjust its appliances and devices based on electricity prices, running them when prices are low and reducing their operation when prices are high.

Smart buildings and homes can generate their own energy through rooftop solar panels or other decentralized energy resources and sell excess energy back to the grid. They can also provide demand response, reducing their energy usage during peak demand periods to help balance the grid. Smart energy systems represent a holistic approach to energy management that can significantly enhance the sustainability and resilience of our cities. They are a testament to the power of technology and innovation in transforming the way we produce, distribute, and consume energy. As we continue to urbanize and grapple with the challenges of climate change, the importance of smart energy systems in urban energy planning will only continue to grow. They hold the promise of a cleaner, more efficient, and more sustainable energy future for all.

11.3.5 Low-Carbon Economic Business Model

Transitioning to a low-carbon economy is a central objective of numerous urban energy planning initiatives. This transition is multifaceted, involving not only technological transformations but also significant shifts in economic structures and policy frameworks. The goal is to reduce greenhouse gas emissions, mitigate climate change, and promote sustainable development.

The transition to a low-carbon economy necessitates the adoption of a range of new technologies. Renewable energy technologies, such as solar PV systems, wind turbines, and hydroelectric power, play a crucial role in reducing carbon emissions. Energy-efficient technologies, including LED lighting, high-efficiency appliances, and electric vehicles, can significantly reduce energy consumption and associated emissions. Moreover, advanced technologies like smart grids and energy storage systems can optimize the use of renewable energy and further enhance energy efficiency.

Alongside technological changes, transitioning to a low-carbon economy also involves substantial economic changes. Traditional industries that are heavily reliant on fossil fuels may decline, while new industries related to renewable energy and energy efficiency may emerge and grow. This shift can create new jobs and stimulate economic growth, but it can also pose challenges, such as job losses in declining industries and the need for workers to acquire new skills. Therefore, strategies for managing this economic transition, such as retraining programs and social safety nets, are an important part of urban energy planning.

Policy changes are another key aspect of the transition to a low-carbon economy. Governments at all levels, from local to national, can implement a variety of policies to promote energy efficiency and renewable energy. For example, feed-in tariffs and power purchase agreements can provide financial incentives to produce renewable energy. Energy performance contracting can encourage energy efficiency improvements in buildings by guaranteeing that the energy savings will cover the cost of the improvements. Green building certification programs, such as leadership in energy and environmental design (LEED) and building research establishment environmental assessment method (BREEAM), can set standards for energy-efficient design and construction and recognize buildings that meet these standards.

The transition to a low-carbon economy can also foster the development of new business models and industries. For instance, the growth of renewable energy can spur the creation of businesses that install and maintain solar panels and wind turbines. The push for energy efficiency can lead to the expansion of industries that produce energy-efficient appliances and vehicles, as well as businesses that offer services like energy auditing and retrofitting. These new business models and industries can create jobs, stimulate innovation, and contribute to economic growth.

11.4 ENERGY TRANSITION IN CITIES

Energy transition in cities is a dynamic process that involves shifting from fossil fuel-based energy systems to more sustainable, renewable energy sources. This transition is driven by the urgent need to mitigate climate change, improve air quality, enhance energy security, and create new economic opportunities.

Cities are at the forefront of this energy transition. As major consumers of energy and emitters of greenhouse gases, cities have a critical role to play in transforming our energy systems. Moreover, with their high population density and concentration of resources, cities are uniquely positioned to pioneer innovative energy solutions.

One of the key aspects of the energy transition in cities is the shift toward renewable energy. Cities around the world are investing in solar, wind, hydro, and other forms of renewable energy. These investments are not only reducing greenhouse gas emissions but also creating jobs, stimulating local economies, and enhancing energy security. Another important aspect is energy efficiency. Enhancing the energy efficiency of buildings, transportation systems, and industrial processes can greatly reduce a city's energy demand. This can be achieved through measures such as stricter building codes, promoting public transportation and active travel, and encouraging the use of energy-efficient appliances.

The energy transition also involves the development of smart grids. These are electricity networks that use digital technology to optimize the production, distribution, and consumption of electricity. Smart grids can accommodate a high share of variable renewable energy, improve the reliability of electricity supply, and empower consumers to manage their energy use. It is not just about technology. It also involves changes in policy, finance, and governance. For example, cities are implementing policies to incentivize renewable energy and energy efficiency, such as feed-in tariffs and energy performance contracting. They are also exploring innovative financing mechanisms to fund energy projects, such as Green Bonds and public–private partnerships (PPPs).

The energy transition is about people. It is about creating a just and inclusive transition that benefits all residents. This means ensuring that the benefits of the energy transition, such as lower energy bills and improved air quality, are shared equitably. It also means engaging residents in the energy transition, for example, through community energy projects and public consultations.

The energy transition in cities is about reimagining and reshaping the energy systems to create a more sustainable, resilient, and equitable future. As we continue to urbanize and grapple with the challenges of climate change, the role of cities in the energy transition will only become more important. They are not only the battlegrounds where the fight against climate change will be won or lost but also the testing grounds for the innovative solutions that will drive this crucial transition forward. The energy transition in cities is not just a necessity but an opportunity to build a better future for all.

While cities are making strides in the energy transition, the reality is that their electric supply is still largely dependent on non-renewable sources. This underscores the pressing need for continued innovation, investment, and policy commitment to accelerate the shift toward 100% renewable energy and realize a truly sustainable future.

11.5 ELECTRIC POWER SUPPLY

The electric power supply in cities typically consists of electric generation, transmission, and distribution, plus demand management to optimize the supply system and minimize losses.

11.5.1 Electric Generation

Electricity generation is the process of converting various forms of energy into electrical energy, which is then distributed and utilized for a wide range of applications. This process can involve traditional power plants that rely on the combustion of fossil fuels, as well as more modern and sustainable approaches that harness renewable energy sources.

11.5.1.1 Traditional Fossil Fuel Power Plants

The most common method of electricity generation has been through the use of traditional power plants that burn fossil fuels, such as coal, natural gas, or oil. In these plants, the chemical energy stored in the fossil fuels is converted into thermal energy, which is then used to power turbines that generate electricity. The basic principle behind this process is the Rankine cycle, where water is heated, converted into high-pressure steam, and then used to spin a turbine connected to an electric generator.

The efficiency of these fossil fuel power plants can vary, with modern facilities employing advanced technologies and systems to improve their overall performance. However, the burning of fossil fuels also comes with significant environmental drawbacks, as it releases greenhouse gases and other pollutants into the atmosphere, contributing to climate change and air pollution.

11.5.1.2 Renewable Energy Sources

In recent years, there has been a growing emphasis on the use of renewable energy sources for electricity generation. These sources, such as wind, solar, hydropower, and geothermal, offer a more sustainable and environmentally friendly approach to electricity production.

11.5.1.2.1 Wind Power

Wind turbines convert the kinetic energy of wind into electrical energy. As the wind blows across the blades of the turbine, it causes them to spin, which in turn rotates a generator, producing electricity. Wind farms, consisting of multiple wind turbines, are often located in areas with consistent and reliable wind patterns, such as coastlines, hilltops, or open plains.

Wind turbines, which convert the kinetic energy of wind into electrical energy, come in various types and designs. Here are some of the most common types:

- *Horizontal Axis Wind Turbines (HAWTs):* These are the most common type of wind turbine. HAWTs have the main rotor shaft and electrical generator at the top of a tower, and they must be pointed into the wind. Small turbines are pointed by a simple wind vane, while large turbines generally use a wind sensor coupled with a servo motor. HAWTs have two or three blades that look like an airplane propeller.
- *Vertical Axis Wind Turbines (VAWTs):* These turbines have the main rotor shaft arranged vertically, meaning they do not need to be pointed into the wind to be effective, which is an advantage on sites where the wind direction is highly variable. VAWTs can harness wind from all directions. There are several types of VAWTs, including the Savonius, Darrieus, and Helical types.
  - Savonius VAWTs are drag-based turbines with scoops that catch the wind. They are often used for low-speed operations.
  - Darrieus VAWTs are lift-based turbines with aerofoil blades. They are more efficient than Savonius turbines but require a starting mechanism to begin operation.
  - Helical Darrieus VAWTs have longer blade areas and are designed to capture more wind energy.
- *The Wind Tree:* This is a decorative urban feature that uses multiple small vertical axis turbines to generate electricity.

- *Vortex Bladeless:* This is a bladeless oscillating mast that harnesses vortices in the wind to generate electricity.
- *Vehicle-Powered Turbines:* These turbines harness the turbulence created by vehicles on highways to generate electricity.
- *Cross-Axis Wind Turbine (CAWT):* This type of turbine combines features of both HAWTs and VAWTs.
- *Cooling Tower Updraft Turbine:* This type of turbine utilizes the exhaust air flow from cooling towers to generate electricity.

Each type of wind turbine has its own advantages and disadvantages, and the choice of turbine depends on the specific requirements of the site, including the wind speed and direction, the available space, and the local environment and wildlife.

The power generated by a wind turbine can be estimated using the following formula:

$$P = \frac{1}{2} \cdot \rho \cdot A \cdot v^3 \tag{11.1}$$

where:

P is the power output in watts.
ρ is the air density in kg/m.
A is the cross-sectional area of the wind in m², which can be calculated as $A = \pi D^2 / 4$ for a turbine with diameter D.
v is the velocity of the wind in m/s.

This formula shows that the power available to a wind turbine is based on the density of the air, the swept area of the turbine blades, and the velocity of the wind. Although wind speed is the most variable input and also the most impactful as it is cubed, the stronger wind speed will also hamper the structural integrity of the wind turbine.

This formula gives the theoretical maximum power output. The actual power output of a wind turbine will be less than this due to various inefficiencies. To calculate the actual power output of a specific turbine, you would need to multiply the theoretical power by the power coefficient of the turbine, which ranges from 0.25 to 0.45. The power coefficient represents the efficiency of the turbine in converting the power in the wind into electrical power.

11.5.1.2.2 Solar Power

Solar power systems, particularly PV panels, convert the energy of sunlight directly into electrical energy. PV cells, made of semiconductor materials, absorb photons from sunlight and generate an electric current, which is then collected and converted into usable electricity. Solar power can be generated on a large scale through utility-scale solar farms or on a smaller scale through rooftop solar installations on residential and commercial buildings.

The power output of a solar panel can be estimated using the formula:

$$P = A \cdot G \cdot \eta \tag{11.2}$$

where:

P is the power output.
A is the area of the solar panel.
G is the solar radiation incident on the panel (also known as solar irradiance, typically measured in watts per square meter).
η is the efficiency of the solar panel.

This is a simplified formula, and actual power output can be influenced by many factors including the angle of the panel, temperature, and spectral distribution of the light. Also, the efficiency of solar panels typically ranges from 15% to 20% for most commercially available panels but can be higher for certain types of panels. It is always best to refer to the manufacturer's specifications for the most accurate information.

11.5.1.2.3 Hydropower

Hydropower, one of the oldest and most reliable sources of renewable energy, harnesses the potential energy stored in water to generate electricity. This energy is typically stored in dams or other types of water reservoirs. As water flows through turbines, it spins generators, which in turn produce electricity. The beauty of hydropower lies in its simplicity and its reliance on the natural water cycle, which continually replenishes the water supply, making it a renewable and sustainable energy source.

The process of generating electricity from hydropower begins with the collection of water in large reservoirs or dams. These structures are strategically located to capture and store water from rivers and other sources. The stored water has potential energy due to its elevation relative to the turbines located at the bottom of the dam. This potential energy is what is converted into electricity. When electricity is needed, gates in the dam are opened to allow water to flow out. The flowing water passes through turbines, causing them to spin. These turbines are connected to generators, and as the turbines spin, so do the generators. Inside the generators, magnets spin within wire coils to produce electricity. The electricity produced is then transmitted through power lines to homes, businesses, and industries.

One of the key advantages of hydropower is its reliability. Unlike solar and wind energy, which are dependent on weather conditions, hydropower can produce electricity consistently as long as there is a sufficient water supply. Moreover, the flow of water can be controlled using gates, allowing production of electricity to be increased or decreased as needed. This makes hydropower a flexible and dependable source of energy. Hydropower is considered a clean energy source as it does not directly emit greenhouse gases. This makes it an attractive option in the fight against climate change. However, the construction of dams and reservoirs can have significant environmental impacts, including changes to ecosystems and displacement of local communities. Therefore, careful planning and management are required to minimize these impacts. There are also small-scale and micro-hydro systems. These systems can be used in remote areas where access to the national grid is not available. They can provide a reliable and sustainable source of electricity for local communities, promoting economic development and improving quality of life.

Technological advancements continue to improve the efficiency and environmental performance of hydropower systems. For example, new turbine designs are being developed to reduce the impact on fish and other aquatic life. There are also ongoing research and development efforts aimed at improving the efficiency of generators and exploring new ways to store and use the energy produced.

The power generated by a hydropower system can be estimated using the formula:

$$P = \eta \cdot \rho \cdot g \cdot h \cdot Q \tag{11.3}$$

where:

- P is the power output.
- η is the efficiency of the turbine and generator (typically between 0 and 1).
- ρ is the density of water (approximately 1000 kg/m³).
- g is the acceleration due to gravity (approximately 9.81 m/s²).
- h is the height of the water column (also known as the hydraulic head).
- Q is the flow rate of the water.

This is a simplified formula, and actual power output can be dictated by many factors including the design and condition of the turbine and generator, as well as environmental factors such as water temperature and silt content. Also, the efficiency of the system can vary depending on the specific technology and design used. It is typically around 90% for modern turbines.

11.5.1.2.4 Geothermal Power

Geothermal power plants are a type of renewable energy source that utilizes the natural heat stored within the Earth's crust and mantle layers. This heat is a result of the radioactive decay of minerals and the original heat from the Earth's formation. The temperature at the Earth's core, which can reach over 5,000°C, is higher than the surface of the Sun. This immense heat continuously flows outward from the core to the crust, creating a virtually inexhaustible supply of energy.

To harness this geothermal energy, wells are drilled into geothermal reservoirs, which are regions of the earth's crust where the heat is particularly concentrated. These reservoirs can be found in regions of volcanic activity, tectonic plate boundaries, or hot springs. The depth of these wells can range from a few hundred meters to over 2 km deep, depending on the location and the temperature of the reservoir. Once the well reaches the reservoir, the heat is accessed in the form of steam or hot water. In some cases, steam can be used directly to drive turbines and generate electricity. In other cases, the hot water is used to heat a secondary fluid with a lower boiling point, which then vaporizes and drives the turbine. This process is known as a binary cycle.

The electricity generated by the turbines is then fed into the power grid for distribution. The steam or hot water, after passing through the turbine, is returned to the reservoir to be reheated and reused, making the process sustainable and environmentally friendly. Geothermal power plants have the advantage of being able to provide a constant and reliable source of power, unlike other renewable sources such as wind or solar, which are dependent on weather conditions. They also have a smaller land footprint compared to other power generation methods and emit very low levels of GHGs. However, they are limited by the availability of suitable geothermal reservoirs and the initial cost of drilling and construction can be high. Despite these challenges, geothermal energy is a promising and growing field in the quest for clean and sustainable energy sources.

Geothermal energy offers several environmental benefits, as shown in Table 11.12.

TABLE 11.12
Environmental Benefits of Geothermal Energy

Low-carbon footprint	Unlike fossil fuels, geothermal energy harnesses earth's natural heat without combustion, meaning fewer GHGs are released into the atmosphere. This makes it a cleaner and more environmentally friendly energy source.
Sustainable and renewable	Geothermal energy is both sustainable and renewable. Earth has been emitting heat for about 4.5 billion years and will continue to emit heat for billions of years into the future because of the ongoing radioactive decay in the earth's core.
Minimal land footprint	Compared to solar farms or wind parks, geothermal power plants require significantly less land. This is beneficial in preserving natural habitats and biodiversity.
Water conservation	Most geothermal facilities have closed-loop water systems, in which extracted water is pumped directly back into the geothermal reservoir after it has been used for heat or electricity production. This reduces the overall water impact of the plants.
Reduction in air pollutants	Geothermal energy does not produce air pollution, making it a healthier alternative to fossil fuels.

11.5.2 Integration of Renewable Energy into the Grid

The global shift toward renewable energy sources like wind, solar, and hydro is a critical step in combating climate change and reducing our reliance on fossil fuels. These renewable sources offer a sustainable way to meet our energy needs while significantly reducing greenhouse gas emissions. However, the integration of these clean energy sources into the existing electrical grid is not without its challenges.

The traditional electrical grid was designed for centralized power generation, typically from fossil fuel power plants. Renewable energy sources, on the other hand, are often decentralized and distributed across various locations. This necessitates a significant upgrade to our grid infrastructure to accommodate these new energy sources. Enhancements to transmission lines, substations, and transformers are required to handle the increased and variable power flow. Renewable energy sources like wind and solar are intermittent by nature – they produce energy only when the wind blows or the sun shines. This variability poses a challenge to maintaining a stable and reliable power supply. To address this, the development of advanced energy storage technologies is crucial. These technologies, such as batteries or pumped hydro storage, can store excess energy produced during peak times and release it when production is low.

The effective management of these variable energy sources requires the implementation of smart grid systems. These systems use digital technology and advanced communication systems to monitor and manage the flow of electricity from various sources in real time. This ensures a more reliable and efficient grid, capable of adjusting to changes in electricity demand and supply. While the adoption of renewable energy sources is a positive step toward a sustainable future, it requires a comprehensive approach to successfully integrate these sources into our existing electrical grid. This includes the development of advanced grid infrastructure, energy storage technologies, and smart grid systems.

11.5.2.1 Advance Grid Infrastructure

The traditional electrical grid, a network of power plants, transformers, and transmission lines, was primarily designed to accommodate centralized power generation. This typically involved large-scale power plants, often powered by fossil fuels such as coal, natural gas, or oil. These plants, usually located away from populated areas due to pollution concerns, generate electricity that is then transmitted over long distances to homes, businesses, and industries. However, the advent of renewable energy sources like wind, solar, and hydro has fundamentally changed the dynamics of power generation. Unlike traditional power plants, renewable energy installations are often decentralized and distributed across various locations. For instance, solar panels can be installed on rooftops of individual homes, wind turbines can be erected in fields, and small-scale hydro plants can be set up in rivers and streams. This decentralization of power generation brings energy production closer to the point of consumption, reducing transmission losses and increasing efficiency.

Yet, this shift also presents a significant challenge. The existing grid infrastructure, designed for a different era of power generation, needs to be substantially upgraded to accommodate these new energy sources. This is a complex task that involves several key components.

First, the transmission lines need to be enhanced to handle the increased and variable power flow. Unlike traditional power plants that provide a steady output, renewable sources can be variable and intermittent, depending on factors like sunlight, wind speed, and water flow. This requires a more flexible and robust transmission network that can handle fluctuations in power generation. Second, substations, which act as nodes in the power grid, converting high-voltage electricity from power plants to lower-voltage electricity for consumers, need to be upgraded or even redesigned. They need to be equipped with advanced technologies to manage the flow of electricity from diverse sources and ensure a stable and reliable power supply. Lastly, transformers, which adjust the voltage of electricity to appropriate levels for transmission and distribution, also need to be enhanced. They need to be capable of handling the variable and sometimes unpredictable power output from renewable sources.

Integrating renewable energy sources requires a comprehensive upgrade of the grid infrastructure, including transmission lines, substations, and transformers. This will enable us to harness the full potential of renewable energy and move toward a more sustainable and resilient energy future.

11.5.2.2 Energy Storage Solutions

Renewable energy sources such as wind and solar have immense potential to meet our energy needs in a sustainable manner. However, they come with their own set of confronts, the most significant of which is their intermittent nature. The power generated by wind turbines and solar panels is heavily dependent on weather conditions. Wind turbines require a certain wind speed to operate efficiently, and solar panels need sunlight. Unfortunately, the wind does not always blow, and the sun does not always shine. This variability can lead to fluctuations in power generation, making it difficult to ensure a consistent supply of electricity. To overcome this challenge, various energy storage technologies have been developed. These technologies can store the excess energy produced during periods of high wind speeds or bright sunlight and release it when the conditions are not favorable for power generation.

One of the most common energy storage technologies is batteries. Batteries, especially lithium-ion batteries, have high energy density and can quickly charge and discharge. This makes them ideal for storing electricity from renewable sources. When the wind is blowing or the sun is shining, the excess electricity generated can be used to charge the batteries. When the wind speed drops or the sun sets, the stored energy can be released back into the grid. Pumped hydro storage is another effective method of storing energy. In this system, excess electricity is used to pump water from a lower reservoir to a higher one. When electricity is needed, the water is released back down, and the potential energy of the falling water is converted back into electrical energy using turbines. Thermal storage is a method that involves storing energy in the form of heat. In concentrated solar power plants, for example, excess thermal energy from the sun can be stored in materials like molten salts. This stored thermal energy can then be used to produce steam and generate electricity even after the sun has set.

These energy storage technologies play a crucial role in balancing supply and demand in the power grid. They ensure that despite the intermittent nature of renewable energy sources, a steady and reliable supply of electricity can be maintained. This not only enhances the reliability of the power grid but also facilitates the wider adoption of renewable energy sources.

11.5.2.3 Smart Grid and Energy Management

Smart grids represent a significant evolution in the technology used to manage and distribute electricity. They leverage digital technology and advanced communication systems to monitor and manage the flow of electricity from various sources, including traditional power plants and renewable energy installations. This ability to handle diverse energy sources is particularly important in the context of the global shift toward renewable energy.

A smart grid is a sophisticated network of sensors, controls, and computers that work together to instantaneously manage the flow of electricity. These components continuously monitor electricity usage across the grid, collecting data on everything from overall demand to the performance of individual pieces of equipment. This data is then analyzed and used to make real-time adjustments to the electricity supply, ensuring that it precisely matches demand. One of the key advantages of smart grids is their ability to adapt to changes in electricity demand and supply in real time. This is a significant improvement over traditional grids, which often struggle to handle sudden changes in demand or supply. For example, on a hot summer day when many people turn on their air conditioners, a traditional grid might struggle to provide enough electricity, leading to blackouts. A smart grid, on the other hand, could anticipate this increase in demand and adjust the electricity supply, accordingly, preventing blackouts.

Smart grids are also better equipped to handle the variable and intermittent nature of renewable energy sources like wind and solar. During periods of high solar or wind production, a smart grid could redirect excess electricity to storage facilities or areas of high demand. This not only prevents waste but also helps to smooth out the fluctuations in renewable energy production. Conversely, when production from these sources is low, a smart grid could draw on stored energy or ramp up output from other sources, ensuring a steady supply of electricity. Smart grids can improve the efficiency and reliability of the electricity supply. By continuously monitoring the performance of the grid, they can quickly identify and address issues, reducing the frequency and duration of power outages. They can also optimize the electricity supply, reducing energy waste and lowering electricity costs.

Smart grids represent a significant advancement in the way we manage and distribute electricity. Smart grids can handle diverse energy sources, adapt to changes in demand and supply in real time, and improve the reliability and efficiency of the electricity supply. As such, they are a crucial component of our transition toward a more sustainable and resilient energy future. In conclusion, the process of electricity generation has evolved significantly, moving beyond traditional fossil fuel-based power plants to incorporate a growing array of renewable energy sources. This shift is driven by the need to address the environmental and sustainability challenges associated with the use of fossil fuels, as well as the ongoing technological advancements in renewable energy technologies and grid integration solutions.

As the world continues to strive for a more sustainable and low-carbon future, the integration of energy-efficient design principles into urban planning and building design, coupled with the continued expansion of renewable energy sources, will play a crucial role in shaping the energy landscape of the future.

11.5.3 Transmission

The process of generating electricity and delivering it to consumers involves several stages. Once electricity is generated at power plants, it needs to be transmitted over long distances to reach the cities and towns where it will be used. This stage of the process, known as power transmission, is a critical component of the electrical power grid. Power plants, which can be powered by a variety of energy sources including coal, natural gas, nuclear energy, or renewables like wind and solar, generate electricity by converting energy from these sources into electrical energy. This electricity is then fed into the power grid, a vast network of interconnected transmission lines and substations that distribute electricity across a region.

The first step in the transmission process involves stepping up the voltage of the electricity generated at the power plant. This is done using a device called a transformer. Transformers work on the principle of electromagnetic induction and can increase (or decrease) the voltage of an electric current. By increasing the voltage for transmission, the current of the electricity is reduced, which in turn reduces the amount of energy lost as heat due to the resistance in the transmission lines. This is why electricity is transmitted at high voltages, often several hundred thousand volts. Once the electricity has been stepped up to the appropriate voltage, it is sent onto the transmission grid. This grid is made up of a network of high-voltage transmission lines that carry electricity over long distances. These lines are typically made of aluminum or copper, both of which are good conductors of electricity, and are mounted on tall towers to keep them well above the ground.

As the electricity travels along the transmission lines, some energy is lost as heat due to the resistance in the lines. However, because the electricity is being transmitted at such a high voltage, these losses are minimized. The design and materials used in the transmission lines, as well as the distance the electricity needs to travel, can all affect the amount of energy lost during transmission. When the electricity reaches the area where it will be used, it must be stepped down to a lower voltage before it can be distributed to homes and businesses. This is done at substations, which are equipped with

TABLE 11.13
The Key Features of High-Voltage Transmission Lines

Voltage levels	• Transmission lines use high voltage (e.g., 132, 220, and 500 kV) to minimize the current and reduce power losses during transmission. • The high voltage is achieved by using step-up transformers at the generation sites, which convert the medium-voltage electricity (e.g., 11 and 22 kV) to the required high-voltage levels.
Conductors	• The conductors used in high-voltage transmission lines are typically made of aluminum or a combination of aluminum and steel (known as ACSR, or aluminum conductor steel reinforced). • The conductors are designed to be lightweight, have high tensile strength, and offer low electrical resistance to minimize losses. • The conductors are arranged in a specific configuration, such as a single-circuit or double-circuit, depending on the required transmission capacity.
Tower structures	• High-voltage transmission lines are supported by tall steel or concrete towers, also known as pylons or transmission towers. • These towers are designed to withstand the weight of the conductors, as well as environmental factors like wind and ice loads. • The tower structures are typically spaced at regular intervals, usually ranging from 300 to 500 m, depending on the terrain and voltage level.
Insulation and clearances	• The high-voltage conductors must be adequately insulated to prevent electric arcing and ensure safe operation. • Sufficient clearance is maintained between the conductors and the ground, as well as between the conductors themselves, to prevent flashovers and ensure public safety.
Substation connections	• High-voltage transmission lines are connected to substations, which are essential hubs in the power grid. • Substations house the necessary equipment, such as transformers, circuit breakers, and switchgear, to manage the flow of electricity and facilitate the transformation of voltage levels.

transformers that can reduce the voltage of the electricity. Once the voltage has been stepped down, the electricity is sent onto the distribution grid, a network of lower-voltage power lines that deliver electricity directly to consumers.

The transmission of electricity from power plants to cities involves stepping up the voltage of the electricity for transmission, sending the electricity over high-voltage transmission lines, and then stepping down the voltage for distribution. Each of these steps is critical to ensuring that the electricity we rely on every day is delivered efficiently and reliably.

High-voltage transmission lines are an essential component of the power grid, responsible for transporting large amounts of electricity over long distances from generation sources to distribution networks. These lines operate at very high-voltage levels, typically ranging from 132 to 500 kV or even higher. The transmission line system is summarized in Table 11.13.

The high-voltage transmission network is the backbone of the electricity grid, enabling the efficient and reliable delivery of power over long distances from generation sources to distribution networks and ultimately to end users. High-voltage transmission lines can have several environmental impacts, as shown in Table 11.14.

The impacts mentioned do exist, but they can be reduced or managed through meticulous planning, the utilization of cutting-edge technologies, and compliance with environmental regulations.

There are several alternatives to traditional high-voltage overhead lines:

1. *Underground Cables:* Underground cables are a viable alternative to overhead transmission lines. They are often used in densely populated areas, or in areas where overhead lines would be visually intrusive or could pose a safety risk. However, the installation cost of underground cables is significantly higher than that of overhead lines.

TABLE 11.14
Possible Environmental Impacts of Transmission Lines

Landscape and aesthetics	High-voltage overhead transmission lines (HVOHL) can cause a visual deterioration of the skyline, reducing its aesthetic appeal. They can disturb forests, wetlands, and other natural areas when placed in undeveloped areas.
Biodiversity	The construction and maintenance of transmission lines can potentially disrupt local ecosystems and harm biodiversity.
Land use	The construction of transmission lines and their access roads can lead to changes in land use, potentially affecting agriculture and other land-based activities.
Electromagnetic fields (EMF)	Transmission lines produce electric and magnetic fields, which have been the subject of research to understand their potential impacts on human health and the environment.
Noise and light impacts	The operation of transmission lines can lead to noise and light pollution, which can affect both humans and wildlife.
Resource depletion	The construction and maintenance of transmission lines involve the use of materials and energy, contributing to resource depletion.
Greenhouse gas emissions	Many high-voltage circuit breakers, switches, and other pieces of equipment used in the transmission and distribution system are insulated with sulfur hexafluoride, which is a potent GHG.

2. *High-Voltage Direct Current (HVDC) Cables:* HVDC technology is used for greater efficiency over longer distances, typically hundreds of miles. It is also used in submarine power cables and in the interchange of power between grids that are not mutually synchronized. An underground HVDC cable with similar capacity can be installed in a much narrower right-of-way (ROW) compared to a high-voltage overhead line.
3. *Micro-Alloyed Copper Conductors:* Micro-alloyed copper conductors offer an interesting alternative to steel-reinforced aluminum conductors for high-voltage overhead lines. They can handle higher currents and have better conductivity than traditional aluminum conductors.
4. *Advanced Conductor Designs:* There are also advanced conductor designs that can reduce energy losses and increase the capacity of overhead lines. For example, multiple subconductors, phase arrangement, conductor arrangement, and surface treatment can all be used to improve the performance of overhead lines.

Each of these alternatives has its own advantages and disadvantages, and the choice between them depends on a variety of factors, including the distance and location of the transmission, the available ROW, the local environment and regulations, and the cost considerations. The advantages and disadvantages in comparison to traditional overhead lines are shown in Table 11.15.

The choice between them depends on a variety of factors, including the distance and location of the transmission, the available ROW, the local environment and regulations, and the cost considerations.

Electricity is typically generated at voltages around 11 kV (or 11000 V) at the power plants. This is a relatively lower voltage, and it is more economical to generate electricity at this level and then step it up for transmission. However, the voltage can be as low as 6 kV or as high as 40 kV depending on the specific design of the power plant. The power output is determined by the amount of steam driving the turbine, which mainly depends on the boiler. The voltage of that power is determined by the current in the rotating winding (i.e., the rotor) of the synchronous generator.

High-voltage transmission lines typically operate at voltages ranging from 69 to 765 kV. However, the voltage can be even higher for certain types of transmission lines. For instance, most HVDC links use voltages between 100 and 800 kV. These are typical values, and the actual voltage can vary depending on the specific requirements of the power system.

TABLE 11.15
Advantages and Disadvantages of Four Alternate Transmission Cables

Methods	Advantages	Disadvantages
Underground cables	Lower transmission losses	More expensive to manufacture
	Less maintenance required	Installation cost is higher compared to overhead lines.
	Can absorb emergency power loads	In case of system failure, it is hard to locate and fix wire breaks.
	No electric field and can be engineered to emit less magnetic field than overhead lines	Underground cables cannot be easily upgraded to expand capacity.
	Requires less land	Susceptible to damage due to ground movement caused by earthquakes
	Less susceptible to severe weather	
	Reduced EMFs, which eliminate potential health issues	
HVDC cables	Efficient for long-distance power transmission	Requires converter stations at both ends
	Can control power flow	Higher costs compared to AC lines
	Suitable for submarine cable transmission	Limitations with fault isolation
	Reduced electrical losses	Requires advanced control systems
Micro-alloyed copper conductors	Better conductivity than traditional aluminum conductors	Heavier and twice as expensive, making it a poor choice for long runs without support
	More durable and less subject to corrosion	Susceptibility to stress corrosion cracking
	Does not expand and contract as much, making many types of connections more secure	
	Higher energy efficiency due to lower electrical resistance and losses	
	Allows for higher operating temperatures and temporary overloads	
Advanced conductor designs	Can carry twice the power capacity	Costs more but provides much greater benefits
	Lose up to 40% less energy when delivering electricity	Installation cost is higher compared to legacy conductors.
	Sag far less than legacy conductors, minimizing ground clearance, fire, and safety concerns	

Electric power plants typically generate electricity at relatively lower voltages, such as 6, 11, or 40 kV. However, for efficient long-distance transmission, the voltage needs to be significantly higher, typically in the range of 69–765 kV. This is because higher voltages reduce power losses during transmission. To achieve this increase in voltage, a device called a step-up transformer is used. The step-up transformer increases the voltage of the electricity from the power plant to the required level for transmission. This is a crucial step in the power distribution process.

Electricity is transmitted at high voltages for several reasons, as shown in Table 11.16.

Transmission of power at high voltages can help achieve lesser power loss and lesser voltage drops, while improving system efficiency and reducing the overall cost of power transmission at the same time.

HVDC transmission has several advantages over traditional high-voltage alternating current (HV-AC) transmission, as shown in Table 11.17.

HVDC transmission can utilize earth return and, therefore, does not require a double circuit while extra high voltage alternating current (EHV-AC) transmission always requires a double circuit. The potential stress on the insulation in the case of a DC system is $1/\sqrt{2}$ times that in an AC system for the same operating voltage. Hence for the same operating voltage less insulation is required. The

TABLE 11.16
The Reasons for High-Voltage Transmission

Reduced power losses	Power loss in transmission lines is proportional to the square of the current flowing through it. By transmitting power at high voltage, the current is reduced, which in turn reduces power losses.
Improved efficiency	Transmitting power at high voltage improves the efficiency of power transmission. This is because the power loss will be higher while transmitted at a lower voltage.
Reduced voltage drop	Voltage drop in the transmission lines is proportional to the current flowing through it and its resistance to the current flow. Hence, stepping up the voltage and reducing the current help to reduce the voltage drop and improve voltage regulation.
Cost-effective	Transmission lines are sized based on the maximum current they must carry. If an amount of power is transmitted at a lower voltage, the size of the conductor required is much higher than the same required for power transmission at higher voltages. Hence, the transmission of power at higher voltages enables a reduction in the conductor required for the same.

TABLE 11.17
HVDC Transmission Advantages over HVAC Transmission

Lower losses over long distances	HVDC losses are three to four times lower than AC over hundreds of miles, enabling economic long-distance transmission.
Asynchronous grid interconnections	HVDC allows power to be transferred between grids that are not synchronized.
Controls power flow	HVDC systems have the ability to control the amount of power that is transmitted, providing greater flexibility and stability.
Submarine cable transmission	HVDC is suitable for s
Efficient long-distance power transmission	HVDC transmission is more efficient for transmitting power over long distances.
Better integration of renewable energy	HVDC can be used to connect remote renewable energy sources, such as offshore wind farms, to the main grid.
Reduced electrical losses	HVDC transmission system needs only two conductors; therefore, the power losses in a DC line are lesser than the losses in an AC line of the same power transfer capability.
No skin effect	There is no skin effect in DC, so there is a uniform distribution of current over the section of the conductor.
Voltage regulation	There is no inductance; hence, the voltage drop due to inductive reactance does not exist in the DC transmission line.
Cheaper in cost	HVDC transmission lines require two-pole conductors while AC system requires three conductors to carry power.

phase-to-phase clearances, phase-to-ground clearances, and tower size are smaller in the case of DC transmission as compared to those for AC transmission. Because of the lesser load on the supporting structures, tower designs are simpler and cheaper. The requirement of "ROW" for a DC line is about 20–40% less than that for an AC line of the same power transmission capability.

EHV-AC transmission needs intermediate substations at an interval of 300 km for compensation, but the HVDC transmission system does not require any such intermediate substation for compensation. However, the DC line terminal equipment because of the high cost of converters is considerably more expensive than AC line terminal equipment. HVDC transmission becomes economical over AC transmission above the break-even distance when the saving in overall costs of conductors, towers, etc. compensates for the additional cost of the terminal equipment (rectifiers and inverters).

Each of these advantages contributes to making HVDC an attractive option for certain power transmission applications.

Here is a simple explanation using electrical engineering principles to prove that HVDC has lower power losses over a long-distance transmission in comparison to HVAC.

The power loss in a transmission line is given by the formula:

$$P_{loss} = I^2R \tag{11.4}$$

The power transmitted is given by the formula:

$$P_{trans} = IV \rightarrow \text{or } I = \frac{P_{trans}}{V}$$

$$\text{Substitute I into } P_{loss} \rightarrow P_{loss} = I^2R = \left(\frac{P_{trans}}{V}\right)^2 R = \frac{P_{trans}{}^2}{V^2} \tag{11.5}$$

As mentioned earlier, the transmission voltage of HVAC systems typically ranges from 33 to 230 kV AC. However, there are also extra high-voltage alternating current (EHV-AC) transmission lines that operate at even higher voltages, such as 220, 400, and 765 kV. These voltages are used to transmit power over long distances with minimal losses.

HVDC transmission systems typically operate at voltages between 100 and 800 kV. However, advancements in technology have led to the development of ultra–high-voltage direct current (UHVDC) transmission systems. For instance, a 1,100 kV HVDC link in China was completed in 2019, which transmitted power over 3,300 km (2,100 mi) with a power capacity of 12 GW. These high voltages allow for efficient long-distance power transmission with minimal losses. The specific voltage used can vary depending on the requirements of the power transmission system.

The above formula shows that the power loss is inversely proportional to the square of the voltage. Therefore, for a given power, increasing the voltage reduces the power loss. As HVDC can transmit higher voltage than HVAC, therefore the HVDC has lower power losses than HVAC.

11.5.4 Distribution

From the transmission line, the high-voltage electricity is then transformed down to lower voltages suitable for use in homes and businesses. This is done through a network of substations and smaller distribution lines. The process of electricity distribution is as follows.

11.5.4.1 Primary Distribution

The primary distribution system is the first stage in the process of delivering electricity to consumers. It involves the transmission of electricity from power plants to substations located in urban areas1. The voltage at this stage typically ranges from 11 to 33 kV.

The primary distribution system serves as the initial stage in the process of delivering electricity to consumers. This system is responsible for the transmission of electricity from power plants, where it is generated, to substations located in urban areas. The power grid is a network designed to deliver electricity from the point of generation to the end consumers. It comprises various components, including power plants, transmission lines, substations, transformers, and distribution lines. The role of the primary distribution system begins here. The electricity produced at power plants is typically at a very high voltage to minimize losses during transmission. However, this high voltage is not suitable for direct consumption as it can be dangerous. Therefore, the voltage needs to be stepped down before it can be distributed to consumers.

This is where substations come into play. Substations are integral parts of the primary distribution system. They are equipped with transformers that reduce the high-voltage electricity from the transmission lines to a lower voltage suitable for the primary distribution system. These substations can be thought of as nodes or junction points in the power grid, directing the flow of electricity to different parts of the city or region. The primary distribution system then carries this lower voltage electricity, typically in the range of 11–33 kV, to different parts of the urban area. The electricity is further stepped down at smaller, local substations or distribution transformers to voltages suitable for residential, commercial, and industrial use.

The design and operation of the primary distribution system require planning and management. Engineers must consider factors such as load demand, peak usage times, and the geographical layout of the area. They must also account for potential issues such as power outages, system faults, and maintenance needs. The primary distribution system is not just about transmitting electricity; it is also about ensuring reliability and quality of service. Utilities companies use sophisticated monitoring and control systems to keep track of the flow of electricity, detect any faults or disruptions, and take corrective action as needed.

High-voltage transmission lines are an essential component of the power grid, responsible for transporting large amounts of electricity over long distances from generation sources to distribution networks. These lines operate at very-high-voltage levels, typically ranging from 132 to 500 kV or even higher.

In recent years, the concept of smart grids has been introduced to make the primary distribution system more efficient and reliable. Smart grids use digital technology and advanced communication systems to monitor and manage the electricity flow. They can automatically detect and respond to changes in electricity demand, integrate renewable energy sources, and even allow consumers to feed electricity back into the grid.

The primary distribution system transmits electricity from power plants to substations in urban areas and stepping down the voltage to safer levels; it ensures that consumers have access to reliable and safe electric power. As technology advances, the primary distribution system continues to evolve, paving the way for a more sustainable and resilient energy future.

Load balancing, also known as demand response, is an aspect of managing urban electricity grids. In the context of electrical power, load balancing refers to the use of various techniques by electrical power stations to store excess electrical power during low-demand periods for release as demand rises. The aim is for the power supply system to have a load factor of 1. Load balancing is essentially about adjusting electricity consumption in relation to the available electrical capacity in the grid. The primary objective is to avoid overloading the grid during periods of high demand and prevent power outages.

In urban areas, where the demand for electricity can vary significantly throughout the day, load balancing is particularly important. For instance, there is often a surge in electricity demand in the morning when people wake up and start using electrical appliances, and in the evening when people return home from work. If not managed properly, these peaks in demand can overload the grid and lead to power outages. To balance the load, utility companies use a variety of strategies. One common approach is demanding a response, where consumers are incentivized to reduce their electricity usage during peak times. This can be achieved through time-of-use pricing, where electricity is more expensive during peak times, encouraging consumers to shift their usage to off-peak times.

Another strategy is the use of energy storage systems, such as batteries, which can store excess power generated during off-peak times and then release it during peak times. This not only helps to balance the load but also makes better use of renewable energy sources, which can be intermittent. With the advent of smart grids, more sophisticated load-balancing strategies are being developed. Smart grids use advanced metering infrastructure to collect real-time data on electricity usage, enabling utility companies to respond more quickly and effectively to changes in demand. For instance, they can use predictive analytics to forecast periods of high demand and adjust their generation and distribution strategies accordingly.

11.5.4.2 Substations

Substations are facilities that contain transformers and other equipment necessary to step down the voltage from the primary distribution level to the secondary distribution level. In urban and suburban areas, substations are often located underground or in small secure plots. Substations are a component of the electrical power distribution system. They serve as nodes within the power grid, facilitating the transfer of electricity from power generation sources to end consumers. The primary function of a substation is to transform the voltage level of the electrical power, making it suitable for transmission over long distances or for safe use by consumers. This transformation is achieved through the use of transformers and other essential equipment housed within the substation.

This is where substations come into play. They are strategically located within the power grid to receive high-voltage electrical power from the transmission lines. The substations contain transformers, which are devices that can step down the voltage from the primary distribution level to the secondary distribution level.

Transformers work on the principle of electromagnetic induction. They consist of two coils of wire, known as the primary and secondary windings, wrapped around a common iron core. When alternating current flows through the primary winding, it creates a changing magnetic field in the core. This changing magnetic field induces a voltage in the secondary winding by adjusting the number of turns in the primary and secondary windings. This transformer can increase or decrease the voltage of the electrical power.

Substations also contain other equipment necessary for the safe and efficient operation of the power grid. This includes circuit breakers and disconnect switches, which can interrupt the flow of electricity in the event of a fault or for maintenance purposes. Protective relays and controls are used to monitor the condition of the electrical power and the equipment, triggering circuit breakers if a problem is detected. Metering devices measure the amount of electricity flowing through the substation, providing valuable data for billing and system analysis.

In urban and suburban areas, space is often at a premium. Therefore, substations are often located underground or in small, secure plots. Underground substations are typically housed in a vault or tunnel, providing protection from the elements and reducing the visual impact on the surrounding area. Similarly, substations located in small, secure plots are often enclosed by a fence or wall and are designed to blend in with the surrounding environment as much as possible.

The substations need to be designed and constructed carefully by considering many factors. Factors such as the expected load demand, the characteristics of the local power grid, and the physical constraints of the site must be considered. Safety is also a paramount concern, with strict standards and regulations governing the design, construction, and operation of substations.

The selection of substation locations in urban areas involves several considerations: Carefully considering these factors allows utility companies to choose substation locations that ensure reliable and efficient power distribution while minimizing impact on the community and the environment. The factors to be considered in the selection of the location of substations are shown in Table 11.18.

The location of electrical substations has been a source of public concern, especially regarding safety aspects. Residents living near substations often express worries about the potential fire hazards associated with these facilities. Additionally, there are ongoing discussions about the potential health impacts of electromagnetic radiation emitted from substation equipment. Communities are increasingly demanding greater transparency and risk assessments from utility companies to ensure that substations are situated in safe locations, away from densely populated areas and critical infrastructure. Addressing these public concerns is essential for building trust and ensuring the safe and reliable operation of the power grid.

TABLE 11.18
Factors to be Considered in the Selection of Substation Locations

Proximity to load centers	Substations should ideally be located as close to the load centers (areas with high electricity demand) as possible. This reduces the length of the distribution lines, which in turn minimizes power losses and improves the efficiency of the power distribution system.
Future load requirements	Utilities companies also need to consider future load requirements when choosing substation locations. Urban areas are often subject to rapid growth and development, which can lead to increases in electricity demand. Therefore, it is important to choose locations that can accommodate potential future expansions.
Space availability	In urban and suburban areas, space is often at a premium. Therefore, substations are often located underground or in small, secure plots. Underground substations are typically housed in a vault or tunnel, providing protection from the elements and reducing the visual impact on the surrounding area.
Safety considerations	Safety is a paramount concern when choosing substation locations. Substations need to be located in areas where they pose minimal risk to the public and where there is easy access for maintenance and emergency crews.
Environmental and aesthetic considerations	Substations should ideally be located in visually screened areas, such as hills, forests, or behind other buildings. This helps to minimize the visual impact of the substations on the surrounding area.
Regulatory and legal factors	There may also be regulatory and legal factors that influence the choice of substation locations. For example, there may be zoning laws or environmental regulations that restrict where substations can be located.
Cost considerations	Cost is a significant factor in the selection of substation locations. This includes not only the cost of the land itself but also the cost of construction, operation, and maintenance of the substation.

Utilities companies or authorities may address public concerns about substation locations through a variety of strategies, as shown in Table 11.19.

Addressing these concerns, utility companies or power authorities can build trust with the community, ensure the successful implementation of the substation project, and ensure the reliable delivery of electricity to the area.

Modular substations, also known as compact substations, are a modern solution to the growing demand for electricity and the need for efficient and flexible power distribution systems. They are essentially a grouping of electrical equipment assembled on a platform or a trailer, integrated and tested in a factory environment to enable energizing upon delivery to the site. Here are some of the key advantages of modular substations for urban areas:

- *Space Efficiency:* Modular substations are designed to be compact, which makes them ideal for urban areas where space is often limited. They can be installed in small, secure plots or even underground, which can save valuable urban space.
- *Reduced Construction Duration:* As modular substations are pre-assembled and tested in a factory, they can be quickly installed and energized. This can significantly reduce the construction duration compared to conventional substations.
- *Improved Safety:* Modular substations are designed with safety in mind. They offer improved security and protection against vandalism or unauthorized access. Also, as they are tested in a controlled factory environment, the risk of on-site installation errors is minimized.
- *Reduced Environmental Impact:* Modular substations have a smaller footprint compared to conventional substations, which can help to reduce their environmental impact. Also, their compact and aesthetically pleasing design can help to minimize their visual impact on the urban landscape.

TABLE 11.19
Strategies to Cope with Public Concerns on Substations

Community engagement	Utility companies often hold community meetings to discuss proposed substation locations and address residents' concerns. They provide detailed information about the project, including the need for the substation, the selection process for the site, and the measures taken to minimize impact on the community.
Transparency	Utilities companies strive for transparency in their operations. They openly share information about the substation project, including the reasons for its necessity, the benefits it will bring to the community, and the potential impacts.
Environmental and health safety	Utilities companies address concerns about the potential health and environmental impacts of substations. They follow strict safety standards and regulations to ensure that substations do not pose a risk to public health or the environment. For example, they ensure that substations do not produce significant external electric fields and that the magnetic fields they generate are within safe limits.
Aesthetic considerations	Utility companies take steps to minimize the visual impact of substations. They may use landscaping, walls, or other screening methods to make the substation less visible. They also consider the aesthetic impact when choosing the location and design of the substation.
Site selection process	Utilities companies follow a rigorous site selection process. They consider a range of factors, including proximity to load centers, future load requirements, space availability, safety considerations, environmental and aesthetic impacts, regulatory and legal factors, and cost considerations.
Addressing property value concerns	Utility companies address concerns about potential impacts on property values. They provide information on studies and data showing the impact of substations on property values, which often show minimal impact.

- *Increased Design Flexibility:* Modular substations offer greater design flexibility. They can be easily expanded or relocated if needed, which makes them a versatile solution for urban power distribution.
- *Enhanced Monitoring and Control:* With the integration of advanced digital technologies, modular substations can offer enhanced monitoring and control capabilities. This can lead to increased reliability and reduced downtime.

Modular substations offer a flexible, efficient, and cost-effective solution for power distribution in urban areas. They address many of the challenges associated with conventional substations, making them an increasingly popular choice for utility companies.

11.5.4.3 Secondary Distribution

The secondary distribution system delivers electricity from substations to individual consumers. The voltage at this stage is typically stepped down to safer levels, usually 230 or 400 V. The secondary distribution system is a component of the electrical power grid, serving as the final stage in the process of delivering electricity to consumers. This system is responsible for the transmission of electricity from substations, where it is transformed to a lower voltage, to individual consumers. The voltage at this stage is typically stepped down to safer levels, usually 230 or 400 V.

The role of the secondary distribution system begins at the substations. Substations are integral parts of the power grid. They are equipped with transformers that reduce the high-voltage electricity from the transmission lines to a lower voltage suitable for the secondary distribution system. The secondary distribution level refers to the lower voltage suitable for consumer use. In addition to transformers, substations also contain other equipment necessary for the safe and efficient operation of the power grid. This includes circuit breakers and disconnect switches, which can interrupt the flow of electricity in the event of a fault or for maintenance purposes. Protective relays and controls are used to monitor the condition of the electrical power and the equipment, triggering circuit breakers if a problem is detected. Metering devices measure the amount of electricity flowing through the substation, providing valuable data for billing and system analysis.

Once the voltage has been stepped down at the substation, the secondary distribution system carries this lower voltage electricity, typically at 230 or 400 V, to different parts of the urban area. The electricity is further stepped down at smaller, local substations or distribution transformers to voltages suitable for residential, commercial, and industrial uses.

The secondary distribution system is not just about transmitting electricity but also about ensuring reliability and quality of service. Utilities companies use sophisticated monitoring and control systems to keep track of the flow of electricity, detect any faults or disruptions, and take corrective action as needed.

Smart grids have been introduced to make the secondary distribution system more efficient and reliable. Smart grids use digital technology and advanced communication systems to monitor and manage the electricity flow. They can automatically detect and respond to changes in electricity demand, integrate renewable energy sources, and even allow consumers to feed electricity back into the grid. The secondary distribution system is indispensable in the electricity supply chain. Transmission of electricity from substations to individual consumers and stepping down the voltage to safer levels ensures that consumers have access to reliable and safe electric power.

During the secondary distribution of electricity to the users, it is possible that the voltage fluctuates. Voltage fluctuations can have several effects on electrical appliances:

- *Overheating and Electrical Stress:* Voltage fluctuations can cause overheating and electrical stress in appliances, leading to reduced lifespan. High voltage can cause appliances to run "too fast and too high," which can shorten their service life.
- *Damage to Components:* If the voltage is too low, the amperage increases, which may result in the components melting or causing the appliance to malfunction.
- *Reduced Efficiency:* Frequent voltage fluctuations can lead to premature wear and tear, reduced efficiency, and increased susceptibility to breakdowns.
- *Performance Issues:* Low voltage can cause lights to dim and heaters to work less effectively, and some smart appliances may report a low voltage error. This type of voltage is also damaging to your appliances.
- *Safety Risks:* High voltage can pose safety risks. Circuit breakers are designed to trip when the incoming voltage is too high for the devices to handle. This protects the appliances, but you may need to manually turn the power back on.

The above issues can vary depending on the specific appliance and the severity of the voltage fluctuation.

The secondary distribution system is one of the essential parts of the electrical power grid, as it delivers electricity from substations to individual consumers. Some of the most important aspects of secondary distribution are shown in Table 11.20.

Within the secondary distribution system, it is possible to have autonomous localized microgrids. Microgrids are localized electric grids that can disconnect from the main grid to operate autonomously. They can vary in size from a single-customer microgrid to a full-substation microgrid, which may include hundreds of individual generators and consumers of power. The secondary distribution system delivers electricity from substations to individual consumers. The voltage at this stage is typically stepped down to safer levels, usually 230 or 400 V. Microgrids fit into this system by providing a localized network of power generation and distribution, often at the secondary distribution voltage levels. Microgrids serve several important purposes as shown in Table 11.21.

TABLE 11.20
Important Aspects of Secondary Distribution System

Voltage levels	The voltage at this stage is typically stepped down to safer levels, usually 230 or 400 V. This is done to ensure the safe use of electricity by consumers.
Transformers	Transformers play a crucial role in the secondary distribution system. They step down the voltage from the primary distribution level to the secondary distribution level.
Distribution lines	These are the lines that carry electricity from the substations to the consumers. They need to be properly maintained and monitored to ensure a reliable power supply.
Load management	Managing the load on the secondary distribution system is crucial. This involves balancing the supply of electricity with the demand from consumers.
Fault detection and protection	The secondary distribution system needs to have mechanisms in place for detecting and isolating faults. This helps to prevent power outages and ensures the safety of the electrical equipment and consumers.
Metering	Metering devices are used to measure the amount of electricity consumed by each customer. This is important for billing purposes and also for monitoring the overall load on the system.
Customer service	Providing reliable and efficient service to customers is a key aspect of the secondary distribution system. This includes handling customer queries and complaints and providing timely and accurate billing.
Infrastructure maintenance	Regular maintenance of the infrastructure, including transformers, distribution lines, and metering devices, is crucial to ensure the reliability and longevity of the secondary distribution system.
Integration with renewable energy sources	With the increasing adoption of renewable energy sources, the secondary distribution system must be capable of integrating these sources and managing the variability in their output.
Adaptability to smart grid technologies	Modern secondary distribution systems need to be adaptable to smart grid technologies. This includes advanced metering infrastructure, demand response mechanisms, and grid automation technologies.

TABLE 11.21
The Purposes of Microgrids

Local control	Microgrids are local electrical grids with defined electrical boundaries, acting as a single and controllable entity. They can operate in grid-connected and island modes, providing local control over power generation and distribution.
Reliability and resilience	Microgrids can operate independently of the main grid, providing a reliable power source during grid outages. This improves the security of supply within the microgrid cell and can supply emergency power.
Integration of distributed energy resources (DERs)	Microgrids can effectively integrate various sources of distributed generation, especially renewable energy sources. This helps to reduce dependence on fossil fuels and promote the use of clean and sustainable energy sources.
Efficiency and cost-effectiveness	Microgrids provide efficient, low-cost, clean energy. They enhance local resiliency and improve the operation and stability of the regional electric grid.
Rural electrification	Stand-alone microgrids offer an option for rural electrification in remote areas and smaller geographical islands. They have their own sources of electricity, supplemented with an energy storage system.
Flexibility	Microgrids provide dynamic responsiveness unprecedented for an energy resource. They can stand on their own or be connected to the larger grid but have the capability of keeping electricity flowing in the case of a power outage.
Cross-sectoral energy usage	An important feature of microgrids is to provide multiple useful energy needs, such as heating and cooling besides electricity. This allows energy carrier substitution and increased energy efficiency due to waste heat utilization for heating, domestic hot water, and cooling purposes.

TABLE 11.22
Roles of Microgrids in Energy Resilience

Local control and independence	Microgrids can operate independently of the main grid. In the event of a grid failure due to a natural disaster, they can disconnect and function autonomously, ensuring a continuous power supply to critical facilities.
Rapid restoration	Microgrids enable faster restoration of electricity after a disaster. They can serve as a backup power source when the utility grid's operations are interrupted due to either high-probability low-impact events like a component failure or low-probability high-impact events – be it a natural disaster or a planned cyberattack.
DERs	Microgrids integrate various DERs, such as solar panels, wind turbines, and energy storage systems. These resources can provide power locally within the microgrid, reducing dependence on the main grid and enhancing resilience.
Optimized power routing	When resources such as solar panels and energy storage are also connected, a microgrid controller can route and optimize power within the grid's footprint. If the larger electrical grid goes down, the distributed energy sources within the microgrid will keep the lights on.
Prosumer-centric networked microgrids	Prosumer-centric networked microgrids integrate DERs, like solar PV panels, wind turbines, battery energy storage systems (BESSs), and battery electric vehicles (BEVs), to form autonomous microgrids capable of operating independently during grid disruptions. Prosumers, acting as both producers and consumers, play a crucial role by generating and sharing electricity within the microgrid. BEVs act as mobile energy storage units during emergencies.
Load management and demand response strategies	Load management and demand response strategies prioritize the energy needs for essential facilities, ensuring uninterrupted operation during adverse weather.

Microgrids enhance the secondary distribution system by providing localized control over power generation and distribution, improving reliability and power quality, and enabling the integration of distributed energy resources. Microgrids are essential in enhancing energy resilience during natural disasters as reflected in Table 11.22.

Unless the power source of microgrids itself is physically destructed by disasters, microgrids enhance the resilience of distribution systems and provide a reliable power source during natural disasters, ensuring uninterrupted power supply and facilitating quicker restoration of electricity.

Microgrids can communicate and coordinate during disasters through the following means, as shown in Table 11.23.

Through advanced control schemes, communication systems, and cybersecurity measures, microgrids can effectively handle communication and coordination during disasters, ensuring a reliable power supply and facilitating quicker restoration of electricity.

11.5.4.4 Underground Distribution

The distribution of electricity in urban areas is a complex process that involves a multitude of factors. The primary method of electricity distribution in these areas is through underground cables. This approach is primarily due to the high population density in urban areas and the need to minimize the impact on the urban landscape. Urban areas are characterized by a high concentration of people, buildings, and infrastructure. This high population density creates a unique set of challenges for the distribution of electricity. Overhead power lines, which are commonly used in less densely populated areas, are not feasible in urban environments due to space constraints and the potential for interference with other structures.

Therefore, the electricity distribution in urban areas is mainly done underground. Underground distribution systems have several advantages over overhead systems. They are less susceptible to

TABLE 11.23
Communication and Coordination During Disasters

Decentralized control	Microgrids can operate without a central controller and its need for communication. This simplifies the setup and recovery of microgrids, making them ideal for quick deployment during emergencies.
Communication-less control scheme	National Renewable Energy Laboratory (NREL) has developed a communication-less control scheme for microgrids. This approach sidesteps the reliance on communications, instead using native controls of battery, solar, and wind systems. This makes it possible to assemble devices into a microgrid without arduous configuration.
Advanced communication systems	Microgrids often incorporate advanced communication systems that allow for real-time monitoring and control of the grid's components. These systems enable optimal performance and reliability, even in the event of a disaster.
Cybersecurity measures	Given that microgrids rely on digital communication networks, they also incorporate cybersecurity measures to protect against cyberattacks. This is particularly important during disasters, when communication networks may be more vulnerable.
Interoperability	Microgrids are designed to be interoperable, meaning they can communicate and coordinate with other microgrids, the main grid, and other infrastructure. This allows for a coordinated response during disasters.

weather-related outages, such as those caused by storms or high winds. They are also less likely to be damaged by trees or wildlife. Underground systems are generally more reliable and require less maintenance than overhead systems.

The decision to place power lines underground is also influenced by aesthetic considerations. Overhead power lines can be visually intrusive and can detract from the overall appearance of the urban landscape. Placing these lines underground allows cities to maintain a cleaner, more attractive skyline. This is especially important in areas with significant architectural or historical value, where overhead lines could detract from the aesthetic appeal of the surroundings.

Underground cables are often housed in common utility ducts. These ducts are shared spaces within the ground where utilities such as electricity, water, and telecommunications are routed. Housing multiple utilities in a common duct can save space and reduce the amount of excavation required when installing or maintaining these services. The use of common utility ducts also allows for easier access to the cables for maintenance or repair. If a problem arises with an underground cable, technicians can access the cable through the utility duct without having to dig up large sections of ground. This can significantly reduce the disruption caused by maintenance or repair work.

However, the use of underground cables and common utility ducts also presents its own set of challenges. For instance, the installation of underground cables can be more expensive and time-consuming than the installation of overhead lines. There is also the risk of damage to other utilities when digging to install or repair underground cables. While underground cables are less susceptible to weather-related damage, they can be more vulnerable to other types of damage, such as corrosion or damage from rodents. They can also be more difficult to repair than overhead lines, as locating and accessing the damaged section of cable can be a complex and time-consuming process.

The use of underground cables and common utility ducts presents certain advantages, and it also comes with its own set of disputes. Despite these challenges, the benefits of underground distribution systems, such as improved reliability, reduced visual impact, and better use of space, make them the preferred choice for electricity distribution in urban areas. As cities continue to grow and evolve, methods and technologies used to distribute electricity will also advance. Innovations in technology and changes in urban planning practices will continue to shape the future of electricity distribution in urban areas.

TABLE 11.24
Considerations for Planning in Urban Electricity Infrastructure

Understanding the current and future energy demand	Cities need to assess their current energy consumption and project future demand. This involves considering factors such as population growth, urbanization trends, and changes in energy use patterns.
Adopting renewable energy sources	Cities are increasingly turning to renewable energy sources such as solar, wind, and hydropower. The integration of these sources into the urban energy mix can reduce dependence on fossil fuels and decrease greenhouse gas emissions.
Implementing smart grids and microgrids	These technologies allow for more efficient distribution of electricity, better integration of renewable energy sources, and improved resilience to disruptions.
Investing in energy efficiency	This can involve upgrading buildings and infrastructure to reduce energy consumption, implement energy-saving technologies, and promote energy-efficient practices among residents.
Engaging stakeholders	Cities need to work with a wide range of stakeholders, including political leaders, regulators, power generators, technology start-ups, manufacturers, and consumers. This collaboration can facilitate innovation and ensure that the energy transition is inclusive and beneficial for all.
Planning for resilience and sustainability	Cities must plan their energy infrastructure to withstand future challenges, such as climate change impacts and specific threats to their infrastructure.
Modernizing regulations	Regulations need to be updated to support new technologies and practices, such as distributed energy generation and energy storage.
Investing in research and development	Continued investment in R&D can lead to new technologies and strategies for energy generation, distribution, and consumption.
Rethinking PPPs	PPPs can be an effective way to finance energy infrastructure projects and bring innovative solutions to market.

Planning for future growth in urban electricity infrastructure is an uneasy process because it is about the prediction of the future. This involves several key steps and considerations as shown in Table 11.24.

Cities can ensure their electricity infrastructure meets future demand sustainably and resiliently by carefully planning substation locations, mitigating fire and electromagnetic risks, engaging the public, and implementing robust safety standards and monitoring protocols to address community concerns.

A head-to-head comparison between underground and overhead electric distribution is shown in Table 11.25.

Both methods have their own advantages and disadvantages. The choice between overhead and underground often depends on a variety of factors, including cost, aesthetics, reliability, and the specific needs and constraints of the area.

11.5.4.5 Energy Efficiency Practices

Urban areas are increasingly implementing energy efficiency practices to reduce energy consumption and environmental impact. This is driven by a growing recognition of the need to transition to more sustainable forms of energy use, in response to the challenges posed by climate change and the ongoing depletion of nonrenewable resources. These practices can include adaptive public lighting, future-ready net-zero buildings, and public transport fleet electrification.

11.5.4.5.1 Adaptive Public Lighting

Adaptive public lighting is a strategy that involves adjusting the intensity of streetlights based on real-time conditions. This can be achieved using sensors and intelligent control systems that monitor

TABLE 11.25
Comparison between Overhead and Underground Electric Distribution

Aspect	Overhead	Underground
Costs	Overhead lines are generally more cost-effective. The materials used for overhead lines, such as conductors and support structures, are inexpensive. Additionally, installing overhead lines is typically less complex, requiring less specialized equipment and labor.	Underground cables are significantly more expensive. The materials used, such as insulated conductors and protective sheaths, are costlier. The installation process involves trenching and the need for specialized equipment, increasing labor and material costs.
Maintenance and durability	Maintenance of overhead lines is relatively straightforward. Routine inspections involve visual assessments of the conductors, insulators, and support structures. Any necessary repairs or replacements can be carried out more easily and quickly.	Underground cables require less maintenance. Once installed, they are protected from environmental factors such as wind, ice, and UV radiation. However, if a fault occurs, locating and repairing it can be challenging and time-consuming.
Accidents and safety	Accidents involving overhead lines can pose risks to the public and maintenance personnel. Contact with live conductors can result in severe injuries or fatalities.	Underground cables are safer as there is less chance for public contact.
Aesthetics	Overhead circuits can be visually intrusive and can detract from the overall appearance of the urban landscape.	Underground cables offer a cleaner look with much less visual clutter.
Reliability	Overhead circuits have more faults, causing more voltage sags, more momentary interruptions, and more long-duration interruptions.	Underground circuits are more reliable with significantly fewer short and long-duration interruptions.
Lifespan	Overhead systems typically have a longer lifespan.	The typical lifespan of an underground transmission system is about half that of an overhead transmission system.

factors such as ambient light levels, weather conditions, and pedestrian or vehicle traffic. When these factors indicate that less lighting is needed, the system can automatically dim the lights to save energy. Conversely, when more lighting is required, the system can increase the intensity of the lights to ensure safety and visibility.

This approach to public lighting can result in significant energy savings. Streetlights are a major source of energy consumption in urban areas, and the ability to adjust their intensity in real time can greatly reduce their energy use. Moreover, adaptive public lighting can also contribute to reducing light pollution, which is a growing concern in many urban areas.

11.5.4.5.2 Future-Ready Net-Zero Buildings

Future-ready net-zero buildings are structures designed to produce as much energy as they consume, resulting in a net-zero energy balance over the course of a year. This is achieved through a combination of energy-efficient design and the integration of renewable energy technologies. Energy-efficient design involves the use of strategies such as passive solar heating, natural ventilation, and high-efficiency insulation to reduce the building's energy needs. This can also include the use of energy-efficient appliances and systems, such as LED lighting and high-efficiency HVAC systems.

The integration of renewable energy technologies typically involves the installation of solar panels or wind turbines to generate electricity. In some cases, it may also involve the use of geothermal systems or biomass energy.

TABLE 11.26
Strategies of Net-Zero Building Design

Financial incentives	Cities can offer financial incentives such as tax credits, rebates, or grants to developers who incorporate net-zero designs into their projects. These incentives can help offset the initial costs of implementing energy-efficient technologies and designs.
Regulatory measures	Cities can implement building codes and standards that require new constructions to meet certain energy efficiency or net-zero standards. This can also include mandatory energy audits or energy performance certificates for buildings.
Capacity building	Cities can invest in capacity development programs to equip architects, engineers, and builders with the skills and knowledge needed to design and construct net-zero buildings.
PPPs	Cities can form partnerships with private sector companies to finance and implement net-zero building projects. These partnerships can leverage the resources and expertise of both the public and private sectors.
Awareness campaigns	Cities can run campaigns to raise awareness about the benefits of net-zero buildings among developers, homeowners, and the public. This can help create demand for net-zero buildings and encourage their adoption.
Demonstration projects	Lead by example by incorporating net-zero designs into public buildings. These projects can serve as a proof of concept and demonstrate the feasibility and benefits of net-zero designs.
Green certifications	Cities can promote green building certifications like LEED, which recognize buildings that meet certain sustainability and energy efficiency criteria.

Future-ready net-zero buildings represent the cutting edge of sustainable architecture and design. They not only reduce energy consumption and greenhouse gas emissions but also can contribute to energy independence by generating their own power.

Cities can incentivize the adoption of net-zero building designs through a variety of strategies, as shown in Table 11.26.

To ensure that financial incentives lead to actual adoption of net-zero building designs, rather than just token compliance, cities need to do the following: (1) Cities should establish clear guidelines and standards for what constitutes a net-zero building. (2) Regulations and standards need to be strictly enforced. (3) Cities can offer performance-based incentives. (4) Cities should promote transparency and accountability. (5) Cities should regularly review and update their incentive programs.

11.5.4.5.3 Public Transport Fleet Electrification

Public transport fleet electrification involves replacing conventional diesel or gasoline-powered public transport vehicles with electric vehicles (EVs). This can include buses, trams, and trains, as well as smaller vehicles used for services like car sharing or taxis. Electric public transport vehicles have several advantages over their conventional counterparts. They are more energy-efficient, meaning they use less energy to travel the same distance. They also produce zero tailpipe emissions, which can significantly improve air quality in urban areas.

Electric public transport vehicles are quieter than conventional vehicles, which can reduce noise pollution in urban areas. They also have lower operating costs, as electricity is generally cheaper than diesel or gasoline, and EVs require less maintenance. However, the transition to electric public transport fleets also presents challenges. These include the need for charging infrastructure, the higher upfront cost of EVs, and issues related to the sourcing and recycling of batteries.

Urban areas are implementing a range of energy efficiency practices to reduce energy consumption and environmental impact. These practices, which include adaptive public lighting, future-ready net-zero buildings, and public transport fleet electrification, represent a significant shift toward more sustainable forms of energy use. They reflect a growing recognition of the need to transition to more sustainable forms of energy use, in response to the challenges posed by climate change and the ongoing depletion of nonrenewable resources. While these practices present their own challenges, they also offer significant opportunities for reducing energy consumption, improving environmental outcomes, and creating more sustainable urban environments.

11.5.4.6 Regulations and Policies

The operation of the electricity distribution system in urban areas is a complex process, influenced by a multitude of regulations and policies. These guidelines are designed to ensure the safety, reliability, and sustainability of the power supply, while also promoting the use of renewable energy and energy efficiency.

Safety standards are a critical aspect of these regulations. They are designed to protect both the workers involved in the operation of the distribution system and the public. These standards cover a wide range of areas, including the design and construction of electrical equipment, the operation and maintenance of the distribution system, and the procedures for dealing with emergencies. For example, there are specific standards for the installation of transformers and power lines, the maintenance of substations, and the response to power outages.

Environmental regulations also play a significant role in the operation of the electricity distribution system. These regulations are designed to minimize the environmental impact of the power supply. They cover areas such as emissions from power plants, the disposal of waste from the power generation process, and the impact of power lines on the local environment. For example, power plants are required to limit their emissions of greenhouse gases and other pollutants, and to dispose of their waste in a safe and environmentally friendly manner.

There are also policies aimed at promoting renewable energy and energy efficiency. These policies are designed to encourage the use of renewable energy sources, such as wind and solar power, and to improve the efficiency of the power generation and distribution process. For example, there may be incentives for power companies to invest in renewable energy projects or requirements for new buildings to be equipped with energy-efficient appliances.

The integration of renewable energy into the existing grid is a key aspect of these policies. This involves the development of technologies and systems that can accommodate the variable and intermittent nature of renewable energy sources. For example, smart grids and energy storage systems can help to balance the supply and demand of electricity and ensure the stability of the grid.

Energy efficiency policies are also important. These policies aim to reduce the amount of energy that is wasted in the power generation and distribution process and to encourage consumers to use electricity more efficiently. For example, there may be regulations requiring power companies to improve the efficiency of their power plants, or programs to educate consumers about energy-saving practices.

The operation of the electricity distribution system in urban areas is governed by a complex set of regulations and policies. These guidelines are designed to ensure the safety, reliability, and sustainability of the power supply, while also promoting the use of renewable energy and energy efficiency. As the demand for electricity continues to grow, these regulations and policies will play a crucial role in shaping the future of the power sector.

The operation of the electricity distribution system in urban areas is a multifaceted process, influenced by a variety of regulations and policies. These guidelines not only ensure the safe and reliable delivery of electricity but also promote the adoption of sustainable practices, paving the way for a greener and more energy-efficient future.

11.5.5 Demand Management

The management and control of the direction and magnitude of power through lines, often referred to as power flow or load flow, is a fundamental aspect of power system operation. This process involves the continuous adjustment of electrical power production and transmission to meet the fluctuating demand of consumers.

Power flow begins at the generation stage, where electricity is produced by various types of power plants, such as coal-fired, nuclear, hydroelectric, and increasingly, renewable energy sources like wind and solar. The generated power is then transmitted over long distances through high-voltage transmission lines. The direction of power flow is typically from the power plants to the load centers, which are areas of high electricity demand, such as urban areas. However, with the increasing integration of distributed generation sources like solar panels and wind turbines, power can also flow in the opposite direction, from the load centers back to the grid. The magnitude of power flow is determined by the amount of electricity being produced by the power plants and the demand for electricity. During periods of high demand, such as during the day when businesses are operating, more power is produced and transmitted. Conversely, during periods of low demand, such as at night, less power is produced.

Controlling the direction and magnitude of power flow is a complex task that requires sophisticated control systems and algorithms. These systems monitor the status of the power grid in real-time, adjusting the output of power plants and the configuration of the transmission network to ensure that supply always matches demand.

One of the key challenges in power flow management is dealing with the variability and unpredictability of both supply and demand. For example, the output of wind and solar power plants can vary depending on the weather conditions, while the demand for electricity can change rapidly due to factors like temperature changes and economic activity. To manage these challenges, power system operators use a variety of strategies, such as demand response, in which consumers are incentivized to reduce their electricity usage during periods of high demand, and energy storage, where excess electricity is stored during periods of low demand and then released back into the grid when demand increases.

Managing and controlling the direction and magnitude of power through lines is a critical aspect of ensuring the reliable and efficient operation of the power system. It involves a delicate balance of adjusting supply to meet demand, while also dealing with the uncertainties and rapid changes that can occur in both. As the power system continues to evolve and become more complex, particularly with the increasing integration of renewable energy sources, the importance of effective power flow management will only continue to grow.

Demand management of electricity, also known as demand-side management (DSM), is a crucial aspect of modern power systems. The key issues in the DSM include the following.

11.5.5.1 Understanding Energy Consumption Patterns

Energy consumption is a dynamic process that fluctuates throughout the day and across different seasons. This variability leads to periods of high demand, known as peaks, and periods of low demand, known as valleys. Understanding these patterns is crucial for effective energy management. During the day, energy consumption typically follows a predictable pattern. In the morning, as people wake up and start their day, there is a surge in energy demand. This demand usually dips during the midday hours when many people are at work or school and then rises again in the evening when people return home and use appliances, lighting, and heating or cooling systems.

Seasonal variations in energy demand are also significant. In colder climates, energy demand tends to peak during the winter months due to increased heating needs. Conversely, in warmer climates, the demand for energy is often highest in the summer due to the use of air conditioning systems. These fluctuations in energy demand present both challenges and opportunities for power system operators. On the one hand, they must ensure that there is enough capacity to meet peak demand. On the other hand, the valleys in energy demand provide opportunities for maintenance and system upgrades.

DSM is a strategy used by power system operators to manage these fluctuations in energy demand. DSM involves understanding consumption patterns and managing energy after it reaches the consumer or "after the meter." DSM strategies can include energy efficiency programs, demand response programs, and time-of-use pricing. Energy efficiency programs aim to reduce overall energy consumption through measures such as improving insulation in buildings or promoting the use of energy-efficient appliances. Demand response programs incentivize consumers to reduce their energy use during peak periods, for example, by offering lower electricity rates during off-peak hours. Time-of-use (ToU) pricing involves charging higher rates during peak demand periods and lower rates during off-peak periods, encouraging consumers to shift their energy use to times when demand is lower.

ToU pricing is a method used by electricity companies to manage the demand and supply of electricity. Some important points about ToU pricing include:

- *Variable Pricing:* ToU pricing refers to electricity prices that vary based on the time of consumption. It involves higher prices during peak demand times and lower prices during off-peak times.
- *Goal:* The goal of ToU pricing is to incentivize customers to consume energy when the cost of generating electricity is low and discourage energy consumption when the cost of generating electricity is high.
- *How It Works:* ToU rates adjust the rate you pay for electricity over the course of the day. When both the cost of generating electricity and the electricity demand are low (i.e., in the middle of the night), the rate paid to use electricity is very low. However, when both the cost of generation and demand for electricity are high (i.e., on the afternoon of a hot summer day), the electricity rate is much higher.
- *Variations:* ToU rates may vary by season, on weekdays versus weekends and holidays, and across multiple periods over an individual day.
- *Necessity:* ToU rates aim to better align the costs that electricity consumers see with the actual cost of producing electricity. By adjusting the rate across the course of the day, week, or month, consumers can better understand the true cost of the electricity they use.

ToU pricing is a strategy used by utilities to manage the demand and supply of electricity, promote energy conservation, and ensure grid stability. Consumers can adjust their behavior to take advantage of ToU rates in several ways:

1. *Shift Usage to Off-Peak Hours:* Use energy-intensive appliances like dishwashers, washing machines, and dryers during off-peak hours when electricity is cheaper.
2. *Charge EVs During Off-Peak Times:* If you own an electric vehicle, consider charging it during off-peak hours to save on electricity costs.
3. *Invest in Energy Storage:* If possible, invest in energy storage systems like home batteries. These can be charged during off-peak hours and used to power your home during peak hours.
4. *Use Smart Appliances and Thermostats:* Smart appliances and thermostats can be programmed to operate during off-peak hours.
5. *Take Advantage of Peak-Hours Rebates:* Some utility companies offer rebates to customers who reduce their energy use during peak hours.

The specific peak and off-peak hours can vary by utility company and region, so it is best for the customers to check with the local provider for the most accurate information.

11.5.5.2 Demand Response Programs

Demand response programs are initiatives that provide an opportunity for consumers, including businesses, to participate in the operation of the electric grid. These programs are designed to reduce or shift electricity usage during peak periods in response to time-based rates or other forms of financial incentives. The concept of demand response is based on the supply and demand equation of electricity. The electricity grid becomes most stressful when there is a high demand for electricity. High demand requires a higher supply of electricity, which both stresses the grid and results in higher prices for all energy users.

Demand response programs work by reducing the demand for electricity during certain periods, thereby cutting prices by reducing the need to run high-cost generators. Instead of supply, or power plants, turning on in response to higher demand, it is demand turning off in response to higher prices and stress on the system. These programs are generally implemented by the utilities themselves or in conjunction with their customers, both residential and commercial. When customers participate in demand response, they are generally incentivized by the utility to adjust their power consumption through reduced electric rates, bill credits, or direct payments.

Methods of engaging customers in demand response efforts include offering time-based rates such as ToU pricing, critical peak pricing, variable peak pricing, real-time pricing, and critical peak rebates. It also includes direct load control programs which provide the ability for power companies to cycle air conditioners and water heaters on and off during periods of peak demand in exchange for a financial incentive and lower electric bills.

The electric power industry considers demand response programs as an increasingly valuable resource option whose capabilities and potential impacts are expanded by grid modernization efforts. For example, sensors can perceive peak load problems and utilize automatic switching to divert or reduce power in strategic places, removing the chance of overload and the resulting power failure. Advanced metering infrastructure expands the range of time-based rate programs that can be offered to consumers. Smart customer systems such as in-home displays or home-area networks can make it easier for consumers to change their behavior and reduce peak period consumption from information on their power consumption and costs. These programs also have the potential to help electricity providers save money through reductions in peak demand and the ability to defer construction of new power plants and power delivery systems, specifically those reserved for use during peak times. One of the goals of the Smart Grid R&D Program is to develop grid modernization technologies, tools, and techniques to utilize demand response and help the power industry design, test, and demonstrate integrated, national electric/communication/information infrastructures with the ability to dynamically optimize grid operations and resources and incorporate demand response and consumer participation.

Energy conservation is a critical aspect of modern life, particularly in the face of increasing energy demand and environmental concerns. Voluntary conservation efforts are one such approach that can significantly contribute to energy sustainability. These efforts are typically activated during critical hours, such as periods of peak energy demand or extreme weather conditions. By altering energy use behavior during these times, it is possible to significantly reduce stress on the energy grid. Peak demand periods refer to those times when the demand for electricity is at its highest. These periods often occur during extreme weather conditions, such as heatwaves or cold snaps, when the need for heating or cooling is greatest. During these times, the energy grid is under significant stress, as it must supply enough electricity to meet this increased demand.

Voluntary conservation efforts are initiatives that encourage individuals and businesses to reduce their energy consumption, particularly during these peak demand periods. These efforts can take various forms from simple actions like turning off unnecessary lights and appliances to more complex strategies like shifting energy-intensive activities to off-peak hours. The activation of these voluntary conservation efforts during critical hours is crucial. By focusing on these periods of high demand, these efforts can have the most significant impact. Even small changes in energy use behavior, when multiplied across many users, can result in substantial reductions in overall energy demand.

Altering energy use behavior involves making conscious decisions about when and how to use energy. For example, instead of running a dishwasher or washing machine during peak hours, these

TABLE 11.27
Strategies of Businesses in Energy Conservation

Energy management systems	Businesses can implement energy management systems that monitor and control energy usage in real time. These systems can automatically adjust energy consumption during peak hours, reducing demand on the grid.
Demand response programs	Businesses can participate in demand response programs offered by utility companies. These programs provide incentives for businesses to reduce their energy usage during peak demand periods.
Energy-efficient equipment	Investing in energy-efficient equipment can significantly reduce a business's overall energy consumption. This includes energy-efficient lighting, HVAC systems, and appliances.
Alternative work schedules	Businesses can consider alternative work schedules, such as flextime or staggered start times, to shift energy demand away from peak hours.
Remote work options	Allowing employees to work from home can reduce the energy demand of office buildings during peak hours.
Energy audits	Regular energy audits can help businesses identify opportunities for energy conservation and efficiency improvements.
Employee education	Businesses can educate employees about the importance of energy conservation and encourage behaviors that reduce energy usage, such as turning off lights and computers when not in use.
Renewable energy sources	Businesses can invest in renewable energy sources, such as solar or wind power, to offset their energy usage and reduce demand on the grid.

tasks could be scheduled for off-peak times when demand is lower. Similarly, adjusting the thermostat by a few degrees during extreme weather conditions can significantly reduce the demand for heating or cooling. These changes in behavior not only reduce stress on the grid but also have other benefits. For instance, they can lead to cost savings for consumers, as energy prices are often higher during peak demand periods. Additionally, by reducing overall energy consumption, these efforts contribute to environmental sustainability by decreasing greenhouse gas emissions associated with energy production. These voluntary conservation efforts can also enhance the resilience of the energy grid. By reducing peak demand, these efforts can help prevent blackouts and other disruptions in energy supply. This is particularly important in the face of increasing extreme weather events due to climate change, which can put additional stress on the energy grid.

Voluntary conservation efforts that activate during critical hours represent a powerful tool for energy sustainability. By encouraging changes in energy use behavior during periods of peak demand, these efforts can significantly reduce stress on the energy grid, contribute to environmental sustainability, and enhance the resilience of our energy systems. Therefore, the promotion and adoption of such efforts should be a key component of our approach to energy management and sustainability. While these efforts are voluntary, their success relies on widespread participation. Therefore, education and awareness-raising about the importance of energy conservation and the impact of our energy use behaviors are crucial. Through collective action, we can all contribute to a more sustainable and resilient energy future.

Businesses can participate and play a significant role in energy conservation during critical hours through various strategies, as shown in Table 11.27.

11.5.5.3 Energy Storage

Energy storage can balance supply and demand in an electricity grid. It allows excess energy produced during periods of low demand to be stored and then released back into the grid when demand increases. This process is vital for maintaining grid stability and ensuring a reliable supply of electricity. During periods of low demand, power plants often generate more electricity than is needed. This excess energy, instead of being wasted, can be stored for later use. Various technologies can be used for this purpose, including pumped hydro storage, batteries, flywheels, and thermal storage.

Pumped hydro storage is the most common form of large-scale energy storage. It involves pumping water uphill to a reservoir when there is excess electricity, then releasing it to generate electricity when demand increases. This process is highly efficient and can quickly respond to changes in demand. Batteries, on the other hand, store energy in a chemical form. They can be charged during periods of low demand and discharged when needed. Batteries are highly flexible and can be used for both short-term and long-term storage. They are particularly useful for integrating intermittent renewable energy sources, like wind and solar, into the grid. Flywheels store energy in a mechanical form. They involve spinning a rotor at high speeds and then using kinetic energy to generate electricity when needed. Flywheels can respond very quickly to changes in demand, making them ideal for providing short-term grid stability. Thermal storage involves storing energy in a thermal form, such as hot water or ice. This stored energy can then be used for heating or cooling when needed. Thermal storage is often used in buildings to shift energy demand from peak to off-peak hours.

When demand increases, the stored energy can be released back into the grid. This process involves converting the stored energy back into electricity, which can then be distributed to consumers. This can help to meet peak demand without needing to build additional power plants. Energy storage not only helps to balance supply and demand but also has several other benefits. It can improve grid reliability by providing backup power in case of outages. It can also reduce greenhouse gas emissions by enabling greater integration of renewable energy sources into the grid.

However, energy storage also confronts several disputes. These include high costs, technical issues, and regulatory barriers. Despite these challenges, energy storage is increasingly important in the future electricity grid.

Energy storage allows excess energy to be stored during periods of low demand and then released back into the grid when demand increases. This process is vital for maintaining grid stability, ensuring a reliable supply of electricity, and facilitating the integration of renewable energy sources into the grid. Despite the challenges, energy storage technologies hold great promise for the future of our energy systems.

Each energy storage technology has different environmental impacts, as reflected in Table 11.28.

These technologies can have environmental impacts, and they also offer significant environmental benefits. Enabling a greater integration of renewable energy sources into the grid can help to reduce greenhouse gas emissions and combat climate change. The environmental impacts of these technologies need to be weighed against their potential benefits.

11.5.5.4 Advanced Metering Infrastructure

Advanced metering infrastructure (AMI) is an integrated, fixed-network system that enables two-way communication between utilities and customers. It is a key technology in the broader evolution of our growing need for efficient, cost-effective, environmentally sustainable energy. AMI systems are smart meters – digital gas meters, electric meters, and water meters that record energy consumption, typically in intervals of an hour or less. These devices typically send data back to the utility company at least once daily. Communication networks serve as the backbone of the two-way communication between smart meters and the AMI head-end system. They can be either wireless or wired, depending on the specific topology of the system. These AMI networks carry data from the smart meters to the head-end system and vice versa, allowing utility companies to send commands to meters.

The data management system is the central repository where all meter data is collected, stored, processed, and analyzed. Here, raw data from smart meters is converted into actionable insights. AMI improves a utility's ability to collect frequent and accurate water usage data to improve billing, leak detection, and water resource management. It supports more frequent billing but more importantly also allows a utility to quickly identify excessive water use that could be the result of leaks.

TABLE 11.28
Environmental Impacts and Costs of Energy Storages

Storage Systems	Environmental Impacts	Costs
Pumped hydro storage	This technology needs large reservoirs. It can impact local ecosystems, including aquatic life, due to changes in water levels and flow rates. However, it does not produce direct emissions during operation.	The initial investment costs can be high due to the need for significant infrastructure, including reservoirs and turbines.
Batteries	Some materials, such as lithium and cobalt, have significant environmental and human rights concerns associated with their mining. Batteries have a limited lifespan and can pose waste management challenges at the end of their life.	Costs of lithium-ion BESSs have dropped 70% since 2012 and are forecasted to drop below the $200/kWh threshold. The total cost can vary based on the size, technology used, and the costs of installation and maintenance.
Flywheels	Flywheels do not require harmful materials and produce no direct emissions during operation. However, the energy used in their manufacture and the materials used in the flywheel and containment system can have environmental impacts.	Flywheels are generally less expensive than batteries and pumped hydro storage, but they are typically used for short-term storage due to energy losses over time.
Thermal storage	This system uses common materials like water or ice, and they do not produce direct emissions during operation. However, if the heat source is nonrenewable, it can contribute to greenhouse gas emissions.	These systems are relatively inexpensive, particularly if they use common materials like water or ice.

Unlike traditional automatic meter readings, AMI's two-way communication model enables more comprehensive data collection and helps companies remotely manage meter functionality. The ability to remotely manage meters not only eliminates the need for manual meter readings but also enables faster response times in the event of a power outage.

AMI can help utility companies collect a range of data, including indicators of tampering, data collected at set intervals, details regarding power outages, and the quality of electricity supplied. It also offers specific advanced capabilities for endpoints used in electric metering. Water utilities, for example, rely on AMI meters to provide comprehensive flow data. If the data points to excessive water use patterns, which might indicate a leak, the company can notify the customer or make the necessary repairs

The emergence of AMI provides utilities with a real opportunity, and it can also be a challenge to determine how to manage the data collected and how to make that data useful to utility customers. AMI can benefit commercial and institutional facilities too. AMI offers frequent (e.g., hourly) water usage data that can be monitored to understand consumption patterns, identify leaks before the bill comes, and improve overall facility water management.

The electric power industry considers demand response programs as an increasingly valuable resource option whose capabilities and potential impacts are expanded by grid modernization efforts. For example, sensors can perceive peak load problems and utilize automatic switching to divert or reduce power in strategic places, removing the chance of overload and the resulting power failure.

AMI refers to systems that measure, collect, and analyze energy usage and communicate with metering devices such as electricity meters, gas meters, heat meters, and water meters, either on request or on a schedule. AMI is a key technology in the broader evolution of our growing need for efficient, cost-effective, environmentally sustainable energy. It enables two-way communication between utilities and customers, collects, stores, and analyzes and presents energy usage data, providing utility companies the ability to monitor electricity, gas, and water usage in real time.

Implementing AMI systems can be challenging due to a variety of factors:

1. *Technology Maturity:* Many AMI solutions are still evolving and are in their infancy compared to more established utility information systems. Rapid growth environments attract companies interested in growth and creating new ideas, new approaches, and high energy. Most rapid growth environments eventually reach a consolidation point, as winners in the market consolidate the smaller players and absorb niche elements of the solution.
2. *Scale Issues:* AMI solutions that work well on a small scale can have issues scaling to larger utilities. Communication networks have little issues with scaling, but the head-end is an area of concern if a technology provider does not have existing clients of large sizes.
3. *Project Management:* Common challenges for AMI projects include failure to meet schedule milestones, failure to meet utility expectations and requirements, poor coordination of necessary implementation tasks, and poor readiness to accept the organizational changes that AMI systems force upon a utility.
4. *Vendor Selection:* An AMI project involves much more than selecting a vendor and waiting for the technology to be deployed. AMI projects require utilities to follow an excellent system implementation and integration process due to challenges related to AMI's inherent complex technology, the lack of depth in many vendors' project services, and the integration points across other enterprise IT systems.
5. *Installation Errors:* AMI deployment problems include hot meter sockets and associated fire hazards, meter registration, batching, and pairing installation errors, meter inventory errors (lost meters and/or inefficient tracking), inefficient processes (for workers, managers, and/or administrators), safety, reliability, or customer-facing process deficiencies.
6. *Data Management:* The emergence of AMI provides utilities with a real opportunity, but it can also be a challenge to determine how to manage the data collected and how to make that data useful to utility customers.

Despite these challenges, AMI systems offer significant benefits, including reduced operational costs and enhanced grid reliability.

11.6 GAS SUPPLY SYSTEMS

This section explores the details of gas supply systems, starting with the diverse supply sources that include natural gas fields and storage facilities. It further explores the vast network of transmission and distribution pipelines that transport gas. The role of pressure regulators and valves in controlling gas flow is discussed, along with the function of meters in measuring usage. The section also highlights the end consumers of the system, the importance of safety requirements, and the critical aspect of system planning.

11.6.1 Supply Sources

The gas supply system begins at a supply source, and it can be multiple supply sources. These sources are diverse and can include natural gas fields, storage facilities, or liquefied natural gas (LNG) ports. Each of these sources plays a unique role in the supply chain, contributing to the efficient and reliable delivery of gas to end consumers. Natural gas fields are the primary source of gas supply. These fields, often located in remote areas, are where natural gas is extracted from the earth. The extraction process involves drilling wells into underground reservoirs to release the gas.

Once extracted, the gas is processed to remove impurities and then transported through pipelines for distribution.

Storage facilities are another crucial part of the gas supply system. These facilities store natural gas during periods of low demand and release it during periods of high demand. This helps to balance the supply and demand of natural gas throughout the year. Storage facilities can be underground, such as depleted gas or oil fields, or above ground, like LNG storage tanks. Natural gas is cooled to a liquid state to form LNG, reducing its volume and making it easier to transport. LNG is often shipped overseas in specialized tanker ships and then re-gasified at the receiving port. This allows natural gas to be transported long distances from countries with abundant natural gas resources to those with high demand.

For natural gas fields, considerations include the environmental impact of extraction, the safety of operations, and the economics of extracting gas from remote locations. For storage facilities, considerations include the safety and integrity of the storage structures, the efficiency of storage operations, and the environmental impact of potential leaks. For LNG ports, considerations include the safety and security of LNG transport, the environmental impact of LNG operations, and the economics of LNG trade. Figure 11.7 depicts the process of natural gas shipping.

The gas supply system has proven to be remarkably resilient and efficient. Advances in technology have made it possible to extract natural gas from previously inaccessible locations, store it safely and efficiently, and transport it across the globe. Regulations and standards have been put in place to ensure the safety and environmental sustainability of operations. Looking ahead, the gas supply system will continue to evolve in response to changes in supply and demand, advances in technology, and shifts in policy. As the world seeks to transition to a low-carbon energy future, the role of natural gas and the infrastructure that supports its supply will undoubtedly be a topic of ongoing discussion and innovation.

FIGURE 11.7 Natural gas shipping.

Source: **Maciej Margas, Wikipedia.**

11.6.2 Transmission and Distribution Pipelines

The gas supply system is a complex network that ensures the safe and efficient transport of gas from supply sources to end consumers. This network comprises two main types of pipelines: high-pressure transmission lines and lower-pressure distribution lines. High-pressure transmission lines are the backbone of the gas supply system. They transport gas from supply sources, such as natural gas fields or LNG ports, to the distribution centers located near the areas of consumption. These pipelines are designed to handle high pressures, typically in the range of 200–1500 pounds per square inch gauge (psig). This high pressure allows for the transportation of large volumes of gas over long distances.

The design and construction of high-pressure transmission lines require careful consideration of several factors. These include the route of the pipeline, the type of terrain it will cross, the materials used in its construction, and the safety measures implemented to prevent leaks and other potential hazards. The pipeline must also be equipped with compressor stations at regular intervals to maintain the pressure and flow of the gas.

Once the gas reaches the distribution centers, it is transferred to lower-pressure distribution lines. These lines deliver the gas to the end consumers, including residential, commercial, and industrial users. Distribution lines operate at significantly lower pressures than transmission lines, typically less than 100 psig. This lower pressure makes the gas safer to handle and use. Distribution lines are further divided into main lines and service lines. Main lines are the larger pipelines that carry gas from the distribution centers to specific neighborhoods or industrial areas. Service lines are the smaller pipelines that branch off from the main lines to deliver gas to individual buildings or homes.

The design and construction of distribution lines need consideration of several factors. These include the layout of the area to be served, the expected demand for gas, and the safety measures implemented to prevent leaks and other potential hazards. The pipeline must also be equipped with pressure regulating stations to reduce the pressure of the gas to a level suitable for use by the end consumers.

During the transmission of gas, several types of stations are required to ensure the safe and efficient transport of gas, as shown in Table 11.29.

These stations are strategically located along the transmission pipeline and are essential for the operation of the gas transmission system.

TABLE 11.29
Stations Required During the Gas Transmission

Compressor stations	These are crucial for maintaining the pressure of the gas in the transmission pipelines. Compressors are housed within these stations, and they work by squeezing or compressing the gas, thus increasing its pressure and driving it down the pipeline.
Metering stations	These stations measure the volume of gas flowing through the pipeline. They are typically located at key points along the pipeline, such as where the gas enters or exits the pipeline.
Mixing stations	These stations are used to mix gases of different compositions or to add odorants to the gas.
Scrubbers and filters	These are generally included in compressor stations. They capture any liquids or solid particles that have condensed out of the gas stream during transport.
Monitoring stations	These stations monitor the condition of the gas and the pipeline. They are often equipped with supervisory control and data acquisition (SCADA) systems for real-time monitoring and control.
Pressure regulating stations	These stations are used to control the pressure of the gas in the pipeline. They reduce the high pressure of the transmission system to a lower pressure suitable for the distribution system.
Emergency shutdown system	This is a safety system designed to quickly isolate sections of the pipeline in the event of an emergency.

11.6.3 Pressure Regulators

Pressure regulators are a critical component in gas supply systems, particularly in high-pressure systems. They play a vital role in controlling the pressure of the gas, reducing it from a high level to a level suitable for delivery to the user. This is crucial for ensuring the safe and efficient operation of the gas supply system.

In a high-pressure gas system, the gas is often transported at pressures that can range from 200 to 1500 psig. This high pressure is necessary for transporting large volumes of gas over long distances. However, such high pressures are not suitable for delivery to the end user. The gas needs to be at a much lower pressure when it reaches the consumer to ensure safety and usability. This is where pressure regulators come into play.

Pressure regulators work by reducing the pressure of the gas as it flows through the system. They do this by using a control valve to restrict the flow of gas, thereby reducing its pressure. The control valve is operated by a diaphragm or piston that reacts to changes in pressure. When the pressure of the gas exceeds a set point, the diaphragm or piston moves, causing the control valve to open and release some of the gas, thereby reducing the pressure. Pressure regulators are designed to maintain a constant outlet pressure, regardless of changes in the inlet pressure or the flow rate of the gas. This is achieved through a feedback mechanism, where the outlet pressure is monitored and used to control the operation of the control valve. If the outlet pressure starts to rise above the set point, the control valve opens to release more gas and bring the pressure back down. Conversely, if the outlet pressure starts to fall below the set point, the control valve closes to restrict the flow of gas and increase the pressure.

Pressure regulators are particularly important in high-pressure gas systems for several reasons. First, they ensure the safety of the system. High-pressure gas can be dangerous if not properly controlled. By reducing the pressure to a safe level, pressure regulators help to prevent accidents and ensure the safe operation of the system. Second, pressure regulators ensure the efficient operation of the system. By maintaining a constant outlet pressure, they ensure that the gas is delivered to the end user at a pressure that is suitable for their needs. This helps to prevent the wastage of gas and ensures that the user gets the most out of their gas supply. Finally, pressure regulators help to protect the equipment in the gas supply system. High-pressure gas can cause damage to equipment if not properly controlled. By reducing the pressure, pressure regulators help to prolong the life of the equipment and reduce maintenance costs.

The pressure regulators control the pressure of the gas, reducing it from a high level to a level suitable for delivery to the user. This ensures the safety, efficiency, and longevity of the system, making pressure regulators an indispensable component in the gas supply chain. However, they can sometimes face issues that affect their performance. Here are some common problems. These issues are shown in Table 11.30.

11.6.4 Valves

Valves are a fundamental component in gas supply systems. They are used to control the flow of gas in the system and can be opened or closed as needed to manage the distribution of gas. This simple sentence encapsulates a complex process that is crucial for the safe and efficient operation of gas supply systems. To understand the role of valves, it is important to first understand the nature of gas flow in a supply system. Gas, like any fluid, flows from areas of high pressure to areas of low pressure. In a gas supply system, the gas is typically stored under high pressure and is transported through pipelines to the end users, who use the gas at a much lower pressure. The flow of gas through the system is driven by this pressure difference.

Valves are devices that are strategically placed within the system to control this flow of gas. They work by restricting or allowing the passage of gas through the pipeline. When a valve is opened, it allows gas to flow through. When it is closed, it prevents gas from flowing. By adjusting the position

TABLE 11.30
Some Common Issues of Pressure Regulators

Incorrect regulator type	It is a common issue in the process. For instance, if we need to control pressure from a source before it arrives at the main process, then we need a pressure-reducing regulator. If we need to control pressure from upstream, then we need a back-pressure regulator.
Pressure variations	Pressure regulators are designed to control system pressure while accounting for changes in system parameters. We might see either a drop or an increase in pressure downstream, which can compromise the quality and safety of our process.
Creep phenomenon	This can occur just after the first commissioning of a regulator when installation debris flows through the system. The contaminants create a fine gap between the regulator's seat and poppet, which can unintentionally allow system media to flow across the seat, resulting in unwanted pressure increases downstream.
High or low water pressure	In the case of water pressure regulators, signs of a bad regulator include very high or low water pressure.
Vibrating/noisy pipes	Vibrating or noisy pipes can also indicate a problem with the pressure regulator.
Water hammers	Water hammers, which are loud banging sounds in the pipes, can be a sign of a bad regulator.
Leaking appliances	Leaking appliances or faucets, even when the water is off, could indicate a problem with the regulator.

of the valve, the flow rate of the gas can be precisely controlled. This ability to control the flow of gas is crucial for several reasons. First, it allows the gas supply to be adjusted to meet the demand. For example, during periods of high demand, more valves can be opened to allow more gas to flow through the system. Conversely, during periods of low demand, some valves can be closed to reduce the flow of gas. Second, valves allow for the isolation of certain parts of the system for maintenance or in the event of an emergency. For example, if there is a leak in a certain section of the pipeline, the valves on either side of that section can be closed to isolate it and prevent gas from flowing into the affected area. Third, valves are used to control the pressure of the gas in the system. Some valves, known as pressure relief valves, are designed to open if the pressure in the system becomes too high. This allows some of the gas to escape, thereby relieving the pressure and preventing potential damage to the system.

The operation of valves may seem simple, but their role in a gas supply system is complex. They are essential for controlling the flow of gas, adjusting the supply to meet demand, isolating parts of the system for maintenance or in emergencies, and controlling the pressure of the gas. Understanding the function of valves is key to understanding the operation of gas supply systems.

There are several types of valves used in gas supply systems, each with its own specific function and application. Some of the most common ones are listed in Table 11.31.

Each type of valve has its own advantages and disadvantages, and the choice of valve depends on the specific requirements of the gas supply system.

11.6.5 Meters

Gas meters are an integral part of the gas supply system. They measure the amount of gas that is delivered to each consumer, allowing the gas company to bill its customers accurately for their usage. This seemingly simple function is a complex process that involves a combination of mechanical engineering, fluid dynamics, and data management. To understand how gas meters work, it is important to first understand the nature of gas flow in a supply system. Gas, like any fluid, flows from areas of high pressure to areas of low pressure. In a gas supply system, the gas is typically stored under high pressure and transported through pipelines to the end users, who use the gas at a much lower pressure. The flow of gas through the system is driven by this pressure difference.

TABLE 11.31
Most Common Types of Valves in Gas Supply Systems

Gate valves	These are commonly used in the industry. They use a gate system to open or close a pipeline entirely. This is an excellent choice if the flow rate needs to be controlled and maintained.
Globe valves	When pressure control is necessary, the globe valve is frequently chosen. It is also often used for open and close operations.
Check valves	These are extremely helpful as they open under pressure and do not allow backflow of fluid or material. Because of the restricted backflow, cross-contamination of the product is prevented.
Plug valves	The plug valve comprises a plug-shaped disk with a horizontal passageway bored through it.
Ball valves	These are used for shut-off and control applications. They provide tight sealing with low torque.
Butterfly valves	These are used for regulating or isolating flow. They are lighter in weight and more compact than other types of valves.
Diaphragm valves	These are used to control the flow of fluids that contain suspended solids.
Solenoid valves	These are electromechanically operated and are used to control the flow of fluids.
Control valves	These regulate the flow or pressure of a fluid.
Relief or pressure relief valves	These are used to control or limit the pressure in a system.
Safety valves	These are used to prevent pressure buildup and potential system failures.
Thermostatic mixing valves	These are used in water systems to ensure a constant, safe shower and bath outlet temperature.
Slide valves	These are typically used in engines and pumps.
Pressure regulating valves	These are used to automatically cut off the flow of a liquid or gas at a certain pressure.

Gas meters are devices that are installed at the point of delivery to the end user. They work by measuring the volume of gas that flows through them. This is typically done using a diaphragm meter, which consists of a series of chambers that fill and empty as the gas flows through them. Each time a chamber fills and empties, it counts as one unit of gas. The total number of units is then recorded on a meter dial, which can be read by the gas company.

The accuracy of the gas meter is crucial for ensuring that customers are billed correctly for their usage. If the meter is not accurate, customers could be overcharged or undercharged for their gas usage. This could lead to disputes and loss of trust between the gas company and its customers. Therefore, gas meters are regularly tested and calibrated to ensure their accuracy. Gas meters also ensure the safety of the gas supply system. They are often equipped with safety devices such as pressure regulators and shut-off valves. The pressure regulator reduces the pressure of the gas to a level that is safe for use by the end user, while the shut-off valve can be used to stop the flow of gas in the event of an emergency. Despite their importance, gas meters are often overlooked in discussions about the gas supply system. However, they are important in ensuring the efficient operation of the system, the accurate billing of customers, and the safety of the end users. As our reliance on natural gas continues to grow, so too will the importance of gas meters in our energy infrastructure.

The calibration process for gas meters is a critical step to ensure their accuracy and reliability. The steps in the calibration process are shown in Table 11.32.

The specific steps and methods used can vary depending on the type of gas meter and the standards followed by the organization performing the calibration. It is always important to follow the manufacturer's instructions and relevant industry standards when calibrating gas meters.

TABLE 11.32
Steps in Gas Meters Calibration

1	*Selection of Calibration Method:* The calibration method depends on the type of gas meter and its intended use. Common methods include reference master meter calibration, primary air by bell/piston prover calibration, critical flow venturi (Sonic nozzle) calibration, and wind tunnel air velocity calibration.
2	*Preparation:* The gas meter is prepared for calibration. This may involve cleaning, initial checks, and setting up the meter in the calibration system.
3	*Reference Master Meter Calibration:* In this method, the meter under test is compared with a duly calibrated master gas flow meter. The readings of the master flow meter and the meter under test are compared using air or standard gas flow rate with a closed-loop calibrated volumetric tank storage system.
4	*Piston/Bell Prover Gravimetric Calibration:* This method involves the use of a piston or bell prover to measure the volume of gas passing through the meter. The meter's readings are then compared with the volume measured by the prover.
5	*Adjustment:* If the meter's readings do not match the reference or expected values, adjustments are made. This could involve tweaking the meter's settings or making physical adjustments to the meter.
6	*Verification:* After adjustment, the meter is tested again to verify that it is now giving accurate readings.
7	*Documentation:* Once the calibration is complete, the results are documented. This documentation serves as proof of calibration and is important for quality control and regulatory compliance.
8	*Regular Recalibration:* Over time, the meter's readings can drift due to various factors such as wear and tear, changes in environmental conditions, or buildup of impurities. Therefore, regular recalibration is necessary to maintain the meter's accuracy.

11.6.6 End Consumers

The end consumers of the gas supply system are a diverse group that includes residential homes, commercial buildings, and industrial facilities. Each of these consumers has a unique set of needs and requirements, and each has a point of connection with building piping. This connection point is where the gas supply system interfaces with the consumer's internal gas system.

Residential homes are perhaps the most familiar type of end-consumer. These homes use natural gas for a variety of purposes, including heating, cooking, and water heating. The connection point for a residential home is typically located near the gas meter, which measures the amount of gas consumed by the home. From this point, the gas is distributed throughout the home via a network of pipes that connect to various appliances and fixtures.

Commercial buildings, such as offices, restaurants, and retail stores, are another type of end-consumer. These buildings often have larger and more complex gas systems than residential homes. They may use natural gas for heating and cooling, cooking, and other purposes. The connection point for a commercial building is usually located in a utility room or other dedicated space. From this point, the gas is distributed throughout the building via a network of pipes that connect to various appliances and systems.

Industrial facilities are a third type of end-consumer. These facilities use natural gas for a wide range of industrial processes, including power generation, manufacturing, and chemical production. The connection point for an industrial facility is typically located at a central plant or utility area. From this point, the gas is distributed throughout the facility via a network of pipes that connect to various equipment and processes.

The connection point between the gas supply system and the consumer's internal gas system is a vital interface. It is here that the pressure and flow of the gas are controlled to meet the consumer's needs. This is typically done using a combination of valves, regulators, and meters. Valves control the flow of gas, regulators control the pressure of the gas, and meters measure the amount of gas consumed.

TABLE 11.33
Differences between Urban and Rural Gas Supply Systems

Aspects	Urban Areas	Rural Areas
Infrastructure	Urban areas typically have a well-established infrastructure for gas supply, including extensive pipeline networks and storage facilities.	Rural areas often lack such infrastructure, making it more challenging to deliver gas.
Population density	Urban areas have a higher population density, which allows for a greater concentration of potential users in a smaller area. This makes it more cost-effective to provide gas supply services.	Rural areas, with their lower population density, may not have the same economies of scale.
Energy access	Despite their greater geographical proximity to grid electricity and other supplies of clean energy, people living in poverty in urban areas also lack energy access.	In rural areas, it is often the lack of electricity delivery infrastructure, such as grid connectivity, that leads to the lack of access.
Energy consumption	Urban families spend at least $30 billion more for energy each year than their country cousins, but each individual urban family spends about $200–$400 less.	Rural areas consume about 95 million Btu per household each year, followed by towns (102 million) and suburbs (109 million).
Energy burden	Lower in urban areas	The percentage of household income spent on energy consumption is often higher in rural areas. This is largely due to higher home heating and cooling costs, and longer commutes requiring more fuel.
Availability of gas networks	There is no rural or urban distinction for gas standing charges.	

The design and operation of the connection point require consideration of several factors. These include the pressure and flow requirements of the consumer, the safety requirements of the gas supply system, and the regulatory requirements of the local jurisdiction. The connection point must be designed and operated in a way that ensures the safe and efficient delivery of gas to the consumer.

Gas supply systems can differ significantly between rural and urban areas due to a variety of factors such as infrastructure, population density, and energy needs. The key differences between the two are shown in Table 11.33.

11.6.7 Safety Requirements

The design of a gas supply system involves numerous safety considerations. One of the key standards that guide this process is ANSI B31.8, also known as ASME B31.8. This standard covers the design, fabrication, installation, inspection, examination, and testing of pipeline facilities used for the transportation of gas. It includes engineering requirements deemed necessary for the safe design and construction of pressure piping.

One of the critical aspects covered by ANSI B31.8 is the guidance for abandoning gas lines. Abandonment refers to the process of permanently taking a pipeline out of service. This could be due to various reasons such as the end of the pipeline's useful life, changes in the gas supply network, or safety concerns.

The abandonment process is carefully regulated to ensure safety and environmental protection. According to the Federal Energy Regulatory Commission (FERC) regulations, which align with ANSI B31.8, criteria for abandonment include compulsory disconnection, purging of gas, and sealing of pipelines at the end of their useful lifetime. Specifically, each pipeline abandoned in place

TABLE 11.34
Right-of-Way Pipelines

Easement	A right-of-way (ROW) agreement between the pipeline company and the property owner is also called an easement. This agreement is usually filed in the public records with property deeds.
Access	The ROW provides access for pipeline workers for inspection, maintenance, testing, or in emergencies.
Restrictions	The ROW identifies an area where certain activities are prohibited to protect public safety and the integrity of the pipeline. This can include restrictions on building structures, planting deep-rooted trees, or digging.
Markers	Above-ground pipeline markers are used along the permanent ROW for easy recognition of the presence of underground pipelines.
Land Acquisition	Pipeline ROWs are acquired from landowners, other utilities, or government entities by obtaining an easement.

must be disconnected from all sources and supplies of gas, purged of gas, in the case of offshore pipelines, filled with water or inert materials, and sealed at the ends. The process of purging involves removing all residual gas from the pipeline, often by flushing it with an inert gas or air. This is crucial to prevent any potential hazards due to leftover gas in the pipeline. Sealing the ends of the pipeline ensures that no gas or other substances can enter or exit the pipeline after it has been abandoned. This is typically done using a mechanical plug or cap.

The ROW for gas transmission pipelines is a strip of land over and around pipelines where some of the property owner's legal rights have been granted to a pipeline company. This ROW allows pipeline companies to construct, operate, inspect, maintain, and protect the pipeline. The width of the ROW can vary depending on the size of the pipeline, the type of terrain, and regulatory requirements. It is typically wide enough to provide a safe working environment for pipeline maintenance and inspections. The important points about the ROW of pipelines are shown in Table 11.34.

Understanding the ROW is crucial for both pipeline companies and landowners. For pipeline companies, it provides the necessary access for pipeline operations and maintenance. For landowners, it is important to know the restrictions in the ROW to ensure they do not inadvertently damage the pipeline or put themselves at risk.

Gas transmission pipelines are a vital part of the energy infrastructure, transporting natural gas from production facilities to distribution systems and end users. The design and installation of these pipelines involve several key requirements to ensure safety, efficiency, and compliance with regulatory standards, as shown in Table 11.35.

The design and installation of gas transmission pipelines involve a set of requirements and considerations. These ensure that the pipelines operate safely and efficiently, delivering natural gas to end users while minimizing environmental impact.

Environmental considerations need to be in place during the installation of gas pipelines. The aspects to be considered in environmental concerns are shown in Table 11.36.

These are general considerations, and the specific environmental impacts and mitigation measures can vary depending on the location and scale of the project.

There are several innovative technologies and practices that can help minimize the environmental impact during construction. These innovative technologies are shown in Table 11.37.

These innovative technologies and practices not only help to reduce the environmental impact of construction but also often result in cost savings and improved project outcomes.

ConcreteDNA is a product developed by Converge, a company specializing in construction technology. It is an AI-powered solution that provides concrete monitoring, data management, and mixed selection. The key features of ConcreteDNA are shown in Table 11.38.

TABLE 11.35
Design and Installation Requirements for Pipelines

Right-of-way (ROW)	The ROW for gas transmission pipelines is a strip of land over and around pipelines where some of the property owner's legal rights have been granted to a pipeline company. The width of the ROW can vary but is typically about 10–50 m wide. However, in some cases, such as in Alberta, Canada, the controlled area of a pipeline is a strip of land 30 m wide on each side of the pipeline. The ROW provides access for pipeline workers for inspection, maintenance, testing, or emergencies.
Depth of pipelines	The depth at which gas pipelines are buried can vary based on local regulations, the type of pipeline, and the specific conditions of the site. In the United States, gas pipelines should be installed 18 to 24 inches (0.45-0.60 m) under the ground, with the main lines being at least 24 inches deep (0.60 m). However, some regulations call for at least 2.5 ft (0.75 m) of cover over the typical oil or gas pipeline.
Pipeline material and diameter	Transmission pipelines are typically made from steel and can range in size from several inches to several feet in diameter.
Pressure	Depending on the product being transported, transmission pipeline systems can be designed to operate from relatively low pressures to over 1,000 pounds per square inch (psi) of pressure.
Safety requirements	The design of the gas supply system includes all safety requirements of ANSI B31.8. This includes guidance for abandoning gas lines, which involves compulsory disconnection, purging of gas, and sealing of pipelines at the end of their useful lifetime.
Inspection and maintenance	Regular inspection and maintenance are crucial to ensure the safety and longevity of gas transmission pipelines. This includes checking for potential leaks, corrosion, and other forms of damage.
Regulatory compliance	Gas transmission pipelines are subject to various local, state, and federal regulations. These regulations cover aspects such as design standards, safety requirements, environmental impact, and ROWs.

TABLE 11.36
Aspects of Environmental Considerations

Route planning	Careful planning of the pipeline route can minimize environmental impact. The route should avoid ecologically sensitive areas and make use of previously disturbed areas or existing infrastructure to minimize habitat destruction.
Erosion control	Construction activities can lead to soil erosion, which can harm local ecosystems and water resources. Erosion control measures, such as silt fences and re-vegetation of disturbed areas, are important.
Habitat protection	Pipeline construction can disrupt habitats and endanger local wildlife. Efforts should be made to minimize habitat loss and fragmentation.
Water protection	Pipelines often cross water bodies, and construction can impact water quality. Measures should be taken to protect water resources, including careful route planning and the use of construction methods that minimize disturbance.
Air and noise pollution	Construction activities can generate air and noise pollution, which can impact local communities and wildlife. Contractors should use equipment that minimizes these impacts and follow work practices that reduce unnecessary noise and emissions.
Waste management	Construction can generate waste, including excavated material, construction debris, and wastewater. A waste management plan should be in place to minimize, manage, and dispose of waste in an environmentally responsible manner.
Restoration	After construction, the land should be restored as close as possible to its original condition. This includes re-vegetation to stabilize the soil and restore habitat.

TABLE 11.36 (Continued)
Aspects of Environmental Considerations

Regulatory compliance	All activities should comply with relevant environmental laws and regulations. This includes obtaining necessary permits, following approved construction methods, and conducting environmental monitoring.
Community engagement	Engaging with local communities and stakeholders can help identify potential environmental issues and develop strategies to address them.
Environmental management plan	An environmental management plan (EMP) should be developed and implemented. The EMP outlines the measures to mitigate the environmental impacts during the construction, operation, and decommissioning phases.

TABLE 11.37
Innovative Technologies to Minimize Environmental Impacts

Sustainable construction practices	These involve using materials and techniques that are environmentally friendly and have a minimal ecological footprint. It prioritizes the efficient use of resources, promotes energy efficiency, and aims to minimize waste and pollution throughout the entire lifecycle of a building.
Energy-efficient design and technologies	Incorporating features such as proper insulation, efficient heating and cooling systems, and the use of renewable energy sources can reduce the energy demand of buildings. This can help mitigate climate change and reduce reliance on fossil fuels.
Water conservation measures	Implementing measures to reduce water consumption, such as installing low-flow fixtures, using rainwater harvesting systems, and implementing efficient irrigation systems, can significantly reduce water usage.
Use of renewable energy sources	Installing solar panels or wind turbines on-site can help generate clean energy for construction activities and further reduce carbon emissions.
Smart construction techniques	Implementing digital technologies, such as ConcreteDNA, can help optimize the construction processes and reduce resource consumption. This includes modularization, 3D printing, robotics, drones, and AI, all of which are helping to decrease project duration and encourage efficiencies.
Waste management plan	A waste management plan should be in place to minimize, manage, and dispose of waste in an environmentally responsible manner.

TABLE 11.38
Key Features of ConcreteDNA

Real-time monitoring	ConcreteDNA uses wireless, embedded sensors to monitor concrete temperature and strength in real time. The data is sent to the cloud, allowing users to access it from any device, anywhere.
Predictive analytics	The platform uses AI to provide predictive analytics on concrete strength and temperature. This allows contractors to make faster, safer decisions on formwork removal, road openings, and more.
Mix selection	ConcreteDNA also assists with mix selection, helping builders construct faster and more efficiently with a lower embodied carbon footprint.
Digital twin integration	ConcreteDNA Pro, a new product tier, integrates concrete sensor data with building information modeling to provide in-depth material performance analytics. This forms the first iteration of a real-time digital twin.
Sustainability	ConcreteDNA Pro has helped companies reduce concrete waste and lower their carbon footprint. For example, Gammon saved over 110 tons of concrete waste by reducing testing on a project in Hong Kong.
Efficiency	The platform helps optimize labor and resources, reduce embodied carbon, and accelerate project timelines. It also helps reduce costs associated with unnecessary lab testing, rework, and delays.

ConcreteDNA can be said to be a comprehensive solution for concrete construction, providing real-time monitoring, predictive analytics, mix selection, and more. It is a significant step toward the digitization of construction, enabling companies to build more efficiently, safely, and sustainably.

11.6.8 System Planning

System planning in gas distribution systems involves carefully balancing multiple factors to ensure economic efficiency, operational safety, and consistent pressure delivery. This complex process requires a thorough analysis of demand, infrastructure, and regulatory requirements to develop an optimal distribution network.

11.6.8.1 Economic Efficiency

The design and operation of a gas distribution system should be economically efficient. This involves careful selection of pipeline routes, materials, and equipment to minimize costs. It also involves optimizing the operation of the system to reduce energy consumption and maintenance costs.

Achieving economic efficiency in gas distribution systems is a critical aspect of system planning. The selection of pipeline routes, materials, and equipment must be carefully considered to minimize upfront capital expenditures and ongoing operational expenses. This includes factors such as the length and diameter of pipelines, the type of materials used, and the choice of compression and storage facilities. The operation of the gas distribution system must be optimized to reduce energy consumption and maintenance costs. This can involve strategies such as load balancing, pressure optimization, and predictive maintenance.

11.6.8.2 Safety

Safety is a paramount concern in the design and operation of gas distribution systems. This involves multiple aspects, starting with ensuring the integrity of the pipelines. The pipelines must be designed and constructed to withstand the pressures and stresses they will encounter during operation. This includes considerations for the materials used, the thickness of the pipes, and the quality of the welds and joints. Regular inspections and maintenance are also crucial to detect and address any potential issues that could compromise the integrity of the pipelines.

The installation of safety devices such as pressure regulators and shut-off valves is another critical aspect of safety. Pressure regulators help maintain the gas pressure within safe limits, preventing overpressure situations that could lead to leaks or ruptures. Shut-off valves allow for the quick isolation of sections of the pipeline in case of an emergency, preventing the spread of gas and minimizing potential damage.

Implementing procedures for emergency response is also vital. This includes having plans in place for dealing with various types of emergencies, such as leaks, fires, or natural disasters. These plans should outline the roles and responsibilities of all personnel, the steps to be taken in response to an emergency, and the resources available for dealing with emergencies. The design of gas distribution systems must adhere to all safety requirements of ANSI B31.8, the standard for gas transmission and distribution piping systems. This standard provides guidelines for the design, construction, inspection, testing, and maintenance of these systems, ensuring they are built and operated to the highest safety standards. Adherence to this standard is not just a regulatory requirement but also a commitment to the safety of the personnel operating the system and the communities it serves.

11.6.8.3 Uniformity of Pressure

Maintaining uniform pressure throughout the gas distribution system is important for its efficient operation. This uniformity ensures that gas is delivered to all consumers at a consistent pressure, regardless of their location in the network. Achieving this uniformity involves a careful design of the pipeline network and the strategic use of pressure-regulating devices.

TABLE 11.39
The Roles of Pressure Monitoring Systems

Real-time monitoring	Pressure monitoring systems provide real-time data on the pressure at various points in the network. This allows operators to quickly identify and address any issues that could disrupt the uniformity of pressure.
Predictive maintenance	By analyzing pressure data over time, these systems can help predict potential issues before they become serious problems. This can prevent disruptions that could affect the uniformity of pressure.
Optimization of network performance	Pressure data can be used to optimize the operation of the network. For example, it can help determine the optimal settings for pressure-regulating devices.
Safety and compliance	Regular pressure monitoring can help ensure the network operates within safe pressure limits and complies with regulatory requirements.
Data-driven decision-making	The data from pressure monitoring systems can inform decisions about network expansion, maintenance scheduling, and other aspects of system planning.
Emergency response	In case of an emergency, such as a leak or a significant drop in pressure, the system can send alerts to operators, enabling a quick response.

The pipeline network should be well-looped within the main area and in all outlying areas whenever practicable and economically feasible. A well-looped system allows gas to flow from multiple directions, which can help maintain uniform pressure and provide redundancy in case of a pipeline failure. However, it is not always practicable or economically feasible to loop a supply line to an outlying area and then back into the main system. In such cases, the negative effects of dead ends can often be mitigated to some extent by looping the line around the area it serves and then back into itself. This can help maintain more consistent pressure throughout the system.

At large stations, it is usually advantageous to extend the high-pressure line to provide two or more connections with the distributing lines. This can help ensure uniformity of pressure and provide assurance of adequate pressure even during periods of high demand. Multiple connections can also provide redundancy, improving the reliability of the system. Maintaining uniform pressure in a gas distribution system is a complex task that requires careful planning and design. It involves a combination of well-designed pipeline networks, strategic use of pressure-regulating devices, and thoughtful planning of pipeline routes and connections. By achieving uniform pressure, gas distribution systems can ensure efficient operation and reliable service to all consumers.

Pressure monitoring systems are indispensable in maintaining uniformity in gas distribution networks. The roles of the pressure-monitoring systems are shown in Table 11.39.

Geographic information systems (GIS) can be an excellent tool for pressure monitoring systems as it can significantly enhance network management in gas distribution systems. The GIS roles in pressure-monitoring systems of gas distribution are summarized in Table 11.40.

The integration of pressure monitoring systems with GIS provides a powerful tool for better network management. It enhances the ability to monitor, analyze, predict, and respond to changes in pressure throughout the gas distribution network.

11.7 RENEWABLE ENERGY IN CITIES

Renewable energy in cities is a rapidly growing field due to the increasing urbanization and the need to mitigate the impacts of climate change. Cities account for 65% of global energy use and 70% of man-made carbon emissions. Therefore, they play a crucial role in the world's transition to a low-carbon economy.

Cities can harness various forms of renewable energy, including solar, wind, hydropower, and bioenergy. Solar panels can be installed on rooftops, parking lots, and other urban spaces to generate

TABLE 11.40
GIS Roles in Pressure-Monitoring Systems

Real-time monitoring and visualization	Pressure monitoring systems provide real-time data on the pressure at various points in the network. When integrated with a GIS, this data can be visualized on a map, making it easier to understand and analyze.
Spatial analysis	GIS allows for spatial analysis, which involves understanding patterns, relationships, and trends in a geographic context. By integrating pressure data, operators can perform spatial analysis to identify areas of high or low pressure, detect leaks, and optimize network performance.
Predictive maintenance	The integration of GIS and pressure monitoring systems can help predict potential issues before they become serious problems. This can prevent disruptions that could affect the uniformity of pressure.
Emergency response	In case of an emergency, such as a leak or a significant drop in pressure, the system can send alerts to operators, enabling a quick response. The location of the issue can be quickly identified on the GIS map, facilitating a faster and more efficient response.
Planning and optimization	The integration of GIS and pressure monitoring systems can support planning and optimization efforts. For example, it can help determine the optimal locations for new pipelines or pressure-regulating devices based on the analysis of pressure data and geographic information.

electricity. Wind turbines, although more commonly associated with rural areas, can also be installed in cities, especially on tall buildings. Hydropower can be harnessed from urban water systems, and bioenergy can be produced from urban organic waste.

As the world continues to contend with the pressing issue of climate change, the role of renewable energy in cities has become increasingly crucial. Cities, which are home to a significant portion of the global population, are major contributors to greenhouse gas emissions and energy consumption. Consequently, the adoption of renewable energy sources in urban areas has emerged as a critical strategy to mitigate the environmental impact of cities and pave the way for a more sustainable future.

One of the primary advantages of renewable energy in cities is its potential to reduce the carbon footprint of urban areas. Traditional energy sources, such as fossil fuels, are a significant contributor to the greenhouse gas emissions that drive climate change. In contrast, renewable energy technologies, such as solar, wind, and geothermal power, produce electricity and heat without releasing harmful emissions. By transitioning to these cleaner energy sources, cities can significantly reduce their carbon footprint and contribute to global efforts to combat climate change.

Renewable energy in cities can enhance energy security and resilience. Reliance on centralized, fossil-fuel-based energy systems can make cities vulnerable to disruptions, such as power outages and supply chain issues. Distributed renewable energy systems, on the other hand, can provide a more decentralized and resilient energy infrastructure, reducing the risk of widespread outages and ensuring a more reliable energy supply for urban communities.

Another key benefit of renewable energy in cities is its potential to create new economic opportunities and job prospects. The renewable energy industry is a rapidly growing sector, and the installation, maintenance, and operation of renewable energy systems can generate a significant number of local jobs. This can contribute to the economic development of urban areas and help to address issues such as unemployment and income inequality. In addition, the integration of renewable energy in cities can also have positive impacts on public health and well-being. By reducing air pollution and noise, renewable energy can improve the overall quality of life for urban residents, leading to better respiratory health, reduced cardiovascular issues, and a more pleasant living environment.

Despite the numerous advantages of renewable energy in cities, several challenges also must be addressed. One of the primary challenges is the availability of suitable space for renewable energy installations, such as rooftop solar panels or wind turbines. Urban areas often have limited land and roof space, and the integration of renewable energy systems must be carefully planned to ensure efficient use of available resources. Another challenge is the need for effective energy storage solutions to address the intermittent nature of some renewable energy sources, such as solar and wind power. Without adequate storage capacity, the integration of renewable energy in cities can be hampered by fluctuations in energy supply and demand. Advances in battery technology and the development of smart grid systems can help to address this challenge and enable a more seamless integration of renewable energy in urban areas.

The adoption of renewable energy in cities may face financial and policy-related barriers. The upfront costs of renewable energy systems can be a deterrent for some urban residents and businesses, and there may be a need for financial incentives, such as tax credits or subsidies, to encourage investment in these technologies. Furthermore, the regulatory and policy landscapes surrounding renewable energy in cities can vary significantly, and the development of clear and supportive policies can be crucial in driving the widespread adoption of these technologies.

Despite these challenges, there are numerous examples of cities around the world that have successfully embraced renewable energy and are reaping the benefits. For instance, the city of Copenhagen, Denmark, has set a goal of becoming carbon-neutral by 2025, and it has made significant strides in the integration of renewable energy sources, such as wind and solar power, into its urban infrastructure. Similarly, the city of Oslo, Norway, has implemented a comprehensive plan to reduce its greenhouse gas emissions, which includes the installation of large-scale solar power systems on municipal buildings and the development of a district heating system powered by geothermal energy.

The integration of renewable energy in cities is inevitable toward a more sustainable and resilient future. By harnessing the power of renewable sources, such as solar, wind, and geothermal energy, cities can significantly reduce their carbon footprint, enhance energy security, create new economic opportunities, and improve public health and well-being. While there are challenges that must be addressed, the potential benefits of renewable energy in urban areas make it a vital component of the global effort to mitigate the effects of climate change and build a more sustainable world.

PROBLEMS AND ISSUES FOR DISCUSSION

1. How can urban energy systems be designed to effectively integrate renewable energy sources while meeting the growing energy demands of urban populations?
2. In what ways does urban energy planning influence the sustainability of cities, and what innovative strategies can be employed to enhance energy efficiency in urban areas?
3. Discuss the role of innovation in facilitating energy transitions within cities, particularly in relation to technological advancements and policy frameworks.
4. What are the challenges and opportunities associated with transitioning to non-conventional electric power sources in urban environments, and how can cities overcome these challenges?
5. Analyze the impact of electric power supply reliability on urban development and economic growth, and propose measures to enhance the resilience of power infrastructure.
6. How does the integration of gas supply into urban energy systems affect the overall energy transition, and what implications does this have for urban planning?
7. Examine the relationship between energy supply and housing affordability in urban areas, and discuss potential policy interventions to address this issue.
8. What role do local governments play in shaping urban energy systems, and how can they foster innovation and collaboration among stakeholders in the energy sector?
9. Evaluate the effectiveness of current urban energy planning practices in addressing climate change and promoting sustainable energy use in cities.
10. How can cities leverage smart grid technology to optimize electric power supply and enhance energy management in urban settings?

11. Discuss the potential for decentralized energy systems in urban areas, and how they can contribute to energy security and sustainability.
12. In what ways can public–private partnerships facilitate the development of innovative urban energy solutions, particularly in the context of energy transitions?
13. Analyze the social implications of energy transitions in cities, focusing on equity and access to clean energy for marginalized communities.
14. How can urban energy planning incorporate considerations for climate resilience, and what strategies can cities employ to adapt to changing environmental conditions?
15. What are the key barriers to implementing non-conventional electric power technologies in urban areas, and how can these barriers be addressed through policy and innovation?
16. Explore the role of community engagement in urban energy planning, and how participatory approaches can enhance the effectiveness of energy transition initiatives.
17. How do energy supply disruptions impact urban economies, and what strategies can cities adopt to mitigate the effects of such disruptions?
18. Discuss the importance of integrating energy efficiency measures into housing policies in urban areas and the potential benefits for residents and the environment.
19. What innovative financing mechanisms can be employed to support the development of sustainable urban energy systems, particularly for low-income communities?
20. Analyze the impact of regulatory frameworks on the adoption of non-conventional electric power sources in cities and suggest reforms to promote innovation.
21. How can cities balance the need for reliable gas supply with the imperative to reduce greenhouse gas emissions in their energy systems?
22. Discuss the potential role of energy storage technologies in enhancing the reliability and flexibility of urban electric power supply systems.
23. In what ways can energy transitions in cities contribute to achieving national and global climate goals, and what challenges must be addressed to ensure success?
24. Explore the relationship between urban design and energy consumption patterns, and how thoughtful planning can lead to more sustainable energy use.
25. How can cities utilize data analytics and smart technologies to improve energy planning and management in urban environments?
26. Discuss the role of education and public awareness in promoting energy transitions in cities, and how informed citizens can drive change.
27. What strategies can urban planners implement to encourage the adoption of EVs, and how does this shift impact urban energy systems?
28. Analyze the potential for integrating renewable energy sources into existing urban infrastructure and the challenges associated with retrofitting traditional systems.
29. How can cities create incentives for businesses to invest in sustainable energy solutions, and what impact might this have on local economies?
30. Explore the role of international cooperation in addressing urban energy challenges, and how knowledge sharing can facilitate innovative solutions.
31. Discuss the implications of energy poverty in urban areas, and what comprehensive strategies can be developed to address this pressing issue.
32. How can urban energy systems be designed to promote social equity and inclusivity, ensuring that all residents have access to clean energy?
33. Analyze the significance of multi-stakeholder collaboration in urban energy planning, and how diverse perspectives can enhance decision-making processes.
34. What role do research and development play in advancing non-conventional electric power technologies, and how can cities support innovation in this area?
35. Discuss the importance of integrating climate adaptation strategies into urban energy planning, particularly in the context of extreme weather events.

36. How can cities leverage technological advancements in energy efficiency to reduce their overall carbon footprint and enhance sustainability?
37. Explore the challenges and opportunities of transitioning from fossil fuels to renewable energy sources in urban gas supply systems.
38. What role does urban agriculture play in promoting sustainable energy practices, and how can cities support the integration of these initiatives?
39. Analyze the impact of energy-efficient housing designs on urban energy consumption, and how these designs can be incentivized through policy.
40. How can cities assess and prioritize their energy needs to ensure a balanced and sustainable energy supply for future generations?
41. Discuss the potential for integrating green infrastructure into urban energy systems, and how this can enhance environmental and social outcomes.
42. What are the implications of emerging technologies, such as blockchain, for urban energy systems and the management of electric power supply?
43. How can cities develop comprehensive energy transition roadmaps that align with both local needs and global sustainability goals?
44. Explore the role of behavioral economics in influencing energy consumption patterns in urban areas, and how this knowledge can inform policy.
45. Discuss the importance of resilience planning in urban energy systems, particularly in the face of natural disasters and climate change impacts.
46. How can cities ensure that their energy transition strategies are adaptable to future technological advancements and changing energy landscapes?
47. Analyze the role of energy cooperatives in promoting community-based energy solutions in urban settings and their impact on local economies.
48. What strategies can urban planners employ to reduce dependency on non-renewable energy sources while ensuring energy security for residents?
49. How can cities facilitate the transition to smart energy systems that enhance the efficiency and reliability of electric power supply?
50. Discuss the importance of integrating health considerations into urban energy planning, particularly in relation to air quality and public health outcomes.

MULTIPLE-CHOICE QUESTIONS

1. What is the primary goal of urban energy planning?
   a. Maximize energy consumption
   b. Ensure energy security and sustainability
   c. Focus solely on fossil fuels
   d. Eliminate all forms of energy
   e. Increase energy prices
2. Which of the following is a non-conventional electric power source?
   a. Coal
   b. Natural gas
   c. Wind energy
   d. Nuclear power
   e. Oil
3. What is a significant challenge in transitioning to renewable energy in urban areas?
   a. Abundance of resources
   b. Infrastructure limitations
   c. Lack of interest from citizens
   d. Excessive funding
   e. Overpopulation

4. Which technology is most associated with smart urban energy systems?
   a. Solar panels
   b. Smart meters
   c. Wind turbines
   d. Biomass boilers
   e. Geothermal heat pumps
5. What role does energy efficiency play in urban housing?
   a. Increases energy costs
   b. Reduces energy consumption
   c. Eliminates the need for energy
   d. Has no impact
   e. Only benefits large corporations
   6. Which of the following best describes a decentralized energy system?
   a. Centralized power plants
   b. Local energy generation
   c. National grid reliance
   d. Fossil fuel dependency
   e. Single-source energy supply
7. What is the main benefit of integrating renewable energy in urban planning?
   a. Increased pollution
   b. Energy independence
   c. Higher energy prices
   d. Decreased quality of life
   e. Limited resource availability
8. Which of the following is an example of an energy transition?
   a. Increasing coal production
   b. Shifting from fossil fuels to renewables
   c. Expanding natural gas usage
   d. Reducing energy efficiency
   e. Focusing solely on nuclear energy
9. What is a significant factor in urban energy demand growth?
   a. Decreasing population
   b. Technological stagnation
   c. Urbanization and economic development
   d. Reduced industrial activity
   e. Increased rural migration
10. Which government action can promote innovation in urban energy systems?
    a. Cutting research funding
    b. Implementing strict regulations
    c. Providing tax incentives for green technologies
    d. Limiting public–private partnerships
    e. Discouraging renewable energy use
11. What is the primary purpose of smart grids in urban energy systems?
    a. To increase energy prices
    b. To enhance energy management and reliability
    c. To eliminate renewable energy sources
    d. To centralize energy production
    e. To reduce energy efficiency

12. Which of the following is a barrier to adopting non-conventional electric power?
    a. High initial investment costs
    b. Abundant resources
    c. Strong public support
    d. Technological advancements
    e. Government incentives
13. What is the impact of energy poverty in urban areas?
    a. Increased access to energy
    b. Enhanced quality of life
    c. Limited access to essential services
    d. Decreased reliance on energy
    e. Improved environmental conditions
14. Which energy source is considered the cleanest for urban electric supply?
    a. Natural gas
    b. Coal
    c. Solar energy
    d. Oil
    e. Nuclear energy
15. How can urban planners encourage electric vehicle adoption?
    a. Reducing charging infrastructure
    b. Increasing parking fees
    c. Expanding charging stations
    d. Limiting public transport options
    e. Promoting fossil fuel vehicles
16. What is the benefit of using energy storage technologies in cities?
    a. Increases energy waste
    b. Reduces energy reliability
    c. Balances supply and demand
    d. Limits renewable energy use
    e. Creates energy monopolies
17. Which factor contributes to the need for innovative urban energy solutions?
    a. Stable energy prices
    b. Climate change challenges
    c. Decreasing population density
    d. Abundant fossil fuel reserves
    e. Lack of technological advancement
18. What is the primary goal of energy transition strategies in cities?
    a. Maintain reliance on fossil fuels
    b. Achieve sustainability and reduce emissions
    c. Increase energy consumption
    d. Limit technological innovation
    e. Focus solely on economic growth
19. Which urban energy planning approach focuses on community involvement?
    a. Top-down planning
    b. Participatory planning
    c. Centralized decision-making
    d. Exclusive stakeholder engagement
    e. Non-transparent processes

20. What is a common feature of smart urban energy systems?
    a. Manual energy management
    b. Real-time data analytics
    c. Limited user engagement
    d. High energy waste
    e. Reliance on outdated technology
21. How can cities improve energy efficiency in housing?
    a. By ignoring building codes
    b. By implementing energy-efficient designs
    c. By increasing energy prices
    d. By using only fossil fuels
    e. By reducing insulation standards
22. What is a significant advantage of gas supply in urban energy systems?
    a. High carbon emissions
    b. Flexibility and reliability
    c. Limited availability
    d. High cost of infrastructure
    e. Environmental degradation
23. Which factor is critical for successful urban energy transitions?
    a. Public opposition
    b. Stakeholder collaboration
    c. Lack of funding
    d. Resistance to change
    e. Decreased awareness
24. What is a potential drawback of relying on non-conventional electric power?
    a. Environmental benefits
    b. Intermittency and reliability issues
    c. Decreased energy independence
    d. High efficiency
    e. Low operational costs
25. How can urban areas address energy supply disruptions?
    a. By ignoring infrastructure maintenance
    b. By diversifying energy sources
    c. By relying solely on fossil fuels
    d. By reducing energy efficiency
    e. By eliminating contingency plans
26. What is the primary environmental benefit of energy transition in cities?
    a. Increased pollution
    b. Reduced greenhouse gas emissions
    c. Higher energy consumption
    d. Greater reliance on fossil fuels
    e. Deterioration of air quality
27. Which policy can incentivize the use of renewable energy in urban areas?
    a. Tax breaks for fossil fuel companies
    b. Subsidies for renewable energy projects
    c. Increased tariffs on solar panels
    d. Limiting renewable energy research
    e. Reducing public awareness campaigns

28. What is the primary function of energy cooperatives in urban settings?
    a. Centralizing energy production
    b. Promoting community-based energy solutions
    c. Increasing energy prices
    d. Reducing public participation
    e. Limiting access to renewable energy
29. How can cities enhance their resilience to energy supply disruptions?
    a. By ignoring climate change impacts
    b. By investing in diverse energy sources
    c. By centralizing energy production
    d. By reducing energy efficiency measures
    e. By limiting public engagement
30. What is a key aspect of sustainable urban housing?
    a. High energy consumption
    b. Energy-efficient designs and materials
    c. Reliance on fossil fuels
    d. Ignoring environmental impacts
    e. Reducing public transport options
31. Which urban planning strategy can reduce energy consumption?
    a. Sprawling development
    b. Mixed-use development
    c. Isolated zoning
    d. Limited public transport
    e. Increasing parking spaces
32. What is a challenge faced by urban energy planners?
    a. Abundance of renewable resources
    b. Balancing energy demand and supply
    c. Excess funding for projects
    d. High public support for energy projects
    e. Low energy consumption rates
33. How can cities promote energy conservation among residents?
    a. By increasing energy prices without communication
    b. By providing education and incentives
    c. By limiting access to information
    d. By ignoring public engagement
    e. By reducing energy efficiency programs
34. What is a benefit of integrating green infrastructure in urban energy systems?
    a. Increased pollution
    b. Enhanced environmental and social outcomes
    c. Higher energy costs
    d. Decreased biodiversity
    e. Limited public engagement
35. Which of the following is a potential outcome of energy-efficient urban design?
    a. Increased energy consumption
    b. Improved quality of life
    c. Greater reliance on fossil fuels
    d. Higher greenhouse gas emissions
    e. Decreased property values

36. What is the role of local governments in urban energy transitions?
    a. To discourage renewable energy use
    b. To facilitate policies and programs that support sustainability
    c. To centralize energy production
    d. To ignore community needs
    e. To increase energy prices
37. Which of the following can enhance urban energy resilience?
    a. Ignoring climate risks
    b. Developing contingency plans
    c. Reducing energy efficiency measures
    d. Limiting stakeholder engagement
    e. Focusing solely on fossil fuels
38. What is a common feature of successful urban energy initiatives?
    a. Lack of community involvement
    b. Strong stakeholder collaboration
    c. Reliance on outdated technology
    d. High energy waste
    e. Centralized decision-making
39. How can cities leverage technological advancements in energy efficiency?
    a. By ignoring new technologies
    b. By adopting smart technologies and practices
    c. By reducing public transport options
    d. By limiting renewable energy use
    e. By maintaining outdated infrastructure
40. What is a significant benefit of community engagement in urban energy planning?
    a. Increased public opposition
    b. Improved trust and collaboration
    c. Limited stakeholder input
    d. Reduced awareness of energy issues
    e. Ignoring community needs
41. Which of the following is a characteristic of energy-efficient buildings?
    a. High energy consumption
    b. Effective insulation and design
    c. Reliance on fossil fuels
    d. Poor air quality
    e. Increased maintenance costs
42. What is a key driver of urban energy innovation?
    a. Public apathy
    b. Climate change pressures
    c. Decreased investment in research
    d. Limited technological advancements
    e. Lack of regulatory support
43. How can cities assess their energy needs effectively?
    a. By ignoring data analysis
    b. By conducting comprehensive energy audits
    c. By reducing public engagement
    d. By relying solely on outdated statistics
    e. By limiting stakeholder collaboration

44. What is a potential drawback of centralized energy systems?
    a. Increased reliability
    b. Vulnerability to supply disruptions
    c. Enhanced energy independence
    d. Lower operational costs
    e. Greater community involvement
45. Which of the following can enhance the integration of renewable energy in urban areas?
    a. Reducing public awareness campaigns
    b. Implementing supportive policies and incentives
    c. Limiting access to renewable technologies
    d. Ignoring community needs
    e. Focusing solely on fossil fuels
46. How can energy cooperatives benefit urban communities?
    a. By centralizing energy production
    b. By promoting community-based energy solutions
    c. By increasing energy costs
    d. By limiting access to renewable energy
    e. By reducing public participation
47. What is a key aspect of energy transition strategies?
    a. Maintaining reliance on fossil fuels
    b. Fostering innovation and sustainability
    c. Reducing energy efficiency
    d. Ignoring climate change impacts
    e. Focusing solely on economic growth
48. What is a potential consequence of energy inefficiency in urban housing?
    a. Lower energy costs
    b. Increased greenhouse gas emissions
    c. Improved indoor air quality
    d. Enhanced occupant comfort
    e. Greater energy independence
49. Which of the following is essential for successful energy transitions in cities?
    a. Public resistance
    b. Strong leadership and vision
    c. Ignoring community feedback
    d. Limiting technological advancements
    e. Focusing solely on economic factors
50. What is the primary environmental goal of urban energy systems?
    a. Increase fossil fuel reliance
    b. Achieve sustainability and reduce emissions
    c. Maximize energy consumption
    d. Decrease energy efficiency
    e. Limit community engagement

ANSWERS TO MULTIPLE-CHOICE QUESTIONS

1	b	11	b	21	b	31	b	41	b
2	c	12	a	22	b	32	b	42	b
3	b	13	c	23	b	33	b	43	b
4	b	14	c	24	b	34	b	44	b
5	b	15	c	25	b	35	b	45	b
6	b	16	c	26	b	36	b	46	b
7	b	17	b	27	b	37	b	47	b
8	b	18	b	28	b	38	b	48	b
9	c	19	b	29	b	39	b	49	b
10	c	20	b	30	b	40	b	50	b

FURTHER READING

Agarwal, A., & Narain, S. (2016). *Urban energy systems: A holistic approach to energy management in cities*. Routledge.

Baker, J. (2020). Energy transitions in cities: The role of local governments. *Journal of Urban Affairs*, 42(3), 456–474. https://doi.org/10.1080/07352166.2019.1667260.

Bauwens, T., & Devine-Wright, P. (2018). The role of community energy in the transition to a low-carbon economy: A review of the literature. *Renewable and Sustainable Energy Reviews*, 81, 152–162. https://doi.org/10.1016/j.rser.2017.07.043.

Bertsch, V., & Luthra, S. (2019). Urban energy planning: The role of policy and technology. *Energy Policy*, 134, 110–121. https://doi.org/10.1016/j.enpol.2019.110121.

Bulkeley, H., & Betsill, M. (2013). *Cities and climate change: Urban sustainability and global environmental governance*. Routledge.

C40 Cities Climate Leadership Group. (2019). *The future of urban energy: A roadmap for sustainable cities*. C40.

Deng, X., & Zhang, W. (2021). *Urban energy transition: The role of local governments in China*. *Sustainability*, 13(4), 2000. https://doi.org/10.3390/su13042000.

Droege, P. (2018). *Urban energy transition: From fossil fuels to renewable energy*. Elsevier.

Edenhofer, O., & Pichs-Madruga, R. (2014). *Climate change 2014: Mitigation of climate change*. Cambridge University Press.

Fischer, C., & Pregernig, M. (2018). The role of innovation in urban energy transitions: A case study of Germany. *Energy Research & Social Science*, 46, 1–10. https://doi.org/10.1016/j.erss.2018.07.014.

Fischer, S., & Newell, R. (2020). Energy supply and housing: The impact of energy efficiency on urban housing markets. *Journal of Housing Economics*, 47, 101–112. https://doi.org/10.1016/j.jhe.2019.101112.

González, M., & Gutiérrez, R. (2019). Urban energy systems and climate change: A review of the literature. *Renewable and Sustainable Energy Reviews*, 99, 1–12. https://doi.org/10.1016/j.rser.2018.10.015.

Hodson, M., & Marvin, S. (2010). *Urban energy transitions: Local action and the challenge of sustainability*. Routledge.

Kivimaa, P., & Virkamäki, V. (2014). The role of cities in energy transitions: A comparative analysis of urban energy planning. *Energy Policy*, 67, 651–661. https://doi.org/10.1016/j.enpol.2013.12.037.

Klein, J., & Ramesh, T. (2017). Non-conventional electric power: Opportunities and challenges in urban settings. *Energy,* 139, 1032–1041. https://doi.org/10.1016/j.energy.2017.08.084.

Lombardi, P., & Mazzarella, F. (2019). Urban energy systems: A framework for sustainable development. *Sustainable Cities and Society*, 49, 101–109. https://doi.org/10.1016/j.scs.2019.101109.

Meyer, M., & Becker, T. (2020). Innovations in urban energy systems: The role of technology and governance. *Environmental Science & Policy*, 107, 1–9. https://doi.org/10.1016/j.envsci.2019.12.007.

Miller, S., & Dyer, M. (2018). Energy transitions in urban areas: A comparative study of policy frameworks. *Urban Studies*, 55(4), 789–803. https://doi.org/10.1177/0042098017707784.

Moreno, J., & Rojas, C. (2020). The role of urban planning in energy transitions: Lessons from European cities. *Cities*, 96, 102–110. https://doi.org/10.1016/j.cities.2019.102110.

Newman, P., & Kenworthy, J. (2015). *Sustainability and cities: Overcoming automobile dependence*. Island Press.

Pérez, C., & Gómez, E. (2019). Smart cities and energy transitions: Innovations in urban energy systems. *Journal of Cleaner Production*, 234, 124–135. https://doi.org/10.1016/j.jclepro.2019.06.032.

Rogers, J., & Hall, S. (2018). Energy supply and urban resilience: A framework for analysis. *Energy Research & Social Science*, 45, 1–11. https://doi.org/10.1016/j.erss.2018.07.005.

Sovacool, B. K. (2018). *The renewable energy transition: A global perspective*. Routledge.

Steffen, W., & Rockström, J. (2016). *Planetary boundaries: Guiding human development on a changing planet*. Stockholm Resilience Centre.

Sullivan, J., & Morrow, C. (2020). Gas supply and urban energy systems: A critical review. *Energy Policy*, 138, 111–120. https://doi.org/10.1016/j.enpol.2020.111120.

Tindale, S., & Smith, R. (2019). The future of urban energy: Transition pathways for cities. Energy Research & Social Science, 49, 1–12. https://doi.org/10.1016/j.erss.2018.10.006.

United Nations. (2020). *World urbanization prospects: The 2018 revision*. United Nations.

Van der Meer, J., & Schipper, L. (2018). Urban energy systems and the role of innovation. *Energy Efficiency*, 11(2), 321–335. https://doi.org/10.1007/s12053-017-9552-5.

Wang, Y., & Zhang, L. (2021). The impact of urban energy planning on sustainable development: A case study of Beijing. *Sustainable Cities and Society*, 66, 102–110. https://doi.org/10.1016/j.scs.2021.102110.

Zheng, Y., & Liu, J. (2019). Non-conventional electric power generation: Trends and challenges. *Renewable Energy*, 132, 116–126. https://doi.org/10.1016/j.renene.2018.08.095.

Zhou, Y., & Zhao, J. (2020). Urban energy transitions: The role of public participation. *Energy Policy*, 137, 111–120. https://doi.org/10.1016/j.enpol.2019.111120.

Zhou, Z., & Wang, H. (2021). Urban energy systems: Integrating renewable energy and energy efficiency. *Energy*, 224, 120–130. https://doi.org/10.1016/j.energy.2021.12013.

12 Information Technology and Communication in Cities

12.1 INTERNET, CONNECTIVITY, AND E-COMMERCE

The internet, a global network of interconnected devices, has revolutionized the way we communicate, access information, and conduct business. It has made connectivity ubiquitous, enabling real-time communication and data exchange across vast distances. This connectivity has given rise to e-commerce, a digital marketplace that transcends geographical boundaries. E-commerce leverages the internet's connectivity to facilitate the buying and selling of goods and services online. It has transformed traditional commerce, making shopping more convenient and accessible. Consumers can browse products from multiple vendors, compare prices, and make purchases from the comfort of their homes. For businesses, e-commerce opens up new markets, reduces operational costs, and provides valuable customer data. However, this digital transformation also brings challenges. Internet connectivity depends on robust and reliable infrastructure, which is not uniformly available, leading to a digital divide. E-commerce transactions require secure platforms to protect against cyber threats and ensure consumer trust. Despite these challenges, the synergistic relationship between the internet, connectivity, and e-commerce continues to drive innovation and growth in the digital economy. As technology advances, solutions to these challenges are being developed, making the digital world more inclusive and secure. Thus, the internet, connectivity, and e-commerce are intertwined, each influencing and shaping the other, collectively transforming the way we live and work.

12.1.1 INTERNET INFRASTRUCTURE

This section covers the physical hardware, transmission media, and other equipment used to provide internet services. It includes data centers, servers, routers, satellites, fiber optic cables, and wireless transmitters. The quality and capacity of this infrastructure can significantly impact the speed and reliability of internet connections.

12.1.1.1 Data Centers

Data centers are the backbone of internet services. They house servers, storage devices, and network connections. These facilities are designed to ensure the smooth operation of IT systems, with features like climate control, fire suppression systems, and redundant power supplies.

12.1.1.1.1 Data Centers: The Backbone of Internet Services

Data centers (Figure 12.1) are the heart and soul of our digital world. They are the physical locations where companies store and access important data and applications. In essence, they are large groups of networked computer servers used for remote storage, processing, or distribution of large amounts

DOI: 10.1201/9781032664613-12

FIGURE 12.1 CERN Data Center.

Source: **Hugo van Meijeren, Wikipedia.**

of data. Data centers are the backbone of our digital age, serving as the vital hubs that power our ever-evolving technological landscape. These specialized facilities are the heart and soul of our interconnected world, where the vast troves of data that drive our digital lives are stored, processed, and distributed.

At their core, data centers are large, purpose-built complexes housing massive arrays of networked computer servers. These servers are meticulously engineered to handle the storage, processing, and distribution of vast quantities of data, catering to the ever-growing demands of businesses, organizations, and individuals alike.

The importance of data centers cannot be overstated. They are the nerve centers that enable the seamless operation of a wide range of digital services, from cloud computing and online banking to social media and e-commerce. Without these vital hubs, the digital revolution that has transformed our lives would simply grind to a halt.

One of the key functions of data centers is remote data storage. Instead of storing data on local devices, companies and individuals can leverage the immense storage capacity of data centers to keep their important files, documents, and applications safe and secure. This not only provides a level of redundancy and reliability that would be challenging to achieve with on-site storage solutions, but also allows for easy access and sharing of data from anywhere in the world. Data centers also do the data processing. These facilities house powerful computing infrastructure, including high-performance CPUs, GPUs, and specialized hardware, that can be leveraged to perform complex computational tasks. This makes data centers an essential resource for industries ranging from scientific research and artificial intelligence (AI) to financial analysis and e-commerce.

The distribution of data is another pivotal function of data centers. By strategically locating these facilities across the globe, companies can ensure that their digital content, services, and applications are delivered to users with lightning-fast speeds and minimal latency. This is particularly important for services that require low latency, such as video streaming, online gaming, and real-time communication platforms.

Data centers are designed with a robust focus on security and reliability. They are equipped with advanced physical and digital security measures, redundant power supplies, and sophisticated cooling systems to ensure the uninterrupted operation of the critical infrastructure they house. This level of resilience is essential for safeguarding the sensitive data and mission-critical applications that are entrusted to these facilities. Data centers are the indispensable hubs that power our digital world. As technology continues to evolve and our reliance on digital services grows, the importance of these specialized facilities will only continue to increase. From remote data storage and processing to content distribution and security, data centers are the heart and soul of our interconnected, technology-driven society.

12.1.1.1.2 Housing Servers, Storage Devices, and Network Connections

At the core of a data center are its servers. These are powerful computers engineered to host applications and data. They come in various forms, including file servers, database servers, mail servers, and web servers, each serving a unique purpose. Storage devices in data centers hold vast amounts of data. These include hard disk drives (HDDs), solid-state drives (SSDs), and newer technologies like storage area networks (SANs) and network-attached storage (NAS). Network connections are the veins and arteries of the data center. They link servers and storage devices together and to the outside world. These connections are typically made via high-speed fiber optic cables that can quickly transmit data over long distances.

12.1.1.1.3 Ensuring Smooth Operation of IT Systems

Data centers are designed to ensure the smooth operation of IT systems. They are equipped with redundant power supplies to prevent outages. If the main power source fails, uninterruptible power supply (UPS) systems and generators kick in to keep the servers running. Climate control is crucial in data centers. Servers generate a lot of heat, and if not properly managed, this heat can lead to equipment failure. Hence, data centers use cooling systems, such as computer room air conditioners (CRACs) and computer room air handlers (CRAHs), to maintain optimal temperatures. Fire suppression systems are also a key feature of data centers. These systems detect and suppress fires while minimizing damage to equipment. They often use clean agent gases, which are safe for people and do not leave residue on sensitive electronics.

The increasing demand for data storage, computing, and analysis has led to the need for hyperscale data centers. Hyperscale data centers represent a significant evolution in data center technology, designed to meet the growing demands of the digital age. These massive facilities are engineered to provide a highly scalable and resilient environment for handling unprecedented volumes of data and computational workloads. As businesses, organizations, and individuals generate and consume ever-increasing amounts of digital information, the need for hyperscale data centers has become paramount. These specialized facilities can seamlessly scale to accommodate the exponential growth in data, offering unparalleled processing power, storage capacity, and redundancy to ensure the continuous availability and reliability of mission-critical digital services. Hyperscale data centers have become the cornerstone of the modern digital infrastructure, enabling the efficient management and processing of the staggering amounts of data that drive our connected world.

12.1.1.1.4 What Is a Hyperscale Data Center?

A hyperscale data center is a massive facility built by companies with vast data processing and storage needs. These data centers are engineered for large-scale workloads with an optimized network infrastructure, streamlined network connectivity, and minimized latency. They can scale up rapidly to meet growing demands.

12.1.1.1.5 *How Do Hyperscale Data Centers Work?*

Hyperscale data centers work on the principle of virtualization, which uses software to create an abstract layer over computer hardware, enabling the division of a single computer's hardware components into multiple virtual machines (VMs). Each VM runs its own operating system and behaves like an independent computer, even though it is running on just a portion of the actual underlying computer hardware.

12.1.1.1.6 *Key Features of Hyperscale Data Centers*

Hyperscale data centers are characterized by their sheer size, often encompassing millions of square feet of space. They house at least 5,000 servers and possibly miles of connection equipment. Redundancy is another important aspect of hyperscale data centers, providing backup measures and/or devices that can automatically activate in the event of equipment failure or power be lost.

12.1.1.1.7 *Who Owns Hyperscale Data Centers?*

The hyperscale market's massive capital requirements dictate that there are only a handful of hyperscale data center operators, namely Amazon, Facebook/Meta, Apple, Google, IBM, and Microsoft. In addition, there are wholesale colocation and data center development companies that cater to them, like Equinix and Digital Realty.

12.1.1.1.8 *Market Worth of Hyperscale Data Centers*

Despite representing less than 10% of the market, more than half of all data processed in 2020 passed through a hyperscale facility. The US hyperscale market alone was valued at $51.5 billion in 2021 and is expected to be worth $63 billion by 2027.

12.1.1.2 Servers

Servers are powerful computers that store, process, and distribute data. They handle requests from clients (like your computer or smartphone), perform computations, and send back the results. Servers can be dedicated to specific tasks, such as web hosting, email, or database management.

12.1.1.2.1 *Servers: The Powerhouses of the Digital World*

In the digital world, servers play a crucial role. They are powerful computers that store, process, and distribute data. They handle requests from clients, perform computations, and send back the results. Servers can be dedicated to specific tasks, such as web hosting, email, or database management.

12.1.1.2.2 *What Is a Server?*

A server is a computer or system that provides resources, data, services, or programs to other computers, known as clients, over a network. In theory, whenever computers share resources with client machines, they are considered servers.

12.1.1.2.3 *How Do Servers Work?*

Servers work on a client-server model where servers host, deliver and manage most of the resources and services to be consumed by the client. This means that the client can be a machine or a process that requests resources or services from a server. When a client device wants to access a file, for example, it sends a request over the network to the server where the file is stored. The server then responds to the request by providing the client with the file. The client can then download the file and use it on their own device.

12.1.1.2.4 *Types of Servers*

Servers can be categorized based on their purpose, as shown in Table 12.1.

TABLE 12.1
Types of Servers

Web servers	These servers deliver web pages to clients who request them. When we type a URL into your browser, our computer sends a request to the server where the webpage is stored. The server then sends the webpage back to our computer for our browser to display.
Database servers	These servers store, retrieve, and manage data in a database. They provide clients with access to database capabilities.
Mail servers	These servers send and receive emails. When you send an email, it goes to the recipient's mail server, which then delivers it to their inbox.
File servers	These servers store files and allow clients to upload and download files. They are often used in businesses to enable employees to share files.
Game servers	These servers are used in online gaming to host multiplayer games. They handle the game logic and keep all players' games in sync.

12.1.1.2.5 Server Hardware and Software

Server hardware is designed to better handle the demanding tasks that servers perform. This includes more powerful processors, more memory, and larger, more reliable storage systems. Server software is also specialized. Server operating systems are designed to handle many connections, run for long periods without rebooting, and secure against attacks.

Servers are the backbone of the internet and many business networks. They store, process, and distribute data, handle requests from clients, and can be dedicated to specific tasks. Without servers, our digital world as we know it would not exist. They are the unsung heroes that work behind the scenes to make sure we can access the digital resources we need when we need them.

12.1.1.3 Routers

Routers (Figure 12.2) are devices that direct data packets along networks. They analyze the destination of a data packet and choose the most efficient route for it. This routing ensures that data travel quickly and accurately from its source to its destination.

12.1.1.3.1 The Traffic Directors of the Digital World

If servers are powerhouses of digital world, routers are the directors. Routers are integral components of any network, large or small. They are devices that direct data packets along networks, ensuring that information travels quickly and accurately from its source to its destination.

12.1.1.3.2 What Is a Router?

A router is a networking device that forwards data packets between computer networks. Routers perform the traffic directing functions on the internet. The most familiar type of routers are home and small office routers that simply forward Internet Protocol (IP) packets between the home computers and the internet.

12.1.1.3.3 How Do Routers Work?

Routers work by analyzing the destination of a data packet and choosing the most efficient route for it. This process is known as routing. When a data packet arrives at a router, the router examines the packet's destination IP address and compares it to a table of available routes, known as a routing table. The router then sends the packet on its way via the best route.

12.1.1.3.4 Routing: Ensuring Efficient Data Travel

Routing is the process that ensures data travels quickly and accurately from its source to its destination. When a data packet is sent from one location to another on the internet, it does not travel in a

FIGURE 12.2 A wireless router.

Source: **Weihao Chiu, Wikipedia.**

TABLE 12.2
Types of Routers

Wireless routers	These are the most common types of routers found in homes and small businesses. They connect devices wirelessly to a network and to each other.
Edge routers	These routers are placed at the edge of a network and connect to the networks of ISPs.
Core routers	These high-capacity routers reside within the body of a network and forward information to other routers.
Virtual routers	These are software-based routers that exist within virtual environments.

straight line. Instead, it is passed from one internet-connected device to another (like routers) until it reaches its destination.

12.1.1.3.5 Types of Routers

There are several types of routers, each designed for specific needs as reflected in Table 12.2.

12.1.1.3.6 Router Hardware and Software

Routers are made up of hardware and software. The hardware includes the physical device and its interfaces for network connections. The software, also known as firmware, includes the operating system of the router and software applications.

Routers analyze the destination of a data packet and choose the most efficient route for it, ensuring that data travels quickly and accurately from its source to its destination. Whether it is a simple email, a web page, or a streaming video, every bit of data you send or receive makes its journey through a series of routers. Without routers, the internet as we know it would not exist.

12.1.1.4 Satellites

Satellites are indispensable in providing internet services to remote areas where it is difficult to lay cables. They receive signals from an earth station and then relay them back to users. Satellite internet can reach places where traditional broadband services cannot. Satellite internet is a technological breakthrough that has revolutionized the way we access the internet, especially in remote and hard-to-reach areas. It operates on a simple yet effective principle: signals are transmitted from an earth station to a satellite, which then relays them back to users on the ground.

The first step in this process involves an earth station, also known as a ground station. This is a terrestrial radio station designed for extraplanetary telecommunication with spacecraft or reception of radio waves from celestial radio sources. The earth station sends the internet data to the satellite in the form of radio signals. This data can be anything from a simple email to high-definition video.

Figure 12.3 illustrates how satellite internet access works by depicting the connection between satellites in orbit and ground-based receivers, providing internet connectivity to remote and underserved areas where traditional infrastructure is unavailable or impractical.

The satellite, positioned in geostationary orbit approximately 36,000 km above the Earth's equator, receives these signals. Geostationary orbit is chosen because the satellite in this orbit appears stationary with respect to a point on the Earth. This means the satellite remains in the same position in the sky, eliminating the need for ground equipment to constantly track the satellite's position.

Once the satellite receives the signals, it amplifies them and sends them back to Earth. This is done using a transponder, which is a device that receives a signal on one frequency and retransmits it at another frequency. This change in frequency helps avoid interference between the incoming and outgoing signals.

The signals, now carrying the internet data, are received by a satellite dish installed at the user's location. This dish is connected to a modem inside the user's premises, which decodes the signals into a form that can be used by devices such as computers, smartphones, and smart TVs.

FIGURE 12.3 Satellite internet access.

Source: Jesster79, Wikipedia.

One of the key advantages of satellite internet is its ability to provide internet services in areas where it is difficult to lay cables. Traditional broadband services rely heavily on physical infrastructure. They require cables to be laid down, either underground or overground, to carry data. This can be challenging in remote areas, hilly regions, islands, and other hard-to-reach places. It is also not cost-effective to lay cables in areas with a sparse population.

Satellite internet, on the other hand, does not have these limitations. As it relies on signals transmitted through space, it can reach virtually any place on Earth that has a clear view of the sky. This makes it an ideal solution for providing internet services in rural areas, remote islands, ships in the ocean, and even aircraft in flight. However, satellite internet does have its challenges. For one, the long distance that signals need to travel can lead to high latency, which is the delay between a user's action and the internet's response. This can affect real-time online activities like video calls and online gaming. Weather conditions can also affect the signal quality.

Despite these challenges, ongoing advancements in technology are continually improving the speed, reliability, and affordability of satellite internet. With the advent of low Earth orbit (LEO) satellite constellations, like SpaceX's Starlink, the future of satellite internet looks promising. These constellations involve deploying thousands of small satellites in lower orbits, reducing latency and increasing internet speeds.

Satellites can be broadly classified into two categories: natural and artificial. The moon is an example of a natural satellite. Artificial satellites, on the other hand, are man-made objects that have been placed into orbit around a celestial body. They serve a variety of purposes and can be classified based on their functions and orbits. Based on their orbit, the types of satellites are shown in Table 12.3 and Figure 12.4.

Each type of satellite and orbit has its own unique set of characteristics and is used for different missions. The choice of satellite type and orbit depends on the intended purpose, the data it acquires, and the services it offers. A geostationary orbit, also referred to as a geosynchronous equatorial orbit (GEO), is a circular geosynchronous orbit that is approximately 35,786 km in altitude above Earth's equator. An object in such an orbit has an orbital period equal to Earth's rotational period, one sidereal day, and so to ground observers it appears motionless, in a fixed position in the sky.

The advantages and disadvantages of geostationary orbit satellites are shown in Table 12.4.

These advantages make geostationary orbit particularly useful for applications such as telecommunications, weather monitoring, and navigation. However, it is worth noting that geostationary orbit also has its disadvantages, such as signal delay due to the long distance the signal needs to travel, and difficulty in providing coverage near the polar regions. These disadvantages need to be considered when choosing the type of orbit for a satellite. However, ongoing advancements in technology are continually improving the performance and reducing the costs associated with geostationary orbits.

TABLE 12.3
Types of Satellites Based on Their Orbits

LEO satellites	These satellites orbit at an altitude of roughly 160–2,000 km above the Earth's surface.
Medium Earth orbit (MEO) satellites	These satellites are positioned at an altitude that is higher than LEO but lower than GEO, that is, 2,000–35,786 km
Geosynchronous equatorial orbit (GEO) satellites	These satellites orbit at an altitude of approximately 35,786 km above Earth's equator.
Sun-synchronous orbit (SSO) satellites	These satellites are placed in a nearly polar orbit to provide coverage of the entire Earth's surface.
Geostationary transfer orbit (GTO) satellites	These satellites are placed in an elliptical orbit that is used to transfer satellites from LEO to geostationary orbit, that is, 600–800 km.

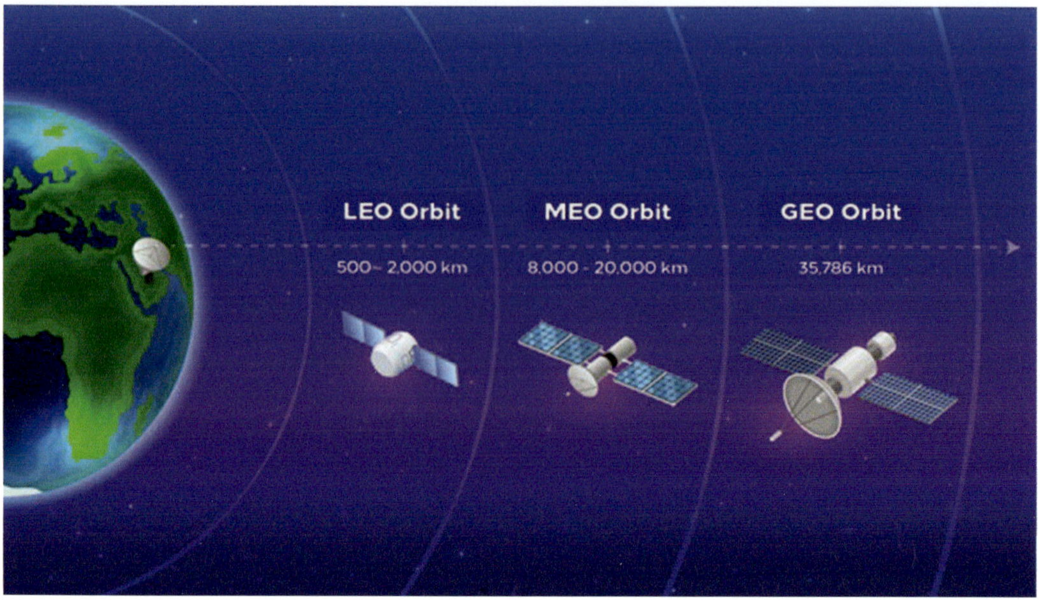

FIGURE 12.4 Orbits of satellite.

Source: **Guerra-Pujol, Wikipedia.**

TABLE 12.4
Advantages and Disadvantages of Geostationary Orbits Satellites

Advantages	Disadvantages
Large Coverage Area: Due to its great height, a satellite in geostationary orbit covers a large geographical area. Only three satellites are required to cover the entire Earth.	*Signal Delay:* The signal requires considerable time to travel from Earth to the satellite and vice versa. The signal travel delay is about 120 ms in one direction. This latency can be problematic for time-critical applications such as real-time voice and video communication.
Continuous Visibility: Satellites in geostationary orbit are visible for 24 hours continuously from a single fixed location on the Earth. This means that ground station antennas do not have to rotate to track them but can be pointed permanently at the position in the sky where the satellites are located.	*Polar Coverage:* Geostationary orbits are located above the equator, making it difficult to provide coverage near the polar regions. This is a significant drawback for locations at high latitudes.
Ideal for Broadcasting and Multi-Point Distribution Applications: Geostationary orbit is ideal for broadcasting and multi-point distribution applications. Communications satellites are often placed in a geostationary orbit so that Earth-based satellite antennas do not have to rotate to track them.	*Signal Strength:* Due to the long transmission distance, the received signal is very weak. This requires better low-noise amplifiers (LNAs) and advanced signal processing algorithms in the satellite modem, increasing the cost of ground station equipment.
No Need for Ground Station Tracking: Ground station tracking is not required as the satellite is continuously visible from Earth all the time from a fixed location.	*High Altitude:* The high altitude of geostationary orbits (approximately 35,786 km above Earth's equator) makes satellite installation quite costly.
No Inter-Satellite Handoff Needed: Inter-satellite handoff is not needed, which simplifies the communication process.	*Limited Coverage at High Latitudes:* Geostationary satellites provide poor coverage at higher latitude places, usually greater than 77 degrees.
Less Complex Receivers: Almost there is no Doppler shift and hence less complex receivers can be used for the satellite communication.	

Notes on Inter-Satellite Handoff: As satellites move quickly across the sky, the connection to a ground station or another satellite can shift from one satellite to another. This shift is called an inter-satellite handoff.

However, LEO also has its own set of challenges. For instance, many satellites are needed to cover the entire region of the Earth, making the installation of such systems costly. Also, LEO satellites move constantly, requiring service to be handed off by each satellite to the next one in the constellation. Despite these challenges, LEO satellites, particularly those in constellations like SpaceX's Starlink, are becoming increasingly popular for providing internet services, especially in remote areas.

12.1.1.5 Fiber Optic Cables

Fiber optic cables (Figure 12.5) are used for high-speed data transmission. They use light signals to transmit data, which allows for faster speeds and higher bandwidth compared to traditional copper cables. Fiber optic cables are commonly used for internet, telephone systems, and cable TV services. Fiber optic cables are a significant advancement in the field of telecommunications and data networking, providing a leap in bandwidth and transmission speed capabilities compared to traditional copper cables. They have become the backbone of the world's internet, telephone systems, and cable TV services.

The fundamental principle behind fiber optic cables is the transmission of data as light signals, rather than electrical signals used in copper cables. This is achieved using optical fibers, which are thin strands of glass or plastic, about the thickness of a human hair. Each optical fiber consists of two parts: the core and the cladding. The core is the inner part where the light signal travels, and the cladding is the outer part that traps the light within the core using a principle of physics called total internal reflection. This allows the light to travel great distances with minimal loss.

The process of data transmission begins with the conversion of electronic data into light signals. This is done by a device known as a transmitter, which uses a light source, usually a laser or an LED, to generate the light signal. The light signal is then injected into the optical fiber. As the light travels

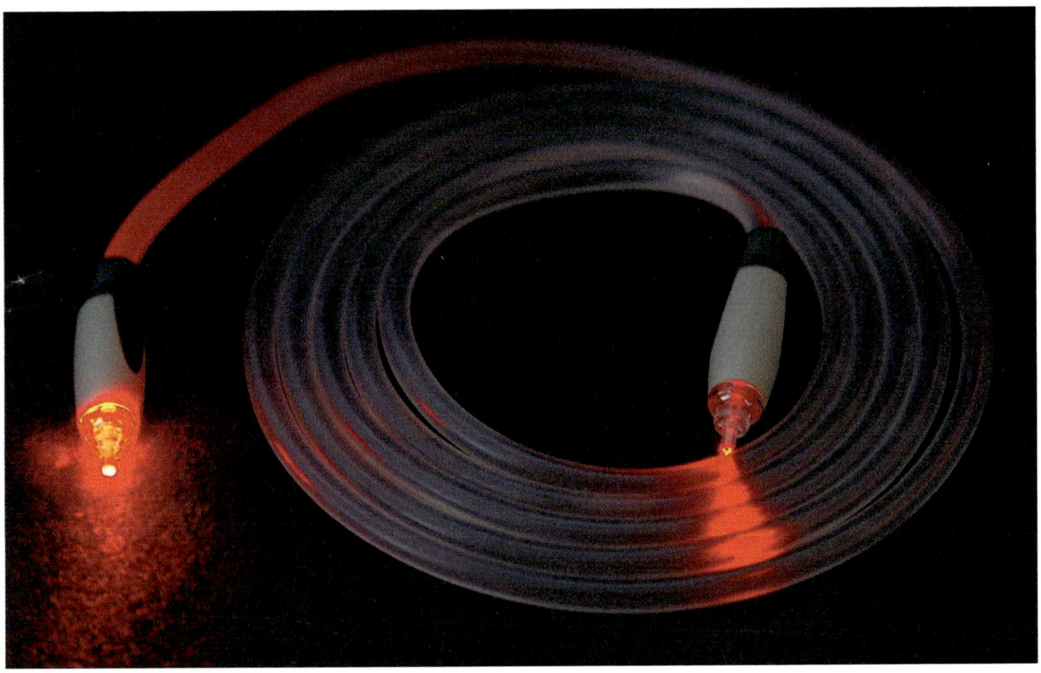

FIGURE 12.5 Fiber optic cable.

Source: **Hustvedt, Wikipedia.**

along the fiber, it bounces off the walls of the core, continually being reflected back into the core due to the difference in refractive indices between the core and the cladding. This allows the light to travel great distances with minimal signal loss. Once the light signal reaches its destination, it is converted back into electronic data by a device known as a receiver. This data can then be used by computers, telephones, televisions, or other electronic devices.

One of the key advantages of fiber optic cables is their high data transmission speed. Because they use light signals, which travel at the speed of light, they can transmit data much faster than copper cables, which transmit data as electrical signals. This makes fiber optic cables ideal for applications that require high-speed data transmission, such as internet services, telephone systems, and cable TV services. Another major advantage of fiber optic cables is their high bandwidth. Bandwidth refers to the amount of data that can be transmitted in each amount of time. Fiber optic cables have a much higher bandwidth than copper cables, allowing them to carry more data at once. This is particularly beneficial for applications that require the transmission of large amounts of data, such as streaming video or downloading large files.

Fiber optic cables also have several other advantages over copper cables. They are less susceptible to interference and signal degradation, allowing for clearer and more reliable data transmission. They are also more secure, as it is difficult to tap into a fiber optic cable without disrupting the signal. Furthermore, because they are made of glass or plastic, they are immune to electrical interference and do not pose a fire hazard.

Despite these advantages, fiber optic cables also have their challenges. They are more fragile than copper cables and require careful handling and installation. They are also more expensive to install, although this cost is often offset by their superior performance and lower maintenance costs.

The fiber optic cables can transmit data as light signals allows for faster speeds and higher bandwidth compared to traditional copper cables, making them an ideal solution for a wide range of applications, including internet, telephone systems, and cable TV services.

Fiber optic cables come in a variety of types, each designed for specific applications. The main types of fiber optic cables are shown in Table 12.5.

TABLE 12.5
Main Types and Configurations of Fiber Optic Cables

	Types
Single mode fiber optic cable	This type of cable has a small core and only allows light to travel in a single path, or mode. It is used for long-distance transmission and can carry signals up to 100 km without the need for amplification.
Multimode fiber optic cable	This cable has a larger core and allows light to travel in multiple paths, or modes. It is used for short-distance transmission, typically less than 2 km.
Plastic optical fiber (POF)	This type of cable uses plastic instead of glass for the core and cladding. It is less expensive and more flexible than glass fiber optic cables, but it has higher signal loss and is typically used for short-distance applications.
	Configurations
Simplex fiber optic cable	This type of cable consists of a single fiber, typically used for applications that require one-way data transfer.
Duplex fiber optic cable	This cable consists of two fibers, allowing for simultaneous, bi-directional data transfer.
Ribbon fiber optic cable	This cable contains multiple fibers organized in a flat, ribbon-like configuration. It is used in applications that require high fiber counts.
Armored fiber optic cable	This cable has an additional layer of protection to prevent damage from harsh conditions, such as rodent attacks, moisture, and extreme temperatures.

Each type of fiber optic cable has unique characteristics and uses, and the choice of cable depends on the specific requirements of the network.

The differences between single-mode and multimode fiber optic cables.

Single-Mode Fiber Optic Cables: These cables are designed to transmit a single beam of light, or mode, through a small diameter core. The technology behind single-mode fiber optic cables dates to the late 1970s when researchers developed the means to produce a fiber with a small enough core for single-mode transmission. Single-mode fiber operates under a narrow range of wavelengths, enabling the transmission of signals over long distances with minimal loss or interference. It's ideal for long-haul signal transmission applications, such as across or between campuses, undersea, or in remote offices. There are essentially no distance limitations.

Multimode Fiber Optic Cables: These cables have a larger core that allows multiple modes or light paths to travel down the fiber at once. This has traditionally limited transmission distance and bandwidth. However, multimode fiber is preferred for shorter distances with high bandwidth needs. Because multiple modes/light paths travel down a multimode fiber cable, it only offers high bandwidth over a short distance. When run over longer distances, modal dispersion (distortion) becomes an issue.

Key differences:

- *Core Size:* Single-mode fiber has a smaller core size compared to multimode fiber. This smaller core size allows single-mode fiber to carry light directly down the fiber, reducing light reflection and lowering attenuation.
- *Transmission Distance:* Single-mode fiber can transmit signals over much longer distances than multimode fiber. This is because multimode fiber experiences more modal dispersion, which can result in distortion and reduced signal quality over long distances.
- *Cost:* Currently, single-mode fiber is typically less expensive than multimode fiber. However, the total cost can depend on other factors, such as the cost of the transceivers.
- *Applications:* Single-mode fiber is typically used for long-distance, high-speed transmissions, while multimode fiber is often used for shorter distances, such as within a building or on a campus.

The choice between single-mode and multimode fiber depends on the specific requirements of the application, including distance, bandwidth, and budget constraints.

The maximum distance of a single-mode fiber link can be estimated by considering the optical budget of the link, the loss per kilometer of the fiber, and the losses introduced by connectors and splices. The formula is shown below.

$$\text{Fiber Length} = \frac{(\text{Optical Budget} - \text{Link Loss})}{\text{Fiber Loss/km}} \tag{12.1}$$

where:

Optical Budget: This is the maximum allowable optical power loss (in dB) that a signal can experience to maintain acceptable performance. It is determined by the difference between the transmitter power and the receiver sensitivity.

Link Loss: This is the total loss (in dB) introduced by connectors, splices, and other components in the link.

Fiber Loss/km: This is the signal loss (in dB) per kilometer of the fiber. It depends on the fiber type and the operating wavelength.

To provide a better picture, here is a hypothetical example to compute the fiber length, given losses:

EXAMPLE 12.1

Suppose we have a fiber optic link with the following characteristics:

- *Optical Budget:* 20 dB (This is the difference between the transmitter power and the receiver sensitivity.)
- *Link Loss:* 2 dB (This is the total loss introduced by connectors, splices, and other components in the link.)
- *Fiber Loss/km:* 0.25 dB/km (This is the signal loss per kilometer of the fiber.)

We can substitute these values into the formula (12.1):

$$\text{Fiber length} = \frac{(\text{Optical budget} - \text{Link loss})}{\text{Fiber loss/km}}$$

$$\text{Fiber length} = \frac{(20\text{ dB} - 2\text{ dB})}{0.25\text{ dB/km}} = 72\text{ km}$$

This means that, under these conditions, the maximum distance that the single-mode fiber link can support is approximately 72 km. This is a simplified calculation, and actual results can vary based on a few factors including the quality of the fiber and connectors, the type and number of network devices in the link, and environmental conditions.

12.1.1.6 Wireless Transmitters

Wireless transmitters are used to send data over radio waves. They are used in Wi-Fi (Wireless Fidelity) networks, mobile networks, and satellite communication systems. Wireless transmitters allow devices to connect to the internet without the need for physical cables. Wireless transmitters are a fundamental component of modern communication systems, enabling the transmission of data over radio waves. They are used in a variety of applications, including Wi-Fi networks, mobile networks, and satellite communication systems. The primary advantage of wireless transmitters is their ability to allow devices to connect to the internet without the need for physical cables.

A wireless transmitter works by converting data into a radio signal, which can then be sent over the airwaves. This process begins with the data to be transmitted, which could be anything from a simple text message to a high-definition video stream. This data is then converted into a format that can be transmitted wirelessly, typically through a process known as modulation. Modulation involves varying one or more properties of a carrier signal, such as its amplitude, frequency, or phase, in accordance with the data to be transmitted. This modulated signal can then be sent over the airwaves using an antenna. The specific frequency of the carrier signal determines the range and speed of the transmission, with higher frequencies generally providing faster speeds but shorter ranges.

Wireless transmitters are used in a variety of applications. In Wi-Fi networks, for example, a wireless transmitter in a router sends data to devices such as laptops, smartphones, and smart TVs. The devices receive the data using a wireless receiver, which demodulates the signal to extract the original data. In mobile networks, wireless transmitters are used in cell towers to send data to and from mobile devices. Each cell tower covers a specific area, known as a cell, and as a device moves from one cell to another, its connection is handed off from one tower to another. This allows for seamless connectivity, even when on the move. In satellite communication systems, wireless transmitters are used to send data to satellites, which then relay the data to receivers on the ground. This allows for global coverage, making satellite communication systems particularly useful in remote areas where traditional wired or wireless networks are not available.

One of the key advantages of wireless transmitters is their ability to allow devices to connect to the internet without the need for physical cables. This provides a great deal of flexibility, as devices can connect from virtually anywhere within range of a transmitter. It also eliminates the cost and complexity of installing and maintaining physical cables. However, wireless transmitters also have their challenges. The range and speed of wireless transmissions can be affected by factors such as physical obstacles, interference from other devices, and the distance between the transmitter and receiver. Security is also a concern, as wireless transmissions can be intercepted by unauthorized devices.

Despite these challenges, ongoing advancements in technology are continually improving the performance and security of wireless transmitters. Techniques such as beamforming, which focuses the wireless signal in a specific direction, and multiple-input multiple-output (MIMO), which uses multiple antennas to improve signal quality, are helping to increase the speed and range of wireless transmissions. Meanwhile, encryption technologies are helping to secure wireless transmissions against interception.

Wireless transmitters enable the transmission of data over radio waves. They provide a flexible and convenient means of connecting devices to the internet, making them an integral part of Wi-Fi networks, mobile networks, and satellite communication systems.

Wireless and wired communication systems are the two primary ways in which data is transmitted and received. Both have their unique characteristics as shown in Table 12.6.

Wired communication involves the use of physical cables and wires to transmit data. The data is typically sent as electrical or optical signals. Wireless communication involves the use of radio waves to transmit data. The data is sent as electromagnetic waves.

Wi-Fi is a wireless networking technology that allows devices to connect to the internet or each other without the need for physical cables. The way Wi-Fi works is shown in Table 12.7.

TABLE 12.6
Wired and Wireless Communications Characteristics

	Wired Communications
Speed	Wired networks generally offer faster data transmission speeds than wireless networks.
Reliability	Wired networks are less prone to interference and offer more stable connections
Security	Wired networks are more secure as they are harder to intercept.
Range	Wired networks are used for short-distance communication because as the length of the wire increases the received signal quality decreases.
Complexity	Implementation of a wired communication system is less complex than a wireless communication system.
	Wireless Communications
Mobility	Wireless networks provide more freedom and flexibility as devices can connect from virtually anywhere within range of a transmitter.
Ease of installation	Wireless networks are quicker to install and offer more layout options.
Range	Wireless networks have a greater range than wired ones.
Interception	Wireless signals can be intercepted by anyone with the right equipment.
Complexity	A wireless communication system is more complex in nature than a wired communication system.

TABLE 12.7
The Works of Wi-Fi

Data conversion	The process begins with data that needs to be transmitted over the network. This data is converted into a radio signal by a device known as a wireless transmitter. This device could be a wireless router, a smartphone, or a computer.
Signal transmission	The wireless transmitter sends the radio signal through the air. This is done using an antenna that is part of the wireless transmitter.
Signal reception	The radio signal is received by a wireless receiver, which could be in another computer, a smartphone, or any other device that is connected to the network.
Data extraction	The wireless receiver converts the radio signal back into data. The device can then use this data for a variety of applications, such as browsing the internet or streaming video.

Wi-Fi operates on a set of wireless communication standards developed by the Institute of Electrical and Electronics Engineers (IEEE). These standards are also known as IEEE 802.11. The specific standard used can affect the speed and range of the Wi-Fi network.

One of the key advantages of Wi-Fi is its flexibility. Devices can connect to a Wi-Fi network from virtually anywhere within range of the wireless transmitter. This makes Wi-Fi an ideal solution for creating networks in homes, businesses, and public places. However, Wi-Fi networks also have their challenges. The range and speed of the network can be affected by factors such as physical obstacles and interference from other devices. Security is also a concern, as unauthorized devices may attempt to connect to the network.

Under the IEEE 802.11, which is the set of standards that define communication for wireless local area networks (WLANs), there are two sets of 802.11s, which are 802.11ac (Wi-Fi 5) and 80211ax (Wi-Fi 6).

802.11ac (Wi-Fi 5): This is a wireless standard that was developed between 2008 and 2013 and finalized in late 2013. It aimed to provide high-throughput connectivity across the 5 GHz band. Wi-Fi 5 improves the earlier 802.11n wireless standard and the MIMO technology introduced by Wi-Fi 4 with a major enhancement named MU-MIMO (Multi-users MIMO). MU-MIMO gives multiple devices the ability to transmit to several clients simultaneously instead of just one or to transfer data to a network client using multiple data streams at once, thus increasing the speed of the transfer.

802.11ax (Wi-Fi 6): This is the most recent wireless standard that went live in 2019. It is set to replace 802.11ac as the de facto wireless standard. Wi-Fi 6 maxes out at 10 Gbps, uses less power, is more reliable in congested environments, and supports better security. It builds on the strengths of 802.11ac, adding efficiency, flexibility, and scalability that allows new and existing networks increased speed and capacity with next-generation applications.

The specific standard used can affect the speed and range of the Wi-Fi network. The choice between these Wi-Fi standards depends on the specific requirements of the network, including factors such as speed, range, and cost.

Upgrading to Wi-Fi 6 (802.11ax) offers several compelling benefits, as shown in Table 12.8.

While upgrading to Wi-Fi 6 can offer these benefits, to take full advantage of Wi-Fi 6, both the router and the devices connecting to it need to support the standard.

12.1.1.7 Network Infrastructure

Network infrastructure is the backbone of any communication system. It comprises all the hardware and software resources that facilitate network connectivity, communication, operations, and management. This infrastructure is crucial for providing a communication path and services between users, processes, applications, services, and external networks or the internet.

TABLE 12.8
Benefits of Wi-Fi 6

Faster speeds	Wi-Fi 6 can theoretically transfer data at up to 10 Gbps, compared with max speeds of around 3.5 Gbps for 802.11ac (Wi-Fi 5). This means you can enjoy faster download and upload speeds, smoother streaming of high-definition content, and more responsive online gaming.
Increased capacity	Wi-Fi 6 is designed to handle more devices at once. It uses technologies like Orthogonal Frequency-Division Multiple Access (OFDMA) and Multi-User Multiple-Input Multiple-Output (MU-MIMO) to manage traffic between devices more efficiently. This is particularly beneficial in environments with many connected devices, such as smart homes.
Improved performance in crowded areas	Wi-Fi 6 performs better in crowded areas with lots of devices, such as apartment buildings and public spaces. It aims to relieve network congestion and provide greater client capacity.
Better battery life	Wi-Fi 6 introduces a feature called Target Wake Time (TWT), which allows devices to determine when they will normally wake up to begin sending and receiving data. This can extend the battery life of mobile devices such as smartphones and tablets, as well as battery-powered smart home devices.
Enhanced security	Wi-Fi 6 provides WPA3 encryption, making it a more secure protocol than its predecessor, Wi-Fi 5.
Greater range	Wi-Fi 6 provides better coverage and stronger signal strength at greater distances.
Futureproofing	As more and more devices start supporting Wi-Fi 6, upgrading your router now can help ensure that you are ready for the future.

TABLE 12.9
Physical Devices of a Network

Routers	These devices connect multiple networks and direct network traffic between them.
Switches	These devices connect multiple devices on a network and use packet switching to receive, process, and forward data to the destination device.
Servers	These are powerful computers that provide services to other computers (clients) over a network.
Firewalls	These devices provide security for the network by controlling incoming and outgoing network traffic based on predetermined security rules.
Wireless access points	These devices allow wireless devices to connect to the network.
Cables	Various types of cables (like Ethernet cables, coaxial cables, and fiber optic cables) are used to connect network devices and transmit data.

12.1.1.7.1 Hardware Resources

The hardware components of a network infrastructure are the physical devices that facilitate the network's operation, as shown in Table 12.9.

12.1.1.7.2 Software Resources

The software components of a network infrastructure are the programs and applications that run on the hardware and enable network functionality. These fundamental software resources are generically shown in Table 12.10.

12.1.1.7.3 Connectivity, Communication, Operations, and Management

The network infrastructure enables connectivity by linking devices and allowing them to communicate. It uses protocols, which are sets of rules that govern data transmission, to ensure effective communication. Operations involve the day-to-day running of the network. This includes managing

TABLE 12.10
Generic Names and Commercial Brands of Required Software

Generic Names of Software	Commercial Brands (Examples)
Network Operating Systems (NOS): These are specialized software that manage network resources, provide network services, and coordinate network activities.	• Microsoft Windows Server • Linux (e.g., Red Hat Enterprise Linux, Ubuntu Server, CentOS) • Cisco IOS (Internetwork Operating System) • Juniper Junos OS • Arista EOS (Extensible Operating System)
Network Management Software: This software helps monitor, control, and manage the network.	• Cisco Prime Infrastructure • VMware vRealize Network Insight • SolarWinds Network Performance Monitor • Nagios Core/Nagios XI • PRTG Network Monitor
Firewall and Security Software: This software helps protect the network from threats and attacks.	• Cisco Adaptive Security Appliance (ASA) • Palo Alto Networks Firewall • Fortinet FortiGate • Check Point Firewall • Juniper SRX Series Services Gateways
Virtualization Software: This software allows the creation of virtual machines, networks, storage, and applications.	• VMware vSphere (ESXi and vCenter) • Microsoft Hyper-V • Citrix Hypervisor (formerly XenServer) • Red Hat Virtualization • Oracle VM VirtualBox

network traffic, maintaining network performance, and ensuring network availability. Management involves overseeing the network's operation to ensure it meets the organization's needs. This includes tasks such as planning, coordinating, and monitoring network activities, as well as troubleshooting network issues.

12.1.1.7.4 Communication Path and Services

The network infrastructure provides the communication path between users, processes, applications, services, and external networks or the internet. This path is facilitated by the network's hardware and software resources. Services provided by the network infrastructure include data storage and retrieval, application hosting, email services, internet access, and many others.

12.1.1.8 Transmission Media

Transmission media, in the context of computer networking, refers to the physical or wireless means through which data is transmitted from one device to another. It serves as the pathway or medium that carries the information from the sender (source) to the receiver (destination).

12.1.1.8.1 Types of Transmission Media

Transmission media can be broadly classified into two categories: guided (wired) and unguided (wireless), as shown in Table 12.11.

12.1.1.8.2 Role of Transmission Media in Data Communication

In the world of data communication, the role of transmission media cannot be overstated. These physical or wireless pathways serve as the critical conduits through which data is transmitted from one device to another, enabling the seamless exchange of information across various networks and applications.

TABLE 12.11
Transmission Media

***Guided Media:* In guided media, signals are transmitted through a physical medium.**	
Twisted pair cable	This is the most common type of guided media. It consists of two conductors twisted together to cancel out electromagnetic interference (EMI) from external sources. Twisted pair cables are used in telephone lines and Ethernet networks.
Coaxial cable	This type of cable has a single copper conductor at its center. A plastic layer provides insulation between the center conductor and a metal shield. Coaxial cables are commonly used in cable television networks and Ethernet networks.
Fiber optic cable	This cable uses light to transmit data. It consists of one or more strands of glass, each only slightly thicker than a human hair. The glass strand functions as a waveguide for light signals. Fiber optic cables offer high bandwidth and can transmit data over long distances.
***Unguided Media:* In unguided media, signals are transmitted wirelessly.**	
Radio waves	These are used in radio and television broadcasting, mobile radio communication, and wireless networking technologies like Wi-Fi and Bluetooth.
Microwaves	These are used in point-to-point communication systems, cellular networks, and satellite communication.
Infrared	These are used in remote control devices and short-range wireless communication such as Infrared Data Association (IrDA) devices.

The choice of transmission media has a profound impact on the overall speed, reliability, cost, and security of the data communication process. Each type of transmission medium comes with its own unique characteristics and trade-offs, making it essential for network designers and administrators to carefully consider the specific requirements of their communication system.

12.1.1.9 Internet Service Providers

Internet Service Providers (ISPs) play a vital role in connecting individuals, households, businesses, and organizations to the global internet infrastructure. These companies are responsible for providing the essential services and infrastructure that enable access to the vast and ever-expanding online world. ISPs can take on various forms and organizational structures, each with its own unique characteristics and approaches to delivering internet services.

12.1.1.9.1 Commercial ISPs

The most common type of ISP is the commercial or for-profit provider. These are private companies that operate as businesses, offering internet access and related services to customers for a fee. Commercial ISPs range from large, nationwide telecommunications giants to smaller, regional or local providers. These commercial ISPs are summarized in Table 12.12.

12.1.1.9.2 Community-Owned ISPs

In some cases, internet services are provided by community-owned or municipal ISPs. These are organizations that are owned and operated by local governments, municipalities, or cooperative entities, with the goal of serving the community's internet needs, as shown in Table 12.13.

12.1.1.9.3 Non-Profit and Charitable ISPs

In addition to commercial and community-owned ISPs, there are also non-profit and charitable organizations that provide internet services. These entities are driven by a mission-oriented approach, aiming to address specific social, educational, or community needs, as shown in Table 12.14.

TABLE 12.12
Commercial ISPs

Large telecommunication companies	• These are major players in the telecommunications industry in the United States, such as Verizon, AT&T, Comcast, and Spectrum. • They often have extensive network infrastructure, including fiber-optic cables, cable TV networks, and wireless towers, which they leverage to provide high-speed internet connectivity to their customers. • Large commercial ISPs typically offer a wide range of internet plans, with varying speeds and data caps, catering to the diverse needs of residential, small business, and enterprise customers.
Regional and local ISPs	• Smaller, regional or local ISPs serve specific geographic areas, often focusing on underserved or rural communities. • These ISPs may have a more limited network infrastructure compared to the larger players, but they can provide more personalized customer service and tailor their offerings to the unique requirements of their local market. • Regional and local ISPs can play a crucial role in bridging the digital divide, ensuring that even remote or underserved areas have access to reliable and affordable internet services.

TABLE 12.13
Public and Community-Owned ISPs

Municipal ISPs	• Municipal ISPs are owned and operated by local governments, often as a public utility or as part of a city's infrastructure. • They can leverage taxpayer funding and community resources to build and maintain their own network infrastructure, providing high-speed internet access to residents and businesses within the municipality. • Municipal ISPs are often driven by the mission of ensuring affordable and equitable internet access for all members of the community, especially in areas where commercial ISPs may not have adequate coverage or investment.
Cooperative ISPs	• Cooperative ISPs are owned and governed by their members, who are typically residents or businesses within a specific community or region. • These ISPs are organized as non-profit entities, with a focus on serving the needs of their member-owners rather than maximizing profits. • Cooperative ISPs can leverage community resources and collective investment to build and operate their own network infrastructure, often in areas where commercial ISPs may have limited presence or investment.

Regardless of their organizational structure or ownership model, ISPs play a crucial role in enabling access to the internet, which has become an essential utility in the digital age. The diverse landscape of ISPs, from large commercial providers to community-driven and non-profit entities, ensures that individuals, businesses, and communities have access to the connectivity and services they need to participate fully in the digital economy and society.

12.1.1.9.4 Services Provided by ISPs

ISPs are key facilitators of internet services, offering a range of services to enable online activities:

1. *Internet Access:* ISPs connect users to the internet, typically offering various plans based on speed and data usage.

TABLE 12.14
Non-Profit and Charitable ISPs

Non-profit ISPs	• Non-profit ISPs are organizations that operate without the goal of generating profits, reinvesting any surplus revenue back into the organization or the community it serves. • They may focus on providing affordable or free internet access to underserved populations, such as low-income households, schools, or community centers. • Non-profit ISPs can be funded through grants, donations, or partnerships with government agencies or philanthropic organizations.
Charitable ISPs	• Charitable ISPs are like non-profit ISPs, but they are specifically organized as charitable entities, often with a broader mission of promoting digital inclusion and bridging the digital divide. • They may offer internet services at no cost or at highly subsidized rates, targeting marginalized communities, rural areas, or populations with limited access to affordable internet. • Charitable ISPs may rely on donations, grants, and other forms of funding to sustain their operations and expand their reach.

2. *Internet Transit:* ISPs can also provide internet transit, which involves routing network traffic between different networks.
3. *Domain Name Registration:* ISPs often offer services to register and manage domain names, which are unique names that identify websites on the internet.
4. *Web Hosting:* ISPs can host websites on their servers, making them accessible to internet users.
5. *Colocation:* ISPs may offer colocation services, where users can house their servers and devices in a secure, ISP-managed facility.

Some ISPs offer value-added services such as email services, allowing users to send and receive emails, and browser packages, providing software for navigating the web. These offerings vary by ISP and are designed to enhance the user's internet experience.

12.1.1.9.5 Access Methods

ISPs provide internet access through various channels, including cable, digital subscriber line (DSL), fiber optics, dial-up, and wireless, with most ISPs offering all options. ISPs offer various methods for connecting users to the internet, each with its own advantages and use cases:

- *Cable:* Cable internet is provided via the same coaxial cable system that's used to deliver cable television. It offers high-speed internet services and is widely available in urban and suburban areas.
- *DSL:* DSL provides internet access via telephone lines but does not interfere with telephone use. Its speed varies based on the distance from the switching station.
- *Fiber Optics:* Fiber optic internet uses light signals to transmit data over thin strands of glass, providing very high-speed internet connectivity. However, its availability is limited as it requires specific infrastructure.
- *Dial-Up:* Dial-up is an older technology that connects users to the internet via telephone lines. It offers slower speeds and ties up the phone line during use.
- *Wireless:* Wireless internet, or Wi-Fi, provides internet access without cables, allowing devices to connect to the internet via a wireless network.

Most ISPs offer multiple options to cater to different customer needs and network conditions. The choice depends on factors like required speed, availability, and cost.

12.1.1.9.6 Charges

ISPs usually charge a fee for their services, which can be based on the speed of the internet connection, the amount of data used, or a combination of both.

12.1.1.9.7 Role in the Internet Economy

ISPs make it possible for their customers to surf the web, shop online, conduct business, and connect with family and friends. Here are how they enable various online activities:

- *Surfing the Web:* ISPs provide internet connectivity that allows users to browse websites, read online articles, watch videos, and more.
- *Online Shopping:* ISPs enable access to online marketplaces and e-commerce websites, allowing users to shop from the comfort of their homes.
- *Conducting Business:* ISPs provide the infrastructure necessary for online business operations, including email communication, video conferencing, cloud services, and more.
- *Connecting with Family and Friends:* ISPs enable social connectivity by providing access to social media platforms, messaging apps, and video calling services.

ISPs act as the gateway to the internet, facilitating a wide range of online activities and making the digital world accessible to their customers.

12.1.1.9.8 Evolution

The evolution of internet access has been transformative. In the early days, internet access was limited to a select few, primarily those with university or government affiliations. This was largely due to the high costs and complex infrastructure required to support internet connectivity. However, advancements in technology and infrastructure, coupled with decreasing costs, have democratized internet access. Today, nearly everyone can access the internet, whether through paid services or free public access points. This has opened a world of information and opportunities, enabling people to learn, work, socialize, and entertain themselves online.

The method of internet access has also evolved significantly. Early internet users relied on dial-up connections, which used existing telephone lines to connect to the internet. While dial-up was revolutionary at the time, it was slow and tied up the phone line. Today, most users have access to high-speed broadband technology, such as DSL, cable, fiber optics, and wireless networks. These technologies provide much faster and more reliable internet access compared to dial-up.

The evolution of internet access has been marked by increasing availability and improving technology. From its origins as a tool for the select few to its current status as a global resource, the internet has truly transformed the way we live, work, and communicate.

Satellite internet providers are instrumental in extending internet access to remote and hard-to-reach areas. In regions where traditional infrastructure like cable or fiber-optic internet is not feasible or cost-effective, satellite internet emerges as a beacon of hope.

Satellite internet works by transmitting data between our device and the internet via satellites orbiting the Earth. Users require a satellite dish installed at their location, which sends and receives signals from a geostationary satellite in space. The satellite transmits data to a network operations center (NOC) on Earth, which is then connected to the internet.

The benefits of satellite internet in remote areas are numerous:

- *Global Coverage:* Satellite internet can reach rural and remote areas where traditional broadband services are unavailable.
- *Improved Connectivity:* For regions with limited or no internet access, satellite internet provides a critical link to the digital world.

- *Reliable Service:* Unlike terrestrial internet connections, satellite internet is less susceptible to outages caused by natural disasters or infrastructural issues.
- *Quick Deployment:* Setting up satellite internet is faster and less intrusive than laying cables or fiber-optic lines.

Satellite internet providers bridge the digital divide, enabling communication, access to information, and participation in the global digital community, regardless of geographical constraints.

The *Digital Divide* refers to the gap between individuals, households, businesses, and geographic areas at different socio-economic levels with regard to their opportunities to access information and communication technologies (ICTs). It includes the imbalance in physical access to technology as well as the imbalances in resources and skills needed to effectively participate as a digital citizen. In essence, it is the difference between those who have effective access to digital and information technology (IT), and those who do not. Factors causing this divide can be classified into three categories: global divide (difference in ICT access between countries), social divide (disparity within each country), and democratic divide (inequality in ICT use).

The digital divide can lead to further socioeconomic disparities as the world becomes increasingly digitized. Those without access are unable to participate in the digital economy, which includes opportunities for jobs, education, and social participation. Addressing the digital divide requires efforts on multiple fronts. Infrastructure needs to be improved in areas lacking access, particularly in rural and underprivileged urban areas. Affordability is another key issue, as the cost of devices and internet services can be prohibitive for many. Digital literacy is also crucial, as simply having access to technology is not enough if users do not have the skills to use it effectively.

The digital divide is a complex issue with significant implications for social equality and economic development. While progress has been made, much work remains to ensure that all individuals have equitable access to digital resources and the skills to use them effectively. This is increasingly important as our world becomes more interconnected and reliant on digital technology.

Understanding the issue of digital divide, we know that bridging the digital divide is crucial for several reasons shown in Table 12.15.

Bridging the digital divide is about ensuring equality of opportunity in a digital age. It is about making sure that everyone can participate fully in our increasingly digital society and economy. Without efforts to bridge the digital divide, we risk exacerbating existing social and economic

TABLE 12.15
Reasons Why Bridging Digital Divide are Important

Economic growth	Access to digital technology can drive economic growth and innovation. It opens new markets, creates jobs, and provides new ways of doing business.
Education	Digital tools can enhance learning and provide access to resources and information that might not be available otherwise. This can lead to improved educational outcomes and skills development.
Social inclusion	The internet allows people to participate in social and cultural activities from which they might otherwise be excluded. It also provides platforms for communication and community building.
Access to services	Many essential services, such as healthcare, banking, and government services, are increasingly being provided online. Bridging the digital divide ensures that everyone can access these services.
Empowerment	Digital literacy can empower individuals by giving them the tools to access information, express their views, and make informed decisions.

TABLE 12.16
Key Indicators of Digital Divide

Internet penetration	The percentage of individuals in each population who have access to the internet. This can be further broken down into urban and rural internet penetration rates.
Device ownership	The percentage of individuals who own a device (like a smartphone, tablet, or computer) that can access the internet.
Broadband access	The percentage of households with access to high-speed internet. This is particularly important as many online activities require a high-speed connection.
Digital literacy rates	The percentage of individuals who have the skills to use digital technologies effectively. This can be measured through surveys or assessments.
Usage statistics	Data on how individuals are using the internet can also provide insight. This might include the percentage of individuals using the internet for different activities (like education, work, socializing, accessing services).
Affordability	The cost of internet access and digital devices as a percentage of average income.

inequalities. Therefore, it's not just a matter of fairness, but also a crucial factor for sustainable development and social cohesion.

Progress in bridging the digital divide can be measured through several key indicators, as shown in Table 12.16.

By tracking these indicators over time and across different populations, we can gain a clearer picture of where progress is being made and where challenges remain. It is also important to consider qualitative factors, such as the impact of digital inclusion on individuals' quality of life. This can be harder to measure but is equally important in understanding the full impact of the digital divide.

12.1.1.10 End-User Equipment

End-user equipment, often referred to as end devices or user devices, are the physical hardware devices that individuals use to access and interact with the internet. These devices include, but are not limited to, desktop computers, laptops, smartphones, tablets, and even smart home devices like Amazon's Echo or Google Home. The primary function of these devices is to provide an interface for users to send and receive data over the internet. This data can be in various forms, including text, images, audio, and video. The devices request information from the internet, process it, and then display it to the user in a human-readable format.

The hardware components of these devices are inevitable in their operation. For instance, the Central Processing Unit (CPU) performs the computations required to run applications and process data. The Random Access Memory (RAM) temporarily stores data that the CPU may need to access quickly. The hard drive or SSD provides long-term storage for data and applications. The network interface card enables the device to connect to the internet, either through a wired connection like Ethernet or a wireless connection like Wi-Fi or cellular data.

The software on these devices is equally important. The operating system, such as Windows, macOS, Android, or iOS, manages the hardware resources and provides services for other software to run. It handles tasks like memory management, process scheduling, and input/output operations. On top of the operating system, there are various applications that users interact with directly. These applications, like web browsers, email clients, and social media apps, make requests to servers on the internet and present the received data to the user.

Web browsers are vital in internet access. They interpret and render HTML, CSS, and JavaScript, the primary languages of the web, into the web pages we see. They also handle user input, like clicking links or submitting forms, and make corresponding requests to web servers.

TABLE 12.17
Security Considerations for End-User Equipment

Anti-virus and anti-spyware software	Install and use anti-virus and anti-spyware software on the computing device, keep software definitions up to date, and run regular scans.
Firewall	Install and enable a hardware and/or software firewall. The primary function of a firewall in computer security is to control and monitor the flow of incoming and outgoing network traffic between a computer or a network and the outside world (the internet or other networks).
Secure file transfer	Use secure methods to securely transfer files to and from the computing device.
Data loss prevention	Implement physical and digital controls for data loss prevention.
Behavior monitoring	Monitor behavior to detect and protect against insider threats.
Data encryption	Encrypt data during transit (email) and rest (device or disk).
Detection and response tools	Use tools for detecting compromised devices and responding to them.
Network Access Control	Implement elevated network access control and device validation.

Smartphones and tablets have additional components that enable internet access. They have cellular radios that can connect to mobile networks, allowing internet access even without Wi-Fi. They also have touchscreens for user input, and various sensors like GPS, accelerometers, and gyroscopes that can provide additional data to applications.

In the context of networking, these end-user devices are often referred to as the "clients" in a client-server model. They initiate requests for data, which are fulfilled by servers. This model is fundamental to how the internet works. When you visit a website, your device (the client) is making a request to a server to send the data for that website.

End-user equipment is a vital component of the internet ecosystem. These devices, equipped with specialized hardware and software, allow users to access, interact with, and contribute to the vast digital landscape of the internet. They are our gateways to the wealth of information and services available online, enabling communication, commerce, entertainment, and much more. As technology continues to advance, we can expect these devices to become even more powerful and versatile, opening up new possibilities for what we can do online.

Security considerations for end-user equipment, also known as endpoint devices, are crucial due to the significant and potentially devastating security risks they can pose. Some primary considerations for security purposes in end-user equipment are reflected in Table 12.17.

These security measures help protect end-user equipment from cyber threats, ensuring the integrity and confidentiality of data. Security is not a one-time task but a continuous process that requires regular updates and monitoring. It is also important to educate users about safe online practices, as human error can often be a weak link in security.

12.1.1.11 IP Address and Domain Names

An IP address is a unique identifier assigned to each device connected to a network. It serves two main purposes: host or network interface identification and location addressing. IP addresses are binary numbers but are most often expressed in human-readable notations, such as **172.16.254.1** (for IPv4), or **2001:db8:0:1234:0:567:8:1** (for IPv6).

There are two versions of IP addresses in use today:

1. *Internet Protocol Version 4 (IPv4):* It is the fourth version of the IP. It uses 32-bit addresses which limits the address space to 4,294,967,296 (2^{32}) possible unique addresses. However, many are reserved for special purposes such as private networks (~18 million addresses) or multicast addresses (~270 million addresses). IPv4 addresses are typically shown in

dot-decimal notation, with four numbers separated by periods, each ranging from 0-255, such as 192.168.1.1.

2. *Internet Protocol Version 6 (IPv6):* It is the most recent version of the Internet Protocol. IPv6 was developed by the Internet Engineering Task Force (IETF) to deal with the long-anticipated problem of IPv4 address exhaustion. IPv6 uses 128-bit addresses, allowing for a total of approximately 3.4×10^{38} unique addresses. IPv6 addresses are represented as eight groups of four hexadecimal digits, each group representing 16 bits (two octets, a group sometimes also called a hextet), separated by colons, such as 2001:0db8:85a3:0000:0000:8a2e:0370:7334.

IP addresses can be either static or dynamic. Static IP addresses are manually set on a device and remain constant until they are manually changed. Dynamic IP addresses are assigned by the network when they connect and can change over time. IP addresses can be either public or private. Public IP addresses are accessible over the internet. Private IP addresses are used within private networks and cannot be directly accessed over the internet.

Public and private IP addresses serve different purposes in the realm of internet connectivity.

Public IP Address:

- A public IP address is a unique IP address assigned to your network router by your ISP and can be accessed directly over the internet.
- It is used to communicate outside the network.
- Public IP addresses are visible to other devices on the internet.
- They are used to interacting and communicating online.
- A public IP address ensures websites, emails, and other online content reach you correctly, acting as a gateway to the internet.
- Public IP addresses can be traced back to the ISP that can easily trace the geographical location.

Private IP Address:

- A private IP address is a unique address that your network router assigns to your device.
- It is used within a private network to connect securely to other devices.
- Private IP addresses are not routed on the Internet, they never leave the local network, and no traffic can be exchanged using private IP addresses.
- Private IP addresses within the same local network must be unique.
- They are only visible to devices that are connected within that network.
- Private IP addresses are more secure than public IP addresses.

The crucial difference between a public and private IP address is that the public IP can be seen by other devices on the Internet, while the private IP cannot[2]. Therefore, public IPs are used to interact and communicate online, while private IPs operate within a local network.

IP addresses are managed globally by the Internet Assigned Numbers Authority (IANA), and by five regional Internet registries (RIRs) responsible in their designated territories for assignment to end users and local Internet registries, such as Internet service providers. The five RIRs are APNIC (Asia-Pacific Network Information Center) with its head quarter location in Brisbane, Australia, RIPE NCC (Réseaux IP Européens Network Coordination Center, Europe), in Amsterdam, the Netherlands, ARIN (American Registry for Internet Numbers, North America), in Chantilly, VA, United States, LACNIC (Latin American and Caribbean Internet Addresses Registry, Latin

America), in Montevideo, Uruguay, and AfriNIC (African Network Information Center, Africa), in Ebène, Mauritius.

Looking at its function, an IP address is a vital component in the infrastructure of any network, including the internet. It allows for devices to communicate with each other, each with a unique identifier, much like a mailing address allows for mail to be sent from one location to another.

There is a strong connection between domain names and IP addresses. This connection is a fundamental aspect of how the internet functions. While IP addresses are the unique numerical identifiers assigned to devices connected to the internet, Domain Names are the human-readable versions of IP addresses, designed to make internet navigation more user-friendly. The assignment and management of IP addresses and domain names are handled by different organizations and authorities. IP addresses are managed and assigned by the IANA, which is the global coordinator for IP address allocation, domain name system (DNS) root zone management, and other internet protocol resources. The DNS and domain name registration are managed by the Internet Corporation for Assigned Names and Numbers (ICANN), a non-profit organization responsible for coordinating the global DNS.

12.1.2 Internet Services and Applications

The internet, a vast network of interconnected devices, has revolutionized the way we live, work, and play. It offers a plethora of services and applications that have become integral to our daily lives and the functioning of modern cities. This includes web browsing, email, video streaming, online gaming, social media, and cloud-based applications. The availability and performance of these services can greatly enhance the quality of life and economic productivity in cities.

12.1.2.1 Web Browsing

Web browsing is one of the most fundamental services offered by the internet. It allows users to access and interact with a vast array of information and services. From reading news articles, researching academic topics, shopping online, to booking tickets for travel or events, web browsing has become an essential part of our daily lives. The performance of web browsing is crucial as it directly impacts the user experience. Faster load times and seamless navigation can significantly enhance the user's experience and productivity.

12.1.2.2 Email

Email is a critical communication tool in both personal and professional settings. It allows for quick and efficient exchange of information, documents, and multimedia files. The performance of email services is vital for timely communication and coordination, especially in business and academic environments.

12.1.2.3 Video Streaming

Video streaming services like Netflix, YouTube, and Amazon Prime have transformed the entertainment industry. They provide on-demand access to movies, TV shows, documentaries, and user-generated content. The performance of these services, particularly the streaming quality and buffering times, significantly affects the viewing experience.

12.1.2.4 Online Gaming

Online gaming is a popular form of entertainment that also fosters social interaction. Games range from simple puzzles to complex multiplayer online games. The performance of online gaming services, particularly latency and server uptime, can greatly impact the gaming experience.

12.1.2.5 Social Media

Social media platforms such as Facebook, Twitter, and Instagram have become central to our social lives. They allow us to connect with friends and family, share experiences, and engage in public discourse. The availability and performance of these platforms are crucial for maintaining social connections and participating in the digital public sphere.

12.1.2.6 Cloud-Based Applications

Cloud-based applications such as Google Docs, Microsoft 365, and Dropbox allow users to create, store, and share documents and files on the internet. These applications enable collaboration and remote work, enhancing productivity in businesses and educational institutions.

The availability and performance of these internet services and applications are not just conveniences; they are essential components of modern urban life. They enable access to information, facilitate communication, provide entertainment, and support economic activities. In cities, where the pace of life is fast and the demand for these services is high, their performance can greatly impact the quality of life and economic productivity.

However, there are also challenges against the provision of these services. The internet requires robust and reliable internet infrastructure, including high-speed broadband connections, data centers, and cloud services. Furthermore, issues such as data privacy and security, digital literacy, and the digital divide also need to be addressed to ensure that these services can be accessed and used safely and equitably.

12.1.2.7 Internet Outages

Internet services and applications are susceptible to outages, which can have a significant impact on users and businesses worldwide. For instance, a major internet outage impacted websites and apps around the world due to a widespread failure in a major content delivery network. Such incidents highlight the importance of robust and reliable internet infrastructure.

12.1.2.8 Research and Development

There is ongoing research and development aimed at improving the performance, security, and usability of internet services and applications. For example, researchers are exploring self-adaptive architectures in Internet of Things (IoT) systems, interoperability between different connected devices, and the analysis of interconnected mobile IoT devices.

12.1.2.9 News and Updates

Staying updated with the latest news and trends related to internet services and applications can provide insights into new features, security threats, and usage trends. This can help users make informed decisions about using these services and applications.

12.1.2.10 Privacy and Security

As internet services and applications often involve the exchange and storage of personal and sensitive data, privacy and security are of paramount importance. Users should be aware of the privacy policies and security measures of the services and applications they use.

The internet shapes urban life and economic productivity. As we continue to embrace digital technologies, it is crucial to ensure the availability and performance of these services, and to address the challenges that come with them, to create connected, resilient, and inclusive cities.

Despite its challenges, internet has also encouraged or generated some emerging trends in its services as shown in Table 12.18.

TABLE 12.18
Some Emerging Trends in Internet Services

AI	AI continues to rank at the top of emerging tech. Generative AI, which can generate text based on a prompt, is gaining attention. Explainable AI, a subset of AI that provides logic and reasoning for its conclusions, is also an emerging trend.
IoT	IoT devices are becoming increasingly used across various industries. The expansion of IoT devices is largely due to the increasing bandwidths we are seeing with 5G.
Web3	Web3, or the decentralized internet, is an emerging trend that aims to create a user-centric online world where users own their data.
Tokenization	Tokenization is the process of converting rights to an asset into a digital token on a blockchain.
Digital–physical fusion	This trend involves the integration of digital and physical elements to create new user experiences.
Human augmentation	This refers to the use of technology to enhance human abilities.
Quantum computing	Quantum computing uses the principles of quantum mechanics to process information. It has the potential to solve complex problems that are currently beyond the capabilities of classical computers.
Blockchain	Blockchain technology provides a decentralized and secure way of recording transactions.

TABLE 12.19
Generations of the World Wide Web

Web1 (1990s)	• The first iteration of the World Wide Web, often referred to as "Web1," was focused on static, read-only web pages. • This phase was characterized by basic HTML websites with limited interactivity and user-generated content.
Web2 (2000s–present)	• The current phase, often called Web2, is the era of the modern internet that we are in now. • Web2 is characterized by the rise of social media, user-generated content, and centralized platforms like Google, Facebook, and Amazon. • In Web2, a small number of tech companies have gained significant control over user data, content, and online experiences.
Web3 (future)	• The term Web3 emerged to describe the next evolution of the internet, aimed at addressing the issues and centralization of Web2. • Web3 envisions a more decentralized, transparent, and user-centric internet, enabled by technologies like blockchain, dApps, and cryptocurrencies.

Amid current centralized internet, in which data is controlled by a few large tech companies, Web3 offers a more decentralized system. The term Web3 originated from the concept of the three different phases or generations of the World Wide Web, as shown in Table 12.19.

The concept of Web3 was first introduced in the early 2010s, particularly within the cryptocurrency and blockchain communities, to describe a new, more decentralized iteration of the internet. The term Web3 was popularized by Ethereum co-founder Gavin Wood in 2014, who used it to refer to a vision of a decentralized internet built on blockchain technology, smart contracts, and decentralized applications (dApps). Since then, the concept of Web3 has gained significant attention and traction, with major tech companies, startups, and the broader crypto community exploring and investing in the development of this next generation of the web. While the specifics of Web3 are still evolving and subject to ongoing debate, the term has become a widely recognized shorthand for the vision of a more decentralized, user-empowered, and blockchain-based internet.

TABLE 12.20
Some Important Aspects of Web3

Decentralization	• Web3 advocates for a shift away from the current centralized model of the internet, where a few large tech giants control and manage vast amounts of data and online services. • The goal is to create a more decentralized internet, where power and control are distributed across a network of individual users and nodes, rather than being concentrated in the hands of a few.
Blockchain and cryptocurrencies	• Web3 heavily relies on blockchain technology, which enables the creation of dApps and the use of cryptocurrencies and digital assets. • Blockchain provides a secure, transparent, and tamper-resistant way to record and validate transactions and data, without the need for a central authority. • Cryptocurrencies, such as Bitcoin and Ethereum, are integral to Web3, as they enable peer-to-peer transactions and the creation of decentralized financial systems.
Ownership and control	• In Web3, the focus is on giving users more control and ownership over their data, digital assets, and online activities. • This includes the ability for users to own and control their digital identities, social media profiles, online content, and other digital possessions, without relying on centralized platforms.
Transparency and trust	• Web3 aims to increase transparency and trust on the internet by leveraging technologies like blockchain, which provide a transparent and tamper-resistant record of transactions and data. • This can help address issues such as data privacy, censorship, and the concentration of power in the hands of a few tech giants.
dApps	• Web3 enables the development of dApps that run on blockchain networks, rather than on centralized servers • These dApps can cover a wide range of use cases, from decentralized finance (DeFi) to social media, gaming, and beyond.

Web3 is a concept that refers to the next evolution of the internet, which aims to decentralize the web and reduce the control and influence of large tech companies and centralized authorities. Some important aspects of Web3 are shown in Table 12.20.

While Web3 is still in its early stages, the concept has gained significant attention and investment from the tech industry, startups, and the broader cryptocurrency and blockchain community. The potential benefits of Web3, such as increased user control, privacy, and transparency, have sparked widespread interest and ongoing development efforts to realize this vision of the internet's future.

The internet and its trends are shaping the future of internet services, offering new possibilities for enhancing efficiency, improving user experiences, and addressing challenges such as privacy and security. However, they also present new challenges and ethical considerations that need to be carefully managed.

12.1.3 E-Commerce

E-commerce, or electronic commerce, refers to the buying and selling of goods and services over the internet. It has revolutionized the way businesses operate and consumers shop, offering greater convenience, choice, and opportunities for both parties.

12.1.3.1 Online Retailers

Online retailers are businesses that sell goods directly to consumers via the internet. These can range from large multinational companies like Amazon and Alibaba, to small independent businesses operating through platforms like Etsy or Shopify. Online retailers offer consumers the convenience of shopping from the comfort of their homes, with a wide range of products available at their fingertips. They also provide businesses with a cost-effective way to reach a global customer base, without the need for physical storefronts.

12.1.3.2 Marketplace Platforms

Marketplace platforms are websites or apps that facilitate e-commerce transactions between buyers and sellers. They provide a platform for sellers to list their products, and for buyers to browse and purchase these products. Examples include eBay, Amazon Marketplace, and Facebook Marketplace. These platforms benefit consumers by offering a wide variety of products from different sellers, often at competitive prices. For businesses, they provide access to a large customer base and tools to manage their online sales.

12.1.3.3 Digital Payment Systems

Digital payment systems are methods of making transactions over the internet. These include credit and debit cards, digital wallets like PayPal and Apple Pay, QR payments and cryptocurrencies like Bitcoin. Digital payment systems offer consumers a fast, convenient, and often secure way to pay for their online purchases. For businesses, they provide a way to accept payments from customers worldwide and can offer features like fraud protection and dispute resolution.

12.1.3.4 Logistics and Delivery Services

Once a purchase is made online, the product needs to be delivered to the consumer. This is where logistics and delivery services come in. These services, which include courier companies, postal services, and delivery apps, ensure that products are picked up from the seller and delivered to the buyer in a timely manner. They are a critical part of the e-commerce process, with fast and reliable delivery often being a key factor in customer satisfaction.

E-commerce offers numerous benefits for consumers. It provides a convenient way to shop, with the ability to browse and purchase products at any time, from anywhere. It offers a wide range of products, often at competitive prices, with detailed product information and customer reviews to aid in decision-making. It also allows for easy price comparison, personalized recommendations, and the ability to shop from international retailers.

For businesses, e-commerce opens up new opportunities for growth and profitability. It allows businesses to reach a global customer base, without the geographical limitations of physical stores. It provides access to customer data, which can be used to personalize marketing efforts and improve products and services. It also allows for cost savings, as businesses can operate with less physical infrastructure and automate various sales and customer service processes. However, e-commerce also presents challenges. These include issues related to privacy and security, the need for reliable internet access and digital literacy among consumers, and the logistical complexities of delivering products to customers worldwide. Businesses also face intense competition in the online marketplace and must continually innovate to attract and retain customers.

E-commerce, with its various components of online retailers, marketplace platforms, digital payment systems, and logistics and delivery services, has transformed the landscape of commerce. It offers consumers greater convenience and choice, while providing businesses with new opportunities for growth and profitability. As technology continues to evolve, e-commerce is likely to continue growing and evolving, shaping the future of buying and selling. Some emerging trends in e-commerce are shown in Table 12.21.

These trends are shaping the future of e-commerce, offering new possibilities for enhancing efficiency, improving user experiences, and addressing challenges such as privacy and security.

TABLE 12.21
Some Emerging Trends in E-commerce

AR shopping experiences	AR enables customers to have immersive shopping experiences from the comfort of their homes. It allows them to visualize products in a real-world context, making informed decisions before making a purchase.
Blockchain for supply chain transparency	Blockchain technology is not new to the e-commerce industry, but its application is evolving. It provides transparency to their shipping clients and improves communication with their contracted owner/operators.
Customized loyalty programs	Personalization has been a hallmark of e-commerce; however, its scope has been historically limited to cross-selling through product recommendations. In 2024, retailers are expected to harness data analytics and AI to offer more personalized content to users and allow users to customize loyalty programs to meet their specific shopping habits and needs.
AI for personalization	AI helps personalize user experience – e-commerce businesses can provide a richer buying experience with this technology.
Hybrid commerce	Hybrid commerce offers a seamlessly connected customer journey – combine your online channels with the brick-and-mortar store experience to improve purchase rate.
CRM	CRM builds a community around your e-Commerce brand – CRM helps you understand your potential customer's needs and behavior to strengthen relationships and improve customer loyalty.
Customer support as a priority	Excellent support positively influences purchasing decisions, cultivates loyalty, promotes a positive brand image, and boosts e-commerce sales.
Social commerce	Strategize social selling to take advantage of its rise – social commerce is cost-effective and encourages in-app purchases.
Subscription models	The subscription model helps customers fulfill their recurring expenses easily – attracts loyal customers, and boosts retention rates and profitability.
Headless commerce	Migrate to headless e-commerce for better ROI – decoupling your e-Commerce site's front and back ends offers more flexibility and the ability to reach customers on any device.
Flexible payment options	Provide multiple payment options to improve user experience – some customers prefer traditional payment methods like bank transfers.

They also present new challenges and ethical considerations that need to be carefully managed. E-commerce, for example, has had a profound impact on traditional commerce in several ways as shown in Table 12.22.

While e-commerce has posed challenges for traditional commerce, it has also provided opportunities for traditional businesses to innovate and reach a larger customer base. The key for traditional businesses is to adapt and find ways to integrate e-commerce into their existing business models. Here are necessary adaptations by traditional retailers to compete with e-commerce, as shown in Table 12.23.

While e-commerce giants pose a significant challenge, there are still many opportunities for traditional retailers to compete and thrive in the digital age.

12.2 TELECOMMUNICATION SYSTEMS

Telecommunication systems are the backbone of modern society, enabling the rapid exchange of information across vast distances. They encompass a wide range of technologies and applications, from traditional telephony to the internet. Telecommunication is the science and practice of transmitting information by electromagnetic means. It involves the transmission of information with an immediacy comparable to face-to-face communication. The term is derived from the Greek prefix "tele-," meaning distant, and the Latin verb "communicare," meaning to share.

TABLE 12.22
Impacts of E-commerce on Traditional Commerce

Increased competition	E-commerce has increased competition as it allows businesses, regardless of their size, to reach a global audience. Traditional brick-and-mortar stores are no longer just competing with other local businesses but with online retailers from around the world.
Changed consumer behavior	With the convenience of shopping anytime and anywhere, consumers have embraced e-commerce, which in turn has changed their shopping habits. Traditional stores have had to adapt by offering more personalized and engaging shopping experiences.
Price transparency	E-commerce has made pricing more transparent. Consumers can easily compare prices online, forcing traditional retailers to be more competitive in their pricing strategies.
Shift in sales strategy	Traditional businesses have had to shift their sales strategies to include online channels. This includes having an online presence through a website or social media and using digital marketing strategies to attract online customers.
Demand for improved customer service	E-commerce platforms provide customer reviews and ratings, which can significantly influence purchasing decisions. This has raised the bar for customer service in traditional commerce as well.
Innovation in delivery and logistics	The growth of e-commerce has led to innovations in delivery and logistics services. Traditional retailers are now offering services like "click and collect," where customers can order online and pick up in-store.
Data-driven decision making	E-commerce generates a wealth of data that can be analyzed to gain insights into customer behavior. This has impacted traditional commerce by making business strategies more data driven.
Reduced operating costs	E-commerce can potentially reduce operating costs by eliminating the need for a physical storefront and reducing manpower costs. This has put pressure on traditional retailers to streamline their operations to remain competitive.

TABLE 12.23
Adaptive Strategies to Compete with E-commerce

Offer something different	Differentiating your store from e-commerce giants can be a key competitive advantage. This could be unique products, superior customer service, or a compelling in-store experience.
Leverage physical stores	Physical stores can offer experiences that online platforms cannot. Retailers can use their stores to provide personal services, product demonstrations, or events.
Create a seamless omnichannel experience	Ensuring that customers have a consistent experience across all channels – online, in-store, mobile – is crucial. Walmart, for example, has done an excellent job of integrating online and in-store sales.
Use data analytics	E-commerce giants excel at using data to understand their customers. Traditional retailers can also use data analytics to gain customer insights and personalize their offerings.
Implement personalized marketing strategies	Personalized marketing can increase customer engagement and sales. This could involve targeted emails, personalized product recommendations, or ads featuring products the customer has shown interest in.
Focus on local	Traditional retailers can use their local knowledge to their advantage. This could involve collaborating with local businesses, using region-specific references in marketing, or offering locally sourced products.
Improve delivery and logistics	Offering fast and reliable delivery can help traditional retailers compete with e-commerce giants. This could involve offering services like "click and collect," where customers order online and pick up in-store.
Provide excellent customer service	Providing excellent customer service can differentiate traditional retailers from their online competitors. This could involve having knowledgeable staff, offering easy returns, or responding quickly to customer inquiries.

A basic telecommunication system consists of three main parts: (1) *Transmitter:* This takes information and converts it into a signal. (2) *Transmission Medium:* Also known as the physical channel, this carries the signal. Examples include wires, optical fibers, and even free space for wireless communication. (3) *Receiver:* This takes the signal from the channel and converts it back into usable information for the recipient.

A telecommunications network is an electronic system of links, nodes, and controls that govern their operations to allow voice and data transfer among users and devices. Examples include telephone networks, computer networks, and the internet. Many transmission media have been used for telecommunications throughout history, from smoke signals and semaphore telegraphs to wires and empty space made to carry electromagnetic signals. Modern technologies generally use electric power and include the telegraph, telephone, television, and radio. Digital transmission is employed to achieve high reliability and because the cost of digital switching systems is much lower than the cost of analog systems. To use digital transmission, however, the analog signals that make up most voice, radio, and television communication must be subjected to a process of analog-to-digital conversion.

Multiplexing is a process that allows for a single medium to transmit several concurrent communications sessions. This increases the efficiency of the communication system by allowing multiple signals to be transmitted over a single communication channel.

Since the 1960s, the proliferation of digital technologies has meant that voice communications have gradually been supplemented by data. The physical limitations of metallic media prompted the development of optical fiber. The internet, a technology independent of any given medium, has provided global access to services for individual users and further reduced location and time limitations on communications.

12.2.1 Fixed Line Telephone

Fixed line telephony, also known as landline telephony, is a method of communication that uses physical media, such as copper or fiber optic cables, to transmit voice and data. Despite the rise of mobile and internet-based communication, fixed line telephony still plays a crucial role in many areas, including business communication, broadband internet access, and emergency services.

12.2.1.1 How It Works

The fixed-line telephone system, also known as the Public Switched Telephone Network (PSTN), is a marvel of human engineering and innovation. It operates on a network of wired connections that span the globe, enabling voice communication between individuals regardless of their geographical location.

The process begins when a person speaks into the telephone's microphone. The microphone is a transducer, a device that converts one form of energy into another. In this case, it converts the mechanical energy of the sound waves produced by the person's voice into electrical energy. This conversion is achieved through a process known as electromagnetic induction. The sound waves produced by the person's voice cause a diaphragm inside the microphone to vibrate. This diaphragm is attached to a coil of wire placed within a magnetic field. As the diaphragm vibrates, it causes the coil to move back and forth within the magnetic field, generating an electrical current. This current is an analog signal, meaning it is a continuous signal that varies in amplitude and frequency in direct proportion to the sound waves.

Once the sound waves have been converted into an electrical signal, they are sent over the network to the recipient's telephone. The network consists of a vast array of wires and cables, switches, and routers that guide the signal to its destination. The signal may be amplified at various points along the way to ensure it reaches its destination without degradation.

TABLE 12.24
Milestones of Telecommunication Systems

Ancient Systems	Before modern telecommunications, people used smoke signals, drums, and homing pigeons for long-distance communication.
Telegraph (1830s)	The first electromagnetic telegraph emerged in the early 1830s, with inventors like Sir William Fothergill Cooke, Sir Charles Wheatstone, and Samuel Morse playing key roles.
Telephone (1870s)	Alexander Graham Bell and Elisha Gray independently designed devices that could transmit speech electrically, marking the birth of the telephone.
Radio and television (late 1800s–early 1900s)	The invention of radio and television enabled wireless transmission of sound and visual information over long distances.
Semiconductor era (mid-1900s)	The invention of semiconductors revolutionized telecommunications, enabling the miniaturization of electronic components and paving the way for digital communications.
First generation (1G) mobile phones (1979)	The Nippon Telegraph and Telephone (NTT) in Japan introduced the 1G of mobile telecommunications, which were still analog.
2G mobile phones (1991)	Launched in Finland, 2G networks introduced digitally encrypted communications, marking the beginning of digital mobile communications.
3G, 4G, and 5G mobile phones (2000s–present)	These generations of mobile telecommunications introduced high-speed internet access, multimedia support, and advanced features like GPS and video conferencing.

Upon reaching the recipient's telephone, the electrical signal is converted back into sound waves. This is achieved using a speaker, which is another type of transducer. The speaker operates in much the same way as the microphone, but in reverse. It receives the electrical signal, which causes a coil within a magnetic field to move back and forth. This movement drives a diaphragm that produces sound waves, which are then heard by the recipient.

The fixed-line telephone system is a testament to the ingenuity of human beings. It is a system that has stood the test of time, evolving and adapting to meet the changing needs of society. Despite the advent of mobile and internet-based communication technologies, it remains an essential part of our global communication infrastructure. Its continued relevance is a testament to the robustness of its design and the fundamental principles on which it operates.

The fixed-line telephone system is a wired connection that enables voice communication between individuals. It operates by converting sound waves into electrical signals, transmitting these signals over a network, and then converting them back into sound waves at the recipient's end. This process, while seemingly straightforward, involves a series of intricate steps and components, each playing a crucial role in ensuring clear and effective communication.

The evolution of telecommunication systems from the telegraph to mobile phones has been a fascinating journey. Some key milestones are reflected in Table 12.24.

This evolution has transformed how we communicate, bringing us closer together and enabling new forms of interaction and collaboration. It is a testament to human ingenuity and the power of technological innovation.

The transition from analog to digital telecommunication marked a significant milestone in the history of communication technology. Analog signals are continuous and vary in time, representing physical measurements. Digital signals are discrete-time signals, formed by digital modulation, representing a sequence of bits. Table 12.25 shows the history of transitions from analog to digital signals.

The transition from analog to digital has brought about numerous benefits, including improved call quality, increased capacity, and the ability to transmit data and voice over the same lines. However, it also presents challenges, particularly for businesses and services that rely on analog

TABLE 12.25
History of Transition from Analog to Digital Signals

Analog systems	Until the early 1980s, the bulk of long-distance transmission was provided by analog systems. In these systems, individual telephone conversations were stacked in 4-kHz intervals across the transmission band, a process known as frequency-division multiplexing (FDM).
Digital systems	With the development of fiber optics, these analog systems were rapidly replaced by digital systems. In digital transmission, telephone signals are first converted from an analog format to a quantized, discrete time format. The signals are then multiplexed together using time-division multiplexing (TDM), a method in which each digitized telephone signal is assigned a specific slot within a fixed time frame.
Codec	A codec, which is a contraction of coder-decoder, converts analog signals into digital signals.
Switch to IP networks	As soon as 2023, analog telephone services in the UK will be switched off and replaced by digital systems using IP technology. Telecommunication suppliers are well on their journey to making this switch to "IP networks."
Analogue switch off 2025	Telecom providers such as British Telecom and Virgin Group UK are in progress of switching off analog communications networks, and this is planned to be completed by 2025.

systems. These organizations need to plan for their migration to digital services, a process that can be resource-intensive and time-consuming

12.2.1.2 Types of Fixed Line Telephone

There are two main types of fixed line telephony: analog and digital. Analog telephony, also known as Plain Old Telephone Service (POTS), uses a continuous signal to represent the voice data. Digital telephony, on the other hand, uses a discrete signal and offers better sound quality and more features.

12.2.1.3 Role in Business Communication

Fixed line telephony has been a great player in business communication. It provides a reliable and secure means of communication, making it ideal for businesses that handle sensitive information. Additionally, fixed line telephony can support advanced features such as conference calling, call forwarding, and automated attendant services.

12.2.1.3.1 Analog Telephony

Analog telephony, also known as POTS, is the original form of telephone service. It uses a continuous signal to represent voice data, which is then transmitted over a network of copper wires. The term "analog" refers to the fact that the signal is analogous to the original sound wave produced by the human voice.

The analog signal is created by converting the sound waves of a person's voice into electrical signals. These electrical signals vary in frequency and amplitude, creating a waveform that is a direct representation of the original sound. This waveform is then transmitted over the telephone line to the receiver, where it is converted back into sound. Analog telephony has been the backbone of telephone communication for over a century. It is simple, reliable, and capable of delivering voice communication over long distances. However, it has its limitations. The quality of the sound can degrade over long distances due to signal loss and interference. Additionally, analog lines can only handle one call at a time, limiting their capacity.

12.2.1.3.2 Digital Telephony

Digital telephony, on the other hand, uses a discrete signal to represent voice data. Instead of a continuous waveform, digital signals consist of binary data – ones and zeros. This binary data is less susceptible to noise and interference, resulting in better sound quality.

TABLE 12.26
Advantages and Disadvantages of Analog and Digital Telephonies

	Advantages	Disadvantages
Analog telephony	• *Simplicity:* Analog systems are generally simpler and easier to use. They do not require complex encoding or decoding processes. • *Reliability:* Analog signals are less susceptible to complete loss, meaning that even in poor conditions, some communication is possible. • *Cost:* Analog devices and infrastructure are typically less expensive than their digital counterparts.	• *Quality:* Analog signals are more susceptible to degradation over long distances and can be affected by physical disturbances like weather and electrical interference. • *Capacity:* Analog lines can only handle one call at a time, limiting their capacity. • *Security:* Analog signals are easier to intercept and eavesdrop on.
Digital telephony	• *Quality:* Digital signals maintain their quality over long distances and are not affected by physical disturbances. They offer clearer voice quality and less noise. • *Capacity:* Digital signals can be compressed, allowing for more calls to be carried over the same line. • *Features:* Digital telephony offers more features like caller ID, call waiting, voicemail, and call forwarding. • *Security:* Digital signals can be encrypted, making them more secure against eavesdropping.	• *Complexity:* Digital systems are more complex due to the need for encoding and decoding processes. • *Cost:* Digital devices and infrastructure can be more expensive than analog ones. • *Power Dependency:* Digital systems rely on a continuous power supply. In case of a power outage, digital phone systems may not work unless they have a backup power source.

The process of converting voice data into digital signals involves sampling the analog signal at regular intervals and then quantizing these samples into a set of discrete values. These values are then encoded as binary data and transmitted over the network. One of the key advantages of digital telephony is its ability to offer more features. Caller ID, call waiting, voicemail, and call forwarding are all made possible because of digital technology. Additionally, digital signals can be compressed, allowing for more calls to be carried over the same line, increasing capacity. Digital telephony also offers better security. As the voice data is encoded as binary, it is much harder to eavesdrop on digital phone calls. Furthermore, digital signals can be encrypted, providing an additional layer of security.

The advantages and disadvantages of both analog and digital telephony are shown in Table 12.26.

In conclusion, while analog telephony has its place, the advantages of digital telephony make it the preferred choice for most modern applications. However, the choice between analog and digital would depend on the specific needs and circumstances of the user.

12.2.1.4 Role in Broadband Internet Access

Fixed line telephony also plays a crucial role in providing broadband internet access. DSL and fiber optic services are types of broadband internet service that use fixed lines to deliver high-speed internet access. These services offer faster speeds and more reliable connections than wireless alternatives.

12.2.1.4.1 Fixed Line Telephony and Broadband Internet Access

Fixed line telephony, which includes both analog and digital phone lines, is important in providing broadband internet access. Broadband internet service refers to high-speed internet access that is always on and faster than traditional dial-up access. It is delivered to homes and businesses through several types of connections, two of which are DSL and fiber optic services.

12.2.1.4.2 Digital Subscriber Line

DSL is a type of broadband internet service that uses existing telephone lines to deliver high-speed internet access. It is called "digital subscriber line" because it converts an analog telephone line into a high-speed digital line. DSL works by utilizing the unused frequency bands on telephone lines to transmit data. This allows for simultaneous voice and data transmission over the same line, meaning you can use the internet and make phone calls at the same time. DSL offers several advantages. It is widely available, as it uses the existing telephone infrastructure. It is also relatively affordable compared to other types of broadband service. However, the speed of DSL service can vary depending on the distance from the service provider's central office. The further you are, the slower the connection.

12.2.1.4.3 Fiber Optic Services

Fiber optic services, on the other hand, use thin strands of glass or plastic, known as optical fibers, to transmit data as pulses of light. Fiber optic services can deliver much higher speeds than DSL, often up to 1 Gbps or more. Unlike DSL, the speed of fiber optic service does not degrade with distance from the service provider's office. This makes it an excellent choice for areas that are far from the central office. Additionally, fiber optic lines can carry more data than traditional copper lines, making them more suitable for today's high-bandwidth applications.

However, fiber optic services are not as widely available as DSL, as they require the installation of new infrastructure. They can also be more expensive than DSL.

12.2.1.4.4 Comparison with Wireless Alternatives

Both DSL and fiber optic services offer faster speeds and more reliable connections than wireless alternatives. While wireless internet services like 4G and 5G offer the convenience of mobility, they can be affected by factors such as signal strength, network congestion, and physical obstructions.

In contrast, fixed line services like DSL and fiber optic provide a dedicated line to each customer, ensuring consistent performance regardless of how many users are on the network. They also offer unlimited data usage, unlike many wireless plans that have data caps.

12.2.1.5 Role in Emergency Services

Fixed line telephony is essential for emergency services. In many countries, emergency calls can be made from a landline without any charge. Furthermore, when a call is made from a landline, emergency services can trace the call to a specific address, enabling them to respond more quickly.

12.2.1.5.1 Fixed Line Telephony and Emergency Services

Fixed line telephony, which includes both analog and digital phone lines, plays a crucial role in emergency services. In case of an emergency, being able to communicate quickly and effectively can be a matter of life and death. This is where fixed line telephony comes in.

12.2.1.5.2 Emergency Calls from Landlines

In many countries, emergency calls can be made from a landline without any charge. This is a critical feature as it ensures that anyone, regardless of their financial situation, can reach out for help in an emergency. The ability to make free emergency calls is not just a feature of fixed line telephony, but a legal requirement in many jurisdictions.

12.2.1.5.3 Call Tracing and Quick Response

One of the key advantages of fixed line telephony in emergencies is the ability to trace the call to a specific address. When a call is made from a landline, the telephone company can identify the physical location of the line. This information can be shared with emergency services, enabling them to respond more quickly. This is particularly important in situations where the caller is unable to speak or does not know their exact location. It also helps in situations where the call gets disconnected. Even if the caller cannot provide their location, emergency services can use the call trace to send help to the right place.

12.2.1.5.4 Reliability of Fixed Line Telephony

Fixed line telephony is also generally more reliable than mobile telephony. Landlines are less likely to suffer from issues like poor signal strength or network congestion that can affect mobile phones. In addition, landlines do not rely on battery power, so they can still be used when the power goes out, provided the phone itself does not require electricity.

The ability to make free emergency calls, the capability to trace calls to a specific location, and the reliability of fixed lines all contribute to their importance. While mobile phones and VoIP services have their place, fixed line telephony remains a critical component of our emergency response infrastructure. As technology continues to evolve, it will be interesting to see how these systems adapt and improve to better serve our emergency communication needs.

12.2.2 Mobile Telephone

This section focuses on mobile phones, which have revolutionized communication by enabling voice calls, text messaging, and internet access on the go. It covers aspects like cellular network infrastructure, mobile devices, and services provided by mobile network operators (MNOs).

12.2.2.1 Mobile Telephony: The Wireless Revolution in Communication

In the span of just a few decades, the world of communication has undergone a profound and far-reaching transformation, and at the heart of this revolution lies the development and widespread adoption of mobile telephony. The humble mobile phone, once a novelty, has now become an indispensable tool that has reshaped the very fabric of how we connect, share information, and access the digital world.

Prior to the emergence of mobile phones, communication was largely tethered to fixed locations – landline telephones, fax machines, and desktop computers. This static model of communication imposed limitations on our ability to stay in touch and access information on the go. However, the rise of mobile telephony has shattered these constraints, ushering in a new era of untethered and ubiquitous communication.

The cellular network infrastructure has enabled the seamless integration of voice, text, and data services across vast geographical areas. By dividing coverage regions into smaller, overlapping cells, each equipped with its own base station, cellular networks have facilitated the dynamic handoff of connections as users move from one location to another. This cell-based architecture, combined with the efficient reuse of radio frequencies (RFs), has allowed mobile networks to accommodate the growing number of users and their increasing demand for connectivity.

Alongside the development of the cellular network, the evolution of mobile devices has been a key driver in the widespread adoption of mobile telephony. From the bulky, analog-based mobile phones of the 1980s to the sleek, feature-rich smartphones of today, the capabilities of these devices have grown exponentially. Modern smartphones are essentially pocket-sized computers, capable of far more than just making voice calls. They can send and receive text messages, access the internet, stream multimedia, and run a vast array of applications that cater to the diverse needs and preferences of users.

The integration of cameras, GPS, and various sensors into mobile devices has further expanded the functionality of these devices, enabling a wide range of mobile services and applications. Whether it is navigating unfamiliar streets, capturing and sharing memorable moments, or accessing real-time information, the capabilities of smartphones have become an integral part of our daily lives.

The proliferation of mobile operating systems, such as iOS, Android, and Windows Phone, has played a crucial role in this transformation. These platforms provide a standardized environment for app development, allowing software developers to create a diverse range of applications that seamlessly integrate with the hardware and connectivity features of mobile devices. This has led to the emergence of a vibrant mobile ecosystem, where users can access a vast array of services and functionalities right at their fingertips.

The impact of this communication revolution extends far beyond personal use. Businesses, organizations, and even governments have embraced the power of mobile technology to enhance productivity, streamline operations, and engage with their stakeholders in new and innovative ways. Mobile devices have become essential tools for remote collaboration, real-time decision-making, and on-the-go access to critical information.

In essence, the development and widespread adoption of mobile telephony has fundamentally transformed the way we connect, share information, and access the digital world. From the cellular network infrastructure to the ever-evolving capabilities of mobile devices, this revolution has liberated communication from the constraints of physical location, ushering in a new era of untethered and ubiquitous connectivity.

12.2.2.2 Cellular Network Infrastructure

The foundation of mobile telephony is the cellular network infrastructure. This is the system of interconnected towers, base stations, and switching centers that allows for seamless voice and data communication across vast geographic areas. The key principles behind cellular networks are frequency reuse and cell-based architecture. Frequency reuse is the ability to use the same RFs in different, non-overlapping geographical areas, known as cells. This efficient use of the finite radio spectrum enables mobile networks to support many users simultaneously.

The cell-based architecture divides the coverage area into smaller, overlapping cells, each with its own base station. As a user moves from one cell to another, their connection is seamlessly handed off between the base stations, ensuring uninterrupted service. This dynamic cell-switching mechanism is a crucial aspect of cellular networks, allowing them to accommodate the mobility of users.

12.2.2.3 Mobile Devices and Capabilities

The evolution of mobile devices has been a driving force behind the widespread adoption of mobile telephony. From the bulky, analog-based mobile phones of the 1980s to the sleek, feature-rich smartphones of today, the capabilities of these devices have grown exponentially. Modern smartphones are powerful pocket-sized computers, capable of far more than just making voice calls. They can send and receive text messages, access the internet, stream multimedia, and run a vast array of applications. The integration of cameras, GPS, and various sensors has further expanded the functionality of these devices, enabling a wide range of mobile services and applications.

The development of mobile operating systems, such as iOS, Android, and Windows Phone, has played a crucial role in the proliferation of smartphones. These platforms provide a standardized environment for app development, allowing software developers to create a diverse range of applications that cater to the needs and preferences of users.

12.2.2.4 Mobile Services and Network Operators

MNOs are the companies that build and maintain the cellular network infrastructure, providing voice, data, and messaging services to their subscribers. These operators invest heavily in network deployment, upgrades, and spectrum acquisition to ensure reliable and high-quality coverage.

One of the primary services offered by MNOs is mobile voice communication. Through the cellular network, users can make and receive calls from anywhere within the coverage area, even across great distances. This has revolutionized personal and business communication, allowing people to stay connected on the go. In addition to voice services, MNOs also offer mobile data services, enabling users to access the internet, stream content, and engage in various online activities through their mobile devices. The evolution of mobile data technologies, such as 3G, 4G, and now 5G, has dramatically increased the speed and capacity of these networks, paving the way for more advanced mobile applications and services. Text messaging, or SMS (Short Message Service), is another core service provided by mobile network operators. This simple yet efficient method of communication has become an integral part of everyday life, allowing users to exchange short text-based messages in real time.

Beyond traditional communication services, MNOs have also expanded into mobile financial services, such as mobile money and mobile banking. These services allow users to perform various financial transactions, like money transfers, bill payments, and even mobile payments, directly from their mobile devices. The role of mobile network operators extends beyond just providing services. They also play a crucial part in the development and deployment of new mobile technologies. MNOs work closely with device manufacturers, software developers, and regulatory bodies to ensure the seamless integration of new features and capabilities into the mobile ecosystem.

12.2.2.5 The Future of Mobile Telephony

As the world continues to embrace the digital age, the future of mobile telephony looks increasingly bright. The ongoing rollout of 5G, the latest generation of cellular networks, promises even faster data speeds, lower latency, and greater network capacity. This will open new possibilities for innovative mobile applications and services, such as immersive virtual reality experiences, real-time remote healthcare monitoring, and autonomous vehicle communication. Additionally, the convergence of mobile technologies with emerging trends like the IoT and AI is expected to further transform the mobile landscape. The integration of IoT devices, such as smart home appliances and wearables, with mobile platforms will enable new levels of connectivity and automation. AI-powered virtual assistants embedded in mobile devices will provide users with personalized and intelligent interactions, enhancing the overall user experience.

The increasing adoption of eSIM (embedded SIM) technology and the rise of mobile-first business models are likely to reshape the traditional mobile service landscape. eSIM technology allows users to easily switch between different mobile network operators without the need for a physical SIM card, providing more flexibility and choice. Mobile-first businesses are leveraging the ubiquity of smartphones to deliver innovative services and experiences, further blurring the lines between physical and digital realms.

Mobile telephony has been a true revolution in the world of communication, transforming the way we connect, access information, and interact with the digital realm. From the foundation of cellular network infrastructure to the ever-evolving capabilities of mobile devices and the diverse services provided by network operators, the mobile telephony landscape continues to evolve, promising even more remarkable advancements in the years to come.

12.2.3 Internet-Based Communication

With the advent of the internet, a variety of communication methods have emerged that use internet protocols to transmit voice, video, and text. This section includes Voice over Internet Protocol (VoIP) services, video conferencing tools, instant messaging (IM) platforms, and social media networks.

12.2.3.1 Internet-Based Communication: Transforming the Way We Connect

The advent of the internet has revolutionized the world of communication, giving rise to a vast array of new methods and platforms that leverage internet protocols to transmit voice, video, and text. This remarkable transformation has ushered in a new era of connectivity, empowering individuals, businesses, and communities to interact and collaborate in ways that were once unimaginable.

12.2.3.2 Voice over IP

At the forefront of this internet-based communication revolution is VoIP technology. VoIP services enable users to make and receive voice calls over the internet, bypassing the traditional PSTN.

VoIP technology works by converting analog voice signals into digital data packets, which are then transmitted over the internet and reassembled on the receiving end. This approach offers several advantages over traditional landline and mobile phone systems, including cost savings, improved call quality, and enhanced features.

VoIP services, such as Skype, Zoom, and WhatsApp, have become increasingly popular for personal and business communication. They allow users to make affordable, high-quality voice and video calls, often with additional features like screen sharing, conference calling, and file sharing. The ubiquity of internet access and the widespread adoption of smartphones have further fueled the growth of VoIP, enabling users to stay connected from virtually anywhere.

12.2.3.3 Video Conferencing Tools

Closely related to VoIP are video conferencing tools, which leverage the power of the internet to facilitate real-time, face-to-face communication across distances. These platforms, such as Zoom, Microsoft Teams, and Google Meet, provide a platform for multi-participant video calls, enabling users to see and hear each other in a virtual meeting environment.

Video conferencing solutions have become increasingly essential in the modern era, especially during the COVID-19 pandemic, when in-person interactions were limited. They have revolutionized the way we conduct business meetings, virtual classrooms, telemedicine consultations, and even personal gatherings, allowing people to collaborate and connect from the comfort of their own homes or offices.

These tools often incorporate additional features like screen sharing, presentation tools, and virtual whiteboards, enhancing the overall collaboration experience. The integration of video conferencing with other productivity tools and cloud-based services has further streamlined the process of remote communication and collaboration.

12.2.3.4 IM Platforms

Alongside VoIP and video conferencing, IM platforms have become a ubiquitous part of internet-based communication. Services like WhatsApp, Facebook Messenger, and iMessage enable users to exchange real-time text messages, share media, and engage in group conversations. The advantages of IM lie in its immediacy, convenience, and asynchronous nature. Users can quickly send and receive messages, often with the ability to see when the recipient is typing or has read the message. This allows for more dynamic and responsive exchanges, facilitating both personal and professional communication.

The integration of IM platforms with mobile devices has further enhanced their accessibility and usage. The ability to seamlessly continue conversations across multiple devices and the incorporation of features like voice and video calls have made IM an indispensable part of modern communication. Moreover, the growing prevalence of rich messaging features, such as stickers, graphics interchange format (GIFs), and emoji, has added an expressive and engaging dimension to IM, allowing users to convey emotions and nuances more effectively.

12.2.3.5 Social Media Networks

The rise of social media networks has profoundly shaped the landscape of internet-based communication. Platforms like Facebook, Twitter, Instagram, and LinkedIn have transformed the way we connect, share information, and engage with our social and professional circles. These social media networks provide a digital space for users to create and curate their online identities, share updates, and interact with others. They enable the exchange of text, images, videos, and other multimedia content, fostering a dynamic and interconnected community.

Beyond personal connections, social media networks have also become powerful tools for businesses, organizations, and influencers to reach and engage with their target audiences. Brands can leverage these platforms to share content, build brand awareness, and foster customer relationships.

The integration of social media with other internet-based services, such as IM and video conferencing, has further enhanced the communication capabilities of these platforms. Users can now seamlessly share updates, participate in group discussions, and collaborate on projects within the same digital ecosystem.

12.2.3.6 The Future of Internet-Based Communication

As the internet continues to evolve and new technologies emerge, the future of internet-based communication promises even more transformative advancements. Emerging trends like the integration of AI and the rise of virtual and augmented reality (VR/AR) are poised to redefine the way we interact and communicate online.

AI-powered chatbots and virtual assistants are already being incorporated into various communication platforms, providing users with intelligent and personalized interactions. As these technologies become more sophisticated, they could enable even more natural and contextual communication experiences, blurring the lines between human and machine interactions. The advancements in VR and AR technologies could revolutionize the way we engage in remote communication and collaboration. Immersive virtual environments could enable users to feel physically present in the same space, facilitating more engaging and intuitive interactions, even across vast distances.

In the years to come, the convergence of internet-based communication, emerging technologies, and evolving user preferences will continue to shape the future of how we connect, share information, and collaborate in the digital realm.

12.2.4 Network Infrastructure and Management

This section deals with the underlying infrastructure that supports all types of telecommunications, including network cables, switches, routers, and servers. It also covers network management aspects such as traffic engineering, network security, and quality of service (QoS).

12.2.4.1 Network Infrastructure and Management: The Backbone of Telecommunications

In the rapidly evolving world of telecommunications, the network infrastructure and its management play a crucial role in enabling the seamless exchange of information, whether it's voice, data, or multimedia. This foundational layer of the communication ecosystem is essential for the reliable and efficient functioning of a wide range of services, from mobile telephony and internet access to enterprise-level collaboration and cloud computing.

12.2.4.2 Base Transceiver Stations

Base Transceiver Stations (BTSs) are a vital part of the telecommunication infrastructure in cities. A BTS, also known as a base station or cell site in the context of mobile communications, is a critical component in wireless communication networks. Its primary function is to facilitate communication between mobile devices (such as cell phones) and the core network infrastructure.

The BTS is responsible for transmitting and receiving wireless signals between mobile devices and the core network. It performs several functions that are critical to the functioning of a mobile communication network. These include radio transmission and reception, call handover, network control, traffic management, and signal strength adjustment.

There are several types of BTS, each with its own characteristics and functions, as shown in Table 12.27.

TABLE 12.27
Types of BTS

Macro BTS	Provides coverage to a large geographical area, typically up to several kilometers.
Micro BTS	Provides coverage to a smaller geographical area, typically up to a few hundred meters.
Pico BTS	Provides coverage to a very small geographical area, typically up to a few tens of meters.
Femto BTS	Provides coverage to a very small area, typically up to a few meters.

BTSs are essential components of cellular networks, as they facilitate communication between mobile devices and the network by transmitting and receiving radio signals. They can be either macrocells, which cover large areas, or microcells, which cover small geographical locations.

BTSs require a stable and reliable power source to operate. This is typically provided by the local power grid, but in case of power outages, backup power sources such as batteries or generators are used.

The BTS is connected to the core network through backhaul links. These links carry both user data and control information between the base station and the network's central elements, such as Mobile Switching Center (MSC) or Base Station Controller (BSC). The backhaul links can be wired (e.g., fiber optic) or wireless (e.g., microwave or satellite links). The choice between wired and wireless backhaul depends on various factors, including the geographical location of the BTS, the available infrastructure, and the cost.

The block diagram of a base station typically includes the following key components:

- *Baseband Processor:* Deals with different communication protocols and interfaces with mobile network infrastructure.
- *Duplexer:* Enables the employment of a single antenna for both transmission and reception.
- *Antenna System:* Transmits and receives RF signals. The antenna system can comprise several antennas to support different frequency bands and technologies.
- *Transceiver:* Manages various modulation and demodulation activities. In transmission and reception, it converts electrical signals to radio waves and vice versa. The transceiver also deals with power control, frequency tuning, and other necessary operations.
- *Power Amplifier:* The RF signals are power amplified before transmission to their destinations for increased signal strength. This is very important for enabling the signals to cover long distances and even penetrate barriers in the communication environment.

In a typical BTS architecture, the transceiver units (TRXs) are connected to the baseband units (BBUs) through digital links. The BBUs handle the processing of digital signals and control multiple TRXs.

The deployment of BTS in cities is a complex process that involves careful planning and coordination between various stakeholders, including city authorities, telecommunication companies, and regulatory bodies.

The first step in the deployment of BTS is planning and site selection. This involves identifying suitable locations for the installation of the base stations. Factors considered during site selection include the coverage area, population density, topography, and existing infrastructure. The aim is to ensure optimal coverage while minimizing interference with other signals.

Before a BTS can be installed, telecommunication companies must comply with various regulations and standards set by regulatory bodies. These may include health and safety standards, environmental impact assessments, and zoning regulations. Compliance with these regulations is crucial to ensure the safety and well-being of the public.

TABLE 12.28
Challenges in Managing BTS Deployment in Cities

Technical knowledge and expertise	There is often a lack of technical knowledge and expertise among the state and local officials who are responsible for deploying the systems. This can make it difficult to understand the complexities of the technology and make informed decisions.
Cost-effectiveness	There is often a lack of quantitative data proving the systems' cost-effectiveness in solving transportation problems. This can make it difficult to justify the investment in BTS deployment.
Funding	There can be a lack of funds, considering other transportation priorities. This can limit the resources available for BTS deployment.
Regulatory challenges	The deployment of BTS often involves navigating complex regulatory landscapes, which can include health and safety standards, environmental impact assessments, and zoning regulations.
Infrastructure challenges	The physical deployment of BTS can be challenging due to factors such as challenging topography and high population density. For instance, the rollout of 5G brings new challenges, requiring small cells to be deployed more densely than previous generations of connectivity, which raises fresh concerns about the aesthetic impact on the urban landscape.
Collaboration challenges	There can be challenges involved in coordinating between various stakeholders, including industry bodies, carriers, power authorities, public works authorities, standards and codes, and government bodies.

Once a site has been selected and all regulatory requirements have been met, the BTS can be installed. This involves the installation of various components, including antennas, transceivers, and power supply units. After installation, the BTS needs to be regularly maintained to ensure optimal performance and reliability.

Historically, cities have played a largely passive role in the implementation of telecoms infrastructure – typically issuing permits and hoping for the best. However, with the advent of smart cities and the increasing demand for digital services, cities are beginning to play a more active role.

Cities are now fostering better collaboration with the telecommunications industry to address the complexities and opportunities around telecom issues. For instance, the rollout of 5G brings new challenges, requiring small cells to be deployed more densely than previous generations of connectivity, which raises fresh concerns about the aesthetic impact on the urban landscape. Moreover, cities are making better use of their assets such as lampposts, street furniture, and buildings to advance the deployment of small cells on shared assets. Some cities have even set up dedicated Telecoms Units as a "one-stop-shop" for operators and infrastructure providers looking to invest or deploy connectivity in the city.

Cities typically face several challenges in managing the deployment of BTSs . These challenges are summarized in Table 12.28.

These challenges highlight the need for cities to foster better collaboration with the telecommunications industry, make better use of their assets, and set up dedicated telecoms units to facilitate the deployment of BTS.

12.2.4.3 Network Cables and Components

Network infrastructure include the physical components that facilitate the transmission of data. This includes a diverse array of cables, such as copper wires, fiber-optic strands, and wireless radio signals, each with its own unique characteristics and applications. Copper cables, traditionally used for landline telephone networks and early internet connections, provide a reliable and cost-effective solution for transmitting data over short to medium distances. However, the growing demand for higher bandwidth and faster speeds has led to the widespread adoption of fiber-optic cables, which can carry significantly more data with lower latency and reduced signal degradation over longer distances. These physical transmission media are complemented by a range of network devices, including switches, routers, and servers, which play crucial roles in directing, managing, and processing the flow of information across the network.

Switches, for instance, act as the gatekeepers of local area networks (LANs), directing data packets to their intended destinations within the same network. Routers, on the other hand, facilitate the exchange of information between different networks, enabling seamless communication between local and wide-area networks, including the internet. Servers, the workhorses of the network infrastructure, provide centralized computing and storage resources, hosting applications, services, and data that are accessed by various clients and devices connected to the network.

12.2.4.4 Network Management and Optimization

Alongside the physical components of the network infrastructure, the management and optimization of these systems are equally important for ensuring reliable, secure, and efficient communication. Network traffic engineering, a key aspect of network management, involves the careful planning, monitoring, and optimization of network resources to accommodate the dynamic and ever-changing demands of users and applications. This includes techniques like load balancing, bandwidth management, and traffic prioritization, which help to ensure that the network operates at peak performance and can handle spikes in traffic without compromising the QoS.

Another critical element of network management is security, as the proliferation of cyber threats, such as hacking, malware, and distributed denial-of-service (DDoS) attacks, poses a constant challenge to the integrity and availability of communication networks. Network administrators employ a variety of security measures, including firewalls, intrusion detection and prevention systems, and virtual private networks (VPNs), to safeguard the network against unauthorized access and malicious activities.

QoS is another important consideration in network management, as it ensures that different types of network traffic, such as voice, video, and data, are prioritized and delivered according to their specific requirements. By implementing QoS policies, network operators can optimize the performance of time-sensitive applications, such as video conferencing and streaming, while still providing adequate resources for less critical data transfers.

12.2.4.5 Emerging Trends in Network Infrastructure

As the world of telecommunications continues to evolve, the network infrastructure is also undergoing transformative changes to keep pace with the growing demands and emerging technologies. One such development is the rise of software-defined networking (SDN), which decouples the control plane from the data plane in network devices, allowing for more flexible and programmable network management. SDN enables network administrators to centrally control and configure the network, making it easier to adapt to changing requirements, deploy new services, and enforce policies across the entire network infrastructure.

Another significant trend is the deployment of edge computing, which brings computing and storage resources closer to the end-users and devices, rather than relying solely on centralized data centers. This approach reduces latency, improves responsiveness, and optimizes the use of network resources, particularly for applications that require real-time processing, such as autonomous vehicles, industrial automation, and IoT systems.

The advent of 5G wireless technology has also transformed the network infrastructure landscape, offering vastly improved speed, capacity, and low latency. 5G networks, with their ability to support a massive number of connected devices and enable ultra-reliable communication, are paving the way for innovative applications in areas like remote healthcare, smart cities, and immersive entertainment. Furthermore, the integration of AI and machine learning (ML) into network management is becoming increasingly prominent. These technologies enable network operators to automate and optimize various aspects of network operations, such as traffic analysis, anomaly detection, and predictive maintenance, leading to enhanced performance, reduced operational costs, and improved user experiences.

12.2.4.6 The Role of Network Infrastructure in the Digital Transformation

As the digital world continues to expand, the underlying network infrastructure has become the backbone that supports the ever-increasing demand for connectivity, communication, and data exchange. From enabling remote work and telemedicine to powering the growth of cloud computing and the IoT, the network infrastructure plays a vital role in the ongoing digital transformation of our society.

Businesses and organizations of all sizes rely on robust and reliable network infrastructure to facilitate collaboration, access cloud-based services, and ensure the efficient flow of information. The ability to seamlessly connect employees, customers, and partners, regardless of their physical location, has become a crucial competitive advantage in the modern business landscape. Moreover, the network infrastructure is essential for the development and deployment of emerging technologies, such as AI, AR, and autonomous systems. These cutting-edge applications often require low-latency, high-bandwidth, and secure network connections to function effectively, further highlighting the importance of a well-designed and managed network infrastructure.

The network infrastructure and its management can be said to be the unsung heroes of the telecommunications industry, providing the foundation upon which a myriad of communication services, digital applications, and innovative technologies thrive. As the demand for connectivity and data exchange continues to grow, the continuous evolution and optimization of network infrastructure will be essential for unlocking new possibilities and driving the digital transformation of our world.

12.2.5 Regulation and Policy

Telecommunication systems are subject to various regulations and policies that govern how they can be used and managed. This section covers topics like spectrum allocation, privacy and data protection laws, and regulatory bodies that oversee telecommunication systems.

Telecommunication systems are vital infrastructure that enable the exchange of information, communication, and connectivity in our modern world. Given their importance and the significant impact, they have on society, these systems are subject to a range of regulations and policies that govern their development, deployment, and usage.

12.2.5.1 Spectrum Allocation

One of the fundamental aspects of telecommunication regulation is the management and allocation of the electromagnetic spectrum. The RF spectrum is a finite and valuable resource that is used for various wireless communication technologies, such as cellular networks, broadcast radio and television, satellite communications, and more. Regulatory bodies, typically national or international in scope, are responsible for allocating and managing the spectrum to ensure efficient and equitable use.

These regulatory bodies, such as the Federal Communications Commission (FCC) in the United States, or the Office of Communications (Ofcom) in the UK, or the International Telecommunication Union (ITU) at the global level, establish rules and procedures for spectrum licensing, allocating frequencies to different services, and coordinating the use of the spectrum across borders. This is crucial to prevent interference between different wireless systems and to ensure that the spectrum is utilized in a way that maximizes its potential for various communication services. Table 12.29 shows some examples of spectrum allocation.

Please note that these frequency bands can vary by country and region and are subject to change based on technological advancements and regulatory decisions.

TABLE 12.29
Spectrum Allocation

Cellular and television spectrums	Cellular systems use various bands, including Frequency Range 1 (FR1), which is 0.41–7.125 GHz, FR3, which is between 7 and 24 GHz, and FR2, which is the high frequency (HF) millimeter spectrum between 24.25 and 71 GHz. For television broadcasting, the very high frequency (VHF) band is used, which ranges from 30 MHz to 300 MHz, and the UHF band, which ranges from 300 MHz to 3 GHz.
Amateur RF allocations	The frequency allocations for amateur radio vary by country and between ITU regions. Some of the frequency ranges include low frequency (2,200 m: 135.7–137.8 kHz), medium frequency (630 m: 472–479 kHz and 160 m: 1.800–2.000 MHz), and HF (ranging from 3 MHz to 30 MHz).
Industrial, scientific, and medical (ISM) bands	The ISM frequency bands are scattered from 6.78 kHz to 245 GHz in the radio spectrum. Some common ISM bands include 6.765–6.795 MHz, 13.553–13.567 MHz, 26.957–27.283 MHz, 40.66–40.70 MHz, 433.05–434.79 MHz, 2.400–2.500 GHz, and 5.725–5.875 GHz.
Unlicensed National Information Infrastructure (U-NII) band	The U-NII band operates over several ranges, including U-NII Low (U-NII-1: 5.150–5.250 GHz), U-NII Mid (U-NII-2A: 5.250–5.350 GHz), U-NII-2B: 5.350–5.470 GHz, U-NII Worldwide (U-NII-2C / U-NII-2e: 5.470–5.725 GHz), U-NII Upper (U-NII-3: 5.725–5.850 GHz), and DSRC/ITS (U-NII-4: 5.850–5.925 GHz).
Aeronautical Mobile Service	The frequency allocations for aeronautical mobile service include 1.164–1.215 MHz and 1.559–1.610 MHz. Other bands used for aeronautical communications include the VHF band and HF band.
Broadcasting service	The broadcasting services use various bands, including longwave AM radio (148.5–283.5 kHz), mediumwave AM radio (520–1700 kHz), and shortwave AM radio (3–30 MHz). For television broadcasting, the VHF band is used, which ranges from 30 MHz to 300 MHz, and the UHF band, which ranges from 300 MHz to 3 GHz.

12.2.5.2 Privacy Data Protection

As telecommunication systems handle vast amounts of sensitive data, including personal information, call records, and internet traffic, they are subject to a range of privacy and data protection laws. These laws aim to safeguard the privacy of individuals by regulating how telecommunication companies collect, store, and use this data.

For example, in the European Union, the General Data Protection Regulation (GDPR) sets strict standards for the handling of personal data, requiring telecommunication providers to obtain explicit consent from users, implement robust security measures, and provide transparent information about their data processing practices. Similarly, in the United States, the Federal Trade Commission (FTC) and state-level regulations, such as the California Consumer Privacy Act (CCPA), establish guidelines for the protection of consumer data in the telecommunication industry.

These privacy and data protection laws not only ensure the protection of individual rights but also help to build trust in telecommunication services and encourage the responsible use of data by service providers.

12.2.5.3 Regulatory Oversight

Telecommunication systems are also subject to oversight by various regulatory bodies that are responsible for ensuring compliance with applicable laws and regulations. These regulatory bodies can have a significant influence on the telecommunication industry, as they have the authority to set standards, issue licenses, and enforce compliance. For example, in the United States, the FCC is the primary regulatory agency responsible for overseeing the telecommunication industry. The FCC is tasked with regulating interstate and international communications by radio, television, wire, satellite, and cable. It sets rules and policies related to licensing, network infrastructure, consumer protection, and more.

Similar regulatory bodies exist in other countries, such as the Ofcom in the UK, the Canadian Radio-Television and Telecommunications Commission (CRTC) in Canada, and the Australian Communications and Media Authority (ACMA) in Australia. These regulatory agencies are vital in shaping the telecommunication landscape, ensuring fair competition, consumer protection, and the overall development of the industry.

12.2.5.4 Implications and Challenges

The regulation and policy framework surrounding telecommunication systems can have significant implications for various stakeholders, including consumers, service providers, and technology developers. For consumers, regulations can provide important safeguards, such as ensuring affordable and accessible services, protecting privacy, and guaranteeing QoS. However, overly restrictive regulations can also limit innovation and the availability of new technologies. For service providers, the regulatory environment can present both opportunities and challenges. Compliance with regulations can be costly and time-consuming, but it also provides a level playing field and can help to build trust in the industry. Regulatory changes can also create new business opportunities, such as the deployment of new technologies or the expansion into new markets. For technology developers, the regulatory landscape can be a significant factor in their product development and deployment strategies. They must navigate a complex web of rules and standards to ensure that their technologies are compatible with the existing infrastructure and meet the requirements set by regulatory bodies.

The rapid pace of technological change in the telecommunication industry can outpace the ability of regulatory bodies to keep up, leading to challenges in adapting policies and regulations to new technologies and market dynamics.

The regulation and policy framework surrounding telecommunication systems is a critical component of the industry, ensuring the efficient and responsible use of this vital infrastructure. The interplay between regulations, privacy, and technological innovation will continue to shape the future of telecommunication systems, with implications for consumers, service providers, and technology developers alike.

12.3 SMART INFRASTRUCTURE AND IOT

The rise of smart infrastructure and the IoT is transforming the way we interact with and manage our built environment. IoT devices and sensors are being integrated into a wide range of infrastructure, from buildings and transportation networks to utilities and public services, enabling the collection and analysis of massive amounts of data. This data is then used to power smart city applications that optimize resource allocation, improve efficiency, and enhance the quality of life for citizens. For example, smart traffic management systems can analyze real-time data on traffic patterns and road conditions to adjust traffic signals and provide commuters with routing recommendations, reducing congestion and emissions. Similarly, smart energy grids can monitor and regulate electricity usage, identifying opportunities for conservation and the integration of renewable energy sources. However, the proliferation of IoT devices and the increasing reliance on data-driven decision-making also raise concerns about security and privacy. Cybersecurity threats, such as hacking and data breaches, can have severe consequences for critical infrastructure, and the collection of personal data by IoT devices raises questions about individual privacy rights and the potential for misuse. Addressing these challenges requires a comprehensive approach to security and data governance, as well as the development of connectivity standards and interoperability protocols that enable seamless integration and communication between the various components of smart infrastructure. As smart cities and IoT-enabled systems become more pervasive, it will be crucial for policymakers, technology providers, and communities to work together to ensure that the benefits of these transformative technologies are realized while safeguarding the safety, security, and privacy of citizens.

12.3.1 IoT Devices and Sensors

The IoT is a network of physical objects embedded with sensors, software, and other technologies to connect and exchange data with other devices and systems over the internet. In the context of smart cities, IoT devices and sensors are the frontline in collecting real-time data to improve city services, enhance sustainability, and improve the quality of life for citizens. As mentioned earlier, the IoT has revolutionized the interaction between us and our built environment, and the proliferation of IoT devices and sensors is a key driver of this transformation. These devices and sensors are the fundamental building blocks of smart city infrastructure, enabling the collection and analysis of vast amounts of data that can be used to optimize resource allocation, improve efficiency, and enhance the quality of life for citizens.

At the most basic level, IoT devices and sensors can include simple environmental monitors, such as those used to measure temperature, humidity, air quality, and other environmental factors. These sensors are often deployed in buildings, public spaces, and infrastructure to provide real-time data on the state of the environment. More advanced IoT devices can include smart meters that track energy and water usage or connected vehicles that can communicate with traffic management systems and other infrastructure.

The range of IoT devices and sensors continues to expand, with innovations in fields like robotics, AI, and ML leading to the development of increasingly sophisticated and versatile devices. For example, some cities are experimenting with IoT-enabled waste management systems that use sensors to monitor trash levels and optimize collection routes, while others are deploying smart streetlights that can adjust their brightness based on the presence of pedestrians or vehicles.

One of the key advantages of IoT devices and sensors is their ability to collect and transmit data in real-time, allowing for the development of dynamic and responsive systems that can adapt to changing conditions. This real-time data can be used to inform decision-making and drive automated responses, such as adjusting traffic signals to alleviate congestion or triggering emergency response protocols in the event of a natural disaster or other crisis. However, the proliferation of IoT devices and the increasing reliance on data-driven decision-making also raise significant concerns about security and privacy. IoT devices can be vulnerable to cyber-attacks, with hackers potentially gaining access to sensitive data or even taking control of critical infrastructure. The vast amounts of personal data collected by IoT devices, such as location, usage patterns, and behavioral information, can pose a threat to individual privacy if not properly safeguarded.

To address these challenges, policymakers, technology providers, and communities must work together to develop robust security protocols and data governance frameworks that protect both the integrity of the infrastructure and the privacy of citizens. This may involve the implementation of encryption technologies, the establishment of clear guidelines for data collection and usage, and the development of interoperability standards that ensure seamless communication between different IoT devices and systems.

The successful deployment of IoT devices and sensors in smart city applications requires a thorough examination of the social and ethical implications of these technologies. For example, the use of connected vehicles and traffic management systems may raise concerns about surveillance and the potential for unequal access to the benefits of these technologies, particularly among marginalized communities. To mitigate these concerns, smart city initiatives must prioritize principles of inclusivity, equity, and transparency, ensuring that the benefits of IoT-enabled infrastructure are distributed equitably and that the decision-making processes are open and accessible to the public.

IoT devices and sensors are the foundation of smart city infrastructure, enabling the collection and analysis of data that can drive innovation, improve efficiency, and enhance the quality of life for citizens. However, the successful deployment of these technologies requires a carefully considered approach that addresses the complex challenges of security, privacy, and social impact. Collaboration and prioritizing the needs and concerns of all stakeholders, policymakers, technology providers, and communities can unlock the full potential of IoT-enabled smart cities and create more sustainable, resilient, and equitable urban environments.

12.3.2 Data Collection and Analysis

The IoT has revolutionized the way we collect, store, and analyze data. In the context of smart cities, this data is invaluable for improving city services, optimizing resource use, and making informed decisions. This essay will delve into the process of data collection and analysis, focusing on technologies like cloud computing, big data analytics, and ML.

The first step in the process is data collection. IoT devices, ranging from simple sensors to complex systems like connected vehicles or smart meters, continuously collect data about their environment. This data can include temperature readings, energy usage, traffic conditions, and much more. The devices then transmit this data to a central system for storage and analysis. The transmission usually happens over the internet, hence the term "Internet of Things".

Once the data is collected, it needs to be stored in a way that facilitates efficient analysis. This is where cloud computing comes in. Cloud computing platforms provide a scalable solution for storing the vast amounts of data generated by IoT devices. These platforms offer virtually unlimited storage capacity and powerful computing resources to process the data. They also provide tools for managing and organizing the data, making it easier to analyze.

With the data collected and stored, the next step is to analyze it and extract meaningful insights. This is typically done using big data analytics and ML techniques. Big data analytics involves examining large and varied data sets to uncover hidden patterns, correlations, and other insights. With today's technology, it is possible to analyze the data in real time, or near real time, which is crucial for smart city applications where timely response is critical. ML, on the other hand, involves training a model to make predictions or decisions without being explicitly programmed to do so. In the context of smart cities, ML models can be used to predict traffic congestion, energy demand, waste generation, and much more. These predictions can then be used to optimize city services and resource use.

The goal of collecting and analyzing data from IoT devices is to gain insights that can inform decision making. For example, data from traffic sensors can be used to optimize traffic light timings, reducing congestion and improving traffic flow. Data from smart meters can provide insights into energy use patterns, informing efforts to improve energy efficiency. In essence, the data serve as the foundation for a feedback loop, where the insights gained inform actions, the results of which are then measured and fed back into the system, leading to continuous improvement.

Through technologies like cloud computing, big data analytics, and ML, these cities can leverage the data to improve services, optimize resources, and make more informed decisions. As technology continues to advance, the possibilities for what can be achieved with this data are only set to grow.

Ensuring data security and privacy in smart cities is a complex task that involves strategies and technologies. Table 12.30 summarized some of the key methods to ensure the security and privacy of the data.

The complex and interdependent nature of smart cities raises significant technical, political, and socioeconomic challenges for those involved in managing these systems.

The relationship between data security and usability can be characterized as a delicate balance. As the level of data openness and accessibility increases to enhance usability for users, the potential risk to data security also rises. Conversely, stricter data security measures can limit the usability and

TABLE 12.30
Methods for Data Security and Privacy

Privacy by design	This principle involves integrating privacy considerations into the design and operation of IoT devices and systems from the outset. It includes measures such as data minimization (collecting only the data that is necessary), purpose limitation (using data only for the purpose it was collected), and ensuring appropriate security measures are in place.
Anonymization "at the edge"	This involves processing data on the IoT device itself (the "edge" of the network) to remove personally identifiable information before it is transmitted to the cloud or other systems. This can help to protect privacy by ensuring that the raw data, which may contain sensitive information, does not leave the device.
Data encryption	Encrypting data both at rest and in transit can help to protect it from unauthorized access. This means that even if a cybercriminal were to intercept the data, they would not be able to read it without the encryption key.
Access controls	Implementing robust access controls can ensure that only authorized individuals or systems can access the data. This might involve techniques such as multifactor authentication, role-based access control, and regular audits of access logs.
Regulatory compliance	Compliance with data protection regulations, such as the GDPR in the European Union, is crucial. These regulations set out strict rules for how personal data can be collected, stored, and used, and non-compliance can result in hefty fines.
Public–private partnerships	In many smart cities, both public and private entities are involved in collecting and managing data. Clear agreements and partnerships can help to define responsibilities and ensure that all parties are taking appropriate steps to protect data.
Citizen engagement	Engaging citizens in decisions about how their data is used can help to build trust and ensure that privacy concerns are addressed. This might involve public consultations, transparency initiatives, and efforts to educate citizens about how their data is used.

accessibility of the data. This creates a trade-off that needs to be carefully managed to strike the right balance between these two competing priorities. The trade-off between data security and usability is a critical aspect of smart city design and operation. This trade-off indicates the inversely proportional relationship that seems to exist between security and usability. The more secure the systems, the less usable they will be, and vice versa.

Usability refers to the ease of use and learnability of a system or product. In the context of smart cities, usability might involve the simplicity of interacting with IoT devices, the intuitiveness of user interfaces, or the accessibility of services across different platforms. High usability can lead to increased adoption of technologies, improved user satisfaction, and more effective utilization of city services.

Security, on the other hand, involves protecting data from unauthorized access or alteration. In smart cities, this could involve securing communication networks, encrypting data, or implementing robust access controls. While strong security measures are crucial for protecting sensitive data and maintaining public trust, they can also make systems more complex and harder to use.

The collection, processing, and transmission of large amounts of data in smart cities can lead to trade-offs between policy efficiency, business profitability, consumer convenience, and personal privacy. For instance, requiring users to go through multiple authentication steps might enhance security but could also lead to frustration and reduced usage. Similarly, while anonymizing data can help protect privacy, it can also make it harder to deliver personalized services.

Balancing security and usability are a challenging task. One approach is to adopt the principle of "privacy by design," which involves integrating privacy and security considerations into the design and operation of systems from the outset. User education is also important, as informed users can make better decisions about their privacy and are more likely to tolerate necessary security measures.

12.3.3 Security and Privacy

As we step into the era of smart infrastructure and IoT, we are witnessing an unprecedented level of connectivity. From smart homes to smart cities, IoT devices are becoming an integral part of our daily lives. However, with this increased connectivity comes the challenge of ensuring security and privacy.

12.3.3.1 Security in IoT

Security in IoT is about protecting the data that these devices collect, transmit, and store. IoT devices are often seen as weak links in the security chain because they can be easier to compromise than traditional computing devices. This is due to a variety of factors, including their ubiquitous nature, the use of insecure communication protocols, and the lack of robust security measures in many devices.

To address these constraints, several measures need to be taken, as shown in Table 12.31.

12.3.3.2 Privacy in IoT

While security focuses on protecting data from unauthorized access, privacy is about protecting the rights of individuals whose data is being collected. With IoT devices collecting vast amounts of data, often of a personal nature, privacy concerns are paramount.

Table 12.32 shows some necessary measures to ensure privacy in the IoT space.

12.3.4 Connectivity and Interoperability

The success and widespread adoption of the IoT rely heavily on the ability of different IoT devices and systems to communicate and work together seamlessly. This concept is known as connectivity and interoperability, and it is a critical aspect of the IoT ecosystem.

TABLE 12.31
Strategies to Cope with Issues in Security in IoT

Secure communication	All data transmitted by IoT devices should be encrypted. This ensures that even if the data is intercepted, it cannot be understood by unauthorized individuals.
Device authentication	IoT devices should have a way to prove their identity before they can connect to a network. This prevents unauthorized devices from joining the network.
Regular updates	IoT devices should receive regular software updates to fix any security vulnerabilities. This requires an efficient and secure update mechanism.
Intrusion detection systems	These systems monitor the network for any unusual activity that could indicate a security breach. They can alert administrators and take corrective action if necessary.

TABLE 12.32
Measures to Ensure Privacy in the IoT Space

Data minimization	IoT devices should only collect the data that is necessary for their operation. This reduces the amount of personal data that is at risk.
User consent	Users should have control over their data. They should be able to decide what data is collected, how it is used, and who it is shared with.
Transparency	Companies should be transparent about their data practices. This includes providing clear and understandable privacy policies.
Anonymization	When data is shared, it should be anonymized to protect the identity of individuals.

TABLE 12.33
Connectivity and Interoperability Components

Data exchange standards	These are agreed-upon methods for devices to format and send data. Standards like JavaScript Object Notation (JSON) and Extensible Markup Language (XML) are commonly used for data exchange. They ensure that the receiving device can interpret the data sent by the transmitting device.
Communication protocols	These are the rules for how data is transmitted and received. IoT devices might use protocols like Message Queuing Telemetry Transport (MQTT), Constrained Application Protocol (CoAP), Hypertext Transfer Protocol (HTTP), etc. The choice of protocol can depend on factors like power consumption, data size, speed, and reliability.
Device management platforms	These are software solutions that allow users to manage their IoT devices, often providing features like device provisioning, firmware updates, security management, and troubleshooting.
Security	This is paramount in any discussion about connectivity and interoperability. As devices connect and share data, they become vulnerable to attacks. Hence, robust security measures like encryption, secure booting, and access control are essential.
Semantic interoperability	This is about ensuring that not only can devices exchange data, but that also the data being exchanged is understood in the same way by all parties. This often involves the use of data models and ontologies.

Connectivity refers to the ability of devices to reach out and communicate with other devices or systems. In the IoT ecosystem, this is crucial as it enables devices to transmit and receive data, thereby facilitating their primary function of collecting, sending, and acting on data. Connectivity can be achieved through various means such as Wi-Fi, Bluetooth, Zigbee, cellular networks, and even wired connections. Interoperability, on the other hand, is the ability of different systems, devices, or applications to work together, even if they are from different manufacturers. It is about the seamless sharing and utilization of information among interconnected devices. Without interoperability, the true potential of IoT cannot be realized, as the power of IoT lies in the ability of devices to interact and collaborate to make intelligent decisions.

The components that enable connectivity and interoperability are shown in Table 12.33.

Connectivity and interoperability ensure that devices can not only connect and communicate with each other, but also work together toward common goals, thereby enabling the creation of smart, responsive, and adaptive systems. However, achieving this requires careful consideration of factors like data standards, communication protocols, security, and semantics. As the IoT landscape continues to evolve, so will the approaches and technologies for achieving connectivity and interoperability. It is an exciting field that is continually pushing the boundaries of what is possible when everything is connected. Some examples of IoT devices that demonstrate good connectivity and interoperability that are currently in operation are given below.

12.3.4.1 Smart Home Ecosystem

- Smart home devices, such as smart thermostats, lighting, and security systems, can connect to and communicate with each other through common protocols like Wi-Fi, Zigbee, or Z-Wave.
- These devices can also integrate with voice assistants like Amazon Alexa or Google Assistant, allowing users to control their smart home with voice commands.
- Smart home platforms, such as Apple HomeKit, Google Home, or Samsung SmartThings, provide a unified interface to manage and automate various connected devices in the home, regardless of their manufacturer.

12.3.4.2 Industrial IoT Systems

- Industrial equipment, sensors, and control systems in factories or production facilities can communicate using standardized protocols like OPC UA (OPC Unified Architecture) or MQTT.
- These protocols enable seamless data exchange and coordination between different industrial components, such as programmable logic controllers (PLCs), SCADA systems, and enterprise resource planning (ERP) software.
- Industrial IoT (IIoT) platforms, like PTC ThingWorx or GE Digital Predix, provide a common interface for monitoring, analyzing, and optimizing industrial operations across multiple sites and equipment manufacturers.

12.3.4.3 Connected Healthcare

- Medical devices, such as wearable fitness trackers, blood pressure monitors, and glucose meters, can connect to mobile apps or cloud-based platforms using Bluetooth, Wi-Fi, or cellular connectivity.
- These devices can share patient data with healthcare providers, enabling remote monitoring, early intervention, and personalized treatment plans.
- Platforms like Apple HealthKit or Google Fit allow users to aggregate data from various connected health devices and share it with their healthcare team, promoting better care coordination.

12.3.4.4 Smart City Infrastructure

- IoT sensors and devices, such as traffic lights, parking meters, streetlights, and environmental monitors, can be integrated into a city's central management system.
- These connected systems can share real-time data, enabling improved traffic flow, efficient resource allocation, and better-informed decision-making by city authorities.
- Open data platforms and Application Programming Interfaces (APIs) allow third-party developers to build applications that leverage the data from these interconnected IoT devices, enhancing the overall smart city experience for residents and visitors.

These examples demonstrate how connectivity and interoperability within the IoT ecosystem can enable seamless data exchange, coordinated control, and integrated experiences across a wide range of applications and industries. As the IoT continues to evolve, the importance of connectivity and interoperability will only grow, driving further advancements and unlocking new use cases.

Some other examples of IoT that demonstrate good connectivity and interoperability are shown below.

- *Audi's connect® System:* Audi offers various connectivity features for their vehicles through the Audi connect® plans. With this IoT system, drivers can receive their EV's charging status, control a vehicle's interior temperature, and review data on their driving behavior – all from their phone.
- *Zubie's IoT Fleet Tracking:* Zubie offers real-time GPS for rent and business fleet tracking while monitoring vehicle health and driver performance. This information can be used to track maintenance, avoid potential accidents, and increase fuel efficiency.
- *Tesla's Connectivity Service:* Tesla's cars come with connectivity features that provide access to features through Wi-Fi only, in addition to basic maps, navigation, and music streaming. Users can purchase Tesla's Premium Connectivity to access all connectivity features through cellular data as well as Wi-Fi.
- *LG Electronics' ThinQ Appliances:* LG Electronics offers home appliances and consumer electronics that are IoT-enabled. These devices can gather environmental information and make appropriate adjustments.

Here is an example of the feature: Audi's connect® System (Box 12.1). The Audi connect® system is an intuitive and intelligent system that provides remote vehicle access and control via the Audi connect® app. It represents the digital connection between the driver, vehicle, and infrastructure.

BOX 12.1 AUDI'S CONNECT SYSTEM

1. *Audi Connect CARE:* This service is included with the purchase of every Audi and provides a suite of peace of mind and remote services. These include emergency call, remote lock/unlock, remote start, stolen vehicle locator, and vehicle status report.
2. *Audi Connect NAV:* This is a paid subscription service that includes additional navigation and infotainment services. It provides real-time traffic, satellite imagery/maps, POI search, traffic information, and natural voice recognition.
3. *Audi Connect PLUS with Full-Speed Wi-Fi:* This service turns your Audi into a Wi-Fi hotspot, providing a full-speed internet connection for up to eight devices at a time. It also includes access to the App Store, which offers apps across Music, Video, Gaming, Navigation, Parking and Charging, Productivity, Weather, and News.
4. *In-Vehicle Trip Planner:* This feature generates routes with access to EV charging stations and ETAs that include charging time estimates.
5. *Remote Climatization:* This allows you to adjust the cabin temperature of your EV from anywhere.
6. *Geofencing:* This feature notifies you when your Audi leaves your set boundaries.

12.4 DATA MANAGEMENT AND ANALYTICS

In the digital age, data management and analytics have become the backbone of decision-making processes in various sectors. It all begins with data collection, the process of gathering and measuring information from different sources. This data is then stored and organized in a structured manner, enabling easy access and efficient processing. The next step is data analysis, where sophisticated algorithms and statistical methods are used to extract meaningful insights from the collected data. These insights are then presented in an understandable format through data visualization and reporting, aiding in the interpretation of complex data and facilitating informed decision-making. However, with the increasing volume and variety of data, data governance has become crucial. It ensures the availability, usability, integrity, and security of data used in an enterprise. Moreover, in an era where privacy has become a major concern, data management also involves the protection of data privacy. It ensures that sensitive information is not misused or accessed without authorization, thereby building trust among data subjects. Thus, data management and analytics encompass a wide range of activities, each of which plays a vital role in turning raw data into valuable insights.

12.4.1 Data Collection

Data collection is the foundational step in the data management and analytics process. It involves gathering information from various sources to be used for further analysis and decision-making. In the context of cities, data collection can involve a wide range of sources, from IoT devices and sensors to social media feeds and public records.

12.4.1.1 IoT Devices and Sensors

IoT devices and sensors are at the forefront of data collection in modern cities. These devices, embedded within the city's infrastructure, collect real-time data on various aspects of urban life.

For instance, traffic sensors monitor vehicle flow and congestion levels, environmental sensors track air quality and temperature, and utility meters measure water and electricity usage. This data is invaluable for city planners and administrators, enabling them to monitor the city's functioning in real-time and make informed decisions.

12.4.1.2 Social Media Feeds

Social media feeds provide a wealth of data on public sentiment and behavior. By analyzing posts, likes, shares, and comments, city administrators can gain insights into the public's opinions on various issues, from public transport and sanitation to safety and community events. This data can be used to gauge public satisfaction with city services, identify areas of concern, and inform policy decisions.

12.4.1.3 Public Records

Public records, such as census data, property records, and crime reports, are another valuable source of data for cities. This data provides a historical perspective, helping city administrators understand long-term trends and patterns. For instance, census data can reveal demographic shifts, property records can shed light on housing trends, and crime reports can help identify areas with high crime rates.

12.4.1.4 Methods and Technologies for Data Collection

The methods and technologies used for data collection have evolved significantly with the advent of digital technology. Traditional methods, such as surveys and manual data entry, are increasingly being supplemented or replaced by automated data collection systems. For instance, IoT devices and sensors typically use wireless communication technologies to transmit data to a central server or cloud-based system. This data is then processed and analyzed using advanced data analytics tools. Similarly, data from social media feeds is collected using web scraping tools and APIs. These tools extract data from social media platforms and convert it into a structured format for further analysis.

Public records, on the other hand, are often digitized and made available through online databases. This data can be accessed and downloaded for analysis using various data extraction tools.

Data collection in the context of cities involves gathering information from a wide range of sources, using a variety of methods and technologies. This data is crucial for understanding the functioning of the city, informing policy decisions, and improving city services. As technology continues to evolve, the scope and scale of data collection are likely to increase, providing even more detailed and timely insights into urban life. However, data collection must be conducted in a responsible and ethical manner, respecting privacy rights and ensuring data security.

12.4.2 Data Storage and Organization

Data storage and organization is a critical aspect of data management and analytics. Once data is collected, it needs to be stored and organized in a way that facilitates analysis. This involves the use of databases, data warehouses, and data lakes, as well as data management systems that ensure data quality and consistency.

12.4.2.1 Databases

A database is a structured set of data. It allows for the storage, retrieval, and manipulation of data. Databases can be relational, where data is organized into tables and relationships are defined between different tables, or non-relational, where data is stored as documents, key-value pairs, wide-column stores, or graph databases. The choice of database often depends on the nature of the data and the specific requirements of the application.

12.4.2.2 Data Warehouses

A data warehouse is a large, centralized repository of data that is used for reporting and data analysis. It integrates data from multiple sources, providing a unified view of the data. Data warehouses are optimized for analytical processing and are typically used for business intelligence (BI) and reporting purposes. They use a different design from operational databases, with data organized around subject areas and following a schema that is optimized for analytics. It is also known as an enterprise data warehouse (EDW), is a system that aggregates data from different sources into a single, central, consistent data store to support data analysis, data mining, AI, and ML. It is a type of data management system designed to enable and support BI activities, especially analytics.

Data warehouses are solely intended to perform queries and analysis and often contain large amounts of historical data. They are designed to handle large amounts of data (petabytes and petabytes) in ways that a standard database cannot. Over time, it builds a historical record that can be invaluable to data scientists and business analysts.

12.4.2.3 Architecture of a Data Warehouse

Data warehouses generally have a three-tier architecture as shown in Table 12.34.

12.4.2.4 Components of a Data Warehouse

A typical data warehouse has four main components, namely (1) A central database for storing and managing data, (2) Extract, Transform, and Load (ETL) tools for preparing the data for analysis, (3) Metadata, which is data about data, providing context and additional information about the data, and (4) Access tools for visualizing and presenting data to business users.

12.4.2.5 Benefits of a Data Warehouse

Data warehouses offer unique benefits that allow organizations to analyze large amounts of variant data and extract significant value from it. They provide a unified view of the data, making it easier for users to access and analyze data. They also allow for historical analysis, enabling organizations to track trends and patterns over time.

Real-time analytics in the context of data warehouses refers to the ability to analyze data as it is created, captured, or updated. Traditional data warehouses are designed for batch processing, where data is loaded and processed at scheduled intervals. However, with the increasing need for real-time insights, data warehouses have evolved to support real-time analytics.

A real-time data warehouse takes the capabilities of a traditional data warehouse a notch higher by enabling the storage of real-time data and allowing you to analyze it nearly instantaneously. This means that the analytics is performed within a short time frame (from milliseconds to minutes) after the new data arrives, depending on the specific business needs and solution complexity.

TABLE 12.34
Architecture of Data Warehouse

Bottom tier	This consists of a data warehouse server, usually a relational database system. It collects, cleanses, and transforms data from multiple data sources through a process known as Extract, Transform, and Load (ETL) or a process known as Extract, Load, and Transform (ELT).
Middle tier	This consists of an online analytical processing (OLAP) server which enables fast query speeds.
Top tier	This is represented by some kind of front-end user interface or reporting tool, which enables end users to conduct ad-hoc data analysis on their business data.

The architecture of a real-time data warehouse involves several key processes:

1. *Data Ingestion:* Real-time data warehouses ingest real-time data with high throughput performance. Depending on the data source type and the physical distance between the data source and the analytics software, data can be ingested into the processing block by several means.
2. *Real-Time Storage:* The real-time storage acts as a buffer that ensures reliable queuing logic, for example, record ordering, scaling resources, delivering messages with minimal latency.
3. *Real-Time Processing and Analytics:* All data processing stages (data ingestion, enrichment, analytics, AI/ML-based analysis) are continuous, run with minimal latency, and enable real-time reporting and ad hoc analytics.

Real-time data warehouses enable organizations to analyze data at the time of creation and extract the value from the data in real-time. This dramatically simplifies storing, understanding, and retrieving this data. However, implementing real-time analytics requires thoughtful planning including the selection of appropriate technologies and the management of data quality and consistency.

Ensuring data consistency in a real-time data warehouse is crucial for maintaining the integrity of the data and ensuring that the same information is used across the organization. Strategies that organizations use to ensure data consistency are shown in Table 12.35.

12.4.2.6 Data Lakes

A data lake is a storage repository that holds a vast amount of raw data in its native format until it is needed. Unlike a data warehouse, which stores data in a structured format, a data lake stores data in an unstructured format. Data lakes are designed to store large volumes of data, including structured, semi-structured, and unstructured data. They are particularly useful for big data and real-time analytics.

Data lakes offer several advantages over traditional databases, making them an increasingly popular choice for data storage and analysis. These advantages are shown in Table 12.36.

Data lake architecture is the foundation for data analytics, reporting, and decision-making. It is designed to store and process data from a variety of sources, changing the way businesses use data. The key components of a data lake architecture are shown in Table 12.37.

TABLE 12.35
Strategies to Ensure Data Consistency

Data standardization	One of the most effective strategies for ensuring data consistency is data standardization. This involves defining a standard data format and structure that is used across all systems. This not only ensures consistency but also simplifies data integration and sharing.
Handling null or invalid values	During the load process, data warehouse developers typically institute design rules to ensure that null or invalid values are properly handled. This is usually accomplished by replacing these values with values that are more meaningful to an end user.
Coordination mechanisms	Achieving strong consistency usually involves using coordination mechanisms like distributed transactions or consensus algorithms to make sure the data stays intact and in sync across the entire distributed system.
Data quality metrics	To test any data quality dimension, you must measure, track, and assess a relevant data quality metric. In the case of data consistency, you can measure the number of passed checks to track the uniqueness of values, uniqueness of entities, corroboration within the system, or whether referential integrity is maintained.

TABLE 12.36
Advantages of Data Lakes in Comparison to Traditional Database

Separation of storage and compute	Unlike traditional systems where computer and storage resources were combined, data lakes enable the separation of computer and storage. This ensures that each resource can be scaled as needed, often reducing costs.
Data integration for better business intelligence and big data analytics	Data lakes are designed to be queryable, meaning they can be easily analyzed using a variety of tools. This makes them ideal for extracting insights from large data sets for purposes such as predictive analytics, machine learning, and data visualization.
Ability to store large volumes of data	Data lakes can store terabytes and petabytes of data from various sources. They are typically both large and inexpensive, making them well suited to the rapid increase in data volumes seen in recent years.
Support for various data sources and data structures	Data lakes can store data from multiple sources and multiple data structures in the same repository. This includes structured, semi-structured, and unstructured data, a diverse approach that would not be possible in a traditional database.
Scalability	Data lakes are designed to handle large volumes of data without compromising performance. This is particularly important as organizations build large, expanding data repositories.
Flexibility	Data lakes offer rapid, flexible data ingestion and storage. They can store any format and size of data, allowing a variety of data types and data sources to be available in one location, which supports statistical discovery.
Cost-effective	Data lakes are often designed for low-cost storage, so they can house a high volume of data at a relatively low price.

TABLE 12.37
Key Components of Data Lake Architecture

Raw data layer	This is the foundational layer of a data lake where data in its raw format is stored. The data can be structured, semi-structured, or unstructured.
Storage layer	At the core of data lake architecture lies a storage layer designed for durability and scalability. It is designed to protect data from corruption or loss and guarantee constant uptime. It also allows for arbitrarily changing voluminous data.
Processing layer	Data lake processing involves one or more processing engines built with the goals of operating on data stored in a data lake at scale.
Data integration layer	This layer is responsible for integrating data from various sources and providing a unified view of the data.
Security and governance layer	This layer ensures data protection and access control. It manages who can access the data and for what uses.

Without proper management and governance, a data lake can turn into a "data swamp" where it is difficult to locate and retrieve information. Ensuring data protection and access control can also become complex because of the volume and diversity of the data.

As discussed earlier, data lakes and data warehouses are both widely used for storing big data, but they are not the same and serve different purposes. The key differences between them are shown in Table 12.38.

12.4.2.7 Data Management Systems

Data management systems are software solutions that enable organizations to manage their data effectively. They provide functionalities for data integration, data quality management, data

TABLE 12.38
Key Differences between Data Lakes and Data Warehouses

Aspect	Data Lakes	Data Warehouses
Data type	Data lakes store raw data, including structured, semi-structured, and unstructured data from various sources.	Data warehouses store processed data that is structured and organized according to specific metrics and attributes.
Data processing	In a data lake, the process is ELT, meaning the data is loaded in its raw format and transformed when it is ready to be used.	In contrast, a data warehouse uses the ETL process, where data is transformed before it is loaded into the warehouse.
Schema	Data lakes use a schema-on-read approach, which means the schema is applied only when the data is read. This offers agility and easy data capture.	Data warehouses, however, use a schema-on-write approach, where the schema is applied before the data is stored. This offers security and high performance.
Users	Data lakes are typically used by data scientists who need in-depth analysis and tools such as predictive modeling.	Data warehouses are more commonly used by business professionals for operations.
Flexibility	Data lakes are highly flexible and can store any type of data.	Data warehouses, however, are less flexible and are designed for specific purposes.
History	Data lakes are a relatively new concept for big data.	The concept of a data warehouse has been around for decades,

governance, and data security. Data management systems ensure that data is accurate, consistent, and secure.

12.4.2.8 Data Quality and Consistency

Data quality refers to the condition of a set of values of qualitative or quantitative variables. It covers various aspects, including accuracy, completeness, consistency, timeliness, validity, and uniqueness. High-quality data is crucial for reliable and meaningful analysis.

Data consistency means ensuring that data is the same across all systems and platforms. It is crucial for maintaining the integrity of the data and ensuring that the same information is used across the organization.

Data storage and organization are vital aspects of data management and analytics. They involve storing and organizing data in a way that facilitates analysis, using databases, data warehouses, and data lakes. Data management systems ensure data quality and consistency. As the volume and variety of data continue to grow, effective data storage and organization will become increasingly important for organizations. All these processes must comply with data privacy and security regulations to ensure that sensitive information is protected.

12.4.3 Data Analysis

Data analysis is a critical process in the data management lifecycle. It involves examining, cleaning, transforming, and modeling data to discover useful information, draw conclusions, and support decision-making. Here are some of the key techniques used in data analysis.

12.4.3.1 Statistical Analysis

Statistical analysis is the collection, interpretation, presentation, and modeling of data to highlight patterns and trends. It involves descriptive statistics (mean, median, mode, range, etc.), inferential statistics (hypothesis testing, regression analysis, etc.), and predictive statistics (forecasting, etc.). It is used to understand the underlying patterns and trends in data, and to make informed decisions.

12.4.3.2 Data Mining

Data mining is the process of discovering patterns and knowledge from large amounts of data. The data sources can include databases, data warehouses, the internet, and other information repositories. Data mining uses a variety of techniques, including association, classification, clustering, prediction, and sequential patterns.

12.4.3.3 Predictive Modeling

Predictive modeling uses statistics to predict outcomes. It involves creating a statistical model, which is then used to predict future behavior. Predictive modeling can be used in various fields, such as finance, marketing, healthcare, and more.

12.4.3.4 Machine Learning

ML is a type of AI that provides systems with the ability to learn and improve from experience without being explicitly programmed. It focuses on the development of computer programs that can access data and use it to learn for themselves. The process of learning begins with observations or data, such as examples, direct experience, or instruction, to look for patterns in data and make better decisions in the future based on the examples that we provide.

12.4.3.5 Artificial Intelligence

AI is the simulation of human intelligence processes by machines, especially computer systems. These processes include learning (the acquisition of information and rules for using the information), reasoning (using rules to reach approximate or definite conclusions), and self-correction. AI is used in a wide range of fields, including healthcare, education, finance, and more.

The goal of data analysis is to turn raw data into actionable information. This involves extracting insights from data and using these insights to make informed decisions. The techniques used in data analysis, such as statistical analysis, data mining, predictive modeling, ML, and AI, all contribute to this goal.

Ensuring the quality of data for analysis is indispensable in data management. High-quality data is essential for making well-informed decisions, performing accurate analyses, and developing effective strategies. Strategies that organizations use to ensure data quality are shown in Table 12.39.

TABLE 12.39
Strategies to Ensure Data Quality

Establish data governance policies	Creating data governance policies ensures uniformity in handling and managing data throughout your organization.
Offer data quality training	Training staff on the importance of data quality and how to maintain it can help improve the overall quality of data.
Keep documentation accurate and up to date	Accurate documentation helps ensure that data is correctly interpreted and used.
Implement data validation techniques	Data validation techniques, such as range checks and consistency checks, can help identify and correct errors in the data.
Implement feedback loops	Feedback loops allow for continuous improvement by identifying errors and implementing corrective actions.
Use data cleansing tools	Data cleansing tools can help identify and correct errors in the data, improving its overall quality.
Monitor data quality metrics	Monitoring data quality metrics can help identify areas for improvement and track the effectiveness of data quality initiatives.

TABLE 12.40
Data Quality Dimensions

Accuracy	This dimension measures the degree to which data represents real-world things, events, or an agreed-upon source. For example, if a prospective employee has an inaccurate interview address, he will not be able to attend the interview until he obtains the accurate address.
Completeness	This dimension checks if all the required data is present. Incomplete data can lead to incorrect analysis and decision-making.
Consistency	This dimension ensures that data is consistent across different datasets and databases. Inconsistent data can lead to confusion and misinterpretation.
Uniqueness	This dimension checks for duplicate entries in the dataset. Duplicate entries can skew the analysis and lead to incorrect conclusions.
Timeliness	This dimension checks whether the data is up to date for its intended use. Outdated data can lead to incorrect analysis and decision-making.
Validity	This dimension checks whether the data conforms to the defined business rules or constraints.
Reliability	This dimension checks if the data is trustworthy and credible.
Currency	This dimension checks if the data is current and relevant.
Conformity	This dimension checks if the data follows the defined standards.
Integrity	This dimension checks the accuracy and consistency of data over its entire lifecycle.
Precision	This dimension checks the level of detail in the data.

The quality of the data can be measured from the dimensions they possess. The data dimensions are measurement attributes of data, which we can individually assess, interpret, and improve. Some common data quality dimensions are shown in Table 12.40.

Metadata management can help to ensure data quality. Metadata provides insights into the structure, context, and interpretation of datasets, which is essential for maintaining data quality. Metadata is data that provides information about other data. It is commonly referred to as “data about data”. Metadata helps describe, organize, and manage various types of data, making it easier to find, understand, and utilize.

Some examples of metadata: Let us say we have a digital photo file on our computer. The actual photo content is the data, but the metadata associated with that photo file could include things such as the following: (1) *File Name:* “family_vacation_2023.jpg”. This provides a descriptive name for the photo file. (2) *File Type:* JPEG. This indicates the file format of the photo. (3) *File Size:* 2.5 MB. This shows the size of the photo file. (4) *Date Created:* May 1, 2024. This is the date the photo was taken, or the file was created. (5) *Camera Make and Model:* Canon EOS Rebel T7. This metadata provides information about the camera used to capture the photo. (6) *GPS Coordinates:* 40.730610, –73.935242. This metadata includes the geographic location where the photo was taken. The data in (1) through (6) above is metadata. Due to this characteristic, metadata is important in data quality. The ways metadata management contributes to data quality are shown in Table 12.41.

12.4.4 Data Visualization and Reporting

In the age of big data, the ability to effectively communicate and interpret the insights derived from data analysis has become increasingly crucial. Data visualization and reporting play a vital role in this process, transforming complex data into easily digestible and impactful representations. Data visualization refers to the graphical representation of information and data. It involves the use of various visual elements, such as charts, graphs, maps, and infographics, to convey complex data in a clear and intuitive manner. The primary goal of data visualization is to facilitate understanding and enable effective decision-making.

TABLE 12.41
Contribution of Metadata Management on Data Quality

Understanding data	Metadata describes the content, format, and context of data, enabling efficient search operations and reducing the time and effort required to find specific information. It helps users understand what the data represents, how it is structured, and how it relates to other data.
Data lineage	By documenting data lineage, metadata enables us to trace the source of the data, identify any transformations it has undergone, and verify its accuracy and reliability. This is particularly important for ensuring data integrity and trustworthiness.
Data standardization	Metadata can help standardize data by defining a common set of terms and definitions. This ensures that everyone in the organization understands the data in the same way, promoting consistency and reducing errors.
Data integration	Metadata plays a key role in data integration. It provides the necessary information for mapping data from different sources, ensuring that the data is correctly aligned and integrated.
Data governance	Metadata is a critical component of data governance. It provides the information needed to establish policies, procedures, and responsibilities for managing data.
Automation	Metadata can be used to automate many data management tasks, such as data validation, data transformation, and data loading. This can improve efficiency and reduce the risk of errors.

When data has been thoroughly analyzed, the results need to be presented in a way that resonates with the intended audience, whether they are executives, policymakers, or the general public. Data visualization techniques can help bridge the gap between the technical complexities of the analysis and the need for clear, actionable insights.

Some common data visualization techniques include the following: (1) *Bar Charts:* Used to compare quantities or frequencies across different categories or time periods. (2) *Line Charts:* Effective for displaying trends and patterns over time. (3) *Scatter Plots:* Useful for identifying relationships and correlations between variables. (4) *Pie Charts:* Illustrate the proportional size of different parts of a whole. (5) *Histograms:* Visualize the distribution and frequency of numeric data. (6) *Heat Maps:* Represent data using a color-coded grid, often used to identify patterns and outliers. (7) *Geographical Maps:* Visualize data with a spatial component, such as regional or demographic information.

The selection of the appropriate visualization technique depends on the nature of the data, the key insights to be communicated, and the intended audience. Effective data visualization not only presents the information but also highlights the most relevant and meaningful aspects, allowing users to quickly grasp the underlying patterns, trends, and relationships.

Beyond static visualizations, reporting tools have emerged as a powerful means of interactively exploring and communicating data insights. These tools provide users with the ability to interact with the data, drill down into specific areas of interest, and customize the visual representation to suit their needs.

Reporting tools often incorporate features such as the following: (1) *Interactive Dashboards:* Consolidated views that bring together multiple visualizations and data sources, allowing users to monitor key metrics and trends. (2) *Filtering and Sorting:* Capabilities that enable users to slice and dice the data based on specific criteria, such as time period, location, or product category. (3) *Data Exploration:* The ability to drill down into the data, uncover underlying details, and discover new insights through intuitive navigation and exploration. (4) *Collaborative Features:* Functionalities that allow users to share reports, annotations, and insights with others, fostering collaboration and collective decision-making. (5) *Automated Reporting:* The generation of scheduled reports or alerts based on predefined thresholds or events, ensuring timely access to critical information.

These reporting tools empower users to take an active role in understanding and interpreting the data, rather than passively consuming static reports. By providing an interactive and user-friendly interface, these tools enable users to explore data, test hypotheses, and make more informed decisions.

The benefits of effective data visualization and reporting extend beyond just improving comprehension. They can also facilitate the following: (1) *Enhance Decision-Making:* By presenting data in a clear and compelling manner, visualizations and reports can help decision-makers quickly identify patterns, trends, and anomalies, leading to more informed and data-driven decisions. (2) *Foster Engagement and Collaboration:* Interactive visualizations and reporting tools can engage stakeholders, encourage dialogue, and facilitate the sharing of insights across an organization. (3) *Improve Communication and Storytelling:* Data visualization and reporting can transform complex data into compelling narratives, making it easier to communicate findings and recommendations to diverse audiences. (4) *Uncover Hidden Insights:* The ability to explore data interactively can reveal unexpected connections, patterns, and insights that might have been overlooked in traditional data analysis. (5) *Increase Data Literacy:* By making data more accessible and understandable, visualization and reporting tools can help improve the overall data literacy of an organization, empowering employees to make better use of data in their decision-making processes.

As the volume and complexity of data continue to grow, the importance of effective data visualization and reporting will only increase. Organizations that invest in developing these capabilities will be better equipped to derive meaningful insights, communicate their findings, and ultimately make more informed and impactful decisions.

Creating effective data visualizations is an art that requires a balance of simplicity, clarity, and accuracy. Table 12.42 shows some best practices for creating effective data visualizations.

12.4.5 Data Governance and Privacy

As the volume and importance of data continue to grow, the need for robust data governance and privacy practices has become increasingly critical. Data governance and privacy are essential

TABLE 12.42
Best Practices for Data Visualization

Keep it simple	Data visualization is all about communicating data insights in a way that is both accurate and easy to understand. When creating visuals, less is almost always more.
Use intuitive visual cues	Visual cues like color, size, and position can help viewers understand the data more quickly and easily.
Avoid distracting elements	Unnecessary decorations or complex designs can distract from the data. Stick to simple and clean designs.
Tell a story	Good data visualizations tell a story. They should have a clear beginning, middle, and end, and guide the viewer through the data.
Use data points carefully	Too many data points can make visualization confusing. Use them sparingly and only when necessary.
Use visual hierarchy effectively	Visual hierarchy helps guide the viewer's eye to the most important parts of the visualization.
Use effective labels	Labels should be clear and concise, and they should accurately represent the data.
Test data visualizations with users	Always test your visualizations with users to ensure they are easy to understand and interpret.
Use color to highlight important information	Color can be used to differentiate or compare data, or to draw attention to key points.
Ensure your data is clean	Dirty or inaccurate data can lead to misleading visualizations. Always clean your data before visualizing it.
Choose the right chart	Different types of data and different messages require different types of charts. Choose the one that best fits your data and your message.

components of the overall data management strategy, ensuring that data are used and handled in a responsible, compliant, and ethical manner.

Data governance refers to the framework of policies, procedures, and accountabilities that define how data is managed within an organization. It establishes the rules, roles, and responsibilities for data ownership, access, security, and usage. Effective data governance helps organizations ensure the quality, consistency, and integrity of their data assets.

The core of data governance lies in the need to address the growing concerns around data privacy. Data privacy encompasses the protection of personal and sensitive information from unauthorized access, use, or disclosure. As individuals and organizations generate and share more data than ever before, the risks of data breaches, misuse, and privacy violations have also increased substantially.

To address these concerns, data governance and privacy practices must be aligned with relevant legal and regulatory requirements. Some examples of the data privacy laws and regulations that organizations must comply with include:

- *GDPR:* Implemented by the European Union, GDPR sets strict standards for the collection, processing, and storage of personal data, with significant penalties for non-compliance.
- *Health Insurance Portability and Accountability Act (HIPAA):* In the healthcare sector, HIPAA mandates the protection of sensitive patient information and establishes guidelines for data security and privacy.
- *CCPA:* This California-specific law grants consumers more control over their personal information and imposes data protection requirements on businesses.
- *Payment Card Industry Data Security Standard (PCI DSS):* This industry standard sets the requirements for the secure handling of payment card data to prevent fraud and data breaches.

In addition to legal compliance, effective data governance and privacy practices also need to address ethical considerations. Organizations must ensure that data is collected, processed, and used in a way that respects individual privacy, promotes transparency, and builds trust with customers, employees, and stakeholders. Key elements of a comprehensive data governance and privacy framework include:

- *Data Policies and Standards:* Clearly defined policies that outline data ownership, access controls, data classification, and acceptable use.
- *Data Security and Risk Management:* Robust security measures, such as encryption, access controls, and incident response plans, to protect data from unauthorized access, modification, or destruction.
- *Data Lifecycle Management:* Processes for the proper handling of data throughout its lifecycle, including data collection, storage, retention, and secure disposal.
- *Data Privacy and Consent Management:* Mechanisms to ensure that personal data is collected and used with the knowledge and consent of individuals, in alignment with privacy regulations.
- *Data Stewardship and Accountability:* Defined roles and responsibilities for data stewards, who are responsible for ensuring data quality, security, and compliance within their respective domains.
- *Data Auditing and Monitoring:* Regular reviews and audits to assess the effectiveness of data governance and privacy practices, as well as the identification and mitigation of potential risks.
- *Employee Training and Awareness:* Comprehensive training programs to educate employees on data governance and privacy best practices, as well as their individual responsibilities in maintaining data integrity and security.

Several benefits can be obtained by organizations through implementing a robust data governance and privacy framework. The benefits include:

- *Regulatory Compliance:* Adherence to relevant data privacy laws and regulations, mitigating the risk of costly fines and legal penalties.
- *Data Security and Risk Mitigation:* Reduced vulnerability to data breaches, unauthorized access, and other security threats, protecting the organization's reputation and assets.
- *Improved Data Quality and Consistency:* Consistent data management practices that ensure the accuracy, completeness, and reliability of data across the organization.
- *Enhanced Trust and Transparency:* Demonstrating a commitment to responsible data management and privacy protection, building trust with customers, employees, and stakeholders.
- *Informed Decision-Making:* Reliable and trustworthy data that can be leveraged to drive more informed and data-driven decision-making.

Organizations that proactively address these critical areas will be better equipped to navigate the complexities of data management, ensure regulatory compliance, and maintain the trust of their stakeholders.

12.5 DIGITAL SERVICES

In the 21st century, digital transformation has revolutionized the way we access and utilize a wide range of services. From public sector to private enterprises, the integration of digital technologies has paved the way for a new era of convenience, efficiency, and accessibility. This digital landscape encompasses a diverse array of services that cater to the evolving needs of individuals, businesses, and communities.

At the forefront of this digital revolution are online public services, which leverage technology to streamline governmental processes and enhance citizen engagement. Through e-government platforms, citizens can access a plethora of services, from filing taxes to obtaining permits and licenses, all with the click of a button. Similarly, digital healthcare services have revolutionized the delivery of medical care, enabling remote consultations, telemedicine, and personalized treatment plans. E-learning services have also transformed the educational landscape, providing learners with unprecedented access to a wealth of knowledge and interactive learning experiences, irrespective of geographical barriers.

The digital ecosystem also encompasses the thriving world of e-commerce and online business services. Consumers can now conveniently browse, purchase, and receive a wide range of products and services from the comfort of their homes, while businesses leverage digital tools to reach new markets, optimize operations, and enhance customer experiences. Furthermore, the digital entertainment and media landscape has expanded, offering a diverse array of digital content, streaming platforms, and immersive experiences that cater to the diverse preferences of modern-day consumers.

12.5.1 Online Public Services

In the digital age, city governments are increasingly embracing the power of technology to enhance the delivery of public services to citizens. Online public services have become a crucial component of modern e-government initiatives, transforming the way individuals interact with their local authorities and access essential resources.

The rise of online public services can be attributed to the growing demand for more efficient, transparent, and accessible government operations. Citizens, accustomed to the convenience and responsiveness of private-sector digital services, have come to expect similar experiences when engaging with their city governments. Online public services address this expectation by leveraging digital platforms and technologies to streamline administrative processes, improve communication, and empower citizens.

At the core of online public services are integrated web-based portals that serve as a one-stop-shop for a wide range of municipal services. These portals allow citizens to perform a multitude of tasks, such as:

1. *Paying Taxes and Utility Bills:* Citizens can conveniently make payments for property taxes, water bills, parking tickets, and other municipal fees through secure online payment gateways, eliminating the need for physical visits to government offices.
2. *Applying for Permits and Licenses:* From building permits and business licenses to pet registrations and event permits, citizens can submit applications, track the status of their requests, and receive approvals through the online portal.
3. *Reporting Issues and Service Requests:* Citizens can use the portal to report public works concerns, such as potholes, abandoned vehicles, or street light outages, enabling city officials to quickly address and resolve these issues.
4. *Accessing Public Records:* Online public service portals often provide citizens with the ability to access and download various public records, such as property deeds, birth certificates, and government meeting minutes, reducing the need for in-person visits to municipal offices.
5. *Engaging with Local Government:* The digital platform can facilitate two-way communication, allowing citizens to submit inquiries, provide feedback, or participate in public consultations, fostering greater civic engagement and transparency.

Beyond the convenience of these core services, online public service portals often integrate additional features that enhance the user experience and the overall efficiency of municipal operations. Some of these advanced capabilities include:

1. *Personalized Dashboards:* Authenticated users can access personalized dashboards that display relevant information, such as upcoming deadlines, payment history, and service request status, enabling them to stay informed and proactive in their interactions with the city.
2. *Mobile-Friendly Design:* Many online public service portals are optimized for mobile devices, allowing citizens to access services and information on-the-go, further enhancing the accessibility and convenience of these digital offerings.
3. *Multilingual Support:* To cater to diverse communities, online public service portals may offer translations or language-switching capabilities, ensuring that all citizens can navigate and utilize the available services.
4. *Integrated Payment Systems:* In addition to standard payment options, online public service portals may integrate with electronic wallets, credit card processors, or other payment gateways, providing citizens with a seamless and secure payment experience.
5. *Integration with Other Municipal Systems:* Online public service portals can be connected to other city databases and information systems, enabling the sharing of data and the streamlining of cross-departmental processes, such as coordinating service requests or updating property records.

The implementation of online public services delivers a range of benefits to both citizens and city governments. For citizens, these digital platforms offer enhanced convenience, improved access to services, and greater transparency in government operations. Residents can interact with their local authorities on their own schedules, without the need for physical visits or prolonged waiting times.

For city governments, online public services enable more efficient and cost-effective service delivery. By automating administrative tasks and reducing the workload on municipal staff, cities can reallocate resources to focus on strategic priorities and improve the overall QoS. Additionally, the data collected through online public service portals can provide valuable insights for data-driven decision-making, allowing city officials to identify trends, address pain points, and optimize service delivery.

Despite the numerous advantages, the implementation of successful online public services emerges with its challenges. Ensuring data privacy and security, maintaining user trust, and ensuring digital inclusivity for all citizens are critical considerations. City governments must invest in robust cybersecurity measures, data governance frameworks, and user-centric design to address these concerns and foster seamless and secure digital experiences. Moreover, the adoption and utilization of online public services may vary among different demographic groups, requiring city governments to implement targeted outreach and digital literacy initiatives to ensure equitable access and participation. This may involve providing training, technical support, and alternative channels for citizens who may not be fully comfortable with or have access to digital technologies.

As the digital transformation continues to reshape the landscape of public service delivery, the role of online public services will only become more critical. City governments that embrace this shift and prioritize the development of user-friendly, secure, and inclusive digital platforms will be better equipped to meet the evolving needs and expectations of their constituents, ultimately enhancing the overall quality of life and civic engagement within their communities.

12.5.2 Digital Healthcare Services

Digital Healthcare Services was partly discussed in Smart Healthcare Systems. However, another part of these systems is discussed in this section. Digital Healthcare Services are transforming the way healthcare is delivered, making it more accessible, efficient, and personalized. This revolution is driven by advancements in digital technology, including telemedicine platforms, online appointment scheduling, electronic health records (EHRs), and mobile health apps.

12.5.2.1 Telemedicine Platforms

Telemedicine platforms allow healthcare professionals to provide medical services remotely, using digital communication tools. These platforms enable real-time interaction between patients and healthcare providers, regardless of their geographical location. Telemedicine can include services like virtual consultations, remote patient monitoring, and digital diagnostic services. It is particularly beneficial for patients in remote areas, those with mobility issues, or during public health crises like the COVID-19 pandemic.

12.5.2.2 Online Appointment Scheduling

Online appointment scheduling systems are digital platforms that allow patients to book, reschedule, or cancel their appointments with healthcare providers. These systems offer convenience and flexibility, as patients can manage their appointments 24/7 from any location. They also improve efficiency for healthcare providers by automating the scheduling process, reducing administrative workload, and minimizing patient no-shows.

12.5.2.3 Electronic Health Records

EHRs are digital versions of patients' health records, which can be shared across different healthcare settings. They contain comprehensive information about a patient's medical history, including diagnoses, medications, treatment plans, immunization dates, allergies, radiology images, and laboratory test results. EHRs improve the quality of care by providing healthcare providers with accurate and up-to-date information, facilitating coordinated care, and reducing medical errors.

12.5.2.4 Mobile Health Apps

Mobile health apps are applications designed to provide health services and manage patient health via mobile devices. These apps can perform various functions, such as tracking physical activity,

TABLE 12.43
Mobile Health Apps Roles

Empowering patients	Mobile health apps empower patients to take control of their health. They offer features that let patients monitor their health metrics, such as physical activity, diet, and medication adherence. This enables better self-management of chronic conditions and allows patients to take immediate action if health metrics increase or decrease.
Enhancing communication	These apps facilitate communication between patients and healthcare providers. They enable remote monitoring and virtual consultations, which can lead to timely interventions and improved health outcomes.
Improving accessibility	Mobile health apps increase accessibility to healthcare services. They are particularly beneficial for people living in remote or rural areas who may lack access to proper healthcare. These apps allow users to book appointments, receive reminders, and even buy medicines online.
Supporting physicians	Healthcare apps can support physicians by providing a faster, more precise approach to diagnosis. Medical professionals can use their phones to view all patient reports, which lets them make quick and correct decisions in emergency cases.
Promoting preventative care	Many health apps focus on preventative care by helping users track daily steps, access exercise and nutrition programs, and manage their health conditions. This can lead to improved health outcomes and reduced healthcare costs.

monitoring vital signs, providing health education, and facilitating communication with healthcare providers. They empower patients to take an active role in managing their health and wellness.

Mobile health apps are indispensable in modern healthcare by empowering patients, improving communication, increasing accessibility, supporting physicians, and promoting preventative care. They hold the potential to significantly improve patient outcomes and transform the healthcare landscape. The ways Mobile health apps can improve patient's outcomes are shown in Table 12.43.

The ubiquity of mobile applications in the digital age has profoundly influenced human life. These compact, yet powerful tools have become indispensable companions, seamlessly integrating into our daily routines. In the healthcare realm, mobile apps have emerged as invaluable resources, empowering users with a wealth of information and capabilities. From fitness trackers to medication management, thousands of mobile applications cater to a wide range of health-related needs. Table 12.44 highlights several successful examples of these mobile health solutions, underscoring the transformative impact of technology on modern healthcare delivery. The convenience and accessibility of these apps have revolutionized the way individuals manage their well-being in the digital era.

AI can be seamlessly integrated into healthcare apps, revolutionizing the delivery of medical services. These integrations can include symptom analysis and diagnosis, personalized treatment recommendations, medication management, and predictive analytics to anticipate and address potential health concerns. Utilizing the power of AI enables healthcare apps to provide more accurate, efficient, and tailored support to users, thereby enhancing overall patient outcomes and well-being. AI can be integrated into healthcare apps in various ways:

1. *Automating Administrative Tasks:* AI can automate mundane activities like administrative work and data entry. This includes maintaining records, authorizing insurance, following up on unpaid bills, scheduling appointments, and managing electronic health records. This automation frees up time for healthcare workers to focus on more value-adding work.
2. *Enhancing Care Delivery:* AI can revolutionize healthcare by improving care outcomes and the productivity and efficiency of care delivery. It can also improve the day-to-day life of healthcare practitioners, letting them spend more time looking after patients.

TABLE 12.44
Examples of Health-Related Mobile Apps

8fit	8fit provides customized workout plans and meal plans to help you get fit	Macros	Calorie counter and meal planner, track your macros and plan your meal.
Charity miles	With Charity Miles, you can earn money for charity whenever you walk, run, or bike.	Noom	Noom is a weight loss app that provides personalized diet plans and tracking tools.
Calorie	Calorie is a food tracking app that provides detailed nutrition information for a wide range of foods.	Pocket yoga	This app provides yoga workouts that you can do at home.
Fitplan	Fitplan offers personalized workout plans to help you achieve your fitness goals.	Shopwell	Shopwell helps you make healthier choices at the grocery store by providing nutrition information for different foods.
Headspace	Headspace is a mental wellness app that provides guided meditation sessions and mindfulness training.	Sleep cycle	This app tracks your sleep patterns and wakes you up during your lightest sleep phase.
MyFitnessPal	This app helps you track your eating and exercise habits and lose weight.	Strava	Strava is a fitness tracking app that provides detailed statistics about your running and cycling activities.

3. *Predictive Analytics and Personalized Treatment Plans:* AI can be used to diagnose, create personalized treatment plans, and even predict patient survival rates. It sifts through large volumes of big data efficiently to identify patterns, anomalies, and trends.
4. *Robotic Surgical Equipment:* AI can be outfitted with robotic surgical equipment to help surgeons better perform surgeries by decreasing their physical fluctuations and providing updated information during the operation.
5. *ML, Deep Learning, Neural Language Processing (NLP), and Robotic Process Automation (RPA):* These are some of the most common forms of AI used within healthcare. They are used to categorize information, predict outcomes, understand human language, interpret documentation, notes, reports, and published research, and automate administrative and clinical workflows.

Integrating AI into healthcare software involves defining your healthcare organization's needs, finding trustworthy developers to implement the solution, choosing the right AI solution, planning for integration, training, testing, and deploying the solution, and monitoring and refining your healthcare AI system. However, integrating AI into healthcare apps comes with challenges, including ensuring data privacy and confidentiality, achieving interoperability with existing healthcare systems, managing scalability and performance issues, navigating the complex regulatory landscape, and prioritizing user experience.

There are some ethical concerns on the use of AI in medical diagnosis as shown in Table 12.45.

These are some of the ethical issues that must be carefully considered as AI becomes more integrated into medical diagnosis and healthcare. Proactive measures are needed to address these concerns.

12.5.3 E-Learning Services

E-Learning Services encompass a wide range of digital platforms used for education and learning. These platforms have revolutionized the way we learn, making education more accessible, flexible,

TABLE 12.45
Ethical Concerns of AI in Medical Diagnosis

Transparency and explainability	AI systems used for medical diagnosis can be complex "black boxes" where the decision-making process is not transparent or easily explainable to patients and doctors. This lack of transparency can make it difficult to understand how the AI reached its conclusions.
Bias and fairness	AI models can learn and perpetuate biases present in the training data, leading to unfair or discriminatory diagnoses for certain demographic groups. Ensuring AI systems are fair and equitable is crucial.
Privacy and data security	Medical data used to train AI models is highly sensitive. There are concerns around data privacy, security, and the potential for misuse of patient information.
Liability and accountability	When an AI system makes a medical mistake, it is unclear who is liable – the healthcare provider, the AI developer, or the hospital. Clear frameworks for accountability are needed.
Overreliance on AI	Over-trusting AI systems could lead to doctors over-relying on the technology and failing to use their own clinical judgment and expertise. This could negatively impact patient care.
Access and equity	The high costs of developing and deploying AI systems could restrict access to only the wealthiest healthcare providers, widening healthcare disparities.
Human oversight	There are concerns that over-automation of medical diagnosis could reduce the role of human doctors, leading to a loss of the human touch and personalized care.
Validation and regulation	Rigorous validation of AI systems for clinical use and clear regulatory guidelines are necessary to ensure safety and efficacy before deployment.

and personalized. They include online courses, virtual classrooms, digital libraries, and learning management systems.

12.5.3.1 Online Courses

Online courses are a form of e-learning where students can study a particular subject or skill over the internet. They are often delivered through a learning management system, which provides a structured environment for course content, assessments, and communication. Online courses can be self-paced, allowing students to learn at their own speed, or instructor-led, where students follow a set schedule. They can cover a wide range of subjects, from academic topics to professional development and personal interest areas.

12.5.3.2 Virtual Classrooms

Virtual classrooms are digital spaces where students and teachers interact in real-time. They replicate the experience of a physical classroom, with features such as live video conferencing, interactive whiteboards, and breakout rooms for group work. Virtual classrooms enable synchronous learning, where students and teachers are online at the same time, facilitating immediate feedback and interaction. They are particularly useful for distance learning, where students and teachers are geographically dispersed.

12.5.3.3 Digital Libraries

Digital libraries are online databases that provide access to a vast array of resources, including books, journals, images, videos, and other multimedia content. They allow users to search, retrieve, and use these resources from anywhere, at any time. Digital libraries can support learning by providing students with a wealth of information at their fingertips. They can also facilitate research, with advanced search capabilities and access to a wide range of academic sources.

12.5.3.4 Learning Management Systems

Learning Management Systems (LMS) are software applications used for the administration, documentation, tracking, reporting, automation, and delivery of educational courses, training programs, or learning and development programs. They provide a centralized platform where educators can create, manage, and deliver educational content, and students can access this content, interact with their peers, and track their progress. LMSs can support a variety of learning approaches, including blended learning, flipped classrooms, and mobile learning.

E-Learning Services have transformed the landscape of education, breaking down geographical barriers, and providing flexible and personalized learning experiences. They leverage digital technology to enhance accessibility, efficiency, and quality of education. E-learning further revolutionizing education, making it more learner-centered and data-driven. However, along with its evolvement, it is necessary to address challenges related to digital literacy, data privacy, and quality assurance to realize the full potential of e-learning.

One of the challenges of the E-learning systems is the quality and effectiveness of online courses. While e-learning offers flexibility and accessibility, ensuring consistent high-quality content, engaging learning experiences, and effective assessment methods can be a significant hurdle. Instructional design, faculty training, and technological infrastructure are crucial factors in overcoming this challenge and delivering impactful online education. The strategies that the educators can apply to cope with these challenges are shown in Table 12.46.

Fostering a sense of community in virtual classrooms is necessary for creating an engaging and effective learning environment. Some strategies that individual educators can apply include:

- *Establishing Digital Community Agreements:* Use a live class session to develop community agreements to guide virtual class interaction. These agreements should outline expectations

TABLE 12.46
Strategies for Successful E-Learning Applications

Clear learning objectives	Each course should have clear and measurable learning objectives. These objectives guide the course design and help students understand what they are expected to learn.
Engaging course activities	Course activities should be engaging and aligned with the learning objectives. They should promote active learning and provide opportunities for students to apply what they have learned.
Effective student–teacher interaction	Regular interaction between students and teachers is crucial for online learning. This can be facilitated through discussion forums, video conferences, and timely feedback on assignments.
Use of assessment data	Assessment data can be used to evaluate teaching effectiveness and student learning outcomes. This data can inform adjustments to teaching methods and course content.
Peer review and external examiners	Peer review and external examiners can provide valuable feedback on the appropriateness and effectiveness of the course.
Student feedback	Student feedback is key to designing an effective course. Regular surveys can be conducted to gather student perceptions of the course quality.
Continuous improvement	Quality online learning requires a commitment to continuous improvement. This involves regularly reviewing and updating course content, teaching methods, and technology tools.
Ensuring accessibility	Online courses should be accessible to all students, including those with disabilities. This involves considering factors like course design, content format, and technology requirements.
Professional development for educators	Educators should be provided with training and support to develop their skills in online teaching. This includes understanding how to use technology effectively and how to engage students in an online environment.

for respectful communication, active participation, and collaborative learning. Revisiting these agreements during every class session reinforces their importance and ensures a cohesive virtual learning environment.

- *Creating a Strong Class Culture:* A strong class culture is essential for a successful virtual classroom. This can be achieved by establishing traditions, such as weekly check-ins or virtual social activities, assigning different roles to students to foster engagement, and encouraging continuous interaction through discussion forums and collaborative projects. A vibrant class culture promotes a sense of belonging and commitment among students.
- *Providing Supplementary Resources and Support:* Offering extra materials, such as video lessons or discussion guides, suggesting relevant readings, and connecting students to academic support services can help foster a sense of community. These supplementary resources demonstrate the instructor's investment in student success and create opportunities for further engagement.
- *Promoting Active Participation:* Encourage active participation by limiting time spent lecturing and instead utilizing video, chat, and other interactive features to engage students. This encourages learners to actively contribute their ideas, ask questions, and collaborate with their peers, fostering a dynamic and engaging virtual classroom experience.
- *Regular Check-Ins and Updates:* Regularly reaching out to students, whether through individual check-ins or dedicated class time, and allowing opportunities for personal and professional updates can help build a sense of community in the virtual classroom. These interactions demonstrate the instructor's care and concern for students' well-being, while also encouraging students to share their experiences and connect with one another on a more personal level. This fosters a supportive and inclusive virtual learning environment.
- *Hosting Digital Advisory Sessions:* Use dedicated digital advisory sessions to discuss academic, social, or career-related issues with students. These one-on-one or small group sessions provide a safe and confidential space for learners to share their concerns, seek guidance, and build stronger relationships with the instructor and their peers. These advisory sessions can significantly enhance the sense of community and support within the virtual classroom.

Building a sense of community in a virtual classroom requires consistent effort and engagement from both educators and students. Instructors must intentionally design and facilitate activities that foster connections, while students must actively participate and contribute to the learning environment. This collaborative approach is essential for cultivating a vibrant and supportive virtual community.

12.5.4 E-Commerce and Online Business Services

The digital transformation of the business world has led to the widespread adoption of e-commerce and online business services. This section explores the various digital platforms and tools that have revolutionized the way companies conduct their day-to-day operations and engage with customers.

12.5.4.1 Online Marketplaces

One of the key components of e-commerce is online marketplaces. These digital platforms connect buyers and sellers, enabling businesses to reach a wider customer base and expand their reach beyond their local or regional markets. Examples of popular online marketplaces include Amazon, eBay, Etsy, and Alibaba. These platforms provide businesses with the infrastructure, tools, and customer base to list and sell their products or services globally. Businesses can leverage the built-in features, such as secure payment gateways, product categorization, and customer reviews, to streamline their online sales process and enhance the customer experience.

12.5.4.2 Digital Payment Systems

Closely tied to the growth of e-commerce is the rise of digital payment systems. These electronic payment platforms facilitate transactions between businesses and their customers, enabling seamless and secure financial transactions. Examples of widely adopted digital payment systems include PayPal, Stripe, Square, and Apple Pay. These services allow businesses to accept a variety of payment methods, including credit cards, debit cards, and digital wallets, without the need for traditional brick-and-mortar point-of-sale (POS) systems. The convenience, security, and global reach of digital payment systems have been instrumental in the expansion of online businesses, as they enable customers to make purchases with ease and confidence.

12.5.4.3 Online Banking and Financial Services

In addition to digital payment systems, the e-commerce landscape has also witnessed the proliferation of online banking and financial services. Businesses can now manage their financial operations, access loans, and monitor their accounts through secure online platforms. Online banking services provided by traditional financial institutions, as well as the emergence of fintech (financial technology) companies, have made it easier for businesses to handle their financial needs digitally. This includes tasks such as invoicing, payroll management, expense tracking, and accessing business loans or lines of credit. The integration of these online financial services with e-commerce platforms and business software has streamlined the financial management processes for companies of all sizes.

12.5.4.4 Business Software as a Service

Another critical component of the e-commerce and online business services ecosystem is the adoption of cloud-based business software, commonly known as Software as a Service (SaaS). These web-based applications are designed to support various business functions, including customer relationship management (CRM), project management, accounting, human resources, and more. Examples of popular SaaS offerings include Salesforce for CRM, Zoho for productivity and collaboration, Intuit QuickBooks for accounting, and Zenefits for HR management. By leveraging SaaS solutions, businesses can access robust software functionalities without the need for on-premise hardware or complex IT infrastructure. These cloud-based tools are typically subscription-based, scalable, and accessible from any device with an internet connection, making them particularly beneficial for small and medium-sized enterprises (SMEs) that may lack the resources for traditional software deployment and maintenance.

12.5.4.5 The Synergy of E-commerce and Online Business Services

The integration and synergy of these various e-commerce and online business services have transformed the way companies operate and interact with their customers. By leveraging the capabilities of online marketplaces, digital payment systems, online banking, and cloud-based business software, companies can streamline their operations, improve efficiency, and enhance the customer experience. For example, a small e-commerce business can use an online marketplace to list its products, accept digital payments, manage its finances through online banking, and utilize SaaS tools for inventory management, customer support, and analytics. This holistic approach to leveraging digital platforms and services allows businesses to scale their operations, reduce costs, and stay competitive in the rapidly evolving e-commerce landscape.

12.5.5 Digital Entertainment and Media Services

In the rapidly evolving digital landscape, the way we consume and interact with entertainment and media has undergone a profound transformation. This section focuses on the digital platforms that have revolutionized the way we experience entertainment and access information, shaping the way we live, learn, and connect with the world around us.

12.5.5.1 Streaming Video and Music Services

The rise of streaming platforms has fundamentally changed the way we consume video and music content. Services like Netflix, Hulu, Amazon Prime Video, and Disney+ have disrupted the traditional cable and satellite television models, offering on-demand access to a vast library of movies, TV shows, documentaries, and original content. These platforms leverage advanced algorithms to provide personalized recommendations, making it easier for users to discover new and relevant content.

Similarly, music streaming services like Spotify, Apple Music, and Amazon Music have transformed the way we listen to music. These services provide users with instant access to millions of songs, playlists, and podcasts, allowing for a more personalized and on-the-go listening experience. The integration of features like curated playlists, artist radios, and personalized recommendations has further enhanced the user experience, making it easier for listeners to discover new music and explore different genres.

12.5.5.2 Online Gaming Platforms

The digital revolution has also had a profound impact on the gaming industry. Online gaming platforms, such as Steam, Epic Games Store, and Xbox Live, have revolutionized the way we access and enjoy video games. These platforms offer a centralized hub for purchasing, downloading, and playing a vast array of games, ranging from blockbuster titles to independent releases.

Beyond just providing access to games, these platforms often include social features, allowing gamers to connect with friends, join online communities, and engage in multiplayer experiences. The integration of features like cloud gaming, cross-platform play, and live streaming has further enhanced the gaming experience, making it more immersive, social, and accessible.

12.5.5.3 Digital News and Magazines

The transition from print to digital has significantly impacted the news and magazine industry. Digital news platforms, such as The New York Times, The Guardian, and CNN, offer users instant access to breaking news, in-depth reporting, and multimedia content. These platforms often leverage advanced technologies like AI and ML to personalize the news experience, providing users with content tailored to their interests and preferences. Digital magazine platforms, such as Apple News+ and Texture, provide users with access to a wide range of digital magazines, newspapers, and periodicals. These platforms offer a seamless and interactive reading experience, often with additional features like video, audio, and interactive content.

The rise of digital news and magazines has not only transformed the way we consume information but has also had a significant impact on the media industry, leading to new business models, revenue streams, and challenges in the realm of journalism and content creation.

12.5.5.4 Social Media Networks

Social media platforms have become an integral part of our digital lives, revolutionizing the way we connect, communicate, and share information. Platforms like Facebook, Twitter (X), Instagram, and TikTok have enabled users to create and share content, engage with others, and access a wealth of information and entertainment.

These social media networks have also become powerful platforms for businesses and organizations to reach and engage with their target audiences. The integration of features like targeted advertising, influencer marketing, and e-commerce capabilities has created new opportunities for brands and content creators to monetize their digital presence. However, the rise of social media has also raised concerns about privacy, data security, misinformation, and the impact on mental health and well-being. Navigating these challenges and ensuring the responsible use of social media platforms is an ongoing effort for both users and platform developers.

12.5.5.5 Challenges and Considerations

The digital transformation of entertainment and media services has brought about numerous benefits, such as increased accessibility, personalization, and the democratization of content creation. However, this shift has also introduced several challenges that must be addressed:

- *Content Curation and Overload:* With the abundance of content available on digital platforms, users can often feel overwhelmed by the sheer volume of options, leading to challenges in content discovery and curation.
- *Data Privacy and Security:* The collection and use of personal data by digital platforms raises concerns about privacy and the potential for misuse or breaches, requiring robust data governance and security measures.
- *Accessibility and Digital Divide:* Ensuring that digital entertainment and media services are accessible to users with diverse abilities and socioeconomic backgrounds is crucial to promoting digital inclusion and equity.
- *Regulation and Content Moderation:* As digital platforms become more influential in shaping our entertainment and information landscape, there is a growing need for effective regulation and content moderation to address issues like misinformation, hate speech, and harmful content.
- *Sustainability and Environmental Impact:* The energy-intensive nature of digital entertainment and media services, particularly in areas like streaming and cloud gaming, raises concerns about the environmental sustainability of these industries.
- *Evolving Business Models and Revenue Streams:* The shift to digital platforms has disrupted traditional business models in the entertainment and media industries, requiring companies to adapt and explore new revenue streams to remain competitive.

The digital transformation of entertainment and media services has profoundly impacted the way we consume, interact with, and experience content. From streaming video and music to online gaming and social media, these digital platforms have revolutionized the entertainment and information landscape, offering unprecedented access, personalization, and opportunities for content creators and users alike. However, this digital revolution has also introduced a range of challenges and considerations that must be addressed to ensure the sustainable and responsible development of these industries. As we navigate this evolving landscape, it is crucial for stakeholders, including platform providers, policymakers, and users, to collaborate and find innovative solutions that prioritize user privacy, content moderation, accessibility, and environmental sustainability.

Businesses can monetize their presence on social media networks in several ways as shown in Table 12.47.

12.6 CYBERSECURITY AND E-GOVERNANCE

Cybersecurity and e-governance are intertwined in the digital era, with cybersecurity measures being a critical component of effective e-governance. E-government services, which provide citizens with access to government services online, must prioritize data privacy and protection to ensure the security of personal information. This involves implementing robust cybersecurity measures, such as encryption, two-factor authentication, and regular security audits. Digital participation and inclusion are also key aspects of e-governance, ensuring that all citizens, regardless of their digital literacy or access to technology, can benefit from e-government services. This requires making these services accessible and user-friendly and providing digital literacy training to those who need it. Finally, regulation and compliance play a crucial role in e-governance and cybersecurity. Governments must establish clear regulations regarding data privacy and protection and ensure that e-government services comply with these regulations. This includes conducting regular compliance

TABLE 12.47
Monetizing the Business on Social Media

Affiliate marketing	Businesses can join affiliate marketing programs and earn a commission for promoting other companies' products or services.
Influencer partnerships	Businesses can partner with influencers who have a large following on social media. These influencers can promote the business's products or services to their followers.
Sponsored posts	Businesses can create sponsored posts that promote their products or services. These posts are typically marked as "sponsored" and can reach a larger audience.
Online courses	Businesses can create online courses related to their industry or expertise. These courses can be sold to their followers.
Online store	Businesses can run online stores and divert traffic to it from their social media accounts.
Consulting services	Businesses can offer their services as consultants. This can be particularly effective for businesses that specialize in a specific area.
Selling leads	Businesses can sell leads to other companies. This involves gathering information about potential customers and selling this information to businesses that can benefit from these leads.
Creator funds	Some social media platforms offer creator funds. These funds provide financial support to creators who produce high-quality content.
Engaging and shareable content	Creating engaging and shareable content can help businesses reach a larger audience. This can lead to increased brand awareness and potential sales.
Influencer marketing	Leveraging influencer marketing can help businesses reach a targeted audience. This involves partnering with influencers who can promote the business's products or services to their followers.
Social media advertising campaigns	Running social media advertising campaigns can help businesses reach a larger audience. These campaigns can be targeted to reach specific demographics, increasing the likelihood of reaching potential customers.
Social media analytics	Using social media analytics can help businesses track and measure performance. This can provide valuable insights into customer behavior and preferences, helping businesses tailor their marketing strategies.

audits and holding service providers accountable for any breaches. In conclusion, cybersecurity and e-governance are mutually reinforcing, with effective cybersecurity measures enhancing the trustworthiness and effectiveness of e-government services, and robust e-governance practices contributing to improved cybersecurity.

12.6.1 Cybersecurity Measures

In the digital age, where technology has become an integral part of our lives, the importance of cybersecurity cannot be overstated. As we increasingly rely on digital systems, networks, and data to power our personal and professional lives, the need to protect these assets from cyber threats has become paramount. This section delves into the various measures that organizations and individuals can take to secure their digital landscapes, ensuring the confidentiality, integrity, and availability of their critical information.

12.6.1.1 Firewalls: The First Line of Defense

Firewalls are the cornerstone of any cybersecurity strategy, serving as the gatekeeper between a network and the internet. These specialized hardware or software systems monitor and control the incoming and outgoing traffic, filtering out potential threats based on pre-defined rules and policies. Firewalls can be configured to block unauthorized access, prevent malicious activities, and safeguard sensitive data from being compromised. Placing firewalls strategically at the network perimeter allows organizations to establish a robust, multilayered defense against cyber-attacks.

12.6.1.2 Encryption: Protecting Data in Transit and at Rest

Encryption is a crucial mechanism for safeguarding data, both in transit and at rest. By converting readable information into an unreadable format, encryption ensures that even if data falls into the wrong hands, it remains unintelligible to unauthorized individuals. This technology is particularly important for protecting sensitive information, such as personal identification, financial data, and confidential business records. Various encryption algorithms, such as AES, RSA, and PGP, are commonly used to secure data, ensuring that only authorized parties with the appropriate decryption keys can access the information.

12.6.1.3 Intrusion Detection and Prevention Systems

Intrusion Detection and Prevention Systems (IDPS) are designed to monitor network traffic and system activities, identifying and responding to potential security breaches in real-time. These systems employ a combination of techniques, including signature-based detection, anomaly-based detection, and behavioral analysis, to identify and mitigate cyber threats. IDPS can detect and prevent a wide range of attacks, such as unauthorized access attempts, malware infections, and denial-of-service (DoS) attacks. By continuously monitoring and analyzing network traffic, IDPS can provide early warning signals, enabling security teams to respond promptly and effectively to minimize the impact of cyber incidents.

12.6.1.4 Secure Software Development Practices

Ensuring the security of software applications is crucial in the fight against cyber threats. Secure software development practices involve incorporating security considerations throughout the entire software development lifecycle, from the design phase to deployment and maintenance. This includes practices such as threat modeling, secure coding techniques, code reviews, and vulnerability assessments. By implementing these measures, organizations can reduce the risk of software vulnerabilities that could be exploited by attackers, thereby enhancing the overall security of their digital systems.

12.6.1.5 Incident Response and Recovery

Despite the best cybersecurity measures, organizations may still face the unfortunate reality of a successful cyber-attack. In such cases, having a well-defined incident response plan is crucial. This plan outlines the steps to be taken during and after a security breach, ensuring a coordinated and effective response. Key elements of an incident response plan include incident detection and analysis, containment and eradication of the threat, recovery of affected systems and data, and post-incident review and lessons learned. By having a comprehensive incident response plan in place, organizations can minimize the impact of a cyber-attack, restore normal operations, and implement measures to prevent similar incidents in the future.

12.6.1.6 Cybersecurity Awareness and Training

Cybersecurity is not solely a technical challenge; it also requires a strong human element. Employees and users play a crucial role in maintaining the overall security of an organization's digital systems. Implementing comprehensive cybersecurity awareness and training programs can help educate users on common cyber threats, such as phishing, social engineering, and password management. Creating a security-conscious culture within the organization enables individuals to become the first line of defense, recognize and report suspicious activities, and adhere to best practices for secure online behavior.

12.6.1.7 Continuous Monitoring and Improvement

Cybersecurity is an ever-evolving landscape, with new threats and vulnerabilities constantly emerging. To stay ahead of these challenges, organizations must adopt a mindset of continuous monitoring and improvement. This involves regularly reviewing and updating security policies, testing the effectiveness of implemented measures, and staying informed about the latest cybersecurity trends and best practices. Continuously monitoring and improving their cybersecurity posture enables organizations to proactively address emerging threats and adapt their security strategies accordingly.

12.6.1.8 Regulatory Compliance and Industry Standards

Many industries and sectors are subject to various regulatory requirements and industry standards related to cybersecurity. These guidelines, such as the GDPR, the HIPAA, and the PCI DSS, outline specific security controls and practices that organizations must adhere to. Compliance with these regulations is not only a legal requirement but also a critical aspect of maintaining the trust and confidence of customers, partners, and stakeholders. Aligning their cybersecurity measures with relevant industry standards and regulatory requirements enables organizations to demonstrate their commitment to data protection and risk management.

12.6.1.9 Leveraging Cybersecurity Frameworks and Best Practices

To enhance their cybersecurity posture, organizations can leverage well-established cybersecurity frameworks and best practices. These include frameworks such as the National Institute of Standards and Technology (NIST) Cybersecurity Framework, the ISO/IEC 27001 standard for information security management, and the Center for Internet Security (CIS) Controls. These frameworks provide a structured approach to identifying, assessing, and mitigating cyber risks, as well as ensuring the implementation of comprehensive security controls. Adopting these proven frameworks and best practices allows organizations to align their cybersecurity strategies with industry-recognized guidelines, ensuring a more robust and effective defense against cyber threats.

12.6.1.10 The Cybersecurity Implications of Cloud Computing

The widespread adoption of cloud computing has significantly transformed the digital landscape, offering organizations numerous benefits in terms of scalability, flexibility, and cost-effectiveness. However, this transition to cloud-based solutions has also introduced a new set of cybersecurity considerations that must be addressed.

One of the primary cybersecurity challenges associated with cloud computing is the issue of data control and sovereignty. When organizations store their data on cloud platforms, they relinquish a certain level of direct control over that data. This can raise concerns about data privacy, as the cloud provider may have access to sensitive information, and the data may be subject to the jurisdiction and regulations of the country where the cloud infrastructure is located. To mitigate these risks, organizations must carefully evaluate the security and privacy policies of their cloud service providers, ensuring that appropriate measures are in place to protect sensitive data.

Another crucial aspect of cloud computing cybersecurity is the shared responsibility model. In this model, the cloud provider is responsible for securing the underlying infrastructure, such as the physical data centers, network, and virtualization layers, while the client organization is responsible for securing its own applications, data, and user access. This shared responsibility can create potential gaps in security if not properly understood and addressed by both parties. Organizations must work closely with their cloud providers to clearly define and implement security controls, access management, and incident response procedures to ensure a cohesive and effective cybersecurity strategy.

The complexity of cloud environments, with their distributed nature and reliance on various interconnected services, can also pose challenges for cybersecurity. Traditional security tools and techniques may not be fully effective in the cloud, as they may not be designed to handle the dynamic and scalable nature of cloud infrastructure. Organizations must adopt cloud-native security solutions and practices, such as cloud access security brokers (CASBs), cloud security posture management

(CSPM), and cloud workload protection platforms (CWPPs), to ensure comprehensive visibility, control, and protection across their cloud environments.

The increased attack surface in cloud computing is another key concern. As organizations move more of their operations and data to the cloud, the potential entry points for cyber threats also expand. Attackers can exploit vulnerabilities in cloud-based applications, APIs, and infrastructure to gain unauthorized access, disrupt operations, or steal sensitive information. To address this challenge, organizations must implement rigorous vulnerability management, secure coding practices, and cloud-specific security monitoring and incident response capabilities.

Finally, the reliance on third-party cloud service providers introduces an additional layer of risk. Cloud providers themselves can become targets of cyber-attacks, and a successful breach of a cloud provider could potentially impact multiple organizations that rely on their services. Conducting thorough due diligence on cloud service providers, including evaluating their security posture, incident response plans, and business continuity measures, is crucial to mitigate the risks associated with third-party dependencies.

To address the cybersecurity challenges posed by cloud computing, organizations must adopt a comprehensive and proactive approach. This includes:

- *Establishing a Robust Cloud Security Governance Framework:* Organizations should define clear policies, roles, and responsibilities for cloud security management, ensuring alignment across the organization and with cloud service providers.
- *Implementing Cloud-Specific Security Controls:* Organizations should leverage cloud-native security solutions, such as CASBs, CWPPs, and CSPM, to gain visibility, control, and protection across their cloud environments.
- *Enhancing Identity and Access Management (IAM):* Robust IAM practices, including multi-factor authentication, role-based access controls, and privileged access management, are crucial in the cloud context to prevent unauthorized access and data breaches.
- *Ensuring Data Protection and Encryption:* Organizations should prioritize the encryption of data at rest and in transit, both within their own cloud environments and when accessing cloud-based services, to safeguard sensitive information.
- *Continuous Monitoring and Threat Detection:* Implementing cloud-specific security monitoring and threat detection capabilities can help organizations quickly identify and respond to emerging cyber threats in their cloud environments.
- *Establishing Incident Response and Business Continuity Plans:* Comprehensive incident response and business continuity plans tailored to cloud environments are essential to ensure the organization's resilience in the face of a cyber-attack or other disruptive events.
- *Fostering Cloud Security Awareness and Training:* Educating employees on cloud security best practices, security responsibilities, and incident reporting procedures can enhance the overall security posture of the organization.

By addressing these considerations and adopting a proactive, cloud-centric cybersecurity approach, organizations can harness the benefits of cloud computing while mitigating the associated security risks. Collaboration with cloud service providers, adherence to industry standards and regulations, and continuous improvement in cloud security measures will be crucial in navigating the evolving cybersecurity landscape in the cloud era.

12.6.2 Data Privacy and Protection

The collection, storage, and processing of personal data have become ubiquitous across various sectors, from healthcare and finance to e-commerce and social media. As the volume and value of personal information continues to grow, the importance of ensuring data privacy and protection has become paramount. This section delves into the policies and practices that organizations must adopt to safeguard the confidentiality of personal data and comply with the evolving regulatory landscape.

At the core of data privacy and protection is the recognition that personal information is an asset that must be treated with the utmost care and respect. Individuals have a fundamental right to privacy, and organizations have a duty to protect the personal data entrusted to them. This includes sensitive information such as names, contact details, financial records, healthcare data, and online behavioral patterns. Failure to adequately protect such data can have severe consequences, ranging from financial penalties and reputational damage to the erosion of consumer trust and the potential for identity theft or other forms of exploitation.

To address these concerns, governments around the world have implemented comprehensive data protection regulations, such as the European Union's GDPR and the CCPA in the United States. These regulations establish clear guidelines and requirements for organizations handling personal data, including obtaining explicit consent for data collection, providing transparent privacy policies, and implementing robust security measures to safeguard the information. Compliance with these regulations is not only a legal obligation but also a critical aspect of maintaining the trust and confidence of customers, partners, and stakeholders.

One of the key principles of data privacy and protection is the concept of data minimization, which states that organizations should only collect and retain the minimum amount of personal data necessary to achieve their intended purpose. This helps to reduce the potential attack surface and mitigate the risks associated with data breaches or unauthorized access. Additionally, the practice of data anonymization, where personal identifiers are removed or obfuscated, can further enhance the privacy and security of data, allowing organizations to leverage the insights and benefits of the information without compromising individual privacy.

Secure data handling practices are another essential component of data privacy and protection. This includes the implementation of robust access controls, encryption of data both in transit and at rest, and the adoption of secure data storage and disposal methods. Organizations should also have clear policies and procedures in place for the handling of personal data, ensuring that employees are trained on the proper protocols and that any third-party service providers or vendors also adhere to the same standards of data protection. In addition to technical safeguards, the human element plays a crucial role in data privacy and protection. Employees must be educated on the importance of data privacy and trained to recognize and report any potential breaches or suspicious activities. Creating a culture of data stewardship, where individuals understand their responsibilities and the consequences of mishandling personal information, can significantly enhance an organization's overall data protection posture.

Incident response and breach notification are also critical aspects of data privacy and protection. In case of a data breach or cyber-attack, organizations must have a well-defined incident response plan that outlines the steps to be taken to contain the incident, mitigate the damage, and notify the affected individuals and relevant authorities in a timely manner. Prompt and transparent communication with stakeholders can help to build trust and mitigate the reputational and financial risks associated with a data breach.

Emerging technologies, such as AI and the IoT, have also introduced new challenges and considerations for data privacy and protection. As these technologies become more prevalent, organizations must carefully evaluate the data collection and processing practices associated with them, ensuring that the appropriate safeguards are in place to protect personal information and comply with regulatory requirements.

To address the evolving landscape of data privacy and protection, organizations must adopt a comprehensive and proactive approach. This includes:

- *Establishing Robust Data Governance and Privacy Policies:* Organizations should develop and regularly review their data privacy and protection policies, aligning them with the relevant regulations and industry best practices.

- *Implementing Data Protection by Design and by Default:* Incorporating privacy and security considerations into the design and development of products, services, and systems can help to mitigate risks from the outset.
- *Conducting Regular Data Protection Impact Assessments:* Evaluating the potential risks and implications of data processing activities can help organizations identify and address privacy concerns.
- *Investing in Data Protection Technologies:* Leveraging encryption, access controls, data masking, and other secure data handling technologies can enhance the overall protection of personal information.
- *Fostering a Culture of Data Stewardship:* Empowering employees to be responsible data stewards, through comprehensive training and awareness programs, can strengthen an organization's data privacy and protection posture.
- *Collaborating with Regulatory Authorities and Industry Partners:* Engaging with regulatory bodies and industry organizations can help organizations stay informed about the latest data privacy and protection requirements and best practices.

Through the adoption of a holistic and proactive approach to data privacy and protection, organizations can not only comply with the evolving regulatory landscape but also build trust and maintain the confidence of their customers, partners, and stakeholders. As the value and sensitivity of personal data continue to grow, the need for a robust and comprehensive data protection strategy will only become more crucial in the years to come.

12.6.3 E-Government Services

E-Government Services, also known as Electronic Government, refers to the use of digital technology to provide government services to citizens. This concept has revolutionized the way governments operate and interact with their citizens, making public services more accessible, efficient, and transparent.

12.6.3.1 Online Portals for Public Services

Online portals for public services are a key component of E-Government Services. These portals are essentially websites or applications that provide a single point of access to various government services. They are designed to be user-friendly, making it easy for citizens to find and use the services they need. These online portals offer a wide range of services. For example, a citizen can use a government portal to pay their taxes. This might involve logging into the portal, entering the relevant tax information, and making a payment. The portal would then provide a receipt or confirmation of the payment. Similarly, a citizen could use a government portal to apply for permits, such as a building permit or a fishing license. The process would typically involve filling out an online application form and paying any required fees. Other services that might be available through a government portal include renewing licenses (such as a driver's license), registering for social security benefits, and accessing healthcare services.

The use of online portals for public services offers several benefits for citizens. First and foremost, it provides convenience. Citizens can access the services they need from the comfort of their own home, at any time of day. This eliminates the need to visit a physical office and wait in line.

In addition, online portals can save citizens' time. The process of using an online portal is often faster than the traditional method of accessing government services. For example, paying taxes through an online portal can be done in a matter of minutes, whereas visiting a tax office to make a payment could take several hours.

Online portals for public services also offer benefits for the government. By providing services online, governments can reduce the administrative burden on their staff. This can result in cost savings, as fewer staff members are needed to process applications and payments. Furthermore, online portals can improve the efficiency of government operations. They allow for the automation of various processes, such as the collection of payments and the processing of applications. This can help to speed up the delivery of services and improve the overall QoS.

Online portals for public services provide a convenient and efficient way for citizens to access a wide range of government services, while also offering benefits for government in terms of reduced administrative burden and improved operational efficiency. As technology continues to develop, we can expect to see further advancements in this area, offering even greater benefits to citizens and governments alike.

12.6.3.2 Digital Identity Systems

Digital identity systems are a crucial aspect of E-Government Services. They provide each citizen with a unique digital identity, which can be used to access government services online. This digital identity is like a digital passport, providing proof of identity in the digital world.

Each citizen's digital identity is unique. This uniqueness is typically ensured using unique identifiers, such as a Social Security number in the United States or Aadhaar number in India. When a citizen registers for a digital identity, they provide certain personal information, which is then linked to their unique identifier. This information might include their name, date of birth, address, and biometric data such as fingerprints or facial recognition data.

Once a citizen has a digital identity, they can use it to access a wide range of government services online. This might include services such as filing taxes, applying for benefits, renewing a driver's licence, or registering to vote. The digital identity serves as proof of identity, allowing the citizen to access these services without needing to provide additional documentation or visit a government office in person.

Digital identity systems typically incorporate robust security measures to protect citizens' personal information. This might include encryption, two-factor authentication, and secure storage of personal data. Encryption ensures that the data is unreadable to anyone who does not have the correct decryption key. Two-factor authentication adds an additional layer of security by requiring the citizen to provide two forms of identification, such as a password and a fingerprint. Secure storage ensures that the personal data is stored in a way that is safe from unauthorized access.

Digital identity systems also streamline the process of identity verification. Without a digital identity system, verifying a citizen's identity can be a time-consuming and complex process, requiring multiple forms of identification and often involving manual checks. With a digital identity system, the process can be automated and completed in a matter of seconds. This not only makes it easier for citizens to access services, but also allows governments to ensure that services are being provided to the right individuals.

Digital identity systems are the core in E-Government Services. They provide a secure and efficient way for citizens to prove their identity and access government services online. However, it's important to note that the implementation of digital identity systems requires careful planning and management to ensure the security of citizens' personal information and the integrity of the system. As technology continues to evolve, we can expect to see further advancements in digital identity systems, offering even greater benefits to citizens and governments alike.

12.6.3.3 Mobile Government Applications

The advent of smartphones has transformed the way we live, work, and interact with the world around us. Recognizing this, many governments have embraced the use of mobile applications to deliver services to their citizens. These applications, often referred to as "m-Government" or "mGov" services, represent a significant evolution in e-Government services.

One of the primary advantages of mobile government applications is their accessibility. As most people carry their smartphones with them at all times, these applications allow citizens to access government services anytime, anywhere. This is a significant improvement over traditional methods, which often require citizens to visit government offices during specific hours or navigate complex websites.

Mobile government applications can offer a wide range of services. For instance, a citizen could use a government mobile application to report a pothole in their street. This not only makes it easier for citizens to communicate with their local government, but also allows the government to respond more quickly and efficiently. Similarly, mobile applications can simplify the process of requesting official documents. For example, a citizen could request a birth certificate through a government mobile application, eliminating the need to visit a government office or mail in a paper application.

In some countries, mobile government applications have even been used to facilitate voting in elections. This can make the voting process more accessible and convenient, potentially increasing voter turnout.

While mobile government applications offer many benefits, they also raise important considerations about security and privacy. Governments must ensure that these applications are secure, and that citizens' personal information is protected. This includes implementing robust authentication mechanisms, encrypting sensitive data, and complying with privacy laws and regulations.

Mobile government applications represent a significant advancement in the delivery of government services. By utilizing the widespread use of smartphones, these applications can make government services more accessible, efficient, and responsive. However, it is important for governments to address the associated security and privacy challenges to ensure the trust and confidence of their citizens. As technology continues to evolve, we can expect to see further innovations in this area, offering even greater benefits to citizens and governments alike.

12.6.3.4 Improving Transparency and Efficiency

E-Government Services also play a significant role in improving the transparency and efficiency of government operations. By digitizing services, governments can collect, process, and analyze data more effectively. This allows for better decision-making and policy formulation. For example, data collected from online tax payments can provide insights into economic trends, which can inform fiscal policy. Moreover, digital technology enables governments to share information with citizens more easily, promoting transparency. Citizens can access information about government budgets, contracts, and performance metrics, fostering accountability and trust in government.

The implementation of e-government services presents both opportunities and challenges for governments. On the positive side, digital technologies can lead to significant improvements in service quality, responsiveness, and cost-effectiveness. E-government initiatives can also enhance citizen satisfaction, trust in public institutions, and overall government performance. However, the successful deployment of e-government services requires thorough planning, significant investments in infrastructure and digital capabilities, and the effective management of complex technological and organizational changes. Governments must address a range of issues, including data security and privacy, digital literacy and inclusion, interoperability between different systems, and the integration of e-government services with legacy public administration frameworks. Moreover, the transition to e-government is not a one-size-fits-all solution. Each country, and even individual government agencies, must tailor their e-government strategies to their unique political, cultural, and socio-economic contexts. This often involves striking a balance between the standardization of digital services and the customization of solutions to meet the diverse needs of citizens.

Despite these challenges, the potential benefits of e-government are substantial. Embracing digital technologies enables governments to enhance the quality, accessibility, and responsiveness of public services, ultimately improving citizens' overall well-being and satisfaction. As the digital transformation of public administration continues to evolve, e-government will play an increasingly important role in shaping the future of government-citizen interactions and the delivery of essential public services.

TABLE 12.48
Constraints in e-Government

Incompatible systems and lack of integration guidance	E-Government initiatives often involve integrating various systems across different departments or agencies. These systems may be incompatible due to differences in technology, data formats, or protocols. Additionally, there may be a lack of clear guidance on how to integrate these systems, leading to inefficiencies and potential failures
Resource constraints	Implementing E-Government services can be resource-intensive, requiring significant investments in technology, infrastructure, and human resources. For many governments, especially those in developing countries, these resources may be scarce.
Lack of management commitment	The success of E-Government initiatives often depends on the commitment and support of top management. Without this commitment, initiatives may lack the necessary resources, face resistance from staff, or fail to align with broader organizational goals
Need for training and cultural change	The shift to E-Government services often requires significant changes in organizational culture and work processes. Staff may need training to develop new skills, and citizens may need to be educated about how to use the new services.
Security and privacy concerns	E-Government services often involve the collection, storage, and processing of sensitive personal data. This raises concerns about data protection, privacy, and cybersecurity. Governments must implement stringent security measures and comply with legal and regulatory requirements to ensure the trust and confidence of citizens.
Ineffective governance and lack of web-related capabilities	Ineffective governance structures and a lack of web-related capabilities can limit the impact of e-government efforts. This includes the ability to manage complex projects, coordinate across different agencies, and develop user-friendly web services.
Reluctance to allow user participation	Some governments may be reluctant to allow user participation in the creation of applications and content. This can limit the effectiveness of e-government services, as user input can help ensure that services are responsive to the needs and preferences of citizens.

E-government services, while transformative and beneficial, do face several constraints and challenges as reflected in Table 12.48.

12.6.4 Digital Participation and Inclusion

Digital participation and inclusion refer to the use of technology to engage citizens in governance and ensure that everyone has equal access to digital resources. This is a crucial aspect of modern governance, as it can enhance democratic processes, improve public services, and reduce social inequalities.

Online platforms for citizen engagement are a key tool for promoting digital participation. These platforms can take various forms, such as websites, social media platforms, or mobile applications. They provide a space where citizens can interact with government officials, participate in policy discussions, provide feedback on public services, and engage in other forms of civic participation. For example, a city might have an online platform where citizens can report issues in their neighborhood, participate in community planning processes, or communicate directly with elected officials.

Digital literacy programs are essential for ensuring digital inclusion. These programs aim to equip citizens with the skills they need to use digital technologies effectively. This might include basic skills like using a computer or smartphone, as well as more advanced skills like navigating the internet, using digital government services, or understanding online privacy and security. Digital literacy programs can be targeted at different groups, such as children, adults, seniors, or marginalized communities, depending on their specific needs.

TABLE 12.49
Aspects to be Considered in Measuring the Impacts of Digital Participations

Use of metrics and qualitative analysis	Governments can measure the impact of digital transformation through metrics and by the qualitative analysis of their business. These metrics can include customer satisfaction scores, revenue growth and retention, and employee productivity
Clearly defined indicators	The best metrics have a clearly defined and defensible causal relationship to a business outcome, work as a leading, not lagging, indicator, address a specific, defined audience, can be understood by a non-IT audience, and drive action when they change from green to yellow to red.
Data analytics	Data analytics, preferably performed by an objective outside consultant, is the best way to measure impact. If that is more cost prohibitive, internal data collection, which will show both positive and negative aspects, can provide a good mirror on how a government is serving the community
Participant-centered approaches	There is a growing recognition of the need for alternative, participant-centered approaches to evaluation. These approaches involve the participants in the evaluation process, ensuring that the evaluation reflects their experiences and perspectives.
Ongoing evaluation	Governments should regularly review and update their evaluation methods to ensure they continue to accurately measure the impact of their digital participation initiatives.

The digital divide refers to the gap between those who have access to digital technologies and those who do not. This divide can be due to various factors, such as income, education, age, geographic location, or disability. Initiatives to bridge the digital divide aim to ensure that everyone has equal access to digital resources. This might involve providing affordable internet access, distributing digital devices, creating accessible digital content, or offering support services for those who struggle with technology.

Digital participation and inclusion involve creating online platforms for citizen engagement, offering digital literacy programs, and implementing initiatives to bridge the digital divide. While these efforts can be challenging, they are crucial for promoting democracy, improving public services, and reducing social inequalities in the digital age. As technology continues to evolve, we can expect to see further innovations in this area, offering even greater benefits to citizens and governments alike.

Measuring the impact of digital participation initiatives is an interesting task that requires a multidisciplinary approach. The aspects to measure the impacts of digital participation are shown in Table 12.49.

12.6.5 Regulation and Compliance

Regulation and compliance refer to the legal and procedural aspects of cybersecurity and e-governance. This involves the creation and enforcement of laws and regulations designed to protect digital information and transactions, as well as the role of regulatory bodies in overseeing these processes.

12.6.5.1 Cybercrime Laws and Regulations

Cybercrime laws and regulations are designed to prevent and punish illegal activities conducted over the internet. This includes activities such as hacking, identity theft, phishing, and the distribution of malicious software. These laws define what constitutes cybercrime, establish penalties for those who commit such crimes, and provide a framework for law enforcement agencies to investigate and prosecute these crimes.

12.6.5.2 Data Protection Laws and Regulations

Data protection laws and regulations are designed to protect the privacy and integrity of personal data. They establish rules for how personal data can be collected, stored, processed, and shared. These laws also give individuals certain rights with respect to their personal data, such as the right to access their data, the right to correct inaccurate data, and the right to have their data deleted.

12.6.5.3 Electronic Transactions Laws and Regulations

Electronic transactions laws and regulations provide a legal framework for conducting transactions over the internet. They define what constitutes a valid electronic contract, establish rules for electronic signatures, and provide protections for consumers who engage in electronic commerce.

12.6.5.4 Role of Regulatory Bodies

Regulatory bodies play a crucial role in overseeing and enforcing these laws and regulations. They monitor compliance, investigate potential violations, and take enforcement action when necessary. These bodies also provide guidance to individuals and organizations on how to comply with the laws and regulations, and they may play a role in resolving disputes related to cybersecurity and e-governance.

Regulation and compliance provide a legal and procedural framework for protecting digital information and transactions, and they play a crucial role in preventing and punishing cybercrime, protecting personal data, and facilitating electronic commerce.

12.7 SUSTAINABLE TECHNOLOGY

Sustainable technology is rapidly transforming the landscape of our digital world, paving the way for a more environmentally friendly future. At the heart of this transformation are energy-efficient data centers that optimize power usage, reducing the carbon footprint of our ever-growing digital data needs. Complementing this, cloud computing offers scalable and efficient computing solutions, minimizing the need for physical infrastructure and thereby conserving resources. Green networking strategies further enhance this conservation effort, optimizing data flow to reduce energy consumption. Meanwhile, e-waste management initiatives ensure responsible disposal and recycling of electronic waste, mitigating harmful environmental impacts. Lastly, the rise of telecommuting and virtual meetings has not only revolutionized our work culture, but also significantly reduced carbon emissions associated with travel. Together, these elements of sustainable technology are leading us toward a more sustainable and eco-friendly digital age.

12.7.1 Energy-Efficient Data Centers

Data centers are facilities that house computer systems and related components, such as telecommunications and storage systems. They are critical for the day-to-day operations of many businesses, providing services such as data storage, processing, and distribution. However, these facilities consume a significant amount of energy, primarily due to the need to power and cool the servers.

12.7.1.1 Advanced Cooling Systems

One of the key strategies for reducing energy consumption in data centers is the use of advanced cooling systems. Traditional data centers often use air conditioning to cool servers, but this can be energy intensive. Advanced cooling systems, such as liquid cooling or free cooling, can be more

TABLE 12.50
Strategies to Improve Data Center Efficiency

Monitoring and managing power in real-time	Real-time monitoring of power usage can help identify inefficiencies and areas for improvement. This can be particularly useful in identifying underutilized servers that can be turned off or repurposed.
Maintaining an accurate asset inventory	Keeping track of all hardware and software assets in a data center can help identify redundant systems and streamline operations.
Automating change management	Automating the process of implementing changes in the data center can reduce errors and improve efficiency.
Tracking efficiency key performance indicators (KPIs)	KPIs such as power usage effectiveness (PUE) can provide valuable insights into the efficiency of a data center.
Reporting on energy consumption	Detailed reports on energy consumption can help identify trends and areas for improvement.
Implementing autonomous optimization solutions	These solutions use artificial intelligence and machine learning to continuously analyze and optimize data center operations.
Optimizing data center space	Data centers built before the advent of server virtualization may be overbuilt for today's equipment needs, enabling further reduction of the necessary space for IT equipment and less IT power.
Optimizing data center cooling	To achieve a minimum level of efficiency, data center managers should ensure the adoption of basic data center cooling best practices

energy efficient. Liquid cooling uses a liquid coolant to absorb heat from the servers, while free cooling uses outside air or water to cool the servers when the outside temperature is lower than the temperature inside the data center.

12.7.1.2 Server Virtualization

Server virtualization is another effective strategy for reducing energy consumption. This involves using software to divide a physical server into multiple virtual servers, each of which can run its own operating system and applications. This allows for more efficient use of server resources, as multiple virtual servers can share the resources of a single physical server. As a result, fewer physical servers are needed, which reduces energy consumption.

12.7.1.3 Efficient Power Supply Units

Efficient power supply units (PSUs) can also help to reduce energy consumption in data centers. PSUs convert the alternating current (AC) power from the mains into direct current (DC) power for the servers. However, this conversion process can be inefficient, resulting in wasted energy. Efficient PSUs are designed to minimize this waste, thereby reducing energy consumption.

There are several strategies that can be employed to improve data center efficiency. These strategies are shown in Table 12.50.

12.7.2 Cloud Computing

Cloud computing is a model of computing where computing resources, such as storage, processing power, and software, are provided over the internet (the "cloud") rather than being hosted locally on a user's own computer or server.

The primary characteristics of cloud computing generally include the aspects as shown in Table 12.51.

TABLE 12.51
Characteristics of Cloud Computing

On-demand access	Cloud computing allows users to access and use computing resources whenever they need them, without the need to manage or maintain the underlying infrastructure.
Scalability and elasticity	Cloud computing resources can be easily scaled up or down based on the user's changing needs, providing flexibility and the ability to handle fluctuations in demand.
Resource pooling	Cloud service providers pool their computing resources, such as storage, processing power, and memory, and share them across multiple users or organizations, effectively utilizing the available resources.
Broad network access	Cloud computing resources are accessible over the internet, allowing users to access them from any location and a variety of devices, such as computers, smartphones, and tablets.
Measured service	Cloud computing services are typically billed on a pay-as-you-go basis, where users only pay for the resources they consume, enabling more efficient and cost-effective use of computing resources.

Cloud computing is a model for delivering IT services where resources are retrieved from the internet through web-based tools and applications, rather than a direct connection to a server. This significantly reduces the need for physical infrastructure, leading to less energy consumption.

12.7.2.1 Sharing of Resources

In cloud computing, resources such as storage, processing power, and memory are pooled together and allocated dynamically to multiple users based on demand. This means that instead of each user having their own dedicated physical resources, they share a pool of resources. This reduces the total amount of physical infrastructure required, which in turn reduces energy consumption.

12.7.2.2 Reduced Energy Consumption

Physical infrastructure like servers, cooling systems, and networks consume a significant amount of energy. Using shared resources in the cloud to reduce the need for this infrastructure can significantly decrease energy consumption. This approach not only saves costs but also promotes environmental sustainability.

12.7.2.3 Efficient Use of Hardware

Cloud computing allows for the efficient use of hardware. In traditional IT setups, hardware often sits idle or is underutilized. In contrast, cloud computing allows for the maximum utilization of hardware as resources can be allocated and de-allocated on demand. This means that hardware rarely sits idle, reducing waste.

12.7.2.4 Reduced Waste

With traditional IT setups, old or outdated hardware often end up as electronic waste. With cloud computing, the responsibility of managing and upgrading hardware falls on the cloud service provider. This means that hardware can be used for as long as it is useful and then properly recycled or disposed of by professionals, reducing electronic waste.

Some examples of cloud computing services are given below:

1. *SaaS:* These are cloud-based software applications that are accessed over the internet.
   - *Salesforce:* A CRM tool that helps businesses manage customer relations and generate sales leads.
   - *Zoom:* A cloud-based software platform for audio and video conferencing.

   - *StreamNative:* A cloud system that supports communities with solutions for accessing data, messaging queues, and event-sourcing.
2. *Infrastructure-as-a-Service (IaaS):* These services provide virtualized computing resources over the internet.
   - *DigitalOcean:* Provides cloud website hosting.
   - *IBM:* Offers bare metal IaaS servers.
3. *Platform-as-a-Service (PaaS):* These services provide a platform allowing customers to develop, run, and manage applications without the complexity of building and maintaining the infrastructure.
   - *AWS:* Offers a multi-language app platform.
   - *Heroku:* Provides a platform for deploying, running, and managing applications.
4. *File Sharing and Data Storage Services:* These services allow users to store files online and access them from any location.
   - *Dropbox:* A popular file-sharing and storage service.
   - *Google Drive:* Another widely used file-sharing and storage service.
5. *Big Data Analysis Services:* These services analyze large and complex volumes of data. Example: Civis Analytics provides customer insight software.
6. *Data Governance Services:* These services help organizations manage their data assets. Example: Carbonite offers data protection and cloud backup.
7. *Cybersecurity Services:* These services protect internet-connected systems from cyber threats. Example: Forcepoint provides a security platform.

These are just a few examples. There are many other cloud computing services available that cater to different needs and use cases.

12.7.3 Green Networking

Green Networking is a concept that aims to achieve energy-efficient networking technologies. It involves the design and operation of networking technologies and systems with minimal environmental impact. The primary goal is to reduce energy consumption and carbon emissions, thereby contributing to environmental sustainability. The need for green networking arises from the fact that ICT infrastructure, including data centers and network systems, consumes a significant amount of energy. This energy consumption contributes to greenhouse gas emissions, which are a major factor in global warming and climate change.

The concept of green networking focuses on three key areas: improving energy efficiency, reducing resource consumption, and leveraging renewable energy sources.

12.7.3.1 Improving Energy Efficiency

One of the primary goals of green networking is to enhance the energy efficiency of networking equipment and infrastructure. This involves optimizing the energy consumption of various networking components, such as routers, switches, and wireless access points.

1. *Hardware Optimization:* Manufacturers of networking devices are increasingly designing their products with power-saving features and energy-efficient components. This includes the use of low-power processors, energy-efficient circuitry, and advanced cooling systems that minimize the energy required for operation.
2. *Software-Driven Energy Management:* Network management software and protocols can be leveraged to intelligently manage the energy consumption of networking devices. This includes techniques like dynamic power scaling, which adjusts the performance of devices

based on traffic load, and sleep mode functionality, which allows devices to enter a low-power state during periods of inactivity.
3. *Network Infrastructure Optimization:* The design and layout of the network infrastructure can also contribute to improved energy efficiency. This may involve strategies like consolidating network equipment, implementing virtualization technologies, and optimizing the placement and utilization of network resources.

12.7.3.2 Reducing Resource Consumption

In addition to energy efficiency, green networking emphasizes the importance of reducing the consumption of other resources, such as materials and physical space.

1. *Miniaturization and Modularization:* Networking equipment is becoming increasingly compact and modular, allowing for more efficient use of physical space and reducing the overall material footprint. This trend is driven by advancements in semiconductor technology and the adoption of smaller, more integrated components.
2. *Recycling and Reuse:* At the end of their lifecycle, networking devices should be properly recycled or repurposed to minimize waste. This involves initiatives like equipment refurbishment, component reuse, and the recovery of valuable materials for future use.
3. *Virtualization and Cloud Computing:* The shift toward virtualization and cloud-based networking infrastructures can lead to significant reductions in physical hardware requirements, as resources can be shared and dynamically allocated based on demand.

12.7.3.3 Leveraging Renewable Energy Sources

To further minimize the environmental impact of networking, green networking approaches aim to utilize renewable energy sources, such as solar, wind, and hydroelectric power, to power networking equipment and infrastructure.

1. *On-Site Renewable Energy Generation:* Networking facilities, such as data centers and telecommunication hubs, can be equipped with on-site renewable energy generation systems, such as solar panels or wind turbines, to directly power their operations.
2. *Renewable Energy Procurement:* In cases where on-site renewable energy generation is not feasible, network operators can procure renewable energy from external sources, such as utility companies or specialized renewable energy providers, to offset the carbon footprint of their networking activities.
3. *Energy-Efficient Network Protocols and Algorithms:* The development and deployment of energy-efficient network protocols and algorithms can help optimize the energy consumption of networking devices and systems, further leveraging the potential of renewable energy sources.

The implementation of green networking practices has the potential to deliver significant environmental benefits, including reduced greenhouse gas emissions, lower energy consumption, and a decreased demand for finite natural resources. Moreover, green networking aligns with broader sustainability initiatives and can contribute to the achievement of global environmental goals, such as those outlined in the United Nations Sustainable Development Goals (SDGs). By embracing green networking, network operators and service providers can demonstrate their commitment to environmental stewardship and position themselves as leaders in the transition toward a more sustainable ICT landscape.

However, the adoption of green networking practices is not without its challenges. Factors such as the upfront costs of upgrading or replacing existing networking infrastructure, the need for specialized expertise and training, and the potential trade-offs between energy efficiency and network performance must be carefully considered. To overcome these challenges, a collaborative effort involving network equipment vendors, service providers, policymakers, and end-users is necessary. Governments and regulatory bodies can play a crucial role by implementing policies and incentives that encourage the adoption of green networking practices, while network operators and manufacturers can invest in research and development to drive further innovation in this field.

Through embracing the principles of green networking, the ICT industry can contribute to a more sustainable future, where the environmental impact of networking technologies is minimized, and the benefits of digital connectivity can be harnessed in an eco-friendly manner. As we navigate the ever-evolving landscape of technological advancement, green networking emerges as a critical component in our collective journey toward a greener, more resilient, and environmentally conscious digital world.

12.7.4 E-Waste Management

Electronic waste, or e-waste, refers to discarded electronic or electrical devices. Used electronics which are destined for refurbishment, reuse, resale, salvage recycling through material recovery, or disposal are also considered e-waste. With the rapid growth of technology, the generation of e-waste has become a significant issue in the field of IT and communication technology.

E-waste contains various toxic and hazardous materials such as lead, mercury, and cadmium, which can be harmful to the environment and human health if not properly managed. Improper disposal of e-waste can lead to serious environmental and health problems. For example, when e-waste is improperly disposed of in landfills, toxic materials can leach into the soil and groundwater, causing pollution.

Sustainable practices in e-waste management involve several key strategies as shown in Table 12.52.

TABLE 12.52
Sustainable Practices in E-Waste Management

Proper e-waste disposal	This involves disposing of e-waste in a manner that does not harm the environment. This can be achieved by using authorized e-waste recycling centers and disposal facilities that follow proper waste management protocols.
Recycling	E-waste contains several valuable materials such as copper, silver, gold, and palladium which can be recovered and reused through recycling. Recycling not only reduces the amount of waste that ends up in landfills but also helps conserve natural resources by reducing the need for mining new materials.
Reuse of electronic components	Many electronic components can be reused in other devices. For example, a hard drive from an old computer can be used as an external storage device. Reusing electronic components not only reduces e-waste but also saves money.
Promotion of green electronics	This involves promoting the use of electronic products that are designed to be environmentally friendly. Green electronics are designed to be energy efficient, use fewer hazardous materials, and be easy to recycle.
Consumer awareness and education	Consumers play a crucial role in e-waste management. By being aware of the issues related to e-waste and how to properly dispose of their old electronic devices, consumers can significantly contribute to sustainable e-waste management.
Government regulations and e-waste legislation	Government regulations and legislation can play a key role in e-waste management. By enforcing strict regulations on e-waste disposal and recycling, governments can ensure that e-waste is properly managed.

Businesses make significant contributions to the quantity of e-waste. Therefore, e-waste practices by businesses will have substantial influence on e-waste issues. Businesses can promote sustainable practices for e-waste management in several ways:

1. *Adopt a Circular Economy Approach:* Businesses can shift from a linear economy (take, make, dispose) to a circular economy by designing products with durability, repairability, and recyclability in mind. Encourage suppliers to provide products that are easily upgradeable or have modular components, extending the lifespan of devices and reducing the frequency of replacements.
2. *Implement E-Waste Recycling Programs:* Establishing efficient recycling systems helps recover valuable materials from discarded devices while minimizing the environmental impact of raw material extraction. Partnering with certified e-waste recyclers ensures the safe and responsible processing of electronics, adhering to strict environmental and data security standards.
3. *Promote Employee Awareness:* Educate employees about the importance of e-waste management and how to properly dispose of electronic devices. This can be done through training programs, workshops, and awareness campaigns.
4. *Establish a Device Reuse Program:* Extending the life of electronic devices through reuse is a sustainable approach. Companies can establish programs for refurbishing and repurposing electronics, making them available to individuals or organizations in need.
5. *Invest in the Future:* Businesses can invest in research and development to create more sustainable and energy-efficient electronic products.
6. *Encourage Repair over Replacement:* Promote the repair of electronic devices instead of replacing them. This not only reduces e-waste but also saves money.
7. *Understanding the Impact of E-Waste:* Electronic waste contains hazardous materials such as lead, mercury, cadmium and various other toxic substances. If not managed properly, these substances can contaminate soil, water and air, posing a serious threat to human health and ecosystems.

Through the adoption of these strategies, businesses are expected to be able to cut down their e-waste and contribute to environmental sustainability, and simultaneously enhance their brand image via corporate social responsibility.

12.7.5 Telecommuting and Virtual Meetings

Telecommuting, also known as remote work or working from home, is a work arrangement in which employees do not commute to a central place of work, such as an office building, warehouse, or store. Instead, they use telecommunication technology, such as the internet and telephone, to work from a remote location, usually their home. Virtual meetings are meetings that take place over the internet using video conferencing software, allowing participants to meet and collaborate from different locations.

Both telecommuting and virtual meetings contribute to saving energy and reducing carbon emissions in several ways:

12.7.5.1 Reduced Commuting

One of the most significant ways telecommuting and virtual meetings save energy is by reducing the need for commuting. Commuting, especially by car, consumes a significant amount of energy and contributes to carbon emissions. By eliminating the need for daily commuting, telecommuting and virtual meetings can significantly reduce energy consumption and carbon emissions.

12.7.5.2 Lower Energy Consumption in Office Buildings

Office buildings consume a significant amount of energy for heating, cooling, and lighting. By reducing the number of people who need to be in the office, telecommuting can significantly reduce the energy consumption of office buildings.

12.7.5.3 Reduced Business Travel

Virtual meetings can reduce the need for business travel, which often involves long-distance flights. Air travel is a significant contributor to carbon emissions, so reducing business travel can have a significant impact on a company's carbon footprint.

12.7.5.4 Efficient Use of Resources

Telecommuting and virtual meetings allow for more efficient use of resources. For example, instead of having a physical meeting room that is only used for a few hours a day, a virtual meeting room can be used by different teams throughout the day, making more efficient use of the company's resources.

12.7.5.5 Promotion of Digital over Physical

Telecommuting and virtual meetings promote the use of digital documents over physical ones, reducing the need for paper and other physical resources. This not only saves energy but also reduces waste.

Telecommuting and virtual meetings are effective strategies for businesses to save energy and reduce carbon emissions. They not only offer environmental benefits but also provide cost savings and increased flexibility for employees. As technology continues to develop, it is expected that more businesses will adopt these practices, leading to even greater energy savings and reductions in carbon emissions.

PROBLEMS AND ISSUES FOR DISCUSSION

1. How has the IoT revolutionized the way we interact with technology in our daily lives? What are the implications of a hyper-connected world where everyday objects enable data collection and automation on a massive scale?
2. Discuss the impact of artificial intelligence on information technology and its role in shaping the future of communication. How do AI-powered automation, machine learning algorithms, and natural language processing technologies influence the way we process information, communicate with each other, and interact with digital devices?
3. How do smart cities leverage technology to improve urban infrastructure and enhance quality of life for residents? What specific technological solutions such as IoT sensors, smart grids, autonomous vehicles, and data analytics are being implemented to address urban challenges related to transportation, energy efficiency, waste management, public safety, and healthcare?
4. What are the potential benefits and challenges of implementing 5G technology in smart cities? How does the ultra-fast and low-latency connectivity of 5G networks enable innovative applications such as real-time data analytics, augmented reality, and autonomous systems in urban environments, while also raising concerns about security, privacy, and digital divide issues?
5. Explore the ethical implications of big data analytics in information technology and its influence on decision-making processes. How do data-driven insights derived from massive datasets impact issues such as privacy, surveillance, bias, discrimination, and accountability in various sectors including healthcare, finance, marketing, law enforcement, and governance?

6. How has cloud computing transformed the way businesses store and access data, and what are the security considerations involved? What are the advantages of cloud-based services such as scalability, cost-efficiency, and remote accessibility, as well as the risks associated with data breaches, data loss, compliance regulations, vendor lock-in, and service availability?
7. Discuss the role of blockchain technology in ensuring secure and transparent transactions in the digital age. How does blockchain's decentralized and tamper-proof ledger system enable trustless peer-to-peer transactions, smart contracts, supply chain traceability, identity verification, voting systems, and tokenized assets, while addressing concerns about scalability, interoperability, energy consumption, and regulatory compliance?
8. What are the key factors that contribute to the success of smart city initiatives around the world? How do factors such as government leadership, public–private partnerships, citizen engagement, digital infrastructure, data governance, interoperability standards, scalability models, sustainability practices, and resilience planning influence the development and adoption of smart technologies in urban contexts?
9. Analyze the impact of social media on communication patterns and interpersonal relationships in the digital era. How do social networking platforms shape public discourse, influence consumer behavior, amplify misinformation, facilitate online activism, foster virtual communities, redefine celebrity culture, blur boundaries between personal and professional identities, and raise concerns about mental health, privacy invasion, cyberbullying, echo chambers, and filter bubbles?
10. How can cybersecurity measures be strengthened to protect sensitive information in an increasingly interconnected world? What are the best practices for securing networks, devices, applications, data storage, cloud services, IoT devices, critical infrastructure, supply chains, and personal information against cyber threats such as malware, ransomware, phishing attacks, social engineering tactics, zero-day vulnerabilities, insider threats, and nation-state espionage?
11. Explore the concept of digital divide and its implications for access to information and communication technologies.
12. Discuss the potential applications of virtual reality and augmented reality in enhancing communication experiences.
13. How can Internet censorship affect freedom of speech and access to information in different parts of the world?
14. Analyze the role of data privacy regulations in shaping the future of information technology and online communication.
15. What are the environmental implications of technology-driven solutions in smart cities, and how can sustainability be prioritized?
16. Discuss the role of 5G technology in enabling faster and more reliable communication networks for smart devices.
17. How do autonomous vehicles contribute to the development of smart transportation systems in urban environments?
18. Explore the concept of digital transformation and its impact on traditional business models in the era of information technology.
19. What are the potential risks associated with the increasing interconnectedness of devices in the IoT ecosystem?
20. How can AI be leveraged to optimize energy consumption and resource management in smart cities?
21. Discuss the role of data analytics in improving decision-making processes within organizations and government institutions.

22. Analyze the impact of social networking platforms on shaping public opinion and influencing political discourse.
23. How can smart city initiatives promote inclusivity and accessibility for individuals with disabilities through technology solutions?
24. Explore the evolution of mobile communication technologies and their influence on global connectivity and collaboration.
25. What are the key considerations for ensuring data security and privacy in a hyperconnected world driven by information technology?
26. Discuss the potential challenges of integrating autonomous systems into existing infrastructure in smart cities.
27. How do emerging technologies such as quantum computing and edge computing redefine the possibilities of information processing?
28. Analyze the role of smart grids in optimizing energy distribution and consumption patterns within urban environments.
29. What are the implications of AI on job automation and workforce dynamics in various industries?
30. Explore the intersection of cybersecurity and national security concerns in an era dominated by digital communication networks.
31. Discuss the role of open data initiatives in promoting transparency and accountability within smart city governance structures.
32. How can blockchain technology be utilized to enhance supply chain management systems and ensure traceability of goods?
33. Analyze the potential impact of 5G technology on healthcare services delivery and patient outcomes in smart cities.
34. What are the ethical considerations surrounding the collection and use of personal data by companies and governments in the digital age?
35. Discuss the role of smart sensors and IoT devices in enabling real-time monitoring and decision-making processes in urban settings.
36. How can smart city solutions address challenges related to traffic congestion and air pollution through data-driven approaches?
37. Explore the concept of digital inclusion and its significance in bridging the gap between those with access to technology and those without.
38. Analyze the role of machine learning algorithms in predicting patterns of behavior and preferences in online communication platforms.
39. What are the potential implications of quantum computing advancements on data encryption methods and cybersecurity protocols?
40. Discuss the challenges of ensuring interoperability among diverse systems and devices within a smart city ecosystem.
41. How do smart buildings leverage IoT technologies to enhance energy efficiency and occupant comfort levels?
42. Analyze the role of digital twins in simulating real-world scenarios for urban planning and infrastructure development.
43. What are the key considerations for safeguarding critical infrastructure systems from cyber threats in a digitally connected environment?
44. Discuss the potential applications of drones in enhancing public safety and emergency response capabilities within smart cities.
45. How can data visualization tools be used to communicate complex information effectively to diverse audiences in smart city projects?

46. Explore the role of 3D printing technologies in revolutionizing manufacturing processes and supply chain management practices.
47. Analyze the impact of smart meters on promoting energy conservation behaviors among residents in smart city environments.
48. What are the challenges associated with ensuring data sovereignty and privacy compliance in cross-border data transfers within smart city networks?
49. Discuss the role of digital identity management systems in enhancing security measures and user authentication processes in online interactions.
50. How can emerging technologies such as Internet satellites expand connectivity options for remote areas and bridge digital divides globally?

MULTIPLE-CHOICE QUESTIONS

1. What does the abbreviation "IT" stand for?
    a. Information Transformation
    b. Internet Technology
    c. Information Technology
    d. Intelligent Telecommunication
    e. Integrated Transmission
2. Which of the following is not a type of network topology?
    a. Star
    b. Mesh
    c. Circle
    d. Bus
    e. Ring
3. What does IP stand for in the context of the internet?
    a. Internet Protocol
    b. Information Provider
    c. Intelligent Program
    d. Interface Port
    e. Internal Protocol
4. Which of the following is not a commonly used programming language?
    a. Java
    b. Python
    c. HTML
    d. RADAR
    e. C++
5. What is the main purpose of a firewall in network security?
    a. Speed up internet connection
    b. Block unauthorized access
    c. Enhance data encryption
    d. Increase network bandwidth
    e. Improve server performance
6. What does IoT stand for in the context of smart cities?
    a. Internet of Things
    b. Intelligent Office Technology
    c. Integrated Online Tools
    d. Information on Traffic
    e. Interactive Operational Task

7. Which communication technology is commonly used for short-range wireless communication between devices?
   a. NFC (Near Field Communication)
   b. DSL (Digital Subscriber Line)
   c. WAN (Wide Area Network)
   d. LTE (Long-Term Evolution)
   e. VPN (Virtual Private Network)
8. What is the purpose of a VPN (Virtual Private Network)?
   a. Speed up internet connection
   b. Securely connect to a private network over the internet
   c. Enhance data encryption
   d. Increase network bandwidth
   e. Improve server performance
9. Which organization is responsible for managing domain names and IP addresses on the internet?
   a. IEEE (Institute of Electrical and Electronics Engineers)
   b. ISO (International Organization for Standardization)
   c. IETF (Internet Engineering Task Force)
   d. ICANN (Internet Corporation for Assigned Names and Numbers)
   e. ITU (International Telecommunication Union)
10. What is the term used to describe the practice of sending fraudulent emails in an attempt to steal sensitive information?
    a. Spoofing
    b. Hacking
    c. Phishing
    d. Spamming
    e. Malware
11. What does the acronym "URL" stand for in the context of the internet?
    a. Universal Resource Locator
    b. Unified Routing Language
    c. User Registration Link
    d. Ultimate Reference Label
    e. User Recognition List
12. Which of the following is not a type of cyber-attack?
    a. Phishing
    b. DDoS (Distributed Denial of Service)
    c. SQL Injection
    d. SMTP (Simple Mail Transfer Protocol)
    e. Ransomware
13. What is the function of a modem in computer networking?
    a. Data storage
    b. Data encryption
    c. Data transmission
    d. Data compression
    e. Data analysis
14. Which technology is used to identify and verify individuals based on their physical or behavioral traits?
    a. RFID (Radio-Frequency Identification)
    b. GPS (Global Positioning System)
    c. OCR (Optical Character Recognition)
    d. Biometrics

   e. QR Code (Quick Response Code)
15. What is the purpose of a DNS (Domain Name System)?
    a. Secure data transmission
    b. Encrypt network traffic
    c. Translate domain names to IP addresses
    d. Manage network hardware
    e. Monitor network performance
16. Which programming language is commonly used for developing web applications?
    a. Java
    b. Swift
    c. Ruby
    d. Pascal
    e. COBOL
17. What is the main function of a router in a network?
    a. Data storage
    b. Data encryption
    c. Data transmission
    d. Data compression
    e. Data analysis
18. Which protocol is commonly used for sending and receiving email messages?
    a. HTTP (Hypertext Transfer Protocol)
    b. FTP (File Transfer Protocol)
    c. SMTP (Simple Mail Transfer Protocol)
    d. TCP (Transmission Control Protocol)
    e. UDP (User Datagram Protocol)
19. What is the purpose of a cache in computing?
    a. Store data for long-term retention
    b. Speed up data access by storing frequently accessed data
    c. Encrypt sensitive information
    d. Monitor network traffic
    e. Analyze system performance
20. Which technology is used to remotely access and control a computer or network?
    a. VPN (Virtual Private Network)
    b. SSH (Secure Shell)
    c. SSL (Secure Sockets Layer)
    d. FTP (File Transfer Protocol)
    e. SNMP (Simple Network Management Protocol)
21. What does the term "IoT" stand for in the context of technology?
    a. Internet of Text
    b. Internet of Telecommunications
    c. Internet of Things
    d. Internet of Transactions
    e. Internet of Techniques
22. Which type of network is characterized by interconnected devices without a central server?
    a. Peer-to-Peer (P2P)
    b. Client-Server
    c. Ring
    d. Bus
    e. Star

23. What is the purpose of a proxy server in a network?
    a. Data storage
    b. Data encryption
    c. Data transmission
    d. Data caching and filtering
    e. Data compression
24. Which technology allows users to access computing resources over the internet on a pay-as-you-go basis?
    a. Virtualization
    b. Cloud Computing
    c. Blockchain
    d. Artificial Intelligence
    e. Machine Learning
25. What is the term used to describe the process of converting human-readable data into a form that can be easily processed by a computer?
    a. Decryption
    b. Encoding
    c. Compression
    d. Hashing
    e. Encryption
26. Which technology is used for wireless communication between devices over short distances?
    a. Bluetooth
    b. LTE
    c. WiMAX
    d. NFC
    e. Zigbee
27. What is the purpose of a modem in networking?
    a. Data storage
    b. Data encryption
    c. Data transmission
    d. Data compression
    e. Data analysis
28. Which of the following is not a commonly used web browser?
    a. Chrome
    b. Firefox
    c. Safari
    d. Thunderbird
    e. Edge
29. What does the abbreviation "AI" stand for in the context of technology?
    a. Advanced Interface
    b. Automated Intelligence
    c. Artificial Intelligence
    d. Augmented Interaction
    e. Adaptive Interface
30. Which technology is used to identify and verify individuals based on their unique physical characteristics?
    a. RFID
    b. GPS
    c. OCR
    d. Biometrics
    e. QR Code

31. What does the term "VoIP" stand for in communication technology?
    a. Voice over Internet Protocol
    b. Video over Internet Protocol
    c. Virtual Office Integration Platform
    d. Voice of Integrated Providers
    e. Visual Output Interface Protocol
32. Which type of software allows users to browse the World Wide Web?
    a. Email client
    b. Web browser
    c. File transfer protocol
    d. Instant messaging application
    e. Video conferencing software
33. What is the purpose of an SSL certificate on a website?
    a. Secure email communication
    b. Encrypt website data
    c. Improve website performance
    d. Block unwanted traffic
    e. Authenticate website ownership
34. Which technology is used to store and manage data in a distributed and secure manner?
    a. Blockchain
    b. Cloud computing
    c. Artificial intelligence
    d. Virtual reality
    e. Augmented reality
35. What is the primary function of a router in a network?
    a. Data storage
    b. Data encryption
    c. Data transmission
    d. Data compression
    e. Data analysis
36. Which technology is used to provide wireless internet access over a large area?
    a. Wi-Fi
    b. Bluetooth
    c. NFC
    d. RFID
    e. Zigbee
37. What is the purpose of a firewall in network security?
    a. Speed up internet connection
    b. Block unauthorized access
    c. Enhance data encryption
    d. Increase network bandwidth
    e. Improve server performance
38. Which of the following is not a common programming language?
    a. Java
    b. Python
    c. HTML
    d. Ruby
    e. DNS

39. What does the term “LAN” stand for in networking?
    a. Local Access Network
    b. Long Area Network
    c. Large Area Network
    d. Link Access Node
    e. Localized Area Node
40. Which technology is used to remotely access and control a computer or network securely?
    a. VPN
    b. FTP
    c. SSH
    d. HTTP
    e. DNS
41. Which of the following is not a commonly used cloud service model?
    a. Infrastructure as a Service (IaaS)
    b. Platform as a Service (PaaS)
    c. Software as a Service (SaaS)
    d. Function as a Service (FaaS)
    e. Database as a Service (DBaaS)
42. What is the purpose of a DNS server in networking?
    a. Secure data transmission
    b. Encrypt network traffic
    c. Translate domain names to IP addresses
    d. Manage network hardware
    e. Monitor network performance
43. Which technology allows users to access computing resources over the internet on-demand?
    a. Virtualization
    b. Cloud Computing
    c. Blockchain
    d. Artificial Intelligence
    e. Machine Learning
44. What is the term used to describe the process of converting data into a code to prevent unauthorized access?
    a. Decryption
    b. Encoding
    c. Compression
    d. Hashing
    e. Encryption
45. Which communication technology is commonly used for short-range wireless communication between devices?
    a. NFC (Near Field Communication)
    b. DSL (Digital Subscriber Line)
    c. WAN (Wide Area Network)
    d. LTE (Long-Term Evolution)
    e. VPN (Virtual Private Network)
46. What is the purpose of a proxy server in a network?
    a. Data storage
    b. Data encryption
    c. Data transmission
    d. Data caching and filtering
    e. Data compression

47. Which of the following is not a common cyber security measure?
    a. Antivirus software
    b. Firewall
    c. Encryption
    d. Virtualization
    e. Intrusion Detection System (IDS)
48. What is the main function of a router in a network?
    a. Data storage
    b. Data encryption
    c. Data transmission
    d. Data compression
    e. Data analysis
49. Which technology is used to remotely access and control a computer or network?
    a. VPN (Virtual Private Network)
    b. SSH (Secure Shell)
    c. SSL (Secure Sockets Layer)
    d. FTP (File Transfer Protocol)
    e. SNMP (Simple Network Management Protocol)
50. What does the term "AI" stand for in the context of technology?
    a. Advanced Interface
    b. Automated Intelligence
    c. Artificial Intelligence
    d. Augmented Interaction
    e. Adaptive Interface

ANSWERS TO MULTIPLE-CHOICE QUESTIONS

1	c	11	a	21	c	31	a	41	d
2	c	12	d	22	a	32	b	42	c
3	a	13	c	23	d	33	e	43	b
4	d	14	d	24	b	34	a	44	e
5	b	15	c	25	b	35	c	45	a
6	a	16	c	26	a	36	a	46	d
7	a	17	c	27	c	37	b	47	d
8	b	18	c	28	d	38	e	48	c
9	d	19	b	29	c	39	a	49	b
10	c	20	b	30	d	40	c	50	c

FURTHER READING

Adegun, A. A., Fonou-Dombeu, J. V., Viriri, S., & Odindi, J. (2024). Ontology-based deep learning model for object detection and image classification in smart city concepts. *Smart Cities*, 7(4), 2182–2207.

Agnew, D., Boamah, S., Bretas, A., & McNair, J. (2024). Network security challenges and countermeasures for software-defined smart grids: A survey. *Smart Cities*, 7(4), 2131–2181.

Anderson, C. (2006). *The long tail: Why the future of business is selling less of more*. Hyperion.

Batty, M., Axhausen, K. W., Giannotti, F., Pozdnoukhov, A., Bazzani, A., Wachowicz, M., ... & Portugali, Y. (2012). Smart cities of the future. *The European Physical Journal Special Topics*, 214(1), 481–518.

Baumann, N. A. (2017). The importance of effective communication: Some food for thought. *The Journal of Applied Laboratory Medicine*, 1(4), 460–461.

Benkler, Y. (2006). *The wealth of networks: How social production transforms markets and freedom*. Yale University Press.

Biesenbach, R. (2018). *Unleash the power of storytelling: Win hearts, change minds, get results*. Eastlawn Media.
Caragliu, A., Bo, C. D., & Nijkamp, P. (2011). An assessment of urban quality of life using a multicriteria decision analytic approach. *International Journal of Urban and Regional Research*, 35(1), 36–48.
Caragliu, A., Del Bo, C., & Nijkamp, P. (2009). Smart cities in Europe. *Journal of Urban Technology*, 18(2), 65–82.
Caragliu, A., Del Bo, C., & Nijkamp, P. (2011). Smart cities in Europe. *Journal of Urban Technology*, 18(2), 65–82.
Carnegie, D. (1936). *How to win friends and influence people*. Simon & Schuster.
Castells, M. (1996). *The rise of the network society*. Blackwell.
Christensen, C. M. (1997). *The innovator's dilemma: When new technologies cause great firms to fail*. Harvard Business Review Press.
Covey, S. R. (1989). *The 7 habits of highly effective people: Powerful lessons in personal change*. Free Press.
Cringely, R. X. (1992). *Accidental empires: How the boys of silicon valley make their millions, battle foreign competition, and still can't get a date*. HarperBusiness.
Dertouzos, M. L. (2001). *The unfinished revolution: Human-centered computers and what they can do for us*. HarperBusiness.
Drucker, P. F. (1999). *Management challenges for the 21st century*. HarperBusiness.
Duggan, M. (2024). *Maps shape our lives–Showing us not just where we are, but who we are*. King's College London.
Friedman, T. L. (2005). *The World is flat: A brief history of the twenty-first century*. Farrar, Straus and Giroux.
Gallo, C. (2016). *The storyteller's secret: From TED speakers to business legends, why some ideas catch on and others don't*. St. Martin's Press.
Gates, B. (1995). *The road ahead*. Viking.
Giddens, A. (1999). *Runaway world: How globalization is reshaping our lives*. Routledge.
Gladwell, M. (2000). *The tipping point: How little things can make a big difference*. Little and Brown.
Green, C., & David, T. (2022). *India vs Australia 3rd T20I highlights: India beat Australia by 6 wickets*. Times of India.
Halegoua, G. R. (2020). *Smart cities*. MIT Press.
Isaacson, W. (2011). *Steve Jobs*. Simon & Schuster.
Johnson, S. (2010). *Where good ideas come from: The natural history of innovation*. Riverhead Books.
Kahneman, D. (2011). *Thinking, fast and slow*. Farrar, Straus and Giroux.
Kelly, K. (1994). *Out of control: The new biology of machines, social systems, and the economic world*. Basic Books.
Kissinger, H., Schmidt, E., & Huttenlocher, D. (2021). *The age of A.I.*. Little and Brown.
Lanier, J. (2010). *You are not a gadget: A manifesto*. Knopf.
Lessig, L. (2004). *Free culture: How big media uses technology and the law to lock down culture and control creativity*. Penguin Press.
Levy, S. (2010). *In the Plex: How Google thinks, works, and shapes our lives*. Simon & Schuster.
Li, F. (2024). *Microsoft CrowdStrike outage: How a single software update was able to cause IT chaos across the globe*. City, University of London.
Makvandi, M., Li, W., Li, Y., Wu, H., Khodabakhshi, Z., Xu, X., & Yuan, P. F. (2024). Advancing urban resilience amid rapid urbanization: An integrated interdisciplinary approach for tomorrow's climate-adaptive smart cities – A case study of Wuhan, China. *Smart Cities*, 7(4), 2110–2130.
McConnell, S. (2004). *Code complete: A practical handbook of software construction*. Microsoft Press.
McLuhan, M. (1964). *Understanding media: The extensions of man*. McGraw-Hill.
Murphy, M. (2024). *Communication theory*. Oxford Academic.
Negroponte, N. (1995). *Being digital*. Knopf.
Norman, D. A. (2013). *The design of everyday things: Revised and expanded edition*. Basic Books.
Parnell, G. S., Specking, E., Pohl, E. A., & Buchanan, R. (2023). Smart cities – A structured literature review. *Smart Cities*, 6(4), 1719–1743.
Patterson, K., Grenny, J., McMillan, R., & Switzler, A. (2002). *Crucial Conversations: Tools for talking when stakes are high*. McGraw-Hill.
Pearce, J. M. (2024). *How better and cheaper software could save millions of dollars while improving Canada's health-care system*. Western University.
Pink, D. H. (2009). *Drive: The surprising truth about what motivates us*. Riverhead Books.

Radu, C., & Smaili, N. (2022). *How hiring more women IT experts improves cybersecurity risk management.* Université du Québec à Montréal (UQAM).
Rigby, D. K., Elk, S., & Berez, S. (2020). *Doing agile right: Transformation without chaos.* Harvard Business Review Press.
Rogers, E. M. (1962). *Diffusion of innovations.* Free Press.
Sanchez Gracias, J., Parnell, G. S., Specking, E., Pohl, E. A., & Buchanan, R. (2023). Smart cities – A structured literature review. *Smart Cities*, 6(4), 1719–1743.
Sinek, S. (2009). *Start with why: How great leaders inspire everyone to take asction.* Portfolio.
Sullivan, J. (2016). *Simply said: Communicating better at work and beyond.* Wiley.
Sullivan, N. P. (2023). *The information age is starting to transform fishing worldwide.* Tufts University.
Tapscott, D. (1996). *The digital economy: Promise and peril in the age of networked intelligence.* McGraw-Hill.
Townsend, A. M. (2013). *Smart cities: Big data, civic hackers, and the quest for a new Utopia.* W. W. Norton.
Turkle, S. (2011). *Alone together: Why we expect more from technology and less from each other.* Basic Books.
Vance, A. (2015). *Elon musk: Tesla, SpaceX, and the quest for a fantastic ruture.* Ecco.
Weinberger, D. (2011). *Too big to know: Rethinking knowledge now that the facts aren't the facts, experts are everywhere, and the smartest oerson in the room is the room.* Basic Books.
Weiser, M. (1991). *The computer for the 21st century.* Scientific American.
West, D. M. (2018). *The future of work: Robots, AI, and automation.* Brookings Institution Press.
Wheeler, T. (2018). *From Gutenberg to Google: The history of our future.* Brookings Institution Press.
Winograd, T., & Flores, F. (1986). *Understanding computers and cognition: A new foundation for design.* Ablex Publishing Corporation.
Zuboff, S. (2019). *The age of surveillance capitalism: The fight for a human future at the new frontier of power.* PublicAffairs.
Zuckerman, E. (2013). *Rewire: Digital cosmopolitans in the age of connection.* W. W. Norton.
Zweig, J. (2015). *Your money and your brain: How the new science of neuroeconomics can help make you rich.* Simon & Schuster.

13 Urban Public Transportation Systems

13.1 GENERIC TYPES OF URBAN TRANSPORTATION SYSTEMS

Urban transportation systems serve as the vital lifelines that sustain and enhance the functionality of cities. Operating akin to a meticulously orchestrated performance, they facilitate the seamless flow of individuals, goods, and services. Envision the dynamic spectacle of subway trains efficiently transporting commuters to their workplaces, buses skillfully maneuvering through city streets, and delivery vehicles deftly navigating intricate urban routes. These intricate networks not only interconnect various city districts but also serve as the crucial link between commercial centers and recreational venues. Whether it be the iconic yellow taxis synonymous with New York City, the highly efficient metro systems characteristic of Tokyo, or the lively tuk-tuks prevalent in Bangkok, each city boasts a distinctive transportation infrastructure that weaves together its urban fabric. Let us delve into these indispensable components that intricately bind our urban landscapes. These systems, akin to the sinews of a metropolis, establish connections between neighborhoods, business centers, and leisure facilities. They are typically classified into several overarching categories:

Metro (Subway) System (Figure 13.1):

- An intricate network of steel and tunnels beneath the city surface, swiftly ferrying passengers through bustling thoroughfares.
- The bustling scenes of rush-hour crowds, the rhythmic hum of trains, and platforms adorned with urban artwork.
- From the iconic yellow lines of New York to the orderly queues in Tokyo, metro systems pulsate with urban vibrancy.

Bus System (Figure 13.2):

- The roar of engines, the presence of vinyl seats, and the familiar jolt as buses commence their journeys.
- Routes intricately crisscrossing the cityscape, linking residential neighborhoods with commercial hubs.
- The bus stop serves as a microcosm of society, encompassing students, seniors, and busy professionals awaiting their respective rides.

Transit Rail System:

- Light rail vehicles gracefully glide along tracks, tracing historic pathways within the city.
- Streetcars resonate with nostalgic charm as they traverse historic districts, their bell chimes reverberating through the streets.

DOI: 10.1201/9781032664613-13

FIGURE 13.1 Metro (subway).

***Source:* DCInfo, Wikipedia.**

- These systems seamlessly blend elements of nostalgia with modernity as they integrate into the urban landscape.

Shuttle System:

- Airport shuttles efficiently transport travelers with their luggage in tow.
- University shuttles facilitate the movement of students from dormitories to academic buildings.
- Though covering short distances, shuttle services play a pivotal role in bridging gaps within the urban transportation network.

Paratransit System:

- Equipped with wheelchair lifts, accommodating drivers, and offering door-to-door services for individuals with mobility challenges.
- Serving as a crucial support system for those in need, ensuring inclusivity and accessibility within urban transit.
- These vehicles carry narratives of resilience, independence, and community support.

Taxi System (Figure 13.3):

- Yellow cabs swiftly navigate through traffic, with their rooftop signs serving as beacons for passengers.
- Uber, Grab, Bolts, and Lyft drivers utilize GPS technology to navigate routes efficiently while earning ratings and gratuities.
- Taxis act as transient companions for hurried travelers and late-night revelers within the urban environment.

FIGURE 13.2 Bus system.

Source: FlushingLocal, Wikipedia.

FIGURE 13.3 New York City taxi.

Source: Henning 48, Wikipedia.

In this symphony of motion, urban transportation systems harmonize efficiency, accessibility, and communal experiences. Each mode of transportation contributes its unique essence, collectively shaping the urban symphony that defines city life.

Given our current lifestyle and heavy reliance on fossil fuels within our transportation systems, coupled with the depletion of easily accessible oil reserves, the necessity for sustainable practices has become increasingly apparent. It is crucial to acknowledge that we have only one Earth, yet our current demands would require the resources of seven Earths to sustain if continued at the present rate. This unsustainable trajectory not only poses a threat to our planet's well-being but also contributes to the degradation of urban air quality over time. Therefore, it is imperative that we transition toward sustainable living practices, with a particular emphasis on fostering sustainability within urban transportation systems. By prioritizing environmentally conscious modes of transport and investing in eco-friendly infrastructure, we can mitigate the negative impact on our environment and work toward ensuring a healthier and more sustainable future for generations to come.

13.2 SUSTAINABLE URBAN TRANSPORTATION SYSTEMS

Sustainable urban transportation systems are at the heart of creating livable, environmentally friendly cities. They are inevitable in reducing greenhouse gas emissions, improving air quality, and promoting healthier lifestyles. These systems encompass a wide range of transport modes, including public transit, cycling, walking, and electric vehicles (EVs), all aimed at minimizing environmental impact while maximizing mobility. The concept of sustainability in transportation goes beyond just the use of clean energy. It also involves efficient use of resources, equitable access, safety, and resilience to changes. Urban planning strategies such as transit-oriented development (TOD) and mixed-use development are key to promoting sustainable transportation by reducing the need for travel and encouraging the use of public transit. Technological advancements, including smart traffic management systems and autonomous vehicles (AVs), also hold great promise for making urban transportation more sustainable. However, achieving sustainable urban transportation is not without challenges. It requires a holistic approach that integrates transportation planning with land use planning, involves community participation, and is supported by appropriate policies and investments. As urban populations continue to grow, sustainable urban transportation systems will become increasingly important for ensuring the livability and sustainability of our cities.

13.2.1 Importance of Sustainability in Urban Transportation

Sustainability in urban transportation is crucial for reducing carbon emissions, improving air quality, and enhancing public health. It promotes the use of eco-friendly transport options, reduces traffic congestion, and fosters more livable, resilient cities for future generations.

13.2.1.1 The Role of Urban Transportation

Urban transportation systems, the arteries and veins of a city, facilitate the flow of people and goods, linking disparate parts of the urban landscape. Yet, these traditional systems, while essential, often contribute to environmental degradation, traffic congestion, and poor air quality. As such, the urgency for sustainable urban transportation has never been more acute. Urban transportation systems are integral to the functioning of cities. They enable the movement of people from their homes to workplaces, schools, shopping centers, and recreational areas. They also facilitate the transport of goods from ports, warehouses, and factories to markets and homes. In essence, these systems connect different parts of the city, making urban life possible. Urban transportation serves as an indispensable element of metropolitan existence, exerting a profound influence on the economic structure, environmental health, and overall quality of life. With the relentless expansion of urban areas, the demand for sustainable urban transportation escalates, becoming a matter of paramount importance.

Urban transportation systems are the essential circulatory networks that allow cities to function and thrive. They facilitate the movement of people, goods, and services between homes, workplaces, businesses, and other key destinations. Effective transportation links different neighborhoods and communities, enabling access to economic opportunities, social services, and shared resources across the urban landscape. However, traditional transportation systems in many cities have struggled to keep pace with growing populations, urbanization, and increasing demands for mobility. Reliance on privately owned vehicles, especially those powered by fossil fuels, has contributed to a range of significant problems. Vehicle emissions are a major source of air pollution, degrading local air quality and posing health risks to residents. The congestion caused by high volumes of cars and trucks on the roads leads to wasted time, economic losses, and further environmental damage. The construction and maintenance of extensive road networks also consumes valuable land area that could otherwise be utilized for housing, green spaces, or other community needs.

13.2.1.2 The Environmental Impact of Conventional Transportation Systems

One of the major drawbacks of traditional transportation systems is their environmental impact. Vehicles powered by fossil fuels emit a range of pollutants, including carbon dioxide, a major greenhouse gas; nitrogen oxides and sulfur dioxide, which contribute to acid rain; and particulate matter, which harms air quality and human health. In addition, the noise generated by traffic can have negative impacts on human well-being and wildlife. The negative impacts of conventional transportation models have made the transition to sustainable urban mobility a pressing priority for city planners, policymakers, and community stakeholders around the world. Sustainable transportation approaches seek to minimize the environmental footprint, improve social equity, and enhance the overall livability of cities.

Transportation is one of the major contributors to global greenhouse gas emissions. Traditional modes of transportation, such as cars and buses, rely on fossil fuels, leading to significant carbon emissions. Sustainable transportation systems, on the other hand, prioritize the use of renewable energy sources, such as EVs and bicycles, which have a lower environmental impact.

One crucial element of sustainable urban transportation is the promotion of active mobility options, such as walking and cycling. By investing in high-quality pedestrian infrastructure, including wide sidewalks, pedestrian crossings, and traffic-calming measures, cities can encourage residents to choose active modes of travel for short trips. Similarly, building extensive networks of protected bike lanes, bike-sharing programs, and bicycle parking facilities can make cycling a safe, convenient, and attractive alternative to driving. Active mobility not only reduces emissions and congestion but also provides opportunities for physical activity, improving public health outcomes.

In addition to active transportation, public transit systems contribute to sustainable urban mobility. Well-designed and efficient bus, light rail, subway, and commuter rail networks can move large numbers of people quickly and efficiently while minimizing the environmental impact per passenger. By providing accessible, affordable, and reliable public transit options, cities can reduce reliance on private vehicles and ensure that all residents, including those without access to cars, can travel to their destinations. Integrating public transit with other modes, such as bike-sharing and micromobility services, can further enhance the convenience and accessibility of sustainable transportation choices.

13.2.1.3 The Social and Economic Impact of Traditional Transportation Systems

The traditional transportation systems, while serving as the backbone of urban mobility, have considerable social and economic implications. The impacts range from traffic congestion to social inequalities, and these effects are felt most acutely in densely populated urban areas.

13.2.1.3.1 Traffic Congestion and Its Implications

Traffic congestion is a pervasive issue in many cities worldwide. It is primarily caused by an over-reliance on private vehicles, inadequate public transportation infrastructure, and poor urban planning. The consequences of traffic congestion are far-reaching, affecting not just the commuters but also the entire urban ecosystem.

From a social perspective, traffic congestion results in wasted time for commuters. The time spent in traffic could have been used productively elsewhere. This lost time can lead to increased stress levels among commuters, negatively impacting their mental health and overall quality of life. From an economic standpoint, traffic congestion leads to decreased productivity. When employees are stuck in traffic, they arrive late at work, leading to lost work hours. According to a study by the Centre for Economics and Business Research, traffic congestion costs economies billions of dollars each year in lost productivity.

13.2.1.3.2 Inequitable Access to Transportation

Traditional transportation systems often fail to serve all city residents equitably. This inequity in access to reliable and efficient transportation is most pronounced among residents living in low-income and peripheral areas. These areas are often poorly served by public transportation, making it difficult for residents to access jobs, education, and other opportunities. This lack of access to transportation can exacerbate existing social inequalities. For instance, if a person cannot easily travel to a job interview, they may miss out on employment opportunities. Similarly, students who cannot easily commute to educational institutions may have limited educational opportunities. Moreover, residents of low-income areas often cannot afford private vehicles, making them reliant on public transportation. When public transportation is inadequate or unreliable, these residents are disproportionately affected.

13.2.1.3.3 Impacts on Economic Development

The social and economic impacts of traditional transportation systems can also hinder economic development. Areas with poor transportation infrastructure may struggle to attract businesses and investments. Without reliable transportation, employees may find it difficult to commute to work, and customers may find it inconvenient to visit businesses. This can lead to decreased economic activity in these areas. Furthermore, the costs associated with traffic congestion – such as fuel costs, vehicle maintenance costs, and the cost of lost time – can drain the economic resources of individuals and communities. These costs can be particularly burdensome for low-income individuals and families, further exacerbating economic inequalities.

Given the economic burden of traffic congestion on low-income individuals, it is imperative to adopt a comprehensive approach for improving public transportation in these areas, thereby alleviating economic inequalities and enhancing community resilience. Improving public transportation in low-income areas is a multifaceted challenge that requires a comprehensive approach. Here are some strategies that cities can adopt:

13.2.1.3.4 Understand the Needs of Different Groups

Enhancing the equity and inclusivity of public transportation requires a nuanced understanding of how different demographic groups utilize these essential services and the unique challenges they face. By collecting and analyzing disaggregated data on transport usership, policymakers and planners can identify underserved populations and tailor solutions to their specific needs. Surveys, for example, can shed light on barriers to access, such as cost or limited-service availability, as well as concerns about personal safety and security. Going beyond quantitative data, community engagement and focus groups can provide a more in-depth, qualitative understanding of the lived experiences of marginalized groups. This holistic approach, combining data-driven insights and direct community input, enables the development of targeted interventions that address the multifaceted mobility needs of all urban residents, ultimately creating a more equitable and accessible public transportation system.

13.2.1.3.5 Improve Coverage, Affordability, Accessibility, Safety, and Security

To truly foster equitable public transportation, cities must take a comprehensive approach that addresses interconnected challenges. By expanding services to underserved neighborhoods, often home to marginalized communities, cities can improve access to vital destinations and opportunities. Ensuring affordability through discounted fares, subsidies, and flexible payment options is essential to make public transit accessible to all, regardless of socioeconomic status. Enhancing physical accessibility through barrier-free design and assistive technologies enables mobility for passengers with disabilities or special needs. Finally, bolstering safety and security measures, from lighting to surveillance, can build rider confidence and encourage broader utilization of public transport networks.

13.2.1.3.6 Promote Active Transportation

Promoting walking, bicycling, and public transit can be particularly beneficial for low-income communities and communities of color. These modes of transportation are often more affordable and can provide healthy physical activity. However, it is important to ensure that these activities are safe and convenient for everyone.

13.2.1.3.7 Introduce On-Demand Shuttles

Bridging the gap between residential areas and major public transportation hubs can be a significant barrier for many underserved urban communities. Introducing on-demand shuttle services to cover this "first and last mile" of the journey can be a game-changer in enhancing access to essential services and opportunities. These flexible, responsive shuttle routes can connect residents directly to bus stops, train stations, and other key transit nodes, making it easier and more convenient for them to utilize the broader public transportation network. This seamless integration of on-demand and fixed-route services can be particularly beneficial for populations with limited mobility, such as the elderly, those with disabilities, or low-income households without access to private vehicles. Addressing this critical last-mile challenge, on-demand shuttles can significantly improve the overall connectivity and accessibility of public transportation, empowering underserved communities to fully participate in the economic and social fabric of the city.

13.2.1.3.8 Reduce Commute Time

The daily commute is a significant factor in determining an individual's ability to access economic and social opportunities, with far-reaching implications for social mobility. A groundbreaking Harvard study found that commute time is the single strongest factor in predicting whether a person can escape poverty, even more so than factors like crime rates or school quality. This underscores the vital importance of efficient and accessible public transportation systems in connecting residents, particularly those in underserved communities, to jobs, education, healthcare, and other essential services. Through optimizing the speed, reliability, and coverage of public transit, cities can reduce commute times, freeing up valuable time and resources for individuals to pursue personal and professional goals. Investing in public transportation that serves the needs of marginalized groups is not only an issue of equity, but also a strategic lever for breaking the cycle of intergenerational poverty and fostering greater social and economic inclusion.

13.2.1.4 The Need for Sustainable Urban Transportation

Sustainability in urban transportation refers to the development and implementation of transportation systems that are environmentally friendly, economically viable, and socially equitable. These systems aim to reduce greenhouse gas emissions, improve air quality, promote economic growth, and enhance the quality of life for all city residents. The need for sustainable urban transportation is more pressing than ever. Sustainable transportation systems aim to minimize environmental impacts, promote social equity, and support economic development. They prioritize the use of clean and renewable energy sources, promote the use of public transportation and active travel modes like walking and cycling, and aim to make transportation infrastructure more resilient to the impacts of climate change. Implementing sustainable transportation systems requires significant investment in infrastructure and technology, changes in policy and planning practices, and shifts in individual behavior. Yet, the benefits – improved air quality, reduced greenhouse gas emissions, enhanced social equity, and economic development – make it a worthwhile endeavor.

Implementing sustainable urban transportation requires a comprehensive and collaborative approach, involving local governments, transportation agencies, urban planners, private sector partners, and community stakeholders. Strategies may include infrastructure investments, policy reforms, incentive programs, public-private partnerships, and public education and engagement campaigns. Prioritizing sustainable transportation initiatives allows cities to not only address environmental concerns but also foster more livable, equitable, and economically vibrant urban environments.

The transition to sustainable urban transportation is a necessary step in creating more sustainable, resilient, and livable cities for the future. Through a combination of innovative approaches, technological advancements, and collaborative efforts, cities can revolutionize the way people and goods move, ultimately improving the quality of life for all urban residents.

13.2.1.5 Emerging Technologies in Urban Transportation

Emerging technologies and innovations are also transforming the landscape of sustainable urban transportation. EVs, powered by renewable energy sources, offer a cleaner and more efficient alternative to traditional gasoline-powered cars. As battery technology continues to improve and charging infrastructure expands, EVs are becoming increasingly accessible and practical for urban residents. Additionally, the rise of shared mobility services, such as car-sharing and ride-hailing, can reduce the need for individual car ownership and encourage the use of pooled transportation options.

Figure 13.4 illustrates an EV charging station, where a white EV is plugged in and charging. The station features a main charging unit with a screen and several cables connected to the car. It is located in a parking area designated for EVs, as indicated by blue markings on the ground. The charging unit displays various instructions and information in Japanese. The station is sheltered by a roofed structure, emphasizing the infrastructure required to support the widespread adoption of eco-friendly transportation options. This setup highlights the importance of accessible and convenient charging facilities in promoting the use of electric vehicles and reducing carbon emissions.

Autonomous and connected vehicle technologies also hold promise for enhancing the sustainability and efficiency of urban transportation. Self-driving cars and trucks can optimize routing, reduce congestion, and improve safety, while connected vehicle systems can enable real-time communication and coordination to improve traffic flow. However, the integration of these emerging technologies must be carefully planned and implemented to ensure equitable access and alignment with broader sustainability goals.

Emerging technologies and innovations are rapidly reshaping the landscape of sustainable urban transportation, offering a wide range of solutions to the environmental, economic, and social challenges posed by traditional mobility models. At the forefront of this transformation are EVs and shared mobility services, which are driving a shift toward cleaner, more efficient, and more accessible transportation options for city residents.

Figure 13.5 features an electric car, a compact, two-seater model, that prominently captures attention with its sleek, modern design. This car painted a striking red with white accents, stands out not only for its aesthetics but also for its significance in sustainable transportation. The vehicle is connected to a charging station labeled "Renewable Energy," underscoring its eco-friendly nature.

FIGURE 13.4 Electric charging station.

Source: **Sakaori, Wikipedia.**

This image highlights the practical aspects of using EVs, which are essential for reducing carbon emissions and dependence on fossil fuels, contributing to a cleaner, greener urban environment.

EVs, powered by renewable energy sources, represent a pivotal step toward decarbonizing urban transportation. Compared to their gasoline-powered counterparts, EVs produce significantly lower greenhouse gas emissions and other air pollutants, contributing to improved local air quality and public health outcomes. As battery technology continues to advance, offering greater driving range and faster charging times, EVs are becoming increasingly practical and accessible for a wider range of urban users.

The expansion of public and private charging infrastructure, such as curbside chargers, parking garage installations, and workplace charging stations, is crucial in making EVs a viable choice for both personal and fleet vehicles. Cities are playing a key role in this transition, implementing policies and incentives to encourage EV adoption, from rebates and tax credits to the establishment of low-emission zones and priority parking for electric cars. Actively promoting the transition to EVs allows municipalities to capitalize on the environmental and public health benefits while also supporting the growth of a sustainable mobility ecosystem.

Complementing the rise of EVs is the proliferation of shared mobility services, which have the potential to reduce the need for individual car ownership and encourage the use of pooled transportation options. Car-sharing programs, where users can access and rent vehicles on-demand, enable more efficient utilization of the existing vehicle fleet and reduce the number of privately owned cars on city streets. Ride-hailing services, such as ride-sharing and microtransit, offer flexible, on-demand transportation, often serving as a first- and last-mile connection to public transit hubs.

The integration of shared mobility with other sustainable transportation modes, such as public transit, cycling, and walking, can create a seamless, multimodal transportation network that enhances overall accessibility and reduces reliance on private vehicles. Making it easier for

FIGURE 13.5 Electric car.

Source: **Felix Muller, Wikipedia.**

residents to access and utilize a range of mobility options enables cities to foster a more sustainable and equitable transportation system. However, the success of shared mobility services in advancing sustainable transportation goals is contingent on careful policy and regulatory frameworks. Cities must ensure that these services are equitably distributed, accessible to underserved communities, and integrated with public transit in a way that complements and enhances, rather than competes with, existing systems. Policies that incentivize the use of shared, electric, and pooled transportation options, while discouraging private vehicle use, can help steer the urban mobility landscape toward greater sustainability.

Beyond EVs and shared mobility, the emergence of autonomous and connected vehicle technologies also holds promise for enhancing the sustainability and efficiency of urban transportation. Self-driving cars and trucks, enabled by advancements in sensors, artificial intelligence (AI), and control systems, have the potential to revolutionize the way people and goods move within cities.

AVs (Figure 13.6) can optimize routing, reduce congestion, and improve safety by eliminating human error and reckless driving behaviors. Seamless communication between AVs and surrounding infrastructure allows for coordinated movements, reduced unnecessary stops and starts, and smoother traffic flows. This results in significant reductions in fuel consumption, emissions, and travel times, fostering more sustainable and livable urban environments.

Connected vehicle technologies, which allow cars, trucks, and public transit vehicles to exchange real-time data with each other and with the transportation network, can further enhance the efficiency and safety of urban mobility. This connectivity enables the coordination of traffic signals, the identification of congestion and hazards, and the provision of routing and navigation information to drivers and passengers. Leveraging these data-driven insights enables cities to better manage and optimize the overall transportation system, improving the user experience and reducing

FIGURE 13.6 Self-driving (autonomous) car.

Source: **GrendelKhan, Wikipedia.**

the environmental impact of urban travel. However, the integration of autonomous and connected vehicle technologies must be carefully planned and implemented to ensure equitable access and alignment with broader sustainability goals. Concerns have been raised about the potential for these technologies to exacerbate existing socioeconomic disparities if they are not deployed in a way that prioritizes the needs of underserved communities.

To address these challenges, cities must adopt a proactive, inclusive, and multifaceted approach to the integration of emerging mobility technologies. This may involve implementing policies and regulations that prioritize public transit, active transportation, and shared mobility options over private, AV use; investing in infrastructure upgrades to support connected vehicle systems; and ensuring that the deployment of AVs and other new technologies is accompanied by robust public education and engagement efforts. Cities must work closely with technology companies, transportation providers, and community stakeholders to develop governance frameworks that align these innovations with broader sustainability and equity goals. This may include establishing data-sharing protocols, setting performance standards, and implementing mechanisms for community input and oversight. Being a holistic and collaborative approach to the integration of emerging mobility technologies, cities can harness the transformative potential of these innovations while mitigating potential risks and unintended consequences. The strategic deployment of EVs, shared mobility services, autonomous and connected vehicle systems, and other sustainable mobility solutions can collectively contribute to more efficient, equitable, and environmentally friendly urban transportation networks.

Some emerging technologies that are revolutionizing urban transportation, making it more efficient, sustainable, and user-friendly, and are potentially shaping the future of urban transportation include:

13.2.1.5.1 Internet of Things and Connectivity

The Internet of Things (IoT) is enabling the creation of smart transportation systems. IoT devices can collect real-time data about traffic conditions, vehicle performance, and user behavior, which can be used to optimize transportation systems. For example, IoT-enabled traffic lights can adjust their timing based on real-time traffic conditions, reducing congestion and improving traffic flow.

13.2.1.5.2 Artificial Intelligence

AI is being used to develop predictive models for traffic patterns, allowing cities to proactively manage congestion and reduce the time commuters spend on the road. AI can also be used to optimize public transit routes and schedules, improve the efficiency of logistics operations, and enhance the safety of AVs.

13.2.1.5.3 Electrification

The electrification of transportation is on the rise. EVs are becoming increasingly common in cities around the world, reducing greenhouse gas emissions and improving air quality. Many cities are also investing in electric public transit vehicles, such as buses and trams, and building the necessary charging infrastructure.

13.2.1.5.4 Shared Mobility

Shared mobility services, such as bike-sharing, car-sharing, and ride-hailing, are changing the way people move around cities. These services can provide flexible and convenient transportation options, reduce the need for private car ownership, and decrease traffic congestion.

13.2.1.5.5 Autonomous Vehicles

AVs have the potential to transform urban transportation. AVs can improve road safety, increase the efficiency of transportation systems, and provide new mobility options for people who are unable to drive. However, the widespread adoption of AVs will require significant advances in technology, infrastructure, and regulation.

13.2.1.5.6 5G and Edge Computing

Technologies like 5G and edge computing are poised to drastically transform public transportation. While 5G networks provide fast connectivity for high-bandwidth data streams, edge computing provides distributed computer power to access, process, and act on insights from that data.

13.2.2 Energy-Efficient Vehicles and Renewable Energy Sources for Transportation

Energy-efficient vehicles and renewable energy sources are transforming the transportation sector, offering a sustainable alternative to traditional fossil fuel-based systems. An in-depth look at these two critical components of sustainable transportation is given below.

13.2.2.1 Energy-Efficient Vehicles

Energy-efficient vehicles, including EVs, hybrid electric vehicles (HEVs), and plug-in hybrid electric vehicles (PHEVs), are revolutionizing the transportation sector. These vehicles use advanced technologies to reduce energy consumption and emissions, making them a more sustainable choice for transportation. EVs operate solely on electricity, producing zero tailpipe emissions. They convert over 77% of the electrical energy from the grid to power at the wheels, compared to conventional gasoline vehicles, which convert about 12–30% of the energy stored in gasoline to power at the wheels. This high energy efficiency, coupled with the potential to use electricity from renewable sources, makes EVs a key component of sustainable transportation.

TABLE 13.1
Strategies to Promote the Adoption of EVs by City

Building public EV charging infrastructure	Cities can invest in public EV charging infrastructure and encourage other stakeholders to do the same. This includes installing charging stations in public parking lots, shopping centers, and other high-traffic areas. Providing convenient and accessible charging options can alleviate 'range anxiety' and make EV ownership more practical.
Offering incentives	Financial incentives, such as tax breaks, can make EVs more affordable. Cities can also offer non-monetary incentives, such as free or priority parking, free charging, and access to bus lanes. These perks can make EVs more attractive to potential buyers.
Encouraging electric car-sharing businesses	Car-sharing businesses can play a crucial role in promoting EV adoption. By including EVs in their fleets, these businesses can expose more people to EVs and help them understand the benefits of electric driving
Enabling faster and cheaper registration for EVs	Reducing the time and cost of vehicle registration for EVs can also encourage their adoption. This can be particularly effective in cities where vehicle registration is typically time-consuming or expensive.
Implementing low- and zero-emission zones	Cities can implement low- and zero-emission zones, where only EVs and other low-emission vehicles are allowed. This can incentivize residents to switch to EVs and help reduce air pollution in city centers.
Raising awareness	Cities can also launch campaigns to educate the public about the benefits of EVs and the availability of charging infrastructure. This can help dispel misconceptions about EVs and encourage more people to consider them as a viable transportation option.

HEVs and PHEVs combine an internal combustion engine with an electric motor and a battery. These vehicles can switch between the electric motor and the engine to optimize fuel efficiency, and in the case of PHEVs, they can also be plugged in to recharge their batteries.

Cities can promote the adoption of EVs through a variety of strategies, as shown in Table 13.1.

13.2.2.2 Renewable Energy Sources for Transportation

Renewable energy sources, such as wind, solar, and biofuels, are increasingly being used to power transportation systems. These energy sources are sustainable, produce little to no greenhouse gas emissions during operation, and can be used to generate the electricity needed to power EVs or produce biofuels.

Solar and wind energy can be harnessed to generate electricity for EV charging stations. This allows for the charging of EVs with renewable energy, further reducing the carbon footprint of transportation. Biofuels, produced from organic materials like plant matter or waste, can be used as a renewable source of fuel for certain types of vehicles. Biofuels can be used in existing vehicles with little to no modification, making them a practical option for reducing emissions from the transportation sector.

13.3 URBAN TRANSPORT PLANNING

Urban transport planning is crucial for developing efficient, sustainable, and accessible transportation systems in cities. It involves strategic design and management of transport infrastructure to meet the needs of urban populations, reduce traffic congestion, and minimize environmental impact. Integrating various modes of transport, such as public transit, cycling, and walking paths, urban planners aim to enhance mobility and connectivity. This plan also addresses social equity, ensuring that all residents, regardless of socioeconomic status, have access to reliable and affordable transportation options. Ultimately, urban transport planning plays a vital role in promoting economic growth, improving quality of life, and fostering sustainable urban development.

13.3.1 The Planning Process: Factors and Considerations

Urban transport planning process involves a multitude of factors and considerations. The process is fundamental in shaping the mobility, sustainability, and livability of cities. In the process, the following issues must be addressed.

13.3.1.1 Understanding the Context

The first step in urban transport planning is understanding the context, which includes the demographic, market, and technology characteristics of society that influence travel behavior. This involves analyzing data on population density, land use, economic activity, and existing transportation infrastructure. Planners also need to consider future trends, such as population growth, urbanization, economic development, and technological advancements.

13.3.1.2 Identifying Problems and Goals

Once the context is understood, planners can identify the problems that need to be addressed and set goals for the planning process. Problems can include traffic congestion, air pollution, lack of accessibility, and safety issues. Goals can include improving mobility, reducing environmental impacts, promoting social equity, and supporting economic development.

13.3.1.3 Developing and Evaluating Alternatives

Planners then develop and evaluate alternative strategies to address the identified problems and achieve the set goals. These strategies can include infrastructure projects, such as building new roads or public transit lines; policy measures, such as implementing congestion pricing or promoting active transportation; and technological solutions, such as adopting intelligent transportation systems.

13.3.1.4 Implementing the Plan

Once a strategy has been chosen, it needs to be implemented. This involves detailed design, securing funding, obtaining necessary approvals, and managing construction. Planners also need to coordinate with various stakeholders, including government agencies, transportation providers, and the community.

13.3.1.5 Monitoring and Evaluation

After the plan has been implemented, it is important to monitor its performance and evaluate its impact. This can involve collecting data on traffic volumes, travel times, air quality, and user satisfaction. The results of this evaluation can then be used to refine the plan and inform future planning efforts.

13.3.1.6 Considerations in the Planning Process

There are several key considerations in the urban transport planning process. One is the need for a long-term perspective. Transport infrastructure is expensive and long-lasting, so it is important to plan for the long term. However, the future is uncertain, so plans need to be adaptable. Another consideration is the need for integration. Transport planning should be integrated with land use planning, as the two are closely interrelated. It should also be integrated with other policy areas, such as environmental protection, economic development, and social equity. A third consideration is the need for public participation. The public should be involved in the planning process, as they are the ones who will be affected by the decisions made. Public participation can help ensure that the plan meets the needs of the community and has public support.

13.3.2 Population Density and Land Use

Urban transport planning is a complex process that requires careful consideration of various factors, including population density and land use. These two factors are interrelated and have a significant impact on the effectiveness and sustainability of urban transportation systems.

13.3.2.1 Population Density and Urban Transport Planning

Population density refers to the number of people living in a specific area. High-density areas typically have more demand for public transportation, walking, and cycling, while low-density areas often rely more on private vehicles. Therefore, understanding population density is crucial for planning transportation systems that meet the needs of different areas.

High population density can support more frequent and diverse public transportation services, as there are more potential passengers to share the costs. However, it can also lead to congestion and increased demand for infrastructure. Therefore, transport planning in high-density areas often involves strategies to manage congestion, such as implementing public transit priority measures, promoting active transportation, and using traffic demand management strategies. In contrast, low-density areas often face challenges in providing efficient public transportation due to the dispersed nature of destinations. In these areas, transport planning may involve strategies to improve the coverage and accessibility of public transportation, promote carpooling, and encourage the use of EV or hybrid vehicles.

13.3.2.2 Land Use and Urban Transport Planning

Land use refers to the human use of land, including residential, commercial, industrial, and recreational uses. Land use patterns significantly influence travel behavior and transportation needs. Mixed land use, which involves integrating different types of land uses in close proximity, can reduce the need for long-distance travel and make walking, cycling, and public transportation more viable. Therefore, promoting mixed land use is often a key strategy in sustainable urban transport planning. On the other hand, segregated land use, where different land uses are separated from each other, often leads to longer travel distances and higher reliance on private vehicles. In areas with segregated land use, transport planning may involve strategies to improve connectivity, such as developing efficient public transportation networks and implementing land use policies to encourage infill development and reduce urban sprawl.

13.3.2.3 The Interplay of Population Density and Land Use

Population density and land use are closely interrelated and together shape the demand for and supply of transportation. High-density, mixed-use areas typically have high demand for public and active transportation and can support a diverse range of transportation options. In contrast, low-density, segregated areas often have high demand for private vehicle travel and face challenges in providing efficient public transportation. Therefore, urban transport planning needs to consider both population density and land use to develop effective and sustainable transportation systems. This involves not only designing transportation infrastructure and services but also implementing land use policies that support sustainable travel behaviors.

Addressing safety concerns related to population density and land use in urban transportation planning needs a comprehensive approach. Some strategies that cities can adopt are shown in Table 13.2.

TABLE 13.2
Cities' Strategies on Safety Concerns

Designing for safety	Cities can incorporate safety features into the design of transportation infrastructure. This can include traffic calming measures, such as speed bumps and roundabouts, to slow down vehicles and reduce the risk of accidents. Pedestrian-friendly design features, such as wide sidewalks, crosswalks, and pedestrian islands, can also enhance safety.
Enhancing public transportation	Public transportation is generally safer than private vehicle travel. Therefore, improving the coverage, frequency, and reliability of public transportation can help reduce the reliance on private vehicles and enhance safety.
Promoting active transportation	Walking and cycling are not only sustainable modes of transportation, but they also have lower accident rates compared to motorized transport. Cities can promote active transportation by providing safe and comfortable infrastructure, such as bike lanes and pedestrian paths.
Implementing traffic demand management strategies	Traffic demand management strategies, such as congestion pricing and carpooling programs, can reduce traffic volumes and improve safety. By reducing the number of vehicles on the road, these strategies can reduce the risk of accidents.
Enforcing traffic laws	Enforcing traffic laws, such as speed limits and drunk driving laws, is crucial for maintaining safety. Cities can use a combination of education, enforcement, and engineering measures to promote compliance with traffic laws.
Using technology	Technology can also play a role in enhancing safety. For example, ITS can provide real-time traffic information, manage traffic flow, and detect and respond to incidents.

13.3.2.4 Equity in Transportation

Equity in transportation is an important issue but frequently overlooked. However, it has gained increasing attention in recent years. The equity in transportation refers to the fair and just provision of transportation services, ensuring that all community members, particularly those identified as underserved, disadvantaged, and overburdened, have access to safe, reliable, comfortable, healthy, and sustainable mobility. For decades, major transportation and infrastructure projects in cities across the United States have contributed to racial and social inequities. Urban freeways and transit infrastructure projects, often paid for in large part by federal transportation funds, have disproportionately displaced and isolated people living in minority neighborhoods, tearing at the fabric of vibrant communities and compounding issues of equity and access to jobs and essential services.

If we effectively incorporate equity into plans, projects, and policies, we are better equipped to create transport alternatives that service society fairly, and that can prioritize the needs of the most disadvantaged members of our communities. Many of these individuals are more likely to be reliant on less costly modes, such as transit, walking, and cycling. The National Academies' Transportation Research Board (TRB) identifies important issues about transportation's impacts on equity, including areas such as access, institutional roles, planning and public participation, and land use and affordable housing. In essence, addressing equity concerns in transportation involves ensuring that all individuals, regardless of their socioeconomic status, have equal access to transportation services, are equally represented in transportation planning and decision-making processes, and equally share in the benefits of transportation investments.

13.3.3 Traffic Patterns and Flow Management

Traffic patterns and flow management are integral aspects of urban transportation planning. They play a crucial role in ensuring the smooth movement of vehicles and reducing congestion, thereby enhancing the overall efficiency of the transportation system.

13.3.3.1 Understanding Traffic Patterns

Understanding traffic patterns is essential in urban transportation planning and management. Traffic patterns refer to the movement and behavior of vehicles on road networks, influenced by various factors such as time of day, day of the week, weather conditions, and special events. For instance, during peak hours, traffic volume typically increases, leading to congestion. By understanding these patterns, transportation planners can implement strategies to manage peak-hour traffic, such as adjusting traffic signal timings or encouraging flexible work hours. Moreover, traffic patterns can provide insights into long-term trends and changes in travel behavior, which can inform transportation planning and policy.

Traffic patterns are also vital for traffic flow management, which involves strategies to optimize traffic flow and reduce congestion. Understanding traffic patterns can help in designing and implementing these strategies. For example, traffic signal optimization involves adjusting the timing of traffic signals based on real-time traffic conditions to improve traffic flow. Similarly, traffic demand management strategies, such as congestion pricing and carpooling programs, can be more effectively implemented with a good understanding of traffic patterns. Furthermore, with the advent of technologies like Intelligent Transportation Systems (ITS) and AI, traffic flow prediction and management have seen significant advancements. These technologies can provide real-time traffic information, manage traffic flow, and detect and respond to incidents, significantly improving the efficiency and safety of transportation systems.

13.3.3.2 Traffic Flow Management

Traffic flow management involves strategies to optimize traffic flow and reduce congestion. These strategies can range from speed control, lane management, traffic signal optimization, to effective signage utilization. Speed control measures, such as speed limits and traffic calming devices, can help regulate traffic flow and improve safety. Lane management strategies, such as dedicated bus lanes or high-occupancy vehicle lanes, can optimize the use of road space. Traffic signal optimization involves adjusting the timing of traffic signals based on real-time traffic conditions to improve traffic flow. Effective signage can guide drivers and help prevent confusion and disruptions that can lead to congestion.

Addressing safety concerns related to traffic patterns in cities involves a comprehensive approach that includes understanding traffic patterns, implementing safety measures, and leveraging technology. Cities can incorporate safety features into the design of transportation infrastructure. This can include traffic calming measures, such as speed bumps and roundabouts, to slow down vehicles and reduce the risk of accidents. Pedestrian-friendly design features, such as wide sidewalks, crosswalks, and pedestrian islands, can also enhance safety. Technology can also enhance safety. For example, ITS can provide real-time traffic information, manage traffic flow, and detect and respond to incidents. Similarly, machine learning algorithms can analyze large amounts of traffic data, learn from patterns and past behaviors, and provide insights into the current and future traffic flow. This can help optimize traffic management, including traffic signal control, route optimization, and demand forecasting.

13.3.4 Urban Transport Policies and Regulations

A policy is a set of ideas or a plan of what to do in particular situations that has been agreed to officially by a group of people, a business organization, a government, or a political party. It serves as a guiding principle helping individuals or organizations make decisions and achieve rational outcomes. Policies are generally adopted by a governance body within an organization and can assist in both subjective and objective decision making. Understanding urban transport policies and regulations is crucial in urban transport planning for several reasons.

First, transport policies and regulations set the framework within which transport systems operate. They define the rules, standards, and procedures that guide the planning, design, operation, and maintenance of transportation infrastructure and services. Understanding these policies and regulations can help ensure that transport plans are compliant and effective. Second, transport policies and regulations reflect the priorities and goals of the government and society. They can aim to promote sustainability, improve mobility, enhance safety, or achieve other objectives. Understanding these policies and regulations can help align transport plans with these goals. Third, transport policies and regulations can influence travel behavior. For example, policies promoting public transit or active transportation can encourage people to use these modes, while regulations such as congestion pricing can discourage car use. Understanding these policies and regulations can help predict and manage travel demand. Lastly, transport policies and regulations can have significant impacts on land use, economic development, social equity, and other aspects of urban life. Understanding these impacts can help integrate transport planning with other areas of urban planning.

Urban transport policies and regulations are crucial in shaping the transportation landscape of cities, influencing how people move, the quality of air, and the overall livability of urban spaces.

13.3.4.1 Urban Transport Policies

Transport policies set the direction for the development and management of transportation systems. They lay the basic framework for downstream planning as well as project identification and prioritization. These policies can aim to promote sustainability, improve mobility, enhance safety, or achieve other objectives. For instance, a city might adopt a policy to promote public transit or active transportation or implement congestion pricing to manage traffic. Policies can also be used to mitigate the environmental impact of urban passenger transport. For example, cities around the world have tried a wide variety of measures, often through regulations or economic incentives, to reduce emissions and improve air quality.

13.3.4.2 Urban Transport Regulations

Regulations, on the other hand, are rules or directives made and maintained by an authority to control the operation of urban transport systems. They define the standards and procedures that guide the planning, design, operation, and maintenance of transportation infrastructure and services. Regulations can cover a wide range of areas, including vehicle emissions standards, speed limits, parking rules, public transit fares, and requirements for accessibility. They can also regulate new forms of mobility, such as ride-hailing services and e-scooters.

13.3.4.3 The Interplay of Policies and Regulations

Policies and regulations often work hand in hand. For example, a policy to promote EVs could be supported by regulations that require the installation of charging stations in public parking lots. Similarly, a policy to improve road safety could be implemented through regulations that enforce speed limits and mandate the use of seat belts.

13.3.4.4 Evaluating Urban Transportation Policies

Evaluating the impact of various urban transportation policies aimed at alleviating traffic congestion and air pollution is essential. This involves measuring crucial quantities, including the marginal external cost of traffic congestion, and evaluating different policies in terms of both efficiency and equity within an integrated framework. For instance, Beijing's municipal government has implemented aggressive demand-side and supply-side policies over the past 15 years, making it an ideal setting for studying urban transportation policies.

13.4 URBAN TRANSPORT DEMAND AND SUPPLY

Transport demand refers to the amount and type of travel that people would choose under specific conditions. The key factors influencing transport demand include:

- *Socioeconomic Factors:* These include income levels, age, gender, occupation, and education level. For instance, higher-income groups might prefer private cars, while students might rely more on public transport.
- *Land Use Patterns:* The spatial distribution of residences, workplaces, educational institutions, and recreational areas affects travel demand. Higher-density areas often have higher public transport demand.
- *Transport Costs:* This includes both monetary costs (fares, fuel costs, tolls) and non-monetary costs (travel time, comfort, safety). Lower costs generally lead to higher demand.
- *Transport Policies:* Policies such as congestion pricing, parking regulations, and subsidies can influence people's travel choices.

Demand analysis involves the use of various models to predict travel behavior under different scenarios. These models consider the above factors to estimate trip generation (how many trips people make), trip distribution (where these trips go), mode choice (what mode of transport is used), and route choice (which path is taken).

The steps for analyzing urban transport demand without involving a complex model, as shown in Table 13.3.

TABLE 13.3
Steps in Demand Analysis

Objective	The objective of demand analysis in the context of urban transportation is to understand and predict the travel behavior of individuals or groups.
Define the scope	Identify the geographical area and the modes of transport you want to analyze. This could be a specific city or region, and could involve public transport, private vehicles, cycling, walking, etc.
Data collection	Gather data related to transport usage in the defined area. This could involve traffic counts, public transport ridership data, surveys of travel behavior, etc. Also collect data on factors that might influence transport demand, such as population density, land use patterns, transport costs, and socioeconomic factors.
Descriptive analysis	Analyze the collected data to understand current travel patterns. This could involve calculating average trip lengths, mode shares (i.e., the percentage of trips made by each mode of transport), peak travel times, etc.
Trend analysis	Look at how travel patterns have changed over time. This could involve analyzing historical data to identify trends in transport demand. For example, has public transport ridership been increasing or decreasing? Has the mode share of private cars been changing?
Comparative analysis	Compare the transport demand in your area with other similar areas. This could provide insights into how transport demand might change in the future or under different conditions.
Impact analysis	Analyze the potential impact of changes in factors that influence transport demand. For example, how might changes in population density, land use patterns, or transport policies affect transport demand?
Reporting	Document your findings in a clear and understandable manner. This could involve creating graphs and charts to visualize travel patterns, writing a report summarizing your findings, or presenting your analysis to stakeholders.

TABLE 13.4
Reasons the Importance of Demand Analysis

Infrastructure planning	By predicting future travel demand, planners can design and implement infrastructure projects that meet these needs. This could involve building new roads, expanding public transit systems, or creating bike lanes.
Policy making	Understanding travel demand can help policymakers create effective transport policies. For example, if a city has high demand for public transit, policies could be implemented to further encourage its use, such as reducing fares or increasing service frequency.
Sustainability	Demand analysis can help identify sustainable transport solutions. If a city has high demand for cycling, for instance, it might invest in bike-sharing programs or dedicated cycling infrastructure.
Efficiency	By matching supply with demand, transport systems can operate more efficiently. This can reduce congestion, improve travel times, and enhance the overall user experience.
Equity	Understanding demand can ensure that all residents have access to transportation, regardless of their location or income level. This is crucial for creating equitable and inclusive cities.

Demand analysis is important to understand and predict the travel behavior of individuals or groups. This understanding is fundamental for several reasons as reflected in Table 13.4.

While the approach of demand analysis can provide valuable insights, it may not be able to capture the complex interactions between different factors influencing transport demand. Therefore, it is typically used as a preliminary analysis or when resources for a full-demand modeling study are not available.

13.4.1 Supply Analysis in Urban Transport

Transport supply refers to the capacity of the transport system to accommodate travel demand. It depends on:

- *Infrastructure:* This includes the quantity and quality of roads, public transit lines, bike lanes, sidewalks, and other transport facilities.
- *Transport Services:* The frequency, speed, reliability, and coverage of public transport services affect their ability to meet travel demand.
- *Vehicles:* The number and capacity of vehicles, whether private cars, buses, trams, or trains, also determine transport supply.
- *Regulations:* Rules regarding vehicle standards, driving behavior, service operation, etc., can impact transport supply.

Supply analysis involves assessing the above elements to determine the transport system's ability to serve travel demand. It also considers how changes in these elements might affect service levels. Here are the steps involved in supply analysis for urban transportation as shown in Table 13.5.

The goal of supply analysis is to understand the capacity of the transport system to meet travel demand, both now and in the future. It provides crucial information for transport planning and policymaking. The supply analysis can also lead to understanding the level of service (LOS) of the transportation system.

Let's consider an example of determining the LOS for a road segment, which is often based on traffic flow characteristics. The Highway Capacity Manual (HCM) provides a standard method for this.

TABLE 13.5
Steps in Supply Analysis

Objective	The main objective of supply analysis in urban transportation is to understand the capacity and efficiency of the existing transportation system and its ability to meet the current and future travel demand.
Identify transport modes	The first step is to identify the modes of transport that are available in the area of study. This could include public transport (buses, trams, metros), private vehicles (cars, motorcycles), and NMT (cycling, walking).
Infrastructure assessment	Evaluate the existing transport infrastructure. This includes the quantity and quality of roads, public transit lines, bike lanes, sidewalks, and other transport facilities.
Service evaluation	Assess the level of service provided by each mode of transport. This could involve measuring factors like frequency, speed, reliability, and coverage of public transport services.
Capacity analysis	Determine the capacity of the transport system. This involves calculating the maximum number of passengers or vehicles that can be accommodated on each mode of transport.
Performance measurement	Measure the performance of the transport system. This could involve metrics like travel speed, delay, reliability, and passenger comfort.
Future projections	Based on trends and planned developments, we will project the future supply of transport. This could involve considering planned infrastructure projects, expected changes in service levels, and anticipated technological advancements.
Gap analysis	Compare the current and projected supply with the current and projected demand to identify any gaps. This will help in identifying areas where the transport supply needs to be increased or improved.
Reporting	Document your findings and present them in a clear and understandable manner. This could involve creating graphs and charts to visualize the data, writing a report summarizing your findings, or presenting your analysis to stakeholders.

The steps for determining the LOS are shown below:

1. *Data Collection:* Collect data on traffic volume (number of vehicles per hour), speed, and roadway conditions (number of lanes, presence of intersections, etc.) for the road segment in question.
2. *Calculate Density:* Density is typically used as the primary measure of effectiveness. It is calculated as the ratio of traffic volume to speed. For example, if there are 200 vehicles per hour and the average speed is 50 km/h, the density would be 200/50 = 4 vehicles per kilometer.
3. *Determine LOS:* Compare the calculated density with the thresholds provided in the HCM to determine the LOS. For a freeway, the HCM might provide the following thresholds:
   - LOS A: Density $<$ 11 vehicles per kilometer
   - LOS B: 11 $\leq$ Density $<$ 18 vehicles per kilometer
   - LOS C: 18 $\leq$ Density $<$ 26 vehicles per kilometer
   - LOS D: 26 $\leq$ Density $<$ 35 vehicles per kilometer
   - LOS E: 35 $\leq$ Density $<$ 45 vehicles per kilometer
   - LOS F: Density $\geq$ 45 vehicles per kilometer

In this example, with a density of 4 vehicles per kilometer, the road segment would be classified as LOS A, indicating free-flow conditions with little or no delay. This is a simplified example. In practice, determining LOS can be more complex and may consider additional factors such as traffic interruptions, pedestrian activity, and driver comfort and convenience. Also, the thresholds can vary

TABLE 13.6
Methods to Improve LOS at Intersections

Optimize signal timing	Adjusting the signal timing can significantly improve traffic flow. This includes optimizing the green time for each phase, coordinating signals along a corridor, and implementing adaptive signal control technologies that adjust signal timings in real time based on traffic conditions.
Improve lane configuration	Adding turn lanes or adjusting lane assignments can reduce conflicts and improve capacity. For example, a dedicated right-turn lane can allow right-turning vehicles to bypass the queue waiting at the red light.
Implement Advanced Traffic Management Systems (ATMS)	ATMS uses technology to improve traffic flow. This could include traffic sensors that provide real-time data on traffic conditions, allowing for dynamic adjustments to signal timings.
Promote alternative modes of transport	Encouraging public transit, cycling, and walking can reduce the number of vehicles at intersections, thereby improving the LOS.
Implement intersection geometric improvements	This could include adding more lanes, implementing roundabouts, or grade separation where feasible.
Traffic demand management (TDM)	TDM strategies such as congestion pricing or high-occupancy vehicle (HOV) lanes can reduce traffic volumes during peak periods.

depending on the specific context and source of the LOS criteria. Always refer to the relevant transportation engineering resources or local guidelines when performing such an analysis.

There are many ways to improve the LOS. Here is an example of what cities can employ to improve the LOS at signalized intersections as shown in Table 13.6. Improving LOS at intersections is essential in urban traffic management.

The appropriate strategies will depend on the specific conditions of each intersection, including its physical layout, traffic volumes, and the surrounding land use. Therefore, a detailed analysis of each intersection is necessary to identify the most effective improvements.

13.4.2 Adaptive Signal Control Technology/Systems

The main purpose of adaptive signal control technology (ASCT) is to manage throughput and improve average travel time, control delay, emissions, and fuel consumption. It adjusts the timing of red, yellow, and green lights to accommodate changing traffic patterns and ease traffic congestion. ASCT is particularly effective on arterials with variable traffic demand.

ASCT works by collecting current demand information, evaluating performance using system-specific algorithms, and then implementing modifications based on the outcome of that evaluation. Traffic sensors collect data, which are then evaluated, and signal timing improvements are developed. Finally, ASCT implements signal timing updates. The process is repeated every few minutes to keep traffic flowing smoothly.

While ASCT can be highly beneficial, it is not an effective solution for every corridor. It does not create capacity; rather, it optimizes green times using a variety of mechanisms. It is therefore most successful in systems near capacity with available green time to reallocate, as well as high traffic volumes to warrant the investment.

13.4.3 Balancing Demand and Supply

Urban transport planning is fundamental in shaping the mobility landscape of cities. The ultimate goal of this planning process is to strike a balance between demand and supply. This means that the transport system should be capable of efficiently and sustainably meeting the travel needs of all urban residents. Achieving this balance involves a combination of managing travel demand and enhancing transport supply.

Travel demand management is a strategic approach that aims to optimize the use of existing transport infrastructure by influencing travel behavior. This is achieved through the implementation of various policies and strategies. One such policy is congestion pricing, which involves charging vehicles for using certain roads during peak hours. The primary objective of congestion pricing is to reduce traffic congestion by encouraging people to alter their travel behavior. This could involve shifting their travel times to off-peak periods, choosing alternative routes, or opting for other modes of transport such as public transit, cycling, or walking. By doing so, congestion pricing can help manage travel demand and improve the efficiency of the transport system.

Promoting non-motorized transport (NMT) is another effective strategy for managing travel demand. NMT modes, such as walking and cycling, are not only sustainable but also efficient for short-distance travel. Cities can encourage the use of NMT by developing pedestrian-friendly infrastructure, implementing bike-sharing programs, and creating safe and accessible cycling lanes. Shifting a portion of travel demand from motorized transport to NMT enables cities to reduce traffic congestion and improve the overall efficiency of the transport system.

Managing travel demand is essential, it is also equally important to enhance the supply side of the transport system. This involves improving the infrastructure and increasing the service frequency.

Infrastructure improvement can take various forms, depending on the specific needs and characteristics of the city. For instance, in cities with high public transit demand, this could involve expanding the transit network, upgrading transit facilities, or implementing bus priority measures. In cities with high levels of cycling, this could involve developing a comprehensive network of cycling lanes or building secure bicycle parking facilities. Improving the infrastructure allows cities to increase the capacity of the transport system and better meet the travel needs of urban residents.

Increasing service frequency, particularly for public transit, is another effective strategy for enhancing transport supply. Higher service frequency can reduce waiting times for passengers, making public transit a more attractive option. It can also increase the capacity of the transit system, allowing it to accommodate higher passenger volumes.

13.4.4 Disequilibrium in Urban Transportation

Disequilibrium, on the other hand, is a state where demand and supply are out of balance. This is often the reality in many urban areas, where rapid growth and changing travel patterns can quickly outstrip the capacity of the transportation system. Disequilibrium can manifest in various ways. One common symptom is congestion, where the demand for road space exceeds the supply, resulting in slow speeds and long travel times. Another symptom is overcrowding on public transit, where the number of passengers exceeds the capacity of the vehicles. Disequilibrium can have significant negative impacts. It can lead to wasted time and increased stress for travelers, increased air pollution due to idling vehicles, and decreased economic productivity due to the inefficiency of the transportation system.

Disequilibrium in urban transportation, a state where demand and supply are out of balance, can be caused by several factors:

- *Rapid Urbanization:* Rapid population growth in cities can lead to an increase in travel demand that outpaces the capacity of the existing transportation system. More people living in larger cities can strain urban transportation systems.
- *Inadequate Infrastructure:* If the transport infrastructure is not sufficient or not well-maintained, it can lead to disequilibrium. The infrastructure supply often struggles to keep up with the pace of mobility growth.

- *Inefficient Public Transport:* If public transport services are infrequent, unreliable, or do not cover all areas of the city, people may resort to using private vehicles, leading to increased demand.
- *Increased Motorization:* The diffusion of the automobile has increased the demand for transport infrastructures. As vehicles spend most of their time parked, motorization has expanded the demand for road infrastructure and parking space.
- *Poor Traffic Management:* Inefficient traffic management and control can lead to congestion, even when there is sufficient infrastructure.
- *Lack of Integrated Planning:* Urban transport systems may not adequately satisfy the requirements of urban mobility if there is a lack of integrated planning.
- *Economic Factors:* Economic growth and increased incomes can lead to higher car ownership and usage, thereby increasing travel demand.
- *Land Use Patterns:* Urban sprawl, where residential areas are spread out, can lead to increased reliance on private vehicles and higher travel demand.

Addressing these causes requires a comprehensive approach that includes improving infrastructure, enhancing public transport, implementing effective traffic management strategies, and promoting sustainable transport modes.

13.4.5 Strategies for Achieving Equilibrium

Given the challenges of disequilibrium, much of urban transportation planning is focused on strategies for achieving or maintaining equilibrium. These strategies can be broadly categorized into two types: demand management and supply enhancement.

Demand management strategies aim to reduce the demand for travel or shift it to less congested times or modes. Examples include congestion pricing, where drivers are charged a fee to use certain roads during peak times, and travel demand management programs, which encourage carpooling, telecommuting, and other alternatives to single-occupancy vehicle travel.

Supply enhancement strategies, on the other hand, aim to increase the capacity of the transportation system. This can involve building new infrastructure, such as roads or transit lines, but it can also involve making better use of existing infrastructure. For example, intelligent transportation systems use technology to improve the efficiency of traffic signal timings, vehicle routing, and other aspects of the transportation system.

13.5 PUBLIC TRANSPORTATION MODES

Public transport, also known as mass transit or public transit, plays a vital role in shaping the mobility landscape of urban areas. It provides an efficient, cost-effective, and environmentally friendly alternative to private vehicle use. The various modes of public transport, each with its unique characteristics and advantages, collectively contribute to the diversity and resilience of urban transport systems. These modes include buses, trams, metros, commuter trains, ferries, and more recently, shared mobility services. Buses, being flexible and requiring relatively less infrastructure, are the most common form of public transport in cities worldwide. Trams and light rail systems offer a higher capacity and speed, making them suitable for medium-demand routes. Metros and commuter trains cater to high-demand routes, offering fast and frequent service over longer distances. Ferries provide an essential mode of transport in cities with suitable water bodies, offering a unique combination of transport and tourism functions. Shared mobility services, such as bike-sharing and ride-hailing, have emerged as important complements to traditional public transport, providing solutions for first- and last-mile connectivity. As cities continue to grow, public transport modes will keep addressing the challenges of congestion, pollution, and urban mobility.

13.5.1 Bus Transit Systems

Bus transit systems form the backbone of public transportation in many cities around the world. They offer a flexible, cost-effective, and environmentally friendly alternative to private vehicle use, playing a crucial role in enhancing urban mobility, reducing congestion, and promoting sustainable development.

13.5.1.1 Components of Bus Transit Systems

A bus transit system comprises several components, each playing a vital role in its operation. These include the buses themselves, which vary in size, capacity, and fuel type; the bus stops, which serve as access points for passengers; the bus routes, which determine the path that buses follow; and the bus schedule, which specifies the frequency of service.

13.5.1.2 Types of Bus Transit Systems

There are various types of bus transit systems, each designed to cater to specific travel needs and urban contexts. These include:

- *Local Bus Service:* This is the most common type of bus service, providing local transportation within a city or town. Buses stop frequently, serving almost all bus stops along their route.
- *Express Bus Service:* Express buses operate on longer routes with fewer stops, offering faster service for passengers traveling longer distances.
- *Bus Rapid Transit (BRT):* BRT systems offer high-frequency, high-capacity service, similar to a metro or light rail system, but at a lower cost. Key features of BRT systems include dedicated bus lanes, off-board fare collection, and priority at traffic signals.

13.5.1.3 Benefits of Bus Transit Systems

Bus transit systems offer numerous benefits. They provide affordable and accessible transportation for all residents, including those without access to private vehicles. Bus transit systems can reduce reliance on private cars, which helps decrease traffic congestion, air pollution, and greenhouse gas emissions. Moreover, promoting compact, TOD through bus transit systems fosters more walkable, livable, and sustainable urban environments.

13.5.1.4 Challenges and Solutions

Despite their benefits, bus transit systems face several challenges. These include traffic congestion, which can lead to unreliable service; insufficient funding, which can limit service quality and coverage; and competition from private cars and ride-hailing services, which can reduce ridership. To address these challenges, cities are implementing various strategies. These include giving buses priority at traffic signals and dedicated bus lanes to improve speed and reliability; investing in cleaner, more efficient buses to reduce environmental impact; enhancing passenger amenities to improve the riding experience; and integrating bus services with other modes of transport to provide seamless mobility.

13.5.1.5 The Future of Bus Transit Systems

Looking ahead, bus transit systems are set to play an even more critical role in urban transportation. Advances in technology, such as electric buses, autonomous driving, and real-time passenger information systems, promise to make bus transit more efficient, sustainable, and user-friendly. At the same time, the growing recognition of public transit's role in addressing urbanization, climate change, and social equity challenges underscores the importance of continued investment in bus transit systems.

13.5.1.6 Electric Bus

Electric buses are increasingly important in reducing emissions from public transportation. They operate without tailpipe emissions, significantly reducing harmful pollutants such as nitrogen oxides (NO_x) and particulate matter. This transformation is particularly crucial for densely populated urban areas, where air quality concerns have been on the rise.

The use of EVs, including buses, is rising in many cities around the world, underpinned by the urgent need to reduce levels of air and noise pollution, and tackle the ever-growing energy-related greenhouse gas (GHG) emissions from conventionally fueled (fossil fuel) vehicles. The Intergovernmental Panel on Climate Change (IPCC) recognizes electrification of short-distance vehicles as a powerful means for reducing emissions.

Electric buses not only help in reducing GHG emissions but also contribute to the reduction of noise pollution in urban areas. They are quieter than conventional diesel buses, contributing to a more pleasant urban soundscape. Moreover, the strong increase of renewable energy sources in electricity generation, observed over the past decade and expected to further grow, maximizes the climate benefits of electric mobility. When eBuses are fueled by electricity that comes from renewable sources, the benefits are even greater.

A recent study reveals that transitioning US transit buses to electric can significantly reduce emissions by 33–65%. This highlights the potential for sustainability and increased accessibility in public transportation systems. However, the transition to electric buses requires careful planning. It should also consider the availability of charging infrastructure, the range of electric buses, the procurement and operational costs, and the source of electricity.

Cities can incentivize private operators to invest in electric buses for public transportation through a variety of strategies as shown in Table 13.7.

A combination of financial incentives, policy support, and infrastructure development can incentivize private operators to invest in electric buses, contributing to a more sustainable public transportation system.

TABLE 13.7
Strategies to Encourage Private Operators to Invest in Electric Buses

Financial incentives	Financial incentives such as grants, tax breaks, reduced vehicle taxes, and subsidies for charging infrastructure development can encourage public transportation agencies and private operators to invest in electric bus fleets. For instance, the FAME-II scheme in India allows a subsidy of Rs. 20,000 per kWh for e-buses with a cap of 40% of the vehicle's cost and a maximum ex-factory price limit of Rs. 2 crores.
Interest rate subvention	Lowering interest rates for loans taken to purchase electric buses can make them more affordable for private operators. For example, a study recommends incentivizing through interest rate subvention at 4–6% for a loan term of seven years for the uptake of e-buses in the private bus sector.
Leasing model	A leasing model can address the high upfront costs related to e-buses. Under this model, private operators can lease electric buses rather than purchasing them outright, reducing the initial investment required.
Policy support	Governments can introduce policies that mandate a certain percentage of buses to be electric. Such policies can create a stable market for electric buses, encouraging private operators to invest in them.
Infrastructure development	Governments can invest in developing charging infrastructure, making it easier for private operators to maintain and operate electric buses.
awareness and capacity building	Governments can also conduct awareness programs and capacity-building workshops to educate private operators about the benefits of electric buses and how to operate and maintain them.

13.5.2 Rail-Based Systems

Rail-based systems are a critical component of urban transportation networks, offering a high-capacity, efficient, and sustainable mode of transport. They include various types of systems such as trams or streetcars, light rail transit (LRT), heavy rail transit (HRT), and commuter rail.

13.5.2.1 Trams or Streetcars

Trams, also known as streetcars, operate on tracks embedded in city streets. They are typically used for short to medium distances and have frequent stops. Trams are often used in downtown areas or historic districts where their slower speed and integration with the streetscape are seen as advantages. They can operate in mixed traffic or have their dedicated right-of-way.

13.5.2.2 Light Rail Transit

Light rail transit (LRT) is a step up from trams in terms of speed and capacity. LRT systems typically have their dedicated right-of-way, either at street level, elevated, or underground, allowing them to avoid traffic congestion. They are often used for medium-to-long distances and have less frequent stops compared to trams. LRT systems can be a cost-effective alternative to heavy rail, offering similar benefits at a lower cost.

13.5.2.3 Heavy Rail Transit

Heavy rail transit (HRT), also known as metro or subway, is designed for transporting large numbers of passengers over long distances. HRT systems operate on fully separated right-of-way, either elevated or underground, allowing them to maintain high speeds and avoid traffic congestion. They have high-frequency service and high-capacity vehicles, making them ideal for serving dense urban areas with high travel demand.

13.5.2.4 Commuter Rail

Commuter rail systems provide service between outlying suburbs or small towns and central city areas. They operate on standard railroad tracks and are designed for longer distances with less frequent stops. Commuter rail systems typically have peak-hour services, catering mainly to commuters working in central city areas.

13.5.2.5 Benefits of Rail-Based Systems

Rail-based systems offer numerous benefits. They provide high-capacity service, making them ideal for serving areas with high travel demand. They are more energy-efficient and environmentally friendly compared to private cars, helping to reduce greenhouse gas emissions and air pollution. Rail-based systems also promote compact, TOD, contributing to more walkable and livable urban environments.

13.5.2.6 Challenges and Solutions

Despite their benefits, rail-based systems also face challenges. These include high capital costs, long construction periods, and the need for careful planning and coordination. However, these challenges can be addressed through sound planning, effective project management, and innovative financing strategies.

The benefits of rail-based systems for urban transportation are shown in Table 13.8.

Despite the numerous advantages of rail-based systems, they also face constraints and challenges such as high capital costs, long construction periods, and the need for careful planning and coordination, as shown in Table 13.9.

A combination of strategic planning, innovative financing, efficient operations, public engagement, and the effective use of technology is expected to be able to cope with existing challenges.

TABLE 13.8
Benefits of Rail-Based Systems

High capacity	Rail-based systems, especially heavy rail transit like metros, can carry a large number of passengers, making them ideal for serving areas with high travel demand.
Efficiency	Trains can transport a large number of passengers quickly and reliably, making them a highly efficient mode of transport.
Reduced traffic congestion	By providing an alternative to private car use, rail systems can help reduce traffic congestion on roads.
Environmental sustainability	Trains, especially electric ones, emit fewer greenhouse gases per passenger-kilometer compared to cars. This makes rail transport a more environmentally friendly option.
Land use efficiency	Rail systems, with their fixed routes and stations, encourage compact, transit-oriented development. This can lead to more efficient land use, with higher-density development focused on transit stations.
Economic development	The presence of a rail system can stimulate economic development and increase property values in the areas around stations.
Accessibility	Rail systems provide a means of transportation for people who do not have access to a car, including the young, the elderly, and the economically disadvantaged.
Safety	Rail travel has a lower accident rate compared to car travel, making it a safer mode of transport.

TABLE 13.9
Constraints and Strategies to Cope with Constraints in Rail-Based Systems

Constraints	Strategies
High capital costs	High capital costs can be addressed through innovative financing mechanisms, such as public–private partnerships, value capture financing, or infrastructure bonds.
Long construction	Long construction periods can be managed through phased construction, where the system is built and opened in segments. This can also help generate revenue earlier, offsetting some of the construction costs.
High operational costs	Operational costs can be reduced through efficient management practices, such as predictive maintenance, energy-efficient operations, and automated systems.
Ridership	Achieving sufficient ridership requires integrated land use and transport planning. This involves focusing development on transit stations (transit-oriented development) and ensuring good connectivity with other modes of transport.
Public support	Gaining public support for rail projects can be facilitated through effective public engagement. This involves communicating the benefits of the project, addressing concerns, and involving the public in the planning process.
Funding	Strong political leadership can help secure funding for rail projects, overcome opposition, and ensure that the project is aligned with broader urban development goals.
Technological challenges	Technological challenges can be addressed through the adoption of proven technologies, staff training, and collaboration with technology providers.
Limited resources and expertise	Forming partnerships with private sector companies, non-profit organizations, and other stakeholders can bring additional resources, expertise, and support to rail projects.

Rail-based systems, including trams, light rail, heavy rail, and commuter rail, offer a unique blend of cost-effectiveness and efficiency compared to other modes of transport.

13.5.2.7 Cost-Effectiveness

In terms of cost, rail transport is often more affordable for long distances due to the high freight capacity and lower fuel consumption. The Association of American Railroads notes that railways handle more than 40% of the long-haul freight in the United States, offering cost advantages for transporting commodities like coal, metals, and agricultural products. However, the initial capital costs for establishing rail infrastructure can be high.

On the other hand, road transportation, including buses and private vehicles, is typically more cost-effective for short distances. Shipping charges are usually lower on shorter routes, making road transportation services less expensive depending on cargo volume and the destination.

13.5.2.8 Efficiency

Rail-based systems are highly efficient for transporting large numbers of passengers or high volumes of goods over long distances. They operate on dedicated tracks, allowing them to avoid traffic congestion and maintain high speeds. This makes rail transport ideal for extensive hauls. Road transportation, while flexible and capable of door-to-door service, can be subject to traffic congestion, which can impact its efficiency. However, road transportation has a fast transit time as trucks can be loaded and unloaded promptly and can deliver goods to their ultimate destination.

13.5.3 Non-Motorized Transport

NMT refers to modes of transportation that do not require an engine or motor. This includes walking, cycling, and other forms of human-powered transport. NMT is gaining popularity, especially in developed countries, due to its low transport externalities such as emissions and traffic congestion, alongside its benefits to physical and mental health. NMT offers basic mobility, affordable transport, and access to public transport. It plays a crucial role in improving the convenience, comfort, and safety of walking and cycling, which reduces the demand for travel by personal motor vehicles. This helps to alleviate the critical traffic challenges facing many cities.

As zero-emission modes, walking and cycling are critical efforts in reducing the harmful local pollution and greenhouse gas emissions. The built environment, geography, weather, and socio-economic factors significantly affect the use of NMT as a travel mode.

Developing a NMT strategy involves several steps. It begins with a discussion of the design principles for NMT facilities, followed by a description of methodologies for assessing the existing NMT environment and institutional capacity. The strategy typically includes components that can be customized for the local context. These may involve specific projects, such as a streetscape redesign or the implementation of a system for managing on-street parking. The strategy can also guide coordination among the multiple agencies that play a role in the design and management of the NMT environment.

In laying out the vision for an enhanced walking and cycling environment, an NMT strategy can complement mobility plans, which offer detailed guidance on specific mobility initiatives. Developing an NMT strategy is essential for promoting and enhancing the use of NMT, and it requires a comprehensive understanding of the local context and effective coordination among various stakeholders. With the right strategy and implementation, NMT can contribute significantly to sustainable urban development and improved quality of life.

Cities can motivate citizens to use NMT through various approaches. These include improving infrastructure, implementing public bicycle systems, promoting pedestrian-oriented development, increasing connectivity, implementing traffic calming measures, and providing education and encouragement programs. These approaches make NMT a more attractive, convenient, and safe option for residents are shown in Table 13.10.

Cities can enhance pedestrian safety through a comprehensive approach. This includes improving infrastructure by developing sidewalks, crosswalks, paths, bicycle lanes, and networks. The construction of bike lanes, sidewalks, and multi-use trails can significantly improve safety for pedestrians and cyclists. Public bicycle systems, such as automated bicycle rental systems, can provide efficient mobility for short, utilitarian urban trips, further promoting pedestrian safety.

Pedestrian-oriented development, which involves the development of land use and building design that encourages walking, can also contribute to pedestrian safety. Increasing road and path

TABLE 13.10
Approach to Motivate Citizens to Use NMT

Improve infrastructure	Develop sidewalks, crosswalks, paths, bicycle lanes, and networks. This includes constructing bike lanes, sidewalks, and multi-use trails.
Public bicycle systems	Implement automated bicycle rental systems designed to provide efficient mobility for short, utilitarian urban trips.
Pedestrian-oriented development	Develop land use and building design that encourages walking.
Increase connectivity	Increase road and path connectivity, with special non-motorized shortcuts.
Traffic calming measures	Implement streetscape improvements, traffic speed reductions, vehicle restrictions, and road space reallocation.
Education and encouragement programs	Implement safety education, law enforcement, and encouragement programs.
Bicycle parking	Provide ample and secure bicycle parking facilities.
Integration with public transport	Integrate bicycles into transit systems, such as providing bike racks in metro stations or on buses.
Address security concerns	Address the security concerns of pedestrians and cyclists to make them feel safe while using NMT.
Policy measures	Implement policy measures such as congestion pricing, vehicle parking policies, and fuel taxes.

connectivity, with special non-motorized shortcuts, can make walking and cycling more convenient and safer. Traffic calming measures, such as streetscape improvements, traffic speed reductions, vehicle restrictions, and road space reallocation, can further enhance pedestrian safety.

Education and encouragement programs, including safety education, law enforcement, and encouragement programs, can raise awareness about pedestrian safety and promote responsible behavior among both pedestrians and motorists. Providing ample and secure bicycle parking facilities can encourage more people to cycle, reducing the number of vehicles on the road and thereby enhancing pedestrian safety.

Integration of bicycles into transit systems, such as providing bike racks in metro stations or on buses, can make cycling a more viable and safe option for commuters. Addressing the security concerns of pedestrians and cyclists can make them feel safe while using NMT. Policy measures, such as congestion pricing, vehicle parking policies, and fuel taxes, can discourage the use of private vehicles and promote walking and cycling.

Deploying vertical bollards or high-visibility paint can alert drivers about upcoming pedestrian crossings. Closing travel lanes during peak hours can reduce traffic and increase pedestrian safety. Implementing pedestrian-only traffic signal phases can stop all vehicular traffic, further enhancing pedestrian safety. Through these strategies, cities can make NMT a more attractive, convenient, and safe option for residents, thereby promoting its use and reaping the associated environmental and health benefits.

13.5.4 Public Bicycle Systems and Bike Lanes

Public bicycle systems, also known as bike share programs or public bike share (PBS) schemes, are shared transport services where bicycles are available for shared use by individuals at low cost. These systems are ideal for short-term urban bicycle rental schemes that enable bicycles to be picked up at any self-serve bicycle station and returned to any other bicycle station, making bicycle-sharing ideal for point-to-point trips. The programs include both docking and dockless systems. Docking systems allow users to rent a bike from a dock, i.e., a technology-enabled bicycle rack, and return it to another dock within the system. Dockless systems offer a node-free system relying on smart technology. In either format, systems may incorporate smartphone web mapping to locate available bikes and docks.

Public bicycle systems have been implemented in approximately 3,000 cities worldwide, including Dubai, New York, Paris, Mexico City, Montreal, and Barcelona. One of the well-known vendors of these systems is PBSC Urban Solutions, which develops bicycle-sharing systems, equipment, parts, and software, and sells its products to cities in Canada, the United States, the United Kingdom, Spain, Brazil, and more.

These systems are designed to provide short-term bicycle rentals for users to travel within a city. They typically involve a network of bike stations or docks, where users can pick up and return bicycles. Public bicycle systems are a significant part of urban transportation, offering an eco-friendly, healthy, and efficient way of commuting.

Cities can enhance cycling safety through a variety of strategies. One such strategy is the creation of bike lanes in car-free zones, where cyclists can safely share the street with pedestrians and other non-motorized road users. The implementation of bicycle boulevards, which are streets designed to prioritize bicycle traffic, can also contribute to cycling safety. The use of pop-up bike lanes, which are temporary bike lanes that separate cars and cyclists, can significantly improve safety. On streets with heavy traffic, bike lanes need sturdy physical barriers to separate cyclists from vehicles, a strategy known as physical segregation.

Adding lanes specifically for cyclists at intersections can increase their visibility, thereby reducing the risk of accidents. Similarly, implementing bicycle signaling at intersections can help prevent accidents. Prohibiting right-hand turns on red for drivers can also reduce accidents involving cyclists.

In some cases, adding tunnels at intersections can be beneficial as it separates cyclists from other traffic, further enhancing safety. Cities can also implement bike racks, lockers, and parking garages to make it easier for people to secure their bikes, a strategy known as bike parking infrastructure. Education and outreach programs can help to promote cycling and improve safety for cyclists.

Bike lane development becomes an essential aspect of modern urban planning and transportation, aiming to create safe, efficient, and sustainable mobility options. The process involves careful planning, design, and implementation, considering various factors such as safety, connectivity, accessibility, and the needs of the community.

There are several types of bike lanes designed to accommodate cyclists and improve safety, as shown in Table 13.11.

Ranking bike lanes can be subjective as it depends on various factors such as the specific needs of cyclists, the volume of traffic, and the surrounding environment. However, from a general perspective considering safety and comfort for cyclists, the ranking could be as shown by Table 13.11. Each type of bike lane offers different levels of protection and separation from vehicles, and the choice of development for bike lane type depends on factors such as traffic volume, speed, and the surrounding environment.

The first step in bike lane development is understanding the needs of the community. This involves assessing the current state of cycling infrastructure, identifying gaps, and understanding the needs and preferences of cyclists and potential cyclists. Surveys, public consultations, and data analysis can provide valuable insights into these aspects. The next step is designing the bike lanes. This involves determining the type of bike lanes (such as protected, buffered, or conventional), their location, and their connectivity to other parts of the city. The design should prioritize safety for cyclists and pedestrians, considering lane setup and protection, managing common conflict zones, and selecting appropriate materials.

The implementation phase involves constructing the bike lanes according to the design. This requires coordination with various stakeholders, including city planners, engineers, contractors, and

TABLE 13.11
Different Types of Bike Lanes (from the Best to the Worst)

Separated bike lanes	Also known as protected bike lanes, these lanes are physically separated from motor vehicle traffic by vertical elements such as bollards, planters, or parked cars.
Buffered bike lanes	These are conventional bicycle lanes paired with a designated buffer space separating the bicycle lane from the adjacent motor vehicle travel lane and/or parking lane.
Contra-flow bike lanes	These are bicycle lanes designed to allow bicyclists to ride in the opposite direction of motor vehicle traffic. They convert a one-way traffic street into a two-way street: one direction for motor vehicles and bikes, and the other for bikes only.
Left-side bike lanes	These are conventional bike lanes placed on the left side of one-way streets or two-way median divided streets.
Conventional bike lanes	These are designated spaces for bicyclists on the roadway, delineated by pavement markings and signage. They are located adjacent to motor vehicle travel lanes and flow in the same direction as motor vehicle traffic.
Sharrows	Also known as shared lane markings, sharrows are markings painted onto streets where bicycles and cars share the same space. Sharrows appear as a picture of a bike with triangular arrows indicating the direction of traffic.

the community. During this phase, it is important to manage potential disruptions to traffic and communicate effectively with the public about the benefits of the new infrastructure.

Once the bike lanes are in place, they need to be maintained and monitored[1]. Regular inspections can identify issues such as damage, debris, or obstructions that need to be addressed. Data collection and analysis can help assess the usage of bike lanes, their impact on traffic and safety, and areas for improvement.

Public engagement is a crucial aspect of bike lane development. This involves ongoing communication and engagement at all stages of the design and implementation of safe bike lanes. Public consultations can provide valuable feedback and foster a sense of ownership and support for the bike lanes among the community.

Bike lane development can have significant benefits for cities. It can promote cycling as a healthy and sustainable mode of transportation, reduce traffic congestion, improve air quality, and enhance the quality of life. However, it requires a comprehensive and thoughtful approach, considering the needs and preferences of the community, the characteristics of the city, and the goal of creating safe, efficient, and sustainable mobility options.

13.5.5 Shared Mobility Services

Shared mobility services refer to transportation services and resources that are shared among users, either concurrently or one after another. This includes public transit; micromobility (bike-sharing, scooter-sharing); automobile-based modes (car-sharing, rides on demand, and micro transit); and commute-based modes or ride-sharing (carpooling and vanpooling). Shared mobility has deep roots, stretching back to 1940s Switzerland, and has expanded over time to include shared micromobility offerings, such as bicycles in the 1960s. The modern manifestation began about a decade ago when the first major ride-hailing player appeared, and the field has grown significantly.

In recent years, new modes and services have emerged, such as pooled ride-sharing with strangers, peer-to-peer car-sharing (driving a stranger's private car), and shared electric scooters, pointing to a sizable potential market in the mobility space. The next big things include autonomous taxis (so-called robo-taxis) and airborne varieties, which have seen huge investment acceleration and traction in recent months.

The shared mobility market accounted for approximately \$130 billion to \$140 billion in global consumer spending in 2019. Out of this, e-hailing accounted for the largest share, \$120 billion to \$130 billion, which is more than 90% of the total market. Taken together, car-sharing and peer-to-peer car-sharing account for less than 10% of this market, which reflects e-hailing's higher convenience. Shared transportation has grown tremendously since 2014 as a result of renewed interest in urbanism and growing environmental, energy, and economic concerns intensifying the need for sustainable alternatives. Simultaneously, advances in electronic and wireless technologies made sharing assets – and data – easier and more efficient. This led to automobile manufacturers, rental car companies, venture-backed startups, and city-sponsored programs springing up with new solutions ranging from large physical networks to mobile applications designed to alter routes, fill empty seats, and combine fare media with real-time arrival and departure information.

Shared mobility services provide a continuum of choices that cover many types of personal trips, and together with a robust public transit system, allow people to get to work, run errands, and get to all the places they need to go in daily life without the need for a personal vehicle. The benefits of shared mobility services are many. They provide more mobility choices, offer last mile and first mile solutions, reduce traffic congestion, mitigate various forms of pollution, reduce transportation costs, create equitable access to jobs and other resources, improve efficiency, identify choices for those who cannot afford to buy and maintain a vehicle, and create accessible mobility options for those with limited physical ability. The shared mobility market could grow rapidly within the next several years. Depending on customer acceptance of shared mobility, regulations in each country, and the progress of technology, spending on shared mobility services could reach \$500 billion to \$1 trillion in 2030.

13.6 PRIVATE TRANSPORTATION

Private transportation, a cornerstone of modern society, is an integral part of our daily lives. It encompasses a wide range of vehicles, from personal cars and motorcycles to private jets and yachts, all of which provide individuals with the freedom to travel according to their own schedules and preferences. This mode of transportation is often associated with convenience, comfort, and time efficiency, as it eliminates the need to adhere to public transit schedules or routes. However, issues such as traffic congestion, environmental impact, and the financial burden of vehicle ownership are significant considerations. Despite these challenges, private transportation continues to evolve, with advancements in technology paving the way for more sustainable and efficient travel options. Innovations such as EVs and autonomous driving systems are transforming the landscape of private transportation, offering promising solutions to some of its inherent problems. As we move forward, the role of private transportation in our society is likely to continue to evolve, shaped by technological advancements, environmental considerations, and changing consumer preferences. The future of private transportation is a fascinating topic, ripe for exploration and discussion. Whether viewed as a necessity, a luxury, or a complex issue requiring solutions, private transportation undeniably plays a crucial role in shaping our modern lifestyle and the urban environments we inhabit. Automobiles are the most prominent modes of private urban transport.

13.6.1 AUTOMOBILES

Automobiles, as a form of private transportation, have significantly shaped the urban landscape and are essential elements of city life. They offer convenience, flexibility, and a sense of personal freedom that public transportation often cannot match. However, the use of automobiles in cities also brings about a host of challenges, including traffic congestion, environmental pollution, and social inequality.

13.6.1.1 Convenience and Flexibility

Automobiles have revolutionized modern transportation, providing individuals with unparalleled convenience and flexibility. Unlike public transportation, which often adheres to fixed schedules and predetermined routes, automobiles offer the freedom to travel at one's own pace and on one's own schedule. This flexibility is particularly beneficial for those who live in areas that are poorly served by public transit. In many suburban and rural communities, public transportation options may be limited or non-existent, making it challenging for residents to access essential services, visit friends and family, or commute to work. With a personal vehicle, individuals can easily navigate these areas, reaching their destinations without the constraints of public transit timetables.

Furthermore, automobiles offer the convenience of traveling at odd hours, which may be necessary for those who work shifts or have irregular schedules. Public transportation often has reduced or limited service during the evenings, weekends, and holidays, leaving those who need to travel at these times with limited options. Automobiles, on the other hand, provide the freedom to travel at any time, day or night, allowing individuals to better accommodate their unique schedules and lifestyle needs.

Automobiles also offer the ability to transport heavy or bulky items, making them an invaluable asset for shopping trips, moving, or transporting large household goods. Public transportation, such as buses or trains, often have limited space and may not be suitable for carrying large or heavy items. Automobiles, with their spacious trunks and rear seats, can easily accommodate these types of loads, allowing individuals to efficiently transport the items they need without the hassle of renting a separate vehicle or relying on delivery services.

The freedom and independence provided by automobiles can have a positive impact on an individual's quality of life. The ability to travel at one's own pace, without the constraints of public transportation, can reduce stress and increase overall satisfaction with one's daily activities. This can be particularly beneficial for those with mobility issues or disabilities, who may find it challenging to navigate public transportation or rely on others for transportation.

Automobiles offer unparalleled convenience and flexibility, allowing individuals to travel at their own pace, access areas with limited public transit, and transport heavy or bulky items. These benefits have made automobiles an essential part of modern life, providing individuals with the freedom and independence to better accommodate their personal and professional needs.

13.6.1.2 Personal Freedom

Owning a car is often viewed as a symbol of personal freedom and independence, providing individuals with the ability to explore the world beyond the confines of public transportation. The car has become more than just a mode of transportation; it has evolved into a reflection of one's identity and social status.

One of the primary benefits of car ownership is the freedom it affords. With a personal vehicle, individuals can venture out to places that may be difficult or time-consuming to reach using public transit. This flexibility allows people to discover new destinations, pursue hobbies and interests, and participate in activities that may be located in areas poorly served by public transportation. Whether it is a spontaneous weekend getaway, a day trip to a remote hiking trail, or a visit to a specialized medical facility, the car empowers individuals to navigate their lives with greater autonomy and control.

The car becomes a personalized space that can be customized to reflect the owner's preferences and values. From the choice of make and model to the selection of interior accessories and personalized decor, the car can serve as an extension of one's personality and style. This personalization not only enhances the driving experience but also contributes to a sense of ownership and pride, as the car becomes a tangible representation of the individual's identity.

The car's status as a symbol of personal identity and social standing cannot be overlooked. In many cultures, the type of car one drives is often perceived as a reflection of their wealth, taste, and social standing. Luxury or high-performance vehicles, for instance, can be interpreted as symbols of success, affluence, and social prestige. Conversely, the choice of a more modest or eco-friendly car may be seen as a statement of environmental consciousness or a rejection of ostentatious displays of wealth. This association between car ownership and social status can have profound implications for individuals. The car can become a powerful tool for self-expression, allowing people to curate their public image and communicate their values and aspirations to the world around them. For some, the car may serve as a means of achieving or maintaining a certain social standing, while for others, it may be a way to differentiate themselves from their peers and establish a unique identity.

The ownership of a car is often closely tied to personal freedom, independence, and the expression of one's identity. The car provides individuals with the ability to explore the world beyond the limitations of public transportation, while also serving as a customizable space that reflects their unique preferences and values. Moreover, the car's status as a symbol of social standing can have a significant impact on how people perceive themselves and how they are perceived by others. As such, the car has become an integral part of modern life, infused with a multifaceted significance that extends far beyond its practical function as a means of transportation.

13.6.1.3 Traffic Congestion

While automobiles offer unparalleled convenience and flexibility, they have also become a significant contributor to the growing problem of traffic congestion in cities around the world. The very benefits that make cars so appealing to individuals – the freedom to travel at one's own pace and the ability to access a wide range of destinations – have also led to an ever-increasing number of vehicles on the roads, ultimately resulting in a less efficient and more frustrating transportation system.

As more people opt for the convenience of private cars, the number of vehicles on city streets has steadily risen, leading to a domino effect of traffic congestion. During peak commuting hours, major thoroughfares and highways become clogged with vehicles, causing travel times to slow to a crawl. This not only leads to increased stress and frustration for drivers but also has a significant impact on overall productivity.

The time lost sitting in traffic jams represents a significant economic cost, as workers are unable to be as productive as they would be if they were able to reach their destinations more quickly. This lost productivity can have ripple effects throughout the economy, as businesses and organizations struggle to maintain efficiency and meet deadlines. Moreover, the additional time spent in traffic can lead to burnout, absenteeism, and other negative consequences for employees, further exacerbating the problem.

Beyond the economic impacts, traffic congestion also contributes to a range of environmental and public health concerns. Idling vehicles emit harmful pollutants, degrading air quality and contributing to the global climate crisis. The increased fuel consumption and emissions associated with stop-and-go traffic also have a detrimental impact on the environment, undermining efforts to promote sustainable transportation and reduce the carbon footprint of urban areas.

Perhaps most concerning, traffic congestion can also pose a significant threat to public safety. Crowded roads increase the risk of accidents, and the additional time spent in traffic can delay emergency response times, potentially putting lives at risk.

To address this challenge, city planners and policymakers must explore a range of strategies to reduce the reliance on private automobiles and promote more sustainable and efficient transportation solutions. This may involve investments in public transit infrastructure, the implementation of congestion pricing schemes, the development of bike-friendly infrastructure, and the promotion of remote work and telecommuting options to reduce the number of vehicles on the roads during peak hours.

While the convenience and flexibility of automobiles will likely continue to be a draw for many individuals, it is clear that the negative impacts of traffic congestion on productivity, the environment, and public safety necessitate a more comprehensive and sustainable approach to urban transportation. Addressing this complex challenge allows cities to balance the benefits of automobile ownership with the need to maintain efficient, safe, and eco-friendly transportation systems.

13.6.1.4 Environmental Impact

Automobiles, while providing unparalleled convenience and mobility, have also emerged as a significant contributor to various forms of environmental pollution, posing a serious challenge to sustainability efforts in urban areas. One of the primary environmental concerns associated with automobiles is their emission of greenhouse gases, such as carbon dioxide (CO_2), methane, and nitrous oxide. These emissions, generated through the combustion of fossil fuels, are a major driver of climate change, contributing to the warming of the planet and the resulting disruptions to the global ecosystem. As the number of vehicles on the road continues to increase, so too does the cumulative impact of these greenhouse gas emissions, making it increasingly difficult to meet the targets set by international agreements aimed at mitigating the effects of climate change.

Automobiles also contribute to the degradation of local air quality through the release of other pollutants, including particulate matter, nitrogen oxides, and carbon monoxide. These pollutants can have significant adverse effects on human health, leading to respiratory issues, cardiovascular problems, and even increased risk of certain types of cancer. The impact is particularly acute in urban areas, where the concentration of vehicles can create a toxic blend of airborne contaminants that pose a serious threat to the well-being of residents.

The noise generated by automobiles, particularly in high-traffic areas, can also have detrimental effects on the environment and human health. Excessive noise pollution can disrupt the natural habitats of wildlife, leading to behavioral changes and decreased reproductive success. For human populations, exposure to prolonged noise from vehicles can contribute to stress, sleep disturbances, and even hearing loss, undermining the overall quality of life in urban areas.

The infrastructure required to support the widespread use of automobiles, such as roads, highways, and parking lots, also places a significant burden on the environment. The construction and maintenance of these facilities often result in the destruction or fragmentation of natural habitats, leading to a loss of biodiversity and ecosystem services. Additionally, the large expanses of paved surfaces associated with this infrastructure can contribute to the urban heat island effect, wherein cities experience higher temperatures than their surrounding rural areas, exacerbating the impacts of climate change and posing additional challenges for urban planning and public health.

To address these environmental concerns, a multi-pronged approach is necessary, incorporating a range of strategies that aim to reduce the reliance on private automobiles and promote more sustainable modes of transportation. This may include investments in public transit systems, the development of pedestrian-friendly and bicycle-friendly infrastructure, the implementation of congestion pricing schemes, and the incentivization of EV and hybrid vehicle adoption. With a comprehensive and collaborative approach, cities can work to mitigate the environmental impact of automobiles and build more livable, sustainable urban environments for current and future generations.

13.6.1.5 Social Inequality

The reliance on automobiles in modern society can have far-reaching implications for social inequality, creating barriers and disadvantages for those who lack access to personal vehicles. One of the primary ways in which automobile-centric transportation systems exacerbate social inequalities is through the issue of access. For individuals who do not own a car, whether due to financial constraints, physical limitations, or other factors, the ability to reach employment opportunities, educational institutions, healthcare facilities, and other essential services can be severely limited. This is particularly true in suburban and rural areas where public transportation options may be scarce or non-existent, effectively isolating those without a car from the economic and social opportunities available to their more affluent, car-owning counterparts.

The financial burden associated with car ownership can also be a significant contributor to social inequalities. The costs of purchasing a vehicle, maintaining it, paying for insurance, and fueling it can place a heavy financial strain on low-income households, leaving less disposable income for other essential needs such as housing, food, and healthcare. This financial pressure can limit the ability of these households to invest in education, save for the future, or build wealth, perpetuating the cycle of poverty and further entrenching social disparities.

The consequences of this unequal access can have far-reaching implications for social mobility and economic opportunity. Individuals without reliable transportation may struggle to secure and maintain employment, as they are unable to consistently reach their workplaces, especially those located in areas not well-served by public transit. This, in turn, can limit their earning potential and their ability to improve their financial situation, ultimately hindering their upward social mobility.

The impact of automobile-centric transportation systems on social inequality is not limited to individual households; it can also have broader societal effects. The disproportionate investment in infrastructure that prioritizes private vehicle use, such as the construction of highways and the expansion of parking facilities, can redirect valuable resources away from the development of more inclusive and equitable public transportation systems. This allocation of resources can further disadvantage low-income communities, widening the gap between those who have access to reliable and affordable mobility options and those who do not.

To address these issues, policymakers and urban planners must adopt a more holistic and equity-focused approach to transportation planning. This may involve increased investment in public transit systems, the development of bike-friendly and pedestrian-friendly infrastructure, the implementation of subsidies or voucher programs to assist low-income households with transportation costs, and the promotion of land use policies that integrate affordable housing with accessible transportation options. Acknowledging and addressing the ways in which automobile-centric transportation systems can exacerbate social inequalities enables communities to work toward building more inclusive and equitable mobility networks that provide equal access to opportunities and resources for all residents, regardless of their socioeconomic status or ability to own a private vehicle.

13.6.1.6 The Future of Automobiles in Cities

As we look toward the future, the role of automobiles in urban landscapes is poised to undergo a significant transformation, driven by the convergence of technological advancements and evolving urban planning strategies. One of the most promising developments in the automotive industry is the rise of EVs. Unlike traditional gasoline-powered cars, EVs produce zero direct emissions, making them a more environmentally friendly mode of transportation. As battery technology continues to improve and charging infrastructure becomes more widely available, the adoption of EVs is expected to accelerate, addressing some of the environmental concerns associated with automobile use. Moreover, the quieter operation and lower maintenance requirements of EVs can contribute to a more pleasant and livable urban environment.

Another emerging technology that holds great promise for the future of automobiles in cities is autonomous driving. Self-driving cars have the potential to significantly improve traffic flow, reduce congestion, and enhance safety by eliminating human error and reckless driving behaviors. AVs can also provide mobility options for those who are unable to drive due to age, disability, or other factors, potentially helping to address issues of social inequality and accessibility. However, the evolution of automobiles in cities is not solely dependent on technological advancements. Urban planners and policymakers are also playing a crucial role in shaping the future of transportation by actively promoting strategies that reduce reliance on private vehicles and encourage the use of alternative modes of mobility.

One such strategy is the development of robust public transportation systems, including buses, trains, and light rail networks. By investing in these systems and making them more accessible, efficient, and reliable, cities can provide residents with viable alternatives to driving, reducing the number of cars on the roads and alleviating issues such as traffic congestion, air pollution, and urban sprawl.

Complementing public transportation initiatives, many cities are also focusing on the expansion of cycling and pedestrian infrastructure. Dedicated bike lanes, bike-sharing programs, and walkable urban design can make it easier and safer for people to commute by bicycle or on foot, reducing the need for automobile use and promoting more sustainable and healthy modes of transportation.

Urban planners are exploring innovative strategies such as car-free zones, congestion pricing, and the integration of land use and transportation planning to discourage private vehicle use and encourage the adoption of alternative mobility options. These approaches aim to create more livable, equitable, and environmentally friendly urban environments that prioritize the needs of people over the dominance of automobiles.

As we look ahead, the future of automobiles in cities will likely involve a delicate balance between technological advancements and comprehensive urban planning strategies. With a holistic approach that addresses the challenges associated with private vehicle use; while harnessing the potential of emerging technologies, cities can strive to create transportation systems that are more sustainable, inclusive, and responsive to the needs of their residents.

13.6.2 Two-Wheelers

Two-wheelers, encompassing bicycles, motorcycles, and electric scooters, are a significant mode of urban transportation. They offer several advantages that make them an attractive choice for city dwellers and are increasingly being recognized for their potential to address some of the challenges associated with urban mobility.

13.6.2.1 Efficiency and Flexibility

Two-wheeled vehicles, such as motorcycles, scooters, and bicycles, have emerged as highly efficient and practical modes of transportation, particularly in the context of congested urban environments. These compact and maneuverable vehicles offer a range of advantages that make them an increasingly attractive choice for commuters and city dwellers.

One of the primary advantages of two-wheelers is their efficient use of space. Compared to larger four-wheel vehicles, two-wheelers require significantly less road and parking space, making them an ideal choice for navigating through the often-crowded streets of modern cities. Their compact size allows them to weave through traffic more easily, avoiding the gridlock that often plagues larger automobiles. This not only benefits the rider but also helps to improve overall traffic flow, reducing congestion and the associated negative impacts on the environment and public health.

The reduced parking requirements of two-wheelers are another significant advantage in urban areas. With limited and often expensive parking spaces available, two-wheelers offer a practical solution by requiring significantly less designated parking area. This can be particularly beneficial for those living in high-density neighborhoods or working in bustling commercial districts where parking availability is scarce and costly.

In addition to their space-saving capabilities, two-wheelers also provide a high degree of flexibility for urban commuters. They can be used for short, point-to-point trips within the city, allowing riders to avoid the delays and inconveniences often associated with public transportation or larger vehicles. This flexibility can be especially valuable for those who need to make multiple stops or travel to destinations not easily accessible by other modes of transport.

Two-wheelers can be seamlessly integrated into multimodal transportation systems. Riders can combine the use of a two-wheeler with public transit, such as trains or buses, to create a more efficient and comprehensive transportation solution. This integration can be particularly beneficial for individuals living in areas with limited public transit coverage or for those who need to travel to destinations not served by public transportation.

The advantages of two-wheelers extend beyond just practical considerations. They can also contribute to a more environmentally friendly and sustainable urban landscape. Many two-wheeled vehicles, particularly electric bicycles and scooters, produce zero direct emissions, helping to reduce the carbon footprint of urban transportation and improve local air quality. Additionally, the promotion of two-wheeler usage can encourage more active and healthy modes of transportation, such as cycling, which can have positive impacts on public health and overall well-being.

13.6.2.2 Affordability

Compared to four-wheeled automobiles, two-wheeled vehicles offer a significantly more affordable alternative for personal transportation. This affordability factor is a key advantage that makes two-wheelers accessible to a wider demographic, including students, low-income individuals, and families.

The initial purchase price of a two-wheeler is generally much lower than that of a car. Even high-end motorcycle or scooter models typically cost a fraction of the price of a new automobile. This lower financial barrier to entry allows individuals with more limited budgets to acquire a reliable mode of transportation, opening up opportunities for education, employment, and personal enrichment that may have been previously inaccessible due to the prohibitive cost of car ownership.

In addition to the initial purchase price, the ongoing costs associated with operating and maintaining two-wheelers are also significantly lower than those of cars. Fuel efficiency is generally much higher for two-wheeled vehicles, resulting in substantial savings on gasoline or electricity expenses. Routine maintenance, such as oil changes and tune-ups, also tends to be less frequent and less expensive for two-wheelers, further reducing the financial burden on owners.

Insurance premiums for two-wheelers are often more affordable as well, particularly for younger or less experienced riders. The lower weight and reduced risk profile of two-wheelers compared to cars can translate into lower insurance costs, making transportation more accessible for those with tighter budgets. This affordability factor is especially beneficial for students, low-income individuals, and families who may be struggling to allocate limited financial resources. By providing a more cost-effective alternative to car ownership, two-wheelers can help these groups overcome transportation barriers and access essential services, educational opportunities, and employment, ultimately improving their quality of life and promoting greater social and economic inclusion.

13.6.2.3 Environmental Impact

The world grapples with the pressing issue of climate change and the urgent need to reduce greenhouse gas emissions. On the other hand, the environmental benefits of two-wheeled modes of transportation have become increasingly evident. Compared to the environmental impact of automobiles, these compact and efficient vehicles offer a significantly more sustainable alternative for personal mobility.

One of the primary advantages of two-wheelers in terms of environmental impact is their greatly reduced emissions of greenhouse gases. Conventional gasoline-powered vehicles are a major contributor to the release of carbon dioxide (CO_2) and other planet-warming pollutants, with the transportation sector accounting for a significant portion of global emissions. In contrast, bicycles and electric scooters, which are powered by human or electric propulsion, produce zero direct emissions, effectively eliminating their contribution to the greenhouse gas problem.

Beyond the direct emissions, two-wheelers also consume significantly less energy than cars. The manufacturing and fuel production processes required for automobiles are highly energy-intensive, whereas the energy demands for the production and operation of two-wheelers are much lower. This reduced energy consumption, combined with the lack of direct emissions, translates into a substantially smaller carbon footprint for individuals who choose to commute by bicycle or electric scooter rather than driving a car.

Two-wheelers contribute far less to air pollution, which is a pressing issue in many urban areas. Automobiles emit a range of pollutants, including particulate matter, nitrogen oxides, and carbon monoxide, which can have detrimental effects on human health and the local environment. In contrast, two-wheelers, particularly electric models, produce minimal to no direct air pollutants, helping to improve the overall air quality in cities and reducing the associated health risks for residents.

The environmental benefits of two-wheelers extend beyond just emissions and energy consumption; they also have a lower impact on noise pollution. The quieter operation of bicycles and electric scooters, especially compared to the engine noise of cars and trucks, can contribute to a more peaceful and serene urban environment, enhancing the overall quality of life for city dwellers.

Choosing to commute by two-wheeler instead of driving a car, individuals can directly contribute to the creation of cleaner, healthier, and more sustainable cities. This individual choice, multiplied across a population, can have a significant impact on reducing the overall environmental footprint of urban transportation and supporting the transition toward an eco-friendlier future.

Cities around the world strive to address the challenges of climate change and environmental degradation, on the other hand, the widespread adoption of two-wheeled modes of transportation can be a powerful tool in achieving these goals. Embracing the environmental benefits of bicycles and electric scooters enables urban residents to shape a more sustainable and livable urban landscape for generations to come.

13.6.2.4 Health Benefits

Cycling has significant health benefits. It is a form of physical activity that can help to improve cardiovascular health, build muscle strength, enhance mental well-being, and promote weight management. Even riding an electric scooter or motorcycle can provide some level of physical activity, particularly for individuals who might otherwise be sedentary.

13.6.2.5 Challenges and Solutions

Despite their advantages, two-wheelers also present some challenges. Safety is a major concern, as two-wheeler riders are more vulnerable to accidents than car occupants. Infrastructure is another issue; many cities lack the necessary infrastructure, such as dedicated bike lanes and secure parking facilities, to support two-wheeler use. However, these challenges are not insurmountable. Cities can take measures to improve the safety of two-wheeler riders, such as implementing traffic calming measures, improving road infrastructure, and promoting safe riding practices. They can also invest in infrastructure to support two-wheeler use, such as building bike lanes and bike parking facilities, and providing bike-sharing programs.

Two-wheelers offer efficiency, flexibility, and affordability, and have a lower environmental impact than cars, and provide health benefits. With the right policies and infrastructure in place, two-wheelers can contribute to more sustainable, livable, and healthy cities. Demand for efficient and sustainable transportation solutions increases the role of two-wheelers in urban mobility is likely to become even more important.

13.6.3 Shared Private Transport

Shared private transport, also known as ride-sharing or ride-hailing, has emerged as a significant development in the transportation landscape, offering an alternative to traditional modes of private and public transportation. This concept involves the shared use of privately owned vehicles, facilitated through digital platforms or applications, to provide transportation services to multiple users simultaneously.

The rise of shared private transport can be attributed to several factors, including the widespread adoption of smartphone technology, the growth of the sharing economy, and the desire for more flexible and efficient transportation options. Companies such as Uber, Lyft, and various regional counterparts have capitalized on these trends, creating digital platforms that connect drivers with passengers, enabling them to share rides and split the costs.

One of the primary benefits of shared private transport is its potential to reduce traffic congestion and environmental impact. Having multiple passengers share a single vehicle reduces the number of cars on the road, leading to decreased fuel consumption and lower greenhouse gas emissions. Additionally, shared private transport can help to alleviate the need for personal vehicle ownership, particularly in urban areas where parking and road infrastructure are limited.

From the perspective of passengers, shared private transport offers several advantages. It provides on-demand transportation, allowing users to request a ride at their convenience and be picked up at their desired location. This flexibility is particularly valuable for individuals who may not have access to their own vehicles or who need to travel at times when public transportation is less readily available. Shared private transport can be more cost-effective than traditional taxi or private hire services, as the costs are shared among multiple passengers. This can make transportation more accessible, especially for those with limited financial resources. In addition to that, the digital platforms used by shared private transport companies often offer features such as real-time tracking, in-app payment, and customer reviews, which can enhance the overall user experience and provide a sense of safety and reliability.

For drivers, shared private transport presents an opportunity to generate additional income by utilizing their personal vehicles during idle or underutilized periods. This can be particularly appealing for individuals seeking flexible work arrangements or supplemental sources of revenue. The platforms provided by shared private transport companies often offer driver-friendly features, such as flexible scheduling, transparent earnings information, and access to a broader customer base.

However, the growth of shared private transport has also raised some concerns and challenges. One of the primary issues is the potential impact on traditional taxi and public transportation services. The increased availability of shared private transport options may lead to a decline in the demand for these traditional modes of transportation, potentially threatening the livelihoods of taxi drivers and public transit operators. This has led to debates around fair competition, regulations, and the equitable provision of transportation services.

Another concern is the potential for shared private transport to contribute to increased vehicle miles traveled and urban sprawl. If shared private transport is perceived as a more convenient and affordable option compared to public transportation, it may encourage individuals to live further from urban centers, leading to longer commutes and a greater reliance on private vehicles. This could have negative consequences for urban planning, public health, and environmental sustainability.

There are concerns regarding the working conditions and employment status of drivers in the shared private transport industry. Some have raised issues related to job security, benefits, and the ability of drivers to collectively bargain or organize. Addressing these concerns requires careful consideration of labor laws and regulations to ensure the fair treatment of workers in this emerging sector.

Despite these challenges, the potential benefits of shared private transport in terms of improved accessibility, reduced environmental impact, and enhanced user experience have led many cities and policymakers to explore ways to integrate and regulate this mode of transportation. Some strategies include the development of partnerships between shared private transport companies and public transportation providers, the implementation of policies to incentivize the use of shared rides, and the establishment of guidelines to ensure the safety and fairness of the industry.

Shared private transport represents a significant development in the transportation landscape, offering both opportunities and challenges. As cities and communities continue to grapple with the complexities of urban mobility, the integration of shared private transport, along with other innovative transportation solutions, will be crucial in shaping the future of sustainable and equitable urban mobility. Carefully balancing benefits and addressing concerns enables policymakers and stakeholders to develop transportation systems that meet the evolving needs of both individuals and communities.

13.6.4 Future of Private Transport

The future of private transportation is poised at the intersection of technology, sustainability, and changing consumer preferences. As we look ahead, several key trends are likely to shape the evolution of private transportation.

13.6.4.1 Autonomous Vehicles

Self-driving or AVs are expected to be a game-changer. AVs can potentially increase safety by reducing human error, provide mobility for those unable to drive, and transform the way we use our time during travel. However, they also pose challenges in terms of regulation, cybersecurity, and societal acceptance.

AVs, or self-driving cars, hold great promise for transforming our transportation systems. However, they also present several significant challenges:

1. *Technological Limitations:* AVs rely on advanced sensors and AI to navigate their surroundings. However, these technologies still have limitations. For example, AI struggles to interpret unusual situations, like a traffic officer waving vehicles through a red light. Similarly, sensors need to reliably detect road features in all conditions, regardless of weather or lighting.
2. *Infrastructure Needs:* AVs require clear lane striping and up-to-date maps to operate effectively. Moreover, they need a robust data storage and processing infrastructure to handle the vast amounts of data they generate.
3. *Cybersecurity Concerns:* As connected devices, AVs are vulnerable to hacking, which could lead to safety issues.
4. *Regulatory Challenges:* The legal framework for AVs is still evolving. Issues around liability, insurance, and traffic laws need to be addressed.
5. *Social Acceptance:* Public acceptance of AVs is uncertain. Concerns around safety, privacy, and job displacement could affect the adoption of this technology.
6. *Business Model:* Finding a profitable business model for AVs is another challenge. The high cost of technology and uncertain demand make it difficult to ensure profitability.
7. *Increased Traffic Problems:* If AVs become available too quickly, that may lead to more personal vehicles on the road, thus disincentivizing more traffic-efficient public transportation options.
8. *Social Equity Concerns:* There are concerns that AVs may reduce affordable mobility options like walking, bicycling, and public transportation.

Despite these challenges, researchers and industry leaders are actively working on solutions. The future of AVs will likely involve ongoing advancements in technology, regulatory adaptations, and shifts in public perception.

The advent of AVs is set to significantly transform urban landscapes, offering new opportunities and challenges for city planners. Some primary impacts of AVs on urban planning include:

1. *Changes in Mobility Conditions:* AVs will significantly affect mobility conditions in the future. The changes in mobility conditions are expected to have an impact on urban development and, more specifically, on location choices, land use organization, and infrastructure design.
2. *Infrastructure Needs:* AVs require clear lane striping and up-to-date maps to operate effectively. Moreover, they need a robust data storage and processing infrastructure to handle the vast amounts of data they generate.

3. *Impact on Urban Form:* Predicting the future impact of AVs on infrastructure and urban form is critical for decision makers at all levels of government. An AV is a disruptive technology in transportation and its impact could resonate through all levels of society and government.
4. *Changes in Land Use:* Several studies suggest that the introduction of AVs could lead to as much as a 90% reduction in vehicle need. There are likely to be fewer, but harder working, vehicles on the road, leading to decreased congestion and increased road space.
5. *Impact on Public Spaces:* 'Mobility corridors' can be created to be shared with pedestrians and cyclists. A mobility corridor is where different modes of transport are brought closer together to share a common route without the need for separation in levels and design.
6. *Changes in Urban Transportation Shift:* We are now faced with the prospect of a new urban transportation shift, given the approaching feasibility of fully AVs. The era of the AVs will likely foster a new and distinct urban model.

The integration of AVs into urban environments necessitates a rethinking of transportation infrastructure, land use, and traffic management. As AVs move from concept to reality, their integration into urban environments necessitates a rethinking of transportation infrastructure, land use, and traffic management.

Redesigning urban spaces to accommodate AVs involves a multi-pronged approach that considers infrastructure, land use, and societal needs. The redesigning urban spaces may involve the following concerns:

1. *Reimagining Road Infrastructure:* AVs require clear lane markings and up-to-date maps. Cities may need to invest in improving road infrastructure and signage to accommodate AVs. Additionally, the advent of AVs could free up parking spaces, transforming them into green spaces, pedestrian zones, or other uses.
2. *Prioritizing Shared Spaces:* The implementation of AVs can liberate urban space and reclaim it for people. This aligns with new urban models such as the superblocks in Barcelona, the 15-minute city in Paris, or tactical urbanism interventions.
3. *Fostering Multimodal Transportation:* AVs should be integrated with other modes of transport to provide a seamless, efficient, and sustainable transport system. This includes public transit, cycling, and walking.
4. *Planning for Pedestrians and Cyclists:* As cities redesign their streets for AVs, they should also consider the needs of pedestrians and cyclists. This could involve creating wider sidewalks, dedicated bike lanes, and safe crossing points.
5. *Addressing Urban Sprawl:* AVs could potentially exacerbate urban sprawl by making commutes less arduous. Urban planners will need to consider how to prevent this and encourage more compact, walkable communities.
6. *Considering Social Equity:* Planners should ensure that the benefits of AVs are accessible to all residents, not just those who can afford to own an AV. This might involve regulations to ensure shared use over ownership.
7. *Leveraging Data:* AVs generate vast amounts of data that can be used to inform urban planning decisions. This data can help cities understand traffic patterns, identify infrastructure needs, and plan for future growth.

The integration of AVs into urban environments necessitates a rethinking of transportation infrastructure, land use, and traffic management. As AVs move from concept to reality, their integration into urban environments necessitates a rethinking of transportation infrastructure, land use, and traffic management.

13.6.4.2 Electric Vehicles

The shift toward EVs is already underway, driven by concerns about climate change and air quality. EVs are not only more environmentally friendly but are also increasingly cost-competitive with internal combustion engine vehicles. The growth of EVs will depend on advancements in battery technology, the availability of charging infrastructure, and supportive government policies.

13.6.4.3 Shared Mobility

The concept of car ownership is being challenged by the rise of shared mobility services like Uber and Lyft. These platforms provide flexible, on-demand transportation without the costs and responsibilities of ownership. Shared mobility can also contribute to reducing traffic congestion and emissions. The future may see an increase in shared ownership models and subscription-based services.

13.6.4.4 Connectivity

The future of private transportation is also likely to be increasingly connected. Vehicles are becoming 'smart' and are being integrated with other devices and systems. This connectivity can enhance the driving experience, improve safety, and enable new services. However, it also raises issues around data privacy and security.

13.6.4.5 Active Transportation

There is a growing recognition of the role of active transportation modes, such as cycling and walking, particularly in urban areas. These modes are not only sustainable but also contribute to health and well-being. The challenge is to create safe and attractive environments for active transportation.

13.6.4.6 Urban Air Mobility

Looking further into the future, urban air mobility (UAM) – the use of air space for personal transportation – could become a reality. Concepts like flying cars and passenger drones are being explored and could potentially alleviate ground-level congestion. However, UAM also presents significant challenges in terms of technology, regulation, and infrastructure.

13.6.4.7 Sustainability

Across all these trends, sustainability is a key theme. There is a growing imperative to reduce the environmental impact of transportation. This is driving innovations in vehicle technology, fuel efficiency, and emissions reduction. It is also leading to a rethinking of transportation systems and urban design to promote sustainable modes of travel.

13.7 LAND USE AND TRANSPORT INTERCONNECTION

Land use and transportation are two fundamental elements of urban living that are inextricably intertwined. The spatial organization of cities, often referred to as land use, determines where people live, work, shop, and engage in recreational activities. This spatial arrangement, in turn, influences the demand for transportation, dictating how, when, and where people travel. Conversely, the availability and quality of transportation infrastructure and services can significantly impact land use patterns by influencing the location decisions of households and businesses. For instance, areas with robust public transit systems often see higher densities and a mix of residential, commercial, and office uses, fostering vibrant, walkable communities. On the other hand, regions dominated by automobile use tend to spread out, leading to sprawling suburban developments that are less efficient and sustainable.

The interconnection between land use and transportation also has profound implications for a range of critical urban issues, including housing affordability, economic development, environmental sustainability, and social equity. Therefore, integrated land use and transportation planning is essential for creating livable, sustainable, and equitable cities. It is a complex, dynamic relationship that urban planners, policymakers, and researchers must understand and consider in their efforts to shape the future of our cities.

In the era of rapid urbanization and technological advancements, the interplay between land use and transportation is evolving, presenting both challenges and opportunities. As we move forward, it is crucial to leverage these changes to promote more sustainable and equitable urban development patterns. This requires innovative approaches, informed decision-making, and collaborative efforts among various stakeholders. The journey is complex, but the potential rewards – healthier, happier, and more sustainable cities – are well worth the effort.

13.7.1 Principles of Land Use and Transport Integration

The integration of land use and transport planning is a fundamental principle that guides the development of sustainable and livable urban environments. This approach recognizes the symbiotic relationship between the way we use land and the transportation systems that serve it and aims to create a more holistic and efficient urban structure.

The integration of land use and transportation planning is crucial for several reasons as shown in Table 13.12.

13.7.1.1 Accessibility

One of the core principles driving the integration of land use and transport is the concept of accessibility. This refers to the ease with which people can reach and engage with the various destinations, services, and opportunities within a given area, such as employment, education, healthcare, and leisure activities. Aligning land use and transport planning enables cities to strive to increase accessibility, ensuring that a diverse range of destinations and amenities are within easy reach of residential areas. This can be achieved through strategies such as mixed-use development, which combines residential, commercial, and recreational spaces in close proximity, as well as the strategic placement of public transportation hubs and the development of pedestrian-friendly infrastructure.

TABLE 13.12
Grounds for Land Use and Transportation Planning Integration

Sustainability	Integrated planning can lead to carbon neutrality by transforming the transport sector, which is a significant contributor to greenhouse gas emissions. It can also combat urban sprawl, promoting compact and mixed-use developments that help preserve natural habitats, reduce the loss of agricultural land, and protect our environment.
Efficiency	It recognizes that transport is about the access and movement of people. Without careful planning, transport can create conflicts with the spaces where we live, work, and play.
Urban form outcomes	Integrated planning was originally mooted to improve urban form outcomes and reduce the negative social impacts of transport systems.
Reduced emissions	It is now also recognized as being a critically important part of reducing transport emissions.
Accessibility	Transport and land-use planning have a significant role in promoting accessibility, which is an important component of urban systems and plays a key role in guiding more sustainable and inclusive land use and transport decisions.
Visionary planning	Integrated land use and transport planning addresses a city or region's longer-term challenges, working toward a shared vision of what a city or region aspires to be in the future.

Improved accessibility not only enhances the quality of life for residents but also supports more sustainable patterns of mobility, reducing the reliance on private vehicles and encouraging the use of active transportation modes like walking and cycling.

13.7.1.2 Sustainability

Sustainability is another guiding principle that underpins the integration of land use and transport. This encompasses environmental, social, and economic considerations, with the goal of creating urban environments that are resilient, equitable, and environmentally responsible.

From an environmental perspective, the integration of land use and transport can help to reduce greenhouse gas emissions, air pollution, and energy consumption associated with urban transportation. By promoting compact, mixed-use development and prioritizing public transit, cycling, and walking, cities can minimize the need for lengthy, automobile-dependent commutes, thereby lowering the carbon footprint of their transportation systems. In addition, the integration of land use and transport can support social sustainability by ensuring that all residents, regardless of their socioeconomic status or mobility capabilities, have access to essential services and employment opportunities. This can be achieved through the strategic placement of affordable housing, the provision of equitable public transportation, and the creation of walkable, inclusive neighborhoods. Moreover, the integration of land use and transport can also contribute to the economic sustainability of urban areas. Enhancing accessibility and connectivity enables cities to attract investment, foster economic development, and support the growth of local businesses, ultimately creating more livable and prosperous communities.

13.7.1.3 Compact City Design

The concept of compact city design is closely linked to the integration of land use and transport. This approach emphasizes the development of dense, mixed-use urban areas that reduce the need for extensive transportation infrastructure and promote the use of sustainable mobility modes.

Compact city design typically involves the following principles:

1. *Densification:* Encouraging the development of higher-density residential and commercial areas, which can support the viability of public transportation and reduce the distances people need to travel.
2. *Mixed-use development:* Integrating a diverse range of land uses, such as housing, employment, shopping, and leisure, within close proximity to one another.
3. *Walkable and Pedestrian-Friendly Environments:* Creating well-designed streetscapes, with ample sidewalks, street-level amenities, and safe crossings, to encourage walking and reduce reliance on private vehicles.
4. *TOD:* Strategically locating high-density, mixed-use developments around public transportation nodes, such as bus stops, train stations, or light rail stations, to maximize accessibility and support the use of public transit.

13.7.1.4 The Integration of Land Use and Transport: A Holistic Approach

The integration of land use and transport planning is a multi-pronged endeavor that requires a comprehensive approach, involving collaboration among various stakeholders, including urban planners, transportation engineers, policymakers, and community representatives.

Through this integrated approach, cities can develop comprehensive strategies that address the diverse needs of their residents, while prioritizing accessibility, sustainability, and the creation of compact, livable urban environments. This may involve the implementation of policies, regulations, and incentives that encourage sustainable land use patterns and mobility choices, as well as the investment in infrastructure and services that support multimodal transportation networks.

13.7.2 Impacts of Land Use on Transport

Land use patterns and transportation are intrinsically linked, shaping the way we live, work, and interact within our urban environments. The interplay between these two elements significantly influences transport demand, modal choice, and overall travel behavior. This discussion will delve into how urban sprawl, mixed-use development, and density impact these aspects and the role of land use policies and transport planning in managing these effects.

13.7.2.1 Urban Sprawl and Transportation

Urban sprawl, characterized by low-density, automobile-dependent development on the outskirts of cities, has significant implications for transportation. Sprawl increases the average travel distances for daily trips, leading to a higher reliance on private vehicles. This dependence results in increased traffic congestion, air pollution, and greenhouse gas emissions. Furthermore, sprawling cities often lack efficient public transit systems, making residents more car dependent. Therefore, policies aimed at curbing urban sprawl often focus on promoting compact, TOD to reduce travel demand and encourage the use of more sustainable transport modes.

13.7.2.2 Mixed-Use Development and Transportation

Mixed-use development, where residential, commercial, and other uses are integrated, can significantly influence transportation patterns. By bringing homes, workplaces, schools, and amenities closer together, mixed-use development can reduce the length and number of daily trips. This proximity makes walking, cycling, and public transit more viable and attractive, reducing reliance on private vehicles. Moreover, mixed-use areas support local businesses by increasing foot traffic. Therefore, land use policies promoting mixed-use development can contribute to more sustainable and efficient transportation systems.

13.7.2.3 Density and Transportation

Density, defined as the number of people or jobs per unit of land area, also plays a crucial role in shaping transportation. Higher densities can support more efficient public transit systems and make walking and cycling more practical. A 10% increase in density typically reduces vehicle miles traveled (VMT) by 0.5–1% as an isolated factor and 1–4% including associated factors such as regional accessibility and mix. However, increasing density must be carefully managed to avoid exacerbating traffic congestion and other urban challenges.

13.7.2.4 Land Use Policies and Transport Planning

Land use policies and transport planning are essential tools for managing the impacts of urban sprawl, mixed-use development, and density on transportation. These policies can guide the development of compact, transit-oriented communities, promote mixed-use development, and manage density to support sustainable transportation. For instance, zoning regulations can encourage mixed-use development, while urban growth boundaries can limit sprawl. Transport planning can complement these policies by prioritizing public transit, walking, and cycling infrastructure in areas with supportive land use patterns.

Equity considerations in land use and transport are essential for creating fair and inclusive urban environments. These considerations ensure that all individuals, regardless of their socioeconomic status, race, age, gender, or ability, have equal access to transportation and the opportunities it provides.

13.7.2.5 Historical Exclusion and Marginalization

Historically, certain groups, including Black, indigenous, and other people of color; lower-income individuals; immigrants; older adults; people with disabilities; women; and lesbian, gay, bisexual, transgender, queer, intersex, asexual+ (LGBTQIA+) individuals, have been marginalized from decision making in land use and transportation. This has resulted in unequal access to transportation, leading to disparities in health, economic opportunities, and overall quality of life.

13.7.2.6 Influence of Land Use Patterns

Land use patterns significantly influence transport access. For instance, urban sprawl often leads to inadequate public transit services in peripheral areas, disproportionately affecting lower-income individuals who rely on these services. On the other hand, compact, mixed-use developments can enhance accessibility, but rising property values in these areas can lead to gentrification and displacement of lower-income residents.

13.7.2.7 Equity in Transportation

Transport equity is about ensuring that all individuals have fair access to transportation services. This includes not only physical access but also affordability, safety, and comfort. For example, research shows that neighborhoods where families of color predominate are less walkable due to higher speed limits, fewer streetlights, and more police stops.

13.7.2.8 Policy Considerations

To promote equity in land use and transport, policies should prioritize the needs of the most disadvantaged members of our communities. This involves acknowledging the history of structural racism in transportation and land use, such as redlining, and ensuring authentic community engagement in decision-making processes. Policies should also aim to provide affordable housing near transit hubs, invest in improving transit services in underserved areas, and ensure safe and comfortable pedestrian and cycling infrastructure.

13.7.3 Transport Infrastructure and Land Development

The development of transportation infrastructure has a profound impact on the way land is used and developed within urban and regional contexts. The strategic placement and design of transport projects can significantly shape the patterns of land use, influence urban growth trajectories, and affect property values in the surrounding areas.

13.7.3.1 Impact on Land Values

One of the most direct ways in which transport infrastructure shapes land development is through its influence on property values. The construction of new transportation projects, such as highways, railways, or transit stations, can dramatically alter the accessibility and desirability of the surrounding land, leading to changes in real estate prices and investment patterns.

Areas with improved accessibility and connectivity, brought about by new transport links, often experience a surge in land and property values. This is because these locations become more attractive for residential, commercial, and industrial development, as they offer better access to employment, amenities, and markets. Developers and investors are typically willing to pay higher prices for land that is well-connected to the transportation network, as it increases the potential for profitable development. Conversely, areas that are bypassed or disconnected from new transport projects may see a relative decline in land values, as they become less desirable for development. This uneven impact on property prices can lead to the phenomenon of "value capture," where the increased land values generated by public investment in transportation are leveraged to help fund the infrastructure project itself or to support other community development initiatives.

13.7.3.2 Urban Growth Patterns

Transport infrastructure not only influences land values but also plays a significant role in shaping the overall patterns of urban growth and development. The placement and design of transportation networks can either encourage or discourage certain types of land use and urban form. For example, the construction of highways and arterial roads often facilitates the development of low-density, automobile-dependent suburban and exurban areas. These areas tend to be characterized by sprawling residential neighborhoods, dispersed commercial centers, and a heavy reliance on private vehicles for transportation. In contrast, the development of public transit systems, such as light rail or metro lines, can encourage the growth of more compact, mixed-use, and TOD around station areas.

The impact of transport infrastructure on urban growth patterns can also have significant implications for the efficiency and sustainability of the transportation system itself. Sprawling, car-dependent development patterns typically result in longer commute times, increased traffic congestion, and higher greenhouse gas emissions, while more compact, mixed-use, and TOD can support the use of sustainable mobility modes like walking, cycling, and public transportation.

13.7.3.3 Spatial Patterns and Urban Form

Beyond influencing the overall patterns of urban growth, transportation infrastructure can also shape the spatial distribution and urban form of land uses within a city or region. The location and design of transportation networks can either reinforce or challenge existing spatial structures and urban morphologies.

For instance, the development of radial transportation networks, such as spoke-and-hub highway systems or radial rail lines, can contribute to the formation of monocentric urban structures, where the central business district (CBD) acts as the primary destination and employment hub. Conversely, the construction of circumferential transportation links, such as ring roads or orbital rail systems, can facilitate the emergence of more polycentric urban forms, with multiple activity centers and employment clusters distributed across the metropolitan area.

The integration of transportation infrastructure with land use planning can also influence the physical form and design of urban environments. For example, the incorporation of pedestrian-friendly streetscapes, bike lanes, and transit-supportive building designs around transportation hubs can foster more walkable, human-centric urban development. In contrast, the prioritization of automobile-oriented infrastructure, such as wide roads and expansive parking lots, can lead to the creation of more car-dependent and less visually appealing urban landscapes.

13.7.3.4 The Importance of Integrated Planning

The profound impact of transportation infrastructure on land development underscores the critical importance of integrated planning approaches that carefully consider the interplay between transportation and land use. Aligning transportation investments with broader urban development goals and strategies enables policymakers and planners to harness the transformative power of transport, shaping communities that are more sustainable, equitable, and livable.

This integrated approach involves collaborative efforts among various stakeholders, including transportation authorities, urban planners, real estate developers, and community representatives. It requires the development of comprehensive plans and policies that coordinate land use decisions, zoning regulations, and infrastructure investments to support the desired patterns of urban growth and development.

Through this integrated planning process, cities and regions can leverage the shaping influence of transportation infrastructure to achieve a range of objectives, such as promoting compact and TOD, preserving natural and agricultural lands, enhancing accessibility and social equity, and fostering more environmentally sustainable mobility choices. Recognizing the profound impact of transportation infrastructure on land development enables policymakers and planners to unlock the potential of integrated planning to create urban environments that are not only well-connected and accessible but also economically vibrant, socially inclusive, and environmentally responsible.

13.8 URBAN MOBILITY

Urban mobility is a dynamic concept that shapes the rhythm of cities worldwide. It encompasses the variety of ways in which people and goods navigate through urban landscapes, from traditional public transit systems and private vehicles to emerging technologies like electric scooters, ride-sharing services, and AVs. The essence of urban mobility lies not just in physical movement but also in the seamless integration of different modes of transport, infrastructure, services, and digital platforms.

The demand for efficient, sustainable, and inclusive urban mobility solutions continue to grow. These solutions are critical in addressing traffic congestion, reducing carbon emissions, improving air quality, and enhancing the overall quality of urban life. They also play a pivotal role in fostering economic growth, social equity, and urban resilience. However, achieving optimal urban mobility is a complex challenge that requires a holistic and innovative approach. It involves understanding the intricate patterns of demand and supply, the interplay between land use and transportation, the impact of socioeconomic factors, and the implications of technological advancements. It also necessitates a strong commitment to safety, accessibility, affordability, and user-centric design.

In the era of digital transformation, urban mobility is also increasingly intertwined with the concept of smart cities. E-governance initiatives, data-driven decision-making, connected infrastructure, and digital participation are all reshaping the way we perceive and experience urban mobility. As we stand on the cusp of this exciting new frontier, the future of urban mobility promises to be as diverse and vibrant as the cities it serves.

13.8.1 Sustainable Urban Mobility

Sustainable urban mobility is a concept that has gained significant attention in recent years, particularly in the context of rapid urbanization, climate change, and technological advancements. It is based on the principle of providing safe, efficient, and accessible transportation options for all city dwellers while minimizing environmental impact, promoting social equity, and supporting economic growth. Here are some key principles and practices of sustainable urban mobility:

13.8.1.1 Active Transportation

Active transportation, such as walking and cycling, is a fundamental aspect of sustainable urban mobility. These modes of transport not only reduce greenhouse gas emissions but also promote physical health and well-being. Cities can encourage active transportation by developing comprehensive networks of sidewalks, bike lanes, and trails, implementing traffic calming measures, and promoting a culture of walking and cycling through education and awareness campaigns.

13.8.1.2 Public Transit

Public transit systems, including buses, trams, and metros, are the backbone of urban mobility. They have the capacity to move large numbers of people efficiently, thereby reducing traffic congestion and air pollution. To enhance the sustainability of public transit, cities can invest in cleaner technologies, improve service quality and coverage, integrate different modes of transport, and adopt fare policies that are affordable for all segments of the population.

13.8.1.3 Shared Mobility

Shared mobility services, such as car-sharing, bike-sharing, and ride-hailing, offer flexible and convenient transportation options that can complement public transit. They can help fill the gaps in the transport network, reduce the need for private car ownership, and facilitate the transition toward a more sustainable and integrated transport system. However, it is crucial to regulate these services to ensure they contribute to the overall sustainability goals of the city.

13.8.1.4 Electric Vehicles

The electrification of transport is a key strategy for reducing carbon emissions and improving air quality in cities. EVs, powered by renewable energy, can significantly reduce the environmental impact of urban mobility. To promote the uptake of EVs, cities can provide incentives for EV purchase, build charging infrastructure, and incorporate EVs into public transit and shared mobility services.

In addition to these practices, sustainable urban mobility also involves land use planning that promotes compact and mixed-use development, thereby reducing the need for long-distance travel. It also requires the use of intelligent transport systems and data analytics to optimize transport operations, improve user information, and enable evidence-based decision-making. Furthermore, sustainable urban mobility emphasizes the importance of public participation and stakeholder engagement in transport planning and decision-making. It recognizes that every city is unique, and therefore, solutions must be tailored to the local context and needs.

13.8.2 Urban Mobility Planning and Policies

Urban mobility planning is a comprehensive approach to designing and managing transportation systems in cities. It involves a wide range of activities, from setting strategic goals and objectives to developing detailed plans and implementing specific projects. Some key steps in the process of urban mobility planning are discussed below.

13.8.2.1 Setting Vision and Goals

The first step in urban mobility planning is to establish a clear vision and set of goals. This involves understanding the unique challenges and opportunities of the city, engaging with stakeholders, and aligning with broader urban development and sustainability goals.

13.8.2.2 Data Collection and Analysis

This involves gathering and analyzing data on current transportation patterns, infrastructure, services, and policies. It includes traffic counts, travel surveys, land use data, and socioeconomic data. This information is used to understand existing conditions, identify problems and opportunities, and inform decision-making.

13.8.2.3 Scenario Planning

This involves developing and evaluating different scenarios for the future of urban mobility. Each scenario represents a different combination of policies, investments, and technologies. Scenario planning helps to explore the potential impacts of different strategies and make informed decisions.

13.8.2.4 Plan Development

Based on the scenario analysis, a detailed urban mobility plan is developed. This includes specific strategies and actions for improving transportation in the city. The plan covers various aspects of urban mobility, including public transit, active transportation, shared mobility, parking, land use, and more.

13.8.2.5 Implementation and Monitoring

Once the plan is developed, it is implemented through specific projects and programs. The performance of the plan is monitored using a set of indicators, and adjustments are made as needed.

Urban mobility planning is guided by a number of policies, including:

Integrated Transport Planning: This policy emphasizes the need for a coordinated and integrated approach to transport planning. It involves considering all modes of transport, integrating land use and transport planning, and coordinating with other sectors such as energy, environment, and health.

TOD: TOD is a development strategy that focuses on creating compact, mixed-use, walkable communities around transit stations. It aims to reduce car dependence, promote active transportation, and create vibrant urban neighborhoods.

Sustainable Mobility Policies: These policies aim to promote sustainable modes of transport, such as walking, cycling, public transit, and EVs. They include a range of measures, from infrastructure investments and service improvements to pricing policies and behavior change programs.

Equity in urban mobility planning is fundamental to ensure that all residents, regardless of their socioeconomic status, have access to safe, affordable, and efficient transportation. To ensure equity in mobility planning cities should adopt the following strategies:

Embed Equity in Planning Processes: Cities can incorporate equity considerations into their transportation planning processes. This includes ensuring that new mobility technologies are distributed across all neighborhoods, providing payment options for residents without bank accounts or smartphones, making services accessible for residents living with disabilities, offering discounted prices for residents with lower incomes, and providing outreach and multilingual informational material.

Prioritize Public Engagement: Engaging with community members throughout the decision-making process is crucial to understanding how they perceive new technologies and what barriers they face in access or use. This ensures that the voices of all residents, especially those from marginalized communities, are heard and their needs are addressed.

Promote Integrated and Inclusive Services: Cities should strive to integrate various modes of transportation and ensure they are inclusive. This includes not only traditional public transit but also emerging mobility services like ride-sharing and bike-sharing.

Implement Flexible Permitting and Proposal Processes: Recognizing that companies are typically willing to negotiate when entering a new market, many cities use a request for proposal (RFP) or permitting process to codify equitable access requirements for companies.

Focus on Affordability: Transportation affordability is a key aspect of equity. Cities can implement fare policies that are affordable for all segments of the population and offer discounted prices for low-income residents.

Ensure Accessibility: It is important to ensure that transportation services are accessible to everyone, including people with disabilities and the elderly.

Monitor and Adjust Policies: Cities should monitor the performance of their transportation systems using equity-focused indicators and make necessary adjustments to ensure that their policies are effectively promoting equity.

13.8.3 Technology and Innovation in Urban Mobility

Technology and innovation are playing a pivotal role in shaping the future of urban mobility. They are transforming the way we travel, making our journeys more efficient, sustainable, and enjoyable. Some aspects on how technology and innovation are reshaping urban mobility are described below.

13.8.3.1 Intelligent Transport Systems

ITS use information and communication technologies to improve transportation systems. They include a wide range of applications, such as traffic management systems, traveler information systems, public transit management systems, and advanced vehicle control systems. ITS can enhance the efficiency, safety, and sustainability of urban mobility by optimizing traffic flow, reducing congestion, improving transit service, and minimizing environmental impact. They also enable real-time data collection and analysis, which can inform decision-making and planning.

13.8.3.2 Mobility as a Service

Mobility as a Service (MaaS) is a new business model that integrates various forms of transport services into a single mobility service accessible on demand. It combines public transit, ride-sharing, bike-sharing, car-sharing, and other services into a unified platform that users can access through a single app. MaaS can offer a seamless, flexible, and user-centric travel experience, reduce the reliance on private cars, and promote sustainable mobility. It also opens up new opportunities for collaboration and innovation among transport operators, technology providers, and other stakeholders.

13.8.3.3 Autonomous Vehicles

AVs, also known as self-driving cars, are vehicles that can navigate without human input. They use sensors, cameras, radar, and AI to perceive their environment, make decisions, and control vehicle functions. AVs have the potential to revolutionize urban mobility by improving road safety, increasing transport efficiency, reducing traffic congestion, and providing mobility for those who are unable to drive. However, they also pose challenges in terms of technology, regulation, infrastructure, and public acceptance, which need to be addressed to realize their full potential.

13.8.3.4 Smart Infrastructure

Smart infrastructure refers to the integration of digital technologies into transportation infrastructure. This includes smart roads equipped with sensors and communication devices, smart parking systems that provide real-time parking information, smart traffic signals that adapt to traffic conditions, and smart charging stations for EVs. Smart infrastructure can enhance the performance, resilience, and sustainability of urban mobility systems. It can also support the deployment of ITS, MaaS, AVs, and other innovative solutions.

In addition to these technologies, there are many other innovations shaping urban mobility, such as EVs, shared mobility services, drone delivery, and hyperloop. Moreover, emerging technologies like blockchain, big data, and the IoT are expected to have a significant impact on urban mobility.

However, technology and innovation are not silver bullets for urban mobility challenges. They need to be complemented by sound policies, robust institutions, and inclusive planning processes. They also need to be guided by a clear vision of sustainable and equitable urban mobility.

Big data are playing a transformative role in shaping urban mobility, providing unprecedented opportunities for understanding and planning urban transportation systems. How big data is reshaping urban mobility is discussed below.

13.8.3.5 Understanding Complex Mobility Systems

Big data analysis and visualization play a key role in understanding complex mobility systems. Analyzing transportation data patterns helps identify inefficiencies in urban transportation systems and propose solutions. This could include optimizing bus routes, enhancing bike-sharing programs, and ensuring that public transportation meets the needs of diverse urban populations.

13.8.3.6 Forecasting and Policy Evaluations

Data has played a role in urban mobility policy planning for decades, especially in forecasting demand, but much less in policy evaluations and assessments. The surge in availability and openness of big data in the last decade seems to provide new opportunities to meet demand for evidence-based policymaking.

13.8.3.7 Combining Big Data with Traditional Data

Research shows that the new types of big data provide new opportunities for evidence-based policymaking; however, they cannot replace traditional data usage (surveys and statistics). Instead, combining big data with survey and Geographic Information System data in ex-ante assessments, as well as in developing decision support tools, is found to be the most effective. This could help policymakers not only to get much more insight from policy assessments, but also to help avoid the limitations of one certain type of data.

13.8.3.8 Demand-Driven Data Use

Future research should engage with policy practitioners to reveal best practices, constraints, and potential of more demand-driven data use in mobility policy assessments in practice. Big data is not just a tool for understanding and planning urban mobility, but a powerful driver of innovation and transformation in the field. It provides new opportunities for policymakers toward evidence-based policymaking. However, big data is just one piece of the puzzle. It needs to be complemented by sound policies, robust institutions, and inclusive planning processes to truly reshape urban mobility for the better.

13.8.3.9 Additional Notes on Big Data

Big data refers to extremely large and diverse collections of structured, unstructured, and semi-structured data that continues to grow exponentially over time. These datasets are so huge and complex in volume, velocity, and variety, that traditional data management systems cannot store, process, and analyze them.

The term "big data" is often described in terms of the "three v's" of big data: volume, velocity, and variety.

- *Volume:* As its name suggests, the most common characteristic associated with big data is its high volume. This describes the enormous amount of data that is available for collection and analysis.
- *Velocity:* This refers to the speed at which new data is generated and the speed at which data moves around. With the growth of the IoT, data is being created every second.
- *Variety:* This refers to the different types of data we can now use. Data today comes in many different formats: structured data, semi-structured data, unstructured data and even complex structured data.

The amount and availability of data are growing rapidly, spurred on by digital technology advancements, such as connectivity, mobility, the IoT and AI. As data continues to expand and proliferate, new big data tools are emerging to help companies collect, process, and analyze data at the speed needed to gain the most value from it.

The three Vs of big data – volume, velocity, and variety – are qualitative concepts that describe the characteristics of big data, but they can be illustrated with some quantitative examples:

Volume: The volume of big data refers to the sheer size of the data sets that need to be processed. For instance, Facebook stores approximately 250 billion images. As of 2016, Facebook had 2.5 trillion posts. These numbers are so large that they are difficult to comprehend, and they are growing every day.

Velocity: Velocity refers to the speed at which new data is generated and needs to be processed. For example, Twitter generates around 5 gigabytes of data per minute or 7 terabytes in a day. Facebook generates 7 gigabytes of data per minute or 10 terabytes of data daily.

Variety: Variety refers to the different types of data available for analysis. This is not typically quantified in the same way as volume and velocity but consider the range of data types that businesses now use: structured numeric data, text documents, email, video, audio, stock ticker data, financial transactions, and so on.

These examples give a sense of the scale of big data in terms of volume, velocity, and variety. However, the specific quantities can vary widely depending on the context and the specific application.

13.9 TRANSPORTATION AND URBAN FORM

Transportation and urban forms are intertwined, each shaping the other in profound ways: the layout of our cities, the density of our neighborhoods, the design of our streets – these are all influenced by our transportation systems. Conversely, the nature of our transportation systems is largely determined by the form of our urban environments.

From the narrow, winding streets of ancient cities designed for foot traffic, to the sprawling suburbs of the automobile era, to the TODs of today, the evolution of urban form has always been closely linked to advances in transportation technology. Each mode of transportation – whether it is walking, cycling, driving, or taking public transit – requires a different kind of urban form to function effectively, but the relationship between transportation and urban form goes beyond mere functionality. It also impacts the quality of life in our cities. It affects how we interact with our environment and with each other. It influences our health, our economic opportunities, our sense of community, and even our identity.

In recent years, there has been a growing recognition of the need to create more sustainable, equitable, and livable urban forms. This has led to a renewed focus on public transit, active transportation, and mixed-use development. It has also sparked interest in new concepts like "complete streets" and "15-minute cities."

As we look to the future, the challenge is to leverage transportation as a tool for shaping urban form in a way that enhances the livability, sustainability, and resilience of our cities. This requires a holistic approach that integrates transportation planning with land use planning, and that considers the social, economic, and environmental impacts of our decisions.

13.9.1 How Transportation Shapes Urbanization

Transportation plays a pivotal role in shaping urbanization, influencing the spatial structure of cities and the patterns of urban growth. How transportation shapes urbanization is discussed below.

13.9.1.1 Influencing Urban Form and Structure

The development and expansion of transportation networks have had a profound impact on the form and structure of cities throughout history. This relationship between transportation and urban development is a complex and dynamic one, with each innovation in transportation technology shaping the growth and evolution of cities in significant ways.

One of the most transformative developments in transportation was the advent of the railroad in the 19th century. The introduction of this new mode of transportation fundamentally altered the spatial organization of cities, leading to the development of suburbs and the decentralization of urban centers. Prior to the railroad, cities were typically compact and densely populated, with most residents living and working within a relatively small geographic area. The arrival of the railroad, however, allowed people to live further from the city center while still being able to commute to their jobs.

This shift in spatial organization was driven by several factors. The railroad enabled rapid and efficient movement of people and goods over longer distances, making it feasible for workers to live outside the city core and commute in for employment. Additionally, the availability of cheap land on the urban periphery, combined with the desire for a more spacious and less crowded living environment, led many middle- and upper-class residents to relocate to the newly emerging suburbs.

The rise of the automobile in the 20th century further accelerated this process of suburbanization and led to the spread of low-density, auto-oriented development. The car allowed for even greater freedom of movement and enabled people to live even farther from the city center, as they were no longer dependent on public transportation or walking to get around. This, in turn, fueled the construction of highways, arterial roads, and sprawling residential developments, often at the expense of more compact, walkable urban areas.

The impact of transportation networks on urban form is not limited to the growth of suburbs. In many instances, the construction of major transportation infrastructure, such as highways or rail lines, has also led to the segregation and isolation of certain neighborhoods or communities within cities. For example, the placement of highways through urban areas often resulted in the displacement of low-income and minority residents, disrupting the social fabric of these communities and contributing to ongoing issues of spatial and economic inequality. Moreover, the reliance on automobile-centric transportation has had significant environmental and public health consequences, including increased greenhouse gas emissions, air pollution, and sedentary lifestyles. In response, many cities are now exploring ways to promote more sustainable and equitable transportation options, such as expanding public transit, improving pedestrian and bicycle infrastructure, and implementing policies to discourage private car use.

The development and expansion of transportation networks have had a profound and lasting impact on the form and structure of cities. From the decentralization of urban centers to the spread of low-density, auto-oriented development, these transportation innovations have fundamentally shaped the way we live, work, and move within the built environment. With the challenges posed by these transportation-driven transformations, it is essential to find ways to create more livable, sustainable, and equitable urban spaces that prioritize the diverse needs and experiences of all residents.

13.9.1.2 Facilitating Economic Growth

Transportation networks facilitate economic growth and development, but they can also contribute to economic disparities if access to transportation is not equitably distributed. Understanding the complex relationship between transportation and the economy is crucial for policymakers and urban planners seeking to create more prosperous and inclusive communities.

The core of the connection between transportation and economic growth is the ability of transportation networks to connect people, goods, and businesses. Efficient transportation systems enable workers to access a wider range of employment opportunities, allowing them to find jobs that better match their skills and preferences. This, in turn, can lead to increased productivity and higher incomes. Similarly, transportation networks facilitate the movement of goods and services, connecting businesses to their suppliers, customers, and markets. This enables economies of scale, specialization, and the agglomeration of related industries – all of which are essential drivers of economic growth.

The agglomeration economies that arise from well-connected transportation networks are particularly important for cities, which have historically been engines of economic activity and innovation. By bringing people, businesses, and resources together in a dense, urban environment, transportation networks enable the exchange of ideas, the sharing of resources, and the development of specialized skills and industries. This clustering of economic activity can lead to significant productivity gains and higher levels of economic output.

However, the benefits of transportation-facilitated economic growth are not always evenly distributed. If access to transportation is unevenly distributed, it can exacerbate economic disparities within a community or region. For example, if public transit systems are inadequate or inaccessible in low-income neighborhoods, residents in these areas may have limited access to job opportunities and struggle to participate fully in the local economy. Similarly, if transportation infrastructure, such as highways or airports, is concentrated in wealthier or more privileged areas, businesses and residents in these locations may enjoy a significant economic advantage over their counterparts in less well-connected areas.

This uneven distribution of transportation access can create a vicious cycle, where economic disparities limit access to transportation, which in turn perpetuates and exacerbates those disparities. This dynamic can lead to the isolation of certain communities, the concentration of poverty and unemployment, and the reinforcement of broader patterns of socioeconomic inequality.

To address these challenges, policymakers and urban planners must take a holistic, equity-focused approach to transportation planning and investment. This may involve prioritizing public transit projects in underserved neighborhoods, ensuring that new transportation infrastructure is distributed equitably across a region, and integrating transportation planning with other economic development and social welfare initiatives. In doing so, communities can harness the economic potential of transportation networks while also ensuring that the benefits of this growth are shared more broadly. This can lead to more inclusive, prosperous, and sustainable local economies, where all residents have the opportunity to participate and thrive.

Transportation networks are essential for facilitating economic growth and development, but they can also contribute to economic disparities if access is not equitably distributed. Policymakers and urban planners must carefully consider the complex relationship between transportation and the economy, and work to create transportation systems that support the economic aspirations of all members of the community.

13.9.1.3 Shaping Land Use Patterns

Transportation infrastructure is the lifeblood of any city. It connects people to jobs, services, and opportunities, dictating how we move, interact, and shape our environment. However, beyond simply getting from point A to point B, investments in roads, railways, and public transit systems have a profound and often underestimated impact on the very fabric of our cities – the land use patterns. This dynamic relationship is a double-edged sword, offering the potential for development and revitalization, yet also carrying the risk of displacement and unintended consequences.

The Power of Accessibility: When a new highway or transit line carves a path through a previously isolated area, it unlocks a treasure trove of potential. Land that was once an afterthought becomes a prime location for development. Businesses are drawn to increased accessibility, offering the promise of easier access to customers, employees, and resources. This can be a transformative force, breathing life into neglected areas and creating new economic opportunities. Imagine a long-forgotten industrial district bypassed by major roads suddenly becoming a bustling hub for creative businesses and startups because of the arrival of a new light rail line. Warehouses that once stood idle can be repurposed into vibrant co-working spaces and art studios, attracting a younger, more dynamic population.

The Corridor Effect in Action: Transportation corridors, the areas flanking a new highway or railway, become magnets for development. Developers see the potential for high visibility and easy access, leading to a surge in construction. This "corridor effect" can be immensely beneficial, creating jobs, boosting the local economy, and fostering a sense of community. However, it is important to recognize that the type of development attracted depends heavily on the planning choices made. Uncontrolled sprawl can quickly replace green spaces with generic strip malls and sprawling parking lots, creating a monotonous, car-dependent environment. On the other hand, well-planned corridor development can prioritize mixed-use projects that integrate residential, commercial, and office spaces. This creates a vibrant, walkable environment where people can live, work, and shop within close proximity, reducing reliance on cars and promoting a more sustainable lifestyle.

The Dark Side of the Coin: While the potential benefits of increased accessibility are undeniable, the story does not always unfold like a utopian dream. Unmanaged development can have negative consequences, particularly for the most vulnerable residents. Rising property values and rents, fueled by increased demand and speculation, can force low-income residents to relocate from their longtime homes. This can disrupt established communities, severing social networks and displacing individuals and families who may struggle to find affordable housing elsewhere. Additionally, gentrification can take hold, transforming once affordable neighborhoods into trendy enclaves. Local shops and cafes that served the community for generations may be forced to close their doors, unable to compete with national chains that cater to a higher-income demographic. This not only erodes the unique character of a neighborhood but also disrupts the social fabric of the community.

Striking a Balance: Planning for Equitable Growth: To maximize the positive outcomes of transportation investments and mitigate the risks, careful planning and a commitment to equitable growth are crucial. Here's where foresight comes into play:

- *Community Involvement at the Forefront:* Engaging residents and stakeholders in the planning process goes a long way in ensuring that development aligns with the needs and aspirations of the community. Open forums, workshops, and surveys can be used to gather input on desired amenities, infrastructure needs, and concerns. This information can then be used to inform zoning regulations that promote affordable housing options and mixed-use development. Protecting historic or culturally significant structures and incorporating design elements that complement the existing architecture can also help safeguard the unique character of a neighborhood.
- *The Power of TOD:* Prioritizing development around transit stations is a powerful tool for promoting sustainable and inclusive growth. TOD principles encourage a mix of residential, commercial, and office spaces within a walkable distance of public transit stops. This not only reduces reliance on cars and congestion but also creates vibrant, pedestrian-friendly environments with a strong sense of community. Imagine a bustling neighborhood where residents can walk to their local grocery store, grab a coffee at their favorite cafe, and catch the train to work without ever having to get behind the wheel.
- *Preserving Green Space and Infrastructure:* Transportation investments should not come at the expense of green spaces or essential infrastructure. Integrating parks, bike lanes, and pedestrian-friendly walkways into development plans ensures a healthy and sustainable environment for everyone. Public spaces provide areas for recreation, socialization, and physical activity, while well-maintained infrastructure like schools, libraries, and healthcare facilities are essential for a thriving community.

13.9.1.4 Creating a Sustainable Future

Transportation investments offer a powerful tool for shaping our cities. Recognizing the impact of land use patterns and prioritizing responsible development can help us create thriving, inclusive communities that serve everyone. This involves prioritizing mixed-use development and fostering walkable environments.

13.9.1.5 Impacting the Environment

Transportation systems have significant environmental impacts, which in turn shape urbanization. They are a major source of air pollution and greenhouse gas emissions. They also consume large amounts of land and can fragment habitats. These environmental impacts can influence where and how cities grow.

13.9.1.6 Driving Technological Innovation

The way we move shapes the way we live. Throughout history, advancements in transportation have fundamentally reshaped our cities. From the cobbled streets traversed by horse-drawn carriages to the sprawling suburbs enabled by the automobile, transportation infrastructure has dictated urban development patterns. Today, we stand on the cusp of another potential revolution – one driven by the rise of ride-sharing services and AVs. These advancements have the potential to disrupt the very fabric of our cities, ushering in new patterns of urbanization with far-reaching consequences.

13.9.1.7 Parking Purgatory No More

One of the most immediate impacts will likely be the decline of the ubiquitous parking lot. Cars are space-hungry beasts, and the vast expanses of asphalt dedicated to parking dominate many urban landscapes. Ride-sharing services, with their on-demand availability and efficient utilization of vehicles, already chip away at the need for personal car ownership. AVs are poised to take this a step further. Imagine a future where cars drop you off at your destination and then autonomously park themselves, freeing up valuable real estate. This could lead to a dramatic reduction in parking lots, transforming them into vibrant public spaces like parks, plazas, or even additional housing.

13.9.1.8 The Rise of the 15-Minute City

Reduced reliance on personal car ownership, coupled with the efficiency of AVs, could pave the way for the rise of the "15-minute city." This concept envisions neighborhoods where all essential amenities – grocery stores, schools, healthcare facilities, workplaces – are within a 15-minute walk, bike ride, or public transit journey from any resident's doorstep. Imagine a world where errands can be run with ease, commutes are minimized, and daily life becomes more convenient and less car centric. This could not only improve public health by encouraging active lifestyles but also foster a stronger sense of community, as residents interact more frequently in shared spaces.

13.9.1.9 The Reinvention of Public Transit

The rise of ride-sharing and AVs does not necessarily spell doom for public transportation. In fact, these technologies have the potential to work in synergy with existing systems. Public transit could become the backbone of the urban mobility network, catering to longer commutes and bulk passenger movement. On the other hand, ride-sharing and AVs could act as "first- and last-mile" solutions, bridging the gap between public transit stations and individual destinations. Imagine seamlessly hopping off a train and hailing an autonomous shuttle to take you directly to your doorstep, eliminating the need for personal car ownership entirely.

13.9.1.10 New Challenges and Considerations

This technological revolution, however, faces its challenges. The ethical considerations surrounding AV safety and data privacy need to be carefully addressed. Ensuring equitable access to these new transportation options is crucial. We must avoid a scenario where only the wealthy have access to AVs, further exacerbating social inequalities. Furthermore, the potential job losses in sectors like taxi driving and parking enforcement need to be addressed through retraining programs and social safety nets.

13.9.1.11 Planning for the Future City

Cities need to be proactive in shaping this future. Here are some key considerations:

- *Infrastructure Investment:* Investing in dedicated lanes for AVs and public transit, as well as robust digital infrastructure to support connected vehicles, will be crucial.
- *Land-Use Planning:* Zoning regulations that encourage mixed-use development and higher-density living can support the 15-minute city concept and efficient public transit networks.

- *Equity and Accessibility:* Subsidies, discounted fares, and ensuring universal design principles in all transportation options can ensure everyone benefits from these advancements.
- *Public Engagement:* Open dialogues with residents are essential to understand their needs and concerns, fostering a sense of ownership over the city's transformation.

The Road Ahead: The rise of ride-sharing and AVs presents a unique opportunity to rethink how our cities function. Adopting these technologies and planning for their impact can help us create cities that are more sustainable, equitable, and ultimately more livable. These advancements can reduce congestion, improve public health, and renew focus on walkable, vibrant communities. The future of our cities is not set in stone – it is a journey we are all embarking on together, and by leveraging the power of transportation technology responsibly, we can build a future where everyone benefits from the freedom of movement.

13.9.1.12 Affecting Quality of Life

Transportation systems are the lifeblood of cities, intricately woven into the fabric of our daily lives. They have a profound impact on the quality of life we experience, influencing everything from how long it takes us to get to work to the air we breathe. Explanations on how transportation shapes our cities and our choices within them are given below.

Commuting times are a major factor influencing quality of life. Efficient public transportation systems, with reliable schedules and widespread coverage, can significantly reduce commute times. This translates to more time spent with family, pursuing hobbies, or simply getting a good night's sleep. Conversely, car-dependent cities with congested roads can lead to frustrating commutes, adding stress and taking a toll on mental and physical well-being. This can influence where people choose to live – opting for neighborhoods closer to work or with better access to public transport to avoid the daily grind of traffic jams.

Accessibility to services and amenities is another crucial factor. A well-connected city ensures residents have easy access to essential services like healthcare facilities, schools, grocery stores, and recreational areas. This is particularly important for individuals who do not own cars or rely on public transport. A city with a robust network of buses, trains, and bike lanes empowers them to lead independent lives, fostering a sense of inclusion and contributing to a vibrant, diverse community. Conversely, cities dominated by car-centric infrastructure can create pockets of inaccessibility, particularly for low-income residents or those with mobility limitations. This can exacerbate social inequalities and limit opportunities.

Traffic safety plays a vital role in shaping how we navigate our cities. Safe and well-maintained roads, dedicated bike lanes, and pedestrian crossings create an environment where people feel comfortable walking, cycling, or using public transport. This not only promotes physical activity but also fosters a sense of community by encouraging interaction in shared spaces. Conversely, congested roads with a high number of accidents can make people feel unsafe, discouraging them from venturing outside their cars. This creates a car-reliant culture, further contributing to congestion and a monotonous urban landscape.

Noise and air pollution generated by car traffic can significantly impact on our quality of life. Cities with efficient public transportation systems and pedestrian-friendly infrastructure experience lower levels of noise and air pollution. This translates to a healthier environment, with reduced respiratory problems and improved overall well-being. On the contrary, cities dominated by car traffic can be noisy and choked with smog, creating a less desirable living environment. This can influence where people choose to live, potentially pushing them toward suburbs with cleaner air, further contributing to urban sprawl.

Transportation systems are not mere infrastructure networks; they are fundamental building blocks of a city. Influencing commute times, accessibility, safety, and pollution levels enables them to shape how we experience our urban environment. These factors, in turn, influence our choices about where to live and work, ultimately shaping the patterns of urbanization and the overall quality of life in our cities. Through prioritizing safe, efficient, and sustainable transportation systems, we can create cities that are not only functional but also healthy, inclusive, and enjoyable places to live.

13.9.2 Spatial Structure of Cities and Transportation Infrastructure

Cities are dynamic entities, constantly evolving and adapting. Their spatial structure, the arrangement and distribution of different functional areas, is a complex interplay of historical, geographical, and economic forces. Transportation infrastructure stands out as a particularly powerful force, shaping how cities develop and influencing the lives of their residents. This essay explores the intricate relationship between spatial structure and transportation infrastructure, examining their mutual influence and implications for urban planning.

13.9.2.1 The Genesis of Spatial Form

The spatial structure of a city is not accidental. It is a product of its history, geography, and economic landscape. Early settlements often arose near natural resources like water bodies or fertile land, dictating their initial layout. Factors like topography, climate, and available building materials also play a role. As cities grow, economic forces come into play. Industries cluster together, creating employment hubs, while residential areas arise within commuting distance or connected by transportation routes. Over time, these factors combined to create one of the main categories of urban spatial structure:

- *Monocentric versus Polycentric:* Traditionally, cities have been categorized as monocentric or polycentric. Monocentric cities, like Paris with its historic core, have a single dominant center where businesses, services, and cultural institutions are concentrated. Transportation infrastructure, like radial roads and public transit lines, typically converge on this central hub. Polycentric cities, on the other hand, have multiple centers of activity, often driven by the emergence of suburban business districts or satellite towns. These structures require a more complex network of transportation infrastructure to connect these dispersed centers.
- *Transportation as a Shaping Force:* Transportation infrastructure does not merely respond to existing spatial patterns; it actively shapes them. Consider the transformative impact of the invention of the automobile. Prior to cars, cities were compact and walkable, with a mix of residential, commercial, and industrial uses within close proximity. The rise of the automobile led to the development of suburbs, sprawling residential areas located outside the city center and heavily reliant on cars for daily commutes. This shift in transportation infrastructure had a profound impact on the spatial structure of cities, creating a more segregated and car-dependent environment.

13.9.2.2 Modern Transportation Trends and their Impact

Contemporary urban planning is witnessing a multifaceted shift in transportation priorities. Public transit systems, once facing decline in some regions, are experiencing a resurgence driven by a collective desire for sustainable and efficient mobility solutions. This renewed interest coincides with a growing emphasis on pedestrian-friendly infrastructure, with walkable neighborhoods and safe cycling facilities becoming central to urban design. Further complicating the landscape is the rise of ride-sharing services, which offer on-demand car access without ownership burdens. The potential disruption of AVs looms large, potentially revolutionizing urban mobility altogether. This confluence of trends presents both challenges and opportunities for urban planners. They must grapple with the integration of these emerging transportation options into existing infrastructure, while ensuring equitable access for all residents. The future of urban design and transportation rests on navigating this complex landscape and fostering a system that prioritizes sustainability, efficiency, and a high quality of life for all inhabitants.

13.9.2.3 Considerations

While these trends offer exciting possibilities, challenges remain. Ensuring equitable access to transportation infrastructure is crucial to avoid creating pockets of social and economic disparity. Additionally, careful planning is needed to integrate new technologies like AVs into existing urban fabric. Infrastructure investments need to consider dedicated lanes for AVs and public transit, alongside robust digital infrastructure to support connected vehicles.

13.9.2.4 Planning for the Future

To create cities that are not just functional but also sustainable and equitable, urban planners need to consider the intricate relationship between spatial structure and transportation infrastructure. Here are some key strategies:

- *Mixed-Use Development:* Zoning regulations can be used to encourage the development of mixed-use neighborhoods with a mix of residential, commercial, and office spaces. This promotes walkability and reduces reliance on cars.
- *Public Engagement:* Open dialogues with residents are essential to understand their needs and concerns regarding transportation options and neighborhood development. This fosters a sense of ownership over the city's transformation.
- *Investing in Public Transit:* Investing in reliable and efficient public transit networks is essential for reducing car dependence and creating more connected cities.
- *Prioritizing Pedestrian and Cycling Infrastructure:* Creating safe and well-maintained sidewalks, bike lanes, and pedestrian crossings encourages walking and cycling, promoting physical activity and fostering a sense of community.

The spatial structure of a city and its transportation infrastructure are not independent entities; they are a dynamic and interdependent system. Understanding their complicated relationship allows us to shape cities that are not only functional but also sustainable, equitable, and enjoyable places to live. This collaborative approach will ensure that our cities evolve.

13.9.3 Urban Form and Transport Sustainability

Urban form and transport sustainability are closely intertwined aspects of urban planning and development. The way cities are designed and organized significantly impacts transportation systems, energy consumption, and overall quality of life for residents. We will explore the relationship between urban form and transport sustainability, highlighting key concepts, challenges, and potential solutions.

13.9.3.1 Urban Form: Shaping Our Cities

Urban form refers to the physical layout, design, and structure of cities. It encompasses various elements, including land use patterns, building density, street networks, public spaces, and architectural styles. Different urban forms emerge based on historical context, cultural preferences, economic activities, and technological advancements. Urban form acts as the blueprint for our cities, dictating how we live, work, and interact with our environment. It is the invisible hand that shapes the physical layout, design, and structure of our urban spaces. This complex tapestry encompasses various elements, weaving together land use patterns, building density, street networks, public spaces, and even architectural styles.

Understanding urban form requires peeling back the layers of history, culture, economics, and technology that influence its development. A city's past often plays a significant role. Early settlements, for instance, might have been clustered around natural resources like water bodies or fertile land, leading to a more compact urban form. Cultural preferences also contribute – think of the labyrinthine alleyways and vibrant squares found in many European cities, contrasting with the grid-like street patterns and sprawling suburbs favored in North America. Economic activities shape urban form as well. Industrial districts with factories and warehouses create a distinct landscape compared to financial centers dominated by towering office buildings.

Technological advancements have also been a powerful force in shaping urban form. The invention of the automobile revolutionized transportation, leading to the rise of suburbs and a more car-dependent urban form. Today, we are witnessing a renewed interest in public transportation and pedestrian-friendly infrastructure. This shift, combined with the potential disruption of AVs, could lead to a transformation of urban landscapes, with a renewed focus on compact development and walkable neighborhoods.

Cities are not static entities frozen in time; they are living, dynamic organisms that constantly adapt and evolve. Through understanding the building blocks of urban form – the land uses, building density, street layouts, public spaces, and architectural styles – we can shape cities that are not just functional, but also sustainable, livable, and adaptable to the needs of their ever-changing populations. Imagine vibrant, walkable communities where homes, shops, and offices coexist in close proximity. Picture public spaces like parks, plazas, and community gardens serving as the city's lungs, offering respite and encouraging social interaction. Imagine architectural styles that not only please the eye but are also energy-efficient, promoting a sustainable future. As new technologies emerge, planning for a sustainable urban form becomes even more crucial. Making informed choices about the way we design and build our cities today ensures our urban landscapes not only survive but thrive for generations to come.

13.9.3.2 Traditional Urban Form

Historically, cities were compact and walkable. Dense neighborhoods with mixed land uses allowed residents to live, work, and access amenities within close proximity. Streets were narrow, and public spaces encouraged social interaction. Examples include medieval European towns like York, England, with their well-preserved city walls and narrow, pedestrian-friendly streets. These towns were often built for defense, with a central market square and a mix of residential and commercial buildings within a compact area. Ancient Asian cities like Kyoto, Japan, with its traditional wooden buildings, narrow lanes, and emphasis on gardens and public spaces, offer another example. These cities were often built around religious centers and prioritized walkability and harmony with nature.

Imagine vibrant, walkable communities like those found in Amsterdam, Netherlands, where homes, shops, and offices coexist in close proximity, connected by an extensive network of canals and bike lanes. Picture public spaces like the sprawling Central Park in New York City, United States or the serene gardens of Suzhou, China, serving as the city's lungs, offering respite and encouraging social interaction. Imagine architectural styles that not only please the eye but are also energy-efficient, promoting a sustainable future, like the innovative, green buildings found in Singapore.

Modern Urban Form: The industrial revolution and the automobile's rise fundamentally reshaped cities. Zoning regulations, a new invention, segregated residential areas from factories and bustling commercial districts. This separation fueled suburban sprawl, as people sought quieter living spaces outside the city center. Cars became essential for daily life, leading to the construction of wide roads, vast parking lots, and an abundance of single-use developments. This car-centric urban form, while initially seeming convenient, had unintended consequences, including increased traffic congestion, environmental pollution, and a decline in walkable, vibrant communities.

Contemporary Urban Form: In recent decades, urban planners have emphasized sustainable urbanism. Concepts like TOD, mixed-use neighborhoods, and pedestrian-friendly streets aim to create vibrant, efficient, and environmentally conscious cities.

Transport Sustainability: Moving Towards a Greener Future. Transportation is a critical component of urban life. However, unsustainable transportation practices contribute to pollution, congestion, and resource depletion. Achieving transport sustainability involves addressing several key challenges:

- *Automobile Dependency:* Car-centric cities lead to traffic congestion, air pollution, and high energy consumption. Encouraging alternatives like walking, cycling, and public transit reduces reliance on private vehicles.
- *Public Transit Systems:* Efficient public transit networks are essential for sustainable cities. Investments in buses, trains, and light rail can reduce greenhouse gas emissions and improve mobility. Integration with land use planning ensures accessibility.
- *Active Transportation:* Promoting walking and cycling enhances health, reduces emissions, and creates vibrant streets. Sidewalks, bike lanes, and safe crossings encourage active transportation.
- *Compact Development:* Dense, mixed-use neighborhoods reduce travel distances. Residents can walk to work, shops, and recreational areas. TODs near train stations minimize car use.
- *Green Infrastructure:* Urban green spaces, parks, and tree-lined streets improve air quality, provide shade, and enhance aesthetics. Green infrastructure also manages stormwater runoff.

Solutions for Sustainable Urban Form and Transport:

- *TOD:* Designing neighborhoods around transit hubs encourages public transport use. High-density residential and commercial areas near stations reduce car trips.
- *Complete Streets:* Streets designed for all users – pedestrians, cyclists, and motorists – promote safety and accessibility. Sidewalks, bike lanes, and tree-lined boulevards create pleasant environments.
- *Mixed-Use Zoning:* Allowing diverse land uses within the same area fosters vibrant communities. Residents can live, work, and shop without extensive travel.
- *Car-Sharing and Bike-Sharing Programs:* Shared mobility services reduce the need for private car ownership. Electric scooters, bike-sharing stations, and carpooling contribute to sustainability.
- *Urban Greening:* Planting trees, creating pocket parks, and developing green roofs enhance urban aesthetics and mitigate heat island effects.

Achieving urban form and transport sustainability requires collaboration among policymakers, urban designers, and citizens. Balancing economic growth, social equity, and environmental stewardship is essential. We can create cities that are both livable and resilient for future generations, if we prioritize sustainable practices.

13.10 CHALLENGES IN URBAN TRANSPORT

Our cities, pulsating hubs of human activity, face a constant struggle: ensuring efficient and sustainable movement of people and goods. Urban transportation, the lifeblood of any metropolis, grapples with a multitude of interconnected challenges. Congestion and inefficiency plague many cities, with gridlocked traffic wasting time, spewing pollutants into the air, and hindering economic productivity. Accessibility and equity are also major concerns. Inadequate or poorly maintained public transport systems can leave residents feeling isolated and dependent on private vehicles, particularly those in underserved areas or with limited mobility. The ever-expanding urban sprawl further complicates the equation, straining existing infrastructure and increasing travel distances for everyone. Safety concerns add another layer of complexity. Pedestrians, cyclists, and other vulnerable users navigate congested roads with constant risk, creating a barrier to active transportation. Finally, the environmental impact of car-centric transportation looms large. Emissions and resource consumption contribute to climate change, highlighting the urgent need for a shift toward sustainable transportation solutions. Finding solutions to these challenges requires a multi-pronged approach, one that prioritizes efficiency, sustainability, and the well-being of all urban residents.

13.10.1 Congestion and Inefficiency

Urban congestion and inefficiency pose significant hurdles for cities worldwide. As populations grow and economic activities intensify, traffic jams, overcrowded public transport, and wasted time become commonplace. The intricate interplay between inadequate infrastructure, car-centric lifestyles, and outdated planning exacerbates the problem. In this essay, we delve into the multifaceted nature of congestion and inefficiency, exploring their causes, consequences, and potential solutions. From gridlocked streets to delayed commutes, the impact reverberates across communities, economies, and the environment. Let us unravel this intricate web and seek pathways toward more efficient and sustainable urban mobility.

Traffic congestion is a pervasive problem in urban areas worldwide. As cities grow and populations increase, the strain on transportation networks intensifies. We will dig into the causes of traffic congestion and its far-reaching consequences, as reflected in Table 13.13.

This overreliance on cars creates a vicious cycle. Traffic congestion, a direct result of too many vehicles on the road, discourages people from using public transport, further increasing car usage and perpetuating gridlock. This leads to effects of the traffic congestion, which not only slows commutes (wasted time) and pollutes the air, but also reduces the overall quality of life in our cities, as shown in Table 13.14.

TABLE 13.13
The Causes of Traffic Congestion

Limited road capacity	• Urban roads often struggle to accommodate the sheer volume of vehicles. Expanding road infrastructure is challenging due to space limitations and existing development. • Bottlenecks, intersections, and narrow lanes exacerbate congestion during peak hours.
High reliance on personal vehicles	• Car ownership remains a popular choice for commuting. The convenience and perceived flexibility of personal vehicles contribute to their widespread use. • As more people own cars, roads become crowded, especially in densely populated areas.
Inefficient public transport	• Public transit systems that are unreliable, infrequent, or poorly connected fail to attract commuters away from private cars. • Lack of investment in public transport infrastructure hinders its effectiveness.
Urban sprawl	• Suburban expansion leads to longer travel distances. Commuters from outlying areas converge on city centers, causing congestion. • Zoning laws that separate residential, commercial, and industrial zones contribute to sprawl.
Traffic incidents and accidents	• Collisions, breakdowns, and road closures disrupt traffic flow. Even minor incidents can cause significant delays. • Emergency services responding to accidents further compound congestion.
Construction and maintenance work	• Road repairs, utility installations, and construction projects create temporary bottlenecks. • Lack of coordination between agencies can prolong disruptions.

TABLE 13.14
Consequences of Traffic Congestion

Wasted time	• Commuters spend hours stuck in traffic, leading to lost productivity and reduced quality of life. • Delays affect work schedules, appointments, and leisure activities.
Air pollution	• Idling vehicles emit pollutants, including nitrogen oxides and particulate matter. • Poor air quality harms public health, exacerbating respiratory conditions.
Economic losses	• Congestion costs businesses and individuals. Time wasted in traffic translates to lost revenue. • Increased fuel consumption and wear on vehicles add to economic burdens.
Stress and mental health impact	• Frustration, anxiety, and stress arise from being stuck in traffic. • Long commutes contribute to mental health issues.
Environmental degradation	• Congestion leads to excess fuel consumption and greenhouse gas emissions. • Noise pollution from honking and engine noise affects residents.
Reduced accessibility	• Congestion limits access to essential services, schools, and healthcare facilities. • Vulnerable populations suffer disproportionately

TABLE 13.15
Mitigating Traffic Congestion

Invest in public transit	• Improve bus and rail networks, enhance frequency, and integrate modes of transport. • Encourage commuters to switch from cars to public transit
Promote active transportation	• Develop pedestrian-friendly infrastructure, cycling lanes, and safe crossings. • Encourage walking and cycling for short trips
Congestion pricing	• Implement tolls or fees during peak hours to discourage unnecessary car use. • Revenue generated can fund public transport improvements.
Smart traffic management	• Use real-time data to optimize traffic flow, adjust signal timings, and reroute vehicles. • Dynamic lane management can ease congestion.
Mixed-use development	• Create compact neighborhoods where residents can live, work, and shop within walking distance. • Reduce reliance on long commutes.

Let us bridge the gap between the consequences of traffic congestion and the strategies to alleviate it. As we explore potential solutions, keep in mind that addressing this issue requires a multipronged approach, involving urban planning, transportation policies, and congestion pricing. The practical steps to mitigate congestion and enhance urban mobility are shown in Table 13.15.

13.10.2 Accessibility and Equity

In a thriving city, efficient and accessible transportation plays a vital role. However, the reality for many urban residents is far from harmonious. This section explores the challenges of ensuring equitable access to reliable and affordable transportation, focusing on inadequate coverage, the lack of disability-friendly infrastructure, and the social and economic consequences of "transportation deserts".

13.10.2.1 The Coverage Gap: When Public Transport Leaves People Behind

One of the most pressing challenges is the uneven distribution of public transportation services. In many cities, sprawling suburbs and low-income neighborhoods often experience limited or even non-existent public transport options. This can be attributed to various factors, including:

- *Cost–Benefit Analysis:* Public transit authorities often prioritize routes with high passenger volumes to ensure financial viability. This can leave sparsely populated areas or those with lower incomes underserved.
- *Urban Sprawl:* The expansion of cities outwards makes it more challenging and expensive to provide comprehensive public transport networks to these outlying areas.
- *Historical Planning:* Legacy infrastructure may not have been designed with public transport in mind, creating additional hurdles in extending services to certain areas.

The consequences of inadequate public transport coverage are far-reaching. Residents in underserved areas become heavily reliant on private vehicles, leading to increased traffic congestion and pollution. This can also create a cycle of poverty, as residents struggle with the financial burden of car ownership and maintenance. Additionally, limited access to public transport can restrict job opportunities, healthcare services, and educational institutions that are only accessible by car.

13.10.2.2 Beyond Barriers: Building Inclusive Infrastructure

Another critical aspect of accessibility is ensuring that public transport infrastructure is inclusive and caters to the needs of all citizens, especially those with disabilities. This includes:

- *Accessible Vehicles:* Low-floor buses, ramps, and wider doorways can significantly improve accessibility for wheelchair users.
- *Accessible Stations:* Elevators, audio announcements, tactile paving, and braille signage are essential for visually impaired and blind passengers.
- *Accessible Pedestrian Infrastructure:* Well-maintained sidewalks, curb cuts, and accessible pedestrian signals are crucial for those with mobility limitations.

The lack of disability-friendly infrastructure can have a profound impact on the lives of people with disabilities. It can restrict their mobility and independence, limiting their access to employment, education, and social activities. Investing in inclusive infrastructure is not just a moral imperative but also a smart economic decision. It allows people with disabilities to participate fully in society and contribute to the workforce.

13.10.2.3 Transportation Deserts: When Mobility Becomes a Luxury

The term "transportation desert" refers to areas with limited access to reliable and affordable public transport options. These areas are often characterized by low-income housing, high car ownership costs, and a reliance on personal vehicles for basic necessities. The consequences of living in a transportation desert are significant:

- *Limited Job Opportunities:* Without access to reliable transportation, residents may struggle to reach job opportunities located outside their immediate vicinity. This can limit their career options and restrict economic mobility.
- *Barriers to Healthcare:* Access to healthcare services can be severely hampered by a lack of public transport, particularly for those who rely on regular medical appointments or specialized care facilities.
- *Educational Disparity:* Children residing in transportation deserts may face challenges getting to school, leading to absenteeism and potentially hindering their educational attainment.
- *Limited Social Interaction:* Limited mobility can isolate residents, particularly seniors and individuals with disabilities, who rely on public transport for social interaction and errands.

13.10.2.4 Building a More Equitable Transportation System

Bridging the gap and ensuring equitable access for all requires a multi-pronged approach:

- *Demand–Responsive Transit:* Implementing innovative solutions like on-demand micro transit services or expanding dial-a-ride programs can provide targeted transport options in underserved areas.
- *Fare Integration:* Integrating fares across different modes of transport can create a seamless and affordable public transport network for all users.
- *Land-Use Planning:* Promoting mixed-use development and prioritizing walkable communities can reduce reliance on cars and improve access to essential services.
- *Infrastructure Investment:* Investing in expanding public transport infrastructure, including new routes, improved stations, and disability-friendly features, is crucial for creating a truly inclusive system.

13.10.2.5 The Road to a More Equitable Future

Ensuring everyone has access to reliable and affordable transportation is not just a social justice issue; it is an economic and environmental imperative. By building a more equitable transportation system, we can create a city that works for everyone, fostering social inclusion, economic opportunity, and a sustainable future for all residents. Imagine a city where everyone, regardless of income, age, or ability, can access jobs, education, healthcare, and social activities with ease. This is the vision we should strive for, prioritizing accessibility and inclusivity in every aspect of urban transportation planning.

13.10.3 Safety and Environmental Impact

Our cities are often a tapestry of movement. Cars, buses, bikes, and pedestrians share the urban landscape, creating a dynamic (and sometimes chaotic) environment. However, for vulnerable users – pedestrians, cyclists, and micromobility users – navigating this environment can be a daily test of nerves. This section explores the safety concerns faced by these users and delves into the broader environmental impact of car-centric transportation.

13.10.3.1 A Constant Struggle: Safety Concerns for Vulnerable Users

Pedestrians, cyclists, and micromobility users like scooter riders face a multitude of safety challenges in urban environments. These risks stem from several factors:

- *Traffic Congestion and Speeding:* Gridlock and excessive speeds create a dangerous environment for vulnerable users. The sheer volume of vehicles, coupled with distracted driving, significantly increases the risk of collisions with pedestrians and cyclists.
- *Inadequate Infrastructure:* Poorly maintained sidewalks, a lack of dedicated bike lanes, and insufficient signage all contribute to safety concerns. Shared roadways without proper separation between cars and vulnerable users create a constant feeling of vulnerability.
- *Blind Spots and Unpredictable Behavior:* Large vehicles often have significant blind spots, increasing the risk of collisions with pedestrians and cyclists, particularly when turning corners. Additionally, unpredictable driver behavior, such as sudden lane changes or failure to yield right of way, adds another layer of danger.
- *Lack of Enforcement:* Inconsistent enforcement of traffic laws can embolden reckless driving behavior, creating a greater threat to vulnerable users.

The consequences of these safety concerns are devastating. Pedestrian fatalities are a major public health concern, with vulnerable users often bearing the brunt of traffic accidents. Injuries from cycling and e-scooter accidents can be severe and life-altering. The constant fear of accidents can deter people from walking and cycling, leading to a decline in physical activity and public health.

13.10.3.2 Beyond Collisions: The Invisible Threat of Pollution

The environmental impact of car-centric transportation extends far beyond safety concerns. Our reliance on private vehicles contributes significantly to:

- *Air and Noise Pollution:* Emissions from vehicles pollute the air we breathe, leading to respiratory problems, cardiovascular disease, and other health issues. The constant hum of traffic creates noise pollution, disrupting sleep and impacting mental health.
- *Greenhouse Gas Emissions:* Transportation is a major contributor to greenhouse gas emissions, fueling climate change and its associated environmental threats like extreme weather events and rising sea levels.
- *Inefficient Use of Urban Land:* Urban landscapes are often dominated by vast parking lots and sprawling roads, consuming valuable land that could be used for parks, green spaces, or housing. This car-centric design fosters urban sprawl, increasing transportation distances and further promoting reliance on vehicles.

13.10.3.3 The Vicious Cycle: Safety Concerns and Environmental Impact

A vicious cycle has trapped our cities in a state of gridlock and pollution. The dominance of cars creates a dangerous environment for pedestrians and cyclists. Wide roads with limited space for walking and cycling infrastructure make it intimidating for people to choose these alternative modes of transport. This leads to a surge in car usage, further exacerbating the problem. Traffic congestion becomes the norm, spewing out pollutants that harm our health and contribute to climate change. The very design that prioritizes car convenience discourages walking and cycling, perpetuating the cycle and trapping us in a web of congestion, pollution, and safety concerns. Breaking free from this cycle requires a multi-pronged approach that prioritizes safety for all users and embraces sustainable alternatives.

13.10.3.4 Building a Safer and Sustainable Future

Shifting toward a more sustainable and equitable transportation system requires a multi-pronged approach:

- *Traffic Calming Measures:* Implementing speed limits, reducing car lanes, and prioritizing traffic signals for pedestrians and cyclists can create a calmer and safer environment.
- *Complete Streets Design:* Designing streets that prioritize all users – pedestrians, cyclists, public transport, and cars – is crucial. This includes dedicated bike lanes, wider sidewalks, and improved pedestrian crossings.
- *Vision Zero:* Embracing the "Vision Zero" approach, which aims to eliminate all traffic fatalities and serious injuries, should be a guiding principle for urban transportation planning.
- *Investment in Public Transport:* Investing in reliable, affordable, and efficient public transport systems provides residents with a safe and attractive alternative to private vehicles.
- *Micromobility Infrastructure:* Encouraging and promoting micromobility options like e-scooters and bicycles require dedicated infrastructure, including charging stations and secure parking facilities.
- *Pedestrian-Friendly Zones:* Creating car-free zones in city centers can significantly enhance pedestrian safety and create spaces for public interaction.

13.10.3.5 A Shared Responsibility: Building a City for Everyone

Ensuring safety and well-being for all the street users goes beyond infrastructure and requires a shared responsibility between city planners, policymakers, and the individuals who navigate these streets.

13.10.3.6 Promoting Street Safety: A Collective Responsibility

The safety and well-being of everyone on our streets are paramount. While city planners and policymakers play a crucial role, ensuring safe streets is a shared endeavor that involves all users. Let us explore how drivers, pedestrians, and cyclists can actively contribute to safer urban environments.

13.10.3.7 Responsible Driving

Drivers, as key participants in traffic, can take several proactive steps to enhance safety:

- *Obey Traffic Laws:* Adhering to speed limits, stop signs, and traffic signals is essential. Reckless driving endangers everyone.
- *Avoid Distractions:* Put away phones, adjust GPS devices before driving, and stay focused on the road.
- *Maintain Safe Speeds:* Speeding increases the severity of accidents. Slow down in residential areas and near schools.
- *Yield to Pedestrians:* Always yield the right of way to pedestrians at crosswalks and intersections.
- *Be Mindful of Vulnerable Users:* Watch out for cyclists, children, and elderly pedestrians. Give them extra space and patience.

13.10.3.8 Pedestrian Safety

Pedestrians, too, have a role to play in their own safety:

- *Obey Traffic Signals:* Wait for the green light at crosswalks and use pedestrian crossings.
- *Stay Visible:* Especially at night, wear reflective clothing or accessories. Make yourself visible to drivers.
- *Look Both Ways:* Before crossing, check for oncoming traffic from both directions.
- *Avoid Distractions:* Put away phones while walking. Being aware of your surroundings is crucial.

13.10.3.9 Cyclist Safety

Cyclists can significantly enhance their safety by following these guidelines:

- *Wear Helmets:* Helmets protect against head injuries. Always wear one when cycling.
- *Use Bike Lanes or Designated Paths:* Stick to bike lanes whenever possible. If not available, ride on the right side of the road.
- *Signal Turns:* Indicate cyclist intentions by using hand signals when turning.
- *Be Defensive:* Assume that drivers may not see the cyclist. Anticipate their actions and be cautious.

Shared Responsibility:

- *Education and Awareness:* Public awareness campaigns can educate all users about their responsibilities and encourage safe behaviors.
- *Infrastructure Improvements:* City planners should design streets with safety in mind – adequate lighting, clear signage, and safe crossings.

- *Community Engagement:* Encourage neighborhood watch programs, community patrols, and reporting unsafe conditions.

Creating safer streets requires collaboration. Every street user must contribute to a culture of responsible behavior, where every user prioritizes safety and well-being. Together, we can build cities where everyone feels secure and protected.

Fostering a culture of shared responsibility is key. When drivers, pedestrians, cyclists, and city planners work together, we can create a safer and more enjoyable urban environment for everyone. Imagine a city where streets are shared with respect and awareness, where the sound of bicycle bells replaces the constant hum of traffic, and where children can walk safely to school. This is the vision we should strive for, where everyone feels comfortable navigating the city, regardless of their chosen mode of transport.

13.11 FUTURE OF URBAN TRANSPORTATION

Cities are on the verge of a transportation revolution. AVs, once relegated to science fiction, are poised to transform urban landscapes. Imagine a future where self-driving cars and electric buses seamlessly navigate city streets, reducing congestion and emissions. Micromobility options like e-scooters and bicycles will be seamlessly integrated into the transportation ecosystem, offering convenient and sustainable last-mile solutions. However, this transformation always comes with its challenges. Infrastructure needs to adapt, regulations need to evolve, and social equity must remain a priority. The future of urban transportation is a tapestry woven with innovation, sustainability, and the promise of a more livable cityscape for all.

13.11.1 Emerging Trends in Urban Transportation

The way we move around our cities is undergoing a profound transformation. Driven by technological advancements, environmental concerns, and a growing emphasis on urban living, a wave of emerging trends is reshaping the landscape of urban transportation. This section discusses some of the most significant trends, analyzing their potential impact and the challenges that lie ahead.

13.11.1.1 The Rise of Micromobility

Cities are witnessing a surge in popularity of micromobility options like bicycles, e-scooters, and e-bikes. These compact, often EVs offer a convenient and sustainable way to navigate short distances within urban areas. Their affordability and flexibility make them attractive alternatives to private cars, particularly for last-mile connections. Dockless bike-sharing programs and electric scooter rentals have further fueled this trend, providing residents with easy access to these vehicles. However, the rapid growth of micromobility has also led to concerns about safety, parking regulations, and integration with existing infrastructure. Cities need to adapt by creating dedicated lanes, improving bike parking facilities, and enforcing safety regulations for both riders and pedestrians.

13.11.1.2 The AV Revolution

Once a futuristic vision, AVs are inching closer to reality. These self-driving cars and buses promise significant benefits, including reduced traffic congestion, improved safety, and increased accessibility for those who cannot drive. AVs have the potential to revolutionize public transportation, offering on-demand services and flexible routes. However, several challenges remain. Regulations need to be developed to ensure the safe operation of AVs on public roads. Cybersecurity concerns need to be addressed to prevent hacking and ensure passenger safety. Also, the ethical implications of AV decision-making in the event of accidents need careful consideration.

13.11.1.3 The Sharing Economy Takes the Wheel

The sharing economy is making inroads into urban transportation, with ride-hailing services like Uber and Lyft becoming ubiquitous in many cities. These platforms offer a convenient and often affordable alternative to traditional taxis. However, concerns have been raised about the impacts on traffic congestion and the working conditions of drivers. Car-sharing services like Zipcar and Turo allow users to rent vehicles by the hour, further reducing the need for car ownership. These services offer greater flexibility than traditional car rentals, catering to short-term needs. However, ensuring equitable access across income levels and integrating these services with public transportation are key considerations.

13.11.1.4 Electrification Takes Center Stage

EVs are rapidly gaining traction as a response to growing environmental concerns. EVs produce zero tailpipe emissions, contributing to cleaner air and a reduction in greenhouse gas emissions. Governments are introducing incentives for EV adoption, and advancements in battery technology are steadily increasing range and reducing charging times. While infrastructure for charging EVs needs to be significantly expanded, the shift toward electrification is a crucial step toward sustainable urban mobility.

13.11.1.5 Integrating Public Transportation for Seamless Journeys

A key trend shaping urban transportation is the focus on seamless integration between different modes of transport. This includes enhancing connections between public transport networks, micromobility options, and ride-hailing services. Multimodal ticketing systems and real-time information sharing are crucial for creating a user-friendly experience that encourages people to utilize a combination of options for their journeys. The integration of public transport with new technologies like on-demand micro transit solutions in underserved areas can expand accessibility and create a more equitable system.

13.11.1.6 Prioritizing Active Transportation

Cities are increasingly prioritizing walking and cycling as modes of transportation. This trend is driven by growing awareness of the health benefits of physical activity and the environmental impact of car-centric lifestyles. Creating pedestrian-friendly zones with wider sidewalks, improved signage, and traffic calming measures are essential for encouraging walking. Investing in dedicated bike lanes, protected intersections, and secure bike parking facilities can make cycling a more attractive and safer option.

13.11.1.7 Data-Driven Transportation Planning

The rise of big data and AI is influencing transportation planning. Real-time traffic data can be used to optimize traffic flow and reduce congestion. Additionally, data on user behavior and travel patterns can inform the development of more efficient and responsive public transport systems. AI can be utilized to optimize bus routes and schedules, ensuring they align with peak demand. However, ensuring data privacy and using AI responsibly will be crucial in implementing these data-driven solutions.

13.11.1.8 Challenges and the Road Ahead

While these emerging trends offer exciting possibilities for the future of urban transportation, significant challenges remain. Cities need to invest in infrastructure upgrades to accommodate new technologies and prioritize safety for all users. Equitable access to new transportation options must be ensured to avoid exacerbating existing inequalities. Regulatory frameworks need to be adapted to address the needs of AVs and micromobility options. Collaboration between city planners, policymakers, technology companies, and public transport agencies is critical for navigating these challenges and reaping the benefits of these emerging trends.

13.11.1.9 A Tapestry of Possibilities

Emerging trends, from the rise of micromobility to the burgeoning potential of AVs, hold the promise of a more sustainable, efficient, and equitable urban mobility experience. Balancing innovation with safety, ensuring equitable access for all, and fostering responsible use of technology will be crucial in shaping a future that works for everyone. Ultimately, the future of urban transportation will likely be a tapestry woven from a combination of these trends, not a single dominant force. Cities will need to adapt and evolve, creating a multimodal transportation ecosystem that seamlessly integrates different modes of transport. Imagine a cityscape where electric buses navigate alongside self-driving cars, while dedicated bike lanes and pedestrian walkways provide safe havens for active commuters. Shared micromobility options like e-scooters seamlessly bridge the gap between public transportation and final destinations.

This vision requires a collaborative effort. Urban planners must create streets that cater to all users – pedestrians, cyclists, public transport, and cars. Technology companies need to prioritize safety and develop solutions that are accessible to everyone. Public transport agencies must embrace innovation and adapt their services to integrate seamlessly with emerging technologies.

The future of urban transportation is filled with possibilities. Through innovation, sustainability and equity, and fostering collaboration, we can create cities that are not only more livable but also more sustainable for generations to come. This future depends on our collective ability to navigate the challenges and harness the potential of the emerging trends that are shaping the landscape of urban mobility.

13.11.2 Impact of AVs and Smart Traffic Management Systems

The way we navigate our cities is on the verge of a seismic shift. AVs, once relegated to science fiction, are poised to transform urban landscapes. Coupled with the rise of smart traffic management systems (STMS), these innovations promise a future of reduced congestion, improved safety, and a more efficient transportation ecosystem. However, this revolution comes with its own set of challenges that need careful consideration.

13.11.2.1 The Promise of AVs

AVs hold immense potential to revolutionize urban transportation. These self-driving cars and buses operate without human intervention, relying on a combination of sensors, cameras, and complex algorithms to navigate roads.

- *Reduced Congestion:* AVs are programmed to follow traffic laws meticulously and react to changing conditions quickly. This can lead to smoother traffic flow and reduced congestion, particularly during peak hours. Traffic jams caused by erratic driving behavior or slow reaction times could become a thing of the past.
- *Enhanced Safety:* Human error is a major contributor to traffic accidents. AVs, with their precise navigation and constant awareness of surroundings, promise to significantly reduce accidents and fatalities on our roads. The potential for a future with zero traffic deaths becomes a possibility.
- *Increased Efficiency:* AVs can be programmed to operate 24/7, optimizing routes and minimizing idle times. This can lead to increased efficiency in public transportation systems and logistics operations.
- *Accessibility for All:* AVs can provide mobility options for those who cannot drive due to age, disability, or other factors. This can significantly increase social inclusion and improve access to employment and essential services.

13.11.2.2 Smart Traffic Management: Orchestrating the Flow

STMS represents another key element in the transformation of urban transport. These systems leverage data and technology to optimize traffic flow and improve overall network efficiency.

- *Real-Time Traffic Management:* STMS collects real-time data on traffic conditions, accidents, and road closures. This information can be used to adjust traffic signals dynamically, reroute traffic around incidents, and warn drivers of potential hazards. Imagine a network that proactively adjusts to minimize delays before they even occur.
- *Improved Infrastructure Utilization:* STMS can optimize the use of existing infrastructure, such as traffic lanes and parking spaces. This can be achieved through dynamic lane allocation or smart parking management systems.
- *Data-Driven Planning:* The data collected by STMS can be used by urban planners to identify bottlenecks, assess future needs, and inform infrastructure development projects. Data become a powerful tool for building a more responsive and adaptable transportation system.

13.11.2.3 A Symbiotic Relationship: AVs and STMS Together

The true transformative potential lies in the synergy between AVs and STMS. When these technologies work in tandem, the benefits are amplified:

- *Enhanced Safety and Efficiency:* AVs integrated with STMS can receive real-time traffic updates and adjust their routes for optimal efficiency and safety.
- *Cooperative Driving:* STMS can connect AVs with each other, creating a "cooperative driving" environment where vehicles "communicate" and coordinate movements to minimize collisions and optimize traffic flow. Imagine a network of vehicles working in harmony for the benefit of everyone on the road.
- *Reduced Emissions:* By promoting smoother traffic flow and optimizing travel times, AVs and STMS can contribute to a reduction in vehicle emissions, benefiting air quality and contributing to a more sustainable urban environment.

13.11.2.4 Challenges on the Road Ahead

Despite the potential benefits, the autonomous future has the following potential challenges and constraints:

- *Infrastructure Investment:* Cities need to invest in upgrading infrastructure to accommodate AVs, such as installing sensors and creating dedicated lanes for self-driving vehicles.
- *Cybersecurity Concerns:* The security of AV systems is paramount. Hacking or software malfunctions could have devastating consequences, requiring robust cybersecurity measures.
- *Regulatory Hurdles:* Regulations need to be developed to establish clear guidelines for the operation of AVs on public roads, addressing issues such as liability in case of accidents.
- *Ethical Considerations:* The ethical implications of AV decision-making in the event of accidents need careful consideration. Who is responsible when an AV makes a mistake?

13.11.2.5 Building a Future for All: Equity and Accessibility

Ensuring equitable access to this future will be crucial. The benefits of AVs should not be limited to those who can afford them. Cities need to consider how to integrate AV technology into public transport systems, ensuring everyone can benefit from the potential for improved efficiency and accessibility.

13.11.2.6 A Collaborative Journey

The transformation of urban transportation toward AVs and smart traffic management systems will be a collaborative journey. Technology companies, policymakers, urban planners, and the public all have a role to play. Fostering collaboration, investing in necessary infrastructure, and addressing the challenges head on enable us to unlock the potential of this revolution and create a future where our cities are:

- *Safer:* With fewer accidents and fatalities, our streets become safer for everyone, pedestrians, cyclists, and drivers alike.
- *More Efficient:* Less congestion translates to faster commutes, increased productivity, and reduced reliance on personal vehicles.
- *More Sustainable:* A future with cleaner-burning electric AVs and optimized traffic flow contributes significantly to lowering urban emissions.
- *More Inclusive:* AVs can bridge the gap for those who cannot drive, fostering greater social inclusion and access to opportunities.

The driverless revolution is not just about technology; it is about creating cities that are livable, sustainable, and accessible for all. Imagine a cityscape where traffic flows seamlessly, air pollution is a distant memory, and people can navigate their environment with ease and safety. This vision is within reach, but it requires a collective effort to navigate the challenges and harness the potential of this transformative technology. As we embark on this journey, let us prioritize safety, sustainability, and equity to build a future of urban transportation that benefits everyone.

13.11.3 MaaS and Future Prospects

The way we navigate our cities is undergoing a fundamental shift. Gone are the days of car ownership being the sole option for urban mobility. MaaS – a revolutionary approach that transforms transportation from a set of individual services into a seamless, integrated ecosystem. This essay explores the core elements of MaaS, its potential impact on urban landscapes, and the exciting prospects it holds for the future.

MaaS represents a paradigm shift in how we approach transportation. Rather than relying solely on private cars, MaaS aims to provide integrated, seamless mobility solutions that cater to individual needs. By combining various modes of transport – such as public transit, ride-sharing, biking, and walking – MaaS promises convenience, efficiency, and sustainability. In this essay, we will explore the concept of MaaS, its current state, and the exciting prospects it holds for the future.

13.11.3.1 Understanding MaaS

MaaS is a user-centric model that prioritizes convenience and accessibility. It aggregates various transportation options – public transport, ride-hailing services, micromobility solutions like bikes and scooters, car rentals, and even carpooling – into a single, unified platform. Imagine a single app where you can plan your entire journey, compare different options in real-time based on cost, travel time, and environmental impact, and seamlessly book and pay for your chosen mode of transport.

13.11.3.2 Key Components of MaaS

- *Multimodal Integration:* MaaS combines public transit, taxis, car-sharing, bike-sharing, and walking into a cohesive system.
- *Digital Platforms:* Mobile apps and websites serve as gateways to MaaS, allowing users to access real-time information, plan routes, and make bookings.
- *Payment Integration:* Users pay for their entire journey through a single account, simplifying transactions.

This user-centric approach offers several key advantages:

- *Convenience and Flexibility:* MaaS eliminates the need for multiple apps and ticketing systems. Users can plan their journeys with ease, opting for the most convenient and efficient option for each leg of their trip.
- *Real-Time Information:* MaaS platforms provide users with real-time information on traffic conditions, arrival times, and available options. This empowers users to make informed decisions about their journeys.
- *Cost-Effectiveness:* MaaS can be more cost-effective than car ownership, particularly for those with infrequent driving needs. Users only pay for the transportation they use, eliminating the burden of car maintenance and insurance.
- *Sustainability:* By promoting multimodal transportation and potentially reducing car usage, MaaS can contribute to a more sustainable urban environment. The shift toward EVs within MaaS options can further amplify this positive impact.

13.11.3.3 Shaping the Future of Urban Mobility

MaaS has the potential to fundamentally reshape urban landscapes:

- *Reduced Traffic Congestion:* By encouraging a shift toward multimodal transportation and optimizing route planning, MaaS can help alleviate traffic congestion, leading to faster commutes and improved air quality.
- *Increased Public Transport Utilization:* A seamless integration of public transport options within MaaS can encourage more people to use these services, contributing to a more sustainable and equitable transportation system.
- *Revitalized City Centers:* Reduced car dependence can lead to the creation of more pedestrian-friendly and vibrant city centers, fostering a more enjoyable urban experience.
- *Data-Driven Planning:* The vast amount of data generated through MaaS platforms can provide valuable insights for urban planners. This data can be used to optimize public transportation networks, identify areas with inadequate coverage, and develop more efficient transportation systems.

13.11.3.4 Challenges and Opportunities

Despite its transformative potential, MaaS faces some key challenges:

- *Infrastructure Investment:* Cities need to invest in infrastructure that caters to different modes of transport, including dedicated bike lanes, expanded public transport networks, and charging stations for EVs.
- *Data Privacy Concerns:* MaaS platforms collect a significant amount of user data. Ensuring data privacy and security is crucial to maintaining user trust.
- *Equity and Accessibility:* MaaS solutions need to be accessible and affordable for everyone, regardless of income or technological literacy. Integrating public transport options and providing user-friendly interfaces are key elements to achieve inclusivity. Ensuring MaaS serves all socioeconomic groups equally is essential.
- *Fragmentation:* Integrating diverse transportation providers remains a challenge due to differing systems, data formats, and regulations.
- *Behavioral Shift:* Convincing car owners to switch to MaaS requires changing deeply ingrained habits.

13.11.3.5 The Road Ahead: Collaboration and Innovation

The success of MaaS hinges on collaboration between various stakeholders:

- *Technology Companies:* Technology companies need to develop user-friendly platforms and ensure data privacy and security.
- *Public Transport Agencies:* Public transport agencies need to embrace integration with MaaS platforms and continue to improve service efficiency and reliability.
- *Urban Planners:* Urban planners need to invest in infrastructure that accommodates multi-modal transportation and supports the MaaS ecosystem.
- *Policymakers:* Policymakers need to develop regulations that address issues like data privacy, competition within the transport sector, and ensuring equitable access to MaaS options.

13.11.3.6 A Glimpse into the Future Possibilities

As MaaS evolves, exciting possibilities lie on the horizon:

- *AVs:* The integration of AVs into MaaS platforms can further enhance efficiency and safety. Imagine a future where self-driving cars and buses are seamlessly integrated into the MaaS ecosystem, offering on-demand transportation options.
- *Personalized Mobility:* MaaS platforms with advanced AI capabilities could personalize transportation options based on user preferences, real-time conditions, and even weather forecasts.
- *First-Mile/Last-Mile Solutions:* Seamless integration of micromobility options like e-bikes and scooters can address "first-mile/last-mile" connectivity challenges, making public transport more accessible.
- *Micromobility Revolution:* Electric scooters, e-bikes, and small EVs are reshaping urban mobility. MaaS platforms will integrate micromobility options, providing last-mile connectivity.

13.11.3.7 Global Adoption of MaaS

- MaaS is gaining traction worldwide, with cities and private companies collaborating to implement MaaS platforms.
- Examples include Helsinki's Whim app, Singapore's Beeline, and Germany's Jelbi.

13.11.3.8 Environmental Impact

- MaaS encourages shared rides and reduces the need for private car ownership.
- By promoting sustainable modes, MaaS contributes to lower emissions and cleaner cities.

13.11.3.9 Urban Air Mobility

- Flying taxis and drones could become part of MaaS networks.
- Imagine booking an air taxi for rush-hour traffic or urgent medical transport.

13.11.4 Role of Technology in Urban Transportation

In cities, pedestrians, cyclists, and vehicles share the urban landscape, creating a dynamic and often chaotic environment due to their movement. Technology, however, is emerging as a powerful force in reshaping this landscape, promising a future of increased efficiency, safety, and sustainability. This section discusses the critical role technology plays in transforming urban transportation, analyzing both its potential benefits and the challenges that need to be addressed.

13.11.4.1 Smart Traffic Management: Orchestrating the Flow

One of the most significant technological advancements is the rise of STMS. These systems leverage data and communication technology to create a more responsive and efficient transportation network.

- *Real-Time Data Analysis:* STMS collect real-time data on traffic conditions, accidents, and road closures through a network of sensors and cameras. This information is then used to adjust traffic signals dynamically, reroute traffic around incidents, and warn drivers of potential hazards. Imagine traffic lights that intelligently adapt to optimize flow based on real-time data, leading to smoother commutes and reduced congestion.
- *Improved Infrastructure Utilization:* STMS can optimize the use of existing infrastructure, such as traffic lanes and parking spaces. Dynamic lane allocation systems, for example, can adjust lane usage based on traffic volume, while smart parking management systems can guide drivers to available parking spots, reducing congestion caused by aimless searching.

13.11.4.2 The Connected Vehicle Revolution

Another game-changer is the rise of Vehicle-to-Everything (V2X) communication technology. This technology allows vehicles to "talk" to each other and with roadside infrastructure, creating a more connected and collaborative transportation ecosystem.

- *Enhanced Safety:* V2X communication enables vehicles to share information about their position, speed, and direction. This allows for collision avoidance systems to react faster and with greater precision, significantly reducing accidents and fatalities on the road.
- *Improved Traffic Flow:* By sharing real-time location data, vehicles can adjust their speed and maintain safe distances, creating a smoother flow of traffic and minimizing bottlenecks. Imagine a future where cars automatically adjust speed based on the car ahead, eliminating stop-and-go situations and reducing congestion.

13.11.4.3 The Rise of Micromobility and Shared Services

Technology is also driving the surge in popularity of micromobility options like e-bikes, scooters, and dockless bike-sharing programs. These compact, often EVs offer a convenient and sustainable way to navigate short distances within urban areas. Smartphone apps facilitate locating, unlocking, and paying for these vehicles, further enhancing their user-friendliness.

13.11.4.4 Reduced Car Dependence

Micromobility options provide a convenient alternative to car travel for short trips, potentially leading to a reduction in car usage and associated traffic congestion. They bridge the gap between public transport stops and final destinations, making public transport more accessible for those in underserved areas.

13.11.4.5 The AV Revolution Looms Large

AVs – self-driving cars and buses – have the potential to revolutionize urban transportation completely. These vehicles rely on a combination of sensors, cameras, and complex algorithms to navigate roads without human intervention.

- *Increased Safety:* Human error is a major contributor to traffic accidents. AVs, with their precise navigation and constant awareness of surroundings, promise to significantly reduce accidents and fatalities.
- *Enhanced Efficiency:* AVs can be programmed to operate 24/7, optimizing routes and minimizing idle times. This can lead to increased efficiency in public transportation systems and logistics operations, particularly for deliveries within urban areas.

13.11.4.6 MaaS: A User-Centric Approach

One of the most exciting advancements is the emergence of MaaS platforms. These platforms aggregate various transportation options – public transport, ride-hailing services, micromobility solutions, car rentals, and even carpooling – into a single, unified app.

Convenience and Efficiency: Users can plan their entire journeys seamlessly, compare different options based on cost, travel time, and environmental impact, and book and pay for their chosen mode of transport all through one platform. This user-centric approach streamlines travel and eliminates the need for multiple apps and ticketing systems.

13.11.4.7 Challenges on the Road Ahead

Despite the transformative potential of technology, several challenges need to be addressed:

- *Infrastructure Investment:* Cities need to invest in upgrading infrastructure to accommodate new technologies. This includes installing sensors for V2X communication, expanding charging stations for EVs, and creating dedicated lanes for bicycles and micromobility options.
- *Cybersecurity Concerns:* The security of connected vehicles and MaaS platforms is paramount. Hacking or software malfunctions could have devastating consequences, requiring robust cybersecurity measures.
- *Data Privacy:* The vast amount of data generated by these technologies raises concerns about data privacy. Clear regulations are needed to ensure that user data collected by STMS, V2X systems, and MaaS platforms is used responsibly and with user consent. User trust hinges on data privacy being protected.
- *Equity and Accessibility:* Technology-driven solutions should not exacerbate existing inequalities. MaaS platforms need to be user-friendly and accessible to those with limited technological literacy or access to smartphones. Subsidies may be necessary to ensure that low-income residents who rely on public transport are not excluded from the benefits of MaaS.
- *Regulation and Policy Development:* Policymakers need to develop regulations that address the unique challenges posed by emerging technologies. This includes establishing clear guidelines for the operation of AVs, regulating data privacy within transportation systems, and ensuring fair competition within the industry.

13.11.4.8 Building a Future for All: Towards a Sustainable and Equitable Urban Mobility

The successful integration of technology into urban transportation requires collaboration between various stakeholders:

- *Technology Companies:* Companies need to develop user-friendly and secure platforms while prioritizing data privacy. They should also work toward making technology accessible for all.
- *Policymakers:* Policymakers need to create a regulatory framework that fosters innovation while addressing safety, privacy, and equity concerns.
- *Urban Planners:* Urban planners need to invest in infrastructure that caters to multimodal transportation and integrates seamlessly with emerging technologies.
- *Public Transport Agencies:* Public transport agencies must embrace innovation and integrate their services with MaaS platforms. They should also strive to improve service reliability and efficiency to compete effectively with new technologies.

13.11.4.9 Looking Ahead

A Glimpse into the Future. As technology continues to evolve, the future of urban transportation holds exciting possibilities:

- *Personalized Mobility:* MaaS platforms with advanced AI capabilities could personalize transportation options based on user preferences, real-time conditions, and even weather forecasts.

Imagine a future where your transportation needs are anticipated and catered to seamlessly, creating a more efficient and stress-free travel experience.

- *The Integration of AVs:* As AV technology matures, its integration into MaaS platforms can further enhance efficiency and provide readily available on-demand transportation options, particularly in underserved areas.
- *Focus on Sustainability:* EVs and micromobility options, coupled with optimized traffic flow through STMS and V2X technology, can significantly contribute to a more sustainable urban environment with reduced emissions and energy consumption.

Technology is not a magic bullet, but it serves as a powerful tool for transforming urban transportation. Through innovation, addressing the challenges, and fostering collaboration, we can create a future of urban mobility that is:

- *Safer:* Reduced accidents and fatalities through AVs and improved V2X communication.
- *More Efficient:* Less congestion through smart traffic management and optimized travel planning.
- *More Sustainable:* Increased use of EVs and optimized traffic flow lead to lower emissions.
- *More Equitable:* Inclusive MaaS platforms and accessible public transport ensure everyone can benefit from technological advancements.

The ultimate goal is to create cities where people, not cars, dictate the flow of movement. Technology has the potential to pave the way for a future where urban transportation is a seamless experience, contributing to a more livable and sustainable future for all.

PROBLEMS AND ISSUES FOR DISCUSSION

1. How does public transportation impact urban congestion?
2. What roles do ride-sharing services play in urban mobility?
3. Discuss the challenges of integrating autonomous vehicles into existing transportation networks.
4. Analyze the impact of bike-sharing programs on urban sustainability.
5. How can cities promote walking as a viable mode of transportation?
6. Compare and contrast bus rapid transit (BRT) systems with traditional bus services.
7. What are the equity implications of fare structures in public transit?
8. Explore the concept of transit-oriented development (TOD) and its benefits.
9. Assess the effectiveness of congestion pricing in reducing traffic.
10. Investigate the role of electric scooters in last-mile connectivity.
11. How can cities improve accessibility for people with disabilities in transportation systems?
12. Discuss the trade-offs between expanding road infrastructure and investing in public transit.
13. Analyze the impact of e-commerce delivery on urban freight transportation.
14. What are the environmental consequences of increased ride-sharing?
15. Explore the concept of complete streets and their impact on urban design.
16. How can cities encourage carpooling among commuters?
17. Evaluate the safety of cycling infrastructure in urban areas.
18. Discuss the role of micro transit in filling gaps in public transportation.
19. What factors influence commuters' mode choice in urban settings?
20. Analyze the impact of ride-hailing services on taxi industries.
21. How can cities balance the needs of pedestrians, cyclists, and motorists?
22. Investigate the feasibility of hyperloop technology for urban transportation.

23. Discuss the challenges of maintaining aging subway systems.
24. Explore the impact of urban sprawl on transportation planning.
25. Assess the effectiveness of park-and-ride facilities.
26. What role do electric buses play in reducing emissions?
27. Analyze the impact of express lanes on highway congestion.
28. How can cities promote active transportation (walking and cycling)?
29. Discuss the social and economic implications of reducing parking spaces.
30. Evaluate the effectiveness of car-sharing programs.
31. What are the benefits of integrated multimodal transportation hubs?
32. Explore the impact of climate change on transportation infrastructure.
33. Assess the role of greenways and bike lanes in urban connectivity.
34. Discuss the challenges of implementing congestion pricing.
35. How can cities improve safety for vulnerable road users (e.g., pedestrians, cyclists)?
36. Analyze the impact of transportation network companies (TNCs) on public transit ridership.
37. What role does urban freight play in supporting local economies?
38. Evaluate the effectiveness of bus lane enforcement.
39. Discuss the potential of aerial gondolas as an urban transportation solution.
40. How can cities address the first-mile/last-mile problem?
41. Analyze the impact of electric trams and light rail systems.
42. What are the implications of reducing parking minimums for new developments?
43. Explore the role of urban logistics centers in efficient goods movement.
44. Discuss the challenges of implementing congestion zones.
45. How can cities encourage car-sharing among residents?
46. Evaluate the impact of transit-oriented design on property values.
47. Analyze the role of smart traffic management systems.
48. What are the social equity considerations in transportation planning?
49. Discuss the impact of e-scooter regulations on safety and accessibility.
50. How can cities promote sustainable freight delivery practices?

MULTIPLE-CHOICE QUESTIONS

1. What does the term "BRT" stand for in the context of urban transportation?
   a. Bus Rapid Transit
   b. Bicycle Route Technology
   c. Bridge Rail Transport
   d. Bullet Rail Transit
   e. Bus Routing Technique
2. Which of the following is an example of a sustainable mode of urban transportation?
   a. Personal car
   b. Helicopter
   c. Electric scooter
   d. Monster truck
   e. Rocket ship
3. What is the purpose of a "park-and-ride" facility in an urban area?
   a. To provide free parking for residents
   b. To encourage cycling
   c. To facilitate seamless transfers between different modes of transportation
   d. To host music concerts
   e. To showcase vintage cars

4. Which type of transportation system operates on dedicated tracks and typically has high capacity and speed?
   a. Tram
   b. Monorail
   c. Cable car
   d. Trolleybus
   e. Segway
5. What does the term "intermodal transportation" refer to?
   a. Using only one mode of transportation (e.g., only buses)
   b. Transporting goods via ships
   c. Seamless movement of passengers or freight across different modes (e.g., combining bus and train travel)
   d. Transporting goods using drones
   e. Transporting goods via horse-drawn carriages
6. Which of the following is an example of a non-motorized mode of transportation?
   a. Subway
   b. Taxi
   c. Bicycle
   d. Motorcycle
   e. Jet ski
7. What is the purpose of a "greenway" in urban planning?
   a. To designate areas for planting trees
   b. To create pedestrian-only zones
   c. To provide space for community gardens
   d. To establish linear parks or recreational corridors
   e. To build underground tunnels for transportation
8. Which transportation mode is characterized by having fixed routes, frequent stops, and low speeds?
   a. High-speed rail
   b. Light rail
   c. Express bus
   d. Hyperloop
   e. Cable car
9. What is the purpose of a "turning lane" at an intersection?
   a. To allow pedestrians to cross safely
   b. To provide space for U-turns
   c. To separate opposing traffic flows
   d. To accommodate left or right turns without disrupting through traffic
   e. To host street performances
10. Which transportation mode is commonly associated with elevated tracks and operates in urban areas?
    a. Commuter rail
    b. Tram
    c. Metro
    d. Cable car
    e. Rickshaw

11. What is the primary purpose of a "feeder bus" system in an urban public transportation network?
    a. To transport livestock within the city
    b. To connect suburban areas to the main transit system
    c. To provide luxury transportation for VIPs
    d. To operate exclusively during rush hours
    e. To transport oversized cargo
12. Which of the following factors can significantly impact traffic congestion in urban areas?
    a. The phase of the moon
    b. Road width and lane configuration
    c. The number of pigeons in the vicinity
    d. The color of traffic signals
    e. The average shoe size of pedestrians
13. What does the term "complete streets" refer to in urban planning?
    a. Streets that have no intersections
    b. Streets with decorative flower beds
    c. Streets designed for all users (pedestrians, cyclists, and motorists)
    d. Streets with synchronized traffic lights
    e. Streets with no parking allowed
14. Which type of transportation system is characterized by small, automated vehicles running on a dedicated track or guideway?
    a. Cable car
    b. Personal rapid transit (PRT)
    c. Hyperloop
    d. Segway
    e. Horse-drawn carriage
15. What is the purpose of a "transit-oriented development" (TOD) near public transportation stations?
    a. To build amusement parks
    b. To encourage car ownership
    c. To create walkable neighborhoods with mixed land uses
    d. To establish helicopter landing pads
    e. To promote horse-drawn carriage rides
16. What is the purpose of a "contraflow lane" in urban traffic management?
    a. To allow pedestrians to cross diagonally
    b. To create a dedicated lane for bicycles
    c. To reverse the direction of traffic flow during specific times (e.g., rush hours)
    d. To encourage synchronized dancing at intersections
    e. To accommodate monster trucks
17. Which of the following is an example of a demand-responsive transit service?
    a. Fixed-route bus system
    b. Subway
    c. Taxi
    d. High-speed rail
    e. Skateboard
18. What does the term "right of way" mean in traffic engineering?
    a. The privilege of honking loudly
    b. The legal right to drive on the left side of the road
    c. The priority given to certain road users (e.g., pedestrians, emergency vehicles)
    d. The obligation to yield to aggressive drivers
    e. The exclusive use of carpool lanes

19. Which transportation mode is associated with the concept of "last-mile connectivity"?
    a. Long-haul freight trains
    b. Cruise ships
    c. Airplanes
    d. Electric scooters
    e. Hot air balloons
20. What is the purpose of a "traffic calming measure" in residential neighborhoods?
    a. To encourage drag racing
    b. To reduce vehicle speeds and enhance safety
    c. To install speed bumps for thrill-seekers
    d. To promote synchronized honking
    e. To organize street food festivals
21. What is the purpose of a "roundabout" (also known as a traffic circle or rotary) in road design?
    a. To host car racing events
    b. To confuse drivers
    c. To facilitate smooth traffic flow and reduce collisions
    d. To showcase public art installations
    e. To serve as a picnic spot
22. Which of the following is an example of a "shared space" concept in urban design?
    a. A designated area for synchronized dance performances
    b. A street where pedestrians, cyclists, and vehicles share the same space without strict segregation
    c. A parking lot exclusively for monster trucks
    d. A tunnel for time-traveling pedestrians
    e. A runway for fashion shows
23. What does the term "modal shift" refer to in transportation planning?
    a. Changing the color of buses
    b. Shifting gears while driving
    c. Encouraging travelers to switch from one mode of transportation to another (e.g., from cars to bicycles)
    d. Installing musical instruments at bus stops
    e. Shifting transportation routes underground
24. Which transportation mode is characterized by using overhead wires to power the vehicles?
    a. Steam locomotives
    b. Electric buses
    c. Horse-drawn carriages
    d. Jet skis
    e. Skateboards
25. What is the purpose of a "transit signal priority" system at intersections?
    a. To give priority to ice cream trucks
    b. To synchronize traffic signals with music beats
    c. To prioritize public transit vehicles (buses, trams, etc.) to reduce delays
    d. To encourage synchronized hand-waving by pedestrians
    e. To display inspirational quotes on traffic lights

26. What is the purpose of a "bus rapid transit (BRT)" system, and how does it differ from a traditional bus service?
    a. BRT systems are designed for leisurely sightseeing tours, while traditional buses focus on speed.
    b. BRT systems use dedicated lanes and stations to provide faster, more efficient service compared to regular buses.
    c. BRT systems exclusively serve rural areas, while traditional buses operate in cities.
    d. BRT vehicles are double-decker buses, while traditional buses are single deck.
    e. BRT systems are only operational during weekends.
27. What roles do traffic impact assessments play in urban development projects?
    a. They evaluate the impact of traffic jams on local coffee shops.
    b. They assess the effect of new developments (e.g., housing complexes, shopping malls) on traffic flow and safety.
    c. They determine the best color for traffic signals.
    d. They measure the noise level of passing vehicles.
    e. They predict the arrival time of the next ice cream truck.
28. What is the purpose of a "transit fare integration" system in a multi-modal transportation network?
    a. To provide free rides to everyone
    b. To ensure that all buses have the same paint color
    c. To allow seamless transfers between different modes of transit using a single fare
    d. To organize synchronized flash mobs at transit stations
    e. To encourage passengers to bring their own music for onboard entertainment
29. Which transportation mode is associated with the concept of "micro-mobility"?
    a. Freight trains
    b. Cruise ships
    c. Electric scooters
    d. Hot air balloons
    e. Monster trucks
30. What does the term "complete trip" refer to in transportation planning?
    a. A journey that includes a stop at a bakery for fresh croissants
    b. A trip that involves multiple modes of transportation (e.g., walking, bus, subway)
    c. A trip taken by synchronized swimmers
    d. A trip where passengers receive complimentary ice cream
    e. A trip to the moon
31. What is the purpose of a "transit-oriented development" (TOD) near public transportation stations?
    a. To build amusement parks
    b. To encourage car ownership
    c. To create walkable neighborhoods with mixed land uses
    d. To establish helicopter landing pads
    e. To promote horse-drawn carriage rides
32. Which transportation mode is commonly associated with elevated tracks and operates in urban areas?
    a. Commuter rail
    b. Tram
    c. Metro
    d. Cable car
    e. Rickshaw

33. What is the purpose of a "greenway" in urban planning?
    a. To designate areas for planting trees
    b. To create pedestrian-only zones
    c. To provide space for community gardens
    d. To establish linear parks or recreational corridors
    e. To build underground tunnels for transportation
34. Which type of transportation system operates on dedicated tracks and typically has high capacity and speed?
    a. Tram
    b. Monorail
    c. Cable car
    d. Trolleybus
    e. Segway
35. What does the term "intermodal transportation" refer to?
    a. Using only one mode of transportation (e.g., only buses)
    b. Transporting goods via ships
    c. Seamless movement of passengers or freight across different modes (e.g., combining bus and train travel)
    d. Transporting goods using drones
    e. Transporting goods via horse-drawn carriages
36. Which of the following is an example of a non-motorized mode of transportation?
    a. Subway
    b. Taxi
    c. Bicycle
    d. Motorcycle
    e. Jet ski
37. What is the purpose of a "turning lane" at an intersection?
    a. To allow pedestrians to cross safely
    b. To provide space for U-turns
    c. To separate opposing traffic flows
    d. To accommodate left or right turns without disrupting through traffic
    e. To host street performances
38. Which transportation mode is characterized by having fixed routes, frequent stops, and low speeds?
    a. High-speed rail
    b. Light rail
    c. Express bus
    d. Hyperloop
    e. Cable car
39. What is the purpose of a "feeder bus" system in an urban public transportation network?
    a. To transport livestock within the city
    b. To connect suburban areas to the main transit system
    c. To provide luxury transportation for VIPs
    d. To operate exclusively during rush hours
    e. To transport oversized cargo
40. Which of the following factors can significantly impact traffic congestion in urban areas?
    a. The phase of the moon
    b. Road width and lane configuration
    c. The number of pigeons in the vicinity
    d. The color of traffic signals
    e. The average shoe size of pedestrians

41. What is the purpose of a “protected bike lane” in urban infrastructure?
    a. To provide a designated area for bike races
    b. To separate cyclists from motor vehicle traffic using physical barriers
    c. To encourage synchronized bell ringing by cyclists
    d. To showcase public art installations
    e. To host unicycling competitions
42. Which transportation mode is characterized by using magnetic levitation to eliminate friction between the vehicle and the track?
    a. Steam locomotives
    b. Maglev trains
    c. Horse-drawn carriages
    d. Skateboards
    e. Pogo sticks
43. What does the term “active transportation” refer to?
    a. Using only electric scooters
    b. Transporting goods via drones
    c. Any mode of transportation that involves physical activity (e.g., walking, cycling)
    d. Transporting passengers in hot air balloons
    e. Transporting goods using catapults
44. What is the purpose of a “traffic calming island” in the middle of a road?
    a. To host beach parties
    b. To slow down vehicle speeds and enhance safety
    c. To display synchronized fountain shows
    d. To encourage synchronized honking by drivers
    e. To serve as a picnic spot
45. Which transportation mode is associated with the concept of “paratransit”?
    a. Luxury limousines
    b. Personal helicopters
    c. Electric scooters
    d. Horse-drawn carriages
    e. Segways
46. What is the purpose of a “bus shelter” at a transit stop?
    a. To provide shade for waiting passengers
    b. To host impromptu dance performances
    c. To display synchronized LED light shows
    d. To encourage synchronized umbrella twirling
    e. To showcase local artwork
47. Which transportation mode is characterized by using a network of cables and pulleys to move vehicles up steep inclines?
    a. Funicular
    b. Jet ski
    c. Electric scooter
    d. Skateboard
    e. Hang glider
48. What is the purpose of a “bicycle-sharing system” in cities?
    a. To provide free bicycles to everyone
    b. To encourage synchronized bell ringing by cyclists
    c. To allow short-term bicycle rentals for commuting or leisure
    d. To organize bicycle races
    e. To showcase antique bicycles

49. Which transportation mode is commonly associated with the concept of "right of way"?
    a. Skateboards
    b. Taxis
    c. Pedestrians
    d. Monster trucks
    e. Unicycles
50. What is the purpose of a "traffic management center" in an urban area?
    a. To coordinate synchronized dance performances at intersections
    b. To monitor and control traffic flow using technology and real-time data
    c. To organize synchronized horn honking events
    d. To display inspirational quotes on electronic billboards
    e. To host traffic-themed art exhibitions

ANSWERS TO MULTIPLE-CHOICE QUESTIONS

1	a	11	b	21	c	31	c	41	b
2	c	12	b	22	a	32	c	42	b
3	c	13	c	23	c	33	d	43	c
4	b	14	b	24	b	34	b	44	b
5	c	15	c	25	c	35	c	45	a
6	c	16	c	26	b	36	c	46	a
7	d	17	c	27	b	37	d	47	a
8	c	18	c	28	c	38	c	48	c
9	d	19	d	29	c	39	b	49	c
10	c	20	b	30	b	40	b	50	b

FURTHER READING

Bertolini, L., & le Clercq, F. (2003). Urban development without more mobility by car? Lessons from Amsterdam, a multimodal urban region. *Environment and Planning A*, 35(4), 575–589.

Cervero, R., & Kockelman, K. (1997). Travel demand and the 3Ds: Density, diversity, and design. *Transportation Research Part D: Transport and Environment,* 2(3), 199–219.

Deakin, E. A. (1971). *The development of transport in modern England.* Routledge.

Gakenheimer, R. (1999). *Urban transport and land use policy in Latin America: A case study approach.* World Bank.

Hall, P. (1988). *Cities of tomorrow: An intellectual history of urban planning and design in the twentieth century.* Blackwell.

Hanson, S. (2010). *The geography of urban transportation* (3rd ed.). Guilford Press.

Hensher, D. A., & Button, K. J. (2001). *Handbook of transport strategy, policy & institutions* (Vol. 5). Elsevier.

Hoel, L. A., & Verhoef, E. T. (2007). *Handbook of transport economics* (Vol. 1). Elsevier.

Hymel, K. M., & Small, K. A. (2007). Pricing and rationing of roads with congestion. *Journal of Urban Economics*, 62(2), 187–203.

Kumar, S., & Barrett, A. G. (2008). An analysis of the environmental impacts of the transport sector in Delhi. *International Journal of Sustainable Development & World Ecology*, 15(4), 376–384.

Levinson, D., & Krizek, K. J. (2008). Accessibility and the journey-to-work: The case of the Atlanta region. *Urban Studies*, 45(1), 205–231.

Litman, T. (2006). *Transportation cost and benefit analysis: Techniques, estimates and implications.* Victoria Transport Policy Institute.

Loukaitou-Sideris, A., & Banerjee, T. (1998). *Urban design downtown: Poetics and politics of form.* University of California Press.

Lucas, K., & Blumenberg, E. (2006). *Stuck in traffic: Urban transport in Africa.* Zed Books.

Marshall, S., & Banister, D. (2000). *An introduction to sustainable transportation: Policy, planning and implementation*. Earthscan Publications.

Meyer, M. D., & Miller, E. J. (2001). *Urban transportation planning: A decision-oriented approach* (2nd ed.). McGraw-Hill.

Newman, P., & Kenworthy, J. (1999). *Sustainability and cities: Overcoming automobile dependence*. Island Press.

Noland, R. B., & Cowart, W. A. (2000). Analysis of metropolitan highway capacity and the growth in vehicle miles of travel. *Transportation Research Part A: Policy and Practice*, 34(9), 719–734.

O'Sullivan, A., Morrall, J., & Shearer, J. (2015). *Walking in cities: Quotidian mobility as urban theory, method, and practice*. McGill-Queen's Press-MQUP.

Pucher, J., & Renne, J. L. (2003). Socioeconomics of urban travel: Evidence from the 2001 NHTS. *Transportation Quarterly*, 57(3), 49–77.

Rodier, C., Givoni, M., & Ryley, T. (2010). *Transport and urban development: Challenges facing the urban world*. Edward Elgar Publishing.

Sustainable Urban Transport Project. (2001). *Sustainable urban transport: Four innovative directions*. Earthscan Publications.

Tolley, R., Turton, B., & Morrissey, J. (2002). *Transport project appraisal in the developing world*. Ashgate.

Vuchic, V. R. (2005). *Urban transit systems and technology*. John Wiley.

Wachs, M., & Kumagai, T. G. (1973). *Physical accessibility as a social indicator for transportation planning: Application to the elderly and children in the Los Angeles metropolitan area*. Transportation Research Record.

Xie, F., Levinson, D., & Zhu, S. (2011). *The fundamental law of road congestion: Evidence from US cities*. Transportation Research Part A: Policy and Practice.

Yanarella, E. J., & Levine, M. G. (2012). *Transforming cities with transit: Transit and land-use integration for sustainable urban development*. Island Press.

14 Smart Cities

14.1 THE KEY COMPONENTS OF SMART CITIES

A smart city is a technologically modern city that uses a network of sensors, data analysis, and communication technology (ICT) to improve the overall function and efficiency of the city. This includes everything from traffic management and public services to public safety and citizen engagement. The city leverages various digital solutions, including information and communication technologies (ICT) and the Internet of Things (IoT), to enhance urban living. Smart city initiatives go beyond merely using technology for better resource utilization and reduced emissions; they aim to create interconnected environments that benefit both residents and businesses. Smart cities integrate data-driven decision-making, connectivity, and innovation to optimize services, infrastructure, and community well-being.

Smart cities are dynamic ecosystems that leverage technology to enhance urban living. These cities integrate various components to create efficient, sustainable, and interconnected environments. The key components of smart cities generally include the aspects as shown in Table 14.1.

Smart cities weave these components together, creating a harmonious blend of technology, sustainability, and quality of life for their inhabitants. Smart cities thrive on interconnected components, enhancing urban life through technology.

14.2 CONCEPTS OF SMART CITIES

The urban landscape is undergoing metamorphosis. Cities are no longer just concrete jungles; they are evolving into interconnected ecosystems powered by information technology (IT) and the internet. This digital transformation fosters the rise of smart cities, where technology seamlessly integrates with infrastructure and services, and smart transportation becomes a cornerstone of efficient and sustainable urban mobility.

14.2.1 The Information Backbone: The Power of IT and the Internet

Imagine strolling through a city where traffic lights adjust seamlessly to optimize flow, trash bins send alerts when they are full, and parking spaces mysteriously indicate their availability. This is not a scene from a futuristic movie – it is the reality of smart cities. These urban environments leverage technology to enhance the lives of residents, improve sustainability, and streamline operations. Let us explore the enchanting world of smart cities, where high-speed internet connectivity weaves together a tapestry of innovation.

DOI: 10.1201/9781032664613-14

TABLE 14.1
Key Components of Smart Cities

Smart infrastructure	• The backbone of a smart city, including roads, bridges, utilities, and public facilities. • Smart infrastructure incorporates sensors and connectivity to monitor performance, detect issues, and optimize maintenance.
Smart buildings	• Energy-efficient structures equipped with automation systems. • Features include smart lighting, HVAC, security, and occupancy sensors. • Buildings interact with the grid, adjusting energy usage based on demand and availability.
Smart transportation	• Enhances mobility and reduces congestion. • *Intelligent Traffic Management:* Adaptive traffic lights, real-time route optimization, and smart parking systems. • *Public Transit Integration:* Seamless connectivity between buses, trains, and other modes. • *Micromobility Solutions:* Electric scooters, bike-sharing, and pedestrian-friendly infrastructure.
Smart energy	• Focuses on efficient energy production, distribution, and consumption. • Includes renewable energy sources, smart grids, and demand-side management. • Smart meters enable real-time monitoring and billing.
Smart healthcare	• Leverages technology for better patient care and resource management. • Electronic health records, telemedicine, and predictive analytics improve health outcomes. • Emergency response systems enhance safety.
Smart technology	• Encompasses IoT devices, data analytics, and connectivity. • Smartphones, wearables, and smart home devices empower citizens. • Data-driven insights inform decision-making.
Smart governance	• Efficient administration through digital platforms. • E-governance, open data, and citizen engagement. • Transparent processes and responsive services.
Smart education	• Technology-enabled learning environments. • Virtual classrooms, personalized content, and lifelong learning. • Education accessible to all, regardless of location.
Smart citizens	• Engaged, informed, and active residents. • Participatory platforms, community initiatives, and social innovation. • Citizens contribute to data collection and problem-solving.

Smart cities lie in their digital nervous system – a network of interconnected devices, sensors, and data streams. This is the way the system works:

- *Internet of Things (IoT):* Imagine lampposts, buses, and waste bins equipped with tiny sensors. These sensors collect specific data – traffic patterns, waste levels, energy consumption – and transmit it wirelessly. The city becomes a living organism, constantly sensing and responding.
- *Real-Time Data Collection:* Pavements, buildings, and vehicles become data sources. Sensors embedded in roads monitor traffic density, while smart meters track water and electricity usage. All this information flows into a central hub, creating a real-time snapshot of the city's pulse.

14.2.1.1 Traffic Management and Efficiency

- *Dynamic Traffic Lights:* Traditional traffic lights follow fixed schedules, leading to congestion during rush hours. In a smart city, traffic lights adjust in real time based on traffic flow. Algorithms optimize signal timings, reducing waiting times and easing gridlock.
- *Smart Parking:* Ever circled endlessly looking for a parking spot? Smart cities have you covered. Parking spaces communicate their availability to your smartphone. No more frustration – just seamless parking.

14.2.1.2 Waste Management and Sustainability

- *Smart Bins:* Trash bins send alerts when they are full. Waste collection routes become more efficient, reducing unnecessary trips and emissions. Cleaner streets benefit both residents and the environment.
- *Energy Efficiency:* Smart grids balance electricity supply and demand. Homes and buildings adjust their energy usage based on real-time rates. Energy-efficient street lighting saves costs and reduces the city's carbon footprint.

14.2.1.3 Informed Decision-Making

- *Data-Driven Insights:* City officials analyze the data pouring in. They spot trends, anticipate needs, and make informed decisions. Should a new park be built? Where should bike lanes go? Data holds the answers.
- *Resource Allocation:* Smart cities allocate resources efficiently. Water distribution, emergency services, and public transportation benefit from real-time data. Residents enjoy better services without waste.

14.2.1.4 Quality of Life and Economic Growth

- *Improved Quality of Life:* Reduced traffic congestion means less stress for commuters. Cleaner air improves health. Smart healthcare systems provide timely medical assistance. The magic of technology enhances well-being.
- *Economic Opportunities:* Smart cities attract businesses. High-tech infrastructure encourages innovation. Job creation follows. The city thrives economically, benefiting everyone.

14.2.2 Smart Transportation: A Symphony of Technology and Efficiency

Smart transportation is the beating heart of a smart city. Technologies like smart traffic management systems (STMS) analyze data on traffic flow, accidents, and road closures in real-time. This allows for dynamic adjustments to traffic signals, rerouting congested areas, and providing real-time traffic updates to drivers. Integration with autonomous vehicles (AVs) further elevates efficiency and safety. AVs, guided by advanced sensors and algorithms, can navigate streets with precision, minimizing human error and potentially reducing accidents.

14.2.3 Micromobility and Multimodality: A Connected Ecosystem

Smart cities are not solely about autonomous cars. Micromobility options like electric bikes and scooters, integrated with the broader transportation network, encourage short-distance trips with minimal environmental impact. Seamlessly dockless bike-sharing programs and on-demand scooter rentals, facilitated by smartphone apps, offer convenient and sustainable "first-mile/last-mile" solutions, connecting users to public transportation hubs. This multimodal approach creates a transportation network that caters to diverse needs, reducing reliance on private cars.

14.2.4 Citizen Engagement: A Two-Way Street

The flow of information in a smart city is not a one-way street. Open data platforms allow residents to access real-time data on air quality, noise levels, and even energy consumption. This transparency empowers citizens to hold city authorities accountable and actively participate in shaping their city's future. Mobile apps facilitate two-way communication, allowing residents to report problems like potholes or malfunctioning streetlights, ensuring a more responsive and efficient urban environment.

14.2.5 Challenges of the Work: Privacy, Security, and Accessibility

The digital work of a smart city faces several challenges. Data privacy is paramount. Robust cybersecurity measures are crucial to protecting sensitive user data from cyberattacks. Ensuring equitable access to technology is essential. Not everyone owns a smartphone or has the digital literacy to navigate these systems. Cities must bridge the digital divide by providing affordable internet access and user-friendly interfaces for all residents.

The intertwining among smart cities, smart transportation, and information technology holds immense potential. With this situation we can observe that there are often overlaps between the three chapters associated with the above issues in this book. On the other hand, the confluence of this intertwining can create a more efficient, sustainable, and livable urban environment for all. It is about creating a symphony where innovation and human well-being are in perfect harmony.

14.2.6 Smart Cities: Where Technology Meets Urban Living

The way we live in cities is on the verge of a significant transformation. Imagine a cityscape where traffic flows smoothly, buildings adjust their energy consumption based on real-time needs, and residents can access essential services with a tap on their smartphones. This is the vision of smart cities – urban environments that leverage information and communication technology (ICT) to create a more efficient, sustainable, and livable future.

14.2.7 The Core of a Smart City: Data as the Driving Force

The heart of a smart city lies in its ability to collect, analyze, and utilize data. Sensors embedded throughout the city – in roads, buildings, and even trash cans – gather real-time information on everything from traffic flow and air quality to energy consumption and noise levels. This data becomes the lifeblood of the smart city, feeding into a central hub for analysis.

14.2.8 Benefits of a Smart City Ecosystem

By harnessing the power of data and technology, smart cities offer a multitude of benefits:

- *Enhanced Efficiency:* Traffic management systems optimize traffic flow, reducing congestion and travel times. Buildings with smart grids adjust energy consumption based on real-time needs, leading to reduced energy waste.
- *Improved Sustainability:* Smart waste management systems optimize collection routes and alert authorities when bins are full. Renewable energy sources can be integrated into the power grid, reducing reliance on fossil fuels.
- *Enhanced Public Services:* Smart city initiatives can improve public services like waste collection, emergency response times, and public safety through real-time data and connected infrastructure.
- *Citizen Engagement:* Open data platforms allow residents to access information and participate in shaping their city's future. Mobile apps facilitate two-way communication between citizens and authorities.

14.2.8.1 Smart Transportation

A key component of smart transportation is a cornerstone of any smart city. This includes:

- *Smart Traffic Management Systems (STMS):* These systems analyze data to dynamically adjust traffic signals, reroute traffic around congestion, and provide real-time information to drivers.

- *Autonomous Vehicles (AVs):* These self-driving vehicles, still in development, promise to revolutionize urban transportation by increasing safety and efficiency.
- *Micromobility Options:* Electric bikes, scooters, and dockless bike-sharing programs offer convenient and sustainable "first-mile/last-mile" solutions, connecting users to public transportation hubs.

14.2.8.2 Challenges of Smart Cities

- *Privacy Concerns:* Data security is paramount. Robust cybersecurity measures are needed to protect sensitive user data.
- *Equity and Accessibility:* Smart city initiatives should be inclusive, ensuring all residents have access to technology and its benefits.
- *Infrastructure Investment:* Cities need to invest in upgrading infrastructure to accommodate smart technologies, such as installing sensors and creating dedicated lanes for AVs or micromobility options.

14.2.8.3 Building a Smarter Future.

The successful development of smart cities relies on collaboration between various stakeholders:

- *Technology Companies:* Developing user-friendly and secure platforms that prioritize data privacy.
- *Policymakers:* Creating regulations that foster innovation while addressing privacy, security, and equity concerns.
- *Urban Planners:* Designing cities that integrate smart technologies seamlessly with existing infrastructure.
- *Citizens:* Actively participating in shaping the vision and implementation of smart city initiatives.

The concept of smart cities holds immense potential to transform our urban landscapes. We can create cities that are not only technologically advanced but also sustainable, and livable, and cater to the needs of all their residents. The future of our cities lies in becoming smarter, not just in terms of technology, but in terms of how they serve the people who live in them. One of the examples of successful smart city projects is shown in Box 14.1.

BOX 14.1 LONDON: PIONEERING SMART CITY INITIATIVES

London, the vibrant capital of the United Kingdom, has been at the forefront of smart city development. Its commitment to leveraging technology for the benefit of its citizens has led to several remarkable initiatives. Here are some key examples:

- *London Development Database (LDD):* The LDD is a collaborative project that tracks construction permits and their various phases in real time. Citizens can access this information through a user-friendly web interface. By providing transparency and real-time updates, the LDD empowers residents, developers, and policymakers alike.
- *Digital Skills Training:* Recognizing the importance of digital literacy, London has implemented programs to equip diverse populations with essential technology skills. These initiatives focus on teaching residents how to use devices like iPads and tablets effectively. By bridging the digital divide, London ensures that everyone can participate in the city's smart ecosystem.

- *Smart Electricity Meters:* London has embraced smart meters, which collect real-time data on electricity consumption. This data-driven approach enables more efficient energy management, reduces waste, and contributes to a greener city. By empowering citizens with insights into their energy usage, London encourages sustainable practices.
- *Transport Innovations:* London's commitment to smart transportation includes projects like the Heathrow pods and driverless cars. The Heathrow pods provide an automated, eco-friendly transport link between the city and the airport. Meanwhile, driverless cars promise safer, more efficient urban mobility. These advancements enhance connectivity and reduce congestion.
- *Energy Efficiency:* Smart buildings play a crucial role in London's sustainability efforts. Real-time monitoring systems ensure structural integrity and prompt maintenance when needed. For instance, sensors detect issues like leaking water pipes, allowing timely repairs. Citizens can also report infrastructure problems, such as potholes, through this integrated system.
- *Urban Resilience:* London's smart city approach extends to disaster preparedness. By leveraging data analytics and IoT, the city enhances its resilience against natural calamities. Whether it is predicting floods, managing emergency responses, or safeguarding critical infrastructure, London's smart systems contribute to a safer urban environment.

London's success story demonstrates that smart cities are not just about technology; they are about improving lives. By integrating data, connectivity, and citizen engagement, London continues to evolve as a model for other cities worldwide. As we look to the future, collaboration, adaptability, and sustainability will remain essential for creating truly smart, resilient urban spaces.

Sources: smartcitystyle.com, stateofgreen.com, sustainabilitymag.com

14.3 SMART INFRASTRUCTURE AND APPLICATIONS

Smart city applications leverage the vast data collected by IoT devices and sensors to enhance the efficiency, sustainability, and quality of life in urban environments. From traffic management and energy optimization to public service delivery and emergency response, these data-driven applications are transforming the way cities operate and enabling more informed decision-making to address the challenges of rapid urbanization. Here are some examples of the applications in smart cities.

14.3.1 Smart Grid for Energy Management

Smart grids are an innovative method of energy management that leverages digital technology to monitor and manage the transport of electricity from all generation sources to meet the varying electricity demands of end users. Smart grids coordinate the needs and capabilities of all generators, grid operators, end-users, and electricity market stakeholders to operate all parts of the system as efficiently as possible, minimizing costs and environmental impacts while maximizing system reliability, resilience, and stability. Discussion in more detail about Smart Grid and Energy Management was given in Urban Energy Supply Systems, and therefore will not be discussed further in this section.

14.3.2 Intelligent Transportation System for Traffic Management

Intelligent Transportation Systems (ITS) are advanced applications that aim to provide innovative services relating to different modes of transport and traffic management. These systems use data analysis, communications, and sensor technologies to improve the efficiency and safety of transportation networks. ITS can help manage congestion, reduce accidents, optimize fuel consumption, and enhance the overall commuter experience. ITS represents a significant leap forward in our ability to manage and optimize transportation networks. Leveraging advanced technologies such as data analysis, communications, and sensors, ITS can boost the efficiency and safety of transportation, delivering numerous benefits to commuters, transport operators, and cities alike.

14.3.2.1 Understanding ITS

ITS is about using technology to solve transportation problems. It involves the collection, processing, and dissemination of information related to transportation systems. This information can be used to manage congestion, reduce accidents, optimize fuel consumption, and improve the overall commuter experience.

14.3.2.2 Components of ITS

ITS is generally composed of several key components. The components are summarized in Table 14.2.

14.3.2.3 Benefits of ITS

ITS are revolutionizing urban mobility through the integration of advanced technologies and real-time data. Monitoring traffic patterns, optimizing traffic signals, and providing personalized commuter information allow ITS to reduce congestion, improve safety, and enhance the overall transportation experience. The benefits of ITS are shown in Table 14.3.

TABLE 14.2
Key Components of ITS

Sensors	These are devices that collect data about the transportation system. They can be installed in vehicles, along roads, or in other strategic locations. Sensors can collect a wide range of data, including vehicle speed, location, and direction; road conditions; weather conditions; and more.
Communications networks	These networks enable the transfer of data between different parts of the ITS. This can include communication between vehicles (V2V), between vehicles and infrastructure (V2I), and between different parts of the infrastructure.
Data analysis tools	These tools take the data collected by the sensors and turn it into useful information. This can involve a wide range of techniques, from simple statistical analysis to complex machine learning algorithms.
User interfaces	These are the tools that present the information to the users of the ITS. This can include everything from traffic light signals and road signs to smartphone apps and in-car navigation systems.

TABLE 14.3
Benefits of ITS

Congestion management	By providing real-time information about traffic conditions, ITS can help drivers avoid congested areas, reducing travel times and improving the overall efficiency of the transportation system.
Accident reduction	ITS can help prevent accidents by providing drivers with timely information about potential hazards, such as icy roads or broken-down vehicles. In addition, ITS can help emergency services respond more quickly when accidents do occur.
Fuel optimization	By helping drivers avoid congestion and choose the most efficient routes, ITS can reduce fuel consumption, saving drivers money and reducing environmental impact.
Improved commuter experience	By providing accurate and timely information, ITS can make commuting less stressful and more predictable. This can improve quality of life for commuters and make public transportation a more attractive option.

14.3.2.4 The Future of ITS

As technology continues to advance, the potential of ITS is only set to grow. Future developments could include the integration of ITS with autonomous vehicles, the use of big data and artificial intelligence (AI) to further improve traffic management, and the development of smart cities where all elements of the urban environment are interconnected and optimized.

Intelligent transportation systems represent a powerful tool for improving the efficiency and safety of our transportation networks. By utilizing advanced technologies, ITS have the potential to transform the way we travel, making our journeys more efficient, safer, and more enjoyable. However, it is important to ensure that these technologies are implemented in a way that is inclusive and benefits all road users.

14.3.3 Smart Buildings for Efficient Resource Use

Smart buildings are structures that use automated processes to control the building's operations, including heating, ventilation, air conditioning, lighting, security, and other systems. A smart building uses sensors, actuators, and microchips to collect data and manage it according to the building's functions and services. This infrastructure helps owners, operators, and facility managers improve asset reliability and performance, reducing energy use, optimizing how space is used, and minimizing the environmental impact of buildings.

Smart buildings represent a significant advancement in the management and operation of building infrastructure. They leverage automated processes, sensors, actuators, and microchips to control various operations, including heating, ventilation, air conditioning (HVAC), lighting, security, and more. This not only improves the performance and reliability of these assets but also optimizes energy use, space utilization, and reduces environmental impact. An example of a smart building project is shown in Box 14.2.

BOX 14.2 CONFLUENCE PARK, SAN ANTONIO, UNITED STATES: A SMART BUILDING BEYOND ENERGY EFFICIENCY

Confluence Park, located in San Antonio, Texas, is a remarkable example of an eco-friendly smart building project. Designed to blend seamlessly with its natural surroundings, Confluence Park serves as an educational and recreational hub while showcasing sustainable practices.

The park's roof features a series of sculptural pavilions that collect rainwater. This harvested water is then filtered and stored in underground cisterns, providing a sustainable water source for irrigation and educational programs. Solar panels integrated into the park's canopies generate renewable energy. These canopies not only provide shade but also power the park's lighting and other electrical needs.

An environmental education center within the park features interactive exhibits. Visitors learn about water conservation, native plants, and sustainable design principles. The building itself serves as a teaching tool, demonstrating energy-efficient construction techniques.

Confluence Park's landscape design includes native plant species that require minimal water and maintenance. The verdant green walls of the atrium and common areas improve air quality and provide a calming ambiance. Smart irrigation systems adjust watering schedules based on weather conditions and soil moisture levels, minimizing water waste and ensuring efficient plant growth.

Throughout the Park, art installations engage visitors and highlight environmental themes. These pieces integrate seamlessly with the natural landscape, emphasizing the importance of sustainability. Confluence Park serves as a gathering place for community events, workshops, and outdoor activities, fostering environmental awareness and appreciation.

Sources: neuroject.com, sustainability-success.com, smartcitystyle.com

14.3.3.1 Components of Smart Buildings

The primary components of a smart building are shown in Table 14.4.

14.3.3.2 Benefits of Smart Buildings

Smart buildings offer numerous benefits, including improved energy efficiency, enhanced comfort and productivity for occupants, reduced operational costs, and a significant contribution to environmental sustainability and urban development, as shown in Table 14.5.

14.3.3.3 The Future of Smart Buildings

Future developments of smart buildings could include the integration of renewable energy sources, the use of AI and machine learning (ML) for predictive maintenance, and the development of buildings that can adapt and respond to their occupants' needs in real time.

Smart buildings offer the potential to improve asset reliability and performance, reduce energy use, optimize space utilization, and minimize environmental impact. However, it is important to

TABLE 14.4
Primary and Auxiliary Components of a Smart Building

	Primary components
Sensors	These are devices that collect data about various aspects of the building, such as temperature, humidity, light levels, occupancy, and energy use.
Actuators	These are devices that perform actions based on the data collected by the sensors. For example, an actuator might adjust the temperature or lighting based on occupancy data.
Microchips	These are used to process the data collected by the sensors and control the actuators.
Control systems	These are systems that manage the overall operation of the building. They use the data from the sensors and the capabilities of the actuators to control the building's systems and optimize their performance.
	Auxiliary components
Integrated Building Management Systems (IBMS)	An IBMS is a centralized, interlinked network of systems that is used to monitor and control the various electrical and mechanical systems in a building. This can include HVAC, lighting, power systems, fire systems, and security systems. An IBMS allows for more efficient operation and management of these systems.
Cybersecurity measures	With the increasing interconnectivity of systems in a smart building, cybersecurity becomes paramount. Protecting the data and systems from cyber threats is an often overlooked but critical component of a smart building.
User interfaces and experience	The interfaces through which occupants interact with the building systems, such as apps or control panels, are a crucial component. These need to be user-friendly and intuitive to ensure occupants can effectively use and benefit from the smart features.
Sustainability features	Beyond energy efficiency, smart buildings can incorporate other sustainability features that are often overlooked. This can include water conservation systems, waste management, use of sustainable materials, and integration with green spaces.
Accessibility features	Smart buildings should also consider accessibility features to ensure they are usable and beneficial for all occupants. This can include features for people with disabilities, such as voice-activated controls, braille signage, and wheelchair-friendly design.

TABLE 14.5
Benefits of Smart Buildings

Improved asset reliability and performance	By continuously monitoring and managing the building's systems, smart buildings can improve the reliability and performance of these assets. This can reduce maintenance costs and extend the life of the equipment.
Reduced energy use	Smart buildings can significantly reduce energy use by optimizing the operation of HVAC, lighting, and other systems. This not only saves money but also reduces the building's carbon footprint.
Optimized space utilization	By monitoring occupancy and usage patterns, smart buildings can optimize the use of space. This can improve the comfort and productivity of occupants and potentially reduce the need for additional space.
Minimized environmental impact	By reducing energy use and optimizing resource utilization, smart buildings can minimize their environmental impact.

ensure that these technologies are implemented in a way that is sustainable and benefits all occupants. This includes considering issues of digital divide and ensuring accessibility in smart building development. As we move forward, smart buildings will be a crucial element in shaping the future of our built environment.

14.3.4 Smart Water System for Water Conservation

Smart water systems are designed to gather meaningful and actionable data about the flow, pressure, and distribution of a city's water supply. Using this data, these systems can identify leaks or inefficiencies, monitor the condition of the infrastructure, and ensure a high quality of water. This leads to more efficient use of water resources, lower operating costs, extended life of the infrastructure, and improved customer service.

Smart water systems are transforming the way municipalities manage and distribute their most precious resource – water. These advanced systems leverage a network of sensors, data analytics, and automated control mechanisms to gather real-time information about the flow, pressure, and quality of a city's water supply, empowering water managers to make more informed and proactive decisions. Smart water systems are the IoT devices and sensors deployed throughout the water infrastructure. These can include flow meters, pressure sensors, water quality monitors, and even smart water meters installed in individual homes and businesses. These sensors collect a wealth of data about the water distribution network, providing a comprehensive view of the system's performance and condition.

The data collected by these sensors is then fed into sophisticated data analytics platforms, where it is processed and analyzed to identify patterns, detect anomalies, and generate actionable insights. For example, by analyzing flow and pressure data, smart water systems can detect the presence of leaks in the distribution system, often before they become visible to the naked eye. This enables water managers to quickly locate and address these issues, reducing water waste and preventing costly infrastructure damage.

Similarly, water quality data can be used to monitor the condition of the water supply, ensuring that it meets the necessary standards for public health and safety. If any irregularities are detected, the system can trigger alerts, allowing water treatment plants to adjust their processes and maintain the high quality of the water.

Beyond just monitoring the system, smart water technologies also enable automated control and optimization. Based on the data collected, the system can dynamically adjust the operation of pumps, valves, and other components to optimize water flow, pressure, and distribution, reducing energy consumption and operational costs.

This level of real-time monitoring and control has significant benefits for both water utilities and the communities they serve. For water utilities, smart water systems can lead to more efficient use of water resources, lower operating costs, and extended infrastructure lifespan. Quickly identifying and addressing issues allows these systems to reduce water losses and minimize the need for costly repairs or replacement of aging pipes and equipment. For consumers, smart water systems can improve service quality, providing more reliable and consistent water pressure and delivery. Additionally, the ability to detect leaks and monitor water usage empowers consumers to better understand and manage their water consumption, potentially leading to cost savings on their utility bills.

One of the primary concerns is the issue of data security and privacy. The vast amounts of data collected by these systems, including personal information from smart water meters, can be vulnerable to cyber threats and data breaches if not properly secured. Robust cybersecurity measures, such as encryption, access controls, and regular software updates, are essential to protect the integrity of the system and the privacy of the data.

Another challenge is the need for seamless integration and interoperability between the various components of the smart water system, including sensors, data analytics platforms, and control systems. Establishing common standards and protocols is crucial to ensure that all the different elements of the system can communicate and work together efficiently. The successful deployment of smart water systems requires significant upfront investment in infrastructure and technology, which can be a barrier for some municipalities, particularly those with limited budgets. To address this, some cities are exploring alternative financing models, such as public–private partnerships or performance-based contracts, to spread the costs and risks across multiple stakeholders. Despite these challenges, the benefits of smart water systems are clear, and their adoption is expected to continue growing as cities and water utilities recognize the value of data-driven, efficient water management. As technology continues to develop and become more accessible, we can expect more innovative applications and solutions emerge, further enhancing the resilience, sustainability, and livability of our urban environments.

The key components of a smart water system typically include the ones shown in Table 14.6.

The integration and seamless operation of these components are crucial for the effective implementation of a smart water system that can deliver the expected benefits of improved efficiency, reduced costs, and enhanced water management.

Smart water systems offer numerous benefits, including efficient use of water resources, reduced operating costs, improved infrastructure lifespan, enhanced water quality, and superior customer service through real-time data and analytics. The benefits are summarized in Table 14.7.

14.3.5 Smart Healthcare

Smart healthcare systems use IoT devices, AI, and data analytics to improve the quality and efficiency of healthcare services. This can include remote health monitoring, telemedicine services, and smart hospitals that leverage technology for patient care and hospital management. For example, wearable devices can monitor a patient's health in real time, providing valuable data to healthcare professionals and alerting them to potential health issues.

14.3.5.1 IoT and AI in Healthcare

The healthcare industry is undergoing a transformative shift, as the integration of emerging technologies like the Internet of Things (IoT), AI, and data analytics are giving rise to smart healthcare systems that are redefining the delivery of medical services. These innovative systems are designed to leverage the power of technology to improve the quality, accessibility, and efficiency of healthcare, ultimately enhancing patient outcomes and experiences.

TABLE 14.6
The Key Components of Smart Water Systems

Sensors and metering devices	• Flow meters: Measure the volume of water flowing through pipes. • Pressure sensors: Monitor the water pressure at various points in the distribution network. • Water quality sensors: Detect parameters like pH, chlorine levels, turbidity, and contaminants. • Smart water meters: Track water usage at the consumer level, providing detailed consumption data.
Communication infrastructure	• Wireless or wired communication networks: Enable the transmission of data from sensors and meters to a central control system. • Cellular, Wi-Fi, or low-power wide-area network (LPWAN) technologies: Facilitate reliable and secure data transfer.
Data management and analytics platform	• Data aggregation and storage systems: Collect, store, and manage the vast amounts of data generated by the sensors and meters. • Advanced data analytics and visualization tools: Identify patterns, detect anomalies, and generate insights from the collected data. • Predictive modeling and optimization algorithms: Forecast demand, identify leaks, and optimize water distribution and treatment processes.
Automated control and optimization systems	• Intelligent control valves and pumps: Dynamically adjust water flow and pressure based on system conditions. • Automated treatment and disinfection processes: Ensure consistent water quality by adjusting chemical dosages and treatment parameters. • Remote monitoring and control capabilities: Allow water managers to monitor and manage the entire system from a central location.
Asset management and maintenance systems	• Condition monitoring and predictive maintenance models: Assess the health of water infrastructure components and predict when maintenance is required. • Geographic Information Systems (GIS): Integrate spatial data to visualize the distribution network and plan infrastructure upgrades or replacements.
Cybersecurity and data privacy measures	• Encryption and access controls: Protect the integrity and confidentiality of the data collected by the smart water system. • Incident response and disaster recovery plans: Ensure the resilience of the system in the face of cyber threats or natural disasters.

TABLE 14.7
The Benefits of Smart Water Systems in Cities

Efficient use of water resources	By providing real-time data on water usage and identifying leaks or inefficiencies, smart water systems can significantly reduce water waste. This leads to more efficient use of water resources and can help conserve this vital resource.
Lower operating costs	By optimizing the operation of the water infrastructure, smart water systems can reduce energy usage and maintenance costs. This can lead to significant cost savings for water utilities and consumers.
Extended infrastructure life	By providing timely data on the condition of the water infrastructure, smart water systems can help detect issues early, before they become major problems. This can extend the life of the infrastructure and reduce the need for costly repairs or replacements.
Improved water quality	Smart water systems can continuously monitor water quality parameters, ensuring that the water supply meets health and safety standards. This can improve public health and customer satisfaction.
Enhanced customer service	By providing real-time data on water usage, smart water systems can provide customers with detailed information about their water consumption. This can help customers better understand their water usage and make more informed decisions about water conservation.

The core of smart healthcare systems is the various IoT devices and sensors that are being deployed in diverse medical settings. These range from wearable health monitors that track a patient's vital signs, activity levels, and other physiological data, to smart hospital beds that can automatically adjust their position to improve patient comfort and prevent bedsores. Continuously collecting and transmitting real-time data, these IoT devices provide healthcare professionals with a more comprehensive and current understanding of a patient's health status, enabling more proactive and personalized care.

The data generated by these IoT devices is then fed into sophisticated AI-powered analytics platforms, which use ML algorithms to identify patterns, detect anomalies, and generate insights that can inform clinical decision-making. For example, an AI-based system might analyze a patient's vital signs and activity data to detect early signs of a medical condition, triggering an alert that allows healthcare providers to intervene before the situation escalates. Similarly, these analytics platforms can be used to optimize hospital operations, such as predicting patient flow, improving resource allocation, and enhancing workflow efficiency.

14.3.5.2 Telemedicine Services

Telemedicine is another significant aspect of smart healthcare. It enables patients to consult with healthcare professionals from the comfort of their homes. This is particularly beneficial for patients living in remote areas or those with mobility issues. Telemedicine services can include virtual consultations, remote patient monitoring, and digital prescriptions. It is one of the transformative applications of smart healthcare systems. Through the utilization of IoT devices and video conferencing technologies, healthcare providers can now offer virtual consultations and monitor patients from the comfort of their own homes. This not only improves access to care, especially for those in remote or underserved areas, but it also reduces the burden on hospital resources and minimizes the risk of exposure to infectious diseases.

14.3.5.2.1 Smart Hospitals

Smart hospitals are emerging as a testament to the power of these technologies. These healthcare facilities are designed to integrate a wide range of IoT devices, from smart beds and monitoring systems to automated medication dispensers and robotic assistants. By automating various administrative and clinical tasks, smart hospitals can improve the efficiency of operations, reduce medical errors, and enhance the overall patient experience.

Smart hospitals take the concept of smart healthcare a step further. They integrate technology into every aspect of hospital operations, from patient care to hospital management. For example, smart beds can automatically adjust to enhance patient comfort, robots can assist in surgeries, and AI can manage hospital resources efficiently.

14.3.5.2.2 Wearable Devices and Remote Health Monitoring

Wearable devices are perhaps the most visible element of smart healthcare. They monitor a patient's health in real-time, providing valuable data to healthcare professionals. For instance, a smartwatch can track a user's heart rate and alert them if any irregularities are detected. Similarly, a glucose monitoring patch can continuously monitor blood sugar levels in diabetic patients, alerting them when insulin is needed.

One of the key applications of smart healthcare is remote health monitoring. Patients no longer need to visit a healthcare facility for routine check-ups. Instead, their health status is continuously monitored via wearable or implantable devices. If any abnormality is detected, healthcare professionals are alerted immediately. This allows for timely intervention, reducing the risk of complications.

However, one of the primary concerns is the issue of data security and privacy. The sensitive nature of medical data, combined with the interconnected nature of these systems, makes them a prime target for cyber threats, such as data breaches and ransomware attacks. Robust cybersecurity measures, including encryption, access controls, and regular software updates, are essential to protect patient information and maintain the integrity of the healthcare system. Another challenge is the need for seamless integration and interoperability between the various components of the smart healthcare system. Ensuring that IoT devices, data analytics platforms, and clinical systems can communicate and share data effectively is crucial for realizing the full potential of these technologies. This requires the development of common standards and protocols, as well as the implementation of comprehensive data governance frameworks.

The successful adoption of smart healthcare systems requires significant upfront investment in infrastructure and technology, which can be a barrier for some healthcare providers, particularly those with limited resources. To address this, some organizations are exploring alternative funding models, such as public–private partnerships or value-based reimbursement schemes, to spread the costs and risks across multiple stakeholders.

Despite these challenges, the potential benefits of smart healthcare systems are substantial, and their adoption is expected to continue growing as healthcare providers recognize the value of data-driven, technology-enabled care.

Smart healthcare is a holistic, technology-driven approach to health management. It essentially utilizes IoT devices, AI, and data analytics to enhance the quality and efficiency of healthcare services. The goal is to provide personalized, timely, and effective patient care while optimizing resources.

Smart healthcare systems are transforming the way medical services are delivered, leveraging IoT devices, AI, and data analytics to improve patient outcomes, enhance operational efficiency, and reduce healthcare costs. These systems are paving the way for a more proactive, personalized, and accessible healthcare landscape through continuously monitoring patient health, optimizing hospital operations, and enabling remote care. As the healthcare industry navigates the challenges of an aging population, the rise of chronic diseases, and the need for more efficient resource management, the deployment of smart healthcare systems will be a critical component in building a more resilient and sustainable healthcare ecosystem.

AI has the potential to revolutionize hospital management, leading to improved efficiency, cost-effectiveness, and patient care by providing data-driven insights and automating various processes. Table 14.8 shows the ways AI can revolutionize hospital management.

TABLE 14.8
The Ways AI Improve Hospital Management

Predictive analytics	AI can analyze vast amounts of data to predict future trends. This can help hospitals anticipate patient influx, manage staff scheduling, and ensure adequate supply of medical equipment and medicines.
Task management	AI can help understand administrative tasks and task completion networks. It can identify patterns and optimize workflows, reducing the administrative burden and increasing efficiency.
Resource allocation	AI can optimize the allocation of resources such as staff, equipment, and facilities. It can help in making the most of available resources, improving patient outcomes, and reducing costs.
Remote monitoring	AI can enable continuous monitoring of patients, reducing the need for hospital visits and freeing up resources.
Automation	AI can automate various operations such as patient flow and scheduling, supply chain management, and managing healthcare facilities. This can build efficiencies and allow healthcare professionals to focus more on patient care.
Customized care	AI can enable personalized care by analyzing individual patient data and predicting specific needs. This can lead to better health outcomes and more efficient use of resources.

14.3.6 Smart Waste Management

In the ever-evolving landscape of sustainability and resource management, smart waste management systems have emerged as a game-changer of the conventional municipal waste management systems. These innovative solutions leverage the power of sensors, data analytics, and technology to transform the way we handle waste, optimize operations, and drive positive environmental impact.

14.3.6.1 Smart Waste Management Systems

Smart waste management systems are a modern approach to dealing with waste in an efficient and sustainable manner. They leverage technology, specifically sensors and data analytics, to optimize waste collection routes, reduce operational costs, and improve recycling rates.

The core of these systems is the use of technology. Sensors installed in waste bins monitor the level of waste in real time. When a bin reaches its capacity, the sensor sends a signal to the waste management service. This real-time monitoring prevents overflow of waste bins, reduces the frequency of collections, and ensures that waste collection is only done when necessary. Data analytics, on the other hand, is used to analyze the data collected by the sensors. It can determine patterns and trends in waste generation and disposal. This information is then used to optimize waste collection routes, ensuring that waste collection trucks take the most efficient route possible. This not only saves time but also reduces fuel consumption and emissions.

Smart waste management systems can significantly reduce operational costs through optimizing waste collection routes and reducing the frequency of collections. They also improve the efficiency of waste management services by ensuring that resources are used where they are most needed. In addition to operational efficiency, smart waste management systems also contribute to sustainability by the improvement of recycling rates and helping reduce the amount of waste that ends up in landfills. They also contribute to the circular economy by ensuring that waste is treated as a resource rather than a problem.

14.3.6.2 Sensors in Waste Management

Sensors are vital devices in smart waste management. They are installed in waste bins and containers to monitor the level of waste. These sensors can detect when a bin is full and send a notification to the waste management service. This helps prevent overflow and reduces the frequency of collections, leading to cost savings.

In smart waste management, sensors are vital devices. These devices are strategically installed in waste bins and containers to monitor the level of waste. The primary function of these sensors is to detect when a bin is full. This is achieved through various technologies such as ultrasonic sensors, which measure the distance between the sensor and the waste, or weight sensors that determine the weight of the waste.

Once the sensor detects that the bin is full, it sends a notification to the waste management service. This notification system is typically facilitated through wireless communication technologies like Wi-Fi or cellular networks. The data transmitted includes information about the bin's location and its fill level. This real-time monitoring and reporting system helps prevent overflow of waste bins. Overflowing bins can lead to various environmental and health issues, including unpleasant odors, littering, and the attraction of pests. By alerting the waste management service when a bin is full, these issues can be mitigated.

Sensors in waste management also contribute to operational efficiency. Traditionally, waste collection trucks follow fixed routes and schedules, leading to unnecessary pickups from half-empty bins. With sensor technology, collections can be optimized based on actual fill levels. This means trucks only need to service bins that are full, reducing the frequency of collections. This reduction in collections leads to significant cost savings. Fewer collections mean less fuel consumption and lower vehicle maintenance costs. It also frees up resources that can be allocated elsewhere, enhancing the overall efficiency of the waste management service.

Sensors can detect different types of waste using various technologies. One of the most common is ultrasonic technology, which measures the fill levels in bins via ultrasonic beams. These sensors can monitor any type of waste, including mixed waste, paper, plastics, glass, clothing, bio-waste, liquids, electronics, and metal, in bins and containers of various types and sizes.

The sensors are robust, water and shock-resistant, and fully functional within a wide temperature range. They are set to measure from 3 cm up to 12 m. The casing is made from recyclable glass fiber-reinforced polyamide, which not only provides an eco-design but also aids in recycling. Once the sensor detects that the bin is full, it sends the data to a cloud-based platform for data-driven daily operations. The data include information about the bin's location and its fill level. This real-time monitoring and reporting system helps prevent overflow of waste bins and contributes to operational efficiency.

While ultrasonic technology is widely used in waste containers, other technologies such as infrared and image-based tech are also used in some applications. Each technology has its advantages and is chosen based on the specific requirements of the waste management system.

Image-based technology for waste detection primarily involves the use of computer vision and deep learning techniques. The process begins with a camera capturing images or videos of the waste. These images are then processed and analyzed using deep-learning algorithms.

One common method involves a Convolutional Neural Network (CNN), a type of deep learning model, on a dataset of waste images. CNN learns to recognize different types of waste, such as cardboard, glass, metal, paper, plastic, and trash. Once trained, the model can classify new waste images into these categories. Some systems use more advanced models like EfficientNet-B0, a compound-scaling based model proposed by Google. This model is pretrained on ImageNet, a large dataset of images, and then fine-tuned on images specific to demographic regions for efficient classification. Such model tuning over transfer learning provides a customized model for classification, highly optimized for a particular region. These image-based waste detection systems can significantly improve the efficiency of waste management processes. They automate and enhance the sorting and recycling of diverse garbage items, reducing the need for manual sorting and thus leading to cost savings and increased recycling rates.

These systems require high-quality, representative images for training, and their performance can be affected by factors such as lighting conditions and image quality. Despite these challenges, image-based waste detection holds significant potential for revolutionizing waste management processes.

Sensors are indispensable in smart waste management. They not only ensure cleaner and healthier environments but also drive operational and cost efficiency. As technology continues to evolve, the role of sensors in waste management is set to become even more significant.

14.3.6.3 Data Analytics in Waste Management

Data analytics is the brain behind smart waste management. It processes the data collected by sensors to derive meaningful insights. For example, it can analyze the waste levels in different bins across a city to determine the most efficient collection routes. This helps save time and fuel, reducing the environmental impact of waste collection.

Data analytics is important in every aspect that involves data, including waste management. Data analytics can unlock the full potential of these innovative systems. As the backbone of smart waste management, data analytics serves as the brain, processing the wealth of information collected by the network of sensors to derive actionable insights and optimize the entire waste management process. Data analytics in smart waste management lies in the ability to collect and analyze real-time data from the sensor network. These sensors, strategically placed in waste bins, dumpsters, and compactors, continuously monitor the fill levels and other relevant metrics, transmitting this information to a central data hub.

The power of data analytics lies in its capacity to transform this raw data into valuable insights that can drive informed decision-making. By applying advanced algorithms and statistical models, data analysts can identify patterns, trends, and anomalies within the waste management system. This allows them to gain a comprehensive understanding of waste generation, collection, and disposal across an entire city or region. One of the primary applications of data analytics in smart waste management is the optimization of collection routes. Analyzing the fill levels of waste receptacles across a given area enables data analysts to determine the most efficient routes for waste collection. This information can then be used to create dynamic, real-time collection schedules, ensuring waste is collected when and where it is needed, without unnecessary trips.

The benefits of optimized collection routes are numerous. First and foremost, it leads to a significant reduction in operational costs. By minimizing the number of collection trips required, waste management companies can save on fuel, labor, and vehicle maintenance expenses. This, in turn, translates to more cost-effective waste management services for the community. The optimization of collection routes directly contributes to the reduction of the environmental impact associated with waste management. With fewer vehicles on the road, the overall carbon footprint of the waste collection process is diminished, aligning with the broader goals of sustainability and environmental stewardship.

Data analytics also enhances recycling rates and improves waste segregation. By analyzing the composition and volume of waste collected, data analysts can identify opportunities to enhance recycling programs, target specific waste streams, and educate the public on better waste sorting practices. This information can be shared with both municipal authorities and citizens, fostering a more informed and engaged community when it comes to waste management.

Beyond route optimization and recycling improvements, data analytics in smart waste management can also be used to predict future waste generation patterns, anticipate potential issues, and proactively address them. Through the analysis of historical data and considering factors such as population growth, urbanization, and seasonal fluctuations, data analysts can develop forecasting models that help waste management companies prepare for and adapt to changing waste management demands.

Data analytics can be the driving force behind the success of smart waste management systems. Processing vast amounts of data collected by sensors enables data analysts to unlock insights that revolutionize waste collection, improve recycling rates, and reduce the environmental impact of waste management. As cities and communities continue to embrace the power of data-driven decision-making, the future of waste management is poised to become more efficient, sustainable, and responsive to the needs of both citizens and the environment.

14.3.6.4 Optimizing Waste Collection Routes

One of the key benefits of smart waste management is the optimization of waste collection routes. Traditional waste collection involves following a fixed route, regardless of whether the bins are full or not. This can lead to inefficiencies and unnecessary fuel consumption. Smart waste management systems, on the other hand, use data analytics to determine the most efficient routes based on the current waste levels. This ensures that only full bins are collected, reducing the number of trips and saving fuel.

14.3.6.5 Reducing Operational Costs

Smart waste management can lead to significant cost savings. By optimizing waste collection routes, it reduces fuel consumption and wear and tear on vehicles. Moreover, preventing overflow reduces the need for costly emergency collections and allows for more efficient use of manpower, as workers can be deployed where they are needed most.

14.3.6.6 Improving Recycling Rates

Smart waste management can also help improve recycling rates. By monitoring the types and quantities of waste, it can provide valuable data for planning recycling programs. For example, if data show that a large amount of recyclable waste is being thrown in regular bins, it could indicate the need for better recycling education or more accessible recycling facilities.

14.3.6.7 Smart Bins

Smart waste management is a revolutionary approach that leverages technology to enhance the efficiency and effectiveness of traditional waste management systems. Smart bins are ordinary waste bins equipped with advanced sensor technology. These sensors are designed to monitor the fill level of the bins. They use various technologies such as ultrasonic waves, infrared, or weight measurement to detect the amount of waste in the bin. Once the waste reaches a certain level, the sensor triggers a notification, thereby indicating that the bin is full. This notification, typically sent via Wi-Fi or cellular networks, is received by the waste management service. The data transmitted include information about the bin's location and its fill level.

The real-time monitoring and reporting system of smart bins helps prevent the overflow of waste. Overflowing bins can lead to various environmental and health issues, including unpleasant odors, littering, and the attraction of pests. Alerting the waste management service when a bin is full mitigates these issues. This ensures cleaner streets and a healthier environment for the community.

Smart bins contribute to operational efficiency. In traditional waste management systems, waste collection trucks follow fixed routes and schedules, leading to unnecessary pickups from half-empty bins. With smart bins, collections can be optimized based on actual fill levels. This means trucks only need to service bins that are full, reducing the frequency of collections. This reduction in collections leads to significant cost savings. Fewer collections mean less fuel consumption and lower vehicle maintenance costs. It also reduces the carbon footprint of the waste management process, contributing to environmental sustainability. Furthermore, it frees up resources that can be allocated elsewhere, enhancing the overall efficiency of the waste management service.

In addition to cost savings, smart bins also contribute to better resource management. By knowing exactly when and where collections are needed, waste management services can plan their routes more effectively. This leads to improved logistics and time management.

The data collected by smart bins can provide information for waste management services. Analyzing this data allows them to identify trends and patterns in waste generation, which can inform future waste management strategies. For example, if certain areas consistently produce more waste, additional resources can be allocated to those areas. Furthermore, the data collected by smart bins can be integrated with advanced data analytics platforms, providing valuable insights into recycling rates, and opportunities for improvement. This data-driven approach empowers waste management authorities to make informed decisions, tailor their services to the specific needs of the community, and drive sustainable waste management strategies.

Beyond the operational benefits, smart bins also contribute to the broader goals of sustainability and environmental stewardship. By minimizing the number of collection vehicles on the road, the carbon footprint associated with waste management is significantly reduced, aligning with the global push for more eco-friendly practices.

Smart waste management systems represent a significant advancement in the way we deal with waste. They promise to make waste management more efficient, cost-effective, and environmentally friendly. However, they also pose challenges, such as the need for reliable sensor technology and secure data handling, which need to be addressed to realize their full potential.

14.3.7 Smart Street Lighting

Smart street lighting systems use sensors and intelligent networks to control city lighting based on various factors such as natural light availability, movement of people and vehicles, and specific events. This can significantly reduce energy consumption and maintenance costs, while improving safety and quality of life for residents.

Smart street lighting systems are an integral part of the smart city infrastructure, designed to enhance energy efficiency, reduce maintenance costs, and improve the quality of life for city residents. These systems use advanced technologies such as sensors, intelligent networks, and data analytics to control city lighting based on various factors. This includes natural light availability, movement of people and vehicles, and specific events.

The core of a smart street lighting system is the intelligent network of sensors installed on each streetlight. These sensors can detect various environmental parameters such as ambient light levels and movement. For instance, light sensors, also known as photocells, can measure the amount of natural light available. When the natural light falls below a certain level, the sensors trigger the streetlights to turn on. Conversely, when the natural light is sufficient, the lights are turned off, saving energy.

Motion sensors, on the other hand, detect the movement of people and vehicles. In areas with low footfall or traffic during certain hours, the lights can be dimmed or turned off completely. When movement is detected, the lights are turned on or brightened. This not only saves energy but also enhances safety by providing light when and where it is needed.

In addition to light and motion sensors, smart street lighting systems can also incorporate other types of sensors to monitor various environmental factors. For example, air quality sensors can measure pollution levels, while noise sensors can monitor noise pollution. This data can be used to gain insights into the environmental health of the city and inform relevant policies and initiatives.

The data collected by the sensors is transmitted via a wireless network to a central management system. This system uses data analytics to process and analyze the data, enabling real-time monitoring and control of the streetlights. For instance, if a light is not functioning properly, the system can immediately detect the issue and alert the maintenance team. This proactive approach can significantly reduce maintenance costs and ensure that issues are addressed promptly, minimizing disruptions to city lighting.

The central management system can use predictive analytics to forecast future lighting needs based on historical data and trends. This can further optimize energy use and reduce costs. For example, if data show that a particular street has little to no activity after midnight, the system can schedule the lights on that street to be dimmed or turned off during those hours.

Smart street lighting systems can also be integrated with other smart city systems to provide additional benefits. For example, streetlights can be synchronized with traffic signals to improve traffic flow. They can also be equipped with charging ports for electric vehicles, contributing to the development of green transportation infrastructure.

In the event of emergencies or special events, the lighting can be remotely controlled to ensure adequate lighting is provided. For instance, in the case of an accident or a public event, the lights in the relevant area can be brightened to enhance visibility.

Intelligent street lighting systems not only result in substantial energy and cost reductions, but also contribute to enhancing safety and the overall well-being of urban inhabitants. As cities worldwide work toward achieving greater sustainability and efficiency, it is undeniable that smart street lighting systems will play a pivotal role in this pursuit.

14.3.8 Smart Parking

Smart parking solutions are an innovative approach to address the growing problem of parking in urban areas. By capitalizing advanced technologies such as sensors, data analytics, and wireless communication, these solutions provide real-time information about available parking spaces. This not only reduces the time spent searching for parking but also results in less traffic congestion and pollution. Smart parking solutions are sensors installed in each parking space. These sensors can detect the presence or absence of a vehicle in space. The most common types of sensors used are ultrasonic and infrared sensors. Ultrasonic sensors work by emitting ultrasonic waves and measuring the time it takes for the waves to bounce back after hitting an object. If a vehicle is present, the waves will bounce back faster, indicating that the space is occupied. Infrared sensors, on the other hand, detect the heat emitted by a vehicle to determine if a space is occupied.

The data collected by the sensors are transmitted via a wireless network to a central management system. This system uses data analytics to process and analyze the data, providing real-time information about the availability of parking spaces. This information can be accessed by drivers through a mobile app or a web portal. The app can guide the driver to the nearest available parking space, significantly reducing the time spent searching for parking.

Other than providing real-time parking information, smart parking solutions can also offer other features such as online payments and reservations. Drivers can reserve a parking space in advance and pay for it online, further enhancing the convenience of parking. Some systems even integrate with other services such as electric vehicle charging stations, providing additional value to drivers.

Smart parking solutions can significantly reduce traffic congestion by minimizing the time drivers spend searching for parking. Research indicates that a substantial portion of urban traffic results from drivers looking for parking spots. Providing real-time parking information allows these solutions to help drivers find parking more quickly, thereby reducing the time they spend on the road.

Less time spent on the road means less fuel consumption and fewer emissions, contributing to a reduction in air pollution. This is particularly important in urban areas, where air quality is often a major concern. Smart parking solutions can play a key role in creating more sustainable cities. Smart parking solutions also offer benefits to parking operators and city authorities. For operators, the real-time data provided by the sensors can help optimize the use of parking spaces, increasing revenue. For city authorities, the data can provide valuable insights into parking patterns and trends, informing urban planning and transportation policies.

Smart parking solutions can be integrated with other smart city technologies to create a more connected and efficient urban environment. For example, they can be linked with traffic management systems to optimize traffic flow, or with public transportation systems to provide integrated mobility solutions.

14.3.9 Smart Public Safety

Smart public safety solutions are a transformative approach to enhancing the capabilities of emergency services. These solutions aim to improve the efficiency, effectiveness, and responsiveness of public safety operations. This includes the use of advanced surveillance systems, predictive policing using data analytics, and improved communication systems for emergency responders.

Advanced surveillance systems form the backbone of smart public safety solutions. These systems use a network of cameras and sensors to monitor public spaces in real time. The data captured by these devices is analyzed using sophisticated algorithms to detect unusual or suspicious activities. For instance, facial recognition technology can be used to identify individuals of interest, while motion detection algorithms can alert authorities to potential incidents. These systems not only enhance the ability of law enforcement agencies to prevent and respond to incidents but also serve as a deterrent to potential offenders.

Predictive policing is another key component of smart public safety solutions. This approach uses data analytics to predict where and when crimes are likely to occur. It involves the collection and analysis of a wide range of data, including crime statistics, demographic data, and social media feeds. ML algorithms are then used to identify patterns and trends in this data, which can inform proactive policing strategies. For example, if data analysis reveals a high incidence of burglaries in

TABLE 14.9
Strategies by Communities to Ensure the Ethical Use of Surveillance Technology

Regulation	Governments should establish clear regulations on the use of surveillance technologies. These regulations should be based on international human rights law and standards. They should define what constitutes legitimate use of surveillance and set limits to prevent abuse.
Transparency	Governments should inform the public about their surveillance activities. This includes disclosing what data is being collected, how it is being used, and who has access to it.
Consent	Whenever possible, surveillance should only be conducted with the informed consent of the individuals being monitored. This means individuals should be aware that they are being monitored and have agreed to it.
Data protection	Robust data protection measures should be in place to prevent unauthorized access to surveillance data. This includes using encryption to protect data in transit and at rest.
Accountability	There should be mechanisms in place to hold those who misuse surveillance technologies accountable. This could include legal penalties for violations and avenues for individuals to seek redress.
Review and oversight	Regular reviews and oversight by independent bodies can help ensure surveillance technologies are being used ethically. These bodies can monitor compliance with regulations, investigate complaints, and recommend changes as needed.
Public participation	The public should be involved in decisions about the use of surveillance technologies. This could include public consultations on proposed surveillance initiatives and opportunities for public input on surveillance policies.
Education and awareness	Communities should be educated about the potential risks and benefits of surveillance technologies. This can help individuals make informed decisions about their own privacy and advocate for ethical surveillance practices.

a particular neighborhood at a specific time, police can increase patrols in that area during those hours. Predictive policing not only enhances the efficiency of law enforcement operations but also helps to prevent crimes before they occur.

Improved communication systems for emergency responders are also a crucial part of smart public safety solutions. These systems ensure that emergency personnel can communicate effectively in real time, both with each other and with command centers. They can include secure mobile networks, wearable devices, and advanced dispatch systems. For instance, GPS-enabled devices can provide real-time location data, helping to coordinate the response to an incident. Meanwhile, advanced dispatch systems can use data analytics to assign responders based on factors such as proximity to the incident and availability of resources. These technologies not only enhance the speed and effectiveness of emergency responses but also ensure the safety of emergency personnel. Smart public safety solutions can also incorporate other technologies to further enhance public safety. For example, drones can be used for aerial surveillance or to deliver emergency supplies. Similarly, AI can be used to analyze social media feeds for potential threats or to monitor crowds during large public events.

Maintaining privacy and promoting the ethical use of surveillance technology is a challenging task that calls for a comprehensive strategy. Some strategies that communities can put into action are shown in Table 14.9.

14.3.10 Smart Education

Smart education systems are a transformative approach to traditional learning methodologies, exploiting digital technology to enhance learning outcomes and improve access to education. These systems encompass a wide range of technologies and methodologies, including online learning platforms, digital classrooms, and AI-powered personalized learning experiences.

Online learning platforms are a key component of smart education systems. These platforms provide a virtual space where students can access educational content anytime, anywhere. This includes recorded lectures, interactive quizzes, discussion forums, and other learning materials. Online learning platforms break down geographical barriers, allowing students from all over the world to learn from the best educators. They also provide flexibility, enabling students to learn at their own pace and according to their own schedules.

Digital classrooms are another important aspect of smart education systems. These classrooms use technology to create an interactive and engaging learning environment. For example, interactive whiteboards can be used to display multimedia content, while student response systems can be used to conduct real-time assessments. Digital classrooms can also incorporate virtual reality or augmented reality technologies to provide immersive learning experiences. For instance, a biology lesson could include a virtual dissection of a frog, providing a hands-on learning experience without the need for actual dissection.

AI-powered personalized learning experiences are perhaps the most revolutionary aspect of smart education systems. AI algorithms can analyze a student's learning patterns, strengths, and weaknesses to provide personalized learning paths. For example, if a student is struggling with a particular concept, the AI system can provide additional resources or exercises to help the student master that concept. On the other hand, if a student excels in a particular area, the AI system can provide more advanced materials to challenge the student and promote further learning.

AI-powered personalized learning experiences are transforming the educational landscape by providing tailored experiences to students. Here are some examples:

- *Adaptive Learning Platforms:* These platforms analyze student data, such as performance on assessments and interactions with learning materials. Based on this analysis, AI provides personalized suggestions and content to help students with areas they are struggling. For instance, if a student is having difficulty with a particular concept, the platform can provide additional resources or exercises to help the student master that concept.
- *Personalized Course Recommendations:* AI can analyze a student's performance, preferences, and profiles to deliver personalized course recommendations. For example, if a student shows a strong interest and aptitude in science, the system might recommend advanced science courses.
- *Individualized Learning Paths:* AI-powered platforms can generate individualized learning paths for students based on their goals, interests, and prior knowledge. Adapting the curriculum and pace to meet the needs of each student allows AI to ensure they receive appropriate challenges and support, maximizing their learning outcomes.
- *Interactive Learning Experiences:* AI can create interactive learning experiences that adapt to the learner's responses. For example, an AI-powered learning game might adjust its difficulty level based on the learner's performance.
- *Predictive Analytics:* AI can use predictive analytics to identify potential learning difficulties before they become significant problems. For example, if a student is consistently struggling with math problems, the system might flag this issue and suggest additional support or resources.

These are just a few examples of how AI is being used to personalize learning experiences. The goal of these technologies is to cater to the unique needs and learning styles of each student, thereby enhancing their learning outcomes and overall educational experience.

Smart education systems also provide benefits beyond improved learning outcomes. For example, they can help reduce the cost of education by eliminating the need for physical textbooks and other learning materials. They can also help reduce the environmental impact of education by reducing the need for physical transportation to and from school. Furthermore, smart education systems can provide valuable data for educators and policymakers. For example, data on student performance can be used to identify areas where students are struggling and develop targeted interventions. Data on the effectiveness of different teaching methodologies can be used to inform teacher training and professional development programs.

There are also constraints in implementation of smart education systems. For example, there are concerns about data privacy and security, as these systems often involve the collection and storage of sensitive student data. There are also concerns about the digital divide, as not all students have access to the necessary technology to participate in smart education systems. Addressing these constraints needs robust data protection measures and initiatives to promote digital inclusion.

In wrapping up our discussion on smart city Applications, these innovations are reshaping our urban landscapes, making them more efficient, sustainable, and enjoyable to live in. They are elevating public services, enriching the quality of life, and fueling economic progress. As we look ahead, the evolution of technology promises to broaden the horizons for smart city applications, laying the groundwork for future advancements in urban development. However, it needs to be remembered that the deployment of these technologies should be carried out in a manner that is inclusive and beneficial to all residents. This means addressing the digital divide and ensuring accessibility are key considerations in the development of smart cities.

14.4 SMART MOBILITY

14.4.1 Smart Mobility: Transforming the Way We Move

In an increasingly connected world, the concept of smart mobility has emerged as a game-changer in the way we approach transportation. Smart mobility is a holistic approach that leverages advanced technologies, data, and innovative solutions to create more efficient, sustainable, and user-centric modes of travel. As the world tackles the challenges of urbanization, congestion, and environmental concerns, smart mobility offers a promising path forward, transforming the way we move and live.

In smart mobility lies the integration of various technologies, including the Internet of Things (IoT), AI, and connected vehicle systems. These advancements enable the seamless flow of information, allowing for real-time monitoring, optimization, and personalization of transportation services. From smart traffic management systems that reduce congestion to shared mobility platforms that provide on-demand access to a variety of transportation options, smart mobility is redefining the urban landscape.

One of the key pillars of smart mobility is the rise of ITS. ITS employs a suite of technologies, such as sensors, communication networks, and data analytics, to enhance the efficiency and safety of transportation infrastructure. Smart traffic signals, for instance, can adjust their timing based on real-time traffic conditions, reducing wait times and minimizing idling, which in turn lowers greenhouse gas emissions and fuel consumption. Similarly, intelligent parking systems can guide drivers to available spots, reducing the time and resources spent searching for parking.

Another integral component of smart mobility is the growth of shared mobility services. Ride-hailing apps, car-sharing platforms, and bike-sharing schemes have become increasingly popular, providing users with alternative options to traditional private vehicle ownership. These services not only offer convenient and on-demand access to transportation, but they also contribute to reducing the environmental impact of transportation by encouraging more efficient utilization of vehicles.

The rise of electric vehicles (EVs) is another significant development in the smart mobility landscape. As concerns over climate change and air pollution continue to mount, EVs have emerged as a cleaner and more sustainable alternative to traditional internal combustion engine vehicles. With advancements in battery technology and the expansion of charging infrastructure, EVs are becoming more accessible and practical for a wider range of users. Integrating EVs into smart mobility solutions, such as shared mobility services and smart charging networks, can further enhance the environmental benefits of this technology.

Autonomous and connected vehicles are also shaping the future of smart mobility. Self-driving cars, enabled by advanced sensors, ML, and vehicle-to-everything (V2X) communication, have the potential to revolutionize the way we travel. These vehicles can navigate roads more efficiently, improve safety by reducing human error, and provide mobility options for those who may not have access to traditional transportation. As the development and testing of autonomous vehicles continue, they are expected to become an integral part of smart mobility ecosystems.

The implementation of smart mobility solutions requires a collaborative effort among various stakeholders, including city planners, transportation authorities, technology companies, and citizens. Developing the necessary infrastructure, such as robust communication networks, smart traffic management systems, and charging stations for EVs, is crucial for the successful deployment of smart mobility initiatives. Policymakers and urban planners must also work closely with the private sector to create regulatory frameworks that support the integration of new mobility technologies and services. Furthermore, the success of smart mobility relies on the active participation and engagement of citizens. Educating the public about the benefits of smart mobility solutions, promoting their adoption, and addressing concerns related to privacy, security, and equity are essential for creating a truly inclusive and sustainable transportation ecosystem.

The need for smart mobility solutions is becoming increasingly pressing. By leveraging cutting-edge technologies, fostering collaboration, and prioritizing sustainability, we can transform the way we move, ultimately creating more livable, efficient, and environmentally friendly cities. The journey toward smart mobility is not without its challenges, but the potential rewards are immense – a future where transportation is seamless, environmentally conscious, and tailored to the unique needs of individuals and communities.

14.4.2 Toward a Sustainable and Equitable Smart Mobility Future

As the world struggles with the challenges of urbanization, climate change, and social inequalities, the need for a holistic approach to transportation has never been more pressing. Smart mobility emerges as a compelling solution, promising to reshape the way we move and live. Integrating advanced technologies, data-driven insights, and innovative mobility services offers a pathway to a more sustainable, efficient, and inclusive transportation ecosystem through smart mobility.

One of the key drivers of smart mobility is the transition to electric vehicles (EVs). As concerns over greenhouse gas emissions and air pollution continue to grow, EVs have become an increasingly attractive alternative to traditional internal combustion engine vehicles. With advancements in battery technology and the expansion of charging infrastructure, EVs are becoming more accessible and practical for a wider range of users. Integrating EVs into smart mobility solutions, such as shared mobility services and smart charging networks, can further enhance the environmental benefits of this technology. However, the successful adoption of EVs is not without its challenges. Ensuring equitable access to charging infrastructure, particularly in underserved communities, is crucial to prevent the widening of existing transportation disparities. Policymakers and urban planners must work closely with the private sector to develop comprehensive strategies that address the needs of diverse populations, ensuring that the benefits of smart mobility are distributed fairly.

Another key aspect of smart mobility is the rise of shared mobility services. Ride-hailing apps, car-sharing platforms, and micro-mobility options, such as e-bikes and e-scooters, have become increasingly popular, providing users with alternative options to traditional private vehicle ownership. These services not only offer convenient and on-demand access to transportation, but they also contribute to reducing the environmental impact of transportation by encouraging more efficient utilization of vehicles.

The integration of these shared mobility services into a seamless, multimodal transportation network is crucial for the success of smart mobility initiatives. Enabling users to easily transition between different modes of transportation, such as public transit, ride-hailing, and micro-mobility, allows smart mobility to enhance the overall accessibility and convenience of the transportation system.

ITSs also play a vital role in smart mobility. These systems employ a suite of technologies, such as sensors, communication networks, and data analytics, to enhance the efficiency and safety of transportation infrastructure. Smart traffic signals, for instance, can adjust their timing based on real-time traffic conditions, reducing wait times and minimizing idling, which in turn lowers greenhouse gas emissions and fuel consumption. Similarly, intelligent parking systems can guide drivers to available spots, reducing the time and resources spent searching for parking.

The development and deployment of autonomous and connected vehicles are also transforming the smart mobility landscape. Self-driving cars, enabled by advanced sensors, ML, and vehicle-to-everything (V2X) communication, have the potential to revolutionize the way we travel. These vehicles can navigate roads more efficiently, improve safety by reducing human error, and provide mobility options for those who may not have access to traditional transportation. As technology continues to evolve, autonomous vehicles are expected to become an integral part of smart mobility ecosystems. However, the integration of autonomous and connected vehicles into smart mobility solutions requires careful consideration of ethical, legal, and security implications. Policymakers and transportation authorities must work closely with technology companies to develop robust frameworks that address concerns related to privacy, data governance, and the societal impact of these technologies.

To achieve a truly sustainable and equitable smart mobility future, a collaborative effort among various stakeholders is essential. City planners, transportation authorities, technology companies, and citizens must work together to create a shared vision and develop the necessary infrastructure, policies, and regulatory frameworks to support the integration of smart mobility solutions.

Engagement and education of the public are also crucial for the success of smart mobility initiatives. Raising awareness about the benefits of smart mobility, promoting the adoption of new technologies and services, and addressing concerns related to accessibility, affordability, and social inclusion can help build a sense of ownership and trust among citizens.

14.4.3 Unlocking the Potential of Smart Mobility: Challenges and Opportunities

As the world becomes increasingly urbanized and the demand for efficient, sustainable, and inclusive transportation grows, smart mobility has emerged as a transformative solution. This holistic approach to transportation leverages advanced technologies, data-driven insights, and innovative mobility services to redefine the way we move and live. However, the path toward a smart mobility future is not without its challenges, and navigating these obstacles is crucial for unlocking the full potential of this transformative concept.

One of the primary challenges lies in the development and deployment of the necessary infrastructure to support smart mobility solutions. The successful implementation of ITS, EV charging networks, and connected vehicle technologies requires significant investments in digital and physical infrastructure. This includes the installation of sensors, communication networks, and supporting systems that enable the seamless flow of data and the integration of various mobility services.

Securing the necessary funding and coordinating the efforts of multiple stakeholders, including government agencies, transportation authorities, and private sector partners, can be a complex and time-consuming process. Policymakers must work to create incentive structures and regulatory frameworks that encourage the development and deployment of smart mobility infrastructure, ensuring that these investments align with the long-term vision for sustainable and equitable transportation.

Another challenge lies in the integration and interoperability of different smart mobility solutions. As the technology landscape continues to evolve, the need for standardized protocols and data-sharing practices becomes increasingly crucial. Ensuring that various mobility services, such as ride-hailing, car-sharing, and micro-mobility, can seamlessly integrate with public transportation systems and other transportation modes is essential for creating a truly multimodal and user-centric mobility ecosystem.

Addressing issues of data governance, privacy, and security is also a critical aspect of smart mobility development. The vast amounts of data generated by smart mobility solutions, including vehicle telemetry, user preferences, and real-time traffic information, must be managed and protected in a way that respects individual rights and earns the trust of citizens. Policymakers and technology companies must collaborate to establish robust data governance frameworks that balance the need for data-driven insights with the protection of personal information.

Equity and accessibility are also significant challenges in the realm of smart mobility. Ensuring that the benefits of these advanced transportation solutions are equitably distributed across diverse communities, including underserved and marginalized populations, is crucial for achieving true social and environmental justice. This may require targeted investments in infrastructure, subsidies for alternative mobility options, and the development of inclusive decision-making processes that incorporate the needs and perspectives of all stakeholders.

Despite these challenges, the opportunities presented by smart mobility are vast and transformative. Harnessing the power of emerging technologies enables smart mobility to enhance the efficiency, sustainability, and resilience of transportation systems, ultimately improving the quality of life for individuals and communities.

One of the most promising opportunities lies in the integration of electric vehicles (EVs) into smart mobility ecosystems. As the world shifts toward decarbonization and the reduction of greenhouse gas emissions, EVs have become an increasingly viable alternative to traditional internal combustion engine vehicles. Integrating EVs into shared mobility services, smart charging networks, and intelligent transportation systems allows smart mobility to significantly contribute to the transition toward a carbon-neutral transportation future.

The rise of shared mobility services, such as ride-hailing, car-sharing, and micro-mobility, also presents exciting opportunities for smart mobility. These services not only provide users with convenient and on-demand access to transportation but also have the potential to reduce private vehicle ownership, lower congestion, and decrease environmental impact. Seamlessly integrating these shared mobility options into multimodal transportation networks allows smart mobility to enhance overall accessibility and support the shift toward more sustainable modes of travel.

The development and deployment of autonomous and connected vehicles are also transforming the smart mobility landscape. These advanced vehicle technologies have the potential to improve road safety, optimize traffic flows, and provide mobility options for individuals who may have limited access to traditional transportation. As technology continues to mature and regulatory frameworks evolve, the integration of autonomous and connected vehicles into smart mobility solutions can unlock new levels of efficiency, accessibility, and personal freedom.

To fully capitalize on these opportunities, stakeholders from the public and private sectors must work collaboratively to address the challenges and create a shared vision for the future of smart mobility. This collaboration must involve policymakers, transportation authorities, technology companies, urban planners, and, most importantly, the citizens who will ultimately benefit from these transformative solutions.

By embracing the potential of smart mobility, we can pave the way for a more sustainable, equitable, and livable future. Through the strategic deployment of advanced technologies, the fostering of multi-stakeholder partnerships, and the prioritization of inclusive and data-driven decision-making, smart mobility can become a catalyst for positive change, reshaping the way we move and live in the years to come.

14.5 SMART LIVING

The boundaries between physical and digital spaces are blurring within the realm of smart living. This burgeoning field leverages advancements in information and communication technology (ICT) to create intelligent homes that adapt to user needs and preferences. Imagine a residence that automatically adjusts lighting and temperature, seamlessly integrates with smart appliances, and boasts a robust connected security system – all managed through a single interface. While smart living promises to enhance comfort, efficiency, and security within our dwellings, key considerations remain. Data privacy concerns, potential security vulnerabilities, and ensuring equitable access to these technologies must be addressed to ensure this revolution truly benefits all stakeholders.

14.5.1 Home Automation

14.5.1.1 Automating Our Life

The fascinating transformation of our home walls from passive structures to intelligent companions is marked by the evolution of home walls from passive to smart. In the past, our home walls served primarily as physical boundaries, providing shelter and privacy. However, with the advent of smart technology, these once inert surfaces have undergone a remarkable metamorphosis. The ways they work are shown below.

14.5.1.2 Intelligent Devices Integration

- Smart devices embedded within our walls now actively participate in our daily lives. These devices include smart thermostats, motion sensors, and voice-activated assistants. They seamlessly blend into the architecture, enhancing both form and function.
- Imagine waking up to a room that adjusts its temperature based on your preferences, because of a smart thermostat that learns your habits. Or consider motion sensors that illuminate the hallway as you walk by, ensuring safety and convenience.

14.5.1.3 Enhanced Convenience

Our walls have become intuitive companions. They respond to our needs without us even asking. For instance:

- *Lighting Control:* Smart lighting systems adjust brightness levels based on natural light, time of day, and occupancy. Dim the lights for movie night or brighten them for a productive work session.
- *Voice Assistants:* Voice-activated devices like Amazon Echo or Google Nest Hub reside discreetly on our walls. They answer questions, play music, and even order groceries – all with a simple command.
- *Interactive Surfaces:* Imagine touch-sensitive walls that display recipes while you cook or turn into a canvas for your child's creativity. These interactive surfaces redefine how we engage with our living spaces.

14.5.1.4 Security Reinforcement

Smart walls contribute significantly to home security:

- *Surveillance Cameras:* Discreetly integrated cameras monitor the premises, alerting us to any unusual activity. Whether we are at work or on vacation, we can check in via our smartphones.
- *Intrusion Detection:* Motion sensors embedded in walls trigger alarms if unauthorized movement occurs. They are like silent sentinels, safeguarding our homes.

- *Access Control:* Smart locks allow us to grant entry remotely. Forgot your keys? No problem – unlock the door with your phone.

14.5.1.5 Energy Efficiency Revolution

Our walls actively participate in energy conservation:

- *Insulation and Climate Control:* Smart walls regulate indoor temperatures efficiently. They adjust heating and cooling based on occupancy patterns and external weather conditions.
- *Solar Integration:* Imagine photovoltaic panels seamlessly integrated into exterior walls, harnessing sunlight to power our homes. These walls become active contributors to sustainability.

14.5.1.6 Personalization and Aesthetics

Smart walls adapt to our moods and preferences:

- *Digital Art Displays:* Transform your living room wall into a rotating art gallery. Display your favorite paintings, family photos, or dynamic digital art.
- *Color-Changing Surfaces:* With smart paint or LED panels, walls can change color to match your mood or the time of day. Morning blue, sunset orange, or calming lavender – it is all at your fingertips.

Our home walls are no longer passive observers; they actively engage with us, anticipate our needs, and enhance our lifestyle. We will witness even more ingenious ways in which our walls become intelligent companions, seamlessly woven into the fabric of our daily existence.

14.5.1.7 Effortless Comfort

Imagine walking into a home where the temperature adjusts automatically to your preference as you unlock the door. Smart thermostats, like Nest or Ecobee, learn your routines and regulate heating and cooling for optimal comfort and energy savings. Similarly, smart lighting systems, controlled via apps or voice assistants like Alexa or Google Assistant, allow you to adjust lights throughout your home, setting the perfect ambiance for any occasion.

14.5.1.8 Enhanced Security

Gone are the days of fumbling for keys. Smart locks, such as those by August or Schlage, allow you to lock and unlock your doors remotely using your smartphone or even voice commands. Additionally, smart security cameras offer real-time monitoring of your home, providing peace of mind whether you are away or tucked in for the night. These cameras can be integrated with smart assistants, allowing you to view live footage or receive alerts on your phone if any unusual activity is detected.

14.5.1.9 Voice is the New Remote

Voice-activated assistants, like Amazon's Alexa or Google's Assistant, have revolutionized how we interact with our homes. These intelligent devices, often embedded in smart speakers like Sonos or Google Home, respond to voice commands, allowing you to control lights, adjust thermostats, play music, or even order groceries, all hands-free.

14.5.1.10 Energy Optimization

Smart devices go beyond convenience; they can also help you save money on your energy bills. Smart plugs, such as those from TP-Link or Kasa, allow you to monitor and control energy consumption of individual appliances. You can schedule lights to turn off automatically when you leave or program your coffee maker to start brewing right before you wake up, eliminating energy waste.

14.5.1.11 A Connected Ecosystem

The beauty of smart devices lies in their ability to work together. Through platforms like Apple HomeKit or Samsung SmartThings, you can create a seamless ecosystem where all your smart devices interact. Imagine setting a "goodnight" routine that dims lights, adjusts the thermostat, and locks the doors – all with a single voice command.

This is just a glimpse into the ever-expanding world of smart home technologies. As these systems continue to evolve and integrate with each other, our homes are poised to become not just sanctuaries but intelligent companions, offering a level of personalized control and convenience that was once unimaginable.

14.5.2 Health and Wellness

In the realm of health and fitness, technology has made significant strides, offering a plethora of tools to help individuals monitor and improve their well-being. This includes wearable fitness trackers, health monitoring apps, and smart scales. These technologies have revolutionized the way we approach health and fitness, making it easier than ever to track physical activity, sleep patterns, and overall well-being.

14.5.2.1 Wearable Fitness Trackers

Wearable fitness trackers are devices worn on the body that measure various aspects of physical activity. They come in various forms, including wristbands, watches, and even clothing. These devices use sensors to track movements, heart rate, and sometimes even body temperature and sweat levels. The data collected can provide insights into the user's level of physical activity, calories burned, distance traveled, and more. Some fitness trackers also include features like GPS for tracking outdoor workouts and alerts to remind users to move if they have been inactive for too long.

14.5.2.2 Health Monitoring Apps

Health monitoring apps are software applications designed to track and record health-related data. They can be installed on smartphones, tablets, or other devices and can monitor a wide range of health metrics. Some apps focus on specific areas like nutrition, mental health, or chronic disease management, while others offer a more comprehensive overview of overall health. These apps often allow users to set goals, track progress over time, and even share data with healthcare providers. Some health monitoring apps can sync with wearable devices for more accurate tracking.

14.5.2.3 Smart Scales

Smart scales are digital scales that do more than just measure weight. They use bioelectrical impedance analysis (BIA) to estimate body composition metrics like body fat percentage, muscle mass, bone mass, and water percentage. Some smart scales can even measure heart rate or provide weather forecasts. The data from smart scales can be synced with health monitoring apps or fitness trackers, providing a more comprehensive view of an individual's health and fitness progress.

14.5.2.4 The Impact of These Technologies

These technologies have made it easier for individuals to take control of their health and fitness. They provide real-time feedback, which can be a powerful motivator for behavior change. For example, seeing the number of steps taken in a day can encourage individuals to move more.

TABLE 14.10
Privacy Concerns of Health Monitoring Apps

Inadequate data protection	Many health apps do not provide sufficient protection for the sensitive health data they collect. This can lead to data breaches and unauthorized access to personal health information.
Data sharing without consent	A significant number of health apps have the potential to share personal data with third parties, such as Google and Facebook. In some cases, this data sharing occurs even when the app's privacy policy states it will not.
Lack of privacy policies	About a third of health apps do not provide access to their privacy policies. This makes it difficult for users to understand how their data is being used and protected.
Violation of privacy policies	Some health apps violate their own privacy policies by collecting and transmitting personal data.
Lack of regulatory oversight	Currently, there are no limitations on what companies can do with the data collected by health apps. This data is not safeguarded by laws like the Health Insurance Portability and Accountability Act (HIPAA), leaving users largely unprotected.
Permanent data	Once health data is collected and shared online, it is available forever. Unlike in Europe, where "right to be forgotten" laws offer some protection, in the U.S., digital health information is "immortal".

Similarly, tracking sleep patterns can highlight the need for better sleep habits. Moreover, these technologies can provide valuable data to healthcare providers, potentially improving the quality of care. For instance, a doctor could adjust a patient's treatment plan based on data from their health monitoring app.

While these technologies can provide valuable insights, they should not replace professional medical advice. They are only tools that can support health and fitness goals, but they do not diagnose or treat conditions.

Wearable fitness trackers, health monitoring apps, and smart scales are powerful tools in the modern health and fitness landscape. They empower individuals to track their physical activity, sleep patterns, and overall well-being, promoting a proactive approach to health and fitness. As these technologies continue to evolve, they will undoubtedly play an increasingly important role in personal health management.

Health monitoring apps, while beneficial, do raise several privacy concerns as shown in Table 14.10.

Before using a health app, it is important for users to understand these potential privacy risks and to review the app's privacy policy. Users should also consider whether the benefits of using the app outweigh these risks. It is always a good idea to consult with a healthcare professional before sharing sensitive health information through an app.

There are several secure alternatives to popular health monitoring apps. The examples are shown in Table 14.11.

It is always important to review their privacy policies and settings to ensure your data are being handled in a way that you are comfortable with. Also, keep in mind that no app can replace professional medical advice. Always consult with a healthcare professional about any health concerns.

14.5.3 Energy Efficiency

In the quest for a more sustainable lifestyle, innovations such as solar panels, smart meters, and energy-efficient appliances play a crucial role. These technologies not only help reduce our carbon footprint but also lead to significant cost savings over time.

TABLE 14.11
Examples of Alternative Health Monitoring Apps

HealthTap	This app allows you to speak with a healthcare professional within 24 hours, through chat or video call. It provides access to primary care video appointments and offers a discount to online therapy appointments with Talkspace.
Microsoft HealthVault	An online service and mobile app where you can keep your own medical records safe and up-to-date and share them securely with healthcare professionals and people.
Apple Health	This is a comprehensive health app that tracks a variety of health data and is integrated with most fitness and health apps.
Noom	This app offers personalized diet plans and tracking, psychological strategies to help you change your habits, and access to a health coach.
Headspace	A popular meditation app that offers guided meditations for a variety of goals, including stress management, focus improvement, and better sleep.
Sleep Cycle	This app tracks your sleep patterns and wakes you up during light sleep.

14.5.3.1 Solar Panels

Solar panels, also known as photovoltaic (PV) panels, are devices that convert sunlight into electricity. They are a key component in the shift toward renewable energy sources, and their popularity has been growing for several reasons.

14.5.3.2 How Solar Panels Work

Solar panels work by allowing particles of light, or photons, to knock electrons free from atoms, generating a flow of electricity. They are made up of many smaller units called photovoltaic cells. Each photovoltaic cell is basically a sandwich made up of two slices of semi-conducting material, usually silicon. When light shines on the cell it creates an electric field across the layers, causing electricity to flow.

14.5.3.3 Environmental Impact

One of the main advantages of solar panels is that they produce clean, green energy. Unlike fossil fuels, they do not emit harmful greenhouse gases or other pollutants during operation. The energy is produced locally, right where it is needed, which reduces the need for complex and costly transportation and distribution networks. This local, decentralized production of electricity can also help reduce losses that occur during electricity transmission.

14.5.3.4 Cost and Affordability

The cost of solar panels has been decreasing over the years, because of technological advancements and economies of scale. This trend is expected to continue, making solar power an increasingly affordable option for many households. In addition, many governments around the world offer incentives for installing solar panels, such as tax credits or feed-in tariffs, further reducing the cost.

14.5.3.5 Maintenance and Lifespan

Once installed, solar panels require little maintenance, especially if they do not use batteries. The only regular service they need is cleaning to remove dust and dirt. Most manufacturers recommend inspecting the system annually just to make sure everything is working properly. Solar panels are also known for their durability and long lifespan. Most manufacturers offer a warranty of 25 years, but the actual lifespan of a solar panel can be much longer, depending on the quality of the panel, the materials used, and the environment where it is installed.

Some of the key formulas associated with solar panels are as follows:

Solar Insolation: Solar insolation (**I**) is the amount of solar radiation incident on a surface area per unit time. It is typically measured in watts per square meter (W/m^2). The average solar insolation at the Earth's surface is about 1000 W/m^2 on a clear day.

Peak Sun Hours: Peak sun hours P refer to the equivalent number of hours per day when solar irradiance averages 1000 W/m^2. For example, six peak sun hours means that the energy received during total daylight hours equals the energy that would have been received had the irradiance for 6 hours been 1000 W/m^2.

Solar Panel Output per Day: The energy output E of a solar panel per day can be calculated by multiplying the power rating P of the panel (in watts), the number of peak sun hours S, and the panel's efficiency η:

$$E = P \cdot S \cdot \eta \tag{14.1}$$

Power Output: The power output **P** of a solar panel is given by the product of the solar cell's efficiency η, the surface area A of the solar cell, and the solar constant G (the amount of solar radiation received per unit area), which is approximately 1361 Watts per square meter.

$$P = \eta \cdot A \cdot G \tag{14.2}$$

Energy Produced: The energy E produced by a solar panel over a certain period of time can be calculated by multiplying the power output P by the time t the solar panel is exposed to sunlight.

$$E = P \cdot t \tag{14.3}$$

Solar Cell Efficiency: The efficiency of a solar cell is calculated as the electrical power output divided by the incident light power. If P_{out} is the power output and P_{in} is the incident light power, the efficiency η can be calculated as:

$$\eta = \frac{P_{out}}{P_{in}} \tag{14.4}$$

Number of Solar Panels Needed: If you know the total amount of energy E you need to produce, the energy E_i produced by a single solar panel, the number n of solar panels needed can be calculated as:

$$n = \frac{E}{E_i} \tag{14.5}$$

Please note that these are simplified formulas. The actual calculations can be more complex and depend on many factors such as the angle and intensity of sunlight, temperature, type of solar cell, and more. Always consult with a solar energy professional for accurate information.

Payback Period: The payback period is the time it takes for a solar panel system to pay for itself through electricity savings. If C is the total cost of the system, E is the annual energy production, and r is the rate per kWh, the payback period T in years is:

TABLE 14.12
Effects of Shading on Solar Panel

Reduction in energy efficiency	Shading reduces the energy efficiency of a PV panel by blocking sunlight. If a shadow is cast on a portion of an array, the cells under the shadow will be unable to produce electricity.
Power loss and uneven current distribution	Shading leads to power loss, uneven current distribution, and reduced system efficiency. Accurate shading analysis during system design helps optimize solar panel placement, select shading-tolerant modules, and validate system performance predictions.
Severe power loss	Shading just one out of 36 cells in a small solar module can reduce total power output by as much as 75%. This is because every PV cell in the cell string has to operate at the current set by the shaded cell. This limitation prevents the unshaded cells from operating at peak performance.
Role of bypass diodes	Bypass diodes can improve solar panel performance under shade conditions by preventing reduced output due to shaded cells. They allow current to bypass the shaded cells, reducing the impact of shading.
Module-level power electronics (MLPEs)	MLPEs, including DC optimizers and microinverters, enhance solar efficiency by enabling independent operation of each panel within an array. This means that even if one panel is shaded, the others can still operate at full capacity.
Importance of regular cleaning and maintenance	Regular cleaning and maintenance practices aimed at reducing energy loss caused by dirt buildup on panel surfaces are also important.

$$T = \frac{C}{E \cdot r} \tag{14.6}$$

Shading can significantly impact the efficiency of solar panels. The effects of shading on the solar panels are shown in Table 14.12.

Solar Irradiance Calculation: To figure out how much solar power we will receive, you need to calculate solar irradiance. This can be calculated using:

$$E = H \times \eta \times A \tag{14.7}$$

Where:

E = energy (kWh)
H = annual average solar radiation ($kWh/m^2/year$)
η = PV panel efficiency (%)
A = area of PV panel (m^2)

PV System Size Calculation: To estimate the size of the PV system required, use:

$$S = \frac{D}{365 \times H \times \eta} \tag{14.8}$$

Where:

S = size of PV system (kW)
D = total energy demand (kWh)
H = average daily solar radiation ($kWh/m^2/day$)
η = PV panel efficiency (%)

Battery Capacity Calculation: If you are planning to include a storage system, calculating the battery capacity is essential. This calculation considers the average daily consumption and desired autonomy (number of days you want your system to operate when there is no sun).

$$C = \frac{D \times N}{V} \tag{14.9}$$

Where:
C = Battery capacity (Ah)
D = Daily energy demand (kWh)
N = Days of autonomy (days)
V = Battery voltage (V)

Inverter Size Calculation: The inverter converts the DC electricity from the panels (and battery if present) into AC electricity for home use. Its size should be at least as large as the PV array output under peak conditions.

$$I = \frac{P}{V} \tag{14.10}$$

Where:
I = Inverter size (kVA)
P = Peak power from the PV array (kW)
V = Voltage at which the system operates (Volts).

Solar panels offer a way for individuals and businesses to reduce their environmental impact, save money, and even achieve a degree of energy independence. As technology continues to advance and the cost of solar panels continues to drop, it is likely that we will see even more widespread use of this renewable energy technology.

14.5.3.6 Smart Meters

Smart meters are revolutionizing the way we manage our energy consumption and the grid that powers our homes and businesses. These digital devices are a crucial component of the smart grid, transforming the way utilities and consumers interact with energy. One of the capabilities of smart meters is their ability to record energy consumption in intervals as short as an hour or less. This granular data provides utilities with a remarkable level of insight into the demand patterns across their service areas. By analyzing this real-time information, utilities can make more informed decisions about managing their resources and optimizing the grid's efficiency.

One of the primary benefits of smart meters is their enhanced communication capabilities. Unlike traditional analog meters, smart meters can transmit consumption data back to the utility on a daily basis, or even more frequently. This near real-time data flow allows utilities to closely monitor usage and quickly respond to changes in demand. This, in turn, enables them to better manage their power generation, distribution, and storage resources, ultimately leading to a more reliable and cost-effective energy system.

The impact of smart meters on utility operations is substantial. With this wealth of data at their fingertips, utilities can identify and address inefficiencies, reduce energy losses, and optimize the overall performance of the grid. For example, by detecting sudden spikes in usage or identifying areas with consistently high demand, utilities can proactively dispatch additional resources or implement targeted load-balancing strategies. This level of granular control and responsiveness is a significant improvement over the more reactive approach of traditional grid management. But the benefits of smart meters extend beyond the utility's operations; they also provide substantial advantages for consumers. One of the most significant benefits is the ability for consumers to track their energy usage in real time. Smart meters often come equipped with in-home displays or companion mobile applications that give consumers a detailed breakdown of their energy consumption, down to the individual appliance level. This visibility empowers consumers to identify and address wasteful energy practices, leading to significant cost savings and a reduced environmental impact.

Some smart meters can provide real-time pricing information to consumers. This allows them to take advantage of dynamic pricing, where electricity rates fluctuate based on the time of day and overall demand on the grid. Consumers can then choose to shift their energy-intensive activities, such as running the dishwasher or charging their electric vehicles, to off-peak hours when electricity is cheaper. This not only reduces their energy bills but also helps utilities smooth out the demand curve, reducing the need for costly peak generation capacity.

The integration of smart meters with the smart grid also opens new opportunities for consumer engagement and participation. For example, some utilities offer incentive programs that encourage consumers to reduce their energy consumption during periods of high demand, often referred to as demand response. By responding to these signals, consumers can earn credits or other rewards, while also helping the utility manage the grid more efficiently. Moreover, smart meters enable utilities to identify and fix problems more quickly. In the past, utilities often relied on customers to report outages. Now, with smart meters, utilities can instantly detect and locate outages, speeding up the restoration process.

For consumers, smart meters offer several benefits. They allow consumers to track their energy usage in real time, helping them identify wasteful practices and reduce energy consumption. This can lead to significant cost savings. For example, by identifying and eliminating unnecessary energy use, a household could potentially save hundreds of dollars per year on their energy bills.

Some smart meters also provide real-time pricing information, allowing consumers to shift their energy use to off-peak times when electricity is cheaper. This is known as demand response. By adjusting their energy use in response to price signals, consumers can save money and help balance the supply and demand of electricity.

As renewable energy sources, such as solar and wind, become more prevalent, smart meters will play a crucial role in facilitating the integration of these intermittent resources into the grid. By providing real-time data on energy generation and consumption, smart meters can help different utilities balance the supply and demand, ensuring a stable and reliable electricity supply even as the grid's generation mix becomes more diverse. Additionally, the rise of EVs presents both challenges and opportunities for the grid. Smart meters can help utilities monitor and manage the impact of EV charging on the grid, ensuring that the additional load is integrated smoothly and efficiently. Furthermore, some smart meters can enable vehicle-to-grid (V2G) technology, allowing EV batteries to serve as distributed energy storage resources, further enhancing the grid's flexibility and resiliency.

Privacy and security concerns have been raised, as the granular data collected by smart meters could potentially be used to infer personal information about consumers' behavior and habits. To address these concerns, robust data privacy and security measures have been developed, including encryption, data anonymization, and strict data access protocols.

Another challenge has been the cost of deploying smart meters, which can be substantial for utilities, especially in large-scale rollouts. However, the long-term benefits of improved grid efficiency, reduced energy losses, and enhanced consumer engagement are often seen as justifying the initial investment. Many governments have also implemented policies and incentives to support the widespread adoption of smart meters, further driving their deployment.

Smart meters are a pivotal component of the smart grid, revolutionizing the way we manage and consume energy, providing real-time data and communication capabilities. These digital devices empower utilities to optimize their operations, while also giving consumers the tools to take control of their energy usage and reduce their environmental footprint. As the energy landscape continues to

TABLE 14.13
The Components of Smart Meters and Their Works

Wireless networks	Most smart meters connect wirelessly. They typically connect to a Home Area Network (HAN) via a low-speed wireless protocol, such as Wireless M-Bus or ZigBee, or they connect to the cellular phone network. The amount of bandwidth needed by each individual meter is low.
Data concentrators	The meters connect to a data concentrator, which pulls together the data from several smart meters and sends it to the central grid. This is typically mounted on a utility pole.
Secure protocols	Smart meters do not connect directly to the internet. Instead, they send their data to a data concentrator. This allows many streams of data to go down one pipe, as it were. This is then sent to a relay and eventually to the utility company's servers where the data for each customer is stored for billing and the aggregate data can be analyzed.
Transmission frequency	Smart meters transmit data in short bursts, not constantly. The exact frequency varies but is typically every fifteen or every thirty minutes. This gives close to real-time readings without constant RF transmission.
Information sent	Only electricity usage information is sent out by the smart meter. The smart meter does not know or have access to any personal information, including other information held by smart appliances.

TABLE 14.14
The Use of Data from Smart Meters for Grid Management

Optimization and control	Smart meters provide real-time data on energy production and consumption. This information helps utilities manage the fluctuating supply from renewable sources, contributing to a more stable and reliable energy distribution system. Utilities can use this data for different power system operations and planning services, such as infrastructure sizing and upgrade, and generation forecasting.
Demand-side management	Smart meter data can be used for demand-side management. This involves adjusting demand for power on the customer's side of the meter. Demand-side management can help reduce peak demand and shift electricity use to off-peak times, thereby saving costs and reducing the need for additional power generation.
Integration of renewable energy sources	Smart meters facilitate the integration of renewable energy sources into the grid. They provide real-time data on energy production and consumption, which helps utility companies manage the fluctuating supply from renewable sources.
Improved grid management	Smart meters enable utilities to monitor grid performance more effectively and respond promptly to fluctuations in demand. This enhanced visibility helps prevent outages, optimize distribution, and integrate renewable energy sources more seamlessly.
Infrastructure modernization	Smart meters are part of the broader trend towards the digitalization of the energy system. They are a key component of smart grids, which use digital and other advanced technologies to monitor and manage the transport of electricity from all generation sources to meet the varying electricity demands of end users.

evolve, with the rise of renewable sources and electric vehicles, smart meters will play an increasingly crucial role in shaping a more sustainable and efficient energy future.

Smart meters communicate with utilities using a combination of wireless networks, data concentrators, and secure protocols. The ways they work are summarized in Table 14.13.

This communication process allows utilities to monitor energy usage in real time, detect and fix problems quickly, and manage resources more efficiently. For consumers, it provides real-time data on energy consumption, helping them identify wasteful practices and reduce energy consumption.

Utilities use the data from smart meters for grid management in several ways, as summarized in Table 14.14. The data from smart meters is invaluable for utilities. It provides them with the information they need to manage their resources more efficiently, respond to outages more quickly, and plan for future infrastructure needs.

14.5.3.7 Energy-Efficient Appliances

Energy-efficient appliances are revolutionizing the way we use energy in our homes and businesses. These appliances are designed to use less energy for the same performance compared to standard appliances. They range from refrigerators and washing machines to light bulbs and televisions. By using less energy, these appliances not only save money on energy bills but also reduce the demand for electricity, leading to fewer greenhouse gas emissions.

The concept of energy efficiency is simple: it is about doing more with less. This means using less energy to perform the same tasks, while cutting energy bills for homes, businesses, and industry. Energy efficiency is a key element of efforts to combat climate change and promote sustainable growth. It is also a cost-effective way to address energy security, air pollution, and energy access challenges.

Energy-efficient appliances are a prime example of how energy efficiency can be achieved in practice. These appliances use advanced technologies and designs to deliver superior performance while using less energy. For example, an energy-efficient refrigerator uses high-efficiency compressors, improved insulation, and more precise temperature and defrost mechanisms to reduce energy use. Similarly, energy-efficient washing machines use innovative designs that reduce the amount of hot water and electricity needed to wash clothes.

The benefits of energy-efficient appliances are numerous. For consumers, the most immediate benefit is the reduction in energy bills. By using less energy, these appliances cost less to operate, saving consumers money. Over time, the savings on energy bills can more than offset the higher upfront cost of energy-efficient appliances. In addition to saving money, energy-efficient appliances also help reduce the demand for electricity. This is particularly important in regions where electricity generation is a major source of greenhouse gas emissions. Energy-efficient appliances help lower greenhouse gas emissions by reducing electricity demand, contributing significantly to the fight against climate change.

Energy-efficient appliances also have benefits for utilities. By reducing the demand for electricity, these appliances can help utilities avoid the need to build new power plants or upgrade existing ones. This can result in significant cost savings for utilities, which can be passed on to consumers in the form of lower electricity rates.

Recognizing the benefits of energy-efficient appliances, the US Environmental Protection Agency and the US Department of Energy run a program called Energy Star. This program certifies appliances that are energy efficient. Consumers can look for the Energy Star label when shopping for appliances to ensure they are buying energy-efficient products. The Energy Star program has been highly successful in promoting energy efficiency. Since its inception, the program has helped consumers save billions of dollars on energy bills and has significantly reduced greenhouse gas emissions. The program has also spurred innovation in the appliance industry, leading to the development of new technologies and designs that further improve energy efficiency.

Energy-efficient appliances are a key tool in the fight against climate change and a major contributor to sustainable growth. They offer significant benefits for consumers, utilities, and the environment. As we continue to grapple with the challenges of climate change and the need for sustainable growth, the role of energy-efficient appliances is likely to grow. They represent a significant step forward in our efforts to create a more sustainable and energy-efficient future.

Some examples of energy-efficient appliances are shown in Table 14.15. Energy-efficient appliances can be seen from the Energy Star label. This label indicates that the appliance meets certain energy efficiency criteria set by the US Environmental Protection Agency and the US Department of Energy.

TABLE 14.15
Examples of Energy-efficient Appliances

Refrigerators	Energy-efficient refrigerators use high-efficiency compressors, improved insulation, and more precise temperature and defrost mechanisms to reduce energy use. For example, the GE Profile Stainless-Steel Smart 4-Door French Door Refrigerator is an energy-efficient appliance that connects to your home Wi-Fi network, allowing you to receive alerts should the internal temperature rise too high.
Dishwashers	Energy-efficient dishwashers use less water and energy. The Whirlpool Monochromatic Stainless-Steel Dishwasher is an example of an energy-efficient dishwasher. It has a quiet 55-decibel operation, cycle-setting memory, quick 1-hour wash cycle, and soil sensor to adjust cleaning power as needed.
Washing machines	Energy-efficient washing machines use less water and electricity to wash clothes. They often have features like load sensing to adjust water levels based on the size of the load.
Dryers	Energy-efficient dryers use less electricity to dry clothes. The Electrolux Titanium Front Load Electric Dryer is one of the top Energy Star appliances. It boasts advanced features such as Predictive Dry that calculates the time each load will need to dry, the Perfect Steam option to release wrinkles and static, and low-heat air dry cycle for more delicate items.
Air conditioners	Energy-efficient air conditioners use less electricity to cool your home or office. They often have features like programmable thermostats and energy-saving modes.
Light bulbs	Energy-efficient light bulbs, such as LED or CFL bulbs, use significantly less energy than traditional incandescent bulbs.
Televisions	Energy-efficient televisions use less energy to operate. They often have features like automatic brightness control and a power-saving sleep mode.

Everyone can contribute to energy-saving efforts in everyday life by doing simple tasks like turning off lights and electronics when not in use, adjusting the thermostat, using energy-efficient appliances, and adopting sustainable transportation options whenever possible. These are shown in Table 14.16.

14.5.3.8 The Impact of Energy-efficient Technologies

Adopting energy-efficient technologies is a crucial step toward a more sustainable future. These technologies, which range from energy-saving appliances to renewable energy systems, can lead to significant energy savings and a reduction in carbon emissions. However, it is important to remember that technology alone is not enough. A truly sustainable lifestyle requires a holistic approach that includes changes in behavior, such as reducing consumption, recycling, and choosing sustainable products.

Energy-efficient technologies are designed to use less energy for the same performance compared to standard technologies. For example, an energy-efficient light bulb uses less electricity to produce the same amount of light as a traditional incandescent bulb. Similarly, an energy-efficient refrigerator uses less electricity to keep food cold compared to a standard refrigerator. These technologies not only save money on energy bills by using less energy but also reduce the demand for electricity, thereby cutting down greenhouse gas emissions.

The benefits of energy-efficient technologies extend beyond individual households. On a larger scale, they can help reduce the overall demand for energy, leading to fewer power plants and less reliance on fossil fuels. This can result in significant reductions in carbon emissions, contributing to efforts to combat climate change. However, while energy-efficient technologies are a key part of the solution, they are not the whole solution. A sustainable lifestyle also requires changes in behavior. This includes reducing consumption, recycling, and choosing sustainable products.

Reducing consumption is perhaps the most fundamental behavioral change required for a sustainable lifestyle. This means using less energy, water, and other resources, and producing less waste. It can involve simple actions like turning off lights when not in use, taking shorter showers,

TABLE 14.16
Everyone's Daily Habit to Save Energy

Replace light bulbs	Traditional incandescent light bulbs consume excessive electricity and do not last as long as energy-efficient alternatives. When shopping for light bulbs, look for the government-backed symbol for energy efficiency, Energy Star.
Use smart power strips	"Phantom energy,“ also known as "standby energy" or "vampire energy," is the electricity used by electronics when turned off or in standby mode. Smart power strips, also known as advanced power strips, eliminate the problem of phantom loads by shutting off the power to electronics when they are not in use.
Install a programmable or smart thermostat	Smart thermostats can help reduce heating and cooling energy use without upgrading your HVAC system. Smart thermostats can automatically turn off or reduce heating and cooling while asleep or away.
Purchase energy-efficient appliances	When purchasing an appliance, you should pay attention to two numbers: the initial purchase price and the annual operating cost.
Use natural light	Whenever possible, use natural light instead of artificial light. Open the blinds or curtains to let in sunlight during the day.
Air seal your home	Sealing cracks, gaps, and leaks and adding insulation can save up to 10% on home heating and cooling costs.
Wash clothes in cold water	Washing clothes in cold water can save a significant amount of energy.
Reduce water heater temperature	Lowering the temperature of your water heater to 120 degrees Fahrenheit can save energy and money.

and buying only what you need. By reducing consumption, we can significantly reduce our environmental footprint.

Recycling is another important aspect of a sustainable lifestyle. By recycling, we can reduce the amount of waste that ends up in landfills and conserve natural resources. Recycling involves separating recyclable materials like paper, plastic, glass, and metal from our waste and ensuring they are sent to facilities where they can be processed and made into new products.

Choosing sustainable products is also crucial. This means choosing products that are made in a way that is environmentally friendly and socially responsible. It can involve buying products made from recycled materials, choosing products with less packaging, and buying from companies that prioritize sustainability.

While adopting energy-efficient technologies is an important step towards a sustainable lifestyle, it is not enough on its own. A truly sustainable lifestyle requires a holistic approach that includes not only technological solutions but also changes in behavior. Reducing consumption, recycling, and opting for sustainable products can enhance the benefits of energy-efficient technologies, making a significant contribution to a more sustainable future.

Talking about sustainability, we cannot ignore the roles of circular economy. The circular economy is a transformative approach that can significantly impact sustainability. It is centered around the idea of designing out waste, promoting the continual use of resources, and regenerating natural systems. In a circular economy, the goal is to reduce material use, redesign materials and products to be less resource intensive, and recapture "waste" as a resource to manufacture new materials and products. This transformative approach moves away from the traditional linear economy's 'take-make-dispose' model to a more sustainable 'reduce-reuse-recycle' model. The circular economy offers several following benefits:

- *Reduced Resource Consumption:* The circular economy concept is all about making better use of natural resources. By keeping products and materials in use for longer, we can significantly reduce the demand for new resources.

- *Economic Growth:* The circular economy can stimulate economic growth by creating new markets and opportunities. It encourages innovation and the development of new business models that can generate economic benefits.
- *Job Creation:* The transition to a circular economy can create jobs in various sectors, including recycling, remanufacturing, and renewable energy.
- *Enhanced Resilience to Shocks:* A circular economy can enhance resilience to shocks by reducing dependence on scarce resources and volatile commodity markets.
- *Decreased Environmental Footprint:* By reducing waste and promoting resource efficiency, the circular economy minimizes greenhouse gas emissions associated with resource extraction, manufacturing, and waste disposal.

However, the transition to a circular economy requires more than just technological innovation. It also requires changes in behavior, such as reducing consumption, recycling, and choosing sustainable products.

The circular economy is a powerful tool for achieving sustainability, and it offers a way to address our environmental challenges while also creating economic opportunities and improving societal well-being. However, to fully realize its potential, we must also embrace changes in our behavior and make sustainable choices in our everyday lives.

14.6 SMART GOVERNANCE

14.6.1 Overview of Smart Governance

A government is the system or group of people governing an organized community, generally a state. It is the political system by which a country or community is administered and regulated. Most of the key words commonly used to describe governments – words such as monarchy, oligarchy, and democracy – are of Greek or Roman origin. Government normally consists of legislature, executive, and judiciary. In more detail, government refers to the exercise of political authority over the actions, affairs, etc., of a political unit, people, etc., as well as the performance of certain functions for this unit or body; the action of governing; political rule and administration. It is the body of people that constitutes the governing authority of a political unit or organization. This includes the officials comprising the governing body of a political unit and constituting the organization as an active agency. The government is responsible for the direction and supervision of public affairs.

On the other hand, governance refers to the processes, structures, and organizational traditions that determine how power is exercised, how stakeholders have their say, how decisions are made, and how decision-makers are held accountable. It encompasses the mechanisms, processes, and institutions through which citizens and groups articulate their interests, exercise their legal rights, meet their obligations, and mediate their differences. Smart governance, consequently, is an extension of traditional governance that incorporates information and communication technology (ICT). It is a system where ICT is integrated with traditional infrastructures, coordinated and managed by human, political, economic, and social processes to promote a sustainable, prosperous and inclusive society.

The key differences between conventional governance and smart governance are:

1. *Technology Integration:* Conventional governance relies on traditional methods of communication and decision-making processes. In contrast, smart governance integrates digital technologies to enhance efficiency and effectiveness.
2. *Data-Driven Decision-Making:* Conventional governance often relies on manual data collection and processing, which can be time-consuming and prone to errors. Smart governance uses data analytics and big data to make informed decisions.

3. *Citizen Engagement:* While conventional governance often involves one-way communication from the government to citizens, smart governance promotes two-way communication. Citizens can provide feedback, participate in decision-making processes, and access services online.
4. *Transparency and Accountability:* Smart governance promotes transparency and accountability through the use of technology. Information is readily available to the public, and the actions of decision-makers can be easily tracked and evaluated.
5. *Sustainability:* Smart governance is more sustainable as it promotes the efficient use of resources. Digital technologies can help reduce paper use, optimize energy consumption, and improve service delivery.

While conventional governance has its merits, the integration of technology in smart governance offers numerous advantages that can lead to more efficient, effective, and inclusive governance. However, the transition to smart governance requires strong political willingness and consideration to ensure that all citizens can benefit from it, and potential challenges such as data security and privacy are adequately addressed.

A democratic government is a system of governance in which power is vested in the people and exercised by them directly or through freely elected representatives. The term 'democracy' is derived from the Greek words 'dēmos' (people) and 'kratos' (rule).

The democratic government are characterized by the following attributes:

1. *Citizen Participation:* In a democracy, the will of the people serves as the basis for collective decisions. Each member of the population expresses their opinion on each issue through voting. As all votes are equal, the opinion held by most members is considered the will of the majority.
2. *Representation:* In most democracies, the needs and wishes of the people are represented by elected lawmakers who are charged with writing and voting on laws and setting policy.
3. *Separation of Powers:* A democratic government provides for the separation of powers between governmental entities. This ensures that no single entity has absolute power and promotes checks and balances.
4. *Protection of Rights:* Democracies ensure the protection of natural rights and civil liberties. They feature competitive elections, freedom of expression, and protection of individual civil liberties and human rights.
5. *Responsibilities:* Democracy is about balancing rights and responsibilities. Citizens can expect the government to do things for them like providing protection, health services, education, and housing[4]. Citizens also have responsibilities like obeying the law and paying tax to the state.

A democratic government is a "government of the people, by the people, for the people". It empowers individuals to exercise political control over the form and functions of their government. Despite its inherent complexities, democracy remains the world's most practiced form of government. By these characteristics, smart governance will not work in undemocratic government, which is no transparency and lack people's participation in the governance.

14.6.2 Digital Government

Digital Government, or e-Government, refers to the use of information and communication technology (ICT), particularly the internet, to enhance the availability and delivery of government services to citizens, employees, businesses, and agencies. It involves the digital interactions between a citizen and their government (C2G), between governments and government agencies (G2G), between government and citizens (G2C), between government and employees (G2E), and between government and businesses/commerce (G2B).

14.6.2.1 Digital Transformation in Government

Digital transformation in government services is a comprehensive and strategic overhaul that goes beyond the mere digitization of existing processes and information. It is about harnessing the power of technology to fundamentally improve how government operates and delivers services to citizens.

14.6.2.2 Enhancing Citizen Experience

In the digital age, citizens expect the same level of service from government as they do from private sector companies. This means services that are accessible, convenient, and user-friendly. Digital transformation helps meet these expectations by enabling services to be delivered online, reducing the need for physical visits to government offices. For example, citizens can file taxes online, apply for permits, or renew their driver's licenses from the comfort of their homes. Moreover, digital transformation allows for personalized services. Leveraging data enables governments to gain insights into individual needs and preferences, allowing them to tailor services accordingly. For instance, a citizen accessing a government health portal could receive personalized health advice based on their medical history.

14.6.2.3 Streamlining Operations

Digital transformation also streamlines government operations, leading to increased efficiency and cost savings. By automating routine tasks, government employees can focus on more complex and value-added work. For example, AI-powered chatbots can handle common citizen queries, freeing up customer service representatives to deal with more complex issues. Furthermore, digital transformation enables better decision-making. Governments generate and collect vast amounts of data. Data analytics can transform raw data into actionable insights, assisting government officials in making well-informed decisions. For example, analyzing traffic patterns can improve the planning of public transportation routes.

14.6.2.4 Delivering Modern Services

Digital transformation paves the way for innovative and modern services. For example, digital IDs allow citizens to prove their identity online, enabling them to access a range of services securely and conveniently. Similarly, blockchain technology can be used for land registries, ensuring transparency and reducing fraud. Digital transformation allows for the integration of various services, providing a seamless experience for citizens. For instance, a single digital platform could allow a citizen to pay taxes, register a business, and apply for a permit.

14.6.2.5 Challenges and the Way Forward

Despite its benefits, digital transformation brings challenges as well. These include data security and privacy concerns, the need for digital literacy among citizens and government employees, and the risk of widening the digital divide. Governments need to invest in infrastructure, enact supportive policies, and foster a culture of innovation. Moreover, they need to put citizens at the center of their digital transformation efforts, involving them in the design and delivery of digital services. Digital transformation in government services is a journey. It is about continually leveraging technology to improve operations and meet changing citizen expectations. While the path may be challenging, the rewards – in terms of improved citizen satisfaction, efficiency, and innovation – are significant.

14.6.2.6 Benefits of Digital Government

The advent of digital technology has revolutionized various sectors, and government is no exception. The concept of digital government, where public administrations leverage digital tools to enhance their services, has gained significant traction worldwide. The benefits of digital government are manifold, impacting both citizens and the government itself.

For citizens, the most apparent benefit of digital government is easier access to information. In the traditional model, obtaining information from government agencies often involved navigating through bureaucratic red tape. However, digital government platforms provide citizens with a one-stop portal where they can access various services and information at their fingertips. This not only saves time but also makes it easier for citizens to understand and comply with regulations. Digital government leads to more efficient public services. Digital tools can automate routine tasks, streamline processes, and reduce paperwork, making public services faster and more efficient. For instance, citizens can now file taxes, apply for permits, or register businesses online, without the need to visit government offices. This convenience is particularly beneficial for those living in remote areas or those with mobility issues.

Greater transparency is another critical benefit for citizens. Digital government platforms can provide real-time updates and allow citizens to track the progress of their applications or complaints. This transparency not only enhances accountability but also builds trust between the government and its citizens. For governments, the digital transformation can lead to significant cost savings. Automating routine tasks and streamlining processes enables governments to reduce operational costs. Additionally, digital platforms minimize the need for physical infrastructure, leading to further savings.

Improved employee productivity is another benefit for governments. Digital tools can free up employees from routine tasks, allowing them to focus on more complex and value-added tasks. This not only improves productivity but also leads to better job satisfaction. Digital government can lead to increased citizen satisfaction. Providing efficient, transparent, and convenient services enables governments to enhance citizen satisfaction and public trust. This is particularly important in today's digital age, where citizens' expectations are shaped by their experiences with private sector digital services. However, the transition to digital government requires significant investment in digital infrastructure, skills training, and change management. Moreover, governments must also address issues related to digital divide and data security to ensure that the benefits of digital government are equitably distributed, and that citizens' data are protected.

14.6.2.7 Challenges in Implementing Digital Government

Despite the potential benefits, implementing digital government is not without challenges. These include data security and privacy concerns, digital divide issues, and resistance to change among employees and citizens. Governments must address these challenges to successfully implement and benefit from digital transformation.

14.6.2.8 Case Studies of Digital Government

In the modern era, the digital revolution has transformed the way governments operate and interact with their citizens. Many governments around the world have made significant strides in implementing digital government, leveraging technology to improve public services, enhance transparency, and foster citizen engagement.

Estonia, a small Baltic nation, is often cited as a leading example of digital government. The country embarked on its digital journey in the late 1990s, and today, virtually all public services are available online. From voting in elections to filing taxes, Estonians can access a wide range of services with just a few clicks. The cornerstone of Estonia's digital government is the X-Road platform, a secure data exchange layer that allows different information systems to communicate and operate together. This has not only streamlined public services but also significantly reduced bureaucratic red tape, saving both time and resources for the government and citizens alike. Moreover, Estonia's digital ID card system is another testament to its digital prowess. Every Estonian citizen and resident have a digital ID card, which serves as a legal travel ID for EU citizens and gives access to all of Estonia's secure e-services. This has fostered a sense of trust and confidence in the digital system among the populace, further driving the adoption of digital services.

On the other side of the world, Singapore's Smart Nation initiative is another excellent example of a digital government. Launched in 2014, the initiative aims to leverage technology to improve living standards, create economic opportunity, and build a closer community. It encompasses several key strategic national projects, including a national digital identity, a cashless society, a smart urban mobility solution, and a digital government service platform.

The Smart Nation initiative is not just about using technology for technology's sake. Instead, it is about harnessing technology to solve real-world problems and make life better for everyone in Singapore. For instance, the initiative has led to the development of smart homes equipped with sensors to monitor the elderly's health, cashless payment solutions to make transactions more convenient, and data analytics tools to improve urban planning and sustainability. The initiative has also created significant economic opportunities. It has also attracted global tech giants, nurtured local start-ups, and created high-value jobs, driving Singapore's economic growth in the digital age.

14.6.2.9 The Future of Digital Government

The future of digital government is a fascinating subject, with the potential to revolutionize the way public services are delivered. The key to this future lies in the ability of governments to be agile, user-centric, and inclusive. This means that governments must be able to adapt quickly to changes, focus on the needs of their citizens, and ensure that everyone has access to digital services.

Emerging technologies such as AI, ML, blockchain, and the Internet of Things (IoT) offer exciting opportunities to further transform government services. AI and ML, for instance, can be used to automate routine tasks, predict trends, and personalize services, making government operations more efficient and responsive. Blockchain, on the other hand, can enhance transparency and security, making it easier for citizens to trust digital services. IoT, meanwhile, can improve the delivery of public services by connecting devices and systems, enabling real-time monitoring and decision-making. However, these technologies also present new challenges, particularly in terms of data security and privacy. As governments collect and process more data, they must ensure that this data is protected from cyber threats. This requires robust cybersecurity measures, including encryption, secure networks, and regular security audits. Governments must also respect the privacy of their citizens. This means that they must have clear policies on data collection, use, and sharing, and they must obtain the consent of their citizens before collecting their data. There are also organizational and cultural challenges. Implementing these technologies requires significant investment in infrastructure and skills training. Governments must also foster a culture of innovation and risk-taking, as the adoption of these technologies often involves trial and error. Furthermore, governments must ensure that their employees are comfortable with these technologies and are willing to use them in their work.

Despite these challenges, the potential benefits of digital government are enormous. For citizens, digital government can provide easier access to information, more efficient public services, and greater transparency. For governments, it can lead to cost savings, improved employee productivity, and increased citizen satisfaction.

Governments must be agile, user-centric, and inclusive, and they must leverage emerging technologies to transform their services. However, they must also address the challenges posed by these technologies, particularly in terms of data security and privacy. By doing so, they can create a digital government that truly serves the needs of its citizens and contributes to a better future for all.

14.6.3 Transparency and Accountability

Transparency and accountability are two fundamental pillars of any governance system, and they become even more critical in the context of Smart Governance. Smart Governance refers to the use of ICT by government bodies to provide public services in a more efficient, convenient, and citizen-centric manner. While the adoption of ICT brings numerous benefits, it also necessitates a renewed focus on transparency and accountability to ensure the trust and confidence of citizens.

Transparency in Smart Governance refers to the openness and clarity with which government bodies operate. It involves making information readily available and accessible to the public, thereby enabling citizens to understand the workings of the government. In the digital age, transparency takes on a new dimension. Governments are not only expected to share information but also to do so in a manner that is easy to understand and navigate. This includes providing real-time updates, visualizing data, and using interactive platforms that allow citizens to explore and analyze information.

Accountability, on the other hand, refers to the government's obligation to answer for its actions. In the context of Smart Governance, accountability involves using ICT to track and report on the performance of government services. This includes setting clear performance metrics, regularly monitoring and evaluating these metrics, and sharing the results with the public. By doing so, governments can demonstrate that they are delivering on their promises and meeting the expectations of their citizens.

Another challenge is ensuring inclusivity. As governments move toward digital services, they must ensure that these services are accessible to all citizens, including those who may not have access to the internet or digital devices. This requires investing in digital infrastructure, providing digital literacy training, and designing services that are easy to use.

Despite these challenges, the potential benefits of transparency and accountability in Smart Governance are enormous. They can enhance citizen trust and satisfaction, improve the quality of public services, and foster a culture of innovation and continuous improvement in government bodies. Moreover, by leveraging these technologies, governments can also drive economic growth and social development. Realizing these benefits requires a strategic and holistic approach that considers not just the technological aspects, but also the organizational, cultural, and policy dimensions of Smart Governance.

Smart Governance can employ blockchain technology. Blockchain technology in smart governance refers to the use of a decentralized, distributed digital ledger system to enhance the efficiency, transparency, and accountability of government operations. This technology is being increasingly integrated into the comprehensive ideology of Smart Cities.

Blockchain technology, a revolutionary concept that has significantly impacted various sectors, including governance, finance, and supply chain, among others, has an intriguing history. The very first appearance of the idea of Blockchain was in 1982 when Cryptographer David Chaum presented a dissertation "Computer Systems Established, Maintained, and Trusted by Mutually Suspicious Groups". Further down the line in 1991, Stuart Haber and W. Scott Stornetta wrote about their work on a cryptographically secured chain of blocks. They wanted to implement a system wherein document timestamps could not be tampered with. In 1992, Haber, Stornetta, and Dave Bayer incorporated Merkle trees into the design, which improved its efficiency by allowing several document certificates to be collected into one block.

The term "blockchain" was not coined until 2008, when an individual (or group of individuals) using the pseudonym Satoshi Nakamoto created a blockchain to serve as the public distributed ledger for Bitcoin cryptocurrency transactions. This implementation of the blockchain within Bitcoin made it the first digital currency to solve the double-spending problem without the need for a trusted authority or central server. The Bitcoin design has inspired other applications and blockchains that are readable by the public and are widely used by cryptocurrencies.

As for why it is called "blockchain", the term comes from the way this technology works. A blockchain is essentially a chain of blocks, where each block is a ledger of publicly available data. This data could be anything from the details of a monetary transaction to medical records to proof of ownership. Each block contains a cryptographic hash of the previous block, a timestamp,

TABLE 14.17
Transparency due to Blockchain Employment

Immutable records	Once data is recorded in a blockchain, it cannot be altered or deleted. This immutability makes blockchain a reliable source of truth, as citizens can trust that the information has not been tampered with.
Transparent transactions	All transactions on a blockchain are visible to all participants, which enhances transparency. This can be particularly useful in areas like public procurement, where blockchain can provide a transparent record of all bids and contracts.
Decentralization	Unlike traditional databases, which are controlled by a single entity, blockchains are decentralized. This means that no single party has complete control over the entire chain, which can prevent misuse of power and enhance trust among citizens.
Auditability	The transparency and immutability of blockchain also make it highly auditable. Governments can use blockchain to track and verify the use of public funds, which can help prevent corruption and misuse of funds.
Smart contracts	Blockchain can also support the use of smart contracts, which are self-executing contracts with the terms of the agreement directly written into code. This can automate and streamline administrative processes, making them more transparent and efficient.

and transaction data. As each block contains information about the previous block, they effectively form a chain, with each additional block linking to the ones before it. Consequently, blockchain transactions are irreversible in that, once they are recorded, the data in any given block cannot be altered retroactively without altering all subsequent blocks.

Blockchain technology plays a fundamental role in enhancing transparency in various sectors, including Smart Governance. Blockchain is a decentralized and distributed digital ledger that records transactions across multiple computers so that any involved record cannot be altered retroactively, without the alteration of all subsequent blocks. In the context of Smart Governance, blockchain can significantly enhance transparency in several ways as shown in Table 14.17.

Blockchain technology has a wide range of applications beyond smart governance and cryptocurrencies. Here are some key areas where blockchain is making a significant impact:

- *Supply Chain Management:* Blockchain can enhance transparency, improve efficiency, and reduce fraud in supply chain management. By using blockchain, supply chain managers can track products from the point of origin to the point of consumption, creating a secure and transparent record of the entire supply chain. This can help reduce waste, improve quality control, and enhance accountability.
- *Healthcare:* In the healthcare sector, blockchain can be used for secure and transparent record-keeping, patient data management, drug traceability, and more. It can also help in research for new drugs and treatments.
- *Digital Identity Verification:* Blockchain can provide a secure and decentralized framework for digital identity verification. This can be particularly useful in areas such as online voting, banking, and e-commerce.
- *Voting Systems:* Blockchain can be used to increase security, ensure accuracy, and facilitate remote voting in voting systems. By using blockchain, voting systems can create a secure and transparent record of votes that is immutable and tamper-proof.
- *Smart Contracts:* Blockchain supports the use of smart contracts, which are self-executing contracts with the terms of the agreement directly written into code. This can automate and streamline administrative processes in various industries such as real estate, legal, and logistics.

- *Internet of Things (IoT):* Blockchain can provide a secure and scalable framework for device-to-device communication in the IoT ecosystem. This can enable real-time data exchange, automated decision-making, and improve system efficiency.
- *Media:* Blockchain can be used to manage intellectual property rights, track digital media ownership, prevent unauthorized content copying, and enable fair compensation for content creators.
- *Government Services:* Many governments are exploring the use of blockchain for various applications, such as land registry, tax collection, and public service delivery. Blockchain can enhance transparency, reduce bureaucracy, and improve public trust in government services.

These are just a few examples. The potential applications of blockchain technology are vast and continue to grow as more industries recognize its benefits.

14.6.4 Data-Driven Decision-Making

Data-driven decision-making is a critical component of smart governance, which refers to the use of ICT by government bodies to provide public services in a more efficient, convenient, and citizen-centric manner. Leveraging data allows governments to make more informed decisions, enhance public services, and improve transparency and accountability.

In the context of smart governance, data-driven decision-making involves collecting, analyzing, and using data to guide policymaking and service delivery. This can include a wide range of data, from demographic data and economic indicators to citizen feedback and social media sentiment. By analyzing this data, governments can gain insights into the needs and preferences of their citizens, the effectiveness of their policies, and the performance of their services.

One of the key benefits of data-driven decision-making is that it allows governments to be more responsive to the needs of their citizens. By analyzing data on citizen feedback, social media sentiment, and service usage, governments can identify areas of concern, understand citizen preferences, and tailor their services accordingly. This can lead to more effective and efficient public services that meet the needs of citizens.

Data-driven decision-making can also enhance transparency and accountability in government operations. Making data publicly available enables governments to provide citizens with a clear view of their operations and decision-making processes. This can enhance trust in government and foster a culture of accountability. Moreover, data-driven decision-making can lead to more effective policymaking. Analyzing data on the impact of past policies enables governments to determine what works and what does not, guiding future policy decisions based on this information. This can lead to more effective policies that deliver better outcomes for citizens. However, implementing data-driven decision-making in smart governance is not without challenges. One of the key challenges is ensuring the quality and reliability of data. Governments need to ensure that the data they collect is accurate, up-to-date, and representative of the population. This requires robust data collection and management practices, as well as the use of advanced analytics tools and techniques.

Another challenge is ensuring data privacy and security. As governments collect and analyze more data, they need to ensure that this data is protected from cyber threats and that the privacy of citizens is respected. This requires robust cybersecurity measures, clear data privacy policies, and strong data governance frameworks. Furthermore, implementing data-driven decision-making requires a cultural shift within government bodies. It requires government officials to embrace data and analytics, and to be willing to make decisions based on data rather than intuition or tradition. This requires training and capacity building, as well as leadership commitment.

AI can play a transformative role in enhancing data-driven decision-making. Its capabilities extend from speeding up data analysis to providing more accurate and predictive insights, enabling

TABLE 14.18
AI Enhances Data-driven Decision-Making in Smart Governance

Speed and accuracy	AI can analyze large datasets without error. This helps teams to focus better on work relevant to their field. AI decision-making allows organizations to make faster, accurate, and consistent decisions by capitalizing on datasets with AI.
Overcoming human limitations	AI allows us to overcome our inherent limitations as human processors–low throughput and cognitive bias–by assigning machines to process the vast volumes of data we produce. This lets us humans apply judgment, culture, value, and context to the decision options machines can produce.
From data-driven to AI-driven	To fully leverage the value contained in data, organizations need to evolve from data-driven to AI-driven workflows. This is not merely an automation play, which is a marginal benefit of integrating AI into decisions.
Enhanced analysis	AI complements Data Analytics by introducing a layer of intelligence that enables more nuanced and sophisticated analysis. It empowers systems to discern patterns, make predictions, and automate decision-making processes.
Strategic decision-making	Effective use of artificial intelligence in strategic decision-making will be one of the biggest determinants for future competitiveness. The interaction between humans and AI, as well as the ability to choose which decisions to delegate to AI, will be among the most important skills for decision-makers.

TABLE 14.19
Contribution of Machine Learning Data-driven Decision-Making

Efficiency and pattern recognition	ML algorithms can analyze large datasets efficiently, identifying patterns that might be difficult for humans to detect. This helps organizations make more informed and unbiased decisions.
Predictive modeling	ML can look at old data and find patterns. Those patterns help guess what will happen next. It is pretty accurate, too. Businesses love this as they can use it to make plans and good choices, like in finance.
Overcoming human limitations	ML allows us to overcome our inherent limitations as human processors–low throughput and cognitive bias–by assigning machines to process the vast volumes of data we produce. This lets us humans apply judgment, culture, value, and context to the decision options machines can produce.
Enhanced analysis	ML complements Data Analytics by introducing a layer of intelligence that enables more nuanced and sophisticated analysis. It empowers systems to discern patterns, make predictions, and automate decision-making processes.
Strategic decision-making	Effective use of artificial intelligence in strategic decision-making will be one of the biggest determinants for future competitiveness. The interaction between humans and AI, as well as the ability to choose which decisions to delegate to AI, will be among the most important skills for decision-makers.

businesses and governments to make better decisions faster. The capability of AI in enhancing data-driven decision-making in smart governance is shown in Table 14.18.

ML can also contribute to enhancing data-driven decision-making. ML is a subset of AI that uses statistical techniques to give computer systems the ability to learn from data, without being explicitly programmed. The ways ML contributes to data-driven decision-making are shown in Table 14.19.

An example on how AI makes decisions based on big data through a process that involves following steps:

Step 1 Data Collection: AI systems first collect vast amounts of data from various sources. This data can be in the form of text, numbers, images, videos, and more. If current data is already in the database, extract the necessary data.

Step 2 Data Preprocessing: The collected data is then preprocessed to remove any noise or inconsistencies. This step is crucial as the quality of data directly impacts the performance of the AI system.

Step 3 Feature Extraction: In this step, relevant features or characteristics are extracted from the preprocessed data. These features are used to train the AI model.

Step 4 Model Training: The AI system is then trained using ML algorithms. During training, the model learns to identify patterns and relationships in the data.

Step 5 Decision-Making: Once the model is trained, it can make decisions based on new data. For instance, an AI model trained to predict fraud can analyze new transactions and decide whether they are likely to be fraudulent.

Step 6 Evaluation and Optimization: The decisions made by the AI system are continuously evaluated for accuracy. The system learns from its mistakes and optimizes its performance over time.

An example of this process in action is the use of AI in healthcare for disease diagnosis. AI systems can analyze medical records, lab results, and imaging data to diagnose diseases and suggest treatment plans. The AI system makes these decisions based on patterns it has learned from analyzing large amounts of historical healthcare data. However, AI can process and learn from vast volumes of data, but it is the humans who apply judgment, culture, value, and context to the decision options machines can produce. This interaction between AI technology and human decision-making is a key aspect of AI-driven decision-making.

Implementing AI systems for decision-making can present several arguments as shown in Table 14.20.

TABLE 14.20
Challenges of Implementing AI in Decision-Making

Data quality and reliability	AI systems heavily rely on high-quality and reliable data for decision-making. Challenges can arise when dealing with incomplete, inaccurate, or biased data, which can lead to flawed insights and decisions.
Infrastructure	For AI systems to function optimally, they need to process large amounts of information in fractions of a second. However, many businesses are still using outdated equipment that is not capable of taking on the challenge of AI implementation.
Integration into existing systems	Incorporating AI in your system is much more than downloading a few plugins. You need to consider whether you have the storage, processors, and infrastructure necessary for the system to function properly.
Lack of understanding and context	AI applies complex algorithms to inputs and data, to produce outputs that its algorithms predict from its training data. However, there can be a lack of human understanding and context.
Ethical considerations	Ethical considerations are another significant challenge in AI implementation. These include issues related to privacy, bias, transparency, and accountability.
Interpretability and explainability	AI systems, particularly those based on complex machine learning models, can sometimes act as a "black box," making it difficult to understand how they arrived at a particular decision.
Overreliance and decision bias	There is a risk that users may over-rely on AI systems or develop a decision bias based on the outputs of the AI system.
Developing "unwarranted" trust	Users might develop unwarranted trust in the AI system, leading to over-reliance on the system and underestimation of its limitations.

Data-driven decision-making is a fundamental component of smart governance. It can lead to more responsive and effective public services, enhance transparency and accountability, and improve policymaking. However, implementing data-driven decision-making requires overcoming several challenges, including ensuring data quality and privacy, and fostering a data-driven culture within government bodies. Despite these challenges, the potential benefits of data-driven decision-making in smart governance are immense, and governments around the world are increasingly embracing this approach to improve their services and decision-making processes.

14.6.5 Citizen Engagement

Citizen engagement is a cornerstone of smart governance, which refers to the use of ICT by government bodies to provide public services in a more efficient, convenient, and citizen-centric manner. Engaging citizens in the decision-making process ensures that government policies and services are responsive to their needs and preferences.

In smart governance, citizen engagement involves a wide range of activities, from soliciting citizen feedback and conducting public consultations to involving citizens in policymaking and service delivery. This can be facilitated through various digital platforms, such as online forums, social media, and mobile apps.

One of the key benefits of citizen engagement is that it allows governments to tap into the collective intelligence of their citizens. By soliciting citizen feedback, governments can gain insights into the needs and preferences of their citizens, identify areas of concern, and tailor their services accordingly. This can lead to more effective and efficient public services that meet the needs of citizens. Citizen engagement can enhance transparency and accountability in government operations. Involving citizens in the decision-making process can provide them with a clear view of their operations and decision-making processes. This, consequently, can enhance trust in government and foster a culture of accountability. Citizen engagement can lead to more inclusive and equitable public services. Engaging all citizens, including marginalized and vulnerable groups, ensures that government services are accessible and address the needs of everyone. Implementing citizen engagement in smart governance faces many challenges. One of the key challenges is ensuring meaningful participation. This requires creating an environment where citizens feel safe to express their views, where their views are considered, and where they can see the impact of their participation. Another challenge is ensuring inclusivity. As governments move towards digital services, they must ensure that these services are accessible to all citizens, including those who may not have access to the internet or digital devices, and this requires investing in digital infrastructure, providing digital literacy training, and designing services that are easy to use.

Governments must also address issues related to data privacy and security. As governments collect and analyze more data, they must ensure that this data is protected from cyber threats and that the privacy of their citizens is respected. This requires robust cybersecurity measures, clear data privacy policies, and strong data governance frameworks. On the other hand, the potential benefits of citizen engagement in smart governance are enormous. They can enhance citizen satisfaction, improve the quality of public services, and foster a culture of innovation and continuous improvement in government bodies. However, realizing these benefits requires a strategic and holistic approach that considers not just the technological aspects, but also the organizational, cultural, and policy dimensions of smart governance.

Citizen's engagement is a core component of smart governance for the sake of transparency and participation. Citizen's engagement refers to the use of ICT by government bodies to provide public services in a more efficient, convenient, and citizen-centric manner. Here are some efficient and effective mechanisms for citizen engagement in smart governance:

- *Digital Platforms for Feedback and Interaction:* Governments can use digital platforms such as online forums, social media, and mobile apps to solicit citizen feedback and conduct public consultations. These platforms allow for real-time interaction and can reach a wide audience.

- *E-Participation:* E-Participation is about fostering civic engagement and open, participatory governance through ICTs. It offers government authorities new possibilities for interaction with citizens and local communities.
- *Open Data Initiatives:* Governments can make data publicly available, providing citizens with a clear view of their operations and decision-making processes. This enhances transparency and allows citizens to make informed decisions.
- *Smart Cities Initiatives:* Smart cities invest in initiatives that enhance citizens' digital literacy, providing training and resources to enable them to effectively participate in digital governance processes.
- *Public–Private Partnerships (PPPs):* Governments can collaborate with private sector companies to develop and implement digital solutions that enhance citizen engagement.
- *Inclusive Design:* Digital services should be designed to be accessible and usable by all citizens, including those with disabilities and those who are not tech-savvy.
- *Trust and Security:* For citizens to engage with digital government services, they need to trust that their data will be handled securely and that their privacy will be respected.

These mechanisms, when implemented effectively, may lead to more responsive and effective public services, enhance transparency and accountability, and improve policymaking. However, implementing citizen engagement in smart governance requires overcoming several constraints, including ensuring data quality and privacy, and fostering a data-driven culture within government bodies.

Citizen's engagement is a fundamental component of smart governance. It is not just a principle to be upheld but also a powerful tool that can drive the digital transformation of government services. As governments around the world continue to embrace smart governance, they must strive to enhance citizen engagement, thereby ensuring that the digital revolution benefits all citizens.

14.7 SMART URBAN ENVIRONMENT

14.7.1 Smart Resource Management

Smart Resource Management (SRM) is one of the components of smart cities, focusing on the effective and efficient use of resources through the application of ICT and IoT devices. This approach aims to ensure sustainability and improve the quality of life for city dwellers while minimizing the environmental impact. The concept of SRM is rooted in the idea that resources, whether they are natural, like water and energy, or man-made, like infrastructure and services, need to be managed intelligently to maximize their value and minimize waste. This is particularly important in urban environments, where the demand for resources is high and the impact on the environment can be significant.

One of the key technologies in SRM is IoT. IoT devices, such as sensors and smart meters, can collect real-time data about the use and condition of resources. This data can then be analyzed to gain insights into usage patterns, predict demand, identify inefficiencies, and make informed decisions about resource allocation. For example, smart meters can provide detailed information about energy use in a building, enabling more efficient use of electricity and reducing carbon emissions.

ICTs provide the infrastructure for collecting, transmitting, storing, and analyzing data from IoT devices for SRM. Moreover, they enable communication and coordination among different stakeholders involved in resource management, including government agencies, service providers, and citizens. For instance, a city council can use ICTs to inform residents about water shortages and encourage them to reduce their water use.

Designing systems and protocols for managing resources is another important aspect of SRM. These systems and protocols need to be flexible and adaptable to accommodate changes in resource availability and demand. They also need to be transparent and accountable to ensure that resources

TABLE 14.21
The Roles of Data Analytics in SRM

Real-time monitoring and control	IoT devices, such as sensors and smart meters, collect real-time data about the use and condition of resources. This data is then analyzed to monitor resource usage and control resource allocation in real-time.
Predictive analysis	Data analytics can be used to predict future resource needs based on historical data and trends. This can help in planning and allocating resources more effectively.
Optimization of resource allocation	Data analytics can help optimize resource allocation by identifying inefficiencies and suggesting improvements. For example, it can help in reducing waste and lowering costs in water and energy distribution.
Improving grid reliability	Data analytics can improve grid reliability by using sensor data to unlock insights about asset performance, enabling utilities to forecast potential equipment failures and intervene long before an outage results.
Enhancing energy efficiency	Data analytics can analyze consumption trends in minute detail, enabling utilities to better predict load requirements and identify times of peak demand.
Cost savings	Data analytics can lead to significant cost savings for utilities in many ways. For example, smart inertia management is a key analytics capability of the best advanced energy management systems (AEMS).
Asset management	With the right smart grid analytics, utilities can translate data into insights about their asset performance management. Those insights can unlock proactive planning and predictive maintenance that led to reduced downtime and optimized asset lifecycles.

are used responsibly and equitably. Furthermore, they need to be integrated with the broader environmental governance of a smart city, aligning resource management with environmental policies and objectives.

Data analytics are essential in SRM. Data analytics provides insights that can lead to more efficient and effective use of resources. The roles of data analytics in SRM are shown in Table 14.21.

Data analytics enhances water management through real-time monitoring, predictive analysis, and resource optimization. It aids in grid reliability, water efficiency, cost savings, and asset management. These methods, detailed in Table 14.22, contribute to sustainable and efficient water use in smart cities.

14.7.2 Smart Environmental Governance

Smart Environmental Governance (SEG) is a transformative approach to environmental management that leverages digital technologies to enforce and implement environmental regulations and policies. This innovative approach marks a significant departure from traditional methods of environmental governance, which are often reactive and less efficient. Instead, SEG promotes proactive, data-driven, and participatory models that are more effective and sustainable.

Traditional environmental governance often relies on manual monitoring and enforcement, which can be slow, labor-intensive, and prone to errors. In contrast, SEG uses digital technologies like sensors, data analytics, and AI to monitor environmental conditions in real-time, predict future scenarios, and automate enforcement processes. This shift not only improves the efficiency and effectiveness of environmental governance but also enables a more proactive approach. Instead of reacting to environmental issues after they occur, SEG allows for the anticipation and prevention of such issues.

Digital technologies are at the core of SEG. Sensors and IoT devices can collect vast amounts of data about various environmental parameters, such as air and water quality, waste generation, and energy consumption. This data can then be analyzed using advanced data analytics and AI

TABLE 14.22
Improving Water Management Offered by Data Analytics

Real-time monitoring and control	Data analytics allows for real-time monitoring and control of water resources. IoT devices, such as sensors, collect real-time data about the use and condition of water resources. This data is then analyzed to monitor resource usage and control resource allocation in real-time.
Predictive analysis	Data analytics can be used to predict future water demand and potential system failures based on historical data and trends. This predictive modeling can help in planning and allocating resources more effectively.
Optimization of resource allocation	Data analytics can help optimize resource allocation by identifying inefficiencies and suggesting improvements. For example, it can help in reducing waste and lowering costs in water distribution.
Enhancing system reliability and efficiency	Data analytics can improve the reliability and efficiency of the water distribution system by monitoring and analyzing real-time data from various sources, such as flow rates, pressure levels, and water quality indicators, potential issues or failures can be identified.
Enhancing water efficiency	Data analytics can improve water efficiency. Real-time data from sensors and smart meters can provide detailed insights into water usage patterns. This information can be used to identify areas of excessive water use and implement measures to reduce wastage.
Cost savings	Data analytics can lead to significant cost savings in water management. By identifying inefficiencies and areas of excessive water use, utilities can implement measures to reduce wastage, leading to substantial cost savings.
Asset management	Assets can include infrastructure like pipes, pumps, and treatment facilities. Data analytics can help manage these assets more effectively by predicting maintenance needs and optimizing their performance

algorithms to gain insights, identify trends, and make predictions. For instance, data analytics can help identify areas with high levels of air pollution, enabling authorities to take targeted actions to improve air quality. One of the key advantages of SEG is its ability to facilitate data-driven decision-making. Providing accurate and timely data enables SEG to help policymakers and environmental managers make evidence-based decisions. This approach leads to more effective policies and strategies tailored to the specific needs and conditions of each area.

SEG also promotes participatory models of environmental governance. Digital platforms can be used to engage citizens in environmental decision-making processes, fostering a sense of ownership and responsibility. Citizens can provide valuable input, report environmental issues, and even participate in environmental monitoring using their own devices. This not only enhances the effectiveness of environmental governance but also promotes environmental awareness and stewardship among citizens.

14.7.2.1 Concept of Smart Environmental Governance

SEG is rooted in the broader concept of smart cities, which use technology to improve the quality of life for their citizens, enhance sustainability, and streamline urban services. In the context of environmental governance, 'smart' refers to the use of digital technologies like IoT devices, AI, and data analytics to monitor environmental conditions, enforce regulations, and engage citizens in environmental stewardship.

The components of SEG are summarized in Table 14.23.

While SEG offers many benefits, it also presents several challenges. These include the need for robust and secure digital infrastructure, the protection of privacy and data security, and the need for digital literacy among citizens and officials. However, these challenges also present opportunities for innovation, collaboration, and capacity building. The benefits and challenges of SEG are presented in Table 14.24.

TABLE 14.23
The Components of SEG

Digital technologies	At the heart of SEG are digital technologies that collect, analyze, and disseminate environmental data. These include sensors that monitor air and water quality, GIS systems that map environmental features, and AI algorithms that predict environmental changes.
Data-driven decision-making	SEG relies on data to make decisions. By analyzing real-time environmental data, policymakers can make informed decisions about resource allocation, policy implementation, and enforcement.
Citizen engagement	SEG also involves citizens. Digital platforms can be used to inform citizens about environmental issues, gather their input, and encourage their participation in environmental conservation.
Policy implementation and enforcement	Digital technologies can also help in the implementation and enforcement of environmental policies. For example, satellite imagery can be used to monitor deforestation, while smart meters can enforce water conservation measures.

TABLE 14.24
Benefits and Challenges of SEG

Benefits	Challenges and Constraints
Improved Environmental Monitoring: Digital technologies can monitor environmental conditions in real-time, providing valuable data for decision-making.	*Technological Infrastructure:* Implementing SEG requires robust and secure digital infrastructure, which can be a challenge for many cities, especially those in developing countries.
Proactive Management: With predictive analytics, SEG allows for proactive environmental management. Instead of reacting to environmental issues, policymakers can anticipate and mitigate them.	*Data Security and Privacy:* With the increased use of digital technologies and data collection, ensuring data security and privacy is a significant challenge.
Increased Transparency and Accountability: Digital platforms can make environmental data accessible to the public, increasing transparency and holding governments accountable.	*Digital Divide:* The digital divide, or the gap between those who have access to digital technologies and those who do not, can hinder the implementation of SEG.
Enhanced Citizen Engagement: By involving citizens, SEG can foster a sense of environmental stewardship and promote sustainable behaviors.	*Technical Expertise:* Implementing and managing digital technologies require technical expertise, which may not be readily available.
	Policy Issue Interdependencies: Policy actors often address complex environmental problems by dividing them into multiple policy issues. This compartmentalization can lead to challenges when these issues are interdependent.
	Public Participation: Engaging citizens in environmental decision-making processes can be challenging, but it is crucial for the success of SEG.
	Regulatory Challenges: The fast pace of technological advancement can outstrip the ability of regulatory frameworks to keep up, leading to potential legal and compliance issues.
	Financial Constraints: The implementation of SEG can be costly, and securing adequate funding can be a challenge.

SEG is essential in creating sustainable cities. By utilizing digital technologies, cities can manage their environmental resources more efficiently, reduce their ecological footprint, and engage citizens in sustainability efforts. SEG represents a new paradigm in environmental management. It offers a proactive, data-driven, and participatory approach to environmental governance. While challenges remain, with continued innovation and investment, SEG holds great promise for creating sustainable and resilient cities.

14.7.3 Smart Community Capabilities

Smart Community Capabilities (SCC) is a concept that emphasizes the importance of community involvement and empowerment in managing environmental resources. It is based on the understanding that a well-informed and educated community can play a significant role in addressing environmental challenges. SCC refers to the collective abilities of a community to manage and protect their local environment. It involves equipping community members with the knowledge, skills, and tools necessary to make informed decisions about environmental resource management. This includes understanding the value of natural resources, the impacts of human activities on these resources, and the strategies for conserving and sustainably using them.

The importance of SCC lies in its potential to drive sustainable environmental practices at the grassroots level. Communities are often the first to experience the impacts of environmental degradation, and they have a vested interest in preserving their local environment. Strengthening SCC enables communities to take ownership of their environment and actively participate in its protection and management.

Achieving SCC requires a multi-faceted approach. First, it involves providing environmental education to community members. This can be done through workshops, seminars, and educational programs that cover various aspects of environmental resource management. Second, it requires building the capacity of community members to apply this knowledge in practice. This can involve training in specific skills, such as waste management or water conservation, and providing tools and resources to support these activities. Finally, it requires creating a supportive environment that encourages community participation in environmental resource management.

Strengthening SCC can have a profound impact on environmental resource management. When community members are knowledgeable about environmental issues and equipped with the skills to address them, they can contribute to the sustainable use of resources, the preservation of biodiversity, and the mitigation of environmental degradation. They can also advocate for environmental policies and practices that reflect their local needs and priorities. SCC can foster a sense of environmental stewardship among community members. This can lead to a shift in attitudes and behaviors toward more sustainable practices, such as reducing waste, conserving water, and protecting natural habitats. Over time, these changes can result in significant improvements in environmental health and sustainability.

Technology can support SCC in several ways, as summarized in Table 14.25.

There are potential risks and constraints for relying heavily on technology for SCC, as summarized in Table 14.26.

Several cities around the world have successfully balanced technology and community engagement for environmental management. They are:

- *London, UK:* London has been recognized as one of the top smart and sustainable cities in the world. It has implemented various digital technologies for environmental management and actively engages its citizens in sustainability efforts.
- *New York, United States:* New York is another city that has successfully integrated technology and community engagement in its environmental management strategies. The city uses data analytics and IoT devices for monitoring environmental conditions and implements policies based on these insights.

TABLE 14.25
The Ways Technology Supports SCC

Digital technologies	State-of-the-art digital technologies are used to create smarter, more intuitive infrastructure, making communities better places to work, play, and invest.
Internet of Things (IoT) and artificial intelligence (AI)	The fusion of IoT with AI-based analysis, also known as AIoT, can help cities reduce noise, improve pedestrian and traffic safety, monitor air or water quality, manage congestion, reduce crime, and streamline billing for city services.
Cloud computing and big data analytics	These technologies enable the processing and analysis of large volumes of data collected from various sources, providing valuable insights for decision-making.
Sensors and sensor networks	These devices collect real-time data about various environmental parameters, which can be analyzed to monitor conditions and make informed decisions.
Shared infrastructure solutions	These offer significant benefits, such as enabling connectivity in areas with limited space and lowering costs.
Smartphones and mobile Apps	Smartphones have become the keys to the city, putting instant information about transit, traffic, health services, safety alerts, and community news into millions of hands.
Data and digital technology	More comprehensive, real-time data gives agencies the ability to watch events as they unfold, understand how demand patterns are changing, and respond with faster and lower-cost solutions.

TABLE 14.26
Potential Risks and Constraints of SCC

Technological infrastructure	Implementing SCC requires robust and secure digital infrastructure, which can be a challenge for many communities, especially those in developing regions.
Data security and privacy	With the increased use of digital technologies and data collection, ensuring data security and privacy is a significant challenge.
Digital divide	The digital divide, or the gap between those who have access to digital technologies and those who do not, can hinder the implementation of SCC.
Technical expertise	Implementing and managing digital technologies require technical expertise, which may not be readily available.
Cybersecurity risks	As cities increase their digital dependency, their potential information and communications technology (ICT) attack surface expands dramatically. This could expose local communities to cyberattacks by a variety of malicious actors.
Reliability of technology	Technology can sometimes fail or malfunction, leading to disruptions in services and potentially causing harm.
Over-reliance on technology	There is a risk that communities may become too dependent on technology, which could lead to a loss of traditional skills and knowledge.
Cost	The implementation of digital technologies can be costly, and securing adequate funding can be a challenge.

- *Paris, France:* Paris has made significant strides in becoming a smart and sustainable city. It uses digital technologies for environmental monitoring and decision-making and encourages citizen participation in environmental conservation.
- *Tokyo, Japan:* Tokyo is known for its advanced use of technology in various sectors, including environmental management. The city uses data-driven approaches for resource management and involves its citizens in environmental initiatives.
- *Santiago, Chile:* Santiago has made significant progress on several green initiatives, reducing the number of days with poor air quality by 70% over the past decade. The city has converted 30% of its bus fleet to electric since 2017.

These cities demonstrate how the combination of technology and community engagement can lead to effective environmental management.

SCC is a powerful approach to environmental resource management that harnesses the power of community involvement and education. SCC can empower communities to take charge of their local environment and contribute to global sustainability efforts. While challenges exist, such as ensuring access to quality environmental education and overcoming barriers to community participation, the potential benefits of SCC make it a worthwhile endeavor.

14.7.4 Smart Faith-based Approach

The faith-based approach is a unique perspective that integrates religious beliefs with environmental stewardship. It is a way of life that considers the environment as an integral part of one's faith, creating a distinct interface between deep ecology and religion. This approach is rooted in the belief that faith can guide individuals and communities towards sustainable practices and environmental consciousness.

Faith, in this context, is not limited to a specific religion or belief system. It encompasses a broad spectrum of spiritual beliefs and practices, all of which can provide a moral and ethical framework for environmental stewardship. The faith-based approach recognizes that every faith tradition has teachings and principles that encourage respect for the natural world and promote sustainable living. The faith-based approach to environmental stewardship begins with the understanding that the environment is a divine creation, deserving of respect and care. This perspective views the natural world not as a resource to be exploited, but as a sacred space to be preserved and protected. It encourages individuals and communities to live in harmony with nature, recognizing that their actions have direct impacts on the health and well-being of the planet.

In the faith-based approach, environmental stewardship is seen as a spiritual duty. It is a way of expressing gratitude for the blessings of nature and acknowledging the divine presence in all living things. This approach encourages individuals and communities to take responsibility for the health of the planet, recognizing that their actions can either contribute to environmental degradation or promote ecological sustainability. The faith-based approach also recognizes the interconnectedness of all life. It understands that the health of the planet is intimately connected to the health of its inhabitants. Therefore, it promotes practices that not only protect the environment but also enhance the well-being of all living creatures.

One of the key strengths of the faith-based approach is its ability to mobilize communities toward environmental action. Faith communities often have strong social networks and shared values, which can be leveraged to promote environmental awareness and action. Through sermons, teachings, and community activities, faith leaders can inspire their congregations to adopt sustainable practices and become advocates for environmental conservation. Moreover, the faith-based approach can provide a sense of hope and resilience in the face of environmental challenges. It can inspire individuals and communities to remain committed to environmental stewardship, even when faced with obstacles and setbacks. Grounding environmental action in faith can infuse a sense of purpose and meaning, driving individuals and communities to persist in their efforts for environmental sustainability.

There are potential constraints and limitations of the faith-based approach to environmental stewardship. These are presented in Table 14.27.

There are several resources available for faith-based environmental education. These resources can help faith communities integrate environmental education into their teachings and practices, fostering a deeper understanding of environmental stewardship and promoting sustainable living. These resources include:

- *Earth Day Network:* The Earth Day Network provides resources linking environmental stewardship and faith. They offer various campaigns and resources for communities of faith,

TABLE 14.27
Potential Constraints of Faith-based Approach

Diverse interpretations	Different faiths have different interpretations of environmental stewardship, and even within a single faith, interpretations can vary widely. This diversity can lead to disagreements and conflicts, hindering collective action.
Religious bias	The faith-based approach may unintentionally favor certain religions over others, leading to bias and discrimination. It is important to ensure that all faiths are equally represented and respected.
Resistance to change	Some religious beliefs and practices may be resistant to change, making it difficult to adopt new environmental practices or adapt to changing environmental conditions.
Religious vs. scientific views	There can be a conflict between religious views and scientific understanding of environmental issues. Some religious beliefs may contradict scientific evidence, making it challenging to reconcile faith with environmental action.
Exclusion of non-religious groups	The faith-based approach may exclude those who do not identify with a particular faith or are non-religious. It is crucial to find ways to engage all individuals, regardless of their religious beliefs.
Overemphasis on divine intervention	Some may interpret the faith-based approach as relying too heavily on divine intervention for environmental solutions, potentially undermining the importance of human action and responsibility.

including Catholic, Buddhist, Muslim, Jewish, Christian, Hindu, Sikh, and interfaith environmental resources.

- *UN Environment Program's Faith for Earth Initiative:* This initiative promotes faith leadership, faith-based organizations, and communities as custodians of far-reaching, value-based perspectives on environmental sustainability. They focus on strengthening partnerships with faith-based organization's leadership for policy impact, greening faith-based organization's investments, operations, and assets, and establishing an accessible knowledge-based support system using faith scripts and scientific evidence.
- *eePRO:* eePRO (Environmental e-Professional) provides resources for environmental leaders on how to approach different faith groups.

It is important to choose resources that align with the specific beliefs and practices of their faith community.

The faith-based approach offers a unique and powerful perspective on environmental stewardship. This approach can inspire individuals and communities to care for the environment as an expression of their spiritual beliefs. It provides a moral and ethical framework for environmental action, promoting sustainable practices and fostering a deep sense of respect for the natural world. As we face the growing challenges of environmental degradation, the faith-based approach offers a hopeful and resilient path toward ecological sustainability.

14.7.5 Global Cooperation, Local Efforts

Global cooperation is a concept that recognizes the interconnectedness of our world and the shared responsibility we all have towards its well-being. It involves considering the environment as a shared common resource and promoting global cooperation to complement local efforts. This approach is rooted in the understanding that environmental issues are not confined to geographical boundaries and that collective action is necessary to address them effectively.

The environment is a shared common resource that sustains all life on Earth. It provides us with the air we breathe, the water we drink, the food we eat, and the natural beauty we enjoy. However, human activities are causing unprecedented damage to the environment, leading to climate change, biodiversity loss, pollution, and other environmental crises. These challenges are global in nature and require a global response.

Global cooperation in environmental stewardship involves countries working together to protect and preserve the environment. This can take various forms, such as international treaties and agreements, joint research and development projects, shared environmental standards and regulations, and collaborative conservation efforts. These initiatives aim to ensure that all countries contribute to environmental protection and that no country bears an undue burden.

Global cooperation also involves recognizing and respecting the diverse ways in which different cultures and societies relate to the environment. It acknowledges that there is no one-size-fits-all solution to environmental challenges and that local knowledge and practices can offer valuable insights for sustainable living. Promoting cultural exchange and learning enables global cooperation to foster a more inclusive and holistic approach to environmental stewardship.

However, global cooperation is not just about governments and institutions. It also involves individuals, communities, businesses, and civil society organizations. Everyone has a role to play in protecting the environment, and global cooperation provides a framework for collective action. Working together allows us to pool our resources, share our knowledge and expertise, and amplify our impact.

Global cooperation also complements local efforts by providing additional resources and support. While local actions are crucial for addressing environmental issues at the grassroots level, they can be limited by various constraints, such as lack of funding, technical expertise, or political will. Global cooperation can help overcome these constraints by providing financial assistance, technical support, policy guidance, and other forms of aid.

Moreover, global cooperation can help address the inequities and injustices associated with environmental issues. Many environmental challenges disproportionately affect marginalized and vulnerable communities, and global cooperation can ensure that their voices are heard, and their rights are protected. Promoting equity and justice enables global cooperation to contribute to a more sustainable and inclusive world.

Global cooperation can promote global and local environmental sustainability, through the following efforts:

- *Addressing Global Challenges:* Environmental issues such as climate change, deforestation, and ocean pollution are global in nature and cannot be effectively addressed by any single country. Global cooperation allows for collective action to tackle these challenges.
- *Sharing Resources and Knowledge:* Through global cooperation, countries can share resources, knowledge, and technologies that can aid in environmental conservation. This can include sharing best practices, research findings, and innovative technologies.
- *Setting International Standards:* Global cooperation can lead to the establishment of international environmental standards and regulations. These can help ensure that all countries adhere to certain environmental practices, reducing global environmental degradation.
- *Promoting Sustainable Development:* Global cooperation can promote sustainable development by ensuring that economic growth does not come at the expense of the environment. This can be achieved through international agreements that balance economic development with environmental conservation.
- *Protecting Biodiversity:* Global cooperation is key to protecting the world's biodiversity. Many species and ecosystems span national borders, and their conservation requires international collaboration.
- *Fostering Peace and Stability:* Environmental degradation can lead to conflicts over resources. Global cooperation in managing shared resources can help prevent conflicts and promote peace and stability.

- *Enhancing Resilience to Environmental Disasters:* Global cooperation can enhance resilience to environmental disasters by facilitating the sharing of information, resources, and best practices.

Global cooperation is essential for promoting environmental sustainability. It allows for collective action, sharing of resources and knowledge, setting of international standards, and enhancing resilience to environmental disasters. It is a powerful tool for environmental stewardship. It recognizes the shared responsibility we all have toward the environment and promotes collective action to address environmental challenges through complementing local efforts, global cooperation can enhance the effectiveness and reach of environmental initiatives.

14.8 SMART CITIZENS

Smart citizens are individuals who leverage technology and data to improve their quality of life, contribute to their communities, and participate in governance. They are a crucial component of smart cities, where digital technologies are used to enhance urban living. A smart city without smart citizens can be likened to a high-tech car without a skilled driver. It may have advanced infrastructure and technology, but without digitally literate and engaged citizens to utilize these resources effectively, the city's potential for efficiency, sustainability, and improved quality of life may remain underutilized or even misused.

Smart citizens are not merely passive consumers of technology. They are active participants in the digital transformation of their cities. They use technology to access information, engage with public services, communicate with their government, and contribute to their communities. They understand the potential of technology to improve their lives and are willing to embrace it.

One of the key characteristics of smart citizens is their ability to use data. They understand the value of data in decision-making and are skilled at using it to solve problems, make informed decisions, and hold their government accountable. They use data to understand their environment, track public services, monitor government performance, and contribute to public debates.

Smart citizens also understand the importance of digital literacy. They are proficient in using digital tools and platforms and are aware of the risks associated with digital technologies, such as privacy breaches and misinformation. They strive to stay informed about the latest technological developments and are proactive in learning new skills. Moreover, smart citizens are active participants in their communities. They use technology to connect with their neighbors, participate in community activities, and contribute to local initiatives. They use digital platforms to share their ideas, voice their concerns, and collaborate with others.

Smart citizens can contribute to smart in governance. They use digital platforms to engage with their government, participate in decision-making processes, and hold their leaders accountable. They use technology to vote, provide feedback on public services, and advocate for their rights. However, being a smart citizen is not just about using technology. It is also about having a civic mindset. Smart citizens care about their communities and are committed to making them better. They are engaged, informed, and proactive. They understand their rights and responsibilities and are willing to take action to improve their cities.

Smart citizens are the driving force behind smart cities. They are digitally savvy, data-driven, and community-oriented. They leverage technology to improve their lives, contribute to their communities, and participate in governance. As our cities become increasingly digital, the role of smart citizens will become even more important. They are the ones who will shape the future of our cities and ensure that technology is used to benefit all citizens.

Involving marginalized communities in digital initiatives is crucial for ensuring inclusivity and equity. Some strategies cities can be employed by the city to cope with this issue include:

- *Digital Literacy Programs:* Conduct training programs to improve digital literacy in marginalized communities. This includes teaching basic computer skills, internet usage, and online safety.
- *Affordable Access:* Work with service providers to ensure affordable internet access in low-income areas. Consider setting up free Wi-Fi zones in community centers.
- *Inclusive Design:* Ensure digital platforms are user-friendly and accessible to people with disabilities. This includes using clear language, large text, and compatibility with assistive technologies.
- *Community Engagement:* Involve community members in the planning and implementation of digital initiatives. This can help ensure the initiatives are relevant and beneficial to the community.
- *Partnerships:* Collaborate with NGOs and community organizations that work with marginalized communities. They can provide valuable insights and help reach out to the community.
- *Targeted Initiatives:* Implement initiatives specifically aimed at increasing digital inclusion in marginalized communities. This could include tech job training programs, providing public computers in community centers, or mobile internet access programs.
- *Representation:* Ensure representation of marginalized communities in decision-making processes related to digital initiatives. Their voices and perspectives are crucial in creating inclusive digital cities.

Through the execution of these strategies, cities have the ability to guarantee that their digital initiatives are not only beneficial to all citizens but also inclusive, particularly for those from marginalized communities. This ensures that everyone, regardless of their background, can participate in and benefit from the city's digital transformation.

PROBLEMS AND ISSUES FOR DISCUSSION

1. What defines a "smart city," and how has this concept evolved over time?
2. How do data analytics and IoT contribute to making cities smarter?
3. Discuss the role of AI in optimizing urban infrastructure.
4. What challenges do smart cities face in terms of privacy and data security?
5. Compare and contrast the implementation of smart city initiatives in developed vs. developing countries.
6. Explore the impact of smart transportation systems on traffic congestion and air quality.
7. How can smart grids enhance energy efficiency and resilience in cities?
8. Investigate the use of sensor networks for waste management and environmental monitoring.
9. Assess the effectiveness of smart parking solutions in reducing urban parking woes.
10. Discuss the role of citizen engagement in shaping smart city policies.
11. What ethical considerations arise when deploying facial recognition technology in public spaces?
12. Analyze the economic benefits of investing in smart city infrastructure.
13. How can blockchain technology improve transparency and trust in municipal services?
14. Explore the concept of "digital twins" and their application in urban planning.
15. Investigate the impact of ride-sharing and micro-mobility services on urban mobility.
16. Discuss the trade-offs between centralized vs. decentralized smart city governance.
17. How can smart buildings enhance energy efficiency and occupant comfort?
18. Examine the role of open data platforms in fostering innovation within cities.
19. What strategies can cities employ to bridge the digital divide and ensure inclusivity?

20. Assess the environmental impact of e-governance and paperless initiatives.
21. Explore the potential of 5G networks in transforming urban connectivity.
22. Discuss the challenges of retrofitting existing cities to become "smart."
23. How can smart street lighting contribute to safety and energy savings?
24. Analyze the role of autonomous vehicles in reshaping urban transportation.
25. Investigate the use of predictive analytics for crime prevention in smart cities.
26. Discuss the impact of smart water management systems on water conservation.
27. How do smart health initiatives improve public health outcomes in cities?
28. Explore the concept of circular economy within the context of smart cities.
29. Assess the scalability of smart city solutions across different urban contexts.
30. What lessons can be learned from successful smart city case studies?
31. Discuss the role of public–private partnerships in funding smart city projects.
32. How can smart agriculture contribute to food security in urban areas?
33. Analyze the social implications of automated decision-making in city services.
34. Explore the potential of smart tourism to enhance visitor experiences.
35. Discuss the role of green infrastructure (e.g., green roofs, vertical gardens) in smart cities.
36. Investigate the impact of smart waste collection systems on sanitation.
37. How can smart sensors improve disaster preparedness and response?
38. Assess the role of mobility-as-a-service (MaaS) platforms in integrated urban transport.
39. Discuss the challenges of interoperability among various smart city technologies.
40. Explore the cultural dimensions of smart city design and implementation.
41. How can smart education initiatives enhance learning outcomes for urban students?
42. Analyze the role of smart street furniture (e.g., benches, kiosks) in enhancing public spaces.
43. Discuss the potential of smart energy districts for sustainable urban living.
44. Investigate the impact of climate change adaptation strategies in smart cities.
45. How do smart waste-to-energy systems contribute to circular economy goals?
46. Assess the role of urban resilience in ensuring long-term smart city success.
47. Explore the use of drones for urban surveillance and emergency response.
48. Discuss the challenges of balancing innovation with affordability in smart city projects.
49. How can smart city initiatives address social inequalities and promote equity?
50. Analyze the role of gender-responsive policies in shaping inclusive smart cities.

MULTIPLE-CHOICE QUESTIONS

1. What does the term "IoT" stand for in the context of smart cities?
   a. Internet of Things
   b. Internet of Transportation
   c. Integrated Operations Technology
   d. Intelligent Optimization Techniques
2. Which of the following is NOT a typical component of a smart city?
   a. Smart traffic lights
   b. Smart waste bins
   c. Smart thieves
   d. Smart energy grids
3. What technology enables real-time monitoring of air quality in urban areas?
   a. Satellite imagery
   b. Smart sensors
   c. CCTV
   d. Morse code

4. Which renewable energy source is commonly integrated into smart city infrastructure?
    a. Thermal energy
    b. Solar energy
    c. Wind energy
    d. Wave energy
5. What is the primary purpose of a smart grid?
    a. To connect all smartphones in the city
    b. To optimize energy distribution and consumption
    c. To create a network of transformers
    d. To distribute electricity
6. Which transportation mode is often associated with smart mobility solutions?
    a. Horse-drawn carriages
    b. Electric scooters
    c. Roller skates
    d. Walk
7. What role does blockchain technology play in smart cities?
    a. Securing Wi-Fi networks
    b. Enhancing data privacy
    c. Predicting weather patterns
    d. Creating virtual reality simulations
8. Which of the following is an example of a smart building feature?
    a. Self-cleaning windows
    b. Talking elevators
    c. Glass roof
    d. Energy-efficient HVAC systems
9. What is the purpose of a digital twin in urban planning?
    a. To create virtual amusement parks
    b. To simulate city growth
    c. To mock-up the planning
    d. To model real-world infrastructure digitally
10. Which of the following technologies is commonly used for smart waste management?
    a. Zero Waste
    b. RFID (Radio Frequency Identification)
    c. Chute disposal system
    d. Sanitary landfill
11. What is the definition of a smart city?
    a. A city with advanced transportation systems
    b. A city with sustainable green initiatives
    c. A city that integrates technology across various domains
    d. A city with efficient waste management
12. Which of the following is an example of a smart innovation?
    a. Traditional street lighting
    b. Self-driving vehicles
    c. Conventional waste disposal methods
    d. Manual traffic management
13. Which indicators help determine whether a city is "smart"?
    a. Advanced technological infrastructure
    b. Sustainable green initiatives
    c. Digital services for city planning
    d. All the above

14. Why is sharing better than owning in a smart city context?
    a. It reduces environmental impact
    b. It optimizes space usage
    c. It is cost-efficient
    d. All the above
15. What does the zero-waste economy aim to achieve?
    a. Minimizing waste generation
    b. Repurposing and recycling products
    c. Designing products for longevity
    d. All the above
16. Which technology is commonly used for monitoring air quality in smart cities?
    a. RFID tags
    b. LiDAR sensors
    c. IoT devices
    d. Satellite imagery
17. What does the term "resilience" refer to in the context of smart cities?
    a. The ability to withstand natural disasters
    b. Efficient public transportation systems
    c. High-speed internet connectivity
    d. Green energy sources
18. Which of the following is an example of a smart grid component?
    a. Traditional power lines
    b. Decentralized energy storage
    c. Manual meter reading
    d. Non-renewable energy sources
19. What role does data analytics play in smart cities?
    a. Optimizing traffic flow
    b. Enhancing public safety
    c. Predicting energy consumption
    d. All the above
20. What is the purpose of a digital twin in urban planning?
    a. Simulating real-world scenarios
    b. Managing waste collection
    c. Monitoring wildlife habitats
    d. None of the above
21. What is the primary goal of a smart transportation system?
    a. Reducing traffic congestion
    b. Increasing fuel consumption
    c. Encouraging single-occupancy vehicles
    d. Promoting public transit
22. Which technology enables real-time monitoring of energy consumption in buildings?
    a. Smart meters
    b. Analog gauges
    c. Manual readings
    d. None of the above
23. What role does blockchain technology play in smart cities?
    a. Enhancing cybersecurity
    b. Managing waste collection
    c. Monitoring air quality
    d. None of the above

24. Which of the following is an example of a smart home feature?
    a. Traditional light switches
    b. Voice-controlled thermostats
    c. Manual door locks
    d. Non-connected appliances
25. What is the purpose of a sensor network in a smart city?
    a. Collecting data for urban planning
    b. Providing free Wi-Fi to residents
    c. Monitoring wildlife habitats
    d. None of the above
26. Which renewable energy source is commonly integrated into smart grids?
    a. Coal
    b. Natural gas
    c. Solar power
    d. Nuclear energy
27. What does the term "citizen engagement" refer to in the context of smart cities?
    a. Involving residents in decision-making
    b. Encouraging passive participation
    c. Ignoring public feedback
    d. None of the above
28. How can smart cities promote sustainable mobility?
    a. Investing in electric vehicle infrastructure
    b. Encouraging private car ownership
    c. Expanding highways
    d. None of the above
29. What is the significance of open data in smart city initiatives?
    a. Enhancing privacy
    b. Fostering transparency
    c. Limiting access to information
    d. None of the above
30. Which organization promotes the development of smart city standards?
    a. United Nations
    b. World Health Organization
    c. International Organization for Standardization (ISO)
    d. None of the above
31. Which technology enables real-time monitoring of water quality in smart cities?
    a. Smart meters
    b. LiDAR sensors
    c. IoT devices
    d. None of the above
32. How can smart cities enhance public safety?
    a. Implementing surveillance cameras
    b. Encouraging community policing
    c. Leveraging data analytics
    d. All the above
33. What role does 5G technology play in smart cities?
    a. Enhancing mobile gaming
    b. Enabling faster internet browsing
    c. Supporting IoT devices and real-time communication
    d. None of the above

34. Which of the following is an example of a smart waste management system?
    a. Traditional landfill disposal
    b. Sensor-equipped trash bins
    c. Manual waste sorting
    d. Non-recyclable materials
35. How can smart cities promote inclusivity?
    a. Providing accessible public transportation
    b. Ignoring diverse cultural backgrounds
    c. Limiting green spaces
    d. None of the above
36. What is the purpose of a mobility-as-a-service (MaaS) platform?
    a. Encouraging private car ownership
    b. Integrating various transportation modes
    c. Expanding highways
    d. None of the above
37. Which organization focuses on sustainable urban development and smart cities?
    a. World Health Organization
    b. United Nations
    c. International Monetary Fund
    d. None of the above
38. How can smart cities address climate change?
    a. Promoting electric vehicle adoption
    b. Encouraging deforestation
    c. Ignoring renewable energy sources
    d. None of the above
39. What is the significance of open-source software in smart city projects?
    a. Enhancing cybersecurity
    b. Fostering collaboration and innovation
    c. Limiting access to information
    d. None of the above
40. What is the purpose of a smart parking system in urban areas?
    a. Reducing traffic congestion
    b. Encouraging car ownership
    c. Expanding parking lots
    d. None of the above
41. Which of the following is an example of a smart building feature?
    a. Manual climate control
    b. Energy-efficient lighting
    c. Non-connected elevators
    d. Traditional fire alarms
42. How can smart cities improve waste management?
    a. Implementing sensor-equipped trash bins
    b. Ignoring recycling programs
    c. Promoting single-use plastics
    d. None of the above
43. What role does edge computing play in smart city infrastructure?
    a. Centralizing data processing in the cloud
    b. Distributing computation closer to data sources
    c. Limiting internet connectivity
    d. None of the above

44. Which technology enables efficient energy distribution in smart grids?
    a. Analog meters
    b. Decentralized energy storage
    c. Non-renewable energy sources
    d. None of the above
45. How can smart cities promote sustainable food systems?
    a. Encouraging local agriculture
    b. Ignoring food waste reduction
    c. Importing food globally
    d. None of the above

ANSWERS TO MULTIPLE-CHOICE QUESTIONS

1	a	10	b	19	d	28	a	37	b
2	c	11	c	20	a	29	b	38	a
3	b	12	b	21	a	30	c	39	b
4	b	13	d	22	a	31	c	40	a
5	b	14	d	23	a	32	d	41	b
6	b	15	d	24	b	33	c	42	a
7	b	16	c	25	a	34	b	43	b
8	d	17	a	26	c	35	a	44	b
9	d	18	b	27	a	36	b	45	a

FURTHER READING

Alawadhi, S., & Alawadhi, T. (2019). Smart city initiatives: A systematic literature review. *Sustainability*, 11(7), 1907.

Albino, V., Berardi, U., & Dangelico, R. M. (2015). Smart cities: Definitions, dimensions, performance, and initiatives. *Journal of Urban Technology*, 22(1), 3–21.

Anthopoulos, L. G. (2015). Understanding the smart city domain: A literature review. *Cities*, 45, 3–12.

Batty, M. (2013). *The new science of cities*. MIT Press.

Bibri, S. E., & Krogstie, J. (2017). Smart sustainable cities of the future: An extensive interdisciplinary literature review. *Sustainable Cities and Society*, 31, 183–212.

Caragliu, A., Del Bo, C., & Nijkamp, P. (2009). Smart cities in Europe: A multicriteria performance analysis. *Journal of Urban Technology*, 16(1), 2–19.

Caragliu, A., Del Bo, C., & Nijkamp, P. (2011). Smart cities in Europe. *Journal of Urban Technology*, 18(2), 65–82.

Caragliu, A., Del Bo, C., & Nijkamp, P. (2011). Smart cities in Europe. *Journal of Urban Technology*, 18(2), 65–82.

Deakin, M., & Al Waer, H. (2011). From intelligent to smart cities. *Intelligent Buildings International*, 3(3), 140–152.

Giffinger, R., Fertner, C., Kramar, H., Kalasek, R., Pichler-Milanović, N., & Meijers, E. (2007). *Smart cities: Ranking of European medium-sized cities*. Centre of Regional Science (SRF).

Glaeser, E. L. (2011). *Triumph of the city: How our greatest invention makes us richer, smarter, greener, healthier, and happier*. Penguin Press.

Gracias, J. S., Parnell, G. S., Specking, E., Pohl, E. A., & Buchanan, R. (2023). Smart cities – A structured literature review. *Smart Cities*, 6(4), 1719–1743.

Hollands, R. G. (2008). Will the real smart city please stand up? *City*, 12(3), 303–320.

Komninos, N. (2002). *Intelligent cities: Innovation, knowledge systems and digital spaces*. Routledge.

Komninos, N. (2006). *The architecture of intelligent cities: Integrating human, collective and artificial intelligence to enhance knowledge and innovation*. Routledge.

Komninos, N. (2008). *Intelligent cities: Variable geometries of spatial intelligence.* Intelligent Buildings International, 1(1), 13–30.

Komninos, N. (2013). *The age of intelligent cities: Smart environments and innovation-for-all strategies.* Routledge.

Komninos, N., Pallot, M., & Schaffers, H. (2013). Special issue on smart cities and the future internet in Europe. *Journal of the Knowledge Economy*, 4(2), 119–134.

Lee, J. H., Hancock, M. G., & Hu, M. (2014). Towards an effective framework for building smart cities: Lessons from Seoul and San Francisco. *Technological Forecasting and Social Change*, 89, 80–99.

Neirotti, P., De Marco, A., Cagliano, A. C., Mangano, G., & Scorrano, F. (2014). Current trends in Smart City initiatives: Some stylised facts. *Cities*, 38, 25–36.

Neirotti, P., De Marco, A., Cagliano, A. C., Mangano, G., & Scorrano, F. (2014). Current trends in Smart City initiatives: Some stylised facts. *Cities*, 38, 25–36.

Neirotti, P., De Marco, A., Cagliano, A. C., Mangano, G., & Scorrano, F. (2014). Current trends in Smart City initiatives: Some stylised facts. *Cities*, 38, 25–36.

Schaffers, H., Komninos, N., Pallot, M., Trousse, B., Nilsson, M., & Oliveira, A. (2011). Smart cities and the future internet: Towards cooperation frameworks for open innovation. *Future Internet*, 3(1), 35–53.

Townsend, A. M. (2013). *Smart cities: big data, civic hackers, and the quest for a new utopia.* WW Norton.

Yigitcanlar, T., & Kamruzzaman, M. (2018). Creating smart cities: A conceptual framework based on the interplay between knowledge, innovation and policy. *Cities*, 81, 145–160.

Zanella, A., Bui, N., Castellani, A., Vangelista, L., & Zorzi, M. (2014). Internet of things for smart cities. *IEEE Internet of Things Journal*, 1(1), 22–32.

15 Urban Greenery and Urban Agriculture

15.1 THE CONCEPT OF URBAN GREENERY

The concept of urban greenery has become increasingly important in modern city planning and design, as it offers a multitude of benefits that contribute to the overall well-being and sustainability of metropolitan areas. Urban greenery, which encompasses parks, gardens, street trees, and other forms of vegetation within the urban landscape, plays a crucial role in enhancing the quality of life for city dwellers. Integrating natural elements into urban environments helps mitigate the effects of heat islands, significantly improving temperature and air quality. The presence of vegetation also aids in the absorption of carbon dioxide and the release of oxygen, contributing to improved air quality and a healthier atmosphere. Moreover, urban greenery serves as a natural sponge, absorbing excess rainwater and reducing the risk of flooding, while also helping to filter pollutants and improve water quality.

Beyond the environmental benefits, urban greenery has a profound impact on the mental and physical health of residents. Studies have shown that access to green spaces can reduce stress, improve mood, and enhance cognitive function. The visual and sensory experience of being surrounded by nature in an urban setting can also promote physical activity and encourage social interaction, fostering a sense of community and well-being.

As cities continue to grow and densify, the integration of urban greenery becomes increasingly crucial, not only for environmental sustainability but also for the overall quality of life and livability of these metropolitan areas. Strategically incorporating green spaces enables policymakers, urban planners, and designers to create vibrant, resilient, and livable cities. These cities blend natural and built environments, offering residents a higher quality of life and a more sustainable future.

15.1.1 Definition and Types of Urban Greenery

Urban greenery refers to the introduction and integration of vegetation in urban environments. It encompasses a wide range of green spaces, including parks, gardens, street trees, green roofs, walls, community gardens, and even urban farms. These green spaces play a crucial role in enhancing the quality of urban life, providing numerous environmental, social, and health benefits.

15.1.1.1 Definition of Urban Greenery

Urban greenery is a term used to describe all types of vegetation found within an urban environment. This includes not only trees and plants but also grass, shrubs, flowers, and other forms of plant life. The purpose of urban greenery is to enhance the aesthetic appeal of urban areas, improve air quality, reduce urban heat island effects, provide habitats for wildlife, and contribute to the well-being of residents.

DOI: 10.1201/9781032664613-15

TABLE 15.1
Types of Urban Greenery

Urban parks	These are green spaces in cities that are set aside for recreation and leisure activities. They are often home to a variety of plant and animal species and provide a natural escape from the urban environment.
Street trees	These are trees planted along the streets in urban areas. They provide shade, improve air quality, and enhance the aesthetic appeal of the city.
Green roofs and walls	These are structures covered with vegetation that provide insulation, reduce heat absorption, and contribute to biodiversity.
Community gardens	These are plots of land in urban areas where residents can grow their own plants and vegetables. They promote community interaction and provide a source of fresh produce.
Urban farms	These are agricultural practices carried out within the city limits. They provide fresh produce and can contribute to food security.
Private gardens	These are green spaces owned by individuals or organizations. They contribute to the overall green cover of the city and provide a personal space for relaxation and enjoyment.
Public squares and plazas	These are open spaces in cities that are often landscaped with trees, plants, and grass. They serve as gathering places for people and host various public events.
River and canal banks	These are green spaces along the water bodies in cities. They provide habitats for various species and recreational spaces for residents.

Urban greenery is an essential component of urban planning and design. It is increasingly recognized for its role in mitigating the impacts of urbanization and climate change. Absorbing carbon dioxide, filtering pollutants, and providing shade allow urban greenery to contribute to climate change mitigation and adaptation. It also enhances biodiversity, provides recreational spaces, and improves mental health.

15.1.1.2 Types of Urban Greenery

Urban greenery, encompassing all vegetation within cities, significantly enhances urban life quality. It includes diverse types such as urban parks, street trees, green roofs and walls, community gardens, urban farms, private gardens, public squares, and green spaces along water bodies. The types of urban greenery are shown in Table 15.1.

Urban greenery provides numerous environmental benefits, which will bring significant improvements to city life. It enhances air quality, reduces urban heat islands, and promotes biodiversity. Furthermore, it aids in stormwater management, noise reduction, and erosion control, making cities more sustainable, livable, and in harmony with nature. These benefits include:

1. *Air Quality Improvement:* Plants absorb pollutants and release oxygen, improving the air quality in urban areas.
2. *Urban Heat Island Mitigation:* Green spaces absorb less heat than concrete and asphalt, helping to lower temperatures in urban areas.
3. *Carbon Sequestration:* Plants absorb carbon dioxide, a greenhouse gas, helping to mitigate climate change.
4. *Habitat Provision:* Urban greenery provides habitats for a variety of wildlife, promoting biodiversity in cities.
5. *Stormwater Management:* Green spaces absorb rainwater, reducing runoff and the risk of flooding.
6. *Noise Reduction:* Vegetation can act as a sound barrier, reducing noise pollution in urban areas.
7. *Erosion Control:* Plant roots stabilize the soil, preventing erosion.

8. *Increased Property Values:* Green spaces can increase the value of nearby properties, leading to higher property taxes for the city.
9. *Energy Savings:* Greenery, especially trees, can provide shade and reduce the need for air conditioning in the summer and act as windbreaks in the winter, reducing heating costs.
10. *Job Creation:* The design, installation, and maintenance of urban green spaces can create jobs and stimulate local economies.
11. *Tourism:* Beautiful parks and gardens can attract tourists, boosting local businesses and generating revenue for the city.
12. *Healthcare Savings:* By improving air quality and providing spaces for exercise, urban greenery can lead to healthier populations and potentially reduce healthcare costs.
13. *Stormwater Management:* Green spaces can absorb rainwater, reducing the need for costly drainage systems.
14. *Business Productivity:* Studies have shown that views of and access to nature can increase productivity in the workplace.

These benefits highlight the economic importance of integrating green spaces into urban planning and development.

15.1.2 The Importance of Urban Greenery

Urban greenery, the integration of vegetation in urban environments, plays a fundamental role in enhancing the quality of urban life and the environment. It provides numerous benefits that are vital for cities, citizens, and the environment itself.

Urban greenery is an essential component of a city. It contributes to the aesthetic appeal of cities, making them more attractive to residents, tourists, and investors. Green spaces, such as parks and gardens, serve as landmarks and give character to urban landscapes. Urban greenery contributes to the economic vitality of cities. It increases property values, leading to higher property taxes. Green spaces also attract tourists, boosting local businesses and generating revenue for the city. Furthermore, the design, installation, and maintenance of urban green spaces create jobs and stimulate local economies. Urban greenery is also essential in urban climate adaptation. It helps mitigate the urban heat island effect, where urban areas become significantly warmer than their rural surroundings. This is particularly important as cities continue to expand, and global temperatures rise due to climate change.

For citizens, urban greenery improves the quality of life in several ways. It provides spaces for recreation and relaxation, contributing to physical health and mental well-being. Access to green spaces has been linked to reduced stress levels, improved mood, and increased life satisfaction. Urban greenery also promotes social cohesion and community building. Parks and community gardens serve as gathering places where people can interact, build relationships, and engage in community activities. They provide opportunities for outdoor learning and environmental education, especially for children. Furthermore, urban greenery can also contribute to food security. Urban farms and community gardens provide fresh produce, contributing to a healthy diet and reducing food costs for urban residents.

From an environmental perspective, urban greenery provides several ecosystem services. It improves air quality by absorbing pollutants and releasing oxygen. It regulates climate by providing shade and reducing the urban heat island effect. It also reduces noise pollution by acting as a sound barrier. Urban greenery is also important for biodiversity conservation. It provides habitats for various species, serving as urban wildlife corridors. This is particularly important in cities where development has led to habitat loss. Urban greenery contributes to stormwater management. It absorbs rainwater, reducing runoff and the risk of flooding. It also prevents soil erosion, protecting the integrity of the urban landscape.

Urban greenery is of paramount importance for cities, citizens, and the environment. It enhances the livability and sustainability of cities, improves the quality of life for citizens, and contributes to environmental health and resilience. As urbanization continues to rise, the integration of green spaces into urban planning and design will be crucial for creating sustainable and resilient cities.

15.1.3 The Role of Urban Greenery in Urban Ecosystems

Urban greenery plays a pivotal role in urban ecosystems, contributing to the environmental, social, and economic well-being of cities. It includes all forms of vegetation within urban areas, such as trees, shrubs, grass, flowers, and other plants.

15.1.3.1 Environmental Role

Urban greenery significantly contributes to the environmental health of urban ecosystems. It improves air quality by absorbing pollutants and releasing oxygen. This process is crucial in urban areas where air pollution is often a major concern. Moreover, urban greenery helps mitigate the urban heat island effect, a phenomenon where urban areas are significantly warmer than their rural surroundings due to human activities. Green spaces absorb less heat than concrete and asphalt, helping to cool urban areas and reduce energy consumption for cooling buildings. Urban greenery also plays a role in water management. It absorbs rainwater, reducing surface runoff and the risk of flooding. It also helps filter pollutants from the water, improving the quality of urban water bodies. Also, urban greenery contributes to biodiversity in cities. It provides habitats for a variety of species, serving as an urban wildlife corridor and helping to maintain ecological balance.

15.1.3.2 Social Role

Urban greenery enhances the social fabric of cities. It provides spaces for recreation and relaxation, contributing to the physical and mental well-being of urban residents. Access to green spaces has been linked to various health benefits, including reduced stress levels, improved mood, and increased physical activity. Urban greenery promotes social cohesion and community building. Parks and community gardens serve as gathering places where people can interact, build relationships, and engage in community activities. They provide opportunities for outdoor learning and environmental education, fostering a sense of environmental stewardship among urban residents.

15.1.3.3 Economic Role

Urban greenery also has economic implications for cities. It increases property values, leading to higher property taxes. It also reduces energy costs by providing shade and reducing the need for air conditioning. Moreover, the design, installation, and maintenance of urban green spaces create jobs and stimulate local economies. Also, urban greenery attracts tourists, boosting local businesses and generating revenue for the city. It also contributes to a city's image and identity, making it more attractive to investors and residents.

Urban greenery contributes to social well-being and stimulates economic growth. As urbanization continues to rise, the integration of green spaces into urban planning and design will be crucial for creating sustainable and resilient cities. Urban greenery is not just an aesthetic feature of cities; it is a vital component of urban ecosystems that contributes to the sustainability and livability of cities. Therefore, it is essential to protect and enhance urban greenery for the benefit of urban ecosystems and the people who live in them.

Given the significance of urban greenery for the city, its residents, and the environment, it is imperative that we advocate for universal access to urban green spaces. This can be achieved by implementing the strategies outlined in Table 15.2.

TABLE 15.2
Strategies for Inclusive and Accessible Urban Greenery

Universal design	Implement universal design principles to ensure that green spaces are accessible to everyone, regardless of age, ability, or socioeconomic status. This includes providing wheelchair-accessible paths, sensory gardens for the visually impaired, and safe, age-appropriate play areas for children.
Location and distribution	Ensure that green spaces are evenly distributed throughout the city so that everyone has access to them. Prioritize areas that currently lack green spaces.
Public transportation	Improve public transportation links to green spaces. This can make it easier for people who do not have private vehicles to access these areas.
Community involvement	Involve local communities in the planning and maintenance of green spaces. This can foster a sense of ownership and ensure that the spaces meet the needs of the community.
Safety and comfort	Make green spaces feel safe and comfortable by providing adequate lighting, clean facilities, and regular maintenance.
Diverse activities	Design green spaces to support a variety of activities, such as walking, picnicking, playing, gardening, and birdwatching. This can attract a wider range of users.
Education and signage	Provide educational materials and clear signage in multiple languages to help people understand and navigate the green spaces.
Affordability	Ensure that access to green spaces is free or affordable to avoid excluding low-income individuals and families.

Implementing these strategies enables cities to develop green spaces that are inclusive, accessible, and advantageous to all residents. This crucial step significantly enhances the quality of urban life and fosters social equity.

15.1.4 Obstacles in Implementing Urban Greenery

Obstacles and constraints in implementing urban greenery may range from technical and financial difficulties to social and political issues.

15.1.4.1 Technical Challenges

One of the primary technical challenges is the lack of suitable space for greenery in densely populated urban areas. High land prices and competing land uses often limit the availability of space for parks and other green spaces. Moreover, the maintenance of urban greenery can be technically challenging. It requires regular watering, pruning, pest control, and other care to keep the plants healthy and the spaces clean and safe. In addition, the selection of suitable plant species that can thrive in the urban environment is also a challenge. Furthermore, the implementation of certain types of urban greenery, such as green roofs and walls, requires technical expertise and can be complex and costly. These structures need to be carefully designed and constructed to ensure they are safe, effective, and durable.

15.1.4.2 Financial Challenges

Implementing and maintaining urban greenery requires significant financial resources. These costs include land acquisition, design and construction, plant materials, irrigation systems, and ongoing maintenance. Moreover, while urban greenery can increase property values and generate economic benefits in the long term, these benefits may not be immediately apparent. This can make it difficult to secure funding and investment for urban greenery projects.

15.1.4.3 Social Challenges

Social challenges include public perception and behavior. While many people appreciate the benefits of urban greenery, others may resist changes to their local environment. For example, some people may oppose the removal of parking spaces to create room for street trees or parks. Ensuring equitable access to urban greenery can be a challenge. In many cities, green spaces are unevenly distributed, with wealthier neighborhoods often having more and better-quality green spaces than poorer ones.

15.1.4.4 Political Challenges

Political challenges include the lack of political will and support for urban greenery. Urban greenery may not be a priority for some city leaders, particularly in cities facing other pressing issues such as housing, transportation, or economic development. The implementation of urban greenery requires coordination among various city departments and stakeholders. This can be challenging in cities where there is a lack of integrated planning and decision-making.

Financial issues are a common problem, particularly in developing countries. It is therefore imperative to provide all possible solutions to cope with this issue. Some considerations for the city to deal with financial issues include the following:

1. *Public–Private Partnerships (PPPs):* Collaborating with private entities can help share the costs and risks associated with urban greening projects. Businesses can contribute funds, resources, or expertise in exchange for benefits such as branding opportunities or improved employee well-being.
2. *Grants and Funding Programs:* Cities can seek grants from government agencies, foundations, and international organizations that support urban greening initiatives. These grants can provide significant funding for green space development and maintenance.
3. *Community Crowdfunding:* Engaging local communities in fundraising can help generate funds for specific projects. This approach not only raises money but also fosters community ownership and involvement in urban greening efforts.
4. *Green Bonds:* Cities can issue green bonds to raise capital for urban greening projects. These bonds are like regular municipal bonds but are specifically used for projects that have environmental benefits.
5. *Value Capture Mechanisms:* Implementing mechanisms such as special assessment districts or tax increment financing can help capture the increased property values resulting from urban greening. The additional revenue can be reinvested in maintaining and expanding green spaces.
6. *Incorporating Green Infrastructure into City Planning:* By integrating green infrastructure into city planning, cities can allocate a budget for green spaces just as they do for other essential city services. This ensures a dedicated funding source for urban greening.
7. *Volunteer Programs:* Engaging volunteers in the maintenance of urban green spaces can help reduce costs. Volunteer programs can also foster community engagement and stewardship of local green spaces.

15.2 THE ROLE OF URBAN AGRICULTURE IN FOOD SECURITY

Urban agriculture contributes to enhancing food security. It provides fresh, nutritious produce directly to urban consumers, reducing reliance on long-distance supply chains. Growing food locally through urban agriculture reduces food deserts – areas where access to fresh food is limited. It promotes biodiversity and resilience against climate change while offering opportunities for income generation and community engagement. This contributes to social inclusion and economic development, making urban agriculture a multifaceted strategy for achieving food security in urban areas.

15.2.1 Understanding Food Security

Food security is defined as a state where all people, at all times, have physical and economic access to sufficient, safe, and nutritious food that meets their dietary needs and food preferences for an active and healthy life. It encompasses four main dimensions:

1. *Physical Availability of Food:* This addresses the supply side of food security and is determined by the level of food production, stock levels, and net trade.
2. *Economic and Physical Access to Food:* An adequate supply of food at the national or international level does not in itself guarantee household-level food security. This dimension focuses on incomes, expenditure, markets, and prices in achieving food security objectives.
3. *Food Utilization:* This is commonly understood as the way the body makes the most of various nutrients in the food. Sufficient energy and nutrient intake by individuals are the result of good care and feeding practices, food preparation, diversity of the diet, and intra-household distribution of food.
4. *Stability of the Other Three Dimensions Over Time:* Even if your food intake is adequate today, you are still considered to be food insecure if you have inadequate access to food on a periodic basis, risking a deterioration of your nutritional status.

Food security, a fundamental aspect of a nation's well-being, is an issue that encompasses various dimensions. It refers to the availability, access, utilization, and stability of food supplies. Availability refers to the sufficient quantity of food of appropriate quality, supplied through domestic production or imports. This involves the consideration of land resources, climate, and farming practices. Technological advancements in agriculture, such as genetically modified crops and precision farming, can enhance food production and thus availability.

Access to food concerns the affordability and allocation of food, as well as the preferences of individuals and households. Economic access is determined by income levels and food prices, while physical access involves the proximity and ease of reaching food sources. Policies ensuring fair trade and regulating food prices, along with initiatives like community gardens and farmers' markets, can improve food access.

Utilization of food relates to the body's ability to ingest and metabolize nutrients, which is influenced by health status, clean water, sanitation, and healthcare access. Education in nutrition and cooking skills also plays a role in the proper utilization of food.

Stability of food supply is crucial, as political unrest, climate change, or economic instability can disrupt food availability and access. Strategies for enhancing stability include diversifying food production, creating buffer stocks, and implementing social safety nets.

Urban agriculture, as mentioned earlier, can contribute significantly to food security by providing fresh produce directly to urban consumers, reducing reliance on long-distance supply chains, and offering opportunities for income generation. However, achieving food security faces constraints and obstacles. One of the major obstacles is climate change. Climate change poses a significant threat, with changing weather patterns potentially affecting crop yields. The global food system's complexity, with its interconnected political, economic, and social factors, also complicates efforts to ensure food security. Moreover, food security is closely linked with other Sustainable Development Goals (SDGs), such as No Poverty (SDG 1), Good Health and Well-being (SDG 3), and Responsible Consumption and Production (SDG 12). Therefore, a holistic and integrated approach is necessary to address food security effectively.

Food security can differ between rural and urban areas due to various factors such as income levels, access to food markets, and the availability of agricultural land.

In rural areas, households often have direct access to agricultural land and can grow their own food, which can contribute to food availability. However, they may face challenges in terms of

TABLE 15.3
The Causes of Food Insecurity in Rural Areas

Natural disasters	Droughts, floods, and climate change can lead to crop failure, affecting the availability of food.
Poor farming practices	Inefficient or unsustainable farming practices can lead to reduced crop yields.
Lack of access to markets	Rural households may face challenges in terms of access to markets, transportation, and storage facilities, which can affect their ability to sell their produce and buy diverse foods.
Land shortage	In densely populated rural areas, there may not be enough land for each household to grow sufficient food.
Lack of infrastructure	The absence of basic infrastructure such as roads, schools, markets, and health facilities can exacerbate food insecurity.
Population growth	Rapid population growth can put pressure on limited resources, leading to food insecurity.
Lack of education	Lack of education can limit people's ability to adopt improved farming techniques or diversify their livelihoods.
Political and economic factors	Issues such as war, political instability, and economic downturns can disrupt food production and distribution.
Health emergencies	Loss of household labor due to health emergencies can also contribute to food insecurity.

access to markets, transportation, and storage facilities, which can affect their ability to sell their produce and buy diverse foods. According to a report from Purdue University, 23% of US rural households are facing food insecurity. This report assesses food security and spending, consumer satisfaction and values, support of agricultural and food policies, and trust in information sources. Food insecurity is a global issue, and the rates can vary significantly between different countries and regions. For instance, in some rural areas of Ethiopia, the food insecurity rate can be as high as 71.6%. Therefore, the situation can vary greatly depending on the specific context of the region.

On the other hand, urban areas often have better access to markets, which can provide a diverse range of food products. However, urban residents are typically reliant on purchasing food, which can be problematic if food prices rise or incomes fall. Urban agriculture can play a role in enhancing urban food security by providing fresh produce directly to urban consumers. Yet, 14% of all households in urban areas are also facing food insecurity. These are general trends, and the actual situation can vary greatly. Addressing food security effectively requires tailored strategies that consider the unique challenges and opportunities in both rural and urban settings.

Food insecurity in rural areas can be caused by a variety of factors, as shown in Table 15.3.

One of the solutions to food insecurity is to implement sustainable agriculture. Sustainable agriculture is a system of farming that aims to provide the resources necessary for present human populations while conserving the planet's ability to sustain future generations. It seeks to integrate three main objectives into its work: a healthy environment, economic profitability, and social and economic equity. Sustainable agriculture is characterized by farming practices that are in harmony with nature, aiming to protect the environment, aid, and expand natural resources, and make the best use of nonrenewable resources. It emphasizes the production of abundant food without depleting the earth's resources or polluting its environment. Practices commonly used in sustainable agriculture include methods to promote soil health, minimize water use, and lower pollution levels on the farm. It also involves the use of polyculture, where multiple crops are grown together, and crop rotation, which can help preserve soil productivity and lessen the need for agricultural chemicals.

Sustainable agriculture is more than a collection of practices. It is also a process of negotiation, balancing the sometimes-competing interests of an individual farmer or of people in a community as they work to solve complex problems about how we grow our food and fiber.

Sustainable agriculture can contribute to addressing food security by contributing to all four pillars of food security – availability, access, utilization, and stability.

- *Availability:* Sustainable agriculture increases food availability by promoting efficient and resilient farming practices that enhance crop yields.
- *Access:* By promoting economic development and increasing incomes for farmers and their families, sustainable agriculture improves access to food.
- *Utilization:* Sustainable agricultural systems offer ecosystem services, such as regulation of soil and water quality, maintenance of soil structure and fertility, which can contribute to the production of nutritious food and its proper utilization.
- *Stability:* Sustainable agriculture contributes to the stability of food supplies by creating an efficient, stable, and resilient food system, enabling us to meet the global demand for food.

Sustainable agriculture practices are environmentally friendly because they promote biodiversity conservation, carbon sequestration, mitigation of greenhouse gas emissions, nutrient cycling, and hydrological services. These practices not only ensure food security but also help in achieving the Sustainable Development Goals (SDGs), including Zero Hunger. Thus, sustainable agriculture is a practical option for achieving global food security while minimizing further environmental degradation.

15.2.2 How Urban Agriculture Contributes to Food Security

Urban agriculture is one of the major factors in enhancing food security, particularly in urban areas in which access to fresh, nutritious food can be limited. The practices of urban agriculture are gaining popularity as a response to the growing concern over food security in urban settings. It increases the availability of fresh food in cities and reduces the distance food travels from farm to plate, ensuring that urban dwellers have access to fresh produce. This is particularly important in "food deserts," urban areas where access to affordable, healthy food options is limited.

Urban farms and other urban cultivation can improve access to food by providing urban dwellers with opportunities to grow their own food. Community gardens, rooftop farms, and other urban agricultural practices empower individuals and communities to take control of their food sources. This not only ensures access to fresh produce but also helps to reduce food costs. It contributes to better nutrition and health by promoting the consumption of fresh, locally grown produce. Fresh fruits and vegetables are rich in essential nutrients, and urban agriculture makes these foods more accessible to urban populations. Moreover, the physical activity involved in urban farming practices can also contribute to improved health.

Urban agriculture can provide economic benefits by creating jobs and providing income for urban dwellers. It can stimulate local economies through the sale of locally grown produce. Furthermore, it can help to reduce food costs, as growing one's own food can be cheaper than purchasing it from a store.

Community gardens and other urban farming initiatives can contribute to community development by fostering social interaction and community cohesion. These spaces can serve as gathering places, promoting social interaction and strengthening community ties. They can also provide educational opportunities, teaching both children and adults about food production and healthy eating. Promoting the efficient use of urban resources, urban agriculture contributes to environmental sustainability. It aids in recycling urban waste, reducing the urban heat island effect, and promoting biodiversity. Additionally, urban agriculture helps lower greenhouse gas emissions by reducing the distance food travels from farm to plate.

TABLE 15.4
Requirements for Urban Agriculture

Availability of space	Urban agriculture requires space for cultivation. This could be in the form of vacant lots, rooftops, balconies, or even indoor spaces for vertical farming.
Access to resources	Adequate access to resources such as water, soil, seeds, and farming equipment is essential.
Supportive policies	Government policies and regulations should support urban agriculture. This includes zoning laws, land tenure rights, and subsidies.
Community engagement	The involvement and commitment of the community are crucial for the success of urban agriculture initiatives.
Training and education	Urban farmers need knowledge and skills in sustainable farming practices. This could be provided through training programs and educational initiatives.
Market access	For urban agriculture to be economically viable, farmers need access to markets to sell their produce.
Waste management	Urban agriculture can contribute to food security by recycling urban waste into compost. However, this requires effective waste management systems.
Safety measures	Urban soils may be contaminated with heavy metals or other pollutants. Therefore, safety measures such as soil testing and the use of clean compost are necessary.
Climate-smart practices	Urban agriculture should employ climate-smart practices to ensure resilience to climate change.
Integration with urban planning	Urban agriculture should be integrated into urban planning to ensure the efficient use of urban resources.

Urban agriculture can contribute to food security under certain conditions or requirements, as summarized in Table 15.4.

15.2.3 Limitations and Obstacles of Urban Agriculture

Urban agriculture, while offering numerous benefits, is bringing limitations and obstacles.

15.2.3.1 Limited Space

One of the most significant challenges for urban agriculture is the limited availability of space. Urban areas are often densely populated, leaving little room for traditional agricultural activities. This lack of open land and suitable space can pose a major obstacle to the growth and development of urban farming initiatives.

In many cities, the scarcity of land is exacerbated by competition with other pressing needs, such as housing, commercial development, and infrastructure. As urban populations continue to grow, the pressure on available land only intensifies, making it increasingly difficult to set aside space for urban agriculture. Despite these challenges, innovative solutions have emerged to help overcome the issue of limited space in urban environments. One such solution is the rise of rooftop gardening and farming. Utilizing the often-underutilized space atop buildings allows urban farmers to create productive growing areas without encroaching on valuable ground-level real estate. Rooftop gardens can take advantage of available sunlight and shelter from harsh weather conditions, providing an ideal setting for cultivating a variety of crops.

Another innovative approach to urban agriculture is vertical farming. This method involves growing crops in stacked, indoor growing systems, often with the aid of artificial lighting and controlled environmental conditions. Vertical farming allows for high-density food production in compact urban spaces, making efficient use of limited square footage. By growing crops vertically, urban farmers can significantly increase the yield per square meter, maximizing their production capacity.

While rooftop gardens and vertical farming offer promising solutions to the space constraint, they often require significant upfront investment and specialized infrastructure. The establishment of these systems can be capital-intensive, requiring the construction of greenhouses, hydroponics systems, or other specialized growing environments. Additionally, the ongoing operational costs, such as energy consumption for lighting and climate control, can pose a challenge for urban farmers, particularly those with limited resources. The feasibility of rooftop gardens and vertical farming may be limited in certain urban settings. The structural integrity of buildings, access to sunlight, and the availability of reliable power sources are all crucial factors that must be considered when implementing these solutions. In some cases, the costs and technical requirements may outweigh the potential benefits, making these approaches less viable for urban farmers.

To address the issue of limited space, some urban agriculture initiatives have turned to alternative growing methods, such as container gardening, small-scale plot farming, and the use of vacant lots or abandoned spaces. These more flexible and lower-cost approaches can be more accessible to a wider range of urban residents, including those with limited resources or access to land. Moreover, the integration of urban agriculture into broader urban planning and development strategies can help to address the space constraints. By incorporating urban farming into the design of new residential, commercial, and mixed-use projects, city planners can ensure that adequate space is allocated for agricultural activities. This holistic approach can help to create more sustainable and resilient urban environments that support local food production.

15.2.3.2 Soil Contamination

Urban soil can be contaminated with a variety of pollutants, including heavy metals, chemicals, and other hazardous substances. This poses a significant challenge for urban agriculture, as the safety and quality of the food produced in these environments can be compromised. Heavy metals, such as lead, cadmium, and arsenic, can accumulate in urban soils due to a range of human activities, including industrial processes, vehicle emissions, and the use of certain pesticides and fertilizers. These toxic substances can be readily absorbed by plants, leading to their presence in the edible portions of crops. Consuming food grown in contaminated soils can have serious health implications, as heavy metals can cause a variety of adverse effects, including neurological damage, kidney problems, and cancer.

In addition to heavy metals, urban soils may also be contaminated with other harmful chemicals, such as polycyclic aromatic hydrocarbons (PAHs), dioxins, and persistent organic pollutants. These substances can originate from sources like industrial activities, waste disposal, and the use of certain pesticides. The presence of these contaminants in the soil can compromise the safety and nutritional value of the food produced, potentially exposing urban residents to significant health risks.

To address the issue of soil contamination, urban farmers and researchers have developed various techniques and strategies. One of the primary approaches is soil testing, which involves analyzing samples of the soil to identify the presence and concentrations of various pollutants. This information can then be used to develop targeted remediation strategies.

Soil remediation methods can include physical, chemical, or biological techniques, depending on the nature and extent of the contamination. For example, techniques such as phytoremediation, where certain plants are used to extract or break down pollutants, can be an effective and environmentally friendly approach. In other cases, more intensive methods, such as soil excavation and replacement, may be necessary to remove heavily contaminated soils. While these remediation techniques can be effective in addressing soil contamination, they often come with significant challenges and limitations. The process of soil testing and analysis can be time-consuming and costly, requiring specialized equipment and expertise. Similarly, the implementation of remediation strategies can be a complex and expensive undertaking, placing a heavy burden on urban farmers and community gardens.

In many cases, the cost and logistical constraints associated with soil testing and remediation may be prohibitive for smaller-scale urban agriculture initiatives or individuals with limited resources. This can create a situation where urban farmers and gardeners are forced to work with contaminated soils, potentially exposing themselves and their communities to health risks. To overcome these constraints, there is a need for more accessible and cost-effective solutions. One approach that has gained attention is the development of "safe growing systems," such as raised beds or container gardens, which allow urban farmers to cultivate crops in clean, uncontaminated soil. These systems can be particularly useful in areas where the underlying soil is known to be heavily polluted, providing a safe and controlled growing environment.

There is a growing emphasis on the importance of education and awareness-raising efforts. Informing urban residents about the risks of soil contamination and the steps to mitigate these risks empowers communities to make informed decisions and adopt safer growing practices. Collaboration between urban farmers, researchers, and local authorities is crucial in addressing soil contamination. Sharing knowledge, resources, and best practices allows stakeholders to develop comprehensive strategies to tackle the unique challenges faced by urban agriculture in their communities.

The issue of soil contamination in urban areas presents a significant challenge for the growth and development of urban agriculture. While techniques such as soil testing and remediation can help address this problem, the associated costs and complexities can be a barrier for many urban farmers and gardeners. Innovative solutions, such as safe growing systems and collaborative efforts, as well as increased public awareness, will be essential in ensuring the long-term sustainability and food safety of urban agriculture initiatives.

15.2.3.3 Access to Resources

Urban farmers often face significant challenges in accessing the necessary resources to support their agricultural activities. These challenges can include limited access to water, difficulties in obtaining high-quality seeds and planting materials, and barriers to acquiring the appropriate farming equipment and tools. One of the primary resource constraints faced by urban farmers is the availability and accessibility of water. In many urban areas, water is a scarce and highly coveted resource, with municipal authorities prioritizing its allocation for residential, commercial, and industrial use. This can leave urban farmers struggling to secure reliable and sufficient water supplies for their crops.

The issue of water scarcity is particularly acute in regions experiencing prolonged droughts or where water infrastructure is aging and overburdened. In these situations, urban farmers may find themselves in direct competition with other water users, leading to conflicts and uncertainties around water rights and allocation. Furthermore, the cost of water for agricultural use can be prohibitively high in urban settings, where water pricing is often structured to favor residential and commercial consumers. This financial burden can make it challenging for urban farmers to maintain viable and productive operations, as a significant portion of their operating budgets may be directed toward water expenses.

Other than water-related constraints, urban farmers may also face difficulties in accessing high-quality seeds, seedlings, and other planting materials. In many urban areas, the availability of specialized agricultural inputs can be limited, as rural-focused seed suppliers and nurseries may not have a strong presence in the city.

The lack of access to diverse and genetically robust planting materials can hinder urban farmers' ability to grow a wide range of crops, adapt to local climatic conditions, and maintain the long-term health and productivity of their land. This, in turn, can limit the variety and quantity of fresh, locally grown produce that urban farmers can bring to their communities.

The acquisition of appropriate farming equipment and tools can present a significant challenge for urban agriculture initiatives. In densely populated urban areas, the space available for storing and maintaining large-scale farming machinery may be scarce or non-existent. This can force urban farmers to rely on smaller-scale, often manual tools and equipment, which may be less efficient and more labor-intensive.

The high cost of specialized farming equipment, such as tillers, cultivators, and irrigation systems, can also be a significant barrier for urban farmers, particularly those with limited financial resources. The lack of access to affordable and well-maintained farming tools can hinder productivity, increase physical strain on the farmers, and limit the overall scale and efficiency of urban agriculture operations.

To address these resource challenges, urban farmers and policymakers have explored a range of innovative solutions and strategies. Some cities have implemented water conservation and recycling programs, as well as subsidies or prioritized water allocations for urban agriculture, in an effort to ensure a more reliable water supply. Similarly, initiatives have emerged to connect urban farmers with seed suppliers and nurseries, providing them with access to high-quality and locally adapted planting materials. These partnerships can help diversify the crop varieties grown in urban settings and strengthen the resilience of urban agriculture systems.

Regarding farming equipment, urban agriculture initiatives have explored the potential of shared or communal equipment pools, where farmers can access and share the use of larger machinery and tools. This collaborative approach can help reduce the financial burden on individual farmers and improve the overall efficiency of urban farming operations.

In addition, urban agriculture organizations and community groups have advocated for increased funding and support from local governments and other stakeholders to address the resource challenges faced by urban farmers. This can include the development of infrastructure, the provision of technical assistance, and the creation of financial incentives or subsidies to help urban farmers overcome the unique obstacles they face.

The limited access to essential resources, such as water, seeds, and farming equipment, poses significant challenges for urban agriculture. These constraints can hinder the productivity, profitability, and overall sustainability of urban farming initiatives. These resource challenges can be addressed through innovative solutions, strategic partnerships, and policy support, allowing urban agriculture to become more resilient and better equipped to serve the needs of urban communities.

15.2.3.4 Regulatory Barriers

Urban agriculture often operates within a complex and challenging regulatory environment. Zoning laws, building codes, and other municipal regulations can present significant barriers to the development and expansion of urban farming initiatives.

One of the key regulatory challenges faced by urban farmers is the issue of zoning. In many cities, land use is strictly zoned, with specific areas designated for residential, commercial, or industrial purposes. The incorporation of agricultural activities within these designated zones can be a point of contention, as urban farming may not align with the intended uses for certain areas. For instance, some cities may have strict zoning regulations that prohibit or severely restrict the establishment of urban farms or community gardens in residential or commercial districts. This can limit the available land that urban farmers can access, forcing them to seek out scarce parcels of land that are zoned for agricultural use, which are often in short supply in densely populated urban areas.

The conversion of land for urban agriculture may require complex and time-consuming approval processes, including zoning variances or amendments. These bureaucratic hurdles can deter potential urban farmers, who may face significant delays and uncertainty in securing the necessary permissions to start or expand their operations. Urban agriculture initiatives can also be subject to a myriad of building codes and other regulations that were not necessarily designed with urban farming in mind. These regulations can create barriers to the construction and operation of essential infrastructure, such as greenhouses, hoop houses, or aquaponics systems.

Building codes may impose strict requirements on the design, materials, and safety features of these structures, which can be costly and time-consuming to comply with. In some cases, urban farmers may find that the regulatory hurdles make it financially or logistically infeasible to invest in the necessary infrastructure to support their agricultural activities.

Another regulatory challenge faced by urban farmers is the issue of livestock and animal husbandry. In many urban areas, the keeping of livestock, such as chickens, goats, or bees, may be prohibited or heavily restricted due to concerns over public health, safety, and nuisance factors. These regulations can limit the ability of urban farmers to integrate animal-based production systems into their operations, which can be an important component of a diversified and sustainable urban agriculture model.

To navigate these complex regulatory environments, urban agriculture advocates and practitioners have employed various strategies. One approach has been to actively engage with policymakers and local authorities to advocate for more favorable and adaptable regulations. This can involve participating in the development of comprehensive urban agriculture plans, pushing for zoning amendments, or lobbying for the creation of specialized urban agriculture zones.

Some cities have recognized the value of urban agriculture and have taken steps to streamline the regulatory process. For example, some municipalities have established dedicated urban agriculture departments or liaison positions to assist urban farmers in navigating the bureaucratic landscape and obtaining the necessary permits and approvals. Furthermore, urban agriculture organizations and community groups have worked to educate and inform local decision-makers about the benefits of urban farming, highlighting its potential contributions to food security, community development, and environmental sustainability. This advocacy work can help build political will and support for regulatory reforms that better accommodate and enable urban agriculture initiatives.

In some cases, urban farmers have also explored innovative ways to work within the existing regulatory frameworks, such as by repurposing abandoned or underutilized urban spaces or by integrating their agricultural activities with other land uses, such as rooftop gardens or vertical farming systems. However, these workarounds can be limited in their scalability and may not be feasible or accessible to all urban farmers, particularly those with limited resources or access to specialized knowledge and expertise.

The complexity of the regulatory environment faced by urban agriculture initiatives presents a significant challenge to the growth and development of this sector. Zoning laws, building codes, and restrictions on livestock and infrastructure can create substantial barriers for urban farmers. Overcoming these regulatory hurdles requires a multifaceted approach, involving collaboration between urban farmers, policymakers, and local authorities to advocate for more supportive and adaptable regulations. Addressing these regulatory challenges makes urban agriculture more accessible, sustainable, and better integrated into the fabric of urban communities.

15.2.3.5 Economic Viability

The economic viability of urban agriculture is another key concern that can hinder the growth and sustainability of this sector. While urban agriculture can provide income and economic opportunities for urban dwellers, the high costs associated with inputs and labor can pose significant challenges. One of the primary economic challenges faced by urban farmers is the high cost of land and infrastructure. In densely populated urban areas, the value of land is often significantly higher than in rural areas, making it difficult for urban farmers to acquire or lease suitable land for their agricultural activities. The need to construct specialized growing structures, such as greenhouses or vertical farming systems, further adds to the upfront capital investment required to establish urban agriculture operations.

The operating costs for urban farmers can be substantial. Urban areas tend to have higher labor costs, as the local cost of living is generally higher compared to rural regions. This can make it challenging for urban farmers to maintain a sufficient workforce to carry out the necessary farming tasks, from planting and harvesting to processing and distribution.

The costs of essential agricultural inputs, such as seeds, fertilizers, and water, can also be higher in urban settings. Due to the limited availability of land and resources, urban farmers may not benefit from the economies of scale that often characterize rural agriculture. This can put them at a disadvantage when it comes to negotiating prices and accessing wholesale suppliers.

Urban farmers may face difficulties in accessing and securing reliable markets for their produce. In many cities, the traditional distribution channels and wholesale markets that serve rural farmers may not be easily accessible or tailored to the needs of urban agriculture. This can make it challenging for urban farmers to find consistent and lucrative outlets for their products.

The lack of access to established marketing channels can force urban farmers to rely more heavily on direct-to-consumer sales, such as through farmers' markets, community-supported agriculture (CSA) programs, or online platforms. While these channels can provide urban farmers with better control over pricing and customer relationships, they may also require significant time and effort to cultivate and maintain, potentially adding to the overall workload and operational complexity.

To address the economic challenges facing urban agriculture, a range of strategies and support systems have emerged. Some urban farming initiatives have explored innovative business models, such as partnering with local restaurants, schools, or institutions to secure stable and predictable revenue streams. Establishing these strategic partnerships allows urban farmers to leverage the demand for locally sourced, fresh produce and ensure a more reliable income stream.

The development of cooperative or collective marketing approaches can help urban farmers overcome the challenges of accessing mainstream distribution channels. Pooling their resources and jointly marketing their products allows urban farmers to achieve greater economies of scale, negotiate better prices, and expand their reach to a wider customer base.

Governments and non-profit organizations have also played a crucial role in supporting the economic viability of urban agriculture. Some cities have implemented policies and incentives, such as land-use zoning, tax credits, or subsidies, to make it more financially feasible for urban residents to engage in agricultural activities. These support mechanisms can help offset the higher costs associated with urban farming and make it a more attractive economic proposition.

Educational and training programs focused on urban agriculture, business management, and marketing can empower urban farmers with the necessary skills and knowledge to operate more efficiently and effectively. Improving their business acumen and understanding of market dynamics enables urban farmers to enhance their ability to navigate economic challenges and identify viable pathways to profitability.

In some cases, urban agriculture initiatives have also explored the integration of value-added processing and product diversification as a means of increasing their economic resilience. Transforming their raw agricultural outputs into value-added products, such as jams, pickles, or baked goods, allows urban farmers to capture a larger share of the consumer market and generate higher profit margins.

Urban agriculture faces complicated obstacles when it comes to economic viability. The high costs of land, infrastructure, labor, and inputs, coupled with the difficulties in accessing reliable markets, can pose significant barriers to the financial sustainability of urban farming initiatives. However, through innovative business models, collaborative marketing approaches, supportive policies, and the development of specialized skills and knowledge, urban agriculture can overcome these economic hurdles and become a more viable and prosperous component of urban communities.

15.2.3.6 Climate Change

Climate change poses a significant threat to urban agriculture, just as it does to traditional agriculture. Urban agriculture, which refers to the practice of growing food within or around cities, is not immune to the impacts of a changing climate. Rising temperatures, changing rainfall patterns, and increased frequency of extreme weather events can all have adverse effects on crop yields and the viability of urban farming.

One of the primary challenges posed by climate change is the rise in global temperatures. As the Earth's average temperature increases, many urban areas are experiencing more frequent and intense heat waves. These high temperatures can stress plants, leading to reduced growth, decreased yields, and even crop failure. In addition, the urban heat island effect, where cities tend to be warmer than surrounding rural areas, can exacerbate the impact of rising temperatures on urban agriculture. This is particularly concerning as many urban farms are in densely populated areas, where the urban heat island effect is more pronounced.

Changing rainfall patterns are another significant threat to urban agriculture. Some regions may experience prolonged droughts, while others may face more intense and frequent rainfall events. Both scenarios can have detrimental effects on crop production. Droughts can lead to water scarcity, making it difficult to maintain healthy plants and sustain urban farming operations. On the other hand, heavy rainfall and flooding can damage crops, erode soil, and create conditions that are conducive to the spread of pests and diseases.

Extreme weather events, such as severe storms, hailstorms, and hurricanes, pose an additional threat to urban agriculture. These events can destroy crops, damage infrastructure, and disrupt the supply chain, making it challenging for urban farmers to maintain consistent food production. For example, a severe hailstorm can batter and destroy delicate leafy greens and other sensitive crops, while a hurricane can flood urban farms, destroy greenhouses, and leave the community without access to locally grown produce.

Beyond the direct impacts on crop yields, climate change can also have indirect effects on urban agriculture. For instance, changes in temperature and precipitation patterns can alter the distribution and behavior of pests and diseases, making it more difficult for urban farmers to manage these threats effectively. Invasive species may also thrive in the changing climate, outcompeting traditional crop plants and further jeopardizing urban food production. Besides, climate change can contribute to water scarcity, which is a significant concern for urban agriculture. Many urban farms rely on municipal water supplies or groundwater sources, both of which can be affected by drought and changing precipitation patterns. As competition for limited water resources increases, urban farmers may face challenges in accessing the water needed to sustain their operations.

To address the threats posed by climate change, urban agriculture practitioners and policymakers must adopt a range of strategies to build resilience and adapt to the changing conditions. These strategies may include:

1. *Diversifying Crop Selection:* Planting a wider variety of crops that are better suited to the shifting climate conditions can help urban farmers mitigate the risks of crop failure.
2. *Implementing Water-efficient Techniques:* Adopting water-saving practices, such as drip irrigation, mulching, and using drought-tolerant plant varieties, can help urban farmers adapt to water scarcity.
3. *Enhancing Soil Health:* Improving soil quality through the use of compost, covering crops, and sustainable soil management practices can help urban farms become more resilient to the impacts of climate change.
4. *Investing in Protective Infrastructure:* Constructing greenhouses, shade structures, and other protective measures can help urban farmers safeguard their crops from extreme weather events and temperature fluctuations.
5. *Fostering Community Engagement:* Collaborating with local residents, urban planners, and policymakers to identify and address the unique challenges faced by urban agriculture in a changing climate can help build more resilient and sustainable food systems.

As the impacts of climate change continue to intensify, the need for robust and adaptable urban agriculture systems will only become more pressing.

15.2.3.7 Lack of Knowledge and Skills

Urban agriculture, the practice of cultivating, processing, and distributing food in or around urban areas, has been recognized as a potential solution to food insecurity in cities. However, one of the significant challenges it faces is the lack of farming knowledge and skills among urban dwellers.

The success of urban agriculture largely depends on the knowledge and skills of the individuals involved. Farming is a complex activity that requires an understanding of various factors such as soil quality, crop selection, pest management, and water conservation. Without proper knowledge and skills, urban dwellers may struggle to establish and maintain productive urban farms.

Urban agriculture often involves unique challenges that are not typically encountered in traditional farming. For instance, urban farmers may need to navigate zoning laws, manage limited space, and deal with potential soil contamination. These challenges require specialized knowledge and skills that many urban dwellers lack.

Training and education programs can play a crucial role in equipping urban dwellers with the necessary knowledge and skills for urban farming. These programs can provide practical training on various aspects of urban agriculture, from basic gardening techniques to more complex topics like vertical farming and hydroponics.

Education programs can also raise awareness about the benefits of urban agriculture, encouraging more people to get involved. They can teach urban dwellers how to make the most of their limited space, how to compost household waste for use in their gardens, and how to grow a variety of crops to ensure a balanced diet.

One of the main obstacles is the lack of resources. Developing and delivering high-quality training programs requires a significant investment. Resources are needed to hire experienced trainers, develop curriculum, and provide hands-on learning opportunities. Another challenge is reaching the target audience. Many urban dwellers may not be aware of the training opportunities available to them or may not see the relevance of these programs to their lives. Therefore, effective outreach strategies are needed to promote these programs and engage the community.

Despite these challenges, there are several strategies that can help make training and education programs more accessible and effective. One approach is to leverage partnerships with local schools, community centers, and non-profit organizations. These institutions often have the facilities and community connections needed to deliver effective training programs. Another strategy is to incorporate urban agriculture into school curricula. This not only equips young people with valuable skills but also fosters an appreciation for agriculture and food production from an early age. Technology can also be a powerful tool in delivering training and education programs. Online courses and tutorials can make learning about urban agriculture accessible to a wider audience. Mobile apps can provide ongoing support and resources to urban farmers, helping them troubleshoot problems and make informed decisions about their gardens.

Training and education programs can help address this gap. However, these programs require significant investment and resources. By leveraging partnerships, incorporating urban agriculture into school curricula, and utilizing technology, we can overcome these challenges and equip urban dwellers with the skills they need to contribute to food security in their communities.

15.2.3.8 Pest and Disease Management

Pest and disease management can be particularly challenging for urban farmers, as the proximity of urban farms to residential areas can limit the use of chemical pesticides. This constraint requires urban farmers to rely more on organic or biological pest control methods, which may be less effective or more costly compared to traditional chemical-based approaches.

In urban pest and disease management, food production with the health and safety of surrounding communities must be in balance. In densely populated urban areas, the application of synthetic pesticides can pose a risk to the local population, particularly to vulnerable groups such as children, the elderly, and those with respiratory conditions. Consequently, urban farmers often face regulatory restrictions or public pressure to avoid the use of chemical pesticides, which can be an essential tool in traditional agricultural pest management.

Without the option of relying on synthetic pesticides, urban farmers must turn to alternative, more environmentally friendly methods of pest and disease control. These organic or biological approaches can include the use of beneficial insects, such as ladybugs or lacewings, to prey on pest populations. Urban farmers may also introduce cover crops, companion planting, or other diversification strategies to create a more balanced ecosystem that discourages the proliferation of pests. However, these alternative pest control methods can be more labor-intensive and less effective than conventional chemical treatments. Beneficial insects, for example, may not be as efficient at reducing pest populations as targeted pesticide applications. In addition, organic pest control methods can be more sensitive to environmental conditions, such as temperature and humidity, making them less reliable in the face of the climatic changes associated with urban heat islands.

Another obstacle in urban pest and disease management is the limited space and fragmented nature of many urban farms. In densely populated urban areas, farms are often small, scattered, and surrounded by residential or commercial properties. This can make it difficult to implement comprehensive pest management strategies, as the close proximity to other land uses can allow pests to easily move between different urban agricultural sites. Urban farms may be more vulnerable to the introduction of pests and diseases from surrounding environments, such as nearby parks, gardens, or even household pets. The increased human activity and traffic in urban areas can also contribute to the spread of pests and pathogens, making it harder for urban farmers to maintain control.

Addressing these constraints, urban farmers must adopt a multi-pronged approach to pest and disease management. This may involve combining various organic and biological control methods, such as the use of beneficial insects, microbial biopesticides, and physical barriers, to create a more comprehensive and resilient pest management strategy. Urban farmers may also need to invest in specialized infrastructure, such as high-tunnels or greenhouses, to create a more controlled environment that can better protect crops from pests and diseases. These structures can help to isolate urban farms from the surrounding urban ecosystem, reducing the risk of pest and disease transmission. In addition, urban farmers may need to collaborate with local authorities, research institutions, and community groups to develop and implement integrated pest management (IPM) strategies that are tailored to the unique challenges of the urban environment. This might involve sharing information, pooling resources, and coordinating pest monitoring and control efforts across multiple urban agricultural sites.

The obstacles of pest and disease management in urban agriculture can be significant, but they are not insurmountable. Embracing a combination of organic and biological control methods, investing in protective infrastructure, and fostering collaborative partnerships enables urban farmers to overcome obstacles and maintain the viability and productivity of their urban farms.

Urban agriculture has the potential to contribute significantly to food security and urban sustainability. Urban agriculture also faces several obstacles that need to be addressed. Overcoming these obstacles will require innovative solutions, supportive policies, and the involvement of various stakeholders, including urban planners, policymakers, and the urban farmers themselves. Despite these challenges, the potential benefits of urban agriculture make it a worthwhile endeavor in our quest for sustainable urban development.

15.3 URBAN GREEN SPACES AND MENTAL HEALTH

Urban green spaces are essential for mental well-being, offering a tranquil escape from the bustling city life. These areas provide a natural environment for relaxation, physical activity, and social interaction. Studies have shown that access to green spaces can reduce stress, anxiety, and depression, promoting a healthier, happier urban population.

15.3.1 The Impact of Urban Green Spaces on Mental Health

Urban green spaces, such as parks, gardens, and forests, play a crucial role in promoting mental health among city dwellers. This essay will delve into the impact of urban green spaces on mental health. Urbanization, while offering numerous benefits, often comes with increased stress, noise, pollution, and social isolation, all of which can negatively impact mental health. Urban green spaces offer a respite from these stressors, providing an oasis of calm and tranquility amidst the hustle and bustle of city life.

One of the primary benefits of urban green spaces is stress reduction. Spending time in nature has been shown to lower levels of cortisol, a hormone associated with stress. The calming effect of natural environments can help individuals relax and unwind, reducing feelings of stress and anxiety.

Urban green spaces can also improve mood. Exposure to nature has been linked to increased levels of serotonin and endorphins, chemicals in the brain that contribute to feelings of happiness and well-being. Furthermore, the beauty and aesthetic appeal of green spaces can evoke positive emotions and enhance mood.

Urban green spaces provide opportunities for physical activity, which is known to have positive effects on mental health. Regular physical activity can reduce symptoms of depression and anxiety, improve mood, and boost self-esteem. Green spaces offer a conducive environment for activities such as walking, jogging, cycling, and playing sports. Green spaces also facilitate social interaction, which is crucial for mental health. They serve as communal hubs where people can meet, socialize, and form connections. This sense of community can alleviate feelings of loneliness and isolation, promoting mental well-being. Spending time in green spaces can foster a connection with nature, which has been associated with improved mental health. This connection can instill a sense of peace and contentment, helping individuals cope with stress and adversity.

Despite their benefits, urban green spaces are often underutilized or neglected. Limited access, safety concerns, and lack of awareness can prevent people from reaping the mental health benefits of these spaces. Therefore, efforts should be made to improve the accessibility, safety, and quality of urban green spaces. Public awareness campaigns can also encourage more people to take advantage of these spaces.

The World Health Organization (WHO) recommends a minimum of 9 m^2 of green space per individual. This guideline is based on the idea that access to green spaces contributes to physical and mental health, providing opportunities for physical activity, relaxation, and social interaction. In addition to this, the WHO also recommends that all people should reside within 300 m of green space, which is roughly equivalent to a 5-minute walk from the houses of urban residents. However, these are only general guidelines, and the ideal amount can vary depending on the specific context of the urban area.

Incorporating more green spaces into densely populated urban areas can be a complex task, but it is not impossible. Urban planners can incorporate the promotion of urban green spaces by implementing these schemes as shown in Table 15.5.

Maintaining urban green spaces in a city requires a comprehensive and sustained approach. First and foremost, regular upkeep is essential. This includes routine tasks such as mowing grass, pruning trees and shrubs, weeding, and litter removal. Ensuring proper irrigation is also crucial, especially in areas with limited rainfall. In addition to these basic maintenance tasks, urban green spaces may require more specialized care. For instance, trees may need to be inspected for disease or pest infestations, and appropriate treatment should be applied as needed. Also, the design of urban green spaces should facilitate easy maintenance. This could include the use of native plants that are adapted to the local climate and thus require less care, or the installation of durable and easy-to-clean amenities.

Community involvement can also play a key role in maintaining urban green spaces. Encouraging local residents to take part in clean-up events or adopt-a-park programs can foster a sense of ownership and responsibility for these spaces. Local authorities should allocate sufficient

TABLE 15.5
Promotion of Urban Green Spaces Amid Limited Urban Land

Utilize vacant spaces	Urban planners can transform vacant lots and abandoned spaces into green spaces. These could be turned into community gardens, pocket parks, or even mini forests.
Rooftop and vertical gardens	Building owners can be encouraged to create rooftop gardens or vertical gardens on the sides of their buildings. This not only adds greenery but also makes efficient use of available space.
Green corridors and street trees	Planting trees along sidewalks and creating green corridors can add greenery and improve the overall aesthetic of the city.
Community involvement	Engaging the community in the planning and maintenance of urban green spaces can ensure their success and sustainability.
Incorporate greenery in building design	Incorporating greenery within building entrances and providing continuous shaded pathways can add greener to city landscapes.
Reclaim urban land	Cities in Europe prioritize reclaiming urban land for green space, implementing nature-based solutions like green roofs and vertical gardens, green schoolyards, green corridors, street trees, pocket parks, community gardens.
Policy support	Supportive policies and regulations can encourage the creation of more green spaces. This could include incentives for property owners to create green spaces or relaxed zoning laws to allow for urban farming.

resources for the maintenance of urban green spaces. This includes not only funding but also training personnel and equipment. Regular audits and inspections can help ensure that maintenance activities are carried out effectively and that urban green spaces continue to provide benefits for all city residents.

15.3.2 The Role of Green Spaces in Stress Reduction

In the fast-paced and often overwhelming urban environment, the importance of green spaces cannot be overstated. As the world becomes increasingly urbanized, with more people living in cities, the need for accessible and well-designed green spaces has become a critical issue for public health and well-being. One of the significant benefits of these natural oases within the concrete jungle is their ability to reduce stress, a growing concern for many city dwellers.

Stress, a common affliction in modern society, can have detrimental effects on both physical and mental health. Prolonged exposure to high levels of stress can lead to a range of negative consequences, including increased risk of cardiovascular disease, weakened immune system, and mental health issues such as anxiety and depression. In urban areas, where the pace of life can be relentless and the built environment can feel oppressive, the availability of green spaces serves as a vital counterbalance, offering a refuge from the stresses of daily life.

The restorative power of nature has been well-documented in scientific research. Studies have shown that spending time in green spaces, such as parks, gardens, or even simply viewing greenery, can have a measurable impact on reducing stress levels. This phenomenon is often referred to as "nature therapy" or "ecotherapy," and it is based on the premise that the human mind and body are inherently connected to the natural environment.

One of the primary ways in which green spaces reduce stress is through their ability to induce a state of mental and physiological relaxation. The presence of lush vegetation, soothing sounds of flowing water, and the absence of urban noise and distractions can trigger a biological response in the body, lowering heart rate, blood pressure, and the production of stress hormones like cortisol. This calming effect helps to counteract the physiological symptoms of stress, allowing individuals to experience a sense of tranquility and rejuvenation.

Green spaces also provide a psychological and emotional respite from the stresses of urban life. The act of simply being in nature, whether it is taking a leisurely stroll through a park or sitting quietly on a bench, can have a profound impact on one's mental state. The natural environment can evoke feelings of awe, wonder, and a sense of connection to something greater than oneself, which can help to put daily worries and concerns into perspective. Green spaces can facilitate social interaction and community engagement, both of which are important factors in stress reduction. Urban parks and public gardens often serve as gathering places for people to connect with one another, engage in physical activity, or simply enjoy the peaceful surroundings. This social interaction and sense of community can help to alleviate feelings of isolation and loneliness, which are known contributors to stress and poor mental health.

To maximize the stress-reducing benefits of green spaces, urban planners and policymakers must prioritize the development and maintenance of high-quality, accessible green infrastructure. This includes not only large, centralized parks but also smaller, neighborhood-level green spaces, such as community gardens, urban forests, and green corridors that connect different parts of the city.

These green spaces should be designed with the needs of the local community in mind, considering factors such as accessibility, safety, and the integration of natural elements. Incorporating features like walking paths, seating areas, and interactive elements can further enhance the restorative and stress-reducing qualities of these urban oases.

Public education and outreach efforts can play a crucial role in encouraging urban residents to take advantage of the stress-reducing benefits of green spaces. Raising awareness about the importance of nature-based activities and providing resources and programming to facilitate their use enables cities to empower their residents to incorporate green space experiences into their daily lives.

As the world continues to urbanize, the role of green spaces in stress reduction will only become more critical. Prioritizing the development and maintenance of these natural havens within the urban landscape allows city planners and policymakers to contribute to the overall health and well-being of their communities, helping to mitigate the negative impacts of stress and promoting a more sustainable and livable urban environment.

The power of green spaces in reducing stress cannot be overstated. These natural oases within the concrete jungle offer a vital respite from the demands of urban life, providing both physical and psychological benefits that can help to alleviate the detrimental effects of stress. Investing in the creation and enhancement of high-quality green spaces enables cities to empower their residents to enjoy the restorative qualities of nature, fostering a more resilient and thriving urban community.

15.3.3 Designing Green Spaces for Mental Health Benefits

As urban populations continue to grow worldwide, the need for well-designed green spaces that promote mental health and well-being has become increasingly crucial. In a world where stress, anxiety, and other mental health challenges are on the rise, the role of green spaces in providing restorative and therapeutic benefits cannot be overstated.

When thoughtfully planned and implemented, green spaces can serve as oases of calm and tranquility within the bustling urban landscape, offering respite from the demands of daily life. These natural havens can have a profound impact on mental health, helping to reduce stress, improve mood, and foster a sense of overall well-being.

One of the key principles in designing green spaces for mental health benefits is to create a sense of restorative experience. This can be achieved by incorporating elements that tap into our innate connection to nature, such as lush vegetation, flowing water, and soothing sounds. Research has shown that exposure to these natural elements can trigger a physiological response in the body, lowering stress hormones and promoting relaxation.

To enhance the restorative qualities of green spaces, urban planners and landscape architects should consider the inclusion of features like meandering paths, serene water features, and shaded seating areas. These design elements not only invite visitors to slow down and immerse themselves in their natural surroundings but also provide a sense of privacy and escape from the outside world.

Another important consideration in designing green spaces for mental health is the concept of biophilic design. Biophilic design principles are rooted in the idea that humans have an innate affinity for the natural world and that incorporating natural elements into the built environment can have positive effects on physical and mental well-being. Incorporating biophilic design strategies, such as the use of natural materials, the integration of natural light, and the creation of visual connections to nature, allows green spaces to become more than just aesthetic amenities. They can become therapeutic landscapes that actively engage the senses and foster a deeper sense of connection to the natural world.

One innovative approach to biophilic design in green spaces is the incorporation of horticultural therapy. This practice involves the use of gardening and other plant-related activities to promote physical, emotional, and social well-being. Engaging in tasks like planting, tending to vegetables, or simply observing the growth and change of plants can have a calming and restorative effect on individuals, particularly those dealing with mental health challenges. In addition to the restorative and biophilic design elements, the layout and accessibility of green spaces are crucial considerations in promoting mental health benefits. Green spaces should be strategically located within urban areas, ensuring that they are easily accessible to a wide range of community members, including those with mobility challenges or limited access to transportation.

The design of these spaces should encourage physical activity and social interaction, both of which are important factors in maintaining good mental health. This can be achieved through the inclusion of walking paths, exercise equipment, and communal gathering spaces that foster a sense of community and social connectedness.

To maximize the mental health benefits of green spaces, it is also important to consider the role of programming and community engagement. Offering a variety of activities and events within these spaces, such as yoga classes, art workshops, or community gardening programs, enables green spaces to become dynamic hubs that cater to the diverse needs and interests of the local population. These programmed activities not only provide opportunities for mental and physical stimulation but also foster a sense of belonging and social inclusion, two key elements in promoting overall well-being.

As cities continue to grapple with the challenges of urbanization and the growing prevalence of mental health concerns, the design of green spaces will become increasingly crucial. Prioritizing the incorporation of restorative, biophilic, and community-oriented design elements enables urban planners and landscape architects to create green spaces that serve as therapeutic sanctuaries, contributing to the overall mental health and well-being of urban residents.

Through collaborative efforts between policymakers, designers, and community stakeholders, cities can develop a comprehensive approach to designing and maintaining green spaces that cater to the diverse mental health needs of their populations. Incorporating these elements, urban planners can not only enhance the livability and sustainability of cities but also empower residents to thrive amidst the unique challenges of contemporary urban life.

15.4 TECHNIQUES IN URBAN AGRICULTURE

15.4.1 Overview of Urban Agriculture Techniques

As the world becomes increasingly urbanized, the need for sustainable and resilient food production systems has become more pressing than ever before. One of the emerging solutions to this challenge is the rise of urban agriculture, a practice that encompasses a wide range of techniques and approaches to cultivating food within or around cities. Urban agriculture offers a multitude of benefits, from improving food security and access to fresh produce to enhancing community engagement and environmental sustainability. Leveraging innovative techniques and the unique characteristics of urban environments enables urban farmers to pave the way for more sustainable and localized food systems.

One of the primary techniques used in urban agriculture is container gardening. This approach involves growing plants in a variety of containers, such as raised beds, planters, or even repurposed items like old tires or buckets. Container gardening is particularly well-suited for urban settings, as it allows for efficient use of limited space and can be easily adapted to balconies, rooftops, or small yards. This technique also provides greater control over soil quality, water management, and pest and disease prevention.

Another popular urban agriculture technique is vertical farming, which involves growing crops in vertically stacked layers or towers. This approach maximizes the use of available space, allowing for high-density food production in even the most cramped urban environments. Vertical farming systems often incorporate advanced technologies, such as LED lighting, hydroponics, or aeroponics, to optimize plant growth and minimize resource consumption.

Aquaponics, a symbiotic system that combines aquaculture (fish farming) and hydroponics (soilless plant cultivation), is another innovative technique gaining traction in urban agriculture. In an aquaponic system, the waste produced by the fish is used to fertilize the plants, while the plants, in turn, help to filter the water for the fish. This closed-loop system is highly efficient, requiring minimal water and no soil, making it well-suited for urban environments with limited resources.

Urban farms can also leverage the concept of intensive gardening, which involves maximizing the productivity of a small space through techniques like companion planting, succession planting, and intensive spacing. Carefully selecting and arranging crops enables urban farmers to significantly increase their yields while minimizing the use of land, water, and other resources.

One of the unique aspects of urban agriculture is the ability to integrate it with the built environment. Rooftop gardens, for example, can transform underutilized urban spaces into productive food-growing areas, while also providing additional benefits such as insulation, stormwater management, and urban heat island mitigation.

Similarly, the integration of urban agriculture with buildings can lead to the development of innovative techniques, such as building-integrated agriculture (BIA), where food production is seamlessly incorporated into the design and construction of urban structures. This approach can include features like living walls, green roofs, and even indoor farming systems within the building itself. In addition to these technical approaches, urban agriculture also encompasses a range of social and community-based practices. Community gardens, for instance, allow residents to come together and collectively cultivate shared plots of land, fostering a sense of community, social cohesion, and educational opportunities.

Urban farms can also serve as hubs for educational and outreach programs, providing hands-on learning experiences for students, offering job training and skill-building opportunities, and engaging the broader community in the process of food production.

Practitioners are constantly exploring new and innovative techniques to address the unique challenges and opportunities presented by the urban environment. This includes the adoption of smart technologies, such as automated irrigation systems, real-time monitoring, and data-driven decision-making, to enhance the efficiency and productivity of urban farms. The integration of urban agriculture with renewable energy systems, such as solar panels or wind turbines, can further contribute to the sustainability and resilience of these urban food production systems.

The wide range of techniques and approaches used in urban agriculture reflects the adaptability and ingenuity of urban farmers and practitioners. Leveraging the unique characteristics of urban environments and embracing innovative solutions, urban agriculture can transform how we think about food production and consumption. This shift contributes to creating more sustainable, resilient, and equitable cities.

As the demand for locally sourced, fresh produce continues to grow, the role of urban agriculture will only become more crucial. Exploring and implementing a diverse array of techniques enables urban farmers to play a pivotal role in building more livable, food-secure, and environmentally conscious communities, paving the way for a more sustainable urban future.

15.4.2 Rooftop Farming and Vertical Gardening

As the world's population continues to urbanize, the need for innovative and space-efficient food production methods has become increasingly pressing. Two emerging urban agriculture techniques that are gaining significant traction are rooftop farming and vertical gardening. These approaches offer promising solutions to the challenges of limited land availability and the growing demand for fresh, locally sourced produce in urban areas.

15.4.2.1 Rooftop Farming

Rooftop farming, also known as urban rooftop agriculture, is the practice of cultivating crops on the roofs of buildings, whether residential, commercial, or industrial. This unique approach to urban agriculture takes advantage of otherwise underutilized space, transforming the often-neglected rooftops into productive green spaces.

One of the primary advantages of rooftop farming is the ability to maximize land usage within dense urban environments. By utilizing vertical space, rooftop farms can significantly increase the amount of arable land available for food production, without requiring additional land footprint at the ground level. This is particularly beneficial in cities where space is at a premium and land prices are high.

Rooftop farms can employ a variety of growing techniques, including raised beds, container gardening, and hydroponic systems. These methods not only allow for efficient use of space but also provide greater control over soil quality, water management, and pest and disease prevention. Also, rooftop farms can incorporate sustainable design features, such as rainwater harvesting systems and renewable energy sources, further enhancing their environmental benefits.

Beyond the practical advantages, rooftop farming also offers a range of ecological and social benefits. The presence of green spaces on rooftops can help mitigate the urban heat island effect, reducing energy demands for cooling and improving air quality. Furthermore, rooftop farms can provide habitat for pollinators and other urban wildlife, contributing to biodiversity and ecosystem health.

From a social perspective, rooftop farming can engage and empower local communities, offering educational opportunities, job training, and community-building activities. The production of fresh, locally grown produce can also improve food security and access to nutritious foods, particularly in underserved urban neighborhoods.

15.4.2.2 Vertical Gardening

Vertical gardening, or vertical farming, is another innovative approach to urban agriculture that maximizes the use of limited space. This technique involves growing crops in vertically stacked layers or towers, often in controlled indoor environments or on the facades of buildings.

One of the primary advantages of vertical gardening is its ability to optimize space utilization, allowing for high-density food production in even the most confined urban settings. Stacking multiple growing levels enables vertical farms to achieve significantly higher yields per square foot of land compared to traditional horizontal farming methods.

Vertical gardening systems often incorporate advanced technologies, such as LED lighting, hydroponics, and automated control systems, to create optimal growing conditions for a wide range of crops. These controlled environments also allow for year-round production, reducing the seasonal dependence and ensuring a more consistent supply of fresh produce. Vertical gardens can also be designed to be highly energy-efficient and environmentally sustainable. Using renewable energy sources, recycling water, and implementing closed-loop systems allows these urban farming techniques to minimize resource consumption and reduce the environmental impact of food production.

One of the unique features of vertical gardening is its ability to be integrated into the built environment, both within and on the exterior of buildings. Indoor vertical farms can be situated in unused or repurposed spaces, such as abandoned warehouses or basements, further enhancing the utilization of urban land. Exterior vertical gardens can be incorporated into building facades, providing a visually striking and functional addition to the urban landscape.

The integration of vertical gardening with the built environment also offers additional benefits, such as improved building insulation, reduced urban heat island effects, and the potential for synergies with other urban systems, like wastewater treatment or renewable energy generation.

As more people move to cities, we need better ways to grow food that are good for the environment. Growing plants on rooftops and in tall, narrow spaces is a great way to do this. These methods can be adjusted to fit different city landscapes. Using these methods, cities can not only ensure they have enough food but also help the environment by lowering air pollution, conserving natural resources, and enhancing the overall livability of urban areas.

We know that urban agriculture practices continue to evolve and gain wider adoption; they hold the potential to transform the way we think about food production and the role of cities in achieving a more sustainable and equitable food system. Through collaborative efforts between urban planners, policymakers, and the local community, the full potential of rooftop farming and vertical gardening can be realized, paving the way for a more resilient and food-secure urban future.

15.4.3 Hydroponics and Aquaponics in Urban Agriculture

We observe that the world's population, particularly in urban areas, continues to grow, and traditional farming methods are struggling to keep up with the increasing demand for food. However, innovative agricultural techniques like hydroponics and aquaponics are providing new solutions for sustainable food production in cities. Hydroponics is a method of growing plants without soil, using mineral nutrient solutions in a water-based environment instead. This approach offers a number of benefits for urban agriculture. Cultivating plants vertically in stacked systems allows hydroponics to maximize the use of limited space in cities. It also requires significantly less water than conventional soil-based farming, making it well-suited for water-scarce urban areas. Also, hydroponic systems can be controlled to provide the optimal growing conditions for plants, leading to faster growth rates and higher yields.

A particularly promising application of hydroponics in urban settings is rooftop farming. Through utilizing the otherwise unused space atop buildings, rooftop hydroponic farms can produce fresh, local produce right in the heart of a city. This helps to reduce the environmental impact and costs associated with transporting food over long distances. Rooftop farms also offer other advantages, such as providing insulation for the building, reducing the urban heat island effect, and even serving as community green spaces.

Hydroponics focuses solely on plant cultivation, aquaponics combines hydroponics with aquaculture, the farming of aquatic organisms like fish. In an aquaponic system, the waste produced by the fish provides nutrients for the hydroponic plants, which in turn help to purify the water for the fish. This symbiotic relationship creates a closed-loop, sustainable ecosystem that requires minimal external inputs. Aquaponics offers several benefits that make it well-suited for urban agriculture. Growing both plants and fish enables the production of a diverse range of food products from a single system. The closed-loop design also minimizes waste and resource consumption, making it a highly efficient use of space and water. Likewise, aquaponic systems can be designed to be modular and scalable, allowing them to be adapted to fit the unique constraints and needs of different urban environments.

One example of successful urban aquaponics is the growing number of commercial operations in cities around the world. In Chicago, for instance, the startup "Farmed Here" operates a large-scale

TABLE 15.6
General Requirements for Hydroponics and Aquaponics

Hydroponics	Water: The pH of the water should be between 5.0 and 7.0.
	Light: Plants need direct sunlight exposure or supplemental lighting for an average of 8–10 hours per day.
	Nutrients: The primary nutrients required for plant growth are nitrogen, phosphorus, and potassium.
	Temperature: Consistent temperatures between 10°C and 21°C for fall plants and 16°C and 27°C for spring plants.
	Oxygen: Supplemental oxygen supply is required for optimal nutrient uptake.
	Structure and Support: Stakes and strings are usually needed to support plants as they grow.
Aquaponics	Dissolved Oxygen: The dissolved oxygen level should be around 5 mg/l.
	pH: The pH of the water should be between 6 and 7.
	Temperature: The water temperature should be between 18°C and 30°C.
	Total Nitrogen: The total nitrogen in the water needs to be monitored and controlled.
	Water Alkalinity: The alkalinity of the water is also an important factor to consider.

indoor aquaponic farm that produces a variety of leafy greens and tilapia fish. Locating their facility within the city enables Farmed Here to provide fresh, locally sourced produce to nearby communities, reducing the environmental impact and costs associated with traditional food supply chains.

Beyond commercial operations, aquaponics is also being embraced by urban communities as a way to improve food security and promote sustainable living. Community-based aquaponic projects, often organized by non-profit organizations or urban farming collectives, provide opportunities for residents to learn about and participate in food production. These initiatives not only increase access to fresh, healthy foods but also foster a sense of community and environmental stewardship.

Despite the promise of hydroponics and aquaponics in urban agriculture, there are still some challenges that need to be addressed. The initial setup costs for these systems can be relatively high, which can be a barrier to widespread adoption, particularly among lower-income urban communities. Additionally, the specialized knowledge and ongoing maintenance required for these systems may present difficulties for some urban farmers.

To overcome these challenges, efforts are underway to make hydroponics and aquaponics more accessible and user-friendly. Research institutions and technology companies are working to develop more affordable and automated systems, reducing the barriers to entry. Educational programs and community-based initiatives are also helping to build the necessary skills and knowledge among urban residents.

Along with the relentless urbanization process, the need for sustainable and efficient food production methods will only continue to grow. Hydroponics and aquaponics offer promising solutions that can help cities become more self-sufficient and resilient in the face of these challenges. Harnessing the potential of these innovative agricultural techniques enables urban communities to improve food security, promote environmental sustainability, and foster social equity.

As a general guideline to implement hydroponics and aquaponics, the conditions required for both are summarized in Table 15.6. Certainly, these guidelines will depend on plants and fish growing.

15.4.4 Innovations and Emerging Techniques in Urban Agriculture

One of the most promising innovations in this field is vertical farming. By cultivating crops in vertically stacked systems, vertical farms can dramatically increase the productivity of limited urban land. These systems often utilize hydroponic or aeroponic growing methods, which eliminate the need for soil and minimize water usage. Some vertical farms are even integrating renewable energy sources, such as solar panels, to create self-sustaining, off-the-grid operations.

An example of a successful vertical farming operation is AeroFarms, a company based in New Jersey. AeroFarms has developed proprietary aeroponic technology that allows it to grow a variety of leafy greens and herbs in a controlled, indoor environment. Precisely controlling factors like temperature, lighting, and nutrient delivery allows AeroFarms to achieve significantly higher yields than traditional farming methods, all while using 95% less water and no pesticides.

Another innovative approach to urban agriculture is the use of aquaponics, which combines aquaculture (fish farming) and hydroponics. In an aquaponic system, the waste produced by the fish provides nutrients for the hydroponic plants, which in turn help to purify the water for the fish. This symbiotic relationship creates a closed-loop, sustainable ecosystem that requires minimal external inputs.

One notable example of urban aquaponics is the work being done by the non-profit organization, Growing Power, in Milwaukee, Wisconsin. Growing Power has established a network of community-based aquaponic farms, where residents can learn about and participate in the production of fresh, local food. These systems not only provide a reliable source of nutritious produce and fish, but also serve as educational hubs and community gathering spaces.

In addition to vertical farming and aquaponics, urban agriculture is also exploring the potential of rooftop gardens and urban greenhouses. By utilizing otherwise unused rooftop spaces, these systems can bring food production right into the heart of the city, reducing the environmental impact and costs associated with transporting food over long distances.

One example of a successful urban greenhouse is the Lufa Farms operation in Montreal, Canada. Lufa Farms has constructed a series of large-scale, commercial greenhouses on the rooftops of industrial buildings, growing a variety of fruits, vegetables, and herbs. Leveraging the controlled environment and year-round growing capabilities of their greenhouses allows Lufa Farms to provide a consistent supply of fresh produce to local communities.

Beyond these physical innovations, urban agriculture is also benefiting from advancements in technology and data analytics. The integration of sensors, automation, and machine learning algorithms is allowing urban farmers to optimize their growing conditions, improve yields, and reduce resource consumption. For instance, startup companies like Agrilyst are developing software platforms that use real-time data and predictive analytics to help urban farmers make more informed decisions about their operations. Tracking factors like soil moisture, nutrient levels, and plant growth allows these platforms to provide personalized recommendations to improve efficiency and productivity.

We are also seeing the emergence of new business models and collaborative approaches. Many urban farming initiatives are being spearheaded by social enterprises, non-profit organizations, and community-based cooperatives, which aim to address issues of food security, social equity, and environmental sustainability. One example of this is the work being done by the non-profit organization, The Urban Farming Guys, in Kansas City, Missouri. The Urban Farming Guys operate a network of community gardens, aquaponic systems, and urban livestock facilities, all of which are managed and maintained by local residents. Empowering community members to participate in food production enables this organization to increase access to fresh, healthy foods while also fostering a sense of civic engagement and environmental stewardship.

The emerging techniques and technologies in urban agriculture hold promise for addressing this challenge, from vertical farming and aquaponics to rooftop greenhouses and data-driven optimization.

15.5 URBAN GREENERY AND CLIMATE CHANGE

Urban greenery, such as parks, gardens, and trees, is essential in reducing the impacts of climate change. The integration of green spaces has become increasingly essential, as cities are developing and expanding. These urban green areas act as natural carbon sinks, absorbing and storing atmospheric carbon dioxide, a major greenhouse gas contributing to global warming. In addition, urban greenery helps regulate local temperatures, providing cooling effects through evapotranspiration and shading, which can offset the urban heat island effect. This can lead to reduced energy demands for cooling buildings, ultimately lowering greenhouse gas emissions from the built environment.

A well-planned urban greenery can help manage stormwater runoff, reducing the risk of flooding and improving water quality. Beyond these environmental benefits, urban green spaces also provide significant social and health benefits, enhancing mental well-being, promoting physical activity, and fostering community cohesion. As the world contends the challenges of climate change, the strategic incorporation of urban greenery in city planning and development has become a critical strategy for building resilient and sustainable urban environments that can adapt to the changing climate and mitigate its effects.

15.5.1 The Role of Urban Greenery in Mitigating Climate Change

Climate change mitigation refers to the actions and strategies aimed at reducing or preventing the emission of greenhouse gases (GHGs) into the atmosphere, as well as enhancing the absorption and storage of these gases. The goal of climate change mitigation is to limit the magnitude and rate of long-term climate change.

The key elements of climate change mitigation are summarized in Table 15.7.

Climate change mitigation efforts aim to limit the long-term impacts of climate change by reducing the amount of GHGs in the atmosphere and enhancing the Earth's natural carbon sinks. These actions are crucial in stabilizing the global climate system and preventing the most severe consequences of climate change.

In addition to climate change mitigation, there is climate change adaptation. These two systems are hand-in-hand to combat climate change. Climate change adaptation refers to the process of adjusting to the actual or expected changes in the climate and their effects. It involves preparing for and responding to the impacts of climate change to reduce vulnerability and increase the resilience of natural and human systems. The elements and corresponding actions of climate change adaptation are shown in Table 15.8.

TABLE 15.7
The Elements and Actions of Climate Change Mitigation

Reducing greenhouse gas emissions	• Transitioning to renewable energy sources (e.g., solar, wind, hydropower, geothermal). • Improving energy efficiency in buildings, transportation, and industrial processes. • Promoting the use of low-emission or zero-emission vehicles. • Reducing deforestation and promoting reforestation and afforestation.
Enhancing carbon sinks	• Preserving and expanding natural carbon sinks, such as forests, wetlands, and oceans. • Developing and implementing carbon capture and storage (CCS) technologies. • Promoting sustainable land use and agricultural practices that increase soil carbon sequestration.
Behavioral and societal changes	• Encouraging sustainable consumption patterns and lifestyles. • Promoting public awareness and education about climate change mitigation. • Implementing policies and regulations to incentivize emissions reductions and sustainable practices.
Technological innovation	• Developing and deploying clean energy technologies (e.g., renewable energy, energy storage, carbon capture). • Improving energy efficiency in various sectors (e.g., transportation, buildings, industry). • Exploring new and emerging mitigation strategies, such as direct air capture, biomass energy with CCS, and innovative carbon removal methods.

TABLE 15.8
Elements and Actions of Climate Change Adaptation

Reducing vulnerability	• Assessing and understanding the risks and vulnerabilities to climate change impacts. • Implementing measures to protect infrastructure, communities, and ecosystems from the effects of climate change, such as sea-level rise, extreme weather events, and changing precipitation patterns.
Building resilience	• Enhancing the ability of natural and human systems to withstand, recover, and adapt to climate-related stresses and disruptions. • Diversifying and strengthening livelihood options and economic activities to reduce dependence on climate-sensitive resources.
Improving preparedness and response	• Developing early warning systems and disaster response plans to mitigate the impacts of climate-related hazards. • Investing in infrastructure and services that can withstand the effects of climate change, such as flood protection, water management systems, and emergency services.
Promoting adaptive capacity	• Increasing knowledge, skills, and resources to enable individuals, communities, and institutions to adapt to climate change. • Fostering collaboration and coordination among stakeholders, including government, private sector, and civil society, to coordinate adaptation efforts.
Mainstreaming adaptation	• Integrating climate change adaptation into development planning, policies, and decision-making processes at various levels (local, regional, national). • Ensuring that adaptation measures are equitable, inclusive, and address the needs of vulnerable populations.

Climate change adaptation aims to minimize the negative impacts of climate change and leverage any potential benefits, enabling natural and human systems to better cope with and thrive in the face of a changing climate.

We believe that the challenges of climate change are escalating if we do nothing, and, therefore, the importance of urban greenery has become increasingly evident. Cities, which are home to a growing number of the global population, contribute to as well as combat the effects of climate change. Urban greenery, encompassing parks, gardens, trees, and other vegetation, has emerged as a crucial component in the strategies aimed at mitigating the impacts of a changing climate.

One of the primary ways in which urban greenery can help mitigate climate change is through carbon sequestration. Plants, trees, and other vegetation act as natural carbon sinks, absorbing and storing atmospheric carbon dioxide, a primary greenhouse gas responsible for global warming. The integration of green spaces becomes essential in counteracting the carbon emissions generated by urban activities, such as transportation and energy consumption. Studies have shown that urban forests alone can offset a significant portion of a city's carbon footprint, making them a valuable asset in the fight against climate change. Beyond carbon sequestration, urban greenery also regulates local temperatures and reduces the urban heat island effect. The urban heat island phenomenon refers to the phenomenon where cities experience higher temperatures compared to their surrounding rural or suburban areas, primarily due to the prevalence of heat-absorbing materials like concrete and asphalt. This effect can lead to increased energy demands for cooling, resulting in higher GHG emissions from power generation.

Urban trees, shrubs, and other vegetation can help mitigate the urban heat island effect through the process of evapotranspiration, where plants release water vapor into the atmosphere, leading to a cooling effect. Additionally, the shading provided by urban greenery can directly reduce the amount of heat absorbed by buildings and paved surfaces. This cooling effect not only improves the overall comfort and livability of urban environments but also reduces the energy required for air conditioning, thereby lowering the associated greenhouse gas emissions. Urban greenery can also manage stormwater runoff and reduce the risk of flooding, which are increasingly compelling issues due to the impacts of climate change. Impervious surfaces, such as roads and buildings, prevent rainfall from infiltrating the ground, leading to rapid stormwater runoff that can overwhelm drainage systems and cause flooding. Urban trees, green roofs, and other green infrastructure can help mitigate this issue by promoting water infiltration, slowing down the rate of runoff, and reducing the overall volume of stormwater. This can not only alleviate the risk of flooding but also improve water quality by filtering out pollutants and reducing the strain on water treatment facilities.

Beyond these environmental benefits, urban greenery also provides significant social and health benefits that contribute to climate change resilience. Well-designed green spaces can improve mental and physical well-being, as discussed earlier, encourage physical activity, and foster community engagement. These factors can enhance the overall adaptive capacity of urban populations, enabling them to better withstand and recover from the impacts of climate change, such as extreme weather events and heat waves.

Competing demands for land use, limited resources, and ongoing maintenance requirements can pose barriers to the widespread adoption of urban green infrastructure. Policymakers and urban planners must navigate these obstacles and prioritize the strategic placement and management of urban green spaces to maximize their climate change mitigation potential. To address these challenges, a holistic approach to urban planning and design is needed. This involves incorporating green infrastructure into the built environment, such as green roofs, living walls, and permeable surfaces, and ensuring that new developments prioritize the preservation and expansion of urban greenery. Additionally, community engagement and education initiatives can help raise awareness and foster public support for the integration of urban greenery, thereby enhancing its long-term sustainability.

The role of urban greenery in mitigating climate change is multifaceted and crucial. Acting as natural carbon sinks, regulating local temperatures, managing stormwater, and providing broader social and health benefits enables urban green spaces to contribute significantly to building resilient and sustainable cities. The strategic integration of urban greenery must be a central component of climate change mitigation strategies, ensuring that cities become part of the solution rather than part of the problem.

15.5.2 Urban Greenery and Carbon Sequestration

Carbon sequestration is the process of capturing, securing, and storing carbon dioxide from the atmosphere. It plays a crucial role in mitigating climate change by reducing the amount of carbon dioxide in the atmosphere. Carbon sequestration occurs both naturally and as a result of anthropogenic activities and typically refers to the storage of carbon that has the immediate potential to become carbon dioxide gas. On the other hand, a carbon sink is any natural or artificial process that absorbs more carbon from the atmosphere than it releases. The largest carbon sinks on Earth are forests, soil, and the ocean. These sinks form an important part of the natural carbon cycle. A carbon sink is a type of carbon pool that has the capability to take up more carbon from the atmosphere than it releases.

Estimating the carbon sink and carbon sequestration of urban greenery involves several steps and calculations. The general approach to estimation is outlined below.

15.5.2.1 Carbon Storage Calculation

We refer, in this case, carbon storage to the total amount of carbon currently stored in the urban greenery. This is often calculated based on the biomass of the trees, which can be estimated using species-specific allometric equations. Allometric equations are mathematical models used in biology to study how biological characteristics change with size. They are derived from empirical data and are often used to uncover scaling relationships among different biological measurements.

The general form of an allometric equation is:

$$Y = aM^b \tag{15.1}$$

where:

Y is a biological variable (such as tree height)
a is a proportionality coefficient
M is a measure of body size (such as body volume or body mass)
b is the scaling exponent

15.5.2.2 Estimation of Carbon Contents of a Tree

To estimate a tree's carbon content using its diameter at breast height (DBH), measure the DBH in centimeters, then calculate the tree's volume as a cylinder. Multiply the volume by the tree's specific gravity to find the biomass, and convert this biomass to carbon content, typically 50% of the dry weight. This provides a metric estimate of carbon storage.

Here is an example of how to calculate the carbon content in a tree using the DBH.

1. *Measure the DBH:* Suppose the DBH of the tree is 30 cm.
2. *Estimate the Tree Height:* Let us assume the tree height is 15 m.
3. *Calculate the Volume:* Treat the tree trunk as a cylinder. The volume of the tree is approximately:

$$\text{Volume} = \pi \times \left(\frac{\text{DBH}}{2}\right)^2 \times \text{Height} \tag{15.2}$$

With DBH of 30 cm = 0.3 m, and height of the tree is 15.0 m, the volume of the tree is:

$$\text{Volume} = \pi \times \left(\frac{0.3\text{ m}}{2}\right)^2 \times 15\text{ m} = 1.06\text{ m}^3$$

4. *Estimate Biomass:*

Assume the density of the tree species is 0.53 ton/m^3:
The biomass: $\text{Volume} \times \text{Density} = 1.06\text{ m}^3 \times 0.53\text{ ton/m}^3 = 0.5618\text{ ton}$

5. *Calculate the Carbon Content:*

The carbon content of some tree species can be found in Table 15.9.
From the table, the carbon content is about 50% of the dry biomass:

$$\text{Carbon content} = 0.5 \times 0.5618\text{ ton} = 0.2809\text{ ton}$$

So, for a tree with a DBH of 30 cm and a height of 15 m, the carbon content would be approximately 0.2809 ton.

15.5.2.3 Carbon Sequestration Calculation

Carbon sequestration refers to the amount of carbon dioxide that is absorbed from the atmosphere by trees annually. This can be calculated using the following steps:

TABLE 15.9
Carbon Content of Some Tree Species

Tree Species	Average Carbon Content
Ulmus laciniata	43.4%
Phellodendron amurense	43.5%
Acer mono	43.8%
Tilia amurensis	44.2%
Populus davidiana	44.5%
Fraxinus mandshurica	44.7%
Juglans mandshurica	44.9%
Quercus mongolica	45.3%
Betulla davurica	45.8%
Betulla platyphylla	46.7%
Picea koreansis	46.9%
Larix gmelinii	47.4%
Pinus koreansis	48.3%
Abies nephrolepis	48.3%

Source: Widagdo et al. (J of ForRes 32:2021, pp. 2545–2556).

Step 1: Calculate the net primary production (NPP), which is the amount of carbon captured by the plant through photosynthesis per unit time.

NPP is the net amount of carbon produced by a plant or group of plants in a given period of time. It can be calculated using the following formula:

$$NPP = GPP - Ra \tag{15.3}$$

where:

NPP is the net primary productivity
GPP is the gross primary productivity, which is the total amount of carbon taken in by the plant
Ra is the amount of carbon lost during respiration

Step 2: Estimate the net ecosystem productivity (NEP), which is the net carbon balance of the system. NEP is a measure of the net amount of carbon uptake in an ecosystem. It can be calculated using the following formula:

$$NEP = GPP - R_t \tag{15.4}$$

where:

NEP is the net ecosystem productivity
GPP is the gross primary productivity, which is the total amount of carbon taken in by the plant
R_t is the rate of all organisms' respiration (total).

Step 3: The difference between NPP and NEP gives the amount of carbon sequestered.

15.5.2.4 Some Notes on the Calculation

Gross primary productivity (GPP) is the total amount of energy, mainly from sunlight, that plants in an ecosystem capture and convert into food via photosynthesis over a certain period. Here are some methods to estimate GPP:

1. *Plant Biomass Production:*
   This method involves measuring the growth of plants over a given period. We can focus on one plant, measure its growth over a week or a month using precise tools like calipers, keep a log, and do our math right.
2. *Carbon Assimilation:*
   This method measures how much carbon dioxide plants "inhale" through photosynthesis over time. The assimilated carbon adds to the plant's mass, contributing to its growth. While this process cannot be measured directly, creative solutions like using carbonyl sulfide (COS) as a proxy are explored to get close estimates.
3. *Resource Utilization:*
   This method is all about how effectively organisms, especially green plants, use available resources such as sunlight and carbon dioxide to create energy through photosynthesis. More sunlight means more energy for photosynthesis, which can increase GPP. Higher levels of carbon dioxide can also boost GPP.
4. *Light and Dark Bottle Method:*
   This method is used in aquatic ecosystems to measure GPP. Two bottles are filled with water containing photosynthesizing organisms. One bottle is exposed to light, and the other is kept in the dark. The light bottle measures NPP, as it allows for both photosynthesis and cellular respiration. The dark bottle measures respiration as it only allows for respiration. The difference in the final and initial values for both bottles gives the GPP.

For the information, the major carbon reservoir on Earth and the approximate size of the carbon storage are shown in Table 15.10.

The above calculations are general guidelines, and the specific calculations may vary depending on the type of urban greenery and the specific conditions of the site. These calculations require a good understanding of botany and environmental science.

TABLE 15.10
Major Carbon Reservoir and Size of Carbon Storage

Source	Size (metric tons)
Rocks: most of Earth's carbon	65.5×10^{12}
Oceans	38.7×10^{12}
Atmosphere	3×10^{9} per year
Plants and soils	Exact amount is difficult to estimate, depending on specific ecosystem
Fossil fuels: coal, oil, and natural gas	Exact amount is difficult to estimate, depending on specific ecosystem
Cryosphere: includes all of Earth's ice and frozen ground	Exact amount is difficult to estimate, depending on specific ecosystem
Earth's outer core	Between 5.5 and 36.8×10^{18}

15.5.3 Urban Heat Island Effect and Green Spaces

The urban heat island (UHI) effect is a phenomenon where urban areas experience significantly higher temperatures compared to their surrounding rural or suburban counterparts. This phenomenon is primarily caused by the replacement of natural, permeable surfaces, such as vegetation and soil, with impervious surfaces like concrete, asphalt, and buildings. These man-made materials absorb and retain heat more efficiently, leading to a noticeable temperature difference between urban and rural areas.

The UHI effect can have significant consequences for both the environment and human well-being. One of the most prominent impacts is the increased energy demand for cooling during the summer months. As urban areas become hotter, the need for air conditioning and other cooling systems rises, leading to higher energy consumption and greenhouse gas emissions. In addition, the UHI effect can exacerbate the effects of heat waves, putting vulnerable populations, such as the elderly and the socioeconomically disadvantaged, at an increased risk of heat-related illnesses and mortality.

The UHI effect also has a negative impact on the local ecosystem. The increased temperatures can disrupt the natural habitats of various plant and animal species, leading to changes in species composition and biodiversity. Furthermore, the UHI effect can contribute to the formation of urban smog, as the increased temperatures can promote the formation of ground-level ozone and other air pollutants.

One effective way to mitigate the UHI effect is through the incorporation of green spaces within urban areas. Green spaces, such as parks, gardens, and urban forests, can provide a cooling effect through evapotranspiration and shading. Vegetation absorbs solar radiation and releases water vapor, which helps to lower the surrounding air temperature. Additionally, the presence of trees and other greenery can provide shade, reducing the amount of heat absorbed by buildings and paved surfaces.

Research has shown that the integration of green spaces in urban areas can lead to significant reductions in local temperatures. A study conducted in Seoul, South Korea, found that the presence of a single large park can lower the surrounding air temperature by up to 3.6°C (6.5°F) during the summer. Similarly, a study in Phoenix, Arizona, observed that neighborhoods with more vegetation had significantly lower surface temperatures compared to those with less greenery.

Beyond their cooling effects, green spaces also offer a wide range of additional benefits to urban communities. They can improve air quality by filtering out particulate matter and other pollutants, and they can also help to reduce noise pollution. Green spaces provide valuable habitats for urban wildlife, supporting biodiversity and promoting ecosystem health. Furthermore, access to green spaces has been linked to improved physical and mental health outcomes, as they offer opportunities for recreation, relaxation, and social interaction.

Despite the well-documented benefits of green spaces, many urban areas around the world are facing challenges in preserving and expanding these vital resources. Rapid urbanization, population growth, and competing land-use demands often result in the loss of green spaces, leading to a decrease in the overall quality of the urban environment. To address this challenge, policymakers and urban planners must prioritize the integration of green spaces into urban design and development strategies.

One approach to integrating green spaces into urban areas is through the implementation of green infrastructure. Green infrastructure refers to a network of natural and semi-natural features, such as parks, urban forests, green roofs, and green walls, that are strategically planned and managed to provide a range of ecosystem services. Incorporating green infrastructure into urban planning creates a more resilient and livable environment, mitigating the urban heat island effect and delivering additional benefits to the community.

Another important strategy is the promotion of urban greening initiatives, which involve the planting and maintenance of trees, gardens, and other vegetation within the urban landscape. These initiatives can be spearheaded by local governments, community organizations, and even individual residents, who can work together to transform their neighborhoods into more verdant and sustainable spaces.

There is an association between urban temperature and urban green spaces. This relationship is often explored using statistical models and spatial analysis techniques. Here are some examples:

15.5.3.1 Linear Mixed-Effects Model and Geographically Weighted Regression

A study conducted in Taipei City, Taiwan, used a linear mixed-effects model and geographically weighted regression to explore the association between pedestrian-level green spaces and ambient temperature. They controlled seasons, land use information, and traffic volume. The results indicated that a higher green vegetation index (GVI) was significantly associated with lower ambient temperatures and temperature differences.

This model is often used to analyze the relationship between the GVI and ambient temperature1. The general form of a linear mixed-effects model is expressed below:

$$Y = X\beta + Z\gamma + \epsilon \tag{15.5}$$

where:

- Y is the dependent variable (e.g., ambient temperature),
- X is the fixed-effects design matrix (e.g., green spaces area)
- β is the fixed-effects parameter vector,
- Z is the random-effects design matrix (e.g., other factors of green spaces)
- γ is the random-effects parameter vector, and
- ϵ is the error term.

15.5.3.2 Multiple Linear Regression (MLR)

MLR is a statistical technique that uses several explanatory variables to predict the outcome of a response variable. The goal of MLR is to model the linear relationship between the explanatory (independent) variables and response (dependent) variables. In the context of this study, MLR can be used to model the relationship between the principal components (obtained from PCA) and land surface temperature (LST). The general form of an MLR model is:

$$Y = \beta_0 + \beta_1 X_1 + \beta_2 X_2 + \ldots + \beta_p X_p + \epsilon \tag{15.6}$$

where:

Y	is the dependent variable (in this case, LST),
β_0	is the y-intercept,
$\beta_1, \beta_2, \ldots, \beta_p$	are the coefficients of the independent variables,
$X_1, X_2, \ldots, X_p$	are the independent variables (in this case, the principal components), and
ϵ	is the error term.

With reliable data, these studies can demonstrate that there is a significant negative correlation between urban green spaces and urban temperatures. However, the exact equations used in these models can be complex and depend on the specific variables and data used in each study.

The urban heat island effect poses a significant challenge that urban areas must address in the face of climate change and ongoing urbanization. Incorporating green spaces and green infrastructure into urban design and development is a proven and effective way to mitigate the UHI effect and deliver a range of additional benefits to urban communities. Policymakers and urban planners can create more livable, sustainable, and resilient cities for the future through incorporating green spaces.

15.6 POLICIES AND REGULATION OF URBAN AGRICULTURE

The growing prevalence of urban agriculture has necessitated the development of comprehensive policies and regulations to govern this increasingly important sector. Policymakers have recognized the need to provide a robust regulatory framework to support and guide urban agricultural practices. These policies must strike a sensitive balance, encouraging the growth of urban farms and gardens while also addressing concerns related to land use, water usage, waste management, and public health. Effective regulation can help to ensure that urban agriculture is integrated seamlessly into the urban landscape, providing tangible benefits to local communities without compromising broader city planning objectives. With thoughtful policy development and implementation, urban agriculture can thrive as a vibrant and integral component of the modern, sustainable city.

15.6.1 Overview of Policies Supporting Urban Agriculture

As mentioned earlier, urban agriculture refers to the cultivation, processing, and distribution of food and other products within cities and their surrounding areas. This includes growing crops, raising livestock and aquaculture, as well as processing and selling these products. Urban agriculture has gained increasing prominence in recent years as a way to address food security, community development, and environmental sustainability in cities.

The practice of growing food in urban environments dates back thousands of years, but it has seen a resurgence in popularity as more people recognize its benefits. Urban agriculture can take many forms, from community gardens and urban farms to rooftop greenhouses and vertical farming operations. It provides fresh, nutritious food for local residents, creates green space, and can even generate income for participants.

Despite its advantages, urban agriculture has historically faced a range of policy and regulatory barriers that have hindered its growth and development. However, in recent years, there has been a growing recognition of the importance of supporting and enabling urban agriculture, and an increasing number of cities and governments have implemented policies and programs to do so.

To promote urban agriculture, policies supporting land use, water management, and financial incentives, among others, are essential. These will foster sustainable city farming practices.

15.6.1.1 Land Use and Zoning Policies

One of the primary challenges facing urban agriculture is access to land. Many cities have zoning laws and land use regulations that either prohibit or severely restrict the ability to grow food within city limits. To address this, some cities have implemented policies that specifically allow and encourage urban agriculture. For example, some cities have created urban agriculture zones or overlays that designate certain areas as suitable for growing food. Others have modified their zoning codes to remove barriers and explicitly permit urban agriculture in residential, commercial, and industrial zones. These policies can help ensure that there is sufficient land available for urban farms, community gardens, and other food production activities.

15.6.1.2 Land Disposition Policies

Policies that permit the acquisition of surplus municipal properties for urban agriculture are fundamental. They provide an opportunity for unused or underutilized public lands to be transformed into productive green spaces. This not only makes it easier for urban farmers to secure land for their operations, but also contributes to the city's sustainability. It encourages local food production, enhances biodiversity, and improves the overall urban environment. These policies, therefore, will significantly contribute to promoting urban agriculture and sustainable city living.

15.6.1.3 Financial Incentives and Funding

Establishing and maintaining urban agriculture initiatives can be financially challenging, particularly for low-income individuals and communities. To address this, some governments and organizations have implemented financial incentives and funding programs to support urban agriculture. These can include tax credits or exemptions for urban farms, grants for community garden projects, and low-interest loans for the purchase of land or equipment. Some cities have also created dedicated urban agriculture funds or budgets to provide ongoing support for these initiatives.

15.6.1.4 Infrastructure and Technical Assistance

Beyond land and financial resources, urban agriculture also requires access to appropriate infrastructure and technical support. This can include things like irrigation systems, greenhouses, processing and storage facilities, and educational resources. Some cities have invested in the development of urban agriculture infrastructure, such as constructing community garden sites with raised beds, water access, and storage sheds. Others have provided technical assistance and training programs to help urban farmers and gardeners develop the skills and knowledge they need to be successful.

15.6.1.5 Procurement and Food Systems Integration

Another important aspect of supporting urban agriculture is ensuring that the food produced can be easily integrated into local food systems and procurement processes. This can involve policies that prioritize the purchase of locally grown urban produce by schools, hospitals, and other public institutions. Some cities have also implemented food hubs or food distribution programs that help connect urban farmers and gardeners with local consumers, restaurants, and other buyers. These initiatives can help create new market opportunities for urban agriculture and make it easier for residents to access fresh, locally grown food.

15.6.1.6 Environmental and Public Health Considerations

Urban agriculture can also play a role in addressing environmental and public health challenges. Policies that support urban agriculture can help mitigate issues such as urban heat island effects, stormwater management, and air quality, as well as promote healthier eating habits and community engagement. For example, some cities have implemented policies that require the integration of urban agriculture into new development projects, such as requiring green roofs or vertical farming systems. Others have developed programs that provide educational resources and support for urban gardeners to adopt sustainable growing practices, such as composting and water conservation.

15.6.2 Regulatory Challenges in Urban Agriculture

Urban agriculture is, to some extent, recognized for its potential to improve food security, promote community development, and contribute to environmental sustainability. However, urban farmers often face a complex and sometimes contradictory regulatory environment that can pose significant challenges to the establishment and expansion of urban agriculture initiatives.

15.6.2.1 Zoning and Land Use Regulations

One of the most significant regulatory challenges for urban agriculture is zoning and land use regulations. These regulations determine what activities can take place in specific areas of a city. In many cities, zoning codes were written before urban agriculture became popular, and as a result, they often do not include provisions for urban farms or gardens. This can create uncertainty for urban farmers, who may not know whether their activities are allowed under local zoning laws. In some cases, urban agriculture may be prohibited in areas zoned for residential or commercial use. Even in areas where urban agriculture is allowed, there may be restrictions on the size of the farm, the types of crops that can be grown, or the use of structures like greenhouses or hoop houses. These restrictions can limit the viability of urban agriculture projects and discourage potential urban farmers.

15.6.2.2 Access to Land

Access to land is another significant challenge for urban agriculture. Vacant or underutilized land in cities is often owned by multiple entities, including private individuals, corporations, and various levels of government. Navigating this complex ownership landscape to secure land for urban agriculture can be a daunting task. Even when land is available, it can be prohibitively expensive, particularly in densely populated urban areas where real estate prices are high. Additionally, urban farmers often face insecure land tenure. They may be able to use a piece of land for a time, but they can be forced to move if the land is sold or redeveloped.

15.6.2.3 Water and Soil Quality Regulations

Urban farmers also have to navigate regulations related to water and soil quality. In many cities, water is a scarce and expensive resource, and urban farmers may face restrictions on their water use. They may also be required to implement water-saving technologies or practices, which can add to their costs. Soil quality is another concern. Urban soils can be contaminated with heavy metals or other pollutants, and urban farmers may be required to test their soil and take steps to remediate any contamination. These regulations are important for protecting public health, but they can also be costly and time-consuming for urban farmers.

15.6.2.4 Food Safety and Sales Regulations

Urban farmers must also comply with food safety regulations, which can be complex and vary from one jurisdiction to another. These regulations may include requirements for handling and processing produce, as well as rules for selling produce to the public. Navigating these regulations can be challenging, particularly for small-scale urban farmers who may not have the resources to hire legal or regulatory experts.

15.6.2.5 Public Perception and Social Acceptance

Urban agriculture can sometimes face resistance from local residents or businesses who may have concerns about aesthetics, noise, smells, or pests. Policies need to address these concerns and promote positive perceptions of urban agriculture.

15.6.2.6 Training and Education

Urban farmers often need training and education in farming techniques, business management, marketing, and regulatory compliance. Policies could support educational programs and resources for urban farmers. The challenge in this issue lies in the fact that urban farming is a multidisciplinary field that requires a wide range of skills and knowledge. Urban farmers need to understand not only farming techniques but also business management, marketing, and regulatory compliance. This can be a significant barrier for individuals who are interested in urban farming but do not have this background. Moreover, urban farming often involves unique challenges that are not present in traditional rural farming, such as space constraints, soil contamination, and water management. Urban farmers need specialized training to navigate these challenges.

15.6.2.7 Integration with Urban Planning

Urban agriculture should be integrated into broader urban planning and development strategies. This includes considering urban agriculture in transportation planning, housing development, and environmental management. Urban agriculture is often seen as a separate or niche activity, rather than a core component of urban development. This can lead to a lack of coordination and support for urban agriculture initiatives.

Urban farms need to be accessible to both the farmers who maintain them and the community members who benefit from them. However, transportation planning may not always consider the needs of urban farms, such as the need for roads or public transit routes that provide access to the farms. Urban agriculture can sometimes be in competition with housing development for land use. As cities grow and develop, open spaces that could be used for urban farms may instead be used for housing. This can make it difficult for urban agriculture initiatives to find and secure land. Urban agriculture can play a significant role in environmental management by providing green spaces, improving biodiversity, and reducing food miles. However, these benefits may not be fully recognized or incorporated into environmental management strategies.

15.6.2.8 Climate Change and Sustainability

Policies should also consider the role of urban agriculture in mitigating climate change and promoting sustainability. This includes supporting practices that reduce greenhouse gas emissions, conserve water and energy, and promote biodiversity. The challenges in considering the role of urban agriculture in mitigating climate change and promoting sustainability are multi-fronts.

The first challenge lies in understanding and quantifying the impact of urban agriculture on climate change and sustainability. This requires comprehensive research and data collection to measure factors such as the reduction in greenhouse gas emissions, water and energy conservation, and biodiversity promotion associated with urban farming practices.

Formulating policies that effectively support sustainable practices in urban agriculture is another challenge. These policies need to be based on sound scientific evidence and must consider local contexts and the diversity of urban agriculture practices.

Once policies are in place, the challenge shifts to implementation and enforcement. This includes providing resources and support to help urban farmers adopt sustainable practices, monitoring compliance with sustainability standards, and acting when these standards are not met.

Urban farmers, policymakers, and the public need to be educated about the benefits of sustainable urban agriculture and the role it plays in mitigating climate change. This can be a significant challenge, particularly in communities where awareness of these issues is low.

Sustainable farming practices often require upfront investment, which can be a barrier for many urban farmers. Policies need to address these economic challenges, for example, by providing financial incentives or subsidies to encourage sustainable practices.

Urban agriculture policies need to balance multiple goals, including food production, environmental protection, and social equity. Achieving this balance can be complex and challenging.

Urban agriculture has the potential to contribute to sustainable urban development. On the other hand, regulatory challenges can pose significant barriers. Policymakers need to work closely with urban farmers, community organizations, and other stakeholders to develop supportive policies and regulations that recognize the unique challenges and opportunities of urban agriculture. This may include revising zoning and land use regulations, improving access to land and water resources, addressing soil contamination issues, and providing clear and consistent food safety regulations. With supportive policies and regulations, urban agriculture can thrive, contributing to food security, community development, and environmental sustainability in cities around the world.

15.6.3 Future Policy Directions for Urban Agriculture

Urban agriculture, the practice of cultivating, processing, and distributing food in or around urban areas, has gained significant attention in recent years. As cities continue to grow and face new challenges, urban agriculture offers a unique solution to improve food security, promote community development, and mitigate environmental impact. However, to fully realize these benefits, there is a need for comprehensive and forward-thinking policy directions.

15.6.3.1 Urban Agriculture: Current State and Challenges

Urban agriculture is a broad term that encompasses a variety of activities related to the production of food within urban environments. This includes community gardens, rooftop farms, and indoor vertical farming, each with its unique characteristics and benefits.

Community gardens are shared spaces where residents can grow their own fruits, vegetables, and herbs. These gardens not only provide fresh produce but also serve as a platform for community engagement and education. They foster a sense of community, promote healthy eating habits, and provide opportunities for physical activity. Rooftop farms make use of underutilized rooftop spaces to grow food. This innovative approach to urban agriculture not only maximizes the use of space but also provides numerous environmental benefits. Rooftop farms can help reduce the urban heat island effect, improve air quality, and manage stormwater runoff. Indoor vertical farming is a high-tech approach to urban agriculture that involves growing crops in vertically stacked layers, often in controlled environments. This method allows for year-round production, efficient use of space, and reduced water usage. It also eliminates the need for pesticides and reduces the carbon footprint associated with transporting food.

We know that urban agriculture contributes to local food systems. It reduces the distance food travels from farm to table, ensuring fresher and more nutritious produce. It also enhances food security by increasing the availability of fresh food in urban areas. In addition, urban agriculture enhances biodiversity by creating green spaces and habitats for various species. It contributes to the preservation of local plant varieties and promotes a balanced ecosystem. Beyond its environmental and food security benefits, urban agriculture also provides numerous social benefits. It fosters community cohesion, provides educational opportunities, and can even contribute to local economies by creating jobs and stimulating local businesses.

Urban agriculture, however, faces several challenges. Limited space in urban areas often restricts the scale of urban farming operations. Soil contamination is another concern, especially in post-industrial cities. Water management is crucial, as urban farming practices must be sustainable and not strain local water resources. Zoning restrictions can also pose challenges. Urban agriculture is often not accounted for in city planning and zoning regulations, making it difficult for urban farmers to secure land and operate their farms.

To overcome these challenges, future policy directions should focus on integrating urban agriculture into city planning processes, promoting sustainable practices, and providing support and resources for urban farmers. This could include revising zoning regulations, providing training and resources for soil testing and remediation, and implementing water-saving technologies.

15.6.3.2 Policy Direction 1: Land Use and Zoning Regulations

One of the primary barriers to urban agriculture is the lack of access to land. To address this, policies should aim to identify and allocate underutilized public and private spaces for agricultural use. This could include vacant lots, rooftops, and even spaces within buildings for indoor farming. Zoning regulations should also be revised to allow for agricultural activities in urban areas.

15.6.3.3 Policy Direction 2: Resource Management

Urban agriculture requires careful management of resources, particularly water and soil. Policies should promote the use of sustainable practices such as rainwater harvesting, composting, and recycling of organic waste. Also, guidelines should be established for soil testing and remediation to ensure safe and productive growing conditions.

15.6.3.4 Policy Direction 3: Education and Training

To ensure the success and sustainability of urban agriculture, there is a need for education and training programs. Policies should support the development of such programs that teach urban residents about sustainable farming practices, food production, and nutrition. This could also include initiatives to foster entrepreneurship and create job opportunities within the urban agriculture sector.

15.6.3.5 Policy Direction 4: Integration into Local Food Systems

Urban agriculture should be integrated into local food systems to enhance food security and resilience. Policies should encourage the establishment of local food hubs that connect urban farms with consumers and businesses. This could also involve supporting farm-to-school programs, farmers markets, and other direct marketing strategies.

15.6.3.6 Policy Direction 5: Research and Innovation

Finally, there is a need for ongoing research and innovation to improve urban agricultural practices and technologies. Policies should facilitate collaborations between universities, research institutions, and urban farmers to drive innovation in areas such as vertical farming, hydroponics, and aquaponics.

Urban agriculture holds great potential for addressing many of the challenges faced by growing urban populations. However, realizing this potential requires comprehensive and forward-thinking policies. By focusing on land use, resource management, education, integration into local food systems, and research, we can pave the way for a sustainable and resilient urban agriculture sector.

Crafting policies that facilitate access to land, provide robust financial incentives, and foster meaningful community involvement, policymakers allow us to unlock the transformative potential of urban agriculture. This, in turn, can contribute to the creation of more resilient, sustainable, and equitable urban communities that are better equipped to meet the evolving needs of their populations.

The policy directions represent a crucial step toward realizing the full benefits of urban agriculture and positioning it as a central component of sustainable urban development. Through these policy directions, cities can harness the power of urban agriculture to build a more food-secure, environmentally conscious, and socially inclusive future, one that is better prepared to address the complex challenges of our rapidly urbanizing world.

15.7 ECONOMIC ASPECTS OF URBAN GREENERY AND AGRICULTURE

Urban greenery and agriculture offer not just environmental benefits, but also economic advantages. Urban agriculture, for instance, can contribute to local economies by creating jobs, stimulating local businesses, and reducing food transportation costs. Green spaces, such as parks and community gardens, can increase property values and attract tourism, providing a boost to the local economy. Furthermore, they can play a role in mitigating the impacts of climate change, potentially saving cities significant amounts in future climate-related costs. However, realizing these economic benefits requires strategic planning and investment. Policymakers must recognize the value of urban greenery and agriculture and integrate them into urban planning and economic development strategies.

15.7.1 Economic Benefits of Urban Greenery

Urban greenery, encompassing parks, gardens, street trees, green roofs, and other forms of vegetation within cities, plays a crucial role in urban sustainability. Beyond their aesthetic appeal, these green spaces provide a multitude of economic benefits that contribute to the prosperity of cities and their residents.

Urban greenery increases property values. Properties that are adjacent to or in close proximity to well-maintained green spaces often have a higher market value compared to those that are not. This phenomenon, often referred to as the "green premium," is backed by numerous studies and real estate market analyses. The "green premium" can be attributed to several factors. Firstly, green spaces improve the aesthetic appeal of an area, making it more desirable to potential buyers. Secondly, these spaces provide recreational opportunities and improve the quality of life for residents, factors that are often reflected in property prices. Lastly, green spaces contribute to environmental sustainability by improving air quality and reducing urban heat, features that are increasingly valued in today's real estate market.

The increase in property values resulting from the "green premium" has significant implications for local economies. Higher property values mean higher property taxes, which can boost local government revenues. These additional funds can be reinvested into public services, contributing to the overall development and well-being of the community. However, the realization of this "green premium" is contingent on the proper maintenance and management of these green spaces. Therefore, urban planning and policies should prioritize the creation and upkeep of urban greenery, recognizing its economic value in addition to its environmental and social benefits.

Urban greenery also contributes to the tourism industry. Parks and gardens are often key attractions for tourists, generating substantial revenue for cities. The presence of urban greenery can enhance the overall appeal of a city, attracting more visitors and boosting local businesses. The planning, installation, and maintenance of urban green spaces create jobs, contributing to local economic development. These range from landscape architects and urban planners to gardeners and maintenance staff. Furthermore, green spaces often stimulate commercial activity in their vicinity, supporting local businesses and fostering vibrant community hubs.

Urban greenery contributes to public health by providing spaces for recreation and physical activity, reducing stress, and improving air quality. These health benefits can be translated into economic values through savings in healthcare costs. For instance, regular physical activity can help prevent chronic diseases, reducing the burden on healthcare systems.

Trees and vegetation can help mitigate the urban heat island effect, reducing the need for air conditioning and resulting in energy savings. These energy savings mean economic benefits. Green spaces can also contribute to stormwater management, potentially saving cities significant costs associated with stormwater infrastructure.

There is growing evidence that access to green spaces can boost employee productivity. Views of nature can reduce stress and increase job satisfaction, leading to improved workplace performance. Businesses located near green spaces may also find it easier to attract and retain employees.

While not a direct economic benefit, urban greenery can foster community cohesion, which in turn can lead to reduced crime rates. Safe and well-maintained green spaces provide venues for community interaction, promoting social ties and shared responsibility for local environments. Reduced crime can enhance property values and business activity, contributing to economic prosperity.

There are several studies that have quantified the impact of nearby greenery on property values.

- A study by Ben et al. (2023) found that home prices, on average, increase by 0.17% if the overall green space accessibility rises by 1%. Furthermore, a 1% increase in the green ratio within a community raises property values by 0.46%.

- The Office for National Statistics in the UK reported that houses and flats within 100 m of public green spaces are an average of £2,500 more expensive than they would be if they were more than 500 m away, representing an average premium of 1.1% in 2016. For detached houses, this premium rises to 1.9%, while flats near green spaces are only 0.6% more expensive than those further away.
- Another source suggests that properties in front of or beside a public park had as much as a 20% increase in value.

These figures illustrate the significant economic benefits of urban greenery. However, the exact percentage increase can vary depending on the specific characteristics of the green space and the local real estate market.

15.7.2 The Economic Impact of Urban Agriculture

Urban agriculture, the practice of cultivating, processing, and distributing food in urban areas, is increasingly recognized as a key element of sustainable urban development. Beyond its environmental and social benefits, urban agriculture has significant economic implications that are often overlooked. This essay explores the economic impact of urban agriculture.

Urban agriculture can be a sizable source of employment and income. From small-scale community gardens to large urban farms, these initiatives create jobs for city dwellers. These range from farming and gardening jobs to roles in processing, packaging, marketing, and distributing the produce. Furthermore, urban agriculture can stimulate entrepreneurship, with individuals starting their own urban farming businesses or related enterprises. Urban agriculture can also contribute to food security by increasing the availability of fresh produce in cities. This can lead to cost savings for urban residents, particularly in low-income neighborhoods in which access to affordable, fresh food can be limited. Growing their own food or purchasing it directly from local urban farms enables residents to save on their food bills. Urban agriculture can stimulate local economic development. Money spent on locally grown produce stays within the community, supporting local farmers and businesses. This can lead to a multiplier effect, where each dollar spent generates additional economic activity in the community.

Like urban greenery, urban agriculture can also enhance property values. Well-maintained urban farms and community gardens can improve the aesthetic appeal of a neighborhood, making it more desirable to potential buyers or renters. This can lead to increased property values, benefiting homeowners and generating additional property tax revenues for cities.

Urban agriculture provides important environmental services that can result in significant cost savings. For example, urban farms and gardens can help manage stormwater runoff, reducing the need for costly stormwater infrastructure. They can also improve air quality and reduce urban heat island effects, leading to potential savings in healthcare and energy costs.

There is a business model in urban agriculture, namely community-supported agriculture (CSA). CSA is a model that connects producers and consumers within the food system more closely. The ways CSA work is summarized in Table 15.11.

CSA offers numerous environmental benefits. By promoting local farming, CSA reduces the need for long-distance transportation of food, thereby decreasing carbon emissions. It also encourages sustainable farming practices, which enhance soil health, biodiversity, and ecosystem resilience. These practices often include organic farming methods, crop rotation, and permaculture, which avoid harmful pesticides and synthetic fertilizers, protecting both land and water resources. CSA helps to preserve farmland and green spaces, which are vital for carbon sequestration and habitat conservation. CSA also raises awareness about the environmental impacts of food production, encouraging more conscious and responsible consumption habits. Therefore, CSA contributes considerably to environmental conservation and the fight against climate change.

TABLE 15.11
CSA Business Model

Mutual support	CSA involves a community of individuals who pledge support to a farm operation. By doing so, the farmland becomes, either legally or spiritually, the community's farm. Both growers and consumers provide mutual support and share the risks and benefits of food production.
How it works	• Consumers become CSA members by paying an agreed amount at the beginning of the growing season. This payment can be made either in one lump sum or in installments. • In return, members receive a share of the farm's harvest throughout the season. This could include fresh produce, herbs, flowers, or other farm products. • CSA members share both the bounty (abundant harvests) and the risks (such as crop failures or weather-related challenges) with the farmers.
Benefits	• Direct Connection: CSA allows consumers to have a direct relationship with the farmers who grow their food. • Local and Seasonal: Members receive fresh, locally grown produce that is in season. • Supporting Small Farms: CSA supports small-scale and sustainable farming practices. • Community Building: It fosters a sense of community and shared responsibility for food production.
Variations	• Some CSAs offer different types of shares, such as vegetable shares, fruit shares, or even meat and dairy shares. • Some CSAs allow members to participate in farm activities, visit the farm, or attend workshops.
History	• The concept of CSA originated in the 1960s in Europe and Japan as a response to industrialized agriculture. • In the United States, the first CSA was established in the 1980s by Jan Vander Tuin and Robyn Van En.
Resources for farmers	• The USDA provides resources for farmers interested in CSA models, including research, guides, and annotated bibliographies. • Educational video series like "CSA Starts Here" help aspiring and beginning CSA farmers learn about best practices.

15.7.3 Cost–Benefit Analysis of Urban Greenery and Urban Agriculture

A cost–benefit analysis is a systematic process used by businesses and organizations to analyze decisions, systems, or projects. It involves comparing the projected or estimated costs and benefits associated with a decision or action to determine whether it makes sense from a business perspective.

Process: The cost–benefit analyst sums up the potential rewards expected from a situation or action and then subtracts the total costs associated with taking that action. Some consultants or analysts also build models to assign a dollar value to intangible items, such as the benefits and costs associated with living in a certain town.

Measurement: A cost–benefit analysis involves measurable financial metrics such as revenue earned or costs saved as a result of the decision to pursue a project. It can also include intangible benefits and costs or effects from a decision, such as employees' morale and customer satisfaction.

Complex Analysis: More complex cost-benefit analysis may incorporate sensitivity analysis, discounting of cash flows, and what-if scenario analysis for multiple options.

Decision Making: All else being equal, an analysis that results in more benefits than costs will generally be a favorable project for the company to undertake. If not, the business should review the project to see if it can adjust either to increase benefits or to decrease costs to make the project viable.

Opportunity Costs: A cost–benefit analysis will also factor the opportunity cost into the decision-making process. Opportunity costs are the alternative benefits that could have been realized when choosing one alternative over another.

Benefits: Urban greenery offers numerous environmental benefits. It improves air quality, mitigates urban heat island effects, enhances biodiversity, and aids in stormwater management. Green spaces also provide recreational and health benefits, promoting physical activity and mental well-being. Furthermore, they contribute to aesthetic values and can increase property values.

Below is a hypothetical example of cost–benefit analysis of an urban greenery project.

Scenario: A city is considering investing in an urban greenery project that includes the development of parks, green roofs, and street trees. The project is expected to cost USD 1 million to implement.

Cost–Benefit Analysis Calculation:

1. *Total Cost:* The total cost of the project is the initial investment, which is $1 million.
2. *Total Benefit:* The total benefit of the project is the sum of the estimated benefits over a certain period. These benefits could include:
   a. *Environmental Benefits:* Improved air quality, reduced urban heat island effect, enhanced biodiversity, and better stormwater management. This can be done by employing environmental valuation.
   b. *Health and Recreational Benefits:* Improved physical and mental health of the residents due to increased opportunities for outdoor activities. These are intangible benefits, and therefore, the calculation of the benefits cannot be done by conventional calculations. This can be done by using environmental valuation.
   c. *Economic Benefits:* Increased property values due to the proximity to green spaces.

By using suitable valuation methods, let us say these benefits are estimated to be worth USD 1.5 million over a period of 5 years. The net benefit is calculated by subtracting the total cost from the total benefit. So, it is USD 1.5 million (total benefit) – USD 1 million (total cost) = USD 500,000. In this case, the net benefit of the project is positive (USD 500,000), which suggests that the project is a good investment from a cost–benefit perspective. However, this is a simplified example, and in real-world scenarios, other factors such as the time value of money, risk, opportunity cost, and intangible benefits and costs may also need to be considered.

Please note that a cost–benefit analysis is only a tool to aid decision-making and should not be the sole determinant of whether to proceed with a project or decision. It is always important to consider other qualitative and quantitative factors as well.

15.7.4 Funding and Financial Challenges in Urban Greenery and Agriculture

This section investigates the funding and financial challenges faced by urban greenery and agriculture projects and discusses potential solutions.

15.7.4.1 Urban Greenery

Urban greenery, which includes parks, gardens, green roofs, and street trees, contributes significantly to the quality of urban life.

15.7.4.2 Funding Challenges

The primary funding challenge for urban greenery projects is the high initial investment required for land acquisition, design, and construction. Land acquisition can be particularly costly in urban areas, where land prices are often high due to demand for residential, commercial, and industrial uses. Design and construction costs can also be substantial, particularly for projects that involve complex landscaping, the installation of irrigation systems, or the construction of amenities such as playgrounds, benches, and pathways. In addition to these initial costs, urban greenery projects also incur ongoing maintenance costs. These can include costs for watering, pruning, mowing, litter removal, and repairs. In some cases, specialized maintenance may be required, such as the care of native plants, the management of stormwater features, or the maintenance of green roofs. These ongoing costs can be substantial and must be factored into the project's budget from the outset. In densely populated urban areas, these costs can be prohibitive. High land prices can make it difficult to secure suitable sites for urban greenery projects. Furthermore, the high density of buildings and infrastructure can limit the space available for such projects. This can lead to smaller, more fragmented green spaces, which may be more costly to design, construct, and maintain on a per-unit-area basis.

To overcome the funding challenges associated with urban greenery projects, innovative financing mechanisms are needed. These could include PPPs, where private entities contribute funding in return for benefits such as branding opportunities or increased property values. Green bonds, which are bonds issued to finance environmentally friendly projects, could also be used. Crowdfunding, where funds are raised from many individuals, typically via the internet, is another potential option.

15.7.4.3 Financial Sustainability

Urban greenery projects generate significant environmental, health, and social benefits. The environmental benefits of these projects are (1) improving air quality, (2) mitigating urban heat island effects, (3) enhancing biodiversity, and aiding in stormwater management. The health benefits include promoting physical activity and mental well-being by providing spaces for recreation and relaxation. The social benefit is, among other things, contributing to community cohesion and can increase property values. However, these benefits often do not translate directly into revenue. For example, while a park may improve air quality and provide recreational opportunities, it does not generate income in the same way a commercial development might.

The financial sustainability of urban greenery projects is often reliant on public funding. This funding can come from various sources, including local, regional, or national government budgets, grants, or public financing mechanisms. However, public funding can be uncertain and subject to budget cuts. This uncertainty can make it difficult to plan for the long-term maintenance and management of urban greenery projects. In addition, urban greenery projects often have ongoing costs, such as maintenance and management costs. These costs can be substantial and must be factored into the project's budget from the outset. Without a reliable source of funding, these ongoing costs can threaten the financial sustainability of the project.

To overcome these challenges, innovative financing mechanisms and strategies are needed. The potential solutions include (1) PPPs, where private entities contribute funding in return for benefits such as branding opportunities or increased property values; (2) green bonds, which are bonds issued to finance environmentally friendly projects, could also be used; (3) crowdfunding, where funds are raised from a large number of individuals, typically via the internet, is another potential option; and (4) policies that recognize and quantify the ecosystem services provided by urban greenery can help justify and secure funding.

15.7.4.4 Urban Agriculture

Urban agriculture involves the cultivation, processing, and distribution of food in urban areas.

15.7.4.5 Funding Challenges

Urban agriculture projects face significant funding challenges, similar to those encountered by urban greenery projects. The initial costs of urban agriculture projects can be high, particularly for land acquisition and infrastructure development. The initial costs of the project include the following: (1) *Land Acquisition:* In densely populated cities, land is scarce and expensive, making it challenging to find suitable space for farming The cost of acquiring land in urban areas can be prohibitive, especially in cities where real estate prices are high (2) *Infrastructure Development:* The development of infrastructure for urban agriculture, such as irrigation systems, greenhouses, and storage facilities, can also be costly. These costs can be particularly high for projects that aim to use advanced farming techniques, such as hydroponics or aquaponics.

In addition to the initial costs, urban agriculture projects also incur ongoing costs. These costs include (1) *Inputs:* the cost of seeds, fertilizers, and other farming inputs. While some urban agriculture projects may aim to minimize these costs using organic farming methods, these methods can also require significant labor and expertise. (2) *Labor:* Labor costs can be substantial, particularly for projects that require skilled workers. In some cases, urban agriculture projects may be able to reduce labor costs by engaging volunteers or providing training and employment opportunities for local residents. (3) *Regulatory Compliance:* Urban agriculture projects may also face costs associated with regulatory compliance. These can include costs related to zoning regulations, health and safety standards, and environmental regulations.

The high initial and ongoing costs of urban agriculture projects can pose significant funding challenges. These projects often rely on a mix of funding sources. These include the following: (1) *Grants and Donations:* While grants and donations can provide important sources of funding, they are often limited and uncertain. Furthermore, the process of applying for grants can be time-consuming and competitive. (2) *Income from Sales:* Income from the sale of produce can help to cover ongoing costs and ensure the financial sustainability of urban agriculture projects. However, urban farmers often face challenges in accessing markets and achieving competitive prices for their produce.

15.7.4.6 Financial Sustainability

Ensuring the financial sustainability of urban agriculture projects can be challenging. One of the ways urban agriculture projects can generate income is through the sale of produce. This income can help cover the costs of inputs such as seeds and fertilizers, labor, and infrastructure, and contribute to the financial sustainability of the project. In addition, urban agriculture can also provide non-food products and a number of functionalities. However, while the sale of produce can generate income, it may not be sufficient to cover all the costs associated with urban agriculture. Furthermore, the income generated from the sale of produce can be uncertain and variable, due to factors such as market competition, price volatility, and the seasonal nature of agricultural production.

Urban agriculture projects often face significant market competition. They may be competing with larger, more established agricultural producers who can benefit from economies of scale and have greater access to markets. This competition can make it difficult for urban agriculture projects to sell their produce at a competitive price and generate sufficient income.

Price volatility is another challenge faced by urban agriculture projects. The prices of agricultural products can fluctuate widely due to factors such as changes in supply and demand, weather conditions, and global market trends. This price volatility can make the income from the sale of produce unpredictable, which can pose challenges for the financial sustainability of urban agriculture projects.

The seasonal nature of agricultural production can also pose financial challenges for urban agriculture projects. Many crops can only be grown during certain times of the year, which can lead to seasonal variations in income. This can make it difficult to cover ongoing costs, such as labor and maintenance, which may need to be paid year-round.

TABLE 15.12
Supports of the Cities for Financial Sustainability

Infrastructure development	Cities can invest in the development of local marketplaces such as farmers markets. These markets provide a platform for local farmers to sell their produce directly to consumers, eliminating the need for middlemen and allowing farmers to retain a greater share of the profits.
Public–private partnerships	Cities can foster public–private partnerships to support local farmers. For example, they can partner with supermarkets and restaurants to source locally grown produce. This not only provides a steady market for farmers but also promotes local economic development.
Education and training	Cities can provide education and training programs for local farmers to enhance their farming and business skills. This can include training on sustainable farming practices, post-harvest handling, quality control, marketing, and business management.
Policy Support	Cities can implement policies that support local farmers. This can include zoning laws that allow for urban farming, tax incentives for businesses that source locally grown produce, and grants or low-interest loans for local farming initiatives.
CSA	Cities can promote CSA models, where consumers pay farmers upfront for a share of the harvest. This provides farmers with a guaranteed market and income, reducing their financial risk.
Promotion of local produce	Cities can launch campaigns to promote the consumption of locally grown produce. This can increase demand for local produce, benefiting local farmers.
Improving access to technology and innovation	Cities can help farmers access modern technologies and innovative farming methods that can increase their productivity and market competitiveness.
Building networks	Cities can help build networks among farmers, consumers, businesses, and institutions. These networks can facilitate information sharing, collaboration, and mutual support.

While urban greenery and urban agriculture projects face significant funding and financial challenges, these can be addressed through innovative financing mechanisms, supportive policies, and community-based models. These initiatives are important in promoting sustainable urban development and deserve continued investment and support.

To support the financial sustainability of urban agriculture activities, cities can assist local farmers in accessing markets through various methods, as shown in Table 15.12.

Cities can support local farmers in accessing markets through a combination of infrastructure development, PPPs, education and training, policy support, promotion of local produce, improving access to technology and innovation, and building networks. These strategies are expected to help local farmers overcome challenges, enhance their market access, and contribute to sustainable urban development.

15.8 FUTURE URBAN GREENERY AND AGRICULTURE INITIATIVES

The future of urban greenery and agriculture is promising for sustainable urban development. Integrating farming and green spaces into urban lifestyles becomes increasingly important. Urban farming is changing the future of agriculture, with new technologies allowing for the growth of food in places where it was previously difficult or impossible. This includes traditional small outdoor community gardens, as well as complex indoor vertical farms that think about growing space in three-dimensional terms. These urban farms can grow more than just fruits and vegetables, they can enhance the natural outlook of a city while reducing its carbon footprint. Urban greenery resulting from these agricultural activities can improve the environment of a metropolitan area. As more than half of the global population now lives in urban areas, a figure expected to rise to 68% by 2050, the role of urban greenery and agriculture in feeding the urban population, improving city climates, and enhancing urban life quality will be more crucial than ever.

15.8.1 Emerging Trends in Urban Greenery

Urban greenery, which includes elements such as parks, gardens, green roofs, and street trees, serves as a fundamental component of urban ecosystems, playing a fundamental role in promoting environmental sustainability, enhancing public health, and fostering social cohesion. With the relentless expansion of cities and the intensification of urbanization, the significance of urban greenery is gaining increased recognition. Concurrently, a number of emergent trends are influencing and shaping the future path of urban greenery, underscoring its critical role in the sustainable development of our urban landscapes. Some current trends associated with urban greenery are shown below. Here are some current trends and the future direction of urban greenery:

15.8.1.1 Current Trends

1. *Increased Focus on Green Infrastructure:* Cities are incorporating more green spaces, such as parks, gardens, green roofs, and urban forests, to provide environmental and social benefits.
2. *Urban Cooling:* Urban greenery can help cool cities, mitigating the urban heat island effect caused by traffic, industrial activity, and concrete buildings. A single healthy tree can have the cooling power of more than ten air-conditioning units.
3. *Air Quality Improvement:* Plants can help reduce pollution through a process known as dry deposition, where harmful pollutants such as microscopic particulate matter become trapped in the leaves.
4. *Multifunctional Green Spaces:* Urban greenery is being designed to serve multiple purposes, such as stormwater management, urban heat island mitigation, air purification, and recreational activities.
5. *Biophilic Design:* The incorporation of natural elements, such as plants, water features, and natural materials, in urban planning and architecture to improve human well-being and connection with nature.
6. *Biodiversity Conservation:* Urban greening can help conserve biodiversity in cities.
7. *Citizen Engagement:* Increased public participation in the planning, maintenance, and stewardship of urban green spaces, fostering a sense of community and ownership.

15.8.1.2 Future Direction

1. *Integrated Green-blue Infrastructure:* The combined use of green (vegetation) and blue (water) elements to create multifunctional, nature-based solutions for urban challenges, such as flood mitigation and urban cooling.
2. *Adaptive and Resilient Urban Greenery:* The design and management of urban green spaces to be adaptable to climate change, increasing extreme weather events, and other environmental stresses.
3. *Technology-enabled Green Spaces:* The use of sensors, data analytics, and automation to optimize the management and maintenance of urban greenery, improving efficiency and responsiveness.
4. *Equitable Access to Urban Greenery:* Ensuring that the benefits of urban greenery are distributed fairly across diverse communities, addressing issues of accessibility and social justice.
5. *Integration with Urban Planning and Design:* The seamless incorporation of urban greenery into the overall urban planning and design process, recognizing its key role in creating sustainable, livable, and resilient cities.
6. *Private and Semi-public Green Areas:* There is potential for the development of new green areas in public, semi-public, and private spaces.

These trends and future directions reflect the growing importance of urban greenery as a critical component of sustainable and livable cities, addressing environmental, social, and economic challenges.

There are several innovative urban greening projects around the world. Here are a few examples:

- *Liverpool, England:* Liverpool has created a living wall in its busy city center. The barren concrete exterior of the St John's Shopping Center was turned into 65 m of leafy habitat. Installed in 2020, Liverpool's living wall has been planted with 14,000 evergreens, perfectly situated to help trap some of the pollution generated by the city's nearby bus station.
- *Curitiba, Brazil:* Curitiba has been working toward a sustainable model since the 1970s and has incorporated natural leisure areas and incentivized recycling as a key part of city planning. The government's afforestation plan has seen the planting of many trees beside roads, in parks, squares, and other public areas.
- *BitaGreen:* BitaGreen develops software and maps that identify how urban green infrastructure helps to reduce flooding and increase the quality of life. They deliver value by screening the most suitable locations combined with cost-benefit assessments to identify the environmental service benefits of green infrastructure.
- *New York City, United States:* In 2016, Brooklyn's Newtown Creek Wastewater Treatment Plant became the site of one of the United States's first biofuel-to-natural-gas distribution projects.
- *London, England:* London encourages its residents to create a rain garden full of plants to reduce the amount of water that goes down the drain and to build a green roof on their shed, bike, or bin store.

These projects highlight the innovative ways cities are incorporating greenery into their urban landscapes to improve the quality of life for their residents and the environment.

15.8.2 Innovations in Urban Agriculture

In recent years, urban agriculture has emerged as a growing trend, driven by the need for more sustainable and localized food production systems. As cities around the world continue to fight for food security, environmental degradation, and the impacts of climate change, innovative approaches to urban agriculture have been gaining momentum. Here we explore some of the key innovations shaping the future of urban farming.

15.8.2.1 Vertical Farming

One of the most notable innovations in urban agriculture is the rise of vertical farming. In response to the limited availability of land in densely populated urban areas, vertical farming leverages technology to grow crops in stacked, controlled environments. These indoor farms, often housed in repurposed industrial buildings or purpose-built facilities, utilize specialized lighting, climate control systems, and hydroponic or aeroponic growing methods to optimize plant growth and maximize yield per square foot. Vertical farming offers several advantages over traditional open-field agriculture. By eliminating the need for soil and reducing water usage by up to 95%, these systems are highly efficient and can operate year-round, independent of seasonal changes or extreme weather events. Moreover, the controlled environment allows for the precise management of factors such as temperature, humidity, and nutrient delivery, leading to increased productivity and the ability to grow a wider variety of crops in urban settings.

Leading companies in the vertical farming industry, such as AeroFarms, Plenty, and Vertical Future, are continuously pushing the boundaries of this technology, developing advanced growing systems and refining their processes to make urban farming more scalable and cost-effective.

15.8.2.2 Urban Aquaponics

Another innovative approach to urban agriculture is the integration of aquaponics systems. Aquaponics combines aquaculture (fish farming) with hydroponics (soilless plant cultivation), creating a closed-loop system where the waste from the fish provides nutrients for the plants, and the plants, in turn, help to purify the water for the fish. Aquaponics systems are well-suited for urban environments, as they can be designed to operate in compact spaces, such as rooftops, abandoned industrial buildings, or shipping containers. The symbiotic relationship between the fish and plants allows for efficient resource utilization, with minimal waste and a reduced environmental footprint.

Startups like Cityfarm and Gotham Greens have pioneered the development of urban aquaponics, demonstrating the potential for these systems to provide fresh, locally grown produce and fish in densely populated areas. Integrating hydroponics with aquaculture provides a sustainable and diverse method for urban food production.

15.8.2.3 Precision Farming and Robotics

As urban agriculture continues to evolve, the integration of precision farming technologies and robotics is becoming increasingly prevalent. Advancements in sensor technology, data analytics, and autonomous systems are transforming the way urban farms are managed and operated. Precision farming techniques, such as the use of sensors to monitor soil moisture, nutrient levels, and plant health, allow urban farmers to optimize their growing conditions and resource usage. This data-driven approach not only improves productivity but also reduces waste and environmental impact. The integration of robotics and automation in urban farming operations is streamlining tasks such as seeding, harvesting, and packaging. Autonomous drones and ground-based robots can perform these labor-intensive activities with precision and efficiency, reducing the need for manual labor and improving the overall economic viability of urban agriculture.

Companies like Iron Ox and Farmshelf are at the forefront of this trend, developing advanced robotic systems and integrated software platforms to enhance the efficiency and scalability of urban farming.

15.8.2.4 Rooftop and Vertical Gardening

While not necessarily a new concept, the growing popularity of rooftop and vertical gardening in urban areas is another notable innovation in urban agriculture. These systems leverage the available vertical space in cities to cultivate a variety of crops, from herbs and leafy greens to fruits and vegetables. Rooftop gardens not only provide a source of fresh produce but also offer additional benefits, such as insulating buildings, reducing urban heat island effects, and improving stormwater management. Vertical gardening systems, such as living walls and modular planter systems, allow urban residents to grow food in limited spaces, including balconies, patios, and even indoor environments.

Companies like Gotham Greens and Urban Farmers are leading the way in the development of innovative rooftop and vertical farming solutions, demonstrating the potential for urban dwellers to engage in food production and enjoy the benefits of locally grown, sustainable produce.

15.8.3 The Role of Technology in Future Urban Greenery and Agriculture

The role of technology in shaping the future of urban greenery and agriculture is pivotal. As urban areas continue to expand, the need for sustainable and efficient methods of food production and green space management becomes increasingly important. Technology offers a range of solutions to these challenges, from vertical farming to smart irrigation systems, and is set to revolutionize the way we approach urban greenery and agriculture.

15.8.3.1 Harnessing Technology to Sustain Urban Greenery

In the face of rapid urbanization, maintaining and expanding urban greenery has become a pressing challenge for city planners and policymakers. However, the integration of cutting-edge technologies offers a promising solution to this dilemma. Leveraging the power of Geographic Information Systems (GIS) and the Internet of Things (IoT) enables urban areas to not only preserve their existing green spaces but also strategically develop new ones, ultimately enhancing the overall sustainability and livability of cities.

Geographic Information Systems (GIS) is a fundamental tool in the strategic planning and management of urban greenery. These advanced spatial analysis tools allow city planners to identify the most suitable locations for new green spaces by considering a variety of factors, such as population density, existing green space distribution, and local climate conditions. GIS technology enables data-driven decision-making, ensuring that urban greenery is strategically placed to maximize its benefits for the local community. For instance, GIS can help identify areas with limited access to green spaces, particularly in densely populated neighborhoods. By mapping this information, city planners can prioritize the development of new parks, gardens, or urban forests in these underserved areas, ensuring that all residents have equitable access to the numerous benefits of urban greenery. GIS can assist in evaluating the overall distribution and connectivity of green spaces, allowing for the creation of a cohesive and accessible network of urban green infrastructure.

In the initial planning phase, GIS can also be employed to monitor and manage existing urban greenery. Integrating spatial data, such as tree inventories, maintenance records, and environmental conditions, allows city authorities to efficiently identify areas needing immediate attention, optimize maintenance schedules, and allocate resources effectively. This data-driven approach to managing urban greenery bolsters the sustainability and resilience of these essential green spaces in the long term.

Complementing the strategic planning capabilities of GIS, the Internet of Things (IoT) offers a powerful tool for real-time monitoring and optimized maintenance of urban greenery. IoT devices, such as sensors and smart meters, can be deployed throughout urban green spaces to collect a wealth of data on various parameters, including soil moisture, light intensity, air quality, and even the health and growth patterns of vegetation.

Transmitting this data continuously to a centralized monitoring system allows IoT technology to help city authorities quickly identify potential issues or stresses affecting urban greenery. This early detection allows for proactive interventions, such as targeted irrigation, pest control, or nutrient management, ensuring the long-term vitality and resilience of urban green spaces.

The integration of IoT with automated irrigation systems and smart watering technologies can significantly improve the efficiency and sustainability of urban greenery maintenance. These systems can precisely monitor and regulate the water needs of plants, reducing wasteful water consumption and ensuring that limited resources are used judiciously.

The benefits of harnessing GIS and IoT technologies for urban greenery extend beyond just maintenance and management. These tools can also facilitate community engagement and citizen science initiatives. Providing residents with access to real-time data on the health and conditions of their local green spaces enables city authorities to foster a sense of ownership and stewardship among the community. This, in turn, can encourage residents to participate in monitoring, reporting, and even volunteer-based maintenance activities, further strengthening the bond between people and their urban green infrastructure.

While the implementation of these technologies may require significant upfront investments, the long-term payoffs in terms of improved sustainability, resilience, and overall quality of life in urban areas are undeniable. Seamlessly integrating GIS and IoT into the planning, management, and maintenance of urban greenery enables cities to ensure their green spaces thrive and provide essential ecosystem services, even in the face of rapid urbanization and environmental challenges.

As cities strive to become more livable, sustainable, and resilient, the strategic use of geospatial and IoT technologies in urban greenery management stands as a shining example of how innovation can help address the pressing challenges of our time. Utilizing these technologies, urban areas can develop vibrant, healthy, and accessible green spaces that enhance the well-being of both people and the planet.

15.8.3.2 Urban Agriculture and the Role of Technology

In an era of rapid urbanization and growing concerns about food security, urban agriculture has emerged as a promising solution to address the challenges of sustainable food production. Technology has played a crucial role in advancing urban agriculture, offering innovative approaches that enhance efficiency, productivity, and sustainability.

One of the most prominent technological advancements in urban agriculture is vertical farming. This method involves growing crops in vertically stacked layers, often within controlled environments. Vertical farming is highly efficient, using up to 95% less water than traditional farming practices. This water-saving technology is particularly valuable in water-scarce regions, where traditional farming methods may not be feasible. Moreover, vertical farming can be conducted year-round, regardless of weather conditions, making it a reliable source of fresh produce even in areas with harsh climates.

Advancements in hydroponics and aeroponics have further expanded the possibilities of urban agriculture. These soilless growing techniques allow plants to be cultivated in nutrient-rich solutions or mist, rather than relying on traditional soil-based methods. This approach offers several advantages, such as the ability to grow a wider variety of crops in limited urban spaces, improved control over growing conditions, and higher yields due to the plants' faster growth rates.

Smart farming technologies have also played a crucial role in enhancing the efficiency and sustainability of urban agriculture. Automated irrigation systems, for instance, can precisely monitor and regulate the water needs of crops, reducing waste and ensuring optimal growing conditions. In addition, the integration of artificial intelligence (AI) and machine learning in pest detection and management has revolutionized urban farming. These technologies can accurately identify and address pest infestations, minimizing the use of harmful pesticides and promoting more sustainable practices.

The benefits of these technological advancements in urban agriculture extend beyond just food production. By locating farms within cities, the need for long-distance transportation and the associated carbon emissions are reduced, contributing to a more environmentally sustainable food system. Furthermore, urban agriculture can provide employment opportunities, foster community engagement, and improve access to fresh, nutritious produce in underserved urban areas. However, the adoption and scaling of these technologies in urban agriculture are not without challenges. The initial investment required for setting up vertical farms, hydroponics systems, and advanced smart farming technologies can be a barrier, particularly for small-scale urban farmers. Moreover, there are concerns about the energy consumption associated with controlled environments and the potential impact on local biodiversity.

To address these challenges, policymakers, urban planners, and technology innovators must work together to create an enabling environment for the widespread adoption of urban agriculture technologies. This may involve developing financial incentives, investing in research and development, and establishing regulatory frameworks that support sustainable urban farming practices.

The integration of technology in urban agriculture has the potential to revolutionize food production in our cities, offering a sustainable solution to the growing demand for fresh, locally sourced produce. By harnessing the power of vertical farming, soilless growing techniques, and smart farming technologies, urban agriculture can contribute to a more resilient and environmentally conscious food system, helping to address the pressing challenges of food security and sustainability in the years to come. While the integration of technology in urban greenery and agriculture presents immense opportunities, it is important to ensure that these technologies are accessible and affordable to all. This will require concerted efforts from policymakers, city planners, technology providers, and communities. With the right approach, technology can truly transform the future of urban greenery and agriculture, making our cities more livable and sustainable.

15.8.4 Vision for Sustainable and Resilient Urban Landscapes

As the world continues to urbanize at a rapid pace, the need for sustainable and resilient urban landscapes has never been more pressing. Cities, which are home to the majority of the global population, are at the forefront of the battle against climate change, environmental degradation, and the growing demand for livable, equitable, and resource-efficient spaces. To address these challenges, a comprehensive vision for the future of urban areas must be embraced, one that seamlessly integrates cutting-edge technologies, forward-thinking urban planning, and a deep commitment to environmental and social sustainability.

At the heart of this vision lies the concept of the "smart and green city" – an urban landscape that harnesses the power of digital innovation to enhance the sustainability and resilience of its infrastructure, public spaces, and natural ecosystems. Through the strategic deployment of technologies such as Geographic Information Systems (GIS), the Internet of Things (IoT), and renewable energy solutions, cities can optimize resource management, improve environmental performance, and foster a higher quality of life for their residents.

GIS, as discussed previously, plays a pivotal role in the strategic planning and management of urban greenery. Utilizing spatial data and advanced analytical tools empowers city planners to pinpoint optimal locations for new parks, urban forests, and green spaces. This ensures an equitable distribution, maximizing both environmental and social benefits. Moreover, GIS can be integrated with real-time IoT monitoring systems to track the health and performance of these green assets, enabling timely interventions and proactive maintenance.

Beyond the management of urban greenery, GIS can also be employed to optimize the broader urban landscape. By mapping infrastructure, transportation networks, and resource flows, GIS can help city authorities identify opportunities for improved efficiency, better resource allocation, and the integration of sustainable technologies. For instance, GIS can assist in the strategic placement of renewable energy generation facilities, such as solar panels and wind turbines, to maximize energy production and minimize the environmental impact.

Complementing the capabilities of GIS, the Internet of Things (IoT) emerges as a transformative technology for achieving sustainable and resilient urban landscapes. IoT-enabled sensors and devices can be deployed throughout the city, collecting real-time data on energy consumption, water usage, air quality, and even the condition of built infrastructure. This wealth of data can then be analyzed and leveraged to optimize resource management, identify areas in need of intervention, and empower citizens to become active participants in the sustainability of their communities.

One prime example of IoT's impact on urban sustainability is the implementation of smart waste management systems. Equipping trash receptacles with sensors allows cities to monitor waste levels and streamline collection routes, thereby lowering fuel consumption and emissions from waste collection vehicles. Furthermore, IoT-enabled waste sorting and recycling facilities can help maximize the recovery and reuse of valuable resources, contributing to a more circular urban economy.

Renewable energy technologies also play a crucial role in the vision for sustainable and resilient urban landscapes. Cities can harness the power of solar, wind, and geothermal energy to transition away from fossil fuel-based energy sources, reducing their carbon footprint and contributing to global climate change mitigation efforts. The integration of smart grids, energy storage systems, and microgrids can further enhance the efficiency and resilience of urban energy infrastructure, ensuring reliable and equitable access to clean energy for all.

Besides the technological aspects, the vision for sustainable and resilient urban landscapes must also include innovative approaches to urban planning and design. It includes the incorporation of nature-based solutions, such as urban wetlands, green roofs, and permeable surfaces, which can mitigate the impacts of climate change, improve stormwater management, and enhance biodiversity.

The concept of the "15-minute city" – where residents can access all their daily needs within a 15-minute walk or bike ride – holds immense promise for creating more livable, inclusive, and environmentally friendly urban spaces. Emphasizing mixed-use development, compact urban forms, and multimodal transportation systems enables cities to reduce dependence on private vehicles, encourage active modes of mobility, and strengthen community ties and social cohesion.

Underpinning this holistic vision for sustainable and resilient urban landscapes is a deep commitment to equity and social justice. As cities strive to become more sustainable, it is essential that the benefits of these initiatives are equitably distributed, ensuring that marginalized communities are not left behind. This may involve targeted investments in green infrastructure, affordable housing, and community-driven sustainability projects, empowering all residents to participate in and reap the rewards of a more livable and sustainable urban environment.

To realize this vision, a collaborative and multidisciplinary approach is required, engaging a diverse array of stakeholders, including city authorities, urban planners, sustainability experts, technology innovators, and community organizations. Encouraging cross-sector partnerships and inclusive decision-making allows cities to tap into collective knowledge, resources, and creativity, transforming their urban landscapes into models of sustainability and resilience.

The journey toward sustainable and resilient urban landscapes is not without its challenges. Issues such as the high upfront costs of technological and infrastructure investments, the complexities of governance and policy coordination, and the need to balance environmental, economic, and social priorities must be addressed. However, the potential benefits of this vision – in terms of improved environmental performance, enhanced quality of life, and the creation of more equitable and livable cities – make it a worthy and necessary pursuit.

As cities continue to develop, the need for a comprehensive, technology-enabled, and community-driven approach to sustainable urban development has never been more pressing. Adopting the vision of a "smart and green city" empowers cities to become innovators, demonstrating how the strategic integration of digital technologies, nature-based solutions, and inclusive planning can transform urban landscapes into models of sustainability, resilience, and livability. This vision not only benefits the residents of today but also safeguards the well-being of future generations, ensuring that our cities remain vibrant, thriving, and in harmony with the natural world.

PROBLEMS AND ISSUES FOR DISCUSSION

1. How do rooftop gardens contribute to sustainable urban living and mitigate the urban heat island effect?
2. What role does vertical farming play in addressing food security and resource efficiency in cities?
3. Discuss the impact of green roofs on urban microclimates and biodiversity.
4. How can urban forests enhance ecological resilience and promote well-being?
5. Explore architectural strategies for integrating green spaces into urban design.
6. Analyze the benefits and challenges of incorporating agriculture into city planning.
7. What policies support the adoption of green infrastructure in urban areas?
8. Compare hydroponics and traditional soil-based agriculture in the context of cities.
9. Investigate the social and economic implications of urban agriculture.
10. How can architects create sustainable urban environments through green design?

11. Examine the ecological benefits of permeable surfaces and green facades.
12. Discuss the potential of agroforestry systems in transforming urban landscapes.
13. What governance structures drive the development of urban green spaces?
14. Evaluate the climate resilience of urban agriculture practices.
15. How do rooftop oases contribute to air quality improvement and mental health?
16. Investigate the feasibility of aquaponics and rooftop beekeeping in cities.
17. Analyze successful case studies of urban agriculture projects worldwide.
18. Discuss the challenges of integrating agriculture into high-rise buildings.
19. How can urban greenery mitigate stormwater runoff and enhance urban aesthetics?
20. Explore the historical evolution of urban green spaces and community gardens.
21. Compare the environmental benefits of green walls and green roofs.
22. Discuss the concept of food sovereignty and localized food production in cities.
23. What innovative technologies support urban agriculture, such as vertical farms?
24. Analyze the economic implications of urban farming for local economies.
25. Examine the role of community gardens in promoting social cohesion and education.
26. How can architects balance aesthetics and functionality in green design?
27. Investigate the impact of urban greenery on property values and well-being.
28. Examine public–private partnerships in advancing urban agriculture initiatives.
29. Discuss the cultural significance of urban green spaces and edible landscapes.
30. How can cities incentivize green roof adoption and sustainable practices?
31. Evaluate the effectiveness of urban agriculture in reducing food miles and waste.
32. Explore the adaptive reuse of vacant lots for community gardens and urban farms.
33. Discuss soil quality management and soil health in urban farming.
34. Analyze the visual impact of vertical farms and rooftop gardens on cityscapes.
35. How can cities promote pollinator-friendly spaces and biodiversity corridors?
36. Investigate the role of urban agriculture in climate change adaptation.
37. Discuss trade-offs between green infrastructure and conventional development.
38. Examine legal and regulatory aspects related to urban farming practices.
39. How do green roofs contribute to energy efficiency and thermal comfort?
40. Explore innovative trends and technologies shaping urban agriculture.
41. Analyze the social equity implications of urban agriculture policies and access.
42. Discuss the role of schools and educational institutions in urban gardening.
43. How can architects integrate edible landscapes into public spaces and parks?
44. Investigate the impact of urban greenery on air pollution reduction.
45. Examine maintenance challenges and best practices for green roofs.
46. Discuss community engagement and participatory design in green spaces.
47. Explore hydroponics and aeroponics as urban farming methods.
48. How can cities balance green infrastructure with other urban needs?
49. Analyze the economic viability and scalability of rooftop agriculture.
50. Discuss future innovations and sustainable practices in urban agriculture.

MULTIPLE-CHOICE QUESTIONS

1. What is the term used to describe the practice of growing food within urban areas?
   a. Urbanization
   b. Rural agriculture
   c. Urban farming
   d. Suburban gardening
   e. City landscaping

2. Which of the following is a benefit of urban green spaces?
   a. Increased air pollution
   b. Decreased biodiversity
   c. Improved mental health
   d. Higher crime rates
   e. Reduced community engagement
3. Which of the following is an example of urban green infrastructure?
   a. Parking lot
   b. Shopping mall
   c. Community garden
   d. Highway overpass
   e. Office building
4. What is the process of covering a roof with vegetation to reduce stormwater runoff and improve energy efficiency called?
   a. Green roofing
   b. Asphalt paving
   c. Concrete sealing
   d. Metal cladding
   e. Bricklaying
5. Which of the following is a common method of vertical farming in urban areas?
   a. Hydroponics
   b. Desert farming
   c. Traditional plowing
   d. Pastoral farming
   e. Rainfed agriculture
6. Which of the following is a technique used in urban agriculture to maximize limited space by growing crops vertically?
   a. Crop rotation
   b. Companion planting
   c. Raised bed gardening
   d. Vertical farming
   e. Seed broadcasting
7. What is the term for the practice of using abandoned or underutilized urban spaces for agriculture?
   a. Rural farming
   b. Urban sprawl
   c. Greenfield development
   d. Brownfield farming
   e. Guerrilla gardening
8. Which of the following is a common method used to compost organic waste in urban settings?
   a. Landfill disposal
   b. Incineration
   c. Vermicomposting
   d. Plastic recycling
   e. Dumping in water bodies

9. What is the term for the practice of integrating food production into urban development projects?
    a. Agroforestry
    b. Permaculture
    c. Edible landscaping
    d. Horticulture
    e. Agribusiness
10. Which of the following is a key benefit of incorporating green spaces into urban planning?
    a. Increased traffic congestion
    b. Higher energy consumption
    c. Improved air quality
    d. Decreased property values
    e. Reduced community engagement
11. What is the term for the process of growing plants without soil, using nutrient-rich water?
    a. Aquaponics
    b. Aeroponics
    c. Soilless farming
    d. Traditional agriculture
    e. Field cultivation
12. Which of the following is a common method of pest control in organic urban farming?
    a. Synthetic pesticides
    b. Chemical fertilizers
    c. Crop rotation
    d. Genetic modification
    e. Soil sterilization
13. What is the term for the practice of using rainwater for irrigation in urban agriculture?
    a. Desalination
    b. Rainwater harvesting
    c. Drip irrigation
    d. Sprinkler system
    e. Water pumping
14. Which of the following is a common practice in community gardening to promote social interaction and cooperation?
    a. Individual plots
    b. High fences
    c. No-trespassing signs
    d. Restricted access
    e. Shared tools and resources
15. What is the term for the process of converting vacant lots or rooftops into green spaces for urban agriculture?
    a. Land speculation
    b. Urbanization
    c. Land banking
    d. Greenfield development
    e. Brownfield reclamation
16. Which of the following is a common type of edible plant used in urban landscaping?
    a. Poison ivy
    b. Tomato plant
    c. Hemlock
    d. Oleander
    e. Foxglove

17. What is the term for the practice of using native plants in urban landscaping to promote biodiversity?
    a. Invasive species planting
    b. Monoculture gardening
    c. Xeriscaping
    d. Native landscaping
    e. Exotic plant cultivation
18. Which of the following is a benefit of green roofs in urban areas?
    a. Increased heat island effect
    b. Reduced energy consumption
    c. Higher air pollution levels
    d. Lower property values
    e. Reduced biodiversity
19. What is the term for the practice of growing food indoors using artificial light sources?
    a. Indoor gardening
    b. Sunless farming
    c. Artificial cultivation
    d. Light-dependent farming
    e. Shade agriculture
20. Which of the following is a common method used in urban agriculture to conserve water?
    a. Overhead watering
    b. Flood irrigation
    c. Mulching
    d. Watering during peak hours
    e. Watering with untreated wastewater
21. Which of the following is a benefit of green roofs in urban areas?
    a. Increased heat island effect
    b. Reduced stormwater runoff
    c. Higher energy consumption
    d. Decreased biodiversity
    e. Improved air pollution
22. What is the term for the practice of using recycled materials in urban gardening projects?
    a. Upcycling
    b. Downcycling
    c. Landfilling
    d. Waste dumping
    e. Incineration
23. Which of the following is a common practice in urban agriculture to improve soil fertility?
    a. Synthetic fertilizers
    b. Tilling
    c. Cover cropping
    d. Soil compaction
    e. Soil erosion
24. What is the term for the practice of growing food indoors using artificial lighting?
    a. Greenhouse farming
    b. Indoor gardening
    c. Controlled environment agriculture
    d. Open field cultivation
    e. Shade farming

25. Which of the following is a common method of preserving harvested crops in urban agriculture?
    a. Freezing
    b. Canning
    c. Dehydration
    d. Fermentation
    e. All of the above
26. What is the term for the practice of using plants to filter pollutants from stormwater in urban areas?
    a. Bioretention
    b. Stormwater diversion
    c. Water treatment plant
    d. Sewage disposal
    e. Chemical oxidation
27. Which of the following is a common practice in rooftop gardening to prevent water runoff and erosion?
    a. Mulching
    b. Overwatering
    c. Pesticide spraying
    d. Soil compaction
    e. Excessive pruning
28. What is the term for the practice of using vertical structures to support climbing plants in urban gardens?
    a. Trellising
    b. Staking
    c. Ground planting
    d. Pruning
    e. Mulching
29. Which of the following is a common method of reducing food waste in urban agriculture?
    a. Composting
    b. Landfill disposal
    c. Incineration
    d. Dumping in water bodies
    e. All of the above
30. What is the term for the practice of growing different types of crops together to maximize space and yield?
    a. Monoculture farming
    b. Polyculture farming
    c. Crop rotation
    d. Companion planting
    e. Intercropping
31. Which of the following is a key benefit of urban agriculture in addressing food security issues?
    a. Increased food prices
    b. Food deserts
    c. Limited access to fresh produce
    d. Overreliance on imports
    e. All of the above

32. What is the term for the practice of using fish and plants together in a recirculating system in urban agriculture?
    a. Aquaculture
    b. Hydroponics
    c. Aquaponics
    d. Vermicomposting
    e. Aeroponics
33. Which of the following is a common method of reducing water usage in urban agriculture?
    a. Overwatering plants
    b. Hand watering only when necessary
    c. Watering during peak sun hours
    d. Using sprinklers for extended periods
    e. Allowing water to run off without absorption
34. What is the term for the practice of growing food in small spaces such as balconies or windowsills in urban areas?
    a. Microgardening
    b. Macrofarming
    c. Miniagriculture
    d. Nanofarming
    e. Urban microfarming
35. Which of the following is a common method of controlling temperature in urban greenhouses?
    a. Natural ventilation
    b. Air conditioning
    c. Heating lamps
    d. Soil sterilization
    e. Overwatering
36. What is the term for the practice of using native plant species to attract beneficial insects in urban gardens?
    a. Companion planting
    b. Monoculture gardening
    c. Exotic landscaping
    d. Invasive species planting
    e. Synthetic fertilization
37. Which of the following is a common method of protecting crops from birds in urban agriculture?
    a. Scarecrows
    b. Chemical repellents
    c. Hunting
    d. Noise pollution
    e. All of the above
38. What is the term for the practice of using raised beds to grow crops in urban gardens?
    a. Container gardening
    b. Ground planting
    c. Vertical farming
    d. Raised bed gardening
    e. Hydroponics
39. Which of the following is a key benefit of incorporating green spaces into urban areas?
    a. Increased air pollution
    b. Decreased biodiversity
    c. Improved mental health
    d. Higher crime rates
    e. Reduced community engagement

40. What is the term for the practice of using plant-based dyes to color urban landscapes?
    a. Natural dyeing
    b. Synthetic dyeing
    c. Artificial coloring
    d. Chemical staining
    e. Paint spraying
41. Which of the following is a common method of reducing water usage in urban landscaping?
    a. Overwatering
    b. Drip irrigation
    c. Watering during peak hours
    d. Watering daily
    e. Watering in the afternoon
42. What is the term for the practice of using recycled materials to create urban art installations?
    a. Upcycling
    b. Downcycling
    c. Landfilling
    d. Waste dumping
    e. Incineration
43. Which of the following is a common method of conserving energy in urban greenhouses?
    a. Natural lighting
    b. Artificial lighting
    c. Heating lamps
    d. Air conditioning
    e. Soil sterilization
44. What is the term for the practice of using reflective surfaces to reduce heat absorption in urban environments?
    a. Heat absorption
    b. Heat reflection
    c. Urban cooling
    d. Albedo effect
    e. Thermal insulation
45. Which of the following is a common method of controlling soil erosion in urban agriculture?
    a. Mulching
    b. Tilling
    c. Overwatering
    d. Soil compaction
    e. Excessive pruning
46. What is the term for the practice of using recycled water for irrigation in urban agriculture?
    a. Greywater recycling
    b. Rainwater harvesting
    c. Drip irrigation
    d. Sprinkler system
    e. Water pumping
47. Which of the following is a common method of reducing pesticide use in urban agriculture?
    a. Integrated pest management
    b. Synthetic pesticides
    c. Chemical fertilizers
    d. Genetic modification
    e. Soil sterilization

48. What is the term for the practice of using sensors and automation to optimize water usage in urban agriculture?
    a. Precision farming
    b. Conventional farming
    c. Industrial agriculture
    d. Traditional farming
    e. Subsistence farming
49. Which of the following is a common method of promoting biodiversity in urban green spaces?
    a. Monoculture planting
    b. Exotic species introduction
    c. Native plant landscaping
    d. Chemical fertilization
    e. Soil sterilization
50. What is the term for the practice of using renewable energy sources to power urban agricultural operations?
    a. Sustainable energy
    b. Fossil fuel energy
    c. Non-renewable energy
    d. Conventional energy
    e. Grid electricity

ANSWERS TO MULTIPLE-CHOICE QUESTIONS

1	c	11	b	21	b	31	e	41	b
2	c	12	c	22	a	32	c	42	a
3	c	13	b	23	c	33	b	43	a
4	a	14	e	24	c	34	a	44	d
5	a	15	e	25	e	35	a	45	a
6	d	16	b	26	a	36	a	46	a
7	e	17	d	27	a	37	a	47	a
8	c	18	d	28	a	38	d	48	a
9	c	19	a	29	a	39	c	49	c
10	c	20	c	30	e	40	a	50	a

FURTHER READING

Alaimo, K., Packnett, E., Miles, R. A., & Kruger, D. J. (2008). Fruit and vegetable intake among urban community gardeners. *Journal of Nutrition Education and Behavior*, 40(2), 94–101.

Andersson, E., Barthel, S., & Ahrné, K. (2007). Measuring social–ecological dynamics behind the generation of ecosystem services. *Ecological Applications*, 17(5), 1267–1278.

Barthel, S., & Isendahl, C. (2013). Urban gardens, agriculture, and water management: Sources of resilience for long-term food security in cities. *Ecological Economics*, 86, 224–234.

Bell, S., & Paskins, J. (1993). The role of urban agriculture in social welfare provision for the elderly. *Landscape and Urban Planning*, 25(1–2), 143–155.

Brown, K. H., & Carter, S. E. (2003). Urban agriculture and community food security in the United States: Farming from the city center to the urban fringe. In N. Bakker, M. Dubbeling, S. Guendel, U. Sabel-Koschella, & H. de Zeeuw (Eds.), *Growing cities, growing food: Urban agriculture on the policy agenda* (pp. 27–42). DSE/ZEL.

Colding, J., & Barthel, S. (2013). The potential of "Urban Green Commons" in the resilience building of cities. *Ecological Economics*, 86, 156–166.

De Bon, H., Parrot, L., Moustier, P., & Alvarez, S. (2010). Urban agriculture in West Africa: Contributing to food security and urban sanitation. In N. Bakker, M. Dubbeling, S. Guendel, U. Sabel-Koschella, & H. de Zeeuw (Eds.), *Growing cities, growing food: Urban agriculture on the policy agenda* (pp. 113–136). DSE/ZEL.

Guitart, D., Pickering, C., & Byrne, J. (2012). Past results and future directions in urban community gardens research. *Urban Forestry & Urban Greening*, 11(4), 364–373.

Haase, D., Larondelle, N., Andersson, E., Artmann, M., Borgström, S., Breuste, J., … & Wurster, D. (2014). A quantitative review of urban ecosystem service assessments: Concepts, models, and implementation. *Ambio*, 43(4), 413–433.

Kortright, R., & Wakefield, S. (2011). Edible backyards: A qualitative study of household food growing and its contributions to food security. *Agriculture and Human Values*, 28(1), 39–53.

Lovell, S. T. (2010). Multifunctional urban agriculture for sustainable land use planning in the United States. *Sustainability*, 2(8), 2499–2522.

McClintock, N. (2010). Why farm the city? Theorizing urban agriculture through a lens of metabolic rift. *Cambridge Journal of Regions, Economy and Society*, 3(2), 191–207.

McClintock, N., Cooper, J., & Khandeshi, S. (2012). Assessing the potential contribution of vacant land to urban vegetable production and consumption in Oakland, California. *Landscape and Urban Planning*, 105(1–2), 240–246.

Mougeot, L. J. A. (2005). *Agropolis: The social, political, and environmental dimensions of urban agriculture*. Earthscan.

Mougeot, L. J. A. (2010). *Growing better cities: Urban agriculture for sustainable development.* IDRC.

Nieuwenhuijsen, M. J. (2003). Urban and transport planning, environmental exposures and health—New concepts, methods and tools to improve health in cities. *Environmental Health*, 2(1), 1–6.

Pearson, L. J., Pearson, L. J., Pearson, L. J., & Pearson, L. J. (2010). Urban agriculture and the alleviation of poverty. *Unasylva*, 61(231), 3–11.

Poe, M. R., McLain, R., Emery, M. R., Hurley, P. T., & Lecompte-Mastenbrook, J. (2013). Urban forest justice and the rights to wild foods, medicines, and materials in the city. *Human Ecology*, 41(3), 409–422.

Smit, J., Ratta, A., & Nasr, J. (1996). *Urban agriculture: Food, jobs, and sustainable cities*. UNDP.

Vitiello, D., & Wolf-Powers, L. (2014). The economic value of urban agriculture: A review of the impacts of vegetable production on property values. *Renewable Agriculture and Food Systems*, 29(4), 386–396.

Wolch, J., Byrne, J., & Newell, J. P. (2014). Urban green space, public health, and environmental justice: The challenge of making cities "just green enough." *Landscape and Urban Planning*, 125, 234–244.

Zezza, A., & Tasciotti, L. (2010). Urban agriculture, poverty, and food security: Empirical evidence from a sample of developing countries. *Food Policy*, 35(4), 265–273.

Index

D

E

F

G

Q

R

S

T

Z

For Product Safety Concerns and Information please contact our EU representative GPSR@taylorandfrancis.com Taylor & Francis Verlag GmbH, Kaufingerstraße 24, 80331 München, Germany